Acta Neurochirurgica
Supplements

Editor: H.-J. Steiger

Brain Edema XII

Proceedings of the 12th International Symposium,
Hakone, Japan, November 10–13, 2002

Edited by
T. Kuroiwa, A. Baethmann, Z. Czernicki,
J. T. Hoff, U. Ito, Y. Katayama, A. Marmarou,
A. D. Mendelow, and H.-J. Reulen

Acta Neurochirurgica
Supplement 86

Springer-Verlag Wien GmbH

T. Kuroiwa
Medical Research Institute,
Tokyo Medical and Dental University,
Tokyo, Japan

A. Baethmann
Institute für Chirurgische Forschung,
Ludwig-Maximilians-Universität,
München, Germany

Z. Czernicki
Department of Neurosurgery,
Polish Academy of Scienes, Warsaw, Poland

J. T. Hoff
Section of Neurosurgery,
University of Michigan, Ann Arbor, USA

U. Ito
Department of Neurosurgery,
Musashino Red Cross Hospital, Tokyo, Japan

Y. Katayama
Department of Neurological Surgery,
Nihon University, Tokyo, Japan

A. Marmarou
Medical College of Virgina,
Richmond, USA

B. A. D. Mendelow
Department of Neurosurgery,
Newcastle Gerneral Hospital, Newcastle-upon-Tyne, UK

H.-J. Reulen
Neurochirurgische Universitätklinik,
Ludwig-Maximilians-Universität,
München, Germany

Originally published by Springer-Verlag / Wien in 2000

Typesetting: Asco Typesetters, Hong Kong

Printed on acid-free and chlorine-free bleached paper

SPIN: 10905760

With partly coloured Figures

Library of Congress Cataloging-in-Publication Data
Brain edema XII : proceedings of the 12th international symposium : Hakone, Japan,
November 10–13, 2002 / edited by T. Kuroiwa ... [et al.].
p. ; cm. – (Acta neurochirurgica. Supplement, ISSN 0065-1419 ; 86)
Includes bibliographical references and index.
ISBN 978-3-211-00919-2 ISBN 978-3-7091-0651-8 (eBook)
DOI 10.1007/978-3-7091-0651-8
1. Cerebral edema–Congresses. I. Title: Brain edema 12. II. Kuroiwa, T. III. Series.
[DNLM: 1. Brain Edema–Congresses. WL 348 B8137 2003]
RC394.E3B7272 2003
616.8–dc21 2003052904

ISSN 0065-1419
ISBN 978-3-211-00919-2

Preface

The 12th International Symposium on Brain Edema and Brain Tissue Injury was held on November 10–13, 2002 in Hakone Japan. This volume is a compilation of the papers presented and discussed in the symposium. The advisory board have edited the papers and summarized their respective sessions. The round table discussion on the third day, a résumé of the scientific essence of the symposium, is also recorded in this volume for readers to have a quick and comprehensive overview of the current status of brain edema research and treatment.

The title of the symposium this time was changed slightly to "Brain Edema and Brain Tissue Injury", as we wanted to emphasize the importance of intercellular and tissue mechanisms as well as intracellular molecular mechanisms in the formation, treatment and resolution of brain edema. Despite rapid advances in diagnostic and therapeutic procedures, brain edema is still a major threat to patients' lives in the neurological/neurosurgical ward. Since brain edema is a multifactorial process associated with most brain tissue injuries, the brain edema symposia have provided a unique opportunity for the exchange of modern laboratory information with clinicians in practice. The symposium consisted of platform sessions, poster sessions and 7 lectures on various topics relating to brain edema. The topics of the symposium ranged from cutting-edge neuroimaging technology, molecular medicine to new therapeutic strategies and ongoing therapeutic trials. Aquaporins and the effects on volume homeostasis were discussed in several papers. Regeneration was another highlight of the symposium. Some manuscripts were dealing with the regeneration of nerve tissue damaged by edema-related processes and the role of progenitor/stem cells on the functional and structural recovery of the neural tissue. There were several reports on the clinical trials on the therapy of intracranial cerebral hemorrhage and decompressive craniectomy.

During the symposium, Hakone Best Presentation Award was selected by a committee chaired by A. D. Mendelow and presented to the best six presentations. The awardees are acknowledged in this volume.

The advisory board has felt that prompt publication is most important to make this volume valuable in the field of rapidly advancing neuroscience. We wish to express our thanks to all authors who enable us to publish this volume approximately 6 months after the symposium. We also express our gratitude to the members of the secretariat for their assistance to have a successful symposium and for prompt publication of this volume.

The Thirteenth International Symposium on Brain Edema will be held in Ann Arbor, USA in 2005 under the chair of J. T. Hoff.

T. Kuroiwa and Editors

Contents

General review

Imaging

Clinical ischemia

Intracerebral and subarachnoid hemorrhage

Listed in Current Contents

General review

Acta Neurochir (2003) [Suppl] 86: 3–6

Concept and pathogenesis of "hypoxic-ischemic encephalopathy"

R. Okeda

Department of Neuropathology, Medical Research Institute, Tokyo Medical & Dental University, Tokyo, Japan

Summary

By experiments of acute carbon-monoxide intoxication, acute nitrogen hypoxia and histotoxic hypoxia using sodium cyanide in cats, and by hemodynamic studies using plastic branch models, the following was elucidated; (1) severe tissue hypoxia, regardless of the underlying cause, and subsequent slight ischemia of the brain due to mild hypotension induce selective involvement of the cerebral white matter and pallidum, these two conditions being necessary and sufficient and this encephalopathy should be separately categorized as "hypoxic-ischemic encephalopathy" in hypoxic brain injuries, (2) the background of the selective involvement of these structures is an enormous development of the cerebrum in the brain, which induces thick white matter resulting in proper and long medullary artery, and especially small diameter ratio of the pallidal perforators to the middle cerebral artery, (3) the long course of the medullary artery produces the blood pressure drop in the deep white matter according to Hagen-Poiseuille's low, and according to that the smaller the diamter ratio, the larger the branching-loss coefficient (energy-loss coefficient), smaller diameter ratio of the pallidal perforator, as compared with that of the putaminal perforator, induces more severe loss of the local blood flow selectively to the pallidum. This state seems to be a failure of compromise between the cardiovascular system and the brain parenchyma.

Keywords: Carbon monoxide intoxication; nitrogen hypoxia; histotoxic hypoxia; sodium cyanide; branching-loss coefficient; hypoxic-ischemic encephalopathy.

Introduction

Aim of this study is to explain the concept and pathogenesis of "hypoxic-ischemic encephalopathy", a phenomenon that is different from the concept of "anoxic-ischemic encephalopathy" introduced by Levine in 1960. Typical "hypoxic-ischemic encephalopathy" can be induced by acute CO intoxication, and characterized by selective, symmetrical involvement of the cerebral white matter with sparing of the subcortical U-fibers. These lesions exhibit patchy myelin loss with axonal swelling and destruction or patchy necrosis. The extracerebral white matter and the gray matter of the cerebral cortex, putamen, and thalamus are generally spared. The pallidum – especially the dorsal portion and the neighboring internal capsule – is frequently involved. Such brain lesions are recognized also in accidents of narcosis or cyanide-intoxication. Following experimental studies using adult cats were performed in order to elucidate the pathogenesis.

Materials and methods

Acute carbon monoxide intoxication [4]

Method. Fig. 1 is a chart of the experiment, in which cats inhaled air that contained 0.3% CO. As the concentration of carboxyhemoglobin in the blood gradually increased, the blood flow in the common carotid artery exhibited a compensatory increase. After 1 hour, however, the blood flow and systemic blood pressure gradually began to decrease, and the inhalation of CO was finished after almost 3 hours, when the systemic blood pressure dropped to about 70 mmHg. Several minutes-interruptions of the CO inhalation was necessary, otherwise the blood pressure would drop progressively, very often resulting in death.

Results. Within a few days these cats presented the characteristic brain lesions shown in Fig. 2. Symmetrical patchy lesions in the cerebral white matter and pallidum occurred. In addition, neuronal ischemic changes or small infarcts were found in the hippocampus and substantia nigra of some cats showing severe blood pressure drop. Local blood flow of the cerebral cortex and white matter, which was measured by hydrogen-clearance method, showed a transient increase and subsequent decrease during CO inhalation, and increased after the end of the inhalation. This blood flow reduction during CO-inhalation seemed to be an important factor for occurrence of cerebral lesions. Therefore, we tried to inhibit the reduction in systemic blood pressure and blood flow of the common carotid artery by frequent interruptions of the CO inhalation, while maintaining the high concentration of carboxyhemoglobin in the blood. In cats in which the blood pressure drop during CO-inhalation was minimally suppressed, the brain lesions were not detected. There is a positive correlation between the severity of the white matter changes and the grade of the blood pressure drop, and the pallidal lesions occurred preferentially in cats that exhibited a relatively severe reduction in systemic blood pressure. This experiment disclosed that in order to produce carbon-monoxide-induced cerebral lesions, the intoxication must generate a concentration of carboxyhemoglobin that is high

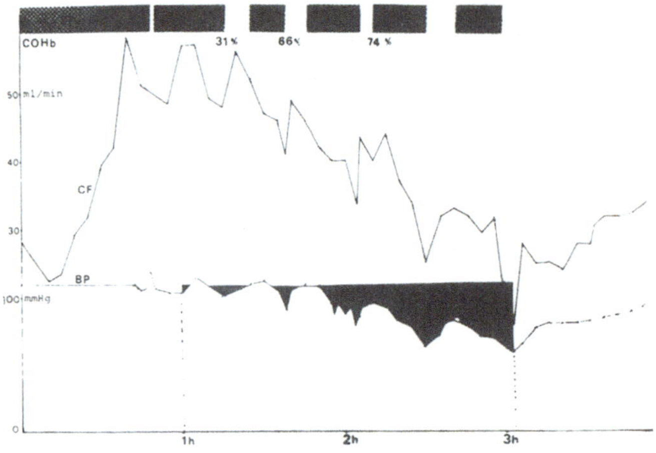

Fig. 1. An experimental chart of acute carbon monoxide poisoning. *COHB* Carboxyhemoglobin, *CF* blood flow of the common carotid artery, *BP* blood pressure of the aorta

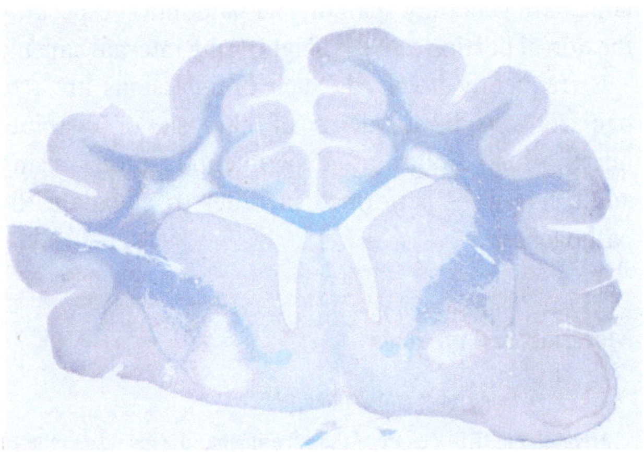

Fig. 2. Pathological changes occur symmetrically in the deep cerebral white matter and the pallidum. Kluever-Barrera's staining

enough to provoke severe hypoxia and blood pressure drop, and that despite the fact that this blood pressure drop was only slight (to about 70 mmHg), the local cerebral blood flow was reduced. This means that in the compensatory phase during CO inhalation, it is likely that almost all of the vessels were maximally dilated, the local blood flow thus became dependent upon systemic blood pressure, and the function of autoregulation was lost. Therefore it appears that even a slight drop in blood pressure causes lowering of the blood flow. Moreover, it is also elucidated that acidosis of the blood had no effect on that blood flow. This raises the question as to whether CO is really necessary for the pathogenesis of these brain lesions, and whether the effects of CO intoxication can be reproduced using other methods, such as nitrogen hypoxia.

Nitrogen hypoxia [5]

Method. The nitrogen hypoxia was induced by gradually raising the nitrogen concentration in the aspirated air over a period of 1.5 hours, and the partial pressure of blood oxygen was reduced to less than 26 mmHg. Then, the systemic blood pressure was artificially lowered to 60–80 mmHg, and this state was maintained for 1 hour.

Results. The brain was involved in cats on which this procedure was carried out, and the pattern of distribution and characteristics of these lesions were the same as those in acute CO-intoxication. Severe hypoxia only, or severe hypotension alone did not cause any pathological changes in the brain. This means that severe hypoxia of the blood and tissue, and the subsequent depletion of cerebral blood flow in response to slight hypotension is common and is essential for the selective involvement of the cerebral white matter and pallidum.

Histotoxic hypoxia by sodium cyanide [2]

Method. A 0.2% solution of sodium cyanide was slowly infused intravenously into a group of cats for more than 2 hours. During the infusion, the systemic blood pressure was spontaneously or artificially lowered to less than 100 mmHg. In another group, the spontaneous reduction in blood pressure observed during the infusion was maintained above 100 mmHg. There was no difference in the severity of blood acidosis between the two groups of animals.

Results. All of the cats in the group with hypotension under 100 mmHg exhibited the pathological changes similar to those observed in acute CO intoxication. Only a few cats from the group with slight hypotension, which was maintained above 100 mmHg, exhibited tiny lesions in limited portions. Therefore, it is concluded that severe tissue hypoxia, regardless of the underlying cause, and subsequent slight ischemia of the brain due to mild hypotension are necessary and sufficient conditions for the selective involvement of the cerebral white matter and pallidum. Such neuropathological changes, therefore, should be separately categorized as "hypoxic-ischemic encephalopathy" in hypoxic brain injuries.

Another issue arose: that is, despite the fact that the entire brain is exposed to this ischemia, why did the resulting lesions occur only in the cerebral white matter and pallidum? Can such a topographical selectivity be explained by the occurrence of especially severe ischemia in such selected portions of the brain? To address this issue, the following experiment was performed.

Measurement of local blood flow in experimental acute CO intoxication in order to analyze the mechanism of selective involvement of the cerebral white matter [6]

Method. The local blood flow in 15 portions including the cerebral and cerebellar white matter of the brain of cats suffering from acute CO intoxication was measured using [^{14}C]iodoantipyrine at the stage just before finishing CO inhalation, when the slight systemic hypotension of 70–80 mmHg occurred.

Results. In general, the local blood flow was higher in the final stage of CO-inhalation, but that of the cerebral white matter did not exhibit such an increase, as compared with the prominent increase in blood flow seen in the cerebellar white matter. This finding means that the local blood flow of the cerebral white matter is reduced more severely at the final stage of CO inhalation than that of the cerebellar white matter. This phenomenon may be explained by a finding from human autopsy cases proved by Dr. Fukazawa [1]. He made arterial casts of autopsy brains and calculated the blood pressure drop along each branch of these casts by measuring the length and size. The cerebral medullary arteries are especially long because the cerebral white matter is especially thick, and, therefore, according to Hagen-Poiseuille's law, the blood pressure of arteries in the deep white matter is lowered more prominently even in this physiological state. This theory is supported by our data about regression lines between the radius and the medial thickness of the arachnoid and cerebral medullary arteries, which were measured morphometrically in normotensive human autopsy cases [8]. The medial thickness of the medullary arteries was significantly thinner than the arachnoid arteries. In other words, the burden of blood pressure is lower there than in the arachnoid arteries. It was concluded that this is the mechanism of selective involvement of the cerebral white matter.

What is the situation in the pallidum? The blood flow to the pallidum and putamen is supported by perforators originating from the middle cerebral artery. However, the pallidum is selectively involved in the pathogenesis of acute CO intoxication or hypoxia induced using nitrogen or histotoxic substances, whereas the putamen is usually spared. Such selectivity cannot be explained angiographically by differences in length or tortuosity of these perforators.

Measurement of the local blood flow of the pallidum and putamen in experimental CO intoxication in cats [7]

Method. In experimental model of acute CO intoxication in cats, the local blood flow of the pallidum and putamen was measured using the hydrogen clearance method. On this occasion, platinum electrodes were inserted into both brain structures, and local blood flow was measured every 30 minutes during CO inhalation.

Results. As the concentration of carboxyhemoglobin increased, the local blood flow of both structures increased, and then, as the blood pressure lowered, it began to decrease. The local blood flow of the pallidum presented an earlier and more severe decrease than that of the putamen, and it was significantly lower than that of the putamen at the terminal stage of CO-inhalation. Therefore, the architectural or geometrical differences between the perforators to both structures, in particular the branching angle and diameter ratio of these perforators to the middle cerebral artery should be considered.

Hemodynamic analysis of effects of branching angles and diameter ratios to the main trunk on the flow of the branch using plastic models [3]

Method. We made plastic branch models comprising a branch to a main trunk at various branching angles and diameter ratios. Branching-loss coefficients at branching sites (which can be called "energy-loss coefficients") were measured under conditions of steady laminar flow. The Reynolds number was 1,500. These coefficients are based on flow disturbances, such as separation of streamlines from the wall, formation of eddies, and complex secondary flow, and they were calculated according to the formula shown in Fig. 3. Practically, under various flow-dividing ratios, that is, the ratio of the flow volume of a branch (Q_2) to that of the main trunk (Q_0), the pressure loss due to branching (ΔP) and velocity at the original portion of the main trunk (V) were measured. This graph of Fig. 3 is the result: the abscissa is the dividing ratio of flow volume, Q_2/Q_0, and the ordinate is the branching-loss coefficient "ξ". "m" is a diameter ratio of a branch to the main trunk. Three branching angles of 45, 90 and 135° were examined.

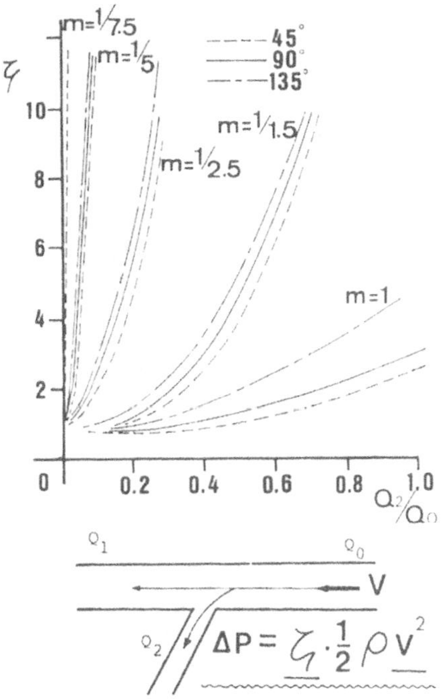

Fig. 3. Relationship between branching-loss coefficients "ζ" and various angles and diameter ratios to the main trunk. Q_0 and Q_2 Flow volumes of the main trunk and the branch, respectively, m: Diameter ratio of the branch to the main trunk. **ΔP** Pressure loss due to branching, ρ viscosity of the fluid, *V* velocity at the original portion of the main trunk

Results. Fig. 3 shows that the larger the branching angle, the larger the coefficient; similarly, the coefficient is larger when the branch is smaller relative to the main trunk. The branching angle of perforators to the putamen and pallidum is nearly 90°, but the diameter of the perforators to the pallidum is from one-third to one-fifth of those to the putamen in cats, monkeys and humans. Therefore, the branching-loss coefficient of the perforators to the pallidum is inherently larger than that of the putamen.

The local flow **F** is generally determined by the formula shown here;

$$\mathbf{F} = (\mathbf{P} - \Delta\mathbf{P})/\mathbf{Rp}$$

Under severe hypoxia, the peripheral resistance **Rp** reaches nearly the minimum level, and the collateral circulation from the surrounding tissues is very poor because these perforators are nearly end arteries. Under this state, the perfusion pressure **P** and pressure loss **ΔP** become important variables. Once hypotension occurs in this state, **ΔP** becomes a more important variable for the local flow **F**. Since the branching-loss coefficient and therefore, **ΔP** of perforators to the pallidum are inherently larger than those to the putamen, even slight hypotension induces more severe lowering of blood flow to the pallidum than to the putamen. This is the mechanism of selective involvement of the pallidum. Since such a difference in the branching-loss coefficient between the pallidal and putaminal perforators is inherently determined in each animal species (including human) having such an arterial structure, these events are destined in such animals.

Discussion

The brains of a rat and cat have very similar volumes to the kidney and heart. However, in human, the brain is about 5 times heavier than the heart, and 8 times heavier than the kidney. This discrepancy is attributable mainly to the enormous development of the cerebrum. Although the cerebral white matter of the cat is thicker as compared with that of the rat, and therefore has proper medullary arteries, in human it is much thicker and therefore the cerebral medullary artery takes an enormously long course, resulting in more severe lowering of the blood flow not only under physiological conditions but also under critical states. The enormous development of the human cerebrum enables us to enjoy a rich life, both materially and spiritually, but conversely it acts disadvantageously in a critical state such as severe hypoxia. This enoumous development of the cerebrum induces not only the long medullary arteries, probably but also an especially small diameter ratio of the pallidal perforators to the middle cerebral artery, because the cerebral development needs big cerebral arteries which promotes small diameter ratio of the perforators. Presumably, hypoxic-ischemic encephalopathy may be an expression of a failure of compromise between the cardiovascular system and parenchyma.

References

1. Fukasawa H (1969) Hemodynamical studies of cerebral arteries by means of mathematical analysis of arterial casts. Tohoku J Exp Med 99: 255–268
2. Funata N, Song S-Y, Okeda R, Funata M, Higashino F (1984) A study of experimental cyanide encephalopathy in acute phase-physiological and neuropathological correlation. Acta Neuropathol 64: 99–107
3. Matsuo T, Okeda R, Higashino F (1989) Hydrodynamics of arterial branching–the effect of arterial branching on distal blood supply. Biorheology 26: 799–811
4. Okeda R, Funata N, Takano T, Miyazaki Y, Higashino F, Yokoyama Y, Manabe M (1981) The pathogenesis of carbon monoxide encephalopathy in the acute phase – physiological and morphological correlation. Acta Neuropathol 54: 1–10
5. Okeda R, Funata N, Song S-J, Higashino F, Takano T, Yokoyama K (1982) Comparative study on pathogenesis of selective cerebral lesions in carbon monoxide poisoning and nitrogen hypoxia in cats. Acta Neuropathol 56: 265–272
6. Okeda R, Matsuo T, Kuroiwa T, Nakai M, Tajima T, Takahashi H (1987) Regional cerebral flow of acute carbon monoxide poisoning. Acta Neuropathol 72: 389–393
7. Song S-Y, Okeda R, Funata N, Higashino H (1983) An experimental study of the pathogenesis of the selective lesion of the globus pallidus in acute carbon monoxide poisoning in cats. With special reference to the chronologic change in the cerebral local blood flow. Acta Neuropathol 61: 232–238
8. Tanoi Y, Okeda R, Budka H (2000) Binswanger's encephalopathy: serial sections and morphometry of the cerebral arteries. Acta Neuropathol 100: 347–355

Correspondence: Riki Okeda, No. 1-5-45, Yushima, Bunkyo-ku, Tokyo, Japan. e-mail: okeda.npat@tmd.ac.jp

Acta Neurochir (2003) [Suppl] 86: 7–10

Pathophysiology of traumatic brain edema: current concepts

A. Marmarou

Division of Neurosurgery, Medical College of Virginia, Virginia Commonwealth University, Richmond, VA, USA

Summary

The generally held concept during the past several decades is that traumatic brain edema is predominately vasogenic emanating from the blood vessels subsequent to blood brain barrier compromise. Much of the experimental data has focused on cryogenic injury models where there clearly is a necrotic lesion surrounded by leaking vessels. However, in closed head injury where brain swelling remains a critical problem, the classification of the type of edema that develops is less clear. Most importantly, studies in the clinical setting have ruled out vascular engorgement as one potential mechanism and these studies have shown that edema and not blood volume is the culprit responsible for brain swelling. We have put forth the notion that traumatic brain edema is a combination of vasogenic and cellular with the cellular component predominating. This article provides an update of our current progress toward supporting this hypothesis and includes an update on the role of aquaporins in traumatic brain edema.

Keywords: Traumatic brain edema; vasogenic; cellular; aquaporin.

Introduction

The brain swelling that accompanies severe head injury accounts for 50% of all deaths. For many years, this swelling was purported to be a result of vascular engorgement and that edema played a minor role in the swelling process. Edema, resulting from a breach of the blood brain barrier, has also been thought to play a role, particularly in contusion where the fluid migration is more easily visualized by computerized tomography (CT). However the rapid expansion of injured brain experienced during surgery further bolstered the blood volume theory and it was posited that the expansion was due to the vasoparalysis of the resistance vessels and subsequent increase in blood volume. Our studies challenged this concept as we were able to measure both blood volume and brain water using non-invasively in head injured patients. The utilization of stable xenon allowed us to measure blood flow and coupled with transit time measures, the absolute blood volume change over the 10 days post injury. Similarly, the MR allowed us to measure brain water in absolute terms of % gm H_2O/gm tissue. Taken in concert these measures allowed us to quantify the respective changes in blood volume and brain water that accompanies severe brain injury in Man. We found that brain edema was responsible for brain swelling and that blood volume was actually reduced after severe brain trauma [4]. Thus, we turned our attention to the development of edema in experimental trauma in order to further characterize the swelling process.

The role of the blood brain barrier in traumatic brain injury

It has long been considered that traumatic brain edema occurs as a result of blood brain barrier breakdown and exudation of intravascular contents into the extra cellular space. This so called "vasogenic" edema has been the dogma of severe brain injury and the terms "vasogenic" edema and "traumatic brain edema" has been used interchangeably. In clinical practice, the concept of a vasogenic edema being responsible for brain swelling was seldom questioned because the exudation of fluid from a site of contusion is clearly visible on the CT scan. However, in diffuse injury the type of edema that contributes to the swelling process is unclear, particularly as there is no visible contrast observed on CT soon after injury. This suggests that the BBB is closed within hours after traumatic brain injury. Moreover, the effects of brain swelling are manifest in the development of refractory ICP generally within 3 to 5 days post injury. If this is the case, then the formation of edema leading to brain swelling must be by other means.

We measured the time of BBB compromise in a rat model of diffuse impact acceleration injury and diffuse injury with secondary insult [1]. The BBB disruption was quantified using T1 weighted magnetic resonance imaging following intravenous administration of the MR contrast agent gadolinium-diethylenetriamine pentaacetic acid (GDPA). We found that in the trauma induced animals, the signal intensity increased dramatically after impact. However, by 15 minutes after impact, permeability decreased exponentially and by 30 minutes was equal to that of the control animals. When secondary insult was introduced, the enhancement was lower than that with trauma alone. This was consistent with reduced blood pressure and blood flow. However, the signal intensity increased dramatically on re-perfusion and establishment of normal blood pressure and was equal to that of controls by 60 minutes post injury. Thus, it seems that the BBB compromise in traumatic brain injury to GDPA is short-lived which infers that the contribution of the vasogenic type of edema to brain swelling is limited. Evidence was mounting that a cellular or "cytotoxic" component of edema must play a pivotal role in traumatic brain injury.

Diffusion weighted imaging for characterizing the type of edema in TBI

The question arose if it were possible to document the temporal course of the vasogenic and hypothetical cellular component in a model of diffuse injury. With the advent of MR, diffusion weighted imaging and the quantification by apparent diffusion coefficient (ADC) made this possible [3]. We applied these techniques in the impact acceleration model in order to establish the contribution of vasogenic and cellular edema [2]. In animals subjected to trauma, we found a significant increase in ADC and brain water content during the first 60 minutes post injury. This was consistent with an increase in the volume of extracellular fluid and vasogenic edema resulting from BBB compromise. This transient increase was followed by a continued decrease in ADC that began 40 minutes post injury and reached a minimum value 7 days post injury. Most importantly, the water content continued to increase over this time indicating the predominance of a cellular edema component. Thus, as ADC represents the algebraic sum of vasogenic (positive ADC component) and cellular (negative ADC component), the net change favors a transient vasogenic edema followed by a predominant cellular edema.

Clinical studies supporting a cellular traumatic edema

The experimental studies strongly supported the concept of a cellular traumatic edema and suggested that the vasogenic component of edema plays less of a role in brain swelling than originally thought. The movement to the clinical setting to provide further information to help clarify the swelling process was of utmost importance. Firstly, MR images of patients studied within 24 hours of injury revealed no evidence of GDPA leakage providing support that the BBB was intact at the time of measurement. Subsequently, several of these patients went on to develop raised intracranial pressure despite barrier closure. An example of refractory ICP development with confirmation of barrier closure in diffuse injury is shown in Fig. 1. Note that the development of raised ICP occurs over several hours reaching a point where cerebral perfusion pressure is reduced to critical levels. This patient did not survive.

Another objective of the clinical studies was to determine in both focal and diffuse injury, the type of edema that developed, vasogenic or cellular. This was done by transferring severely brain injured patients, after stabilization, to the MR suite for diffusion

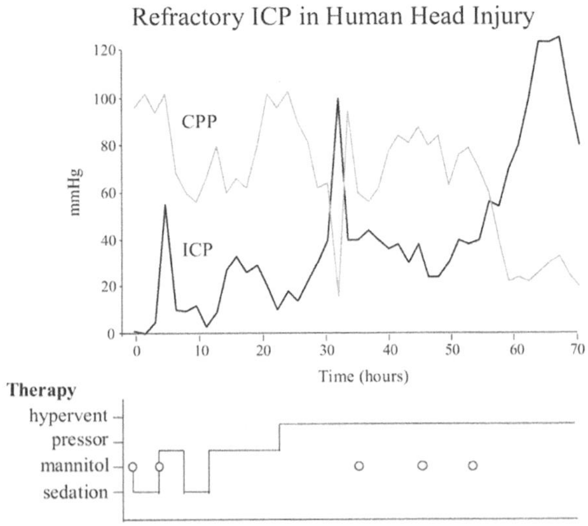

Fig. 1. The development of refractory ICP in a severely head injured patient. Studies of MR images fail to reveal opening of the BBB within the first 24 hours post injury. However, brain swelling ensues reinforcing the concept of a predominant cellular edema in severe trauma

weighted imaging studies. At the completion of the MR studies, the patients were transferred to the CT suite for measurement of CBF utilizing stable xenon. This was important as it was necessary to determine if ischemia played a role in the development of cellular edema. We observed, with focal injury, that ADC was increased with concomitant increase in brain water in the region of contusion (unpublished data). However, distal to the lesion site, the ADC was reduced signifying a cellular edema. Importantly, the CBF studies indicated that CBF was above ischemic threshold at time of measurement. Certainly, it is possible that the patient suffered an ischemic insult prior to measurement. Nevertheless, swelling and increased ICP occurred following these measures and thus we concluded that the process of swelling was ongoing with a predominant cellular component not due to reduced CBF. With diffuse injury, the water content was increased as expected, however the ADC was reduced, again signifying the development of a predominant cellular edema.

Summary of supportive evidence for a traumatic cellular edema

Summarizing these studies, firstly, it was determined that with experimental injury the BBB opening was shortlived and subsequent swelling was due to a cellular component of edema. This was confirmed by diffusion weighted imaging studies in rats which showed that after a transient rise in ADC within 60 minutes, the ADC thereafter reduced and was sustained for several days along with increased water content. Similar observations were made in the clinical setting in severely head injured patients. Specifically, that MR images confirmed that BBB in patients was intact in the presence of continued swelling and increased ICP. Secondly, that in areas of focal and diffuse injury, ADC was reduced in the absence of ischemic levels of CBF.

Taken in concert, these studies provide compelling evidence that the predominant form of edema in traumatic brain injury is cellular and not vasogenic.

Movement of sodium and obligate water into traumatically brain injured tissue

With the predominance of cellular edema, our laboratories focused on the problem of identifying how sodium and obligate water entered the cell. It was posited that with barrier closure and the absence of increased water in the interstitium that the pathway of excess sodium was passing through the astrocytic endfoot. Two experiments were conducted in an attempt to provide supportive evidence. Firstly, radio-active sodium was administered i.v. and its uptake by a microdialysis electrode placed in tissue was measured.

Role of Astrocytic End Foot in Passage of Na and Obligatory Water

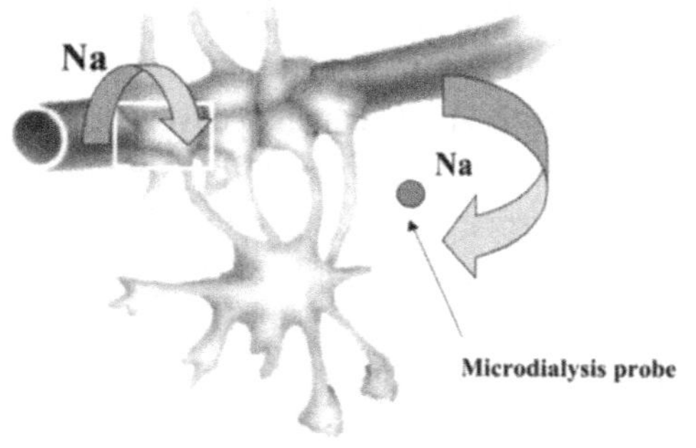

Fig. 2. Studies are now focused on identifying the pathway for sodium and obligatory water increase in traumatically injured brain, either through the vessel wall or alternate pathways. Experiments utilizing microdialysis probes in the tissue measuring uptake of Na^{22} show no uptake suggesting that sodium may enter via astrocytic endfeet

Although tissue sodium increased following TBI in experimental animals, Na^{22} was not observed in the microdialysis fluid suggesting that the Na^{22} was not passing from blood to the extra cellular space. In the second experiment, we studied the inactivation of a very specific astrocytic endfoot water channel protein, Aquaporin 4 (AQP4) via Protein Kinase C (PKC) phosphorilation. (See Amorini *et al.*) We used two well-known potent PKC activators: Phorbol 12–13 dybutirate (PDB) and Phorbol 12 myristate 13 acetate (PMA). We found that water content and sodium were reduced after intrathecal application of PMA providing support to the notion that astrocytic endfeet provide the pathway for sodium and obligatory water in the edematous process.

Mechanisms for a traumatic cellular edema and potential neuroprotectants

The proposed mechanisms for the development of a cellular edema include ischemia, mitochondrial impairment, membrane breakdown and ionic dysfunction. Having shown that swelling occurs in the absence of ischemia, our focus has been in studying the mitochondrial impairment that is associated with TBI. We conducted studies which measured both N-Acetyl-Aspartate and ATP in animals subjected to the impact acceleration model [7]. We found that NAA is reduced synchronously with ATP, which recovers to baseline values in moderate brain injury but remains consistently low with severe brain injury. It is believed excess calcium is implicated in mitochondrial dysfunction. An essential point of post-traumatic Ca^{++} overloading is the generation of the mitochondrial permeability transition pore (MPTP), a transmembrane protein on the inner membrane which permeabilizes the membrane to solutes, uncouples oxidative phosphorylation and causes swelling of the organelle. As recently demonstrated, the MPTP phenomenon is blocked in vitro and in vivo by Cyclosporin A (CsA), an immunosuppressive drug that specifically inhibits the opening of the pore by unbinding mitochondrial matrix cyclophilin from the pore [5, 6]. Studies are now in progress (See Fukui this issue) to evaluate the effect of CsA in TBI. Thus far, protection has been achieved at levels of 35 mg/kg administered intravenously. More importantly, clinical trials are now in progress to assess the safety and efficacy of CsA in severe traumatic brain injury.

In conclusion, the notion that cellular edema plays a major role in the swelling process requires that alternate mechanisms, such as mitochondrial impairment, and new approaches to therapy be considered if we are to improve outcome from traumatic brain injury.

References

1. Barzo P, Marmarou A, Fatouros P *et al* (1996) Magnetic resonance imaging-monitored acute blood-brain barrier changes in experimental traumatic brain injury. J Neurosurg 85: 1113–1121
2. Barzo P, Marmarou A, Fatouros P *et al* (1997) Contribution of vasogenic and cellular edema to traumatic brain swelling measured by diffusion weighted imaging. J Neurosurg 87: 900–907
3. Ito J, Marmarou A, Barzó P *et al* (1996) Characterization of edema by diffusion-weighted imaging in experimental traumatic brain injury. J Neurosurg 84: 97–103
4. Marmarou A, Fatouros PP, Barzo P *et al* (2000) Contribution of edema and cerebral blood volume to traumatic brain swelling in head-injured patients. J Neurosurg Aug 93(2): 183–193
5. Okonkwo DO, Povlishock JT (1999) An intrathecal bolus of cyclosporin A before injury preserves mitochondrial integrity and attenuates axonal disruption in traumatic brain injury. J Cereb Blood Flow Metab 19: 443–451
6. Scheff SW, Sullivan PG (1999) Cyclosporin A significantly ameliorates cortical damage following experimental traumatic brain injury in rodents. J Neurotrauma 16: 783–792
7. Signoretti S, Marmarou A, Tavazzi B, Lazzarino G, Beaumont A, Vagnozzi R (2001) N-Acetylaspartate reduction as a measure of injury severity and mitochondrial dysfunction following diffuse traumatic brain injury. J Neurotrauma 18: 977–991

Correspondence: Anthony Marmarou, Ph.D., Division of Neurosurgery, Medical College of Virginia, Virginia Commonwealth University, P.O. Box 950508, Richmond, Virginia 23298-0508, U.S.A. e-mail: marmarou@hsc.vcu.edu

Acta Neurochir (2003) [Suppl] 86: 11–15

Brain edema from intracerebral hemorrhage

J. T. Hoff and **G. Xi**

Department of Neurosurgery, University of Michigan Health System, Ann Arbor, Michigan, USA

Summary

Sequential changes in brain parenchyma surrounding an intracerebral hemorrhage are described here. Re-bleeding occurs within the first several hours after the initial hemorrhage in about 30% of cases. The coagulation cascade is activated as soon as blood encounters tissue. Perihematomal brain edema develops in response to clot retraction, thrombin formation, erythrocyte lysis, hemoglobin toxicity, complement activation, mass effect, and blood-brain barrier disruption. Early hematoma evacuation interrupts edema formation. The toxicity of extravasated blood in brain parenchyma has not been studied well in traumatic injury or in hemorrhagic tumor models yet, but similar mechanisms of edema formation are likely to occur in these conditions.

Keywords: Intracerebral hemorrhage; thrombin; edema; coagulation cascade; hemoglobin; complement; blood-brain barrier.

Introduction

Extravasation of blood into brain parenchyma results from a variety of lesions. Often, the hemorrhage causes immediate death. If the patient survives the ictus, the resulting hematoma within brain parenchyma triggers secondary insults and additional mortality. The sequential changes in brain that follow the intracerebral hemorrhage (ICH) are described here.

Early hematoma enlargement

Most patients stop bleeding shortly after a spontaneous ICH. However, rebleeding after the ictus accelerates neurological deterioration. Hemorrhage recurs within the first 24 hours in about 30% of patients with spontaneous ICH [4].

Clot formation and retraction

The coagulation cascade is activated as soon as blood enters the brain parenchyma. Both the extrinsic and intrinsic pathways for clot formation are involved. Prothrombin produces the serine protease, thrombin. Thrombin converts fibrinogen to fibrin, forming an unretracted fibrin clot. At the same time, erythrocyte aggregation starts and a platelet plug forms at the bleeding site. When the hematoma forms, clot retraction begins and continues for several hours [14]. When clot retraction is complete, the hematocrit of the hematoma is about 90%, whereas the normal hematocrit of whole blood is half that amount. Both isolated viable brain tissue and necrotic brain debris can be found within the clot. Serum from the clot accumulates around it and probably contributes to perihematomal edema by formation of an oncotic gradient [11].

Mechanism of brain edema after ICH

Edema develops during the acute and subacute stage following ICH. Several mechanisms are responsible for edema development. They include hydrostatic pressure on parenchyma surrounding the hematoma, clot retraction, the coagulation cascade and thrombin formation, erythrocyte lysis and hemoglobin toxicity, complement activation, mass effect, ischemia, and blood-brain barrier (BBB) disruption [23, 40, 42, 43].

Hydrostatic pressure and clot retraction

Early brain edema around the clot is due to hydrostatic pressure and clot retraction [35, 43]. Early CT scans indicate that the hypodensity rim around the clot is due to clot retraction [19]. Hydrostatic and oncotic pressure gradients between the hematoma and the surrounding tissue also contribute to early perihematomal brain edema [1, 35].

Coagulation cascade and thrombin formation

The coagulation cascade plays an important role in early edema formation following ICH. In an experimental model, heparinized autologous blood failed to produce perihematomal edema within 24 hours [42]. The same phenomenon occurs in humans [9, 10].

Thrombin is an essential component of the coagulation cascade and is responsible for early brain edema formation following ICH. Thrombin-induced brain edema is partly due to opening of the BBB. Hirudin, an anticoagulant and thrombin inhibitor, inhibits edema formation in a rat ICH model [22]. Recently, Hamada *et al.* tested antithrombin therapy in patients with ICH and found reduced brain edema following treatment with argatroban, a specific thrombin inhibitor [12]. While quantity and length of time for thrombin production in the clot is not known, it is known that 1-ml of whole blood produces about 260 to 360 units of thrombin. Thrombin is probably produced in the clot continuously until prothrombin is depleted [22]. After the BBB breaks down, more prothrombin may pass through the barrier into the brain where it is converted into thrombin.

Erythrocyte lysis and hemoglobin toxicity

Erythrocytes within a clot preserve their normal biconcave configuration for a few days after an ICH [14]. Thereafter, they lose their normal shape and start to lyse [5, 6, 40, 42]. Hemoglobin release from lysis of red cells in a human parenchymal clot increases during the first few days after ictus [38].

Erythrocyte lysis results from either depletion of intracellular energy reserves, formation of Membrane Attack Complex (MAC) after activation of the complement system, or both. Erythrocytes, which lack mitochondria, use glucose as fuel to produce adenosine triphosphate (ATP) in order to maintain intact membranes. Cells begin to lyse when glucose is depleted. At that same time, the complement cascade is activated and MAC forms in the brain [15, 16].

Prior to erythrocyte lysis, hemoglobin degrades into several forms including oxyhemoglobin, deoxyhemoglobin, and methemoglobin. Immediately after ICH, the hematoma contains oxyhemoglobin because the blood is well oxygenated. Several hours later, loss of oxygen converts the oxyhemoglobin to deoxyhemoglobin, which remains for several days. Deoxyhemoglobin later denatures to methemoglobin and ferrous iron oxidizes to the ferric form [3]. Hemosiderin and bilirubin are additional breakdown products of hemoglobin [21]. Eventually, the hematoma is replaced by a fibroglial scar or a cavity [8].

Edema around a parenchymal clot reaches its peak between 3 and 7 days following whole blood infusion [43]. Thrombin-induced brain edema peaks earlier, however (i.e., 1 to 2 days). A comparison of edema formation produced by either ICH or thrombin infusion suggests there may be delayed edema formation triggered by lysis of red blood cells and hemoglobin release [40]. In contrast to edema formation following thrombin injection, infusion of erythrocytes causes edema to form maximally 3 days later, suggesting a reason for the difference in responses between the ICH and thrombin. Delayed brain edema in the second or third week after onset in ICH patients is associated with significant midline shift on serial CT scans [31]. This delayed edema is probably due to erythrocyte lysis and hemoglobin-induced brain injury.

Infusion of lysed erythrocytes into the basal ganglia of rats results in marked brain edema formation by 24 hours [17]. The edema appears to be hemoglobin-mediated, since an infusion of rat methemoglobin at concentrations found in erythrocytes results in marked increases in brain water content. Other studies also indicate that hemoglobin has deleterious effects on the brain [20, 30].

The adverse effects of hemoglobin vary with its chemical form. Oxyhemoglobin is a spasminogen that has been implicated in cerebral vasospasm [24]. In ICH, however, oxyhemoglobin in the hematoma is present for only a few hours following the hemorrhage [3]. It is unlikely, therefore, that oxyhemoglobin plays a primary role in ICH-induced edema formation, a suggestion supported by the fact that ICH fails to reduce cerebral blood flow (CBF) in the rat [29]. Intracerebral infusion of methemoglobin mimics the effects of erythrocytes on edema formation and fails to produce arterial vasospasm as well [24].

Complement activation

The brain is separated from the general circulatory system by the BBB, which accounts for its relatively isolated immunological status. After ICH, plasma proteins, including complement components, pass directly into brain parenchyma contributing to brain edema. Complement activation is triggered by ICH brain injury and edema follows. N-acetylheparin, a heparin congener without anticoagulant properties, inhibits complement activation and attenuates brain edema following ICH [15, 16].

Complement-related brain injury may be due to MAC formation and the classic inflammatory response. MAC consists of C5b-9 complement forms which are assembled following complement activation [2]. Recent investigations have demonstrated that MAC not only causes cell lysis, but also modulates cellular functions such as the release of cytokines, eicosanoids, oxygen radicals, and matrix proteins [13]. In addition, MAC insertion occurs in neurons and endothelial cells, causing neuronal death and BBB leakage. MAC is assembled after ICH, and clusterin, an inhibitor of MAC formation, is upregulated in the brain parenchyma [15, 16].

Anaphylatoxin complement C5a is generated following complement activation. C5a is also a potent chemoattractant for polymorphonuclear leukocytes and contributes to inflammatory cell injury. Inflammatory cells react to nanomolar concentrations of C5a by chemotaxis, up-regulation of adhesion molecules, and release of oxygen radicals. Complement C5a can be detected around the clot. Finally, systemic complement depletion by cobra venom factor, a non-toxic protein derived from cobra venom, forms a stable C3/C5 convertase which leads to complement depletion and reduces perihematomal edema in the rat ICH model [39, 40, 42].

Mass effect

Although erythrocytes are responsible for the majority of the mass effect created by the hematoma, packed red cells alone fail to generate more brain edema than sham-treated animals. On the other hand, an intracerebral infusion of plasma causes rapid edema formation [41]. Further, non-clotting blood produces minimal perihematomal edema in both rat and porcine models of ICH. In the human, perihematomal edema is scant around non-clotting blood despite a fairly large mass [22]. These results indicate that mass causes brain injury by mechanical force and increased intracranial pressure, especially in large hematomas. However, brain edema which follows ICH in patients who survive the ictus for more than a few hours is not simply from mass effect, but also from reduced CBF produced by the clot [28, 32, 34].

Ischemia

CBF adjacent to a hematoma decreases [25]. Although intracranial pressure rises transiently around the hematoma, an ischemic penumbra is not apparent in the first 5 hours after ICH, and an energy deficit does not occur around the clot unless it is very large. Despite the presence of severe brain edema around the hematoma at all times, ATP levels stay normal and brain phosphocreatine contents gradually increase during the first 8 hours [37]; thus, an energy deficit does not occur around the hematoma [29]. While thrombin contributes to edema formation following ICH, an intracerebral injection of thrombin does not significantly reduce CBF in the vicinity of the clot [23]. If the hematoma is large, however, CBF will fall as intracranial pressure rises.

Blood-brain barrier disruption

BBB dysfunction occurs following ICH and it contributes to brain edema formation [1]. After ICH, the BBB remains intact to large molecules for the first several hours. Eight to 12 hours later, however, BBB permeability in the perihematomal region increases markedly. BBB disruption following ICH is related to thrombin formation since thrombin alone, in amounts produced by the hematoma, causes marked increase in BBB leakage [43].

Treatment

Randomized trials have been accomplished, but clot evacuation by surgery remains controversial [27]. Recently, thrombolysis with tissue plasminogen activator and aspiration under CT-guidance to reduce the hematoma volume has been effective [26]. Infusion of tissue plasminogen activator (tPA) directly into the hematoma prior to clot aspiration has been used in humans [18, 33]. In a rat model, aspiration of clot with tPA reduced both clot volume and brain injury [7]. Recently, Wagner *et al.* infused tPA into hematomas in a pig model at 3 hours after induction and aspirated the liquified clots 1 hour later. Clot removal following tPA treatment resulted in a reduction in hematoma volume compared to unevacuated controls. Clot removal also reduced brain edema volume, BBB disruption, and improved cerebral tissue pressure [36].

Intracerebral hemorrhage from trauma or tumor

Extravasation of blood into brain parenchyma, whether from a spontaneous hypertensive hemorrhage, trauma, or tumor, initiates a sequence of pathological events including mass effect, brain edema, and inflammation. Resolution of the hematoma as well as the edema and inflammation which accompany the hemorrhage, follows days to weeks later provided the patient survives the initial insult. The toxicity of extravasated blood in brain parenchyma has not been studied well in traumatic injury or in tumors that tend to hemorrhage spontaneously. This area of investigation awaits exploration.

References

1. Betz AL, Iannotti F, Hoff JT (1989) Brain edema: a classification based on blood-brain barrier integrity. Cerebrovasc Brain Metab Rev 1: 133–154
2. Bhakdi S, Tranum-Jensen J (1984) On the cause and nature of C9-related heterogeneity of terminal complement complexes generated on target erythrocytes through the action of whole serum. J Immunol 133: 1453–1463
3. Bradley Jr WG (1993) MR appearance of hemorrhage in the brain. Radiology 189: 15–26
4. Brott T, Thalinger K, Hertzberg V (1986) Hypertension as a risk factor for spontaneous intracerebral hemorrhage. Stroke 17: 1078–1083
5. Darrow VC, Alvord EC Jr, Mack LA, Hodson WA (1988) Histologic evolution of the reactions to hemorrhage in the premature human infant's brain: a combined ultrasound and autopsy study and a comparison with the reaction in adults. Am J Pathol 130: 44–58
6. Enzmann DR, Britt RH, Lyons BE, Buxton JL, Wilson DA (1981) Natural history of experimental intracerebral hemorrhage: sonography, computed tomography and neuropathology. Am J Neuroradiol 2: 517–526
7. Figueroa BE, Keep RF, Betz AL, Hoff JT (1997) Surgical management of basal ganglia hemorrhage in rat: evidence for reduce brain edema with early hematoma evacuation. Surg Form 48: 539–541
8. Garcia JH, Ho KL, Caccamo D (1994) Intracerebral hemorrhage: pathology of selected topics. In: Kase CS, Caplan LR (eds) Intracerebral Hemorrhage. Butterworths, Boston, pp 45–72
9. Gebel JM, Brott TG, Sila CA, Tomsick TA, Jauch E, Salisbury S, Khoury J, Miller R, Pancioli A, Duldner JE, Topol EJ, Broderick JP (2000) Decreased peri-hematomal edema in thrombolysis-related intracerebral hemorrhage compared with spontaneous intracerebral hemorrhage. Stroke 31: 596–600
10. Gebel JM, Sila CA, Sloan MA, Granger CB, Mahaffey KW, Weisenberger J, Green CL, White HD, Gore JM, Weaver WD, Califf RM, Topol EJ (1998) Thrombolysis-related intracranial hemorrhage: a radiographic analysis of 244 cases from the GUSTO-1 trial with clinical correlation. Global utilization of streptokinase and tissue plasminogen activator for occluded coronary arteries. Stroke 29: 563–569
11. Gomori JM, Grossman RI, Goldberg HI, Zimmerman RA, Bilaniuk LT (1985) Intracranial hematomas: imaging by high-field MR. Radiology 157: 87–93
12. Hamada R, Matsuoka H, Wagner KR, Broderick JP, de Courten-Myers GM, Xi GH, Keep RF, Hoff JT (2000) Antithrombin therapy for intracerebral hemorrhage. Stroke 31: 794–795
13. Hansch GM, Shin MI (1998) Complement attack phase. In: Rother K, Till GO, Hansch GM (eds) The Complement System. Springer, Berlin Heidelberg New York Tokyo pp 115–145
14. Hayman LA, Pagani JJ, Kirkpatrick JB, Hinck VC (1989) Pathophysiology of acute intracerebral and subarachnoid hemorrhage: applications to MR imaging. Am J Roentgenol 153: 135–139
15. Hua Y, Xi G, Keep RF, Hoff JT (2000) Complement activation in the brain after experimental intracerebral hemorrhage. J Neurosurg 92: 1016–1022
16. Hua Y, Xi G, Keep RF, Hoff JT (2000) N-acetylheparin, a nonanticoagulant, diminishes brain edema after experimental intracerebral hemorrhage. Stroke 31: 339
17. Huang FP, Xi G, Hua Y, Nemoianu A, Hoff JT (2002) Brain edema after experimental intracrebral hemmorhage: role of hemoglobin degradation products. J Neurosurg 96: 287–293
18. Kaufman HH, Schochet SS, Koss W, Herschberger J, Bernstein D (1987) Efficacy and safety of tissue plasminogen activator. Neurosurgery 20: 403–407
19. Kendall BE, Radue EW (1978) Computed tomography in spontaneous intracerebral hematomas. Br J Radiol 51: 563–573
20. Koening ML, Meyerhoff JL (1997) Neurotoxicity resulting from prolonged exposure to hemoglobin. Soc Neurosci Abstr 23: 1935
21. Koeppen AH, Dickson AC, McEvoy JA (1995) The cellular reactions to experimental intracerebral hemorrhage. J Neurol Sci 134: 102–112
22. Lee KR, Colon GP, Betz AL, Keep RF, Kim S, Hoff JT (1996) Edema from intracerebral hemorrhage: the role of thrombin. J Neurosurg 84: 91–96
23. Lee KR, Kawai N, Kim S, Sagher O, Hoff JT (1997) Mechanisms of edema formation after intracerebral hemorrhage: effects of thrombin on cerebral blood flow, blood-brain barrier permeability, and cell survival in a rat model. J Neurosurg 86: 272–278
24. MacDonald RL, Weir BK (1991) A review of hemoglobin and the pathogenesis of cerebral vasospasm. Stroke 22: 971–982
25. Mendelow AD, Bullock R, Teasdale GM, Graham DI, McCulloch J (1984) Intracranial hemorrhage induced at arterial pressure in the rat. Part 2: short-term changes in local cerebral blood flow measured by autoradiography. Neurol Res 6: 189–193
26. Montes JM, Wong JH, Fayad PB, Awad IA (2000) Sterotactic computed tomographic-guided aspiration and thrombolysis of intracerebral hematoma: protocol and preliminary experience. Stroke 31: 834–840
27. Morgenstern LB, Frankowski RF, Shedden P, Pasterur W, Grotta JC (1998) Surgical treatment for intracerebral hemorrhage (STICH): a single-center, randomized clinical trial. Neurology 51: 1359–1363
28. Nath FP, Kelly PT, Jenkins A, Mendelow AD, Graham DI, Teasdale GM (1987) Effects of experimental intracerebral hemorrhage on blood flow, capillary permeability, and histochemistry. J Neurosurg 66: 555–562
29. Qureshi AL, Wilson DA, Traystman RJ (1999) No evidence for an ischemic penumbra in massive experimental intracerebral hemorrhage. Neurology 52: 266–272
30. Regan RF, Panter SS (1993) Neurotoxicity of hemoglobin in cortical cell culture. Neurosci Lett 153: 219–222

31. Ropper AH, King RB (1984) Intracranial pressure monitoring in comatose patients with cerebral hemmorhage. Arch Neurol 41: 725–728
32. Ropper AH, Zervas NT (1982) Cerebral blood flow after experimental basal ganglia hemorrhage. Ann Neurol 11: 266–271
33. Schaller C, Rohde V, Meyer B, Hassler W (1995) Stereotactic puncture and lysis of spontaneous intracerebral hemorrhage using recombinant tissue-plasminogen activator. Neurosurgery 36: 328–333
34. Suzuki J, Ebina T (1980) Sequential changes in tissue surrounding ICH. In: Pia HW, Longmaid C, Zierski J (eds) Spontaneous intracerebral hematomas. Springer, Berlin Heidelberg New York Tokyo pp 121–128
35. Wagner KR, Xi G, Hua Y, Kleinholz M, de Courten-Myers GM, Myers RE, Broderick JP, Brott TG (1996) Lobar intracerebral hemorrhage model in pigs: rapid edema development in peri-hematomal white matter. Stroke 27: 490–497
36. Wagner KR, Xi G, Hua Y, Zuccarello M, de Courten-Myers GM, Broderick JP, Brott TG (1999) Ultra-early clot aspiration after lysis with tissue plasminogen activator in a porcine model of intracerebral hemorrhage: edema reduction and blood-brain-barrier protection. J Neurosurg 90: 491–498
37. Wagner KR, Xi G, Kleinholz M, de Courten-Myers GM, Myers RE (1998) Early metabolic alterations in edematous perihematomal brain regions following experimental intracerebral hemorrhage. J Neurosurg 88: 1058–1065
38. Wahlgren NG, Lindquist C (1987) Haem derivatives in the cerebrospinal fluid after intracranial hemorrhage. Eur Neurol 26: 216–221
39. Xi G, Hua Y, Keep RF, Younger JF, Hoff JT (2002) Brain edema after intracerebral hemorrhage: the effects of systemic complement depletion. Acta Neurochir (Supp) 81: 253–256
40. Xi G, Keep RF, Hoff JT (1998) Erythrocytes and delayed brain edema formation following intracerebral hemorrhage in rats. J Neurosurg 89: 991–996
41. Xi G, Wagner KR, Hua Y, de Courten-Myers GM, Myers RE (1996) Plasma but not packed red blood cells infused in pig cerebral white matter induces rapidly developing, marked edema. Soc Neurosci Abstr 22: 1424
42. Xi G, Wagner KR, Keep RF, Hua Y, de Courten-Myers GM, Broderick JP, Brott TG, Hoff JT (1998) The role of blood clot formation on early edema development following experimental intracerebral hemorrhage. Stroke 29: 2580–2586
43. Yang GY, Betz AL, Chenevert TL, Brunberg JA, Hoff JT (1994) Experimental intracerebral hemorrhage: relationship between brain edema, blood flow, and blood-brain barrier permeability in rats. J Neurosurg 81: 93–102

Correspondence: Julian T. Hoff, M.D., Department of Neurosurgery, University of Michigan Health System, 1500 E. Medical Center Drive, Room 2128TC, Ann Arbor, Michigan 48109-0338, USA. e-mail: jhoff@umich.edu

Acta Neurochir (2003) [Suppl] 86: 17–20

Hydrostatic brain edema: basic mechanisms and clinical aspect

K. Shima

Department of Neurosurgery, National Defense Medical College, Tokorozawa, Japan

Summary

The term hydrostatic brain edema results from unfavorable hydrtostatic pressure gradiernts between blood vessels and brain tissue. Arterial hypertension combined with decompressive craniectomy produces extensive brain edema in the arterial boundary zone of the decompressive area. The increased hydrostatic pressure gradient enhances tissue damage and causes the biphasic opening of the blood-brain barrier (BBB). The hydrostatic pressure alone is capable of causing the initial BBB opening and this induced edema results in derangement of cerebral microcirculation and metabolism. With morphological opening and metabolic damage of the vascular wall, the second BBB opening is elicited by the amplified hydrostatic pressure gradient, which is similar to that in vasogenic edema. In the clinical aspect, the beneficial effect of decompressive craniectomy in the treatment of uncontrollable ICP and brain edema remains controversial. External decompression may have adverse effects on severe brain edema and swelling. In this regards, control of driving force for the formation of brain edema could be the treatment of choice as an initial step. In addition, recent reports provided by MR imaging indicated new information on the pathophysiological features of the patients with acute hypertension. In the patients with hypertension due to reversible posterior leucoencephalopathy (RPL) syndrome, MR images show reversible signal abnormalities in the bilateral occipital lobes, suggesting hydrostatic brain edema.

Keywords: Hydrostatic brain edema; hypertension; decompressive craniectomy; blood-brain barier; reversible posterior leucoencephalopathy.

Introduction

The term hydrostatic brain edema was introduced by Ishii *et al.* [1]. It results from the unfavorable hydrostatic pressure gradients between blood vessels and the surrounding brain tissue. The present study focuses on hydrostatic brain edema caused by the two major hydrostatic factors such as arterial hypertension and decompressive craniectomy on the base of our experimental and clinical studies.

Experimental aspect

Previous studies from our laboratory have shown that arterial hypertension combined with decompressive craniectomy produces extensive brain edema in the arterial boundary zone of the decompressive area [1–3].

Experiment 1 [1, 2]

Hypertension in anesthetized cats was induced by inflating a balloon catheter placed immediately distal to the left subclavian artery. A sustained hypertension ranging 200 to 250 mmHg for 30 minutes was easily obtained. Craniectomy was made by removing the right frontoparietal bone (2.5 × 2.0 cm). In order to determine the permeability changes of macromolecules through BBB, Evans blue was injected as a tracer. Ten minutes following the dye injection, each animal was subjected to hypertension for 30 minutes, or craniectomy, or craniectomy followed by the hypertension for 30 minutes. Animals were allowed to survive for 60 minutes. Induced hypertension in the normal cat produced multifocal cortical lesions stained with Evans blue, predominantly at the posterior part of the suprasylvian gyrus. Craniectomy also resulted in similar multifocal cortical edema in both hemispheres, though more marked on the side of the craniectomy. When animals subjected to both hypertension and craniectomy, extensive brain edema with a shift of midline structures developed on the side of craniectomy 48 hours after the insult. In hypertension or craniectomy alone, there was a significant increase in water content 1 hour after the procedure, but not 48 hours later. In contrast, when hypertension combined with craniectomy, the increase of tissue water was highly significant until 48 hours later.

Experiment 2 [3]

Hydrostatic edema was induced by bolus injection of autologous heparinized blood through the left common carotid artery in Sprague-Dawley rats. The area of craniectomy was made by removing the left parietal bone. The ICP in craniectomy group was significantly lower than that of non-craniectomy group. The extravasation of Evans blue was observed in all animals of both caraniectomy and non-craniectomy groups and recognized as patchy spots predominantly at the watershed areas. As compared with non-craniectomy group, the extravasated Evans blue spots were usually more dense and widely distributed in craniectomy group (Fig. 1 A and B). Forty-eight hours after the insult, diffuse and homogeneous staining was found. In contrast to non-craniectomy group, in the craniectomized animals in which Evans blue was injected at 24 hours after the hypertensive insult and sacrificed at 48 hours, extravasated Evans blue spots were confined to the area posterior to the craniectomy portion (Fig. 1 C). Brains from craniectomy group showed a pronounced increase in water content which persisted 48 hours later.

The regional blood flow and glucose metabolism measured by quantitative autoradiographic methods using [^{14}C] iodoantipyrine and [^{14}C] deoxyglucose. In craniectomy group, misery perfusion observed at 24 hours after the insult and the cerebrovascular dynamics and metabolism were significantly decreased after 48 hours. Under this condition, reopening of BBB was recognized at the occipital lobes of the arterial boundary zone.

These experimental results have demonstrated that the hydrostatic pressure alone is capable of causing the initial BBB opening and this edema results in derangement of cerebral microcirculation and successive metabolic damage. The hydrostatic edema is prolonged or aggravated and accompanied by cytotoxic factors. With morphological and metabolic damage of the vascular wall, the second BBB opening is elicited by the amplified hydrostatic pressure gradient, which is similar to that in vasogenic edema.

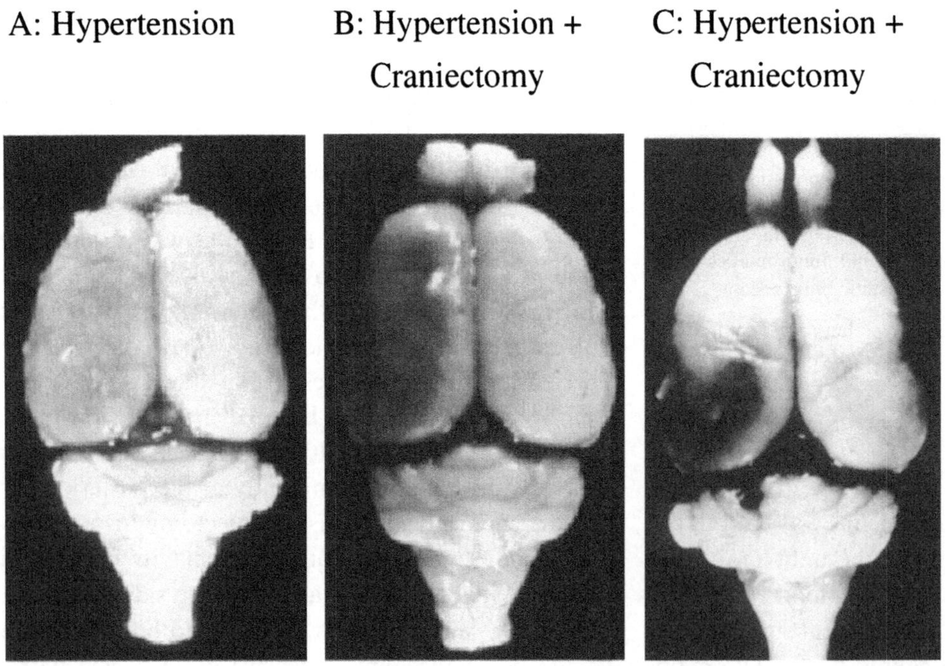

Fig. 1. Extravasation of Evans blue at 48 hours following induced hypertension. (A) Hypertension alone. (B and C) Combination of hypertension and decompressive craniectomy. Evans blue injection was given together with the hypertensive insult (A and B) or 24 hours after the hypertensive insult (C)

Clinical aspect

Recent reports provided by MR imaging indicated new information on the pathophysiological features of the patients with acute hypertension. In 1996, Hinchey *et al.* [4] defined RPL syndrome as a clinicoradiological category characterized by headache, confusion, visual disturbance, seizures in the setting of acute rise of blood pressure, eclampsia, or following immunosuppressive treatment. Lesions associated with this syndrome are typically hyperintense on T2-weighted MR images. Sakai *et al.* [5] recently reported findings from a prospective clinical study in which T2-weighted MR imaging was used for the evaluation of RPL syndrome. Between January 2001 and April 2002, 82 patients with pregnant induced acute hypertension were admitted to our hospital and treated. Among 20 patients entering this study, 5 patients, 25%, showed RPL syndrome associated with hyperintense signals on T2-weighted MR images predominantly in the bilateral occipital lobes (Fig. 2). Follow-up MR images returned to normal within 12 to 92 days. It has been also reported recently on hypertensive encephalopathy and RPL that perfusion image shows decreases of blood flow and apparent diffusion coefficient (ADC) map shows predominantly hyperintense areas of free motion of water, as seen in the vasogenic edema [6]. However, there were any signs of cytotoxic edema or subsequent infarction. These findings are reversible changes with no residual morphological changes. MR findings suggest that the spectrum of RPL ranges from an initially reversible phase of vasogenic edema.

From 1995 to 2002, 24 patients with uncontrolled intracranial hypertension were admitted to our hospital and underwent decompressive craniectomy. Of 24 patients, 25% had died (all patients except for one were aged more than 50 years) and 45.8% survived with an unfavorable outcome (GOS 2, 3). Beneficial effect of decompressive craniectomy remains controversial. External decompression may have adverse effects on severe brain edema and swelling. It is still unknown whether there is any chemical substance, which may affect to the development of brain edema after craniectomy. We investigated the relationship brain edema and natriuretic peptides in 61 patients with subarachnoid hemorrhage [7]. The serum ANP (atrial natriuretic peptide) and BNP (brain natriuretic peptide) levels in patients with brain edema were significantly higher than those in non-edema patients between the days 4 and 14. After craniectomy, the high levels of serum BNP progressed along with brain edema, but disappeared soon after cranioplasty. Brain edema might simulate an excessive secretion of natriuretic peptides.

A: at admission

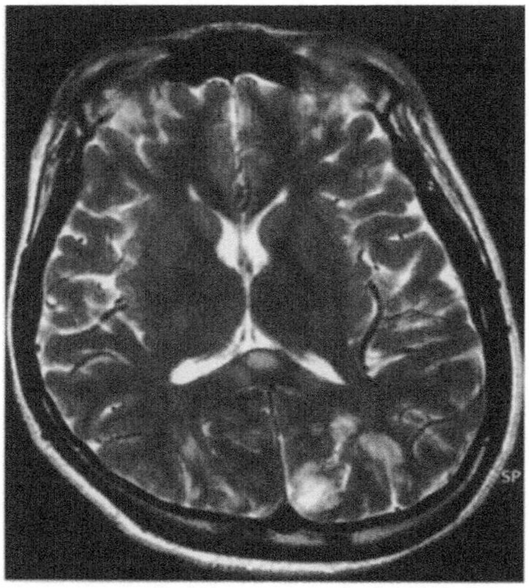

B: Day 15

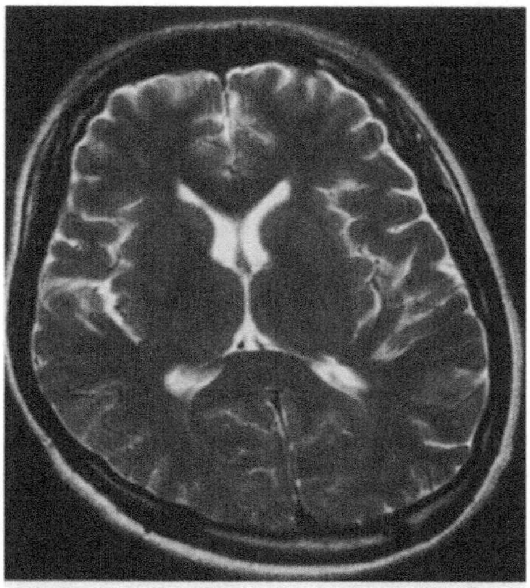

Fig. 2. T2-weighted MR images in the patients with reversible posterior leucoencephalopathy syndrome at admission (A) and Day 15 (B). A 31-year-old and 33 weeks' pregnant woman was admitted with a history of headache and visual disturbances. At admission, the patient had hypertension (170/120 mmHg), proteinuria and pedal edema

Discussion

With craniectomy and hypertension, tissue pressure decreases, resulting in an increase in perfusion pressure, even though the blood pressure remains unchanged. Then, the transmural hydrostatic pressure gradient may increase. Hydrostatic pressure gradient of large magnitude will accelerate extravasation of water, ions, macromolecules and other chemical substances, at least in its initial stage, from structurally normal vessel walls into metabolically normal brain tissue because of hydrostatic or hydrodynamic forces. Hydrostatic brain edema is a variant of vasogenic edema, but the developmental pattern may be sufficiently specific to warrant the separate classification of it.

Arterial hypertension produces focal edema, hydrostatic brain edema, even in normal brain, suggesting an important role for hydrostatic pressure in the development and exacerbation of brain edema. In the treatment of uncontrollable intracranial hypertension, control of the transmural hydrostatic pressure gradient, which is driving force for the formation of edema, could be the treatment of choice as an initial step.

References

1. Ishii S, Koike J, Hatashita S (1983) Brain edema in hypertensive intracerebral hemorrhage. In: Mizukami M *et al* (eds) Hypertensive intracerebral hemmorrhage. Raven Press, New York
2. Hatashita S, Koike J, Sonokawa T, Ishii S (1985) Cerebral edema associated with craniectomy and arterial hypertension. Stroke 16: 661–668
3. Umezawa H, Shima K, Chigasaki H, Ishii S (1993) Local cerebral blood flow and glucose metabolism in hydrostatic brain edema. Neurol Med Chir (Tokyo) 33: 139–145
4. Hinchey J, Chaves C, Appignani B, Breen J, Pao L, Wang A, Pessin MS, Lamy C, Mas J, Caplan LR (1996) Reversible posterior leucoencephalopathy syndrome. N Engl J Med 334: 494–500
5. Sakata I, Tokumaru AM, Matsuda H, Kaji T, Fujii M, Kusano S (2002) MRI findings of brain changes in severe preeclampsia. Neuroradiology 44 [Suppl] (in press)
6. Sundgren PC, Edvardsson B, Holtas S (2002) Serial investigation of perfusion disturbances and vasogenic oedema in hypertensive encephalopathy by diffusion and perfusion weighted imaging. Neuroradiology 44: 299–304
7. Nawashiro H, Fukui S, Otani N, Katoh H, Ishihara S, Tsuzuki N, Ohnuki A, Miyazawa T, Shima K (2002) Focal brain edema and natriuretic peptides in patients with subarachnoid hemorrhage. In: Kuroiwa T *et al* (eds) Brain edema XII. Springer, Berlin Heidelberg New York Tokyo

Correspondence: Katsuji Shima, Department of Neurosurgery, National Defense Medical College, 3-2 Namiki, Tokorozawa, Saitama 359-8513, Japan. e-mail: shima@me.ndmc.ac.jp

Acta Neurochir (2003) [Suppl] 86: 21–27

Non-invasive imaging methods for the characterization of the pathophysiology of brain ischemia

K.-A. Hossmann

Max-Planck-Institute for Neurological Research, Department of Experimental Neurology, Cologne, Germany

Summary

Non-invasive imaging methods are increasingly used to study the evolution and therapy of brain diseases under both clinical and experimental conditions. In the animal experiment, these methods can be supplemented by invasive tissue assays to allow precise characterization of the underlying pathophysiology. Based on such an approach, this review evaluates the importance of in vivo nuclear magnetic resonance (NMR) and positron emission tomography (PET) for the understanding of the pathophysiology of brain ischemia.

Keywords: Global ischemia; focal ischemia; nuclear magnetic resonance; positron emission tomography; injury prediction.

Introduction

With few exceptions, brain diseases are focal. This is most evident for stroke, brain trauma or tumors, but even more diffuse alterations such as neurodegenerative disorders or global hypoxic or ischemic disturbances also result in topical injury patterns which, depending on the specific vulnerability, affect different structures in different ways.

Heterogeneity causes considerable problems for the analysis of a detailed brain pathophysiology, and can only be overcome by using advanced multiparametric imaging techniques. Such methods combine a multitude of pictorial hemodynamic, biophysical, metabolic and molecularbiological assays that can be applied to the same experiment. Most of these assays require the preparation of tissue sections and are therefore restricted to animal experiments and to one timepoint in the evolution of the pathological process. However, the combination with non-invasive methods such as in vivo nuclear magnetic resonance (NMR) or positron emission tomography (PET) provides the opportunity to follow – though with a limited number of accessible parameters – the dynamics of the process and, equally important, to compare the experimental data with investigations in patients.

Obviously, these non-invasive imaging methods have to be carefully validated under clinically relevant experimental conditions to avoid misinterpretations. The following overview will focus on the application of NMR and PET for the study of the pathophysiology of brain ischemia.

Pathophysiology of global brain ischemia

Complete interruption of cerebral blood flow in the normothermic animal, as induced by cardiac arrest, results in the rapid breakdown of all metabolic and electrophysiological functions of the brain. Within a few seconds, tissue oxygen stores are used up and anaerobic glycolysis is stimulated for maintenance of energy metabolism. However, due to the limited reserves of glucose and glycogen and the greatly reduced ATP yield of anaerobic glycolysis, this pathway also rapidly ceases, leading to breakdown of the brain tissue energy state and of all endergonic metabolic processes within 8 to 10 minutes after onset of circulatory arrest [4].

The metabolic disturbances are reflected by the progressive failure of electrophysiological functions. Spontaneous EEG activity is shut down within less than 10 seconds, synaptical activity and evoked potentials disappear within 5–6 minutes, followed shortly afterwards by the anoxic depolarization of cell membranes and the equilibration of extra/intracellular ion gradients [17]. If blood flow – and hence oxygen and glucose delivery – is homogeneously restored, aerobic energy metabolism recovers within minutes after normothermic circulatory arrest of up to one hour, leading to the restoration of ion exchange pumps and the gradual restitution of electrophysiological functions. Under conditions of slow or inhomogeneous reperfusion (no-reflow phenomenon), anaerobic rather than oxidative metabolism is stimulated, resulting in the progression of lactacidosis, tissue hyperosmolality, brain swelling and eventually secondary circulatory arrest. With increasing ischemia time, an increasingly higher blood reperfusion pressure has to be built up to prevent no-reflow [6]. This is feasible in specially designed experimental models of global brain ischemia, but not after cardiac failure. No-reflow is therefore the most important limitation of brain resuscitation after cardiac arrest [7].

Another pathophysiologically relevant hemodynamic disturbance is post-ischemic hypoperfusion. This disturbance is due to the uncoupling between brain metabolism and flow regulation, which under conditions of increasing metabolic activity may cause a progressive mismatch between oxygen supply and metabolism [18].

Disturbances in signal transduction pathways, and here in particular molecular injury cascades that result in the progressing deterioration of mitochondrial and endoplasmic reticulum (ER) function, are greatly dependent on the individual "pathoclitic" sensitivity of brain cells to ischemia [27, 32]. In selectively vulnerable brain regions such as CA_1 sector of hippocampus, a depolarization time of 16 min suffices to trigger the activation of lethal injury pathways (B. Momeni and G. Mies, in preparation). In contrast, other parts of the brain are able to tolerate depolarization times of much longer duration which, however, is of little pathophysiological relevance as long as no-reflow cannot be prevented.

Non-invasive imaging of global brain ischemia

The most important pathophysiological variables of global brain ischemia, i.e. blood flow and metabolic activity, can be addressed by various non-invasive imaging methods. PET permits quantitative measurement of cerebral blood flow, glucose and oxygen utilization and, therefore, provides solid information about the viability of the tissue [34]. An important variable that can be addressed by PET, is the glucose/oxygen uptake ratio. Independent measurements by arterio-venous sampling techniques revealed that under critical conditions of post-ischemic hypoperfusion this ratio sharply rises, reflecting secondary stimulation of anaerobic glycolysis which, if not immediately remedied, may lead to secondary brain energy failure [30].

More direct information on brain energy state is obtained by NMR. The combination of ^{1}H and ^{31}P spectroscopy allows semiquantitative measurements of ATP, PCr, lactate and intracellular pH, although with limited regional resolution [4]. An important aspect of MR spectroscopy is the reliable assessment of suppressed brain function because instrumental failure may prevent acquisition of MR spectra, but it cannot cause artificial shifts of spectral lines. Thus, the shift of the phosphorus resonances from ATP and CrP to the inorganic phosphate peak is an unequivocal sign of energy failure and, hence, of impeding brain death.

An indirect, but highly valuable MR parameter of ischemic injury is the apparent diffusion coefficient (ADC) of water [25]. ADC closely correlates with the extracellular fluid volume and is therefore a sensitive marker of the disturbances of water and ion homeostasis. During global ischemia both extracellular fluid volume and ADC sharply decline in response to anoxic depolarization, and both return to normal as soon as ion exchange pumps are reactivated [21]. As these pumps are energy-dependent, ADC mapping faithfully reflects the regional restoration of the energy state of the brain (Fig. 1). Similarly, absence of metabolic recovery or secondary energy failure are mirrored by the decline of ADC. This rule is only violated under conditions of severe vasogenic brain edema or after necrotic dissociation of cell membranes, where interstitial fluid volume and ADC rise independently of the viability status of the brain tissue.

For the evaluation of cell viability in selectively vulnerable brain regions, imaging of the PET-detectible benzodiazepine receptor ligand [^{11}C] flumazenil could be of interest. This ligand binds only to intact cells [36] and can therefore be used to follow the evolution of delayed cell death. An alternative method is the measurement of N-acetyl-acetate (NAA) by localized proton MR spectroscopy. NAA is a specific marker of viable neurons that can be recorded in combination with other pathophysiologically relevant variables in order to establish possible injury mechanisms.

Finally, neuroimaging is suitable for monitoring the suppression and return of neuronal functions. In fact, functional MR imaging performed in rats submitted to 15 min cardiac arrest revealed a close correlation of the amplitude of evoked potentials and of the neuro-

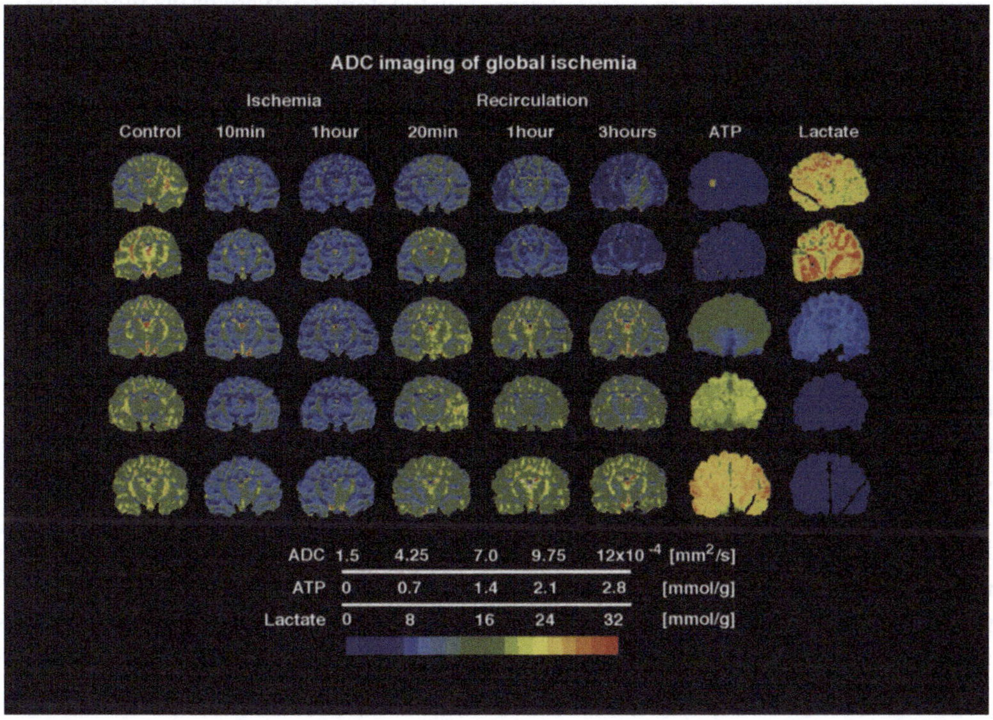

Fig. 1. ADC and biochemical maps of cat brain submitted to 60 min circulatory arrest. Comparison of 5 animals, 2 of which (upper rows) failed to recover after ischemia. Note close correlation of ADC with the biochemical status of the brain (modified from 21)

logical deficit score with the signal intensities of blood-oxygen level dependent (BOLD) T_2 *- or perfusion-weighted imaging (PWI) [35]. Measurements of blood flow and glucose utilization by PET could provide the same information, but the better temporal and spatial resolution favours MR imaging for this purpose.

The recent development of molecular imaging methods opens new perspectives for the analysis of the molecular pathophysiology of ischemic cell injury. Examples are the PET-detectible tracers [^{11}C] PK11195 for the visualization of activated microglia [10], [^{18}F] fluoro-DOPA and [^{11}C] nemonapride for the detection of dopaminergic neurons [22, 37], [1^{-11}C] octanoate for imaging phospholipid metabolism [26] or [^{8}F] FDDNP for tracing Alzheimer-type neurodegeneration [1]. $\alpha_v\beta_3$-targeted nanoparticles can be detected by MRI and have been used for the study of angiogenesis [2], and both PET and MR methods are available for imaging cells labelled in vitro for tracing cell migration [15] or gene transfer [9, 28]. These methods have not yet been applied to global brain ischemia, but could be readily adapted for addressing specific issues of molecular injury.

Pathophysiology of focal brain ischemia

Occlusion of a major brain artery causes a gradient of flow values that decline from the peripheral to the more central parts of the vascular territory. According to the threshold concept of brain ischemia, the flow decline produces a gradient of increasingly severe functional and metabolic disturbances [19]. At flow levels of about 50% of control, cerebral protein synthesis ceases and the amplitudes of spontaneous EEG and evoked cortical potentials begin to decline. At about 30% of control stimulation of anaerobic glycolysis results in lactacidosis, and at flow levels below 15–20% of control, energy metabolism fails and cell membranes depolarize. The range between the beginning functional impairment and the threshold of energy failure is the penumbra, which according to the original definition by Symon and his colleagues is a region of impaired function, but preserved structural integrity [3]. Anatomically, the penumbra occupies the peripheral part of the occluded vascular territory, and it borders on one side to the normally perfused intact tissue and on the other to the infarct core, in which blood flow has declined below the threshold of energy failure.

With ongoing ischemia time the core region of impaired energy metabolism expands, leading to the gradual infarction of the surrounding penumbra [16]. The main reason for this aggravation is the mismatch between oxygen supply and oxygen requirements of the penumbral tissue, notably in response to peri-infarct depolarizations which produce a sharp increase of metabolic activity [20]. Contributing injury factors are various molecular injury cascades which result, among others, in the activation of apoptotic pathway and the progressing dysfunction of mitochondria and endoplasmic reticulum [27, 32].

Restoration of blood flow after focal brain ischemia salvages the penumbra, but in the core of the ischemic territory, complex pathophysiologies are initiated which differ, depending on the duration of ischemia and the quality of metabolic recovery. In brain areas with a mismatch between restoration of energy metabolism and ongoing disturbances of cerebral protein synthesis, secondary injury may evolve. The "maturation" interval of this delayed damage increases inversely with the duration of vascular occlusion [23]. After 30 min ischemia, this interval may be as long as 3 weeks [5], but after 60 min ischemia, secondary energy failure occurs between 3 and 6 hours [11]. If the duration of ischemia is further extended, primary recovery of energy metabolism begins to fail until, after 3–6 hours ischemia, the outcome of transient and permanent ischemia is the same [24].

Non-invasive imaging of focal brain ischemia

A fundamental requirement of all non-invasive stroke imaging methods is the differentiation between infarct core and penumbra. Using PET, the infarct core corresponds to the region in which oxygen consumption ceases, whereas in the penumbra some residual metabolism persists [12]. This differentiation can be refined by measuring the extraction of blood oxygen. In the penumbra, oxygen extraction is increased because in viable tissue the decline of arterial oxygen supply is compensated by a higher oxygen extraction rate. In contrast, oxygen extraction ceases as soon as tissue is irreversibly damaged, even if some residual blood flow persists [12]. Measurement of glucose utilization characterizes the penumbra as a region of increased (anaerobic) glycolysis, whereas in the core metabolism fails. However, glucose utilization also varies with functional activity or the infiltration of inflammatory cells, both of which may lead to misinterpretations.

An indirect approach to identify the penumbra is the quantitative measurement of blood flow. According to the threshold concept of brain ischemia, penumbra corresponds to a flow range between approximately 15–20% and 50% of control [19]. However, these values apply only to acute vascular occlusion because with increasing duration of ischemia, the threshold of energy failure rises. After prolonged ischemia, blood flow may also improve without recovery of brain tissue. Flow measurements for the differentiation between core and penumbra should therefore be interpreted with caution.

An unequivocal biochemical criterion for the differentiation between core and penumbra is the tissue energy state which can be measured by localized ^{31}P MR spectroscopy. By definition, energy-rich phosphates are preserved in the penumbra, whereas in the core the brain energy state collapses. The penumbra can be demarcated from the normal non-ischemic tissue by the presence of lactacidosis, as detected by ^{31}P and ^{1}H spectroscopy [8]. However, as the spatial resolution of MR spectroscopy is limited, such measurements provide only a rough estimate of the actual situation.

A regionally far better resolved MR approach is diffusion-weighted imaging, which in combination with perfusion-weighted imaging is widely used to identify penumbral tissue in patients. This method is based on the assumption that DWI signal intensity increases only in the core of the infarct, whereas blood flow declines throughout the territory of the occluded vessel. The penumbra can thus be identified as a mismatch between reduced blood flow and undisturbed DWI signal intensity [38]. However, quantitative measurements of the apparent diffusion coefficient in experimental brain infarcts demonstrate that the assumption of an undisturbed diffusion in the penumbra is incorrect. ADC begins to decline at flow values well above the threshold of energy failure, probably because of the lactate-induced rise of intracellular osmolality [29]. After anoxic depolarization, ADC further declines due to the associated fluid shift between extra- and intracellular compartment, but this decline is gradual rather than stepwise, preventing a sharp demarcation between core and penumbra on the one hand, and between penumbra and the intact peri-infarct tissue, on the other. Calculated ADC threshold values obtained from correlation studies with invasive

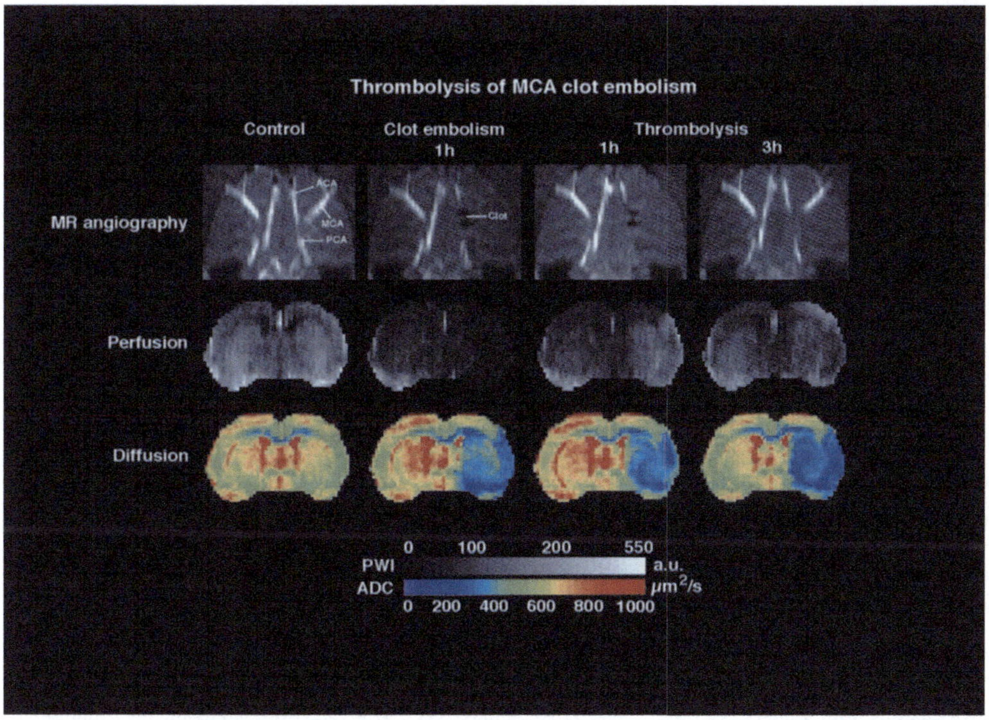

Fig. 2. Thrombolytic recanalization of rat middle cerebral artery following embolization with homologous blood clots. Comparison of MR angiograms (horizontal maximum intensity projection maps) with perfusion and diffusion changes. Note successful recanalization after clot lysis and restitution of blood flow, but failure of ADC recovery. The dissociation is due to molecular injury mechanisms which at the present cannot yet be addressed by MR imaging methods (modified from 14)

ATP and pH measurements are 76% of control for the transition from core to penumbra, and 86% of control for the transition from penumbra to the normal tissue [31]. However, with the development of vasogenic edema, ADC may again rise irrespective of the metabolic state of the tissue (pseudonormalisation) [39]. ADC thresholding for demarcation between core and penumbra is therefore only possible during the acute phase of focal ischemia.

PET and MR methods are particularly suited for monitoring the reversal of ischemic injury after transient focal ischemia, notably during thrombolytic reperfusion of clot embolism [14]. The combination of MR angiography with diffusion and perfusion imaging provides a powerful tool to follow the relationship between vascular recanalization, tissue reperfusion and functional recovery (Fig. 2). Similarly, PET imaging of blood flow and metabolism is successfully used to document the outcome of thrombolytic treatment in stroke patients [13].

An important safety concern of thrombolytic therapy is the exclusion of irreversible injury. As the application of thrombolytic drugs bears the risk of intracerebral bleedings, treatment should be restricted to patients in which brain tissue is still viable. Non-invasive indicators of undamaged tissue are normal metabolic rate of oxygen, normal ATP, pH and lactate, or normal ADC. Signatures of necrotic tissue are the combination of low ADC with increased T_1 or T_2 [40], or the loss of receptor interactions, such as the binding of flumazenil to benzodiazepine receptors [12]. A more difficult task is the prediction of recovery of still viable tissue which has suffered ischemic breakdown of energy metabolism and terminal depolarization. In fact, complete loss of energy-rich phospates, metabolic suppression or the decline of ADC to less than 60% of control are potentially reversible [21]. However, the chances of recovery decline if metabolic failure is aggravated by post-ischemic complications, such as edema formation or reperfusion deficits. As the incidence of such complications varies under different pathophysiological conditions, predictions can only be made in standardized experimental models [33].

In conclusion, advanced non-invasive imaging of basic pathophysiological and pathobiochemical variables of brain ischemia provides information on the dynamics of injury evolution and recovery with high temporal and spatial resolution. The application of such methods to both animal experiments and patients allows the design of disease-relevant animal models and provides the opportunity of supplementing missing data by additional analytical studies, which can only be performed in the animal experiment. It is this combination which promises further progress in the understanding of the pathophysiology of brain ischemia.

References

1. Agdeppa ED, Kepe V, Liu J, Flores-Torres S, Satyamurthy N, Petric A, Cole GM, Small GW, Huang SC, Barrio JR (2001) Binding characteristics of radiofluorinated 6-dialkylamino-2-naphthylethylidene derivatives as positron emission tomography imaging probes for beta-amyloid plaques in Alzheimer's disease. J Neurosci 21: RC189
2. Anderson SA, Rader RK, Westlin WF, Null C, Jackson D, Lanza CM, Wickline SA, Kotyk JJ (2000) Magnetic resonance contrast enhancement of neovasculature with alpha(v)beta(3)-targeted nanoparticles. Mag Res Med 44: 433–439
3. Astrup J, Symon L, Siesjö BK (1981) Thresholds in cerebral ischemia – the ischemic penumbra. Stroke 12: 723–725
4. Behar KL, Rothman DL, Hossmann K-A (1989) NMR spectroscopic investigation of the recovery of energy and acid-base homeostasis in the cat brain after prolonged ischemia. J Cereb Blood Flow Metab 9: 655–665
5. Du C, Hu R, Csernansky CA, Hsu CY, Choi DW (1996) Very delayed infarction after mild focal cerebral ischemia: A role for apoptosis? J Cereb Blood Flow Metab 16: 195–201
6. Fischer EG, Ames A (1972) Studies on mechanisms of impairment of cerebral circulation following ischemia: effect of hemodilution and perfusion pressure. Stroke 3: 538–542
7. Fischer M, Hossmann K-A (1995) No-reflow after cardiac arrest. Intensive Care Medicine 21: 132–141
8. Franke C, Brinker G, Pillekamp F, Hoehn M (2000) Probability of metabolic tissue recovery after thrombolytic treatment of experimental stroke: a magnetic resonance spectroscopic imaging study in rat brain. J Cereb Blood Flow Metab 20: 583–591
9. Gambhir SS, Herschman HR, Cherry SR, Barrio JR, Satyamurthy N, Toyokuni T, Phelps ME, Larson SM, Balatoni J, Finn R, Sadelain M, Tjuvajev J, Blasberg R (2000) Imaging transgene expression with radionuclide imaging technologies. Neoplasia 2: 118–138
10. Gerhard A, Neumaier B, Elitok E, Glatting G, Ries V, Tomczak R, Ludolph AC, Reske SN (2000) In vivo imaging of activated microglia using [C-11] PK11195 and positron emission tomography in patients after ischemic stroke. Neuroreport 11: 2957–2960
11. Hata R, Maeda K, Hermann D, Mies G, Hossmann K-A (2000) Evolution of brain infarction after transient focal cerebral ischemia in mice. J Cereb Blood Flow Metab 20: 937–946
12. Heiss WD (2000) Ischemic penumbra: evidence from functional imaging in man. J Cereb Blood Flow Metab 20: 1276–1293
13. Heiss WD, Grond M, Thiel A, von Stockhausen HM, Rudolf J, Ghaemi M, Lottgen J, Stenzel C, Pawlik G (1998) Tissue at risk of infarction rescued by early reperfusion: a positron emission tomography study in systemic recombinant tissue plasminogen activator thrombolysis of acute stroke. J Cereb Blood Flow Metab 18: 1298–1307
14. Hilger T, Niessen F, Diedenhofen M, Hossmann K-A, Hoehn M (2002) Magnetic resonance angiography of thromboembolic stroke in rats: indicator of recanalization probability and tissue survival after recombinant tissue plasminogen activator treatment. J Cereb Blood Flow Metab 22: 652–662
15. Hoehn M, Küstermann E, Blunk J, Wiedermann D, Trapp T, Föcking M, Arnold H, Hescheler J, Fleischmann BK, Bührle C (in press) Monitoring of implanted stem cell migration in vivo: a highly resolved in vivo magnetic resonance imaging investigation of experimental stroke in rat. Proceedings of the National Academy of Sciences USA
16. Hoehn-Berlage M, Norris DG, Kohno K, Mies G, Leibfritz D, Hossmann K-A (1995) Evolution of regional changes in apparent diffusion coefficient during focal ischemia of rat brain: the relationship of quantitative diffusion NMR imaging to reduction in cerebral blood flow and metabolic disturbances. J Cereb Blood Flow Metab 15: 1002–1011
17. Hossmann K-A (1971) Cortical steady potential, impedance and excitability changes during and after total ischemia of cat brain. Exp Neurol 32: 163–175
18. Hossmann K-A (1997) Reperfusion of the brain after global ischemia – hemodynamic disturbances. Shock 8: 95–101
19. Hossmann K-A (1994) Viability thresholds and the penumbra of focal ischemia. Ann Neurol 36: 557–565
20. Hossmann K-A (1996) Periinfarct depolarizations. Cerebrovasc Brain Metab Rev 8: 195–208
21. Hossmann K-A, Fischer M, Bockhorst K, Hoehn-Berlage M (1994) NMR imaging of the apparent diffusion coefficient (ADC) for the evaluation of metabolic suppression and recovery after prolonged cerebral ischemia. J Cereb Blood Flow Metab 14: 723–731
22. Huang S, Yu D, Barrio JR, Grafton S, Melega WP *et al* (1991) Kinetics and modeling of L-6-[18F]fluoro-DOPA in human positron emission tomographic studies. J Cereb Blood Flow Metab 11: 898–913
23. Ito U, Spatz M, Walker JT, Jr, Klatzo I (1975) Experimental cerebral ischemia in mongolian gerbils. I. Light microscopic observations. Acta Neuropathol 32: 209–223
24. Kaplan B, Brint S, Tanabe J, Jacewicz M, Wang X-J, Pulsinelli W (1991) Temporal thresholds for neocortical infarction in rats subjected to reversible focal cerebral ischemia. Stroke 22: 1032–1039
25. Kucharczyk J, Mintorovitch J, Asgari HS, Moseley M (1991) Diffusion/perfusion MR imaging of acute cerebral ischemia. Magn Res in Med 19: 311–315
26. Kuge Y, Kawashima H, Minematsu K, Hasegawa Y, Yamaguchi T, Miyake Y, Hashimoto T, Imanishi M, Shiomi M, Tamaki N, Hashimoto N (2000) [1-C-11]Octanoate as a PET tracer for studying ischemic stroke: Evaluation in a canine model of thromboembolic stroke with positron emission tomography. Biol Pharmaceut Bulletin 23: 984–988
27. Lipton P (1999) Ischemic cell death in brain neurons. Physiol Rev 79: 1431–1568
28. Louie AY, Huber MM, Ahrens ET, Rothbacher U, Moats R, Jacobs RE, Fraser SE, Meade TJ (2000) In vivo visualization of gene expression using magnetic resonance imaging. Nature Biotech 18: 321–325
29. Matsuoka Y, Hossmann K-A (1982) Brain tissue osmolality after middle cerebral artery occlusion in cats. Exp Neurol 77: 599–611
30. Nemoto EM, Hossmann K-A, Cooper HK (1981) Post-ischemic hypermetabolism in cat brain. Stroke 12: 666–676

31. Olah L, Wecker S, Hoehn M (2000) Secondary deterioration of apparent diffusion coefficient after 1-hour transient focal cerebral ischemia in rats. J Cerebral Blood Flow Metab 20: 1474–1482
32. Paschen W, Doutheil J (1999) Disturbances of the functioning of endoplasmic reticulum: a key mechanism underlying neuronal cell injury? J Cerebral Blood Flow Metab 19: 1–18
33. Pillekamp F, Grune M, Brinker G, Franke C, Uhlenkuken U, Hoehn M, Hossmann K-A (2001) Magnetic resonance prediction of outcome after thrombolytic treatment. Mag Res Imaging 19: 143–152
34. Rudolf J, Ghaemi M, Haupt WF, Szelies B, Heiss WD (1999) Cerebral glucose metabolism in acute and persistent vegetative state. J Neurosurg Anesthesiol 11: 17–24
35. Schmitz B, Hoehn-Berlage M, Kerskens CM, Bottiger BW, Hossmann K-A (1998) Recovery of the rodent brain after cardiac arrest: A functional MRI study. Magn Res Med 39: 783–788
36. Sette G, Baron JC, Young AR, Miyazawa H, Tillet I, Barre L, Travere JM, Derlon JM, Mackenzie ET (1993) In vivo mapping of brain benzodiazepine receptor changes by positron emission tomography after focal ischemia in the anesthetized baboon. Stroke 24: 2046–2057
37. Shimada Y, Uemura K, Ardekani BA, Nagaoka T, Ishiwata K, Toyama H, Ono K, Senda M (2000) Application of PET-MRI registration techniques to cat brain imaging. J Neurosci Meth 101: 1–7
38. Staroselskaya IA, Chaves C, Silver B, Linfante I, Edelman RR, Caplan L, Warach S, Baird AE (2001) Relationship between magnetic resonance arterial patency and perfusion-diffusion mismatch in acute ischemic stroke and its potential clinical use. Arch Neurol 58: 1069–1074
39. Warach S, Gaa J, Siewert B, Wielopolski P, Edelman RR (1995) Acute human stroke studied by whole brain echo planar diffusion-weighted magnetic resonance imaging. Ann Neurol 37: 231–241
40. Welch KMA, Windham J, Knight RA, Nagesh V, Hugg JW, Jacobs M, Peck D, Booker P, Dereski MO, Levine SR (1995) A model to predict the histopathology of human stroke using diffusion and T-2-weighted magnetic resonance imaging. Stroke 26: 1983–1989

Correspondence: Prof. Dr. Konstantin-A. Hossmann, Max-Planck-Institute for Neurological Research, Department of Experimental Neurology, Gleueler Str. 50, D-50931 Cologne, Germany. e-mail: hossmann@mpin-koeln.mpg.de

Acta Neurochir (2003) [Suppl] 86: 29–32

Cécile & Oskar Vogt: The significance of their contributions in modern neuroscience

I. Klatzo

Gaithersburg, Maryland, USA

The last decades of the past century have brought striking progress in molecular biology, opening a dramatic elucidation of the molecular genetic mechanisms involved in normal development, genetic diseases, and programmed cell death (apoptosis). The possibility of control and manipulation of such mechanisms has opened new and exciting horizons for *gene therapy* and *eugenics.*

In the frenzy of new discoveries, the names of *Cécile* and *Oskar Vogt* have been practically forgotten, although over one hundred years ago they envisaged these developments and laid the foundation for modern, genetically oriented neuroscience. Thus, the main purpose of this paper is to bring from obscurity the contributions of the Vogts who, in a visionary way, laid the foundation for the most spectacular progress now unfolding in the neurological sciences.

The curriculum vitae of *C&O Vogt* can be briefly outlined as follows:

Cécile Mugnier, was born in 1875 in Annecy, Haut-Savoie, France, whereas Oskar Vogt was born 1870 in Husum, Schleswig-Holstein, Germany. They met in 1897 in Paris and, in spite of vastly different cultural backgrounds, promptly recognized the unique closeness of their scientific interests and life's objectives. Both being of a rebellious nature and in confrontation with established dogmas and the academic establishment, they decided, after their marriage in 1899, to strike out on their own by starting independent brain research, depending solely on their earnings from private practice which, fortuitously, included the Krupp family, one of the most powerful in Germany [2].

With the unshakable support of Friedrich Alfred Krupp, the head of the industrial Krupp empire, Oskar Vogt was able, in spite of fierce opposition from academia, to become the director of the Kaiser Wilhelm Brain Research Institute, whereas Gustav Krupp von Bohlen und Halbach, the successor of Friedrich Alfred, protected the Vogts from persecution and annihilation by the Nazi regime. When in 1933 Adolf Hitler became the absolute dictator of Germany, the position of Oskar Vogt as the head of K-W Brain Research Institute became untenable, and Gustav Krupp provided funds for the opening of a private Brain Research Institute in Neustadt, Schwarzwald where the Vogts were able to continue their scientific careers until the end of their lives.

The development and formation of their ideas in brain research can be traced as follows:

Perhaps the hardships and struggles the Vogts were exposed to in their youth inspired them to search obsessively for a comprehensive *elucidation of basic workings of the human mind.* Accepting the brain as the organ in charge of all *nervous functional activities*, they arrived at the supposition that the latter *must be related to different brain structures.*

To validate this assumption they undertook monumental *cytoarchitectonic studies* describing about 200 areas in the human cerebral cortex which distinctly differed in neuronal cytology and arrangement. In later studies, the Vogts were able to correlate some of these areas with specific functional responses, using *electrical stimulation* [4].

Oskar revealed his intense interest in nature during his high-school years, while exploring the marshes around Husum, and he became especially fascinated by the *Variability* within species which he could observe, particularly in insects.

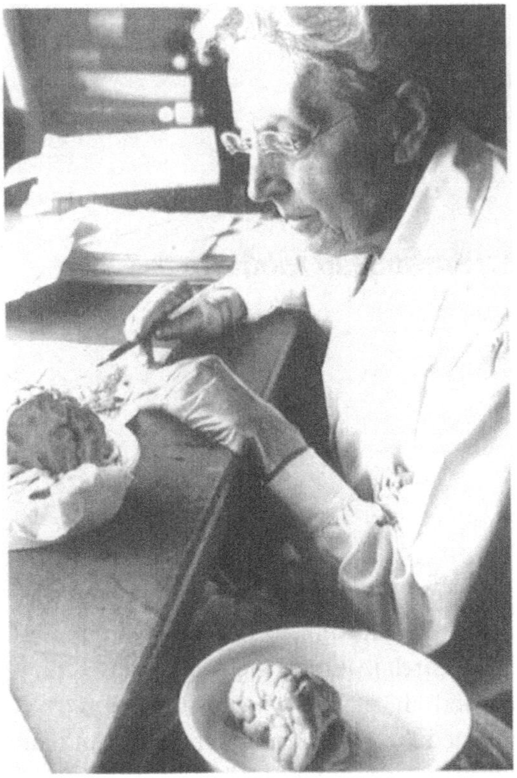

Fig. 1. Cécile at work. Neustadt Institute (1947)

Fig. 2. Oskar. Drawing by Gustav A. Rieth Courtesy of Hedwig Rieth

Later, when working with Cécile in Berlin, they came to the assumption that the *variability* observed in insects and in the structure of the human cerebral cortex might be related to similar mechanisms of *genomic mutations* induced by environmental changes, and this eventually led the Vogts to the theory of *pathoclisis* [3].

The significance and relevance of the pathoclisis theory, the Vogts' greatest contribution to modern Neuroscience, lies in relating their cytoarchitectonic observations to *molecular genetics*, a field which plays a dominant role in current neuroscience research. The assumption that specific differences between cytoarchitectonic (topistic) regions are based on *specific differences in molecular* (physicochemical) composition of the cells, provides the basis for the *selective vulnerability* observed in various regions. According to the Vogts, *Pathoclisis* (clisis = predisposition, in ancient Greek) can be defined as *genomically-determined excessive variability, reaching in intensity the degree of a pathological change.*

In sublethal injury (such as the regions of penumbra in cerebral ischemia), pathoclisis is expressed in the activation of two types of genomic molecular responses, i.e., those *accelerating cell death*, or those inducing a variety of defensive mechanisms which ameliorate damage and *stimulate cell recovery*. The confrontation of these forces thus resembles a battle between the forces of good and evil, determining whether the cell will live or die, with the outcome of the struggle depending on the balance of genomic responses characteristic of each topistic unit. Our observations by fluorescent TUNEL staining on apoptotic death in cerebral ischemia [1], have indicated that positive Tunel staining, when confined to the cell nucleus, does not indicate the point of no return. Such neurons appear to be capable of recovery as long as cytoplasmic DNA can still be active.

In sublethal injury, the induced *genomic reactivity* varies according to the *selective vulnerability of various topistic units.* Thus, in cerebral ischemia, the fluctuating genomic reactivity may play a significant role in designing therapeutic *strategies*, with a proper timing of their application, which would be *enhance regenerative expressions and suppress harmful ones.*

In an attempt to localize the anatomical site of various psychic phenomena the Vogts included in their cytoarchitectonic studies the brain of people with some outstanding intellectual qualities. Their studies on "elite brains" led to Oskar Vogt receiving an invitation from the Soviet Government to study Lenin's brain in the Brain Institute in Moscow [2] which would be established expressly for this purpose. The acceptance of this invitation led to five years (1925–1930) of Oskar Vogt's commuting between the institutes in Berlin and Moscow, and brought some controversial findings, mainly due to the severe damage of Lenin's brain caused by three consecutive strokes and due to political pressures. It also provided a source for attacks by the Nazis following the Vogts' return to Berlin.

The last studies of the Vogts, related to pathoclisis, were carried out at their Institute in Neustadt and were focused on *aging* [5, 6]. These studies revealed that, according to the principles of pathoclisis, cell populations in various topistic regions demonstrate *different*, genomically regulated *aging schedules*.

One of the main conclusions of the Vogts was that "*inactivity accelerates and activity delays the aging of the nerve cells*". They believed that the effect of inactivity must be similar to that which occurs in the transsynaptic degeneration of healthy neurons, a process by which neurons are deprived of synaptic stimulation from interconnected but severely damaged neuronal centers.

Otherwise, in the "elite brains" of their brain collection, the Vogts found that the topistic areas related to some *outstanding ability* were associated with rather *delayed neuronal aging*.

Relevant to the subject of this Symposium would be the interpretation of some mechanisms operative in *brain edema*, drawing on the Vogts' theory of pathoclisis. Thus, the intensity and extent of *cytotoxic edema* is significantly influenced by the *selective vulnerability* of involved topistic units.

Since, according to the Vogts, pathoclisis does not have to be confined to neurons, and, in some instances, may also involve glia or vasculature [7], it can be assumed that selective, topistic changes in vascular permeability may affect the dynamics of post-ischemic *vasogenic brain edema*.

The greatness of the Vogts is based both on their penetrating *vision for the Future of Neuroscience and on the sterling quality of their character, which sustained a "test of fire" during the years of Nazi persecution in Germany.* They never waivered in their convictions nor compromised their integrity, neither in the time of the Kaiser's monarchy, during the shaky Weimar Republic, nor in the terror-stricken Third Reich. They repeatedly risked their lives to save some of their Jewish friends from death, and, sailing through the rough seas of intolerance and terror, they maintained a straight course towards reaching their scientific aims. Their lives, enmeshed in struggles and pursuing visionary brain research under the most trying of circumstances, should serve as an inspiration to a new generation of neuroscientists.

Summary

A) By introducing and propagating the use of *multidisciplinary approaches*, Cécile and Okar Vogt played a decisive role in *establishing brain research as a distinct branch of science*, and thus laid the foundations for *modern Neuroscience.*

B) The Vogts' and Brodmann's *monumental work on brain cytoarchitectonics*, supplemented with *electrical stimulation* of the exposed cerebral cortex, demonstrated *correlations between evoked functional responses and the topography of brain regions endowed with distinct cytoarchitectonic structure (topistic units).* This opened the field of *functional brain mapping,* a subject hotly pursued at the present time.

C) One of the most important aspects of *Pathoclisis* is that the Vogts based their theory on *molecular genetics.* They envisaged that in any topistic unit the neurons had a different specific, *molecular* ("physicochemical") composition, accounting for differences observed in the dynamics of *neuronal development* and *involution*.

In sublethal injury, the induced *genomic reactivity* varies according to the *selective vulnerability of various topistic units.* Thus, in cerebral ischemia, the fluctuating genomic reactivity may play a significant role in designing therapeutic *strategies*, which would include *enhancing regenerative expressions and suppressing harmful ones.*

D) The pathoclisis theory may be of assistance in analyzing the nature of various *genetic disorders*, which could be considered for the application of *gene therapy*.

References

1. Gordon WC, Colangelo V, Bazan NG, Klatzo I (1999) Aspects of the maturation phenomenon observed by the TUNEL method. Maturation phenomenon in cerebral ischemia III. Springer, Berlin Heidelberg New York Tokyo
2. Klatzo I (2002) Cécile and Oskar Vogt: The visionaries of modern neuroscience. Acta Neurochir (Wien) [Suppl] 80

3. Vogt C&O (1922) Erkrankungen der Grosshirnrinde im Lichte der Topistik – Pathoklise und Pathoarchitektonik. J Psychol Neurol 28: 9–171
4. Vogt C&O (1926) Die vergleichend-architektonische und die vergleichend reizphysiologische Felderung der Grosshirnrinde unter besonderer Berücksichtigung der menschlichen. Naturwissensch 14: 1190–1194
5. Vogt C&O (1946) Aging of nerve cells. Nature 157: 304
6. Vogt C&O (1947) Über Wesen und Ursache des Alterns der Hirnzellen. Forschungen und Fortschritte 21/23: 4–6
7. Vogt C&O (1951) Importance of neuroanatomy in the field of neuropathology. Neurology 1: 205–218

Correspondence: Igor Klatzo, Gaithersburg, Maryland, USA. e-mail: klatzo@aol.com

Imaging

Acta Neurochir (2003) [Suppl] 86: 35–37

Estimating blood-brain barrier opening in a rat model of hemorrhagic transformation with Patlak plots of Gd-DTPA contrast-enhanced MRI

J. D. Fenstermacher[1], **R. A. Knight**[2,3], **J. R. Ewing**[2,3], **T. Nagaraja**[1], **V. Nagesh**[2], **J. S. Yee**[2], and **P. A. Arniego**[2]

[1] Department of Anesthesiology, Henry Ford Health Sciences Center, Detroit, MI, USA
[2] Department of Neurology, Henry Ford Health Sciences Center, Detroit, MI, USA
[3] Department of Physics, Oakland University, Rochester, MI, USA

Summary

Patlak plot processing [2, 3] of Gd-shifted T_1 relaxation-time images from a rat model of hemorrhagic transformation yielded estimates and maps of the blood-to-brain influx rate constant of Gd-DTPA (K_1). The Patlak plots also produced a heretofore unrecognized parameter, the distribution space of the intravascular-Gd-shifted protons (V_p), an index of blood-to-tissue transfer of water. The K_1 values for Gd-DTPA were very high for the regions of blood-brain barrier (BBB) opening and were similar to those of ^{14}C-sucrose concurrently obtained by quantitative autoradiographic (QAR) analysis. In these same ROI's, V_p was five-fold greater than normal, which suggests that the permeability of the BBB to water was also increased. The ^{14}C-sucrose space of distribution in the ischemic ROI's was around 8%, thus indicating a sizable interstitial space. The spatial resolving power of Gd-DTPA-ΔT_1 imaging was rather good, although no match for ^{14}C-sucrose-QAR. This study shows that quantitative ΔT_1-MRI estimates of regional blood-brain transfer constants of Gd-DTPA and water distribution are possible when Patlak plots are employed to process the data. This approach may be useful for tracking the time-course of BBB barrier function in both animals and humans.

Keywords: Blood-brain barrier; Gd-DTPA imaging; magnetic resonance imaging; hemorrhagic transformation; water distribution; quantitative autoradiography; cerebral edema.

Introduction

One of the possible consequences of reperfusion after several hours of cerebral ischemia (stroke) is blood-brain barrier (BBB) opening. This opening usually results in considerable cerebral edema and in some cases may be so severe as to lead to bleeding within the brain (hemorrhagic transformation or HT), coma, and death. The goal of this study was to test the proposition that magnetic resonance imaging (MRI) can detect moderate to large BBB opening. A rat model of reperfused middle cerebral artery (MCA) occlusion that often leads to petechial parenchymal hemorrhages by 24 hr and earlier opening of the BBB was used for this testing. Blood-to-brain transfer constants (K_1) for Gd-DTPA were calculated from Patlak plots [2, 3] of MRI-T1 data sets and were compared to quantitative autoradiographic (QAR) measurements of K_1 for ^{14}C-sucrose, thereby confirming or denying the accuracy and reliability of the MRI method. If successful in establishing its validity, then this minimally invasive MRI technique should be useful in predicting sites and severity of BBB opening following stroke and other neuropathologic conditions that might involve microvascular damage. It could also be useful in staging various human brain diseases and in selecting and monitoring treatment.

Methods

Transient focal ischemia was induced in male Wistar rats (n = 7, 275–300 g) by intraluminal suture occlusion of the MCA. Reperfusion was achieved by gently withdrawing the filament after 3 hour of occlusion. Some 12–24 hours later, petechial hemorrhages appear in the preoptic area and anterior hypothalamus, the caudate-putamen (mostly the lateral part), globus pallidus, and the piriform, parietal, and insular cortices in about 65% of the animals.

A series of MRI measurements were made during the occlusion and reperfusion period with a 7 Tesla, 20 cm bore Magnex magnet interfaced to a SMIS console. The following MRI parameters were assayed:

- T1-weighted images: spin-echo sequence (TR/TE = 1000 ms/20 ms).
- T2 measurement: multiecho sequence (TR/TE = 3000 ms/30, 60, 90, 120 ms).
- Apparent diffusion coefficient (ADC): pulsed gradient spin-echo sequence (TR/TE = 1500 ms/40 ms, b-values = 10, 400 and 800 s/mm^2).
- Cerebral blood flow (CBF) measurement: gradient-echo arterial spin tagging sequence (TR/TE = 20 ms/6 ms, 64 × 128, Nave = 96, 1 slice, 2 mm thick).

– T1 measurement: Look-Locker sequence (TR = 80 ms, TE = 4 ms, 64 × 64, 23 echoes, 5 slices, 1.8 mm thick).

T1-, T2- and diffusion-weighted images (T1WI, T2WI and DWI, respectively) were acquired in multislice mode (13 contiguous slices, 1 mm slice thickness, 32 mm field of view, 128 × 128 matrix). T2, ADC and CBF measurements were obtained during and after ischemia to identify the ischemic lesion.

The assessment of BBB barrier function was begun by injecting Gd-DTPA (0.2 mmol/Kg, i.v. bolus) after 2.5 hr of reperfusion. The "blood-tissue distribution" of Gd-DTPA-T_1-shifted protons in a 2.0 mm thick central slice was serially imaged (T1WI) every 3–4 min for 20–30 minutes. Baseline T1WI were obtained prior to Gd-DTPA injection.

The value of K_1 was calculated from the serial changes in the relaxation time (ΔT_1, which is inversely proportional to Gd-DTPA concentration) after the bolus via the model-independent Patlak plot [2, 3]. The Patlak plot yields not only K_1 (the slope of the linear part of the uptake curve) but also the distribution volume of the protons (water) with relaxation times shortened by plasma-contained Gd (V_p). The latter is given by the y-intercept of the line and is an index of water transfer across the BBB.

Immediately after MR imaging, the animals were removed from the magnet and infused with 100 µCi of 14C-sucrose over a 20-minute period using a stepped down procedure that maintained a relatively constant blood sucrose level. A series of timed arterial blood samples were collected throughout the study, and the rats were decapitated immediately after taking the 20-minute sample. The heads were rapidly frozen in isopentane cooled to −40 °C. The frozen brain with meninges, blood, and CSF in place was removed from the head and, while still frozen, cut into a series of coronal brain sections (20 µm thick) taken at 400 µm intervals through the entire brain. These sections and a set of ^{14}C-standards were exposed on x-ray films for approximately 25 days. Adjacent sets of brain sections were also cut and stained with cresyl violet and H&E for histologic evaluation and measuring brain swelling (edema). Regions of interest (ROI's) were selected in brain areas that demonstrated ischemic injury (from DWI and T2WI images), Gd-DTPA enhancement, and ^{14}C-sucrose extravasation; sets of spatially matched ROI's were also selected on the contralateral side.

The MR images were processed offline using a SUN workstation to generate maps of the various transfer numbers and to identify ROI's. Patlak plots were made of the Gd-DTPA-ΔT_1 data from brain and the sagittal venous sinus blood. Coronal maps of K_1 for Gd-DTPA and V_p were constructed from these data.

The radioactivity in the arterial plasma samples was assayed by beta counting and in the tissue sections by an MCID image analysis system. ^{14}C-sucrose distribution spaces (V_{dist}) and influx rate constants (K_1) were determined. The single-time point equation was used to calculate K_1. Because the MRI maps were for a single *2.0 mm* thick coronal piece of tissue but the QAR images were for *0.02 mm* thick sections, the QAR data for all of the sections that lay within the MRI map were "averaged" to produce a composite value of sucrose K_1 for the MRI-selected regions of interest.

Results

Maps of ΔT_1, K_1 of Gd-DTPA, and V_p were constructed, and regions of interest (ROI's) selected in areas that showed BBB opening. In all instances, these ROI's closely matched the areas of sucrose leakage observed on the ^{14}C-autoradiograms. The mean Gd-DTPA transfer constant (±SD) in the reperfusion lesion was 4.1 (±0.8) × 10^{-3} ml g^{-1} min^{-1}, which is 200-fold larger than normal [2]. On the contralateral side, the slope of the Patlak plot line was not appreciably different than zero; as is well known, a longer period of measurement is needed to assess the transfer of a molecule with very low BBB permeability such as DTPA and sucrose. This suggests that barrier function was still normal in the non-ischemic hemisphere after 3 hr of unilateral MCA occlusion plus 3 hr of reperfusion. The mean K_1 of sucrose in the ischemic ROI's was 6.4 (±1.7) × 10^{-3} ml g^{-1} min^{-1} or about 50% greater than that of Gd-DTPA. Such a difference was expected due to the several small physicochemical and physiological differences between them.

Among the seven rats and ROI's, the individual K_1 values varied about two-fold for Gd-DTPA and slightly more for sucrose. A linear regression of K_1 of Gd-DTPA versus K_1 of ^{14}C-sucrose indicated that the two transfer constants were well correlated. In other words, for an ROI where K_1 of ^{14}C-sucrose was relatively high, so was K_1 of Gd-DTPA, and for an ROI where the former was relatively low so was the latter.

Mean V_p (±SD) from the Patlak plot was 0.071 (±0.004) ml g^{-1} for the ipsilateral ROI's but only 0.015 (±0.016) ml g^{-1} for the matching contralateral areas. This difference indicates more ready movement of protons from blood to brain in the ischemic regions than in the non-ischemic ones and an increase in water transfer across an ischemia-reperfusion altered BBB. The mean spaces of ^{14}C-sucrose distribution were 0.078 (±0.023) and 0.005 (±0.002) ml/g tissue for the ischemic and contralateral ROI's, respectively. The latter sucrose space equals the volume of microvascular plasma in normal rat brain [1], whereas the former undoubtedly includes both the plasma and some or all of the extracellular space in the ischemic ROI's.

The histologies showed considerable ventricular distortion and tissue swelling on the ischemic side. Our three-dimensional analysis of the serial histological images to date suggested that the ipsilateral hemisphere was 10–14% larger than the contralateral hemisphere and was, therefore, edematous after 3 hr of MCA occlusion plus 3 hr of reperfusion.

Discussion

This study shows for this model of reversible ischemia that quantitative ΔT_1-MRI estimates of blood-brain transfer constants of Gd-chelates are possible when Patlak plots are employed for data processing and that the Gd-DTPA MRI approach is nearly as sensitive to BBB opening in this model as is the ^{14}C-sucrose QAR method, one of the gold standards for such determinations. In addition, the Patlak plots also indicated an acute increase in the blood-to-tissue distribution of the inverted protons with plasma-Gd shortened relaxation times; V_p was about five-fold higher in the ischemic ROI's than in the contralateral ones.

Water very rapidly exchanges between plasma and red cells; accordingly the inverted protons – whether Gd-shifted or not, whether in the ischemic areas or not – virtually instantaneously move from plasma into red cells. The water space of microvascular-contained blood in most brain areas is around 0.01 ml/g [1]. On the contralateral side, V_p was somewhat larger than 0.01 ml/g; if this difference is real, then the plasma Gd-shifted protons may quickly reach a steady-state with not only the red cells but also the capillary wall. Again, if true, then the water barrier of the BBB may be at the abluminal membrane of the capillary endothelial cell.

Rather surprisingly, the mean distribution space of sucrose and the plasma Gd-shifted protons were both around 0.075 ml/g in the ipsilateral ROI's. The extracellular space in normal brain is about 0.18 ml/g, i.e., 18% [4]. During ischemia and reperfusion, the astrocytes and the neuronal and glial processes of the neuropil swell (cytotoxic edema), and neuronal cell bodies may shrink, swell, or not change volume. The result of these volume (water) shifts is generally believed to reduce the size of the interstitium. With a leaky BBB, the next barrier to Gd-DTPA and possibly to water may be at the neuronal and glial cell membranes; the similarity in these two distribution spaces is consistent with this suggestion. Be that as it may, these large volumes of distibution indicate that the interstitium has not been obliterated and is patent to small molecules and bulk flow of edema fluid.

The transfer constants in the areas of BBB opening were much larger than normal, e.g., around 200 times so for Gd-DTPA. This indicates a huge impairment in BBB barrier function at this time for this model of HT. In view of this, we will next investigate if this early opening of the BBB is predictive of sites where HT later occurs and cerebral edema forms and examine the role of thrombolytic agents in BBB disruption and these other neuropathologic processes.

The lower limit of sensitivity of T_1-MRI for the detection of BBB opening was not determined in the present study. It may well be that small openings that can be determined accurately by QAR-measurements – for instance, a ten-fold increases in K_1 – can not be detected by MRI and Gd-shifting of proton T_1 relaxation times. We caution that this new technique and the present "validation" cannot be used to rule out BBB opening for all pathologies. Ascertainment of small, subtle changes in BBB-barrier function may require a different methodology or sizable technical improvement, e.g., better spatial resolution, in MR imaging.

Acknowledgments

This work was supported by NIH grant NS38540 and a Bugher Award from the American Heart Association to RAK and by Fund for Henry Ford grants to JDF and JRE. The superb technical assistance of Ms. Kelly Ann Randall and Ms. Jun Xu is gratefully acknowledged.

References

1. Bereczki D, Wei L, Otsuka T, Hans F-J, Acuff V, Patlak C, Fenstermacher J (1993) Hypercapnia slightly raises blood volume and sizably elevates flow velocity in brain microvessels. Am J Physiol 264: H1360–H1369
2. Blasberg RG, Fenstermacher JD, Patlak CS (1983) The transport of alpha-aminoisobutyric acid across brain capillary and cellular membranes. J Cereb Blood Flow Metab 3: 8–32
3. Patlak CS, Blasberg RG, Fenstermacher JD (1983) Graphical evaluation of blood-to-brain transfer constants from multiple-time uptake data. J Cereb Blood Flow Metab 3: 1–7
4. Patlak CS, Fenstermacher JD (1975) Measurements of blood-brain transfer constants by ventriculocisternal perfusion. Am J Physiol 229: 877–874

Correspondence: Dr. Joseph D. Fenstermacher, Department of Anesthesiology, M-312, Henry Ford Hospital, 2799 West Grand Boulevard, Detroit, MI 49202, USA. e-mail: anjdf@neuro.hfh.edu

Acta Neurochir (2003) [Suppl] 86: 39–40

Recent development of MR imaging technique for the investigation of brain function

O. Takizawa

Medical Solution Marketing Division, Siemens Asahi Medical Technologies Ltd., Tokyo, Japan

Summary

Recently, a variety of MR techniques have been developed to investigate brain functions, such as high resolution brain images and various functional brain images. The BOLD fMRI technique opened the new field of non-invasively investigating local brain activities.

As analyses are becoming more and more detailed, we frequently meet the problem of limitations with regard to S/N and the physiologic motions of the subjects to be measured. One solution for higher S/N is the introduction of high field MR systems. Improvement can also be achieved via the instrumental side. The Magnetom MR systems from Siemens introduced a high performance gradient system and flexible pulse sequence structure. Also great advances were made in fMRI measurements with real-time motion correction by introduction of PACE (prospective acquisition correction) technique.

To apply such sophisticated technique in the clinical setting, the concept of in-line evaluation technique is employed, which means real-time data processing. As soon as scanning is terminated, the processed images can be obtained to be used immediately for diagnoses.

Keywords: 3T MR system; fMRI; iPAT; PACE; diffusion.

Introduction

Recently, a variety of MR techniques have been developed to investigate brain functions. MR imaging can generate high resolution brain images and diffusion analyses to reveal the relation of the anatomy and its function. BOLD fMRI technique opened the new field of non-invasively investigating local brain activities.

Since the analyses are becoming more and more sophisticated, we frequently meet the problem of limitations with regard to S/N and physiological motion of the subjects to be measured. Another common problem is the volume to be processed after scanning. If the resulting images are to be used for clinical diagnoses, we need to obtain the results immediately after the scan.

To obtain higher S/N also the instrumental side must be considered. MR systems with higher magnetic fields above 1.5T have been developed. In addition, various techniques such as PACE are utilized to suppress the effect of physiologic motion.

Discussion

The series of MR systems from Siemens, Magnetom Harmony 1T, Symphony, Sonata 1.5T, and 3T Trio and Allegra system, offer the high performance gradient system and flexible pulse sequence structure.

The 3T MR systems, Magnetom Trio and Allegra from Siemens, with high performance gradient systems could improve the functionality of human brain studies. The use of a higher field strength MR system, although it is still limited to investigational purposes in Japan (above 1.5T), offers significant improvement of S/N. The unique feature of the 3T Allegra system can facilitate visual stimulation for fMRI experiments. The 3T whole body system, called Trio, allows imaging of the entire region of the body. Improvement of the RF coil and the data receiving system, called integrated parallel acquisition technique (iPAT), shortens scanning time and echo time (TE) for echo planer imaging, which will then significantly improve image quality especially for diffusion imaging.

Great advances were made in fMRI measurement with real-time motion correction by introducting PACE (prospective acquisition correction) technique. This technique allows image realignment against brain motion during data acquisition. Knowing the displacement of the brain in real time, the subsequent measurements will be performed for the renewed slice position. This method can avoid instability of signal intensity resulting from the difference of spin history of

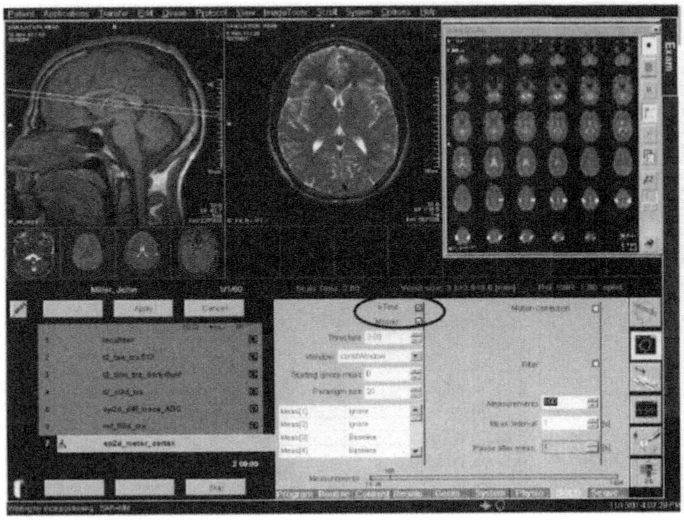

Fig. 1. Task card for Real-time fMRI imaging and display

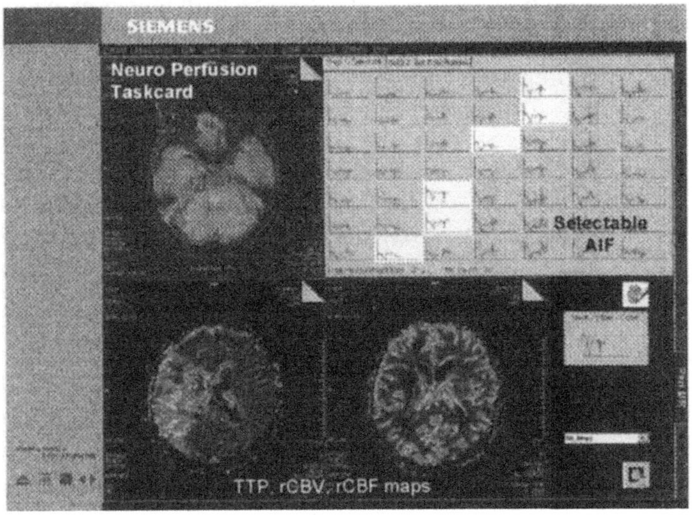

Fig. 2. Task card for Neuro Perfusion imaging and evaluation

each proton within the measuring pixel in relation to RF excitation.

The real-time data processing system introduced by the Magnetom software offers a processed image after sophisticated analyses which allows to make diagnoses immediately after the measurement. This real-time functionality, called in-line technology, is especially important for clinical emergencies requiring urgent diagnostic decisions.

By combining these new techniques with the excellent performance of the gradient and the RF system, brain functional studies using MR can be expanded for both investigational and clinical tasks.

Correspondence: Osamu Takizawa, Ph.D., M.R., Medical Solution Marketing Division, Siemens Asahi Medical Technologies Ltd., Takanawa Park Tower, 20-14, Higashi Gotanda 3-chome, Shinagawa-ku, Tokyo 141-8644, Japan. e-mail: osamu.takizawa@siemens.co.jp

Acta Neurochir (2003) [Suppl] 86: 41–44

Energy metabolism and cerebral blood flow during cytotoxic brain edema induced by 6-aminonicotinamide

D. Kurita[1], **M. Haida**[2], and **Y. Shinohara**[1]

[1] Department of Neurology, Tokai University School of Medicine, Kanagawa, Japan
[2] Department of Physiology, Tokai University School of Medicine, Kanagawa, Japan

Summary

We investigated the progression of cytotoxic brain edema induced by 6-aminonicotinamide (6-ANA), a potent antimetabolite of nicotinamide, by measuring the time courses of changes in brain tissue water state (with MRI), histology (with H&E staining), energy metabolism (with ^{31}P-NMR), brain hemoglobin concentration (with near-infrared spectroscopy; NIRS), cerebral blood flow and volume (CBF, CBV), mean arterial blood pressure (MABP), and brain activity (with EEG) up to 10 hours (h). Change in cerebrovascular autoregulation was also investigated. 6-ANA (120 mg/kg) was administered intraperitoneally to 30 male Wistar rats (250–350 g). After 10 h, the T2-weighted signal intensity was increased ($p < 0.05$), and H&E staining showed severe vacuolation of glial cells. ATP production/consumption and intracellular pH were well maintained up to 10 h, while the intensity of the phosphomonoesters (PME) signal was significantly increased ($p < 0.05$). Oxygen consumption gradually decreased from 4 to 10 h. CBF and MABP were all significantly increased (by 2.5-fold for CBF) ($p < 0.05$). Theta and delta wave amplitudes were reduced at 10 h. In summary, 6-ANA (120 mg/kg) induced cytotoxic brain edema from 4 to 10 h. Energy balance and brain activity were well maintained up to 10 h, though cerebrovascular autoregulation was impaired.

Keywords: 6-aminonicotinamide; energy metabolism; CBF.

Introduction

Application of nondestructive and noninvasive techniques, which do not influence the cerebral state, should be an effective approach to study the progression of brain edema. 6-Aminonicotinamide (6-ANA) is known to induce cytotoxic brain edema due to its effects on 6-phosphogluconate dehydrogenase and nicotinamide phosphoribosyl transferase [4]. To examine energy balance, brain activity and cerebrovascular autoregulation in a progressive cytotoxic brain edema model induced with 6-ANA, we investigated the time courses of changes in brain tissue water state, histology, energy metabolism, brain hemoglobin concentration, cerebral blood flow and volume (CBF, CBV), mean arterial blood pressure (MABP), and brain activity by using MRI (T2-weighted imaging), H&E staining, ^{31}P nuclear magnetic resonance (^{31}P-NMR) spectroscopy, near-infrared spectroscopy (NIRS), laser-Doppler flowmetry, a blood pressure amplifier, and EEG up to 10 hours (h).

Materials and methods

Experimental animals

Thirty male Wistar rats (250–350 g) were anesthetized intraperitoneally with alpha-chloralose (50–75 mg/kg) and urethane (500–750 mg/kg). Cytotoxic brain edema model [8]: 6-ANA was administered intraperitoneally at the dose of 120 mg/kg.

Apparatus

MRI (MR Vectra Fast, GE Yokogawa, Japan) T2-weighted images were taken with a spin echo method, TE 130 msec, TR 3200 msec, 20 acquisitions, slice thickness 3 mm, FOV 9 cm, matrix 160 × 256, and axial to anterior-posterior bregma, with 10 slices.

^{31}P-NMR (BEM 250/80, Phospho-Energetics Inc., USA) spectra were taken with a single 90 degree pulse of 20 μsec, TR 10 sec, 120 acquisitions, and a surface coil of 2.0 mm diameter. The coil was centered at 1.5 mm caudal to the bregma. The area of each peak of the spectrum was determined by the least-squares procedure using a Gaussian function. PCr index, pHi (intracellular pH), and the relative phosphomonoesters (PME) are given by the following equations (see ref. 7).

$$PCr\ index = PCr/(PCr + Pi) \quad (1)$$

$$pHi = 6.77 + \log(\delta - 3.29)/(5.68 - \delta) \quad (2)$$

$$Relative\ PME = PME/\beta\text{-}ATP \quad (3)$$

where δ is the chemical shift difference between the *PCr* and *Pi* resonances. NIRS (OM-100A, Shimadzu, Japan) were taken at wavelengths of 780, 805, and 830 nm. NIRS probes used quartz fibers serving as a 2.5 mm diameter light source and a 1 mm diameter detector. Two burr holes of 2–3 mm diameter were made at 2 mm lateral and 5 mm posterior, and 2 mm lateral and 3 mm anterior to the bregma. The probes were fixed in the burr holes with an epoxy adhesive. Relative changes in oxygenated hemoglobin (oxy-Hb), deoxygenated hemoglobin (deoxy-Hb), and total hemoglobin (total-Hb) were given by the following equations (see ref. 5):

$$Oxy\text{-}Hb = -3.6132 * \Delta abs1 + 1.1397 * \Delta abs2 + 3.0154 * \Delta abs3 \quad (4)$$

$$Deoxy\text{-}Hb = 3.7837 * \Delta abs1 - 0.7833 * \Delta abs2 - 2.5679 * \Delta abs3 \quad (5)$$

$$Total\ Hb = 0.1705 * \Delta abs1 + 0.3564 * \Delta abs2 + 0.4475 * \Delta abs3 \quad (6)$$

where $\Delta abs1$, $\Delta abs2$, and $\Delta abs3$ are the changes in the absorbance readings at 780, 805, and 830 nm, respectively.

Laser-Doppler flowmetry (FLO-C1, Neuroscience Inc., USA) was conducted at 780 nm with a 1 mm glass fiber probe serving as a light source and a detector. The skull was thinned by making a 2–3 mm burr hole 3 mm lateral and 3 mm anterior to the bregma. The probe was fixed in the burr hole with an acrylic resin cement. CBF and CBV are given by the following equations (3).

$$CBF = K_V \int_{f_1}^{f_2} f \times P(f)\,df / \langle I^2 \rangle \quad (7)$$

$$CBV = K_M \left\{ -\ln\left[1 - \int_{f_1}^{f_2} P(f)\,df / \langle I^2 \rangle \right] \right\} \quad (8)$$

where I is the intensity of scattered light, f is the frequency of the Doppler shift, $P(f)$ is the power spectrum, K_V and K_M are constants, and f_1 and f_2 are the signal bandwidths, set between 30 Hz and 30 kHz.

The EEG (AB-601G, Nihon Kohden, Japan) frequency spectrum was taken by FFT with monopolar recording, sampling time 10 msec, time constant 0.3 sec, and 2^{18} data points. Silver pellets (2 mm in diameter) were used as the electrodes. The skull was thinned by making two 2–3 mm burr holes 3 mm lateral and 3 mm anterior, and 1.5 mm lateral and 10 mm anterior to the bregma. The active and the indifferent electrodes were fixed in the burr holes with an acrylic resin cement. The measurements were divided into three groups in order to avoid interaction. Measurement group I: Brain tissue water state and histology were examined by T2-weighted imaging and H&E staining. Data from the 6-ANA model at 4 h and 10 h were compared with those for the sham-operated control. Measurement group II: Energy metabolism and brain hemoglobin concentration were continuously measured by using ^{31}P-NMR and NIRS up to 10 h. Measurement group III: CBF, CBV, and brain activity were continuously measured by using laser-Doppler flowmetry and EEG up to 10 h. The mean arterial blood pressure (MABP) was monitored simultaneously. In groups II and III, data were also obtained pre-injection of 6-ANA, as a control.

Statistics

Data are reported as means ± SD. Statistical analyses were done by AVOVA with the post-hoc Tukey-Kramer test in measurement group I and Fisher's PLSD test in measurement groups II and III. A P value < 0.05 was considered statistically significant.

Results 1 (measurement group I)

T2-weighted images of rat brain of the sham-operated control and the 6-ANA model at 10 h showed no visually apparent difference between the cortical regions under the two conditions. Therefore, we used 0.2% $CuSO_4$ solution as a reference to compare them. The relative signal intensity was calculated by dividing the signal intensity in the ROI of the cortical region by that of the reference in the control, and the 4-h and 6-h 6-ANA models. The relative signal intensity in the 10-h 6-ANA model was significantly increased in comparison with the control. Glial cells were severely vacuolated in axial HE-stained sections in the second layer of the parietal lobe of the 10-h 6-ANA model, as shown in the photograph (magnification: 400 times).

Results 2 (measurement group II)

The changes in PCr index and pHi in the control, and at 4 h, and 10 h after 6-ANA injection are shown in Fig. 1. The PCr index and pHi showed no significant differences under the three conditions. The relative PME was increased significantly over the control at 10 h in the 6-ANA model. In typical time-courses of relative changes in oxy-Hb, deoxy-Hb, and total Hb of the rat brain obtained by using NIRS, total Hb and oxy-Hb gradually increased during 4 to 10 h after in-

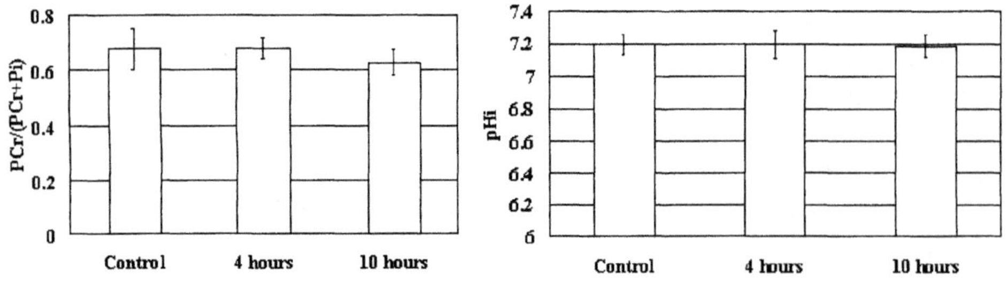

Fig. 1. Changes in PCr index (left) and pHi (right) of rat brain

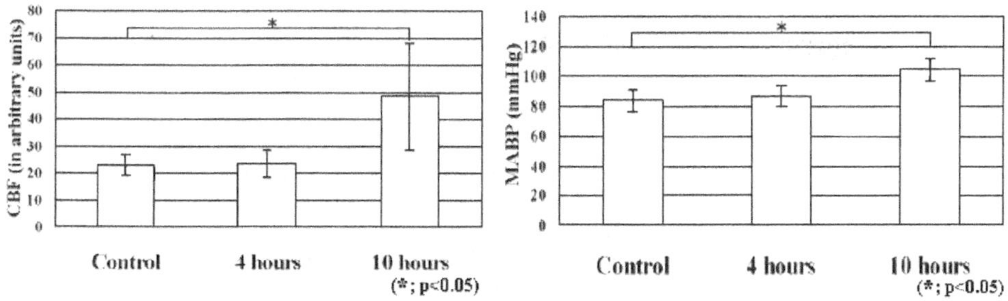

Fig. 2. Changes in CBF (left) and MABP (right) of rat brain

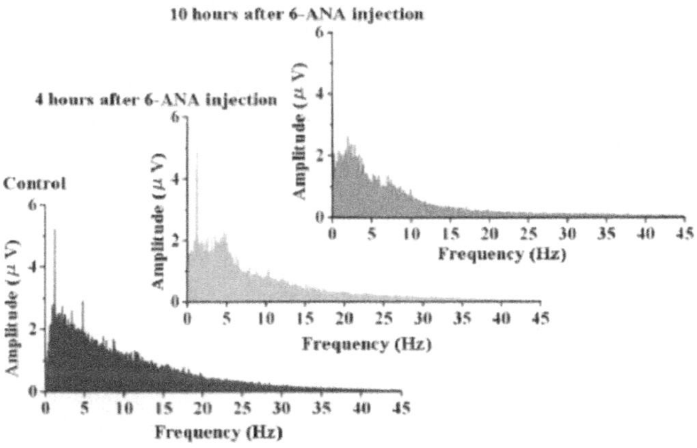

Fig. 3. Changes in EEG frequency spectrum of rat brain

jection of 6-ANA at 0 hour. The change in deoxy-Hb was a mirror image of that in oxy-Hb.

Results 3 (measurement group III)

The changes of CBF and MABP in the control, and at 4 h, and 10 h after 6-ANA injection are shown in Fig. 2. CBF and MABP were significantly increased at 10 h in the 6-ANA model compared with the control, while CBV was significantly decreased.

Typical EEG frequency spectra of the rat brain before 6-ANA injection (control), and at 4 h and 10 h after 6-ANA injection are shown in Fig. 3. The control shows theta and delta waves, while the amplitude of the theta waves was reduced at 4 h, and that of both waves was reduced at 10 h in the 6-ANA model. The physiological status before 6-ANA injection is summarized in Table 1.

Table 1. *Physiological parameters in control rats*

Arterial blood pH	Arterial blood PCO_2 (mm Hg)	Arterial blood PO_2 (mm Hg)	Rectal temperature (°C)	Blood sugar (mg/dl)
7.37 ± 0.02	37.3 ± 5.7	82 ± 5	37.7 ± 1.0	97 ± 9

Discussion

The T2-weighted image signal intensity increase and the vacuolation observed with H&E staining at 10 h after 6-ANA indicate that cytotoxic brain edema was induced by 6-ANA at the dose of 120 mg/kg. The blood-brain barrier function appeared to be maintained since metabolic alkalosis was not apparent from the pHi up to 10 h [2]. The increase of PME may reflect accumulation of fructose-1,6-bisphosphate due to inhibition of 6-phosphogluconate dehydrogenases by 6-ANA, and increase of phosphorylcholine and phosphorylethanolamine in response to glial cell membrane destruction by 6-ANA [6]. The increases of CBF and MABP, and the decrease of CBV were considered due to the progression of brain edema from 4 to 10 h after 6-ANA injection. Neurons were little affected by 6-ANA, since ATP production/consumption as reflected by the PCr index and pHi remained constant, total Hb and oxy-Hb increased, brain activity in EEG did not disappear, and no damage was observed in H&E

staining. There may be different metabolic pathways operating in glial and neural cells. Cerebrovascular autoregulation was impaired since CBF increased about 2.5 times, with an increase of MABP [1].

References

1. Harper SL, Bohlen HG, Rubin MJ (1984) Arterial and microvascular contributions to cerebral cortical autoregulation in rats. Am J Physiol 246: H17–H24
2. Hubesch B, Marinier DS, Hetherington HP, Twieg DB, Weiner MW (1989) Clinical MRS studies of the brain. Invest Radiol 24(12): 1039–1042
3. Kashima S, Oka S, Ishikawa J, Ohsawa T, Hiki Y (1994) Measurement of tissue blood volume in a model system and in the canine intestine by dynamic light scattering. Lasers in the Life Sci 6(2): 79–90
4. Kuchiwaki H, Inao S, Yoshida K, Sugita K (1995) Effect of intracarotid administration of 6-aminonicotinamide on cerebral blood flow in cats. Stroke Mar 26(3): 473–478
5. Kutsuzawa T, Shioya S, Kurita D, Haida M, Yamabayashi H (2001) Effects of age on muscle energy metabolism and oxygenation in the forearm muscles. Med Sci Sports Exerc 33: 901–906
6. Mahmood U, Street JC, Matei C, Ballon D, Martin DS, Koutcher JA (1996) In vivo detection by ^{31}P NMR of pentose phosphate pathway block secondary to biochemical modulation. NMR Biomed May 9(3): 114–120
7. Petroff OA, Prichard JW, Behar KL, Alger JR, den Hollander JA, Shulman RG (1985) Cerebral intracellular pH by ^{31}P nuclear magnetic resonance spectroscopy. Neurology 35: 781–788
8. Sasaki S (1982) Brain edema and gliopathy induced by 6-aminonicotinamide intoxication in the central nervous system of rats. Am J Vet Res 43(9): 1691–1695

Correspondence: Daisaku Kurita, Department of Neurology, Tokai University School of Medicine, Isehara-shi, Kanagawa 259-1193, Japan. e-mail: kurita@is.icc.u-tokai.ac.jp

Acta Neurochir (2003) [Suppl] 86: 45–48

PET neuroreceptor imaging as predictor of severe cerebral ischemic insult

T. Nariai[1], **Y. Shimada**[4], **K. Ishiwata**[3], **T. Nagaoka**[1], **J. Shimada**[5], **T. Kuroiwa**[2], **K.-I. Ono**[4], **K. Hirakawa**[1], **M. Senda**[3], and **K. Ohno**[1]

[1] Department of Neurosurgery, Tokyo Medical and Dental University, Japan
[2] Department of Neuropathology, Tokyo Medical and Dental University, Japan
[3] Positron Medical Center, Tokyo Metropolitan Institute of Gerontology, Japan
[4] Department of Veterinary Clinical Pathobiology, University of Tokyo, Japan
[5] Pharmaceutical Research Laboratories, Kyowa Hakko Kogyo Co., Ltd., Japan

Summary

Measurement of the adenosine A_1 receptor (A1-R) with positron emission tomography (PET) using a newly developed positron ligand, [1-*methyl*-^{11}C]8-dicyclopropylmethyl-1-methyl-3-propylxanthine (MPDX), were performed in a cat middle cerebral artery (MCA) occlusion and reperfusion. Eighteen adult cats underwent PET measurement of; 1) cerebral blood flow (CBF), 2) A1-R, 3) central benzodiazepine receptor (BDZ-R) and 4) glucose metabolism with ^{15}O labeled water, MPDX, ^{11}C-flumazenil (FMZ) and ^{18}F-fluorodeoxyglucose (FDG), respectively. The CBF, A1-R, BDZ-R and FDG uptake were serially measured after 60 min occlusion of MCA in this order. MPDX binding and FMZ binding, but not CBF and FDG uptake, were significantly reduced in the groups with severer ischemic insult than in the groups with no or milder insults. Of the two receptor ligands, the reduction rate of the MPDX binding to A1-Rs was larger in a group that caused fatal ischemic insult. The newly developed PET in vivo imaging technique using MPDX was suitable in evaluating the function of adenosine and A1-Rs in relation to cerebral ischemia.

Keywords: Positron emission tomography; cerebral ischemia; adenosine A_1 receptor; cat; MCA occlusion; blood flow; glucose metabolism; benzodiazepine receptor.

Introduction

Adenosine is present in large amounts in mammalian brain and plays a role as an endogenous modulator of synaptic functions in the central nervous system. The effect is mediated by two major subtypes of receptors: adenosine A_1 receptors that exhibit a higher affinity for adenosine and inhibit adenylyl cyclase, and A_2 receptors which exhibit a lower affinity for adenosine and stimulate adenylyl cyclase. It is known to play an important role in the process of ischemic cerebral damage, as well as in its protection [16, 17]. Our laboratory has proposed several radioligands to image the adenosine receptor system in the living brain with positron emission tomography (PET) [3, 9, 10, 14, 15, 18, 20]. Of these, [1-*methyl*-^{11}C]8-dicyclopropylmethyl-1-methyl-3-propylxanthine (MPDX) is proposed as a promising PET ligand for adenosine A_1 receptors (A1-Rs) in the central nervous system. In this report, we evaluated the A1-R in the middle cerebral artery (MCA) occlusion and reperfusion model of cats with MPDX and animal PET scanner to clarify the significance of A1-R quantification in relation to cerebral ischemia.

Materials and methods

Animal care and preparation for experiments

Nineteen adult cats weighing from 3 to 4 kg were purchased from IEFA CREDO (IOC cats, Lyon, France) and were used for experiments. The cats were intubated following injection of thiopental sodium (15 mg/kg i.v.) and atropine sulphate (0.05 mg/kg s.c.). During the experimental procedure, the anesthesia was maintained with isoflurane (1.5–2.0%).

Six cats were examined without treatment and were used as normal controls. PET measurement was performed with a model SHR-2000 PET camera designed for animal scanning (Hamamatsu Photonics Co. Hamamatsu, Japan).

^{15}O labeled water was injected intravenously into the cat and scanned for 3 min to image blood flow. Ten min later, [^{11}C]MPDX was injected and scanned for 60 min in a dynamic mode. Sixty min after the baseline measurement with [^{11}C]MPDX, the second PET measurement with [^{11}C]MPDX was started was started after injection of DPCPX(8-Cyclopentyl-1,3-dipropylxanthine). Then approximately 60–80 min later, the cat was injected with [^{11}C]FMZ and scanned for 60 min.

In thirteen cats, the right MCA was exposed with transorbital approach and was ready to be occluded. The MCA was reopened after 60 min of occlusion. Animals were subjected to serial measurements of PET. CBF of the cat was measured three times with intravenous injection of $H_2{}^{15}O$: 15 min before the MCA occlusion, 5 min after the occlusion and 5 min after reperfusion. Then, the cat underwent a series of PET measurement A1-R with ^{11}C-MPDX, BDZ with ^{11}C-FMZ, and glucose metabolism with ^{18}F-fluorodeoxyglucose (^{18}F-FDG). The injected dose was 200 MBq for [^{15}O]water, [^{11}C]MPDX and for [^{11}C]FMZ, and 100 MBq for ^{18}F FDG.

After the experiment, a licensed veterinarian, who was one of authors (Y. S.), observed the neurological and systemic physiological parameters of the cats. Four cats determined as having severe systemic and neurological conditions by cerebral ischemia were euthanized by an i.v. overdose of pentobarbital immediately after completing the PET scans without withdrawing anesthesia. Animals that were determined to be in good condition were continuously monitored by a veterinarian based on the postoperative management protocol of the veterinary hospital. Two months after the initial experiment, the cats were evaluated by T2 weighted MR images with an animal MRI system (4.7 Tesla, Unity plus SIS 200/330, Varian, PaloAlto, CA, USA). PET data analyses were carried out on UNIX workstations (Silicon Graphics Inc., Mountain View, CA, USA) using the Dr. View image analysis software system (Asahi Kasei Joho System, Tokyo, Japan). All the PET images were co-registered with the MRI of each animal with the semi-automatic registration method as described previously [19].

All experimental procedures were approved by the Animal Care and Use Committee of the Tokyo Metropolitan Institute of Gerontology (Protocol No. 99064).

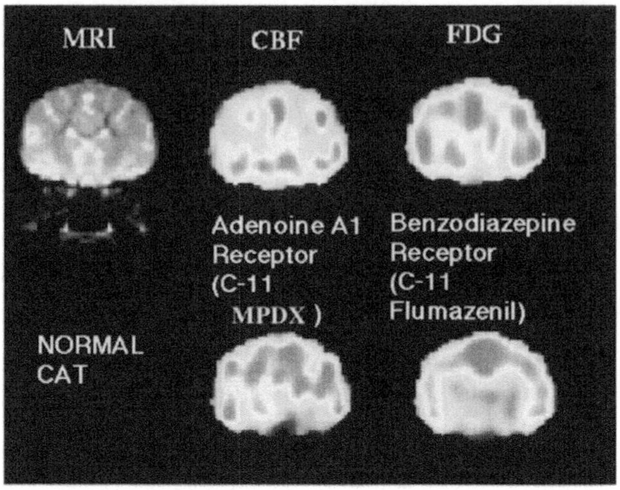

Fig. 1. PET images in normal control, which express CBF, ^{18}F-FDG uptake, cBZD-R binding and A1-R binding were displayed. All the PET images were co-registered with the animal's own MRI to obtain correct anatomical configuration. ^{11}C-MPDX distributed among most brain regions including basal ganglia and brain stem. ^{11}C-FMZ binding was observed mainly in cerebral cortex but sparse in basal ganglia and brain stem

Results

PET images of control animal

PET images in normal control, which express CBF, ^{18}F-FDG uptake, cBZD-R binding and A1-R binding were displayed in Fig. 1. All the PET images were co-registered with the animal's own MRI to obtain correct anatomical configuration. ^{11}C-MPDX distributed among most brain regions including basal ganglia and brain stem. ^{11}C-FMZ binding was observed mainly in cerebral cortex but sparse in basal ganglia and brain stem. Distribution of ^{11}C-MPDX among whole brain region makes the evaluation of ischemic event, evaluation of which is difficult by ^{11}C-FMZ imaging.

PET imaging in ischemia

Four cats did not survive the 1st day and were euthanized because of severe neurological condition and were classified as the group with the most severe ischemic insult (Group D). Three cats survived and had infarction both in the cortex and striatum (Group Cor). The other six had infarcted lesion only in the striatum (Group Str). Data were compared with that of normal control (Group N). The CBF during occlusion of MCA was significantly reduced in all three ischemic groups compared to the normal group. However, no statistically significant difference in the reduction rate was found among the three ischemic groups. The right/left ratio of CBF and FDG uptake after reperfusion did not show any statistically significant difference among all groups. Both the binding of MPDX and FMZ to A1-Rs and BDZ-R, respectively, were significantly less in Group D than in groups N and Str. Of two receptor ligands examined, the reduction rate in the binding of MPDX in group D was larger than that of FMZ binding in the same group. The larger reduction in binding of MPDX and FMZ to A1-Rs and BDZ-Rs, respectively, was always a sign that severer ischemic damage has occurred. On the other hand, in terms of CBF and glucose metabolism, not only reduction but also increase may be a sign of severer damage. As shown in Fig. 2, the widespread and close to zero reduction in the binding of MPDX to the A1-Rs (arrow) was a sign that subsequent severe ischemic damage is occurring; this animals did not survive. In this animals, reduction of the BDZ-R and glucose metabolism was not marked as MPDX binding. The CBF after reperfusion was increased in the ischemic side than in the contralateral side (hyperemia).

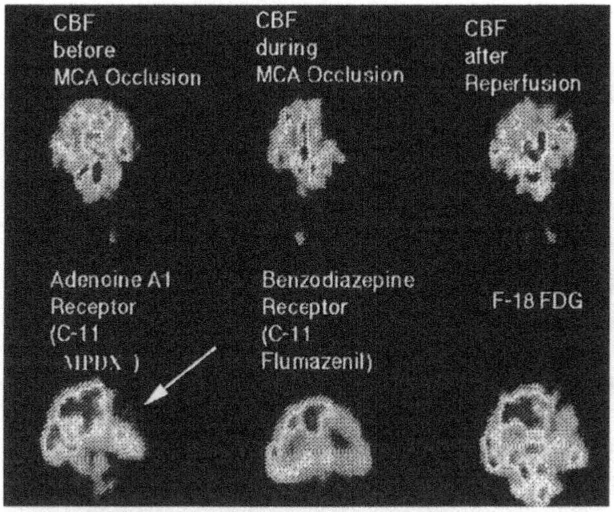

Fig. 2. PET image to demonstrate the CBF before, during and after the reperfusion, A-R binding with ^{11}C-MPDPA, and BZD-R binding with ^{11}C-FMA and the glucose metabolism with ^{18}F-FDG. The widespread and close to zero reduction in the binding of MPDX to the A1-Rs (arrow) was a sign that subsequent severe ischemic damage is occurring; this animals did not survive. In this animals, reduction of the BDZ-R and glucose metabolism was not marked as MPDX binding. The CBF after reperfusion was increased in the ischemic side than in the contralateral side (hyperemia)

Discussion

In this paper, we have demonstrated that use of ^{11}C-MPDX was a premising PET imaging method to measure the binding potential of A1-R. Use of this tracer in cerebral ischemia may be useful to estimate the degree of subsequent ischemic cerebral damage. Of the other parameters simultaneously analyzed, the binding potential of FMZ to BZD-Rs was also a good indicator for early detection of ischemic lesions.

It has already been reported that BZD-R as a marker of neuronal integrity is a superior method for early and accurate detection of cerebral damage in acute stages of ischemia [6, 7]. Our present results of BZD-R imaging also showed its superiority to CBF and CMRGlc imaging. The A1-R imaging in the present study indicated that it is also a superior method for detection of ischemic damage. Moreover, the reduction rate of A1-R binding in severe ischemia was larger than that of BZD-R. There is possibility that the A1-R imaging can evaluate the pathophysiology of cerebral ischemia on the different theoretical background from the BZD-R imaging.

It is recognized that the ischemic condition initiates a marked elevation of extracellular adenosine [4, 11, 12]. In this condition, competition between intrinsic adenosine and an extrinsically injected tracer might cause a decreased binding of MPDX to A1-Rs in the ischemic area, where adenosine is released in the extracellular space much more than in the non-ischemic area. If this hypothesis stands, the A1-R imaging by PET may provide a new method to monitor not only the extent of already damaged neuronal integrity, but also the severity of the on-going cerebral ischemic insult.

As we have already completed PET A1-R imaging in the pre-clinical condition in the primate brain [8], application of this method in the human brain may be possible in the near future. As the neuroprotective role of adenosine is postulated not only in cerebral ischemia, but also in neurotrauma and epilepsy [2, 5, 13, 21], the PET A1-R imaging have good potential to become a powerful new technique for evaluation and treatment of various neurological disorders.

References

1. Ardekani BA, Braun M, Hutton BF, Kanno I, Iida H (1995) A fully automatic multimodality image registration algorithm. J Comput Assist Tomogr 19: 615–623
2. Dunwiddie TV (1999) Adenosine and suppression of seizures. Adv Neurol 79: 1001–1010
3. Furuta R, Ishiwata K, Kiyosawa M, Ishii S, Saito N, Shimada J, Endo K, Suzuki F, Senda M (1996) Carbon-11-labeled KF15372: a potential central nervous system adenosine A1 receptor ligand. J Nucl Med 37: 1203–1207
4. Hagberg H, Andersson P, Lacarewicz J, Jacobson I, Butcher S, Sandberg M (1987) Extracellular adenosine, inosine, hypoxanthine, and xanthine in relation to tissue nucleotides and purines in rat striatum during transient ischemia. J Neurochem 49: 227–231
5. Headrick JP, Bendall MR, Faden AI, Vink R (1994) Dissociation of adenosine levels from bioenergetic state in experimental brain trauma: potential role in secondary injury. J Cereb Blood Flow Metab 14: 853–861
6. Heiss WD, Graf R, Fujita T, Ohta K, Bauer B, Lottgen J, Wienhard K (1997) Early detection of irreversibly damaged ischemic tissue by flumazenil positron emission tomography in cats. Stroke 28: 2045–2051; discussion 2051–2052
7. Heiss WD, Kracht L, Grond M, Rudolf J, Bauer B, Wienhard K, Pawlik G (2000) Early [(11)C]Flumazenil/H(2)O positron emission tomography predicts irreversible ischemic cortical damage in stroke patients receiving acute thrombolytic therapy. Stroke 31: 366–369
8. Ishiwata K, Nariai T, Kimura Y, Oda K, Kawamura K, Ishii K, Senda M, Wakabayashi S, Shimada J (2002) Preclinical studies on [^{11}C]MPDX for mapping adenosine A1 receptors by positron emission tomography. Ann Nucl Med 16: 377–382
9. Ishiwata K, Noguchi J, Wakabayashi S, Shimada J, Ogi N, Nariai T, Tanaka A, Endo K, Suzuki F, Senda M (2000) 11C-labeled KF18446: a potential central nervous system adenosine A2a receptor ligand [in process citation]. J Nucl Med 41: 345–354

10. Ishiwata K, Takai H, Nonaka H, Ishii S, Simada J, Senda M (2001) Synthesis and preliminary evaluation of a carbon-11-labeled adenosine transporter blocker [(11)C]KF21562. Nucl Med Biol 28: 281–285
11. Matsumoto K, Graf R, Rosner G, Shimada N, Heiss WD (1992) Flow thresholds for extracellular purine catabolite elevation in cat focal ischemia. Brain Res 579: 309–314
12. Melani A, Pantoni L, Corsi C, Bianchi L, Monopoli A, Bertorelli R, Pepeu G, Pedata F (1999) Striatal outflow of adenosine, excitatory amino acids, gamma-aminobutyric acid, and taurine in awake freely moving rats after middle cerebral artery occlusion: correlations with neurological deficit and histopathological damage. Stroke 30: 2448–2454; discussion 2455
13. Nilsson P, Hillered L, Ponten U, Ungerstedt U (1990) Changes in cortical extracellular levels of energy-related metabolites and amino acids following concussive brain injury in rats. J Cereb Blood Flow Metab 10: 631–637
14. Noguchi J, Ishiwata K, Furuta R, Simada J, Kiyosawa M, Ishii S, Endo K, Suzuki F, Senda M (1997) Evaluation of carbon-11 labeled KF15372 and its ethyl and methyl derivatives as a potential CNS adenosine A1 receptor ligand. Nucl Med Biol 24: 53–59
15. Noguchi J, Ishiwata K, Wakabayashi S, Nariai T, Shumiya S, Ishii S, Toyama H, Endo K, Suzuki F, Senda M (1998) Evaluation of carbon-11-labeled KF17837: a potential CNS adenosine A2a receptor ligand. J Nucl Med 39: 498–503
16. Rudolphi KA, Schubert P, Parkinson FE, Fredholm BB (1992) Adenosine and brain ischemia. Cerebrovasc Brain Metab Rev 4: 346–369
17. Rudolphi KA, Schubert P, Parkinson FE, Fredholm BB (1992) Neuroprotective role of adenosine in cerebral ischaemia. Trends Pharmacol Sci 13: 439–445
18. Shimada Y, Ishiwata K, Kiyosawa M, Nariai T, .Oda K, Toyama H, Suzuki F, Ono K, Senda M (2002) Mapping adenosine A(1) receptors in the cat brain by positron emission tomography with [^{11}C]MPDX. Nucl Med Biol 29: 29–37
19. Shimada Y, Uemura K, Ardekani BA, Nagaoka T, Ishiwata K, Toyama H, Ono K, Senda M (2000) Application of PET-MRI registration techniques to cat brain imaging. J Neurosci Methods 101: 1–7
20. Wakabayashi S, Nariai T, Ishiwata K, Nagaoka T, Hirakawa K, Oda K, Sakiyama Y, Shumiya S, Toyama H, Suzuki F, Senda M (2000) A pet study of adenosine A1 receptor in anesthetized monkey brain [In Process Citation]. Nucl Med Biol 27: 401–406
21. Winn HR, Welsh JE, Rubio R, Berne RM (1980) Changes in brain adenosine during bicuculline-induced seizures in rats. Effects of hypoxia and altered systemic blood pressure. Circ Res 47: 568–577

Correspondence: Dr. Tadashi Nariai M.D., Department of Neurosurgery, Tokyo Medical and Dental University, 1-5-45 Yushima, Bunkyo-ku, Tokyo 113-8519, Japan. e-mail: nariai.nsrg@tmd.ac.jp

Acta Neurochir (2003) [Suppl] 86: 49–52

PET investigation of post-traumatic cerebral blood volume and blood flow

N. Hattori[1], **S.-C. Huang**[1], **H.-M. Wu**[1], **W. Liao**[1], **T. C. Glenn**[2], **P. M. Vespa**[2], **M. E. Phelps**[1], **D. A. Hovda**[1,2], and **M. Bergsneider**[2]

[1] Department of Molecular and Medical Pharmacology, David Geffen School of Medicine at UCLA, Los Angeles, CA, USA
[2] Brain Injury Research Center, Division of Neurosurgery, David Geffen School of Medicine at UCLA, Los Angeles, CA, USA

Summary

Hemodynamic changes following traumatic brain injury (TBI) may reflect cellular damage leading to secondary injury. The purpose of this study was to investigate the regional hemodynamic parameters acutely after TBI among regions in and around contusions. Sixteen patients (11 male, 5 female) showing evidence of contusion on CT and 18 normal volunteers (12 male, 6 female) underwent positron emission tomography (PET) with O-15 CO and O-15 H2O to estimate cerebral blood volume (CBV) and cerebral blood flow (CBF), respectively. A flow to volume ratio (FVR = CBF/CBV) was also calculated as an index of vasodilatation. The hemodynamic parameters were compared among contusion, pericontusion, and remote areas. Globally, hemodynamic parameters did not differ between patients and normal volunteers, and did not correlate with intracranial pressure (ICP). Regionally, contusional and pericontusional areas showed significantly lower CBF and FVR compared with normal volunteers, while CBV did not differ significantly. The correlation between CBF and CBV was significant ($r = 0.37$, $p < 0.01$). Remote areas did not show a significant difference in any of the PET parameters. In conclusion, regional brain edema is likely to occur in contusion and pericontusion areas, while some of the contusional tissue may show vascular engorgement.

Introduction

Acute hemodynamic changes following severe traumatic head injury may induce vascular dilatation with hyperemia or compression of the microvasculature causing ischemia. In either case, cerebral edema may result and reflect reversible/irreversible cellular damage. Measurement of hemodynamic parameters acutely after injury may help to estimate the possibility of secondary injury [4]. Clinical medical imaging permits in-vivo assessment of hemodynamic parameters including cerebral blood flow (CBF) and cerebral blood volume (CBV). In general, prior studies have demonstrated that in severe traumatic brain injury (TBI), CBF is reduced during the initial week of injury [1, 5, 9]. No similar trend or clear pattern has been seen with respect to post-traumatic changes of CBV. The absolute value of CBV has been reported to be increased [11] or decreased after traumatic brain injury [5].

Functional imaging using positron emission tomography (PET) can produce parametric images of hemodynamic parameters in conjunction with CT and MRI studies. The combination of O-15 water and O-15 carbon monoxide PET studies permit in-vivo measurement of cerebral blood flow (CBF) and cerebral blood volume (CBV), respectively. Using these parameters, post-traumatic hemodynamic disturbances have been classified as follows: vascular dilatation as an increase in CBV relative to CBF; cerebral edema as a proportional reduction in CBF and CBV, and cerebral ischemia as CBF < 20 ml/100g/min [10]. The goal of this study was to investigate these regional hemodynamic changes acutely after TBI comparing contusional, pericontusional and remote areas.

Materials and methods

Subjects

Sixteen patients (11 male, 5 female) showing clear evidence of a cerebral contusion on CT scan were included in this study. Epidemiological and physiological data are summarized in Table 1. A control group of 18 normal volunteers (12 male, 6 female; mean age was 34 ± 8 years, range 21 to 47 years) were similarly studied. All study procedures were performed with informed consent approved by the UCLA Institutional Review Board and the UCLA Radiation Safety Office.

Study protocol

For the TBI cohort, routine clinical CT scans were obtained at admission as well as within 48 hours of the PET study. Patients were

Table 1. *Cinical characteristics of the patients*

No. of patients	16 (5 women)
Age (range)	38 ± 17 (17–81) years
Median GCS score (range)	11 (3–15)
Time after injury (range)	64 ± 26 (17–107) hours
ICP at the time of PET (range)*	17 ± 4 (12–29) mmHg
Papillary abnormality	
– Both	5
– One side	1
– None	10
Cause of injury	
– MVA	8
– Fall	4
– GSW	1
– Other	3
Medication at the time of PET	
– Midazolam	7 cases
– Morphine	6 cases
– Neuromuscular blocker	5 cases
– Mannitol	1 case
– Phenytoin	1 case
– Propofol	1 case

GCS Glasgow Coma Scale; *ICP* Intra Cranial Pressure; *MVA* Motor Vehicle Accident; *PET* Positron Emission Tomography; *GSW* Gun Shot Wound.

* ICP was vailable in 15/16 patients.

admitted to the intensive care unit (ICU) after initial stabilization or surgical evacuation of intracranial hematomas causing excessive mass effect, and were treated in accordance with the Guidelines for Treatment of Severe Traumatic Brain Injury [2]. The contusions and surrounding areas analyzed had not been surgically disturbed. The PET scanning was performed at the earliest possible time after injury.

Positron emission tomography

Each patient underwent a single PET session that consisted of two separate PET scans: O-15 labeled CO (for cerebral blood volume) followed by O-15 H2O (for cerebral blood flow). A transmission study was obtained for attenuation correction. Dynamic acquisition of PET and serial arterial blood sampling were started immediately after a single breath inhalation of CO_2, or a bolus intravenous injection of H2O, respectively. Administered doses were 555 MBq for CO_2 and H2O. The acquisition protocol consisted of 6 frames (6×60 sec) for CO_2 and 26 frames (6×5 sec, 9×10 sec, 6×30 sec and 5×60 sec) for H2O, respectively.

All PET data were acquired with a Siemens/CTI ECAT HR+ (Knoxville, TN) in three-dimensional acquisition mode (collimating septa retracted). To minimize the effect of scatter originating from out of field-of-view activities, acquisition was done after manually installing a lead shield annulus (CTI Neuro-Insert) with 9 mm thickness and 35 cm aperture diameter. The scanner produces 63 planes covering 15.5 cm of axial field of view. The intrinsic spatial resolution of the scanner is 4.3 mm full-width at half-maximum resolution (FWHM) at center of the central axis of the gantry.

Acquired image sets were reconstructed with a filtered back-projection method using a Hanning filter (cutoff frequency of 0.3 cycle per projection element) to generate 128×128 matrices. Spatial resolution of the resulted images was ~5.5 mm FWHM. All dynamic scans were corrected for dead-time, attenuation (using a 10 minute transmission scan obtained immediately before the dynamic studies), scatter and decay prior to image reconstruction.

Image analysis

In order to anatomically localize contusions on the PET images, CT (obtained within 48 hours of the PET study) images were first rotated along a three-dimensional axis to coregister them to the PET images. The coregistration algorithm was based on maximization of the mutual information between two images. Powell's multidimensional search algorithm was then applied to search for the set of transformation parameters that maximizes the mutual information [7, 8]. CT images were rigidly transformed using the parameters to match the PET image. The coregistered CT images were then resliced into 63 planes to produce the same planes with the images of PET.

Regions of interest (ROIs) were placed on the coregistered CT. Each ROI was manually drawn tracing the area of contusion, a pericontusional zone, and remote cortex. The pericontusional area was defined as the low-density area on CT located around the contusion (visually determined). ROIs were drawn on frontal and parietal cortex for the normal control subjects. Values for regional CBV and CBF were obtained after generating parametric images of CBV (in ml/100g) and CBF (in ml/100g/min). A flow-to-volume ratio (FVR = CBF/CBV) was calculated as an index of vasodilatation.

Statistical analysis

Difference of means between the 2 groups was investigated using students' t-test. Correlation of two continuous variables was investigated using Pearson's correlation coefficient. Level of significance was set to be $p < 0.05$, while $p < 0.01$ was considered to show a trend.

Results

Globally, patients with TBI had slightly higher CBV (6.4 ± 2.6 ml/100g) compared with normal volunteers (5.4 ± 1.3 ml/100g, p = n.s.) whereas CBF was similar for patients (38.8 ± 9.8 ml/100g/min) and normal volunteers (39.8 ± 5.3 ml/100g/min). As a result, FVR was slightly lower in patients but did not differ significantly (8.7 ± 3.1 vs. 9.6 ± 2.1, p = n.s.). Correlations between global PET parameters versus ICP showed no significant trend of correlations.

Table 2 summarizes the regional PET parameters with comparison to the values in the cortex of normal volunteers. Regionally, hemodynamic parameters demonstrated a clear difference among contusional, pericontusional, and remote areas. The contusional and pericontusional areas showed significantly lower CBF compared with those of normal volunteers, while CBV did not differ significantly. Figure 1 shows a plot of the CBF versus CBV following TBI to analyze the correlation between the two hemodynamic variables. There was a significant trend of positive correlation between the two variables ($r = 0.37$, $p < 0.01$), although the plot showed a scattered distribution of CBV in some of the contusional and pericontusional regions. The absolute CBF values were

Table 2. *Summary of regional PET parameters*

	Patients			Volunteers
	Contusion n = 16	Pericontusion n = 16	Remote n = 16	Normal cortex n = 18
CBF (ml/100g/min)	12.3 ± 8.8*	27.8 ± 13*	46.5 ± 15.3	47.7 ± 7.5
CBV (ml/100g)	3.6 ± 2.0	4.3 ± 1.9	5.0 ± 1.8	4.5 ± 1.3
FVR	5.0 ± 4.4*	8.2 ± 4.4†	11.4 ± 3.8	12.5 ± 3.3

CBF Cerebral Blood Flow, *CBV* Cerebral Blood Volume, *FVR* Flow to Volume Ratio.
* $p < 0.001$, † $p < 0.01$, ‡ $p < 0.05$ vs. Normal volunteers.

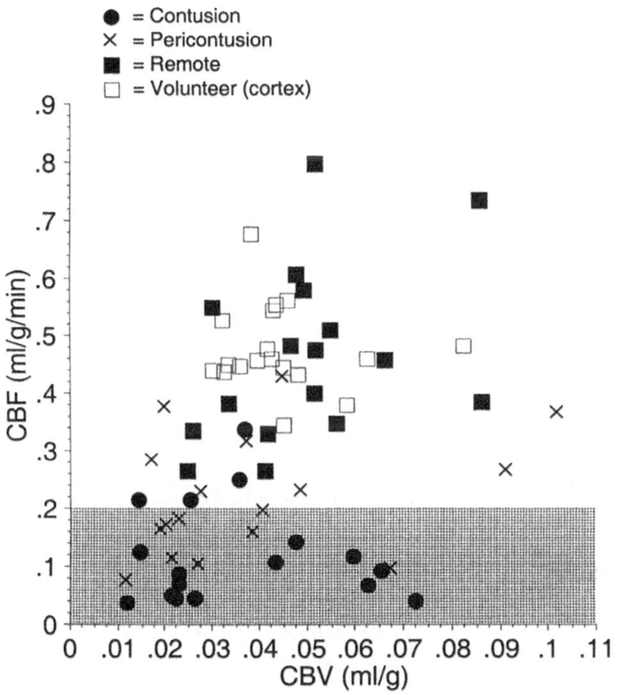

Fig. 1. Scatter plot showing the correlation between cerebral blood flow (*CBF*) and cerebral blood volume (*CBV*). Shaded zone represents ischemia

below 20 ml/100g/min in 9/16 (56%) of pericontusional and 12/16 (75%) of contusional areas. Regional FVR was significantly lower than that of normal cortex in contusional and pericontusional areas. Remote areas did not show a significant difference compared with those of normal volunteers in any of the PET parameters (CBF, CBV and FVR).

Discussion

The nature of cerebral edema following TBI, particularly that surrounding intracerebral contusions, remains unclear. Some have suggested that local increases in intracranial pressure cause compression of microvasculature with resultant ischemia and cytotoxic edema [3, 6]. In this study, we indeed observed a significantly reduced CBF in both contusional and pericontusion areas. Reduction of CBF was profound in the area of contusion showing 75% of cases had CBF value less than 20 ml/100g/min, while it was in 56% of pericontusional areas, and none of remote areas. These findings generally agree with the previous study by Schroeder *et al.* despite CBF being measured using a different modality (Xe-CT) in their study [9].

Although regional hemodynamic parameters showed a significantly reduced CBF in and around the area of contusion, CBV did not change significantly. Furthermore, in the pericontusional areas, we observed that the reduced CBF was coupling with reduced CBV. Our results did not suggest vascular compression as the sole etiology of the pericontusional edema, since FVR, an index of vascular dilatation, was significantly reduced in pericontusional and contusional areas. Although brain edema is the likely mechanism of these regions, our findings suggest that vascular dilatation may also contribute to the edematous change seen on CT imaging.

Hemodynamic changes in contusional areas were more profound than pericontusional areas. Some of the contusional areas showed a high CBV relative to CBF, suggesting vascular engorgement. Hemodynamic parameters of the remote areas did not differ significantly from those of normal volunteers.

Global hemodynamic parameters did not change significantly acutely after traumatic brain injury. The global hemodynamic parameters may show different values if the similar analysis were made on the population showing clear evidence of brain swelling. Our present findings on global CBF generally agree with previous investigations reporting slight decrease in 1–5 days after injury. On the other hand, we did not observe a reduction in global CBV as reported by Marmarou *et al.* [5]. The discrepancy may be explained by difference in severity of injury between the two studies (ours with a higher GCS). As a result, our study might not have included as many patients with severe global brain swelling (as confirmed with well controlled ICP in the current study population, Table 1). The narrow distribution of ICP may also explain the lack of correlation between ICP and hemodynamic parameters in the present population.

In conclusion, regional hemodynamic parameters differ significantly among contusion, pericontusion and remote areas. Regional brain edema with vasodilatation is likely to occur within contusion and pericontusion areas since some of the contusional and pericontusional tissue appear to show vascular engorgement. In patients without brain swelling, global and remote areas do not show abnormal hemodynamic parameters.

Acknowledgment

This study was supported by the UCLA Brain Injury Research Center, NINDS 30308 and DoE DE FC0387-ER630615.

References

1. Alexander MJ, Martin NA, Khanna R, Caron M, Becker DP (1994) Regional cerebral blood flow trends in head injured patients with focal contusions and cerebral edema. Acta Neurochir (Wien) [Suppl] 60: 479–481
2. Bullock RM, Chestnut RM, Clifton GL, Ghajar J, Marion DW, Narayan RK, Newell DW, Pitts LH, Rosner MJ, Walters BC, Wilberger JE (2000) Guidelines for the management of severe head injury. Brain Trauma Foundation, American Association of Neurological Surgeons, Joint section on neurotrauma and critical care. J Neurotrauma 17: 449–627
3. Di X, Goforth PB, Bullock R, Ellis E, Satin L (2000) Mechanical injury alters volume activated ion channels in cortical astrocytes. Acta Neurochir (Wien) [Suppl] 76: 379–383
4. Hovda DA, Becker DP, Katayama Y (1992) Secondary injury and acidosis. J Neurotrauma 9 [Suppl] 1: S47–60
5. Marmarou A, Fatouros PP, Barzo P, Portella G, Yoshihara M, Tsuji O, Yamamoto T, Laine F, Signoretti S, Ward JD, Bullock MR, Young HF (2000) Contribution of edema and cerebral blood volume to traumatic brain swelling in head-injured patients. J Neurosurg 93: 183–193
6. Marmarou A, Portella G, Barzo P, Signoretti S, Fatouros P, Beaumont A, Jiang T, Bullock R (2000) Distinguishing between cellular and vasogenic edema in head injured patients with focal lesions using magnetic resonance imaging. Acta Neurochir (Wien) [Suppl] 76: 349–351
7. Powell MJD (1964) An efficient method for finding the minimum of a function of several variables without calculating derivatives. Comput J 7: 155–162
8. Press WH, Teukolsky SA, Vetterling WT, Flannery BP (1992) Numerical recipes in C 2 edn. Cambridge University Press, Cambridge University Press
9. Schroder ML, Muizelaar JP, Bullock MR, Salvant JB, Povlishock JT (1995) Focal ischemia due to traumatic contusions documented by stable xenon-CT and ultrastructural studies. J Neurosurg 82: 966–971
10. Schroder ML, Muizelaar JP, Kuta AJ, Choi SC (1996) Thresholds for cerebral ischemia after severe head injury: relationship with late CT findings and outcome. J Neurotrauma 13: 17–23
11. Tenjin H, Ueda S, Mizukawa N, Imahori Y, Hino A, Yamaki T, Kuboyama T, Ebisu T, Hirakawa K, Yamashita M, Nakahashi H (1990) Positron emission tomographic studies on cerebral hemodynamics in patients with cerebral contusion. Neurosurgery 26: 971–979

Correspondence: Naoya Hattori, M.D., Ph.D., Box 956948, Los Angeles, CA 90095-6948, USA. e-mail: nhattori@mednet.ucla.edu

Acta Neurochir (2003) [Suppl] 86: 53–55

Quantitative evaluation of cerebral vascular permeability using multi-slice dynamic CT

K. Nambu[1,2], T. Nariai[3], and T. Terada[4]

[1] Medical Engineering Research & Development Center, Medical Systems Company, TOSHIBA Corporation, Japan
[2] Institute of Advanced Biomedical Engineering & Science, Tokyo Womens Medical University, Japan
[3] Department of Neurosurgery, Tokyo Medical and Dental University, Japan
[4] Department of Neurosurgery, Wakayama Medical University, Japan

Summary

We developed a method to measure regional permeability of cerebral capillary vessels, extracellular fluid space volume, and vascular bed volume using multi-slice dynamic CT with iodinated contrast medium. We implemented the method as a novel software "BBB study", and evaluated it in four cases of brain tumor. The examination time was about 20 minutes, and the resulting functional maps had quality sufficient to distinguish area of irregular permeability enhancement.

Keywords: Vascular permeability; extracellular fluid space; dynamic CT; multi-slice CT.

Introduction

Cerebral vascular permeability, extracellular fluid space volume, and vascular bed volume indicate the histopathological condition of a region, and they are important considerations for diagnosis and treatment of brain tumor and edema. We have developed a mathematical model which explain regional contrast enhancement process of iodinated contrast medium in cerebral tissue, including parameters corresponding to permeability, extracellular fluid space volume, and vascular bed volume. In normal brain tissue, BBB (Blood Brain Barrier) inhibits leakage of the contrast medium from vessels. When BBB breaks down, the medium can leak to extracellular fluid space by free diffusion, but still it can not enter into cells. Therefore:

$$C(t) = VeCe(t) + VpCp(t)$$

$$dCe/dt = Kb(Cp(t) - Ce(t))$$

$$C(0) = Cp(0) = Ce(0) = 0$$

where

t: time
C(t): enhancement of CT number at a region
Ce(t): concentration of contrast medium in extracellular fluid space
Cp(t): concentration of contrast medium in intravascular plasma
Ve: rational volume of extracellular fluid space
Vp: rational volume of intravascular plasma
Kb: capillary permeability (P-S product)

and we also can calculate

$$Ki = VeKb$$

as an index of capillary leakage of contrast medium from capillary vessels.

Based on this model, we have developed a method to measure the permeability and other parameters using dynamic CT, and we have verified it by more than a hundred cases of brain tumor [1]. The method could measure regional Ki, Kb, Ve, Vp, and Ce(t), from given C(t) and Cp(t). However, clinical application of the method was limited, because the method required a CT scanner to repeat scanning in a few seconds cycle, so actually, only one slice position could be examined in an examination.

We now can use the multi-slice CT scanner to acquire volumetric dynamic CT, and it became possible to examine the spatial distribution of capillary vessels of irregular permeability in one examination. So, we implemented the method as a novel software "BBB study" for a multi-slice CT scanner, and examined it preliminarily in clinical cases.

Table 1. *Delayed dynamic CT protocol: scanning timing after injection (seconds)*

5 7 9 11 13 15 17 19 21 23 25 27 29 31 33 35 37 39 41 43 45
65 95 125 155 185 215 245 305 455 605 755 905 1055 1205

Materials and methods

We examined a case of pituitary adenoma, and three cases of glioblastoma. By careful explanation about the merits and demerits of the examination, informed consent was obtained in all cases. After a bolus injection of contrast medium (Iopamidol 42–50 mL, 3–5 mL per second) via the cubital vein, we applied a delayed dynamic CT protocol (Table 1) for four slice positions of each 8 mm thickness, with technique of 120 kV, 100 mA, and 1 second per scan, using a multi-slice CT scanner "Aquilion/Multi 4 slice system" (Toshiba, Tokyo). We measured the arterial contrast enhancement Cp(t) at MCA. Then the "BBB study" software calculated maps of the following parameters by a least square curve fitting algorithm: Ki (index of capillary leakage [%/sec]), Kb (capillary permeability [1/sec]), Vp (plasma volume [%]), and Ve (extracellular fluid space volume [%]). The maps were displayed in gray scale and color scale.

Results

The examination took about 20 minutes for CT scan, and several minutes to calculate the maps for four slice positions. The total radiation dose was 52 cGy for each slice. We obtained clear maps showing an increase in permeability and extracelluler fluid space volume in the case of an adenoma, and two cases of glioblastoma (Fig. 1). However, one of the cases of glioblastoma did not show an explicit increase of those parameters (Fig. 2).

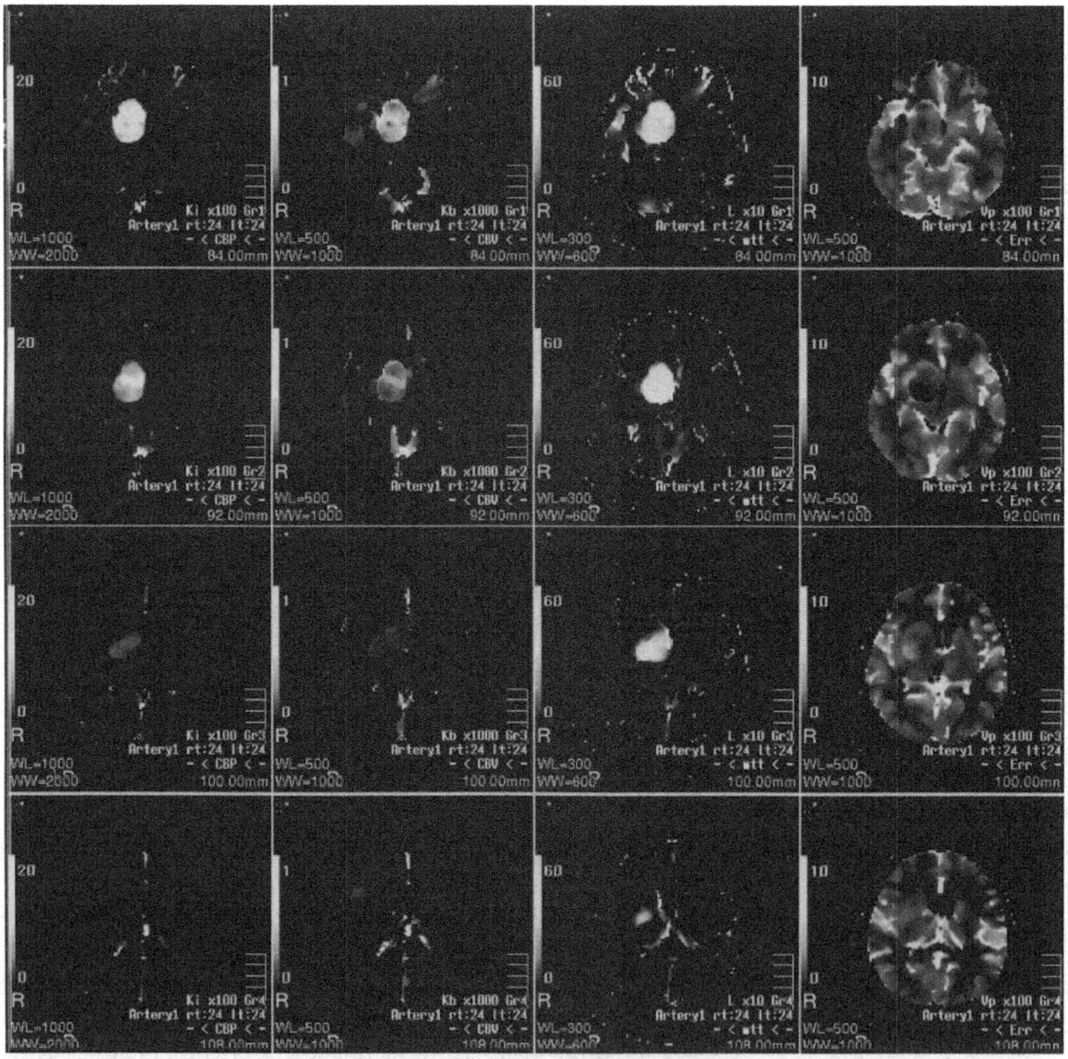

Fig. 1. Ki, Kb, Ve, Vp map (left to right) of four slices in a case of pituitary adenoma

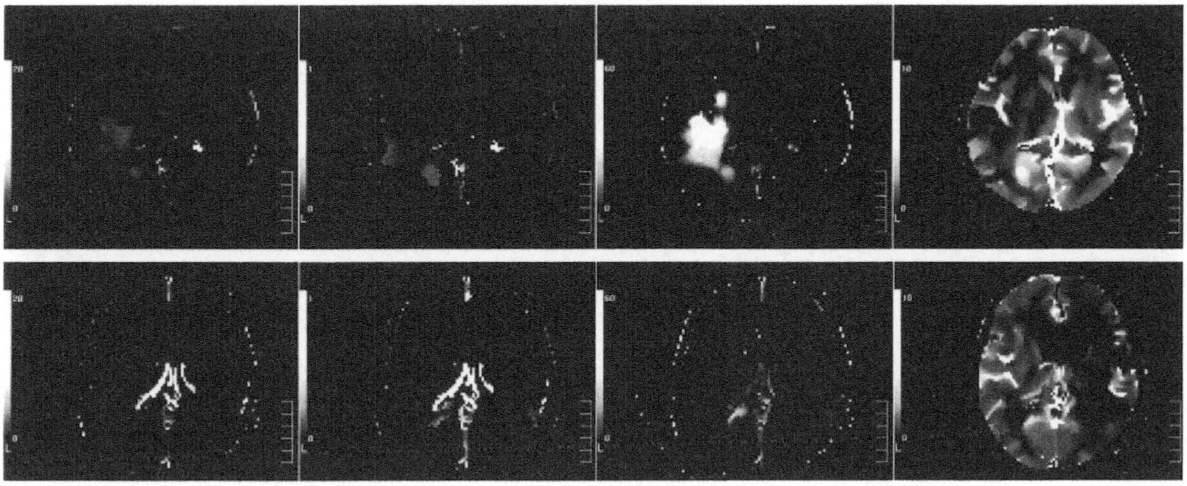

Fig. 2. Ki, Kb, Ve, Vp map (left to right) of two cases of glioblastoma. A case shows explicit leakage and extension of extracellular fluid space (upper row), but another case does not (lower row)

Discussion

Our dynamic CT scanning protocol consists of two parts: early scans (period of 40 seconds after injection) and delayed scans. The delayed scans are necessary to detect the slow leakage process of contrast medium from capillary vessels, and the early scans detect the first pass of bolus to separate the contrast enhancement effect of vessels, from the contrast enhancement in the extracellular fluid space.

Our method requires only a CT scanner and iodinated contrast medium, so it is possible to execute in most institutions. Because the examination takes about 20 minutes, the patient should be fixed carefully. In the cases of glioblastoma, the maps demonstrated various degree of BBB breakdown. Thus, the method can be applied for planning of chemotherapy.

Since the image quality of the maps were good, we are testing the performance of our method in reduced radiation dose, by decreasing the number of scans and X ray technique. Furthermore, the protocol of contrast agent administration and the early part of the dynamic CT scanning are fully compatible with examination to measure cerebral vascular blood perfusion using "CBP study" [2]. So it is possible to examine perfusion or permeability in one examination, and it will save time and X ray dose especially in acute cases.

References

1. Terada T, Nambu K, Hyotani G *et al* (1992) A method for quantitative measurement of cerebral vascular permeability using X-ray CT and iodinated contrast medium. Neuroradiology 34: 290–296
2. Nambu K, Takehara R, Terada T (1996) A method of regional cerebral blood perfusion measurement using dynamic CT with an iodinated contrast medium. Acta Neurol Scand [Suppl] 166: 28–31

Correspondence: Kyojiro Nambu, Medical Engineering R&D Center, 1385 Shimo-Ishigam, Otawara, Tochigi, 324-8550 Japan. e-mail: nambu@mel.nasu.toshiba.co.jp, k.nambu@nifty.com

Acta Neurochir (2003) [Suppl] 86: 57–62

Acetazolamide vasoreactivity evaluated by transcranial harmonic perfusion imaging: relationship with transcranial Doppler sonography and dynamic CT

T. Shiogai[1], **M. Koshimura**[1], **Y. Murata**[1], **H. Nomura**[2], **A. Doi**[2], **M. Makino**[3], **T. Mizuno**[3], **K. Nakajima**[3], and **H. Furuhata**[4]

[1] Department of Clinical Neurosciences, Kyoto Takeda Hospital, Kyoto, Japan
[2] Department of Radiology, Takada Hospital, Kyoto, Japan
[3] Department of Neurology, Kyoto Prefectural University of Medicine, Kyoto, Japan
[4] ME Lab, Tokyo Jikei University School of Medicine, Tokyo, Japan

Summary

To establish the reliability and clinical significance of transcranial ultrasonic harmonic perfusion imaging (HPI), we evaluated HPI's relationships with transcranial Doppler (TCD) and with dynamic CT (DCT), during acetazolamide (ACZ) vasoreactivity tests.

Methods. The subjects were 12 neurological patients. Time-averaged maximum velocity (TAVMX) in the middle (MCA) and posterior cerebral arteries was measured by TCD. Time-intensity (-density) curves of HPI (DCT) after bolus intravenous contrast injections were created in 3 regions of interest (ROI) on the axial plane involving the temporal lobe, basal ganglia, and thalamus on both sides. Assessments of vasoreactivity were based on comparisons conducted before and after ACZ administration in terms of: a) relative changes (%Δ) of the TCD TAVMX, b) HPI contrast area enlargement, c) %Δ of calculated cerebral blood volume and flow of the HPI and DCT.

Results. 1) TCD vasoreactivity decrease in the left MCA tended to correlate with lower frequency of HPI contrast area enlargement on the left side. 2) HPI and DCT vasoreactivity tended to be disturbed in the same side ROIs.

Conclusions. Transcranial HPI achieves repeatable non-invasive bedside evaluation of cerebrovascular reserve capacity through qualitative and quantitative measurements of brain tissue perfusion, and will have clinical value in pathophysiological follow-up and therapeutic effectiveness determination of neurointensive care patients.

Keywords: Acetazolamide vasoreactivity; transcranial harmonic perfusion imaging; ultrasound contrast agents; transcranial Doppler sonography; dynamic CT

Introduction

Recently, transcranial harmonic imaging utilizing ultrasound contrast agents and color duplex sonography has been introduced for repeatable non-invasive bedside measurements of brain tissue perfusion [12, 14–16], particularly for ischemic stroke patients [3, 6, 13]. However, quantitative reliability of the transcranial harmonic perfusion imaging (HPI) has not yet been established, mainly due to bone- and depth-dependent attenuation of the ultrasound signals [12, 17].

Vasoreactivity tests utilizing transcranial Doppler (TCD) and cerebral blood flow (CBF) measurements have been introduced for evaluating cerebrovascular reserve capacity in ischemic cerebrovascular diseases, especially extracranial carotid occlusive lesions [5, 9–11]. Some correlation of vasoreactivity between TCD and CBF measurements has been pointed out, although superiority of the modalities has not been established [2, 9–11]. Furthermore, techniques for CBF measurements, such as single photon emission CT (SPECT) and xenon-enhanced CT (Xe-CT), are complex, labor-intensive, and impossible to use during treatments at bedside, particularly for neuro-intensive care patients.

To overcome the problem relating to depth dependant attenuation of ultrasound signals, we have introduced vasoreactivity tests for qualitative and quantitative evaluation in almost the same cerebral regions by HPI. The objective of this study is to establish the reliability and clinical significance of transcranial HPI, and to this end we evaluated HPI's relationships with TCD and dynamic CT (DCT) during ACZ vasoreactivity tests.

Table 1. *Clinical characteristics of the patients**

n	12
Age	
mean ± SD (range)	56 ± 19 (29–81)
Primary diagnosis	
cerebral infarction	
lacunae	2
hemispheric	2
angioma (cavernous and venous)	2
Moyamoya disease	1
cerebral hemorrhage	1
subarachnoid hemorrhage	1
cerebral contusion	1
others (headache and epilepsy)	2
*Side of cerebral and/or vascular lesions based on imaging studies**	
right hemisphere	6
left hemisphere	4
diffuse or none	2
Days between HPI and DCT	
mean ± SD (range)	6 ± 7 (1–21)

* Cerebral lesions involving temporal lobe, basal ganglia, and thalamus diagnosed by CT, MRI and stenotic lesions in the internal carotid artery diagnosed by MRA and/or color duplex sonography.

Table 2. *TCD and HPI findings before and after ACZ administration*

	ACZ administration	
	Before	After
TCD	TAVMX (cm/s)*	TAVMX increase (%Δ)*
MCA (R)	60 ± 21	24 ± 22
MCA (L)	63 ± 19	13 ± 18
PCA (R)	35 ± 14	18 ± 22
PCA (L)	43 ± 18	17 ± 24
HPI (% of cases)	positive contrast area	contrast area enlargement
R hemisphere	12 (100%)	9 (75%)
L hemisphere	12 (100%)	5 (42%)

* *TAVMX* Time-averaged maximum velocity, mean ± SD.

Patients and methods

The subjects were 12 various neurological patients (aged 29–81 years, mean, 56) with a stable clinical condition (Table 1). Utilizing a SONOS 5500 ultrasound system with an S4 transducer (Philips), 62 integrated backscatter images taken by transient response harmonic imaging (2-sec intervals) after 7-ml bolus intravenous injections of ultrasound contrast agent (Levovist®, 2.5 g) were evaluated on the axial plane involving the temporal lobe (TL), basal ganglia (BG), and thalamus (Th) via temporal windows. The HPI was evaluated before and after (15 and 30 minutes on the right and left sides, respectively) injection of ACZ (Diamox® 500 mg) (Fig. 1). Time-averaged maximum velocity (TAVMX) in the middle cerebral and posterior cerebral arteries (MCA, PCA) was measured by TCD just before HPI. DCT was performed before and 15 minutes after ACZ administration. Sixty scans (1-sec intervals) were analyzed at the same axial planes of HPI after a 45 ml contrast medium (Iopromide®) bolus intravenous injection. Time-intensity curves (TIC) of HPI were created in 4 regions of interest (ROI: TL, BG, Th, and MCA) on both sides (Fig. 2). Time-density curves (TDC) of DCT in the 4ROI (TL, BG, TH, and anterior cerebral artery) were also measured. Assessments of vasoreactivity were based on comparisons conducted before and after ACZ administration in terms of a) relative changes (%Δ) of the TCD TAVMX b) enlargement of HPI contrast area, c) relative changes (%Δ) of the TIC (TDC) parameters of HPI and DCT, i.e., peak intensity (PI) or peak height (PH), area under curve (AUC), mean transit time (MTT), calculated cerebral blood volume (CBV = AUC in the parenchymal ROIs/AUC in the arterial ROIs), and calculated cerebral blood flow (CBF = CBV/MTT) [7].

Utilizing a Student t-test, statistical significance was set at p values less than 0.05.

Results

a) TCD vasoreactivity (Table 2): In TAVMX values before ACZ administration, lateral differences were not obvious in the MCA, but the right PCA value was slightly lower. TAVMX increase (%Δ) tended to decrease in the left MCA (13 ± 18%) in comparison with the right MCA (24 ± 22%), however there was no obvious laterality in the PCA.

b) Contrast area enlargement of HPI was observed in 42% and 75% of cases on the left and right sides, respectively (Table 2).

c) HPI and DCT vasoreactivity (Table 3): In comparison with DCT values before ACZ administration, HPI values of calculated CBV increase caused by AUC increase and MTT prolongation were obvious. In HPI, CBV and CBF in the left side ROIs were lower than those in the right side ROIs. However, lateral differences of DCT parameters were not apparent.

After ACZ administration, %Δ of all PHI parameters in the right TL decreased in comparison with those in the left TL. The same tendency of DCT parameters in the TL was observed. Lateral differences of HPI parameters in the BG and Th ROIs were tended to be the same as those of the DCT parameters.

There was no statistical significance between HPI and DCT parameters.

Discussion

Vasoreactivity evaluated by TCD and HPI

In our series, a relationship between lower frequency of HPI contrast area enlargement and decreased %Δ was identified only in the MCA. This result is probably due to wider vascular distribution in the MCA than in

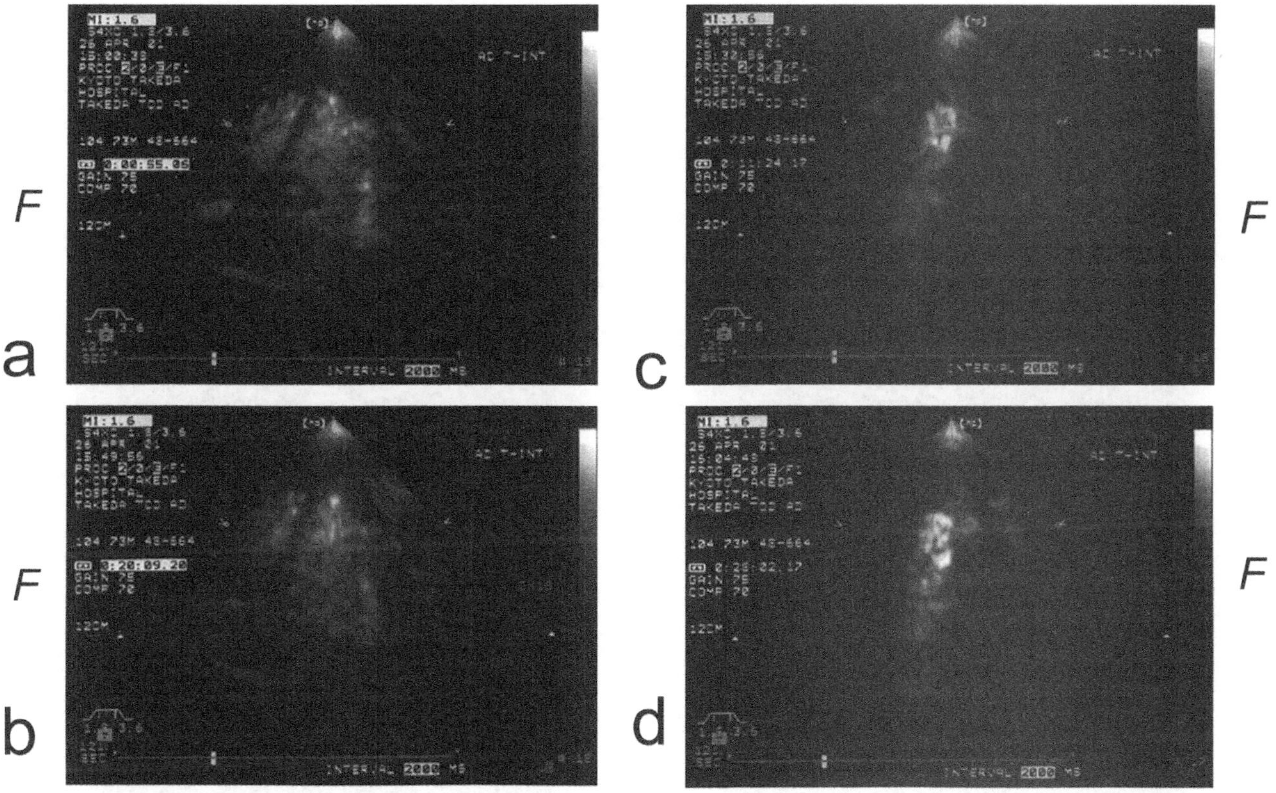

Fig. 1. HPI of a 72-year-old patient with a tight stenosis of the left internal carotid artery. A1 portion of the left anterior cerebral artery as a collateral circulation was not visualized by magnetic resonance angiography. F indicates the frontal side. LV indicates bilateral anterior horn of the lateral ventricle. Post-contrast HPI via right (a and b) and left (c and d) temporal windows demonstrated before (a and c) and after (b and d) ACZ injection. Preserved vasoreactivity indicated by increased high intensity areas after ACZ injection (b and d) was obvious in the right side (b) involving the TL, BG and Th in comparison with the left side (d), which involved mainly BG

the PCA [19]. Quantitative correlation of vasoreactivity evaluated by TCD and CBF measurements based on SPECT or Xe-CT are not always high ($r = 0.39$–0.67) in normal subjects and stroke patients [9, 10, 18]. There was no correlation between flow velocity observed by TCD and CBF in normal subjects [2] and patients with carotid occlusive lesions [11]. These discrepancies of measured vasoreactivity between TCD and HPI are probably due to topographical differences between the main trunks of MCA and PCA and small regions of interest in the TL, BG, and Th. From a clinical point of view, the easier qualitative evaluation of vasoreactivity by HPI, with the same usefulness as SPECT etc, is superior to DCT.

Vasoreactivity evaluated by DCT and HPI

In our previous series of quantitative relationship between DCT and HPI, we did not obtain close linear correlations of TIC (TDC) parameters between DCT and HPI [17]. However, the correlation in MTTs tended to be higher than those in PI and AUC [17]. In resting values before ACZ administration, as compared to DCT, a higher ratio of parenchymal AUC and arterial AUC resulted in higher values of calculated CBV. Furthermore, MTTs in HPI also prolonged in comparison with MTTs in DCT. In these results, AUC increase and MTT prolongation derived from bolus HPI TIC is probably caused by bubble saturation due to a non-linear relationship between microbubble concentration and image intensity [1]. In this regard, it may be necessary to visualize using images with a wide dynamic range such as a conventional B-mode, instead of an integrated backscatter image. Recently, other parameters in relation to transit time derived from bolus TIC analysis, such as positive gradient [4] or time to peak intensity [8] have been introduced. However, further clarification is inevitably

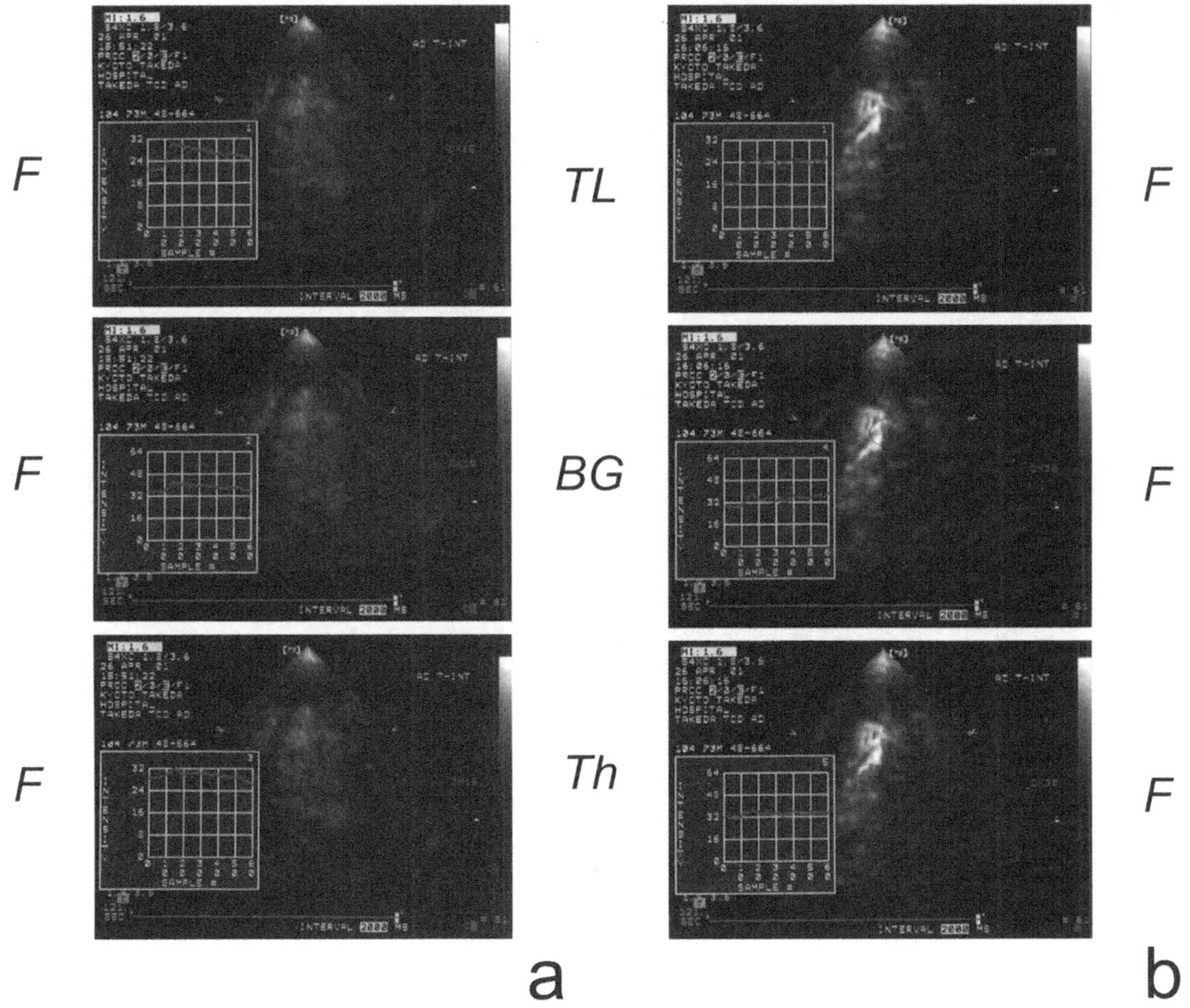

Fig. 2. Time-intensity curve analysis of HPI. Time-intensity curve analyses in the three regions of interest (*ROI*) in the temporal lobe (*TL*), basal ganglia (*BG*), and the thalamus (*Th*) were performed via a right temporal window before (a) and after (b) ACZ injection

required before the clinical use of such vasoreactivity studies.

There were some correlated tendencies between DCT and HPI vasoreactivity. However, a high variation of measured HPI parameters was observed in comparison with DCT parameters. This probably related to unstable fixation of ultrasonic probe during HPI measurements and a different slice of the visualized image before and after ACZ administration. In other relevant points, differing patients' conditions, such as blood pressure, PaCO2, pulse rates etc., between HPI and DCT measurements should be considered [17]. Lateral differences of HPI vasoreactivity are probably related to background pathological lesions. However, temporal differences between right and left measurements of HPI after ACZ administration should also be taken into account. Continuing studies are necessary for improving data variability and establishing clinical reliability of HPI.

Conclusions

Transcranial HPI allows repeatable non-invasive bedside evaluation of cerebrovascular reserve capacity by qualitative and quantitative measurements of brain tissue perfusion, and will have clinical value in the

Table 3. *HPI and Dynamic CT (DCT) findings before and after acetazolamide (ACZ) administration**

HPI	ACZ administration						
	Before					After	
	PI (dB)	AUC (dB × s)	MTT (s)	CBV	CBF	% ΔCBV	% ΔCBF
TL (R)	12.5 ± 5.3	889 ± 558	57.1 ± 7.5	0.82 ± 0.72	0.84 ± 0.68	22 ± 127	24 ± 116
TL (L)	10.9 ± 5.0	728 ± 401	56.4 ± 9.8	0.59 ± 0.14	0.65 ± 0.22	43 ± 115	28 ± 80
BG (R)	10.2 ± 4.1	703 ± 409	55.1 ± 12	0.56 ± 0.23	0.62 ± 0.28	1.9 ± 35	13 ± 66
BG (L)	9.84 ± 3.6	562 ± 265	54.3 ± 10	0.49 ± 0.21	0.55 ± 0.21	49 ± 101	38 ± 85
Th (R)	5.13 ± 3.4	358 ± 338	48.8 ± 15	0.36 ± 0.46	0.40 ± 0.44	40 ± 96	46 ± 87
Th (L)	3.53 ± 2.3	142 ± 109	45.5 ± 16	0.12 ± 0.08	0.16 ± 0.09	166 ± 317	60 ± 127
MCA (R)	18.3 ± 5.4	1321 ± 566	56.7 ± 10				
MCA (L)	16.7 ± 6.2	1241 ± 607	60.0 ± 6.9				
DCT	PH (HU)	AUC (HU × s)	MTT (s)	CBV	CBF	% ΔCBV	% ΔCBF
TL (R)	9.21 ± 3.3	116 ± 41	21.4 ± 3.8	0.22 ± 0.17	0.63 ± 0.44	14 ± 31	17 ± 32
TL (L)	9.92 ± 3.3	121 ± 45	20.8 ± 3.4	0.21 ± 0.12	0.62 ± 0.34	36 ± 58	38 ± 59
BG (R)	7.13 ± 2.8	93 ± 43	21.1 ± 3.6	0.15 ± 0.08	0.43 ± 0.19	27 ± 35	28 ± 35
BG (L)	7.12 ± 1.6	90 ± 23	20.8 ± 3.2	0.16 ± 0.09	0.47 ± 0.28	30 ± 58	32 ± 58
Th (R)	7.96 ± 2.0	102 ± 33	21.6 ± 4.2	0.17 ± 0.09	0.46 ± 0.44	37 ± 47	46 ± 59
Th (L)	7.6 ± 2.4	95 ± 33	20.7 ± 3.4	0.23 ± 0.19	0.46 ± 0.28	50 ± 66	46 ± 59
ACA	55.3 ± 26	701 ± 307	20.2 ± 3.4				

* *TL* Temporal lobe, *BG* basal ganglia, *Th* thalamus, *R* right, *L* left, *MCA* middle cerebral artery, *ACA* anterior cerebral artery, *PI* peak intensity, *PH* peak height, *AUC* area under curve, *HU* Housfield's units, *MTT* mean transit time, *CBV* cerebral blood volume, *CBF* cerebral blood flow.

pathophysiological follow-up of neurointensive care patients and in the determination of therapeutic effectiveness.

References

1. Claassen L, Seidel G, Algermissen C (2001) Quantification of flow rates using harmonic grey-scale imaging and an ultrasound contrast agent: an in vitro and in vivo study. Ultrasound Med Biol 27: 83–88
2. Dahl A, Russell D, Rootwelt K, Nyberg-Hansen R, Kerty E (1995) Cerebral vasoreactivity assessed with transcranial Doppler and regional cerebral blood flow measurements. Dose, serum concentration, and time course of the response to acetazolamide. Stroke 26: 2302–2306
3. Federlein J, Postert T, Meves S, Weber S, Przuntek H, Buttner T (2000) Ultrasonic evaluation of pathological brain perfusion in acute stroke using second harmonic imaging. J Neurol Neurosurg Psychiatry 69: 616–622
4. Harrer JU, Klotzsch C (2002) Second harmonic imaging of the human brain: the practicability of coronal insonation planes and alternative perfusion parameters. Stroke 33: 1530–1535
5. Heiss WD, Podreka I (1993) Role of PET and SPECT in the assessment of ischemic cerebrovascular disease. Cerebrovasc Brain Metab Rev 5: 235–263
6. Meairs S, Daffertshofer M, Neff W, Eschenfelder C, Hennerici M (2000) Pulse-inversion contrast harmonic imaging: ultrasonographic assessment of cerebral perfusion. Lancet 355: 550–551
7. Meier P, Zierler LK (1954) On the theory of the indicator-dilution method for measurement of blood flow and volume. J Appl Physiol 6: 731–744
8. Meves SH, Wilkening W, Thies T, Eyding J, Holscher T, Finger M, Schmid G, Ermert H, Postert T (2002) Comparison between echo contrast agent-specific imaging modes and perfusion-weighted magnetic resonance imaging for the assessment of brain perfusion. Stroke 33: 2433–2437
9. Muller M, Voges M, Piepgras U, Schimrigk K (1995) Assessment of cerebral vasomotor reactivity by transcranial Doppler ultrasound and breath-holding. A comparison with acetazolamide as vasodilatory stimulus. Stroke 26: 96–100
10. Piepgras A, Schmiedek P, Leinsinger G, Haberl RL, Kirsch CM, Einhaupl KM (1990) A simple test to assess cerebrovascular reserve capacity using transcranial Doppler sonography and acetazolamide. Stroke 21: 1306–1311
11. Pindzola RR, Balzer JR, Nemoto EM, Goldstein S, Yonas H (2001) Cerebrovascular reserve in patients with carotid occlusive disease. Stroke 32: 1811–1887
12. Postert T, Muhs A, Meves S, Federlein J, Przuntek H, Buttner T (1998) Transient response harmonic imaging: an ultrasound technique related to brain perfusion. Stroke 29: 1901–1907
13. Postert T, Federlein J, Weber S, Przuntek H, Buttner T (1999) Second harmonic imaging in acute middle cerebral artery infarction: preliminary results. Stroke 30: 1702–1706
14. Seidel G, Algermissen C, Christoph A, Claassen L, Vidal-Langwasser M, Katzer T (2000) Harmonic imaging of the human brain. Visualization of brain perfusion with ultrasound. Stroke 31: 151–154
15. Shiogai T, Tsunezuka C, Koshimura M. Takayasu N, Nagakane Y, Nakase T, Makino M, Nakajima K, Furuhata H (2000)

Comparative visualization of brain tissue perfusion in various neurological disorders utilizing contrast-enhanced harmonic imaging in integrated backscatter and power Doppler methods. Cerebrovasc Dis 10 [Suppl 1]: 44

16. Shiogai T, Tsunezuka C, Ohara T, Imai K, Makino M, Nakajima K, Furuhata (2001) Quantitative measurements of brain tissue perfusion by transcranial contrast-enhanced harmonic imaging: Correlation with dynamic CT. Cerebrovasc Dis 11 [Suppl 3]: 13
17. Shiogai T Tsunezuka C, Ohara T, Imai K, Makino M, Nakajima K, Furuhata (2001) Clinical significance of transcranial contrast-enhanced harmonic perfusion imaging (HPI) as a quantitative method of measuring cerebral blood flow. Neurotraumatology 24: 77–82 (In Japanese)
18. Sorteberg W, Lindegaard KF, Rootwelt K, Dahl A, Nyberg-Hansen R, Russell D, Nornes H (1989) Effect of acetazolamide on cerebral artery blood velocity and regional cerebral blood flow in normal subjects. Acta Neurochir (Wien) 97(3–4): 139–145
19. Tatu L, Moulin T, Bogousslavsky J, Duvernoy H (1998) Arterial territories of the human brain cerebral hemispheres. Neurology 50: 1699–1708

Correspondence: Toshiyuki Shiogai, M.D., Ph. D., Department of Clinical Neurosciences, Kyoto Takeda Hospital, Minamikinutacho 11, Nishinanajo, Shimogyo-ku, Kyoto 600-8884, Japan. e-mail: shiogait@pop11.odn.ne.jp

Acta Neurochir (2003) [Suppl] 86: 63–64

Functional brain imaging of optical topography

T. Kawaguchi

Application Development Office, Hitachi Medical Corporation, Japan

Summary

Hitachi medical corporation developed and produced a new modality called Optical Topography System "ETG-100". This system uses near infrared light as in spectroscopy. ETG-100 measures the oxy hemoglobin, deoxy hemoglobin and total hemoglobin volume changes in cerebral cortex.

This optical topography system has various features, i.e., displaying the hemoglobin volume change in 2 dimensional images, non-invasiveness, compactness, low cost and less patient restraint.

This system is now being used for the research of higher-order brain functions, for example language function, language recognition. Research of brain function disease of new born babies, cerebral blood flow infarction detection are expected.

Keywords: Hemoglobin optical; topography near; infrared spectroscopy functional; image.

The optical topography (OT) system ETG-100 developed by Hitachi Medical corporation, is the first commercial topographic imaging system using near infrared spectroscopic technique. This system measures cerebrocortical blood volume change.

The optical topography system measures the transition of oxy and deoxy hemoglobin volumes in the cerebral cortex applying the hemoglobin's characteristic absorption of near infrared light which has high permeation in the human body. When measuring, this system irradiates two wave length laser light to the patient's head via optical fiber. The optical topography system displays the time-course graph and topography image of the oxy and deoxy hemoglobin change. These are calculated from the detected signals of the light reflected and scattered in the brain. The activated area and dynamic changes of oxygen metabolism can be measured with optical topography images that express the changes of cerebrocortical hemoglobin volume. It delivers two dimensional images and shows the regional cerebral cortex activation with 2 dimensional motion pictures. ETG-100 OT system can display the topographic images on the MR brain surface image as three dimensional image. The picture of ETG-100 is shown in Fig. 1.

Specifications of this system are:

Measuring items:
- oxy-hemoglobin concentration change
- deoxy-hemoglobin concentration change
- total hemoglobin concentration change

Measurement Channels: 24 points maximum
Sampling Time: 0.1 seconds minimum
Probe Configurations:
- 3 × 3 for left and right temporal lobe
- 4 × 4 for occipital or frontal lobes

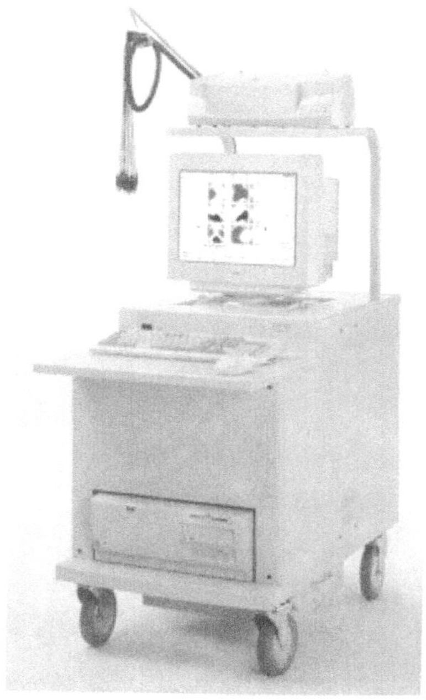

Fig. 1.

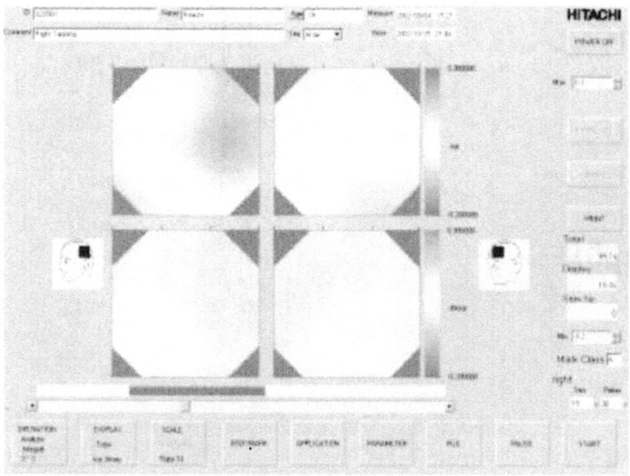

Fig. 2.

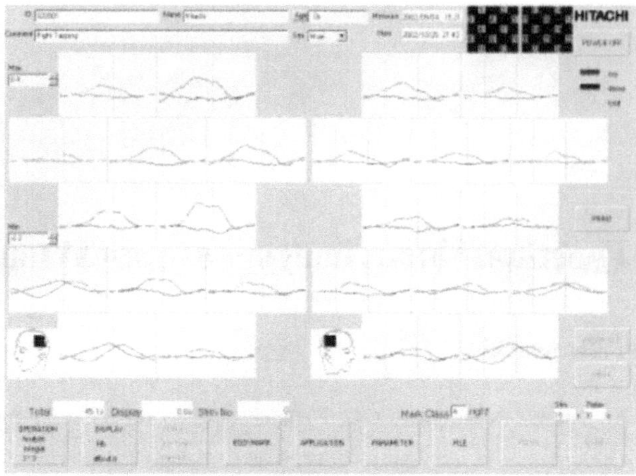

Fig. 3.

Data Format:

- Hemoglobin graphs
- 2D video of hemoglobin density

Right data are motor area hemoglobin volume change images with finger tapping.

Fig. 2 shows one picture of 2D video of hemoglobin density change. In hatched timing that is the finger tapping task time, right motor's oxy hemoglobin volume increase.

Fig. 3 shows the same task data displayed in hemoglobin graph. This graph shows the measuring point and hemoglobin change with time.

This optical topography system offers the following features:

(1) non-invasive procedure
(2) compact unit and low cost
(3) less patient restraint.

This system is now being used for research of higher-order brain functions. For example language function, language recognition. Research of brain function diseases of new born babies and cerebral blood flow infarction detection are expected.

Correspondence: Tsuneaki Kawaguchi, Application Development Office, Hitachi Medical Corporation, 2-1 shintoyofuta kashiwa-city chiba, Japan zip277-8678. e-mail: kawaguchi@kf.hitachi-medical.co.jp

Experimental global ischemia

Acta Neurochir (2003) [Suppl] 86: 67–70

Significant shrinkage of extracellular space during global cerebral ischemia: differences in gray and white matter ischemia

E. Kumura[1], **C. Dohmen**[3], **R. Graf**[3], **T. Yoshimine**[2], and **W.-D. Heiss**[3]

[1] Department of Neurosurgery, Hanwa Memorial Hospital, Osaka, Japan
[2] Osaka University Graduate School of Medicine, Osaka, Japan
[3] Max-Planck Institute for Neurological Research, Cologne, Germany

Summary

Brain extracellular space (ECS) provides an important microenvironment for neurons and glial cells. In the present study, we investigated differences in ischemic changes of ECS in gray and white matter during global ischemia in cats (n = 8). Diffusion capacity of tetramethylammonium ion (TMA^+, 74.1Da) was evaluated by measuring extracellular TMA^+ concentration applied via microdialysis. In both gray and white matter, TMA^+ concentration significantly increased within 10 minutes after the induction of ischemia, and sustained during 120 min of ischemia. The increase, however, was slower and smaller in white matter than in gray matter. In other two animals, volume fraction and tortuosity of ECS were determined by iontophoresis during global ischemia. After 30 min of ischemia induction, volume fraction was decreased and tortuosity was increased in both gray and white matter. In white matter, decrease in volume fraction and increase in tortuosity were smaller than in gray matter. The present study demonstrates that cerebral ischemia induces not only significant shrinkage of ECS volume but also restricts molecular diffusion within. Smaller changes of diffusion capacity within white matter ECS during ischemia may be relevant for lower ischemic vulnerability of the region.

Keywords: Cerebral ischemia; white matter; extracellular space; diffusion; microdialysis; iontophoresis.

Introduction

Brain extracellular space (ECS) provides an important microenvironment for neurons and glial cells [9]. ECS contains mainly CSF and additional matrix of proteoglycans and glycosaminoglycans. Although precise physiological role of ECS is not elucidated, shrinkage of ECS has been reported during cerebral ischemia or anoxia. In 1980, Hansen *et al.* reported that brain ischemia caused nearly 50% of shrinkage of ECS [2]. Based on molecular diffusion analysis, Nicholson *et al.* proposed two properties of ECS, which were volume fraction, as a relative amount of ECS in the tissue, and tortuosity, a measure of hindrance of cellular obstructions within ECS [6]. Vorísek *et al.* demonstrated that decrease in volume fraction of ECS was associated with increase in tortuosity during anoxia [14].

Ischemic ECS shrinkage is considered a consequence of massive transmembrane ionic fluxes due to rapid energy depletion. The ionic fluxes are inevitably associated with redistribution of water across the membrane resulting in swelling of cells and shrinkage of ECS compensatory. Until recently, studies about ischemic changes in extracellular ions, especially sodium and potassium ions, were focused on gray matter [3]. Another important constituent of the brain, white matter has been less studied with regard to ischemic ion shifts. We demonstrated that changes in extracellular sodium, potassium, calcium and hydrogen ions were slower and smaller in white matter than in gray matter during global ischemia [4, 5]. According to the difference in ischemic ion shifts between gray and white matter, shrinkage of white matter ECS is expected to be smaller than that of gray matter, while dynamic changes of white matter ECS during global cerebral ischemia are not fully elucidated.

The structure of brain ECS resembles foam, and diffusing small molecules within ECS execute random movements that affect their concentration distribution. Therefore, size and geometry of ECS can be estimated by measuring concentration of small molecule delivered into ECS from a point source [8, 9]. In the present study, we investigated differences in ischemic changes of ECS of gray and white matter during 120 min of global cerebral ischemia. Because gray matter and white matter differ functionally, metabolically and structurally, ischemic changes of ECS in those regions may differ. Changes in diffusion capacity of tetramethylammonium ion (TMA^+, 74.1Da) in ECS was evaluated by measuring extracellular TMA^+ concentration applied via either microdialysis or iontophoresis before and after the induction of ischemia.

Materials and methods

Eight adult cats were anesthetized with α-chloralose (60 mg/kg) under artificial ventilation of 70% N_2O and 30% O_2. Intravenous infusion of 5 mg/kg/h gallamine triethiodide for muscle relaxation was maintained throughout the experiment. Both common carotid arteries and both proximal subclavian arteries were exposed. To form snare ligatures, nylon sutures were placed around each exposed artery [4]. Arterial blood pressure was monitored continuously. Arterial blood gases were measured intermittently, and P_aCO_2 was kept at 30–35 mmHg using an expiratory end-tidal gas monitor. Two burr holes were made above the temporal auditory cortex and above the frontal cortex. Assemblies of a microdialysis probe, a TMA^+ selective microelectrode and a platinum electrode were inserted at the middle ectosylvian gyrus site 1 mm deep into cortex (gray matter), and at the marginal gyrus site 9 mm deep into the corona radiata (white matter) [11]. Thermocouples were inserted adjacent to the assemblies. Both burr holes were filled with gelfoam soaked in CSF. Body and brain temperatures were controlled separately at 37.0 °C with two independent feedback temperature control systems. Platinum iridium electrodes were used to measure cerebral blood flow by hydrogen clearance method. Electrocorticogram was recorded from reference electrode of TMA^+ sensitive microelectrode placed at gray matter. Microdialysis probes (CMA/12, Carnegie Medicin, Sweden) were perfused continuously with artificial CSF containing 1 mM of TMA^+-chloride at a flow rate of 2 µl/min. Extracellular concentration of TMA^+ ($[TMA^+]_o$) was measured continuously by TMA^+ selective microelectrode placed at 1.5 mm distance from microdialysis probe in both gray and white matter. Double-barrel theta glass was used for TMA^+ selective microelectrodes. The ion-sensing barrel was filled with a liquid ion exchange (Coring 477317) with backfilling solution of 150 mM TMA^+-chloride. Reference barrel was filled with 150 mM NaCl. TMA^+ selective microelectrodes were calibrated at known concentrations of TMA^+-chloride solutions at 37 °C. After 2 hours of stabilization of $[TMA^+]_o$, global cerebral ischemia was initiated by occluding the 4 extracranial arteries and kept for 120 min, and $[TMA^+]_o$ were recorded in gray and white matter simultaneously.

In additional two animals, diffusion parameters of TMA^+ were measured using iontophoretic application of TMA^+-chloride (1 M). Assemblies of TMA^+-chloride iontophoresis electrode and TMA^+ sensitive microelectrode placed with tips distance 150 µm were calibrated under free-diffusion condition (0.3% agar) at 37 °C [7]. Then the assemblies were inserted in gray and white matter in the above described manner. TMA^+-chloride was injected by iontophoresis at 100 nA with 60 sec duration followed by 90 sec of observation period, and repeated. $[TMA^+]_o$ was measured continuously and time-concentration curves of $[TMA^+]_o$ were obtained at each iontophoresis. From time-concentration curves of $[TMA^+]_o$, volume fraction and tortuosity were calculated [7]. Global cerebral ischemia was initiated by occluding the extracranial arteries and kept for 120 min. Throughout the experiment, iontophoretic TMA^+ injection was repeated, and time-concentration curves of $[TMA^+]_o$ were recorded in gray and white matter simultaneously. Each 4 successive curves starting at 30 min, 60 min and 120 min were analyzed to calculate volume fraction and tortuosity. All experimental data are expressed as means ± standard deviation (SD). The significance of differences at $p < 0.05$ was tested using analysis of variance (ANOVA) with multiple *post hoc* comparisons.

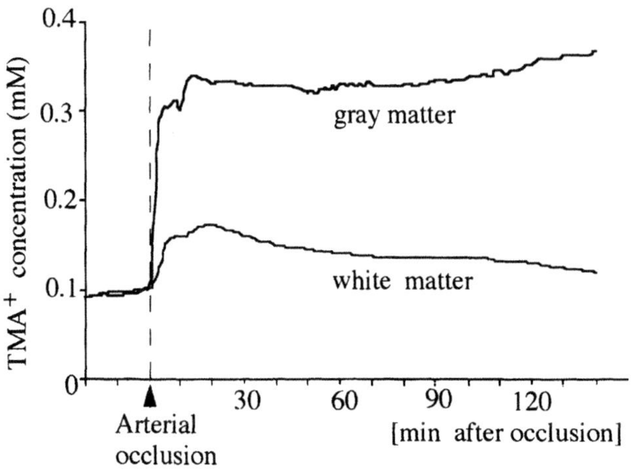

Fig. 1. Example of changes in tetramethylammonium ion (TMA^+) concentration measured with combination of microdialysis and TMA^+ selective microelectrode in gray and white matter during global cerebral ischemia

Results

Figure 1 shows representative $[TMA^+]_o$ recording with TMA^+ application via microdialysis in gray and white matter before and after multiple extracranial arterial occlusion. Immediately after the arterial occlusion, all cats showed significant blood flow reduction (<3 ml/100 g/min) in both gray matter and white matter, and flat electrocorticogram. In gray matter, $[TMA^+]_o$ increased in two steps: First steep increase about 10 min after occlusion was interrupted by rather gradual increase. At 10 min after the occlusion, $[TMA^+]_o$ increased rapidly reaching 315 ± 28% to that of pre-ischemia levels ($p < 0.01$). During the remaining course of the experiment, $[TMA^+]_o$ in gray matter sustained the levels reaching 342 ± 32% of pre-ischemia levels at 120 min after occlusion ($p < 0.01$). In white matter, similar change of $[TMA^+]_o$ were observed after occlusion, although the speed of increase was slower than that in gray matter. Initial increase at 10 min after occlusion reached 156 ± 22% to that of pre-ischemia levels ($p < 0.01$), and which continued at 30 min after occlusion reaching 177 ± 20% of pre-ischemia levels ($p < 0.01$). During the remaining course of the experiment, the levels sustained in white matter (Fig. 2). In additional two animals, analyses of time-concentration curves of $[TMA^+]_o$ applied with

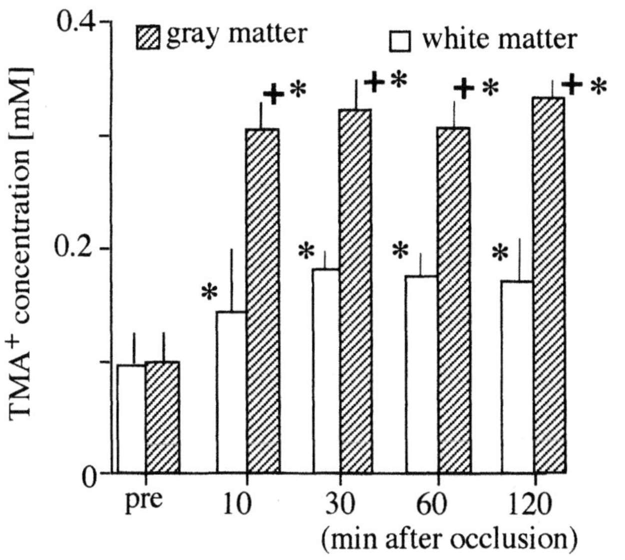

Fig. 2. Changes in tetramethylammonium ion (TMA^+) concentration measured with combination of microdialysis and TMA^+ selective microelectrode in gray and white matter during global cerebral ischemia. (*; $p < 0.001$ vs. pre, +; $p < 0.001$ vs. white matter, pre: pre-ischemia)

iontophoresis revealed pre-ischemia values of volume fraction and tortuosity in gray matter were 0.20 ± 0.02 and 1.5 ± 0.1 respectively, and 0.22 ± 0.04 and 1.5 ± 0.1 in white matter. In gray matter, volume fraction was decreased to 0.05 ± 0.02 at 30 min after the occlusion ($p < 0.01$ vs pre-ischemia), and the levels were kept at 60 min and 120 min. Tortuosity of gray matter ECS was increased to 2.0 ± 0.2 at 30 min after the occlusion ($p < 0.01$ vs pre-ischemia), and which was sustained during occlusion. In white matter, volume fraction was decreased to 0.07 ± 0.02 at 30 min after the occlusion ($p < 0.01$ vs pre-ischemia), and the levels were sustained thereafter. Tortuosity of ECS in ischemic white matter was increased to 1.7 ± 0.2 at 30 min after the occlusion ($p < 0.01$ vs pre-ischemia), and sustained thereafter. The changes of these diffusion parameters of white matter were much smaller than that of gray matter.

Discussion

In the present study, we demonstrated that ischemic shrinkage of ECS of white matter was smaller than that in gray matter during ischemia. To evaluate ischemic changes of ECS by small molecule diffusion within, we measured changes in TMA^+ concentration continuously administered via microdialysis in the first experiment. Since administered TMA^+ stays within ECS with little uptake [7], increase in $[TMA^+]_o$ reflects shrinkage of ECS due to water escape from ECS to intracellular space. During initial 10 min after occlusion, $[TMA^+]_o$ increased steeply in both gray and white matter. The increase was smaller and slower in white matter than in gray matter. During this acute period of cerebral ischemia, energy depletion causes failure of sodium-potassium pump, which release potassium into ECS. In turn, sodium ion enters neurons and glial cells accompanied with water. As ischemia-related extracellular ions shifts were smaller and slower in white matter compared to gray matter [4, 5], which may explain slow and small changes in $[TMA^+]_o$ of white matter. After 30 min of occlusion, $[TMA^+]_o$ continued high levels in both gray and white matter. During the late period of ischemia, ion homeostasis was completely lost and acidosis reached its plateau [4, 5]. These conditions could keep $[TMA^+]_o$ to be high.

By measuring extracellular TMA^+ concentration applied via microdialysis, real-time changes of ECS were estimated as shown in Fig. 1. Application of TMA^+ via microdialysis is based on the dialysis principle in which a membrane permeable to water and small solutes is continuously perfused on one side with a solution containing TMA^+ whereas the other side faces brain ECS. Continuous flow through the dialysis probe carries TMA^+ to the measuring site by a concentration gradient. When concentration of TMA^+ reached equilibrium between inside and just outside of dialysis probe's membrane, TMA^+ concentration measured at fixed distance from the probe is proportional to the concentration of TMA^+ applied and $(\text{tortuosity})^2/(\text{volume fraction})$ of ECS [1, 10]. Therefore, increase in TMA^+ concentration observed after the induction of ischemia could indicate changes in volume fraction, tortuosity or both.

To evaluate tortuosity and volume fraction of ECS in the present animal model, we analyzed time-concentration curves of TMA^+ by iontophoresis in other two animals. The results showed that pre-ischemic values of volume fraction did not show regional differences between gray matter and white matter, and the values were almost similar as reported elsewhere [9, 12, 14]. At 30 min of induction of ischemia, volume fraction was significantly decreased both in gray and white matter, and the degree of volume fraction changes was smaller in white matter than in gray matter. Because volume fraction directly related with the size of ECS, the data indicated that ischemic shrinkage of ECS was smaller in white matter than that in gray matter. Tortuosity is another definitive factor of small molecule diffusion in ECS. In the present study, tortuosity increased after 30 min of occlusion in both gray and white matter, and increase in tortuosity was smaller in white matter than in gray matter. Since tortuosity is a measure of hindrance of cellular obstructions within ECS, the difference indicated that restriction of diffusion was stronger in gray matter than in white matter. The reason for the difference in tortuosity between gray and white matter can not be explained yet, and cyto-architectural differences between the two regions may be involved. In the brain, anisotropy of diffusion was reported in several regions. Anisotropy is known as differences in tortuosity of three-dimensional planes [13]. We measured only in x-axis, and did not evaluate anisotropy of diffusion in the present study.

Using microdialysis and iontophoresis technique, we demonstrated that cerebral ischemia induces not only significant shrinkage of ECS volume but also restricts molecular diffusion within. Shrinkage of ECS is a passive process by the water movement from ECS to intracellular space, which may aggravate ischemic changes in ions and excitotoxin such as glutamate and hinder clearance of those substances from ECS. Therefore, smaller changes of ischemic shrinkage of white matter ECS during ischemia may be relevant for lower ischemic vulnerability of the region.

References

1. Benveniste H, Hansen AJ, Ottosen NS (1989) Determination of brain interstitial concentrations by microdialysis. J Neurochem 52: 1741–1750
2. Hansen AJ, Olsen CE (1980) Brain extracellular space during spreading depression and ischemia. Acta Physiol Scand 108: 355–365
3. Hansen AJ (1985) Effect of anoxia on ion distribution in the brain. Physiol Rev 65: 101–148
4. Kumura E, Dohmen C, Rosner G, Graf R (1997) Ischemic ion shifts in white matter are smaller and slower than in gray matter. J Cereb Blood Flow Metab [Suppl] (1) 17: S784
5. Kumura E, Graf R, Dohmen C, Rosner G, Heiss WD (1999) Breakdown of calcium homeostasis in relation to tissue depolarization: comparison between gray and white matter ischemia. J Cereb Blood Flow Metab 19: 788–793
6. Nicholson C (1985) Diffusion from an injected volume of a substance in brain tissue with arbitrary volume fraction and tortuosity. Brain Res 333: 325–329
7. Nicholson C (1991) Quantitative analysis of extracellular space using the method of TMA^+ iontophoresis and the issue of TMA^+ uptake. Can J Physiol Pharmacol 70: S314–S322
8. Nicholson C (1993) Ion-selective microelectrodes and diffusion measurements as tools to explore the brain cell microenvironment. J Neurosci Meth 48: 199–213
9. Nicholson C, Syková E (1998) Extracellular space structure revealed by diffusion analysis. Trends Neurosci 21: 207–215
10. Parsons LH, Justice JB Jr (1994) Quantitative approaches to in vivo brain microdialysis. Crit Rev Neurobiol 8: 189–220
11. Reinoso-Suarez F (1961) Topographischer Hirnatlas der Katze. E Merck AG, Darmstadt
12. Syková E, Svoboda J, Polák J, Chvátal A (1994) Extracellular volume fraction and diffusion characteristics during progressive ischemia and terminal anoxia in the spinal cord of the rat. J Cereb Blood Flow Metab 14: 301–311
13. Rice ME, Okada YC, Nicholson C (1993) Anisotropic and heterogeneous diffusion in the turtle cerebellum: implications for volume transmission. J Neurophysiol 70: 2035–2044
14. Vorísek I, Syková E (1997) Ischemia-induced changes in the extracellular space diffusion parameters, K^+, and pH in the developing rat cortex and corpus callosum. J Cereb Blood Flow Metab 17: 191–203

Correspondence: Eiji Kumura, M.D., Ph.D., Department of Neurosurgery, Hanwa Memorial Hospital, 7-11-11, Karita, Sumiyoshi-ku, Osaka, Osaka 558-0011, Japan.

Acta Neurochir (2003) [Suppl] 86: 71–74

NTP and PCr responses to hypoxia by hypothermic and normothermic respiring, superfused, neonatal rat cerebrocortical slices: an NMR spectroscopy study at 14.1 Tesla

L. Litt[1], K. Hirai[1], V. J. Basus[2], and T. L. James[2]

[1] Department of Anesthesia, University of California, San Francisco, San Francisco, California, U.S.A.
[2] Department of Pharmaceutical Chemistry, University of California, San Francisco, San Francisco, California, U.S.A.

Summary

Although mechanisms of hypothermic neuroprotection during oxygen deprivation have long been investigated, further characterizations of various molecular mechanisms are appropriate. Anticipating future studies of hypothermia and hypoxia/ischemia, we investigated the extent to which our ex vivo, NMR-based, superfused brain slice model might be helpful. (Slices are ≈350 μm thick, with 18 slices per 8 mm NMR tube.) ^{31}P NMR spectroscopic measurements were made of hypothermia-induced changes in high energy phosphates, while simultaneously monitoring and controlling tissue temperature, using ^{1}H NMR, the high spectroscopic resolution available at 14.1 Tesla (600 MHz for protons), and a recently published protocol. NTP and PCr concentrations in healthy, well-oxygenated slices decreased to (55 ± 15)% and (66 ± 30)% of their respective values at 28.0 °C when warmed to 38.0 °C, in approximate agreement with earlier in vivo studies by others. During 30 min hypoxia NTP and PCr decreased to non-observable values, regardless of temperature. After reoxygenation, NTP and PCr recoveries as percentages of respective prehypoxia values were (63% ± 16%; 70% ± 5%) for hypothermic slices (28.0 °C), and (46% ± 13%; 41% ± 21%) for normothermic slices (38.0 °C), confirming hypothermic neuroprotection during oxygen deprivation in this model, which appears suitable for use in further studies.

Keywords: Nuclear magnetic resonance (NMR); NTP (nucleotide triphosphates); ATP; PCr (phosphocreatine); brain slice; hypoxia.

Introduction

Brain temperature, an extremely important variable in studies of hypoxia and ischemia, can be determined to ±0.5 °C from accurate ^{1}H NMR determinations of the chemical shift for the resonance frequency of water protons, measured in parts per million (ppm) relative to resonance frequencies of standard NMR reference compounds [3, 13]. The temperature dependence water's chemical shift (≈0.11 ppm per 10 °C towards the right, towards the upfield region) has long been known, and calibration curves have been established by different groups [3, 13]. In our experiments, where respiring brain slices are superfused in an NMR tube inside a radiofrequency (RF) coil in an NMR spectrometer, temperatures were obtained from chemical shift difference between the water proton resonance and the resonance for methyl protons in lactate, which was used as an internal reference for chemical shifts and was taken as 1.32 ppm at 37 °C. Using ^{31}P NMR spectroscopy, we compared NTP (nucleotide triphosphates; includes ATP [≈80%], GTP, UTP, and CTP) and PCr (phosphocreatine) concentrations before and after 30 min of hypoxia, both at 28 °C and at 38 °C.

Materials and methods

General principles underlying NMR temperature determinations

Our use of NMR spectroscopy to determine temperature was based on three convenient circumstances. The first is that high resolution NMR spectrometers are routinely capable of detecting and monitoring the NMR resonance frequency of the deuteron in D_2O, heavy water. It is common for high resolution NMR spectrometers to lock all frequencies to that resonance frequency. The second is that there is no toxicity from D_2O when it is added to tissue perfusates at a low concentration (≈3% in our case). The third is that the chemical shift in ppm for the D_2O deuterium resonance has the same temperature dependence as the ^{1}H chemical shift for the H_2O proton resonance. On a practical level, the method worked as follows. Temperature-induced changes in the D_2O resonance frequency caused the spectrometer to make compensating adjustments in its primary magnetic field (B_o). Although such B_o changes kept the ppm assignment for the chemical shift of the water resonance at a fixed value, they changed the ppm values for all other chemical shifts. In the case of a 1 °C temperature increase, the spectrometer continued to assign the same chemical shift (≈4.7 ppm) to the water resonance

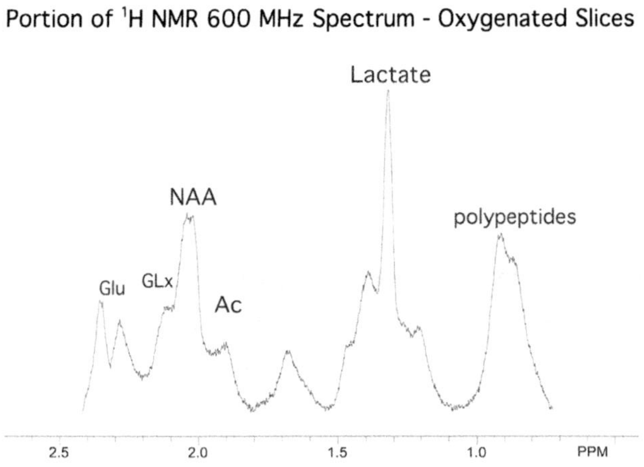

Fig. 1. Portion of a proton NMR spectrum at 600 MHz (14.1 Tesla). The actual frequencies of the resonance peaks that are labeled do not change with temperature. However, a temperature increase causes the resonance frequency for water to decrease, as described in the text. The water resonance (not shown but to the left side of the x axis) thus moves closer to the labeled peaks. However, the spectrometer is set up to always produce the same ppm reading for water. Therefore the ppm values assigned to the resonances shown are increased

while giving a .011 ppm increase to the lactate chemical shift from 1.32 ppm to 1.33 ppm. A 1 °C temperature increase also increased all other ^{1}H and ^{31}P chemical shifts by .011 ppm. Confirmation that chemical shift changes arose from a temperature change came from observing that the chemical shift for CH_3 lactate never changed relative to the chemical shift for CH_3 protons in N-acetyl-aspartate (NAA), which is 2.01 ppm at 37 °C (Fig. 1). We also verified using ^{31}P NMR that there were no pH changes. An approximate formula for the chemical shift change vs. temperature, consistent with earlier studies is:

$$\text{(Water ppm)} - \text{(lactate ppm)} = -.011\ (\text{ppm}/°\text{C}) \times (\text{Temp in }°\text{C}) + (3.757\ \text{ppm}) \qquad \text{[Equation 1]}$$

Note that because the deuterium resonance locked onto by the spectrometer is for D_2O in the perfusate, it is really perfusate temperature that is being measured, as opposed to tissue temperature, even though the lactate and NAA reference signals come from intracellular compartments. However, the ACSF volume in an 8 mm NMR tube is approximately 10 times the volume of slices, with each slice weighing ≈ 30 mg. Thus we are safe in stating that slice temperature is identical to ACSF temperature. Further details of our temperature method are described elsewhere [11].

Slice preparation and experimental design

The protocol for obtaining brain slices was approved by the UCSF Committee on Animal Research. In each experiment, 18 respiring cerebrocortical slices, each 350 μm thick, were obtained from nine 7 day old (10–12 g) Sprague-Dawley rats, and loaded into a 7 inch long, 8 mm diameter NMR tube (Wilmad-Labglass, Buena, New Jersey, http://www.wilmad.com) where they were superfused, as described previously [6, 11]. As explained in a recent publication [11], ACSF temperature in the NMR tube could be adjusted by ±0.5 °C.

Hypoxia

Thirty minute ex vivo hypoxia experiments were carried out after verifying from ^{1}H NMR spectroscopy (using Eq. 1 as described above) that the brain slice temperature was either (38.0 ± 0.5) °C or (28.0 ± 0.5) °C. Each experiment was repeated 3 times. After three hours of metabolic recovery from decapitation, a thirty minute hypoxia period was induced without stopping flow by switching at the remote reservoir to hypoxic ACSF ($PO_2 \leq 15$ mm Hg, $PCO_2 =$ 40 mmHg, pH = 7.4) whose preparation involved 30 minutes of vigorous bubbling with a 95% N_2/5% CO_2 gas mixture. The oxygen content of hypoxic ACSF was checked with blood gas and O_2 saturation measurements. Hypoxia was concluded by switching back to oxygenated ACSF at the remote reservoir, which was then continued for 1.5 hrs after hypoxia. NMR data were acquired continuously before, during, and after hypoxia.

Ex vivo ^{1}H and ^{31}P NMR spectroscopy

NMR studies were performed in the UCSF Magnetic Resonance Laboratory (http://picasso.nmr.ucsf.edu) using a 14.1 Tesla (600 MHz) Varian INOVA spectrometer with a customized, multinuclear Z-SPECT 8 mm radiofrequency probe that was optimized for this project (3NG600-8, Nalorac Corporation, Martinez, CA 94553). D_2O (3%) was added at the last moment to the ACSF to provide a chemical shift reference for the frequency lock. Basic RF pulsing sequences were used. ^{1}H spectra at 599.92 MHz were acquired with 4 acquisitions, using a jump-and-return pulse sequence with a 2 sec acquisition time and a 4 sec relaxation delay. ^{31}P spectra at 242.86 MHz were acquired with 800 acquisitions using a 90° tip angle, 8000 points, a 1 sec acquisition time and a 1 sec relaxation delay. With this repetition rate, the NTP peaks were still fully relaxed, but not the PCr peak, which was reduced to ≈50% of its fully relaxed value. ^{1}H and ^{31}P spectra acquisitions were interleaved, with acquisition times being approximately one minute for the former and 10 minutes for the latter. Relative NMR signal intensities for PCr and NTP were obtained from areas of curves fit to resonance peaks with commercial software (MacNuts Software, Acorn NMR Inc., Livermore, CA, U.S.A.). Lactate intensities were taken as peak heights relative to a broad polypeptide peak just upfield. Intracellular pH (pHi) was determined as in earlier studies [4, 11] from the chemical shift difference, δ, between phosphoethanolamine (PE) and PCr, using the formula: $\text{pHi} = 5.63 + \log[(\delta - 3.19)/(6.95 - \delta)]$.

Results

^{1}H NMR spectroscopy

Figure 1 shows the upfield region of a typical 1-minute spectrum obtained with 4 acquisitions. The sharp 1.32 ppm lactate peak for methyl protons sits among other resonances on a broad peak that is to the left (downfield) of a second (upfield) similarly broad peak. The broad peaks, once speculated upon as coming from lipids, have been studied in animal and human studies and identified for healthy brain tissue as polypeptides and macromolecules [1, 5, 9]. However, a previous study in slices did find a fatty acyl methylene peak associated with lipolysis as a somewhat sharp narrow resonance at 1.28 ppm [5], just to the right of

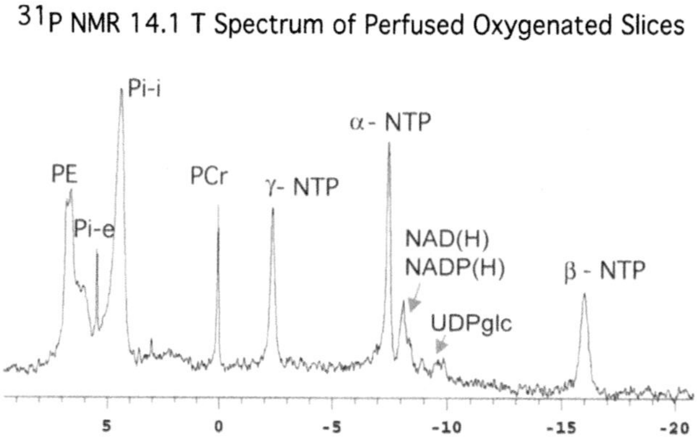

Fig. 2. Low noise ^{31}P pre-hypoxia spectrum obtained by summing NMR spectra for 12 ten minute pre-hypoxia runs, each of which assigns zero to the PCr chemical shift. Identified resonances include NTP, phosphoethanolamine (PE), and inorganic phosphate: Pi-i (intracellular) and Pi-e (extracellular). NTP consists primarily of ATP, but GTP and other nucleotide triphosphates have the same chemical shift. NTP was fully relaxed. PCr, however, was approximately 50% of its fully relaxed value

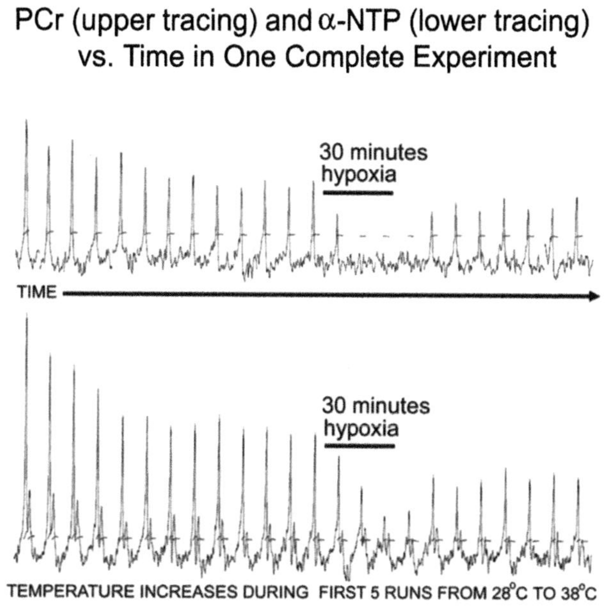

Fig. 3. Sequential PCr peaks (top) and α-NTP peaks (bottom) from successive 10-min ^{31}P spectra obtained during one hypoxia experiment. Brain slice temperature, which was 28 °C for the two left-most peaks, reached 38 °C by the end of the fifth run. The y-axis scales are slightly different for the two tracings. A solid bar representing 30 min is above data taken during hypoxia

lactate as a shoulder on the broad peak, which in Fig. 1 corresponds to a region just to left of a shoulder peak we see at 1.22 ppm, a frequency corresponding to a different resonance also found in earlier studies [5, 9]. Unambiguously evaluating the smaller overlapping peaks in one's spectra commonly requires the use of 2D NMR studies and extract studies. At this time we have not yet evaluated the smaller peaks, because we are sure from hypoxia perturbations in separate ex vivo and extract ^{13}C NMR experiments about identifying lactate. However, as pointed out in the earlier study [5], it is feasible to consider tracking the acyl methylene peak during hypoxia/ischemia studies as a potential measure of injury. Although we found small variations in overall lactate signal intensities from one slice ensemble to the next, successive data acquisitions in a single experiment, where one makes repeated measures of the same slice ensemble, showed almost no lactate fluctuations under steady state conditions.

NMR spectroscopy measures of PCr and NTP

Figure 2 shows the sum of twelve, sequential, ten-minute, normoxic, normothermic ^{31}P spectra from one experiment, obtained after slices had 3 hrs of metabolic recovery from decapitation. Typical plots showing the time course of metabolite quantifications and pHi values, as well as typical associated histological and immunohistochemical findings for such data are shown elsewhere [6]. Figure 3 shows the time course for each of two ^{31}P resonances, α-NTP and PCr, during the course of an entire experiment. The three 30 minute perioc s of hypoxia are indicated by a horizontal line above the corresponding peaks. Each ^{31}P spectrum from which the two spectral regions were excised represents 11 minutes of experimental time, because a one minute ^{1}H spectrum was acquired in between 10 minute ^{31}P spectra. Decreases both in NTP and PCr during slice warming can be seen at the left of each tracing to have been completed by the time of the fifth run. Quantitatively from all experiments, NTP and PCr concentrations in healthy, well-oxygenated slices decreased to (55 ± 15)% and (66 ± 30)% of their respective values at 28.0 °C when warmed to 38.0 °C. During 30 min hypoxia NTP and PCr decreased to non-observable values, regardless of temperature. After reoxygenation, NTP and PCr recoveries as percentages of prehypoxia values were, respectively, (63% ± 16%; 70% ± 5%) for hypothermic slices (28.0 °C), and (46% ± 13%; 41% ± 21%) for normothermic slices (38.0 °C).

Discussion

Our results confirm well known findings regarding hypothermic neuroprotection [2, 10, 12]. At colder temperatures high energy phosphate concentrations are higher, and hypoxia-induced losses are smaller. Biochemical mechanisms of hypothermic neuroprotection during oxygen deprivation originate with the Arrhenius Equation, which connects statistical mechanics with the rate constant, K, of a chemical reaction: $K = Ko\exp[-(\text{activation energy})/kT]$, where k is Boltzmann's constant, and T is absolute temperature. Because activation energies can involve many chemical bonds, they are temperature dependent and can change impressively and differently over a few degrees centigrade. Thus hypothermia can differentially modulate biochemical reactions. It is only recently that efforts have begun at characterizing differential hypothermic influences on apoptotic and necrotic processes such as mitochondrial release of cytochrome c, caspase activation, and ROS/NOS free radical injury [7, 8, 14]. Mitochondria, once investigated primarily because they produce the body's key energy substrate, ATP, are now also known for the life and death messages they can send to their cellular hosts. We believe that bioenergetic measurements such as those in this study can be used to precisely define insults and critical energy thresholds which, when combined with appropriate immunohistological studies, will be useful in characterizing mechanisms of hypothermic protection.

Acknowledgments

This study was supported by the National Institutes of Health via R01 GM34767 (LL).

References

1. Behar KL, Ogino T (1993) Characterization of macromolecule resonances in the ^{1}H NMR spectrum of rat brain. Magn Reson Med 30: 38–44
2. Chopp M, Knight R, Tidwell CD, Helpern JA, Brown E, Welch KM (1989) The metabolic effects of mild hypothermia on global cerebral ischemia and recirculation in the cat: comparison to normothermia and hyperthermia. J Cereb Blood Flow Metab 9: 141–148
3. Corbett RJ, Laptook AR, Tollefsbol G, Kim B (1995) Validation of a noninvasive method to measure brain temperature in vivo using ^{1}H NMR spectroscopy. J Neurochem 64: 1224–1230
4. Espanol MT, Litt L, Yang GY, Chang LH, Chan PH, James TL, Weinstein PR (1992) Tolerance of low intracellular pH during hypercapnia by rat cortical brain slices: A ^{31}P/^{1}H NMR study. J Neurochem 59: 1820–1828
5. Gasparovic C, Berghmans K (1998) Ca^{2+} and Mg^{2+} modulated lipolysis in neonatal rat brain slices observed by one- and two-dimensional NMR. J Neurochem 71: 1727–1732
6. Hirai K, Sugawara T, Chan PH, Basus VJ, James TL, Litt L (2002) Cytochrome c associated apoptosis during ATP recovery after hypoxia in neonatal rat cerebrocortical slices. J Neurochem 83: 309–319
7. Horiguchi T, Shimizu K, Ogino M, Yamaguchi N, Suga S, Inamasu J, Kawase T (2002) Neuroprotection role of adenosine under hypothermia in the rat global ischemia involves inhibition of not dopamine release but delayed postischemic hypoperfusion. Brain Res 952: 222
8. Inamasu J, Suga S, Sato S, Horiguchi T, Akaji K, Mayanagi K, Kawase T (2000) Postischemic hypothermia attenuates apoptotic cell death in transient focal ischemia in rats. Acta Neurochir (Wien) [Suppl] 76: 525–527
9. Kauppinen RA, Kokko H, Williams SR (1992) Detection of mobile proteins by proton nuclear magnetic resonance spectroscopy in the Guinea pig brain ex vivo and their partial purification. J Neurochem 58: 967–974
10. Laptook AR, Corbett RJ, Sterett R, Garcia D, Tollefsbol G (1995) Quantitative relationship between brain temperature and energy utilization rate measured in vivo using ^{31}P and ^{1}H magnetic resonance spectroscopy. Pediatr Res 38: 919–925
11. Litt L, Hirai K, Basus VJ, James TL (2002) Temperature control of respiring rat brain slices during high field NMR spectroscopy. Brain Res Protocols (in press)
12. Williams GD, Dardzinski BJ, Buckalew AR, Smith MB (1997) Modest hypothermia preserves cerebral energy metabolism during hypoxia-ischemia and correlates with brain damage: a ^{31}P nuclear magnetic resonance study in unanesthetized neonatal rats. Pediatr Res 42: 700–708
13. Yablonskiy DA, Quirk JD, Nehra D, Bretthorst GL, Ackerman JJH (2002) On the role of blood flow in brain temperature regulation. Proc Intl Soc Mag Reson Med, vol 10, p 1370
14. Yenari MA, Iwayama S, Cheng D, Sun GH, Fujimura M, Morita-Fujimura Y, Chan PH, Steinberg GK (2002) Mild hypothermia attenuates cytochrome *c* release but does not alter Bcl-2 expression or caspase activation after experimental stroke. J Cereb Blood Flow Metab 22: 29–38

Correspondence: Lawrence Litt, Ph.D., M.D., Department of Anesthesia, University of California, San Francisco, Box 0648, 521 Parnassus Avenue, Room C455, San Francisco, CA 94143-0648, U.S.A.

Acta Neurochir (2003) [Suppl] 86: 75–78

Matrix metalloproteinases are not involved in early brain edema formation after cardiac arrest in rats

F. Xiao[1], **T. Arnold**[1], **S. Zhang**[1], **N. Imtiaz**[1], **A. Khan**[1], **J. S. Alexander**[2], **S. Conrad**[1], and **D. Carden**[1]

[1] Department of Emergency Medicine, Louisiana State University Health Science Center, Shreveport, USA
[2] Department of Physiology, Louisiana State University Health Science Center, Shreveport, USA

Summary

Introduction. Resuscitation from cardiac arrest (CA) often results in a poor neurological outcome possibly due to an incomplete understanding of the pathophysiology of brain injury following CA-induced global cerebral ischemia. Brain edema is an important manifestation after CA and is associated with significant morbidity and mortality. The matrix metalloproteinases (MMPs) contribute to brain edema formation following focal cerebral ischemia. The objective of this study was to investigate the role of an MMP inhibitor, GM6001, in CA-elicited brain edema.

Methods. Eighteen rats were subjected to normothermic (37.5 $\pm$ 0.5 °C) CA induced by eight minutes of asphyxiation and assigned to a CA-control group (CA), an alcohol-placebo group (CA + ETOH), or a GM6001-treated group (CA + GM6001). GM6001 in 100% alcohol or a vehicle was given i.v. before CA to achieve a whole blood concentration of 10 μM. Animals were resuscitated with CPR, ventilation and epinephrine. Brain edema was determined by brain wet-to-dry weight ratio at one hour after resuscitation.

Findings. Brain wet-to-dry weight ratio was 4.86 $\pm$ 0.09 in CA, 4.76 $\pm$ 0.12 in CA + ETOH ($p = 0.30$ vs. CA), and 4.72 $\pm$ 0.03 in CA + GM6001 ($p = 0.17$ vs. CA and 0.42 vs. CA + ETOH).

Interpretation. MMPs are not involved in brain edema formation one hour following CA.

Keywords: Cardiac arrest; brain-blood barrier; matrix metalloproteinase; brain edema; cerebral ischemia and reperfusion.

Introduction

Cardiopulmonary resuscitation (CPR) attempts outside of special care units have achieved suboptimal results. Improved neurological recovery in humans after normothermic sudden cardiac arrest (CA) of up to eight minutes would represent an important breakthrough since urban-suburban ambulance response times of eight minutes are feasible. Even when restoration of spontaneous circulation (ROSC) is achieved by current treatments, about 30% of survivors suffer permanent brain damage, including motor and cognitive deficits [19]. The search for brain-damage mitigating treatments has been disappointing, primarily due to an incomplete understanding of the pathophysiological alterations following CA.

The pathophysiological mechanisms of brain damage caused by CA and resuscitation are complicated and multifactorial [28], including the primary insult of CA (complete temporary *global* brain ischemia) and secondary inflammatory responses during and after resuscitation, also known as the post-resuscitation syndrome. A characteristic feature of this syndrome is an increased microvascular permeability that ultimately leads to interstitial edema [33]. We have recently documented that experimental CA of eight-minute asphyxiation results in brain edema following one hour of ROSC [35]. It has also been documented that brain edema formation correlates with morphological neuronal injury, including endothelial pinocytosis and opening of the endothelial cell-to-cell tight junctions [4, 5]. Diffuse brain edema has been reported in CA patients using CT and MRI scanning and predicts a poor neurological outcome [7, 36]. Improved understanding of the mechanisms responsible for brain edema formation may facilitate the development of the effective therapeutic interventions following CA.

The mechanisms underlying brain edema formation remain unclear. The matrix metalloproteinases (MMPs), a family of proteolytic enzymes that are capable of cleaving basal membrane proteins, are now understood to contribute to an increased cerebral microvascular permeability, resulting in brain edema, after focal cerebral I/R [6, 8]. The objective of this study was to examine the role of an MMP inhibitor, GM6001, in brain edema formation after CA and resuscitation in rats.

Methods and material

Experimental model and study design

The rat model used in this study, e.g., eight minutes of asphyxial CA, is reproducible, clinically relevant and permits evaluation of the effects of resuscitative interventions including mild hypothermia [34, 35]. This study was approved by the Institutional Animal Care and Use Committee and followed the National Guidelines for the care and use of animals. Eighteen rats, subjected to normothermic ($37.5 \pm 0.5\,^{\circ}C$) CA induced by eight minutes of asphyxiation, were assigned to a CA-control group (CA, $n = 6$), an alcohol-placebo group (CA + ETOH, $n = 6$), or a GM6001-treated group (CA + GM6001, $n = 6$).

Study protocol

All rats had free access to food and water before each experiment. Anesthesia was induced with ketamine (50 mg/kg) and xylazine (7 mg/kg), intraperitoneally (i.p.). After tracheostomy, controlled intermittent positive pressure ventilation (IPPV) (Harvard Rodent Ventilator Model 683) was started with tidal volume of 7 ml/kg and humidified 100% O_2, positive end-expiratory pressure (PEEP) of 2 cm H_2O, and a frequency of 40 inflations per minute. Ketamine and xylazine were given as needed to facilitate surgery and during the whole experiment. The femoral artery and vein were cannulated with PE 50 catheters via sterile cutdown in the right groin for blood pressure (BP) and administration of GM6001 or a vehicle of alcohol, respectively. Rectal temperature probe was placed for core temperature monitoring. BP and temperature were monitored and recorded using a Data Recording & Analysis System (PowerLab).

After baseline measurements of BP and rectal temperature, another dose of ketamine and xylazine was given. GM6001 (Chemicon International, Temecula, CA) in 100% alcohol or alcohol alone was given i.v. before CA to achieve 10 µM in total blood volume. This dose was determined based on the company's recommendation (10–25 µM) and a study in which GM6001 at an average plasma concentration of 0.15 µM improves cardiovascular and hepatocellular functions in an animal model of hemorrhagic shock [32]. Then the rat was ventilated with IPPV of 100% O_2 for one minute followed by room air for four minutes. Asphyxia was induced by stopping IPPV and clamping the ventilator tubing at the end of exhalation. Asphyxiation was maintained for eight minutes. CA, expected to occur after three minutes of asphyxia, was defined as systolic blood pressure ≤ 25 mmHg and pulsatile pressure ≤ 10 mmHg. Thus, eight minutes of apnea was expected to result in about 5 minutes of CA (no flow).

After apnea and airway obstruction of eight minutes, IPPV was restarted with 100% O_2 and a frequency of 60 breaths per minute, and external cardiac massage was performed using sternal compression by two fingers at a rate of 200 per minute. After external CPR of 10 to 15 seconds, epinephrine at a dose of 0.01 mg/kg was given by IV push without interruption of CPR. ROSC was defined as a spontaneous systolic BP > 60 mmHg. Hypotension of systolic BP < 80 mmHg was treated with normal saline 0.5 ml, repeated to a maximum of 5 ml.

At one hour of ROSC, another dose of ketamine and xylazine was given, and the animal was exsanguinated. The brain was removed in whole and weighed immediately as wet weight. The brain then was placed in a scintillation vial and put into a laboratory oven with a temperature of $105\,^{\circ}C$ for slow evaporation. The brain was again weighed after 24 hours as dry weight. Brain edema was determined by brain wet-to-dry weight ratio.

Data analysis

To determine the sample size needed in each group, a power analysis was performed based on previously published results using the same model [35] with $\alpha = 0.05$ and the effect size = 1.15. The analysis suggested a sample size of 6 in each group with a power of 0.88. All data obtained from this study were entered into a statistical software of SPSS 11.0 for Windows. One-way ANOVA with a post-hoc analysis of Scheffe's procedure was used for individual group comparisons. The Student's t-test was used for specific p values. Significant differences were defined as $p < 0.05$.

Results

There were no statistically significant differences in animal body weight, or mean arterial pressure between groups. The mean volume of normal saline administration was 1.45 ± 0.3 ml in the CA-control animals (CA); 0.8 ± 0.2 ml in the alcohol-placebo animals (CA + ETOH, $p = 0.148$ vs. CA); and 0.2 ± 0.2 ml in the GM6001-treated animals (CA + GM6001, $p = 0.01$ vs. CA and 0.06 vs. CA + ETOH).

Brain wet-to-dry weight ratio was 4.85 ± 0.09 in the CA-control animals; 4.76 ± 0.12 in the alcohol-placebo animals; and 4.72 ± 0.03 in the GM6001-treated animals ($p > 0.05$, Fig. 1).

Discussion

Brain edema, defined as "an abnormal accumulation of fluid associated with volumetric enlargement of the brain" [15], is a serious complication of brain injury following brain trauma, ischemic or hypoxic insults, including CA [7, 15, 20, 35]. Brain edema can be localized either intracellularly (cytotoxic edema) or interstitially (vasogenic edema) [15]. Because of the major effect of hypoxia on cells, cerebral ischemia re-

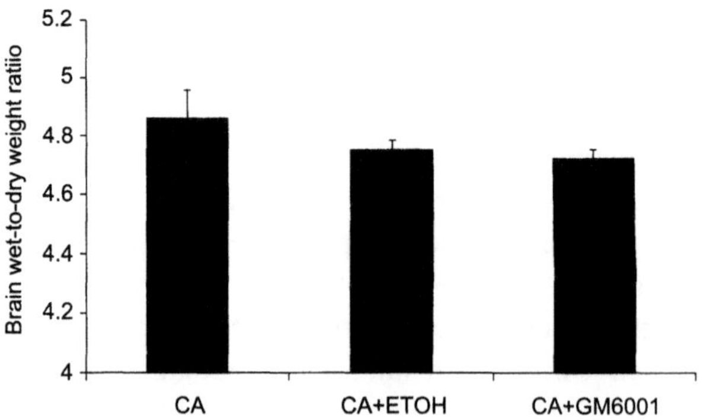

Fig. 1. Brain edema formation at one hour of ROSC following cardiac arrest in rats. There were no differences between groups ($p > 0.05$). *CA*-Cardiac arrest-control group; CA + ETOH – alcohol-placebo group; and CA + MG6001 – GM6001-treated group

sults primarily in cytotoxic edema. After reperfusion, vasogenic edema plays a critical role in brain edema formation. Vasogenic edema is important because of the increased risk for hemorrhage from the damaged vessel [5].

Vasogenic brain edema is a consequence of damage to the blood-brain barrier (BBB). Cerebral capillaries are the major site of the BBB. The extracellular matrix (ECM) participates in the BBB through the formation of a basal lamina (basement membrane) between the mesenchymal endothelial cells and astrocytes.

BBB disruption, as determined by horseradish peroxidase (HRP) permeability, is biphasic in rats undergoing CA induced by cardiac vessel bundle occlusion of 3.5 to 10 min [23]. The first increase appears within two minutes of ischemia and lasts through the first post-insult hour. Permeability returns to normal three hours following the insult. However, BBB opening reappears six hours after I/R and remains open for 24 hours. Another study in piglets using asphyxial CA reveals a significant increase in sodium fluorescein permeability measured at two hours of reperfusion [9].

The mechanisms by which cerebral I/R elicits BBB disruption remain unclear. One recently described pathway is through the formation and/or release of a family of proteases known as the matrix metalloproteinases (MMPs) [3, 6, 12, 13]. Recent in vitro evidence has shown that the increased vascular permeability in human umbilical vein endothelial cell induced by protein tyrosine phosphatase inhibitor was attenuated by a MMP inhibitor, 1,10-phenanthroline [31]. Increased MMP synthese/activity has also been observed in vivo following focal cerebral I/R and is believed to contribute to the breakdown of BBB integrity [8, 11, 12, 24, 26]. MMP-9 is synthesized and activated within 3–4 hours of reperfusion, which is consistent with the timing of barrier dysfunction determined by the Evans blue extravasation [8, 12]. In a recent study, Planas *et al.* reported that MMP9 is induced progressively from four hours to four days after focal cerebral ischemia [22]. The contribution of MMPs to elevated BBB permeability is also supported by studies in which an MMP inhibitor, BB-1101, reduced MMP activation, brain water/sodium content and brain permeability to 14C sucrose after experimental intracranial hemorrhage and lipopolysaccharide-induced brain injury [21, 27]. BBB disruption, as determined by Evans blue leakage, is also significantly attenuated in MMP9 knock-out mice compared with wild-type controls [2]. Based on the assumption that MMPs primarily target components of the matrix, it has been suggested that altered barrier function reflects disturbances in the basal lamina. However, recent evidence that MMP2 is expressed in astrocytic processes and MMP9 in microglial and endothelial cells suggests that MMPs may disrupt BBB through damage to several components of the BBB [2, 25]. Although several studies indicate that MMPs play important roles in brain I/R injury elicited by focal cerebral ischemia [6, 8, 24, 26], comparatively little is known about the role of MMPs in brain edema that accompanies CA. Interventional strategies that inhibit MMP activity have been shown to be protective in various animal models of brain injury [1, 14, 16, 21, 24, 26, 27, 30].

GM6001 (Ilomastat), a hydroxemate MMP inhibitor, has been shown to inhibit intimal hyperplasia of iliac artery after stent placement in rabbits [18], reverse experimental autoimmune encephalomyelitis by restoring the disrupted BBB in mice [10], reduce TNFa level and improve the survival rate in mice subjected to endotoxin [29], improve cardiovascular and hepatocellular function in rats subjected to hemorrhagic shock [32], and attenuate neuronal injury in bacterial meningitis in rats [17]. However, there has been no study investigating the effect of GM6001 on brain edema formation following CA.

In this study, we examined the brain edema-reducing effect of GM6001 in rat CA models. Decreased use of normal saline in GM6001-treated animals is indirect evidence that GM6001 may have a beneficial hemodynamic effect. However, at one hour of ROSC, the MMP inhibitor did not provide a protective effect on brain edema formation following global cerebral ischemia induced by asphyxial CA.

Conclusion

This study suggests that MMPs are not involved in the brain edema formation one hour after global cerebral ischemia elicited by cardiac arrest in rats. Further studies are warranted to determine the role of MMP in the later phase of brain edema formation following cardiac arrest.

References

1. Asahi M, Asahi K, Jung JC, del Zoppo GJ, Fini ME, Lo EH (2000) Role for matrix metalloproteinase 9 after focal cerebral ischemia: effects of gene knockout and enzyme inhibition with BB-94. J Cereb Blood Flow Metab 20: 1681–1689

2. Asahi M, Wang X, Mori T, Sumii T, Jung JC, Moskowitz MA, Fini ME *et al* (2001) Effects of matrix metalloproteinase-9 gene knock-out on the proteolysis of blood-brain barrier and white matter components after cerebral ischemia. J Neurosci 21: 7724–7732
3. Chan VT, Zhang DN, Nagaravapu U, Hultquist K, Romero LI, Herron GS (1998) Membrane-type matrix metalloproteinases in human dermal microvascular endothelial cells: expression and morphogenetic correlation. J Invest Dermatol 111: 1153–1159
4. Dietrich WD, Busto R, Halley M, Valdes I (1990) The importance of brain temperature in alterations of the blood barrier following cerebral ischemia. J Neuropathol Exp Neurol 49: 486–497
5. Dietrich WD, Halley M, Valdes I, Busto R (1991) Interrelationships between increased vascular permeability and acute neuronal damage following temperature-controlled brain ischemia in rats. Acta Neuropathol 81: 615–625
6. Fujimura M, Gasche Y, Morita-Fujimura Y, Massengale J, Kawase M, Chan PH (1999) Early appearance of activated matrix metalloproteinase-9 and blood-brain barrier disruption in mice after focal cerebral ischemia and reperfusion. Brain Res 842: 92–100
7. Fujioka M, Okuchi K, Sakaki T, Hiramatsu K, Miyamoto S, Iwasaki S (1994) Specific changes in human brain following reperfusion after cardiac arrest. Stroke 25: 2091–2095
8. Gasche Y, Fujimura M, Morita-Fujimura Y, Copin JC, Kawase M, Massengale J, Chan PH (1999) Early appearance of activated matrix metalloproteinase-9 after focal cerebral ischemia in mice: a possible role in blood-brain barrier dysfunction. J Cereb Blood Flow Metab 19: 1020–1028
9. Gidday JM, Park TS, Gonzales ER, Beetsch JW (1997) CD18-dependent leukocyte adherence and vascular injury in cerebral circulation after ischemia. Am J Physiol 272: H2622–2629
10. Gijbels K, Galardy RE, Steinman L (1994) Reversal of experimental autoimmune encephalomyelitis with a hydroxamate inhibitor of matrix metalloproteases. J Clin Invest 94: 2177–2182
11. Heo JH, Lucero J, Abumiya T, Koziol JA, Copeland BR, del Zoppo G (1999) Matrix metalloproteinases increase very early during experimental focal cerebral ischemia. J Cereb Blood Flow Metab 19: 624–633
12. Hirohata S, Kusachi S, Murakami M, Murakami T, Sano I, Watanabe T, Komatsubara I *et al* (1997) Time dependent alterations of serum matrix metalloproteinase-1 and metalloproteinase-1 tissue inhibitor after successful reperfusion of acute myocardial infarction. Heart 78: 278–284
13. Huang ZG, Xue D, Preston E, Karbalai H, Buchan AM (1999) Biphasic opening of the blood-brain barrier following transient focal ischemia: effects of hypothermia. Can J Neurol Sci 26: 298–304
14. Jiang X, Namura S, Nagata I (2001) Matrix metalloproteinase inhibitor KB-R7785 attenuates brain damage resulting from permanent focal cerebral ischemia in mice. Neurosci Lett 305: 41–44
15. Klatzo I (1967) Presidental address. Neuropathological aspects of brain edema. J Neuropathol Exp Neurol 26: 1–14
16. Lapchak PA, Chapman DF, Zivin JA (2000) Metalloproteinase inhibition reduces thrombolytic (tissue plasminogen activator)-induced hemorrhage after thromboembolic stroke. Stroke 31: 3034–3040
17. Leib SL, Leppert D, Clements J, Tauber MG (2000) Matrix metalloproteinases contribute to brain damage in experimental pneumococcal meningitis. Infect Immun 68: 615–620
18. Li C, Cantor WJ, Nili N, Robinson R, Fenkell L, Tran YL, Whittingham HA, Tsui W, Cheema AN, Sparkes JD, Pritzker K, Levy DE, Strauss BH (2002) Arterial repair after stenting and the effects of GM6001, a matrix metalloproteinase inhibitor. J Am Coll Cardiol 39: 1852–1858
19. Longstreth WT Jr, Inui TS, Cobb LA, Copass MK (1983) Neurologic recovery after out-of-hospital cardiac arrest. Ann Intern Med 98(5 Pt 1): 588–592
20. Morimoto Y, Kemmotsu O, Kitami K, Matsubara I, Tedo I (1993) Acute brain swelling after out-of-hospital cardiac arrest: pathogenesis and outcome. Crit Care Med 21: 104–110
21. Mun-Bryce S, Rosenberg GA (1998) Gelatinase B modulates selective opening of the blood-brain barrier during inflammation. Am J Physiol 274: R1203–1211
22. Planas AM, Sole S, Justicia C (2001) Expression and activation of matrix metalloproteinase-2 and -9 in rat brain after transient focal cerebral ischemia. Neurobiol Dis 8: 834–846
23. Pluta R, Lossinsky AS, Wisniewski HM, Mossakowski MJ (1994) Early blood-brain barrier changes in the rat following transient complete cerebral ischemia induced by cardiac arrest. Brain Res 633: 41–52
24. Romanic AM, White RF, Arleth AJ, Ohlstein EH, Barone FC (1998) Matrix metalloproteinase expression increases after cerebral focal ischemia in rats: inhibition of matrix metalloproteinase-9 reduces infarct size. Stroke 29: 1020–1030
25. Rosenberg GA, Cunningham LA, Wallace J, Alexander S, Estrada EY, Grossetete M, Razhagi A, Miller K, Gearing A (2001) Immunohistochemistry of matrix metalloproteinases in reperfusion injury to rat brain: activation of MMP-9 linked to stromelysin-1 and microglia in cell cultures. Brain Res 893: 104–112
26. Rosenberg GA, Estrada EY, Dencoff JE (1998) Matrix metalloproteinases and TIMPs are associated with blood-brain barrier opening after reperfusion in rat brain. Stroke 29: 2189–2195
27. Rosenberg GA, Navratil M (1997) Metalloproteinase inhibition blocks edema in intracerebral hemorrhage in the rat. Neurology 48: 921–926
28. Safar P (2000) On the future of reanimatology. Acad Emerg Med 7: 75–89
29. Solorzano CC, Ksontini R, Pruitt JH, Auffenberg T, Tannahill C, Galardy RE, Schultz GP, MacKay SL, Copeland EM 3rd, Moldawer LL (1997) A matrix metalloproteinase inhibitor prevents processing of tumor necrosis factor alpha (TNF alpha) and abrogates endotoxin-induced lethality. Shock 7: 427–431
30. Sumii T, Lo EH (2002) Involvement of matrix metalloproteinase in thrombolysis-associated hemorrhagic transformation after embolic focal ischemia in rats. Stroke 33: 831–836
31. Wachtel M, Frei K, Ehler E, Fontana A, Winterhalter K, Gloor SM (1999) Occludin proteolysis and increased permeability in endothelial cells through tyrosine phosphatase inhibition. J Cell Sci 112 (Pt 23): 4347–4356
32. Wang P, Ba ZF, Galardy RE, Chaudry IH (1996) Administration of a matrix metalloproteinase inhibitor after hemorrhage improves cardiovascular and hepatocellular function. Shock 6: 377–382
33. Xiao F (2002) Brain edema and cerebral resuscitation: the present and future. Acad Emerg Med 9: 933–946
34. Xiao F, Safar P, Radovsky A (1998) Mild protective and resuscitative hypothermia for asphyxial cardiac arrest in rats. Am J Emerg Med 16: 17–25
35. Xiao F, Zhang Z, Arnold TC, Alexanderl JS, Huang J, Carden DL, Conrad SA (2002) Mild hypothermia induced before cardiac arrest reduces brain edema formation in rats. Acad Emerg Med 9: 105–114

Correspondence: Feng Xiao, M.D., Department of Emergency Medicine, Louisiana State University Health Sciences Center, Shreveport, 1501 Kings Highway, Shreveport, LA 71130. e-mail: fxiao@lsuhsc.edu

Acta Neurochir (2003) [Suppl] 86: 79–82

Regional N-acetyl-aspartate level and immunohistochemical damage in the hippocampus after transient forebrain ischemia in gerbils

K. Konaka, H. Ueda, M. Nakano, J.-Y. Li, M. Matsumoto, S. Sakoda, and **T. Yanagihara**

Department of Neurology, Osaka University Graduate School of Medicine, Osaka, Japan

Summary

We investigated changes in regional N-acetyl-aspartate (NAA) levels in the vulnerable CA1 and resistant CA3 areas of the hippocampus after transient forebrain ischemia in gerbils. Under light ether anesthesia, bilateral common carotid arteries of adult male Mongolian gerbils (60–80 g) were occluded for 5 min and reperfused for 7 days. Brains from experimental and control gerbils (n = 4 each) were frozen in situ, and frozen sections (20 μm) were prepared using cryostat (−20 °C). After overnight lyophilization, the CA1 and CA3 areas were dissected out separately, weighed (50–200 μg), and the supernatant of the perchloric acid extract was used for assay of NAA using HPLC. Adjacent 10 μm-thick sections were used for immunohistochemical analysis using antiserum against microtubule-associated protein I and II. The preischemic NAA levels were not significantly different between CA1 and CA3 areas. After transient ischemia, a significant ($P < 0.01$) decrease in the NAA level was observed in the CA1 area, but not in the CA3 area of the hippocampus. Immunohistochemical ischemic damage evolved only in the CA1 area. Thus, the decrease of the regional NAA level was associated with development of immunohistochemical neuronal damage.

Keywords: Selective vulnerability; microtubule-associated proteins; cerebral ischemia; immunohistochemistry; neuronal death.

Introduction

N-acetyl-aspartate (NAA) exists almost exclusively in the nervous system, and is the second most abundant amino acid next to L-glutamate [1]. Although its function is unclear, recent studies revealed that NAA is mainly localized to the neuronal compartment and thus could be used as an index of tissue viability in various disease states [10]. We have developed a method for simultaneous assessment of regional metabolite levels and histological alterations [11], and showed that the regional NAA level decreased in the CA1 area of the hippocampus, the most vulnerable area in the cerebral ischemia model in gerbils [4]. We also found that the extent of decrease was closely associated with the extent of neuronal damage, as evaluated by conventional histology, after preconditioning [4] or after pretreatment with a subtype of glutamate receptor antagonist [5]. These results supported the notion that NAA could be used as an index of neuronal viability in experimental ischemia. However, morphological changes as revealed by conventional histology such as hematoxylin and eosin (H-E) or cresyl-violet are not specific for neuronal damage.

In the past, we have shown that the immunohistochemical procedure using antibodies against microtubule-associated proteins (MAPs) could detect neuronal damage after ischemic insults (for review see [12]), and loss of the immunohistochemical reaction for mixed MAPs (MAP I and II) was the most sensitive indicator to show cytoplasmic and dendritic damages in the CA1 area of the hippocampus after global ischemia in gerbils [13]. In the present study, we therefore combined the immunohistochemical procedure for mixed MAPs and measurement of NAA levels with HPLC in vulnerable CA1 and resistant CA3 areas before and after ischemia to explore the mechanism for decrease of NAA in areas with neuronal damage.

Materials and methods

Adult male Mongolian gerbils (*Meriones unguiculatus*) weighing 60–80 g were used. We followed the guidelines of the National Institutes of Health (Guide for the Care and Use of Laboratory Animals) and our procedures have been approved by the Animal Facility Committee of the Osaka University Medical School. Under ether anesthesia, both common carotid arteries were occluded with miniature aneurysmal clips through the midline skin incision in the neck [2]. After ischemia for 5 min, gerbils were reperfused for 7 days. Brains from control (non-operated) and experimental gerbils (n = 4 each) were frozen in situ, isolated in a cold box (−30 °C), trimmed, mounted on a tissue table, and sectioned into 10 or 20 μm with a cryostat (−20 °C) as described [11].

Each 10 μm thick section was mounted on a chilled slide glass, fixed in chilled acetone solution (−20 °C) for 10 min and air-dried. The previously established immunohistochemical procedure [2] was used with an antiserum against MAP I and II [13] in combination with peroxidase-antiperoxidase complex. The final chromogenic reaction was accomplished with diaminobenzidine and hydrogen peroxide. Counterstaining with hematoxylin was carried out to visualize cell nuclei.

Each 20 μm thick section was lyophilized overnight. The CA1 and CA3 areas were dissected out separately under a stereozoom microscope equipped with a transilluminator and weighed (50–200 μg) using a CAHN-32 microbalance (CAHN Instruments, Inc., Cerritos, CA, USA) in a room with constant temperature (20 °C) and humidity (50%). After homogenization by sonication in 0.5 N perchloric acid (PCA) solution, the soluble fraction was separated by centrifugation, neutralized by addition of 1 N potassium hydroxide, and used for analysis of NAA.

The NAA level was determined by HPLC as described previously [4], using an anion-exchanger, DEAE-2SW column (TOSOH, Tokyo Japan) equilibrated with 80 mM sodium phosphate buffer, pH 6.0. At a flow rate of 0.8 ml/min, NAA was eluted at 15.0 min, which was monitored at 210 nm with the absorbance unit full scale of 0.01–0.005. The NAA level was expressed as pmol per μg dry weight (mean ± SD).

The difference in NAA levels between two areas and that between preischemic and postischemic periods in each area were analyzed using one-factor ANOVA followed by Bonferroni's test. *P*-value of less than 0.05 was taken as significant. A commercial software Stat-View 4.1 (Abacus Concepts Inc., Berkeley, CA, USA) was used for analyses.

Results

All gerbils became flaccid soon after induction of ischemia. After reperfusion, they resumed normal behaviour within several hours and no gerbils developed seizures. Loss of the immunohistochemical reaction for MAPs developed in the whole CA1 area of the hippocampus after transient ischemia in all gerbils (Fig. 1A, B). A higher magnification revealed normal architecture in the CA1 (Fig. 1C) and CA3 areas (Fig. 1D) in control gerbils; disintegration of cytoskeletal architecture was evident in the neuronal parenchyma and dendrites in the CA1 area after ischemia (Fig. 1E). In contrast to the CA1 area, however, no abnormality was observed in the CA3 area even at a higher magnification (Fig. 1F).

Figure 2 shows changes in regional NAA levels in each area before and after ischemia. The preischemic

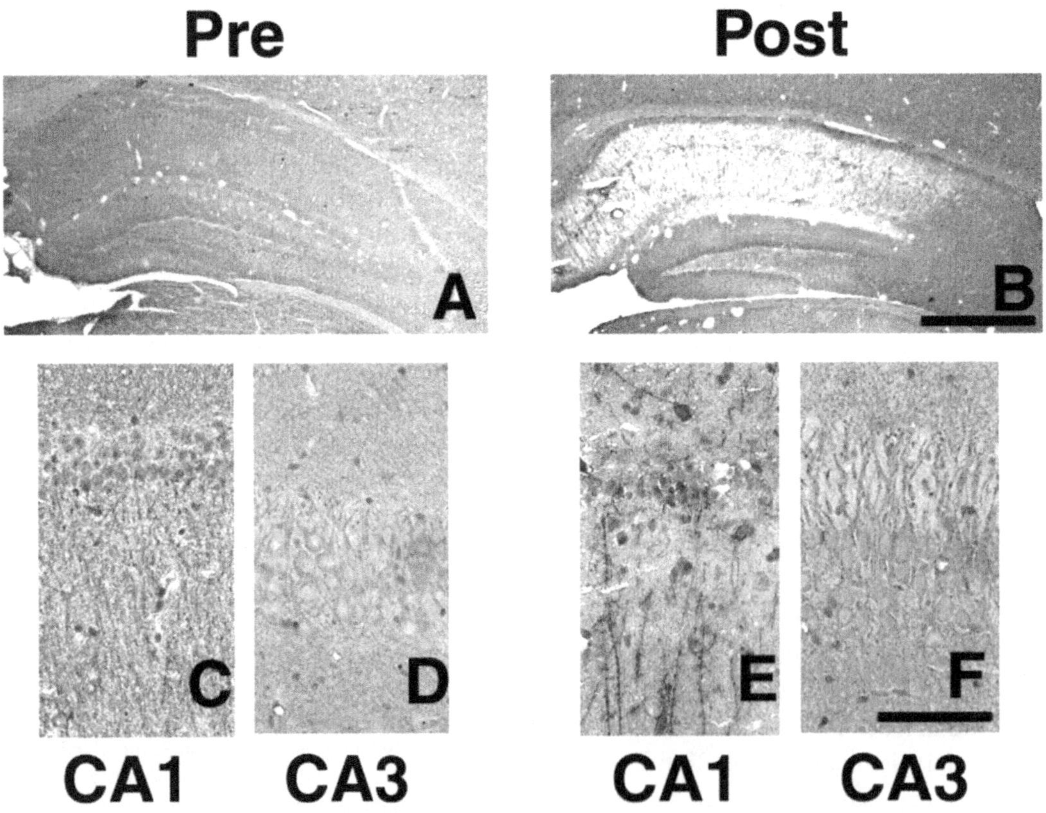

Fig. 1. Lower (A, B) and higher (C–F) magnification of the immunohistochemical reaction for mixed MAPs before (A, C, D) and after (B, E, F) ischemia for 5 min and subsequent reperfusion for 7 days. Note extensive loss of the immunohistochemical reaction in the whole CA1 area in panel B. Higher magnification in the CA1 area revealed extensive damage in the cytoplasm and dendrites after ischemia (E). Bars in panel B and F represent 1000 μm and 100 μm, respectively

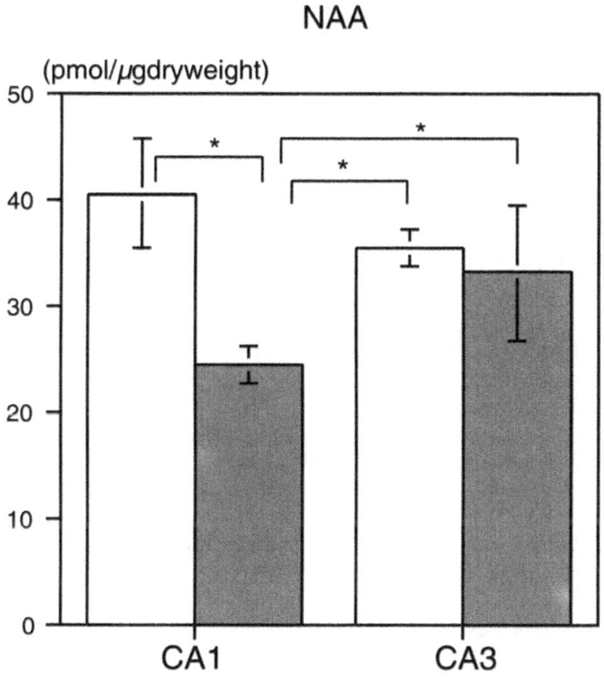

Fig. 2. Regional NAA levels in the CA1 and CA3 areas of the hippocampus before (open bars) and after (filled bars) ischemia and subsequent reperfusion. Note a significant decrease of regional NAA in the CA1 area, but not in the CA3 area after ischemia. $*P < 0.01$

NAA levels (pmol/µg dry weight) were 40.5 ± 5.1 in the CA1 area, and 35.6 ± 1.7 in the CA3 area, respectively. There was no regional difference in preischemic NAA levels. After ischemia for 5 min and subsequent reperfusion for 7 days, a significant ($P < 0.01$) decrease was observed in the CA1 area (24.5 ± 1.8) as compared with the preischemic level. In contrast, the NAA level in the CA3 area after ischemia (33.1 ± 6.3) was not significantly different from the preischemic level. The difference between the preischemic level in the CA3 and postischemic CA1 level was also significant ($P < 0.01$). These data indicated that a decrease of the regional NAA level seemed to be closely associated with the immunohistochemical neuronal damage in the hippocampus.

Discussion

In the past, we showed that the NAA level in the CA1 area of the hippocampus decreased in proportion to the extent of dead neurons after transeint forebrain ischemia in gerbils [4, 5]. In these studies, surviving neurons were evaluated using conventional histology such as H-E [5] or cresyl-violet [4], where neurons with a sharply delineated nucleus and a clearly distinguished nucleolus were counted. The present study showed that such neurons had normal immunohistochemical reaction for MAPs, while damaged neurons had either loss or disturbed patterns of immunoreactivity in the cytoplasm and dendrites (Fig. 1E). Therefore, the present study showed that the decrease of NAA was also associated with cytoplasmic damages in neurons.

The extent and distribution of neuronal damage in the hippocampus in the present study was consistent with our previous reports [2, 12, 13] and others [3] using the same ischemia model. The whole CA1 area was diffusely and extensively damaged; i.e., more than 90 percent of neurons in the pyramidal layer were dead or damaged, but all neurons in the CA3 were intact (Figs. 1A–1F). The reason for such differential vulnerability after global ischemia is uncertain, but the involvement of glutamate toxicity is probable [5, 7].

The preischemic NAA level in the CA1 area was also similar to those from our previous reports [4, 5] and from others [8]. The extent of decrease in NAA after reperfusion was also comparable to that in the previous reports [4, 5, 8]. To our knowledge, this is the first study dealing with NAA in the CA3 area. It appears that the preischemic NAA level in the CA3 area was the same as that in the CA1 area and the level was not altered after ischemia.

The mechanism for the decrease of the NAA level in the ischemic neuronal damage is still uncertain. Although enzymatic degradation of NAA has been proposed [8], our hypothesis is that the release of intracellular NAA to the extracellular space as the result of membrane damage is the cause. The present study may support our hypothesis, since damaged neurons had disintegrated cytoskeletal architecture as indicated by derangement of the immunohistochemical reaction for MAPs. Previous reports also showed impairment of the immunohistochemical reaction for creatine kinase-BB [2] and spectrin [6] with this ischemia model. Our electron microscopic study also revealed extensive cytoplasmic and plasma membrane damage [9].

In summary, the present study showed that the regional NAA level seemed to be a sensitive indicator of neuronal damage in the whole hippocampal subregions and that the decrease of NAA seemed to be closely associated with disintegration of cytoskeletal proteins. Further studies will be necessary to verify the reliability of the regional NAA level as tissue viability outside the hippocampus and to clarify the mechanism for decrease of NAA in ischemic damage.

References

1. Birken DL, Oldendorf WH (1989) N-acetyl-L-aspartic acid: a literature review of a compound prominent in 1H-NMR spectroscopic studies of brain. Neurosci Biobehav Rev 13: 23–31
2. Hatakeyama T, Matsumoto M, Brengman JM, Yanagihara T (1988) Immunohistochemical investigation of ischemic and postischemic damage after bilateral carotid occlusion in gerbils. Stroke 19: 1526–1534
3. Kirino T (1982) Delayed neuronal death in the gerbil hippocampus following ischemia. Brain Res 239: 57–69
4. Nakano M, Ueda H, Li JY, Matsumoto M, Yanagihara T (1998) Measurement of regional N-acetylaspartate after transient global ischemia in gerbils with and without ischemic tolerance: an index of neuronal survival. Ann Neurol 44: 334–340
5. Nakano M, Ueda H, Li JY, Matsumoto M, Yanagihara T (2001) A potent AMPA/kainate receptor antagonist, YM90K, attenuates the loss of N-acetylaspartate in the hippocampal CA1 area after transient unilateral forebrain ischemia in gerbils. Life Sci 69: 1983–1990
6. Roberts-Lewis JM, Savage MJ, Marcy VR, Pinsker LR, Siman R (1994) Immunolocalization of calpain I-mediated spectrin degradation to vulnerable neurons in the ischemic gerbil brain. J Neurosci 14: 3934–3944
7. Rothman SM, Olney JW (1986) Glutamate and the pathophysiology of hypoxic-ischemic brain damage. Ann Neurol 19: 334–340
8. Sager TN, Topp S, Torup L, Hanson LG, Egestad B, Moller A (2001) Evaluation of CA1 damage using single-voxel 1H-MRS and un-biased stereology: can non-invasive measures of N-acetyl-asparate following global ischemia be used as a reliable measure of neuronal damage? Brain Res 892: 166–175
9. Tomimoto H, Yanagihara T (1992) Immunoelectron microscopic study of tubulin and microtubule-associated proteins after transient cerebral ischemia in gerbils. Acta Neuropathol 84: 394–399
10. Tsai G, Coyle JT (1995) N-acetylaspartate in neuropsychiatric disorders. Prog Neurobiol 46: 531–540
11. Ueda H, Tagawa K, Furuya E, Matsumoto M, Yanagihara T (1999) A combined analysis of regional energy metabolism and immunohistochemical ischemic damage in the gerbil brain. J Neurochem 72: 1232–1242
12. Yanagihara T (1990) Immunohistochemical assessment of cerebral ischemia. In: Schurr A, Rigor BM (eds) Cerebral ischemia and resuscitation. CRC Press, Boca Raton, pp 389–412
13. Yanagihara T, Brengman JM, Mushynski WE (1990) Differential vulnerability of microtubule components in cerebral ischemia. Acta Neuropathol 80: 499–505

Correspondence: H. Ueda, Department of Neurology, D-4, Osaka University Graduate School of Medicine, 2-2 Yamada-Oka, Suita, Osaka, 565-0871, Japan. e-mail: ueda@neurol.med.osaka-u.ac.jp

Acta Neurochir (2003) [Suppl] 86: 83–85

Extracellular superoxide dismutase following cerebral ischemia in mice

S. Fukui[1], **H. Nawashiro**[1], **T. Ookawara**[2], **K. Suzuki**[2], **N. Otani**[1], **H. Ooigawa**[1], and **K. Shima**[1]

[1] Department of Neurosurgery, National Defense Medical College, Tokorozawa, Japan
[2] Department of Biochemistry, Hyogo College of Medicine, Nishinomiya, Japan

Summary

We describe the changes in extracellular superoxide dismutase (EC-SOD) following cerebral ischemia in mice. Mice were subjected to transient forebrain ischemia and reperfusion. The measurements of EC-SOD using ELISA showed increased brain EC-SOD after 24 h of reperfusion. The immunohistochemical examination showed that EC-SOD immunoreactivity in cortical and striatal capillary wall was conspicuous after 3 h. EC-SOD immunoreactivity was also noted in cortical neurons after 24 h. Northern blot analysis showed an increased EC-SOD mRNA expression in the brain after 24 h. In situ hybridization study demonstrated no mRNA expression of EC-SOD following ischemia and reperfusion in the capillary wall. These findings suggest that serum EC-SOD might accumulate on brain endothelial cells, while cortical neurons produce EC-SOD themselves after cerebral ischemia with reperfusion.

Keywords: Extracellular Superoxide Dismutase (EC-SOD); cerebral ischemia; reperfusion; Blood-Brain Barrier (BBB); oxidative stress; mice.

Introduction

Recent evidence [7–9] has demonstrated that the third isozyme, extracellular SOD (EC-SOD) isozyme, also plays an important role in defining the severity of ischemic brain damage in mice. EC-SOD is the predominant isozyme in arteries and serum [4, 10], where it contributes up to 70% of the total SOD activity [6, 10]. However, EC-SOD and EC-SOD mRNA are only slightly expressed in the normal brain [1, 5], whereas either Cu/Zn-SOD or Mn-SOD is abundantly expressed in the normal brain. We have, therefore, been interested in the role and kinetics of EC-SOD in an ischemic brain damage.

Most serum EC-SOD originates from the endothelia or smooth muscle cells of systemic arteries, and is circulating in blood [3, 10]. In the present study, we determined the topographical localization and the quantitative change in both EC-SOD and its mRNA expression following cerebral ischemia in mice.

Materials and methods

Animals and forebrain ischemia

C57BL/6NCrj (C57BL/6) mice, weighing from 20 to 27 g were used. The procedures involving laboratory animals were approved by the committee on animal experiments of the National Defense Medical College. General anesthesia was induced with 2.5% isoflurane and maintained with 1.5% isoflurane in a mixture of 70% nitrous oxide and 30% oxygen by means of an open-face mask. The rectal temperature was controlled at between 36.5 °C and 37.5 °C by a heating lamp. The bilateral common carotid arteries were occluded with small aneurysmal clips (Vessel Clip No. 16; Kono, Tokyo, Japan) for 20 minutes and reperfused.

ELISA

At 3 and 24 hours after a BCCAO or a sham operation, the blood was collected from the right ventricle of the heart and the brains were perfused with phosphate-buffered saline (PBS). The brains were then removed and divided into two sections (rostral and caudal sections) by a coronal cut with a blade at the level of 5 mm caudal from the frontal pole. To quantify EC-SOD protein in the brain and serum using an enzyme-linked immunosorbent assay (ELISA), brain samples were obtained by homogenization of the rostral section of the brain, and serum samples were obtained from the blood by centrifugation. The procedure for the quantification of mouse EC-SOD using an ELISA has been described previously [5].

Immunohistochemistry

The procedure for the immunostaining for mouse EC-SOD has been described previously [5]. The lungs of normal mice were used as positive controls. As negative controls, the lungs of normal mice were incubated with nonimmune sera instead of the affinity-purified anti-mouse EC-SOD IgG. Bound antibodies were detected with HRP-labeled donkey anti-rabbit IgG polyclonal antibody. Counterstaining was carried out using hematoxylin.

For the detection of immunoglobulin extravasation, an immunohistological study for immunoglobulin was also performed with goat polyclonal antibody against mouse immunoglobulin.

Northern blot analysis

The brain tissue for a Northern blot analysis had neither perfusion nor post-fixation. To quantify EC-SOD mRNA in brains, brain samples were obtained by the homogenization of the rostral section of the brain. Total RNAs in the brain samples were isolated from each tissue using RNAzol™ B (TEL-TEST, INC., Friendswood, TX). The procedure of the Northern blot analysis for EC-SOD mRNA has been described previously [5].

In situ hybridization

All procedures were performed according to the manufacturer's instructions (RNA Labeling and Detection Kit; Roche Diagnostics, Mannheim, Germany) at room temperature unless a different temperature is stated. The mouse EC-SOD cDNA clone (pBSMEC1212) was isolated from a mouse kidney cDNA library (Ookawara, T., *et al.* unpublished data), and then was excised with Pst I to be a 400 bp DNA fragment, corresponding to nucleotides 497–896 [2]. The DNA fragment was inserted into a pSPT19 vector (Roche Diagnostics) and was used as a template for the synthesis of digoxigenin-labeled antisense and sense RNA probe (DIG RNA labeling kit, Roche Diagnostics).

Statistical analysis

All values reported here are expressed as means ± SD. A quantitative analysis of either EC-SOD using ELISA or EC-SOD mRNA was analyzed by one-way ANOVA followed by Fisher's post hoc analysis. A p value of less than 0.05 was considered to be statistically significant.

Results

ELISA data

The measurements of EC-SOD using an ELISA revealed the increase of the brain EC-SOD after cerebral ischemia and indicated the concomitant reduction of the serum EC-SOD. After 24 h of reperfusion, the brain EC-SOD increased significantly ($p < 0.002$, ANOVA with post hoc Fisher's test) compared to the sham (S)-24 h group. The levels of brain EC-SOD (ng/mg protein; mean ± SD) were 19.45 ± 1.52 in the S-3 h group ($n = 10$), 20.73 ± 1.37 in the ischemia (I)-3 h group ($n = 11$), 19.20 ± 0.91 in the S-24 h group ($n = 10$), and 22.20 ± 3.32 in the I-24 h group ($n = 10$). The levels of serum EC-SOD (ng/mg protein; mean ± SD) were 103.5 ± 5.0 in the S-3 h group ($n = 10$), 92.7 ± 13.6 in the I-3 h group ($n = 11$), 102.5 ± 15.4 in the S-24 h group ($n = 10$), 95.3 ± 22.2 in the I-24 h group ($n = 10$).

Increased EC-SOD immunoreactivity after transient forebrain ischemia

In the I-3 h group ($n = 4$), the EC-SOD immunoreactivity in the cortical capillary vessel walls was conspicuous compared with that in the S-3 h groups ($n = 4$). The staining in the capillary vessel walls was relatively uniformly distributed within the capillary vessel walls. There was no other change in the EC-SOD immunoreactivity after 3 h of reperfusion. In the I-24 h group ($n = 4$), the EC-SOD immunoreactivity appeared in the cortical neurons, whereas no EC-SOD immunoreactivity was observed in the cortical neurons of either the normal or S-24 h group ($n = 4$). Some small cortical cells that obtained the EC-SOD immunoreactivity after 24 h of reperfusion might be astrocytes or microglias. The increase in the EC-SOD immunoreactivity in the capillary vessel walls remained after 24 h of reperfusion.

Northern blot analysis for EC-SOD mRNA

Northern blot analysis ($n = 49$) revealed that the EC-SOD mRNA expression in brains increased after 24 h of reperfusion. The levels of brain EC-SOD mRNA normalized to those of GAPDH increased significantly in the I-24 h group, compared with the control, I-1 h, I-12 h, and S-24 h group ($p < 0.02$,

$p = 0.0005$, $p < 0.005$, $p < 0.01$, respectively, ANOVA with post hoc Fisher's test).

In situ hybridization for EC-SOD mRNA

The specificity of the probes was confirmed by in situ hybridization using a sense probe, in which decreased specific signals of EC-SOD mRNA were noted. In situ hybridization studies showed no clear changes either in the distribution of the EC-SOD mRNA or in the level of the mRNA expression of EC-SOD after cerebral ischemia.

Discussion

EC-SOD is a secretory tetrameric Cu and Zn containing glycoprotein, which is produced and secreted to the extracellular space mainly by smooth muscle cells or endothelia. This enzyme is mainly located in the vascular walls at a higher level to suppress oxidative stress under physiological conditions.

We herein describe the kinetics of EC-SOD in cerebral ischemia and reperfusion. After 3 h of reperfusion, the EC-SOD brain/serum ratio and EC-SOD immunoreactivity in the brain capillary vessel walls increased significantly, but no mRNA expression of EC-SOD was found in the brain capillary walls. These findings strongly suggest that EC-SOD accumulates on brain capillary vessels following cerebral ischemia.

After 24 h of reperfusion, EC-SOD immunoreactivity was found in the cortical neurons, whereas no EC-SOD immunoreactivity was detected in the cortical neurons in neither the control nor the I-3 h group. In addition, there was no immunoreactivity of immunoglobulin through the brains in any of the groups (data not shown). The in situ hybridization study revealed that cortical neurons have abundant EC-SOD mRNA. Moreover, a Northern blot analysis showed an increased mRNA expression of EC-SOD in the I-24 h group. These findings suggest that after 24 h of reperfusion, the cortical neurons produce EC-SOD themselves after cerebral ischemia.

References

1. Folz RJ, Crapo JD (1994) Extracellular superoxide dismutase (SOD3): tissue-specific expression, genomic characterization, and computer-assisted sequence analysis of the human EC-SOD gene. Genomics 22: 162–171
2. Folz RJ, Guan J, Seldin MF, Oury TD, Enghild JJ, Crapo JD (1997) Mouse extracellular superoxide dismutase: primary structure, tissue-specific gene expression, chromosomal localization, and lung in situ hybridization. Am J Respir Cell Mol Biol 17: 393–403
3. Fukai T, Galis ZS, Meng XP, Parthasarathy S, Harrison DG (1998) Vascular expression of extracellular superoxide dismutase in atherosclerosis. J Clin Invest 101: 2101–2111
4. Karlsson K, Sandstrom J, Edlund A, Edlund T, Marklund SL (1993) Pharmacokinetics of extracellular-superoxide dismutase in the vascular system. Free Radic Biol Med 14: 185–190
5. Ookawara T, Imazeki N, Matsubara O, Kizaki T, Oh-Ishi S, Nakao C, Sato Y, Ohno H (1998) Tissue distribution of immunoreactive mouse extracellular superoxide dismutase. Am J Physiol 275: C840–847
6. Oury TD, Day BJ, Crapo JD (1996) Extracellular superoxide dismutase in vessels and airways of humans and baboons. Free Radic Biol Med 20: 957–965
7. Sheng H, Bart RD, Oury TD, Pearlstein RD, Crapo JD, Warner DS (1999) Mice overexpressing extracellular superoxide dismutase have increased resistance to focal cerebral ischemia. Neuroscience 88: 185–191
8. Sheng H, Brady TC, Pearlstein RD, Crapo JD, Warner DS (1999) Extracellular superoxide dismutase deficiency worsens outcome from focal cerebral ischemia in the mouse. Neurosci Lett 267: 13–16
9. Sheng H, Kudo M, Mackensen GB, Pearlstein RD, Crapo JD, Warner DS (2000) Mice overexpressing extracellular superoxide dismutase have increased resistance to global cerebral ischemia. Exp Neurol 163: 392–398
10. Stralin P, Karlsson K, Johansson BO, Marklund SL (1995) The interstitium of the human arterial wall contains very large amounts of extracellular superoxide dismutase. Arterioscler Thromb Vasc Biol 15: 2032–2036

Correspondence: Hiroshi Nawashiro, M.D., D.M.Sc., Department of Neurosurgery, National Defense Medical College, 3-2 Namiki, Tokorozawa, Saitama 359-8513, Japan. e-mail: nawa1957@me.ndmc.ac.jp

Acta Neurochir (2003) [Suppl] 86: 87–91

Peroxynitrite and caspase-3 expression after ischemia/reperfusion in mouse cardiac arrest model

K. Dohi[1], **H. Ohtaki**[2], **R. Inn**[2], **Y. Ikeda**[2], **H. S. Shioda**[2], and **T. Aruga**[1]

[1] Department of Critical Care and Emergency Medicine, School of Medicine, Showa University, Tokyo, Japan
[2] Department of Anatomy, School of Medicine, Showa University, Tokyo, Japan

Summary

NO is a putative neurotransmitter and neuromodulator in the brain. NO is not functioning as a direct neurotoxin. NO with the superoxide radical product peroxynitrite (ONOO-) is much more cytotoxic under tissue impairment conditions. Caspase-3, a potent effector of apoptosis that is triggered via several different signaling pathways, may play a very important role in neuronal cell death caused by various brain injuries. The relationship between mouse caspase-3 and peroxynitrite remains unclear. In the present study, we examined the in vivo expression of 3-nitrotyrosine (a metabolite of peroxinitrite) and caspase-3 after cerebral ischemia produced in a global ischemia model using mice (i.e., a cardiac arrest model). 3-nitrotyrosine immunoreactivity was detected in neuronal cells in the hippocampal dentate nucleus, and cortical regions starting at 12 hrs after ischemia. In particular, numerous neuronal cells were highly immunoreactive for 3-nitrotyrosine in the cortical regions. In hippocampal CA1 pyramidal neurons, 3-nitrotyrosine immunoreactivity was detected from 24 hrs. Caspase-3 immunopositive cells were observed in approximately the same area in which the positive reaction to the anti-nitrotyrosine antibody was observed. These results provide direct evidence for the induction of 3-nitrotyrosine and caspase-3 expression in vivo in an ischemia model using mice. The present findings suggest that peroxynitrite generated by cerebral ischemia/reperfusion was strongly cytotoxic and induced neuronal cell death (apoptosis) mediated by caspase-3.

Keywords: Peroxynitrite; ischemia; NO; apoptosis; free radicals; cytokine; IL-1; mouse; model; nitrotyrosine.

Introduction

The generation of free radicals is closely related to the development of several nervous diseases [9, 3]. Various free radicals have also been reported to be involved in the tissue impairment accompanying cerebral ischemia. One free radical scavenger (MCI-816) has been shown to prevent neuronal death after ischemia via a nitric oxide (NO) scavenging process [1, 16]. Although NO is a putative neurotransmitter in the brain, it is not likely to be a direct neurotoxin by itself [7]. When combined with the superoxide radical product peroxynitrite (ONOO-), NO is much more cytotoxic under conditions of tissue impairment. Whereas oxidative stress and the deleterious effects of NO are of current interest, only a limited number of papers have investigated the role of peroxynitrite (ONOO-) in mice. We therefore conducted morphological observations of 3-nitrotyrosine (NO2-Tyr) derived from peroxynitrite (ONOO-) using a mouse cardiac arrest model.

The caspase family of cysteine proteases plays a role in several steps of the process of apoptosis. Currently, at least 14 members of the caspase family have been identified; these various caspases have been designated by numbers as caspase-1 to caspase-14. Although the caspases have different cellular and molecular substrates, all of them possess processing target sites containing aspartate [15]. Caspase-3 is a potent effector of apoptosis, which is triggered via several different signaling pathways [2]. Activation of caspase-3 also contributes to apoptosis in CA1 pyramidal neurons after ischemia and traumatic brain injury [4–6, 14]. Thus, caspase-3 may play an important role in neuronal cell death caused by various brain injuries. However, the interaction between caspase-3 and peroxynitrite remains unclear. In the present study, we studied the expression of 3-nitrotyrosine (a metabolite of peroxinitrite) and of caspase-3 in a new global ischemia mouse model (i.e., a cardiac arrest model).

Materials and methods

Animals

Male BALB/c mice aged 8 to 10 weeks (20–30 g) were obtained from the Saitama Experimental Animal Center (Saitama, Japan) and were housed in a temperature- (22 °C) and light-controlled room (lights on at 0600 and lights off at 1800). The animals were supplied with standard laboratory chow and tap water. Animal care and experiments complied with the Principles of Laboratory Animal Care listed in the *Guidelines for the Care and Use of Laboratory* (National Institutes of Health publication 86-23, 1985).

Mouse model of ischemia/reperfusion

All animals were intubated under anesthesia with 3.5% halothane and anesthesia was maintained with 2% sevoflurane in 70% N_2O and 30% O_2. Core temperature was continuously monitored with a flexible rectal thermometer and was maintained at 37 °C. A polyethylene catheter was inserted into the left femoral artery. Blood pressure and electrocardiograms were monitored during all of the procedures. Global cerebral ischemia was induced according to the rat cardiac-arrest model. With the anesthetized mouse in the supine position, an incision of about 0.5 cm in length was made along in the midline of the chest. The occluding device was gently inserted into the mediastinum at the level of the second intercostal segment. The occluding device was manipulated such that its distal end moved along the dorsal wall of the thorax. The distal end was then twisted 45° in order to position it under the bundle of the superior aortic artery. A complete interruption of circulation due to compression of the superior aortic artery was accomplished by lifting the occluding device and by simultaneously applying external finger pressure (Fig. 1A). After 5 minutes of transient global ischemia, and under artificial ventilation (tidal volume, 2.5 ml at 200 times per minute; 100% O_2), heart massage was performed in order to facilitate resuscitation (Fig. 1B). After a recovery period of 30 minutes, the animals were disconnected from the respirator. The animals were then housed in cages with free access to food pellets and water.

Tissue preparation. Following recirculation periods of 0, 12, and 24 hours, the animals were sacrificed. Under pentobarbital anesthesia (50 mg/kg, i.p.), the brains were perfusion-fixed with 4% paraformaldehyde and 0.4% glutaraldehyde in 0.1 M phosphate buffer (PB) in preparation for immunohistochemical staining. After the brains were removed from the animals, they were postfixed in the same fixative overnight at 4 °C. Brains were cryoprotected in PB containing 20% sucrose for 2 days prior to all stainings.

Immunohistochemistry

The tissue sections were immunohistochemically stained with monoclonal antibody against 3-nitrotyrosine and caspase-3 (Cell Signaling Technology, Beverly, MA, USA). The frozen sections were preincubated with 10% normal horse serum. The sections were then incubated with the antibodies overnight at 4 °C. The sections were then incubated with the biotinylated second antibody for 1 h at room temperature, followed by incubation with the avidin-biotin-peroxidase complex for 30 min at room temperature.

Results

In non-ischemic hippocampal tissue, 3-nitrotyrosine immunoreactivity was not detected (Fig. 2A). Twelve hours after ischemia, low levels of 3-nitrotyrosine were detected in neuronal cells in the hippocampal dentate nucleus, and cortical regions (data not shown). At 24 hours after ischemia, low levels of 3-nitrotyrosine were detected in neuronal cells in the hippocampal CA1 region. 3-nitrotyrosine immunoreactivities were increased in the hippocampal dentate nucleus, and cortical regions at 24 h (Fig. 2B, C). In particular, a large number of neuronal cells were highly immunoreactive for 3-nitrotyrosine in the cortical regions (Fig. 2D). The reaction was intensified in a time-course manner (data not shown). In neighboring slices, caspase-3-positive cells were observed in approximately the same areas as was the positive reaction to the anti-nitrotyrosine antibody. (Fig. 2F, G, H).

Discussion

The present study employed a cardiac-arrest model that was first introduced in the rat [12]. This cardiac-arrest model is advantageous over other models in that it is less invasive, easier to perform, and produces a more uniform global ischemia [10]. The following morphological features of ultrastructural apoptotic change were observed: nuclear and cytoplasmic condensation, cell shrinkage, and nuclear fragmentation, all of which have previously been observed in the rat model [6]. We have previously reported that mice lacking Interleukin-1 (IL-1) alpha/beta (double knockout mice) exhibited markedly reduced neuronal loss and apoptotic cell death in a mouse cardiac-arrest model [13]. The inducible nitric oxide (iNOS) immunoreactivity observed after global ischemia was lower in IL-1 knockout mice than in wild-type mice. The levels of nitrite (NO(2) (−)) and nitrate (NO(3) (−)) in the hippocampus of wild-type mice increased with time after ischemia-reperfusion, whereas this increase was significantly inhibited in IL-1 knockout mice [13]. These data strongly suggest that endogenous IL-1 contributes to ischemic brain damage; this influence may act through the release of nitric oxide by iNOS. NO is not likely to be a direct neurotoxin, whereas its conversion to peroxinitrite has been associated with cell death [7]. Peroxinitrite, which is formed by the reaction of NO with superoxide, activates the nuclear enzyme PARS (poly (ADP-ribose) synthetase). The present study revealed that 3-nitrotyrosine and caspase-3 immunoreactivities were observed in the same regions of neurons. It appears that peroxynitrite generated by cerebral ischemia/reperfusion is strongly cytotoxic and induces neuronal cell death (apoptosis) mediated by caspase-3. Caspase-3 activity is known to contribute to delayed neuronal death (apoptosis) after transient ischemia in the last stage of apoptotic pathways [4, 8, 11]. These findings indicate that the neurotoxicity of IL-1 after

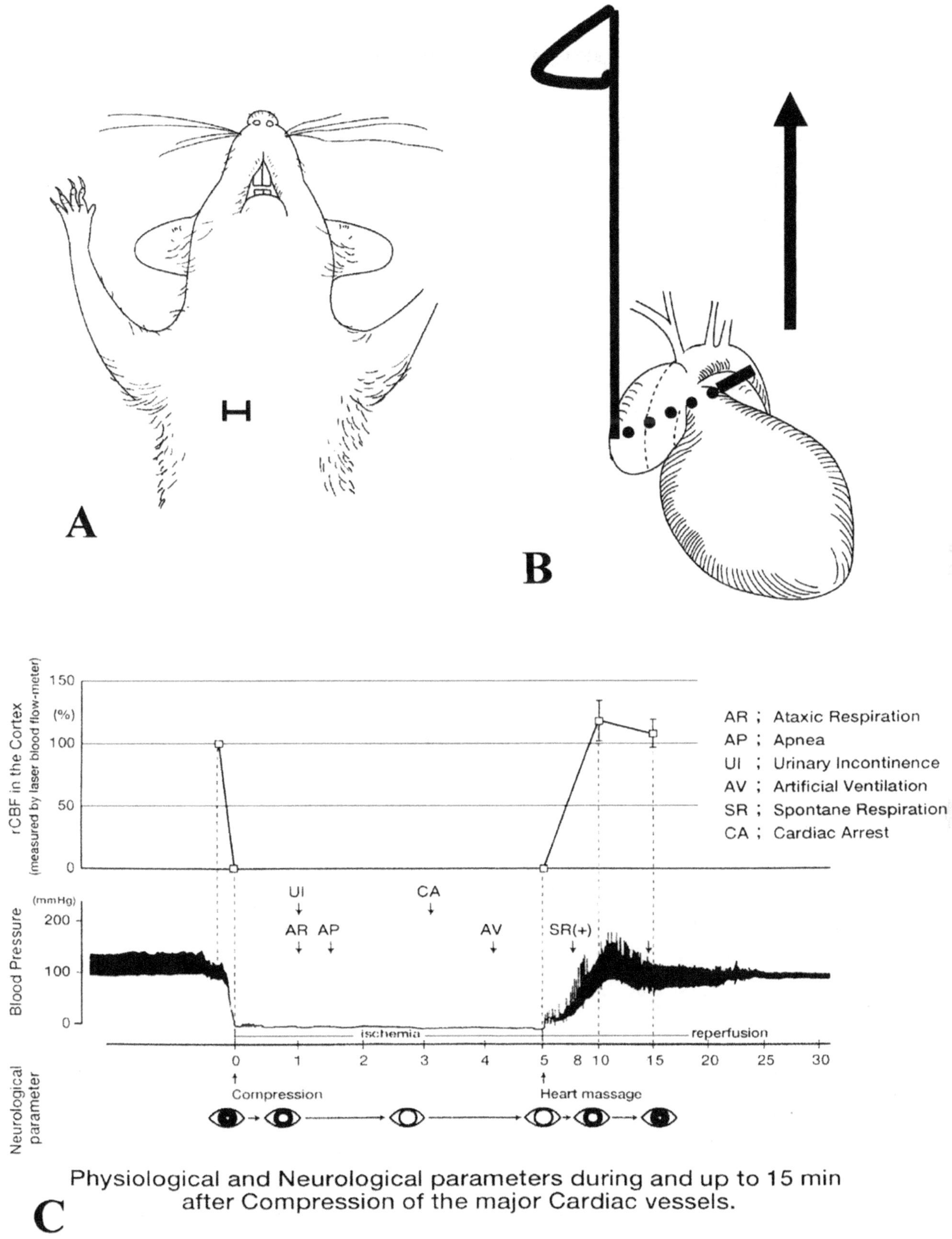

Fig. 1. The surgical procedures and the occluding device used in this model. Skin incision (A). Scheme for the surgical procedure. The major cardiac vessels are lifted and occluded by the occluding device. Regional CBF (upper panel; n – 5) and blood pressure (middle panel) recorded during cardiac arrest and recirculation. Time course of the neurological parameters (middle and lower panels) during the operation (B)

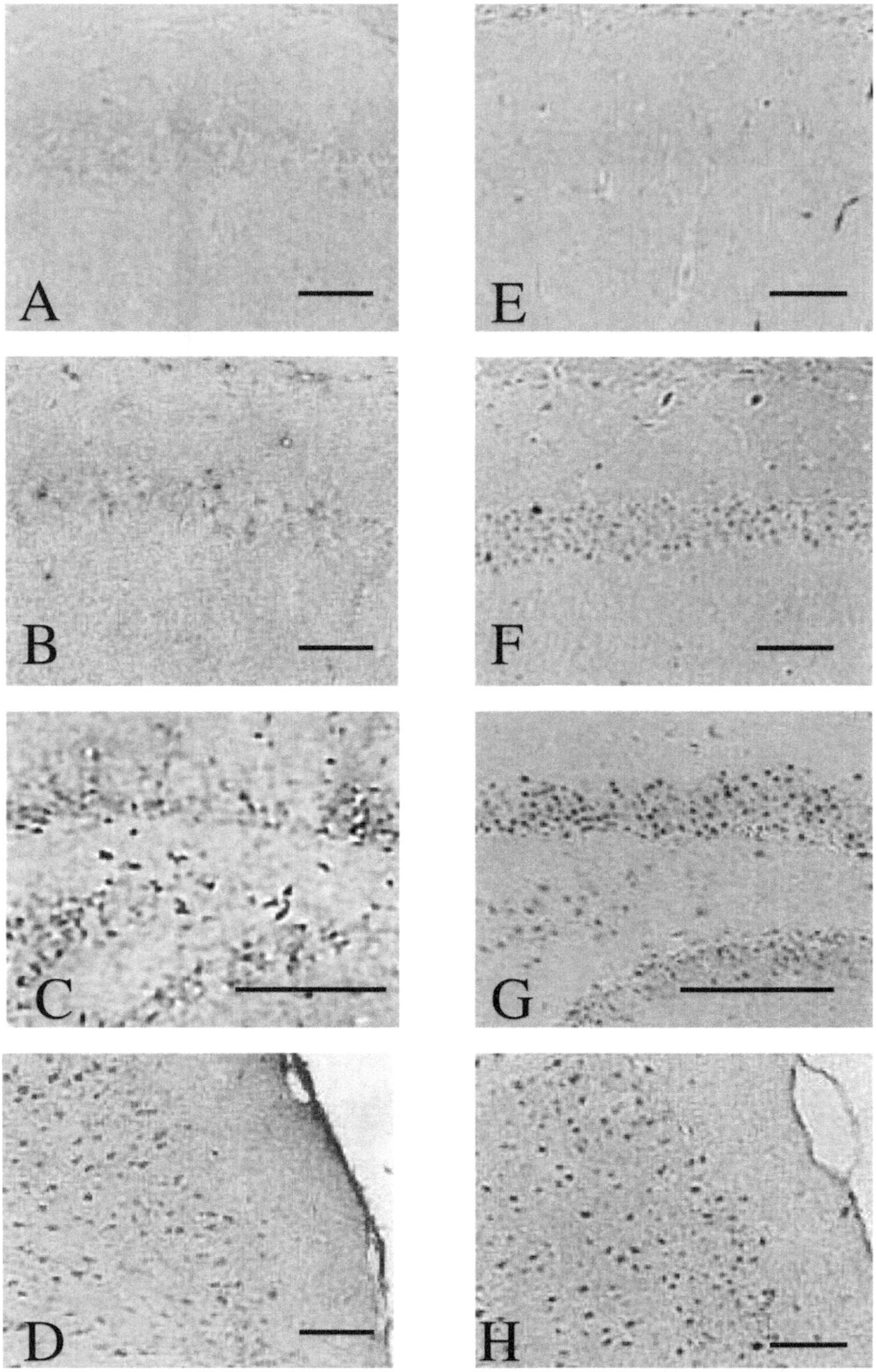

Fig. 2. No 3-nitrotyrosine immunoreactivity was observed in the hippocampal CA1 area (A) of the sham control group. Epxression of 3-nitrotyrosine immunoreactivity in neurons in the hippocampal CA1 (B), dentate gyrus (C), and cortex (D) at 24 hours after ischemia. No caspase-3 immunoreactivity was observed in the hippocampal CA1 area (E) in the sham control group. Caspase-3 immunopositive neurons in the hippocampal CA1 (F), dentate gyrus (G), and cortex (H) at 24 hours after ischemia. (Bars 100 μm)

ischemia is strongly related to the apoptotic pathway through iNOS activation, NO and peroxinitrite production, PARS activation, and caspase-3 activation. However, it should be noted that the present study is highly preliminary, and further studies will be needed to determine the role played by free radicals in cases of brain ischemia.

In the present study, we demonstrated that 3-nitrotyrosine and caspase-3 were markedly upregulated in the cortex, hippocampal CA1, and dentate gyrus in a mouse global ischemia model (i.e., a cardiac arrest model). This is the first report demonstrating that 3-nitrotyrosine (a marker of peroxinitrite) is expressed after ischemia in mice. Our results strongly suggest that peroxinitrite mediates ischemic neuronal cell death (apoptosis) caused by IL-1.

References

1. Abe K, Yuki S, Kogure K (1988) Strong attenuation of ischemic and postischemic brain edema in rats by a novel free radical scavenger. Stroke 19: 480–485
2. Alnemri ES, Livingston DJ, Nicholson DW, Salvesen G, A TN, Wong WW (1996) Human ICE/CED-3 protease nomenclature. Cell 87: 171
3. Chan PH (2001) Reactive oxygen radicals in signaling and damage in the ischemic brain. J Cereb Blood Flow Metab 21: 2–14
4. Chen J, Nagayama T, Jin K, Stetler RA, Zhu RL, Graham SH, Simon RP (1998) Induction of caspase-3-like protease may mediate delayed neuronal death in the hippocampus after transient cerebral ischemia. J Neurosci 18: 4914–4924
5. Clark RS, Kochanek PM, Watkins SC, Chen M, Dixon CE, Seidberg NA, Melick J, Loeffert JE, Nathaniel PD, Jin KL, Graham SH (2000) Caspase-3 mediated neuronal death after traumatic brain injury in rats. J Neurochem 74: 740–753
6. Dohi K, Shioda S, Mizushima H, Homma H, Ozawa H, Nakai Y, Matsumoto H (1998) Delayed neuronal cell death and microglial cell reactivity in the CA1 region of the rat hippocampus in the cardiac arrest model. Med Electron Microsc 31: 85–93
7. Eliasson MJ, Huang Z, Ferrante RJ, Sasamata M, Molliver ME, Snyder SH, Moskowitz MA (1999) Neuronal nitric oxide synthase activation and peroxynitrite formation in ischemic stroke linked to neural damage. J Neurosci 19: 5910–5918
8. Enari M, Sakahira H, Yokoyama H, Okawa K, Iwamatsu A, Nagata S (1998) A caspase-activated DNase that degrades DNA during apoptosis, and its inhibitor ICAD [see comments] [published erratum appears in Nature 1998 May 28; 393(6683): 396]. Nature 391: 43–50
9. Ikeda Y, Long DM (1990) The molecular basis of brain injury and brain edema: the role of oxygen free radicals. Neurosurgery 27: 1–11
10. Kawai K, Nitecka L, Ruetzler CA, Nagashima G, Joo F, Mies G, Nowak TS Jr, Saito N, Lohr JM, Klatzo I (1992) Global cerebral ischemia associated with cardiac arrest in the rat: I. Dynamics of early neuronal changes. J Cereb Blood Flow Metab 12: 238–249
11. Kondratyev A, Gale K (2000) Intracerebral injection of caspase-3 inhibitor prevents neuronal apoptosis after kainic acid-evoked status epilepticus. Brain Res Mol Brain Res 75: 216–224
12. Korpatchev W, Lysenkov S, Thieliz P (1982) Modielirowanlie klinczeskoj smerti l postreanimationznoj bolezni u krys. Palol Fiziol Exp Ter 3: 78–80
13. Mizushima H, Zhou CJ, Dohi K, Horai R, Asano M, Iwakura Y, Hirabayashi T, Arata S, Nakajo S, Takaki A, Ohtaki H, Shioda S (2002) Reduced postischemic apoptosis in the hippocampus of mice deficient in interleukin-1. J Comp Neurol 448: 203–216
14. Ni B, Wu X, Su Y, Stephenson D, Smalstig EB, Clemens J, Paul SM (1998) Transient global forebrain ischemia induces a prolonged expression of the caspase-3 mRNA in rat hippocampal CA1 pyramidal neurons. J Cereb Blood Flow Metab 18: 248–256
15. Salvesen G, Dixit V (1997) Caspases: intracellular signaling by proteolysis. Cell 91: 443–446
16. Satoh K, Ikeda Y, Shioda S, Tobe T, Yoshikawa T (2002) Edarabone scavenges nitric oxide. Redox Rep 7: 219–222

Correspondence: Kenji Dohi, Department of Emergency and Critical Care Medicine, School of Medicine, Showa University, Tokyo, Japan, 1-5-8 Hatanodai, Shinagawa-ku, Tokyo, 142-8555, Japan. e-mail: kdop@med.showa-u.ac.jp

Acta Neurochir (2003) [Suppl] 86: 93–96

Expression of tumor necrosis factor α (TNFα) following transient cerebral ischemia

L. Yin[1,3], H. Ohtaki[1,3], T. Nakamachi[1], K. Dohi[1], Y. Iwai[1], H. Funahashi[1], R. Makino[2], and S. Shioda[1,3]

[1] Department of Anatomy, Showa University School of Medicine, Tokyo, Japan
[2] Clinical Research Laboratory, Showa University School of Medicine, Tokyo, Japan
[3] The Core Research for Evolutional Science and Technology (CREST) of Japan Science and Technology Corporation (JST), Tokyo, Japan

Summary

It has been considered that tumor necrosis factor α (TNFα) is participated in the Alzheimer's, and Parkinson's diseases, brain injury and brain ischemia. However, expression of TNFα after brain ischemia has not been demonstrated in detail. Therefore we examined the cellular expression of TNFα during and after transient middle cerebral artery occlusion (tMCAO) in mice by use of reverse transcriptase-polymerase chain reaction and immunohistochemical technique. TNFα mRNA expression was gradually increased in the neocortex of the ipsilateral hemisphere during ischemia and peaked at 1 hour after reperfusion. Then, the mRNA expression decreased and peaked again at 24 hours after reperfusion. TNFα-like immunoreactivities were observed in the process such as dendrite of neuron slightly before ischemia, and markedly increased in neurons in addition to the process of the ipsilateral hemisphere at 1 and 24 hours after ischemia. The results suggest that the expression of TNFα is up-regulated in the neurons after tMCAO. TNFα may induce ischemic neuronal cell death during ischemic insult.

Keywords: Tumor necrosis factor α (TNFα); middle cerebral artery occlusion; mice; RT-PCR; immunohistochemistry.

Introduction

Increased levels of tumor necrosis factor α (TNFα) in brain, cerebrospinal fluid and plasma have been found in several disorders of the central nervous system (CNS), including Alzheimer's disease, multiple sclerosis, Parkinson's disease, cerebral malaria, and cerebral ischemia [11]. TNFα may contribute to ischemic brain damage through a variety of proinflammatory effects, such as alteration of blood-brain barrierpermeability [9], activation of microglia and astrocytes, induction of cellular adhesion molecule expression [7]. In response to brain ischemia, the relation of TNFα and brain damage is still unknown. In the present study, we investigated the induction of TNFα and its expression mRNA in the mice brain after transient cerebral ischemia. Our results suggest that TNFα induces ischemic neuron cell death in the tMCAO model.

Materials and methods

Mice and production of transient middle cerebral artery occlusion (tMCAO)

All experimental procedures involving animals were approved by the Institutional Animal Care and Use Committee of Showa University (#02099). Adult male BALB/c strain mice were purchased from Charles River Japan (Kanagawa, Japan). Male mice weighing 24–28 g (8–10 weeks) were subjected to focal cerebral ischemia. The anesthesia was induced with inhalation of 2.0% sevoflurane in N_2O/O_2 (70/30%), and body temperature was maintained by a heating blanket. The brain ischemia was induced by occlusion of the left middle cerebral artery (MCA) using the intraluminal filament technique. Left common carotid artery, external carotid artery (ECA), internal carotid artery (ICA) and pterygopalatine artery were exposed and the distal part of the ECA was occluded. Then, 7-0 round tip nylon suture was introduced into the ICA via the ECA, advanced 8- to 10-mm distal to the carotid bifurcation and occluded MCA. At 1 hour after ischemia, the mice were reanesthetized and the suture was withdrawn through the ECA.

Reverse transcriptase-polymerase chain reaction (RT-PCR) for TNFα

Brains were removed from the sham-operated controls (0 hour) and other animals at 0.5 and 1 hour during ischemia and at 1, 3, 6, 12, 24, and 48 hours after reperfusion (each time point 3 animals). The neocortex of the ipsilateral hemisphere was immediately separated on the plates cooled with ice, and it was snap frozen using liquid nitrogen. The samples were stored at −80 °C until assayed. Total mRNA was extracted using the acid phenol-guanidinium thiocianate method [3]. Total RNA was subjected to RT-PCR analysis. The cDNA was synthesized from RNA. The PCR amplification of cDNA with primers specific for TNFα (Forward; 3′-GCG ACG TGG AAC TGG CAG AAG A-5′, Reverse; 3′-GGT ACA ACC

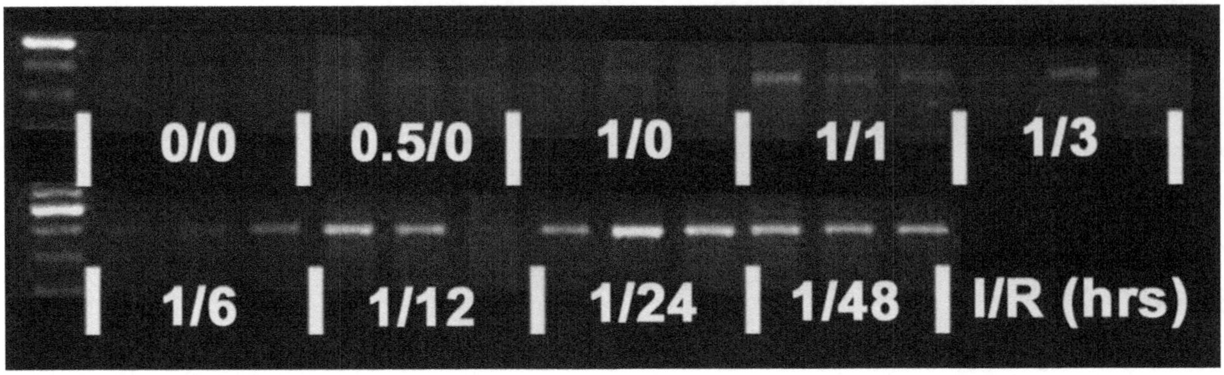

Fig. 1. Induction of TNFα mRNA in the ipsilateral cortex of mice brain following tMCAO, as assesses by RT-PCR. Each value (n = 3) was mean the time (hrs) of ischemia/reperfusion (I/R). Gel lanes are numbered according to the following order: 0/0, 0.5/0 and 1/0 hourduring ischemia and 1/1, 1/3, 1/6, 1/12, 1/24 and 1/48 hours after tMCAO for amplified RNAs from the ipsilateral brain cortex

CAT CGG CTG GCA-5′, 384 bp) and GAPDH (Forward; 3′-AAT GTG TCC GTC GTG GAT CT-5′, Reverse; 3′-TGT TGC TGT AGC CGT ATT CA-5′, 253 bp) as a house-keeping gene was carried out. The number of cycles ensuring non-saturating PCR conditions was established in the preliminary experiments. The PCR products were analyzed by electrophoresis on a 2% agarose gel containing ethidium bromide.

Immunohistochemistry for TNFα

Mice, which had been subjected to 1 and 24 hours after tMCAO, were anesthetized with sodium pentobarbital (50 mg/kg) and then perfused transcardially with 0.9% saline followed by 2% paraformaldehyde in 50 mM phosphate buffer. Brains were then removed, postfixed and cryoprotected in 20% sucrose. The sections were cut on a cryostat at a thickness of 8-µm.

The non-specific binding of the primary antiserum was blocked using 10% normal goat serum in PBS for 60 minutes and then, the sections were incubated overnight at 4 °C with polyclonal goat anti-TNFα antibody (Santa Cruz, CA). Then, the sections were incubated with biotinylated horse anti-goat IgG (1:200, Vector, Burlingame, CA) and were incubated in an avidin-biotin complex solution (Vector) followed by diaminobenzidine (DAB; Vector) as a chromogen. The negative control was performed using 10% normal goat serum instead of the primary antibody. The sections were observed with a fluorescant microscope (Olympus AX-70; Olympus, Tokyo, Japan).

Results

Expression of TNFα mRNA after tMCAO

As shown in the Fig. 1, induction of TNFα mRNA was hardly detected in the non-ischemic and ipsilateral brain cortex during ischemia. The expression of TNFα mRNA in the cortex of ipsilateral hemisphere was clearly increased at 1 hour after tMCAO. At 3 to 6 hours after tMCAO, the expressions were once decreased. The mRNA expression for TNFα in the ipsilateral cortex was dramatically increased again at 12 hours and was peaked at 24 hours after tMCAO. The expression of TNFα mRNA was sustained to 48 hours after tMCAO.

Immunohistochemistry for TNFα

Because the expression of TNFα mRNA was recognized at 1 and 24 hours after tMCAO in the ipsilateral cortex, we determined the TNFα protein at the time of high levels by use of immunohistological technique. Very weak TNFα-like immunoreactivities (TNFα-LI) were found in the normal and sham-operated mice. TNFα-LI was observed in the process like as dendrite in processes (Fig. 2A). After tMCAO, the TNFα-LI was induced in the ipsilateral cortex. The TNFα immunoreactive reaction products were noted in the cytoplasm of neurons and especially abundant around nucleus as well as dendritic process at 1 hour after tMCAO (Fig. 2B). After 24 hours tMCAO, the TNFα-LI in the core of infarction area a low level of immunoreactivity became weak. However, the TNFα-LI was intense around the infarction, the penumbra. The TNFα-LI was not seen in sections incubated without the primary antibody or with pre-absorbed antiserum incubated with the recombinant mouse TNFα, indicating that TNFα was expressed specifically in the brain following tMCAO (data not shown).

Discussion

The present results have clearly demonstrated that TNFα is expressed in neuronal cells of the ipsilateral cortex following transient cerebral ischemia. While

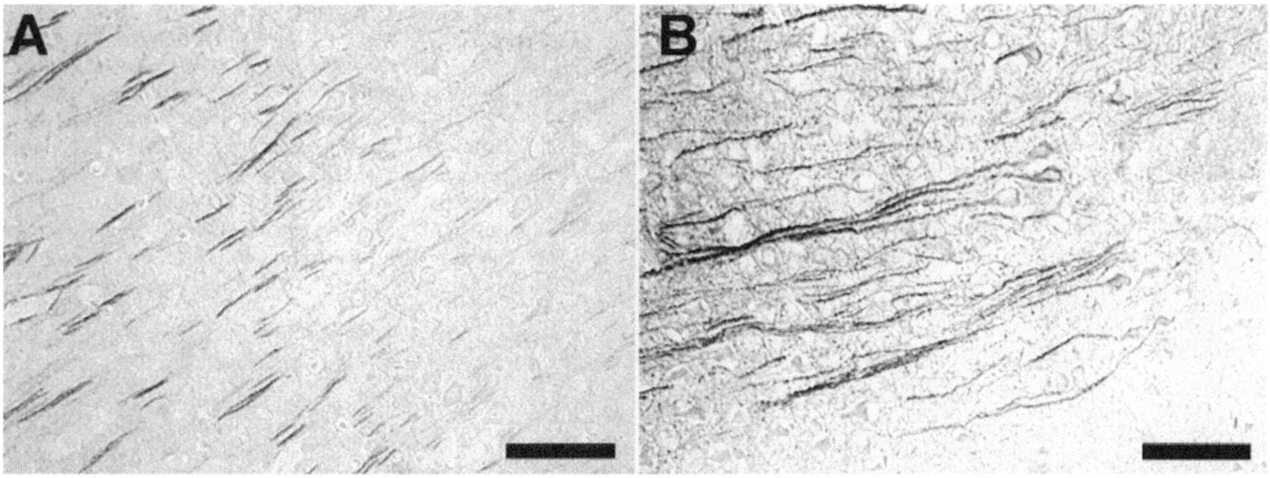

Fig. 2. Expression of TNFα immunoreactivity after transient middle cerebral artery occlusion (*tMCAO*). Photographs show results for sham-operated control mice (A) and animals studied 1 hour (B) after tMCAO in the cortex of the ipsilateral hemisphere

only a few levels of TNFα-like immunoreactivity were found in the sham-operated mice. Using RT-PCR and immunohistochemistry, the expression of TNFα induced by tMCAO revealed a early transient profile with a peak increase in the immunoreactive cells taking place by 1 hour after tMCAO and a late transient profile occurring at 24 hours following tMCAO. The role of TNFα in ischemic brain injury has not yet been fully elucidated. Administration of recombinant TNFα into the brain has been reported to induce neuronal cell death, and to produce a significant neutrophil adherence and an accumulation in capillaries and small blood vessels during ischemia [6]. Since reducing circulating neutrophil levels or their adhesion to blood vessels can significantly reduce ischemic damage, these data suggest that TNFα contributes to neurodegeneration. TNFα has been partivipeted to oligdendroglial injury, to cytotoxicity of lipopolysaccharide-stimulated astrocytes, and to astrocytic proliferation in vitro [8]. TNFα is shown to be roduced mainly by activated macrophages, but is also also produced by microglia and astrocytes in culture [5]. In addition, the evidence is accumulated that glial cells contributed to a cytokine network in the central nervous system [10]. Although cerebral ischemic injury promotes glial proliferation, its effects on production of cytokines have not been fully elucidated. However the cellular source of TNFα has not been identified in vitro after cerebral ischemia. Further studies are warranted to clarify exactly the cell sources of TNFα following tMCAO.

However in vitro, experiments that it is reported the addition of TNFα into the cultured neuron was diminished the neuronal cell death after excitotoxic and oxidative insults [1]. TNFα stimulates the expression of nerve growth factor in variety of cell types including fibroblasts, glioblastoma cells, and astrocytes [4], which may contribute to neuronal survival in and around the ischemic area. Cell expression of TNFα receptors is closely regulated in response to ischemia and injury. When mice rendered genetically deficient in both type of TNFα receptors (TNFR-KO) were subjected to tMCAO, the neuronal damage was exacerbated compared to the wide-type ones, which indicates that TNFα serves a neuroprotective function in vivo [2]. It is likely that the actions of TNFα during cerebral ischemia may be related to cell types, which deliver and accept TNFα signals and induce brain pathology. In the CNS, cultured microglia and astrocytes have been shown to produce TNFα upon several kind of stimulation, and TNFα immunoreactivity was detected in neurons following brain injury and cerebral ischemia.

In the present study, we studied the induction of TNFα mRNA and its expression after cerebral ischemia. The mechanism of TNFα in ischemic injury may relate to cell types which deliver/or accept TNFα signals. Further studies are warranted to clarify the cell sources of TNFα and expression of the TNF receptor. Preventing TNFα expression or blocking the signal transduction in some types of cells will be an effective treatment in cerebral ischemia.

Acknowledgments

This work was supported in part by a Showa University Grant-in-Aid for Innovative Collaborative Research Projects and a Special Research Grant-in-Aid for Development of Characteristic Education from the Japanese Ministry of Education, Culture, Sports, Science and Technology (S.S.), and a Showa University High-Technology Research Center Project from the Ministry of Education, Science, Sports and Culture of Japan (H.O. and S.S.).

References

1. Barger SW, Hoster D, Furukawa K, Goodman Y, Krieglatein J, Mattson J (1995) TNFα and TNFβ protect hippocampal neurons against amyloid β-peptide toxicity: evidence for involvement of a NB-κB binding factor and attenuation of peroxide and Ca^{2+} accumulation. Proc Natl Acad Sci USA 92: 9328–9332
2. Benveniste EN (1992) Inflammatory cytokines within the central nervous system: sources, function, and mechanism of action. Am J Physiol 263: C1–C16
3. Chomczynski P, Sacchi N (1987) Single-step method of RNA isolation by acid guanidium thiocyanate-phenol-chlorlform extraction. Anal Biochem 162: 156–159
4. Hattori A, Tanaka E, Murase K, Ishida N, Chatani Y, Tsujimoto M, Hyashi K, Kohno M (1993) Tumor necrosis factor stimulates the synthesis and secretion of biologically activity nerve growth factor in non-neurons cells. J Bio Chem 268: 2577–2582
5. Liebrman AP, Pitha PM, Shin HS, Shin ML (1989) Production of tumor necrosis factor and other cytokines by astrocytes stimulated with lipopolysaccharide or neurotropic virus. Proc Nan Acad Sci USA 86: 6348–6352
6. Liu T, Clark PK, Mcdonnel PC, Yong PR, White MS, Barone FC, Feuerstein GZ (1994) Tumor necrosis factor α expression ischemic neurons. Stroke 25: 1481–1488
7. Robbins DS, Shiraji Y, Drysdale BE, Lieberman A, Shin HS, Shin ML (1987) Production of cytotoxic factor for oligdendrocytes by stimulated astrocytes. J Immunol 139: 2593–2597
8. Sharief MK, Thompson EJ (1992) In vivo relationship of tumor necrosis factor α to blood-brain barrier damage in patients with active multiple sclerosis. J Neuroimmunol 38: 27–33
9. Tarlow MJ, Jenkins R, Comis SD, Osborne MP, Stephens S, Stanley P, Crocker J (1993) Ependymal cells of the choroid plexus express tumour necrosis factor-α. Neuropathol Appl Neurobiol 19: 324–328
10. Uno H, Matsuyama T, Akita H, Nishimura H, Sugita M (1997) Induction of tumor necrosis factor α in the mouse hippocampus following transient forebrain ischemia. J Cereb Blood Flow Metab 17: 491–497
11. Wesselingh SL, Gough NM, Finlay-Jones J, McDonald PJ (1990) Detection of cytokine mRNA in astrocyte cultures using the polymerase chain reaction. Lymphokine Res 9: 177–185

Correspondence: Dr. Seiji Shioda, Department of Anatomy, Showa University School of Medicine, 1-5-8 Hatanodai, Shinagawa-Ku, Tokyo, 142-8555, Japan. e-mail: shioda@med.showa-u.ac.jp

Acta Neurochir (2003) [Suppl] 86: 97–100

Evaluation of neuronal cell death after a new global ischemia model in infant mice

H. Ohtaki[1,3], S. Mori[2,3], T. Nakamachi[1], K. Dohi[1], L. Yin[1,3], S. Endo[1], Y. Okada[2,3], and S. Shioda[1,3]

[1] Department of Anatomy, Showa University School of Medicine, Tokyo, Japan
[2] Department of Cell Physiology, National Institute for Physiological Sciences, Okazaki, Japan
[3] The Core Research for Evolutional Science and Technology (CREST) of Japan Science and Technology Corporation (JST), Tokyo, Japan

Summary

For the first time we set up a new model for global ischemia in the infant mice, and time-dependent changes of the blood-brain barrier (BBB) disruption and neuronal cell death were investigated in detail. Infant C57/Bl6 mice (postnatal 13 days) were anesthetized with inhalation of sevoflurane in N_2O/O_2 (70/30%) and were subjected to global ischemia by bilateral common carotid artery occlusion (CCAO) for 25 minutes. Disruption of BBB was noted at 4 hours and increased up to 24 hours after the injection of 2% Evan's Blue in the transient CCAO (tCCAO) model. Evaluation of neuronal cell death was determined with toluidine blue staining. Morphological changes of neurons after tCCAO were clearly observed in the hippocampal CA1 region but were slightly detected in the CA3 region. However, there were no morphological changes in the hippocampal dentate gyrus, the neocortex, the striatum and the hypothalamus. The number of survival neurons in the CA1 was significantly decreased at 2 days and sustained up to 4 days after tCCAO. These data indicate that this method is very useful to induce selective vulnerability in mouse hippocampus, and it provides a reliable ischemic model in infant mice.

Keywords: Global ischemia; blood-brain barrier; toluidine blue staining; neuronal cell death; hippocampus; infant mice.

Introduction

Recent progress in experiments using transgene and gene deficient mice incited us to develop the model of ischemia in mice. In many papers the use of global [7–9, 12] and focal ischemia [1–4, 10] model has been reported and these studies reveal a role of many factors during the process of neuronal cell death. Meanwhile, studies on cultured neurons and/or slices of brains have clearly shown the mechanism of induction of neuronal cell death and cell death factors. However, there are still plenty of discrepancies between these studies in animals, and also in cultured cell and brain slices. Most of the ischemic models in animals have shown neuronal cell death in the adult mice. Since neuronal cultured cells prepared from adult mice are more vulnerable than those of fetal and infant mice, studies using primary cultures of neuronal cells and slices of brain widely use brains of the fetal and infant mice. Moreover, the ischemic model in cultured systems such as glucose deprivation and hypoxia is not necessarily the same as in pathophysiological condition.

The purpose of this study is to determine a new model of global ischemia in infant mice, which is useful for studies ex vivo, and to evaluate the disruption of the blood-brain barrier (BBB) and neuronal cell death.

Materials and methods

Animals and production of transient global ischemia

All experimental procedures involving animals were approved by the Institutional Animal Care and Use Committee of Showa University (#02095). C57BL/6 infant mice (postnatal 7 or 8 days) with their mother were purchased from Charles River Japan (Kanagawa, Japan). Thirteen days after birth, the mice were subjected to transient global ischemia. Induction of transient forebrain ischemia in mice was based on the model of transient common carotid artery occlusion (tCCAO, ref; 7, 12). Anesthesia was induced by inhalation of 1.5% sevoflurane in N_2O/O_2 (70/30%) and the body temperature was maintained at 37.5 °C with heat blanket. The common carotid artery was carefully exposed and isolated through a midline incision. Then, the bilateral common carotid arteries were occluded with clips (Zen temporary clip, Oowa-tusho, Tokyo, Japan) for 25 minutes and reperfused by removing the clip. After tCCAO, the mice and their mothers were kept for one night in a chamber with room temperature maintained at 35 °C to keep the body temperature uniform.

Detection of BBB disruption

After 0, 4, 8, and 24 hours tCCAO, the mice were anesthetized with inhalation of 1.5% sevoflurane in N_2O/O_2 (70/30%), and were infused 2% Evans blue into the right jugular vein 30 minutes before sacrifice. The mice were then anesthetized with inhalation of diethyl ether, and sacrificed by decapitation. Thereafter, the brain were removed, and immediately postfixed into 10% buffered-formalin. Thirty minutes after fixation, the brains were photographed using a digital camera.

Evaluation of morphology

After 0, 1, 2, 3, and 4 days tCCAO, the mice were anesthetized with inhalation of diethyl ether, and were sacrificed by decapitation. Then, the brains were immediately removed, and postfixed into 10% buffered-formalin. Subsequently they were paraffin-embedded on the hippocampal region and the 4-μm-thick coronal sections were applied to toluidine blue staining. After staining, the sections were observed and the number of cells counted with a light microscopy (Olympus AX-70; Olympus, Tokyo, Japan).

The number of cells were counted separately in four areas (two areas per each hemisphere) in the hippocampal CA1 and CA3 regions per one section. One area was 260 μm (W) × 180 μm (H) and the number of cells was shown as the average per one mouse.

The data was expressed as mean ± standard error. Statistical comparisons were made with Bornferroni's t-test. $P < 0.05$ was considered statistically significant.

Results

Mouse tCCAO model and the BBB disruption

There was almost no animal that died during the 25 minutes ischemia. Moreover, the percentage of survival 3 days after tCCAO was 90% or more. The body weight of the mice subjected to tCCAO was decreased remarkably at 1 day after tCCAO. However, the body weight of the mice had recovered at 2 days and increased after 3 days following tCCAO.

Evaluation of the BBB disruption after tCCAO was determined by penetration of Evans blue into the brain intravenously. The surface of the brain was slightly dyed blue 4 hours after tCCAO, and the penetration of Evans blue progressively increased up to 24 hours after tCCAO. These results suggest that the BBB was disrupted after tCCAO, and its disruption progressed with time.

Evaluation of neuronal cell death after tCCAO

The extent of neuronal cell death was compared in the hippocampus, neocortex, striatum and hypothalamus in mice brain subjected to tCCAO (Fig. 1). Four days after tCCAO, severe cell death of neurons was clearly observed in the CA1 region of the hippocampus, and a slight neuronal cell death was seen in the CA3 region of the hippocampus. However, neuronal cell death was difficult to recognize in the dentate gyrus of hippocampus, neocortex, striatum, and hypothalamus.

Subsequently, cell death of neurons during four days after tCCAO was evaluated morphologically and the number of survival neurons were counted in the CA1 and the CA3 regions of the hippocampus (Fig. 2). As shown in Fig. 2A, the pyramidal neurons in the CA1 of the hippocampus were noted with large round nuclei and clear nucleolus. Moreover, the cell layer of the CA1 was regularly arranged. In the pyramidal neurons of the CA1 at 1 day after tCCAO we sometimes observed neurodegeneration such as showing the nuclei deeply dyed with coloring matter and shrunk. However, there was little neuronal cell death as a whole (Fig. 2B). The degenerating cells were dramatically increased at 2 days and sustained up to 4 days after tCCAO. The nuclei of the pyramidal neurons in the hippocampus CA1 region appeared to clearly shrink and the pyramidal layer exhibited an irregular arrangement (Fig. 2C and D). The number of intact neuronal cells in the CA1 of the hippocampus were significantly decreased at 2 to 4 days after tCCAO. Although the number of neuronal cells of the CA3 area decreased gradually, there was no significant difference during the experiment (Fig. 2E).

As a result, we developed a global ischemia model in infant mice and observed delayed neuronal cell death, demonstrated specifically in the hippocampus, especially in the CA1 region.

Discussion

Recent advances in experiments in animal models of ischemia, the cultured neuron and brain slices determined the black box in the central nervous system and the neuronal cell death. However, neither the models inducing the neuronal cell death, nor the age of animals used had common standards. Therefore, some discrepancies arose between the experiments in animals and cultured cell and brain slices. To rule out these discrepancies, we tried to establish a new model of ischemia in infant mice.

Ischemia is induced by transient common carotid artery occlusion for 25 minutes. This model has become well known for induction of delayed neuronal cell death in the gerbil [5, 6, 11] and has recently been reported in adult mice [7, 9, 12]. Since this model of ischemia can be produced with a comparatively easy surgical operation, it appears suitable for the ischemic model in infant mice. In the present study, using infant mice, we found that neuronal cell death after tCCAO

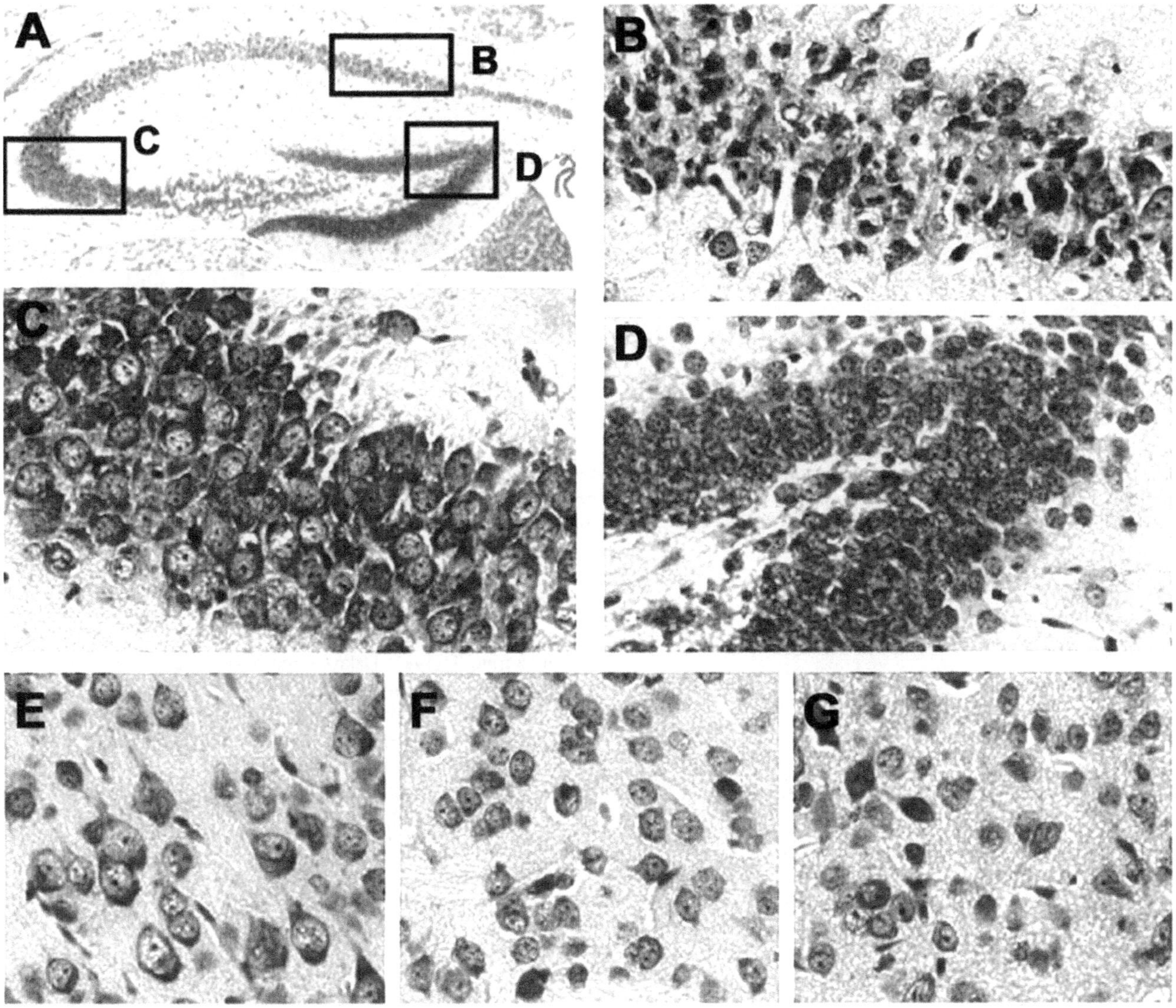

Fig. 1. Representative image of each brain areas 4 days after tCCAO. The photographs show the hippocampus (A, low magnification), the CA1 region (B), the CA3 region (C) and the dentate gyrus (D) of the hippocampus (high magnification) and the neocortex (E), the striatum (F) and the hypothalamus (G)

was above all severely induced in the CA1 region of the hippocampus, and also mildly induced in the CA3 region of the hippocampus, but not the other regions such as the hippocampal dentate gyrus, neocortex, striatum, and hypothalamus. The neuronal cell death induced by tCCAO was not caused after 1 day tCCAO, but was induced after 2 days tCCAO. Therefore, this neuronal cell death is regarded as delayed neuronal cell death. With the permeability of the BBB to the Evans blue, this ischemic model was accompanied by BBB disruption. Disruption of BBB carries out transportation of a substance into the brain. This model may hence be suitable for the estimation of intravenously injected substances.

More recently, we determined that the hippocampal slices prepared from infant mice subjected to tCCAO took Calcein, a calcium indicator, into the neuronal cells ex vivo and bound to calcium. This model can be adapted to experiment systems both in vivo and ex vivo (un-published data).

In summary, we developed a new global ischemia model, that can bc adapted also for ex vivo settings using infant mice. This model of brain ischemia caused disruption of BBB and specifically induced

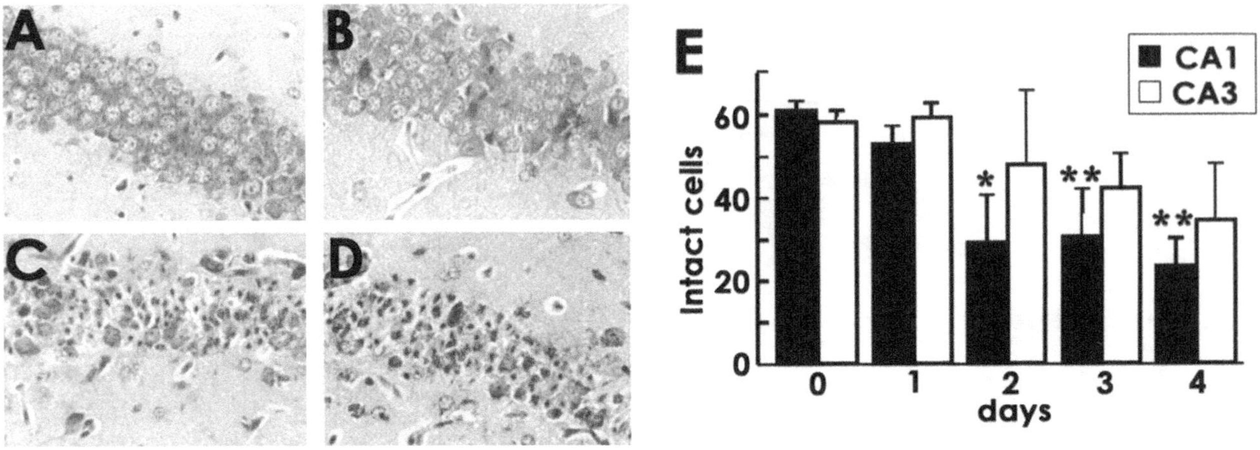

Fig. 2. Time-dependent changes of the pyramidal layer of CA1 region in hippocampus after tCCAO. Representative image of the hippocampus 0 (A), 1 (B), 2 (C), and 4 (D) days after tCCAO. (E) The number of un-damaged cell by staining toluidine blue. Each value is shown as the mean ± SE (n = 5 mice for each time point). *: $P < 0.05$, **: $P < 0.01$ compared with day 0

delayed neuronal cell death in the CA1 region of hippocampus.

Acknowledgments

This work was supported in part by a Showa University Grant-in-Aid for Innovative Collaborate Research Projects and a Special Research Grant-in Aid for development of Characteristic education from the Japanese Ministry of Education, Culture, Sports, Science, and Technology (S.S.). This study was also supported in part by grants from a Showa University High-Technology Research Center Project from the Japanese Ministry of Education, Culture, Sports, Science, and Technology (H.O. and S.S.).

References

1. Bruce AJ, Boling W, Kindy MS, Peschon J, Kraemer PJ, Carpenter MK, Holtsberg FW, Mattson MP (1996) Altered neuronal and microglial responses to exitotoxic and ischemic brain injury in mice lacking TNF receptors. Nat Med 2: 788–794
2. Connolly Jr ES, Winfree CJ, Stern DM, Solomon RA, Pinsky DJ (1996) Procedural and strain related variables significantly affect outcome in a murine model of focal cerebral ischemia. Neurosurgery 38: 523–532
3. Hara H, Huang PL, Panahian N, Fishman MC, Moskowitz MA (1996) Reduced brain edema and infarction volume in mice laking the neuronal isoform of nitric oxide synthase after transient MCA occlusion. J Cereb Blood Flow Metab 16: 605–611
4. Huang Z, Huang PL, Panahian N, Dalkara T, Fishman MC, Moskowitz MA (1994) Effect of cerebral ischemia in mice deficient in neuronal nitric oxide synthase. Science 265: 1883–1885
5. Ide T, Takada K, Qiu JH, Saito N, Kawahara N, Asai A, Kirino T (1999) Ubiquitin stress response in postischemic hippocampal neurons under nontolerant and tolerant conditions. J Cereb Blood Flow Metab 19: 750–736
6. Kirino T (1982) Delayed neuronal death in the gerbil hippocampus following ischemia. Brain Res 239: 57–69
7. Kitagawa K, Matsumoto M, Yang G, Mabuchi T, Yagita Y, Hori M, Yanagihara T (1998) Cerebral ischemia after bilateral carotid artery occlusion and intraluminal suture occlusion in mice: evaluation of the patency of the posterior communicating artery. J Cereb Blood Flow Metab 18: 570–579
8. Mizushima H, Zhou CJ, Dohi K, Horai R, Asano M, Iwakura Y, Hirabayashi T, Arata S, Nakajo S, Takaki A, Ohtaki H, Shioda S (2002) Reduced postischemic apoptosis in the hippocampus of mice deficient in interleukin. 1. J Comp Neurol 448: 203–216
9. Murakami K, Kondo T, Kawase M, Chan PK (1998) The development of a new mouse model of global ischemia: focus on the relationships between ischemia duration, anesthesia, cerebral vasculature, and neuronal injury following global ischemia in mice. Brain Res 780: 304–310
10. Schneider A, Martin Villalba A, Weih F, Vogel J, Wirth T, Schwaninger M (1999) NF kB is activated and promotes cell death in focal cerebral ischemia. Nat Med 5: 554–559
11. Tanaka K, Shirai T, Nagata E, Dembo T, Fukuuchi Y (1997) Immunohistchemical detection of nitrotyrosine in postischemic cerebral cortex in gerbil. Neurosci Lett 235: 85–88
12. Yang G, Kitagawa K, Matsushita K, Mabuchi T, Yagita Y, Yanagihara T, Matsumoto M (1997) C57BL/6 strain is most susceptible to cerebral ischemia following bilateral common carotid occlusion among seven mouse strains: selective neuronal death in murine transient forebrain ischemia. Brain Res 752: 209–218

Correspondence: Hirokazu Ohtaki, Department of Anatomy, Showa University School of Medicine, 1-5-8 Hatanodai, Shinagawa-ku, Tokyo 142-8555, Japan. e-mail: taki@med.showa-u.ac.jp

Acta Neurochir (2003) [Suppl] 86: 101–103

Hippocampal heme oxigenase-1 in a murine cardiac arrest model

Y. Imaizumi[1,2], H. Mizushima[2], K. Dohi[1,2], H. Ohtaki[2,3], H. Funahashi[2], and S. Shioda[2,3]

[1] Department of Neurosurgery, Showa University School of Medicine, Tokyo, Japan
[2] Department of Anatomy, Showa University School of Medicine, Tokyo, Japan
[3] The Core Research for Evolutional Science and Technology (CREST) of Japan Science and Technology (JST), Japan

Summary

Heme oxygenase-1 (HO-1) is a member of the heat shock protein family (HSP-32). It responds to thermal stress in cultured glial cells. To our knowledge, nothing is known about the expression and response of the HO-1 in cerebral ischemia. Therefore, we show here the induction of HO-1 in the brain of mice after global cerebral ischemia. HO-1-like immunoreactivity was detected at 12, 24, and 48 hours after ischemia recirculation. The HO-1-like immunoreactive cells were observed in astrocytes in the hippocampal dentate gyrus and CA1. The peak level of HO-1-like immunoreactivity was found 48 hours after the recirculation. HO-1-like immunoreactivity was observed in GFAP-positive astrocytes by use of a double immunostaining method. These results provide direct evidence for the induction and localization of HO-1 immunoreactivity in vivo in a mouse cerebral ischemia. We suggest that HO-1, produced in astrocytes after ischemia-recirculation, may directly affect neurons to protect from cell death.

Keywords: Heme oxigenase-1; cytokines; cerebral ischemia; cardiac arrest; astrocytes; mouse.

Introduction

Three heme oxigenase (HO) proteins have been identified until now. HO-1 is a stress-induced protein and its isoforms, HO-2 and HO-3, are constitutively expressed [6, 10]. HO-1 is a 32-kDa protein, and is identified as heat shock protein-32 (HSP-32) [12, 13]. HO-1 is reported to be induced by ischemia and hemorrhage in the rat central nervous system [9, 13]. The one most important function of HO is oxidation of the heme molecule into biliverdin, carbon monoxide, and iron [6]. Bilirubin, which is metabolized by biliverdin reductase, is shown to be a powerful antioxidant. Induction of the HO-1 may protect from cell death from oxidative stress by increasing bilirubin production and by fostering the sequestration of catalytic iron with ferritin. Recent physiological and morphological data have indicated that the HO-1 is synthesized and secreted in several kinds of cells including microglia, astrocytes, and neurons in the central nervous sytem [7, 8]. These results suggest that endogenous HO-1 plays an important role in protecting from oxidative damage after ischemia recirculation. The cardiac arrest model has recently been developed to improve defects of the four-vessel occlusion model [1, 3, 5]. The purpose of the present study is to determine first, the expression of HO-1 in brain and second, to identify cells that produce HO-1 in brain after ischemia recirculation in the mouse.

Materials and methods

Animals

Male adult BALB/c mice (25–30 g) were housed on a 12–24 h light-dark cycle and given food and water ad lib. The animals were anesthetized with 3.5% halothane and maintained with 2% sevoflurane in 70% N_2O and 30% O_2. A 24-gauge polyethylene catheter was inserted into the left femoral artery. Blood pressure and the electrocardiogram were monitored during all of the procedures. Arterial blood samples were analyzed at 37 °C for pH, pCO_2, and pO_2 10 min before cardiac arrest and 10 min after reperfusion. Global cerebral ischemia was induced according to the rat cardiac-arrest model [1]. Briefly, an anesthetized mouse was placed supine and an incision was made in the midline of the chest. The occluding device was gently inserted in the mediastinum at the level of the 4th intercostal segment. The device was manipulated to the dorsal wall of the thorax. Then the distal end was positioned under the major cardiac blood vessels. A complete interruption of the circulation was accomplished by lifting the occluding device and, at the same time, applying gentle downward pressure with the fingers on the sternum. After 5 min of transient global ischemia, under artificial ventilation with a tidal volume of 2.5 ml at 200 times per min (100% O_2), heart massage was started for resuscitation. After a recovery period of 30 min, the animals were disconnected from the respirator. The animals were housed in cages with free access to water and food pellets.

Immunohistochemistry

The animals were anesthetized with pentobarbital and then they were perfused for 30 min with a fixative that consisted of 400–500 ml of 2% paraformaldehyde and 0.1% glutaraldehyde (Electron Microscopy Sciences, Cherry Hill, NJ) in 0.1 M phosphate buffer (pH 7.2) for 20–30 min. Twenty-μm sections from the brains were cut on a cryostat and picked up on gelatin-coated slides. Endogenous peroxidase activity was blocked with 0.01 M phosphate buffer that contained 0.5% hydrogen peroxide and 2% normal goat serum. Sections were rinsed with PBS, preincubated with 10% normal goat serum in PBS, and then incubated with rabbit anti-HO-1 antibody (Oncogene Research Products, Cambridge, MA), at a dilution of 1:100–1:1,000. After incubation, tissue sections were immunostained by the avidin-biotin complex method (Vector Laboratories, Burlingame, CA). They were subsequently developed with 3,3′-diaminobenzidine-4HCl (Vector Laboratories) and 0.05 ml of a 5% solution of hydrogen peroxide in Tris-HCl buffer. Sections were dehydrated in a graded ethanol series and coverslipped with DPX.

For double immunostaining, coronal sections were incubated with the solution mixed with both anti HO-1 antibody and mouse anti-GFAP antibody (Sigma, St. Louis, USA) overnight at 4 °C. These sections were incubated with Cy3-labeled donkey anti-goat IgG antibody (1:1000, Amersham International Plc, Buckinghamshire, UK) for 2 h at room temperature and incubated with Alexa 488-labeled goat anti mouse IgG antibody (Molecular Probes, Eugene, OR, USA) for 2 h at room temperature. The sections were mounted and examined with an Olympus AX-70 fluorescence microscope (Olympus Co., Ltd., Tokyo, Japan). To verify the specificity of staining, the sections were (1) incubated with non-immune serum instead of the primary antibody, (2) incubated with the primary antibody without the second and third antibodies, and (3) incubated with the primary antibody that had been reabsorbed with the antigen.

Results

Blood pressure was immediately dropped to zero after cardiac arrest. With commencement of resuscitation after 5 minutes of total ischemia, the blood pressure rose promptly. Pre-operative blood gas data (n = 4) at 10 min before cardiac arrest were pH 7.33 ± 0.02, PaO_2 121.20 ± 7.44 mmHg, $PaCO_2$ 40.12 ± 1.91 mmHg. Post-operative data (n = 4) at 10 min after cardiac arrest were pH 7.23, PaO_2 164.08 ± 19.13 mmHg, $PaCO_2$ 44.49 ± 2.76 mmHg. Blood pH temporarily decreased after global ischemia.

In the nonischemic hippocampal tissue, low levels of HO-1-like immunoreactivity were observed. At 12 hours after ischemia, HO-1-like immunoreactivity was increased in the glial cells in the hippocampal region. At 24 hours after ischemia, many glial cells showed HO-1-like immunoreactivity. By 48 hours after ischemia, amoeboid microglia and astrocytes were increased in number and distributed close to the hippocampal fissure and they were highly immunoreactive for HO-1. The increase of HO-1-like immunoreactivity seemed to be specifically found in astrocytes and microglia in the hippocampal region.

In the non-ischemic dentate gyrus, low levels of HO-1-like immunoreactivity were detected. At 12 hours after ischemia, HO-1-like immunoreactivity was increased in glial cells in the dentate gyrus. At 24 hours after ischemia, glial cells showing HO-1-like immunoreactivity were increased in number. By 48 hours after ischemia, amoeboid microglia and astrocytes were highly immunoreactive to HO-1. The increase in the HO-1-like immunoreactivity in the dentate gyrus seemed to be more evident than in the hippocampal CA-1.

The number of microglial cells and astrocytes gradually increased during 12–48 hours after ischemia-reperfusion. These glial cells were small, with elongated cell bodies, and had one or more thick and thin processes. Some of them had numerous short secondary branches along their length, suggesting glial cells. Therefore we used double-labeling immunohistochemistry to determine the cell type of HO-1-like immunoreactivity at 48 hours after ischemia reperfusion.

Discussion

We used the cardiac-arrest model first introduced by Korpatchev *et al.* in the rat [5]. The operative mortality is relatively low (less than 10%), and the resuscitated animals remain physiologically stable and available for extended observations for indefinite periods of time [1, 3, 14].

The present study revealed that the peak level of HO-1 mRNA expression and HO-1-like immunoreactivity was 12 hours and 24 hours after ischemia recirculation, respectively. It has been reported that the intense gene expression of HO-1 mRNA is detected 12 hours after ischemia in rat brain [4, 18]. Paschen *et al.* [11] have reported that the HO-1 mRNA level increases consistently up to 24 hours in the cortex and striatum, while it peaks at 4 hours and thereafter decreases in the hippocampus after transient global ischemia recirculation. As to HO-1 induction in brain ischemia, Matsuoka *et al.* [7] have reported that HO-1 is expressed early but transiently in pyramidal neurons and later but continuously in the microglia in the CA1 subfield. Singhal *et al.* [15] have reported that HO-1 immunoreactive microglial cells peaked at 24 hours in the ischemic region. We found that HO-1 immunoreactive microglial cells peaked at 24 to 48 hours in the hippocampal region and dentate gyrus, and its expression was detected in astrocytes after brain ischemia recirculation by double-labeling immunohistochemistry. It seems that HO-1 immunoreactivity appears earlier in astrocytes than microglial cells in hippocampal tissue and dentate gyrus after ischemia, but further studies are needed to clarify this detail.

We found a marked increase in the HO-1 expression in mouse hippocampus and dentate gyrus after ischemia-recirculation. HO-1 expression was detected in astrocytes by use of double-labeling immunohistochemistry. Despite the observation that the apoptotic cell death occurs in the CA1 and CA2 pyramidal neurons and dentate gyrus granule neurons [14], the HO-1 expression was not altered in neurons in these areas. Instead, HO-1-like immunoreactivity was increased in astrocytes after ischemia recirculation. There have been many studies showing a marked increase in HO-1 mRNA expression in the rat brain following cerebral ischemia [4, 18]. It is suggested that the HO-1 mRNA expressing cells are microglia or astrocytes [4].

The most important function of HO is oxidation of the heme molecule in accordance with NADPH-cytochrome P-450 reductase followed by cleavage of heme into biliverdin, carbon monoxide, and iron [6]. It is also reported that bilirubin, which is metabolized by biliverdin reductase, is a powerful antioxidant [16, 17]. Carbon monoxide, that activates guanylyl cyclase, acts as a putative neural messenger and carbon monoxide has the ability to relax smooth muscle and block platelet aggregation [19]. In contrast to these functions of HO, HO-1 also produces free iron which may act as a source of oxidative stress in diseases of the CNS in humans [2].

In summary, we have provided evidence for the increase in HO-1 and its transcript in the hippocampus and dentate gyrus after ischemia-recirculation. Furthermore, we have found that increase in HO-1-like immunoreactivity is localized selectively in astrocytes. HO-1, produced in the astrocytes, may play a very important role in protecting from neuronal damage.

References

1. Dohi K, Shioda S, Mizushima H, Homma H, Ozawa H, Nakai Y, Matsumoto K (1998) Delayed neuronal cell death and microglial cell reactivity in the CA1 region of the rat hippocampus in the cardiac arrest model. Med Electron Microsc 32: 85–93
2. Gelman BB (1995) Iron in CNS disease. J Neuropathol Exp Neurol 54: 477–486
3. Kawai K, Nitecka L, Ruetzler CA, Nagashima G, Joo F, Mies G, Nowak TS, Saito N, Lohr JM, Klatzo I (1985) Global cerebral ischemia associated with cardiac arrest in the rat: I. Dynamics of early neuronal changes. J Cereb Blood Flow Metab 12: 238–249
4. Koistinaho J, Miettinen S, Keinanen R, Vartiainen N, Roivainen R, Laitinen JT (1996) Lomg-term induction of haem oxygenase-1 (HSP-32) in astrocytes and microglia following transient focal brain ischemia in the rat. Eur J Neurosci 8: 2265–2272
5. Korpatchev WG, Lysenkov SP, Thieliz PS (1982) Modielirowanlie klinczeskoj smerti l postreanimationznoj bolezni u krys. Palol Fiziol Exp Ter 3: 78–80
6. Maines MD (1988) Heme oxigenase: function, multiplicity, regulatory mechanisms, and clinical applications. FASEB J 2: 2557–2568
7. Matsuoka Y, Kitamura Y, Ozaki M, Sakata M, Tsukahara T, Taniguchi T (1998) Induction of heme oxygenase-1 and major histocompatibility complex antigens in transient forebrain ischemia. J Cereb Blood Flow Metab 18: 824–832
8. Matsuoka Y, Ozaki M, Zhao H, Asai S, Ishikawa K, Kitamura Y (1999) Phosphorylation of c-jun and its Localization with heme oxygenase-1 and cyclooxygenase-2 in CA1 pyramidal neurons after transient forebrain ischemia. J Cereb Blood Flow Metab 19: 1247–1255
9. Matz P, Turner C, Weinstein PR, Massa SM, Panter SS, Sharp FR (1995) Heme-oxygenase-1 induction in glia throughout rat brain following experimental subarachnoid hemorrhage. Brain Res 713: 211–222
10. McCoubrey WK Jr, Huang TJ, Maines MD (1997) Isolation and characterization of a cDNA from the rat brain that encodes hemoprotein heme oxygenase-3. Eur J Biochem 247: 725–732
11. Paschen W, Uto A, Djuricic B, Schmitt J (1994) Heme oxygenase expression after reversible ischemia of rat brain. Neurosci Lett 180 (1): 5–8
12. Shibahara S, Muller RM, Taguchi H (1987) Transcriptional control of rat heme oxygenase by heat shock. J Biol Chem 262: 12889–12892
13. Shibahara S, Muller R, Taguchi M, Yoshida T (1985) Cloning and expression of cDNA for rat heme oxygenase. Proc Natl Acad Sci USA 82: 7865–7869
14. Shioda S, Zhou CJI, Dohi K, Mizushima H, Funahashi H, Ohtaki H, Nakajo S, Arata S, Hirabayashi T, Horai R, Asano M, Iwakura Y, Takaki A (2000) Reduced postischemic apoptosis of mice deficient in interleukin-1. Soc Neurosci Abstr 30: 1810, 670.16
15. Singhal AB, Wang X, Sumii T, Mori T, Lo EH (2002) Effects of normobaric hyperoxia in a rat model of focal cerebral ischemia-reperfusion. J Cereb Blood Flow Metab 22: 861–868
16. Stocker R, Glazer AN, Ames BN (1987) Antioxidant activity of albumin-bound bilirubin. Proc Natl Acad Sci USA 84: 5918–5922
17. Stocker R, Yamamoto Y, McDonagh AF, Glazer AN, Ames BN (1987) Bilirubin is an antioxidant of possible physiological importance. Science 235: 1043–1046
18. Takeda A, Onodera H, Sugimoto A, Itoyama Y, Kogure K, Shibahara S (1994) Increased expression of heme oxygenase mRNA in rat brain following transient forebrain ischemia. Brain Res 666: 120–124
19. Verma A, Hirsch DJ, Glatt CE, Ronnett GV, Snyder SH (1993) Carbon monoxide: a putative neural messenger. Science 259: 381–384

Correspondence: Seiji Shioda, Department of Anatomy, Showa University School of Medicine, 1-5-8 Hatanodai, Shinagawa-ku, Tokyo 142-8555, Japan. e-mail: shioda@med.showa-u.ac.jp

Acta Neurochir (2003) [Suppl] 86: 105–111

Calcineurin and cyclophilin D are differential targets of neuroprotection by immunosuppressants CsA and FK506 in ischemic brain damage

H. Uchino[1], **N. Ishii**[1], and **F. Shibasaki**[2]

[1] Department of Anesthesiology, Tokyo Medical University Hachioji Medical Center, Tatemachi, Hachioji, Japan
[2] Department of Molecular Cell Physiology, Tokyo Metropolitan Institute of Medical Science, Honkomagome, Bunkyo-ku, Tokyo, Japan

Summary

The search for an effective treatment for global ischemia following cardiac arrest has proved to be very difficult. However, studies by Uchino *et al.* [1, 2] show that the immunosuppressant cyclosporin A (CsA), when administered in such a way that the drug can bypass the blood brain barrier (BBB), dramatically reduces ischemic damage in rat forebrain preparations. An alternative immunosuppressant, FK506, is apparently less efficacious [3, 4]. Both CsA and FK506 are specific inhibitors of immunophilins, (CsA inhibits cyclophilins, FK506 inhibits FKBPs), and of calcineurin, a type 2B Ser/Thr phosphatase that is abundant in the central nervous system [5]. The superiority of CsA may be partly attributable to its selective amelioration of mitochondrial damage, as assayed in vivo and in vitro. Our results suggest that pathways involving calcineurin and cyclophilins, particularly mitochondrial cyclophilin D, play pivotal roles in the development of ischemic brain damage. The present findings may inform the search for new drugs in the treatment of global ischemic damage to the brain, and in other organs.

Keywords: Brain ischemia; cyclosporin A; FK506; calcineurin; MRI; cyclophilin D; Mitochondrial Permeability Transition (MPT); bad; and cytochrome *c*.

Introduction

Brief periods of brain ischemia can lead to specific forms of damage at the cellular level that are accompanied by a cascade of biochemical events. The influx of calcium into cells is a key event among several factors in the pathogenesis of ischemic brain damage. The influx is due to the opening of calcium channels on the cell membrane, which in turn is due to presynaptic release of glutamate resulting in membrane depolarization. The rise in free cytosolic calcium concentration triggers a further cascade of reactions. These include lipolysis, proteolysis and phosphorylation/dephosphorylation of regulatory proteins, which include kinases and phosphatases. Transient global ischemia may cause severe brain damage with serious social, medical, and economic implications.

In the past, the effectiveness of neuroprotective treatment following ischemia has been at best mild to moderate. Recently, interest has focused on the use of selective immunosuppressants for the treatment of ischemia. The neuroprotective effects of immunosuppressants on ischemic brain damage were first reported for Tacrolimus (FK506) in rats subjected to focal ischemia [11]. Pre-treatment with cyclosporin A (CsA) also dramatically reduces forebrain ischemic damage in rats, provided that the drug can be delivered to the brain without being blocked by the blood-brain barrier (BBB) [15–17]. Although no direct comparison has been made between the two agents, the effects of FK506 in this model of ischemia seem less pronounced than those of CsA [3].

Ischemia is one of the principal causes of acute organ failure following cardiac arrest. The acute effects may involve a mitochondrial dysfunction, due to activation of the mitochondrial permeability transition (MPT) pore [8]. The MPT pore is a complex found at sites connecting the inner and outer mitochondrial membrane. It is comprised of several mitochondrial proteins including VDAC, ANT, and cyclophilin D (CyPD) [1]. CsA and FK506 are specific inhibitors of immunophilins, *i.e.* prolyl *cis/trans* isomerases, and of calcineurin, a type 2B Ser/Thr phosphatase that is abundant in the central nervous system [9]. Recently, calcineurin was found to regulate cell death [12–14], partially through dephosphorylation of the pro-apoptotic protein Bad [19].

On the basis of these findings, we sought to evaluate the anti-ischemic effects of CsA and FK506. Our hypothesis was that the neuroprotective action of these agents involved the inhibition of calcineurin and/or cyclophilins such as CyPD, and ultimately the function of the MPT pore.

Materials and methods

Rat models of transient forebrain ischemia

Wistar rats (8- to 10-weeks-old) weighing 300–350 g were anesthetized and global ischemia was induced by bilateral common carotid artery occlusion/exsanguination for 10 min, as described previously [16]. One week prior to the ischemic insult, a Hamilton syringe (outer diameter 450 mm) was introduced into the hippocampus on one side under stereotactic guidance. This procedure, termed needle penetration, permitted CsA to enter the brain and was performed under anesthesia. Both drugs were then administrated intraperitoneally 3 days prior to the ischemic insult at doses of 10 mg/kg for CsA, or 1 or 2 mg/kg for FK506, respectively. The injections were continued once per day until the day of the ischemic insult.

Magnetic resonance imaging

Diffusion-weighted magnetic resonance imaging (DWI) was performed using a 4.7-T superconducting animal imaging system (Unity INOVA, Varian, Palo Alto, CA). The image data were then preprocessed using Application Visualization System 5 (AVS5, Advanced Visual Systems, Waltham, MA, USA). Signals of non-neural origin were eliminated by density slicing. Pixel intensity was rendered in pseudo-color with 256 color levels. The area of interest was selected manually using the crop module of AVS5.

Biochemical assay

Samples from hippocampal region CA1, CA3 and DG were dissected and prepared at fixed time points following the onset of recirculation, as described previously (XX-please insert reference). The extracted tissue was then separated into three samples. For the assay of calcineurin activity, frozen tissue (10 mg) from the three regions was homogenized in 10 volumes (ml)/g wet tissue of lysis buffer A. Following fractionation with a filter unit, 150 ng of protein was used to determine calcineurin activity, as described previously [18]. For the analysis of Bad, calcineurin, and cytochrome c (Cyt c), immunoblotting was performed using standard procedures. Swelling of isolated mitochondria isolated from male Wistar rats (300 g) was estimated from changes in light scattering at a wavelength of 540 nm measured in mitochondrial suspensions (0.5 mg of protein in 2 ml), as described previously [7].

Results

Pretreatment with CsA provides marked neuroprotection in rat forebrain ischemia

In vehicle-treated animals that had been subjected to the needle penetration (NP) procedure, mean damage to the CA1 region was 91.9 $\pm$ 5.8% (Fig. 1a[1], column 1). Nearly total suppression of cell damage (3.7 $\pm$ 3.8%, $p < 0.001$ vs. vehicle) was observed in animals pretreated with NP + 10 mg/kg CsA (column 2). In contrast, pretreatment with 1 mg/kg FK506 without NP (column 3) partially decreased damage to 58.0 $\pm$ 9.9% ($p < 0.001$ vs. vehicle). A similar suppression of cell damage (51 $\pm$ 12%) was observed in rats pretreated with 1 mg/kg and NP (column 4). However, suppression of cell damage in rats with NP + 2 mg/kg FK506 (column 5, 72 $\pm$ 7%) was less than in rats pretreated with only 1 mg/kg FK506 ($p < 0.05$ vs. column 3). Typical examples of neuronal damage in CA1 are shown in Fig. 1b.

We further performed DWI to evaluate the neuroprotective effects and the time course of ischemic damage in the whole brain (Fig. 1c and d). Following 7 days of recovery time, areas of hyperintensity were prominent in the hippocampal CA1 region as well as parietal cortical regions. These areas of hyperintensity were totally suppressed by CsA, but not by FK506. No inter-hemispheric differences in the neuroprotective effect of CsA were observed.

Increase of calcineurin activity and dephosphorylation of Bad in CA1 following ischemia

As shown in Fig. 2a[2], calcineurin activity increased by 300–450% relative to the control following 1 h

[1] Reprinted from Neurobiology of Disease, Vol. 10, Uchino *et al.* "Differential neuroprotection by cyclosporin A and FK506 following ischemia corresponds with differing abilities to inhibit calcineurin and the mitochondrial permeability transition", Pages 219–233, Copyright (2002), with permission from Elsevier Science.

[2] This figure is also reprinted from Uchino *et al.*, Copyright (2002), with permission from Elsevier Science.

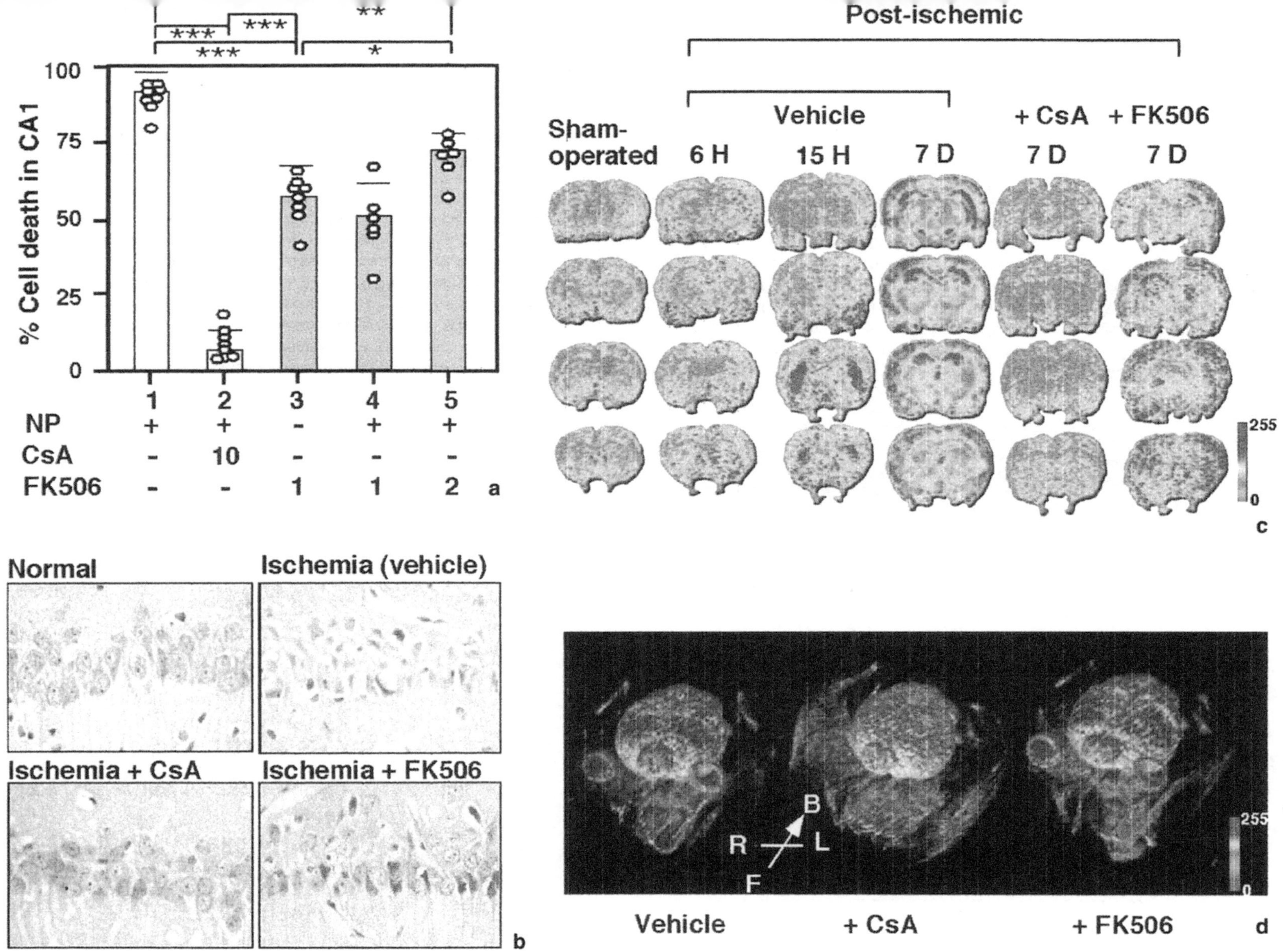

Fig. 1. The neuroprotective effect of CsA and FK506 in forebrain ischemia. (a) Histopathological outcome in the CA1 region of the hippocampus following 7 days recovery time from a 10 min. ischemic insult. Figure shows data from the vehicle-injected animals (column 1; n = 8), from animals that had been pre-treated with 10 mg/kg CsA (column 2; n = 8), 1 mg/kg FK506 (column 3; n = 8 and column 4; n = 6) or 2 mg/kg FK506 (column 5; n = 6) (*** $p < 0.001$, ** $p < 0.01$, * $p < 0.05$; one-way ANOVA with post-hoc Scheffé's test). A needle penetration had been performed 3 days before the ischemic insult in all animals except for those that received 1 mg/kg FK506 (column 3). (b) Examples of histopathological findings in hippocampal CA1 neurons. (c) Magnetic resonance images taken from a sham-operated rat, and from a vehicle-treated rat following 6 hours (6H), 15 hours (15H), and 7 days (7D) recovery time. (d) 3D-imaging by volumetric reconstitution. Each reconstruction represents 24 DWI slices from the whole brain of each rat. The panels show frontal images of rats pretreated with vehicle (left), CsA (middle), or FK506 (right) following 7 days of recovery time. This figure is reproduced and modified with permission from Uchino *et al.* [18] and Elsevier Science (see footnote in text)

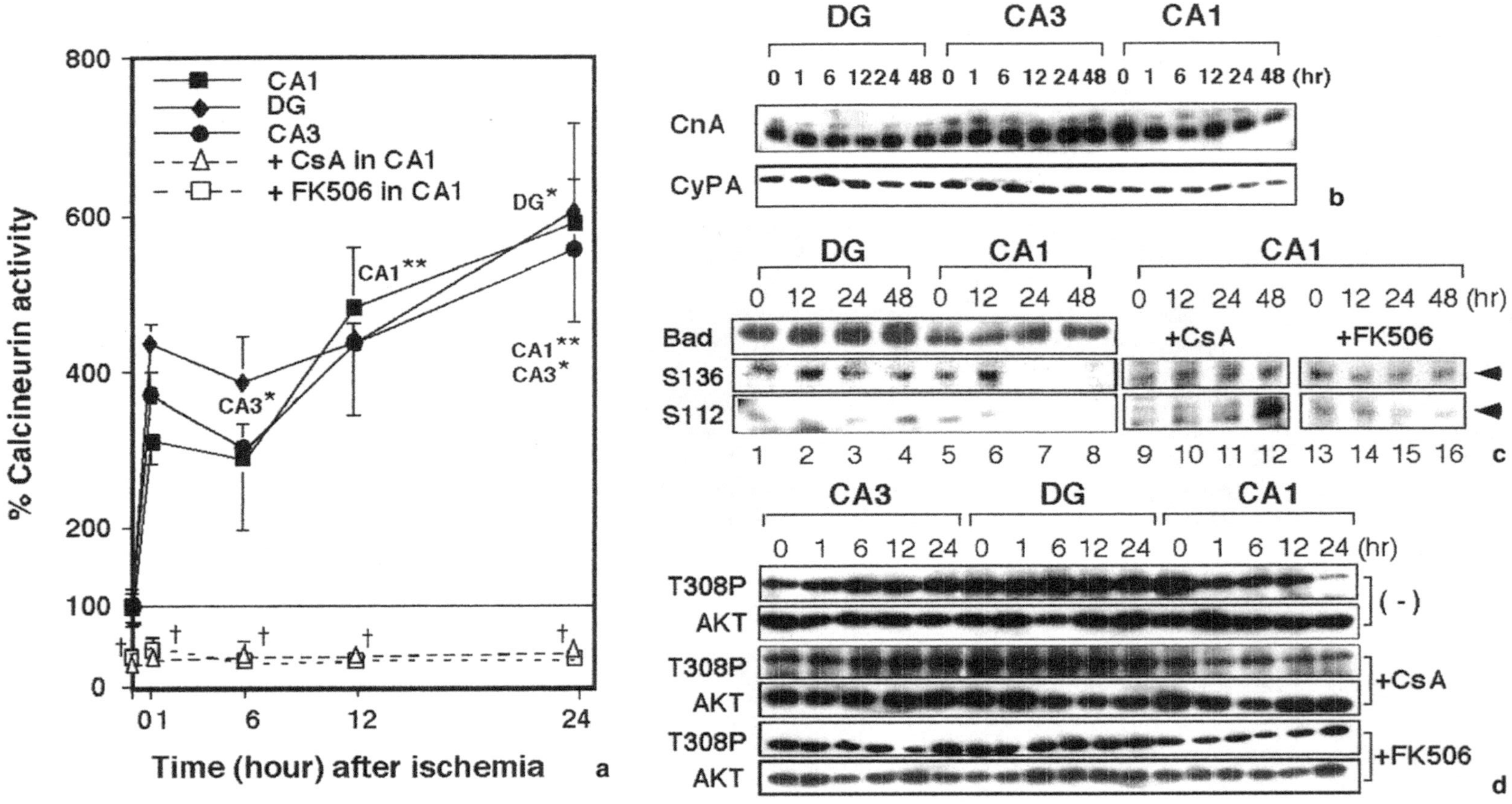

Fig. 2. Bad dephosphorylation by calcineurin correlates with cell death in CA1, but not in DG or the CA3 region of hippocampus. (a) Calcineurin activity in hippocampal CA1, CA3 and DG regions following transient forebrain ischemia. The supernatant (150 ng of protein) prepared from CA1 (closed squares), CA3 (closed circles) and DG (closed diamonds) was assayed for calcineurin activity at the time indicated. Calcineurin activity is shown in vehicle-injected animals as a percentage of the activity in each hippocampal region. To evaluate the effect of CsA and FK506, the supernatant from the CA1 region of rats pretreated with CsA (open triangles) or FK506 (open squares) was determined (** $p < 0.01$, * $p < 0.05$, † $p < 0.01$; one-way ANOVA with post-hoc Scheffé's test). (b) Immunoblot analysis of calcineurin A (CnA) and cyclophilin A (CyPA) in preparations of whole hippocampus by anti-calcineurin A or anti-cyclophilin A. Immunoblots prepared from 50 mg samples of the total lysate and separated on 10% SDS-PAGE from rat brains sampled at 0, 1, 6, 12, and 24 h following ischemia. Data from pretreated and non-pretreated animals are shown, with Cyclophilin A as an internal control. (c) Each 100 mg sample of total lysate from CA1 or DG, obtained at the time indicated, was separated on 15% SDS-PAGE, followed by immunoblotting. Samples were stained with anti-Bad antibody (Bad) or anti-phosphoserine 112-Bad (Bad-S112) and anti-phosphoserine 136 (Bad-S136) antibodies (lanes 1 to 8). Lysates from the CA1 region pretreated with CsA were also analyzed with both anti-phospho-Bad antibodies (lanes 9 to 12). (d) Immunoblots of the above lysates with (+) or without (−) CsA were probed with anti-phosphothreonine 308 of Akt/PKB (T308P) or anti-Akt/PKB (AKT) antibodies. All immunoblot results are representative of at least three separate experiments. Reproduced with permission from Uchino *et al.* [18] and Elsevier Science (see footnote in text)

of recirculation. A secondary increase was observed at 12 h ($p < 0.01$ vs. values at 1 h) and 24 h, peaking at approximately 600% (CA1, 591 ± 126%; CA3, 557 ± 89.2%; DG, 606 ± 74.9%). Calcineurin induces activity-dependent cell death through Bad, a pro-apoptotic protein of the Bcl-2 family [19]. To further explore the relationship between calcineurin and Bad, we examined Bad phosphorylation with site- and phospho-specific antibodies and compared the results to those with an antibody that recognizes the Bad protein.

Dephosphorylation of Bad on Ser 112 and Ser 136, the key phosphorylation sites, was observed in the CA1 region, but not in CA3 or DG. This dephosphorylation was inhibited by pre-treatment with either CsA or FK506 (Fig. 2c), and was preceded by an increase in Bad phosphorylation 12–24 h following ischemia. Since the degree of phosphorylation of Bad in the CA1 region may depend on the balance between calcineurin phosphatase activity and Akt/PKB kinase activity, we examined the activity of Akt/PKB, using a phospho-specific antibody to Akt/PKB Thr 308. Phosphorylation of this site reflects kinase activity. Phosphorylation of Akt/PKB at Thr 308 was markedly reduced following 24 h recovery time in the CA1 region in animals pretreated with either CsA or FK506 (Fig. 2d). The effect was not seen in other hippocampal regions. These findings, along with the fact that FK506 and CsA inhibit both the dephosphorylation of Bad and calcineurin activity, suggest that Bad dephosphorylation is mediated by calcineurin. The differences in anti-ischemic effects between CsA and FK-506, however, could not be explained in terms of their differential effects on calcineurin activation, or in terms of Bad phosphorylation.

Suppression of Cyt c release and mitochondrial swelling in CA1 by CsA but not FK506

Following 24 h recovery time from ischemia, release of Cyt c from mitochondria was observed exclusively in the CA1 region. CsA blocked or strongly reduced Cyt c release, whereas FK506 exerted, at best, a weak effect on Cyt c release [18]. Electron microscopic analysis was performed to assess morphological changes to mitochondria following ischemic reperfusion. During the recirculation/recovery period, mitochondrial morphology in the CA1 region of control animals advanced from partial swelling (24 h) (Fig. 3a[3]) to massive swelling associated with disruption of the mitochondrial cristae (48 h) (Fig. 3b). Animals treated with CsA had contracted mitochondria that were morphologically normal. In contrast, the mitochondria from the FK506-treated animals were partially swollen with some disruption of the mitochondrial cristae (data not shown).

Suppression of calcium-induced swelling of isolated brain mitochondria by CsA

The fact that CyPD, but not other immunophilins such as FKBP, regulates the MPT pore is confirmed by results that show that CsA, but not FK506, suppresses mitochondrial swelling and pore assembly in vitro [5]. CsA reduces the rate of Ca^{2+}-induced mitochondrial swelling in a dose-dependent fashion, while FK506 had no inhibitory effect on mitochondrial swelling.

Discussion

Differences between the effects of CsA and FK506 following ischemic brain damage have been reported previously. However, these differences may be due to the fact that the two drugs were evaluated using different animal models. It is possible that the mechanisms that mediate neuronal cell death following blood vessel occlusion may not be the same in each animal model. In addition, the interpretation of the results from previous work is made difficult by the low penetration of CsA through the BBB compared to FK506. We have demonstrated that CsA provides potent neuroprotection when administered in conjunction with ipsilateral needle penetration, a procedure that allows CsA to breach the BBB. In fact, tissue concentration of CsA is increased by a factor of 3–5 following needle penetration [17]. In contrast, tissue concentration of FK506 readily reaches the therapeutic range even without the needle penetration procedure. The difference between the permeability of CsA and FK506 is due to an active efflux mechanism via the p-glycoprotein that acts on CsA but not on FK506 [10].

Recently, calcineurin has been reported to induce cell death through dephosphorylation of Bad, a pro-

---

[3] This figure is also reprinted from Uchino *et al.*, Copyright (2002), with permission from Elsevier Science.

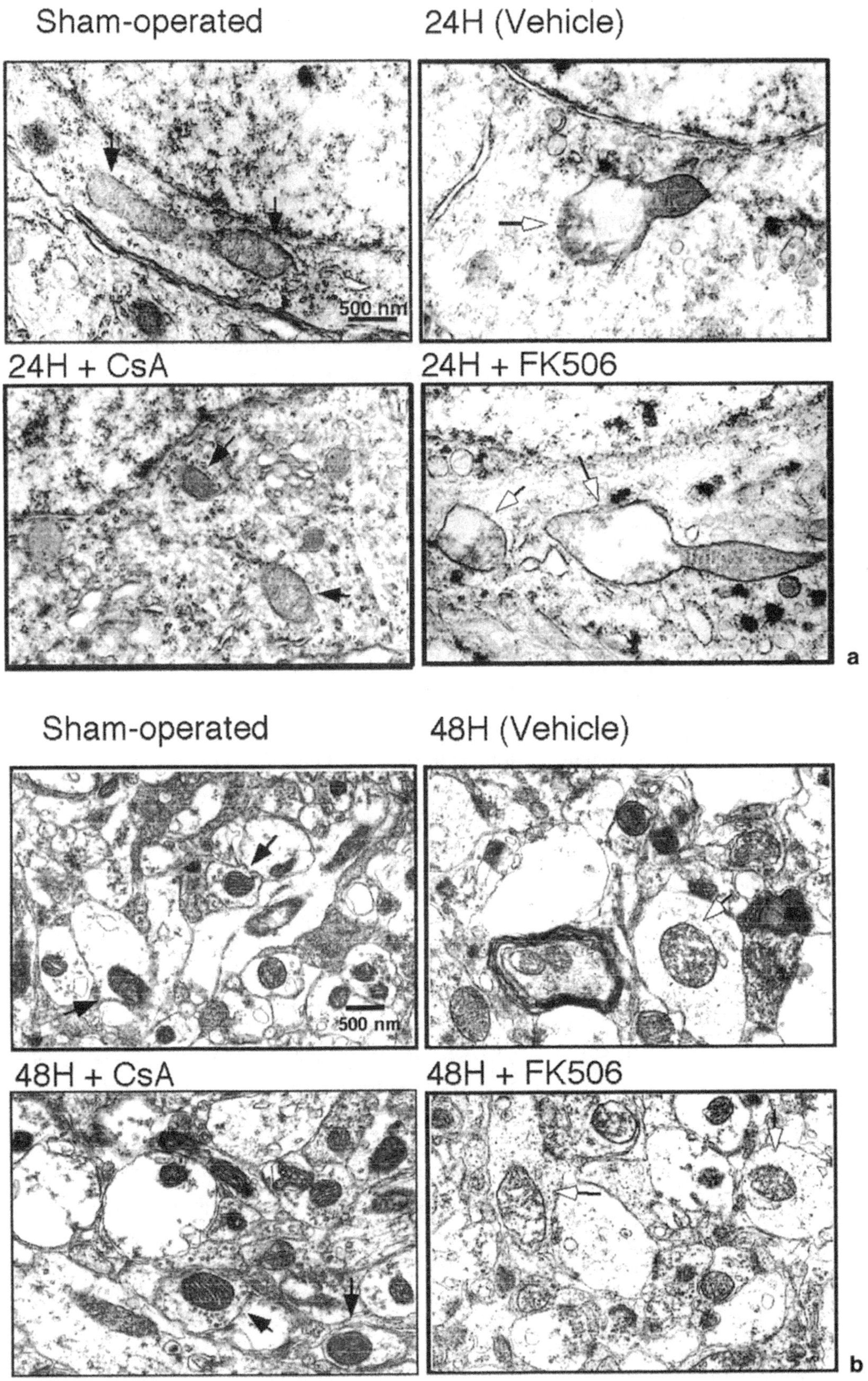

Fig. 3. Evaluation of the effects of CsA and FK506 on mitochondrial swelling by electron microscopy. (a) Ultrastructural changes at 24 h of reperfusion in pyramidal cell bodies from the CA1 region. The specimens for electron microscopy were prepared from the brains of sham-operated animals or ischemic animals pretreated with CsA, FK506 or vehicle. Black and white arrows show normal and swollen mitochondria, respectively. (b) Ultrastructural changes at 48 h of reperfusion in the molecular layer superjacent to the crest in the CA1 region. The specimens prepared for electron microscopy were from the brains of sham-operated animals (left upper) or ischemic animals pretreated with CsA (left lower) or FK506 (right lower) or vehicle (right upper). All pictures were taken at a magnification of ×15,000. Reproduced with permission from Uchino *et al.* [18] and Elsevier Science (see footnote in text)

apoptotic member of the Bcl-2 family [19]. Dephosphorylation of Bad is first observed approximately 24 h following ischemia. Interestingly, this dephosphorylation is only observed in the CA1 region, and not in CA3 or DG, suggesting that apoptotic cascades are initiated with Bad in the CA1 region. This protein may act as a pro-apoptotic mediator by activating mitochondrial depolarization related to MPT pore formation.

Despite the fact that calcineurin is the enzyme responsible for the dephosphorylation of Bad, we have found that calcineurin is activated in all three hippocampal regions. It is therefore likely that the activity of Bad kinases, such as Akt/PKB [2], is reduced by ischemia specifically in CA1. In fact, we show that phosphorylation of threonine 308 of Akt/PKB, an indicator of kinase activation, is selectively reduced in CA1 24 h following ischemia. There was no corresponding change in Akt/PKB expression.

Our results indicate that both CsA and FK506 can prevent the downregulation of Akt/PKB phosphorylation that would otherwise follow from ischemic damage. We sought to explore the mechanism underlying this result. Examining tissue sampled 24 h after the ischemic insult, we characterized the changes in mitochondrial morphology at the electron microscopic (EM) level and prepared western blots to monitor the release of cytochrome c. The results indicate that there is complete suppression of mitochondrial swelling and cytochrome c release 24 h following ischemic damage in the CsA-treated animals. These effects were, at best, only partial in the FK506-treated animals.

CsA has been shown to block MPT pore assembly by binding to the CyPD component of the MPT complex [4]. In vitro experiments using isolated mitochondria from brain (Fig. 3) strongly suggest that the differences between the levels of neuroprotection afforded by CsA and FK506 are due to CsA's selective suppression of mitochondrial swelling, via inhibition of CyPD. These results suggest that pathways involving calcineurin and cyclophilins, particularly mitochondrial cyclophilin D, play pivotal roles in ischemic brain damage.

References

1. Crompton M (1999) The mitochondrial permeability transition pore and its role in cell death. Biochem J 341: 233–249
2. Datta SR, Dudek H, Tao X *et al* (1997) Akt phosphorylation of BAD couples survival signals to the cell intrinsic death machinery. Cell 91: 231–241
3. Drake M, Friberg H, Boris Moller F *et al* (1996) The immunosuppressant FK506 ameliorates ischemic damage in the rat brain. Acta Physiol Scand 158: 155–159
4. Friberg H, Connern C, Halestrap AP, Wieloch T (1999) Differences in the activation of the mitochondrial permeability transition among brain regions in the rat correlate with selective vulnerability. J Neurochem 72: 2488–2497
5. Friberg H, Ferrand Drake M, Bengtsson F *et al* (1998) Cyclosporin A, but not FK506, protects mitochondria and neurons against hypoglycemic damage and implicates the mitochondrial permeability transition in cell death. J Neurosci 18: 5151–5159
6. Kochi S, Takanaga H, Matsuo H *et al* (1999) Effect of cyclosporin A or tacrolimus on the function of blood brain barrier cells. Eur J Pharmacol 372: 287–295
7. Kristián T, Gertsch J, Bates TE, Siesjö BK (2000) Characteristics of the calcium triggered mitochondrial permeability transition in nonsynaptic brain mitochondria: effect of cyclosporin A and ubiquinone O. J Neurochem 74: 1999–2009
8. Lemasters JJ, Nieminen AL, Qian T *et al* (1998) The mitochondrial permeability transition in cell death: a common mechanism in necrosis, apoptosis and autophagy. Biochim Biophys Acta 1366: 177–196
9. Liu J, Farmer JD, Jr, Lane WS *et al* (1991) Calcineurin is a common target of cyclophilin cyclosporin A and FKBP FK506 complexes. Cell 66: 807–815
10. Sakata A, Tamai I, Kawazu K *et al* (1994) In vivo evidence for ATP dependent and P glycoprotein mediated transport of cyclosporin A at the blood brain barrier. Biochem Pharmacol 48: 1989–1992
11. Sharkey J, Butcher SP (1994) Immunophilins mediate the neuroprotective effects of FK506 in focal cerebral ischemia. Nature 371: 336–339
12. Shibasaki F, McKeon F (1995) Calcineurin functions in Ca^{2+}-activated cell death in mammalian cells. J Cell Biol 131: 735–743
13. Shibasaki F, Hallin U, Uchino H (2002) Calcineurin as a multifunctional regulator. J Biochem (Tokyo) 131: 1–15
14. Shibasaki F, Kondo E, Akagi T, McKeon F (1997) Suppression of signalling through transcription factor NF AT by interactions between calcineurin and Bcl2. Nature 386: 728–731
15. Siesjö BK, Elmer E, Janelidze S *et al* (1999) Role and mechanisms of secondary mitochondrial failure. Acta Neurochir Suppl (Wien) 73: 7–13
16. Uchino H, Elmer E, Uchino K *et al* (1995) Cyclosporin A dramatically ameliorates CA1 hippocampal damage following transient forebrain ischemia in the rat. Acta Physiol Scand 155: 469–471
17. Uchino H, Elmer E, Uchino K *et al* (1998) Amelioration by cyclosporin A of brain damage in transient forebrain ischemia in the rat. Brain Res 812: 216–226
18. Uchino H, Minamikawa, Tachino R, Kristián K *et al* (2002) Differential neuroprotection by cyclosporin A and FK506 following ischemia corresponds with differeing ability to inhibit calcineurin and the mitochondrial permeability transition. Neurobio Dis 10: 213–233
19. Wang HG, Pathan N, Ethell IM *et al* (1999) Ca^{2+} induced apoptosis through calcineurin dephosphorylation of BAD. Science 284: 339–343

Correspondence: Hiroyuki Uchino, Department of Anesthesiology, Tokyo Medical University Hachioji Medical Center, Tatemachi 1163, Hachioji 193-0998, Japan. e-mail: huchino@hotmail.com

Acta Neurochir (2003) [Suppl] 86: 113–116

FK506 attenuates the post-ischemic perturbation of protein kinases and tyrosine phosphorylation in the gerbil hippocampal CA1 sectors

K.-I. Katsura[1], **J. Kurihara**[2], **M. Watanabe**[1], **K. Takahashi**[1], and **Y. Katayama**[1]

[1] The Second Department of Internal Medicine, Nippon Medical School, Tokyo, Japan
[2] Department of Pharmacology, Faculty of Pharmaceutical Sciences, Teikyo University, Kanagawa, Japan

Summary

To explore effects of Immunosuppressant FK506 on signal transduction pathway, we studied changes in subcellular distribution of protein kinase Cγ (PKCγ), CaM kinase II (CaMKII), as well as changes of tyrosine phosphorylation levels after ischemia.

Male Mongolian gerbils were divided into 3 groups; FK506 (1 mg/kg, 3 mg/kg) and vehicle. FK506 was administered intravenously after 5 min ischemia. At the designated time points (0 time, 5 min ischemia, 1 hour, or 24 hour recovery), heads were frozen and samples were taken from CA1 subfield of hippocampus. Western blot analysis was carried out with specific antibodies for PKCγ, CaMKII, and phosphotyrosine.

FK506 administration significantly decreased translocation of PKCγ and CaMKII at 24 h of recovery ($p < 0.05$, ANOVA followed by Student-Newman Keuls' test) in P2 fraction. The levels of tyrosine phosphorylated p160, p140, p100, p90, and p80 in P2 fraction were also significantly decreased with FK506 treatment at 24 h of recovery.

The persistently elevated PKCγ and CaMKII level in P2 fraction which may be related to cell death, are attenuated with FK506 treatment. FK506 may contribute to recover calcium homeostasis in the post ischemic phase and promote cell survival.

Keywords: FK506; Tacrolimus; PKC; CaMKII; phosphotyrosine.

Introduction

The immunosuppressant FK506 (Tacrolimus), binding to a member of the immunophilins named FK binding protein 12 (FKBP12), inhibits calcineurin, a calcium and calmodulin-dependent protein phosphatase. Sharkey and Butcher (1994) [13] showed that FK506 has a neuroprotective effect in a rat focal ischemia model, when given 1 min after the arterial occlusion. We have shown protective effects of FK506 against focal ischemia model using filament [1]. FK506 administration has been shown to increase the number of surviving hippocampal pyramidal neurons in the transient global ischemia [7]. However, the protective mechanisms by which FK506 ameliorates ischemia-reperfusion injury have not been elucidated. Besides, we have shown the recovery of translocated CaM kinase II (CaMKII), which shares calmodulin with calcineurin, in surviving CA1 neurons with ischemic preconditioning [10], and persistent translocated protein kinase Cγ (PKCγ) after exaggerated fatal ischemia with enhanced acidosis [9].

To explore the effects of FK506 on the signal transduction pathway, we studied the changes in the subcellular distribution of PKCγ, CaMKII as well as the changes of phosphotyrosine levels after ischemia in hippocampal CA1 sector.

Materials and methods

Male Mongolian gerbils, fasted overnight with free access to tap water, were anesthetized with 3% halothane and N_2O/O_2 (70:30). The ischemic insult was given by bilaterally occluding common carotid arteries. Rectal and temporal muscle temperatures were maintained close to 37 °C [4]. At the designated periods (0 time, end of 5 min ischemia, 1-hour and 24-hour reperfusion), gerbils were immersed into liquid nitrogen. In the 1 h and 24 h reperfusion groups, animals were allowed to recover and warmed by heat lamp to maintain close to 37 °C. Animals were divided into 3 groups; FK506 (1 mg/kg, 3 mg/kg) treated and vehicle treated group. FK506 was administered intravenously at 1 min after recirculation. Frozen heads were coronally sectioned at the levels of hippocampus. Samples (~3 mg) were taken from the CA1 subfield of the hippocampus. Western blot analysis was carried out [10].

Subcellular fractionation, electrophoresis and immunoblotting

The tissue was sonicated in ice-cold homogenization buffer. The homogenate was spun at 800 g for 10 min at 4 °C. The supernatant was collected and was spun at 9200 g for 15 min at 4 °C. The resulting pellet was regarded as a crude synaptosomal fraction (P2). The supernatant was further spun at 176.000 g for 1 hour at 4 °C to obtain the final supernatant (cytosolic fraction, S3). Samples

(15 μg protein) were subjected to electrophoresis on 7.5% SDS-polyacrylamide gel. Separated proteins were electrotransferred onto PVDF membranes. The membranes were incubated with a monoclonal antibody against γ isoform of protein kinase C, or monoclonal antibody against CaMKII alpha subunit, or monoclonal antibody against phosphotyrosine. Finally, the membranes were visualized using an enhanced chemiluminescent detection system and images were captured by lumino CCD camera. The density of the bands was quantified by ATTO densitograph system [10].

Results

Effects of FK506 on PKCγ distribution

As shown in Fig. 1, the changes in subcellular distribution of PKCγ in the vehicle treated group were similar to the previous results in rats [10]. The PKCγ levels in the FK506 administered groups did not differed from vehicle treated group at 1 h of recovery. In particulate fraction, FK506 treatment accelerated the recovery of translocated PKCγ at 24 h of recirculation (1 mg/kg; $p < 0.05$, 3 mg/kg; $p < 0.01$, post hoc Student Newman Keuls test). Changes in the cytosolic PKCγ level were not affected by FK506 administration.

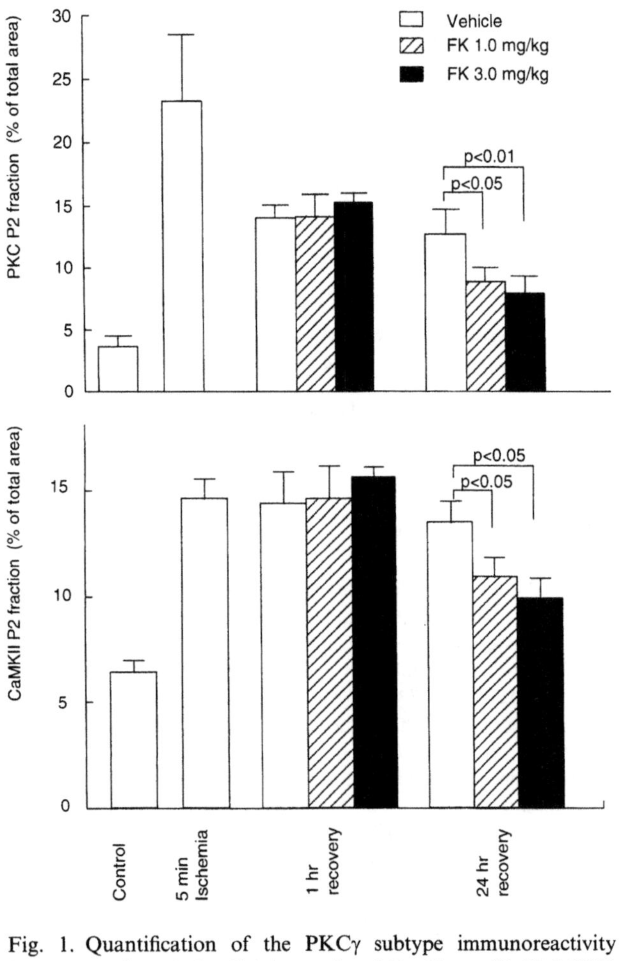

Fig. 1. Quantification of the PKCγ subtype immunoreactivity (upper panel) and the Calcium calmodulin kinase II (CaMKII) immunoreactivity (lower panel) after 5 min of ischemia in synaptosomal (P2) subcellular fractions. FK506 were administered intravenously at 1 min after the end of ischemia. Open bar shows the group without FK506 treatment. Hatched or filled bar shows FK506 treated groups with the dose of 1 mg/kg or 3 mg/kg, respectively. In P2 fraction at 24 hours of recovery, FK506 treatment significantly decreased the PKCγ and CaMKII levels (2 factor ANOVA followed by Student-Newman-Keuls' test)

Effects of FK506 on CaMKII distribution

As shown in Fig. 1, the changes in CaMKII subcellular distribution in the vehicle treated group were similar to the previous results [10], thus, a persistent translocation of CaMKII was observed following the second lethal ischemia.

A significant recovery of synaptosomal CaMKII level toward pre-ischemic level were observed at 24-hour reperfusion in FK506 treated groups. In the cytosolic fraction, no significant effect of drug administration on the CaMKII levels was observed.

Effects of FK506 on phosphotyrosine levels

Five protein bands with approximate molecular weights of 80 (p80), 90 (p90), 100 (p100), 140 (p140), and 160 (p160) showed significant reduction with FK506 treatment at 24 h of recovery. FK506 had no significant effects at 1 h of recovery.

Discussion

It has been shown that the extent of PKC translocation to the cell membrane during ischemia correlates well with the extent of cell injury. Thus, intra-ischemic hypothermia, which provides a significant neuroprotection against cell death, causes less PKC to translocate during ischemia [3]. On the contrary, systemic hyperglycemia or hypercapnia, which aggravates ischemic cell death [8], enhances the PKCγ translocation [9].

Our results showed the significant recovery of PKCγ translocation after administration of FK506. As previously reported with the exaggerated acidosis model, persistently elevated PKCγ level in crude synaptosomal fraction may be related to cell death [9]. Since PKC works in opposite direction as calcineurin, the present results suggest that PKC may be downregulated through feed back mechanisms by increased

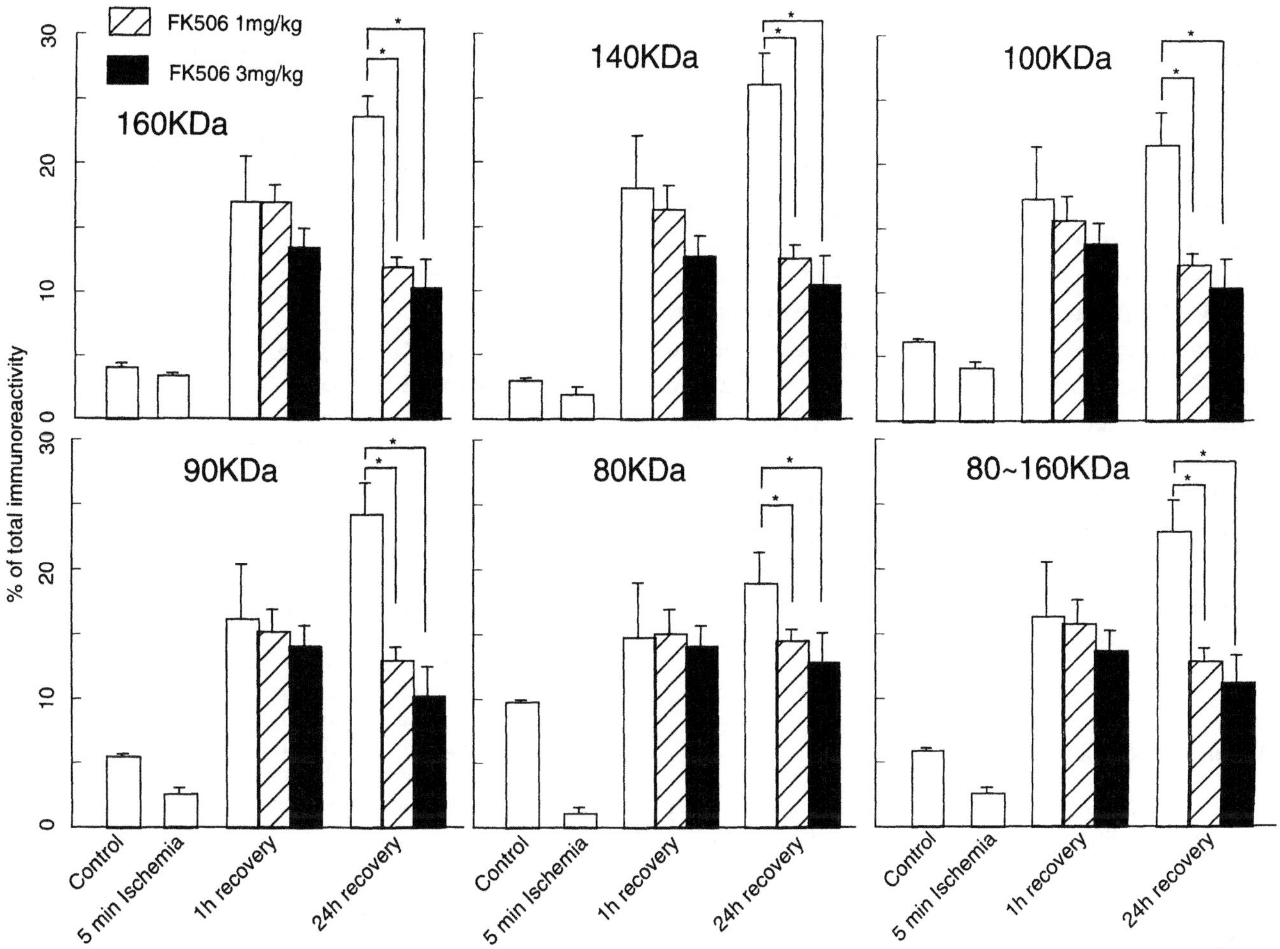

Fig. 2. Changes in the level of tyrosine-phosphorylated proteins p160, p140, p100, p90, p80, and total of p80~p160 in crude synaptosomal fraction of the CA1 sector of gerbil hippocampus. The data are expressed as % of total immunoreactivity of bands on each Western blot. Open column; vehicle treated group. Hatched column: FK506 1 mg/kg treated group, filled clolumn; FK506 3 mg/kg group. Results from 4 experiments are indicated as mean $\pm$ SE. *P $<$ 0.05 denote significant differences from vehicle treated group at 24 h recovery (2 factor ANOVA followed by Student-Newman Keul's test)

phosphorylated materials after inhibition of calcineurin with FK506 post-treatment.

CaMKII is a multifunctional protein kinase, which plays both physiological and pathological roles in neurons [11]. Similarly to PKCγ, translocation of CaMKII seems to correlate with ischemic cell death. Hu and Wieloch (1995) [6] have shown that translocation of CaMKII is transient in the neocortex and CA3/DG, two areas relatively resistant against ischemia, but persists up to 24 hours of reperfusion in the vulnerable CA1 region, which results in delayed neuronal death. Translocation of CaMKII to the crude synaptosomal fraction partly recuperated at 24-hour recovery in the preconditioned group, but not in the non-preconditioned group [10]. These studies suggest that the preconditioning ischemia enhances the recovery of elevated CaMKII in particulate fraction which may be detrimental to cell survival, leading to protection of cell damage in CA1 region. In the present study, CaMKII was translocated significantly to the crude synaptosomal fraction and persisted for 24 hours without drug treatment. FK506 post-treatment showed significant reduction of translocated CaMKII in P2 fraction.

In view of the protecting effects of FK506 administration, persistent translocation of PKCγ and CaMKII in P2 fraction may be related to cell death and the protective mechanisms of FK506 treatment may involve downstream signal transduction pathway below PKC and CaMKII activation.

Tyrosine-phosphorylation of some synaptosomal proteins, in particular the tyrosine phosphorylation of p120 and p180, is enhanced during reperfusion [5]. In this experiment, those proteins could correspond to the p100 and p160, respectively. As shown in the present investigation, the phosphorylation of these proteins is significantly reduced with FK506 treatement. It is most likely that p180 (p160) is the NMDA receptor subunit NR2A/2B [12]. Tyrosine phosphorylation of NMDA receptor by src family of tyrosine kinases is known to enhance the NMDA-evoked responses. Therefore a reduction in tyrosine phosphorylation of the NR2A/2B could contribute to decrease damage by attenuating intracellular calcium levels. It is reported that TNF-receptor and IL-receptor could be observed within p80 fraction. Inflammatory responses are implicated in the pathophysiology of ischemic brain injury [2] and anti-inflammatory drugs with neutrophil inhibition have been shown to be cerebroprotective. In this context, it is possible to speculate that FK506 might affect neuronal protein kinases by inhibiting the post-ischemic inflammatory or immune responses.

Acknowledgements

This study was supported by the Ministry of Education of Japan, Grant-in Aid for Scientific Research (C2), 13670671, 2001.

References

1. Arii T, Kamiya T, Arii K, Ueda M, Nito C, Katsura K, Katayama Y (2001) Neuroprotective effect of immunosuppressant FK506 in transient focal ischemia in rat. Neurol Res 23: 755–760
2. Barone FC, Feuerstein GZ (1999) Inflammatory mediators and stroke: new opportunities for novel therapeutics. J Cereb Blood Flow Metab 19: 819–834
3. Cardell M, Boris-Möller F, Wieloch T (1991) Hypothermia prevents the ischemia-induced translocation and inhibition of protein kinase C in the rat striatum. J Neurochem 57: 1814–1817
4. Hiraide T, Katsura K, Muramatsu H, Asano G, Katayama Y (2001) Adenosine receptor antagonists cancelled the ischemic tolerance phenomenon in gerbil. Brain Res 910: 94–98
5. Hu B-R, Wieloch T (1994) Tyrosine phosphorylation and activation of mitogen-activated protein kinase in the rat brain following transient cerebral ischemia. J Neurochem 62: 1357–1367
6. Hu B-R, Wieloch T (1995) Persistent translocation of Ca2+/Calmodulin-dependent protein kinase II to synaptic junctions in the vulnerable hippocampal CA1 region following transient ischemia. J Neurochem 64: 277–284
7. Ide T, Morikawa E, Kirino T (1996) An immunosuppressant, FK506, protects hippocampal neurons from forebrain ischemia in the mongolian gerbil. Neurosci Lett 204: 157–160
8. Katsura K, Kristián T, Smith M-L, Siesjö BK (1994) Acidosis induced by hypercapnia exaggerates ischemic brain damage. J Cereb Blood Flow Metab 14: 243–250
9. Katsura K, Kurihara J, Siesjö BK, Wieloch T (1999) Acidosis enhances translocation of protein kinase C but not Ca2+/Calmodulin-dependent protein kinase II to cell membranes during complete cerebral ischemia. Brain Res 849: 119–127
10. Katsura K, Kurihara J, Kato H, Katayama Y (2001) Ischemic preconditioning affects the subcellular distribution of protein kinase C and calcium/calmodulin-dependent protein kinase II in the gerbil hippocampal CA1 neurons. Neurol Res 23: 751–754
11. Schulman H, Hanson PI (1993) Multifunctional Ca2+/calmodulin-dependent protein kinase. Neurochem Res 18: 65–77
12. Shamloo M, Wieloch T (1999) Changes in protein tyrosine phosphorylation in the rat brain after cerebral ischemia in a model of ischemic tolerance. J Cereb Blood Flow Metab 19: 173–183
13. Sharkey J, Butcher SP (1994) Immunophilins mediate the neuroprotective effects of FK506 in focal cerebral ischaemia. Nature 371: 336–339

Correspondence: Ken-Ichiro Katsura, M.D., Ph.D., The Second Department of Internal Medicine, Nippon Medical School, 1-1-5 Sendagi, Bunkyo-Ku, Tokyo 113-8603, Japan. e-mail: k-katsur@nms.ac.jp

Acta Neurochir (2003) [Suppl] 86: 117–122

Blood-brain barrier dysfunction and amyloid precursor protein accumulation in microvascular compartment following ischemia-reperfusion brain injury with 1-year survival

R. Pluta

Department of Neuropathology, Medical Research Centre, Polish Academy of Sciences, Warsaw, Poland

Summary

This study examined the late microvascular consequences of brain ischemia due to cardiac arrest in rats. In reacted vibratome sections scattered foci of extravasated horseradish peroxidase were noted throughout the brain and did not appear to be restricted to any specific area of brain. Ultrastructural investigation of leaky sites frequently presented platelets adhering to the endothelium of venules and capillaries. Endothelial cells demonstrated pathological changes with evidence of perivascular astrocytic swelling. At the same time, we noted C-terminal of amyloid precursor protein/β-amyloid peptide (CAPP/βA) deposits in cerebral blood vessels, with a halo of CAPP/βA immunoreactivity in the surrounding parenchyma suggested diffusion of CAPP/βA out of the vascular compartment. Changes predominated in the hippocampus, cerebral and entorhinal cortex, corpus callosum, thalamus, basal ganglia and around the lateral ventricles. These data implicate delayed abnormal endothelial function of vessels following ischemia-reperfusion brain injury as a primary event in the pathogenesis of the recurrent cerebral infarction.

Keywords: Blood-brain barrier; amyloid precursor protein; beta-amyloid; platelets; brain ischemia; cardiac arrest; Alzheimer's disease; rats.

Introduction

The profile of early neuropathology that is observed in an experimental rat model of global brain ischemia produced by cardiac arrest shares a commonality with neurodegeneration processes in Alzheimer's disease [4, 8, 11, 12]. Recent findings propose an important role for vascular pathological factors (e.g. ischemia-reperfusion) contributing to the neurodegenerative processes [2, 3, 12]. These factors include the appearance of a variety of vasospastic events within the cerebral blood vessel wall [15], increased numbers of endothelial microvillar formations, cellular invaginations, microthrombus formation [7] and focal increase in vessel permeability for cellular [9] and uncellular [10] blood components. Above-mentioned pathological changes seriously complicate the recovery state. More recent studies have been performed in this rat model designed to focus on the question of whether or not the neuropathological picture observed in long-term postresuscitation stages differs from that observed in early phases of Alzheimer's disease. The results of such studies suggested that rats who survived 6–12 months following ischemia-reperfusion brain injury developed cerebral abnormalities including continuous neuronal loss, gliosis, brain atrophy and accumulation of different fragments of amyloid precursor protein (APP) [11, 12, 14]. As in Alzheimer's disease, there were multifocal and widespread diffuse amyloid plaques [11, 12]. This reaction is indicative of an active, slowly progressing pathological process. Ischemia-reperfusion injury has been identified as a risk factor for Alzheimer's disease [2, 3], and as such, similarities exist between the pathology found after ischemia-reperfusion brain injury and the earliest stages of pathology in Alzheimer's disease [12]. Earlier findings in these rats suggested a focal blood-brain barrier (BBB) disturbance was observed throughout the brain in gray and white matter associated capillaries and arterioles following 1–24 hours of ischemia [10]. This model closely mimics the condition of early Alzheimer's disease pathology and because of the long-term survival of animals; it offers the opportunity to understand the mechanism(s) of neurodegeneration in Alzheimer's disease. Our focus here will be to present a brief description of major pathological events that occur within the microvascular compartment (BBB, platelets and APP changes) as a consequence of ischemia-reperfusion brain injury with long-term survival.

Materials and methods

Female Wistar rats ($n = 6$; 3 months old) weighing 150–180 g housed under diurnal lighting conditions and allowed food and water ad libitum were anesthetized during 10-min cardiac arrest [7] by diethyl ether and room air. As controls ($n = 5$) sham operated rats in due time were sacrificed. Tissue samples from brains for light and electron microscope studies were obtained one year later. After blood washout with Ringer's solution, fixation was performed by transcardiac perfusion with a solution containing 2% paraformaldehyde and 1% glutaraldehyde in 0.1 M sodium cacodylate buffer at pH 7.3.

Horseradish peroxidase (HRP) (Sigma type VI) served as an indicator of BBB changes. It was infused into the femoral vein (50 mg HRP in 1 ml PBS/rat), and was allowed to circulate for 30 min.

One hemisphere was cut at 50–80 μm in the coronal plane with a vibratome. For the demonstration of HRP, all tissue slices were incubated in a filtered solution of 3,3′-diaminobenzidine-tetra HCl [10]. After mounting slices on microscope slides, they were examined by light microscope. Some sections were selected for ultrastructural studies. These sections were then post-fixed with 1% osmium tetroxide for 1 h at 4 °C, washed several times in S/C buffer, dehydrated in a graded ethanol series, cleared in propylene oxide and embedded in Spurr plastic. Electron microscopy was performed on either a Hitachi H-7000 or Philips EM-420 T electron microscopes.

For immunocytochemistry, we used monoclonal antibody (mAb) 22C11 against the N-terminal of APP, mAb 6E10 recognizing 1–17 aa residues of βA, polyclonal antibodies (pAb) SP 28 recognizing 1–28 aa residues of βA, and pAb RAS 57 against the C-terminal of APP [8, 11]. Control studies included omission of the primary antibodies.

Results

BBB examination

At one year after cardiac arrest brain vibratome sections contained single sites of extravasated HRP (Fig. 1A–F). Light microscopic investigation commonly revealed mild diffuse and focal immunoreactivity within cortical layers (Fig. 1A–C,E). Penetrating vessels also displayed HRP staining of the vessel wall (Fig. 1F). HRP was occasionally seen in endothelium but without parenchymal extravasations. In other areas, including hippocampus, thalamus, basal gan-

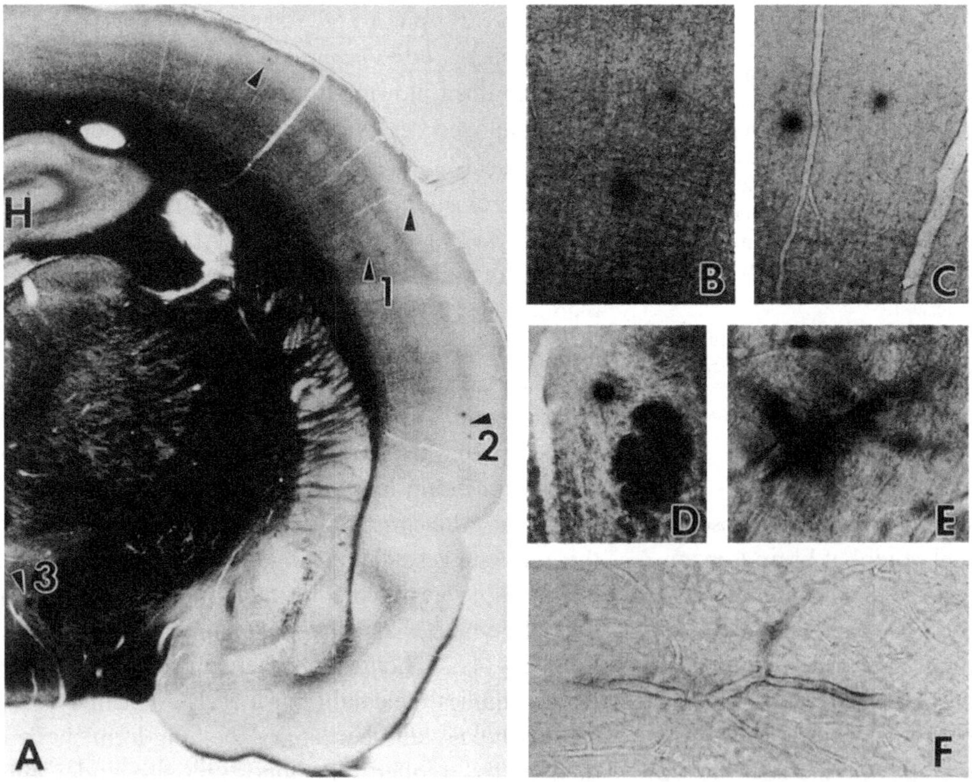

Fig. 1. Vibratome sections reacted for horseradish peroxidase (*HRP*) histochemistry. (A) In reacted section scattered foci of extravasated HRP (arrowheads) are seen throughout the brain. Atrophy of hippocampus (*H*) is noted. ×10. (B) High magnification of Fig. 1A area 1. ×40. (C) High magnification of Fig. 1A area 2. ×40. (D) High magnification of Fig. 1A area 3. ×40. (E) Evidence for platelets and red cells stasis of cortical vessels with HRP extravastion is also apparent. ×400. (F) The wall of cortical arteriole is outlined by HRP. ×100

glia, cerebellum and white matter, isolated and scatter sites of HRP permeability were seen. In addition to HRP extravasations, residual platelets with red cells were occasionally seen in some brain sections in columns indicating capillary congestion (Fig. 1E). These vascular alterations were noted throughout the brain and did not appear to be restricted to any specific part of brain. There was, however, apparent relationship between platelets stasis and HRP leakage (Fig. 1E). Generally post-ischemic rats demonstrated random BBB alterations. These alterations were spotty. Peroxidase extravasations involved arterioles, veins, venules and microvessels surrounded by perivascular spaces. Staining for HRP around leaking vessels looked like "puff of smoke". At the same time we observed brain atrophy particularly of hippocampus (Fig. 1A). In sham operated rats sites of extravasated HRP were not seen in vibratome sections.

Platelets examination

In some sections vessels appeared unremarkable except for luminal platelets with occasionally trapped between them red blood cells that partially obstructed circulation network. Platelets, sometimes occurring in clusters were noted adhering to the endothelial lining of venules and capillaries. As effect of these changes, we observed increase in the numbers of vessels completely obstructed by platelets (Fig. 2A) or platelets with their membranous remnants. Occasionally we found platelets on the perivascular-abluminal side of some vessels (Fig. 2B). We also noted recent aggregates of platelets and thrombi, which presented varying degrees of degranulation. In the most severely affected samples showing evidence of infarction, the neuropil appeared pale and edematous with vacuolation (Fig. 2A). In these areas, astrocytic cell bodies and perivascular astrocytic endfeet were severely swolen (Fig. 2A). Luminal platelets were identified in venulles and capillaries. Platelets commonly contained pseudopodia, which in some cases made contact with the endothelial surfaces. At sites of platelets-endothelial adhesion, junctional complexes between the two cells were observed. Endothelial microvillus projections were also seen extending toward the luminal platelets. Endothelial cells were frequently vacuolated and commonly contained plasma membrane blebs and discontinuities. These changes occurred in arterioles, venules and capillaries. Blood vessels appeared well perfused with no evidence of intraluminal platelets stasis in control group.

APP examination

All rats revealed increased brain immunoreactivity to the C-terminal of APP as well as to the βA (Fig. 3A,B). The expression of both was observed mainly extracellularly. Multiple, abundant, extracellular CAPP/βA immunoreactive deposits embraced or adjoined the blood vessels, mainly capillaries spreading multifocally outward into the parenchyma (Fig. 3A,B). The deposits were of various sizes, irregular in shape and frequently well delineated. Perivascular amyloid deposits surrounded blood vessels, forming perivascular cuffs (Fig. 3A,B). Diffuse, broad, but faintly CAPP/βA-positive perivascular zones like "puff of smoke" were also seen (Fig. 3A). Vascular lumens and pericytes as well as the inner and outer side of capillary walls were labeled also. Amyloid immunoreactivity inside cerebral blood vessels, with a halo of CAPP staining in the surrounding parenchyma (Fig. 3A), suggested diffusion of CAPP/βA out of the vascular compartment. In some cases, mostly in larger arterioles and veins, penetration of the CAPP/βA was limited to perivascular spaces with little or no leakage into the brain parenchyma. In other cases, CAPP/βA penetrated broadly to the surrounding tissue. Perivascular and vascular deposits were present in the hippocampus, cerebral and entorhinal cortex, corpus callosum and/or around the lateral ventricles. None of the control rats revealed the presence of this type of deposits.

Discussion

Our data document the presence of random and focal abnormalities inside cerebral vessels and their walls one year after cardiac arrest. The late pathological changes documented here after ischemia-reperfusion injury include: 1) HRP within cerebral endothelial cells and extravasated into surrounding brain parenchyma, 2) platelets accumulation within cerebral vessels and platelets in perivascular space, 3) CAPP and βA inside cerebral microvessels and within endothelial cells and leakage of these proteins into surrounding brain parenchyma, 4) perivascular swelling of astrocytic endfeet with continuous reinfarctions.

The occurrence of perivascular swelling one year after cardiac arrest with surrounding vacuolated pa-

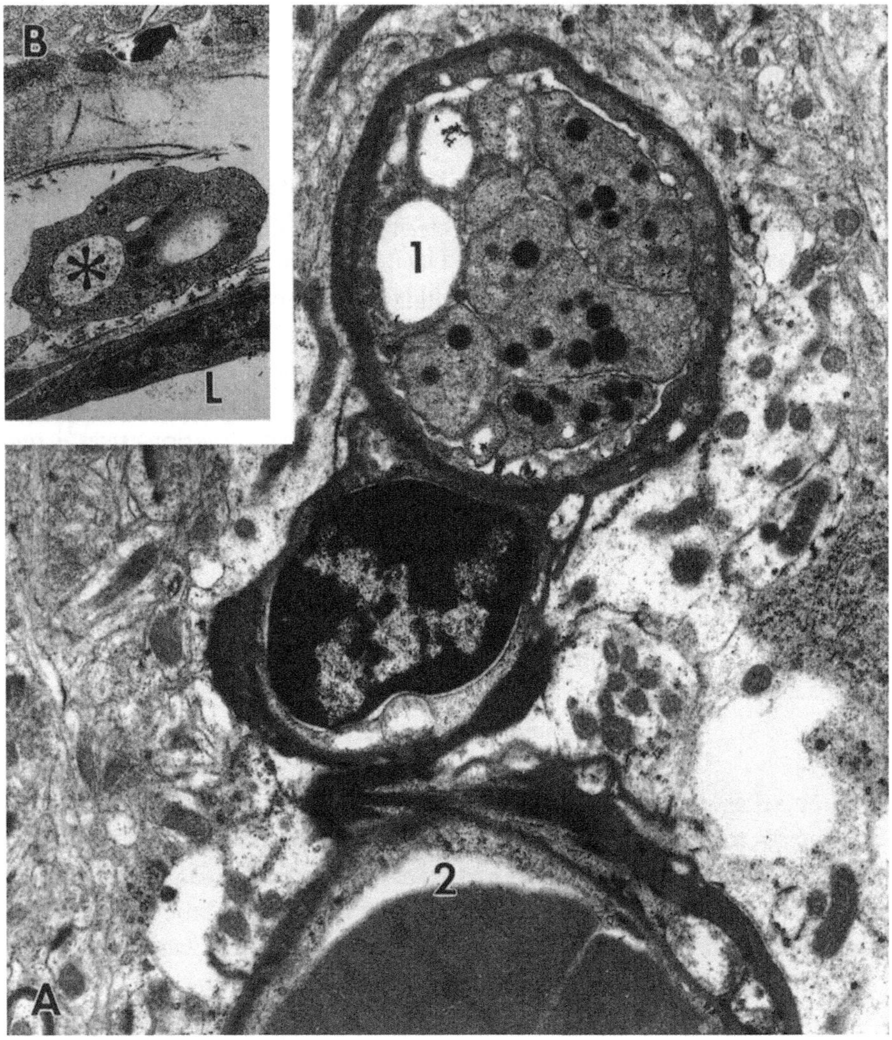

Fig. 2. (A) Vessel bifurcation is shown both branches of which are occluded by platelets (*1*) and a red blood cells (*2*). (cerebral cortex). ×15000. (B) A platelet (*) is shown in the perivascular space of a vessel (*L*), (thalamus) ×18000

renchyma represents vasogenic edema. Random HRP extravasations into diffuse brain areas in the present study are consistent with our previous finding in cardiac arrest with short-term survival [10]. Although there were scattered areas of HRP extravasations into brain tissue, most of the HRP was found within endothelial cells. Taken together, these findings support a transcellular mechanism of HRP transport across the BBB in this study [10].

An important finding of this study was the presence of platelets aggregates inside vessels and in perivascular space one year after cardiac arrest episode. The presence of platelets in microvessels is essential in implicating them in both early microvascular injury and late reperfusion cerebral blood flow disturbances [9]. Their presence in areas of endothelial damage suggests that they are involved in the pathogenesis of these changes. This suggests that ischemia is required for platelets-endothelial cell interactions observed in this study [9]. Our data of pseudopodial projections by both platelets and endothelial cells directed towards each other provides biochemical evidence for endothelial-platelets cell adhesion, attachment and movement. This may have an important functional role in transendothelial platelets passage. Microinfarcts associated with both endothelial alterations and the presence of platelets reflects a "no reflow phenomenon" in these areas [1]. Platelets can reduce or completely stop cerebral blood flow either by plugging vessels or by causing vasoconstriction through the release of TXA_2. Alternatively, the presence of new microinfarcts in our rat model might be explained by variability in cerebral

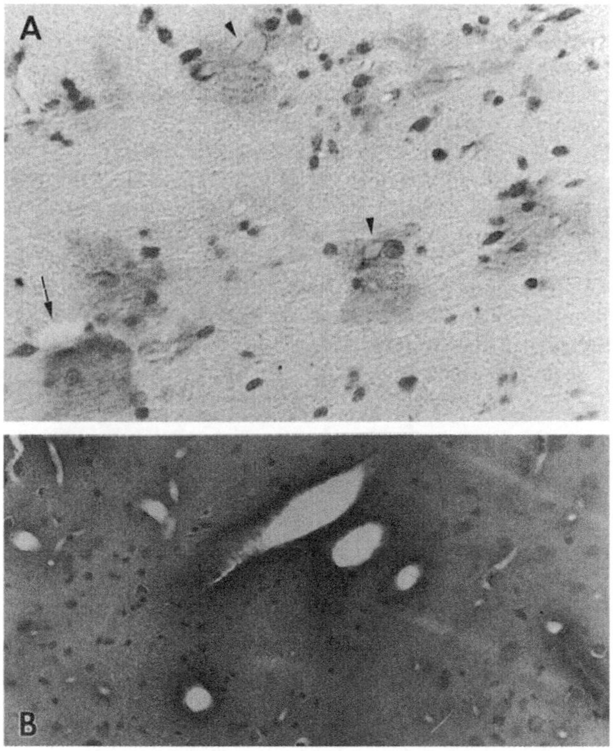

Fig. 3. (A) CAPP-positive material inside (arrowheads) and surrounding (arrowheads) or in contact (arrow) with the capillaries. Note moderate immunostaining for astrocytes and oligodendrocytes. ×400. (B) Perivascular β-amyloid peptide-positive material. Note strong staining for astrocytes and oligodendrocytes. ×400, (corpus callosum)

blood flow in late stages following cardiac arrest. More recently, it has become recognized by us that neuropathological processes continue well beyond the acute stage [12, 14].

An increase in BBB permeability may allow for entry CAPP/βA from blood into endothelial cells or into brain parenchyma which could then change vascular tone or have direct pathological effects on brain tissue itself. Platelets have been found to contain relatively large amounts of APP. Because APP is a potent protease inhibitor and possesses growth factor activity, these results also implicate APP in wound repair [6, 14]. The resultant protein would more likely escape from platelets and enter the brain tissue across a highly permeable BBB in areas of their attachment. CAPP/βA accumulation and platelet aggregation correlated very well with BBB dysfunction. Additionally perivascular platelets can be a source of APP products and enter directly the brain parenchyma [9]. Platelets with own proteins cannot be excluded as a potential pathogenic factor affecting the brain parenchyma. This idea is supported by recent findings presenting APP in neuronal bodies after ischemia [5] and in brain microvessels after cardiac arrest [13]. Release of APP to the brain tissue from pathologically behaving platelets after brain ischemia could have significant clinical implications for ischemic dementia and dementia of the Alzheimer's type. Platelet-endothelial interactions and platelets in the perivascular space after ischemia-reperfusion brain injury may be partly involved in the late injurious process leading, over a lifetime, to cumulative, repeated, silent, focal and continuous cerebral reinfarctions, which possibly cause a slowly progressing postischemic encephalopathy as well as other neurodegenerative diseases, i.e. Alzheimer's disease.

Acknowledgments

The author is indebted to the following founding sources: EC contract QLG3-CT-2000-00934, SPUB and MRC. He is grateful to Dr E. Kida for her help in immunohistochemistry and S. Januszewski for technical assistance.

References

1. Ames A III, Wright RL, Kowada M, Thurston JM, Majno G (1968) Cerebral ischemia. II. The no-reflow phenomenon. Am J Pathol 52: 437–453
2. Etiene D, Kraft J, Ganju N, Gomez-Isla T, Hyman B, Hedley-Whyte ET, Wands JR, Monte SM de la (1998) Cerebrovascular pathology contributes to heterogeneity of Alzheimer's disease. J Alzheimer Dis 1: 119–134
3. Kalaria RN (2000) The role of cerebral ischemia in Alzheimer's disease. Neurobiol Aging 21: 321–330
4. Kida E, Pluta R, Lossinsky AS, Golabek AA, Choi-Miura NH, Wisniewski HM, Mossakowski MJ (1995) Complete cerebral ischemia with short-term survival in rat induced by cardiac arrest. II. Extracellular and intracellular accumulation of apolipoprotein E and J in the brain. Brain Res 674: 341–346
5. Lin B, Ginsberg MD, Busto R (2001) Hyperglycemic but not normoglycemic global ischemia induces marked early intraneuronal expression of β-amyloid precursor protein. Brain Res 888: 107–116
6. Nihashi T, Inao S, Kajita Y, Kawai T, Sugimoto T, Niwa M, Kabeya R, Hata N, Hayashi S, Yoshida J (2001) Expression and distribution of beta amyloid precursor protein and beta amyloid peptide in reactive astrocytes after transient middle cerebral artery occlusion. Acta Neurochir (Wien) 143: 287–295
7. Pluta R, Lossinsky AS, Mossakowski MJ, Faso L, Wisniewski HM (1991) Reassessment of a new model of complete cerebral ischemia in rats. Method of induction of clinical death, pathophysiology and cerebrovascular pathology. Acta Neuropathol 83: 1–11
8. Pluta R, Kida E, Lossinsky AS, Golabek AA, Mossakowski MJ, Wisniewski HM (1994) Complete cerebral ischemia with short-term survival in rats induced by cardiac arrest. I. Extracellular accumulation of Alzheimer's β-amyloid protein precursor in the brain. Brain Res 649: 323–328
9. Pluta R, Lossinsky AS, Walski M, Wisniewski HM, Mossakowski MJ (1994) Platelet occlusion phenomenon after short-

and long-term survival following complete cerebral ischemia in rats produced by cardiac arrest. J Brain Res 35: 463–471

10. Pluta R, Lossinsky AS, Wisniewski HM, Mossakowski MJ (1994) Early blood-brain barrier changes in the rat following transient complete cerebral ischemia induced by cardiac arrest. Brain Res 633: 41–52
11. Pluta R, Barcikowska M, Kida E, Zelman I, Mossakowski MJ (1997) Late extracellular deposits of β-amyloid precursor protein in ischaemic rat brain show different immunoreactivity to the N- and C-terminal. Alzheimer's Res 3: 51–57
12. Pluta R (2000) The role of apolipoprotein E in the deposition of β-amyloid peptide during ischemia-reperfusion brain injury. A model of early Alzheimer's disease. Ann NY Acad Sci 903: 324–334
13. Pluta R (2000) No effect of anti-oxidative therapy on cerebral amyloidosis following ischemia-reperfusion brain injury. Folia Neuropathol 38: 188–190
14. Pluta R (2002) Glial expression of the beta amyloid peptide in cardiac arrest. J Neurol Sci in press
15. Wisniewski HM, Pluta R, Lossinsky AS, Mossakowski MJ (1995) Ultrastructural studies of cerebral vascular spasm after cardiac arrest-related global cerebral ischemia in rats. Acta Neuropathol 90: 432–440

Correspondence: Ryszard Pluta, M.D., Ph.D., Department of Neuropathology, Medical Research Centre, Polish Academy of Sciences, 02-106 Warsaw, Pawińskiego 5 Str., Poland. e-mail: pluta@medres.cmdik.pan.pl

Acta Neurochir (2003) [Suppl] 86: 123–127

Neuroprotective effect from ischemia and direct free radical scavenging activity of Choto-san (kampo medicine)

K. Dohi[1], **K. Satoh**[2], **Y. Ikeda**[3], **H. Ohtaki**[4], **S. Shioda**[4], and **T. Aruga**[1]

[1] Department of Emergency and Critical Care Medicine, School of Medicine, Showa University, Tokyo, Japan
[2] Department of Medicinal Information, School of Pharmaceutical Sciences, Showa University, Tokyo, Japan
[3] Department of Neurosurgery, School of Medicine, Showa University, Tokyo, Japan
[4] Department of Anatomy, School of Medicine, Showa University, Tokyo, Japan

Summary

Choto-san is a kampo medicine that is widely used in patients with cerebral infarction, but the details of its mechanism of action remain unclear. We examined the neuroprotective effects of Choto-san using an experimental cerebral ischemia model (i.e., a rat cardiac arrest model). We also investigated the ability of Choto-san to eliminate or inhibit the activity of free radicals. It was found that Choto-san significantly prevents delayed neuronal cell death after ischemic reperfusion. Electron spin resonance demonstrated that the formation of hydroxyl- and superoxide-DMPO spin adducts were inhibited by Choto-san. The results of this study indicated that Choto-san prevents delayed neuronal cell death in the hippocampal CA1 region after ischemia. Direct free radical scavenging activity is among the pharmacological effects of Choto-san.

Keywords: Kampo medicine; free radical; delayed neuronal cell death; apoptosis; rat; model; electron spin resonance; Choto-san; ischemia; NO.

Introduction

It has recently been reported that there is a strong relationship between various neuronal diseases and ROS (reactive oxygen species) [6].

Choto-san is a kampo medicine that is widely used in patients with cerebral infarction and chronic headache, but the details of its mechanism of action remain unclear. We examined the neuroprotective effects of Choto-san using an experimental cerebral ischemia model (i.e., a rat cardiac arrest model). In addition, we investigated the free radical scavenging ability of Choto-san.

Materials and methods

Analysis of Choto-san by 3D-high performance liquid chromatography (3D-HPLC) (Fig. 1)

Choto-san (1 g) was extracted with 20 mL of methanol under ultrasonication for 30 min followed by centrifugation. Analysis of the supernatant solution was performed by high performance liquid chromatography (HPLC).

Ischemic reperfusion model (cardiac arrest model)

Male Sprague-Dawley rats (250–300 g) were intubated under anaesthesia with 3.5% halothane and were maintained with 2% sevoflurane in 70% N_2O and 30% O_2. Global cerebral ischemia was induced according to the rat cardiac-arrest model [3, 9, 11]. With the anaesthetized rat in the supine position, the occluding device was manipulated such that its distal end moved along the dorsal wall of the thorax. A complete interruption of circulation due to compression of the superior aortic artery was accomplished by lifting the occluding device and by simultaneously applying external finger pressure. After 5 minutes of transient global ischemia, with the rat under artificial ventilation using a tidal volume of 2.5 ml at 200 per minute (100% O_2), heart massage was performed in order to facilitate resuscitation. After a recovery period of 30 minutes, the animals were disconnected from the respirator. Animal care and experiments complied with the Principles of Laboratory Animal Care listed in the Guidelines for the Care and Use of Laboratory (National Institutes of Health publication 86-23, 1985).

Histological analysis

Following recirculation periods of 7 days, the animals were sacrificed by perfusion/fixation under pentobarbital anaesthesia (50 mg/kg, i.p.). The brains were perfusion-fixed with 4% paraformaldehyde and 0.4% glutaraldehyde in 0.1 M phosphate buffer (PB) in preparation for staining with the terminal deoxynucleotidyl transferase

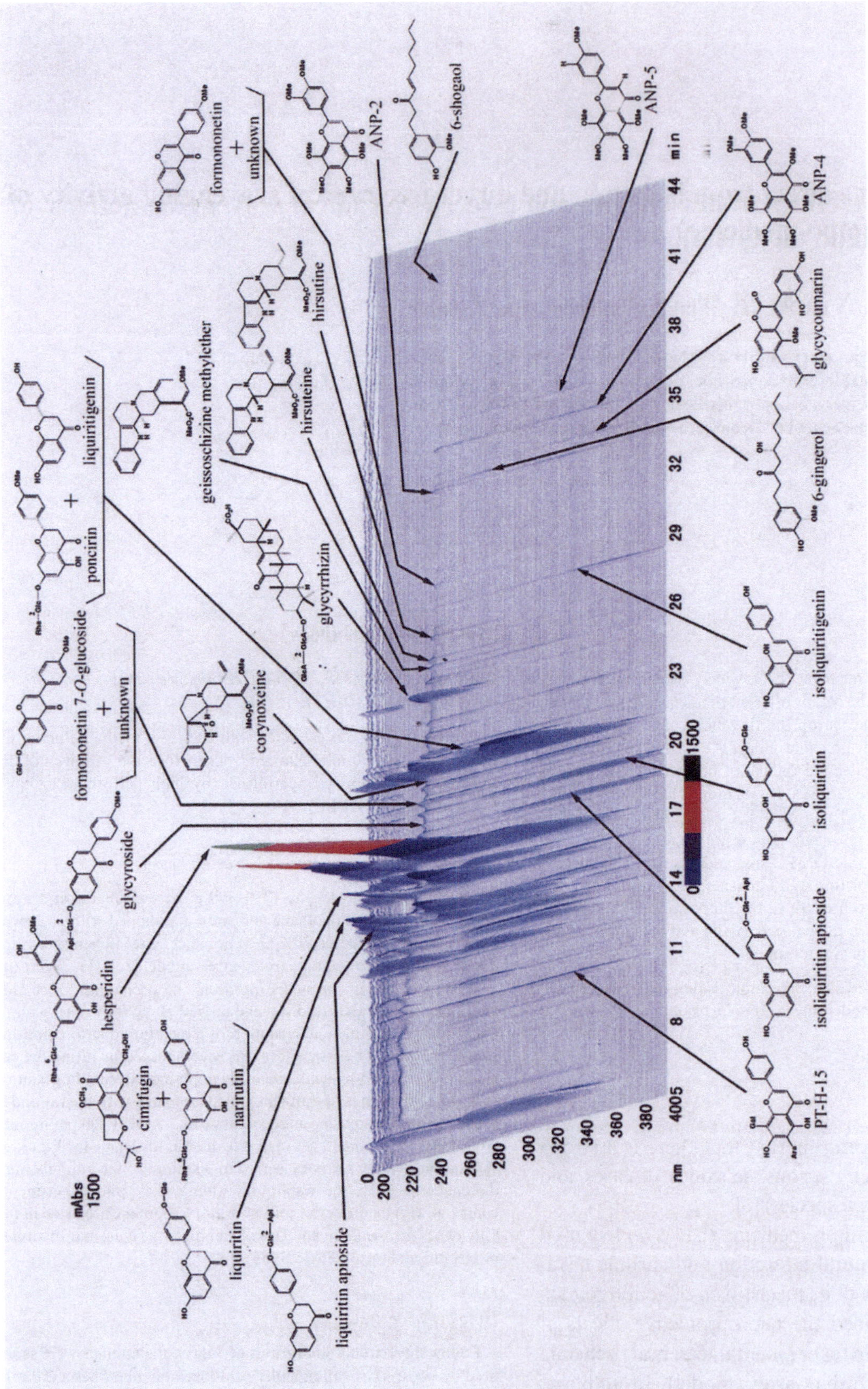

Fig. 1. 3D-high performance liquid chromatography (3D-HPLC) profile of Choto-san

(TdT)-mediated dUTP nick-end labelling (TUNEL) method. TUNEL-stained hippocampal sections at 7 days after ischemia-reperfusion (n = 5 in each group) were used for the quantitative analysis. In two sections (240 (W) × 100 (H) m^2) in the CA1 area of the hippocampus (on both the left and right sides), the TUNEL-positive pyramidal cells were counted as apoptotic neurons.

Direct free radical scavenging activities of Choto-san

The superoxide and hydroxyl radical scavenging activity of Choto-san was measured in vitro by electron-spin resonance spectrometry using 5,5-dimethyl-1-pyrroline-1-oxide (DMPO) as a spin trap, which was obtained from Dojin Chemical (Tokyo, Japan). Electron spin response detection of the spin adduct was performed at room temperature using a JESREIX X-band spectrometer (JEOL, Tokyo, Japan).

Electron spin resonance measurement conditions

Measurement conditions for ESR were as follows: magnetic field, 335.6jt5.0 mT; microwave power, 8.0 mW; sweep time, 2 mm/0.03 s; modulation amplitude, 0.1 mT.

Results

1. Many TUNEL-positive pyramidal cells were found in the CA1 region bilaterally 7 days after recirculation in the control group (Fig. 2A). In the group administered Choto-san, a few TUNEL-positive cells were apparent in the CA1 region. Choto-san prevents delayed neuronal cell death in the hippocampal CA1 region after ischemia (Fig. 2B).
2. ESR study demonstrated that the formation of a superoxide radical DMPO spin adduct was significantly inhibited by 10 μg/ml Choto-san and was completely inhibited by 1 mg/ml Choto-san (Fig. 3A). ESR study also demonstrated that the formation of a hydroxyl radical-DMPO spin adduct was significantly inhibited by 100 μg/ml Choto-san and was completely inhibited by 2 mg/ml Choto-san (Fig. 3B).

Discussion

Choto-san is a kampo medicine used to treat elderly patients with chronic headaches, a heavy feeling in the head and dizziness associated with hypertension. Choto-san is also effective in patients with vascular dementia. In cases of vascular dementia, Choto-san is thought to act as a Ca+ channel blocker, an agonist for the 5-HT receptor, and an inhibitor of lipid peroxidation [8, 12, 13]. Choto-san consists of 11 crude in fixed proportions: Uncaliae Ramurus et Uncus is thought to be the most crucial of those components. Uncaliae Ramurus et Uncus has several beneficial ef-

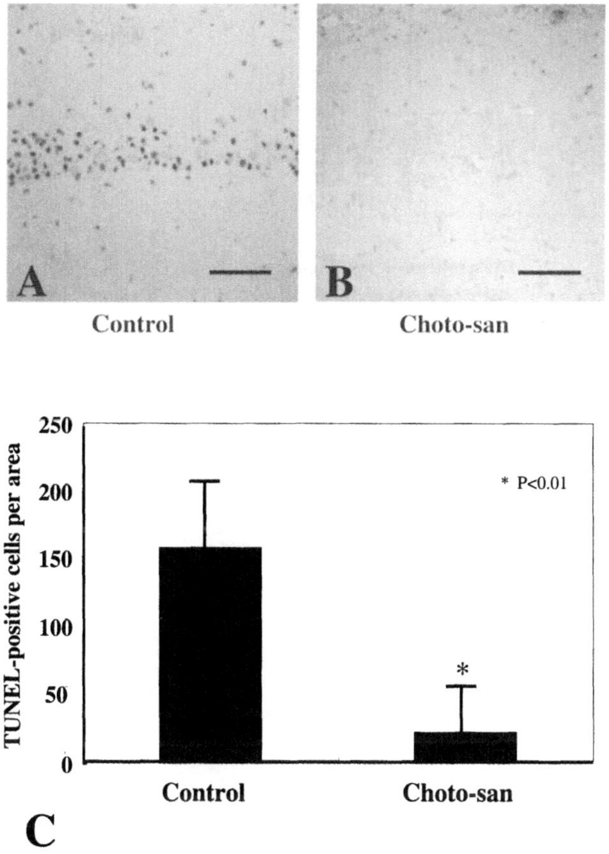

Fig. 2. TUNEL staining in the hippocampal CA1 region 7 days after ischemia-reperfusion. A large number of TUNEL-positive neurons were observed in the control group (A). Choto-san suppressed delayed neuronal cell death after ischemia

fects on areas of neuronal damage: it acts as a calcium antagonist, an agonist for the 5-HT receptor, and as an inhibitor of lipid peroxidation. It is thought that these pharmacological properties are the basic mechanisms of action of Choto-san in the treatment of vascular dementia.

The production of reactive oxygen species (ROS) was found to increase after ischemia in an experimental ischemic model. Recent studies have furthermore suggested that free radicals may be involved in the pathogenesis of brain injury and brain edema, and that free radical scavengers can reduce brain edema due to trauma or ischemia [2, 6]. MCI-186 (3-methyl-1-phenyl-2-pyrazolin-5 one), a free radical scavenger, significantly reduced the area of cerebral damage 24 hrs after MCA occlusion and it also significantly ($P < 0.05$) reduced hydroxyl radical levels. MCI-186 attenuates brain edema by suppressing the production of ROS [1]. In clinical settings, MCI-186 has been shown to exert neuroprotective effects in patients with cerebral infarction [5]. Indomethacin, a non-steroidal

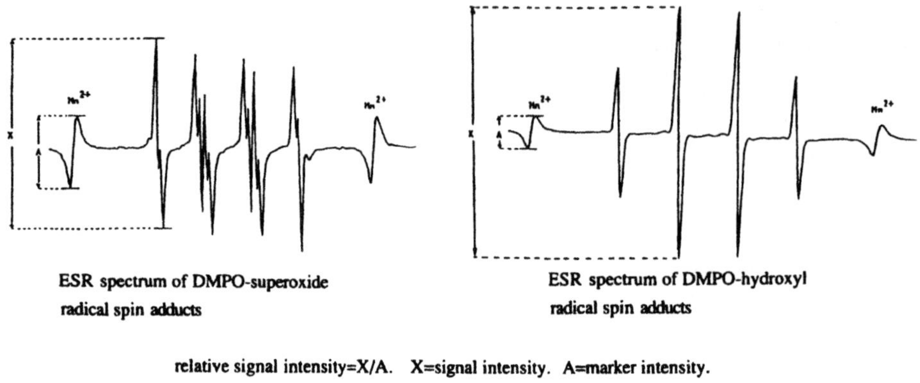

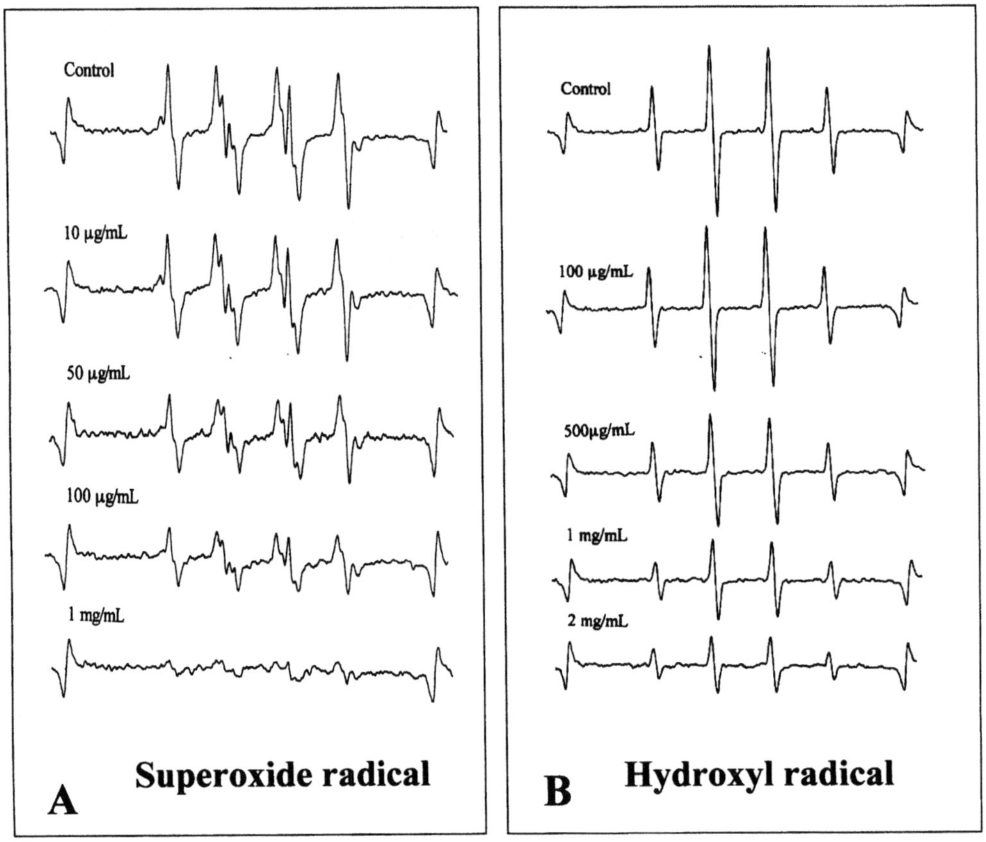

Fig. 3. Superoxide radical scavenging activity of Choto-san (A). Hydroxyl radical scavenging activity of Choto-san (B)

anti-inflammatory drug, exhibits anti-inflammatory, analgesic, and antipyretic activity. Indomethacin also prevented delayed neuronal death in the CA1 area after ischemia [10]. Although the exact mechanism of action is unknown, the reversible inhibition of cyclooxygenase, the enzyme that catalyzes the formation of prostaglandin precursors from arachidonic acid, is considered to be responsible for these effects [4]. However, it has been reported that indomethacin carries out various ROS scavenging activities [7]. The antioxidative effect of indomethacin may be its important pharmacological property. In the future, antioxidative agents are expected to be incorporated into the major neuroprotective drugs.

In this study, we reported the neuroprotective effects and free radical scavenging activities of Choto-san. Choto-san prevents delayed neuronal death (apoptosis) in the CA1 area of a rat cardiac arrest model; it is possible that free radical scavenging was the major mechanism of therapeutic action in the present study. However, it should be noted that this study was strictly preliminary; further studies will be necessary in order to determine the role of free radicals in patients with brain ischemia.

Acknowledgments

The authors are grateful to Dr. Iwao Sakakibara and Dr. Naoko Hattori (Kampo and Pharmacognosy Laboratories, Tsumura) for their technical support in the 3D HPLC experiments.

References

1. Abe K, Yuki S, Kogure K (1988) Strong attenuation of ischemic and postischemic brain edema in rats by a novel free radical scavenger. Stroke 19: 480–485
2. Chan PH (2001) Reactive oxygen radicals in signaling and damage in the ischemic brain. J Cereb Blood Flow Metab 21: 2–14
3. Dohi K, Shioda S, Mizushima H, Homma H, Ozawa H, Nakai Y, Matsumoto H (1998) Delayed neuronal cell death and microglial cell reactivity in the CA1 region of the rat hippocampus in the cardiac arrest model. Med Electron Microsc 31: 85–93
4. Harrigan MR, Tuteja S, Neudeck BL (1997) Indomethacin in the management of elevated intracranial pressure: a review. J Neurotrauma 14: 637–650
5. Houkin K, Nakayama N, Kashiwada T (1998) Neuroprotective effect of the free radical scavenger MCI-186 in patients with cerebral infarction: clinical evaluation using magnetic resonance imaging and spectroscopy. J Stroke Cerebrovasc Dis 7: 315–322
6. Ikeda Y, Long DM (1990) The molecular basis of brain injury and brain edema: the role of oxygen free radicals. Neurosurgery 27: 1–11
7. Ikeda Y, Matsumoto K, Satoh K (2001) Direct superoxide scavenging activity of nonsteroidal anti-inflammatory drugs: determination by electron spin resonance using the spin trap method. Headache 41: 138–141
8. Itoh T, Shimada Y, Terasawa K (1999) Efficacy of Choto-san on vascular dementia and the protective effect of the hooks and stems of Uncaria sinensis on glutamate-induced neuronal death. Mech Ageing Dev 111: 155–173
9. Kawai K, Nitecka L, Klatzo I (1992) Global cerebral ischemia associated with cardiac arrest in the rat: I. Dynamics of early neuronal changes. J Cereb Blood Flow Metab 12: 238–249
10. Kondo F, Kondo Y, Ogawa N (2000) Delayed neuronal death in hippocampal CA1 pyramidal neurons after forebrain ischemia in hyperglycemic gerbils: amelioration by indomethacin. Brain Res 853: 93–98
11. Korpatchev W, Lysenkov S, Thieliz P (1982) Modielirowanlie klinczeskoj smerti l postreanimationznoj bolezni u krys. Palol Fiziol Exp Ter 3: 78–80
12. Sugimoto A, Goto K, Miyamoto KI (2000) Effect of Choto-san, a Kampo medicine, on the cerebral blood flow autoregulation in spontaneously hypertensive rats. Jpn J Pharmacol 83: 135–142
13. Yuzurihara M, Goto K, Terasawa K (1999) Effect of Choto-san, a kampo medicine, on impairment of passive avoidance performance in mice. Phytother Res 13: 233–235

Correspondence: Kenji Dohi, Department of Emergency and Critical Care Medicine, School of Medicine, Showa University, Tokyo, Japan, 1-5-8 Hatanodai, Shinagawa-ku, Tokyo, 142-8555, Japan. e-mail: kdop@med.showa-u.ac.jp

Experimental focal ischemia

Acta Neurochir (2003) [Suppl] 86: 131–135

Temporal profile of experimental ischemic edema after threshold amount of insult to induce infarction – ultrastructure, gravimetry and Evans' blue extravasation

U. Ito[1,3], **T. Kuroiwa**[2], **S. Hanyu**[3], **Y. Hakamata**[3], **E. Kawakami**[1], **I. Nakano**[3], and **K. Oyanagi**[1]

[1] Department of Neuropathology, Tokyo Metropolitan Institute for Neuroscience, Tokyo, Japan
[2] Department of Neuropathology, Medical Research Institute, Tokyo Medical and Dental University, Tokyo, Japan
[3] Department of Neurology, Jichi Medical School, Tochigi, Japan

Summary

When a threshold amount of temporary ischemic insult to induce focal infarction was given to the unilateral cerebral hemisphere of gerbils, a small focal infarct surrounded by a wide penumbra developed in the rostral portion of the cerebral cortex. During the first 5 hours following recirculation, whole astrocytic cell bodies and processes in the ischemic hemisphere were swollen, with an increase in the number of glycogen granules and in number and size of mitochondria. This swelling was an active reaction of astrocytes for neuronal protection, scavenging potassium, glutamate, and other neuronal metabolic products, and for generating fuels·for neurons (cyto-reactive edema). This reactive astrocytic swelling continued in the penumbra, but some dead neurons were found disseminated among the surviving neurons. Whereas, at 12~48 hours, focal infarction developed in which all cell membranes lost their Gibbs-Donnan's equilibrium due to energetic failure of their membranous Na^+/K^+ ATPase. This is the cytotoxic edema (cyto-necrotic edema). In the infarct focus, when pericapillary astrocytic end-feet were damaged, the capillary BBB was broken; and thus vasogenic edema was superimposed on the cytotoxic edema.

Keywords: Cerebral ischemia; ischemic brain edema; penumbra; infarction; cytotoxic edema; vasogenic edema; cyto-reactive edema; cyto-necrotic edema; disseminated selective neuronal necrosis (DSNN); glycogen granules; astrocytic mitochondria; extracellular space; intracellular space.

Introduction

We have suggested that in the maturation phenomenon in cerebral ischemia, severe ischemic insult induces a rapidly developing large infarction [5]. In order to reduce the rate of maturation of ischemic injuries, we gave a threshold amount of ischemic insult to induce a cerebral infarction, and obtained a slowly developing large ischemic penumbra around a small focal infarct in the cerebral cortex [4, 8]. In this model, initially disseminated selective neuronal necrosis (DSNN) develops, followed by evolution of a gradually enlarging infarcted focus surrounded by a large penumbra in which DSNN still develops [4, 8]. Using this model, we studied the ultrastructure, gravimetry, and Evans' blue permeability in the penumbra as well as in the focal infarction of the cerebral cortex.

Materials and methods

1. Under short anesthesia with 2% halothane, 70% nitrous oxide and 30% oxygen, we occluded the left carotid artery of Mongolian gerbil twice, at a 5 hr interval, for 10 min each time. Ischemia-positive animals were selected according to stroke index during the first occlusion. Animals were sacrificed at 15 min, and 5, 12, 24, 48, 72 hr and 7 day after the second occlusion.

2. *Ultrastructural and histological study:* The whole brain was fixed by trans-cardiac perfusion with cacodylate buffered glutalaldehyde fixatives for electron microscopy and phosphate buffered formaldehyde fixative for light microscopy. The fixed brains were cut coronally at the levels of the chiasma (Face A: focal infarction in the penumbra) and infundibulum (Face B: penumbra only). In both faces, ultrathin sections including the 3rd~5th cortical layers were prepared from the left ischemic hemisphere at the mid-point between the interhemispheric fissure (IF) and the rhinal fissure (RF). Alternate sections were double-stained with uranyl acetate and lead citrate solution, and observed under an electron microscope. Paraffin sections were separately stained with hematoxylin-eosin (HE), periodic acid fuchsin Schiff (PAS), and glial fibrillary acidic protein (GFAP).

3. *Gravimetry and Evans' blue permeability:* One hour before decapitation, 2% Evans' blue was injected intravenously. The brain was sectioned coronally at the levels of the chiasma and infundibulum. We measured the specific gravity of the cortical slice including the middle 1/3 segment between IF and RF, in each face [2].

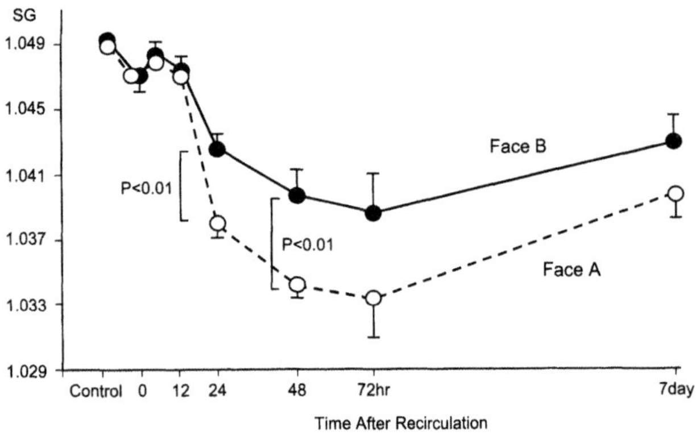

Fig. 1. Temporal profile of specific gravity in Faces A and B

Results

Face A (focal infarction evolving after the initial DSNN)

Fifteen minutes after the second 10-min period of ischemia, the specific gravity of the ischemic cortex had decreased slightly, but significantly, from the control values ($P < 0.05$, Fig. 1). Under EM observation, slight swelling of the astrocytic cell processes and perivascular end-feet was recognized, without an increase in the number of glycogen granules. By 5 to 12 hr, however, the specific gravities were significantly lower than the control values ($P < 0.05$). Astrocytic cell bodies (Fig. 2B) and cell processes (Fig. 2A) swelled continuously, with an increase in their cytoplasmic volume, number of glycogen granules, and mitochondrial number and size. By 12 hr, small foci of infarcts developed in the cerebral cortex where glycogen granules had disappeared, and the rest of the cortex showed progressive DSNN. The disseminated dying neurons showed different grades of electron density of their cell body surrounded by remarkably swollen astrocytic cell processes (Fig. 2C). At 24 hr, the specific gravity decreased further compared with control values ($P < 0.01$) and continued to decrease up to 48 hr (Fig. 1). A focal infarction developed by the enlargment and fusion of small foci of infarction, while DSNN progressed in the surrounding cortical tissues. In the infarct focus, all neurons were necrotic and the astrocytic processes were fragmented, showing cystic degeneration (Fig. 3A). At 24~48 hr, Evans' blue extravasated only in the middle third of the cortical segment, between the IF and the RF, corresponding to the infarct foci. EM of the infarct foci showed that most of the perivascular end-feet had disappeared, although some had remained attached to the vascular wall but showed cystic degeneration (Fig. 3B). Extravasated fluid was

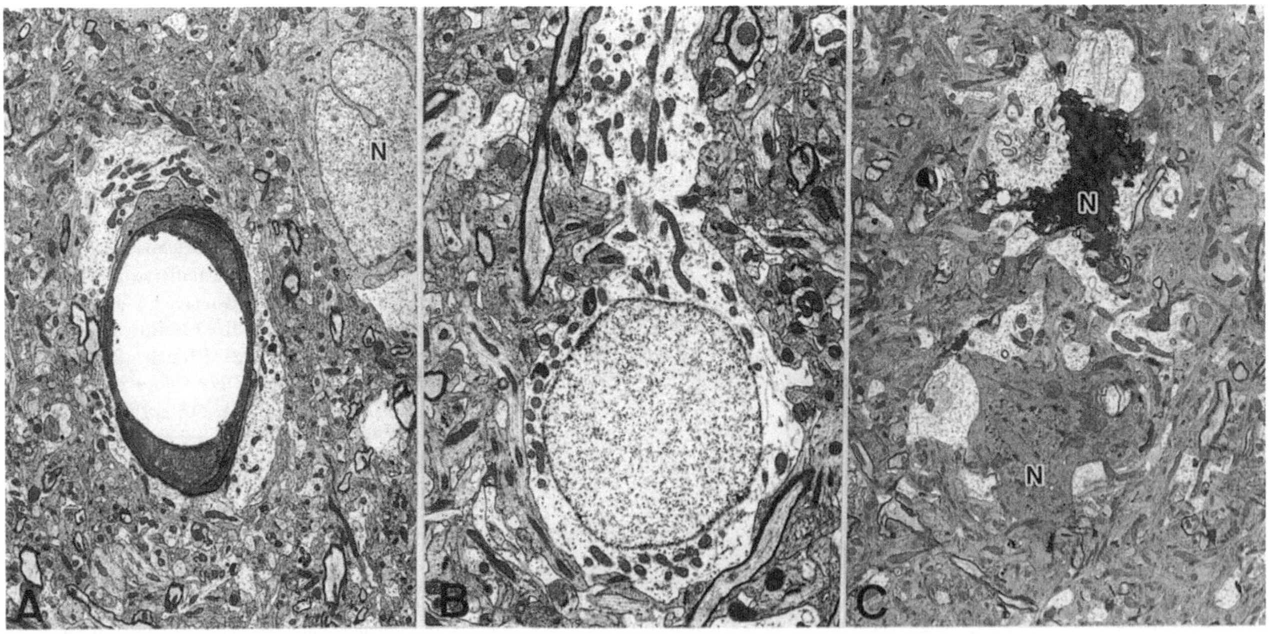

Fig. 2. EM of penumbra at 5 hr after ischemia. (A) Perivascular astrocytic foot processes and cell processes are swollen, with increases in the numbers of glycogen granules and mitochondria (×5,900), *N* neuron. (B) Astrocytic cell body and cell processes are swollen and contain increased numbers of glycogen granules and mitochondria (×7,800). (C) Dissimilated dying neurons with various degrees of electron density (*N*) surrounded by remarkably swollen astrocytic cell processes (×4,600)

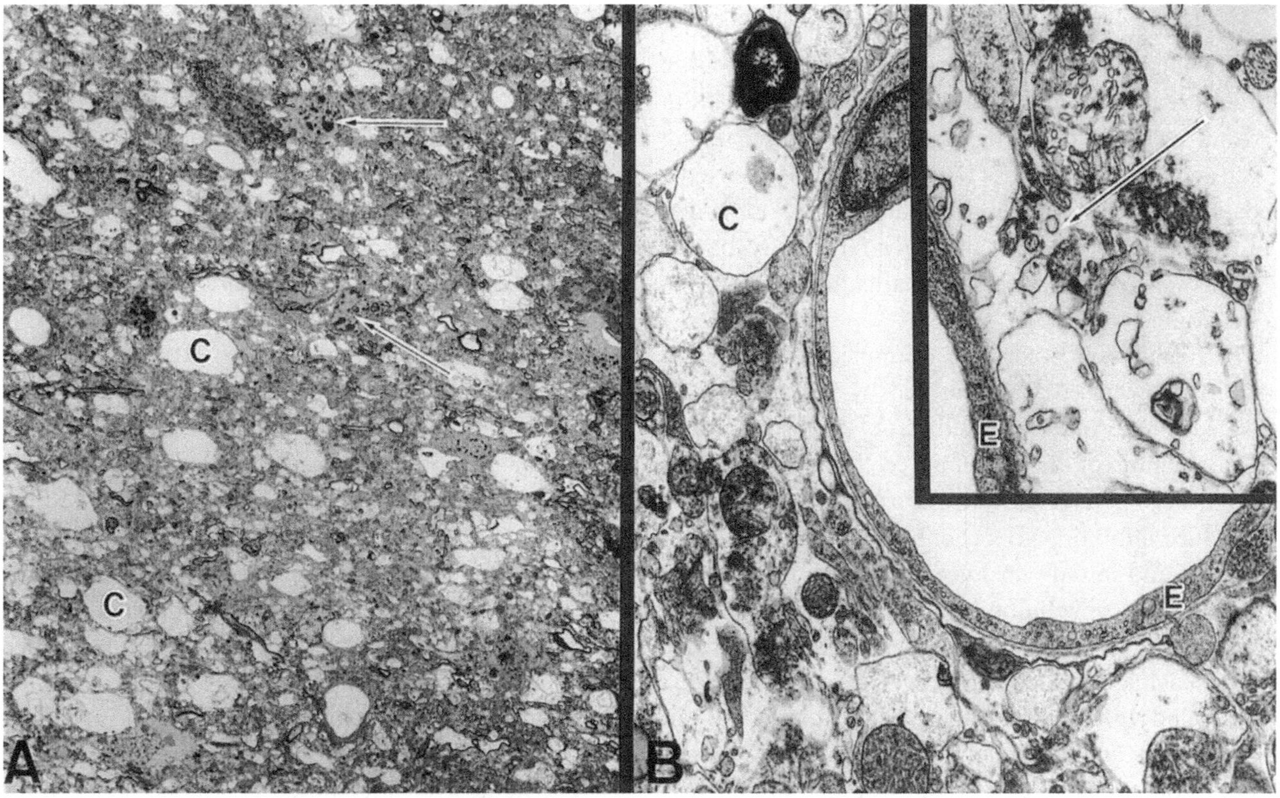

Fig. 3. EM of infarct. (A) 12 hr after ischemia. Cystically degenerated astrocytic cell processes (*C*) and other neuropils are scattered among the degenerated neurons (arrows) (×1,200). (B) 24 hr after ischemia. Perirvascular astrocytic foot processes have disappeared, and astrocytic cell processes and other neuropils are dilated (*C*) (×11,000). Some of them have ruptured into (arrow) the widened extracellular space by extravasated edema fluid (inset: ×22,600), *E* capillary endothel

recognized in the extracellular space adjacent to the vascular wall. At 48 hr, cystically degenerated astrocytic cell processes were found disrupted into the extracellular spaces, which had been further widened by vasogenic edema (Fig. 3B). Brain edema subsided over 3 to 7 days, and the remaining infarct cavity became filled with sweeper cells and protein-rich fluid. In other post-ischemic cerebral cortex tissue, especially that surrounding the infarct foci, there was an increasing number of reactive astrocytes with the cytoplasm rich in glial fibers.

Face B (only DSNN develops)

EM findings were the same as those for Face A until 5 hr (Fig. 2A,B). By 12∼48 hr, no infarction had occurred, and only DSNN developed, as evidenced by an increasing number of the electron-dense dying neurons surrounded by swollen astrocytic cell processes (Fig. 2C). Astrocytic cell bodies and cell processes swelled continuously, showing increased cytoplasmic volume and number of glycogen granules and increased number and size of mitochondria. By 72 hr∼7 days, the rate of development of DSNN slowed but DSNN still occurred; and glial fibers were increased in number in the astrocytic cell bodies and processes. The specific gravities were significantly lower than the control values but significantly higher than the values for Face A (Fig. 1). Evans' blue did not extravasate at any time, nor did macrophages appear.

Discussion

The brain edema that follows temporary ischemia has been thought to consist of cytotoxic edema followed by vasogenic edema, commencing after restoration of blood flow [6, 10].

Previously, using a unilateral, carotid occlusion model in Mongolian gerbils, we found differences in the temporal profiles of ischemic brain edema by measuring water, sodium, potassium, and radioactive iodine labeled serum albumin (RISA) content during and after reperfusion following temporary ischemia for different lengths of time [6]. Following 30 min of ischemia, slight brain edema occurred but was arrested and gradually reduced thereafter, after reperfusion. Following 1 hr of ischemia and then reperfusion, more brain edema occurred but the rate of increase in brain edema dropped by 12 hr, and then increased along with an increase in RISA content, indicating vasogenic edema [6]. Following more than 3 hr of ischemia, in which the brain was already infarcted, vasogenic edema intensified immediately after the start of recirculation.

In the present study, we examined the temporal profile of brain edema measured by gravimetry and Evans' blue permeability, and that of the histology and ultrastructure of the ischemic cerebral cortex, after administering a threshold ischemic insult required to induce infarction. Here, only DSNN developed in Face B, and initially in Face A and was followed in Face A by the evolution of a gradually enlarging infarct focus [8]. Gravimetric data indicated that more severe edema developed in Face A than in Face B [2]. We found that cyto-reactive edema developed, together with active swelling of viable astrocytic cell bodies, processes, and their end-feet, and an increase in glycogen granules, in association with DSNN [7]. When the infarction progressed, there was necrotic edema and passive swelling of all astrocytic processes and some of the neuronal processes, owing to failure of the energy-dependent mechanism in the cell membranes. Then, vasogenic edema evidenced by Evans' blue permeability occurred in the infarcted foci, expanding the extracellular spaces, which were further distended by the rupture of the cystically degenerated astrocytic and neuronal processes that resulted in a connection between the intra- and extracellular spaces.

Under physiological conditions, cyto-membranous ionic homeostasis is maintained by Na^+/K^+-ATPase activity expelling the intracellularly accumulated cations. In cytotoxic edema, astrocytic swelling after an ischemic insult had previously been considered to be detrimental to cells, and it was thought to be due to disruption of the Gibbs-Donnan equilibrium of the cell membrane caused by an energy disturbance of Na^+/K^+-ATPase activity in the membrane [10]. However, investigations using cultured astrocytes and rat C6 glioma cells revealed that neither hypoxia nor ouabain, which inhibit Na^+/K^+-ATPase activity, caused astrocytic swelling [11, 20]. On the other hand, viable astrocytes can swell spontaneously under a number of circumstances. Astrocytes may accumulate Na^+ and expel H^+, of which increase in the astrocytic cytoplasm occurrs due to elevated intra- and extracellular lactacidosis and respiratory products [22], via ionchannels in the cellmembrane, such as the Na^+/H^+ antiporter, Na^+/HCO_3^- co-transporter, and Na^+-dependent Cl^-/HCO_3^- exchanger [20]. These ionchannels are especially abundant in the end-feet around the capillaries. Secondly, astrocytes may accumulate Na^+ and glutamate which increase at the synapse, due to the function of the glutamate/$2Na^+$ co-transporter [16, 18]. Finally, astrocytes may accumulate Na^+ and K^+ that have been increased in concentration extracellularly by neuronal excitation, via the function of the $Na^+/K^+/2Cl^-$ co-transporter [12, 17, 20].

In our previous study using the same experimental model as in the present study [14, 15], the ATP content and succinic dehydrogenase (SDH) activity did not decrease, but were sustained at normal levels following the temporary ischemia causing progressive DSNN without infarction. Similar levels were maintained after temporary ischemia sufficient to induce infarction, and after focal infarction evolved, the levels decreased drastically.

The number of mitochondria is rather small in the normal astrocyte [3]. In the previous study, we found that astrocytic mitochondria increased in both number and size, and that no astrocytic cell death occurred in association with DSNN in the penumbra of the cerebral cortex or in the infarct area before evolution of infarction. However, once infarction evolved, both neurons and astrocytes degenerated [9]. The increase in both number and size of astrocytic mitochondria showing intensified electron-dense matrices and swelling of the cristae, as well as the swelling and glycogen accumulation in the astrocytic cell processes, suggest an activated astrocytic compartmentation of the cortical energy metabolism [1, 3], due to mitochondrial heterogeneity [1, 21]. Such activation is used for housekeeping of extracellular neuronal products such as K^+, glutamate, and asparate, and for processing to generate fuel molecules that can be used by the neighboring neurons [3, 11–13, 16, 17].

Conclusion

1) After ischemic insult at the threshold required to induce infarction, in penumbra of the cerebral cortex in which only DSNN progresses, *cyto-reactive edema* develops together with active swelling of viable astrocytic cell bodies, processes and end-feet, as an astrocyte-mediated neuronal protection mechanism [11,

19]. The astrocytes act as a buffer against various metabolites that have increased in level following the metabolic derangement arising from the ischemia. 2) After creation of threshold amount of the insult to induce focal infarction in the cerebral cortex, *cytoreactive edema* occurs first, associated with progressive DSNN, but as the infarction evolves later, *cytonecrotic edema* develops in the infarction, with passive swelling of all of the astrocytic and some of the neuronal processes due to energetic failure of the cell membranes. Then, *vasogenic edema* ensues, widening the extracellular spaces, which are further distended by the rupture of cystically degenerated astrocytic and neuronal processes that connect the intra- and extracellular spaces.

References

1. Coles JA (1995) Glial cells and the supply of substrates of energy metabolism to neurons, Neuroglia. In: Kettenmann H, Ransom B (eds), Oxford University Press, New York Oxford, pp 793–804
2. Hakamata Y, Hanyu S, Kuroiwa T, Ito U (1997) Brain edema associated with progressive selective neuronal death or impending infarction in the cerebral cortex. Acta Neurochir (Wien) [Suppl] 70: 20–22
3. Hamprecht B, Dringen R (1995) Energy metabolism. Neuroglia. In: Kettenmann H, Ransom B (eds), Oxford University Press, New York Oxford, pp 473–487
4. Hanyu S, Ito U, Hakamata Y, Yoshida M (1995) Transition from ischemic neuronal necrosis to infarction in repeated ischemia. Brain Res 686: 44–48
5. Ito U, Spatz M, Walker J Jr, Klatzo I (1975) Experimental cerebral ischemia in mongolian gerbils. I. Light microscopic observations. Acta Neuropathol 32(3): 209–223
6. Ito U, Ohno K, Nakamura R, Suganuma F, Inaba Y (1970) Brain edema during ischemia and after restoration of blood flow. Measurement of water, sodium, potassium content and plasma protein permeability. Stroke 10: 542–547
7. Ito U, Hanyu S, Hakamata Y, Nakamura M, Arima K (1997) Ultrastructure of astrocytes associated with selective neuronal death of cerebral cortex after repeated ischemia. Acta Neurochir (Wien) [Suppl] 70: 46–49
8. Ito U, Hanyu S, Hakamata Y, Arima K, Oyanagi K, Kuroiwa T, Nakano I (1999) Temporal profile of cortical injury following ischemic insult just-below and at the threshold level for induction of infarction – light and electron microscopic study. Maturation phenomenon in cerebral ischemia III. Springer, Berlin Heidelberg New York Tokyo, pp 227–235
9. Ito U, Kuroiwa T, Hanyu S, Hakamata Y, Ito S, Nakano I, Oyanagi K (2000) Ultrastructure and morphometry of astroglial mitochondria following temporary ischemia. Maturation phenomenon in cerebral ischemia IV. In: Bazan NG, Ito U, Marcheselli VL, Kuroiwa T, Klatzo IE (eds), Springer, Berlin Heidelberg New York Tokyo, pp 253–259
10. Katzmann R, Classen CR, Klatzo I, Meyer JS, Waltz AG (1977) Report of joint committee for stroke resources IV. Brain edema in stroke. Study group on brain edema in stroke. Stroke 8: 513–540
11. Kempski O, Volk C (1996) Glial protection against neuronal damage. Maturation phenomenon in cerebral ischemia II. Springer, Berlin Heidelberg New York Tokyo
12. Kimelberg HK, Rutledge E, Goderie S, Charniga C (1995) Astrocytic swelling due to hypotonic or high K^+ medium causes inhibition of glutamate and aspartate uptake and increases their release. J Cereb Blood Flow Metab 15(3): 409–416
13. Kraig RP, Lascola CD, Caggiano A (1995) Glial response to brain ischemia. Neuroglia. In: Kettenmann H, Ransom B (eds), Oxford University Press, New York Oxford, pp 964–976
14. Kuroiwa T, Mies G, Hakamata Y, Hanyu S, Okeda R, Ito U (1999) Mitochondrial dysfunction and maturation phenomenon in ischemic gerbil cortex. Maturation phenomenon in cerebral ischemia III ed. In: Ito U, Orzi F, Kuroiwa T, Fieschi C, Klatzo IE (eds), Springer, Berlin Heidelberg New York Tokyo, pp 237–241
15. Kuroiwa T, Mies G, Hermann D, Hakamata Y, Hanyu S, Ito U (2000) Regional differences in the rate of energy impairment after threshold level ischemia for induction of cerebral infarction in gerbils. Acta Neuropathol 100: 587–594
16. Martin DL (1995) The role of glia in the inactivation of neurotransmitters. Neuroglia. In: Kettenmann H, Ransom B (eds), Oxford University Press, New York Oxford, pp 732–745
17. Newman EA (1995) Glial cell regulation of extracellular potassium. Neuroglia. In: Kettenmann H, Ransom B (eds), Oxford University Press, New York Oxford, pp 717–731
18. Noble LJ, Hall JJ, Chen S, Chan PH (1992) Morphologic changes in cultured astrocytes after exposure to glutamate. J Neurotrauma 9: 255–267
19. Rosenberg GA, Aizeman E (1989) Hundred-fold increase in neuronal vulnerability to glultamate toxicity in astrocyte-poor cultures of rat cerebral cortex. Neuroscience (Lett) 103: 162–168
20. Siesjo BK (1992) Pathophysiology and treatment of focal cerebral ischemia. Part I: pathophysiology. J Neurosurg 77: 169–184
21. Sonnewald U, Hertz L, Schousboe A (1998) Mitochondrial heterogeneity in the brain at the cellular level. J Cereb Blood Flow Metab 18(3): 231–237
22. Staub F, Baethmann A, Peters J, Weight H, Kempski O (1990) Effects of lactacidosis on glial cell volume and viability. J Cereb Blood Flow Metab 10: 866–876

Correspondence: Umeo Ito, MD, Ph.D., 4-22-24, Zenpukuji, Suginami-ku, Tokyo 167-0041, Japan. e-mail: umeo-ito@nn.iij4u.or.jp

Acta Neurochir (2003) [Suppl] 86: 137–140

Prediction of tissue survival after stroke based on changes in the apparent diffusion of water (cytotoxic edema)

M. Sakoh[1], **T. Ohnishi**[1], **L. Ostergaard**[2], and **A. Gjedde**[2]

[1] Department of Neurological Surgery, Ehime University School of Medicine, Ehime, Japan
[2] Center of Functionally Integrative Neuroscience, Aarhus University, Denmark

Summary

This study tested the hypothesis that the estimate of the apparent diffusion coefficient (ADC; an index of cytotoxic edema) of water is a reliable pathophysiological index of the viability of ischemic brain tissue. CBF, the partition volume of water (PVW; an index of vasogenic edema), cerebral metabolic rate of oxygen ($CMRO_2$), and of glucose (CMR_{glc}) were measured before and after permanent middle cerebral artery occlusion (MCAO) or reperfusion with positron emission tomography (PET) in pigs. Then, the ADC was measured by diffusion-weighted magnetic resonance imaging (DW-MRI) and was compared with physiological variables obtained by PET and with histological findings. Both after permanent MCAO and reperfusion, the ADC was significantly correlated to the $CMRO_2$ and CMR_{glc}. The sequential decrease of ADC was also correlated to the sequential reduction of $CMRO_2$ in relation to the residual CBF after permanent MCAO. The infarction coincided with a threshold $CMRO_2$ of 50% of the contralateral value and a threshold ADC of 75% of the contralateral value, respectively. The PVW was decreased in relation to the magnitude of residual CBF after MCAO, whereas it was transiently increased in the severely ischemic lesions below 75% of the contralateral ADC value after reperfusion. Thus, the ADC is a reliable pathophysiological index which allows therapy to be tailored to the condition of individual patients in clinical practice.

Keywords: Diffusion weighted imaging; apparent diffusion coefficient; ischemia; positron emission tomography; cerebral metabolic rate of oxygen.

Introduction

The window of possible treatment of brain tissue struck by ischemia varies from person to person [11, 12]. A diagnostic procedure that assesses the viability of brain tissue would allow therapy to be tailored to the conditions of individual patients suffering from acute ischemic stroke. Positron emission tomography (PET) shows that ischemic injuries can be distinguished on the basis of severity as either completed ischemia in which the metabolic decline has been rapid and the damage remains irreversible after a few hours, or partial and still evolving ischemia (i.e. the penumbra of stroke-in-evolution) in which the metabolic decline occurs slowly and the damage remains potentially reversible for several hours [8, 9, 13]. However, in clinical practice, the viability of acutely ischemic tissue is usually not measurable by PET. It is well-known that diffusion-weighted magnetic resonance imaging (DW-MRI) provides a rapid means of manifesting cellular injury as cytotoxic edema in evolving ischemic lesions [3, 7, 10]. The diffusion abnormalities, manifested by reduced apparent diffusion coefficient (ADC) of water, correlate with the perfusion deficit and metabolic derangement of ischemic lesions after permanent middle cerebral artery occlusion (MCAO) [2, 10].

In the present study, we investigated these processes by multitracer PET, DWI, and pathohistology, applied to a newly-developed model of MCAO and its reperfusion in pigs [9]. The objective of this study was to test the hypothesis that the estimate of the ADC of water is a reliable pathophysiological index of the viability of ischemic brain tissue. Moreover, we investigated the correlation between cytotoxic edema and vasogenic edema in order to study the relation between brain edema and the reversibility of ischemic brain tissue after stroke.

Materials and methods

The research project was approved by the Danish National Committee for Animal Research Ethics (DaNCARE). Twenty seven female country-bred Yorkshire pigs weighing 38 to 45 kg were divided into two groups; (1) permanent MCAO (n = 18) and (2) reperfusion after 2 hour-MCAO (n = 9). Focal cerebral ischemia was transorbitally induced by occlusion of the left proximal MCAs. PET studies were performed at baseline and every hour up to 6 hours after permanent MCAO. In the reperfusion experiments, PET scans were obtained before and 1 hour after MCAO and every hour up to 6 hours after reperfusion. CBF, the partition volume of water (PVW; an index of vasogenic edema), and the cerebral metabolic rate of oxygen ($CMRO_2$) were measured in 18 pigs. Cerebral metabolic rate of glucose (CMR_{glc}) was measured at the end of the PET scans, i.e. once in each pig. Just after the PET scans, the ADC was measured with DW-MRI after permanent MCAO or reperfusion. Then, angiography and histological examination were performed. Moreover, the ADC was measured at baseline and every hour up to 6 hours after permanent MCAO in 9 pigs.

Positron emission tomography

CBF was estimated by i.v. bolus injection of 800 MBq $H_2{}^{15}O$ in the scanner (Siemens/CTI ECAT EXACT HR). The PVW was calculated as the regional ratio (K_1/k_2) for $H_2{}^{15}O$ of the unidirectional clearance rates from vascular bed to brain (K_1) and from brain to vascular bed (k_2). $CMRO_2$ was measured by single-breath inhalation (1 L) of 1200 MBq ^{15}O followed by 10-second breath holding. CMR_{glc} was measured by i.v. bolus injection of 300 MBq ^{18}F-fluorodeoxyglucose (FDG).

Diffusion weighted-magnetic resonance imaging

Imaging was performed using a GE Signa Horizon 1.0 Tesla Imager (GE Medical Systems, Milwaukee, WI, U.S.A.). After a sagittal scout, a T1-weighted three-dimensional spoiled gradient echo sequence (time of repetition [TR] = 8 ms, time of echo [TE] = 1.5 ms, 20-degree flip angle) was acquired for later coregistration of MRI and PET data. Subsequently, multi-slice DWI scans were performed by single-shot, spin-echo echo-planar-imaging (SE-EPI). The pigs were studied with a trace DWI sequence by acquiring an unweighted image S_0 with a b-factor of 0 s/mm^2 before each diffusion weighted image in three orthogonal directions, S_x, S_y, and S_z. Diffusion weighting was performed with a b-factor of 1000 s/mm^2. TR/TE was 5000/109 ms. Diffusion weighted images were collected in 6 slices. Slice thickness was 5 mm with an interslice gap of 2 mm, field-of-view was 14 × 14 cm, and the matrix size was 96 × 96, leading to an in-plane resolution of 1.46 × 1.46 mm^2.

ADC maps were calculated according to the equation [4]:

$$ADC = -\left[\ln\left(\frac{S_x}{S_0}\right) + \ln\left(\frac{S_y}{S_0}\right) + \ln\left(\frac{S_z}{S_0}\right)\right] \Big/ 3b$$

Statistical analysis was carried out by one-way repeated measures analysis of variance (ANOVA) in order to determine whether $CMRO_2$ and ADC differed significantly. For the assessment of correlation between ADC and physiological variables (CBF, $CMRO_2$ and CMR_{glc}), linear least squares regression was performed. A p-value of 0.05 was used for statistical significance throughout this report. The values were expressed as the mean ± SD.

Results

Comparison of histological damage, diffusion abnomality, and $CMRO_2$ deficit

The infarcted lesions were clearly delineated as hypointense areas with ADC below 75% of the contralateral value and $CMRO_2$ below 50% of the contralateral value after the MCAO, despite reperfusion.

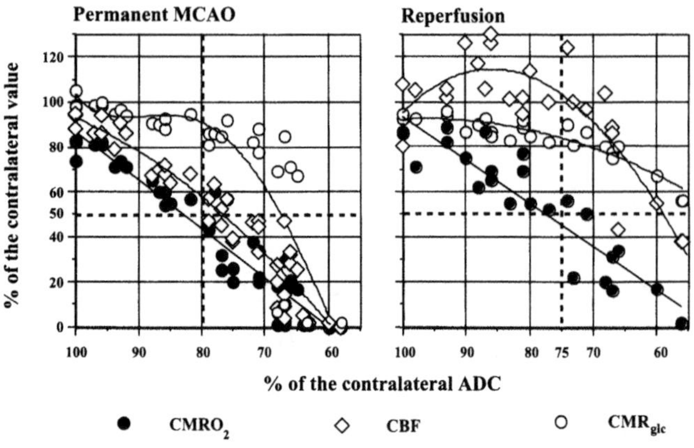

Fig. 1. Regional correlation between ADC and physiological variables (CBF, $CMRO_2$, and CMR_{glc}) 6 hours after permanent MCAO and reperfusion. The ADC was significantly correlated to $CMRO_2$, CMR_{glc}, and CBF after MCAO, whereas it failed to correlate with CBF after reperfusion

Regional correlation between ADC and physiological variables (CBF, $CMRO_2$, and CMR_{glc})

The regional correlations between ADC and CBF, $CMRO_2$, and CMR_{glc} are illustrated in Fig. 1. The ADC was significantly correlated to CBF, $CMRO_2$, and CMR_{glc} after permanent MCAO. In contrast, the ADC was well correlated to $CMRO_2$ and CMR_{glc}, whereas it was unrelated to CBF after reperfusion.

After permanent MCAO, the ischemic $CMRO_2$ threshold of 50% of the contralateral value reflected an ADC value of 80% of the contralateral value. After 2 hours-transient MCAO, the $CMRO_2$ threshold corresponded to an ADC value of 75% of the contralateral value.\

Sequential changes of ADC and $CMRO_2$ after permanent MCAO

The sequential changes of ADC and $CMRO_2$ after permanent MCAO are illustrated in Fig. 2. The ADC was well correlated to $CMRO_2$ in relation to the magnitude of residual CBF.

Regional correlation between Partition Volume of Water (PVW), CBF, and ADC

The changes of PVW related to the residual CBF are illustrated in Fig. 3. The control value of PVW was 0.80 ± 0.02 at baseline. During 6 hours after MCAO, the PVW was sequentially decreased in relation to the magnitude of residual CBF. In contrast, after reperfu-

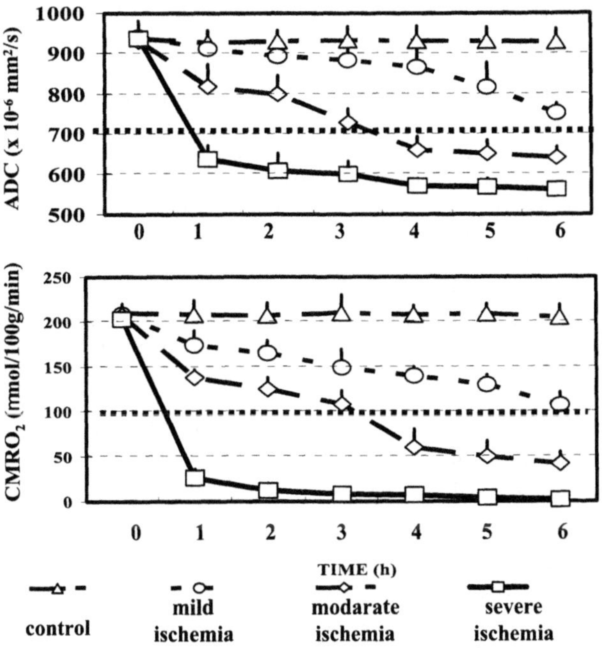

Fig. 2. Sequential changes in $CMRO_2$ and ADC before and after MCAO. The ADC was sequentially correlated to $CMRO_2$ in relation to the magnitude of residual CBF

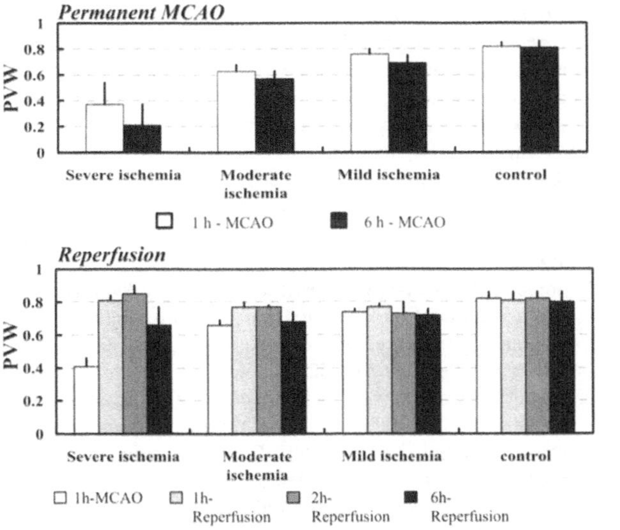

Fig. 3. Changes of PVW related to the residual CBF after MCAO with and without reperfusion. During 6 hours after MCAO, the PVW was sequentially decreased in relation to the degree of residual CBF. In contrast, after reperfusion, it was transiently increased in the severely ischemic tissue

sion, it was transiently increased in the severely ischemic tissue.

The regional correlations between PVW and ADC showed that the PVW was severely decreased at ADC values below 75% of the contralateral value after permanent MCAO, whereas it was increased at ADC values below 75% of the contralateral value 2 hours after reperfusion.

Discussion

Positron emission tomography (PET) has enabled researchers to follow the dynamics of regional pathophysiology leading to infarction after the onset of stroke, but its applicability is limited in clinical practice and frequently unavailable in acute situations [8, 13]. DWI, although still incompletely understood, rapidly provides a simple map of the metabolic state [2, 4, 7, 10]. In recent clinical practice, this technique is increasingly used prior to specific therapy of individual patients with acute stroke. Such assessment of cerebral ischemia is of value for tailoring reperfusion therapies to individual patients with evolving focal ischemia of the brain, provided the prediction of tissue viability is reliable [11, 12]. In the present study, we tested the reliability of the ADC as a pathophysiological index of brain tissue viability.

The initial decrease of the ADC in ischemic stroke reflects cellular swelling caused by the failure of water to pass from the extracellular space into the intracellular space after rapid loss of ion homeostasis [3, 7]. The ADC decrease follows the failure of the P-type Na^+-K^+-ATPase and the ensuing lactate accumulation in ischemic brain cells [2]. The present study demonstrated that the decrease of ADC correlates reliably with the reduction of $CMRO_2$ and CMR_{glc} in ischemic regions, after both permanent MCAO and reperfusion [10]. As a guide to therapy, the close linear relation between the decreases of ADC and $CMRO_2$ makes the ADC a good pathophysiological index, because the decline of $CMRO_2$ is the most reliable indicator of ischemic tissue viability [8, 9]. Thus, the practical threshold for the viability of ischemic tissue would be an ADC decrease to 75% of the contralateral value after the onset of stroke.

It is generally believed that cerebral ischemia is potentially curable in its early phase, when the edema is still of the cellular subtype [3]. The acute decrease of water diffusibility does not reflect a change of the water content of the ischemic tissue in the first few hours because of the absent increase of signal intensity on T_2-weighted and proton density imaging [3, 7]. Instead, the diffusion abnormality reflects cellular edema, i.e. cytotoxic edema [3]. Vasogenic edema develops over time as a result of breakdown of the blood-brain barrier causing excessive water accumulation and tissue swelling as infarction evolves [5]. The correlation between cytotoxic and vasogenic edema was examined during the cytotoxic phase of ischemic cerebral edema, which occurred within 6 hours of the onset of stroke [3]. The PVW indicates an apparent extravascular distribution volume of water and, as such, is an index of vasogenic edema, because the PVW was measured as the regional ratio of the unidirectional $H_2{}^{15}O$ clearance rates between the vascular bed and the brain tissue (K_1/k_2). In the present study, the water content of the ischemic region decreased in relation to the magnitude of residual CBF after the MCAO. Therefore, vasogenic edema did not evolve in parallel with cytotoxic edema at the stroke onset. On the other hand, the PVW was transiently increased at ADC values below 75% of the contralateral value after reperfusion. The ADC value would be the critical point for the viability of ischemic brain tissue, because the generation of vasogenic edema might participate in that of malignant infarction. Thus, the present findings indicate that the continuous water supply to ischemic regions after reperfusion might ameliorate the generation of vasogenic edema in the ischemic lesions with ADC values above 75% of the contralateral value.

Conclusion

This study showed that the ADC is a practical and reliable index of ischemic tissue viability. We found that DWI lesions with ADC values above 80% of the contralateral value are reversible as late as 6 hours after MCAO and DWI lesions with ADC values below 75% of the contralateral value might be irreversible as early as 2 hours after the MCAO.

Acknowledgments

This study was funded by the Scandinavia-Japan Sasakawa Foundation, the Danish Medical Research Council grants 9305246, 9305247, 9601888, and 9802833, and the Institute of Experimental Clinical Research, University of Aarhus, Denmark.

References

1. Davis D, Ulatowski J, Eleff S *et al* (1994) Rapid monitoring of changes in water diffusion coefficients during reversible ischemia in cat and rat brain. Magn Reson Med 31: 454–460
2. Kohno K, Hoehn-Berlage M, Mies G *et al* (1995) Relationship between diffusion-weighted MR images, cerebral blood flow, and energy state in experimental brain infarction. Magn Reson Imaging 13: 73–80
3. Kuroiwa T, Nagaoka T, Ueki M *et al* (1998) Different apparent diffusion coefficient. Water content correlations of gray and white matter during early ischemia. Stroke 29: 859–865
4. Le Bihan D, Breton E, Lallemand D *et al* (1988) Separation of diffusion and perfusion in intravoxel incoherent motion MR imaging. Radiology 168: 497–505
5. Loubinoux I, Volk A, Borredon J *et al* (1997) Spreading of vasogenic edema and cytotoxic edema assessed by quantitative diffusion and T2 magnetic resonance imaging. Stroke 28: 419–427
6. Miyabe M, Mori S, van Zijl PCM *et al* (1996) Correlation of the average water diffusion constant with cerebral blood flow and ischemic damage after transient middle cerebral artery occlusion in cats. J Cereb Blood Flow Metab 16: 881–891
7. Moseley ME, Cohen Y, Mintorovitch J *et al* (1990) Early detection of regional cerebral ischemia in cats: comparison of diffusion- and T2-weighted MRI and spectroscopy. Magn Reson Med 14: 330–346
8. Powers WJ, Grubb RL, Darriet D *et al* (1985) Cerebral blood flow and cerebral metabolic rate of oxygen requirements for cerebral function and viability in humans. J Cereb Blood Flow Metab 5: 600–608
9. Sakoh M, Ostergaard L, Rohl L *et al* (2000) Relationship between residual cerebral blood flow and oxygen metabolism as predictive of ischemic tissue viability: sequential multitracer positron emission tomography scanning of middle cerebral artery occlusion during the critical first 6 hours after stroke in pigs. J Neurosurg 93(4): 647–657
10. Sakoh M, Ostergaard L, Gjedde A *et al* (2001) Prediction of tissue survival after middle cerebral artery occlusion based on changes in the apparent diffusion of water. J Neurosurg 95: 450–458
11. Sorensen AG, Copen WA, Østergaard L *et al* (1999) Hyperacute stroke: Simultaneous measurement of relative cerebral blood volume, relative cerebral blood flow, and mean tissue transit time. Radiology 210: 519–527
12. Ueda T, Sakaki S, Yuh WTC *et al* (1999) Outcome in acute stroke with successful intra-arterial thrombolysis and predictive value of initial Single-Photon Emission-Computed Tomography. J Cereb Blood Flow Metab 19: 99–108
13. Wise RJ, Bernardi S, Frackowiak RS *et al* (1983) Serial observations on the pathophysiology of acute stroke. Brain 106: 197–222

Correspondence: Dr. Masaharu Sakoh, Department of Neurological Surgery, Ehime University School of Medicine, Shigenobu-cho, Onsen-gun, Ehime, 791-0204, Japan. e-mail: sakoh@m.ehime-u.ac.jp

Acta Neurochir (2003) [Suppl] 86: 141–145

Recovery of apparent diffusion coefficient after embolic stroke does not signify complete salvage of post-ischemic neuronal tissue

Y. Tanaka[1], **T. Kuroiwa**[2], **N. Miyasaka**[3], **F. Tanabe**[3], **T. Nagaoka**[4], and **K. Ohno**[1]

[1] Department of Neurosurgery, Tokyo Medical and Dental University, Tokyo, Japan
[2] Department of Neuropathology, Tokyo Medical and Dental University, Tokyo, Japan
[3] Department of Comprehensive Reproductive Medicine, Tokyo Medical and Dental University, Tokyo, Japan
[4] Department of Physiology and Biophysics, Albert Einstein College of Medicine, New York, USA

Summary

Background and purpose. It has been shown in previous studies that recovery of apparent diffusion coefficient (ADC) of water is not necessarily associated with a reversal of tissue damage. We examined the relationships between temporal ADC change and histological outcome after embolic stroke in rat using an experimental MRI scanner.

Subjects and methods. Male Wistar rats underwent embolization of the right middle cerebral artery with autologous clot. ADC and relative cerebral blood flow (rCBF) were measured consecutively for 3 hours after embolization. The brains were prepared for histological examination.

Results. Transient decline of ADC was observed during the initial 2 hours after embolization. In these areas, serial rCBF measurement revealed incomplete recovery. In some other areas, ADC decrease without rCBF recovery was observed during 3 hours postischemia. Histological examination revealed infarction in areas with persistent ADC decrease. Scattered neuronal death was noticeable in areas with a transient – in some areas less than 0.5 hour – ADC decrease.

Conclusion. Consecutive MRI analysis demonstrated postischemic transient ADC decrease in an embolic stroke model of rat. The normalization of ADC does not signify normalization of postischemic tissue, since the areas often evolve postischemic neuronal death at a later time.

Keywords: Cerebral ischemia; embolus; magnetic resonance imaging; neuronal death; rat.

Introduction

Diffusion weighted imaging (DWI) is a powerful tool that can detect early ischemic changes in both experimental models and human stroke patients. An initial decrease of the apparent diffusion coefficient (ADC) of water is reversible if reperfusion is established shortly after the onset of ischemia. However, it has been shown in previous studies, mostly using an intra-luminal thread occlusion model, that the recovery of ADC is not necessarily associated with the reversal of tissue damage. In this study, we examined the relationships between temporal ADC change and histological outcome after embolic stroke in rat using an experimental magnetic resonance imaging (MRI) scanner.

Materials and methods

Animal preparation

The animal experiments were performed in accordance with our institutional guidelines for animal research. Six male Wistar rats weighing 250 to 350 g were anesthetized with 1.5% isoflurane under intubation. Body temperature was maintained at 37–38 °C throughout the experiment using a water heating system. The left femoral artery and vein were cannulated with a PE-50 catheter for continuous monitoring of blood pressure and drug injection, respectively. The inspired oxygen (FiO_2) and expiratory carbon dioxide ($EtCO_2$) concentrations were monitored continuously, and the tidal volume was adjusted to produce $EtCO_2$ of 35–40 mm Hg.

Preparation of the embolus

The method used to prepare the embolus was adapted and modified from Busch *et al.* [1] Autologous blood was collected 24 hours before surgery and retained for 2 hours at 20 °C, and for 22 hours at 4 °C. Then the blood clot was passed through a PE-10 catheter (inner diameter 0.28 mm) and cut into small cylindrical segments with a length of 1.5 mm. Six of these clots (around 0.4 μl) were transferred into a PE-50 catheter filled with saline and stored until embolization.

Animal surgery

Right common carotid artery (CCA), internal carotid artery (ICA), and external carotid artery (ECA) were exposed. A PE-50 catheter, containing the blood clots, was inserted into the ECA, with its tip being located close to the carotid bifurcation. Then the clots were injected over a period of 1 minute during which the CCA was temporarily occluded.

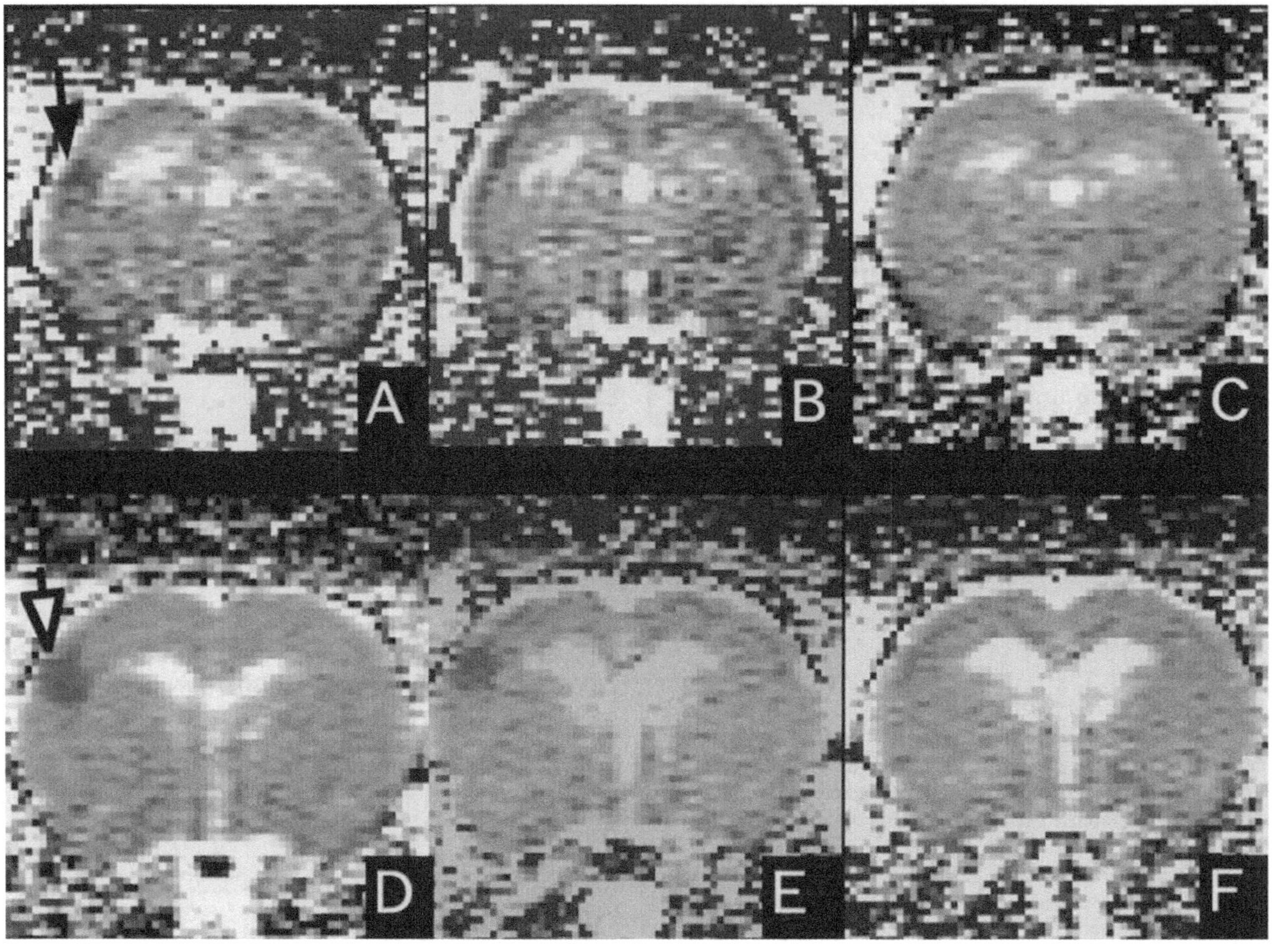

Fig. 1. ADC maps of the representative rats in group A (A–C) and B (D–F). A–C, ADC maps show a transient reduction of ADC within 30 minutes (arrow) and complete reversal without secondary reduction. (A) Initial ADC map after embolization. (B) 30 minutes (C) 180 minutes. D–F, ADC declined area existed over 60 minutes (open arrow), but finally disappeared. (D) initial ADC map after embolization (E) 60 minutes (F) 180 minutes

MRI measurement

The MR imaging studies were performed using a 4.7 tesla superconducting MRI with a 33-cm horizontal bore magnet and a 65-mT/m maximum gradient capability (Unity INOVA, Varian, Palo Alto, CA). A 20 mm diameter surface coil was tuned to 200 MHz for radiofrequency excitation and MR signal reception. After the surgical procedure, each rat was fixed firmly on the surface coil. Diffusion-weighted MR imaging was performed using a multisection, spin-echo sequence with parameters of 1500/80/1 (TR/TE/excitations), a matrix of 128 × 64, a field of view of 35 × 35 mm, and a section thickness of 2 mm without an intersection gap. The diffusion gradients were applied along the three orthogonal directions (x-, y-, and z-axes). The resulting values for the gradient factor b were 0 or 1200 s/mm^2 [8]. Gradient echo serial measurement was used to perform dynamic contrast enhanced perfusion weighted imaging (PWI) for measuring relative CBF. A 2 mm single slice was acquired with parameters of 10/5 (TR/TE), a field of view of 35 × 35 mm, and a matrix of 64 × 64. A total of 80 scans were obtained. 0.2 ml gadopentate dimeglumine was injected after acquisition of the 17th image.

MRI data analysis

Magnetic resonance image analysis of apparent diffusion coefficient (ADC) and cerebral blood flow (CBF) map sections were performed using image-processing software MR Vision (MR Vision Co.).

Ischemic lesions, where ADC values decreased to less than 80% of contralateral hemisphere, were defined from the ADC map obtained immediately after injection of blood clots. In these lesions, ADC values were measured consecutively for 180 minutes. One region of interest (ROI) at the center of the ischemic lesion (10 × 10 pixels on CBF maps) and another in the homologous region of the contralateral hemisphere were selected to measure the CBF index (CBFi). The CBFi ratio was calculated by dividing the ipsilateral CBFi values by the contralateral CBFi values.

Histological investigation

After the final MRI measurement, rats were transcardially perfused with hepalinized saline and 10% formalin under deep anesthesia. The brain was immersed in formalin solution, after which the

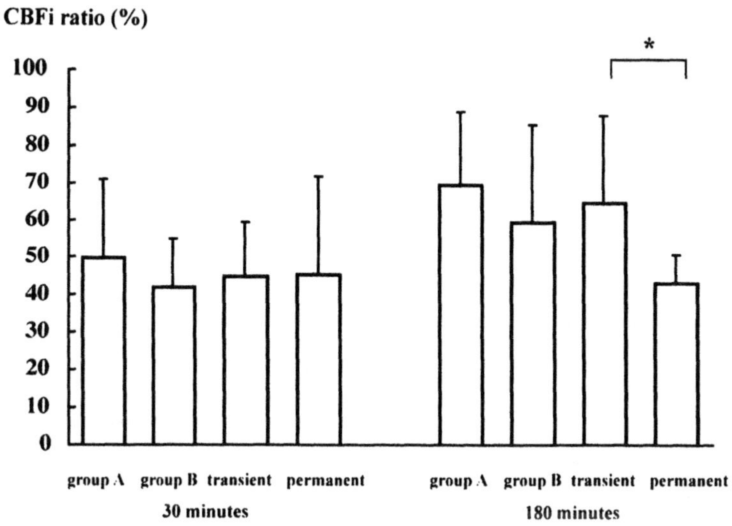

Fig. 2. Changes of CBFi ratio in each group or lesion. Transient includes both lesions of group A and group B. There was significant difference between Transient and Permanent lesions ($^*P < 0.05$)

brain was cut and stained with Hematoxylene-Eosin (H&E) for evaluation of ischemic cell damage.

Statistical analysis

All data are presented as mean ± SD. The Student's *t*-test was used to analyze parametric data. Differences with P values of less than .05 were considered significant.

Results

MRI findings

All the rats had both temporarily ADC declined lesions and permanently ADC declined lesions. In all the transient lesions, the ADC values did not decrease again after recovery throughout the period of the experiments. The rats were divided into the following two groups: group A rats (n = 3) had transient lesions which showed recovery of the ADC values within 30 minutes (Fig. 1), and group B rats (n = 3) had transient lesions which showed recovery of the ADC values from 30 minutes to 120 minutes. The mean ADC value in the transient lesions for group A was $4.28 \pm 0.57 \times 10^{-4}$ mm^2/s and group B was $4.34 \pm 0.56 \times 10^{-4}$ mm^2/s (measured immediately after embolization), and there was no significant difference between the two groups.

In the transient lesions including groups A and B, serial CBFi measurement revealed incomplete recovery from 45.91 ± 17.97% of the control at 0.5 hour after embolization to 64.54 ± 23.41% at 3 hours (Fig. 2). In permanent lesions, an ADC decrease without rCBF recovery (45.20 ± 26.64% at 0.5 hour, 42.95 ± 7.64% at 3 hours) was observed during 3 hours postischemia.

Histological findings

The histological findings of transient lesions of group A showed that some neurons shrank mildly to moderately with ischemic changes (cytoplasmic eosinophilia and pyknotic nuclei), and others were intact. In group B, there was some variation in each rat, but most neurons were moderately to severely shrunken, and a few intact neurons were observed (Fig. 3).

Discussion

Acute ADC reduction during the early phase of ischemia and complete reversal of ADC following reperfusion has been reported by many authors [2, 5–8], but most of them used the thread occlusion model or the hypoxia/ischemia model. In this study, we produced the transient ischemic lesion using a new embolic model. This model was very easy to create, without any using special technique or instrument.

Overgaard *et al.* used thrombin-added autologous clot to prevent spontaneous recanalization and reached a 100% infarction rate in their embolic model [9]. They also demonstrated a smaller clot volume provided smaller infarctions, and one animal showed no infarction. Moreover, some other authors reached a lower infarction rate using a plain autologous clot without adding thrombin or any other agents [3, 4]. These studies suggested that a low-volume, plain autologous clot was not enough to pro-

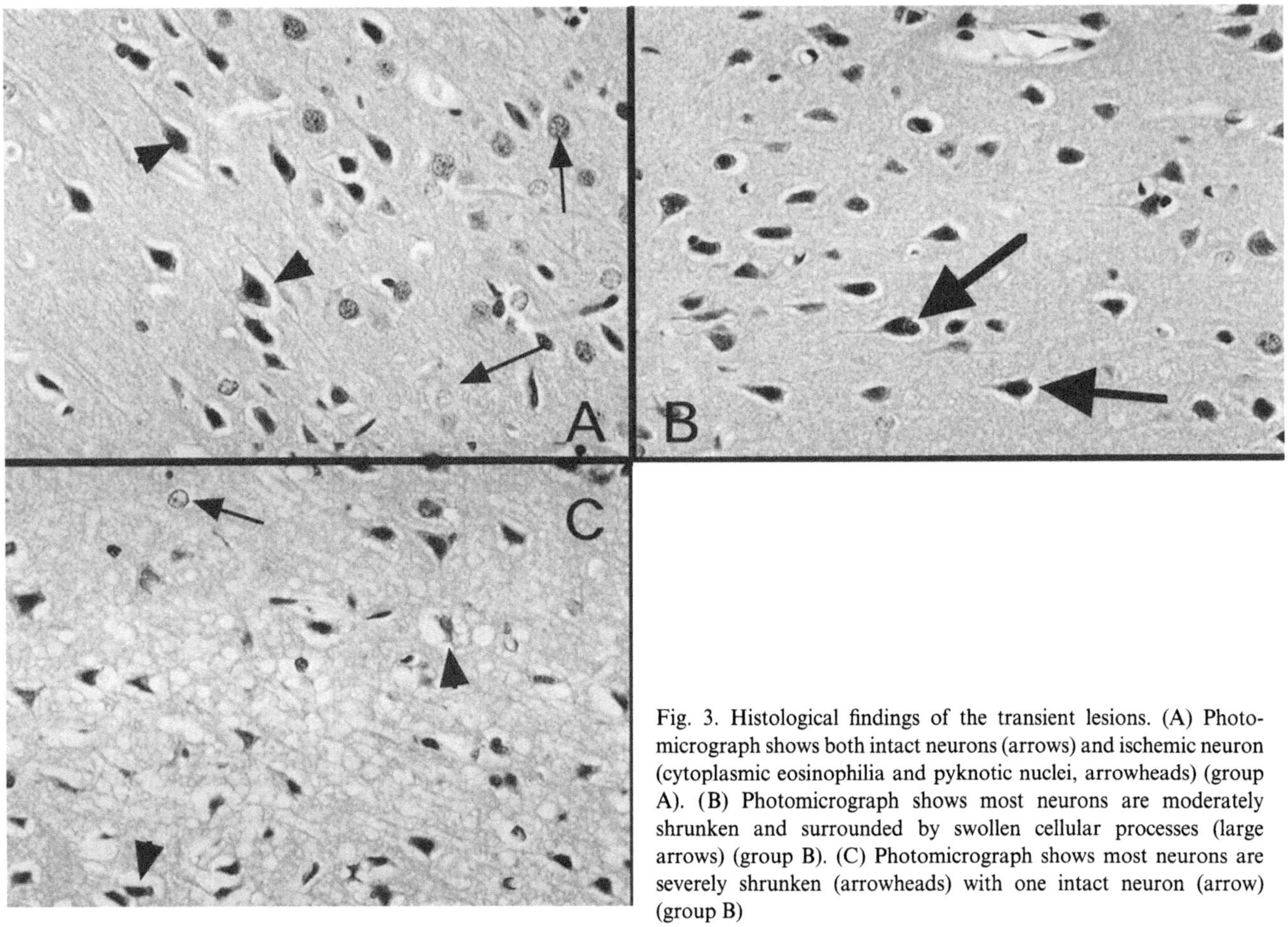

Fig. 3. Histological findings of the transient lesions. (A) Photomicrograph shows both intact neurons (arrows) and ischemic neuron (cytoplasmic eosinophilia and pyknotic nuclei, arrowheads) (group A). (B) Photomicrograph shows most neurons are moderately shrunken and surrounded by swollen cellular processes (large arrows) (group B). (C) Photomicrograph shows most neurons are severely shrunken (arrowheads) with one intact neuron (arrow) (group B)

duce permanent ischemia because of the hemolysis system or some other reason, such as collateral flow. So, we thought that our model could produce transient lesion because a low-volume, plain autologous clot was used. The improvement of regional CBF strongly supported this hypothesis. Our study showed that regional CBF of the transient ADC decline lesion increased incompletely (41.30%–96.05%), but all of the lesions showed improvement. On the other hand, permanent lesions demonstrated no improvement of CBF. In this model, the embolic source was autologous clots, so the reason why regional CBF improved in transient lesions may be that the hemolysis system was activated and the embolic clot moved to the distal side of the middle cerebral artery or was completely resolved.

Histological analysis demonstrated different findings between the lesions of groups A and B. Selective neuronal death after a short period of focal ischemia has been documented in previous studies [5, 7, 8]. In this study, group A exhibited this change. On the other hand, group B showed moderate to severe neuronal injury, indicative of cerebral infarction.

According to these results, we thought that even quick reversal of ADC reduction does not signify the complete salvaging of post-ischemic neuronal tissue. Moreover, the degree of neuronal damage is dependent on the duration of ADC reduction. Therefore, we speculate that in the treatment with cerebral ischemia, both rapid recovery of CBF and neuroprotective therapy may be important to prevent neuronal injury.

Conclusion

Consecutive MRI analysis demonstrated a postischemic transient ADC decrease in an embolic stroke model of rat. The normalization of ADC does not signify normalization of postischemic tissue, since the areas often evolve postischemic neuronal death at a later time.

References

1. Busch E, Krueger K, Hossmann K-A (1997) Improved model of thromboembolic stroke and rt-PA induced reperfusion in the rat. Brain Res 778: 16–24
2. Hasegawa Y, Fisher M, Latour LL *et al* (1994) MRI diffusion mapping of reversible and irreversible ischemic injury in focal brain ischemia. Neurology 44: 1484–1490
3. Kaneko D, Nakamura N, Ogawa T (1985) Cerebral infarction in rats using homologous blood emboli: development of a new experimental model. Stroke 16: 76–84
4. Kudo M, Aoyama A, Ichimori S, Fukunaga N (1982) An animal model of cerebral infarction. Stroke 4: 505–508
5. Li F, Han SS, Tatlisumak T, *et al* (1999) Reversal of acute apparent diffusion coefficient abnormalities and delayed neuronal death following transient focal cerebral ischemia in rats. Ann Neurol 46: 333–342
6. Li F, Liu K-F, Silva MD, *et al* (2000) Transient and permanent resolution of ischemic lesions on diffusion-weighted imaging after brief periods of focal ischemia in rats. Stroke 31: 946–954
7. Minematsu K, Li L, Sotak CH, *et al* (1992) Reversible focal ischemic injury demonstrated by diffusion-weighted magnetic resonance imaging in rats. Stroke 23: 1304–1311
8. Miyasaka N, Nagaoka T, Kuroiwa T, *et al* (2000) Histopathologic collerates of temporal diffusion changes in a rat model of cerebral hypoxia/ischemia. AJNR 21: 60–66
9. Overgaard K, Sereghy T, Boysen G, *et al* (1992) A rat model of reproducible cerebral infarction using thrombotic blood clot emboli. J Cereb Blood Flow Metab 12 (3): 484–490

Correspondence: Yoji Tanaka, 1-5-45 Yushima, Bunkyo-Ku, Tokyo, Japan. e-mail: tanaka.nsrg@tmd.ac.jp

Acta Neurochir (2003) [Suppl] 86: 147–151

Temporal evolution of apparent diffusion coefficient and T2 value following transient focal cerebral ischemia in gerbils

I. Yamada[1], **T. Kuroiwa**[2], **S. Endo**[3], and **N. Miyasaka**[4]

[1] Department of Radiology, Tokyo Medical and Dental University, Tokyo, Japan
[2] Department of Neuropathology, Medical Research Institute, Tokyo Medical and Dental University, Tokyo, Japan
[3] Animal Research Center, Tokyo Medical and Dental University, Tokyo, Japan
[4] Department of Obstetrics and Gynecology, Tokyo Medical and Dental University, Tokyo, Japan

Summary

We examined the time course of apparent diffusion coefficient (ADC) and T2 values in gerbils subjected to transient focal cerebral ischemia and compared them with histopathologic changes. Ten gerbils were subjected to two times 10-min occlusions of the left common carotid artery and examined with diffusion-weighted and T2-weighted MR imaging, at the interocclusion period and 1 hour, 2 hours, 1 day, 2 days, 4 days, and 7 days after the second occlusion. Immediately after the last MR imaging, their brains were examined histopathologically. ADC values decreased 1 hour after the second occlusion and continued to decrease up to 1 or 2 days later. ADC values remained lowered up to 4 days and slightly recovered at 7 days. T2 values were normal at 1 and 2 hours and began to increase at 1 day. T2 values began to recover at 4 days. Histopathologically, infarction was confined in the fronto-parietal cortex, dorsolateral caudate nucleus and dorsolateral thalamus, and neuronal necrosis was found in the pyramidal cell layer of the hippocampus. This study indicates that sequential ADC and T2 studies of gerbils subjected to transient focal cerebral ischemia provide a useful tool for evaluating temporal evolution of ischemic brain injury and edema, including cytotoxic and vasogenic edema.

Keywords: Cerebral ischemia; gerbil; transient ischemia; MR imaging; diffusion-weighted image; ADC; T2 value.

Introduction

Transient cerebral ischemia is followed by the development of tissue infarction and/or selective neuronal death, depending on the depth and duration of the ischemic insult [6]. Although tissue infarction in severe ischemia evolves rapidly and becomes apparent within several hours after the onset of ischemia, the infarction process may only become apparent as late as a week after circulation has been restored. This particularly occurs when the ischemic insult is mild and of the threshold level for the induction of infarction [2, 4]. Therefore, a long time observation is necessary in cases of transient ischemia of the threshold level for infarction.

Diffusion-weighted MR imaging allows one to assess the early stages of cerebral ischemia by detecting decreased water diffusion that may be related to cytotoxic edema [10]. However, from the intermediate to late stages, diffusion-weighted imaging shows complex temporal profiles [7, 12]. These temporal profiles may reflect the cascade of ischemic evolution itself and T2 shine-through effects.

In this study, we examined the time course of apparent diffusion coefficient (ADC) and T2 values in gerbils subjected to transient focal cerebral ischemia. We used a model of transient ischemia of the threshold level for infarction in order to induce the slow evolution [6]. Our purpose was to elucidate the temporal profiles of ADC and T2 value changes in the evolution of ischemic brain injury and edema and to compare them with histopathologic changes.

Materials and methods

Experimental protocol

Adult Mongolian gerbils (n = 10; weight, 60 to 80 g) were used in this study. The animals were anesthetized with 5% isoflurane. Anesthesia was subsequently maintained with 1.5% isoflurane in a mixture of 30% oxygen and 70% nitrous oxide that was supplied via a nose cone, and the gerbils were allowed to breathe spontaneously throughout the study. The left common carotid artery was occluded via a midline incision using a small vascular clip, then the animals were allowed to recover from anesthesia, and blood flow was restored by releasing the clip after 10 min of occlusion. The second 10-min period of ischemia was similarly induced 5 hours later. After the second period of ischemia, the animals were returned to their cages and allowed free access to water and food.

Stroke symptoms were evaluated according to a stroke index (SI) of 0–25 points reported previously [9, 13], in which hair roughed up to tremor, paucity of movement, hypesthesia of the ear, head cocked, eyes fixed open, ptosis, a splayed-out hind limb, circling, seizures, and extreme weakness were scored 1, 1, 1, 3, 3, 1, 3, 3, 3, and 6 points, respectively. Animals manifesting a SI score of more than 10 during the initial ischemia were selected for use as ischemia-positive animals [13].

The two-occlusion procedure of the left common carotid artery was initiated, and ischemia-positive animals (n = 10) were selected for the following imaging experiments. MR imaging was performed at the interocclusion period and 1 hour, 2 hours, 1 day, 2 days, 4 days, and 7 days after the second period of occlusion. Brain specimens were examined histopathologically immediately after the final MR imaging study for all of the gerbils.

MR imaging examination

All MR images were obtained using a 4.7-T superconducting system with a 33-cm horizontal bore magnet and a 67 mT/m maximum gradient capability (Unity INOVA; Varian, Palo Alto, Ca, USA), and a circular surface coil of 4-cm diameter was used. The animal was placed in a supine position on a thermal water blanket maintained at 37.5 °C, and an MR compatible stereotaxic frame was used to prevent motion artifacts.

Diffusion-weighted imaging was performed with a multisection, spin-echo sequence which included a repetition time of 1,500 msec, an echo time of 80 msec, a matrix of 128 × 64, a field of view of 30 × 30 mm, a section thickness of 2 mm without intersection gap, and one signal acquired. The diffusion gradients were applied along the three orthogonal directions (x-, y-, and z-axes), and the resulting values for gradient factor b were 0 or 1,200 sec/mm^2. The acquisition time for one set of diffusion-weighted images was 6 minutes and 24 seconds.

For T2 calculation, multisection spin-echo imaging without diffusion gradient was performed with a repetition time of 1,500 msec and three echo times of 20, 50, and 80 msec. Other imaging parameters included a matrix of 128 × 64, a field of view of 30 × 30 mm, a section thickness of 2 mm without intersection gap, and one signal acquired. The acquisition time for one set of T2-weighted images was 4 minutes and 48 seconds.

Data analysis

All analyses were performed using a Sun Sparc 10 workstation (Sun Microsystems, Ca, USA) and image analyzing software (XDS software; Davis Bioengineering). ADC maps were calculated on a pixel-by-pixel basis using the following equation: $ADC = \ln(So/S)/(b - bo)$, where So and S are the signal intensities of the two diffusion-weighted images that represent the average of three values in three orthogonal planes, i.e. the trace of the diffusion tensor, and bo and b are 0 and 1,200 sec/mm^2, respectively.

T2 maps were calculated on a pixel-by-pixel basis using the following equation: $SI = K\exp(-TE/T2)$, where the repetition time $\gg$ T1, SI is the signal intensity, and K is a proportionality constant. With data obtained at three different echo times (20, 50, and 80 msec), the T2 relaxation time was calculated from the equation using a regression analysis.

Histological examination

Immediately after the final MR imaging acquisition, the animals were removed from the MR unit under deep anesthesia. Their brains were fixed by perfusion buffered formalin via an indwelling left ventricular catheter for up to 30 min. The brains were then removed and cut into five serial coronal sections 2 mm in thickness using a tissue slicer. The sections were processed for hematoxylin-eosin and Kluever-Barrera staining.

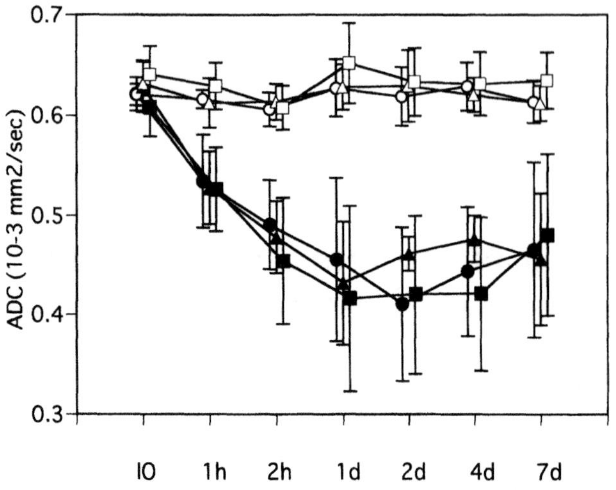

Fig. 1. The temporal profiles of ADC values in gerbils. The ipsilateral cortex (closed circle), caudate nucleus (closed triangle), and thalamus (closed square) showed a significant decrease of ADC one hour after the second occlusion, and the ADCs in these regions continued to decrease up to one or two days after the second occlusion. The ADCs in these regions slightly recovered seven days after the second occlusion, although they were still lower than those in the contralateral hemisphere. In the contralateral cortex (open circle), caudate nucleus (open triangle), and thalamus (open square), no ADC changes were observed. *Error bars* Standard deviations, *IO* interocclusion period

Results

Time-courses of ADC changes

Figure 1 displays the temporal profiles of ADC data for the gerbils. The cortex, caudate nucleus, and thalamus in the ipsilateral cerebral hemisphere showed a significant decrease of ADC one hour after the second occlusion, and the ADCs in these regions continued to decrease up to one or two days after the second occlusion. The lowest ADCs persisted during the period from one to four days. The ADCs in these regions slightly recovered seven days after the second occlusion, although they were still lower than those in the contralateral hemisphere. In the contralateral cortex, caudate nucleus, and thalamus, however, no ADC changes were observed throughout the experimental period (Fig. 1).

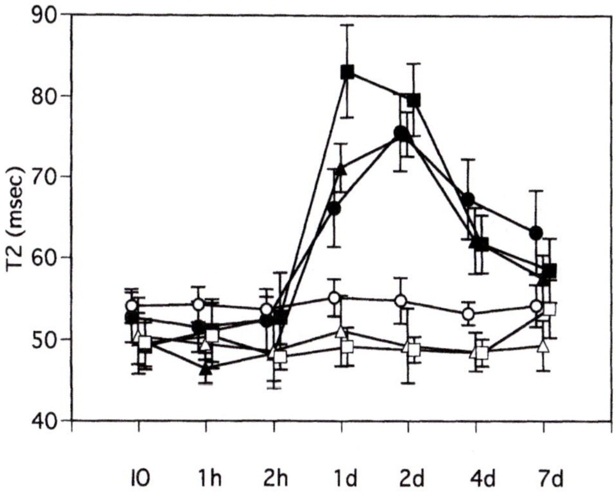

Fig. 2. The temporal profiles of T2 values in gerbils. The ipsilateral cortex (closed circle), caudate nucleus (closed triangle), and thalamus (closed square) showed a significant increase of T2 values one day after the second occlusion. The highest T2 values persisted during the period of one to two days after the second occlusion. Then, T2 values in these regions significantly recovered during the period of four to seven days after the second occlusion. In the contralateral cortex (open circle), caudate nucleus (open triangle), and thalamus (open square), no T2 changes were observed. *Error bars* Standard deviations, *IO* interocclusion period

Time-courses of T2 changes

Figure 2 displays the temporal profiles of T2 data for the gerbils. There was no change of T2 values one and two hours after the second occlusion. The cortex, caudate nucleus, and thalamus in the ipsilateral cerebral hemisphere showed a significant increase of T2 values one day after the second occlusion, and the T2 values in the cortex and caudate nucleus further increased up to two days after the second occlusion. The highest T2 values persisted during the period of one to two days after the second occlusion. Then, T2 values in these regions significantly recovered during the period of four to seven days after the second occlusion. In the contralateral cortex, caudate nucleus, and thalamus, however, no T2 changes were observed throughout the experimental period (Fig. 2).

Histopathologic findings

Histologically, infarction defined as the necrosis of neuronal and glial components with inflammatory cell infiltration was observed in the ipsilateral cerebral cortex, caudate nucleus, and thalamus (Fig. 3). There was extensive infiltration of macrophages with capillary proliferation in the infarction. Neuronal necrosis was found in the pyramidal cell layer of the ipsilateral hippocampus. No infarction was noticeable in

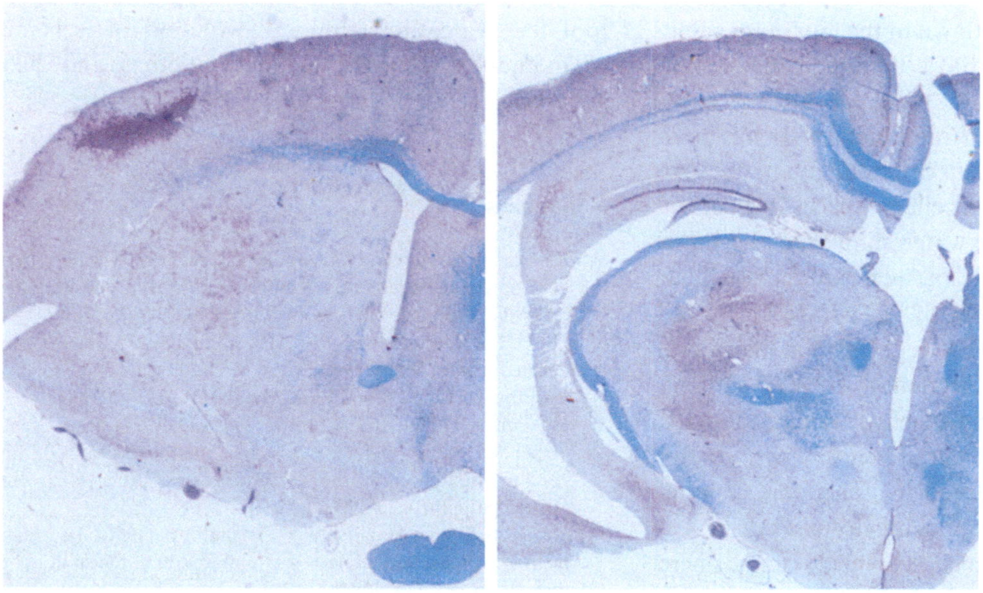

Fig. 3. Photomicrographs of coronal sections with Kluever-Barrera staining show areas of infarction in the ipsilateral cerebral cortex, the dorsolateral part of the caudate nucleus, and thalamus. Extensive infiltration of macrophages with capillary proliferation in the infarction is noted

the contralateral cerebral hemisphere in any of the gerbils.

Discussion

In the present study, focal cerebral ischemia was induced in gerbils by two times 10-min occlusions of the left common carotid artery, separated by 5-h intervals. This transient ischemia has the following characteristics: (a) the evolution of infarction was slow and therefore suitable for analyzing the temporal profile of its development and accompanying metabolic disturbances, (b) the distribution of the lesion (areas of infarction and selective neuronal death) was highly reproducible, and (c) the mortality rate was negligible [6]. Although cerebral infarction often develops rapidly in clinical cases, experimental as well as clinical data show that the injury process may take several days to weeks after the initial ischemic insult. The induction of focal cerebral ischemia by two times 10-min occlusions was suitable for the analysis of the temporal and spatial profiles of postischemic tissue injury, since both infarction and selective neuronal death evolved slowly in various regions of the postischemic brain [4, 6].

Our data demonstrate that the cortex, caudate nucleus, and thalamus in the ipsilateral cerebral hemisphere showed a significant decrease of ADC one hour after the second occlusion, and that the ADCs continued to decrease up to one or two days after the second occlusion. Thus, the ADC values decreased gradually for one to two days down to the minimum value, which lasted for several days. Although the mechanisms responsible for ADC reduction are not completely understood, the extracellular space has been considered to be the main determinant of effective ADC [1, 11, 14]. The extracellular spaces in ischemic tissue become smaller as water shifts into the intracellular space as a result of increased intracellular osmolarity and impairment of ion pumps in the cell membrane [8]. Kuroiwa *et al.* reported that in cerebral regions that developed infarction, tissue ATP content, in parallel with the succinic dehydrogenase activity, fell to their lowest levels over a four-day period after circulation had been restored [6]. Therefore, our results suggest that in this model, cytotoxic edema occurs gradually accompanied by energy impairment and mitochondrial enzymatic dysfunction.

In our results, ADC values remained lowered up to four to seven days after circulation had been restored. There was no pseudonormalization of ADC values during the experimental period of one week. In this regard, histopathologic examinations demonstrated that there was extensive infiltration of macrophages in infarction areas, and that the cellular density was increased at later phases. The high cellular density of macrophages in infarction areas may be related to the finding that ADC values remained lowered for a longer period.

Furthermore, our data demonstrate that the cortex, caudate nucleus, and thalamus in the ipsilateral cerebral hemisphere showed a significant increase of T2 values one day after the second occlusion, and that the T2 values in the cortex and caudate nucleus further increased up to two days after the second occlusion. Then, T2 values in these regions significantly recovered during the period of four to seven days after the second occlusion. Thus, T2 values showed a late peak at 24 to 48 hours after the occlusion. Hakamata *et al.* reported that at 24 and 48 hours in the infarcted focus, the blood-brain barrier was disrupted and vasogenic edema developed [3]. Ito *et al.* also reported that the blood brain barrier injury occurred in relatively late phases in the reperfusion ischemia model [5]. Therefore, our results suggest that the T2 map well depicts the delayed occurrence of vasogenic edema and its gradual disappearance in the reperfusion ischemia model.

In conclusion, sequential diffusion-weighted and T2-weighted MR imaging studies of gerbils subjected to transient focal cerebral ischemia provide a useful tool for evaluating temporal evolution of ischemic brain injury and edema, including cytotoxic and vasogenic edema.

References

1. Benveniste H, Hedlund LW, Johnson GA (1992) Mechanism of detection of acute cerebral ischemia in rats by diffusion-weighted magnetic resonance microscopy. Stroke 23: 746–754
2. Du C, Hu R, Csernansky CA, Hsu CY, Choi DW (1996) Very delayed infarction after mild focal cerebral ischemia: a role for apoptosis? J Cereb Blood Flow Metabol 16: 195–201
3. Hakamata Y, Hanyu S, Kuroiwa T, Ito U (1997) Brain edema associated with progressive selective neuronal death or impending infarction in the cerebral cortex. Acta Neurochir (Wien) [Suppl] 70: 20–22
4. Hanyu S, Ito U, Hakamata Y, Yoshida M (1995) Transition from ischemic neuronal necrosis to infarction in repeated ischemia. Brain Res 686: 44–48
5. Ito U, Go KG, Walker JT Jr, Spatz M, Klatzo I (1976) Experimental cerebral ischemia in Mongolian gerbils: III. Behaviour of the blood-brain barrier. Acta Neuropathol (Berl) 34: 1–6

6. Kuroiwa T, Miles G, Hermann D, Hakamata Y, Hanyu S, Ito U (2000) Regional differences in the rate of energy impairment after threshold level ischemia for induction of cerebral infarction in gerbils. Acta Neuropathol (Berl) 100: 587–594
7. Matsumoto K, Lo EH, Pierce AR, Wei H, Garrido L, Kowall NW (1995) Role of vasogenic edema and tissue cavitaion in ischemic evolution on diffusion-weighted imaging: comparison with multiparameter MR and immunohistochemistry. Am J Neuroradiol 16: 1107–1115
8. Matsuoka Y, Hossmann KA (1982) Brain tissue osmolarity after middle cerebral artery occlusion in cats. Exp Neurol 77: 599–611
9. McGraw CP (1977) Experimental cerebral infarction: effects of pentobarbital in Mongolian gerbils. Arch Neurol 34: 334–336
10. Miyasaka N, Kuroiwa T, Zhao FY, Nagaoka T, Akimoto H, Yamada I, Kubota T, Aso T (2000) Cerebral ischemic hypoxia: discrepancy between apparent diffusion coefficients and histologic changes in rats. Radiology 215: 199–204
11. Moseley ME, Kucharczyk J, Mintorovitch J *et al* (1990) Diffusion-weighted MR imaging of acute stroke: correlation with T2-weighted and magnetic susceptibility-enhanced MR imaging in cats. Am J Neuroradiol 11: 423–429
12. Neumann-Haefelin T, Kastrup A, de Crespigny A, Yenari MA, Ringer T, Sun GH, Moseley ME (2000) Serial MRI after transient focal cerebral ischemia in rats: dynamics of tissue, injury, blood-brain barrier damage, and edema formation. Stroke 31: 1965–1973
13. Ohno K, Ito U, Inaba Y (1984) Regional cerebral blood flow and stroke index after left carotid artery ligation in the conscious gerbil. Brain Res 297: 151–157
14. Sevick RJ, Kanda F, Mintorovitch J *et al* (1992) Cytotoxic brain edema: assessment with diffusion-weighted MR imaging. Radiology 185: 687–690

Correspondence: Ichiro Yamada, M.D., Department of Radiology, Faculty of Medicine, Tokyo Medical and Dental University, 1-5-45 Yushima, Bunkyo-ku, Tokyo 113-8519, Japan. e-mail: yamada.crad@med.tmd.ac.jp

Acta Neurochir (2003) [Suppl] 86: 153–157

Prediction of malignant infarction: perifocal neurochemical monitoring following prolonged MCA occlusion in cats

S. Toyota[1,2], **R. Graf**[1], **M. Valentino**[1], **T. Yoshimine**[2], and **W.-D. Heiss**[1]

[1] Max-Planck Institut für neurologische Forschung, Köln, Germany
[2] Department of Neurosurgery, Osaka University Medical School, Osaka, Japan

Summary

Neurochemical monitoring in the ischemic core predicts malignancy in focal ischemia in cats. Since perifocal regions are more suitable for clinical microdialysis (MD) applications, we tested whether monitoring at such site predicts also malignancy. – Laser Doppler (LD) probes, pressure microsensors, and MD/HPLC measured cerebral blood flow (CBF), intracranial pressure (ICP), and extracellular glutamate (Glu), respectively. The middle cerebral artery was occluded (MCAO) for 3 hours followed by 6 hours reperfusion. Additionally, LD measured CBF in ischemic core. – MCAO reduced CBF in the core in all below 25% of control. In animals exhibiting malignancy (eye dilatation during reperfusion), MCAO decreased CBF in the perifocal site to around 35%. CBF primarily recovered following recirculation but decreased thereafter as ICP rose due to progressive edema formation. Glu increased concomitantly. In cats exhibiting a benign course, MCAO decreased CBF in the perifocal site to around 55%. Recirculation normalized CBF, and Glu did not increase. During MCAO, Glu differences between both groups were not significant. – Glu determinations in perifocal sites taken during MCAO do not predict fatal outcome. This contrasts with determinations in the core. After reperfusion, Glu elevation in perifocal sites may serve as a rather late predictor of malignancy.

Keywords: Cerebral ischemia; focal; malignant stroke; perifocal; glutamates; intracranial pressure; cats.

Introduction

Malignant, space-occupying brain edema is a relevant, serious complication in various clinical situations including large hemispheric stroke [6, 11]. In the search for monitoring tools that permit early identification of malignant edema in large hemispheric stroke, microdialysis (MD) has recently been introduced in intensive patient care [1, 2, 12]. With intracerebral microdialysis, excessive release of Glu into extracellular space has been documented in various animal models of experimental stroke [4, 9, 13–15]. One important issue for the application of such invasive technique is the site of monitoring [3]. In clinical situations, microdialysis probes are mostly inserted into regions peripheral to the ischemic core, because such perifocal regions are non-eloquent and routinely used for neurosurgical insertion of invasive devices. Regarding the distance to core regions, however, these predetermined areas are often not well defined. Therefore, to date, the role of neuroactive substances in perifocal regions remains obscure. Neurochemical monitoring predicts malignancy in the ischemic core in cat focal ischemia [15]. We tested now in the same model whether perifocal neurochemical monitoring possesses a similar prognostic value in malignant experimental stroke.

Methods

In halothane-anesthetized cats (n = 10), the left middle cerebral artery (MCA) was occluded transorbitally for 3 hours followed by 6 hours reperfusion. Microdialysis probes (diameter: 250 μm; length of the active membrane: 1 mm, perfusion rate: 1 μl/min) were inserted into a perifocal site (marginal gyrus) of the MCA territory in the left cerebral cortex. Concentrations of glutamate (Glu) and other amino acids in dialysate were analyzed by HPLC. Adjacent to microdialysis probes, a laser Doppler probe measured regional CBF (LDF-CBF), a strain-gauge MicroSensor measured ICP, and a thermocouple measured regional brain temperature. Another laser Doppler probe was placed in the ischemic core (ectosylvian gyrus). After completion of the preparation, the skull was sealed.

Variables (ICP, CBF measured by LDF, MABP, brain temperature, end-tidal CO2) were continuously recorded with the use of a PC-based data acquisition system (DASY LAB). After a stabilization period of at least 2 hours, the MCA was occluded for 3 hours and then reopened. The observation was followed 6 hours into the reperfusion phase.

The experiments were terminated by perfusion fixation with 4% paraformaldehyde solution. Brains were removed and additionally immersion fixed for at least 2 weeks. After paraffin embedding, the brains were stained with hematoxylin-eosin or with a combination of

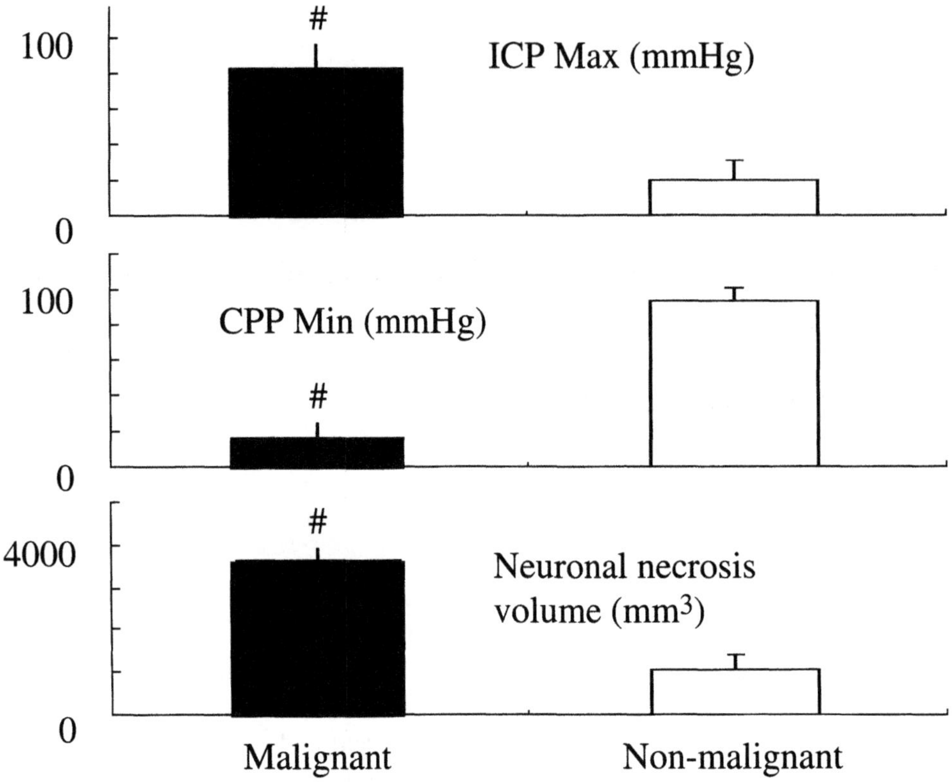

Fig. 1. Maximum value of ICP, minimum value of CPP, and neuronal necrosis volume in groups exhibiting a non-malignant (white bars; n = 5) and a malignant course (black bars; n = 5), respectively. #$P < 0.05$, significant difference between groups

Luxol fast blue and cresyl violet. Brain damage was macroscopically and microscopically determined on individual brain sections.

Results

MCAO reduced LDF-CBF in all animals below 25% of control in the ischemic core. Five of 10 cats that showed pupil dilation in the course of reperfusion were defined as the malignant group, and the other five were defined as the non-malignant group.

MABP did not change throughout MCAO but increased slightly in the non-malignant group and decreased in the malignant group following reperfusion reaching final levels of <100 mm Hg. These levels were significantly lower than those obtained in the non-malignant group. Similar to MABP, ICP did not change during MCAO but started to increase in both groups almost immediately after reopening of the MCA. Already in the early phase of reperfusion, this increase was much more pronounced in the malignant group. ICP continued to rise in this group throughout the observation period and reached final values of >70 mm Hg. As a consequence of MABP and ICP alterations, CPP did not change during the occlusion phase, and in the non-malignant group, it remained almost unaltered throughout reperfusion. In the malignant group, in contrast, CPP started to fall immediately after reopening of the MCA, and this decrease continued until values of <30 mm Hg were reached indicating severe global disturbance of cerebral perfusion. The maximum value of ICP during reperfusion in the malignant group was significantly higher than that in non-malignant group (Fig. 1). The minimum value of CPP during reperfusion period in the malignant group was significantly lower than that in non-malignant group (Fig. 1). The volume of neuronal necrosis in the malignant group was significantly larger than that in the non-malignant group (Fig. 1).

In the malignant group, LDF-CBF decreased during MCAO in the perifocal region to around 35% of control and recovered upon recirculation. During the further course of the reperfusion period, LDF-CBF decreased as edema formation proceeded (Fig. 2). Glu tended to increase in the perifocal region during

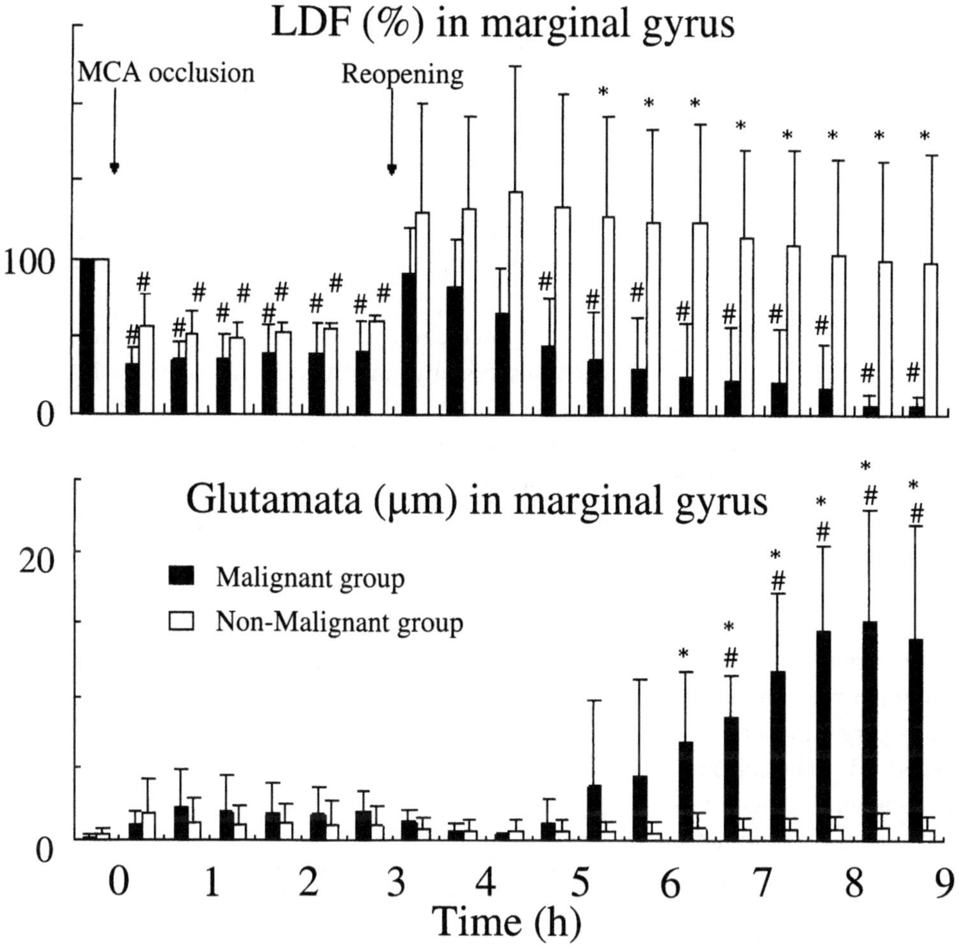

Fig. 2. Changes in LDF-CBF and glutamate in a perifocal region (marginal gyrus) in the 2 groups exhibiting a non-malignant (white bars; n = 5) and a malignant (black bars; n = 5) course of infarction as a function of time. Mean ± SD values are plotted every half hour. *P < 0.05, significantly different from preischemic control; #P < 0.05, significantly different from group exhibiting a malignant course

MCAO, but this increase was not significant (Fig. 2). During reperfusion, Glu increased in the perifocal region as LDF-CBF decreased. In the non-malignant group, LDF-CBF decreased during MCAO in the perifocal region to around 55% of control. Recirculation resulted first in hyperperfusion and thereafter in a normalization of LDF-CBF. Glu did not increase during reperfusion. During MCAO, Glu did not differ significantly between both groups (Fig. 2). After reperfusion, Glu had a threshold-like relationship with CPP. It was elevated when CPP decreased below about 50 mmHg (Fig. 3).

Discussion

Various parameters monitored in the course of ischemia and reperfusion provide good correlates and are therefore good markers of a malignant course. The increase of ICP and decrease of CPP along with the delayed, secondary glutamate rise during reperfusion in core regions of focal ischemia have such marker function [15]. The present study documents that the secondary rise of Glu is not only observed in core but also in border regions of focal ischemia. This result is not surprising since ICP elevation and resulting decline of CPP affect the whole brain and are therefore also recognized in brain regions outside the ischemic focus. Glu elevation after reperfusion in the investigated perifocal regions was caused by the drastic decrease of CPP below 50 mmHg resulting from excessive edema formation.

In view of the critical condition that is generated by malignant edema formation, the more relevant question is, however, whether monitored variables have a

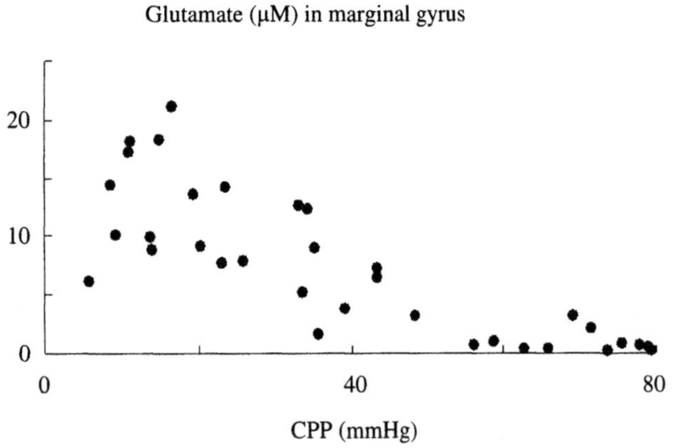

Fig. 3. Relation between CPP and glutamate elevation in the group exhibiting a malignant course of infarction. Data were obtained from the phase after reperfusion that followed the primary recovery. Thirty-minute data were plotted. Note the LDF threshold for extracellular glutamate elevation at a CPP of 40 to 50 mm Hg

prognostic value to predict the malignant course as early as possible. It has for example been shown in a rat model, that decompressive surgery is most effective if started early after reperfusion [5]. Neurochemical monitoring in core regions has shown that such early predictor function is provided best by a pronounced Glu rise (and rise of other amino acids) obtained in an early stage already during ischemia. This elevation of Glu was significantly higher in cats that exhibited a malignant course in the later phase of the reperfusion period compared to cats with a non-malignant course [15].

Thus, we think that neurochemical monitoring in perifocal regions presumably does not provide predictive patterns at early time points. Early prediction seems to be achieved only, if probes can be located within the ischemic core. Such locations would have the additional advantage that traumatization of remote regions by implantation of the probes could be avoided. In view of clinical conditions, however, early monitoring with microdialysis is difficult, because several criteria such as a clear indication by CT, or the agreement of patients or of relatives are prerequisites for the timely application of this invasive technique. The decision on the site of microdialysis probe implantation in patients seems also problematic. In the few microdialysis studies in stroke patients performed so far, remote locations have been chosen because these regions are non-eloquent and routinely used. An original case report with fatal stroke showed increased glutamate and glycerin and lactate/pyruvate ratio levels in the contralateral hemisphere 24 hours before ICP rise and subsequent brain death [1]. In a more recent study on ten patients that applied microdialysis in the immediate vicinity of the infarct, results were more variable, and the time lag between substrate elevation and appearance of herniation was rather short or did not exist [12].

As evident from positron emission tomography experiments, studies applying modern imaging techniques would presumably provide best requirements for early prediction of fatal stroke [15]. Attempts have been successfully undertaken by applying diffusion-weighted MRI [10] and early cerebral CT scanning for assessment of attenuated corticomedullary contrast [7], and flumazenil positron emission tomography [8]. Timely application of neuroimaging techniques is often difficult, and sequential measurements are not readily performed in the clinic. Thus, a combination with invasive monitoring techniques like ICP measurement or microdialysis seems favorable, particularly, if implantation of invasive techniques is guided by early neuroimaging.

References

1. Berger C, Annecke A, Aschoff A, Spranger M, Schwab S (1999) Neurochemical monitoring of fatal middle cerebral artery infarction. Stroke 30: 460–463
2. Berger C, Schabitz WR, Georgiadis D, Steiner T, Aschoff A, Schwab S (2002) Effects of hypothermia on excitatory amino acids and metabolism in stroke patients: a microdialysis study. Stroke 33: 519–524
3. Bullock R, Zauner A, Woodward J, Young HF (1995) Massive persistent release of excitatory amino acids following human occlusive stroke. Stroke 26: 2187–2189
4. Enblad P, Frykholm P, Valtysson J, Silander HC, Andersson J, Fasth KJ, Watanabe Y, Langstrom B, Hillered L, Persson L (2001) Middle cerebral artery occlusion and reperfusion in primates monitored by microdialysis and sequential positron emission tomography. Stroke 32: 1574–1580
5. Engelhorn T, Doerfler A, Kastrup A, Beaulieu C, de Crespigny A, Forsting M, Moseley ME, Faraci FM (1999) Decompressive craniectomy, reperfusion, or a combination for early treatment of acute "malignant" cerebral hemispheric stroke in rats? Potential mechanisms studied by MRI. Stroke 30: 1456–1463
6. Hacke W, Schwab S, Horn M, Spranger M, De Georgia M, von Kummer R (1996) 'Malignant' middle cerebral artery territory infarction: clinical course and prognostic signs. Arch Neurol 53: 309–315
7. Haring HP, Dilitz E, Pallua A, Hessenberger G, Kampfl A, Pfausler B, Schmutzhard E (1999) Attenuated corticomedullary contrast: An early cerebral computed tomography sign indicating malignant middle cerebral artery infarction. A case-control study [see comments]. Stroke 30: 1076–1082
8. Heiss W-DD, C Schneweis, S Staub, F Grond, M Graf, R (2001) Neuromonitoring in patients with large hemispheric stroke: Microdialysis in combination with PET-Imaging. J Cereb Blood Flow Metab 21: S107
9. Hillered L, Hallstrom A, Segersvard S, Persson L, Ungerstedt U

(1989) Dynamics of extracellular metabolites in the striatum after middle cerebral artery occlusion in the rat monitored by intracerebral microdialysis. J Cereb Blood Flow Metab 9: 607–616
10. Oppenheim C, Samson Y, Manai R, Lalam T, Vandamme X, Crozier S, Srour A, Cornu P, Dormont D, Rancurel G, Marsault C (2000) Prediction of malignant middle cerebral artery infarction by diffusion-weighted imaging. Stroke 31: 2175–2181
11. Rudolf J, Grond M, Stenzel C, Neveling M, Heiss WD (1998) Incidence of space-occupying brain edema following systemic thrombolysis of acute supratentorial ischemia. Cerebrovasc Dis 8: 166–171
12. Schneweis S, Grond M, Staub F, Brinker G, Neveling M, Dohmen C, Graf R, Heiss WD (2001) Predictive value of neurochemical monitoring in large middle cerebral artery infarction. Stroke 32: 1863–1867
13. Shimada N, Graf R, Rosner G, Wakayama A, George CP, Heiss WD (1989) Ischemic flow threshold for extracellular glutamate increase in cat cortex. J Cereb Blood Flow Metab 9: 603–606
14. Taguchi J, Graf R, Rosner G, Heiss WD (1996) Prolonged transient ischemia results in impaired CBF recovery and secondary glutamate accumulation in cats. J Cereb Blood Flow Metab 16: 271–279
15. Toyota S, Graf R, Valentino M, Yoshimine T, Heiss WD (2002) Malignant infarction in cats after prolonged middle cerebral artery occlusion: glutamate elevation related to decrease of cerebral perfusion pressure. Stroke 33: 1383–1391

Correspondence: Dr. Rudolf Graf, Max-Planck Institut für neurologische Forschung, Gleueler Strasse 50, 50931 Köln, Germany. e-mail: rudolf.graf@pet.mpin-koeln.mpg.de

Acta Neurochir (2003) [Suppl] 86: 159–162

A comparison of long-term neurological symptoms after two different focal ischemic models in Mongolian gerbils

S. Ishibashi[1], **T. Kuroiwa**[2], **S. Endo**[3], **R. Okeda**[2], and **H. Mizusawa**[1]

[1] Department of Neurology and Neurological Science, Graduate School of Medicine, Tokyo Medical and Dental University, Tokyo, Japan
[2] Department of Neuropathology, Graduate School of Medicine, Tokyo Medical and Dental University, Tokyo, Japan
[3] Animal Research Center, Graduate School of Medicine, Tokyo Medical and Dental University, Tokyo, Japan

Summary

The aim of the present study was to elucidate the relationship between specific tasks and the responsible ischemic lesions after experimentally induced cerebral hemispheric ischemia in Mongolian gerbils. We used the elevated body swing test (EBST) to evaluate asymmetry motor behavior, the bilateral asymmetry test (BAT) to evaluate sensory dysfunction, and the T-maze test to assess cognitive deficit during 4 weeks after either 10-min single unilateral carotid artery occlusion (sCCAO) or 10-min repeated unilateral occlusion (rCCAO). rCCAO produced persistent sensorimotor and severe cognitive deficits. Infarction was confined to the ipsilateral cerebral cortex, hippocampus, dorsolateral nucleus of thalamus and caudate nucleus. sCCAO did not induce sensorimotor abnormalities, but did produce mild cognitive deficits; these animals also showed increased locomotor activity during the early post-ischemic period. In sCCAO, neuronal death was confined to the ipsilateral CA1 sector of the hippocampus. Thus, unilateral CA1 neuronal death was sufficient for evolution of cognitive deficits in this model of experimental focal ischemia. Ischemic lesions extending to the ipsilateral cerebral cortex, basal ganglia, and hippocampus produced persistent and severe sensorimotor impairment accompanying severe cognitive deficits. These findings regarding region-specific behavioral tasks in cerebral ischemia will facilitate improved assessment of stroke therapy.

Keywords: Stroke; gerbils; behavior; cognition; CA1; locomotor activity.

Introduction

Animal models have provided valuable insight into the pathophysiology of stroke and have played an influential role in the development of treatment strategies. Many behavioral testing methods have been used to assess the long-term neurological functions in ischemic rodent models. With the recent advance for stroke models, it has become more important to investigate the relationship between region-specific behavioral tasks and the extent and severity of ischemic regions to evaluate therapeutic potential effectively. However, it is difficult to establish the region-specific behavioral tasks in focal ischemic rodent models because of the complex pattern of sensorimotor and cognitive dysfunctions resulting from widespread induced ischemic lesions. Comparing the behavioral outcomes of different durations of cerebral artery occlusion likely will be important to identify region-specific behaviors of focal ischemic models.

In Mongolian gerbils, depending on the duration of ischemia, the extent of an ischemic lesion after temporary carotid artery occlusion varies from selective neuronal death of the CA1 sector to extensive hemispheric infarction [3]. Using a unilateral carotid occlusion model, we aimed to elucidate the relationship between the distribution of ischemic tissue injury and characteristics of the associated neurological dysfunctions.

Materials and methods

In this experiment, we used 65 Mongolian gerbils of both sexes (age, 16 to 20 weeks; weight, 60 to 72 g). The animals first were divided into two groups: common carotid artery-occluded (CCAO) and sham-operated animals. For the induction of ischemia, the animals were anesthetized with 2% isoflurane, the left common carotid artery was occluded with a mini vascular clip, the animals were allowed to recover from anesthesia, and the clip was removed after 10 min of occlusion. Stroke symptoms were evaluated according to a stroke index (SI) of 0–25 points as reported previously [5]. Animals manifesting a SI score of more than 10 were selected as "postischemic animals," which we then allocated into two groups receiving single sCCAO or repeated rCCAO. The rCCAO animals underwent a second, similarly induced 10-min period of ischemia 5 h after the first occlusion. Sham-operated animals were operated the same except for left carotid artery occlusion.

sCCAO, rCCAO, and sham-operated animals underwent a series of behavioral tests, which were conducted before surgery and on days 4, 7, 14, 21, and 28 post-stroke. The researcher conducting the behavioral testing was blind to the experimental condition.

The elevated body swing test (EBST) was used to evaluate asymmetric motor behavior. Animals were held by the base of the tail and elevated about 5 cm from the tabletop. The direction of body swing, defined as an upper body turn of >10 degrees to either side, was recorded for 1 minute during each of three trials per day. The numbers of left and right turns were counted, and the percent of turns made to the side contralateral to the stroked hemisphere (% right-biased swing) was determined.

The bilateral asymmetry test (BAT) is a test of unilateral sensory dysfunction. Two small pieces of adhesive-backed paper dots (of equal size 60 mm^2 each) were used as bilateral tactile stimuli occupying the distal-radial region on the wrist of each forelimb. The time (maximum, 3 min) to remove each stimulus from the forelimb (the "removal time") was recorded for each of three trials per day.

The T-maze spontaneous alternation task is a method to test exploratory behavior and working memory [2]. Animals were allowed to alternate between the left and right goal arms of a T-shaped maze, which measured 60 (stem) × 25 (arm) × 10 (width) cm, throughout a 15-trial continuous alternation session. Once they have entered a particular goal arm, a door was lowered to block entry to the opposite arm. The door was re-opened only after animals returned to the start arm, thus allowing a new alternation trial to be started. The spontaneous alternation rate was calculated as the ratio between the alternating choices and total number of choices (random choice, 50%; alternation at every trial, 100%; no alternation, 0%). We also recorded the time to complete the 15 trials (complete time).

At the end of the observation period, animals were deeply anesthetized with diethylether, sacrificed, and fixed by perfusion with 4% paraformaldehyde. The brain was cut into 6 serial coronal sections (thickness 2.0 mm) from the level of the anterior pole of the caudate nucleus to the posterior pole. The sections were immersion-fixed in buffered formalin for 1 day and embedded in paraffin. Each of the 6 coronal sections was sliced in thickness of 6 μm and stained with Klüver-Barrera (Luxol fast blue and cresyl violet) for evaluation of the ischemic lesions.

The data were analyzed using repeated measures analysis of variance (ANOVA) with the independent variables as treatment group and day of testing. Statistical comparison at each measurement day between groups was made by the unpaired *t*-test. The level of statistical significance was set at $P < 0.05$. All values are presented as the mean ± SD.

Results

sCCAO (n = 9), rCCAO (n = 10) and sham-operated animals (n = 10) were selected for behavioral testing and histological analysis.

Behavioral tests

In the EBST, the % right-biased swing was significantly higher in rCCAO animals compared with sCCAO and sham-operated animals throughout the 4 weeks postischemia. rCCAO animals showed a strong and persistient tendency to turn their upper bodies toward the side contralateral to the stroked hemisphere, whereas sCCAO and sham-operated animals showed no bias (Fig. 1A).

In the BAT, the removal time for the contralateral forelimb was significantly longer in rCCAO animals compared with sCCAO and sham-operated animals throughout the 4 weeks post-stroke. rCCAO animals showed a longer latency to remove a tape of the side contralateral to the stroked hemisphere (Fig. 1B). sCCAO and sham-operated animals showed no difference in behavior from that before occlusion.

Before occlusion, animals tended to choose a goal arm alternatively and chose left or right with equal frequency in the T-maze. However, the spontaneous alternation rate in sCCAO animals was significantly lower than the rate in the sham-operated animals, whose rate was approximately random (50%; Fig. 1C). In rCCAO animals, this rate was further decreased below the ratio of random choice. rCCAO tended to choose only the left goal arm after ischemia. In addition, the time to complete 15 trials was significantly shorter in sCCAO animals than in rCCAO and sham-operated animals (Fig. 1D).

Histology

Sham-operated controls showed no ischemic change. In sCCAO animals, selective neuronal death was confined only to the ipsilateral CA1 sector of the hippocampus (Fig. 2A). No infarction was detectable. In rCCAO animals, infarction was confined to the ipsilateral cerebral cortex, hippocampus, dorsolateral nucleus of thalamus and caudate nucleus (Fig. 2B). In the rostral cortex, infarction involved the entire cortical layer, whereas in the caudal cortex, it was laminar and confined to the middle cortical layer. A zone of selective neuronal death surrounded the infarction.

Discussion

Our experiment revealed important differences in the behavioral changes that follow sCCAO and rCCAO in gerbils. To summarize: (1) locomotor hyperactivity was inducible by unilateral CA1 injury, (2) sensorimotor deficits were induced by the additional ischemic lesion of the cortex and basal ganglia, and (3) cognitive dysfunction was associated with both hippocampal and non-hippocampal lesions.

Indicative of their postischemic locomotor hyperactivity, sCCAO animals required significantly less time to complete 15 trials in a T-maze than did sham-

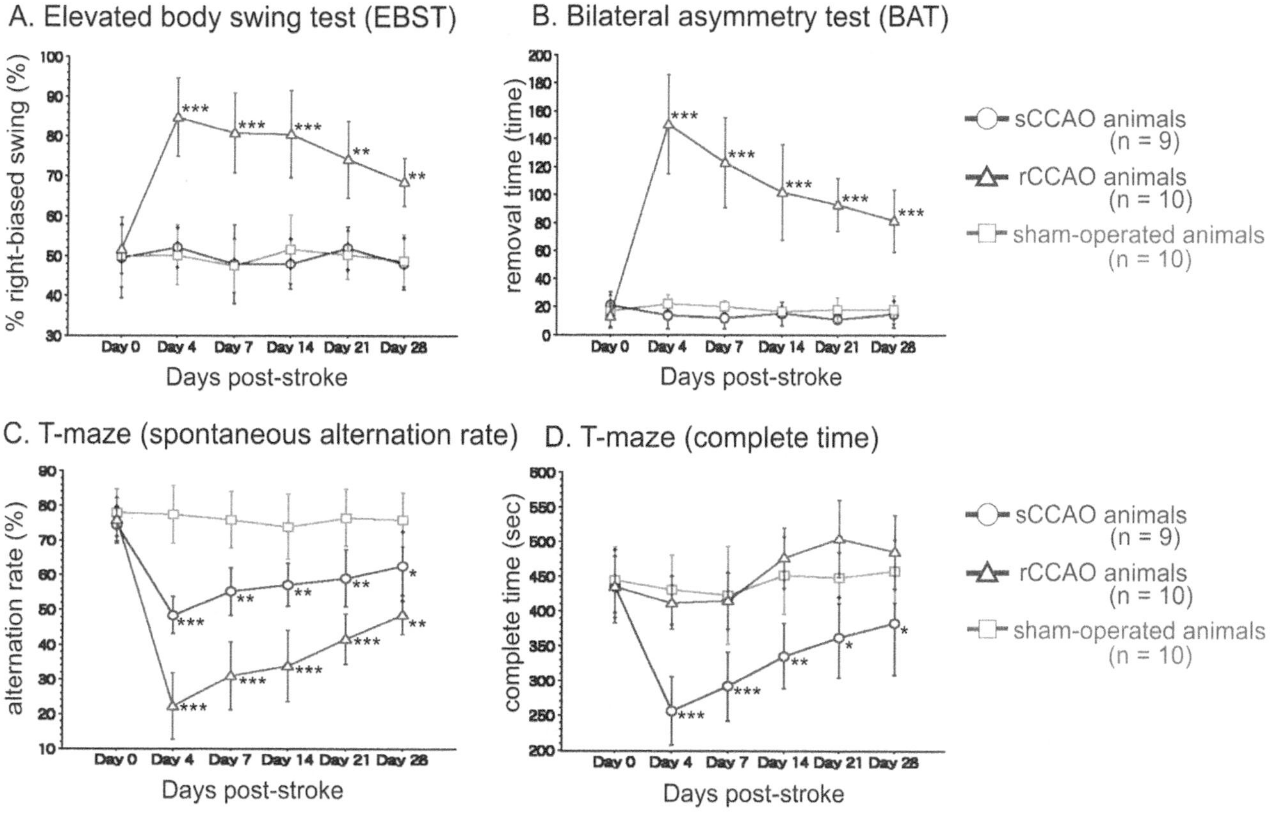

Fig. 1. Time course of behavioral tests from sham-operated, sCCAO, and rCCAO animals for 28 day period. (A) % right biased swing (Elevated body swing test). rCCAO animals showed a strong and persisting tendency to turn their upper bodies to the side contralateral to the stroked hemisphere. (B) Removal time (Bilateral asymmetry test). rCCAO animals took a longer latency to remove a tape of the side contralateral to the stroked hemisphere. (C) Spontaneous alternation rate (T-maze). The spontaneous alternation rate in sCCAO animals was decreased at the random alternation choice. rCCAO animals showed further decreased spontaneous alternation rate. (D) Complete time (T-maze). Complete time was significantly shorter in sCCAO animals than rCCAO and sham-operated animals

operated and rCCAO animals. Postischemic locomotor hyperactivity previously has been reported to occur in several animal models of cerebral ischemia [4, 6]. Results of a previous study on the histological changes after bilateral common carotid artery occlusion in gerbils indicated that CA1 neuronal loss is sufficient for the induction of locomotor hyperactivity [4]. The mechanism might be related to the excitotoxicity of postischemic neural tissue, because the CA1 sector after transient global ischemia has a high extracellular glutamate level [1]. However, locomotor hyperactivity was not observed in our rCCAO animals despite severe ischemic injury in the CA1 pyramidal layers. Multiple cerebral regions therefore are responsible for postischemic locomotor activity. Evidence in the rat after focal ischemia indicates the influence of laterality on postischemic locomotor hyperactivity [6]. We also observed a substantial influence of laterality on the postischemic locomotor hyperactivity in the present model. Further study is needed to elucidate the mechanism of postischemic locomotor hyperactivity.

The spontaneous alternation rate in the T-maze has been reported to reflect exploratory behavior and spatial memory, which enable animals to acquire information about novel places and things [2]. The hippocampus is widely recognized to have an important role in learning and memory. It is therefore reasonable to assume that animals with intact CA1 function have high spontaneous alternation ratios, reflecting good recent memory, and animals with CA1 injury show an alternation ratio of approximately 50% because of the recent memory disturbances [2]. In fact, animals with CA1 lesions in a previous study showed a decrease in the spontaneous alternation rate, to approximately the 50% level, as did our sCCAO group. However, in our rCCAO animals, the spontaneous alternation ratio decreased below 50%. At present, we have no additional findings to explain this interesting change,

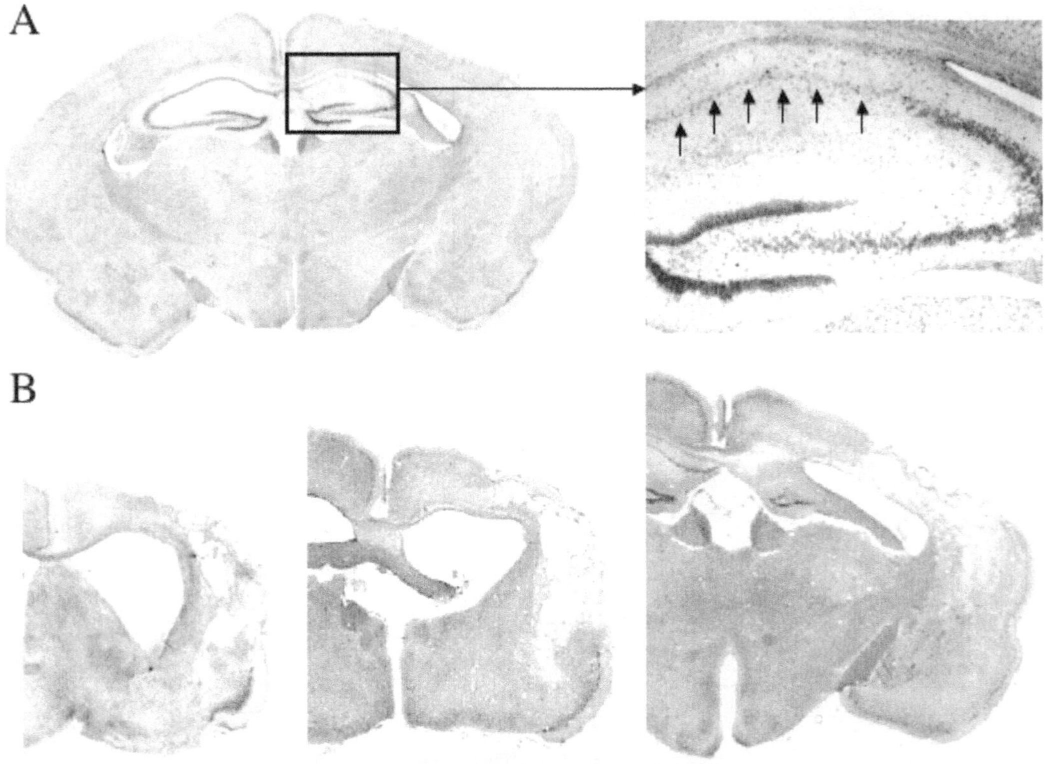

Fig. 2. Photographic display of representative K.B.-stained coronal brain sections taken from sCCAO (A) and rCCAO (B) animals 28 days after ischemia. (A) Selective neuronal death in the CA1 sector of unilateral hippocampus (arrows). (B) Infarction was confined to the ipsilateral cerebral cortex, hippocampus, dorsolateral nucleus of thalamus, and caudate nucleus

and further investigation is necessary to elucidate the mechanisms of this reduction.

Overall, our present findings regarding the long-term neurological deficits and changes in region-specific behavioral tasks after cerebral ischemia likely will be important for improved assessment of stroke therapy.

References

1. Drejer J, Benveniste H, Diemer NH, Schousboe A (1985) Cellular origin of ischemia induced glutamate release from brain tissue in vivo and in vitro. J Neurochem 45: 145–151
2. Gerlai R (1998) A new continuous alternation task in T maze detects hippocampal dysfunction in mice: A strain comparison and lesion study. Behav Brain Res 95: 91–101
3. Ito U, Spatz M, Walker JT, Klatzo I (1975) Experimental cerebral ischemia in Mongolian gerbils. Acta Neuropath (Berl) 32: 209–223
4. Kuroiwa T, Bonnekoh P, Hossmann KA (1991) Locomotor hyperactivity and hippocampal CA1 injury after transient forebrain ischemia of gerbils. Neurosci Lett 122: 141–144
5. Ohno K, Ito U, Inaba Y (1984) Regional cerebral blood flow and stroke index after left carotid artery ligation in the conscious gerbil. Brain Res 297: 151–157
6. Robinson RG (1979) Differential behavioral and biochemical effects of right and left hemispheric cerebral infarction in the rat. Science 205: 707–710

Correspondence: Dr. T. Kuroiwa, Department of Neuropathology, Tokyo Medical and Dental University, 1-5-45, Yushima, Bunkyo-ku, Tokyo 113-8519, Japan. e-mail: t.kuroiwa.npat@mri.tmd.ac.jp

Acta Neurochir (2003) [Suppl] 86: 163–166

Thrombin exacerbates brain edema in focal cerebral ischemia*

Y. Hua[1], J. Wu[1], R. F. Keep[1,2], J. T. Hoff[1], and G. Xi[1]

[1] Department of Neurosurgery, University of Michigan, Ann Arbor, Michigan, U.S.A.
[2] Department of Physiology, University of Michigan, Ann Arbor, Michigan, U.S.A.

Summary

Thrombin contributes to edema formation after intracerebral hemorrhage. Recent studies suggest that thrombin may also play a role in ischemic brain damage. In the present study, adult male Sprague-Dawley rats were anesthetized with pentobarbital. Middle cerebral artery (MCA) was occluded using the suture method. We found that brain thrombin activity was elevated after permanent MCA occlusion as was prothrombin messenger RNA expression. Intracerebral injection of a thrombin inhibitor, hirudin, reduced neurological deficits following cerebral ischemia. In contrast, intracerebral administration of exogenous thrombin (at a dose that is non-toxic to normal brain), markedly exacerbated brain edema after transient focal cerebral ischemia. These results indicate that extravascular thrombin inhibition may be a new therapeutic target for cerebral ischemia.

Keywords: Thrombin; brain edema; cerebral ischemia; neurological deficits; rats.

Introduction

Cerebral ischemia kills about 160,000 people a year in the United States and is also the major cause of neurological disability. There are currently no effective treatments except thrombolytic recanalization with tissue plasminogen activator (tPA) within three hours of the onset of symptoms [19]. Many different mechanisms have been implicated in the secondary brain injury after cerebral ischemia [9]. Thrombin contributes to edema formation after intracerebral hemorrhage [13, 21], and recent evidence suggests that it may play a role in ischemic brain injury [17]. Thus, thrombin receptors are upregulated in brain slices after oxygen-glucose deprivation and hirudin, a thrombin inhibitor, reduces infarct volume in gerbils. The cerebrovasculature becomes permeable to plasma proteins within several hours after ischemia [1] and this would allow the entry of prothrombin into brain parenchyma although the brain itself may be an alternate source.

The present study investigates whether focal cerebral ischemia increases brain thrombin concentrations, whether a thrombin inhibitor can reduce ischemic brain damage and whether ischemic brain is more susceptible to thrombin-induced injury than normal brain.

Materials and methods

Animal preparation

Adult male Sprague-Dawley rats (275–325 g) were anesthetized with pentobarbital (40 mg/kg, i.p.). After anesthesia was achieved, a polyethylene catheter (PE-50) was inserted into the right femoral artery in order to monitor arterial blood pressure and to obtain blood for analysis of blood gases, blood pH, hematocrit, blood glucose concentration. Body temperature was maintained at 37.5 °C by using a feedback-controlled heating pad.

Animal groups

Middle cerebral artery (MCA) was occluded using the suture method. Three sets of experiments were performed. In the first set, rats underwent 4 or 24 hrs of permanent occlusion and the cortex was then divided in core, intermediate, and outer zone samples (referenced to the MCA territory [5, 15]). A thrombin-specific chromogenic substrate, S2238, was used to assay thrombin activity while prothrombin mRNA was semi-quantitated using reverse transcription-polymerase chain reaction (RT-PCR). In the second set of experiments, rats underwent 60 minutes of transient MCA occlusion. Hirudin (5 U) was injected into the ipsilateral caudate (coordinates: 0.2 mm anterior, 5.5 mm ventral, and 4.0 mm lateral to the bregma) just after MCAO. Control rats received saline injection. Neurological deficits (forelimb placing) were evaluated three days later. In the third set of experiments, rats received an intracerebral infusion of 1 U of thrombin or saline while undergoing either 30 minutes of transient MCA occlusion or no occlusion. Brain edema was determined after 24 hours.

* Supported by Grant-In-Aid 0255721N (GX) from American Heart Association and grants NS-17760 (JTH), NS-39866 (GX) from the National Institutes of Health.

Middle cerebral artery occlusion

The MCA was occluded using the suture method of Zea Longa [14]. Briefly, the bifurcation of the common carotid artery was exposed, the external carotid artery is ligated distally. A 3-0 monofilament nylon suture, its tip rounded by heating, was introduced into the internal carotid artery lumen through the stump of the external carotid artery and gently advanced into the internal carotid artery 20–22 mm past the common carotid artery bifurcation to block the origin of MCA.

Thrombin activity measurement

For thrombin activity measurements, rat brains were perfused transcardially with saline. The brains were then removed, the hemisphere was divided, and the brain stem, hippocampus, and thalamus were discarded. The remaining telencephalic tissues ipsi- and contralateral to the MCAO were placed flat on a nonabsorbent surface and the portion inferior to the rhinal fissure was removed. Three samples (ipsi- or contralateral) were taken from the remaining tissue using 5- and 7-mm cork borers [5, 15]. The core region was taken from the cortex and striatum underlying the MCA immediately distal to the region of occlusion. The intermediate region was a ring of tissue surrounding the core zone, whereas the outer region consisted of the remainder of the cortical tissue. Brain samples were homogenized and thrombin activities were measured using the thrombin-specific chromogenic substrate S2238 (Chromogenix). The final concentration of S2238 was 0.3 mmol/L in PBS and the absorption at 405 nm of the supernatant was after measured one hour. A calibration curve of thrombin activity was made for the thrombin activity quantitation [16]. Phenyl-propyl-arginyl-chloromethyl ketone (PPACK) was used for thrombin inhibition.

Reverse Transcription-Polymerase Chain Reaction (RT-PCR) analysis

Prothrombin mRNA was determined using semiquantitative RT-PCR [11]. Prothrombin sense 5′-TCCTCGCTTGGTGTCATTC; prothrombin anti-sense 5′-CAAGCACTCCAGAACCAGA (Operon Technologies Inc, Alameda, CA). Brain total RNA was extracted with Trizl reagent and cDNA was synthesized by reverse transcription (RT). PCR product was electrophretically separated in a 1% agarose gel, stained with ethidium bromide, visualized under UV light, and photographed. PCR products were normalized using glyceraldehyde-3-phosphate dehydrogenase (GAPDH) PCR product. Data are expressed as pixels.

Forelimb placing test

The vibrissae-elicited forelimb-placing test [10] was used to assess ischemic brain damage. Animals were held by their torsos allowing forelimb to hang free. While holding the animal, the experimenter made gentle up and down movement in space prior to place testing. This facilitated muscle relaxation and eliminated any struggling movement. Independent testing of each forelimb was induced by brushing the respective vibrissae on the edge of a table top once per trial for 10 trials. Intact animals place the forelimb of both sides quickly onto the countertop. A score of one was given watch time the rat placed its forelimb on the edge of the tabletop in response to the vibrissae stimulation. Percent successful placing responses were determined [10].

Brain water content

Rats were sacrificed by decapitation under deep anesthesia. Tissue samples were weighed on an electronic analytical balance to obtain the wet weight (WW). The tissue was then dried in a gravity oven at 100 °C for more than 24 hours to determine the dry weight (DW). Tissue water contents (%) were calculated as ((WW − DW)/WW)*100.

Statistical analyses

All values were reported as the mean ± SD. Data were analyzed by ANOVA with a Scheffe's multiple comparisons test or by Student t test. Differences were considered significant at the $p < 0.05$ level.

Results

All physiological variables including mean arterial blood pressure, pH, PaO_2, $PaCO_2$, hematocrit, and blood glucose were within normal ranges.

Twenty-four hours after permanent MCA occlusion, thrombin activity was increased in the ipsilateral core zone (0.6 ± 0.2 vs. 0.3 ± 0.1 U/gram brain tissue in the contralateral, $n = 6$, $p < 0.05$). Prothrombin mRNA expression was also increased in the ipsilateral core zone after 4 hours of permanent occlusion and in the ipsilateral intermediate and outer zones after 24 hours of occlusion.

Sixty minutes of transient MCAO caused marked forelimb placing deficits at day three after MCA occlusion (Fig. 1). These deficits were reduced in animals treated with the thrombin inhibitor, hirudin (Fig. 1).

In sham-occluded rats, infusion of saline or 1 U of thrombin did not cause significant edema formation as assessed by the difference in water content between ipsilateral and contralateral sides. Thirty minutes of

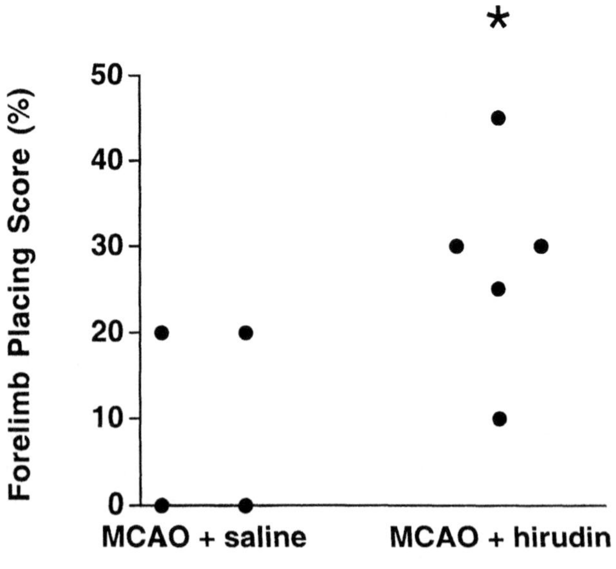

Fig. 1. Forelimb placing score (percentage of successful placements by the impaired limb) at day 3 after 60 minutes transient MCAO with or without hirudin treatment; *$p < 0.05$ vs. MCAO + saline

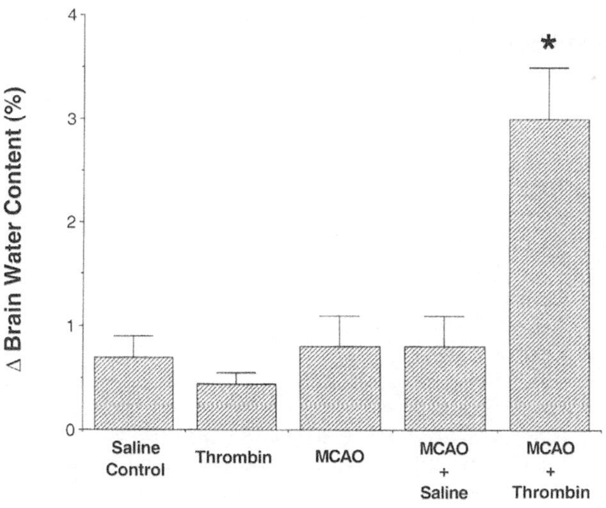

Fig. 2. The difference of brain water content (%) between the ipsilateral and the contralateral basal ganglia. In the first and second groups, the rats had received an intracerebral injection of either 50-μl saline (saline control) or one unit thrombin in 50 μl saline (thrombin) into the left basal ganglia. In the third, fourth and fifth groups, the rats had either 30 min of transient middle cerebral artery occlusion (MCAO), 30 min MCAO plus 50 μl saline intracerebral infusion (MCAO + Saline), or 30 min MCAO plus one unit thrombin infusion (MCAO + Thrombin). All rats were killed at 24 hours for brain edema measurement. Values are expressed as mean ± SD. n = 6. *$p < 0.05$ vs. the other groups. Contralateral water contents were not different among the five groups

transient MCA occlusion with saline injection also did not result in significant edema at 24 hours (the difference = 0.8 ± 0.3%, n = 6). In contrast, rats that underwent 30 min of transient MCA occlusion and received 1 U of thrombin had marked edema formation (the difference = 3.0 ± 0.5%; n = 6, $p < 0.05$ vs. other groups; Fig. 2).

Discussion

Our present study demonstrates that brain thrombin activity and prothrombin gene expression are upregulated after MCA occlusion in the rat. This, along with our previous results on the role of thrombin in hemorrhagic brain injury [13, 21], suggests that thrombin might have a role in ischemic brain damage. This hypothesis is supported by our finding that a thrombin inhibitor attenuates neurological deficits after transient MCA occlusion, and that exogenous thrombin can exacerbate ischemic brain.

Thrombin is a serine protease and an essential component in the coagulation cascade. It is produced immediately in the brain after intracerebral hemorrhage, brain trauma or blood-brain barrier breakdown following many kinds of brain injury [8]. We have demonstrated that thrombin is responsible for early brain edema formation following intracerebral hemorrhage [13, 21]. There is growing direct evidence suggesting that thrombin may play crucial roles in brain injury after cerebral ischemia. Intraventricular infusion of hirudin, a specific inhibitor of thrombin, increases the survival of hippocampal CA1 neurons after global cerebral ischemia in gerbils [17]. This hypothesis is supported by our data demonstrating that hirudin reduced MCA occlusion-induced neurological deficits.

Thrombin, at high concentrations kills neurons and astrocytes in vitro [12]. However, the thrombin concentrations found during ischemia do not appear to reach those levels. A potential reason for this discrepancy is that much lower concentrations of thrombin may cause cell death in stressed cells. Thus, Weinstein *et al.*, examining Ad12 HER 10 cells, found that thrombin concentration as low as 1.2×10^{-3} U/ml caused cell death in hypoglycemic (0.1 mM glucose) but not normoglycemic conditions [20]. In this study, we demonstrate that a thrombin injection that would cause no edema in a normal brain causes marked edema in a brain that has been subjected to a brief period of ischemia.

Although the primary role of thrombin in hemostasis is through cleaving fibrinogen to fibrin and inducing platelet aggregation, other important cellular activities of thrombin may be related to thrombin receptor activation. Three protease-activated receptors (PARs), PAR-1, PAR-3 and PAR-4, have been identified as thrombin receptors [4]. PARs are markedly upregulated in hippocampus after oxygen-glucose deprivation in vitro [18].

In animal models at least, glutamate-induced excitotoxicity is an important mechanism in ischemic brain injury [2]. For example, MK 801, the N-methyl-D-aspartate (NMDA) antagonist, reduces brain injury in many studies of focal cerebral ischemia. Activation of PAR-1 potentiates NMDA receptor responses in CA1 pyramidal cells and the effect of thrombin can be mimicked by a thrombin receptor agonist (SFLLRN). Thrombin-induced NMDA receptor potentiation is reduced in PAR-1 knockout mice [7].

The brain as well as blood may be a site of thrombin production. In vitro studies have shown that prothrombin mRNA is expressed in the cells of the nervous system [6]. Prothrombin mRNA is upregulated after spinal cord injury [3]. Our present study indicates that there is an increase of prothrombin expression in the brain after cerebral ischemia. These results suggest that thrombin may be formed and cause brain injury even if the blood-brain barrier is intact.

In summary, both brain thrombin activity and brain prothrombin messenger RNA are elevated after permanent MCA occlusion. Hirudin can reduce ischemia-induced neurological deficits and cells stressed by ischemia appear to be more susceptible to thrombin-induced injury. In combination, these results suggest that thrombin is involved in ischemic brain damage and that modulating extravascular thrombin using serine protease inhibitors or using specific thrombin inhibitors may form novel strategies for stroke treatment.

References

1. Betz AL, Dietrich WD (1998) Blood-brain barrier dysfunction in cerebral ischemia. In: Ginsberg MD, Bogousslavsky J (eds) Cerebrovascular disease: pathophysiology, diagnosis, and management. Blackwell Science, pp 358–370
2. Choi DW (1997) The excitotoxic concept. In: Welch KMA, Caplan LR, Reis DJ *et al* (eds) Primer on cerebrovascular diseases. Academic Press, San Diego, pp 187–190
3. Citron BA, Smirnova IV, Arnold PM, Festoff BW (2000) Upregulation of neurotoxic serine proteases, prothrombin, and protease-activated receptor 1 early after spinal cord injury. J Neurotrauma 17: 1191–1203
4. Coughlin SR (2000) Thrombin signalling and protease-activated receptors. Nature 407: 258–264
5. Dickinson LD, Betz AL (1992) Attenuated development of ischemic edema in vasopressin-deficient rats. J Cereb Blood Flow Metab 12: 681–690
6. Dihanich M, Kaser M, Reinhard E, Cunningham D, Monard D (1991) Prothrombin mRNA is expressed by cells of the nervous system. Neuron 6: 575–581
7. Gingrich MB, Junge CE, Lyuboslavsky P, Traynelis SF (2000) Potentiation of NMDA receptor function by the serine protease thrombin. J Neurosci 20: 4582–4595
8. Gingrich MB, Traynelis SF (2000) Serine proteases and brain damage – is there a link? Trends in Neurosciences 23: 399–407
9. Ginsberg MD, Bogousslavsky J (1998) Cerebrovascular disease: pathophysiology, diagnosis, and management. Blackwell Science
10. Hua Y, Schallert T, Keep RF, Wu J, Hoff JT, Xi G (2002) Behavioral tests after intracerebral hemorrhage in the rat. Stroke 33: 2478–2484
11. Hua Y, Xi G, Keep RF, Wu J, Jiang Y, Hoff JT (2002) Plasminogen activator inhibitor-1 induction after experimental intracerebral hemorrhage. J Cereb Blood Flow Metab 22: 55–61
12. Jiang Y, Wu J, Hua Y, Keep RF, Xiang J, Hoff JT, Xi G (2002) Thrombin-receptor activation and thrombin-induced brain tolerance. J Cereb Blood Flow Metab 22: 404–410
13. Lee KR, Colon GP, Betz AL, Keep RF, Kim S, Hoff JT (1996) Edema from intracerebral hemorrhage: the role of thrombin. J Neurosurg 84: 91–96
14. Longa EZ, Weinstein PR, Carlson S, Cummins R (1989) Reversible middle cerebral artery occlusion without craniectomy in rats. Stroke 20: 84–91
15. Martz D, Beer M, Betz AL (1990) Dimethylthiourea reduces ischemic brain edema without affecting cerebral blood flow. J Cereb Blood Flow Metab 10: 352–357
16. Stoop AA, Lupu F, Pannekoek H (2000) Colocalization of thrombin, PAI-1, and vitronectin in the atherosclerotic vessel wall: a potential regulatory mechanism of thrombin activity by PAI-1/vitronectin complexes. Arteriosclerosis, Thrombosis & Vascular Biology 20: 1143–1149
17. Striggow F, Riek M, Breder J, Henrich-Noack P, Reymann KG, Reiser G (2000) The protease thrombin is an endogenous mediator of hippocampal neuroprotection against ischemia at low concentrations but causes degeneration at high concentrations. Proc Natl Acad Sci USA 97: 2264–2269
18. Striggow F, Riek-Burchardt M, Kiesel A, Schmidt W, Henrich-Noack P, Breder J, Krug M, Reymann KG, Reiser G (2001) Four different types of protease-activated receptors are widely expressed in the brain and are up-regulated in hippocampus by severe ischemia. Eur J Neurosci 14: 595–608
19. The NINDS rt-PA stroke study group (1995) Tissue plasminogen activator for acute ischemic stroke. The National Institute of Neurological Disorders and Stroke rt-PA Stroke Study Group [see comments]. New Engl J Med 333: 1581–1587
20. Weinstein JR, Lau AL, Brass LF, Cunningham DD (1998) Injury-related factors and conditions down-regulate the thrombin receptor (PAR-1) in a human neuronal cell line. J Neurochem 71: 1034–1050
21. Xi G, Wagner KR, Keep RF, Hua Y, de Courten-Myers GM, Broderick JP, Brott TG, Hoff JT (1998) The role of blood clot formation on early edema development following experimental intracerebral hemorrhage. Stroke 29: 2580–2586

Correspondence: Guohua Xi, M.D., R5550 Kresge I, University of Michigan, Ann Arbor, Michigan 48109-0532. e-mail: guohuaxi @umich.edu

Acta Neurochir (2003) [Suppl] 86: 167–168

Infiltration of tissue plasminogen activator through cerebral vessels: evaluation using a rat thromboembolic stroke model

T. Kano, T. Harada, and **Y. Katayama**

Department of Neurological Surgery, Nihon University School of Medicine, Tokyo, Japan

Summary

In recent years, evidence has been accumulating that tissue plasminogen activator (tPA) possesses neurotoxic effects. However, such deleterious effects have been attributed to endogenously generated tPA. In the present study, focusing on exogenously administered tPA for the purpose of fibrinolysis, we evaluated the extent and the degree of extravasated tPA in a rat model of thromboembolic stroke. Even after early recanalization of occluded cerebral vessels, significant infiltration of tPA occurred through the cerebral vessels. It is assumed that exogenous tPA also exerts neurotoxic effects in the ischemic brain tissue.

Keywords: Rat; thromboembolic stroke; reperfusion; tPA infiltration; neurotoxic effects.

Introduction

In patients with acute ischemic stroke, it has been shown that early recanalization of occluded vessels by administration of tissue plasminogen activator improves clinical outcome [4]. Contrary to the beneficial effect of tPA through its fibrinolytic activity, recent experimental studies have indicated that tPA aggravates ischemic neuronal damage through its proteolytic activity [2, 5, 6]. Such deleterious activity of tPA has been examined by focusing on endogenous tPA, not on exogenous tPA. In the present study, using an established rat model of thromboembolic stroke [3], we investigated the extent and the degree of infiltration of exogenous tPA through the cerebral vessels into the brain tissue following administration of tPA for the purpose of fibrinolysis.

Materials and methods

Male Sprague-Dawley rats (250–300 g) were employed in a thromboembolic stroke model [3]. Animals were anesthetized with halothane (1–1.5%) under spontaneous respiration in an air-oxygen mixture. Rectal temperature were maintained between 36.5 °C to 37.5 °C with a heating lamp. Reperfusion was induced by injecting tPA (Alteplase; 10 mg/kg, 1 mg/ml in saline) intravenously over 30 min beginning at 30 min after the onset of ischemia. At 2 h after the beginning of tPA administration, animals were killed with a lethal overdose of sodium pentobarbital and decapitated. Then, the extent of tPA infiltration was evaluated by immunohistochemistry using anti-human tPA antibody ($n = 5$). Quantification of the infiltrated tPA was made by enzyme-linked immunosorbent assay (ELISA) methods ($n = 8$). In another subset of animals ($n = 10$), histological examinations by Hematoxylin-Eosin staining (H-E staining) were performed to evaluate ischemic neuronal injury at 24 h after the onset of ischemia. To confirm induction of ischemia and reperfusion after the administration of tPA, cerebral blood flow (CBF) was evaluated by laser doppler flowmetry (LDF) at the ischemic core of this model. All experiments were performed following an approved protocol by our institution for the care and use of laboratory animals. Student *t* tests were used to compare results and *P* values of <0.05 were considered significant.

Results

Just after the induction of ischemia, CBF reduced to around 15% compared to baseline value. After the administration of tPA, CBF recovered gradually. When CBF at 2 h after the beginning of tPA administration was below 50%, those animals were excluded from the study judging that recanalization of occluded vessels was not induced enough. Among animals included in the study, CBF just after the induction of ischemia and CBF at 2 h after the beginning of tPA administration was 15.9 ± 11.3 and $95.4 \pm 18.4\%$ (average $\pm$ SD), respectively, indicating that favorable recanalization was achieved.

Striatum and somatosensory cortex from the ischemic hemisphere were stained with anti-human tPA that was administered exogenously. The regions corresponded to the ischemic territory of this model. In the non-ischemic hemisphere, tPA staining was not

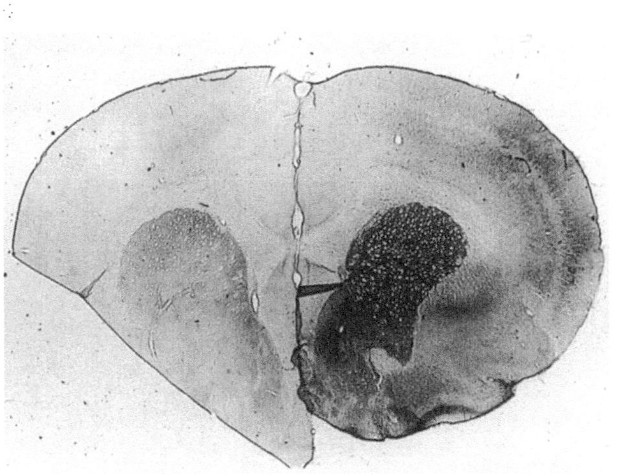

Fig. 1. Immunohistochemistry using anti-human tPA antibody. Striatum and somatosensory cortex are stained

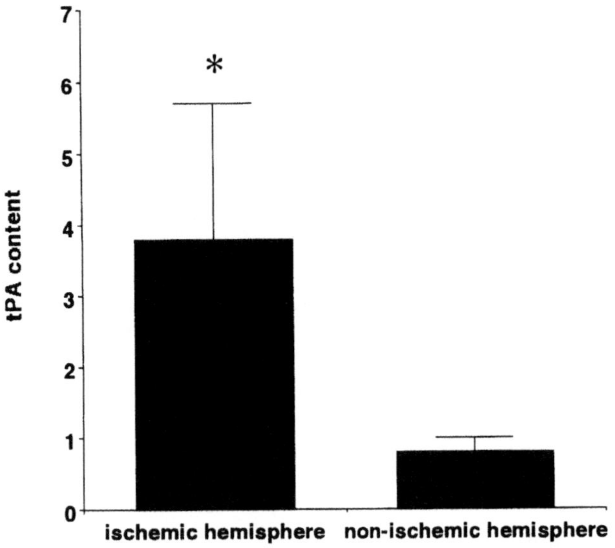

Fig. 2. tPA content in ischemic hemisphere and in non-ischemic hemisphere. Values are ng/ml of homogenized brain in saline (average ± SD). *$P < 0.01$, compared to non-ischemic hemisphere

revealed (Fig. 1). tPA contents were significantly higher in the ischemic hemisphere compared to the non-ischemic hemisphere: 3.8 ± 1.9 and 0.8 ± 0.2 ng/ml of homogenized brain in saline (average ± SD), respectively ($P < 0.01$) (Fig. 2). On histological preparations by H-E staining, necrotic neuronal cells were observed scattered within the ischemic regions only. Confluent infarction was not observed.

Discussion

The present results demonstrate that a remarkable amount of tPA infiltrates into the ischemic brain tissue through the cerebral vessels. When tPA is administered at delayed time point after brain tissue has already sustained irreversible ischemic damage, tPA would naturally pass through disrupted blood brain barrier (BBB). However, in the present study, we administered tPA at an early time when the BBB usually appeared, by electronmicroscopy, to be preserved (data not shown).

In the reperfusion period after transient cerebral ischemia, it is assumed that production of oxygen radicals increases in the small vessels of the brain. Superoxide radicals have been shown to play an important role in the development of vasogenic brain edema by increasing permeability of the BBB [1]. Accordingly, augmented permeability of tPA through the cerebral vessels may be caused by the effects of superoxide radicals.

As long as exogenously administered tPA stays within the cerebral vessels, tPA should promote the desirable (thrombolytic) effect of restoring CBF. Contrary to this, once tPA goes into brain tissue, tPA is supposed to exert an undesirable (neurotoxic) effect. One possible cascade by exogenous tPA is that tPA/plasmin degrades the extracellular matrix and sensitizes neurons to cell death [2]. Further study is required to determine whether the concentration of tPA detected in this study is enough to amplify ischemic neuronal cell injury.

References

1. Chan PH, Longar S, Fishman RA (1987) Protective effects of liposome-entrapped superoxide dismutase on posttraumatic brain edema. Ann Neurol 21: 540–547
2. Chen Z-L, Stickland S (1997) Neuronal death in the hippocampus is promoted by plasmin-catalyzed degradation of laminin. Cell 91: 917–925
3. Kano T, Katayama Y, Tejima E, Lo EH (2000) Hemorrhagic transformation after fibrinolytic therapy with tissue plasminogen activator in a rat thromboembolic model of stroke. Brain Res 854: 245–248
4. NINDS rt-PA Stroke Study Group (1995) Tissue plasminogen activator for acute ischemic stroke. N Engl J Med 333: 1581–1587
5. Tsirka SE, Rogove AD, Bugge TH, Degen JL, Stickland S (1997) An extracellular proteolytic cascade promotes neuronal degeneration in the mouse hippocampus. J Neurosci 17: 543–552
6. Wang YF, Tsirka SE, Strickland S, Stieg PE, Soriano SG, Lipton SA (1998) Tissue plasminogen activator (tPA) increase neuronal damage after focal cerebral ischemia in wild-type and tPA-deficient mice. Nature Med 4: 228–231

Correspondence: Tsuneo Kano, M.D., Ph.D., Department of Neurological Surgery, Nihon University School of Medicine, 30-1 Oyaguchi Kamimachi, Itabashi-ku, Tokyo 173-8610, Japan. e-mail: tuneok@med.nihon-u.ac.jp

Acta Neurochir (2003) [Suppl] 86: 169–172

Mice deficient in cytosolic phospholipase A_2 are less susceptible to cerebral ischemia/reperfusion injury

S. Tabuchi[1], N. Uozumi[2], S. Ishii[2], Y. Shimizu[1], T. Watanabe[1], and T. Shimizu[2]

[1] Department of Neurosurgery, Institute of Neurological Sciences, Tottori University School of Medicine, Yonago, Japan
[2] Department of Biochemistry and Molecular Biology, University of Tokyo Faculty of Medicine, Tokyo, Japan

Summary

To determine the role of cytosolic phospholipase A_2 ($cPLA_2$) in infarct development, wild-type and $cPLA_2$ knock-out mice were subjected to focal cerebral ischemia for 75 min by occluding the middle cerebral artery using nylon filament and subsequent reperfusion by withdrawing the filament. The neurological deficit severity was evaluated by a modified 4-point scale. After the reperfusion period (72 h), mice were killed, and the brains were cut into four 2 mm coronal sections using a rodent brain matrix. Sections were stained with 2% 2,3,5-triphenyltetrazolium chloride (TTC). The infarct volume was 87.19 ± 27.54 mm^3 (mean $\pm$ SD, n = 11) in the wild-type mice and 48.20 ± 31.32 mm^3 (n = 10; $P < 0.01$ vs. wild-type) in the knock-out mice. Less severe functional neurological deficits were observed in knock-out mice at 72 h after ischemia when compared with wild-type. Thus, disruption of $cPLA_2$ resulted in significant reduction of infarct area and neurological deficit severity in the MCA occlusion model. These data indicate a critical role for $cPLA_2$ in the pathogenesis of cerebral ischemia/reperfusion injury.

Keywords: Focal cerebral ischemia; reperfusion injury; mice; $cPLA_2$.

Introduction

Previous studies suggest that transient cerebral ischemia elicits an inflammatory response that is augmented by reperfusion [2, 12, 13]. Phospholipase A_2 is a key mediator of the production of inflammatory elements, such as the arachidonic acid metabolites, eicosanoids and platelet activating factor (PAF) [4, 5, 18]. The cytosolic phospholipase A_2 ($cPLA_2$) subtype has substrate specificity to arachidonic acid [14]. Thus, the goal of the current study was to investigate the contribution of $cPLA_2$ in cerebral ischemia/reperfusion injury using $cPLA_2$-deficient mice established in the laboratory of Shimizu [21].

Materials and methods

Animal model

Wild-type ($cPLA_2$+/+) and $cPLA_2$-deficient ($cPLA_2$−/−) mice generated by gene targeting [21] were maintained at the animal experiment facility of Tottori University. All animals were derived from a mixed C57BL6J/129-Ola strain. cPLA2-deficient mice underwent a series of backcrosses to C57BL/6J, resulting in a highly homogeneous genetic background. Genotypes were determined by PCR using 50–100 ng of genomic DNA obtained from tail biopsies, as previously described [21]. Experiments were carried out in accordance with the National Institutes of Health Guide for the Care and Use of Laboratory Animals-Revised 1996. All efforts were made to minimize the number of animals used and maintain humane care. Adult male mice (18 to 23 g) were anesthetized with 1.0% halothane and maintained with 0.5% halothane in 70% N_2O and 30% O_2 by inhalation mask. Body temperature was maintained at 37 ± 1 °C by a blanket controlled by a rectal temperature probe. The right femoral artery was cannulated to determine arterial blood pressure and sample arterial blood gases. Mean arterial blood pressure was recorded before ischemia, during ischemia, and after reperfusion with a computerized data acquisition system (MacLabs). Eighty-microliter samples were withdrawn into heparinized capillary tubes, and arterial blood gases were measured with a blood gas analyzer (OPTI1, AVL Medical Instruments Japan, Tokyo). Because repeated blood withdrawal from mice was not compatible with further ischemia experiments, the samples for physiological data were taken from a separate set of animals (n = 3 each). Transient focal cerebral ischemia was induced by occlusion of the MCA using an intraluminal filament technique, as described previously [11]. After the 75 min occlusion period, the mice were reanesthetized, and reperfusion was accomplished by withdrawing the intraluminal filament.

Assessment of cerebral blood flow

A laser-Doppler flowmeter (ALF1100, Advance, Japan) with a 0.7 mm probe (Advance) was employed to determine changes in regional cerebral blood flow (rCBF). The skull was exposed through a midline sagittal incision, and the probe tip was placed on the skull surface, 4 mm lateral to midline and 1 mm posterior to the bregma. Saline was circulated around the probe during measurements. Data are presented as percentage of the preischemic rCBF.

Neurological deficit scoring

The neurological deficit severity was determined at 30 min after onset of ischemia and 72 h after reperfusion using a modified 4-point scale (0 = no deficit, 1 = flexion of contralateral forepaw and torso upon lifting by the tail, 2 = circling to one side, 3 = spinning or unable to walk spontaneously).

Detection and quantification of cerebral infarction

After the reperfusion period, mice were killed with a lethal dose of pentobarbital (150 mg/kg IP). The brains were immediately removed, and four 2 mm coronal sections were cut with a rodent brain matrix (RBM-2000C, Activational System, MI). The brain sections were stained with 2% 2,3,5-triphenyltetrazolium chloride (TTC) in saline at room temperature for 15 min and fixed in 10% formalin. The digitized image of each brain section and infarct area was measured using NIH Image software. The infarct volume was calculated by the sum of the infarct area of different brain slices times the thickness of the slices.

Statistical analysis

Data are presented as means ± SD. In most experiments, when comparisons were made between knock-out and wild-type mice, a one-way ANOVA was used, followed by the Student's t test using Statview 4 software. The Mann-Whitney U analysis was used to compare neurological deficit scores between groups. Significance was noted when $P < 0.05$.

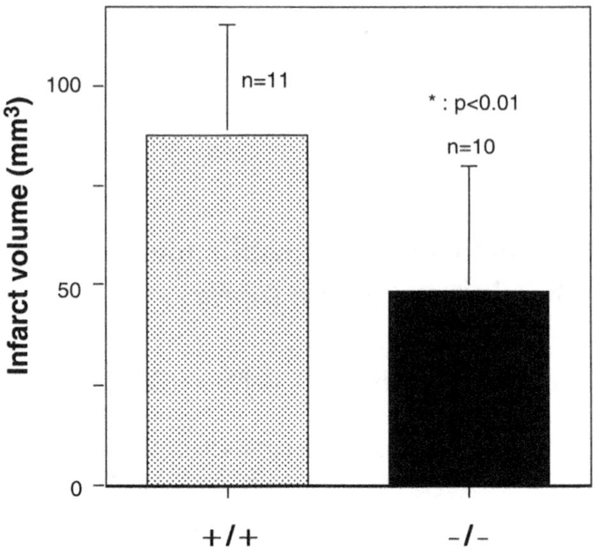

Fig. 1. Total infarction volumes after 75 min of transient focal cerebral ischemia followed by 72 h reperfusion in wild-type and knock-out mice. Data are expressed as mean ± S.D. Knock-out mice had significantly smaller infarction volume compared with wild-type (n = 11 and 10 for wild-type and knock-out mice, respectively). *$P < 0.01$ compared with wild-type, using Students t test

Results

Regional CBF (rCBF) was diminished during the period of MCA occlusion and was restored during the reperfusion period without differences in knock-out and wild-types groups (data not shown). Further, no differences were noted in mean arterial blood pressure or arterial pO_2, pCO_2 and pH between the genotypes (data not shown). The infarct volume was 87.19 ± 27.54 mm^3 (mean ± SD, n = 11) in the wild-type mice and 48.2 ± 31.32 mm^3 (n = 10, $P < 0.01$ vs. wild-type) in knock-out mice (Fig. 1). Attenuation of injury was more evident in the dorsolateral cerebral cortex (i.e., ischemic penumbra) (Fig. 2). Less severe neurological deficits were observed in knock-out mice at 72 h after ischemia when compared with wild-type (Fig. 3). Mortality rates of the wild-type and knock-out mice during this 72 h experimental period were 26.67% and 0%, respectively.

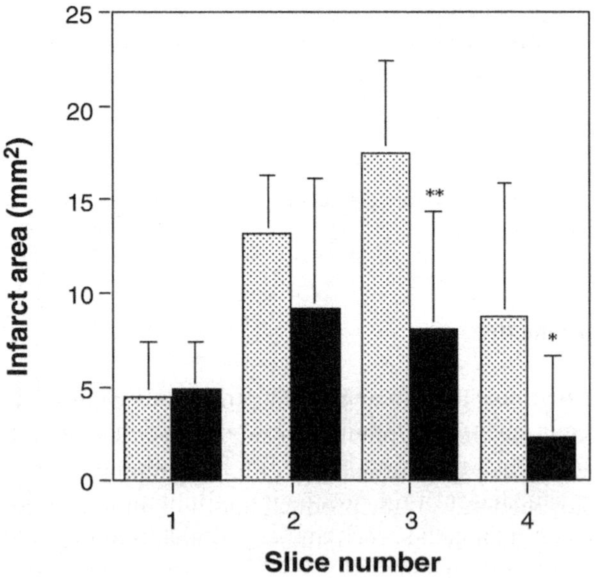

Fig. 2. Quantitative measurements of infarct area in four coronal slices. The brain was cut coronally, 1 mm, 3 mm, 5 mm and 7 mm distal to the frontal pole (slice number 1, 2, 3, 4, respectively) by using the mouse brain matrix. Data are expressed as means ± S.D. The reduction of infarct area is more evident in the posterior coronal slices in knock-out mice. **$P < 0.01$, *$P < 0.05$ compared with wild-type. ▨ +/+, ■ −/−

Discussion

Experimental evidence supports a role for phospholipase A_2 in ischemia and reperfusion injury. Phospholipase A_2 hydrolyzes the ester bond of phospholipids at the sn-2 position, thereby producing arachidonic acid, oxygenated metabolites (eicosa-

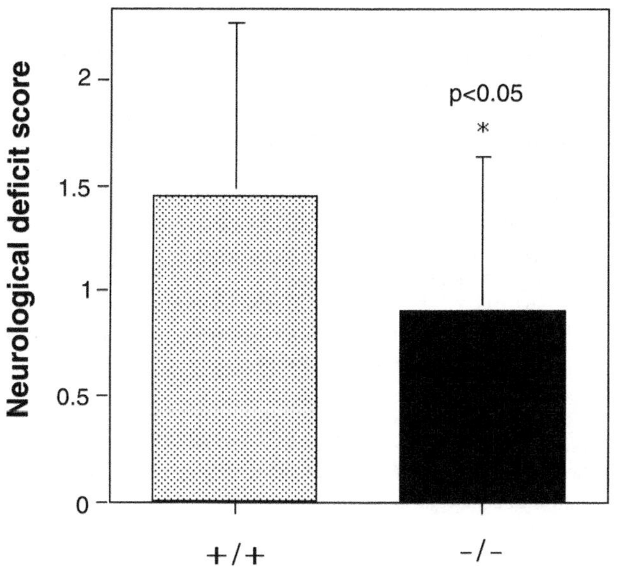

Fig. 3. Neurological deficit scores in wild-type and knock-out mice at 72 h after reperfusion. Data are expressed as means ± S.D. Knock-out mice showed milder neurological deficit compared to wild-type. $^*P < 0.05$ compared with wild-type, using Mann-Whitney *U* test

noids) and PAF. These end-products have been implicated in the pathogenesis of cerebral ischemia and reperfusion injury [4, 5, 18]. Cytosolic PLA_2 ($cPLA_2$) is present in the brain [7] and is the only PLA_2 subtype that selectively liberates arachidonic acid [14]. Further, previous studies have demonstrated increased activity of $cPLA_2$ during cerebral ischemia and reperfusion [5, 17]. Also, inhibitors of cyclooxygenase (COX) and 5-lipoxygenase (5-LO) can ameliorate stroke injuries in animal models [3, 8, 15, 16, 20], and $cPLA_2$ acts upstream of these end-products.

There is abundant evidence that an acute inflammatory reaction associated with reperfusion contributes to the development of secondary brain damage [2, 12, 13]. In addition to $cPLA_2$-derived products, it has been postulated that oxygen free radicals, adhesion molecules and cytokines are also involved in this reaction [2, 12, 13]. Considering that most stroke patients receive medical attention more than 6 h after onset of symptoms, blocking this secondary damage might be an important therapeutic strategy.

In this study, the disruption of $cPLA_2$ significantly reduced the infarction volume and neurological deficit in the MCA occlusion model. This result is consistent with the previous report [6] that employed mice with a different genetic background (129/Sv mouse). As null-mutation effects in mice are dependent on the background genotype [10, 19] and mouse strain differences in susceptibility to cerebral ischemia have been reported [1], the genetic background of the animals must be taken into consideration when generalizing to other models. Among mice strains, C57BL/6J is particularly susceptible to cerebral ischemia [9]. As our mice are a hybrid of two mouse strains (C57BL/6J and 129/0la), we used a backcross strategy to the C57BL/6J strains and examined the effect with a nearly pure C57BL/6J genetic background.

The results of the present study indicate a critical role for $cPLA_2$ in the pathogenesis of cerebral ischemia/reperfusion injury. Anti-inflammatory therapies involving inhibition of $cPLA_2$ may play an important therapeutic role in the clinical management of stroke.

References

1. Barone FC, Knudsen DJ, Nelson AH, Feuerstein GZ, Willette RN (1993) Mouse strain differences in susceptibility to cerebral ischemia are related to cerebral vascular anatomy. J Cereb Blood Flow Metab 13: 683–692
2. Barone FC, Feuerstein GZ (1999) Inflammatory mediators and stroke: New opportunities for novel therapeutics. J Cereb Blood Flow Metab 19: 819–834
3. Baskaya MK, Hu Y, Donaldson D, Maley M, Rao AM, Prasad MP, Dempsey RJ (1996) Protective effect of the 5-lipoxygenase inhibitor AA-861 on cerebral edema after transient ischemia. J Neurosurg 85: 112–116
4. Bazan NG, Allan G, Rodriguez de Turco EB (1993) Role of phospholipase A2 and membrane-derived lipid second messengers in membrane function and transcriptional activation of genes: implications in cerebral ischemia and neuronal excitability. Kogure K, Hossman KA, Siesjo BK (eds) Progress in brain research, vol 96. Elsevier, pp 247–256
5. Bonventre JV, Koroshetz WJ (1993) Phospholipase A_2 (PLA_2) activity in gerbil brain: Characterization of cytosolic and membrane-associated forms and effects of ischemia and reperfusion on enzymatic activity. J Lipid Med 6: 457–471
6. Bonventre JV, Huang Z, Taheri MR, O'Leary E, Li E, Moskowitz MA, Sapirstein A (1997) Reduced fertility and postischemic brain injury in mice deficient in cytosolic phospholipase A_2. Nature 390: 622–625
7. Clemens JA, Stephenson DT, Smalstig EB, Roberts E, Johnstone ED, Sharp JD, Little SP, Kramer RM (1996) Reactive glia express cytosolic phospholipase A_2 after transient global forebrain ischemia in the rat. Stroke 27: 523–535
8. Cole DJ, Patel PM, Reynolds L, Drummond JC, Marcantonio S (1993) Temporary focal cerebral ischemia in spontaneously hypertensive rats: The effect of Ibuprofen on infarct volume. J Pharmacol Exp Ther 266: 1713–1717
9. Connolly ES, Winfree CJ, Stern DM, Solomon RA, Pinsky DJ (1996) Procedual and strain-related variables significantly affect outcome in a murine model of focal cerebral ischemia. Neurosurgery 38: 523–532
10. Gerlai R (1996) Gene-targeting studies of mammalian behavior: is it the mutation or the background genotype? Trends Neurosci 19: 177–181
11. Huang Z, Huang PL, Panahian N, Dalkara T, Fishman MC,

Moskowitz MA (1994) Effects of cerebral ischemia in mice deficient in neuronal nitric oxide synthase. Science 265: 1883–1885
12. Iadecola C, Alexander M (2001) Cerebral ischemia and inflammation. Curr Opin Neurol 14: 89–94
13. Jean WC, Spellman SR, Nussbaum ES, Low WC (1998) Reperfusion injury after focal cerebral ischemia: The role of inflammation and the therapeutic horizon. Neurosurgery 43: 1382–1397
14. Kramer RM, Stephenson DT, Roberts EF, Clemens JA (1996) Cytosolic phospholipase A_2 ($cPLA_2$) and lipid mediator release in the brain. J Lipid MediatCell Signal 14: 3–7
15. Mabe H, Nagai H, Suzuka T (1990) Role of brain tissue leukotriene in brain oedema following cerebral ischaemia: effect of a 5-lipoxygenase inhibitor, AA-861. Neurol Res 12: 165–168
16. Nagayama M, Niwa K, Nagayama T, Ross ME, Iadecola C (1999) The cyclooxygenase-2 inhibitor NS-398 ameliorates ischemic brain injury in wild-type mice but not in mice with deletion of the inducible nitric oxide synthase gene. J Cereb Blood Flow Metab 19: 1213–1219
17. Saluja I, Song D, O'Regan MH, Phillis JW (1997) Role of phospholipase A_2 in the release of free fatty acids during ischemia-reperfusion in the rat cerebral cortex. Neurosci Lett 233: 97–100
18. Shimizu T, Wolfe LS (1990) Arachidonic acid cascade and signal transduction. J Neurochem 55: 1–15
19. Sibilia M, Wagner EF (1995) Stain-dependent epithelial defects in mice lacking the EGF receptor. Science 14: 234–238
20. Ting P (1990) Indomethacin attenuates early postischemic vasogenic edema and cerebral injury. Adv Neurol 52: 119–126
21. Uozumi N, Kume K, Nagase T, Nakatani N, Ishii S, Tashiro F, Komagata Y, Maki K, Ikuta K, Ouchi Y, Miyazaki J, Shimizu T (1997) Role of cytosolic phospholipase A_2 in allergic response and parturition. Nature 390: 618–622

Correspondence: Sadaharu Tabuchi, Department of Neurosurgery, Institute of Neurological Sciences, Tottori University School of Medicine, 36-1 Nishi-machi, Yonago 683-8504, Japan. e-mail: stabuchi@gj8.so-net.ne.jp

Acta Neurochir (2003) [Suppl] 86: 173–175

Regional distribution of potassium and phosphorus in ischemic brain tissue of rats with X-ray fluorescence analysis

T. Mituhashi, S. Hatashita, and **I. Ogino**

Neurosurgery, Juntendo University Urayasu Hospital, Urayasu, Japan

Summary

Focal cerebral ischemia was induced by occlusion of the left middle cerebral artery (MCAO) for 1 and 3 hours in 14 rats. Five sham-operated rats served as controls. Regional cerebral blood flow (rCBF) was assessed by ^{14}C-iodoantipyrine radioluminography. The potassium and phosphorus contents in brain tissue were assessed using energy-dispersive X-ray fluorescence spectrometry. In the inner periphery within the ischemic tissue where rCBF was severely reduced to 6.61 ± 0.95 ml/100 g/min 1 hour after MCAO, the potassium content decreased slightly to 12.82 ± 0.15 mg/g. Potassium content decreased further to 10.61 ± 0.24 mg/g over 3 hours after occlusion, similar to that in the ischemic core. In contrast, in the outer periphery where rCBF was moderately reduced, the potassium fell slightly to 12.78 ± 0.46 mg/g 3 hours after MCAO. Tissue phosphorus content decreased to 8.28 ± 0.24 mg/g only in the ischemic core. These findings demonstrate that potassium and phosphorus in brain tissue depend primarily on the extent, magnitude and duration of ischemia. Ion homeostasis of cell membranes may not be completely disturbed in the periphery around ischemic tissue for a short time after the onset of ischemia, even though blood flow is markedly reduced.

Keywords: Cerebral blood flow; cerebral ischemia; potassium; phosphorus; X-ray fluorescence.

Introduction

We recently demonstrated that X-ray fluorescence (XRF) analysis is a new method for determining the spatial distribution of ionic and elemental concentrations in brain tissue [3]. The present study evaluates the relationships between cerebral blood flow, potassium and phosphorus contents in adjacent brain tissues made variably ischemic.

Materials and methods

Adult male Sprague-Dawley rats, weighing 300–400 g, were anesthetized with ketamine hydrochloride (40 mg/kg) and xylazine (10 mg/kg). Focal cerebral ischemia was induced by occlusion of the left middle cerebral artery (MCA) by a transretro-orbital approach for 1 and 3 hours in 14 rats. Five sham-operated rats served as controls.

Levels of potassium and phosphorus were simultaneously measured in the coronal brain tissue by energy-dispersive X-ray fluorescence (XRF) spectrometry [3] with a Micro XRF Element Monitor having a two-dimensional mapping function and operated at an accelerating voltage of 50 kV, beam current of 1 mA and diameter of 180 μm. (SEA 2010, Seiko Ins. Inc., Japan). Regional cerebral blood flow (rCBF) in the adjacent sections was assessed by ^{14}C-iodoantipyrine radioluminography [2].

Five areas were measured at and around the MCA territory in a coronal section through the caudal caudate-putamen of the hemisphere with MCA occlusion and through homologous sites in the contralateral hemisphere. The inner and outer peripheries were identified in the forelimb area of the frontal cortex adjacent to the MCA territory and in the hindlimb area, respectively. The core was the central area of the parietal cortex. The non-ischemic cortical tissue was area 1 of the frontal cortex [1].

All data are presented as mean ± standard deviation. Parametric data at different points were compared using analysis of variance followed by Scheffe's test.

Results

Brain tissue potassium

A map of potassium concentration in coronal section of brain clearly showed depletion of potassium in the MCA territory at 1 and 3 hours after MCA occlusion (Fig. 1). This potassium map corresponded to a gradient change of blood flow in the ischemic brain tissue.

Blood flow in the core of the parietal cortex was reduced to 1.70 ± 0.43 ml/100 g/min 1 hour after MCA occlusion. Tissue potassium concentration decreased to 11.26 ± 0.26 mg/g in the same ischemic core, significantly different from controls (15.14 ± 0.38 mg/g). Brain tissue potassium did not fall further over the next 3 hours.

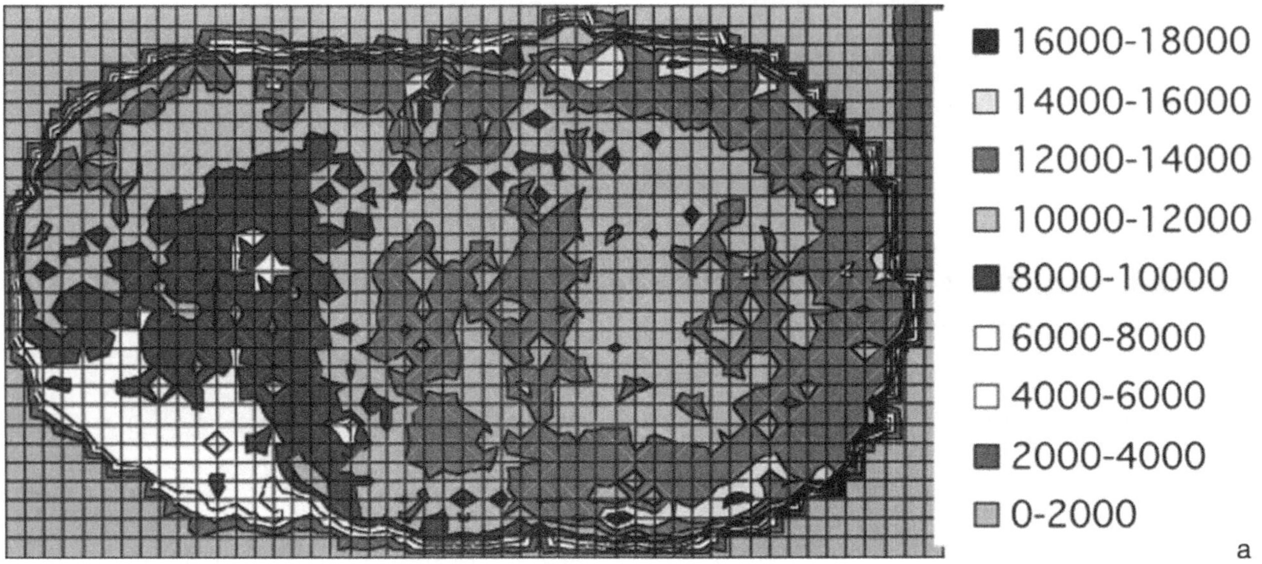

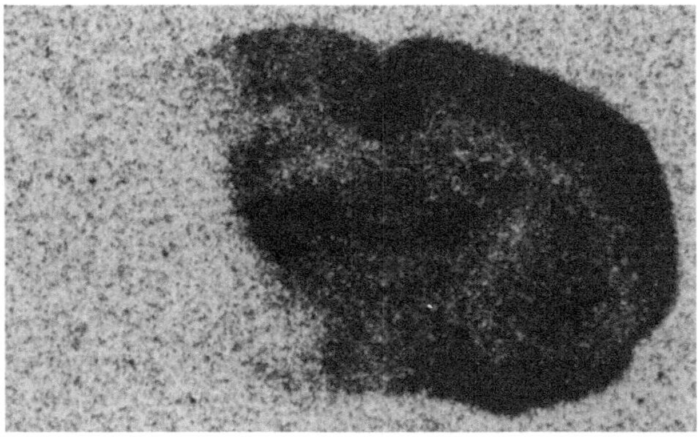

Fig. 1. Coronal sections of brain on ^{14}C-iodoantipyrine radioluminogram (a) and the potassium mapping with XRF (b) 1 hour after MCA occlusion

In the inner peripheral zone within the ischemic tissue where rCBF was severely reduced to 6.61 ± 0.95 ml/100 g/min 1 hour after MCA occlusion, the potassium concentration decreased slightly to 12.82 ± 0.15 mg/g. The extent of this depletion in the inner periphery was large on the potassium map. Three hours after MCA occlusion, the potassium content in the same area decreased further to 10.61 ± 0.24 mg/g, similar to that in the ischemic core.

In the outer peripheral zone surrounding ischemic tissue where rCBF was reduced to 24.15 ± 1.44 ml/100 g/min 1 hour after MCA occlusion, the potassium content was 14.53 ± 0.35 mg/g, not significantly different from controls. Three hours after MCA occlusion, potassium in the same area fell slightly to 12.78 ± 0.46 mg/g, although rCBF did not differ from that at 1 hour after MCA occlusion (Fig. 2).

Brain tissue phosphorus

The phosphorus map showed a decrease of concentration only in the ischemic core 3 hours after MCA occlusion. Phosphorus content was 8.28 ± 0.24 mg/g in the ischemic core of the parietal cortex, significantly different from controls (9.62 ± 0.28 mg/g). There was no reduction of phosphorus concentration in the outer or inner periphery at 1 and 3 hours after MCA occlusion.

Blood flow, brain tissue potassium and phosphorus

The relationship between blood flow and potassium in brain tissue in sham-operated control animals and animals with MCA occlusion is shown in Fig. 3. There was a significant correlation between blood flow and tissue potassium concentration (r = 0.868, n = 76,

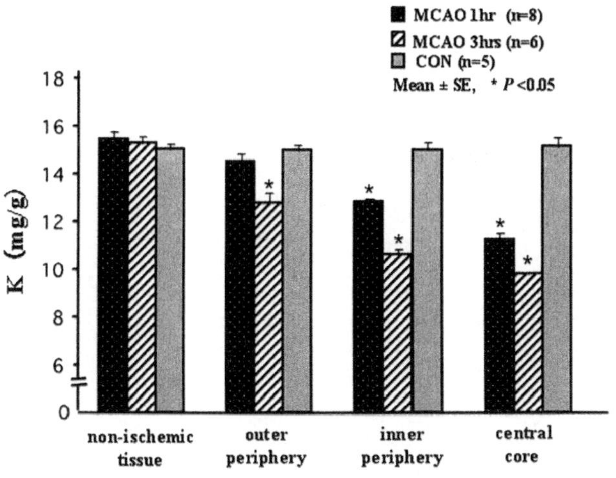

Fig. 2. Changes of the potassium contents in the brain tissue 1 and 3 hours after middle cerebral artery occlusion (*MCAO*) and sham operation (con). *P*: Statistical significance from values with control

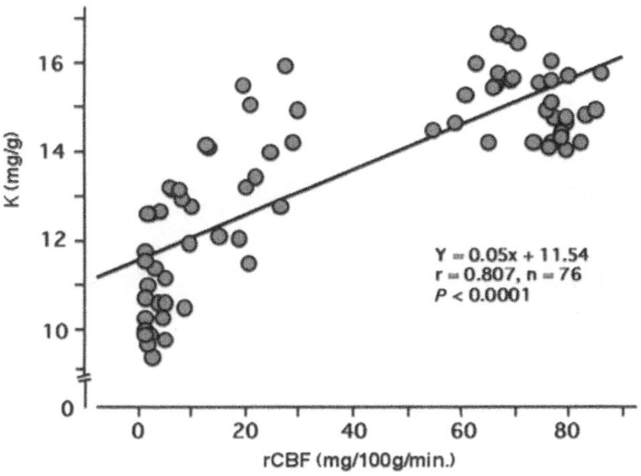

Fig. 3. Scatterplot of blood flow and potassium in brain tissue in sham-operated control animals and animals with middle cerebral artery occlusion

$P < 0.001$). In the peripheral area surrounding ischemic tissue, the relationship between blood flow and tissue potassium concentration was more variable, depending on the duration of ischemia. In contrast, the potassium concentration did not correlate with phosphorus in animals 1 and 3 hours after MCA occlusion and in sham-operated controls ($r = 0.336$, $n = 76$).

Discussion

We evaluated potassium and phosphorus in ischemic brain tissue with energy-dispersive X-ray fluorescence (XRF) spectrometry. The X-ray fluorescence analysis had been applied for measurement of major and trace elements in biological samples for a decade. Recently, the energy dispersive XRF instrument with a collimated X-ray beam has been developed. This instrument can scan serially with a small X-ray beam and demonstrate two-dimensional mapping of the element concentration. We recently demonstrated that energy-dispersive XRF is a new method for determining the spatial distribution of ionic concentrations, including potassium and phosphorus, in brain tissue.

Disturbed ion homeostasis under ischemic conditions has been well established. Intracellular potassium is released into the extracellular space when blood flow is decreased below a critical level. The potassium is removed from the tissue and uptake of sodium and water occurs. Tissue potassium depletion, therefore, is a sensitive indicator of ischemic tissue damage.

We have demonstrated that potassium and phosphorus contents decrease in ischemic brain tissue, depending on the duration and the depth of the ischemia. Depletion of the tissue potassium corresponds to a gradient of blood flow more than that of tissue phosphorus.

In the present study, the tissue potassium was somewhat decreased in the inner peripheral zone within the ischemic tissue where blood flow was severely reduced 1 hour after occlusion. This reduction of the tissue potassium in the inner periphery was smaller than that in the core. This implies that the ion homeostasis of cell membrane may be not completely disturbed in the periphery around ischemic tissue for a short time after the onset of ischemia, even though blood flow was reduced severely.

In contrast, tissue phosphorus decreased only in the ischemic core where blood flow was reduced severely 3 hours after occlusion. Depletion of the phosphorus is primarily attributed to the nucleic acid and phospholipid contents of brain tissue severely damaged with ischemic injury.

References

1. Hatashita S, Hoff JT (1990) Brain edema and cerebrovascular permeability during cerebral ischemia in rats. Stroke 21: 582–588
2. Hatashita S, Tajima A, Ueno H, Ishimaru S, Sato H, Takahashi S (1997) Blood volume and flow velocity through parenchymal microvessels in ischemic brain edema of rats. Acta Neurochir (Wien) [Suppl] 70: 23–26
3. Takahashi S, Hatashita S, Taba Y, Sun XZ, Kubota Y, Yosida (2000) Determination of the spatial distribution of major elements in the rat brain with X-ray fluorescence analysis. J Neurosci Meth 100: 53–62

Correspondence: Shizuo Hatashita, M.D., Neurosurgery, Juntendo University Urayasu Hospital, 2-1-1 Tomioka, Urayasu 279-0021, Japan. e-mail: nsjura@olive.ocn.ne.jp

Acta Neurochir (2003) [Suppl] 86: 177–180

Changes of free cholesterol and neutral lipids after transient focal brain ischemia in rats

H. Kamada, K. Sato, M. Iwai, K. Ohta, I. Nagano, M. Shoji, and **K. Abe**

Department of Neurology, Graduate School of Medicine and Dentistry, Okayama University, Japan

Summary

We investigated temporal and spatial changes of free cholesterol (FC) and neutral lipids (NLs) after brain ischemia with filipin complex staining to detect mainly FC and Nile Red staining for NLs such as cholesteryl ester (CE) and triacylglyceride (TAG). Filipin stanining decreased during 1 to 7 d and lost at 21 d after transient middle cerebral artery (MCA) occlusion in the ischemic core, but did not change in the penumbra. Nile Red positive droplets reached the maximum at 7 d after transient MCA occlusion (MCAO) and gradually decreased in the core, while the peak time delayed in the penumbra. Most Nile Red positive droplets were double positive for ED1 in the core, and were localized within GFAP positive cells in the penumbra. The present study suggests that changes of FC and NLs are different temporally and spatially between the ischemic core and the penumbra in relation to degenerative and regenerative neural processes following brain ischemia. Macrophages and astrocytes are presumed to play important roles in lipid metabolism for neural reorganization of the ischemic brain injury in the ischemic core and the penumbra, respectively.

Keywords: Cholesterol; lipid; brain ischemia; rat.

Abbreviations

ApoE apolipoprotein E; *APP* amyloid precursor protein; *CBF* cerebral blood flow; *CE* cholesteryl ester; *CNS* central nervous system; *FC* free cholesterol; *FFAs* free fatty acids; *MCA* middle cerebral artery; *MCAO* middle cerebral artery occlusion; *NLs* neutral lipids; *PB* phosphate buffer; *PBS* phosphate-buffered saline; *TAG* triacylglyceride.

Introduction

Neural membranes are mainly composed of lipids (cholesterol and phospholipids) and many species of proteins. Similar to other somatic cells, neurons maintain cholesterol homeostasis by regulating supply and efflux of cholesterol [3]. Under physiological condition, majority of free cholesterol (FC) is provided by de novo synthesis from acetyl-CoA or by hydrolytic release from cholesteryl ester (CE) within the CNS, while little FC is imported from systemic circulation. Triacylglyceride (TAG) plays an important role in providing free fatty acids (FFAs) for phospholipid synthesis [14]. Changes in cholesterol balance across cells affect both the rate of processing of β-amyloid precursor protein (APP) and the level of Apolipoprotein E (ApoE). ApoE-knock out mice showed slower clearance of infarct tissue after brain ischemia than wildtype mice [8]. Thus, cholesterol metabolism plays an important role in the pathological conditions including Alzheimer's disease and brain ischemia. However, cholesterol and neutral lipids (NLs) metabolisms after brain ischemia have not been well clarified. In the present study, therefore, possible temporal and spatial changes of FC and NLs were examined in rat brain after transient MCAO (tMCAO).

Materials and methods

Adult male Wistar rats at age 11 weeks (body weight; 250–280 g) were utilized. The right MCA was occluded by an insertion of nylon thread with silicon coat through common carotid artery under halothane anesthesia [7]. After 90 min of MCAO, the blood flow was restored by removal of the nylon thread. Sham control animals were treated in the same way without MCAO. Rats were perfused through the heart with ice cold heparinized saline followed by paraformaldehyde in 0.1 M phosphate buffer (PB, pH 7.4) at 1, 7, 21, and 56 d (n = 6 in each) after the reperfusion. Rat brains were removed, post-fixed immersion in the same fixative and quickly frozen. For histochemistry, frozen coronal sections were cut at 20 μm on a cryostat.

Histological changes were studied with filipin complex staining to detect mainly FC [11], Nile Red staining for neutral lipids such as CE and TAG [4]. Double staining for Nile Red plus ED1, MAP2 or GFAP was performed. The sections were examined with fluorescence microscopy for filipin signals and Nile Red signals, and with light microscopy for ED1, MAP2, or GFAP.

The number of staining cells in 0.25 mm^2 of brain area were counted, and categorized into 6 grades as follows: no, a few (1–3), a small (4–50), moderate (51–100), large (101–200), and larger (201–500) number of stained cells into (−), (±), (+), (2+), (3+), and (4+), respectively.

Results

Filipin staining

Filipin complex staining showed a fuzzy bright blue halo marginating neural cells with a diffuse darker blue background in sham control brain. The neural cell bodies represented black spots in the microscopic field. In the ischemic core, the number of filipin positive cells greatly reduced without intensity change at 1 d after tMCAO, further decreased with reducing the staining intensity at 7 d, and disappeared at 21 and 56 d with a dark black background. On the other hand, the filipin signal did not change in the number and intensity in the ischemic penumbra until 56 d after tMCAO.

Nile Red staining

The staining for Nile Red was weakly present in the sham control brain. In the ischemic core, Nile Red positive droplets appeared at 1 d, which markedly accumulated at 7 d, distributing in the whole ischemic core. At 21 d, the number of Nile Red positive droplets slightly decreased. Although the number of droplets greatly decreased at 56 d in the ischemic core, the Nile Red staining formed small granular aggregates. In the ischemic penumbra, Nile Red positive droplets increased at 7 d, reached the maximum at 21 d, and decreased in the number at 56 d. In the penumbra, each staining was small and did not form strong circular nor granular aggregates.

Double staining for Nile Red and ED1

Double staining for Nile Red plus ED1 demonstrated that the number of double positive cells peaked already at 7 d, decreased slightly at 21 d, and greatly at 56 d in the ischemic core (Fig. 1). However, in the ischemic penumbra, majority of Nile Red positive droplets were not colocalized with ED1 positive cells (Fig. 1). The ratio of double positive cells among Nile Red single positive droplets or ED1 single positive cells was much smaller in the ischemic penumbra than the core.

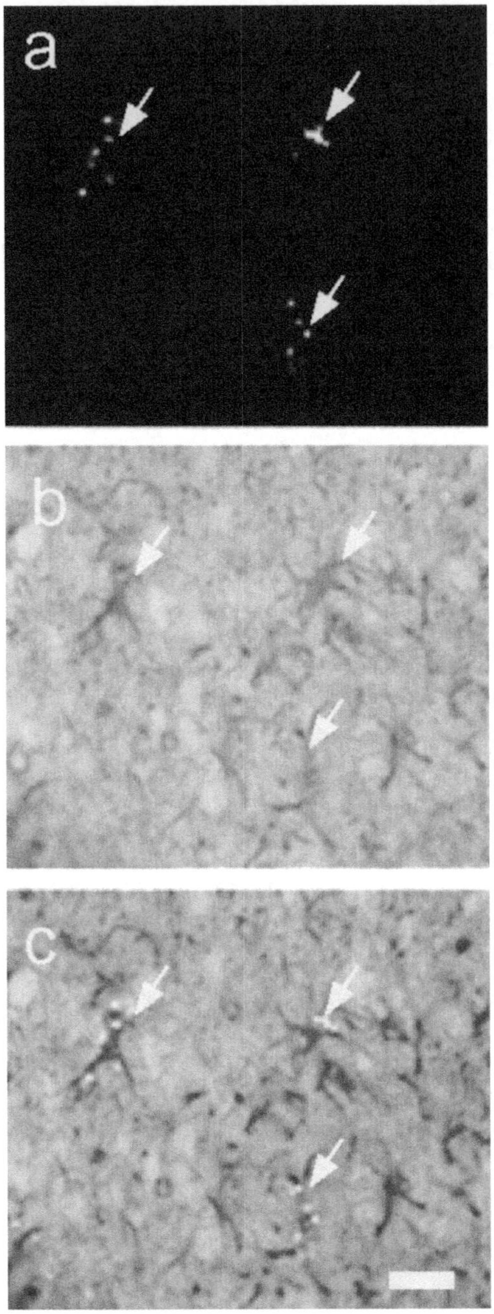

Fig. 1. Single Nile Red (a; arrows), single GFAP (b; arrows) and the merged image (c) in the ischemic penumbra at 21 d after 90 min of transient MCAO. Scale bar = 20 μm

Double staining for Nile Red plus MAP2 or GFAP

In the ischemic core, no Nile Red positive droplets were colocalized with MAP2 or GFAP positive cells. In the penumbra, however, majority of Nile Red positive droplets were localized within GFAP positive cells, while some Nile Red positive droplets were adjacent to MAP2 immunopositive dendrites.

Discussion

Filipin staining principally detects FC and the fuzzy bright halos marginating neural cells represent neural cell membrane. The diffuse darker background suggests a presence of low level FC in the neuropil of normal brain. Therefore, the progressive decrease of filipin staining in the core suggests a decrease of FC level in plasma membrane with acute necrotic cell death after tMCAO. On the other hand, no change of filipin staining in the penumbra indicates that most neural cells in the penumbra survived without losing detectable FC from their cell membrane.

NLs such as CE and TAG accumulate in the ischemic core with a peak at 7 d. CE and TAG are the primary component of NLs droplets in macrophage [12]. Therefore, the accumulation of Nile Red positive droplets in the core suggests an incorporation of FC into migrating macrophage and conversion into CE. FFAs derived from membrane phospholipids may be temporally stored as CE and TAG in the macrophage after brain ischemia, and could be reutilized for cholesterol or phospholipid synthesis.

Nile Red positive droplets in the ischemic core once formed a circle at 7–21 d, but formed granular aggregates at 56 d. This change suggests that incorporation of FC into macrophage persisted for a couple of weeks, and small segmentation occurred at around 8 w after the ischemia. Macrophage uptakes oxidized lipoprotein and stores CE and TG without limit [2]. On the other hand, each droplet was smaller and did not aggregate in the penumbra, because majority of droplets accumulated in astrocytes, which had a regulation system to inhibit influx of FC.

The present results indicate that macrophages and phagocytic microglia appear later at 7–21 d after brain ischemia associated with clearance of necrotic tissue. Activated microglia in the penumbra invaded the core, and changed into macrophage [6]. The number of ED1 positive cells decreased at 21 d in the core when that became a peak in the penumbra, suggesting a moving of NLs bearing macrophage from the core into the systemic circulation. Most ED1 positive cells (probably macrophages) in the core may be full of CE converted from FC, and microglia in the penumbra may not be necessary to be activated with positive ED1 staining for incorporating FC to convert into CE [1].

In the penumbra, intense MAP2 immunostaining were found in dendrites and soma for 7 to 56 d after brain ischemia. Majority of the Nile Red positive droplets were localized in astorcytes, and some were adjacent to MAP2 positive dendrites in the penumbra. Our study suggests that neuronal remodeling occurs in the penumbra after MCAO and that astrocytes are involved in lipid redistribution for neural reconstruction after brain ischemia, as were observed in rat models of peripheral nerve injury [10] and entorhinal cortex lesion [5]. Although the mechanism of transporting cholesterol or other lipids in CNS is not still clear, ApoE that is expressed in macrophage or astrocyte may play an important role in the transport and recycle of FC, CE, and TAG after brain ischemia [8, 9, 13].

In conclusion, the present study suggests that changes of FC and NLs are different temporally and spatially between the ischemic core and the penumbra in relation to degenerative and regenerative neural processes following brain ischemia taking a couple to several weeks. Macrophages and astrocytes are presumed to play important roles in lipid metabolism for reorganization of the ischemic brain tissue in the ischemic core and the penumbra, respectively.

Acknowledgments

This work was partly supported by Grant-in-Aid for Scientific Research (B) 12470141 and (Hoga) 12877211 and National Project on Protein Structural and Functional Analyses from the Ministry of Education, Science, Culture and Sports of Japan, and by grants (Itoyama Y, Kimura I and Kuzuhara S) from the Ministry of Health and Welfare of Japan.

References

1. Brierley JB, Brown AW (1982) The origin of lipid phagocytes in the central nervous system: the intrinsic microglia. J Comp Neurol 211: 397–406
2. Chait A, Heinecke J (1994) Lipoprotein modification cellular mechanisms. Curr Opin Lipidol 5: 365–370
3. Dietschy JM, Truly SD (2001) Cholesterol metabolism in the brain. Curr Opin Lipidol 12: 105–112
4. Fowler SD, Greenspan P (1985) Application of Nile Red, a fluorescent hydrophobic probe, for the detection of neutral lipid deposits in tissue section: comparison with oil red O. J Histochem Cyotchem 33: 833–836
5. Jurevics H, Bouldin TW, Toews AD, Morell P (1998) Regenerating sciatic nerve does not utilize circulating cholesterol. Neurochem Res 23: 401–406
6. Kato H, Kogure K, Liu XH, Araki T, Itoyama Y (1996) Progressive expression of immunomolecules on activated microglia and invading leukocytes following focal cerebral ischemia in the rat. Brain Res 734: 203–212
7. Kitagawa H, Sasaki C, Sakai K, Mori A, Mitsumoto Y, Mori T, Fukuchi Y, Setoguchi Y, Abe K (1999) Adenovirus-mediated gene transfer of glial cell line-derived neurotrophic factor prevents ischemic brain injury after transient middle cerebral artery occlusion in rats. J Cereb Blood Flow Metab 19: 1364–1344

8. Kitagawa K, Matsumoto M, Kuwabara K, Ohtsuki T, Hori M (2001) Delayed, but marked, expression of apolipoprotein E is involved in tissue clearance after cerebral infarction. J Cereb Blood Flow Metab 21: 1199–1207
9. Mahley RW (1988) Apolipoprotein E: Cholesterol transport protein with expanding role in cell biology. Science 240: 622–630
10. Mauch DH, Nagler K, Schumacher S, Goritz C, Muller EC, Otto A, Pfrieger FW (2001) CNS synaptogenesis promoted by glia-derived cholesterol. Science 294: 1354–1357
11. Mori T, Paris D, Town T, Rojiani AM, Sparks L, Delledonne A, Crawford F, Abdullah LI, Humphrey JA, Dickson DW, Mullan MJ (2001) Cholesterol accumulates in senile plaque of Alzheimer disease patients and in transgenic APPsw mice. J Neuropathol Exp Neurol 60: 778–785
12. Nishikawa K, Arai H, Inoue K (1990) Scavenger receptor-mediated uptake and metabolism of lipid vesicles containing acidic phospholipids by mouse peritoneal macrophages. J Biol Chem 265: 5226–5231
13. Poirier J, Baccichet A, Dea D, Gauthier S (1993) Cholesterol synthesis and lipoprotein reuptake during synaptic remodeling in hippocampus in adult rats. Neuroscience 55: 81–90
14. Sun GY, MacQuarrie RA (1989) Deacylation-reacylation of arachidonoyl groups in cerebral phospholipids. Ann NY Acad Sci 559: 37–55

Correspondence: Hiroshi Kamada, M.D., Department of Neurology, Graduate School of Medicine and Dentistry, Okayama University, 2-5-1 Shikata-cho, Okayama 700-8558, Japan. e-mail: h_kamada@cc.okayama-u.ac.jp

Acta Neurochir (2003) [Suppl] 86: 181–185

Osteopontin infusion into normal adult rat brain fails to increase cell proliferation in dentate gyrus and subventricular zone

K. A. Sailor, V. K. Dhodda, V. L. Raghavendra Rao, and **R. J. Dempsey**

Department of Neurological Surgery, University of Wisconsin, Madison, Wisconsin

Summary

In the first week after focal ischemia in adult brain, the basal level of neurogenesis increases dramatically in two distinct areas: The dentate gyrus (DG) of the hippocampus and the subventricular zone (SVZ) of the lateral ventricles. It is possible that this remotely induced neurogenesis is the result of a proliferation inducing factor, or factors, diffusing from the infarction to the neurogenic regions. The secreted protein osteopontin (OPN) is a possible factor. In this study, OPN mRNA levels were measured in the cerebral infarction of adult rats that underwent 1 hour of middle cerebral artery occlusion (MCAO). OPN mRNA levels increased 36.0, 55.0 and 46.7 fold at 6, 24 and 72 hours reperfusion respectively. We also determined whether OPN alone could be responsible for this ischemia-induced neurogenesis. OPN (2.4 μg/day) was infused into the lateral ventricles of the brain in non-ischemic adult male rats, continuously over three days. Bromodeoxyuridine (BrdU) immunohistochemistry was performed and the total BrdU positive ($BrdU^+$) cells were counted. OPN, compared to aCSF infusion, decreased $BrdU^+$ cells in DG and had no significant effect on cell proliferation in the SVZ. This study indicates that osteopontin *alone* does not increase cell proliferation in the normal adult brain.

Keywords: Osteopontin; neurogenesis; focal cerebral ischemia; bromodeoxyuridine; spontaneously hypertensive rat; integrin.

Introduction

It is now clear that the adult brain constantly undergoes neurogenesis [10–12]. This neurogenesis primarily occurs in to two distinct areas of the brain: The dentate gyrus (DG) of the hippocampus and the subventricular zone (SVZ) of the lateral ventricles. Various studies have demonstrated that focal cerebral ischemia further upregulates this neurogenesis via an unknown mechanism as a possible repair response [14, 15, 27]. In the standard models of focal ischemia, the cortex becomes infarcted without damaging the DG; yet, the ipsilateral DG cell proliferation is significantly upregulated. It is possible that this remote upregulation is a result of a proliferation-inducing factor, or factors, diffusing from the ischemic core and/or penumbra to the neurogenic regions, triggering an increase in cell proliferation.

Osteopontin (OPN), a secreted protein with diverse functions, has been demonstrated to be highly upregulated in the ischemic infarct. Its primary roles are involvement in bone mineralization, inflammation, wound repair, angiogenesis, cell survival, inhibiting inducible nitric oxide synthase, tumor proliferation and development of the cerebellum and brainstem [16, 18, 26]. In the first week after transient focal and global cerebral ischemia, OPN has been demonstrated to increase significantly over control levels [17, 28]. OPN binds to certain variant forms of CD44 and numerous integrins, specifically the $\alpha_v\beta_3$, $\alpha_v\beta_5$, $\alpha_v\beta_1$ and $\alpha_8\beta_1$ integrins, which are surface adhesion proteins with numerous α and β subunits that combine to form more than 20 different known receptors [3, 13, 19]. In brain development, the integrin receptors are thought to play a vital role in neurite outgrowth, migration, differentiation, and plasticity [2]. The DG granule cell layer of the adult brain expresses the $\alpha_v\beta_5$ integrin [21] and the hilar region of the DG expresses the $\alpha_8\beta_1$ integrin [9]. For these reasons, we tested whether OPN is responsible for this ischemia-induced neurogenesis by determining if exogenous OPN infused into the normal adult brain can increase neurogenesis.

Materials and methods

Middle cerebral artery occlusion

One hour MCAO was performed in spontaneously hypertensive adult rats (Charles River Laboratories, Wilmington, MA), 300 grams in weight, under halothane anesthesia, as previously described [8]. Sham animals underwent the procedure without the induction of

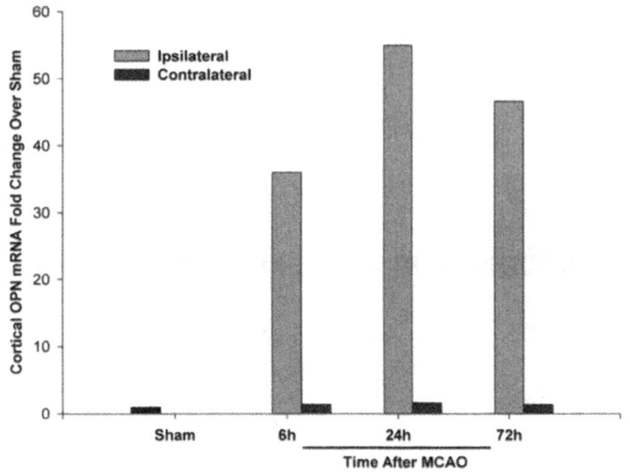

Fig. 1. Real-time PCR osteopontin mRNA fold changes versus sham from 1 h MCAO rat cortex, ipsilateral and contralateral to injury at 6 h, 24 h and 72 h reperfusion

ischemia. The animals were housed and cared for in accordance with the *Guide for the Care and Use of Laboratory Animals*, NIH publication 85-23, revised 1985. The animal use and procedures were approved by the Research Animal Resources and Care Committee of the University (RARC) of the University of Wisconsin-Madison.

Tissue preparation and real-time PCR

The rats were killed with halothane overdose and decapitation at 6 h (n = 3), 24 h (n = 3) and 72 h (n = 3) and sham animals were killed at 72 h (N = 3). The ipsilateral and contralateral cortex was removed and the homologous region was dissected from sham cortex. Total RNA was extracted from the samples and real-time PCR was performed as described earlier [24, 25]. The OPN primer was designed using the Primer Express Software (Applied Biosystems) based on the Gene Bank accession number M14656. The primer sequences for OPN were GCCAGCCAAGGACCAACTAC and CTGCCAAACTCAGCCACTTTC.

Pump preparation and implantation

A separate group of animals were implanted with Alzet osmotic minipumps (model 1003D, 1 μl/hr for 3 days) attached to a brain infusion kit II catheter (DURECT Corp., Cupertino, CA) that was placed into the left lateral ventricle (coordinates: lateral, −1.5 mm; anterior-posterior, −0.8 mm; and dorsal-ventral, −4.8 mm). Pumps were prepared as described previously [22]. Briefly, mouse recombinant OPN (Sigma-Aldrich, St. Louis, MO) was dissolved into aCSF (10 ng/μl) containing (in mM) 119 NaCl, 3.1 KCl, 1.2 $CaCl_2$, 1 $MGSO_4$, 0.5 KH_2PO_4, 25 $NaHCO_3$, 5 D-glucose, 2.2 urea, pH 7.4, 0.22 μm filtered, loaded into pumps, attached to the brain infusion kit and surgically implanted. 2.4 μg of OPN was infused in each experimental animal per day (n = 4) with control animals receiving aCSF only (n = 4). Animals were injected with BrdU (Sigma-Aldrich) 50 mg/kg i.p. twice per day at twelve-hour intervals (see Fig. 2).

Immunohistochemistry and cell quantification

Animals were killed with halothane overdose and decapitation at 72 h. Brains were removed, flash frozen in −20 °C isopentane, cryosectioned at 20 μm and mounted on slides. BrdU Immunohistochemistry was performed based on the methods by Zhou *et al.* [30]. Briefly, the sections were placed in 0.3% hydrogen peroxide in 100% methanol for 10 min. After washing in TBS, DNA was denatured by microwave irradiation. The sections were incubated in 10% horse serum for 30 min at 37 °C, and then in mouse anti-BrdU antibody (Oncogene, La Jolla, CA; 1:200 dilution) 1 h. After washing the sections were incubated in rat pre-adsorbed, biotinylated horse anti-mouse IgG (Vector Labs, Burlingame, CA) for 1 hour at 37 °C. The sections were then incubated for 1 h in avidin-boitin solution (Vector) at RT, washed and reacted with diaminobenzidine (DAB, Vector Labs). The sections were dehydrated and cover-slipped in permanent mounting medium. The slides were digitally imaged using an Olympus IX70 inverted microscope (Olympus USA, Melville, NY) with a Spot camera (Diagnostic Instruments, St. Sterling Heights, MI). BrdU-positive cells reacted with DAB were quantified using 8 sections from each experimental group, at the level of the DG, 3.3 mm caudal and SVZ, 0.3 mm rostral from the bregma. For quantification, the DG and SVZ area and $BrdU^+$ cell counts were measured using ImageJ software (NIH, Bethesda, MD, USA).

Results

Osteopontin mRNA levels after MCAO

OPN mRNA levels increased dramatically in the ipsilateral cortex compared to sham operated 36.0, 55.0 and 46.7 fold at 6 h, 24 h and 72 h respectively (Fig. 1). The increase is thought to be primarily due to macrophage infiltration into the ischemic core synthesizing and secreting OPN. Wang *et al.* used *in situ* hybridization of OPN with antibodies against the ED-1 antigen, which is specific for macrophages and microglia, and noticed a high degree of colocalization with macrophages [28]. We used SHR rats in this study, which Wang *et al.* also demonstrated to have a greater OPN increase after ischemia than normotensive rats [28].

Osteopontin infusion and $BrdU^+$ cell quantification

To determine whether OPN could be responsible for ischemia-induced neurogenesis, we infused OPN into the left lateral ventricle of normal, non-ischemic SHR rats and injected BrdU i.p. to label dividing cells. BrdU was injected 6 times at 12-hour intervals immediately after cannula implantation and the animals were sacrificed at 72 hours (Fig. 2). BrdU immunohistochemistry was performed and images were acquired. In the DG, the $BrdU^+$ cells were primarily located in the subgranule cell layer of the DG as indicated by the dark cells (Fig. 2a–d). In the lateral ventricle sections,

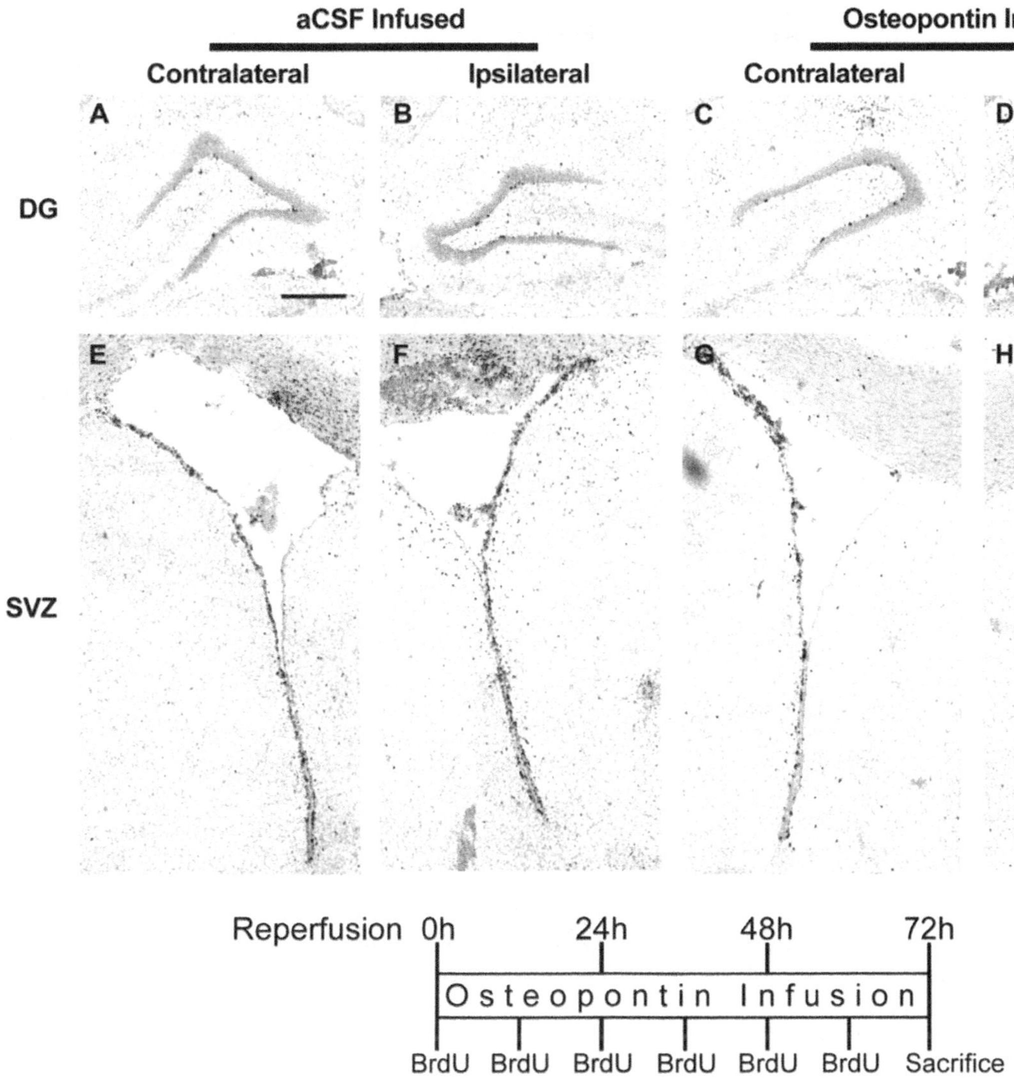

Fig. 2. BrdU immunohistochemistry of DG (A–D) and SVZ (E–H) at 72 hours of infusion, using the injection schedule below the sections. Note the darkly stained BrdU^{+} cells in the subgranule cell layer of the DG and lateral lining of the ventricles. Bar = 500 μm

the cells counted were located in the SVZ, the lateral lining of the ventricles, as indicated by the layer of dark BrdU^{+} cells (Fig. 2e–h). ANOVA analysis was performed to determine significance of the cell counts. In the DG, the BrdU^{+} cells decreased with OPN infusion, ipsilateral to the implantation, compared to the DG BrdU^{+} cells with aCSF infusion, ipsilateral to the implantation (Fig. 3a, $p < 0.05$). Contralateral DG BrdU^{+} cell counts were not statistically different from each other or as compared to the ipsilateral groups. There was no statistical change in any of the SVZ groups (Fig. 3b). These results indicate that OPN alone does not have a direct role, or possibly has an inhibitory role in adult neurogenesis.

Discussion

OPN has multiple roles in the various organs of the body. OPN is primarily involved in bone formation and remodeling, yet it is involved in other functions including: Controlling kidney stone formation [29], angiogenesis and smooth muscle formation and migration, high expression in carcinomas [4] and esophageal tumors [5], inhibition of nitric oxide synthase, reduction of cell peroxidase levels, protection from hypoxic injury [7], and it is highly upregulated in cerebral ischemia [28]. One role that we attempted to determine was whether the tremendous OPN induction after focal ischemia could be one of

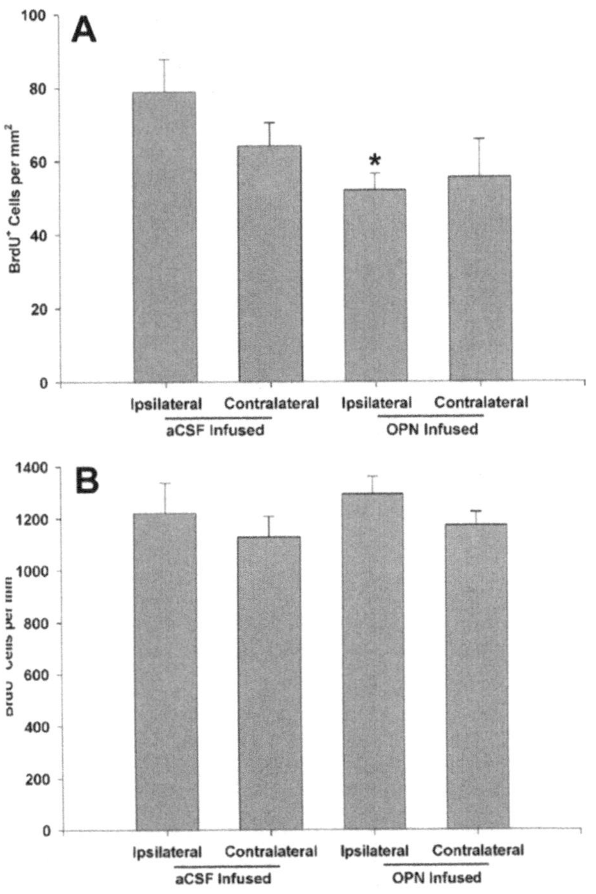

Fig. 3. $BrdU^+$ cells in the DG (A) and SVZ (B) per mm^2 with OPN or aCSF infusion cell counts ipsilateral or contralateral to the implantation. The number of DG $BrdU^+$ OPN ipsilateral cells were significantly lower (* $p < 0.05$) than that of the ipsilateral aCSF infused DG. SVZ $BrdU^+$ cell counts were not significantly different from each group

the factors responsible for ischemia-induced neurogenesis.

The results indicate that exogenous infusion of OPN into the normal adult brain had little, or possibly inhibitory effects on basal neurogenesis. The aim of this study was to screen the possibility of OPN's neurogenic potential. After focal cerebral ischemia, a cascade of biochemical events ensues which could cloud results. OPN was infused into the normal adult brain as a means to determine the role of OPN alone. It may be possible that OPN has a synergistic role with some other factor that is upregulated after ischemia, or it may have a role in adult neuronal progenitor cell migration and differentiation. Chavis *et al.* have demonstrated that the β_3 integrin subunit may be essential for hippocampal excitatory synapse maturation [6]. The β_3 integrin subunit only forms heterodimers with the α_{IIb} or the α_V subunit. The granule cell layer of the DG has been demonstrated to contain the α_V integrin subunit. OPN binds to the $\alpha_V\beta_3$ integrin and it is possible that the hippocampal β_3 subunit, which is associated with synaptic maturation, may be part of the $\alpha_V\beta_3$ integrin. OPN's possible role in neuronal migration and synaptic maturation needs further study.

Most likely, the primary role of OPN after ischemia is in macrophage recruitment. It could be through this mechanism that OPN indirectly has an effect on neurogenesis. O'Donnell *et al.* have demonstrated that in the ischemic infarct, insulin-like growth factor-1 (IGF-1), a known adult neurogenic factor [1], is secreted from macrophages [20]. OPN has also been shown to be a strong promoter of macrophage infiltration [23]. By increasing macrophage infiltration, OPN increases IGF-1 secretion and may indirectly promote ischemia-induced neurogenesis. Furthermore, Nam *et al.* have demonstrated that OPN has a synergistic effect on IGF-1 by enhancing its mitogenic properties [19]. When the combination of IGF-1 and OPN was administered to smooth muscle cells, the cell proliferation was significantly higher than with IGF-1 administration alone.

Further studies using higher doses of OPN as well as OPN knockdown or knockout studies, in conjunction with ischemia will be needed to further determine OPN's role in ischemia-induced neurogenesis.

References

1. Aberg MA, Aberg ND, Hedbacker H, Oscarsson J, Eriksson PS (2000) Peripheral infusion of IGF-I selectively induces neurogenesis in the adult rat hippocampus. J Neurosci 20: 2896–2903
2. Anton ES, Kreidberg JA, Rakic P (1999) Distinct functions of alpha3 and alpha(v) integrin receptors in neuronal migration and laminar organization of the cerebral cortex. Neuron 22: 277–289
3. Brandenberger R, Schmidt A, Linton J, Wang D, Backus C, Denda S, Muller U, Reichardt LF (2001) Identification and characterization of a novel extracellular matrix protein nephronectin that is associated with integrin alpha8beta1 in the embryonic kidney. J Cell Biol 154: 447–458
4. Brown LF, Papadopoulos-Sergiou A, Berse B, Manseau EJ, Tognazzi K, Perruzzi CA, Dvorak HF, Senger DR (1994) Osteopontin expression and distribution in human carcinomas. Am J Pathol 145: 610–623
5. Casson AG, Wilson SM, McCart JA, O'Malley FP, Ozcelik H, Tsao MS, Chambers AF (1997) ras mutation and expression of the ras-regulated genes osteopontin and cathepsin L in human esophageal cancer. Int J Cancer 72: 739–745
6. Chavis P, Westbrook G (2001) Integrins mediate functional pre- and postsynaptic maturation at a hippocampal synapse. Nature 411: 317–321

7. Denhardt DT, Lopez CA, Rollo EE, Hwang SM, An XR, Walther SE (1995) Osteopontin-induced modifications of cellular functions. Ann NY Acad Sci 760: 127–142
8. Dogan A, Baskaya MK, Rao VL, Rao AM, Dempsey RJ (1998) Intraluminal suture occlusion of the middle cerebral artery in Spontaneously Hypertensive rats. Neurol Res 20: 265–270
9. Einheber S, Pierce JP, Chow D, Znamensky V, Schnapp LM, Milner TA (2001) Dentate hilar mossy cells and somatostatin-containing neurons are immunoreactive for the alpha8 integrin subunit: characterization in normal and kainic acid-treated rats. Neuroscience 105: 619–638
10. Eriksson PS, Perfilieva E, Bjork-Eriksson T, Alborn AM, Nordborg C, Peterson DA, Gage FH (1998) Neurogenesis in the adult human hippocampus. Nat Med 4: 1313–1317
11. Gage FH (2002) Neurogenesis in the adult brain. J Neurosci 22: 612–613
12. Gould E, Tanapat P, McEwen BS, Flugge G, Fuchs E (1998) Proliferation of granule cell precursors in the dentate gyrus of adult monkeys is diminished by stress. Proc Natl Acad Sci USA 95: 3168–3171
13. Hynes RO (1992) Integrins: versatility, modulation, and signaling in cell adhesion. Cell 69: 11–25
14. Jiang W, Gu W, Brannstrom T, Rosqvist R, Wester P (2001) Cortical neurogenesis in adult rats after transient middle cerebral artery occlusion. Stroke 32: 1201–1207
15. Jin K, Minami M, Lan JQ, Mao XO, Batteur S, Simon RP, Greenberg DA (2001) Neurogenesis in dentate subgranular zone and rostral subventricular zone after focal cerebral ischemia in the rat. Proc Natl Acad Sci USA 98: 4710–4715
16. Lee MY, Choi JS, Lim SW, Cha JH, Chun MH, Chung JW (2001) Expression of osteopontin mRNA in developing rat brainstem and cerebellum. Cell Tissue Res 306: 179–185
17. Lee MY, Shin SL, Choi YS, Kim EJ, Cha JH, Chun MH, Lee SB, Kim SY (1999) Transient upregulation of osteopontin mRNA in hippocampus and striatum following global forebrain ischemia in rats. Neurosci Lett 271: 81–84
18. Mazzali M, Kipari T, Ophascharoensuk V, Wesson JA, Johnson R, Hughes J (2002) Osteopontin – a molecule for all seasons. QJM 95: 3–13
19. Nam TJ, Busby WH Jr, Rees C, Clemmons DR (2000) Thrombospondin and osteopontin bind to insulin-like growth factor (IGF)-binding protein-5 leading to an alteration in IGF-I-stimulated cell growth. Endocrinology 141: 1100–1106
20. O'Donnell SL, Frederick TJ, Krady JK, Vannucci SJ, Wood TL (2002) IGF-I and microglia/macrophage proliferation in the ischemic mouse brain. Glia 39: 85–97
21. Pinkstaff JK, Detterich J, Lynch G, Gall C (1999) Integrin subunit gene expression is regionally differentiated in adult brain. J Neurosci 19: 1541–1556
22. Rao VL, Dogan A, Todd KG, Bowen KK, Kim BT, Rothstein JD, Dempsey RJ (2001) Antisense knockdown of the glial glutamate transporter GLT-1, but not the neuronal glutamate transporter EAAC1, exacerbates transient focal cerebral ischemia-induced neuronal damage in rat brain. J Neurosci 21: 1876–1883
23. Singh RP, Patarca R, Schwartz J, Singh P, Cantor H (1990) Definition of a specific interaction between the early T lymphocyte activation 1 (Eta-1) protein and murine macrophages in vitro and its effect upon macrophages in vivo. J Exp Med 171: 1931–1942
24. Song G, Cechvala C, Resnick DK, Dempsey RJ, Rao VL (2001) GeneChip analysis after acute spinal cord injury in rat. J Neurochem 79: 804–815
25. Song G, Dhodda VK, Blei AT, Dempsey RJ, Rao VL (2002) GeneChip analysis shows altered mRNA expression of transcripts of neurotransmitter and signal transduction pathways in the cerebral cortex of portacaval shunted rats. J Neurosci Res 68: 730–737
26. Takagi H, Suzuma K, Otani A, Oh H, Koyama S, Ohashi H, Watanabe D, Ojima T, Suganami E, Honda Y (2002) Role of vitronectin receptor-type integrins and osteopontin in ischemia-induced retinal neovascularization. Jpn J Ophthalmol 46: 270–278
27. Takasawa K, Kitagawa K, Yagita Y, Sasaki T, Tanaka S, Matsushita K, Ohstuki T, Miyata T, Okano H, Hori M, Matsumoto M (2002) Increased proliferation of neural progenitor cells but reduced survival of newborn cells in the contralateral hippocampus after focal cerebral ischemia in rats. J Cereb Blood Flow Metab 22: 299–307
28. Wang X, Louden C, Yue TL, Ellison JA, Barone FC, Solleveld HA, Feuerstein GZ (1998) Delayed expression of osteopontin after focal stroke in the rat. J Neurosci 18: 2075–2083
29. Yasui T, Fujita K, Asai K, Kohri K (2002) Osteopontin regulates adhesion of calcium oxalate crystals to renal epithelial cells. Int J Urol 9: 100–108
30. Zhou SS, Gospe SM (1998) Double labeling of proliferating neurons with anti-BrdU and anti-neun: an improved immunohistochemical technique utilizing microwave irradiation. J Histotech 21: 201–204

Correspondence: Robert J. Dempsey, M.D., University of Wisconsin-Madison, Department of Neurological Surgery, 600 Highland Ave, K4/818 CSC, Madison, WI 53792. e-mail: Dempsey@neurosurg.wisc.edu

Acta Neurochir (2003) [Suppl] 86: 187–189

Dissociative increase of oligodendrocyte progenitor cells between young and aged rats after transient cerebral ischemia

K. Ohta, M. Iwai, K. Sato, N. Omori, I. Nagano, M. Shoji, and **K. Abe**

Department of Neurology, Graduate School of Medicine and Dentistry, Okayama University, Japan

Summary

Oligodendrocyte progenitor cells (OPCs) mature to oligodendrocyte and remyelinate axons. OPCs express neuron-glial antigen 2 (NG2) chondroitin sulfate proteoglycan. We stained NG2 to investigate the effect of aging on the OPCs after cerebral ischemia. NG2 positive cells were examined at 1, 3, 7 days after 90 min of transient middle cerebral artery occlusion in young and aged brains. The number of NG2 positive cells increased in the ischemic penumbra at 3 and 7 days after reperfusion. At 7 day, the number of NG2 positive cells was significantly greater in the young than the aged brains, and the processes of NG2 positive cells enlarged and were more highly branched in the young than the aged brains. These results suggest that the young brain shows a higher potential of proliferation and process branching of OPCs than the aged brain.

Keywords: Transient ischemia; aging; remyelination; oligodendrocyte; oligodendrocyte progenitor cell; neuron-glial antigen 2 (NG2).

Introduction

Brain ischemia causes cessation of oxidation and reduction of energy metabolism, thus damaging the function and structure of brain cells [1–4]. It is reported that aged brain shows a reduced biological plasticity to meet emergency conditions, and ischemic damage is more pronounced with advancing age [1, 5]. Although stroke in humans usually afflicts the elderly, young-adult animals have been used in most experimental studies on cerebral ischemia [6]. Only a few works have been reported with aged animals.

Oligodendrocyte progenitor cells (OPCs) make up 5–8% of the glial cell population in the CNS with satellite-shape body and small processes. Their processes are in contact with nodes of Ranvier and synapses [7]. OPCs are present during development in rat brain for differentiation into oligodendrocytes [8]. But, they remain in adult rat brain. The number of OPCs increases after cerebral ischemia [6], mechanical brain injury [9], inflammation [10], and kainate injection. Because OPCs have potentials to mature into oligodendrocyte and to remyelinate of injured axons [7, 8], OPCs membranes express neuron-glial antigen 2 (NG2) chondroitin sulfate proteoglycan, platelet-derived growth factor-α receptor (PDGFαR), and O4 antigen [8]. We examined the difference of NG2 positive OPCs between young and aged brain after ischemia.

Materials and methods

Adult male Wistar rats (SLC, Hamamatsu, Japan) at age 11 weeks (body weight 240–300 g, young-adult group) and at age 55–70 weeks (body weight 390–540 g, aged-adult group) were anesthetized with an intraperitoneal injection of pentobarbital (10 mg/250 g rat). A burr hole with a diameter of 2 mm for measurement of cerebral blood flow (CBF) was made in the skull using an electric drill to avoid traumatic brain injury. The location of the burr hole was 3 mm dorsal and 4 mm lateral to the right from bregma, which is located in the upper part of the middle cerebral artery (MCA) territory. Dura mater was preserved.

On the next day, the rats were anesthetized by inhalation of a nitrous oxide/oxygen/halothane (69%:30%:1%) mixture during surgical preparation. The right MCA was occluded by an insertion of a 4-0 nylon thread through the common carotid artery according to our previous report [12–14]. The nylon thread for aged rats was 2 mm longer than that for young rats and they were the same thickness. Body temperature was maintained at $37 \pm 0.3\,^{\circ}C$ during the surgical procedure for MCA occlusion (MCAO). After 90 min occlusion, CBF was restored by removing the thread. CBF of the right frontoparietal cortex was measured using Omega Flow FLO-C1 (Neuroscience Inc. Tokyo, Japan) before, during and after MCAO. Age-matched sham control animals were treated in the same way without MCAO in the young and aged-adult groups ($n = 4$ in each). The animals recovered at ambient temperature (21–24 °C) until sampling. This experiment was approved by the Animal Committee of Graduate School of Medicine and Dentistry, Okayama University.

For immunohistochemical staining, rat forebrains were removed and quickly frozen in powdered dry ice at 1, 3 and 7 days ($n = 5$ in each) after reperfusion. Coronal sections at the caudate level were cut on a cryostat at −20 °C (10 μm thickness), and collected on glass slides coated with polylysine. The sections were fixed in 4% paraformaldehyde and quenched with 0.3% H_2O_2 in methanol for 30 min to block endogenous peroxidase action. They were incubated for 60 min in phosphate-buffered saline (PBS, pH 7.4) containing 5% bovine serum albumin (Sigma. St. Louis, MO) in order to block non-specific binding of the antibody. The immunofluorescence staining for NG2 was carried out using rabbit anti-NG2 chondroitin sulfate proteoglycan polyclonal antibody (diluted 1:800, Chemicon, Temecula, CA) overnight at 4 °C. The sections were incubated with rhodamine-labeled goat anti-rabbit IgG (1:500, Chemicon) for 1 hr.

Staining was examined under the microscope (Olympus BX51, Olympus Optical Co., Ltd. Tokyo, Japan), equipped with a rodamine filter set. Images (1024 × 1024 pixels) were obtained by Aquacosmos System (Hamamatsu Photonics, Hamamatsu, Japan). Immunopositive cells in the ipsilateral cerebral cortex (ischemic core and penumbra) and contralateral cerebral cortex in 0.15 mm^2 of each animal were counted in a blind manner and averaged. The intensity of immunoreative staining for NG2 was analyzed using Adobe Photoshop 5.0J. The differences between them were statistically analyzed by two-way ANOVA (analysis of variance) followed by Fisher's Protected Least Significant Difference (Fisher's PLSD) with $p < 0.01$ as significant.

Results

Right hemisphere CBF was not different between the young- and aged-adult rats before, during and after MCAO as we previously reported (data not shown).

In sham control brains of both young and aged-adult groups, NG2 positive cells were detected in every part of the brain including cerebral cortex, caudate and corpus callosum. The NG2 positive cells were small in size, and round or oval in shape (Fig. 1A, E, arrowheads). In sham control animals, the number of NG2 positive cells was not different between the young- and aged-adult groups.

In young-adult animals the number of NG2 positive cells began to decrease at 1 day after reperfusion in the ischemic core, and greatly decreased at 3 and 7 days. In the ischemic penumbra, the number of NG2 positive cells showed a gradual increase with significance at 3 and 7 days after the reperfusion (Figs. 1B–D). Although NG2 positive cells had similar size and shape at 1 day (Fig. 1B) to the sham control (Fig. 1A), they began to have small-branched processes at 7 days (Fig. 1D, arrowheads). In the contralateral cerebral cortex, no significant change was observed in the number or shape of NG2 positive cells.

In aged-adult animals, the number of NG2 positive cells also decreased at 1 day after reperfusion in the ischemic core, and greatly decreased at 3 and 7 days. In the ischemic penumbra, the number of NG2 positive cells increased with significance at 3 and 7 days after reperfusion (Figs. 1F–H). Although NG2 positive cells had similar size and shape at 1 day (Fig. 1F) to the sham control (Fig. 1E), they began to have small and branched processes at 7 days (Fig. 1H, arrowheads). In the contralateral cerebral cortex, no change was observed in the number or shape of NG2 positive cells.

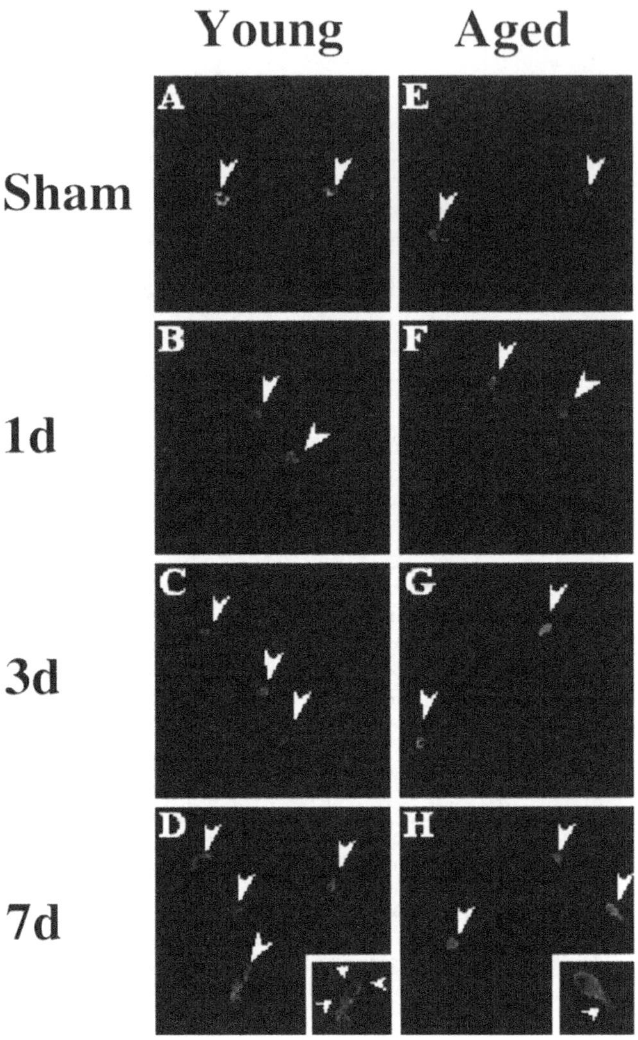

Fig. 1. Immunofluorescence staining for NG2 in the penumbra of sham control (*SC*) and 1, 3 and 7 days after the reperfusion. Young brains in panels A–D, and aged brains in E–H. Note the increase of NG2 positive cell number after 3 days of reperfusion, especially in the young brain (C, D). Also note multiple and branched processes at 7 day in the young brain (D, yellow arrowheads)

Although changes in the number of NG2 positive cells were similar between young- and aged-adult animals in both the ischemic core and the penumbra, the number was significantly greater in the young animals than the aged at 7 days in the penumbra. In addition, the number of processes and branches were much smaller in the aged animals (Fig. 1H) than the young (Fig. 1D).

Discussion

The number of NG2 positive cells greatly decreased in the ischemic core of rat brain at 3 and 7 days. Many previous reports showed that most neuronal cells die at an early stage of reperfusion in the ischemic core [2, 15]. Thus, the early decrease of NG2 positive cells in the ischemic core may be the result of oligodendrocytic cell death.

The number of NG2 positive cells increased in the ischemic penumbra of rat brain at 3 and 7 days in both young and aged animals. This is consistent with previous reports which show that OPCs have dividing ability in the adult brain after ischemia [6, 7, 16]. Because OPCs have low ability to migrate, the increased number of NG2 positive cells can be attributed to new proliferation, but not migration, from another place such as the subventricular zone of the lateral ventricle. In the contralateral hemisphere, the number and shape of NG2 positive cells did not change, probably because there was no ischemic stress on that side.

The present study found that the number of NG2 positive cells was larger in the young brain than the aged at 7 days (Figs. 1D, H), although the right CBF was not different between the two groups. It is reported that remyelination is performed by premyelinating oligodendrocytes in multiple sclerosis lesions, and these oligodendrocytes are considered to differentiate from OPCs. Therefore, after the ischemic injury, proliferation of NG2 positive cell in the penumbra suggests rearrangement of axonal myelination. Such a difference between young and aged brains indicates stronger plasticity and recovery of damaged areas in the young brain.

Enlarged number and process elongation of NG2 positive cells were previously reported after mechanical brain injury [9], inflammation [10], and kinate injection. However, the mechanism proliferation enhancement of NG2 positive cells after brain insults are poorly understood. It is not clear why this enhancement occurs between the young brain and aged one. The number of processes and branches were much smaller in the aged animals (Fig. 1H) than the young (Fig. 1D) in this study. Because processes elongation of NG2 positive cells reflects one stage of oligodendrocyte differentiation, the age-dependent decrease of the processes elongation in the present study (Fig. 1D, H) also suggests less plasticity and recovery in the aged brain. An unknown factor could be present in the aged brain that inhibits proliferation of new oligodendrocytes.

In previous reports, some OPCs expressed a neuronal marker (microtubule associated protein 2) [19] and an astrocytic marker (glial fibrillary acidic protein) [20] in vitro. The function of NG2 positive cells is not only to contribute to repair of damaged tissue by supplying myelin-forming cells [7] but also to have stem cell like function. Therefore, NG2 positive OPCs could provide neuronal and astroglial cells in vivo. Further study may be required to learn the role of OPCs in the recovery of ischemic brain.

Acknowledgments

This work was partly supported Grant-in-Aid for Scientific Research (B) 12470141 and (Hoga) 12877211 and National Project on Protein Structural and Function Analyses from the Ministry of Education, Science, Culture and Sports of Japan, and by grants (Itoyama Y, Kimura I and Kuzuhara S) from the Ministry of Health and Welfare of Japan.

References

1. Abe K (2000) Therapeutic potential of neurotrophic factors and neural stem cells against ischemic brain injury. J Cereb Blood Flow Metab 20: 1393–1408
2. Abe K, Aoki M, Kawagoe J, Yoshida T, Hattori A, Kogure K, Itoyama Y (1995) Ischemic delayed neuronal death: a mitochondrial hypothesis. Stroke 26: 1478–1489
3. Abe K, Tanzi RE, Kogure K (1991) Selective induction of Kunitz-type protease inhibitor domain-containing amyloid precursor protein mRNA after persistent focal ischemia in rat cerebral cortex. Neurosci Lett 125: 172–174
4. Kondo T, Raff M (2000) Oligodendrocyte precursor cells reprogrammed to become multipotential CNS stem cells. Science 289: 1754–1757
5. Levine JM (1994) Increased expression of the NG2 chondroitin-sulfate proteoglycan after brain injury. J Neurosci 14: 4716–4730
6. Levine JM, Reynolds R, Fawcett JW (2001) The oligodendrocyte precursor cell in health and disease. Trends Neurosci 24: 39–47
7. Levine JM, Stincone F, Lee YS (1993) Development and differentiation of glial precursor cells in the rat cerebellum. Glia 7: 307–321
8. Litten L, Raff M (1990) Differentiation signals in the CNS: Type-2 astrocyte development in vitro as a model system. Neuron 5: 111–119
9. Nishiyama A, Yu M, Drazba JA, Vincent KT (1997) Normal and reactive NG2 + glial cells are distinct from resting and activated microglia. J Neurosci Res 48: 299–328
10. Tanaka K, Nogawa S, Ito D, Suzuki S, Dembo T, Kosakai A, Fukuuchi Y (2001) Activation of NG2-positive oligodendrocyte progenitor cells during post-ischemic reperfusion in the rat brain. Neuroreport 12: 2169–2174

Correspondence: K. Abe, M.D., Ph.D., Department of Neurology, Graduate School of Medicine and Dentistry, Okayama University, 2-5-1 Shikata-cho, Okayama 700-8558, Japan. e-mail: shinnai@cc.okayama-u.ac.jp

Acta Neurochir (2003) [Suppl] 86: 191–194

Suppression of oxidative stress after transient focal ischemia in interleukin-1 knock out mice

H. Ohtaki[1,4], **A. Takaki**[2], **L. Yin**[1,4], **K. Dohi**[1], **T. Nakamachi**[1], **M. Matsunaga**[1], **R. Horai**[3], **M. Asano**[3], **Y. Iwakura**[3], and **S. Shioda**[1,4]

[1] Department of Anatomy, Showa University School of Medicine, Tokyo, Japan
[2] Department of Integrative Physiology, Graduate School of Medical Sciences Kyushu University, Fukuoka, Japan
[3] The Institute of Medical Science, Laboratory of Animal Research Center, The University of Tokyo, Tokyo, Japan
[4] The Core Research for Evolutional Science and Technology (CREST) of Japan Science and Technology Corporation (JST), Tokyo, Japan

Summary

Interleukin-1 (IL-1) contributes to ischemic neurodegeneration. However, the mechanisms regulating action of IL-1 are still poorly understood. In order to clarify this central issue, mice that were gene deficient both IL-1α and β (IL-1 KO) and wild-type mice were subjected to 1 hour transient middle cerebral artery occlusion (tMCAO). The concentration of 8-hydroxy deoxyguanosine (8OHdG) which is considered to be a reliable oxidative DNA damage by superoxide anion, in brain and of total nitric oxide (NO) in plasma were determined by use of HPLC. Twenty-four hours after tMCAO, the ratio of 8OHdG to dG in the ipsilateral hemisphere of wild-type mice were 2.24×10^{-3} and 4.41×10^{-3} in the neocortex and striatum, respectively. The concentration of 8OHdG in the ipsilateral hemisphere of the wild-type mice was higher than that of the IL-1 KO mice. The concentration of total NO in the plasma of IL-1 KO mice was also lower than that of the wild-type 24 hours after tMCAO. These results strongly suggest that IL-1 is participated in generating reactive oxygen spices and it aggravates and induces the ischemic neuronal cell death. (183 words)

Keywords: Interleukin-1 (IL-1); focal brain ischemia; 8-hydroxy deoxyguanosine (8OHdG); nitric oxide; oxidative stress.

Introduction

Interleukin-1 (IL-1), a proinflammatory cytokine, has diverse actions in brain and there is considerable evidence implicating that it plays an important role in neurodegeneration [3, 10, 12]. The neuronal cell death of IL-1 knockout (IL-1 KO) mice is reported to decrease neurodegeneration as compared with that of wild-type mice after transient middle cerebral artery occlusion [3] and we have recently reported the similar results in the transient global ischemia induced by using heart arrest model [10]. However, the mechanisms regulating the action of IL-1 are still poorly understood.

One hypothesis is that the cellular damages involving reactive oxygen species-mediated oxidative damage evoke neurodegeneration. It is considered that the reactive oxygen species are mass-produced at the time of pathological condition, such as infection, inflammation, and ischemia, and they play a critical role in neuronal cell damage [1, 9, 14]. Previous reports have implicated that IL-1 induces reactive oxygen species for tissue damage. The addition of IL-1β into the cultured human microglia and rat microvascular endothelial cells stimulates inducible NOS (iNOS) mRNA expression and generates NO [2, 5]. A free radical inducer, cyclooxygenase-2, is induced by IL-1β during inflammatory periods in the hippocampal neuronal cultures [13]. These results suggest that reactive oxygen species formation may be associated with the neurodegenerative mechanism induced by IL-1 after ischemia/reperfusion. However, the relationship between IL-1 and reactive oxygen species formation after brain ischemia is still unclear. Therefore, the purpose of the present study is to determine whether IL-1 induces the reactive oxygen species after transient focal ischemia by using IL-1 KO mice.

Materials and methods

Mice

Mice with homozygous disruption of both IL-1α genes β (IL-1 KO) have been described previously [7]. IL-1 KO mice that had been backcrossed for seven to nine generations into BALB/c strain were used. Wild-type mice were generated from the same chimeric founder. All experimental procedures involving animals were approved by the Institutional Animal Care and Use Committee of Showa University (#02097).

Production of transient middle cerebral artery occlusion (tMCAO)

Adult male mice weighing 24 to 27.5 g (8 to 12 weeks) were subjected to focal cerebral ischemia. Anesthesia was induced by inhalation of 2.0% sevoflurane in N_2O/O_2 (70/30%) and body temperature was maintained by a heating blanket. Ischemia was induced by occlusion of the left middle cerebral artery (MCA) using the intraluminal filament technique. Left common carotid artery, external carotid artery (ECA), internal carotid artery (ICA) and pterygoparatine artery were exposed and the distal part of ECA was occluded. Then, a 7-0 round tip nylon suture was introduced into the ICA via the ECA, advanced 8- to 10-mm distal to the carotid bifurcation and occluded MCA. At 1 hour after ischemia, the mice were reanesthetized and the suture was withdrawn through ECA. Twenty-four hours after tMCAO, the animals were deeply anesthetized and blood and brain samples were collected. The blood samples were immediately heparinized and the plasma samples were obtained. The brain samples were immediately separated into cortex and striatum of both hemispheres on the plates cooled with ice and then they were snap frozen using liquid nitrogen. The samples were stored at −80 °C until assayed.

Measurement of brain 8-hydroxy deoxyguanosine (8OHdG) and deoxyguanosine (dG) levels

The 8OHdG levels in brain were measured by HPLC system including the electro-chemical detector. In brief, each brain was homogenized and centrifuged at 15000 rpm for 5 minutes. Then, the supernatant was collected and deproteinized by 50% methanol, centrifuged again and applied to microdialysis. The percent of collection by microdialysis was approximately 50%. These samples were analyzed by use of HPLC for detection of 8OHdG and dG. All data were shown by the ratio of 8OHdG to dG (8OHdG/dG).

Measurement of plasma NO_2^- and NO_3^- (total NO) levels

The total NO levels were measured by automatically HPLC system obtained from Eicom (Kyoto, Japan). In brief, the plasma samples after brain ischemia were deproteinized by 50% methanol, centrifuged and applied to HPLC. The HPLC was composed from separation of NO_2^- and NO_3^-, the reduction from NO_3^- to NO_2^-, and the measurement of NO_2^-. The colorimetric assay was based on the modified Griess's methods. The absorbance of the color of the product dye at 540 nm was measured. Total NO was calculated from the sum of NO_2^- and NO_3^-.

Results

Mouse tMCAO model

In our preliminary experiments we found that mice weighing less than 24.0 g were more likely to suffer from hemorrhage after reperfusion, while very few mice weighing in excess of 28.0 g clearly showed neurological symptoms during the tMCAO. Therefore, only mice weighing in the range of 24 to 27.5 g (7–14 weeks of age) were subjected to focal cerebral ischemia in this experiment. It should be noted that approximately 15% of mice in this body weight range that were subjected to tMCAO did not display the expected neurological symptoms, or hemorrhaged after reperfusion. These mice were excluded from the analysis on the basis of technical error. The duration of ischemia (1 hour) was chosen because, while it induced injury, ischemia of longer duration (for 2 hours) was associated with a high mortality (approximately 50% at 24 hours). The level of mortality due to severe infarction after tMCAO was 26% within 24 hours of tMCAO. In the results presented here, the data from approximately 60% of animals that originally underwent tMCAO were used in the final analysis. There was no significant difference in the mortality between the two groups at 24 and 48 hours after tMCAO.

Brain 8OHdG levels

The 8OHdG is oxidative products of deoxyguanosine (dG), and is considered to be a reliable oxidative DNA damage by superoxide anion (O_2^-) and hydrogen peroxide (H_2O_2). Usually, oxidative dG (8OHdG) is cut off from DNA and is excreted into urine, and DNA is repaired. However, it is considered that in the pathological condition, excess production of 8OHdG at oxidative stress is accumulated in cells and causes oxidative neuronal damage. Thus, we

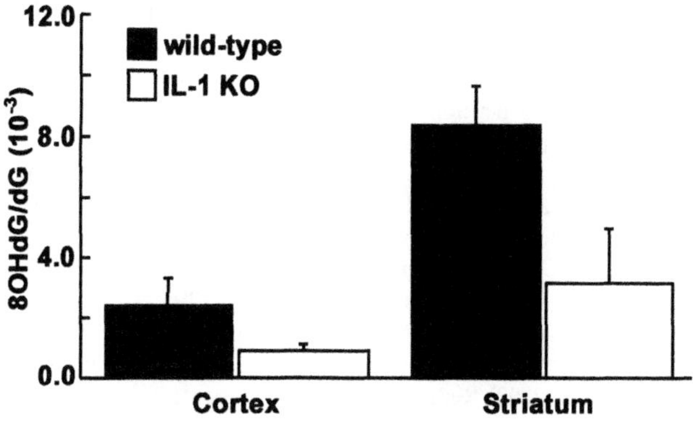

Fig. 1. 8OHdG/dG ratio of the cortex and the striatum at 24 hours after tMCAO in the wild-type and IL-1β KO mice. Brain 8OHdG/dG ratio of wild-type mice was higher than that of IL-1 KO mice at 24 hours after tMCAO. Each value are shown the mean ± SD (n = 3 to 4)

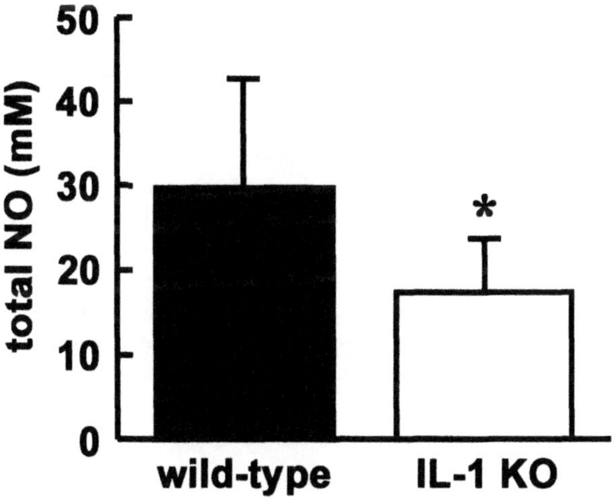

Fig. 2. Plasma total NO level at 24 hours after tMCAO in the wild-type and IL-1β KO mice. Plasma total NO level of wild-type mice (n = 7) was significantly higher than that of IL-1 KO mice (n = 7; * $P < 0.05$) at 24 hours after tMCAO. Each value is shown the mean ± SD

measured the levels of 8OHdG in brain to clarify whether IL-1 after transient focal ischemia increases the oxidative stress by O_2^- and H_2O_2 or not. Twenty-four hours after tMCAO, the ratio of 8OHdG/dG in the ipsilateral hemisphere of the wild-type mice was 2.24×10^{-3} and 4.41×10^{-3} in the cortex and the striatum, respectively (Fig. 1, n = 3–4). The concentration of 8OHdG in the ipsilateral hemisphere of the wild-type mice was higher than that of the IL-1 KO mice.

Plasma total NO levels

Moreover, we compared the relationship of NO and IL-1 in plasma. (n = 7 in each group, Fig. 2). The plasma total NO level of the wild-type mice (29.6 ± 13.1 mM) was significantly higher than that of the IL-1 KO mice (17.3 ± 6.28 mM, $p < 0.05$) after 24 hours after tMCAO (Fig. 2).

Discussion

A number of previous reports have implicated that IL-1 plays an important role in neurodegeneration associated with ischemic insult [3, 10, 12]. However, the mechanisms underlying this are not well clarified in detail. For this reason, we made this experiment to clarify the effects of IL-1 and the basis of its role in neurodegeneration by reactive oxygen species using IL-1 KO mice.

There are several reports showing that the reactive oxygen species are induced during ischemia and are deteriorated neuronal cell damage [1, 9, 14]. Homma *et al.* [6] have used immunohistochemical techniques to show that three NOS subtypes are increased at different periods in the ipsilateral hemisphere after permanent ischemia in rat. Furthermore, the experiments in transgenic mice over-expressing superoxide dismutase and NOS knockout were revealed that reactive oxygen species play very important roles in brain damage [4, 8].

Recently, we have reported that the immunoreactivities of iNOS in the IL-1 KO mice were weaker than those of wild-type mice in the hippocampal region following forebrain ischemia, and that NO levels of IL-1 KO mice in the hippocampal region were lower than those of wild-type mice [10]. Moreover, we reported the number of immunopositive cells for 3-nitrotyrosine (3-NT), which is a reactive oxidant between L-tyrosine and peroxynitrite ($ONOO^-$), in the wild-type and IL-1 KO mice after tMCAO. In the paper, the number of 3-NT positive cells in the wild-type mice is significantly higher than that of IL-1 KO mice [11]. In the present study, the generation of plasma NO and of 8OHdG in the IL-1 KO mice was down-regulated at 24 hours after tMCAO as compared with wild-type mice. These and our recent results strongly suggest that IL-1 regulates reactive oxygen species including NO, O_2^- and $ONOO^-$ and it aggravates neuronal cell death.

Acknowledgments

This work was supported in part by a Showa University Grant-in-Aid for Innovative Collaborate Research Projects and a Special

Research Grant-in Aid for development of Characteristic education from the Japanese Ministry of Education, Culture, Sports, Science, and Technology (S.S.). This study was also supported in part by grants from a Showa University High-Technology Research Center Project from the Japanese Ministry of Education, Culture, Sports, Science, and Technology (H.O. and S.S.).

References

1. Bolaños JP, Almeida A (1999) Role of nitric oxide in brain hypoxia-ischemia. Biochim Biophys Acta 1411: 415–436
2. Bonmann E, Suschek C, Spranger M, Kolb-Bachofen V (1997) The dominant role of exogenous or endogenous interleukin-1 beta on expression and activity of inducible nitric oxide synthase in rat microvascular brain endothelial cells. Neurosci Lett 230: 109–112
3. Boutin H, LeFeuvre RA, Horai R, Asano M, Iwakura Y, Rothwell NJ (2001) Role of IL-1α and IL-1β in ischemic brain damage. J Neurosci 21: 5528–5534
4. Chan PH, Kawase M, Murakami K, Chen SF, Li Y, Calagui B, Reola L, Carlson E, Epstein CJ (1998) Overexpression of SOD1 in transgenic rats protects vulnerable neurons against ischemic damage after global cerebral ischemia and reperfusion. J Neurosci 18: 8292–8299
5. Ding M, St Pierre BA, Parkinson JF, Medberry P, Wong JL, Rogers NE, Ignarro LJ, Merrill JE (1997) Inducible nitric-oxide synthase and nitric oxide production in human fetal astrocytes and microglia. J Biol Chem 272: 11327–11335
6. Homma H, Mizushima H, Arai Y, Dohi K, Matsumoto K, Shioda S, Nakai Y (1997) Effect of focal cerebral ischemia on nitric oxide synthase expression in rats. Med Electron Microsc 30: 55–62
7. Horai R, Asano M, Sudo K, Kanuka H, Suzuki M, Nishihara M, Takahashi M, Iwakura Y (1998) Production of mice deficient in genes for interleukin (IL)-1α, IL-1β, IL-1α/β, and IL-1 receptor antagonist shows that IL-1β is crucial in turpentine-induced fever development and glucocorticoid secretion. J Exp Med 187: 1463–1475
8. Huang Z, Huang PL, Panahian N, Dalkara T, Fishman MC, Moskowitz MA (1994) Effects of cerebral ischemia in mice deficient in neuronal nitric oxide synthase. Science 265: 1883–1885
9. Iadecola C (1997) Bright and dark sides of nitric oxide in ischemic brain injury. Trends Neurosci 20: 132–139
10. Mizushima H, Zhou CJ, Dohi K, Horai R, Asano M, Iwakura Y, Hirabayashi T, Arata S, Nakajo S, Takaki A, Ohtaki H, Shioda S (2002) Reduced postischemic apoptosis in the hippocampus of mice deficient in interleukin-1. J Comp Neurol 448: 203–216
11. Ohtaki H, Funahashi H, Dohi K, Oguro T, Horai R, Asano M, Iwakura Y, Yin L, Matsunaga M, Goto N, Shioda S. Suppression of oxidative neuronal damage after transient middle cerebral artery occlusion in mice lacking interleukin-1. Nerosci Res (in press)
12. Rothwell NJ (1999) Cytokines–killers in the brain? J Physiol 514: 3–17
13. Samad TA, Moore KA, Sapirstein A, Billet S, Allchorne A, Poole S, Bonventre JV, Woolf CJ (2001) Interleukin-1β-mediated induction of COX-2 in the CNS contributes to inflammatory pain hypersensitivity. Nature 410: 471–475
14. White BC, Sullivan JM, DeGracia DJ, O'Neil BJ, Neumar RW, Grossman LI, Rafols JA, Krause GS (2000) Brain ischemia and reperfusion: molecular mechanisms of neuronal injury. J Neurol Sci 179: 1–33

Correspondence: Hirokazu Ohtaki, Department of Anatomy, Showa University School of Medicine, 1-5-8 Hatanodai, Shinagawa-ku, Tokyo 142-8555, Japan. e-mail: taki@med.showa-u.ac.jp

Acta Neurochir (2003) [Suppl] 86: 195–198

Mild hypothermia enhances the neuroprotective effects of a selective thrombin inhibitor following transient focal ischemia in rats

T. Kamiya, C. Nito, M. Ueda, K. Kato, S. Amemiya, A. Terashi, and **Y. Katayama**

Second Department of Internal Medicine, Nippon Medical School, Tokyo, Japan

Summary

The aim of this study is to determine whether a selective thrombin inhibitor, Argatroban, would prevent neuronal cell death and whether extra-mild hypothermia (35 °C) would enhance the neuroprotective effect of a selective thrombin inhibitor following transient focal ischemia in rats.

Sprague-Dawley rats were subjected to MCAo using an intraluminal suture technique for 2 hrs. The rats were reperfused for 24 h and decapitated for infarct and edema analysis. Argatroban-treated animals received a continuous injection of argatroban (3.0 mg/kg) for 24 hrs after onset of ischemia, while vehicle-treated groups received same dose of vehicle. During ischemia, temporal muscle and rectal temperatures were monitored and maintained at 37 °C in the normothermic animals and at 35 °C in the hypothermic animals.

Argatroban ameliorated the cortical ischemic damage significantly ($p < 0.05$). Moreover, argatroban with mild hypothermia decreased the cortical infarct or edema volume significantly compared with those of groups I and III ($p < 0.05$). Argatroban improved neurological symptoms significantly and also improved survival rate.

These results demonstrate that extra-mild hypothermia (35 °C) enhances neuroprotective effects of a selective thrombin inhibitor, argatroban, suggesting that this combined therapy may be a new therapeutic strategy for the treatment of acute stroke.

Keywords: Thrombin; apoptosis; hypothermia; focal ischemia; rat.

Introduction

A novel agent, ((2*R*, 4*R*)-4-methyl-1,2,3,4-tetrahydro-8-quinolinesulfonyl)-L-arginyl-2 piperidine-carboxylic acid monohydrate [argatroban], which is a derivative of arginine that binds to the active site of thrombin competitively, has a strong property in that it inhibits thrombin induction selectively. Argatroban (molecular weight 526.66) with a half-life of 30 minutes, had an immediate anticoagulant effect after intravenous injection [7], ameliorated microcirculatory disturbances by inhibiting microthrombin formation in microvessels [5], and reduced the size of infarcts in a rat experimental ischemia model [9].

Recent experimental studies have demonstrated that hypothermia is a powerful tool in neuronal protection following cerebral ischemia [3, 8]. Hypothermic treatment has also been used clinically as one of the strategies for reducing ischemic brain injury, however extreme hypothermia potentially has systemic adverse effects, such as arrythmia induction and respiratory insufficiency. Thus, mild hypothermia management to avoid severe adverse effects has clinically strongly been necessary.

The aim of this study is, therefore, to determine whether extra-mild hypothermia (35 °C) would enhance the neuroprotective effect of a selective thrombin inhibitor, argatroban following transient focal ischemia in rats.

Materials and methods

Twenty male Sprague-Dawley rats weighing 250–300 g were used. The rats were fasted overnight before surgery with free access to tap water. Anesthesia was initially induced with 5% halothane, and then was maintained with 1% halothane in a mixture of 70% N_2O and 30% O_2 under spontaneous breathing. Blood gases (pH, $PaCO_2$, PaO_2) and blood glucose levels were measured just before and during cerebral ischemia (60 and 120 minutes after ischemia induction). Rectal and temporal muscle temperatures were maintained during the surgical procedures. Focal cerebral ischemia was induced by inserting a silicone rubber-coated 4-0 nylon thread through the left internal carotid artery for 2 hours, and reperfusion was achieved by withdrawing the thread for 24 hours [1].

Animals were randomly divided into the following four groups (each, $n = 6$): (I) vehicle-treated normothermic group (control) at 37 °C (rectal and temporalis muscle temperatures); (II) argatroban-treated normothermic group at 37 °C; (III) vehicle-treated hypothermic group at 35 °C; (IV) argatroban-treated hypothermic group at 35 °C. Temporal muscle and rectal temperatures were maintained during ischemia at 37 ± 0.2 °C (normothermic groups) or 35 ± 0.2 °C (hypothermic groups).

The osmotic pumps infused with vehicle and argatroban solution, were implanted i.p. immediately after MCA occlusion in the vehicle-

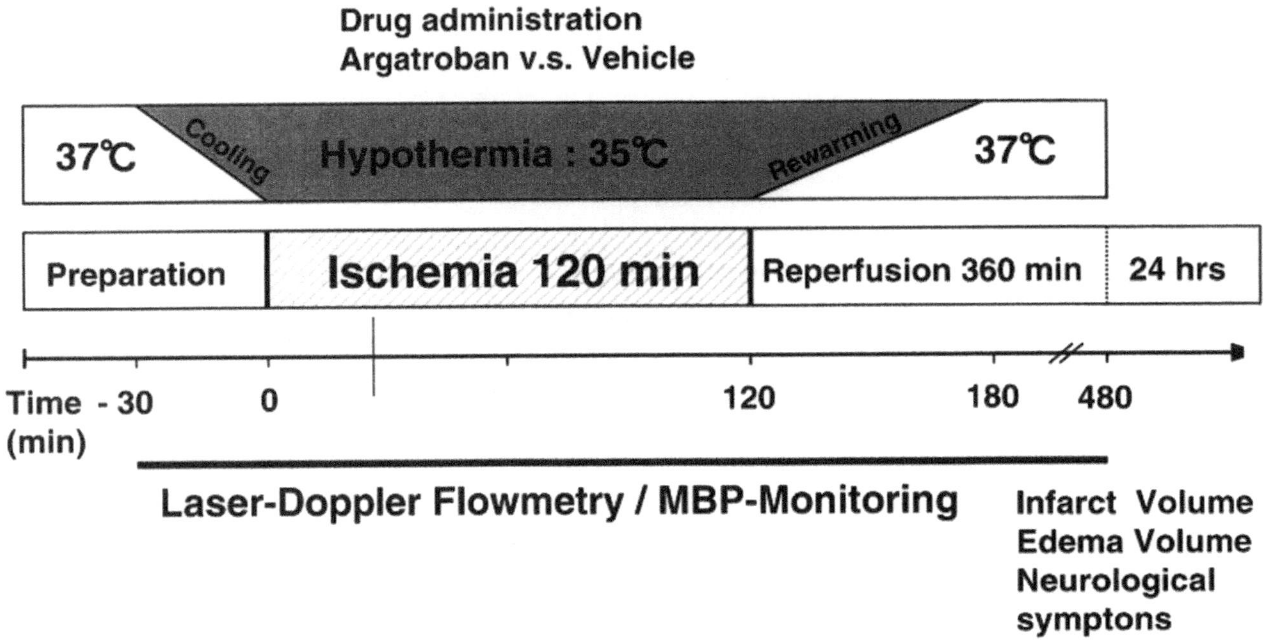

Fig. 1. Study flow diagram

and argatroban-treated groups, respectively, and the i.p. administration was continuously performed until the end of experiment. The rats were reperfused for 24 hours and decapitated for infarct and edema analysis [6]. Immediately before decapitation, all animals were neurologically examined by an investigator blinded to the various treatment groups using an established scoring system [2]. A protocol of the experiment is presented in Fig. 1.

Statistical analyses were performed with the use of Stat View 5.0 (SAS Institute Inc.). Infarct and edema volumes were analyzed with one-way ANOVA followed by Fisher's post hoc test. Values are presented as mean $\pm$ SD, and differences were considered significant at the $p < 0.05$ level.

Results

Infarct volume

The cortical infarct volumes in the vehicle-treated normothermic (control), vehicle-treated hypothermic, argatroban-treated normothermic, and argatroban-treated hypothermic groups were 205 ± 55, 170 ± 27, 162 ± 28, and 114 ± 27 mm^3, respectively (Fig. 2). There were no statistically significant differences in the cortical infarct volume among the vehicle-treated normothermic group and the vehicle-treated hypothermic groups. The cortical infarct volume was significantly reduced in the argatroban-treated normothermic group compared with the vehicle-treated normothermic group, and the reduction was 83% of those in the control group. The striatal infarct volumes in the vehicle-treated normothermic, vehicle-treated hypothermic, argatroban-treated normothermic and argatroban-treated hypothermic groups were 95 ± 18, 92 ± 8, 97 ± 16, and 83 ± 15 mm^3, respectively (Fig. 2). There were no statistically significant differences in the striatal infarct volume among all groups.

Edema volume

The cortical edema volumes in the vehicle-treated normothermic, vehicle-treated hypothermic, argatroban-treated normothermic and argatroban-treated hypothermic groups were 68 ± 23, 52 ± 13, 61 ± 16, and 29 ± 11 mm^3, respectively (Fig. 3). Although there were also no statistically significant differences in the cortical edema volumes among the argatroban-treated normothermic and vehicle-treated groups, the volume was significantly smaller (−57%) in the argatroban-treated hypothermic group compared with the vehicle-treated normothermic group. The striatal edema volumes in the vehicle-treated normothermic, vehicle-treated hypothermic, argatroban-treated normothermic and argatroban-treated hypothermic groups were 38 ± 6, 33 ± 2, 43 ± 17, and 33 ± 9 mm^3, respectively (Fig. 3). There were no statistically significant differences in the striatal infarct volume among all groups.

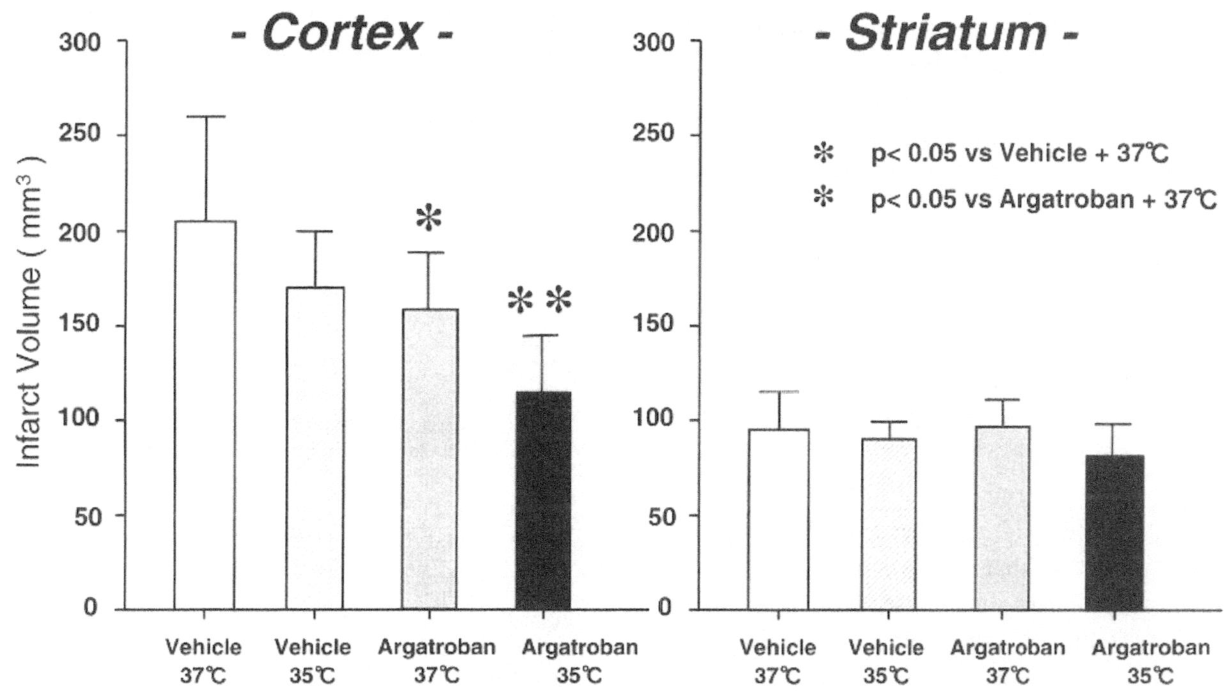

Fig. 2. Effects of single and combination treatment on infarct volume. Rats received mild hypothermia (35 °C), argatroban, and argatroban-plus-mild hypothermia (35 °C). Argatroban dose was 16.7 μg/kg/min either single or in combination with mild hypothermia. Argatroban alone significantly reduced the infarct volume in the cortex. A combination of argatroban and mild hypothermia (35 °C) ameliorated the infarct volume in the cortex significantly compared with that in animals treated by argatroban only, whereas mild hypothermia (35 °C) alone was unaffected. There were no statistically significant differences in the striatal infarct volume among all groups

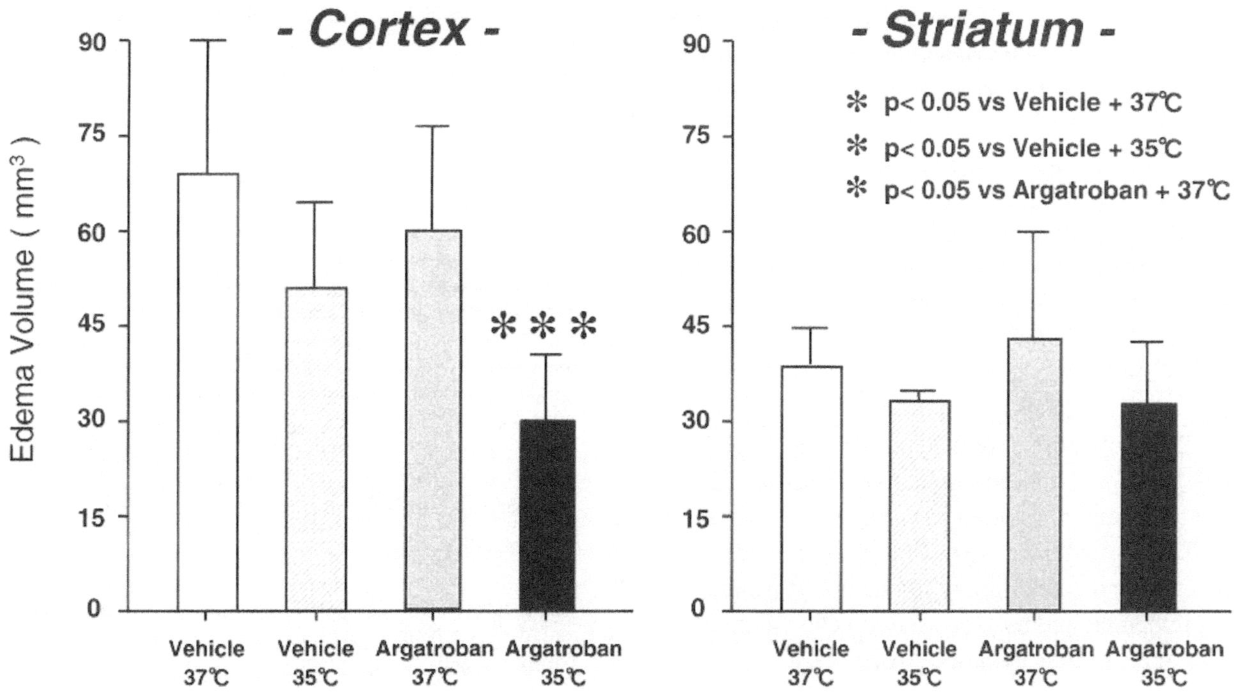

Fig. 3. The edema volumes in the vehicle-treated normothermic, vehicle-treated hypothermic, argatroban-treated normothermic and argatroban-treated hypothermic groups. Although there were also no statistically significant differences in the cortical edema volumes among the argatroban-treated normothermic and vehicle-treated groups, the volume was significantly smaller (−57%) in the argatroban-treated hypothermic group compared with the vehicle-treated normothermic group. There were no statistically significant differences in the striatal infarct volume among all groups

Neurological symptons

Argatroban improved posture of neurological symptons significantly, and also improved hemiplegia of neurological symptoms, but there was no statistically significant value. Argatroban also increased survival rate.

Discussion

Under normothermic conditions, argatroban significantly reduced the infarct volume in the cortex with no effect on edema volume in the cortex, but not in the striatum, suggesting that argatroban has maximal effects on the ischemic penumbra, but not the ischemic core, under normothermic conditions. A combination of argatroban and mild hypothermia (35 °C) ameliorated the infarct volume in the cortex significantly compared with that in animals treated by argatroban only, whereas mild hypothermia (35 °C) alone was unaffected. These results demonstrate that a selective thrombin inhibitor, argatroban, prevents neuronal cell death in the penumbra, and that mild hypothermia enhances neuroprotective effects of argatroban.

Obvious neurological symptoms, such as posture or hemiplegia was seen 24 hours after MCA occlusion. Argatroban significantly improved these neurological symptoms, posture, and also improved hemiplegia. The appearance of neurological deficits in the posture test has been reported to well correlate with infarct area [2, 4].

Brain temperature is an important determinant of neuronal injury following cerebral ischemia [3, 8]. Although the protective mechanisms of hypothermia are complex, several reports have shown that mild hypothermia reduces neuronal apoptosis directly in vitro [10] and in vivo [11].

In this study, the temperature of 35 °C enhances neuroprotective effects of argatroban without the adverse effects of more severe hypothermia on physiological functions. Indeed, mild hypothermia (35 °C) did not affect any physiological variables in the present study. The present data shows that this combination is an effective and safe strategy for neuronal protection during ischemia, and may be useful in clinical medical care for acute human ischemic stroke.

In conclusion, argatroban reduced the infarct area, especially the cortex seemed to be a penumbra zone, and improved neurological deficit such as posture 24 hours after occlusion, when it was administered just after ischemia. Moreover, mild hypothermia enhances neuroprotective effects of argatroban, suggesting that this combined therapy may be a new therapeutic strategy for the treatment of acute stroke.

References

1. Arii T, Kamiya T, Arii K, Ueda M, Nito C, Katsura K, Katayama Y (2001) Neuroprotective effect of immunosuppressant FK506 in transient focal ischemia in rat. Neurol Res 23: 755–760
2. Bederson JB, Pitts LH, Tsuji M, Nishimur MC, Davis RL, Bartowski H (1986) Rat middle cerebral occlusion: Evaluation of the model and development of a neurologic examination. Stroke 17: 472–476
3. Busto R, Dietrich WD, Globus MY, Valdes I, Scheinberg P, Ginsberg MD (1987) Small differences in intraischemic brain temperature critically determine the extent of ischemic neuronal injury. J Cereb Blood Flow Metab 7: 729–738
4. Hakim AM, Hogan MJ, Carpenter S (1992) Time course of cerebral blood flow and histological outcome after focal cerebral ischemia in rats. Stroke 23: 1138–1144
5. Hara T, Iwamoto M, Ishihara M, Tomikawa M (1994) Preventive effect of argatroban on ellagic acid induced cerebral thromboembolism in rats. Haemostasis 24: 351–357
6. Jacewicz M, Tanabe J, Pulsinelli WA (1992) The CBF threshold and dynamics for focal cerebral infarction in spontaneously hypertensive rats. J Cereb Blood Flow Metab 12: 359–370
7. Jeske W, Walenga JM, Lewis BE, Fareed J (1999) Pharmacology of argatroban. Exp Opin Invest Drugs 8: 625–654
8. Karibe H, Chen J, Zarow GJ, Graham SH, Weinstein PR (1994) Delayed induction of mild hypothermia to reduce infarct volume after temporary middle cerebral artery occlusion in rats. J Neurosurg 80: 112–119
9. Kawai H, Yuki S, Sugimoto J, Tamao Y (1996) Effects of a thrombin, argatroban, on ischemic brain damage in the rat distal middle cerebral artery occlusion model. J Pharmacol Exp Ther 278: 780–785
10. Xu L, Yenari MA, Steinberg GK, Giffard RG (2002) Mild hypothermia reduces apoptosis of mouse neurons in vitro early in the cascade. J Cereb Blood Flow Metab 22: 21–28
11. Yenari MA, Iwayama S, Cheng D, Sun GH, Fujimura M, Morota Fujimura Y, Chan PH, Steinberg GK (2002) Mild hypothermia attenuates cytochrome *c* release but does not alter *Bcl2* expression or caspase activation after experimental stroke. J Cereb Blood Flow Metab 22: 29–38

Correspondence: Dr. Tatsushi Kamiya, Second Department of Internal Medicine, Nippon Medical School, 1-1-5, Sendagi, Bunkyo-ku, Tokyo, 113-8603, Japan. e-mail: kamiya@nms.ac.jp

Acta Neurochir (2003) [Suppl] 86: 199–203

The neuroprotective effect of a free radical scavenger and mild hypothermia following transient focal ischemia in rats

C. Nito, T. Kamiya, S. Amemiya, K. Katoh, and **Y. Katayama**

Division of Neurology, Second Department of Internal Medicine, Nippon Medical School, Tokyo, Japan

Summary

Edaravone, a novel free radical scavenger, has been reported to reduce ischemic damage in rats subjected to transient focal ischemia. The aim of this study is, therefore, to investigate the effect of a combined therapy with edaravone and mild hypothermia of 35 °C. Sprague-Dawley rats were subjected to MCA occluding an intraluminal suture technique for 2 hrs. The rats were reperfused for 24 h and decapitated for infarct and edema analysis. Animals were randomly devided into four groups: (I) vehicle + normothermia (control) (II) vehicle + mild hypothermia (III) Edaravone + normothermia (IV) Edaravone + mild hypothermia. Mild hypothermia alone had no reduction of the brain damage. The edaravone alone significantly reduced edema volume. The combined treatment with edaravone and mild hypothermia reduced both infarct and edema volume. In addition, this treatment provided for the best functional outcome. These results demonstrate that free radical scavenger, edaravone attenuates brain edema and that the combined therapy with edaravone and mild hypothermia significantly reduces not only edema but also infarct on transient focal cerebral ischemia in rats. The neuroprotective effects seen in this study may be due to the combined interaction of antiedema activity between edaravone and mild hypothermia, suppressing free radical production.

Keywords: Free radical scavenger; hypothermia; neuroprotection; focal ischemia; rat.

Introduction

Edaravone (3-methyl-1-phenyl-pyrazolin-5-one) is a novel antioxidant, and inhibits both hydroxyl radicals and iron-induced peroxidative vascular endothelial cell damage [13]. Furthermore, edaravone has been tested in various different experimental models for evaluation of its protective effects in cerebral ischemia/reperfusion and myocardial ischemia/reperfusion [1, 14]. Although any free radical scavengers have been developed for clinical use, edaravone is one of the most expecting of these agents. Therefore, it has been clinically applied for the treatment of acute stroke in Japan since June 2001.

Recent experimental studies have demonstrated that hypothermia is a powerful tool in neuronal protection following cerebral ischemia [5]. Studies on the effect of mild or moderate hypothermia on cerebral ischemia were performed in various models of global or focal cerebral ischemia [4, 7]. Multiple mechanisms for hypothermia-induced neuroprotection have been identified, such as reduced metabolic rate and energy depletion, decreased excitatory transmitter release, decreased generation of free radicals, improved ion homeostasis, and reduced vascular permeability, blood-brain barrier disruption, and edema [6, 18]. Recently, experimental studies suggest that the combination of mild hypothermia with neuroprotective drugs is more effective than each of these alone [15, 16]. There are no data available with regard to a combined treatment of free radical scavenger, edaravone and hypothermia.

The present study was designed to evaluate the singular and combined effects of edaravone and/or mild hypothermia of 35 °C on ischemic damage in rats subjected to transient focal ischemia.

Materials and methods

Sprague-Dawley rats weighing 250–300 g were used. Anesthesia was initially induced with 1% halothane in a mixture of 70% N_2O and 30% O_2 under spontaneous breathing. Blood gases and blood glucose levels were measured just before and during cerebral ischemia. Animals were subjected to 2 hours MCA occluding followed by 24 hour reperfusion using an intraluminal suture technique [2]. Animals were randomly divided into the following four groups (each, n = 6): (I) vehicle-treated, normothermic group (control) (II) Edaravone-treated, normothermic group (III) vehicle-treated, mild hypothermic (35 °C) group (IV) Edaravone-treated, mild hypothermic (35 °C) group. Temporal muscle and rectal temperatures were maintained during ischemia at 37 ± 0.2 °C (normothermic groups)

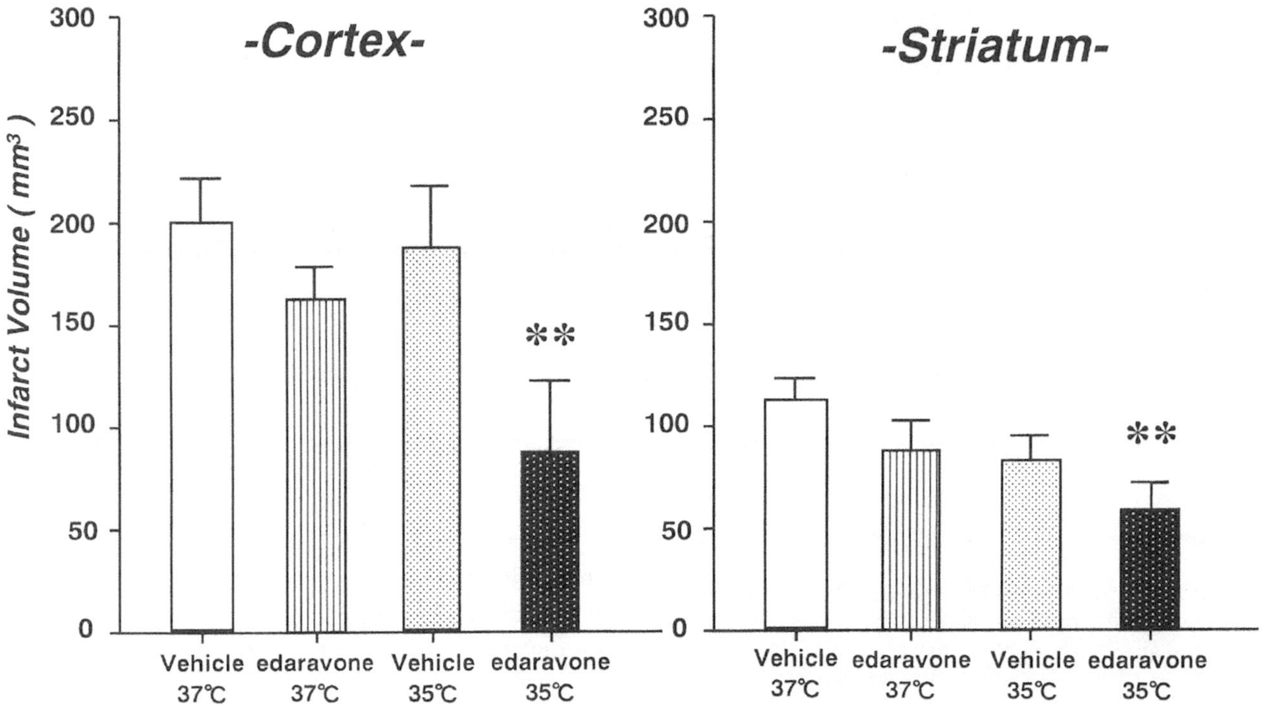

Fig. 1. The cortical or striatal infarct volumes after 2 hour middle cerebral artery occlusion followed by 24 hour reperfusion. Values (mm^3) are expressed as mean ± SD. ** $p < 0.001$ vs vehicle-treated normothermic group by one way ANOVA

or 35 ± 0.2 °C (hypothermic groups). Animals were intravenously administered 3.0 mg/kg of edaravone or vehicle, provided by Mitsubishi Pharma Corporation (Tokyo, Japan), just prior to the reperfusion. Twenty four hours after reperfusion, the animals were decapitated and removed. Twenty coronal sections were cut and were stained with hematoxylin and eosin. The cortical and striatal areas (mm^2) from each slice were summed and multiplied to obtain an infarct volume (mm^3). A volume of edema (mm^3) was determined by subtracting the non-ischemic cortical volume from the ischemic cortical volume [9].

Neurological symptoms of each animal were examined blindly 24 hours after reperfusion using an established scoring system [3]. In these test, the postural reflex test and hemiparesis of animals were observed.

Statistical analyses were performed with the use of StatView 5.0. Physiological data for each time point, infarct volumes, edema volumes and neurological function scores were analyzed with one-way ANOVA. Values are presented as mean ± SD, and differences were considered significant at $p < 0.05$ level.

Results

There were no statistically significant differences in the mean arterial blood pressure, pH, $PaCO_2$, PaO_2 or blood glucose levels among the groups.

Infarct volume

The cortical infarct volumes in the vehicle-treated normothermic, vehicle-treated hypothermic, edaravone-treated normothermic, and edaravone-treated hypothermic groups were 201 ± 37, 181 ± 44, 163 ± 22, and 90 ± 37 mm^3, respectively. The cortical infarct volume was significantly reduced in the edaravone-treated hypothermic group compared with the other groups, and the reduction was 55% of those in the vehicle-treated normothermic group (Fig. 1). The striatal infarct volumes in the vehicle-treated normothermic, vehicle-treated hypothermic, edaravone-treated normothermic, and edaravone-treated hypothermic group were 108 ± 12, 91 ± 10, 96 ± 15, and 70 ± 15 mm^3, respectively. The combination with edaravone and mild hypothermia significantly reduced the striatal infarct volume (−35%) compared with the vehicle-treated normothermic group (Fig. 1).

Edema volume

The cortical edema volumes in the vehicle-treated normothermic, vehicle-treated hypothermic, edaravone-treated normothermic, and edaravone-treated hypothermic group were 82 ± 18, 59 ± 16, 42 ± 13, and 23 ± 17 mm^3, respectively. In

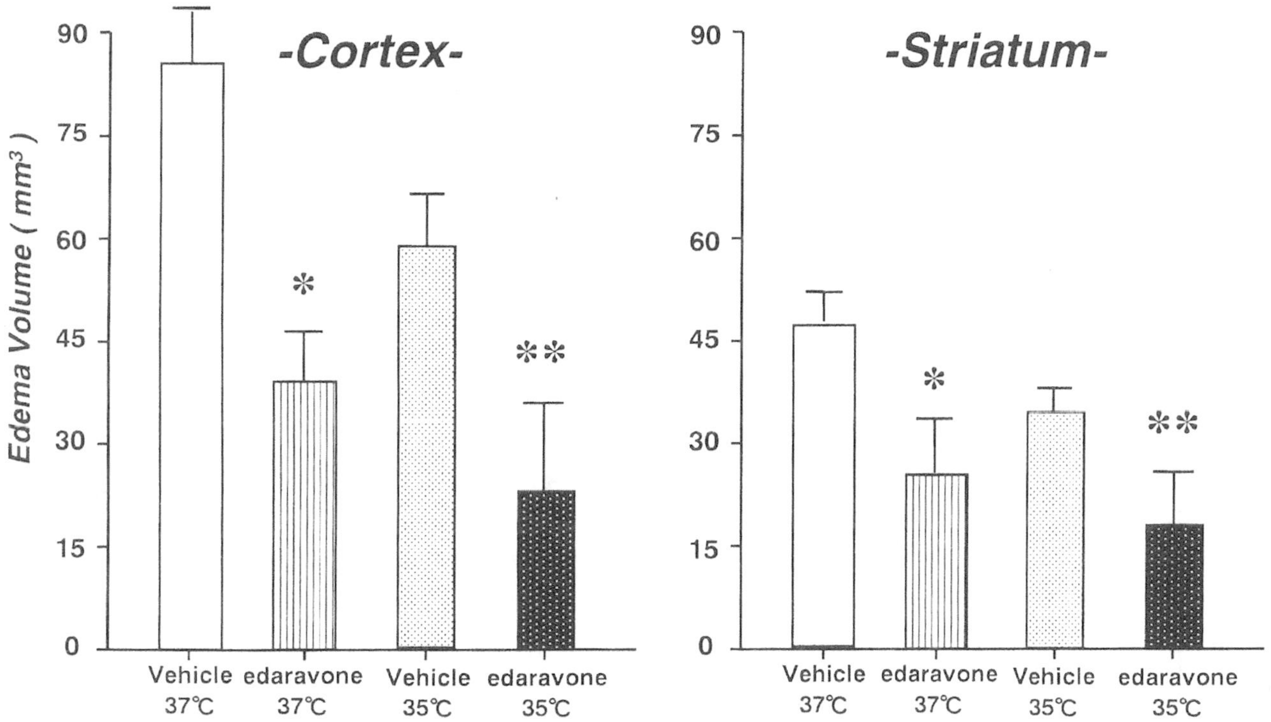

Fig. 2. The cortical or striatal edema volumes after 2 hour middle cerebral artery occlusion followed by 24 hour reperfusion. Values (mm^3) are expressed as mean $\pm$ SD. ** $p < 0.001$, * $p < 0.01$ vs vehicle-treated normothermic group by one way ANOVA

edaravone-treated normothermic group, the cortical edema volume was significantly less than those in vehicle-treated normothermic group (−72%). Moreover, edaravone with hypothermic group reduced the cortical edema volume significantly compared with those in vehicle-treated normothermic group (−49%) (Fig. 2). The striatal edema volumes in the vehicle-treated normothermic, vehicle-treated hypothermic, edaravone-treated normothermic, and edaravone-treated hypothermic group were 46 ± 10, 34 ± 3, 27 ± 10, and 21 ± 8 mm^3, respectively. The striatal edema volumes in edaravone-treated normothermic or edaravone-treated hypothermic group were significantly smaller than those in vehicle-treated normothermic group (−41%, −54%, respectively) (Fig. 2).

Neurological symptoms

The median neurological score was 2.5 ± 0.5 in the control group. The combination treatment of Edaravone and hypothermia significantly improved the neurological symptoms (hemiplegia) compared with the control group ($p < 0.05$) (Fig. 3).

Discussion

With use of the suture model in the rat MCA, the present study demonstrated that a low dose of edaravone (3.0 mg/kg) combined with mild intraischemic hypothermia (35 °C) significantly reduces infarct and edema volumes in the cortex and the striatum. However, mild intraischemic hypothermia (35 °C) alone failed to reduce postischemic brain damage. Under normothermic conditions, the protective effects of edaravone were observed in the edema, but not in the infarct.

Taken together, these results suggest that the combination of edaravone administration and mild intraischemic hypothermia expands the therapeutic time window for edaravone-alone. Moreover, the combination reduced both infarct and edema volumes, suggesting that the combination protects not only ischemic penumbra but also ischemic core.

Brain temperature is an important determinant of neuronal injury following cerebral ischemia [5]. Hypothermic treatment has also been used clinically as one of the strategies for reducing ischemic brain injury,

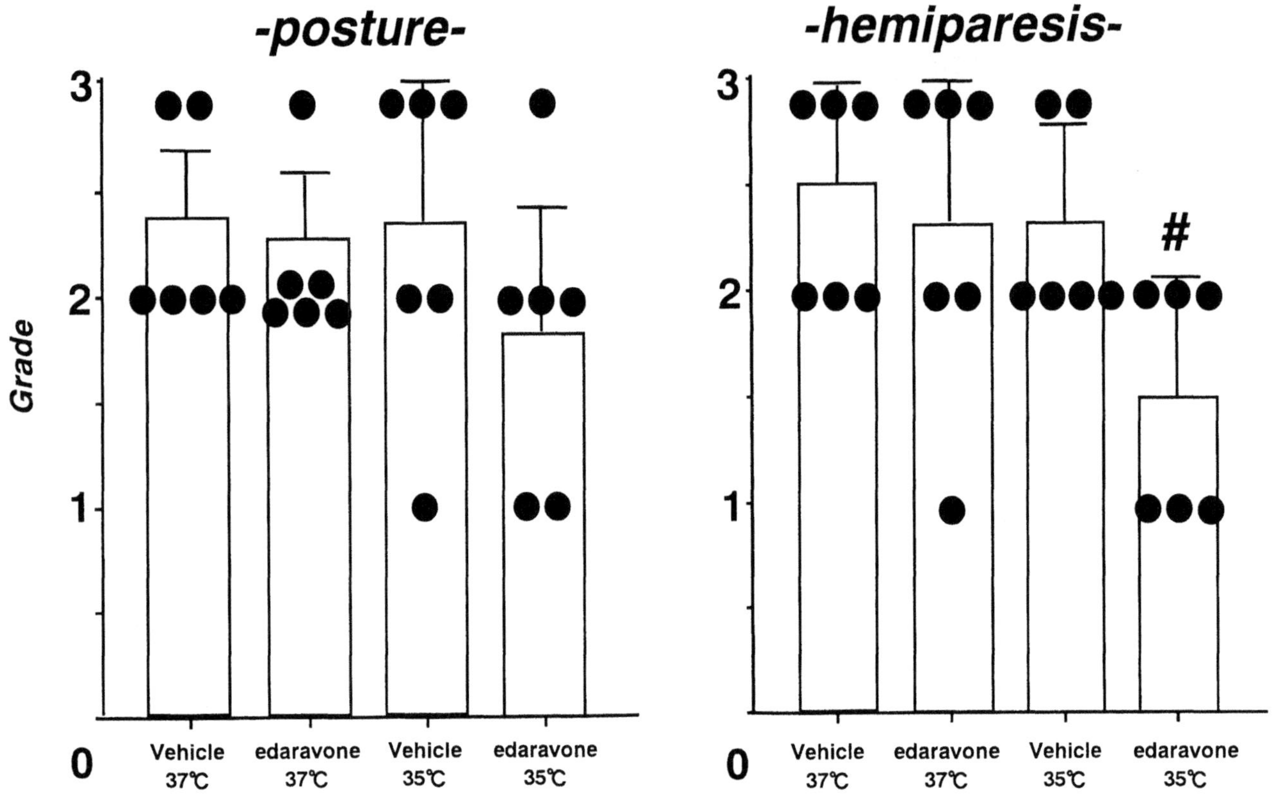

Fig. 3. The neurological score after 2 hour middle cerebral artery occlusion followed by 24 hour reperfusion. Each symbol depicts the individual neuro-score of a single animal. # $p < 0.05$ vs vehicle-treated normothermic group by one way ANOVA

however the usefulness of deep therapeutic hypothermia for cerebral ischemia has been limited by adverese effects such as myocardial arrhythmia, hypotension, respiratory insufficiency, blood hypercoagulability, and hemodynamic instability [17]. Severe hypothermia is known to have these side effects, however, mild hypothermia is much safer than the classical deep hypothermia [10]. We chose mild hypothermia of 35 °C because clinical application and continuity can be accomplished both easily and safely, without severe complication. Indeed, mild hypothermia (35 °C) did not affect any physiological variables in the present study. The temperature of 35 °C may improve edaravone's effect without the adverse effects of more severe hypothermia on physiological functions.

Although the mechanisms of ischemic brain damage have not been clearly determined, accumulated experimental evidence suggests that production of free radicals is possibly one of the major factors involved. Several reports have demonstrated that free radicals generated during ischemia play an important role in the development of neuronal damage [11, 12]. Although the protective mechanisms of hypothermia are complex, one has been identified as decreased generation of free radicals from microglia [8, 19]. Microglial cells produces superoxide anions and nitric oxide, and hypothermia potentially inhibits the production. Accordingly, the neuroprotective effects seen in this study may be due to the combined interaction of antiedema activity between edaravone and mild hypothermia, suppressing free radical production.

The present study demonstrates that free radical scavenger, edaravone attenuates brain edema and that a combined therapy with edaravone and mild hypothermia significantly reduces not only edema but also infarct on transient focal cerebral ischemia in rats. These data show that this combination is an effective and safe strategy for neuronal protection during ischemia, and may be useful in clinical medical care for acute human ischemic stroke.

References

1. Abe K, Yuki S, Kogure K (1988) Strong attenuation of ischemic and postischemic brain edema in rats by a novel free radical scavenger. Stroke 19: 480–485

2. Arii T, Kamiya T, Arii K, Ueda M, Nito C, Katsura K, Katayama Y (2001) Neuroprotective effect of immunosuppressant FK506 in transient focal ischemia in rat. Neurol Res 23: 755–760
3. Bederson JB, Pitts LH, Tsuji M, Nishimur MC, Davis RL, Bartowski H (1986) Rat middle cerebral occlusion: Evaluation of the model and development of a neurlogic examination. Stroke 17: 472–476
4. Bering EA Jr (1974) Effects of profound hypothermia and circulatory arrest on cerebral oxygen metabolism and cerebrospinal fluid electrolyte composition in dogs. J Neurosurgery 40: 199–205
5. Busto R, Dietrich WD, Globus MY, Valdes I, Scheinberg P, Ginsberg MD (1987) Small differences in intraischemic brain temperature critically determine the extent of ischemic neuronal injury. J Cereb Blood Flow Metab 7: 729–738
6. Dietrich WD, Busto R, Globus MY, Ginsberg MD (1996) Brain damage and temperature: cellular ans molecular mechanisms. Adv Neurol 71: 177–194
7. Goto Y, Kassell NF, Hiramatsu K, Soleau SW, Lee KS (1993) Effects of intraischemic hypothermia on cerebral damage in a model of reversible focal ischemia. Neurosurgery 32: 980–985
8. Ho YK, Jing Z, Claude AP (1996) Brain temperature alters hydroxyl radical production during cerebral ischemia/reperfusion in rats. J Cereb Blood Flow Metab 16: 100–106
9. Jacewicz M, Tanabe J, Pulsinelli WA (1992) The CBF threshold and dynamics for focal cerebral infarction in spontaneously hypertensive rats. J Cereb Blood Flow Metab 12: 359–370
10. Kader A, Brisman MH, Maraire N, Huh JT, Solomon RA (1992) The effect of mild hypothermia on permanent focal ischemia in rat. Neurosurgery 31: 1056–1061
11. Kinouchi H, Epstein C, Mizui T, Carlson E, Chen SF, Chan PH (1991) Attenuation of focal cerebral ischemia injury in transgenic mice overexpressing CuZn superoxidedismutase. Proc Natl Acad Sci USA 88: 11158–11162
12. Kitagawa K, Matsumoto M, Oda T, Niinobe M, Hata R, Handa N, Fukunaga R, Isaka Y, Kimura K, Kamada T (1990) Free radical generation during brief period of cerebral ischemia may trigger delayed neuronal death. Neuroscience 35: 551–558
13. Murota S, Morita I, Suda N (1990) The control of vascular endothelial cell injury. Ann NY Acad Sci 598: 182–187
14. Nishi H, Watanabe T, Sakurai H, Yuki S, Ishibashi A (1989) Effect of MCI-186 on Brain Edema in Rats. Stroke 20: 1236–1240
15. Nito C, Kamiya T, Ueda M, Arii T, Terashi A, Katayama Y (2001) Mild hypothermia enhances neuroprotective effects of immunosuppressant FK506 following transient focal ischemia in rats. J Cereb Blood Flow Metab 21(Suppl 1): S405
16. Schmid-Elaesser R, Hungerhuber E, Zausinger S, Baethmann A, Ruelen HJ (1999) Combination drug therapy and mild hypothermia. Stroke 30: 1891–1899
17. Schubert A (1995) Side effect of mild hypothermia. J Neurosurg Anesthesiol 7: 139–147
18. Schwab S, Schwarz S, Spranger M, Keller E, Bertram M, Hacke W (1998) Moderate hypothermia in the treatment of patients with severe middle cerebral artery infarction. Stroke 29: 2461–2466
19. Si QS, Nakamura Y, Kataoka K (1997) Hypothermic suppression of microglial activation in culture: inhibition of cell proliferation and production of nitric oxide and superoxide. Neuroscience 81: 223–229

Correspondence: Chikako Nito, MD, Division of Neurology, Second Department of Internal Medicine, Nippon Medical School, 1-1-5 Sendagi, Bunkyo-ku, Tokyo 113-8603, Japan. e-mail: cnito @sk9.so-net.ne.jp

Acta Neurochir (2003) [Suppl] 86: 205–207

Therapeutical efficacy of a novel non-peptide bradykinin B_2 receptor antagonist on brain edema formation and ischemic tissue damage in focal cerebral ischemia

S. Zausinger[1], D. B. Lumenta[2], D. Pruneau[3], R. Schmid-Elsaesser[1], N. Plesnila[2], and A. Baethmann[2]

[1] Department of Neurosurgery, Klinikum Grosshadern, Ludwig-Maximilians-University, Munich, Germany
[2] Institute for Surgical Research, Klinikum Grosshadern, Ludwig-Maximilians-University, Munich, Germany
[3] Centre de Recherche, Laboratoires Fournier, Daix, France

Summary

Objective. Bradykinin has been identified as a mediator of secondary brain damage in acute insults. We currently studied neuroprotective properties of a bradykinin B_2 receptor antagonist (LF16-0687 Ms) in transitory focal cerebral ischemia to assess infarct formation and the development of brain edema.

Material and methods. 55 Rats were subjected to 90 min of MCA-occlusion. The receptor antagonist was administered at two dose levels, given from 30 min prior to ischemia over two days after ischemia. Ischemic tissue damage was quantified at day 7 after MCA-occlusion together with assessment of brain edema in separate experiments. Neurological recovery was studied daily.

Results. Animals receiving treatment (low dose) had a better functional recovery, particularly at days 3 and 4 ($P < 0.05$). Infarct formation was significantly attenuated in these animals in both total and cortical brain tissue by 50, or 80%, respectively. Postischemic brain swelling was significantly lowered, i.e. by 62%.

Conclusions. Our findings provide further support for a mediator role of bradykinin in ischemic brain damage including edema formation, obviously by ligand binding to the bradykinin B_2 receptor. The availability of a receptor antagonist may afford opportunity for translation of this experimental treatment into stroke patients.

Keywords: Cerebral ischemia; focal; bradykinin B_2 receptor antagonist; neuroprotection; brain edema.

Introduction

Bradykinin is an active peptide formed by a kallikrein-induced enzymatic cleavage of large precursors, the kininogens. Its effects are mediated by two types of G-protein coupled receptors, the B_1 (inducible) and the B_2 (constitutive) receptor, while the B_2 receptor accounts for the majority of acute physiological effects of bradykinin. In peripheral organs bradykinin induces vasodilation, increases vascular permeability, and stimulates immune cells and primary afferent neurons to produce pain. All components of the kallikrein-kinin system have been identified in the brain [1] where kinins have been found to stimulate the production and release of eicosanoids, cytokines, NO, free radicals and excitatory amino acids [3]. Furthermore, kinins open the blood-brain barrier and dilate extra- and intraparenchymal cerebral arteries [15]. Ischemic brain damage is associated with an activation of the kallikrein-kinin system, resulting in a sustained production of kinins [3]. The purpose of the present study was to examine potential neuroprotective effects of LF 16-0687 Ms, a most potent non-peptide antagonist of human B_2 receptors [9] on ischemic brain damage and swelling after focal cerebral ischemia in rats.

Materials and methods

Animal preparation and monitoring

55 male Sprague-Dawley rats (250 to 300 g body weight) were intubated and mechanically ventilated with 0.8% halothane in a mixture of 70% N_2O and 30% O_2 to maintain normal arterial blood gases [16]. A feedback-controlled heating lamp and pad were used to maintain temporalis muscle and rectal temperature at 37.0 °C. The tail artery and left femoral vein were cannulated for blood sampling, monitoring of arterial blood pressure, and administration of fluids and drugs. Serum glucose, blood gases, lactate, hemoglobin, and hematocrit were measured before, during, and after ischemia.

Middle cerebral artery occlusion

All rats were subjected to transient MCA occlusion by insertion of a silicone-coated 4-0 nylon monofilament via the external carotid artery as previously described [13]. The filament was advanced via the external carotid artery until LDF showed a sharp decrease of ipsilateral LCBF to 20–30% of baseline, indicating adequate occlusion of the MCA.

Drug administration and treatment arms

Study I. Animals were randomly assigned to one of three treatment arms (n = 11 each): (a) vehicle-treated (NaCl 0.9%) controls, (b) animals receiving LF 16-0687 Ms s.c. (low dose: 12.0 mg/kg/day), or (c) LF 16-0687 Ms (high dose: 36.0 mg/kg/day) 30 min before ischemia, at onset of reperfusion, and 6 hours after reperfusion on day 1 in three portions. On days 2 and 3 after ischemia two additional doses (each half daily dose) were administered s.c. with an 8 hour interval.

Study II. Formation of ischemic brain edema was evaluated in a subsequent study in animals receiving the more effective low-dose regimen of LF 16-0687 Ms (see results) as described above in comparison with that of a vehicle-treated control group (n = 8 per group).

Quantification of ischemic damage

After the ischemia insult, the neurological function of all animals was evaluated daily, using a 6-point scale [17]. Seven days after cerebral ischemia animals were perfused transcardially with 2% paraformaldehyde for fixation of the brain. The brain was embedded in paraffin, cut into 4-μm-thick coronal sections at 400-μm intervals and the brain slices were stained with hematoxylin and eosin. 24 slices per brain containing the entire infarct were analyzed for infarcted area.

Determination of brain swelling

All animals of study II were sacrificed 48 hours after ischemia. After removal of the brains, the hemispheres were separated along the interhemispheric plane. Both hemispheres were weighed for wet weight. Then, the hemispheres were dried for 24 hours at 110 °C for determination of dry weight. Based on the gravimetrical differences between both hemispheres, swelling of the ischemic hemispheres in both ischemic and non-ischemic hemispheres was obtained.

Statistical analysis

The physiologic variables of each time point were analyzed with one-way ANOVA. Infarct volumes and neurological function scores were analyzed with Kruskal-Wallis ANOVA on ranks. If multiple comparisons were necessary, Dunnett's test or Dunn's test for neurological function scores were applied. Brain swelling and hemispheric water content were analyzed with Mann-Whitney rank sum test. Differences were considered significant at $P < 0.05$.

Results

The experimental groups did not differ with respect to the pre-, intra-, or postischemic blood pressure, hemoglobin, hematocrit, or serum glucose or arterial blood gases.

Study I

Functional outcome. Low dose LF 16-0687 Ms afforded a significant improvement of the neurological recovery from postischemic day 3 to 4 as compared with the vehicle-treated controls ($P < 0.05$).

Infarct volume. Low dose LF 16-0687 Ms significantly attenuated total and cortical infarction by 50% and 80.1%, respectively ($P < 0.05$). Low dose treatment with LF 16-0687 Ms was found to be significantly effective ($P < 0.05$) in limiting cortical infarction. Animals receiving high dose LF 16-0687 Ms had a cortical infarct volume lowered by 54%, without attaining statistical significance vs. the untreated controls. Treatment with LF 16-0687 Ms at both dose levels tended to reduce infarction of the basal ganglia.

Study II

Ischemic brain swelling. Swelling of the ischemic hemispheres 48 hours after ischemia was markedly inhibited by 62% ($P < 0.05$) in animals receiving the bradykinin B_2 receptor antagonist. Accordingly, gravimetry of the ischemic hemispheres revealed an increase in weight of only $5.9 \pm 3.2\%$ (mean $\pm$ SD) compared to the weight increase of $15.5 \pm 7.9\%$ in the vehicle-treated controls.

Discussion

Antagonization of bradykinin action at the B_2 receptor, utilizing the selective antagonist LF 16-0687 Ms, significantly reduced ischemic brain swelling, improved the postischemic neurological recovery, and inhibited infarction predominantly in brain cortex in rats subjected to transient MCA occlusion. LF 16-0687 Ms appears to be a safe drug since major physiological variables were not changed by the treatment.

LF 16-0687 Ms is one of the most potent antagonists of the human B_2 receptor described so far [9]. LF 16-0687 Ms appears to be safe, since blood pressure and blood gases were not adversely influenced according to former observations [7, 8, 14]. The compound was observed to attenuate vasogenic brain edema in rats subjected to a cryogenic lesion of the exposed brain cortex [7]. In both rats and mice with closed head injury, LF 16-0687 Ms was found to improve the functional neurological outcome with reduction of posttraumatic brain edema by 50 to 70% [8]. Administration of LF 16-0687 Ms after a controlled cortical impact injury in rats significantly decreased the CSF level of taurine, an amino-acid considered to play a role in the cell volume regulation of the brain [14].

In accordance with our findings, Plesnila *et al.* [7] have obtained beneficial results when using a lower dose of LF 16-0687 Ms, while higher doses failed to affect brain edema from induction of a focal lesion. The failure to afford neuroprotection by administration of high doses of a B_2 receptor antagonist was previously confirmed by experiments on the selective peptide-based B_2 receptor antagonist, CP-0597, in rats with reversible MCA-occlusion [10]. Previous work has succeeded to distinguish the role of the B_1 and B_2 receptor in focal cerebral ischemia in rats, suggesting that blocking of the B_2 receptor is protective, while antagonism of the B_1 receptor may be detrimental for the functional recovery of the brain [11, 12]. From a pharmacological point of view, it is conceivable that complete blocking of B_2 receptors by high dose would prevent potential beneficial effects of bradykinin, e.g. dilation of arterioles in the penumbra with an impaired delivery of oxygen and substrate as well as clearance of harmful metabolites. The marked attenuation of postischemic brain swelling by LF 16-0687 Ms draws attention to bradykinin-related mechanisms of brain edema evolution. Consumption of kininogens as an indicator of kinin formation, or directly, an increase of the bradykinin concentration in the interstitial space of brain tissue in concentrations sufficient to induce vasodilation, extravasation, and edema formation have been found in cerebral trauma and ischemia in animals and humans [2, 5, 6]. Our data corroborate earlier experimental evidence that pharmacological blockade of bradykinin generation [6] or action by a B_2 receptor antagonist [4] inhibits ischemic brain edema formation and neuronal damage, making it more likely that the pharmacological effects of LF 16-0687 Ms are mediated by the inhibition of B_2 receptors located on vascular endothelial cells. The marked neuroprotective efficacy of the non-peptide agent LF 16-0687 Ms in combination with its lacking evidence of adverse effects make it a promising candidate for clinical evaluation.

Acknowledgment

This work was supported by the Deutsche Forschungsgemeinschaft (Schm1067/3-1), Friedrich Baur Stiftung (to N. Plesnila), and Laboratoires Fournier, Daix, France. Authors wish to thank Veronika Bischoff for excellent technical assistance.

References

1. Chen E-Y, Emerich DF, Bartus RT, Kordower JH (2000) B_2 bradykinin receptor immunoreactivity in the rat brain. J Comp Neurol 427: 1–18
2. Ellis EF, Chao J, Heizer ML (1989) Brain kininogen following experimental brain injury: evidence for a secondary event. J Neurosurg 71: 437–442
3. Francel PC (1992) Bradykinin and neuronal injury. J Neurotrauma 9: S27–S45
4. Kamiya T, Katayama Y, Kashiwagi F, Terashi A (1993) The role of bradykinin in mediating ischemic brain edema in rats. Stroke 24: 571–576
5. Kasuya H, Shimizu T, Okada T, Takahashi K, Summerville T, Kitamura K (1988) Activation of the coagulation system in the subarachnoid space after subarachnoid haemorrhage: serial measurement of fibrinopeptide A and bradykinin of cerebrospinal fluid and plasma in patients with subarachnoid haemorrhage. Acta Neurochir Wien 91: 120–125
6. Maier-Hauff K, Baethmann A, Lange M, Schürer L, Unterberg A (1984) The kallikrein-kinin system as mediator in vasogenic brain edema. Part 2: studies on kinin formation in focal and perifocal brain tissue. J Neurosurg 61: 97–106
7. Plesnila N, Schulz J, Stoffel M, Eriskat J, Pruneau D, Baethmann A (2001) Role of bradykinin B_2 receptors in the formation of vasogenic brain edema in rats. J Neurotrauma 18: 1049–1058
8. Pruneau D, Chorny J, Benkowitz V (1999) Effect of LF 16-0687 Ms, a new nonpeptide bradykinin B_2 receptor antagonist in a rat model of closed head trauma. J Neurotrauma 16: 1057–1065
9. Pruneau D, Paquet JL, Luccarini JM, *et al* (1999) Pharmacological profile of LF16-0687, a new potent non peptide bradykinin B_2 receptor antagonist. Immunopharmacology 43: 187–194
10. Relton JK, Beckey VE, Hanson WL, Whalley ET (1997) CP-0597, a selective bradykinin B_2 receptor antagonist, inhibits brain injury in a rat model of reversible middle cerebral artery occlusion. Stroke 28: 1430–1436
11. Relton JK, Beckey VE, Rothwell NJ, Whalley ET (1997) Bradykinin B_2 receptor antagonist CP-0597 reduces infarct volume after focal cerebral ischaemia in the rat: comparison with HOE-140, NPC-17731 and MK801. Br J Pharmacol 120: 75P (Abstract)
12. Relton JK, Rothwell NJ, Whalley ET (1997) Blockade of the bradykinin B_1 receptor reverses the neuroprotective effect of B_2 receptor antagonism after focal cerebral ischaemia in the rat. Br J Pharmacol 120: 76P (Abstract)
13. Schmid-Elsaesser R, Zausinger S, Hungerhuber E, Baethmann A, Reulen HJ (1998) A critical reevaluation of the intraluminal thread model of focal cerebral ischemia. Evidence of inadvertent premature reperfusion and subarachnoid hemorrhage in rats by Laser-Doppler flowmetry. Stroke 29: 2162–2170
14. Stover JF, Dohse NK, Unterberg A (2000) Significant reduction in brain swelling by administration of nonpeptide kinin B_2 receptor antagonist LF 16-0687 Ms after controlled cortical impact in rats. J Neurosurg 92: 853–859
15. Wahl M, Whalley ET, Unterberg A, *et al* (1996) Vasomotor and permeability effects of bradykinin in the cerebral microcirculation. Immunopharmacology 33: 257–263
16. Zausinger S, Baethmann A, Schmid-Elsaesser R (2002) Anesthetic methods in rats determine outcome after experimental focal cerebral ischemia: Mechanical ventilation is required to obtain controlled experimental conditions. Brain Res Protocols 9: 112–121
17. Zausinger S, Hungerhuber E, Baethmann A, Reulen HJ, Schmid-Elsaesser R (2000) Neurological impairment in rats after transient middle cerebral artery occlusion: a comparative study under various treatment paradigms. Brain Res 863: 94–105

Correspondence: Dr. Stefan Zausinger, Department of Neurosurgery, Ludwig-Maximilians-University, Klinikum Grosshadern, Marchioninistr. 15, 81377 Munich, Germany. e-mail: Stefan.Zausinger@nc.med.uni-muenchen.de

Acta Neurochir (2003) [Suppl] 86: 209–212

Iodoamphetamine (IMP) uptake in the brain is increased after experimental cerebral venous hypertension in the rat

M. Yamada, I. Yuzawa, and K. Fujii

Department of Neurosurgery, Kitasato University School of Medicine, Kanagawa, Japan

Summary

Cerebral venous hypertension (VH) is a pathological condition associated with arteriovenous malformation, and is an important determinant factor for clinical symptom and outcome. Using a rat whole brain VH model created by a carotid-jugular fistula and contralateral jugular vein ligation we measured superior sagittal sinus pressure (SSSP) and cerebral blood flow (CBF) by the ^{14}C-iodoamphetamine (IMP) indicator fractionation technique under isoflurane anesthesia with controlled ventilation. SSSP was significantly elevated in the fistula group (15 ± 1 mmHg) (mean ± SEM) compared to the vessel occlusion (5 ± 1) and sham operated groups (5 ± 1) ($P < 0.05$). Mean absolute CBF in the fistula (103 ± 3 ml/min/100 g) and vessel occlusion (90 ± 4) groups were elevated compared to the sham group (77 ± 3) ($P < 0.05$). Physiological parameters (mean blood pressure, heart rate, and blood gas analysis) before measurement were not different. In this VH model we found increases in SSSP and IMP uptake in the brain. We assume the mechanism of IMP uptake increase in VH animal may be slow circulation with a possible change of first pass extraction of IMP.

Keywords: Cerebral blood flow; arteriovenous malformation; venous hypertension; iodoamphetamine.

Introduction

Cerebral venous hypertension (VH) is a pathological condition associated with cerebral and dural arteriovenous malformations. It is considered to be an important determinant factor for clinical symptoms and outcome [5, 7–9]. However, the pathophysiology of that condition is not well understood. We evaluated a rat whole brain VH model to elucidate the cerebral circulation under that condition.

Materials and methods

The animals in this study received humane care and treatment according to the "Guide for the Care and Use of Laboratory Animals" published by the National Institutes of Health (NIH publication 85-23, revised 1985). Animals were allowed free access to food and water.

Creation of the cervical AV fistula

S-D rats (200–400 g) were used. In all animals, anesthesia was induced with 2.5% halothane and maintained at 1.0% with spontaneous breathing. Surgery was performed under sterile conditions. The right common carotid artery (CCA) and the right external jugular vein (EJV) were exposed by a midline neck incision. Rats were devided into three groups: fistula, ligation, and sham. In the fistula group, a CCA-EJV fistula was constructed by proximal CCA and distal EJV end-to-end anastomosis (Fig. 1). In the ligation group, both vessels were ligated. In the sham group, they were exposed only. After surgery, 1% lidocaine was infiltrated at the site and the skin was closed. Halothane was discontinued and the animals were awakened.

In a pilot study, fistula-created animals who underwent left EJV ligation 2 days later did not survive (n = 2). Therefore contralateral vein ligation was delayed. Five days after the first procedure, all the animals were reanesthetized with halothane. Using the same skin incision the left EJV was exposed. In both fistula and ligation groups, the EJV was ligated. In the sham group no more intervention was added.

General preparation for pressure measurements

Anesthesia was induced with 2–2.5% and maintained with 1.5% isoflurane in 70% N_2O and 30% O_2. Rats were tracheostomized, placed in the stereotaxic frame and ventilated artificially. End-tidal CO_2 was monitored by a CO_2 analyzer (OIR7101, Nihon Kohden, Tokyo, Japan). Both femoral arteries and a jugular vein were cannulated with polyethylene catheters (PE-50, Intramedic, Beckton Dickinson) for continuous arterial blood pressure and heart rate monitoring, blood withdrawal, and isotope administration. Arterial blood gas and pH were analyzed at baseline in each experiment. Rectal temperature was maintained at approximately 37 °C with a thermostatically controlled mat.

The superior sagittal sinus (SSS) near the confluence was exposed by a small craniectomy. The SSS was punctured with a 24-gauge needle to monitor SSS pressure (SSSP). After physiological stabilization, SSSP was recorded for at least 10 min.

Absolute CBF measurement

After physiological stabilization and SSSP measurement, CBF was determined using an indicator fractionation technique [1, 6]. Briefly, arterial blood was withdrawn continuously from the femoral artery at a rate of 0.6 mL/min (Harvard pump). Five microcuries

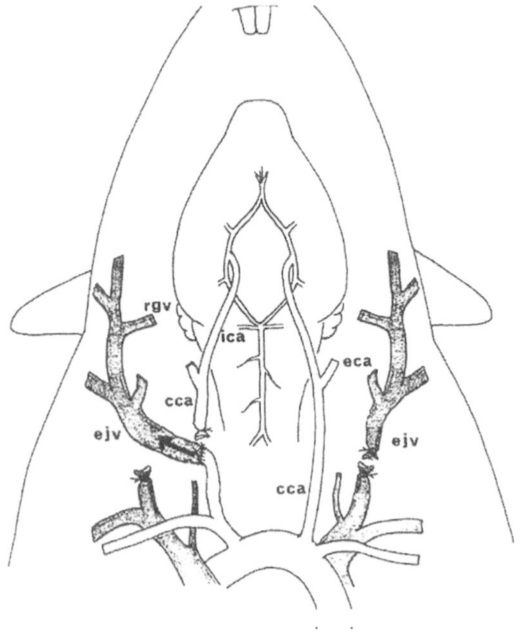

Fig. 1. Carotid-jugular arteriovenous fistula. An end-to-end anastomosis was completed between the proximal common carotid artery (*CCA*) and distal external jugular vein (*EJV*). Arrow indicates the direction of blood flow. Rgv indicates retroglenoid vein

of *N*-isopropyl-[methyl 1,3-^{14}C]-*p*-iodoamphetamine (American Radiolabeled Chemicals Inc, St. Louis, MO) dissolved in 0.2 mL saline was injected into the jugular vein as a bolus (<1 second). Twenty seconds after injection, the animal was decapitated and the blood withdrawal terminated simultaneously. The brain was removed quickly and frozen in isopentane chilled with dry ice. Brain was cut, divided, and weighed. After adding H_2O_2 and Solvable (Packard) and incubating (50 °C overnight), scintilation fluid was added. Twelve hours after shaking, radioactivity in brain and blood were measured by liquid scintilation spectrometry (RackBeta 1209, LKB). CBF was calculated according to the method of Van Uitert and Levy [6] and Betz and Iannotti [1]. The equation is as follows;

$$CBF_{(ml/100g/min)} = \frac{0.6 \times \text{brain } ^{14}\text{C count} \times 100}{\text{blood } ^{14}\text{C count} \times \text{brain weight(g)}}$$

CBF correction

Fifty microcuries of ^{3}H-rat serum albumin dissolved in 0.2 mL saline was administerd intravenously four min before CBF measurements in some animals (sham and fistula groups, n = 4 each). In order to exclude apparent CBF increase by extracerebral plasma-related IMP, we corrected the brain ^{14}C count by subtraction of the cerebral plasma-related ^{14}C count, which was calculated by counts of brain and blood ^{3}H and blood ^{14}C. The equation is as follows;

$$\text{corrected brain } ^{14}\text{C} = \text{brain } ^{14}\text{C} - \text{cerebral plasma-related } ^{14}\text{C}$$
$$= \text{brain } ^{14}\text{C} - \left(\text{brain } ^{3}\text{H} \times \frac{\text{blood } ^{14}\text{C}}{\text{blood } ^{3}\text{H}}\right)$$

Statistical analysis

Data are presented as mean ± SEM. Differences were compared by one-way factorial ANOVA followed by Fisher's protected least-squares difference test (physiologic parameters and absolute CBF study) and by unpaired two-tailed student's *t*-test (CBF correction). The software StatView 4.0 (Abacus Concepts, Inc., Berkeley, CA, USA) was used for statistical analysis. *P* values of <0.05 were considered significant.

Results

Physiological variables are shown in Table 1. There were no within-group differences in mean arterial blood pressure, heart rate, and blood gas analysis before measurements.

Effects of fistula creation and vessel ligation on SSSP

SSSP was significantly elevated in the fistula group (15 ± 1 mmHg) compared to the vessel occlusion (5 ± 1) and sham groups (5 ± 1) ($P < 0.05$) (Table 1).

Effects of cerebral venous hypertension on CBF

Reginal absolute CBF was measured separately in the cortex, caudoputamen, frontal lobe, and occipital lobe. In each area regional CBF in the fistula group tended to be higher and regional CBF in the sham group lower (Fig. 2). In three areas (right frontal lobe and both occipital lobes) within-group differences between the sham and fistula groups were significant ($P < 0.05$).

Mean absolute CBF in the fistula (103 ± 3 ml/min/100g) and occlusion groups (90 ± 4) were elevated

Table 1.

Group	SSSP (mmHg)	MBP (mmHg)	HR	pH	pO2 (mmHg)	pCO2 (mmHg)
Sham (n = 8)	5 ± 1	97 ± 2	296 ± 16	7.45 ± 0.01	151 ± 7	38 ± 1
Occlusion (n = 4)	5 ± 1	99 ± 7	292 ± 17	7.46 ± 0.01	161 ± 16	38 ± 2
Fistula (n = 11)	15 ± 1*	93 ± 2	317 ± 6	7.44 ± 0.01	154 ± 7	40 ± 1

Values are mean ± SEM. *SSSP* superior sagittal sinus pressure, *MBP* mean arterial blood pressure, *HR* heart rate. (**P* < 0.0001, one-way factorial ANOVA, Fisher's protected least-squares difference test).

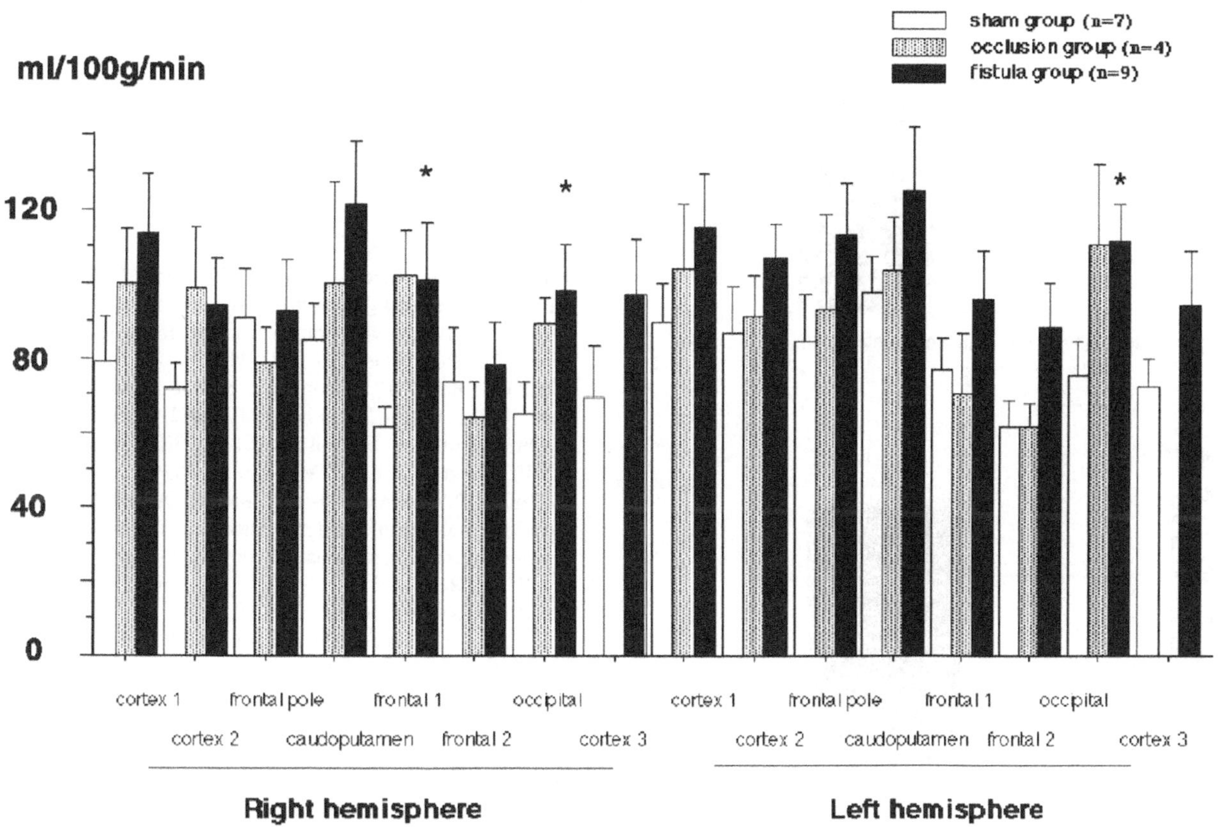

Fig. 2. Absolute cerebral blood flow was measured in eight areas of both hemispheres in three groups. Data are expressed as mean ± SEM. *$P < 0.05$ as compared with sham group

compared to the sham group (77 ± 3), where within-group differences were also significant ($P < 0.05$) (Fig. 3a). Even after correction by subtraction of plasma related IMP in the brain, mean CBF was elevated in the fistula group ($P < 0.05$, unpaired *t*-test) (Fig. 3b).

Discussion

In the patient with a dural AVM, symptoms are closely related to the pattern of venous drainage [5, 7]. Angiographical cortical venous reflux and engorged pial veins are important as risk factors for pathological manifestations [8, 9]. PET studies of patients with dural AVMs presenting with neurological deficits showed reduced CBF, increased cerebral blood volume, and increased oxygen extraction fraction in brain areas corresponding to areas of retrograde venous drainage on angiograms [2]. SPECT studies using IMP showed low perfusion area in the same manner [3]. Venous congestion or hypertention is considered to be the underlying pathology, but is not well understood.

In this rat VH model we found an increase in brain IMP uptake. The mechanism of IMP uptake into brain includes a pH difference between intra- and extracellular fluid and binding to non-specific high volume amine binding sites (4.10). By using a plasma tracer, we tried to exclude the possibility of an apparent CBF increase by increased extracerebral blood volume and found that was not the case. To make the correction more accurate, cerebral hematocrit should be included in the formula. In the occlusion group SSSP was not elevated, but CBF was higher than that in the sham group. Therefore, the mechanism of IMP uptake increase in a VH animal is possiblly due to alteration of first pass extraction of IMP by slow circulation and not by IMP behaviour change due to VH. Reasons for alteration of the IMP cerebral blood distribution coefficient after VH are unknown. Other methods of CBF measurement may help to interpret these data correctly. When comparing these results with patient data, we should remember there is a difference in the cerebral circulation between whole brain VH in this

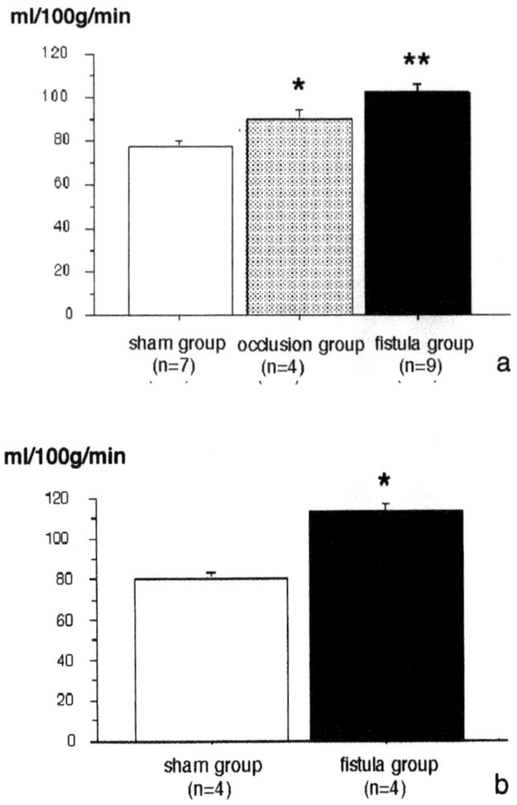

Fig. 3. (a) Mean absolute cerebral blood flow before correction. Data are expressed as mean ± SEM. There were differences among three groups (*$P < 0.05$, one-way factorial ANOVA followed by Fisher's protected least-squares difference test). (b) Mean absolute cerebral blood flow after correction by subtraction of plasma related IMP in the brain. (*$P < 0.05$, unpaired t-test)

model and localized VH in patients. There are also considerable methodological differences even though the same tracer is used.

References

1. Betz AL, Iannotti F (1983) Simultaneous determination of regional cerebral blood flow and blood-brain glucose transport kinetics in the gerbil. J Cereb Blood Flow Metab 3: 193–199
2. Iwama T, Hashimoto N, Takagi Y, Tanaka M, Yamamoto S, Nishi S, Hayashida K (1997) Hemodynamic and metabolic disturbances in patients with intracranial dural arteriovenous fistulas: positron emission tomography evaluation before and after treatment. J Neurosurg 86: 806–811
3. Kawaguchi T, Kawano T, Kaneko Y, Tsutsumi M, Ooigawa H, Kazekawa K (1999) [Dural arteriovenous fistula of the transverse-sigmoid sinus with intraventricular hemorrhage: a case report]. No Shinkei Geka 27: 1133–1138
4. Kuhl DE, Barrio JR, Huang SC, Selin C, Ackermann RF, Lear JL, Wu JL, Lin TH, Phelps ME (1982) Quantifying local cerebral blood flow by N-isopropyl-p-[123I]iodoamphetamine (IMP) tomography. J Nucl Med 23: 196–203
5. Lasjaunias P, Chiu M, ter Brugge K, Tolia A, Hurth M, Bernstein M (1986) Neurological manifestations of intracranial dural arteriovenous malformations. J Neurosurg 64: 724–730
6. Van Uitert RL, Levy DE (1978) Regional brain blood flow in the conscious gerbil. Stroke 9: 67–72
7. Vinuela F, Fox AJ, Pelz DM, Drake CG (1986) Unusual clinical manifestations of dural arteriovenous malformations. J Neurosurg 64: 554–558
8. Willinsky R (1994) Venous congestion: an MR finding in dural arteriovenous malformations with cortical venous drainage. AJNR 15: 1501–1507
9. Willinsky R (1999) Tortuous, elongated pial vein in intracranial dAVF: corelation with presentation, location, and MR findings in 122 patients. AJNR 20: 1031–1036
10. Winchell HS, Horst WD, Braun L, Oldendorf WH, Hattner R, Parker H (1980) N-isopropyl-[123I] p-iodoamphetamine: single-pass brain uptake and washout; binding to brain synaptosomes; and localization in dog and monkey brain. J Nucl Med 21: 947–952

Correspondence: Masaru Yamada, M.D., Department of Neurosurgery, Kitasato University School of Medicine, 1-15-1 Kitasato, Sagamihara, Kanagawa 228-8555 Japan. e-mail: yamadam@med.kitasato-u.ac.jp

Acta Neurochir (2003) [Suppl] 86: 213–217

Vasogenic edema and VEGF expression in a rat two-vein occlusion model

R. Kimura[1], **H. Nakase**[1], **T. Sakaki**[1], **T. Taoka**[2], and **T. Tsuji**[3]

[1] Department of Neurosurgery, Nara Medical University, Nara, Japan
[2] Department of Radiology, Nara Medical University, Nara, Japan
[3] Diagnostics Research Support, Preclinical Development Department, Nihon Schering K.K.

Summary

Vasogenic edema plays an important etiologic role in the pathogenesis of cerebral venous circulation disturbances (CVCDs). Since vascular endothelial growth factor (VEGF) is a major mediator in angiogenesis and vascular permeability, including induction of vasogenic edema, the present study was undertaken to investigate whether it has any relevance to CVCDs.

Male Wistar rats (n = 15) were used. Two adjacent cortical veins were occluded photochemically using rose bengal dye and fiberoptic illumination, with evaluation 24 hours thereafter by magnetic resonance imaging (MRI). Each brain was removed from the skull immediately after MRI and processed for hematoxylin-eosin staining (H&E staining) of sections for histopathology and comparison with MRI data. VEGF expression as demonstrated immunohistochemically appeared to coincide with vasogenic edema, diagnosed as high intensity areas on apparent diffusion coefficient of water (ADCw) maps. On the basis of these data, we conclude that VEGF is related to formation of vasogenic edema in the acute stage of CVCD.

Keywords: Cerebral vein occlusion; magnetic resonance imaging; vasogenic edema; vascular endothelial growth factor; rat.

Introduction

The pathogenesis of cerebral venous infarction remains controversial. However, considerable evidence indicates that several major factors are involved, including extension of venous thrombi, venous ischemia, brain edema and venous hypertension. Recent animal studies have shown that cytotoxic edema generally occurs in acute venous infarction, preceding the onset of vasogenic edema [3].

Vascular endothelial growth factor (VEGF), involved in angiogenesis, promotes vascular permeability [2] and thus contributes to vasogenic edema. It is known to be induced by hypoxia caused by brain ischemia. Previous studies have demonstrated that VEGF enhances vasogenic edema production and tissue damage in the acute stages of arterial infarction. The present study was undertaken to investigate the relevance of VEGF to vasogenic edema formation in the presence of cerebral venous circulation disturbance (CVCD).

Materials and methods

This present animal experimentation was conducted according to the guidelines approved at the 80th general assembly of the Japan Science Council (1980).

Animal preparation

We used 15 male Wistar rats weighing 230–300 g. After premedication with 0.5 mg atropine sulfate, chloral hydrate was administered by intraperitoneal injection (36 mg/100 g wt) for induction of anesthesia. Anesthesia was maintained with 0.5–1.0% halothane in 50% N2O and 50% O2 during surgery. Spontaneous ventilation was maintained during the procedures. In all animals, the body (rectal) temperature was kept close to 37.0 °C. Polyethylene catheters were placed in the tail artery. The arterial line provided continuous registration of arterial blood pressure and blood gas analysis. PaO2, PaCO2, and arterial pH were measured using a blood gas analyzer (ABL 330, Radiometer, Copenhagen, Denmark). Blood pressure was monitored via the intra-arterial catheter connected to a pressure transducer (Polygraph system RM-600, Nihon Koden, Tokyo, Japan). Each rat was mounted on a stereotactic frame (SR-6, Narishige Inc., Japan) and a 1.5 cm midline skin incision and a cranial window (4.5 mm × 6 mm) were made over the left parietal region using a high-speed drill and an operating microscope (Zeiss, Wetzlar, Germany). A continuous flow of physiologic saline cooled the drill tip during the craniotomy. The dura was left intact, and the left parietal cortex was exposed.

Cortical vein occlusion by the photochemical thrombotic technique

Two adjacent cortical veins (100 μm) were occluded photochemically using rose bengal dye and fiberoptic illumination. Rose bengal was injected slowly without decreasing the systemic pressure below 50 mg/kg. Care was taken to avoid illumination of tissues and other vessels near the target veins. The fiber was introduced into the targeted location of the vein for 10 min for the first vein and another

10 min with an additional injection of a half dose of rose bengal for the second vein. The resected bone flap was then repositioned and the skin wound closed. The rats were returned to individual cages and allowed free access to water and food.

Magnetic resonance imaging

The rats underwent brain MRI 24 hours after surgery to monitor infarct evolution. The time point of 24 hours after vein occlusion was chosen based on earlier data from our laboratory [6]. Animals were re-anesthetized (2.0% halothane in 100% O2) and placed in a supine position with their heads inside birdcage radiofrequency coils. The MRI experiments were performed with a 4.7 T Varian Unity Inova MRI system (Varian Inc., Palo Alto, California, USA) equipped with shielded gradients. Diffusion-weighted images (DWI) and T2-weighted images (T2WI) were acquired.

A diffusion-weighted spin-echo image was obtained by applying a motion-probing gradient of 0 and 4 G/cm. The b-values were 0 and 419.859 s/mm^2 for DWI. From these images, pixel-by-pixel maps of the apparent diffusion coefficient (ADC) of water were calculated. Imaging parameters were as follows: 256 × 256 matrix; field of view, 60 mm; repetition time, 1200 ms; echo time, 60 ms; 2 average; 8 coronal slices; slice thickness, 1.5 mm; and gapless. Total imaging time was <15 minutes per image set. MRI scans were generated and post-processed with the use of MRVision (superscript: TM) version 1.5.7 (L.A. Systems Inc.).

Histological preparation

After MRI measurements, the rats were immediately sacrificed. Each brain was removed from the skull and fixed in 10% buffered formaldehyde solution for 2 days, then cut into several 2-mm thick coronal forebrain sections using a brain slicer. The sections were embedded in paraffin and stained with hematoxylin and eosin.

Immunohistochemistry

A primary rabbit polyclonal antibody to VEGF was used at a concentration of 10 μg/mL diluted with PBS (phosphate buffer solution) for VEGF staining. The streptavidin-biotin method was performed with a Histofine Max-PO (R) kit (Nichirei Co) for polyclonal antibodies. Colour development was with 300 μg/mL of DAB (simple stain DAB solution: 3,3′-diaminobenzidine tetrahydrochloride: Nichirei Co) in 0.003% hydrogen peroxide for 1 minute. The sections were finally counterstained lightly with hematoxylin.

Statistical analysis

Data were expressed as means ± SD for physiological variables such as blood gases and mean arterial blood pressure (MABP). Statistical significance was concluded at an error probability of $P < 0.05$. All statistical comparisons were performed with SigmaStat software (Jandel Scientific, Erkrath, Germany).

Results

Physiological variables

There were no significant changes in physiological parameters of blood gases (PaO2 and PaCO2) and MABP before and after venous occlusion.

MRI evaluation

T2WI and DWI revealed high intensity biconvex areas centered on the occluded veins in all rats. ADCw maps revealed high intensity areas in deep white matter around infarctions, indicative of vasogenic edema (Fig. 1).

Histological findings

All 15 rats demonstrated various degrees of parenchymal damage such as multiple petechial hemorrhage surrounding the dilated capillaries and infarction in the white and gray matter (Fig. 2a).

Comparison between MRI and histological findings

Infarct areas were coincident with the high intensity areas on DWI. ADCw maps showed vasogenic edema as high intensity areas, and light microscopy revealed only dilated spaces surrounding the capillaries in the lesions.

Immunohistochemistry

VEGF was expressed exclusively in the areas of vasogenic edema, evident as high intensity areas on the ADCw maps (Fig. 2b).

Discussion

Venous infarction has received relatively little experimental and clinical attention. However, there is much interest in brain injury following CVCDs as a result of the increasing number of neurosurgical operations for aged patients and the development of skull base neurosurgery [12]. Features vary with the etiology, and the severity of disturbance depends on the degree of vasogenic edema and collateral circulation.

Vasogenic edema develops due to an increase in vascular permeability (VP) and breakdown of the blood-brain barrier (BBB) [4], and is especially important with CVCDs. Vascular leakage with brain injury caused by cerebral ischemia [11], physical trauma [7] and central nervous system infections [9] also results in vasogenic edema. This is known to be induced by VEGF, a secreted mitogen associated with angiogenesis which is also a potent vascular permeability factor. VEGF was first purified from ascitic fluid of guinea pigs implanted with hepatocarcinomas by Sen-

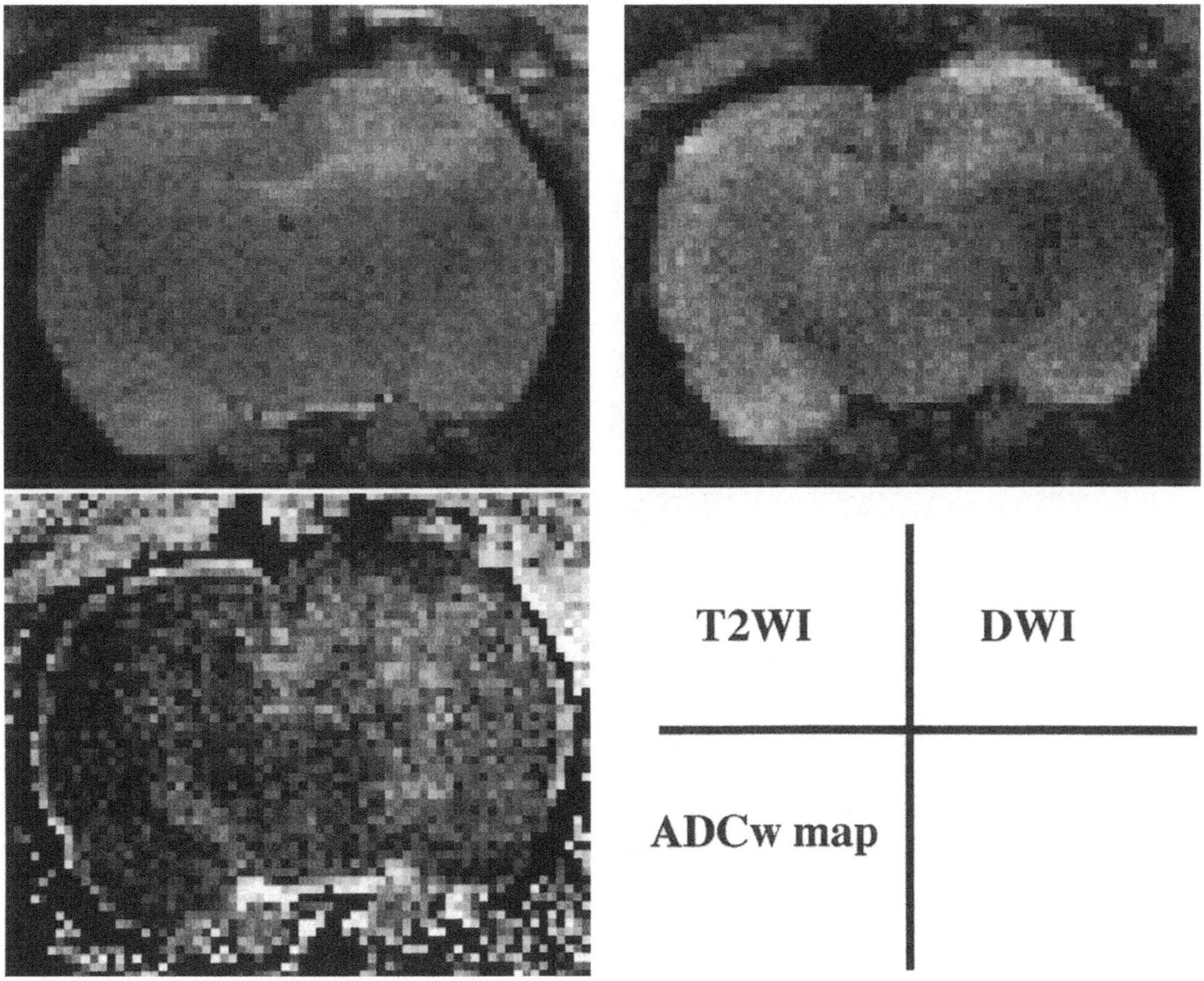

Fig. 1. A coronal view magnetic resonance T2-weighted image (*T2WI*), diffusion-weighted image (*DWI*), and apparent diffusion coefficient of water (*ADCw*) map 24 hours after 2 vein occlusion in a typical experiment (No. 7). T2WI shows a heterogeneous high intensity area extending from the surface of cortex into the deep white matter and corpus callosum in the left fronto-parietal cortex. DWI reveals a high intensity area only on the surface of the left fronto-parietal cortex. The ADCw map demonstrates a high intensity area (vasogenic edema) in the deep white matter and a low intensity area (infarction), corresponding with the DWI high intensity area

ger *et al.* in 1983 [13], who named it vascular permeability factor (VPF).

Pichiule *et al.* reported VEGF upregulation in transient global ischemia in rat brain induced by cardiac arrest and resuscitation [10]. Cobbs *et al.* detected increases in both VEGF mRNA and protein expression in the ischemic border zone 24 hours after occlusion of the middle cerebral artery in rats. They argued that the factor could have a role in the pathophysiology of recovery from ischemia [1]. VEGF can markedly enhance angiogenesis in the ischemic brain and reduce neurological deficits during stroke recovery so that it may pave the way for "therapeutic angiogenesis" of cerebrovascular disease. Such treatment may conceivably be useful to prevent stroke in patients with repeated transient ischemic attacks in whom there is a surgically inaccessible stenosis of a major intracerebral artery [5].

Although VEGF enhances cerebral microvascular perfusion and protects the ischemic border zone, it also increases blood-brain barrier (BBB) leakage in the acute stages of stroke. Zheng *et al.* reported that BBB leakage results in vasogenic edema which then can cause secondary injury in ischemic brain [15]. Ferrara *et al.* used an antibody against VEGF to reduce vasogenic edema formation and tissue damage

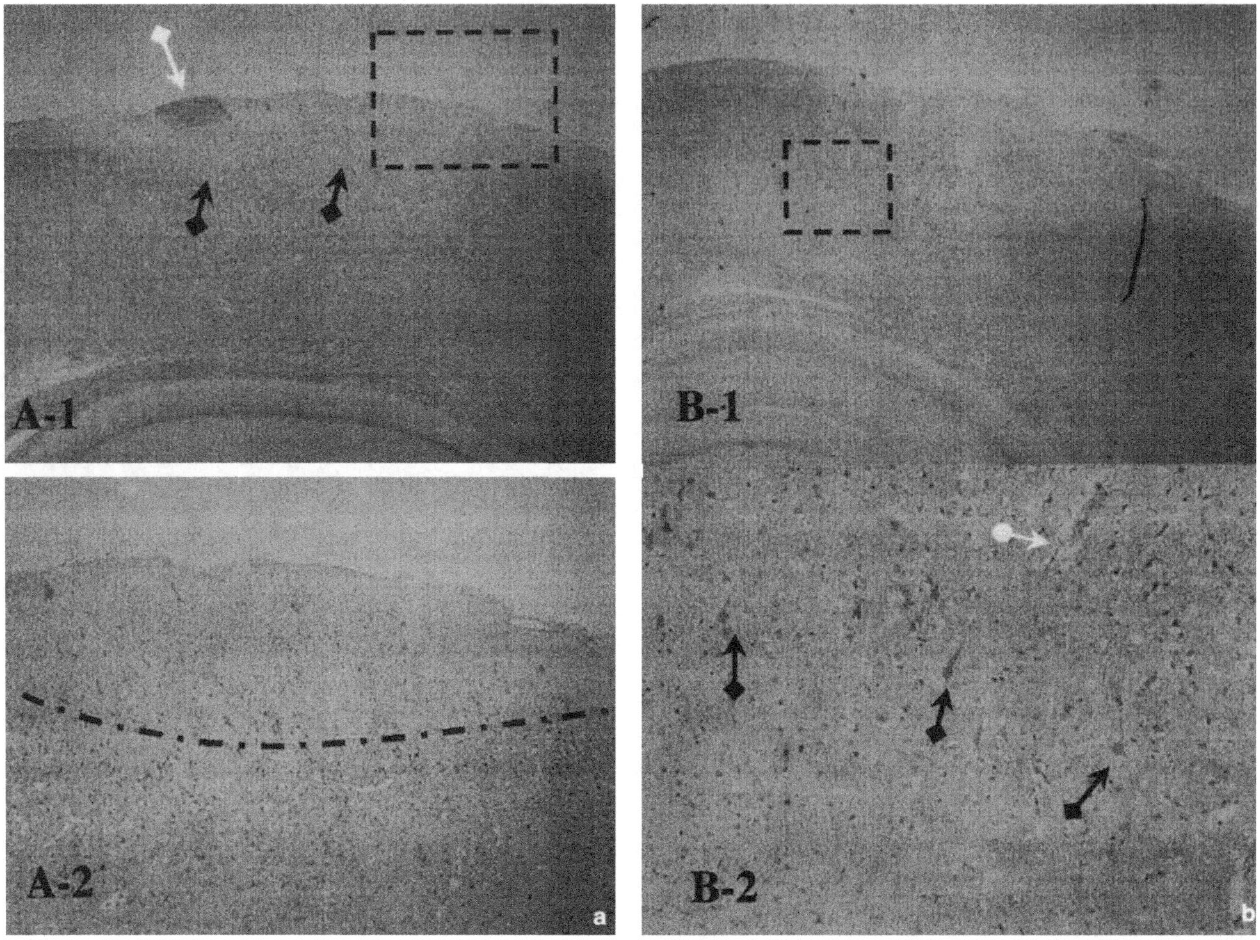

Fig. 2. (a) Photomicrographs of a coronal section stained with hematoxylin and eosin (*HE*) in the same experiment. (A-1) Black arrows indicate the infarction and the white arrow the subarachnoid hemorrhage resulting from CVCD. The dotted square indicates the territory magnified in Fig. A-2. (original magnification ×40). (A-2) The dotted line indicates the border between the infarction and vasogenic edema. No remarkable change can be detected in the area of vasogenic edema (original magnification ×100). (b) Immunohistochemical staining for anti-VEGF in the same experiment. (B-1) The dotted square indicates the territory magnified in Fig. B-2. (original magnification ×40). (B-2) The white arrow shows the dilated space surrounding the capillary in the area of vasogenic edema. VEGF is expressed exclusively (black arrows) in the area of vasogenic edema. (original magnification ×200)

after ischemia/reperfusion injury in the mouse brain [14]. Robert *et al.* also reported that Src kinase regulates VEGF-mediated vascular permeability in the brain following stroke as well as suppression of Src activity decreasing vascular permeability, thereby minimizing brain injury [8]. Therefore, inhibition of VEGF expression or activity may become a new hopeful therapy for acute stages of stroke.

In the present study, we detected VEGF expression in areas of vasogenic edema caused by CVCDs in a rat two cerebral cortical vein occlusion model [6]. We propose that inhibition of VEGF expression or activity may reduce vasogenic edema formation and tissue damage under these conditions.

References

1. Cobbs CS, Chen J, Greenberg DA, Graham SH (1998) Vascular endothelial growth factor expression in transient focal cerebral ischemia in the rat. Neurosci Lett 249: 79–82
2. Ferrara N, Davis-Smyth T (1997) The biology of vascular endothelial growth factor. Endocr Rev 18: 4–25
3. Forbes KP, Pipe JG, Heiserman JE (2001) Evidence for cytotoxic edema in the pathogenesis of cerebral venous infarction. AJNR Am J Neuroradiol 22: 450–455
4. Go KG (1997) The normal and pathological physiology of brain water. Adv Tech Stand Neurosurg 23: 47–142
5. Kovas Z, Ikezaki K, Samoto K, Inamura T, Fukui M (1996) VEGF and Flt Expression Time Kinetics in Rat Brain Infarct. Stroke 27: 1865–1873
6. Nakase H, Kakizaki T, Miyamoto K, Hiramatsu K, Sakaki T (1995) Use of local cerebral blood flow monitoring to predict brain damage after disturbance to the venous circulation: corti-

cal vein occlusion model by photochemical dye. Neurosurgery 37: 280–286
7. Patel S, Takahashi JL, Kilty DW (1997) Role of vascular endothelial growth factor in blood-brain barrier breakdown and angiogenesis in brain trauma. J Neuropathol Exp Neurol 56: 912–921
8. Paul R, Zhang ZG, Eliceiri BP, Jiang Q, Boccia AD, Zhang RL, Chopp M, Cheresh DA (2001) Src deficiency or blockade of Src activity in mice provides cerebral protection following stroke. Nat Med 7: 222–227
9. Pfister HW, Scheld WM (1997) Brain injury in bacterial meningitis: therapeutic implications. Curr Opin Neurol 10: 254–259
10. Pichiule P, Chavez JC, Xu K, LaManna JC (1999) Vascular endothelial growth factor upregulation in transient global ischemia induced by cardiac arrest and resuscitation in rat brain. Brain Res Mol Brain Res 74: 83–90
11. Rosenberg GA (1999) Ischemic brain edema. Prog Cardiovasc Dis 42: 209–216
12. Sekhar L, Chanda A, Morita A (2002) The preservation and reconstruction of cerebral veins and sinuses. J Clin Neurosci 9: 391
13. Senger DR, Galli SJ, Dvorak AM, Perruzzi CA, Harvey VS, Dvorak HF (1983) Tumor cells secrete a vascular permeability factor that promotes accumulation of ascites fluid. Science 219: 983–985
14. van Bruggen N, Thibodeaux H, Palmer JT, Lee WP, Fu L, Cairns B, Tumas D, Gerlai R, Williams SP, van Lookeren Campagne M, Ferrara N (1999) VEGF antagonism reduces edema formation and tissue damage after ischemia/reperfusion injury in the mouse brain. J Clin Invest 104: 1613–1620
15. Zhang ZG, Zhang L, Jiang Q, Zhang R, Davies K, Powers C, Bruggen N, Chopp M (2000) VEGF enhances angiogenesis and promotes blood-brain barrier leakage in the ischemic brain. J Clin Invest 106: 829–838

Correspondence: Dr. Ryota Kimura, Department of Neurosurgery, Nara Medical University, 840 Shijo-cho, Kashihara, 634, Nara, Japan. e-mail: rkimura@nmu-gw.naramed-u.ac.jp

Acta Neurochir (2003) [Suppl] 86: 219–222

Astroglial swelling in the neuronal depolarization ensemble

M. Tomita, N. Tanahashi, H. Takeda, M. Takao, Y. Tomita, T. Amano, and **Y. Fukuuchi**

Department of Neurology, School of Medicine, Keio University, Tokyo, Japan

Summary

The coupling between neuronal depolarization and astroglial swelling was examined. First, previous in vitro data for the swelling of cultured neurons (N18; rat neuroblastoma) and astroglia (C6; rat astroglyoma) upon exposure to a hypoosmotic solution were reappraised. Neurons swelled rapidly, forming blisters, and easily burst, whereas astroglia resisted swelling and slowly assumed a large full-moon shape. The time constant of swelling was 35.2 ± 7.8 s for N18 and 594.8 ± 554.0 s for C6. The glial plasmic membrane was found to be much stronger than the neuronal one, presumably due to a well-developed cytoskeleton. To overcome such neuronal membranous weakness, strong astroglial processes need to cover the neurons including the cell body and synapses, as demonstrated electron-microscopically. Next, *in situ* astroglial swelling was investigated in rats. During K^+-induced cortical spreading depression, increases or decreases of a wave-ring spread of light (550 nm) transmission through a 1 mm-thick cerebral cortical layer was observed. The moving local optical density decrease in the cortex was attributable to local vascular bed compression induced by astroglial swelling, since concomitant occurrence of colocated dynamic capillary flow stall was confirmed by a hemodilution technique. Astroglial swelling may occur in an ensemble acting during neuronal depolarization, suggesting that neurons and astroglia behave like a unit complex.

Keywords: Cytotoxic edema; depolarization; hyperemia; light transmission; neuroglial unit complex; cortical spreading depression.

Introduction

Neuronal depolarization is a major component of both functional activation and pathologies of the brain. Recovery or deterioration of neurons may depend on the efficiency of the flow response. One unknown factor is the role of the astroglia, whose processes surround not only the vasculature including capillaries, but also the neurons, including synapses. In cerebral ischemia where neuronal depolarization occurs, the astroglia swell overwhelmingly, whereas the neurons tend to shrink. Cortical spreading depression (CSD) may provide an appropriate model, since neuronal depolarization and coupled flow changes take place within a limited time and space. Van Harreveld and Ochs [9] reported rapid brain cell swelling during CSD, based on recordings of electrical conductivity.

This communication attempts first to reappraise previous data on brain cell swelling, and then to examine the roles of astroglia during K^+-induced CSD, since CSD has been reported to display "moving" neuronal depolarization and associated vascular events.

Materials and methods

The materials comprised I) in vitro data from our previous experiments in which the differential behavior of cultured glial cells (C6; astroglioma) and neuronal cells (N18; rat neuroblastoma) exposed to a hypoosmotic solution was examined [5], and II) cortical spreading depression (CSD) was induced by K^+ microinjection under α-chloralose-urethane anesthesia in 6 rats.

In the in vitro experiment, images of cell swelling recorded on videotape employing VEC-DIC microscopy were acquired with a Scion frame grabber at every 15 s. The difference in cell swelling speed was compared between the neurons and astroglia during the hypoosmotic shock.

In the in vivo experiment, a transillumination method was applied to the rat sensorimotor cortex as described elsewhere [3]. The transmitted light at $\lambda = 550 \pm 10$ nm (one of the isosbestic points of hemoglobin) in a dark room was associated with changes in blood volume [8]. Images of the transilluminated cortex in a 2 mm × 2 mm ROI were recorded continuously on videotape and fed into a personal computer through a Scion LG-3 frame grabber card at an acquisition rate of 5 frames/s. The light transmission (LT) changes in the frames were automatically represented logarithmically on a 256 gray scale. Subtractions were performed between the first image (control) and each subsequent image. Concentrated K^+ solution was microinjected into the cortical parenchyma at the center of the ROI (Fig. 2). During continuous video-recording of the ROI after CSD induction, 25 μl of saline (a negative blood indicator) was injected as a spike into the internal carotid artery. In total, 2500 hemodilution curves were obtained from the hemodilution processes in individual flow pixels of a 50 × 50 matrix in the ROI and were analyzed with Matlab software (The MathWorks). The mean transit time (MTT)

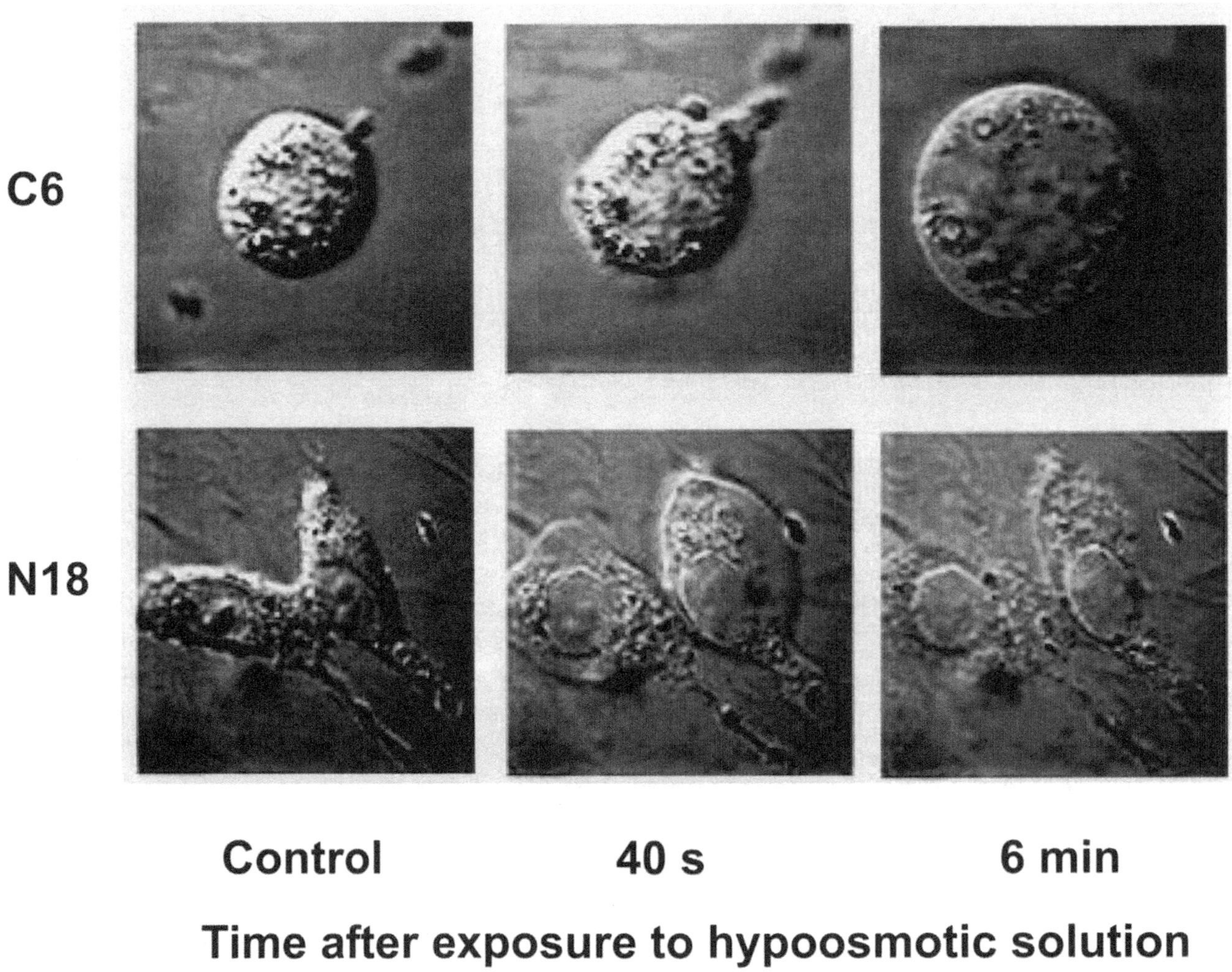

Fig. 1. Upper row: Swelling of astroglia (C6). The cell swelled slowly taking a time of up to 6 min to become like a full moon. Lower row: Swelling of neurons (N18). The cell swelled rapidly, forming blisters, and burst easily at 40 s after exposure

in the individual pixels was calculated by employing a customized algorithm ($MTT = \int t \cdot \Delta LT(t)\,dt / \int \Delta LT(t)\,dt$) and was arranged at an appropriate address to construct a two-dimensional (2-D) flow map. The LT map and corresponding 2-D flow map were then merged, and flow values at identical locations were plotted against the corresponding LT values on the abscissa.

Results and comments

In vitro *experiment*

Upon exposure to the hypoosmotic solution, both N18 and C6 swelled. However, they differed in their speed and manner of swelling. Neurons swelled rapidly forming blisters and easily burst (Fig. 1, lower row), whereas astroglia resisted swelling and slowly assumed a large full-moon shape (Fig. 1, upper row). The time constant of the swelling (mean $\pm$ SD, $n = 5$) was

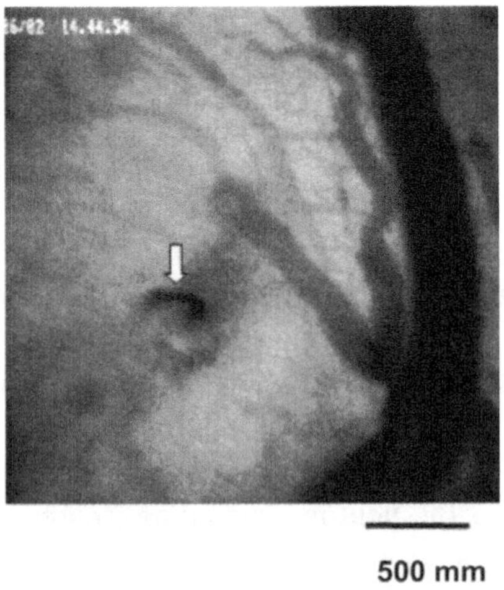

Fig. 2. Control image of the cerebral cortex in the ROI (1 mm thick and 2 mm × 2 mm wide) transilluminated with 550 nm light, which was continuously video-monitored. The arrow indicates the site of K^+ injection

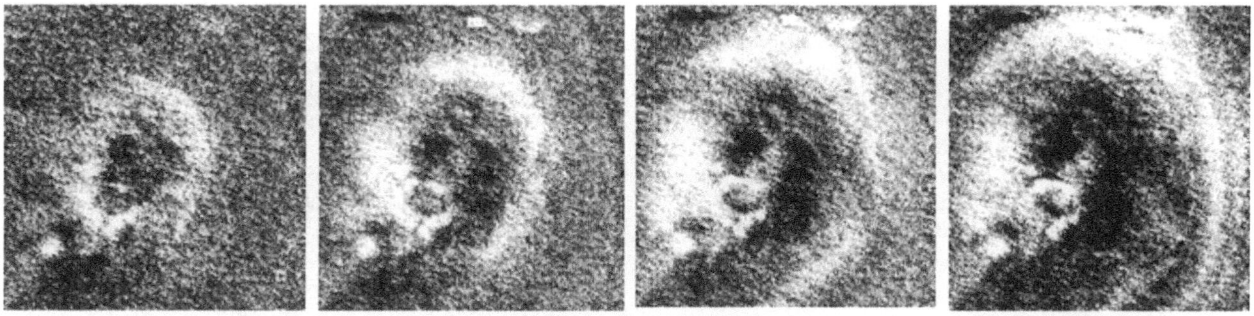

Fig. 3. Sequential subtraction images at 4, 7, 10, and 12 s after induction of cortical spreading depression. Note the wave-ring spread of a light ring (ischemia) at 2.5 mm/s speed followed by a dark band (hyperemia)

35.2 ± 7.8 s for N18 and 594.8 ± 554.0 s for C6. This implied that the plasmic membrane of glia was more than 10 times stronger than that of neurons. The latter membrane appeared to be made up of numerous receptors with a scanty cytoskeleton, whereas the glial membrane was strengthened with a well-developed cytoskeleton. To compensate for weakness of the neuronal membrane, astroglial processes need to cover the entire neuronal cell body, as demonstrated by electron-microscopic pictures.

In vivo *experiment*

K^+ injection into the cerebral cortex (Fig. 2) produced a repeating concentric light wave-ring spread as shown in the subtraction pictures at 4, 7, 10, and 12 s after CSD induction (Fig. 3). Since a green color filter (550 nm, one of the isosbestic points of hemoglobin) was used, the increase in LT of the light ring could be due to a CBV decrease, and therefore an RBC decrease (plasma skimming) in the local capillaries, and the subsequent decrease in LT or dark band, due to an RBC increase. The 2-D flow maps also showed a wave-ring spread of flow changes (ischemia and hyperemia). Plot profiles of the flow along a line passing through the injection site revealed that the flow changes with CSD were very rapid, with an average time constant of 0.53 ± 0.21 s ($n = 5$).

Discussion

From the thermodynamic standpoint [6], depolarization of the plasmic membrane means sodium influx, coupled transport of water with the sodium influx [2], and swelling of the cell. This results in loss of ionic gradient, loss of electrical gradient, and loss of osmotic gradient across the membrane. Depolarizing neurons *per se* can also swell initially. For example, Tasaki *et al.* [4] reported transient swelling of neurons during synaptic transmission in the sympathetic ganglion. Such neuronal depolarization is immediately transmitted to the glia. Synaptic transmission of glutamate may be one of the signalling processes [1]. K^+, a depolarizer, could be another candidate, because it effluxes upon neuronal depolarization. We found a rapid decrease in local CBV and CBF with CSD. This was attributed to transient astroglial swelling, since minimal local cell volume increase in the tissue would cause compression of neighboring capillaries. If such ischemia is not compensated by a flow increase, astroglial swelling could continue to elevate intracranial pressure [7]. Astrocyte collaboration with neuronal depolarization in an ensemble constitutes a neuro-glial complex which may contribute to capillary flow regulation.

References

1. Magistretti PJ, Pellerin L, Rothman DL, Shulman RG (1999) Energy on demand. Science 283: 496–497
2. Nagasawa M, Tasaka M, Tomita M (1986) Coupled transport of water and ions through membranes as a possible cause of cytotoxic edema. Neurosci Lett 66: 19–24
3. Schiszler I, Tomita M, Fukuuchi Y, Tanahashi N, Inoue K (2000) New optical method for analyzing cortical blood flow heterogeneity in small animals: validation of the method. Am J Physiol 279: H1291–H1298
4. Tasaki I, Nakaye T, Byrne PM (1985) Rapid swelling of neurons during synaptic transmission in the bullfrog sympathetic ganglion. Brain Res 331: 363–365

5. Tomita M, Fukuuchi Y, Terakawa S (1994) Differential behavior of glial neuronal cells exposed to hypotonic solution. Acta Neurochir (Wien) [Suppl] 60: 31–33
6. Tomita M, Gotoh F (1992) Cascade of cell swelling (cytotoxic edema): thermodynamic potential discharge of brain cells following membrane injury. Am J Physiol 262: H603–H610
7. Tomita M, Gotoh F, Kobari M (1988) Colloid osmotic pressure of cat brain homogenate separated from autogenous CSF by a copper ferrocyanide membrane. Brain Res 474: 165–173
8. Tomita M, Gotoh F, Sato T, Amano T, Tanahashi N, Tanaka K, Yamamoto M (1978) Photoelectric method for estimating hemodynamic changes in regional cerebral tissue. Am J Physiol 235: H56–H63
9. Van Harreveld A, Ochs S (1957) Electrical and vascular concomitants of spreading depression. Am J Physiol 189: 156–166

Correspondence: Prof. Minoru Tomita, M.D., Department of Neurology, School of Medicine, Keio University, 35 Shinanomachi, Shinjuku-ku, Tokyo 160-8582, Japan. e-mail: mtomita@sc.itc.keio.ac.jp

Acta Neurochir (2003) [Suppl] 86: 223–226

An Na^+/H^+ exchanger inhibitor suppresses cellular swelling and neuronal death induced by glutamate in cultured cortical neurons

S. Yamamoto[1], **Y. Matsumoto**[1,2], **Y. Suzuki**[2], **T. Tsuboi**[1], **S. Terakawa**[1], **N. Ohashi**[3], and **K. Umemura**[2]

[1] Photon Medical Research Centre, Hamamatsu University School of Medicine, Hamamatsu, Japan
[2] Department of Pharmacology, Hamamatsu University School of Medicine, Hamamatsu, Japan
[3] Research Division, Sumitomo Pharmaceuticals Co. Ltd., Osaka, Japan

Summary

We examined the effects of a selective Na^+/H^+ exchanger inhibitor, SM-20220 (*N*-(aminoiminomethyl)-1-methyl-1*H*-indole-2-carboxamide methanesulfonate), on neuronal death induced by glutamate in rat cortical neurons. Morphological changes in neurons were observed with a differential interference contrast microscope, and cellular swelling was analysed. Neuronal death was assessed by staining the cell with propidium iodide. The intracellular calcium concentration ($[Ca^{2+}]_i$) and the intracellular pH were measured by fluorescence imaging with fluo-3/AM as an indicator for $[Ca^{2+}]_i$ and BCECF/AM for pH, respectively. SM-20220 (0.3 to 30 nM) dose-dependently attenuated glutamate (300 μM)-induced neuronal death in a dose-dependent fashion over 6 hours, and inhibited acute cellular swelling following glutamate (100 μM) exposure. SM-20220 suppressed the persistent $[Ca^{2+}]_i$ increase following glutamate (500 μM) exposure, and inhibited intracellular acidification induced by glutamate (1 mM). The activation of the Na^+/H^+ exchanger system may enhance the progress of cerebral damage and oedema following glutamate exposure. SM-20220, a Na^+/H^+ exchanger inhibitor, suppressed neuronal death and cellular swelling induced by glutamate through inhibition of both Ca^{2+} influx and acidification in neurons.

Keywords: Glutamate; Na^+/H^+ exchanger; calcium overload; cellular swelling; intracellular pH.

Introduction

In neurons, ischaemia causes the release of the excitatory amino acid, glutamate, which activates the receptor-dependent Ca^{2+} channel, resulting in Ca^{2+} overload [1]. Glutamate also produces intracellular acidosis in culture neurons [3, 8]. The Na^+/H^+ exchanger (NHE) is essential for intracellular pH (pH_i) regulation under both physiological and pathological conditions [9]. The intracellular acidosis activates the Na^+/H^+ exchange system, resulting in Na^+ influx that extrudes protons from the cell. This then causes cellular swelling and Ca^{2+} overload that exchanges Na^+ by reversing Na^+/Ca^{2+} exchange. Inhibition of Na^+/H^+ exchange therefore should prevent not only Na^+-mediated cellular swelling but also Ca^{2+} overload. We investigated the effects of a newly synthesized NHE inhibitor [4], SM-20220 (*N*-(aminoiminomethyl)-1-methyl-1*H*-indole-2-carboxamide methanesulfonate), on glutamate-induced neuronal death in rat cortical cultures.

Materials and methods

Cortical neurons were prepared from one-day-old Wistar rats, and cultured on collagen-coated dishes (Iwaki Glass Base Dish, Asahi Techno Glass Co., Tokyo, Japan) as described previously [6]. The cells were maintained for 10 to 14 days at 37 °C in a 5% CO_2-containing atmosphere, and then used for experiments. To observe the morphological changes of the neurons, a cultured dish containing cells was placed on the stage of an inverted differential interference contrast (DIC) microscope (Axiovert 10, Zeiss, Oberkochen, Germany). The image was obtained with a charge-coupled device (CCD) camera (C6489, Hamamatsu Photonics K.K., Hamamatsu, Japan), and contrast-enhanced digitally by using an image processor (ARGUS 20, Hamamatsu Photonics K.K.). Throughout the experiment, the temperature of the medium was maintained at 34 ± 2 °C. To induce glutamate excitotoxicity in primary cultures, they were incubated at 37 °C for 6 hours with artificial cerebrospinal fluid (aCSF) containing glutamate, and then the cultures were incubated at 37 °C for 15 minutes with aCSF containing propidium iodide (PI, 3 μM), which does not penetrate viable cells but enters injured cells. $[Ca^{2+}]_i$ and pH_i were measured by fluorescence imaging with fluo-3 and with BCECF, respectively. Neurons were incubated with aCSF containing fluo-3/AM (4.4 μM, Dojindo, Kumamoto, Japan) for 20 minutes, or containing BCECF/AM (1 μM, Dojindo) for 10 minutes, and were observed with a microscope (IX 70, Olympus, Tokyo, Japan) equipped with a confocal scanner of a microlens-attached Nipkow-disk (CSU-10, Yokokawa Electric Co., Tokyo, Japan). Comparison of groups was made by one-way analysis of variance (ANOVA) followed by least significant difference test. A P value < 0.05 was considered significant.

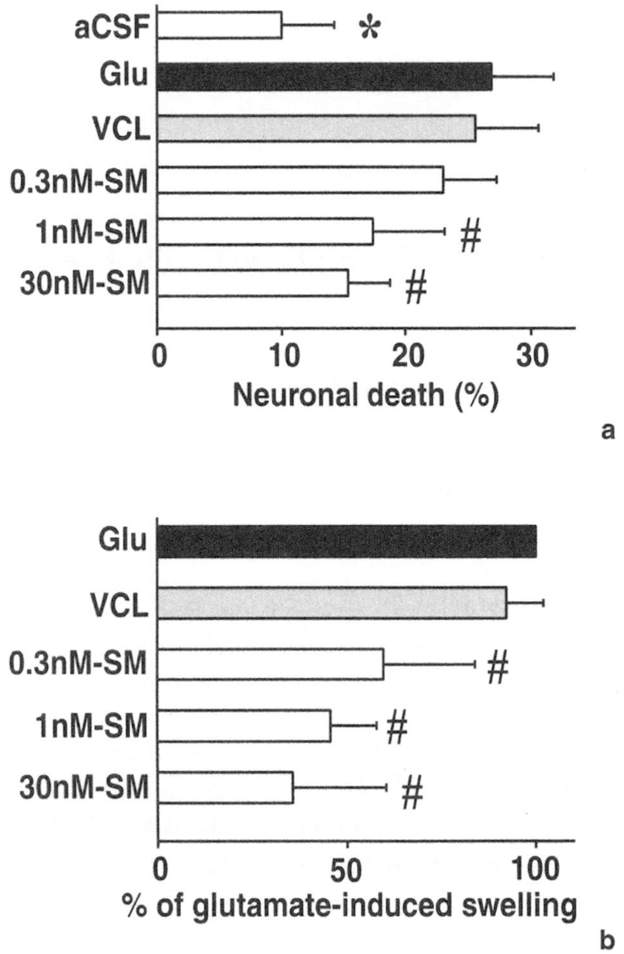

Fig. 1. (a) Effects of SM-20220 on glutamate-induced neuronal death. Neuronal death was analysed 6 hours after an application of aCSF, 500 μM glutamate (Glu), Glu with vehicle (VCL, 8% polyethyleneglycol 400), and Glu with SM-20220 (SM). The death rate was expressed as a percentage of the total number of neurons. SM significantly protected against neuronal death in a dose-dependent manner. * $P < 0.01$ different from Glu. # $P < 0.05$ different from VCL. Data = mean ± SD, n = 4 cultures in different preparations. (b) Effects of SM-20220 on cellular swelling 10 min after glutamate exposure. SM significantly inhibited the cellular swelling induced by glutamate in a dose-dependent manner. # $P < 0.05$ different from VCL. Data = mean ± SD, n = 6

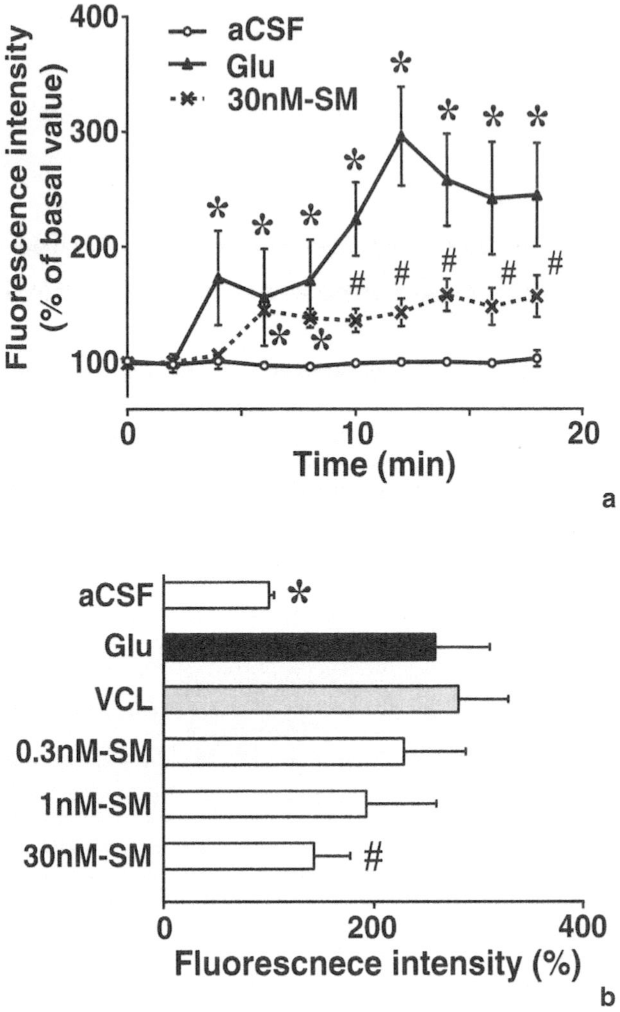

Fig. 2. (a) Time course of $[Ca^{2+}]_i$ change in neurons following glutamate exposure. Fluorescence images were recorded following an application of aCSF, 500 μM glutamate (Glu), and Glu with 30 nM SM-20220 (SM). Note that Glu induced an initial peak at 4 min followed by persistent increase. SM markedly suppressed the persistent increase in $[Ca^{2+}]_i$. * $P < 0.05$ different from aCSF. # $P < 0.05$ different from Glu and aCSF. Values = mean ± SD. n = 4. (b) Effects of SM-20220 on $[Ca^{2+}]_i$ 15 min after glutamate exposure. Fluorescence images were recorded 15 min after an application of aCSF, 500 μM glutamate (Glu), Glu with vehicle (VCL), and Glu with SM-20220 (SM). SM significantly inhibited the increase in $[Ca^{2+}]_i$ induced by glutamate in a dose-dependent manner. * $P < 0.01$ different from Glu. # $P < 0.01$ different from VCL. Data = mean ± SD, n = 4

Results

1. SM-20220 decreases neuronal death and cellular swelling induced by glutamate

SM-20220 (0.3 to 30 nM) attenuated glutamate (300 μM)-induced neuronal death in a dose-dependent fashion over 6 hours (Fig. 1a). Following glutamate (100 μM) exposure, the area of the neuron gradually increased to 117.7 ± 8.0% (n = 6) of the baseline in 10 minutes (Fig. 1b). SM-20220 (0.3 to 30 nM) inhibited the cellular swelling.

2. SM-20220 suppresses glutamate-induced increase in $[Ca^{2+}]_i$

Fig. 2a shows the sequential percent change in fluorescence intensity of fluo-3. Glutamate (500 μM) alone produced the initial rise followed by persistent increase in $[Ca^{2+}]_i$, while SM-20220 (30 nM) significantly suppressed persistent $[Ca^{2+}]_i$ increase. At 15 min after glutamate exposure, SM-20220 suppressed a rise in $[Ca^{2+}]_i$ dose-dependently (Fig. 2b).

3. SM-20220 improves glutamate-induced acidification

Glutamate (1 mM) gradually evoked a decrease in the fluorescence intensity of BCECF indicating a reduction in pH_i. SM-20220 (30 nM) partially but significantly suppressed it after glutamate exposure (Fig. 3a). Fig. 3b shows SM-20220 (0.3 to 30 nM) dose-dependently inhibited the decrease in pH_i at 15 min by a dose-dependent response.

Discussion

In this study, we examined the effects of 0.3, 1.0, and 30 nM SM-20220 on neuronal death. This is based on results indicating the IC_{50} for inhibition of the NHE in neurons was 5 nM [5]. For optimum imaging, we used different concentrations of glutamate in each experiment of the present study. Since the effects of glutamate, even in different concentrations, were completely inhibited by N-methyl-D-aspartate (NMDA) receptor antagonist (unpublished observation), the glutamate-induced events, including neuronal death, cellular swelling, and change in $[Ca^{2+}]_i$ and pH_i, must be mediated through NMDA receptor. These results justify our conclusion even in the experiments conducted using different concentrations of glutamate.

Our results showed that glutamate-induced cellular swelling was inhibited by SM-20220. This observation provides evidence indicating that the NHE activation is involved in the processes of glutamate-induced cellular swelling. In this study, we also observed that glutamate induced intracellular acidosis in neurons. Intracellular acidosis activates the Na^+/H^+ exchange system to extrude protons from the cell, resulting in Na^+ influx. The accumulation of Na^+ increases intracellular osmolarity resulting in water influx and then cell volume increases. Na^+ is mainly counteracted by energy-consuming ATP-dependent Na^+/K^+ adenosine triphosphatase. Conceivably, the energy failure mediated by mitochondrial dysfunction induced by glutamate cannot counteract Na^+ and leads to further cellular swelling. Neuronal swelling leads to the rupture of the cell membrane eventually, then neuronal death. Thus, suppression of cellular swelling mediated by NHE inhibitor can be neuroprotective.

Ca^{2+} overload has been thought to be important in causing glutamate-induced neuronal death [2, 7]. In the present study, we observed that SM-20220 suppressed an increase in $[Ca^{2+}]_i$ following glutamate exposure. Suppression of Ca^{2+} influx, therefore, must be one of the mechanisms of neuroprotection provided by NHE inhibitor. The initial rise can be induced by NMDA receptor activation because NMDA receptors are very permeable to Ca^{2+} among glutamate re-

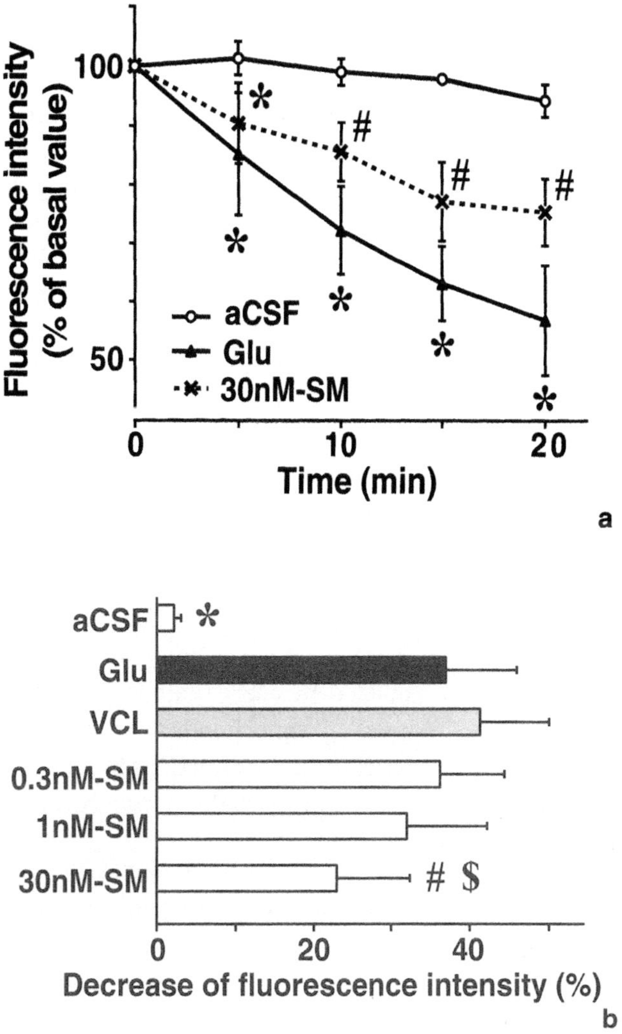

Fig. 3. (a) Time course of intracellular acidification in rat cultured cortical neurons. Neurons were treated with aCSF alone, 1 mM glutamate (Glu), and Glu with 30 nM SM-20220 (SM). SM attenuated the decrease in pH_i induced by the glutamate. * $P < 0.05$ different from aCSF. # $P < 0.05$ different from Glu and aCSF. Values = mean ± SD. n = 5. (b) Effects of SM-20220 on intracellular acidification 15 min after glutamate exposure. Fluorescence images were recorded 15 min after an application of aCSF, 1 mM glutamate (Glu), Glu with vehicle (VCL), and Glu with SM-20220 (SM). SM significantly attenuated the decrease in pH_i induced by glutamate in a dose-dependent manner. * $P < 0.01$ different from Glu. # $P < 0.05$ different from VCL. $ $P < 0.05$ different from aCSF. Data = mean ± SD, n = 5

ceptors and induces Ca^{2+} influx [2]. It is unlikely that SM-20220 inhibited the initial Ca^{2+} influx because SM-20220 has little effect against various channels including Na^+, K^+, and Ca^{2+} (L and N types) channels and glutamate receptors [5]. As discussed above, intracellular acidosis activates the NHE, resulting in the Na^+ influx. Probably it reverses the Na^+/Ca^{2+} exchange and then causes Ca^{2+} overload. The NHE inhibitor can decrease this Ca^{2+} overload and can be neuroprotective.

The NHE is essential for pH_i regulation in the brain under both physiological and pathological conditions [9]. It is expected that the NHE inhibitor may worsen acidosis following glutamate exposure. However, in our present study the NHE inhibitor did not worsen, but rather, improved acidification in neurons following glutamate application. The reason why NHE inhibitor improves intracellular acidification is not clear. Possible explanations are: 1) Glutamate-induced acidification is mediated by energy failure [8]; 2) energy failure is caused by mitochondrial dysfunction through Ca^{2+} overload; and 3) NHE inhibitor decreased Ca^{2+} overload and partially improved intracellular acidification as a result.

In conclusion, SM-20220, an NHE inhibitor, suppressed neuronal death and cellular swelling induced by glutamate through inhibition of both Ca^{2+} influx and acidification in neurons. Activation of the NHE system may enhance the progress of cerebral damage and oedema following cerebral ischaemia. Blockade of NHE is likely to be of therapeutic potential in the treatment of ischaemic stroke.

Acknowledgments

This study was supported by the unrestricted gift of SM-20220 from Sumitomo Pharmaceuticals, Osaka, Japan.

References

1. Choi DW (1988) Calcium-mediated neurotoxicity: relationship to specific channel types and role in ischaemic damage. Trends Neurosci 11: 465–469
2. Choi DW (1992) Excitotoxic cell death. J Neurobiol 23: 1261–1276
3. Hartley Z, Dubinsky JM (1993) Changes in intracellular pH associated with glutamate excitotoxicity. J Neurosci 13: 4690–4699
4. Kitano M, Kojima A, Nakano K, Miyagishi A, Noguchi T, Ohashi N (1999) Synthesis and biological activity of *N*-(aminoiminomethyl)-1*H*-indole carboxamide derivatives as Na^+/H^+ exchanger inhibitors. Chem Pharm Bull (Tokyo) 47: 1538–1548
5. Kuribayashi Y, Itoh N, Kitano M, Ohashi N (1999) Cerebroprotective properties of SM-20220, a potent Na^+/H^+ exchange inhibitor, in transient cerebral ischaemia in rats. Eur J Pharmacol 383: 163–168
6. Nagai N, Yamamoto S, Tsuboi T, Ihara H, Urano T, Takada Y, Terakawa S, Takada A (2001) Tissue-type plasminogen activator is involved in the process of neuronal death induced by oxygen-glucose deprivation in cultures. J Cereb Blood Flow Metab 21: 631–634
7. Randall RD, Thayer SA (1992) Glutamate-induced calcium transient triggers delayed calcium overload and neurotoxicity in rat hippocampal neurons. J Neurosci 12: 1882–1895
8. Wang GJ, Randall RD, Thayer SA (1994) Glutamate-induced intracellular acidification of cultured hippocampal neurons demonstrates altered energy metabolism resulting from Ca^{2+} loads. J Neurophysiol 72: 2563–2569
9. Yao H, Ma E, Gu XQ, Haddad GG (1999) Intracellular pH regulation of CA1 neurons in Na^+/H^+ isoform 1 mutant mice. J Clin Invest 104: 637–645

Correspondence: Seiji Yamamoto, M.D., Ph.D., Photon Medical Research Centre, Hamamatsu University School of Medicine, 1-20-1 Handayama, Hamamatsu, 431-3192, Japan. e-mail: seijiy@hamamed.ac.jp

Acta Neurochir (2003) [Suppl] 86: 227–230

Akt phosphorylation and cell survival after hypoxia-induced cytochrome *c* release in superfused respiring neonatal rat cerebrocortical slices

K. Hirai[1], **T. Hayashi**[2], **P. H. Chan**[2], **V. J. Basus**[3], **T. L. James**[3], and **L. Litt**[1]

[1] Department of Anesthesia, University of California, San Francisco, San Francisco, California, U.S.A.
[2] Departments of Neurosurgery, Neurology and Neurological Sciences, and Program in Neurosciences, Stanford University School of Medicine, Stanford, California, U.S.A.
[3] Department of Pharmaceutical Chemistry, University of California, San Francisco, San Francisco, California, U.S.A.

Summary

Phosphorylation of Akt before hypoxia (30 min) and during reoxygenation (4 h) was evaluated in superperfused neonatal rat cerebrocortical slices (350 μm, P7, Sprague-Dawley). Cytosolic cytochrome *c* intensities in Western blots, which were increased at the end of hypoxia, were decreased during reoxygenation. Western blot intensities of phosphorylated Akt (phospho-Akt), nearly undetectable at the end of hypoxia, recovered quickly during reoxygenation, in a trend opposite that for cytochrome *c*. At 1.5 h and 4 h after hypoxia they became larger or the same as before hypoxia. Total Akt was unchanged by hypoxia and reoxygenation. Phosphocreatine (PCr) and nucleotide triphosphates (NTP) were measured in parallel 14.1 Tesla ex vivo ^{31}P NMR superfused brain slice studies. PCr and α-NTP were nearly undetectable at the end of hypoxia. Although they recovered quickly after hypoxia, they were lower than before hypoxia. Reductions in phospho-Akt during hypoxia were consistent with the general unavailability of basic high energy phosphates. Preferential phosphorylation of Akt after hypoxia suggested that substantial reductions in intracellular energy, as indicated by PCr and NTP, might be tolerated by processes important for generating phospho-Akt. Additionally, the post-hypoxia increase in phospho-Akt might have contributed to concomitant reductions in cytosolic cytochrome *c*.

Keywords: Akt; brain slice; cytochrome *c*; hypoxia; NMR (nuclear magnetic resonance); NTP (nucleotide triphosphates); phosphorylation.

Introduction

Phosphorylation of the serine-threonine kinase Akt (also known as protein kinase B) provides anti-apoptiotic enzyme activation that can critically alter the balance between survival and apoptosis. Akt activation, which can be initiated by insulin and various growth and survival factors [4, 13], is known to occur near the cell membrane, where a conformational change permits phosphorylation at Thr308 and Ser473. Phospho-Akt returns to the cytosol and cell nucleus, where it promotes cell survival by phosphorylating and thereby inhibiting pro-apoptotic proteins, such as BAD [5, 6], glycogen synthase kinase-3 (GSK-3) [3], Forkhead family members [1, 12] and caspase-9 [2]. Certain unphosphorylated pro-apoptotic proteins can oligomerize in outer mitochondrial membranes, forming pores that permit release of cytochrome *c* to the cytosol. Thus an important anti-apoptotic action of phospho-Akt is the prevention of mitochondrial release of cytochorome *c*, and thereby the prevention of "intrinsic pathway" apoptosis [11, 15]. We recently reported that hypoxia-induced cytochrome *c* apoptosis was clearly visible in our brain slice model [9]. In this study we measured levels of total Akt and phospho-Akt, in order to test the hypothesis that decreasing levels of cytosolic cytochrome *c* would require adequate levels of NTP and PCr, and be associated with increasing levels of phospho-Akt, a protective protein.

Materials and methods

Slice preparation and experimental design

The protocol for obtaining brain slices was approved by the UCSF Committee on Animal Research. In each experiment, 18 respiring cerebrocortical slices, each 350 μm thick, were obtained from nine 7 day old (10–12 g) Sprague-Dawley rats, as described previously [7]. After decapitation under isoflurane anesthesia two cerebrocortical slices were obtained rapidly from the right and left lateral surfaces by sliding the brain past a blade fixed 350 μm above a flat lubricated surface. Slices so obtained were transferred to a 15 ml conical tube which was used as a chamber to superfuse slices (Becton Dickinson, Franklin Lakes, NJ, USA). Flowing through the chamber was oxygenated artificial cerebrospinal fluid (oxy-ACSF), which consisted of a modified Krebs balanced salt solution containing 124 mmol/L

NaCl, 5 mmol/L KCl, 1.2 mmol/L KH_2PO_4, 1.2 mmol/L $MgSO_4$, 1.2 mmol/L $CaCl_2$, 26 mmol/L $NaHCO_3$ and 10 mmol/L glucose. The bicarbonate buffer for oxy-ACSF, maintained by continuous bubbling with a 95% O_2/5% CO_2 gas mixture, had constant values of PCO_2 (40 mmHg), PO_2 (600–650 mmHg) and pH (7.4). The tube was immersed in a water bath whose temperature was kept at 37 °C. Slices were superfused with fresh oxy-ACSF at a flow rate of 10 ml/min and allowed to recover metabolically for 3 h.

A 30-minute period of hypoxia began after 3 h metabolic recovery, at a time defined as t = −0.5 h. Hypoxia was induced by stopping the ACSF flow. At the end of hypoxia (t = 0 h), slice superfusion with oxy-ACSF was resumed and continued for 4 h. Slice sampling was done at each of five predetermined times: just before and just after hypoxia (t = −0.5 h and t = 0 h, respectively), and at 0.5, 1.5 and 4 h after reoxygenation following hypoxia.

Western blot analysis of total Akt, phospho-Akt (Ser473), and cytochrome c

Whole cell fractions for each time point were obtained from three slices that were homogenized with an ultrasonicater in 400 µl of cold suspension buffer (20 mmol/L HEPES-KOH [pH 7.5], 250 mmol/L sucrose, 10 mmol/L KCl, 1.5 mmol/L $MgCl_2$, 1 mmol/L EDTA, 1 mmol/L EGTA, 1 mmol/L dithiothreitol, 0.1 mmol/L phenylmethylsulfonyl fluoride, 2 µg/mL aprotinin, 10 µg/mL leupeptin, 5 µg/mL pepstatin, and 12.5 µg/mL N-acetyl-leu-leu-norleucinal) containing phosphatase inhibitor cocktail I and II (1 : 100 dilution for each, Sigma, St. Louis, MO, USA). Homogenates were centrifuged at 8000 g for 20 min at 4 °C and supernatants were used as the whole cell fraction. Protein concentrations were determined by the Bradford method (Bio-Rad, Hercules, CA, USA). Twenty micrograms of protein from each sample were separated by SDS-PAGE on a 10–20% Tris glycine gel (Invitrogen, Carlsbad, CA, USA) and transferred to a polyvinylidine difluoride membrane (Invitrogen). The primary antibodies were: a 1 : 1000 dilution of a polyclonal anti-Akt antibody (Cell Signaling Technology, Beverly, MA, USA), which detects total Akt; and a 1 : 1000 dilution of polyclonal anti-phospho-Akt (Ser473) antibody (Cell Signaling Technology), which were reacted overnight at 4 °C. Western blots were performed with horseradish peroxidase-conjugated anti-rabbit Ig (Amersham Pharmacia Biotec, Piscataway, NJ, USA) using the chemiluminescence method.

Extractions of cytosolic fractions were performed with only minor modifications to procedures described previously [8]. Three slices from each time point were homogenized by gentle pounding (35 strokes) in a glass tissue grinder (Wheaton, Millville, NJ, USA) that contained 200 µl of cold suspension buffer. Homogenates were centrifuged at 750 g for 10 min at 4 °C. Resulting supernatants were centrifuged at 8000 g for 20 min at 4 °C. 8000 g supernatants were further centrifuged at 100000 g for 60 min at 4 °C and resulting supernatants were used as the cytosolic fraction. Eight micrograms of protein from each sample were loaded on a 10–20% Tris glycine gel. The primary antibody was a 1 : 1000 dilution of a monoclonal anti-cytochrome *c* antibody, which was reacted overnight at 4 °C. Western blots were performed as described above.

Ex vivo *^1H and ^{31}P NMR spectroscopy*

Ex vivo NMR studies of superfused, respiring slices were performed using an 8 mm NMR tube (Wilmad Glass, Buena, NJ, USA) in a 14.1 Tesla (600 MHz) Varian INOVA spectrometer having a customized, multinuclear Z-SPECT radiofrequency probe (3NG600-8, Varian-Nalorac Corporation, Martinez, CA, USA). Twenty slices were superfused in the NMR tube with 4 ml/min of oxy-ACSF flow and hypoxia were induced by stopping the ACSF flow after 3 h metabolic recovery. Basic RF pulsing sequences were used. ^1H spectra at 599.92 MHz were acquired with 4 acquisitions, using a jump-and-return pulse sequence, 3000 points, a 2 sec acquisition time and a 4 sec relaxation delay. ^{31}P spectra at 242.86 MHz were acquired with 200 acquisitions using a 45° tipping angle, 8000 points, a 1 sec acquisition time and a 1 sec relaxation delay. ^1H and ^{31}P spectra acquisitions were interleaved, with acquisition times being approximately one minute for the former and 10 minutes for the latter.

Results

Western blot analysis of total Akt and phospho-Akt (Ser473)

Total Akt immunoreactivity in the whole cell fraction was seen as a single band at a molecular mass of 60 kDa and showed no significant change throughout an experiment (Fig. 1, upper lane). In contrast, phospho-Akt (Ser473) immunoreactivity, which was also seen at 60 kDa, became nearly undetectable at the end of hypoxia (t = 0 h) and was increased after hypoxia (Fig. 1, middle lane). At t = 0.5 h phospho-Akt was seen as a faint band, although subsequently its intensity became the same or larger than before hypoxia. The same trend was found in three independent experiments.

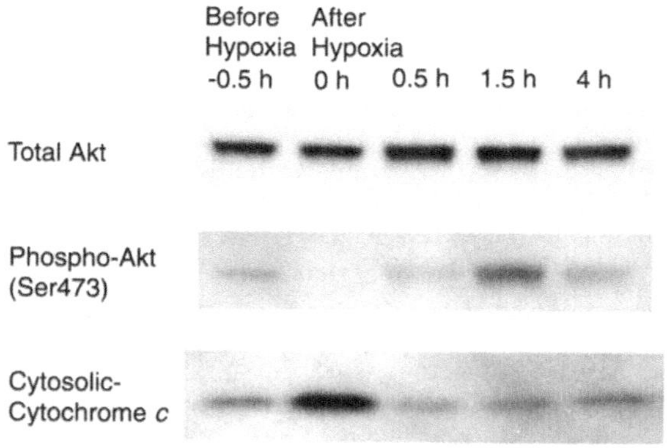

Fig. 1. Representative Western blots of total Akt (upper lane) and phospho-Akt (Ser473) (middle lane) in the whole cell fraction and cytochrome *c* in the cytosolic fraction (lower lane) before (t = −0.5 h) and after (t = 0, 0.5, 1.5, and 4 h) hypoxia. Total Akt immunoreactivity is evident at 60 kDa and did not change throughout an experiment. While phospho-Akt (Ser473) immunoreactivity, evident at 60 kDa, almost disappeared at the end of hypoxia (t = 0 h), it reappeared quickly after hypoxia (t = 0.5 h), and 1.5 and 4 h after hypoxia had intensities that were the same or larger than those before hypoxia. Cytochrome *c* immunoreactivity is evident at 15 kDa. It showed strong intensity at the end of hypoxia (t = 0 h) and became weak again during reoxygenation (t = 0.5, 1.5, and 4 h)

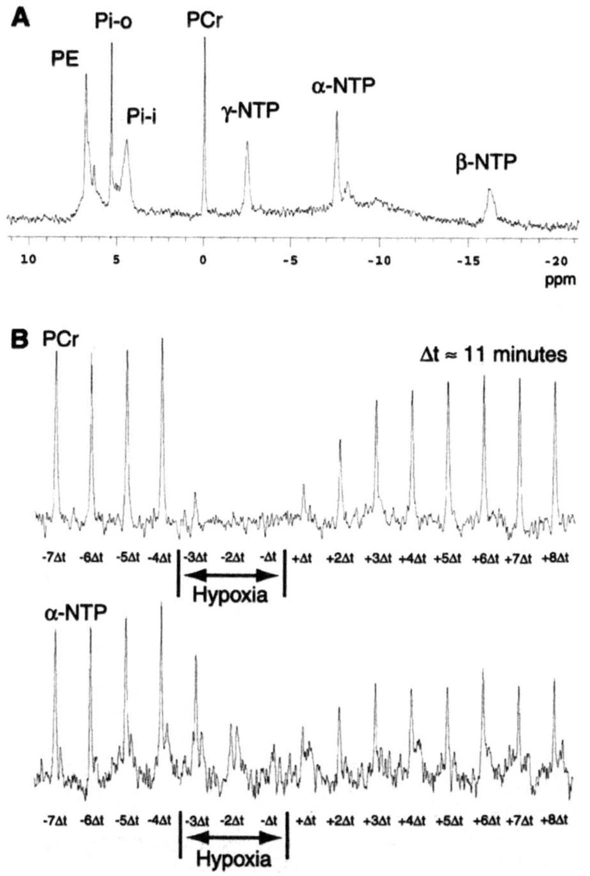

Fig. 2. (A) Representative high resolution 14.1 Tesla ^{31}P NMR spectrum of twenty well oxygenated, respiring slices, superfused in an 8 mm NMR tube. PCr is used to define the chemical shift origin for the ppm scale. Prominent metabolite resonances identifiable in the figure include: phosphoethanolamine (*PE*), extracellular inorganic phosphate primarily in the perfusate (Pi-o), intracellular inorganic phosphate (Pi-i), PCr, and the three phosphate resonances of NTP. (B) Segments for particular resonances from sequential NMR spectra, displayed horizontally from left to right, show the time course of two NMR metabolites (PCr and α̃NTP) in a typical experiment. Prior to hypoxia PCr and α̃NTP are tall. PCr is nearly undetectable in the second run after starting hypoxia, whereas α̃NTP is only modestly decreased and becomes nearly undetectable in the third run after starting hypoxia. The figure demonstrates metabolic recovery to steady state values within five acquisition periods following hypoxia. With this hypoxic insult, however, metabolic recovery is incomplete

Western blot analysis of cytochrome c

Cytochrome *c* immunoreactivity in the cytosolic fraction was seen as a single band at a molecular mass of 15 kDa (Fig. 1, lower lane). Although the intensity of cytochrome *c* was faint before hypoxia, it was increased at the end of hypoxia (t = 0 h), but was again decreased after hypoxia, during reoxygenation (t = 0.5, 1.5, and 4 h). The same trend was found in three independent experiments.

NMR spectroscopy measures of PCr and NTP

Figure 2A shows a composite ^{31}P spectrum, made from summing the data for 12 sequential acquisitions in the two hour period immediately before hypoxia. The sharp NTP and PCr peaks indicate high metabolic integrity. Figure 2B shows the time course of representative PCr and α-NTP resonances throughout an experiment. It shows that PCr and α-NTP became nearly undetectable at the end of hypoxia and recovered quickly but incompletely after hypoxia.

Discussion

In this study we investigated changes in phospho-Akt levels during and after hypoxia to assess its importance as a cell survival factor. Because total Akt levels did not change during or after hypoxia, the disappearance of phospho-Akt during hypoxia and its subsequent recovery during reoxyenation are consistent with the Akt phosphorylation state being dependent on adequate levels of PCr and NTP, the primary intracellular sources of high energy phosphates. Association of a rapid disappearing and quick reappearing of phospho-Akt with hypoxia and reoxygenation have been reported previously [10, 15]. However, in our brain slice model PCr and α̃NTP were lower during reoxygenation than before hypoxia, whereas phospho-Akt increased to equal or higher levels compared to those preceding before hypoxia. On the average, cells in the slices seemed to exhibit a higher priority for Akt phosphorylation than for total energy restoration. There have been reports showing regional differences in phospho-Akt after ischemia [14, 17], possibly due to regional differences in the severity of the stress [14]. Our data motivate inquiries about the extent to which phospho-Akt is needed for cell survival after hypoxia. That is, if NTP recovers substantially but phospho-Akt does not, or *vice-versa*, will the extent of cytochrome *c* apoptosis be substantially different?

It is also important to ask if there are discernable relationships between cytosolic concentrations of cytochrome *c* and the likelihood of cell survival. In the protocol for this study, cytosolic cytochrome *c* exhibited a trend that was the reverse of that shown by phospho-Akt. When one was increased, the other was decreased, and *vice-versa*. This is consistent with – but does not prove – a causal relationship, such as phospho-Akt inhibition of cytochrome *c* release from mitochondria by mechanisms involving Bcl-2 family proteins [5, 6, 16, 18].

Speculation has occassionally occured regarding critical energy levels and thresholds for apoptosis. The divergence seen in our data between the extent of NTP recovery and the extent of phospho-Akt recovery might suggest modulations in critical energy hypotheses. If one can demonstrate very influential roles for specific anti-apoptotic proteins, such as Akt, that must be phosphorylated in order to be protective, then critical phosphorylation energies might exist separately for different compounds, and depend on how far apoptosis has progressed. For example, failure to have phospho-proteins that inhibit cytochrome *c* release will have greater consequences at the stage prior to release, but not at a later stage, when it is also important to have phosphorylation capabilities for inhibiting initiator and effector caspases. Because mitochondria produce the energy to sustain a cell's life while at the same time controlling the signals of its death, experiments should be possible that elucidate key releationships between these two opposite processes.

Acknowledgments

This study was supported by National Institutes of Health Grants R01 GM34767 (LL), P50 NS14543 and R01 NS25372 (PHC).

References

1. Brunet A, Bonni A, Zigmond MJ, Lin MZ, Juo P, Hu LS, Anderson MJ, Arden KC, Blenis J, Greenberg ME (1999) Akt promotes cell survival by phosphorylating and inhibiting a Forkhead transcription factor. Cell 96: 857–868
2. Cardone MH, Roy N, Stennicke HR, Salvesen GS, Franke TF, Stanbridge E, Frisch S, Reed JC (1998) Regulation of cell death protease caspase-9 by phosphorylation. Science 282: 1318–1321
3. Cross DA, Alessi DR, Cohen P, Andjelkovich M, Hemmings BA (1995) Inhibition of glycogen synthase kinase-3 by insulin mediated by protein kinase B. Nature 378: 785–789
4. Datta SR, Brunet A, Greenberg ME (1999) Cellular survival: a play in three Akts. Genes Dev 13: 2905–2927
5. Datta SR, Dudek H, Tao X, Masters S, Fu H, Gotoh Y, Greenberg ME (1997) Akt phosphorylation of BAD couples survival signals to the cell-intrinsic death machinery. Cell 91: 231–241
6. del Peso L, Gonzalez-Garcia M, Page C, Herrera R, Nunez G (1997) Interleukin-3-induced phosphorylation of BAD through the protein kinase Akt. Science 278: 687–689
7. Espanol MT, Litt L, Yang GY, Chang LH, Chan PH, James TL, Weinstein PR (1992) Tolerance of low intracellular pH during hypercapnia by rat cortical brain slices: A 31P/1H NMR study. J Neurochem 59: 1820–1828
8. Fujimura M, Morita-Fujimura Y, Murakami K, Kawase M, Chan PH (1998) Cytosolic redistribution of cytochrome c after transient focal cerebral ischemia in rats. J Cereb Blood Flow Metab 18: 1239–1247
9. Hirai K, Sugawara T, Chan PH, Basus VJ, James TL, Litt L (2002) Cytochrome c associated apoptosis during ATP recovery after hypoxia in neonatal rat cerebrocortical slices. J Neurochem 83: 309–319
10. Hu BR, Fux CM, Martone ME, Zivin JA, Ellisman MH (1999) Persistent phosphorylation of cyclic AMP responsive element-binding protein and activating transcription factor-2 transcription factors following transient cerebral ischemia in rat brain. Neuroscience 89: 437–452
11. Kennedy SG, Kandel ES, Cross TK, Hay N (1999) Akt/Protein kinase B inhibits cell death by preventing the release of cytochrome c from mitochondria. Mol Cell Biol 19: 5800–5810
12. Kops GJ, Burgering BM (1999) Forkhead transcription factors: new insights into protein kinase B (c-akt) signaling. J Mol Med 77: 656–665
13. Lawlor MA, Alessi DR (2001) PKB/Akt: a key mediator of cell proliferation, survival and insulin responses? J Cell Sci 114: 2903–2910
14. Noshita N, Lewen A, Sugawara T, Chan PH (2001) Evidence of phosphorylation of Akt and neuronal survival after transient focal cerebral ischemia in mice. J Cereb Blood Flow Metab 21: 1442–1450
15. Ouyang YB, Tan Y, Comb M, Liu CL, Martone ME, Siesjo BK, Hu BR (1999) Survival- and death-promoting events after transient cerebral ischemia: phosphorylation of Akt, release of cytochrome C and Activation of caspase-like proteases. J Cereb Blood Flow Metab 19: 1126–1135
16. Yang E, Zha J, Jockel J, Boise LH, Thompson CB, Korsmeyer SJ (1995) Bad, a heterodimeric partner for Bcl-XL and Bcl-2, displaces Bax and promotes cell death. Cell 80: 285–291
17. Yano S, Morioka M, Fukunaga K, Kawano T, Hara T, Kai Y, Hamada J, Miyamoto E, Ushio Y (2001) Activation of Akt/protein kinase B contributes to induction of ischemic tolerance in the CA1 subfield of gerbil hippocampus. J Cereb Blood Flow Metab 21: 351–360
18. Zha J, Harada H, Yang E, Jockel J, Korsmeyer SJ (1996) Serine phosphorylation of death agonist BAD in response to survival factor results in binding to 14-3-3 not BCL-X(L). Cell 87: 619–628

Correspondence: Lawrence Litt, Department of Anesthesia, University of California, San Francisco, Box 0648, 521 Parnassus Avenue, Room C455, San Francisco, CA 94143-0648, U.S.A. e-mail: llitt@itsa.ucsf.edu

Acta Neurochir (2003) [Suppl] 86: 231–234

The spinal GDNF level is increased after transient spinal cord ischemia in the rat

J. Tokumine[1], **K. Sugahara**[1], **O. Kakinohana**[2], and **M. Marsala**[2]

[1] Department of Anesthesiology, University of the Ryukyus, Japan
[2] Department of Anesthesiology, University of California, San Diego

Summary

Glial cell line-derived neurotrophic factor (GDNF) is known as the most potent neurotrophic factor against injury. We have characterized spinal GDNF changes after ischemia to clarify its possible physiological role against ischemic damage.

Spinal ischemia in the rat was produced by cross-clamping of the thoracic aorta together with systemic hypotension. The spinal tissue GDNF level was measured by enzyme-linked immunoabsorbent assay (ELISA) and the localization of GDNF in the tissue was examined by immunohistochemistry.

GDNF was increased reaching two peaks after ischemia. The first peak was at 2 hrs after onset of recirculation derived from alpha motor neurons. The second GDNF peak was at 72 hrs provided by astrocytes.

These data suggest a necessity of GDNF to increase to protect against ischemic damage, and that activated astrocytes may have an important role in maintaining the GDNF level.

Keywords: GDNF; spinal ischemia.

Introduction

Glial cell line-derived neurotrophic factor (GDNF), a novel member of the transforming growth factor-beta family, is known to support survival, development, and maintenance of specific functions in different populations of neurons during development and following injury [1, 3, 5, 6, 12]. Recent reports have suggested that GDNF is a more efficient survival factor for spinal alpha motor neurons than the other neurotrophic factors [8].

To confirm the therapeutic efficacy of GDNF against ischemic damage, three conditions must inevitably be satisfied. First, extrinsic administration of GDNF should protect neurons against ischemic damage [4, 10, 14]. Second, decreasing GDNF levels in the tissue should enhance neuronal damage in ischemia [13], and third, GDNF should be up-regulated in respective conditions to protect neurons in cerebral ischemia [9, 15]. While some conditions are satisfied experimentally, so far a protective function of GDNF may only be assumed in cerebral ischemia, although little is known about the dynamics of GDNF upregulation. The objective of the present study was, therefore, to characterize the changes in spinal cord expression of GDNF in cord ischemia in rats.

Material and methods

The work described in this study was performed according to protocols approved by the Institutional Animal Care and Use Committee of the University of California, San Diego.

Animal preparation and induction of spinal cord ischemia [11]

Male Sprague-Dawley rats (weight 350–400 g) were anesthetized with 3% halothane in an oxygen – room air mixture and maintained with 1.5% halothane delivered by an inhalation mask. Body temperature was monitored with a rectal probe and paravertebral muscle needle-probe, and controlled in a range of 37.5–38.5 °C by a heating pad. The tail artery was cannulated for monitoring distal blood pressure (DAP). The left common carotid artery was cannulated with a 20-gauge polytetrafluoroethylene catheter for monitoring proximal arterial perfusion pressure (PAP) and withdrawal of blood. Spinal cord ischemia was induced by inflation of an intra-aortic balloon and systemic hypotension (40 mmHg) during occlusion. For the experiments, we choose an ischemia period of 6 min. This ischemia period allows for full recovery of the motor function at 24 hrs recirculation in all operated rats. After ischemia, the balloon was deflated, and blood withdrawn for arterial hypotension was re-infused during a 60-second period. Thereafter, 4 mg of protamine sulfate were administered subcutaneously. Recovery and stabilization of the arterial blood pressure was monitored for an additional 10 min, after which the arterial lines were removed and the surgical wounds closed.

For sham operation, the same procedure was performed without inflating the aortic balloon nor collecting blood for inducing hypotension. All physiological data obtained during operation were subjected to statistical analysis.

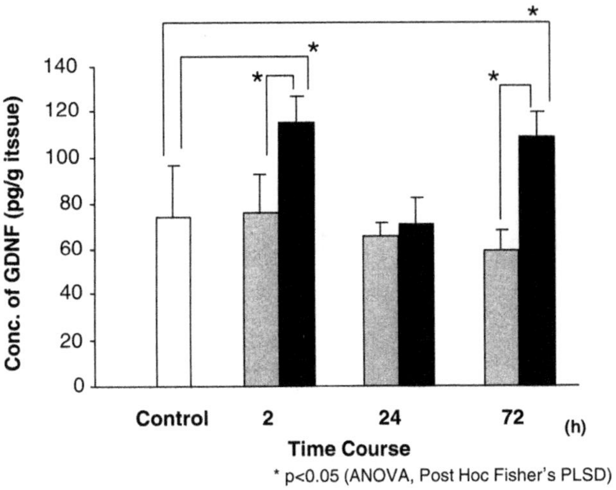

Fig. 1. Changes of the tissue GDNF level at 2, 24, and 72 h after spinal cord ischemia in rats. ■ Ischemia; ▨ Sham

ELISA

After 2, 24, and 72 hrs survival, rats (n = 6 in each group) were sacrificed with an overdose of anesthetics and decapitated. Fourteen rats were studied as non-treated controls. Spinal cords were obtained by the hydro-extrusion technique. Then, the lumbar enlargement was cut immediately, and its weight was measured for the following calculation. The removed spinal cord was put into a 1.5 ml microcentrifuge tube with 150 ml of ice-cold homogenizing buffer (137 mM NaCl, 20 mM Tris, pH 8.0, 1% NP40, 10% glycerol, 1 mM PMSF, 10 μg/ml aprotinin, 1 μg/ml leupeptin, and 0.5 mM sodium vanadate). Then, it was homogenized completely with a tube pestle. 600 μl of extraction buffer (Dulbecco's phosphate-buffered saline, pH 7.35, containing 645 μM $MgCl_2$, and 905 μM $CaCl_2$) was added to the homogenized sample, and mixed completely with a vortex mixer. The extract was centrifuged at 10,000 g for 15 min. A supernatant of 600 μl was obtained and 30 μl of 1N HCl were added (<pH 2) [7]. After incubation for 15 min at room temperature (RT), 30 μl of 1N NaOH were added to neutralize the extract sample. Then, the sample was centrifuged again at 10,000 g for 15 min. The obtained supernatant was kept at −20 °C before using for ELISA. The samples were preserved no longer than one month.

The ELISA plates were coated with an anti-GDNF monoclonal antibody diluted in carbonate buffer, pH 8.2 (1 : 1000), and incubated overnight at 4 °C. After discarding the content of the wells, each well of the plate was blocked with 200 μl of Block & Sample buffer for 6 hrs at RT.

GDNF standards ranging from 0 to 250 pg/ml were prepared, and administered to the two columns on the plate. 100 μl of the prepared extract sample were administered per well. 50 μl of an arbitrary chosen sample and 50 μl of an internal standard (125 pg/ml) were mixed and diluted in the same manner as the GDNF standards. All samples and standards were administered in duplicate. The samples and standards were incubated overnight at 4 °C with mild shaking. After discarding the content of the wells, the plate was washed vigorously five times with 350 μl of Tris-buffered saline-Tween 20 (TBST: 20 mM Tris-HCl, pH 7.6, 150 mM NaCl, 0.05% v/v Tween-20) on a shaker for 10 min for each washing. Then, the plates were incubated with chicken anti-human GDNF polyclonal antibody (1 : 500) for 24 hrs at 4 °C. After washing the plates again with TBST, they were incubated with horseradish peroxidase-conjugated anti-chicken antibody (1:4000) with shaking for 2 hrs at RT. After washing the plates again, 0.02% hydrogen peroxide and tetramethylbenzidine were added for 15 min at RT. To stop the enzyme substrate reaction, 100 μl of 1 M phosphoric acid was added to each well. The absorbance of the samples and standard was measured at 450 nm. The tissue GDNF concentrations were calculated from the measurement of the sample, and the sample added with internal standard using a calibration curve under consideration of the weight of the spinal cord tissue sample.

Histology

After 2, 24, and 72 hrs, two rats each were sacrificed with an overdose of pentobarbital, and transcardially perfused with 150 ml of ice-cold heparinized saline followed by 4% paraformaldehyde in phosphate buffer, pH 7.4. Then, the spinal cord was removed and postfixed for 6 hrs. The tissue obtained was cryoprotected in sucrose solution (30%), and snap-freezed with petroleum ether at −70 °C. Spinal cord sections (10 μm) were cut with a cryostat and mounted on glass slides.

Immunohistochemical staining was performed for GDNF, and glial fibrillary acidic protein (GFAP) as specific marker of astrocytes. 4′-6-diamidino-2-phenylindole (DAPI) was used for staining of the cell nucleus.

The sections were first incubated for 5 min with 0.1 M phosphate buffered-saline, pH 7.4 (PBS). Then, they were incubated for 3 hrs with 5% goat serum in PBS, 0.2% Triton X-100 (5% Serum/PBS-TX) for blocking non-specific binding. We used a mouse anti-human GDNF monoclonal antibody (1:400, Promega co.), and a rabbit anti-GFAP polyclonal antibody (1:200, Chemicon Int.) with 5% Serum/PBS-TX as primary antibodies. The tissues were incubated with the primary antibodies overnight at 4 °C with shaking. After washing 3 times for 5 min with PBS-TX, anti-rabbit IgG (Molecular Probes) and anti-mouse IgG (Molecular Probes) were used for secondary antibodies, respectively. Then, staining by DAPI was performed. Thereafter, the tissues were washed with PBS 3 times for 5 min, and were dried at RT. The fluorescence was analyzed with a fluorescence microscope.

A negative immunohistochemical control was performed by primary antibody omission from the otherwise regular immunolabeling protocol.

Statistical analysis

Statistical analysis was performed by one-way ANOVA followed by post hoc Fisher's PLSD. A value of $P < 0.05$ was considered significant. Data were expressed as mean ± S.D.

Results

There was no significant difference between the ischemia group and the sham-operated group in physiological variables before and during surgery. The figure shows the changes of the tissue GDNF level after recirculation. In contrast to the normal level in the sham-operated group, an increase of the concentration of GDNF in the spinal cord was recognized in the ischemia group at 2 and 72 hrs after recirculation as compared to the sham-operated group or non-operated controls ($P < 0.05$).

To ensure the types of cells expressing GDNF, immunohistochemistry was performed. Whereas alpha motor neurons in tissue of sham operated animals showed a minor immunoreactivity, many clearly GDNF positive granules were observed in the cell bodies of motoneurons in the ischemic tissue.

In contrast to the GDNF staining of astrocytes in animals of the sham-operated group, astrocytes of the ischemic tissue were stronger positive for GDNF (data not shown). Furthermore, these cells obviously had a hypertrophic and hyperplastic morphology.

Discussion

The present study demonstrates two GDNF peaks of up-regulation after transient spinal cord ischemia in rats. In comparison to rabbit spinal cord ischemia [10], there were some differences as to the pattern of GDNF up-regulation, neurological outcome after recirculation, and histology of the spinal cord. In our own experiments, the neurological outcome and histological findings of α motor neurons after 24 hrs recirculation were similar with that in the control group, whereas the findings of experiments with rabbits were indicating progressive neuronal dysfunction and delayed death of alpha motor neurons. Further, in rabbits with spinal cord ischemia up-regulation of GDNF was recognized in alpha motor neurons, but not in astrocytes. Therefore, the pattern of GDNF up-regulation might be closely related with the neurological outcome. A recent report suggested a role of GDNF for the protection of neurons. Accordingly, GDNF gene delivery by adenovirus was found to inhibit the delayed death of alpha motor neurons after spinal cord ischemia in rabbits [10]. Taken together, we speculate that GDNF is an essential factor with protective properties against ischemia, and that astrocytes have a pivotal role in this context against ischemic damage by producing GDNF.

Also in transient focal cerebral ischemia two peaks of up-regulation of GDNF m-RNA were reported in rats by Wei G *et al.* [15]. They showed that the first peak was caused by neurons, while the second peak by astrocytes. These data agree with the results of the present study, even though different types of ischemia were studied. Whereas Wei G *et al.* observed an up-regulation of GDNF by measuring mRNA [15], we found GDNF up-regulation at the protein level.

We did not study other neurotrophic factors in the current experiments. However, in our preliminary investigations, the concentration of GDNF under normal conditions was quite low as compared to that of NT-3 or BDNF (73.9 ± 23.1, 585.4 ± 154.5, 3156.9 ± 405.6 pg/g tissue, respectively). NT-3 and BDNF but not GDNF[2] have been reported to have a functional role in the normal state.

Our present and preliminary studies suggest that up-regulation of GDNF is more specific in ischemia than up-regulation of NT-3 or BDNF, and that it has an important role for the reorganization of damaged tissue.

We conclude that GDNF is an essential protective factor for neurons against ischemic damage. Maintenance of the GDNF level in damaged tissue should be critical for repair.

References

1. Beck KD, Valverde J, Alexi T, Poulsen K, Moffat B, Vandlen RA, Rosenthal A, Hefti F (1995) Mesencephalic dopaminergic neurons protected by GDNF from axonotomy-induced degeneration in the adult brain. Nature 373: 339–341
2. Eide FF, Lowenstein DH, Reichardt LF (1993) Neurotrophins and their receptors – current concepts and implications for neurologic disease. Exp Neurol 121(2): 200–214
3. Henderson CE, Phillips HS, Pollock RA, Davies AM, Lemeulle C, Armanini M, Simpson LC, Moffet B, Vandlen RA, Koliatsos VE, Rosenthal A (1994) GDNF: a potent survival factor for motoneurons present in peripheral nerve and muscle. Science 266: 1062–1064
4. Kitagawa H, Sasaki C, Sakai K, Mori A, Mitsumoto Y, Mori T, Fukuchi Y, Setoguchi Y, Abe K (1999) Adenovirus-mediated gene transfer of glial cell line-derived neurotrophic factor prevents ischemic brain injury after transient middle cerebral artery occlusion in rats. J Cereb Blood Flow Metab 19(12): 1336–1344
5. Li L, Wu W, Lin LF, Lei M, Oppenheim RW, Houenou LJ (1995) Rescue of adult mouse motoneurons from injury-induced cell death by glial cell line-derived neurotrophic factor. Proc Natl Acad Sci USA 92: 9771–9775
6. Lin LH, Doherty DH, Lile JD, Bektesh S, Collins F (1993) GDNF: a glial cell line-derived neurotrophic factor for midbrain dopaminergic neurons. Science 260: 1130–1132
7. Okragly A, Haak-Frendscho M (1997) An acid-treatment method for the enhanced detection of GDNF in biological samples. Exp Neurol 145: 592–596
8. Saarma M, Sariola H (1999) Other neurotrophic factors: glial cell line-derived neurotrophic factor (GDNF). Microsc Res Tech 45(4–5): 292–302
9. Sakurai M, Hayashi T, Abe K, Yaginuma G, Meguro T, Itoyama Y, Tabayashi K (1999) Induction of glial cell line-derived neurotrophic factor and c-ret porto-oncogene-like immunoreactivity in rabbit spinal cord after transient ischemia. Neurosci Lett 276(2): 123–126
10. Sakurai M, Abe K, Hayashi T, Setoguchi Y, Yaginuma G, Meguro T, Tabayashi K (2000) Adenovirus-mediated glial cell line-derived neurotrophic factor gene delivery reduces motor neuron injury after transient spinal cord ischemia in rabbits. J Thorac Cardiovasc Surg 120(6): 1148–1157
11. Taira Y, Marsala M (1996) Effect of proximal arterial perfusion pressure on function, spinal cord blood flow, and histopatho-

logic changes after increasing intervals of aortic occlusion in the rat. Stroke 27: 1850–1858
12. Tomac A, Lindqvist E, Lin LF, Ogren SO, Young D, Hoffer BJ, Olson L (1995) Protection and repair of the nigrostriatal dopaminergic system by GDNF in vivo. Nature 373: 335–339
13. Tomac AC, Grinberg A, Huang SP, Nosrat C, Wang Y, Borlongan C, Lin SZ, Chiang YH, Olson L, Westphal H, Hoffer BJ (2000) Glial cell line-derived neurotrophic factor receptor alpha1 availability regulates glial cell line-derived neurotrophic factor signaling: evidence from mice carrying one or two mutated alleles. Neuroscience 95(4): 1011–1023
14. Wang Y, Lin SZ, Chiou AL, Williams LR, Hoffer BJ (1997) Glial cell line-derived neurotrophic factor protects against ischemia-induced injury in the cerebral cortex. J Neurosci 17(11): 4341–4348
15. Wei G, Wu G, Cao X (2000) Dynamic expression of glial cell line-derived neurotrophic factor after cerebral ischemia. Neuroreport 11(6): 1177–1183

Correspondence: Joho Tokumine, Department of Anesthesiology, Faculty of Medicine, University of the Ryukyus, 207 Uehara, Nishihara, Okinawa, Japan 903-0215. e-mail: tokumine@otc.ne.jp

Clinical ischemia

Acta Neurochir (2003) [Suppl] 86: 237–240

Identification of malignant brain edema after hemispheric stroke by PET-imaging and microdialysis

W.-D. Heiss, C. Dohmen, J. Sobesky, L. Kracht, B. Bosche, F. Staub, S. Toyota, M. Valentino, and **R. Graf**

Max-Planck-Institut für neurologische Forschung, Köln, Germany

Summary

Cerebral blood flow (CBF) and extent of irreversible tissue damage as well as the time course of extracellular concentration of amino acids, substrates of energy metabolism, and purine metabolites, intracranial pressure and tissue oxygen tension were assessed in 34 patients with large strokes covering more than 50% of the MCA territory. The results were compared to findings in the experimental model of transient (for 3 hours) MCA occlusion in cats.

In the experimental model as well as in the clinical setting development of malignant brain infarcts (due to formation of space occupying brain edema) was predicted by the size of critically hypoperfused tissue and the volume of irreversibly damaged tissue. The course of malignant infarcts was characterized by progressive increase in concentrations of excitatory amino acids, lactate, pyruvate, glycerol, hypoxanthine and in intracranial pressure, while cerebral perfusion pressure and tissue oxygen tension decreased. These results clearly differentiate a malignant from a benign course of large hemispheric infarction. The methods can be used to identify patients at risk for formation of space occupying edema and to select patients who could benefit from invasive therapeutic strategies.

Keywords: Malignant brain infarction; MCA stroke; PET; microdialysis; intracranial pressure; tissue oxygen tension; excitatory amino acids; lactate; cerebral blood flow; flumazenil.

Introduction

Malignant brain infarcts due to space occupying brain edema develops in about 10% of patients with ischemic stroke in the middle cerebral artery (MCA) territory and carries a high mortality [1]. Invasive treatment strategies – decompressive hemicraniectomy and hypothermia [3, 4] – might improve mortality and morbidity of these patients. The selection of patients who can benefit and the determination of the best time for these interventions however are still controversial and require a better assessment of the extent of ischemic damage and the course of pathophysiological changes within and in the surrounding ischemic territory. Therefore, a study was performed in patients at risk of malignant infarction in whom PET was applied to determine regional cerebral blood flow and extent of irreversible ischemic damage. Multimodal neuromonitoring was used to follow regional changes of physiologic variables and biochemical substrates in brain tissue. These results were compared to findings in experimental MCA occlusion obtained by repeated PET studies, continuous microdialysis and intracranial pressure monitoring.

The experimental model: transient MCA occlusion in cats

Methods

In 11 halothane anesthetized cats, the left middle cerebral artery (MCA) was occluded for 3 hours followed by 6 hours of reperfusion. Pressure sensors measured simultaneously mean arterial blood pressure and intracranial pressure (ICP), and cerebral perfusion pressure (CPP) was calculated. The extracellular concentration of various biochemical compounds was determined by microdialysis and HPLC (for details see Toyota *et al.* 2002, in these proceedings); the concentration of glutamate was followed for this study only. PET determined regional changes of CBF or CMR_{glc} in relation to alterations of CPP by using ^{15}O-water or ^{18}F-fluorodeoxyglucose (bolus i.v. injections) on a CTI/Siemens ECAT EXACT HR PET scanner. Regional analysis of PET images was performed stepwise in a pixel based approach. First, hemispheres were marked on control CMR_{glc} images. Second, core and border zone regions were determined on CBF images taken during ischemia. Third, the size of both of these regions was related to the size of the affected hemisphere.

Results

Out of 11 animals that were exposed to transient MCA occlusion, 3 developed a malignant course (malignant group). This included excessive rise of ICP and drop of CPP, secondary deterioration of CBF following primary recovery after recirculation, and symptoms of herniation including dilatation of pupils. Sequential CBF measurements at the end of the 3 hour MCA occlusion revealed severe ischemia extended over 55–75% and moderate ischemia in 5–10% of the ipsilateral hemisphere. Reopening of the MCA induced hyperperfusion in large portions of the ischemic volume which turned into hypoperfusion after 1–2 hours and led finally to severe global ischemia with transtentorial herniation and a cerebral perfusion pressure below 50–60 mmHg.

In the 8 animals with a non-malignant course, severe ischemia covered only 10–45% and moderate ischemia 15–30% of the hemisphere at the end of MCA occlusion. Recirculation induced immediate hyperperfusion followed by hypoperfusion in regions which had been severely ischemic. CBF reduction, however, did not spread into regions outside the ischemic focus. Perfusional and metabolic deficits at the end of the observation period were marked only in the formerly ischemic regions.

Intracranial pressure did not change during MCA occlusion. It started to increase in both groups almost immediately after reopening of the MCA and remained at that level in the non-malignant group. It continued to rise in the malignant group throughout the observation period reaching final values above 70 mmHg. As a consequence, cerebral perfusion pressure remained almost unaltered in the non-malignant group, but started to fall immediately after recirculation and continued to fall to values below 30 mmHg in the malignant group.

Extracellular concentration of glutamate increased significantly in the core of ischemia in the malignant group during MCA occlusion and in later periods after reperfusion, when decreased CPP indicated development of severe ischemia which later on also spread to the borderzone of the MCA territory.

Comment

The results from the experimental model demonstrate that a large ischemic core, in conjunction with a small ischemic borderzone, favors progression into a malignant course presumably caused by excessive edema formation. This progression can be predicted by early CBF determinations and goes along with progressive increase in glutamate concentration and intracranial pressure.

The clinical problem: large hemispheric strokes

Methods

The clinical study comprised 34 patients suffering from severe middle cerebral artery (MCA) syndrome with hemiparalysis, conjugate gaze palsy and a hypoattenuation on early (<12 h) CCT exceeding 50% of the MCA territory. Neurologic deficit was assessed by use of the National Institute of Health Stroke Scale (NIHSS) at admission, after 24 hours and after 10 days, and by use of the modified Rankin Scale after 3 months. CT scans were performed on admission, 6 to 12 hours later and at day 5. All patients were treated in the neurological intensive care unit and received antiedematous therapy with mannitol and 30° elevation of the upper body. Clinical course was regarded malignant when patients developed signs of uncal herniation with a unilaterally dilated pupil and when the follow-up CT showed space-occupying brain edema with midline shift. Patients were included in the study, which was approved by the Ethics Committee of the Medical Faculty of the University of Cologne, after written consent from the next of kin. Neuromonitoring was started within 12–34 hours after stroke onset using a microdialysis probe (CMA/70), an ICP measuring device (Codman) and an oxygen sensing probe (Licox) inserted through one burr hole into the parenchyma of the frontal lobe of the infarcted hemisphere. Dialysates were analyzed immediately at the bedside for extracellular concentrations of glutamate, lactate, pyruvate, and glycerol using the CMA 600 Microdialysis Analyzer. Samples collected in vials over periods of 120 min were analyzed post hoc for several other amino acids, and purine catabolites. ICP, partial tissue oxygen pressure (ptO_2) as well as mean arterial blood pressure and arterial oxygen saturation were continuously monitored on the mainframe computer.

PET studies were performed on an ECAT EXACT

HR scanner (Siemens/CTI, Knoxville, USA) in two- or three-dimensional data acquisition mode providing 47 contiguous 3-mm slices at 5 mm FWHM in-plane reconstructed resolution [6]. Twenty mCi (740 MBq) [^{11}C]-flumazenil (FMZ) was injected intravenously. The distribution and accumulation of this tracer was followed for 60 min by serial scanning. Tracer distribution within 2 min after injection represents regional cerebral blood flow (CBF), which can be quantified by a reference method [5]. Regions with critically decreased perfusion were operationally set to 50% of early FMZ distribution relative to the mean of the contralateral hemisphere. A moderately hypoperfused zone suggesting penumbra was defined by early FMZ concentration between 50 and 70% of the contralateral mean. Later FMZ uptake at steady state reflects binding to central benzodiazepine receptors and is a reliable marker of neuronal integrity. As an indicator of irreversible ischemic tissue damage an FMZ uptake threshold of 3.4 times the mean value of white matter was used [2].

All results are expressed as mean $\pm$ standard deviation. Spearman's correlation coefficient was used to compare measured values with patient outcome assessed by the mRS after 3 months. Differences between the patient groups with malignant and benign clinical course were analyzed by student's t-test for unpaired samples using $P < 0.05$ as the level of significance.

Results

Sixteen of the 34 patients with ischemic changes extended over more than 50% of the MCA territory in early (<12 h) CT suffered a malignant course and developed space occupying brain edema. Ten of these patients were treated conservatively, 9 of those died due to transtentorial herniations, one survived with severe neurological defects. Five of the patients with malignant course underwent hemicraniectomy and survived with moderate handicap and one died. The other 18 patients did not develop malignant brain edema (benign group) and survived with mild to moderate handicap, 3 died due to non-neurological reasons.

PET with i.v. administration of FMZ was performed within 3–28 h (mean 17 h) after onset of clinical symptoms in 18 patients. The volume of critically hypoperfused tissue (CBF < 50% of contralateral mean) was larger in the malignant (n = 8) than in the benign (n = 10) group. Additionally, CBF within the critically hypoperfused volume was significantly lower in the malignant group. Finally, the volume of irreversible neuronal damage (FMZ accumulation < 3.4 times mean of white matter) was significantly larger in the malignant than in the benign group. The "penumbra zone" defined by flow reduced to 50–70% of contralateral mean and preserved FMZ binding was smaller in the malignant than in the benign group. The volume of irreversible neuronal damage and the volume of severe ischemia correlated significantly with the clinical outcome of the patients (expressed by the mRS after 3 months), whereas the correlation between the mean CBF within the hypoperfused tissue and the volume of tissue at risk with the mRS was less strong.

The impact of metabolic alterations assessed by continuous neuromonitoring on the clinical course was analyzed by correlating the peak values of the various variables to the outcome of the patients. The mean peak values of patients with benign course for glutamate (20.5 $\pm$ 16.5), aspartate (2.1 $\pm$ 1.3) and GABA (1.5 $\pm$ 2.2 mol/L) were significantly ($P < 0.05$) lower than for patients with malignant course (glutamate: 306 $\pm$ 194.6; aspartate: 34.5 $\pm$ 38.1; GABA 18.0 $\pm$ 20.3 mol/L). Similarly, lactate, pyruvate, lactate/pyruvate ratio, hypoxanthine, glycerol and ICP increased significantly in the malignant group. However, no significant differences in the concentrations of structural amino acids were observed. CPP and ptO_2 were significantly lower in patients with a malignant course.

Comment

This complex study combining continuous neuromonitoring with early PET revealed significant differences in substrate concentrations, ICP and ptO_2 values as well as the extent of critical CBF reduction and early neuronal damage between patients who are developing malignant brain edema and those who do not. It is important in this context that PET – especially the volume of critically hypoperfused tissue and that of reduced uptake of a marker of neuronal integrity – predicts the subsequent changes in biochemical markers of ischemia, in increase of intracranial pressure and decrease of tissue oxygen pressure, which accompany the clinical course and define the critical stage of deterioration. The combination of these methods helps to identify patients at risk for formation of space occupying edema and may be useful for the selection of patients who might benefit from invasive interventional therapeutic strategies. The results from the clinical study are in accordance with findings in the experimental model where large critically ischemic volumes predicted a malignant course characterized by progressive deterioration, extension of hypoperfusion, increase of intracranial pressure, and increase of extracellular glutamate.

References

1. Hacke W, Schwab S, Horn M, Spranger M, De Georgia M, von Kummer R (1996) 'Malignant' middle cerebral artery territory infarction: clinical course and prognostic signs. Arch Neurol 53: 309–315
2. Heiss W-D, Kracht LW, Thiel A, Grond M, Pawlik G (2001) Penumbral probability thresholds of cortical flumazenil binding and blood flow predicting tissue outcome in patients with cerebral ischaemia. Brain 124: 20–29
3. Schwab S, Schwarz S, Spranger M, Keller E, Bertram M, Hacke W (1998) Moderate hypothermia in the treatment of patients with severe middle cerebral artery infarction. Stroke 29: 2461–2466
4. Steiner T, Ringleb P, Hacke W (2001) Treatment options for large hemispheric stroke. Neurology 57 [Suppl] 2: S61–S68
5. Thiel A, Löttgen J, Grond M, Pietrzyk U, Heiss W-D (2001) Estimation of regional cerebral blood flow levels in ischemia using [^{15}O]water or [^{11}C]flumazenil PET without arterial input function. J Comput Assist Tomogr 25: 446–451
6. Wienhard K, Dahlbom M, Eriksson L, Michel C, Bruckbauer T, Pietrzyk U, Heiss W-D (1994) The ECAT EXACT HR: Performance of a new high resolution positron scanner. J Comput Assist Tomogr 18: 110–118

Correspondence: Prof. Dr. W.-D. Heiss, Max-Planck-Institut für neurologische Forschung, Gleueler Str. 50, 50931 Köln, Germany. e-mail: wdh@pet.mpin-koeln.mpg.de

Acta Neurochir (2003) [Suppl] 86: 241–246

Cerebral blood flow index image as a simple indicator for the fate of acute ischemic lesion

H. Igarashi[1,2], M. Hamamoto[1,2], H. Yamaguchi[2], S. Ookubo[1], J. Nagashima[3], H. Nagayama[1], S. Amemiya[1], K. Arii[1], M. Sakamaki[1], and Y. Katayama[1]

[1] The Second internal Medicine, Nippon Medical School, Tokyo, Japan
[2] Neurology, Tama Metropolitan Geriatric Hospital, Japan
[3] Radiology, Tama Metropolitan Geriatric Hospital, Japan

Summary

Background and purpose. To evaluate the feasibility of utilizing cerebral blood flow (CBF) index images, we attempted to investigate 1) whether CBF index images can reveal the resulting infracted area, 2) whether the CBF index can correlate other modality (SPECT).

Methods. DWI and DPI were obtained in 17 patients within 12 hours of stroke onset and follow up MRI. On three DPI delivered images, namely relative regional cerebral blood volume (rrCBV), uncorrected mean transit time (MTTu) and CBF index images, correlations between initial lesion volume of and follow up infarction volume of three images and rCBF images delivered with singular value decomposition (SVD) methods were assessed. Then ^{99m}Tc-ECD SPECT was taken immediately after MRI to correlate to MRI data.

Results. Among the three images, lesion volume of CBF index images against follow up infarct volume had the highest correlation ($r = 0.995$) to a linear fit and the slope was closest to 1.0 (0.91) and had identical accuracy to the regression coefficient of rCBF images. CBF index well correlated to SPECT delivered CBF.

Conclusion. CBF index images can accurately predict final infarct volume. Evaluating CBF index images together with DWI can guide the initial assessment in the acute stage of cerebral ischemia.

Keywords: Cerebral ischemia; magnetic resonance imaging; diffusion-weighted; magnetic resonance imaging; perfusion-weighted; cerebral blood flow; stroke; acute.

Introduction

In acute stage of cerebral ischemia, many calculated maps from DPI, i.e. Cerebral Blood Volume (CBV), Cerebral Blood Flow (CBF), MTT, CBF index, Time to Peak (TTP), were used to assess the mal-perfusion area. However, in the clinical setting, there is no consensus as to which parameter can best predict tissue outcome. In the initial MR examination for acute cerebral ischemia, the assessment of the ischemic lesion must be made as quickly as possible. Thus representative DPI delivered images should fulfill the following conditions.

1. Images must be obtained quickly.
2. It should be easy to distinguish the mal-perfusion area.
3. The parameter should have a distinct threshold for the outcome of ischemic lesion (infarcted or survival for follow up study).
4. The extent of the apparent mal-perfusion area should represent the resulting infracted area.
5. Images must be produced automatically without any interpretation of the raw DPI images, such as the selection of the region of interest for the arterial input function, or the manual selection of the DPI images to be calculated.

Among three simple DPI delivered images, namely relative regional cerebral blood volume (rrCBV), uncorrected MTT (MTTu), which is obtained from first moment of time to the *non* deconvolved contrast concentration curve in tissue, and CBF index images, which can be calculated automatically and are widely used in clinical perfusion studies. CBF index image [2], (determined by dividing rrCBV with MTTu based on the central volume principle), is thought to reflect CBF, although there are limitations to the interpretation of this parameter. CBF index image is a likely candidate for fulfilling the former conditions. In the acute stage of ischemia, tissue outcome in the ischemic lesion depends strongly on CBF. From this point of view, we attempted to investigate, 1) whether the CBF index has a threshold for the tissue outcome, 2) whether CBF index images represent the resulting infarcted area better than rCBV and MTT images, and if so, 3) whether the predictive ability of CBF index image is comparable to the rCBF image delivered from singular value decomposition (SVD) deconvolution methods, which are regarded as most accurate in predicting final infarct area [6].

Subjects and methods

Patient selection

17 consecutive patients who arrived at Tokyo Metropolitan Tama geriatric hospital during the acute stage of cerebral infarction and satisfied the following inclusion criteria were studied (Table 1).

1. MRI taken within 12 hours of the time last observed without neurological symptoms.
2. Neither past history of cerebral infarction, nor old vascular accident was seen on MRI.
3. Large artery occlusion (carotid artery or main trunk of middle cerebral artery) was identified with MRA.
4. Had not received intravenous or intraarterial thrombolytic therapy.
5. Neither large trunk stenosis nor hypoperfusion area was seen in contra lateral side, with carotid ultrasound, MRA, or SPECT.
6. No apparent clinical event was seen between the initial and follow up MRI.

All of them were cardiogenic embolic infarctions. The duration between the onset and the scan are determined from the time when the patient was last known to have no obvious neurological deficits. To assess the natural course of infarct progression, patients taking thrombolytic therapy were excluded from the study, although most of them were taking anticoagulant and/or osmotic therapy.

MR I

MR imaging was performed with a 1.5 tesla MR imaging unit (Magnetom Vision, Siemens, Godchingen, Germany) with optional shielded gradient. Images were taken as follows: gradient recall sagital images (localizer), fast T2 images, DWI, MRA, DPI. The entire protocol required approximately 20 minutes. Follow up MR imaging was performed between 8 and 20 days of onset.

Diffusion weighted imaging

DWI was performed using multi-slice single-shot, spin-echo EPI sequence. To eliminate anisotropy of the diffusion image, three orthogonal diffusion images were sampled with two level (40 sec/mm^2 and 1000 sec/mm^2) of B-value, and trace diffusion images and trace apparent diffusion coefficient images were computed. Imaging parameters were as follows: repetition time: 6 Sec, echo time: 108 msec, field of view: 24 × 24 cm, image matrix: 256 × 256, slice thickness: 6 mm, intersection gap: 2 mm. Each image was added three times to improve the signal to noise ratio. Each set of 16 slices on the transverse plane was acquired for 18 seconds per one B-value, which resulted in 144 seconds in a total imaging time. Fourier transformed imaging files were then transferred to a workstation (SPARCstation, Sun Microsystems, Milpitas, California, USA), and trace diffusion images and trace ADC images were computed by MRI image manipulation software (MR vision, MR vision co. Menro Park, CA, USA).

Dynamic perfusion imaging

Perfusion imaging was performed using bolus tracking with a multi-slice, single-shot, gradient-echo EPI sequence with a temporal resolution of 2 seconds. 11 image slices were obtained every 2 seconds for 120 seconds (total 11 × 60 = 660 images), and for the sixth scan, 0.1 m mol per kilogram of body weight gadodiamide (Omniscan, Nycomed, Oslo, Norway) was injected manually via cubital vein for 5 seconds followed by saline flush. Imaging parameters were as follows; repetition time: 2 sec, echo time 54 msec, field of view: 24 × 24 cm, image matrix: 128 × 128, slice thickness: 6 mm, and intersection gap: 2 mm. 11 slices, which covered a thickness of 86 mm in the axial direction, were selected to include both cerebral hemispheres.

Fourier transformed images were transferred to a Workstation, and perfusion parameters were obtained from the following calculation with MR vision software. First, a time-relaxibility curve of each pixel was calculated with equation (1). The first five images were omitted to archive saturation.

$$\Delta R2^* = -\left[\ln \frac{So}{St}\right] \Big/ TE \tag{1}$$

where **So** was the baseline signal intensity, **St** was the signal intensity of each time point and **TE** was the echo time. Then, numerical fitting of the $\Delta R2^*$ curve using a gamma variate function was performed (equation 2).

$$R2^*(t) = \kappa(t - ta)^{\alpha} \ell^{[-(t-ta)]/\beta'} \quad (2)$$

This fitting was performed to avoid the effects of recirculation, and to obtain numerical estimations for hemodynamic parameters by determining a starting (T_a) and ending (T_e) point concentration curve. In this study, rrCBV, CBF index, and MTTu were used to assess perfusion parameters. rrCBV and MTTu were calculated according to the indicator dilution theory.

Then, the CBF index was calculated according to the central volume principle (equation 5).

$$CBF\ index = \frac{rrCBV}{MTTu} \quad (3)$$

For imaging presentation, each parameter was calculated in pixel-by-pixel manner.

To calculate deonvolved rCBF maps, we used the nonparametric singular-value decomposition (SVD) method described by Østergard *et al.* [4, 5]. This method involves deconvolution of the tissue concentration-time curve on a pixel-by-pixel basis with an arterial input function (AIF). AIF was determined manually from five to ten pixels over the proximal middle cerebral artery of the ischemic hemisphere to avoid underestimation of rCBF due to delay or dispersion of ischemic region [7]. rCBF map was determined from the peak height of the deconvolved curve.

Follow up MRI

Follow up T2 weighted fast spin-echo images were obtained between the seventh to twentieth day (average 13.5th day) from the onset. Imaging parameters were as follows; repetition time: 4 Sec, echo time: 96 msec, field of view: 24 × 24 cm, image matrix: 256 × 256, slice thickness: 6 mm, intersection gap: 2 mm. These images were transferred to the workstation for lesion volume analysis described in the data analysis section.

HMPAO-SPECT

CBF imaging was performed with a triple detector system SPECT scanner (Multispect3; Siemens, Godchingen, Germany). Each patient received 740 MBq of ^{99m}Tc-HMPAO i.v. in a darkened room with their eyes closed. The CBF imaging was started 5 minutes after injection of ^{99m}Tc-HMPAO. The image slices were set parallel to the orbitomeatal line. The resolution was 2.89 mm in-plane and 5.78 mm slice thickness.

Data analysis

Comparison of the lesion volume

On each scan, the lesion volume from rrCBV, MTTu, CBF index and rCBF images were assessed quantitatively. The lesions were outlined in a blinded manner by five neurologists who use DWI and PI in a daily clinical situation. Each of them outlined lesions twice per image and averaged a total of ten for each image. Determining final infarct volume was done semi-automatically on MR vision.

Correlation between CBF index, rCBF and HMPAO SPECT

SPECT image and DWI were co-registered using software (ICON TM; Siemens, Godchingen, Germany) on Power Macintosh G3 personal computer (Apple computer, Cupertino, CA. USA). Eleven of 15 × 15 mm square shape ROIs in a ischemic hemisphere and the mirror regions in contra-lateral hemisphere were placed on the SPECT CBF images and corresponding CBF index images. Then correlation between the ratios of ischemic ROI and contra-lateral ROI from SPECT CBF image and CBF index image was evaluated.

Statistical analysis

The Statistical Package for the Social Sciences (SPSS Inc. Chicago, IL, USA) was used for the statistical analysis. Comparison of the two correlation coefficients was assessed using the standardized value of the difference between the normalizing transform of correlation coefficients.

Results

Correlations between initial lesion volumes and follow up infarction volume

Mean intra-observer differences between paired measurements were 7.23% for MTT, 4.36% for rCBV, 5.87% for CBF index and 6.52% for rCBF. For inter-observer differences, the values were 9.24% for MTT,

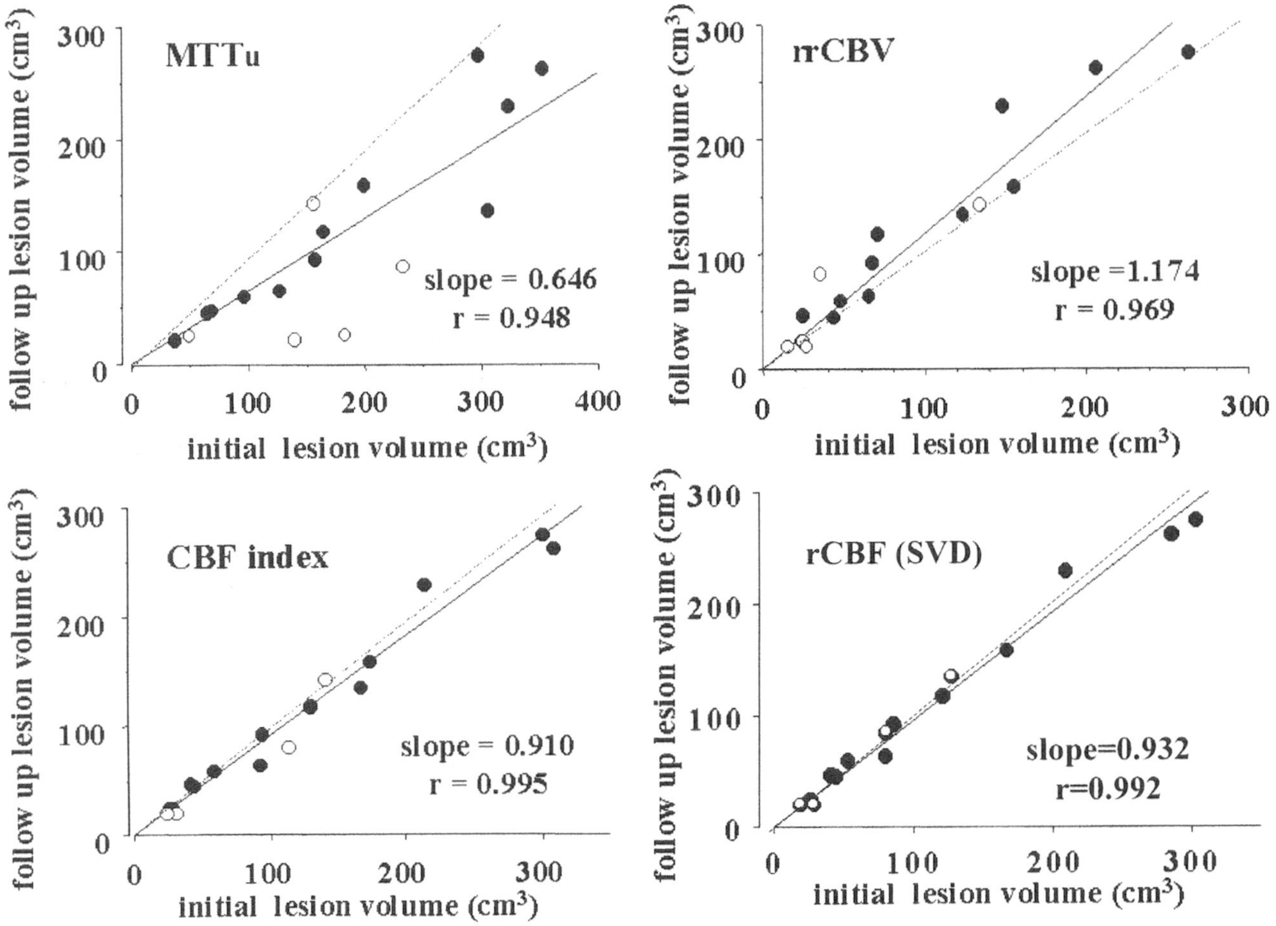

Fig. 1. Scatter plots of initial lesion volume against follow-up infarct volume. Among the three images, CBF index and rCBF have significantly higher correlation to a linear fit and slope closest to 1. Open circles show cases with IC obstruction

5.32% for rrCBV, 8.71% for CBF index and 9.21% for rCBF. All intra- and inter-observer reliabilities were $r > 0.93$ (0.932–0.996). Among the correlations between each DPI delivered lesion volume and final infarct volume, rCBF and CBF index showed the best correlation statistically (103% and 108% versus mean final infarct volume respectively, Figure 1). MTTu lesion volume tended to overestimate the infarct volume (157% versus mean final infarct volume), especially in cases with I.C. occlusion (Open circle, Fig. 1), whereas rrCBV lesion tended to underestimate the infarct volume (87% versus mean final infarct volume). The coefficients of linear regression against final infarct volumes were highest in CBF index images and rCBF images ($r = 0.992$ and 0.995 respectively, no statistical difference was seen) compared with MTTu images and rrCBV images.

Correlation between CBF index and HMPAO SPECT

Figure 2 shows a scatter plot of the CBF ratio between the ischemic hemisphere and the contra lateral (non-ischemic) hemisphere from DPI versus HMPAO-SPECT. SVD delivered rCBF and SPECT delivered rCBF had a strongest correlation, however still there were strong correlation between both DPI delivered data and SPECT delivered CBF ratio.

Discussion

The CBF index image reflects CBF but does not represent the actual CBF. Strictly speaking, since the CBF index is calculated from rrCBV and MTTu, there may be some limitations in assessing the cerebral blood flow with the CBF index images. The nature of

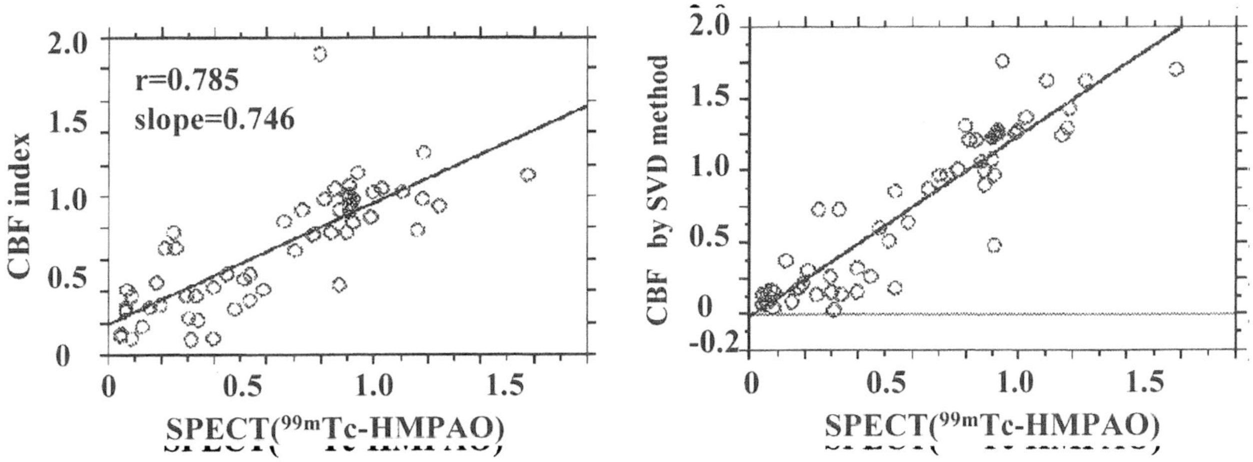

Fig. 2. Correlations between ^{99m}Tc-HMPAO SPECT delivered CBF ratio (ratio to non-ischemic mirror region) and two DPI delivered CBF ratio

MTTu must be taken into consideration. Since the MTTu is affected by the transit time in the input artery, a delay in arrival of the contrast agent and/or broadening of the tracer concentration curve (dispersion) could occur if stenosis of the input artery exists. Moreover, as Robert *et al.* [8] claim, MTT itself obtained from the indicator dilution theorem (MTTidt) has some limitations for quantification, since MTTidt is based on the assumption that all the vascular structures in the images are identical. Thus MTTidt is not a "true" MTT but an approximation. Although, Robert *et al.* did acknowledge that MTTidt is valuable for assessing regional differences within an individual. Thus the CBF index delivered from rrCBV and MTTu must be affected by the limitations of MTTu. Nevertheless the CBF index image is simple to deliver and fulfills the conditions described in the introduction. Furthermore CBF index well correlated with the relative rCBF delivered from SPECT, as well as with stable xenon CT [1] and autoradiography studies [9]. These results indicate that the CBF index is feasible for accessing regional changes in cerebral blood flow semi quantitatively.

During the initial therapeutic assessment of the acute stage of ischemia, time is limited. Thus, the DWI and DPI combined study needs to be assessed very shortly after imaging. Among the three automatically calculated images, namely MTTu, rrCBV and CBF index images, CBF index images have an apparent threshold for assessing the outcome of ischemic lesion and the extent of the low CBF index area predicts the resulting infarcted area. From this perspective, we summarized that CBF index images together with DWI can guide the initial assessment of the acute stage of cerebral ischemia.

References

1. Hagen T, Bartylla K, Piepgras U (1998) Correlation of regional cerebral blood flow values obtained by perfusion-MRI and stable xenon computed tomography. Rivista di Neuroradiologica 11 [Suppl] 2: 176–178
2. Hamberg LM, Macfarlane R, Tasdemiroglu E, Boccalini P, Hunter GJ, Belliveau JW, Moskowitz MA, Rosen BR (1993) Measurement of cerebrovascular changes in cats after transient ischemia using dynamic magnetic resonance imaging. Stroke 24: 444–451
3. Haraldeth O, Jones RA, Muller TB, Fahlvik AK, Oksendal AN (1996) Comparison of dysprosium DTPA BMA and superparamagnetic iron oxide particles as susceptibility contrast agent for perfusion imaging of regional cerebral ischemia. J Magn Reson Imaging 6: 714–717
4. Østergaard L, Sorensen AG, Kwong KK, Weisskoff RM, Gyldensted C, Rosen BR (1996) High resolution measurement of cerebral blood flow using intravascular tracer bolus passages. Part II: Experimental comparison and preliminary results. Magn Reson Med 36: 726–736
5. Østergaard L, Weisskoff RM, Chesler DA, Gyldensted C, Rosen BR (1996) High resolution measurement of cerebral blood flow using intravascular tracer bolus passages. Part I: mathematical approach and statistical analysis. Magn Reson Med 36: 715–725
6. Parsons MA, Yang Q, Barbar PA, Darby DG, Desmond P, Gerraty RT, Tress BM, Davis SM (2001) Perfusion magnetic resonance imaging maps in hyperacute stroke: relative cerebral blood flow most accurately identifies tissue destined to infarct. Stroke 32: 1581–1587
7. Sorensen G, Riemer P (2000) How do I postprocess perfusion images? Cerebral MR perfusion imaging. In: Sorensen G, Reimer P (eds) GeorgeTieme Verlag, Stuttdart, pp 43–51
8. Weisskoff RM, Chesler D, Boxerman JL, Rosen BR (1993) Pitfalls in MR measurement of tissue blood flow with intravascular tracers: which mean transit time? Magn Reson Med 29: 553–558

9. Wittlich F, Kohno K, Mies G, Norris DG, Hoehn-Berlage M (1995) Quantitative measurement of regional blood flow with gadolinium diethylenetriaminepentaacetate bolus track NMR imaging in cerebral infarcts in rats: validation with the iodo[14C]antipyrine technique. Proc Natl Acad Sci 92: 1846–1850

10. Zieler KL (1962) Theoretical basis of indicator dilution methods for determining blood flow and volume. Circ Res 10: 393–447

Correspondence: Hironaka Igarashi M.D., The Second Internal Medicine, Nippon Medical School, 1-1-5 Sendagi, Bunkyo-ku, Tokyo, 113-8603, Japan. e-mail: higara@nms.ac.jp

Acta Neurochir (2003) [Suppl] 86: 247–249

Transient elevation of serum bilirubin (a heme oxygenase-1 metabolite) level in hemorrhagic stroke: bilirubin is a marker of oxidant stress

K. Dohi[1], **Y. Mochizuki**[2], **K. Satoh**[3], **H. Jimbo**[2], **M. Hayashi**[2], **I. Toyoda**[2], **Y. Ikeda**[2], **T. Abe**[2], and **T. Aruga**[1]

[1] Department of Emergency and Critical Care Medicine, Showa University, School of Medicine, Tokyo, Japan
[2] Department of Neurosurgery, Showa University, School of Medicine, Tokyo, Japan
[3] Department of Medicinal Information, School of Pharmaceutical Sciences, Showa University, Tokyo, Japan

Summary

Bilirubin (Bil) is the end product of heme catabolism. The production of Bil reflects heme oxygenase-1 expression in response to oxidative stress in various diseases. To assess the role of Bil as a marker of oxidative stress in cases of brain damage, we measured serum Bil concentrations in patients with hemorrhagic stroke. Serum levels of total Bil were measured in 20 subarachnoid hemorrhage patients with symptomatic vasospasms and in 23 patients with intracerebral hemorrhage; concentrations were measured every day for 14 consecutive days. Serum Bil levels were significantly elevated in the early phases in both groups. Moreover, transient elevation was observed on the day prior to the observation of clinical manifestations of symptomatic vasospasm after SAH. Bil, known to be a powerful antioxidant, was induced after hemorrhagic stroke, reflecting the intensity of oxidative stress. Plasma Bil concentrations might serve as a useful marker of oxidative stress in hemorrhagic stroke patients.

Keywords: Heme oxygenase-1; hemorrhagic stroke; bilirubin; free radical; subarachnoid hemorrhage; intracerebral hemorrhage; oxidative stress; antioxidant.

Introduction

Heme oxygenases (Hos) are rate-limiting enzymes in heme degeneration. They catalyze the cleavage of the heme ring to form ferrous iron, carbon monoxide, and biliverdin. Biliverdin is rapidly metabolized to bilirubin (Bil), which is known to possess powerful antioxidant properties [11, 12]. Heme oxygenase-1 (HO-1) is an inducible protein activated under systemic inflammatory conditions by oxidative stress. Induction of HO-1 may be associated with cellular protection against oxidative stress that is exerted by Bil [7]. It is already known that HO-1 is induced in various types of organ injury and that it protects against oxidative stress [3, 8]. Bil generated after upregulation of the HO-1 pathway is cytoprotective against oxidative stress [2]. Recent reports have shown that HO-1 is also expressed in various central nervous system diseases, including Alzheimer's disease and stroke [2, 6, 9]. To investigate the role of HO-1 in patients with cerebrovascular disease, we measured the serum Bil concentration in cases of hemorrhagic stroke. We also investigated the clinical usefulness of Bil as an oxidative stress marker in the diagnosis of neuronal damage and vasospasm after subarachnoid hemorrhage.

This study protocol was approved by the The Committee for the Ethics of Human Study of Showa University. Informed consent was obtained from all participants.

Patients and methods

Between 1997 and 1999, 20 subarachnoid hemorrhage (SAH) patients with symptomatic vasospasm and 23 patients with hypertensive intracerebral hemorrhage (ICH) were studied. Bil concentrations in the serum were measured for 14 consecutive days by radioimmunoassay. Symptomatic vasospasm after SAH was diagnosed by neurological examination and transcranial doppler (TCD) or cerebral angiography. Patients with complicated liver dysfunction were excluded from this study.

Statistical analysis

All values are presented as the mean $\pm$ standard deviations (SD). Differences between groups of serum Bil levels were analyzed by repeated ANOVA. Serum Bil concentrations at different time points were compared with the 2-tailed Student's t test. Statistical significance was determined at $P < 0.05$.

Results

Bil concentrations were significantly elevated in the early phase of disease in both groups. In the SAH group, the Bil concentration was highest on Day 1, and decreased gradually until Day 12 (Fig. 1). A transient significant increase in the T-Bil concentration was observed one day prior to the onset of ischemic symptoms with symptomatic vasospasm after SAH ($P < 0.01$; Fig. 2). In the ICH group, the Bil concentration was highest on Day 2 ($P = 0.085$), then declined with time (Fig. 1). The mean Bil concentrations for a period of 14 days were higher in the ICH group (0.65 ± 0.21 mg/dl) than in the SAH group (0.52 ± 0.14 mg/dl) ($P < 0.01$).

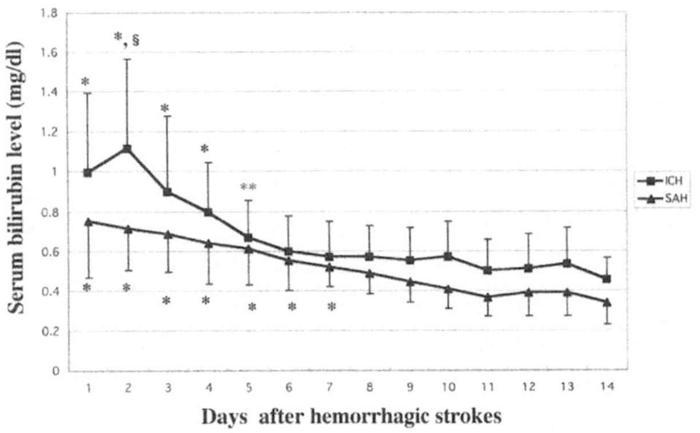

Fig. 1. Time-course of serum total-bilirubin *T-Bil* concentrations after stroke. *ICH* intracerebral hemorrhage group, *SAH* subarachnoid hemorrhage group. * $P < 0.01$, § $P < 0.05$

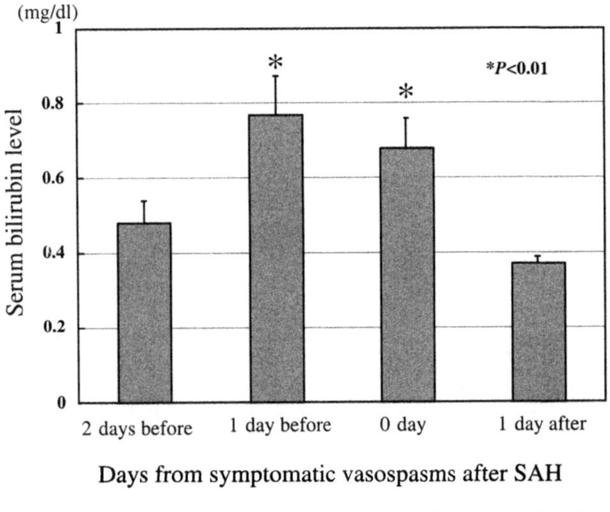

Fig. 2. Time course of serum total-bilirubin *T-Bil* concentrations in patients with symptomatic vasospasms after *SAH* subarachnoid hemorrhage. * $P < 0.01$

Discussion

The mechanism of HO-1 expression following episodes of cerebrovascular disease remains to be clarified. The present study demonstrated that serum Bil concentrations were significantly increased in SAH and ICH patients. Our findings indicate that the production of Bil in the serum is attributable to systemic and local oxidative stress caused by vascular and brain damage.

The time course of the serum T-Bil concentrations of the SAH group differed from that of the ICH group. These differences between the two groups could be due to differences in the respective mechanisms of vascular damage and brain damage. While SAH is caused by bleeding from major vessels, ICH is caused by bleeding from small vessels. SAH may cause severe diffuse primary brain damage at the time of onset, while ICH typically leads to space-occupying lesions. The transient elevation of the Bil concentration on Day 2 in the ICH group may suggest secondary brain damage caused by perifocal ischemia associated with intracerebral hematoma. In the SAH group, a transient Bil elevation was observed one day prior to the onset of vasospasm. These findings appear to reflect oxidative stress caused by vasospasm. The transient elevation in Bil concentrations in the absence of liver dysfunction in SAH patients may indicate the existence of vasospasm.

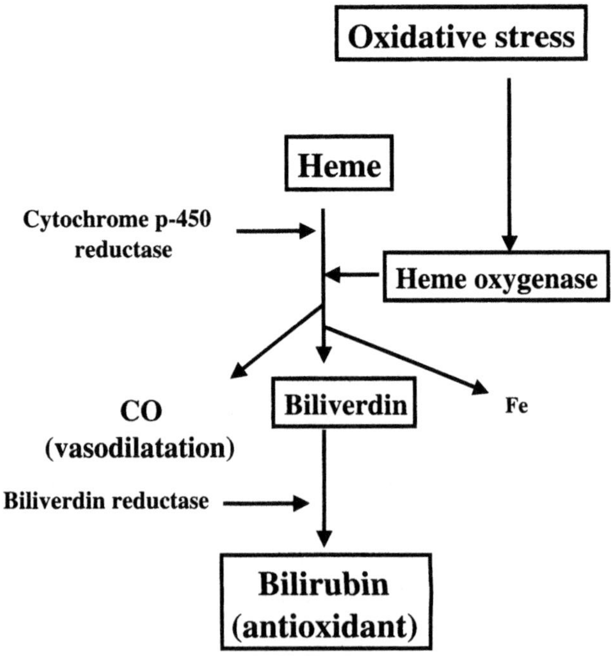

Fig. 3.

The pathophysiological effects of Bil include protection against oxidative stress [2, 11]. Elevated serum Bil concentrations are associated with a decreased risk for early familial coronary artery disease [4, 10]. The other heme metabolite mediated by HO-1, carbon monoxide (CO), is involved in vasodilatation and is thought to perform a protective role in vessel remodeling based on the inhibition of endothelin-1 and platelet-derived growth factor-β (PDGF-β) [1, 5]. HO-1 directly reduces vasoconstriction and inhibits cell proliferation during vascular injury [3]. Elevations in the level of Bil in patients with hemorrhagic stroke and vasospasm after SAH in this series may reflect protection from systemic inflammatory conditions due to oxidative stress. Further studies will be necessary in order to clarify the role of induction of Bil in patients with hemorrhagic stroke. The present preliminary investigation indicates that serum Bil concentrations, which are easily measured, provide a useful marker of oxidative stress caused by vascular and brain damage in patients with hemorrhagic stroke.

References

1. Brouard S, Otterbein LE, Soares MP (2000) Carbon monoxide generated by heme oxygenase 1 suppresses endothelial cell apoptosis. J Exp Med 192: 1015–1026
2. Clark JE, Foresti R, Motterlini R (2000) Dynamics of haem oxygenase-1 expression and bilirubin production in cellular protection against oxidative stress. Biochem J 348: 615–619
3. Duckers HJ, Boehm M, Nabel EG (2001) Heme oxygenase-1 protects against vascular constriction and proliferation. Nat Med 7: 693–698
4. Hopkins PN, Wu LL, Williams RR (1996) Higher serum bilirubin is associated with decreased risk for early familial coronary artery disease. Arterioscler Thromb Vasc Biol 16: 250–255
5. Kourembanas S, Morita T, Christou H (1997) Mechanisms by which oxygen regulates gene expression and cell-cell interaction in the vasculature. Kidney Int 51: 438–443
6. Matz PG, Weinstein PR, Sharp FR (1997) Heme oxygenase-1 and heat shock protein 70 induction in glia and neurons throughout rat brain after experimental intracerebral hemorrhage. Neurosurgery 40: 152–160; discussion 160–162
7. Oguro T, Yoshida T (1998) [New role of stress protein heme oxygenase – its relationship with physiopathology]. J Toxicol Sci 23: App 41–53
8. Otterbein LE, Kolls JK, Choi AM (1999) Exogenous administration of heme oxygenase-1 by gene transfer provides protection against hyperoxia-induced lung injury. J Clin Invest 103: 1047–1054
9. Schipper HM, Liberman A, Stopa EG (1998) Neural heme oxygenase-1 expression in idiopathic Parkinson's disease. Exp Neurol 150: 60–68
10. Schwertner HA, Jackson WG, Tolan G (1994) Association of low serum concentration of bilirubin with increased risk of coronary artery disease. Clin Chem 40: 18–23
11. Stocker R, Yamamoto Y, Ames BN (1987) Bilirubin is an antioxidant of possible physiological importance. Science 235: 1043–1046
12. Wu TW, Fung KP, Weisel RD (1996) Antioxidation of human low density lipoprotein by unconjugated and conjugated bilirubins. Biochem Pharmacol 51: 859–862

Correspondence: Kenji Dohi, M.D., Department of Neurosurgery, Showa University, 1-5-8 Hatanodai, Shinagawa-ku, Tokyo. e-mail: kdop@med.showa-u.ac.jp

Acta Neurochir (2003) [Suppl] 86: 251–253

Relation between brain oxygen metabolism and temperature gradient between brain and bladder

A. Sakurai, K. Kinoshita, T. Atsumi, T. Moriya, A. Utagawa, and **N. Hayashi**

Department of Emergency and Critical Care Medicine, Nihon University School of Medicine, Tokyo, Japan

Summary

Brain temperature is slightly higher than systemic core temperature normally. Fluctuations of a temperature gradient between brain and core body have recently been reported after a severe brain insult. The pathophysiological significance of the gradient fluctuations is unclear. This study aims to identify the gradient fluctuations between brain and core temperatures after a brain insult.

Temperature gradient (brain temperature minus bladder temperature: °C) was measured in 11 patients (125 points) with severe brain injury (4 patients with subarachnoid hemorrhage, 4 with cerebral hemorrhage, and 3 with traumatic brain injury). Cerebral perfusion pressure (CPP; mmHg) and jugular venous blood saturation (SjO_2; %) was also measured.

The average gradient was 0.29 ± 0.285 °C when CPP was above 50 mmHg. SjO_2 was inversely related to the temperature gradient in a significant manner ($r = 0.472$; $P < 0.0001$). Temperature gradients tended to increase and then decrease when CPP < 50 mmHg.

This study demonstrates that increased temperature gradient has a significant inverse correlation with SjO_2 at CPP > 50 mmHg. The current results suggest that the fluctuations in temperature gradient in critical conditions reflect brain ischemia.

Keywords: Brain temperature; bladder temperature; SjO_2; cerebral ischemia.

Introduction

Clinical studies have shown that therapeutic hypothermia is effective in patients with traumatic brain injury (TBI) [8] and out-of-hospital cardiac arrest [2, 12], while hyperthermia may worsen neurological outcome after TBI [6] or stroke [7]. It is important to control the temperature to protect injured brain. A gradient between brain and core body temperatures is frequently observed in brain-injured patients [5, 9]. One clinical study [11] reported that the difference in temperature between brain and body fluctuated when cerebral perfusion pressure (CPP) was 50 mmHg or less. However, the significance of this temperature gradient is not known. This study aimed to identify the significance of the temperature gradient.

Materials and methods

Brain and bladder temperatures were continuously measured in 11 patients (125 points) with severe brain injury (4 patients with subarachnoid hemorrhage, 4 with cerebral hemorrhage, and 3 with traumatic brain injury). The probe for measuring brain temperature was designed for pressure-temperature monitoring with ventricular drainage (Integra NeuroScience^{TM}, Camino, San Diego, USA). Bladder temperatures were obtained using a temperature-sensing Foley catheter (Bardex I.C.^{TM}, C.R. Bard, Inc., Georgia, USA.). All patients received controlled ventilation (keeping $PaCO_2$ at 35 to 40 mmHg), cerebrospinal fluid drainage, sedation, neuromuscular blocking agents, and mannitol to control intracranial pressure (ICP). Cerebral perfusion pressure (CPP) was calculated as mean arterial blood pressure minus ICP. A jugular vein catheter (Opticath P540-H^{TM}, Abbott Laboratories, North Chicago, USA) was inserted in the right jugular bulb, appropriate placement of which was confirmed with skull radiographs, and jugular venous blood saturation (SjO_2; %) was measured concomitantly. All SjO_2 values were obtained by re-calibrated data for blood samples drawn through the catheter.

Patients were divided according to CPP into two groups, CPP > 50 mmHg or more and CPP < 50 mmHg, retrospectively, because we speculated that the fluctuation of temperature gradients between brain and bladder could be observed in patients with brain injury.

Statistical analysis

Correlation coefficient and simple linear regression analysis were used to describe the relationships between temperature differences (brain-bladder temperature: °C) and SjO_2 and between temperature differences and CPP. Statistical significance was defined as $P < 0.05$.

Results

Demographic data for these 11 patients are summarized in Table 1.

Table 1. *Demographic characteristics of the 11 patients*

Age (yr)	58.7 ± 21.0
Gender	
– Male	5
– Female	6
Diagnosis	
– SAH	4
– ICH	4
– TBI	3
GCS after resuscitation	
– 3–5	8
– 6–8	3
Outcome	
– Alive	7
– Dead	4

SAH Subarachnoidal hemorrhage, *ICH* intracranial hemorrhage, *TBI* traumatic brain injury, *GCS* Glasgow Coma Scale, * Mean ± SD.

At CPP > 50 mmHg, the differences in brain-bladder temperature averaged 0.29 ± 0.285 °C. SjO_2 was inversely and moderately related to the temperature gradient (r = 0.472, $P < 0.0001$: Fig. 1a) when CPP was >50 mmHg. There was no significant correlation between the temperature gradient and CPP.

At CPP < 50 mmHg (n = 5), the temperature differences showed an abnormal temperature gradient (bladder > brain temperature) in 2 cases. These patients had uncontrollable ICP and the their final outcome was death. Transient, but markedly increased temperature gradients (1.2 and 1.3 °C) were observed in 2 patients. One patient showed a temperature gradient of 0.2 °C with uncontrollable ICP, but finally demonstrated an abnormal temperature gradient (bladder > brain temperature).

Discussion

Many clinical studies have reported that therapeutic hypothermia was effective in patients having a variety of neurological insult [2, 8, 12], while patients with hyperthermia had a poor outcome in TBI [6] or stroke [7]. Some authors have reported that small differences of brain temperature in intraischemic brain conferred a protective effect [3], but a small increase in brain temperature has a significant effect after cerebral ischemia in rats [10]. Therefore, physicians should control the brain temperature strictly in neurological surgery and the neuro-intensive care unit. Although brain temperature appears to be a key determinant of neurologic outcome, clinical monitoring of brain temperature is not routine or substituted for core body temperature. While fluctuations of gradient between brain and core body temperatures are frequently observed in brain-injured patients [1, 5, 9, 11], the significance of this temperature gradient is unclear. Current studies

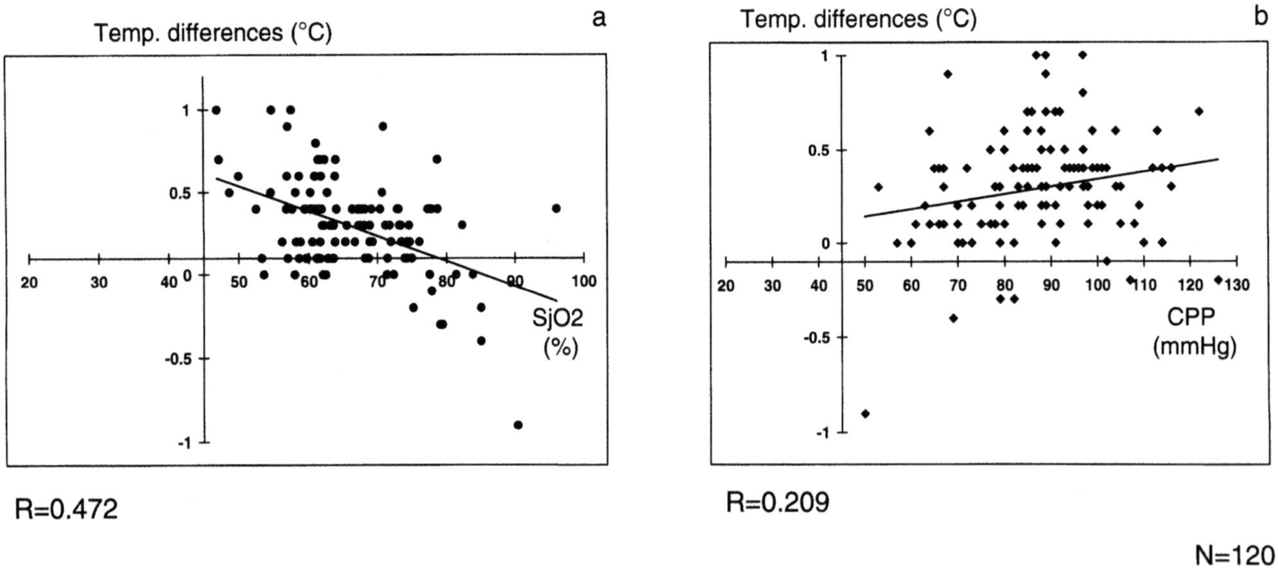

Fig. 1. Relationship between SjO_2 and temperature differences (a), and between CPP and temperature differences (b) at CPP > 50 mmHg. Temperature differences had a significant inverse correlation with SjO_2 (r = 0.472, $P < 0.0001$), and there was no significant correlation between the temperature differences and CPP. Temp. differences: temperature differences between brain-bladder; *SjO₂* jugular venous blood saturation; *CPP* cerebral perfusion pressure

from a limited number of patients suggest that fluctuations in temperature gradient, overshooting normal values, in critical brain conditions may indicate brain ischemia. This hypothesis is important in the detection of secondary ischemic brain injury after TBI or stroke. Further study is needed to clarify the relation between the temperature gradient and brain ischemia.

Brain temperature seems to be determined by 3 major factors: 1) the production of local heat by metabolic processes in the brain; 2) the rate of local cerebral blood flow; and 3) arterial blood temperature [1]. In clinical studies, brain temperature is usually higher than core body temperature [9]. The experimental study suggested that this gradient did not change even when the brain temperature changed from 33 to 37 °C in normal conditions [1]. However, Hayashi N *et al.* found that brain temperature rose with intracranial hypertension in severely brain-injured patients, while systemic core temperature did not increase significantly [5]. Rumana *et al.* reported that a brain-rectal temperature difference tended to increase when CPP was 50 to 20 mmHg and decreased significantly at CPP $<$ 20 mmHg [11]. They suggested that an adverse temperature gradient (rectum $>$ brain temperature) might indicate misery perfusion. In the same manner, our limited data showed that the temperature differences between brain and bladder indicated dynamic changes, such as extremely increased or an adverse temperature gradient, at CPP less than 50 mmHg.

Few studies have attempted to identify the significance of the temperature gradient between brain and core temperature at CPP $>$ 50 mmHg. Our study demonstrates that the differences in temperature gradient between the brain and the bladder had a significant inverse correlation with SjO_2 at CPP $>$ 50 mmHg. Therefore, our study using SjO_2 suggests that the temperature gradient might reflect the cerebral blood flow and metabolism at CPP $>$ 50 mm Hg. Another study reported that differences between jugular bulb and bladder temperature increased when moderate hyperventilation was performed [4]. Excess hyperventilation vasoconstricts brain vessels and decreases SjO_2 resulting from reduced CBF. Both results suggest that the fluctuation of temperature differences overshooting normal values between brain and core body temperature may occur in brain ischemia.

Conclusion

This study demonstrates that increased temperature gradient differences have a significant inverse correlation with SjO_2 at CPP $>$ 50 mmHg. The current results from a limited number of patients suggest the fluctuations in the temperature gradient in critical brain conditions may reflect in brain ischemia.

References

1. Ao H, Moon JK, Tanimoto H *et al* (2000) Jugular vein temperature reflects brain temperature during hypothermia. Resuscitation 45(2): 111–118
2. Bernard SA, Gray TW, Buist MD, Jones BM, Silvester W *et al* (2002) Treatment of comatose survivors of out-of-hospital cardiac arrest with induced hypothermia. N Engl J Med 346(8): 557–563
3. Busto R, Dietrich WD, Globus MY *et al* (1987) Small differences in intraischemic brain temperature critically determine the extent of ischemic neuronal injury. J Cereb Blood Flow Metab 7(6): 729–738
4. Clausen T, Rieger A, Roth S *et al* (2000) Cerebrovenous blood temperature-influence of cerebral perfusion pressure changes and hyperventilation: evaluation in a porcine study and in man. J Neurosurg Anesthesiol 12(1): 2–9
5. Hayashi N, Hirayama T, Udagawa A *et al* (1994) Systemic management of cerebral edema based on a new concept in severe head injury patients. Acta Neurochir (Wien) [Suppl] 60: 541–543
6. Jiang JY, Gao GY, Li WP *et al* (2002) Early indicators of prognosis in 846 cases of severe traumatic brain injury. J Neurotrauma 19(7): 869–874
7. Kammersgaard LP, Jorgensen HS, Rungby JA *et al* (2002) Admission body temperature predicts long-term mortality after acute stroke: the Copenhagen Stroke Study. Stroke 33(7): 1759–1762
8. Marion DW, Penrod LE, Kelsey SF *et al* (1997) Treatment of traumatic brain injury with moderate hypothermia. N Engl J Med 336(8): 540–546
9. Mellergard P (1994) Monitoring of rectal, epidural, and intraventricular temperature in neurosurgical patients. Acta Neurochir (Wien) [Suppl] 60: 485–487
10. Minamisawa H, Mellergard P, Smith ML *et al* (1990) Preservation of brain temperature during ischemia in rats. Stroke 21(5): 758–764
11. Rumana CS, Gopinath SP, Uzura M *et al* (1998) Brain temperature exceeds systemic temperature in head-injured patients. Crit Care Med 26(3): 562–567
12. The Hypothermia after Cardiac Arrest Study Group (2002) Mild therapeutic hypothermia to improve the neurologic outcome after cardiac arrest. N Engl J Med 346(8): 549–556

Correspondence: Atsushi Sakurai, M.D., Department of Emergency and Critical Care Medicine, Nihon University School of Medicine, 30-1 Oyaguchi-Kamimachi, Itabashi-ku, Tokyo, 173-8610, Japan. e-mail: sakurai@med.nihon-u.ac.jp

Experimental trauma

Acta Neurochir (2003) [Suppl] 86: 257–260

Neuropeptide release influences brain edema formation after diffuse traumatic brain injury

R. Vink[1,3], A. Young[2], C. J. Bennett[2], X. Hu[2], C. O. Connor[1], I. Cernak[3], and A. J. Nimmo[2]

[1] Department of Pathology, The University of Adelaide, Adelaide, Australia
[2] School of Pharmacy and Molecular Sciences, James Cook University, Townsville, Australia
[3] Department of Neuroscience, Georgetown University, Washington, DC, USA

Summary

The mechanisms associated with edema formation after traumatic brain injury (TBI) have not been fully elucidated. In peripheral tissue injury, the neurogenic component of inflammation plays a significant role in increased vascular permeability and edema formation. However, few studies have examined the role of neuropeptide induced neurogenic inflammation following TBI. Adult male Sprague-Dawley rats were either left untreated, or pre-treated with capsaicin (125 mg/kg s.c.) or equal volume vehicle, and injured 14 days later using the 2-meter impact-acceleration model. Subgroups of animals were assessed for blood brain barrier (BBB) permeability (Evans Blue), brain edema (wet weight/dry weight) and functional outcome (Barnes maze and Rotarod) for up to 2 weeks post-trauma. Increased BBB permeability was present in untreated animals between 3 and 6 h after injury but not at later time-points. Edema was maximal at 5 h after trauma, declined and then significantly increased over the 5 days post-trauma. In contrast, capsaicin pre-treated, neuropeptide-depleted animals exhibited no significant increase in BBB permeability or edema compared to vehicle treated animals after injury. Notably, motor and cognitive impairments were significantly reduced in the capsaicin-pretreated animals. We conclude that neurogenic inflammation contributes to the development of edema and posttraumatic deficits after diffuse TBI.

Keywords: Neurotrauma; edema; neuropeptides; capsaicin; outcome; inflammation; neurogenic.

Introduction

The mechanisms associated with edema formation after traumatic brain injury have not been fully elucidated. Studies of peripheral tissue injury have demonstrated that increased vascular permeability and edema formation develop through a process known as neurogenic inflammation [15]. Neurogenic inflammation is a neurally elicited reaction that has the typical characteristics of an inflammatory response including vasodilation, protein extravasation and tissue swelling. It is initiated by stimulation of sensory C-fibers, which results in the release of neuropeptides. These neuropeptides then facilitate plasma protein extravasation from blood vessels and associated edema formation [15]. Extensive studies have now shown that these sensory C-fibers can be activated by capsaicin that will induce the release of the neuropeptides, including substance P and calcitonin gene related peptide [5]. Repeated administration of capsaicin to adult animals results in a transient depletion of neuropeptide stores and subsequent inhibition of neurogenic inflammation lasting up to 3 weeks [10]. Such depletion provides an opportunity to examine the relationship between neuropeptide release and edema formation after traumatic brain injury. The current study therefore characterizes the temporal profile of increased vascular permeability and edema formation after traumatic brain injury, and examines the relationship between neuropeptides and such edema formation.

Materials and methods

Adult male Sprague Dawley rats (n = 77; 400 ± 25 g) were fed and watered *ad libitum* before being anesthetized with sodium pentobarbital (60 mg/kg i.p.) and subjected to impact-acceleration induced brain injury as previously described in detail elsewhere [6, 9]. Briefly, the skull was exposed by a midline incision and a 10 mm stainless steel disc (3 mm depth) was affixed centrally between lambda and bregma using a polyacrylamide adhesive. The animal was then placed on a 10 cm deep foam bed and injury induced by dropping a 450 g weight a distance of 2 m onto the stainless steel disc. Rectal temperature was maintained at 37 °C throughout using a thermostatically controlled heating pad. A further 14 animals served as uninjured controls (shams). A final group of 40 animals was equally subdivided and administered either 125 mg/kg capsaicin in 10% alcohol, 10% TWEEN 80 and 80% saline by subcutaneous injection over three days (25 mg/kg; 50 mg/kg; 50 mg/kg), or equal volume vehicle alone over the same three day period. This capsaicin treatment has previously been shown to induce a depletion of neuropeptides lasting for up to 3 weeks [10]. At 14 days after capsaicin or vehicle treatment, animals were injured as described above.

After injury, a time course for blood brain barrier (BBB) permeability and edema formation in the cortex was established in non-treated animals. For assessment of BBB integrity, Evans blue dye (0.3 ml of 2% solution in phosphate buffered saline) was injected intravenously 30 minutes prior to decapitation at various time points postinjury (n = 7/timepoint). The BBB permeability to Evans blue dye was assessed by measuring light absorbance in transverse sections of the brains using the Image Pro Plus image analysis computer program. For quantitation of edema formation, animals were killed at various time points (n = 7/timepoint) by decapitation under pentobarbital anesthesia and their brains rapidly removed, the cortex dissected and weighed. Cortices were then dried for 48 h at 100 °C and reweighed. Water content was then calculated and expressed as a percentage of total cortex weight. All sham (uninjured) animals were assessed for BBB permeability and edema formation at 4 h after surgery.

In capsaicin and vehicle treated animals, BBB permeability and edema formation was determined at their maximum points established above (n = 7 per group). Furthermore, functional outcome (n = 6/group) was assessed daily using the Rotarod and Barnes Maze for motor and cognitive outcome, respectively. Briefly, the rotarod test as used in previous TBI studies [8, 9] requires an animal to walk on a motorized rotating assembly of 18 rods, each 1 mm in diameter. The rotational speed of the assembly is increased from 0 to 30 revolutions per minute (rpm) in intervals of 3 rpm every 10 s. The duration in seconds at the point at which the animal either completed the 2 minute task, fell from the rods, or gripped the rods and spun for 2 consecutive revolutions rather than actively walking, was recorded as the task score. Finally, the Barnes Maze learning and memory paradigm [1, 7] incorporates animals being placed under a cover in the center of an elevated 1-meter diameter board containing 19 holes around the periphery. One of the holes is the entrance to a darkened escape tunnel that was not visible from the surface of the board. After activating a series of bright lights and an aversive sound, the cover was lifted and the latency in seconds for the animal to locate and enter the darkened escape tunnel was recorded. For functional outcome tests, animals were pre-trained on the tasks for 5 days prior to injury and assessed for posttraumatic functional deficits for 6 days after injury.

All data are expressed as mean and SEMs. Statistical differences were determined using one way analysis of variance followed by individual Student-Newman-Keuls post-hoc tests. A p value of 0.05 was considered significant.

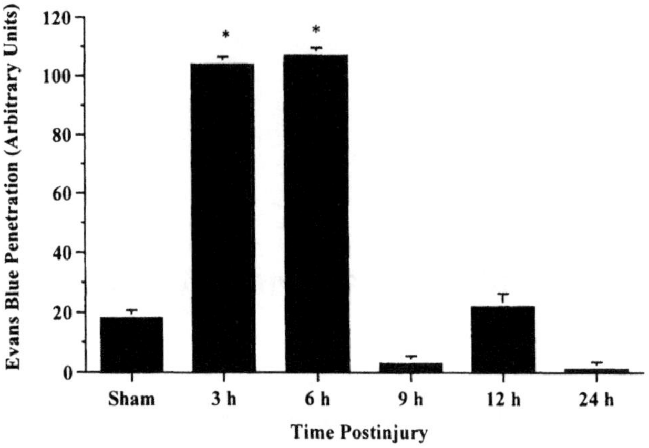

Fig. 1. Time course of blood brain barrier permeability to Evans Blue in the rat cortex following diffuse traumatic brain injury (n = 7/group). * p =< 0.05 versus sham

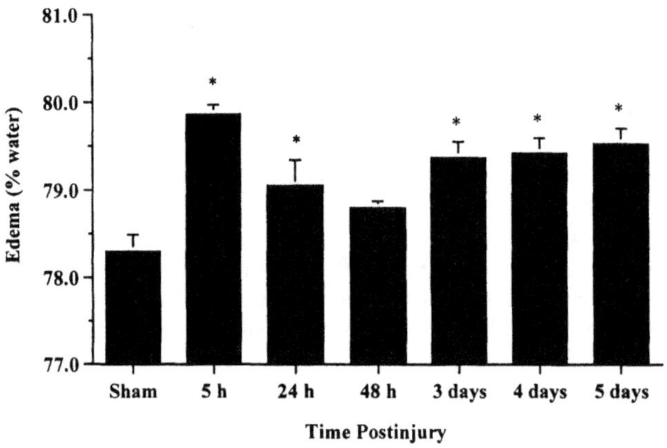

Fig. 2. Temporal profile of edema formation in rat cortex following diffuse traumatic brain injury (n = 7/group). * = $p < 0.05$ versus sham

Results

A significant ($p < 0.05$) increase in cortical BBB permeability to Evans blue was observed between 3 and 6 h after diffuse trauma, but not at any other time point (Fig. 1). Indeed, only residual amounts of Evans blue could be detected in the cortex at 9, 12 or 24 h, and these were no different from that found in sham (uninjured) animals. Presumably, this was vascular in origin. Notably, no significant Evans blue was detected in either the cerebellum or subcortical layers. Cortical water content in sham (uninjured) animals was 78.30 ± 0.19% (Fig. 2). After injury, there was a significant ($p < 0.05$) increase in water content in the cortex within the first 24 h, a decrease by 48 h, and then a gradual significant increase over the ensuing 3 days. At 5 h, which was selected to correspond to the established maximal BBB permeability after injury, the edema formation was the greatest of any time point after injury (79.87 ± 0.11%).

For capsaicin studies, we selected the 5 h time point that would normally reflect maximal BBB permeability and edema formation after trauma. Prior depletion of neuropeptides with capsaicin pre-treatment resulted in a significant ($p < 0.05$) attenuation of both post-traumatic, cortical BBB permeability and edema formation compared to vehicle treated controls (Fig. 3). Indeed, there was no significant difference between

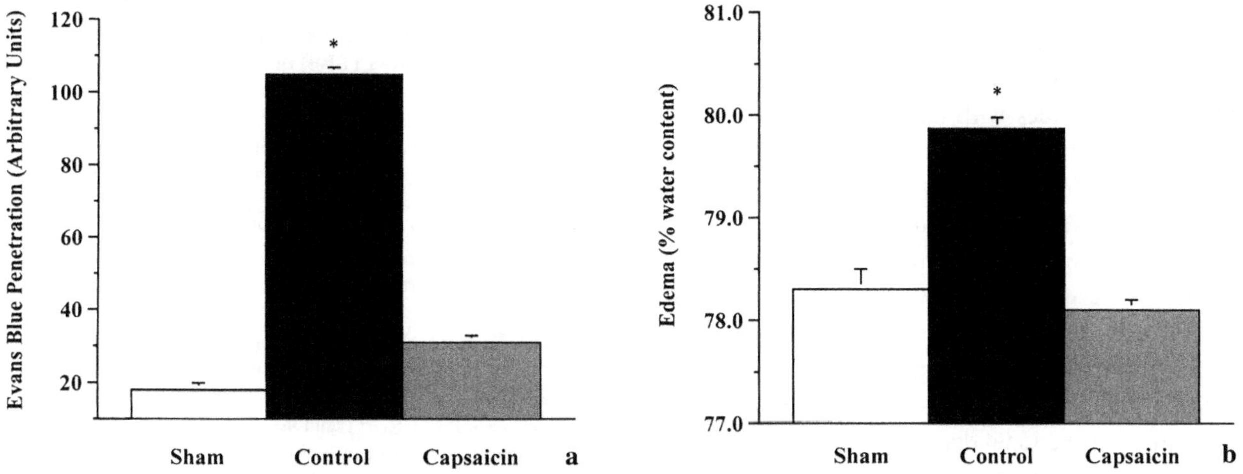

Fig. 3. Effects of capsaicin pretreatment on (a) cortical blood brain barrier permeability and (b) cortical edema, at 5 h following diffuse traumatic brain injury in rats (n = 7/group). * = $p < 0.05$ versus shams and capsaicin

capsaicin treated animals and shams. Furthermore, capsaicin treated animals that were assessed for functional outcome demonstrated a significant attenuation of both motor and cognitive deficits after trauma. In the rotarod test, injury to vehicle treated controls resulted in a decline from 107 ± 2 secs to 23 ± 6 secs at 24 h posttrauma ($p < 0.05$). With repeated exposure to the task, this motor performance improved at a rate of 16 secs per day to preinjury levels by 6 days after trauma. In contrast, capsaicin treated animals did not demonstrate a motor deficit after trauma, averaging 101 ± 2 secs after injury, which was similar to the 111 ± 2 secs in sham (uninjured) animals over the same 6 day period. In the cognitive test, latency to escape the aversive stimulus in vehicle treated controls increased from 13 ± 3 secs to 186 ± 20 secs at 24 h after injury ($p < 0.05$). With repeated exposure to the task, cognitive performance improved at a rate of 32 secs per day such that by day 6 after injury, cognitive function was similar to preinjury values. Capsaicin treated animals did not demonstrate any cognitive deficit after trauma, recording values that averaged 10 ± 1 sec after injury. This value was not significantly different from sham (uninjured) animals that recorded a value of 14 ± 2 secs over the same period.

Discussion

Previous studies have demonstrated that posttraumatic edema formation is a mixture of both vasogenic and cytotoxic phases following diffuse traumatic brain injury [2]. Our present results are consistent with these previous findings. We demonstrate that diffuse brain injury causes an increase in BBB permeability that is maximal between 3 and 6 h after trauma. The increased BBB permeability was associated with an early edema peak, which then transiently subsided before increasing over the next 3 days. The temporal association between the early peak in edema and the increased BBB permeability suggests that this early edema may be vasogenic in origin. Moreover, the subsequent increase in edema over the following days, in the absence of BBB permeability, supports the view that this second phase is most likely cytotoxic in origin [2]. We specifically examined the vasogenic phase of edema in view of the fact that BBB opening has been shown to be permissive to subsequent cytotoxic edema [3].

Virtually all blood vessels of the body are surrounded by sensory nerve fibers that contain neuropeptides. Cerebral arteries, in particular, appear to receive a dense supply of these neuropeptide-containing neurons, and studies of migraine [14] have demonstrated that these neuropeptides are a therapeutic target to reduce vascular permeability. The involvement of neuropeptides and neurogenic inflammation in peripheral edema has been well characterized [15], and it is therefore consistent that these neurons may play a role as mediators of neurogenic inflammation following brain injury. There is some previous evidence in central nervous system tissue that neurogenic inflammation may play a role in edema formation. In brain ischemia, activation of neuropeptide receptors in the endothelium has been shown to contribute to edema formation [13]. However, no studies have examined a potential role in traumatic brain injury. In the present study, we have demonstrated that prior depletion of C-fiber neuropeptides with capsaicin inhibits the development of BBB permeability and edema following traumatic brain injury. Moreover, the inhibition of increased BBB permeability and edema formation was associated with a profound attenuation of functional deficits. Our results therefore support the hypothesis that neurogenic inflammation may be associated with the acute vasogenic edema formation following diffuse traumatic brain injury, and also in the development of posttraumatic functional deficits.

Various neuropeptides are depleted by capsaicin treatment [5, 10], with substance P and calcitonin gene related peptide having been identified as causing increased microvascular permeability and potent vasodilation, respectively [11]. Moreover, antagonists to the receptors of these neuropeptides have been characterized [4, 12]. Accordingly, there exists an opportunity to inhibit neurogenic inflammation by posttraumatic interventions and identify the roles of the individual neuropeptides in the edema process.

Acknowledgments

Supported, in part, by an Australian National Health and Medical Research Council Grant to RV and AJN.

References

1. Barnes CA (1979) Memory deficits associated with senescence: a neurophysiological and behavioral study in the rat. J Comp Physiol Psychiat 93: 74–104
2. Barzo P, Marmarou A, Fatouros P, Hayasaki K, Corwin F (1997) Contribution of vasogenic and cellular edema to traumatic brain swelling measured by diffusion-weighted imaging. J Neurosurgery 87: 900–907
3. Beaumont A, Marmarou A, Hayasaki K, Barzo P, Fatouros P, Corwin F, Marmarou C, Dunbar J (2000) The permissive nature of blood brain barrier (BBB) opening in edema formation following traumatic brain injury. Acta Neurochir (Wien) [Suppl] 76: 125–129
4. Doods H (2001) Development of CGRP antagonists for the treatment of migraine. Curr Opin Investig Drugs 2: 1261–1268
5. Dray A (1992) Neuropharmacological mechanisms of capsaicin and related substances. Biochem Pharmacol 44: 611–615
6. Foda MAA, Marmarou A (1994) A new model of diffuse brain injury in rats; Part II: morphological characterization. J Neurosurg 80: 301–313
7. Fox GB, Fan L, LeVasseur R, Faden AI (1998) Effect of traumatic brain injury on mouse spatial and nonspatial learning in the barnes circular maze. J Neurotrauma 15: 1037–1046
8. Hamm RJ, Pike BR, O'Dell DM, Lyeth BG, Jenkins LW (1994) The rotarod test: an evaluation of its effectiveness in assessing motor deficits following traumatic brain injury. J Neurotrauma 11: 187–196
9. Heath DL, Vink R (1995) Impact acceleration induced severe diffuse axonal injury in rats: characterization of phosphate metabolism and neurologic outcome. J Neurotrauma 12: 1027–1034
10. Kashiba H, Ueda Y, Senba E (1997) Systemic capsaicin in the adult rat differentially affects gene expression for neuropeptides and neurotrophin receptors in primary sensory neurons. Neuroscience 76: 299–312
11. Newbold P, Brain SD (1995) An investigation into the mechanism of capsaicin-induced oedema in rabbit skin. Brit J Pharmacol 114: 570–577
12. Rupniak NM, Kramer MS (1999) Discovery of the antidepressant and anti-emetic efficacy of substance P receptor (NK1) antagonists. Trends Pharmacol Sci 20: 485–490
13. Stumm R, Culmsee C, Schafer MK, Krieglstein J, Weihe E (2001) Adaptive plasticity in tachykinin and tachykinin receptor expression after focal cerebral ischemia is differentially linked to gabaergic and glutamatergic cerebrocortical circuits and cerebrovenular endothelium. J Neurosci 21: 798–811
14. Williamson DJ, Hargreaves RJ (2001) Neurogenic inflammation in the context of migraine. Microsc Res Tech 53: 167–178
15. Woie K, Koller M, Heyeraas KJ, Reed RK (1993) Neurogenic inflammation in rat trachea is accompanied by increased negativity of interstitial fluid pressure. Circ Res 73: 839–845

Correspondence: Robert Vink, Ph.D., Department of Pathology, University of Adelaide, Adelaide, SA 5005, Australia. e-mail: Robert.Vink@adelaide.edu.au

Acta Neurochir (2003) [Suppl] 86: 261–263

Modulation of aquaporin-4 water transport in a model of TBI

A. M. Amorini, J. G. Dunbar, and **A. Marmarou**

Division of Neurosurgery, Medical College of Virginia, Virginia Commonwealth University, Richmond, VA, USA

Summary

Our Laboratory has pursued the hypothesis that traumatic brain edema is predominantly cellular and recent supportive evidence has been obtained indicating a non-extracellular route for sodium and water entering brain. The aim of this study was to investigate if astrocytic endfeet are involved in this passage, using a potent activator of Protein Kinase C (phorbol ester) to modify and closing the Aquaporin 4 (AQP4), a water channel specific for astrocytic endfoot.

Anaesthetized Sprague-Dawley rats were subjected to an intracerebroventricular bolus of phorbol ester (50 pmol/4 μl) or vehicle, in the right hemisphere and after 30 minutes they were exposed to the well-established cortical contusion model (3 mm depth at 6 m/sec) on the same side. After trauma, they were subjected to 5 hours of drug continuous infusion, then sacrificed. Water content measurements for both right (injured) and left (uninjured) hemispheres were calculated using the wet weight/dry weight technique.

Results of these experiments showed a significant decrease in water content in injured phorbol treated animals, underlying that AQP4 regulation plays an important role in brain edema following stroke, and supporting the concept of cellular formation for edema via astrocytic foot processes.

Keywords: Brain edema; protein kinase C; aquaporin-4; phorbol ester; astrocytic endfeet.

Introduction

Several studies have shown the importance of ions and consequentially water transport through the blood brain barrier and its increase like an explanation for the development of brain edema in ischemic and traumatized brain tissue [1, 4, 6]. Recent studies [3] have also pointed out that brain tissue water content and swelling of pericapillary astrocytic foot processes in Aquaporin-4 deficient mice were significantly reduced after acute water intoxication and ischemic stroke providing compelling evidence of an alternate route for water transfer and indicating a substantial involvement of the water channels in brain volume homeostasis.

The aim of this study was to elucidate if Aquaporin-4, which is very specific for astrocytic endfeet, plays an important role in brain swelling in a model of head injury. For this reason we chose to fit an in vitro study for cell swelling reduction based on the inactivation of the AQP-4 via Protein Kinase C (PKC) phosphorilation [2]. We used a phorbol ester (PMA) as a potent PKC activator.

Material and methods

All animals received human care in compliance with the VCU Institutional Animal Care and Use Committee regulations, according to NIH standards and guidelines. Adult male Sprague-Dawley rats (300–400 g) were placed in a chamber attached to an anesthesia machine (70% O_2, 30% N_2O, 4% halothane) for several minutes and then intubated and ventilated (70% O_2, 30% N_2O, 1.5% halothane). The femoral artery and vein were cannulated and animals were placed in a stereotactic frame and a midline incision made into the scalp. A 1.5 mm burr hole was drilled above the dural surface (+0.5 lateral, −0.5 mm) and a 25 g cannula placed into the ventricle. Placement was confirmed using a slow infusion of saline within an increased fluid-pressure system. When the tip of the catheter passed through brain tissue and into the ventricular cavity, the release of measured pressure confirmed desired location.

Animals (n = 6) were first subjected to a bolus of PMA (50 pmol/4 μl), or vehicle (n = 6) in the right ventricle; after 30 minutes the ventricular catheter was removed and a right hemisphere craniotomy (from bregma – lamda-below the temporal ridge) was then performed, keeping the underlying dura intact. Animals were then placed under the piston of a pneumatic impactor specially designed for delivery of a blow to the exposed rat brain. The impactor was set at 6 m/s and a depth of 3 mm.

After trauma, the ventricular catheter was replaced in the injured hemisphere for continuous infusion of PMA (5 h) or vehicle at the rate of 2 μl/min established to maintain the regular turnover of CSF. Control experiments were performed in animals (n = 6) subjected only to cortical contusion, keeping the needle inserted in brain for 5 hours post injury without any flow.

Arterial blood pressure, body temperature, and arterial blood gases were monitored for a period of 5 hours post injury. The brain was then removed for the determination of brain water content in

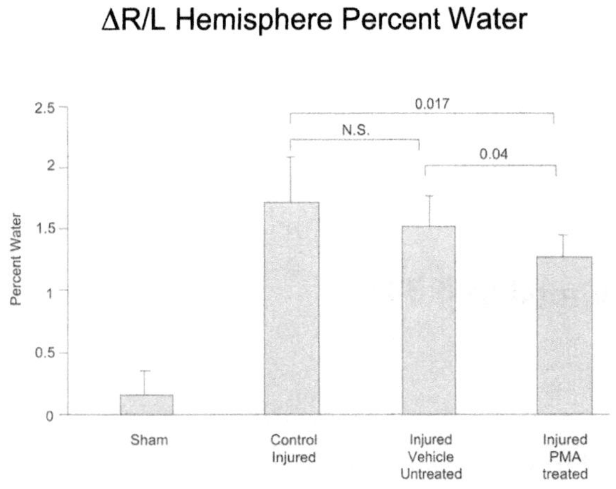

Fig. 1. Percent of water content change in injured animals treated and non-treated by Phorbol Ester (*PMA*)

both right (injured) and left (uninjured) hemispheres by using the wet weight/dry weight technique and for sodium concentration calculated using the flame photometry method. Additional animals (n = 3, sham), which did not undergo any surgical and traumatic procedure, were sacrificed to check the normal water values.

Results

All results are expressed as percent of tissue water relative to the average difference between right (injured) and left (contralateral) hemisphere ± SD.

The statitistical significance of difference between group means was evaluated by Student's unpaired two-tail t-test. A p value of less than 0.05 was considered significant.

Figure 1 shows a significant decrease ($p < 0.05$) in delta % water content [PMA-treated group (mean value = 1.25 ± 0.17) with the injured vehicle-untreated group (mean value = 1.52 ± 0.23)]. The difference between PMA-treated group and control-injured group (mean value = 1.72 ± 0.37) was also significant ($p < 0.05$). No significant difference was found between control-injured group and injured vehicle-untreated group.

Discussion

Many studies in the last two decades were conducted to explain the development of brain edema in ischemic and contused brain tissue [1, 4, 6]; our laboratory has continued to pursue the hypothesis that traumatic brain edema is predominantly cellular and recent supportive evidence has been obtained (data non published) indicating a non-extracellular route for Na and obligatory water entering brain. In this study our aim

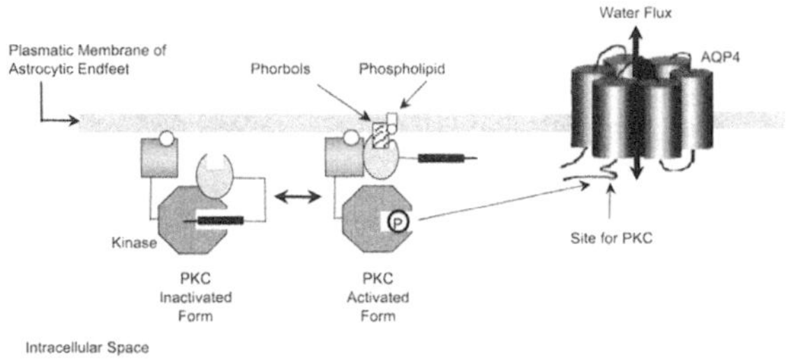

Fig. 2. Scheme of the activation mechanism of Protein Kinase C by Phorbol Ester (*PMA*) and consequent modulation of Aquaporin 4 functionality

was to show if astrocytic endfeet are involved in this movement, accomplished by utilizing a phorbol ester (PMA) to block Aquaporin-4, the water channel specific of astrocytic endfeet. Phorbol esters are natural products isolated from plants used as preferred pharmacological tools to study the activation of PKC: they mimic the action of a cellular lipid second messenger in a more powerful way because they are not readily metabolized into the cells, resulting in a constitutive activation of the enzyme [7]. This activation consists in increasing the affinity of the protein for membrane: only in this form it can transfer phosphate from ATP to specific cellular protein acceptors, like Aquaporin-4 (fig. 2). It is also well known that the specific AQP-4 owns two PKC phosphorylation sites [5]. In this work we performed for the first time experiments utilizing phorbols to activate PKC and consequentially modify AQP-4 structure thereby blocking water flux into contused rat brain. Results prove clearly this gating exists in vivo, as shown by a significant decrease in water accumulation during the five hours post cortical contusion in animals treated by phorbol ester. These results are supportive of an intracellular pathway for water and an involvement of astrocytic endfeet.

Further experiments are in progress to assess the role of phorbols in the movement of ions in the contused brain model.

Acknowledgments

We thank Prof. Giuseppe Lazzarino, Ph.D., for his constant biochemical support, and Dr. Shinji Fukui, M.D., for the helpful discussions.

References

1. Betz AL, Keep RF, Beer ME, Ren X (1994) Blood-brain barrier permeability and brain concentration of sodium, potassium, and chloride during focal ischemia. J Cereb Blood Flow Metabolism 14: 29–37

2. Han Z, Wax MB, Patil RV (1998) Regulation of Aquaporin-4 water channels by Phorbol Ester-dependent protein Phosphorylation. J Biological Chem 273: 6001–6004
3. Manley GT, Fujimura M, Ma T, Noshita N, Filiz F, Bollen AW, Chan P, Verkman AS (2000) Aquaporin-4 deletion in mice reduces brain edema after acute water intoxication and ischemic stroke. Nat Med 6: 159–163
4. Menzies SA, Betz AL, Hoff JT (1993) Contribution of ions and albumin to the formation and resolution of ischemic brain edema. J Neurosurg 78: 257–266
5. Neely JD, Christensen BM, Nielsen S, Agre P (1999) Heterotetrameric Composition of Aquaporin-4 Water Channels. Biochemistry 38: 11156–11163
6. Soares HD, Cloherty K, McIntosh (1992) Development of prolonged focal cerebral edema and regional cation changes following experimental brain injury in the rat. J Neurochem 58: 1845–1852
7. Newton AC (2001) Protein Kinase C: Structural and Spatial Regulation by Phosphorylation, Cofactors, and Macromolecular Interactions. Chem Rev 101: 2353–2364

Correspondence: Anthony Marmarou, Ph.D., Division of Neurosurgery, Medical College of Virginia, Virginia Commonwealth University, P.O. Box 950508, Richmond, Virginia 23298-0508, U.S.A. e-mail: marmarou@abic.vcu.edu

Acta Neurochir (2003) [Suppl] 86: 265–266

Correlation of lesion volume and brain swelling from a focal brain trauma*

J. Eriskat[2], M. Fürst[1], M. Stoffel[3], and A. Baethmann[1]

[1] Institute for Surgical Research, Klinikum Großhadern, Ludwig-Maximilians-Universität, München, Germany
[2] Department of Neurosurgery, University Würzburg, Würzburg, Germany
[3] Department of Neurosurgery, University Bonn, Bonn, Germany

Summary

Brain edema and secondary growth of a traumatic brain tissue necrosis are important manifestations of secondary brain damage and of prognostic significance in severe head injury. Aim of the current study was to analyze the interdependency of the resulting brain swelling from the size of the focal traumatic lesion.

Male Sprague-Dawley rats were intubated and mechanically ventilated. A trephination was made over the left parietal cortex for induction of a cold lesion. Different injury severities were achieved by variing the contact time of the cooled copper-cylinder and the exposed cortex. Animals were randomized into 12 experimental groups. Hemispheric brain swelling was measured in groups A1–A6 ($n = 4$–8) by gravimetry 24 hrs after lesions of six increasing severity levels. Correspondingly, in animals of groups B1–B6 ($n = 5$–7) the volume of necrosis was planimetrically assessed in histological serial sections of the brain obtained 24 hrs after trauma of different severity.

In groups A1–A6, hemispheric brain swelling (increase in weight) was growing with increasing contact duration of the cold probe with the exposed cerebral cortex, i.e. from $7.7 \pm 0.4\%$ (5 s) to a maximum of $9.9 \pm 0.5\%$ (25 s). Longer contact periodes (30 s) were not further effective to increase hemispheric brain swelling. The contact times and extent of swelling were linearly correlated between 5 s and 25 s ($r = 0.47$; $p < 0.01$). The volume of necrosis in groups B1–B6 increased from 35.7 ± 3.7 mm^3 (5 s) to 106.3 ± 10.3 mm^3 (30 s). There was again a linear correlation between the duration of contact of the cold probe (i.e. injury severity) with the brain cortex and the volume of necrosis ($r = 0.77$; $p < 0.01$).

Accordingly, the lesion volume could be increased in a reproducible manner from 35.7 up to 106.3 mm^3 by extending the contact times of the cooling device and cerebral cortex. Hemispheric swelling, predominantly due to vasogenic brain edema, was expanding in relationship with the volume of necrosis.

Keywords: Secondary brain damage; brain edema; cold injury; necrosis.

Introduction

Brain edema and secondary growth of a traumatic brain tissue necrosis are important manifestations of secondary brain damage [1, 2], and of prognostic significance in severe head injury [3]. Aim of the current study was to analyze the interdependency of the resulting brain swelling from the size of a focal traumatic lesion.

Material and methods

Male Sprague-Dawley rats (270–330 g) were intubated and mechanically ventilated with halothane/N_2O/O_2. Body temperature was monitored by a rectal probe and kept between 37.0 and 37.5 °C by a feedback controlled heating pad. A polyethylene catheter was inserted into the tail artery for measurement of arterial blood pressure and drawing of blood samples. Following fixation of the skull in a stereotactic frame a trephination was made over the left parietal cortex without opening of the dura mater. A highly standardized cortical lesion was induced by focal freezing of the brain utilizing attachment of a computer-controlled metal cylinder (∅ 5 mm, −80 °C) filled with dry ice/acetone [2]. Animals were randomly assigned to 12 experimental groups for analysis of hemispheric swelling and volume of necrosis after different injury severities.

Animals of groups A1–A6 ($n = 4$–8) were subjected to brain lesions of increasing severity. Different injury severities were achieved by varying the contact time of the freezing probe with the exposed cortex from 5 s in group A1 to 30 s in group A6. Animals were allowed to wake up after skin suture. Hemispheric swelling was measured by gravimetry 24 hrs after lesion [5]. For that purpose animals were anesthetized again, sacrificed by exsanguination, and the brains were quickly removed. The hemispheres were separated exactly in the median plane and weighed (fresh weight). After 24 h drying at 110 °C the hemispheres were weighed again (dry weight). Swelling of the traumatized hemisphere was calculated as percent increase of fresh weight over that of the contralateral uninjured hemisphere.

In groups B1–B6, the volume of necrosis was assessed planimetrically in histological serial sections obtained 24 hrs after trauma. As in groups A1 to A6, animals of groups B1 to B6 were exposed to a freezing lesion with increasing contact duration of the cold probe with brain cortex from 5 s (B1) to 30 s (B6). 24 hrs after trauma animals were anesthetized again. Brains were quickly frozen in isopentane (−50 °C). Frozen coronal serial sections of 20 μm thickness were cut with a cryostat and stained according to NISSL. The areas

* Supported by BMBF-Verbund Neurotrauma FKZ 90 30 911

of necrosis were assessed planimetrically. From this data the volume of necrosis was calculated according to Cavalieri's principle.

Values are given as mean ± SEM. Data was examined by linear regression analysis for interdependencies between injury severity and resulting lesion size as well as the developing hemispheric brain swelling.

Results

The focal lesion had no influence on systemic parameters, such as arterial blood pressure, blood gases, and body temperature. In groups A1–A6 hemispheric brain swelling was increasing with prolonged contact times of the probe with brain cortex from 7.7 ± 0.4% (5 s) to a maximum of 9.9 ± 0.5% (25 s). Longer contact times (30 s) were not effective to further increase hemispheric brain swelling. The contact times and extent of swelling were linearly correlated between 5 s and 25 s ($r^2 = 0.47$, $p < 0.01$). The volume of necrosis in groups B1–B6 increased from 35.7 ± 3.7 mm^3 (5 s) to 106.3 ± 10.3 mm^3 (30 s). There was again a linear correlation between contact time (i.e. injury severity) and volume of necrosis ($r^2 = 0.77$; $p < 0.01$). The histomorphometrically assessed hemispheric brain swelling in groups B1–B6 ranged from 8.6 ± 1.0% to 9.1 ± 0.7%, yet without a statistical relationship to the contact time ($r^2 = 0.03$).

Discussion

As shown, the secondary expansion of a traumatic brain tissue necrosis is a function of the severity of the primary insult – the longer the freezing exposure, the larger the resulting volume of necrosis. Apparently, the volume of tissue necrosis is an important determinant of the resulting hemispheric brain swelling, albeit up to an "upper limit" of the lesion size, where even greater degrees of injury severity do not induce additional swelling of the exposed cerebral hemispheres. Such a "ceiling effect" may have various mechanisms. Among others the geometry of the lesion, the distance of the brain tissue necrosis to the cerebral ventricels or other factors might have played a role.

References

1. Baethmann A, Maier-Hauff K, Kempski O, Unterberg A, Wahl M, Schürer L (1988) Mediators of brain edema and secondary brain damage. Crit Care Med 16: 972–978
2. Eriskat J, Schürer L, Kempski O, Baethmann A (1994) Growth kinetics of a primary brain tissue necrosis from a focal lesion. Acta Neurochir (Wien) [Suppl] 60: 425–427
3. Katayama Y, Tsubokawa T, Miyazaki S, Kawamata T, Yoshino A (1990) Oedema fluid formation within contused brain tissue as a cause of medically uncontrollable elevation of intracranial pressure: the role of surgical therapy. Acta Neurochir (Wien) [Suppl] 51: 308–310
4. Klatzo I, Piraux, Laskowski EJ (1958) The relationship between edema, blood-brain barrier and tissue elements in a local injury. J Neuropathol Exp Neurol 17: 548–564
5. Unterberg A, Dautermann C, Baethmann A, Müller-Esterl W (1986) The kallikrein-kinin system as mediator in vasogenic brain edema. Part 3: inhibition of the kallikrein-kinin system in traumatic brain swelling. J Neurosurg 64: 269–276

Correspondence: Jörg Eriskat, Department of Neurosurgery, University Würzburg, Josef-Schneider-Str. 11, D-97080 Würzburg, Germany. e-mail: eriskat.j@nch.uni-wuerzburg.de

Acta Neurochir (2003) [Suppl] 86: 267–269

A role of glial fibrillary acidic protein in hippocampal degeneration after cerebral trauma or kainate-induced seizure

N. Otani, H. Nawashiro, N. Nomura, S. Fukui, N. Tsuzuki, S. Ishihara, and **K. Shima**

Departments of Neurosurgery, National Defense Medical College, Tokorozawa, Japan

Summary

Astrocytes perform a variety of functions in the adult central nervous system (CNS). Recent evidence suggests the robust upregulation of glial fibrillary acidic protein (GFAP) after CNS insult. However, little is known about the role of GFAP in the hippocampal degeneration after brain injury. We herein compared the GFAP knockout (KO) and wild type (WT) mice on the histological and behavioral outcome in response to cerebral trauma or kainic acid (KA)-induced seizure. Although all KO mice showed hippocampal CA3 neuronal degeneration, WT mice did not show any neuronal degeneration in CA3 subfield at 72 hrs after trauma. Thereafter, KO mice showed a higher susceptibility to KA-induced seizures and an increased number of pyknotic CA3 neurons 72 hrs after KA administration. These results indicate that GFAP plays a crucial role in the hippocampal neurodegeneration after CNS insult.

Keywords: Astrocytes; glial fibrillary acidic protein; hippocampus; excitotoxicity; kainate; seizures; traumatic brain injury.

Introduction

Glial fibrillary acidic protein (GFAP), a member of the family of intermediate filament structural proteins, is found predominantly in astrocytes of the central nervous system (CNS). Recent evidence suggest the potential use of GFAP as biological markers of neurotoxicity after CNS insult.

There were close relationships between inflammation, cytokine production, and astrogliosis. Reactive astrocytes induce the expression of a variety of molecules such as neurotrophins and growth factor families. In addition, GFAP-null mice were observed to be more sensitive to traumatic brain injury or cerebral ischemia [7]. We herein examined the role of GFAP for neurotoxicity using traumatic brain injury or kainic acid (KA) administration model to compare the behavioral and histological outcome produced in GFAP knockout (KO) mice with that produced in wild type (WT) mice.

Materials and methods

GFAP-KO mice (B6, 129-G*fap*) were obtained from the Laboratory of Gene Engineering (University of Gunma, Maebashi, Japan). Mice of mixed B6, 129 genetic background heterozygous for the GFAP-KO allete were crossed to generate the homozygous KO mice and WT littermates used for these experiments. Impact acceleration head injury was delivered onto the vertex using a weight drop device as previously described [7]. After trauma, the mice remained motionless for some time and manifested loss of righting reflex; the time required for recovery from these deficits was measured.

Kainic acid (KA), dissolved in phosphate buffered saline, was prepared for induction of limbic seizures. Animals were subcutaneously injected with KA in a dose of 10 ($n = 6$) or 30 mg/kg ($n = 6$). The behavior of animals was continuously observed after KA administration for 5 hrs. Severity of seizures was classified using a modified behavioral scale as previously described [4].

Seventy-two hours after cerebral trauma or KA administration, animals were perfused transcardially with normal saline followed by 4% buffered paraformaldehyde under intraperitoneal anesthesia. The brains were embedded in paraffin. Serial 5 μm thick coronal sections were prepared, and were stained with haematoxylin and eosin. The severity of neuronal damage in the hippocampal CA3 subfield was evaluated by the number of the damaged neurons with appearance of pyknotic changes.

All data are expressed as the mean ± SD. The data was statistically analyzed using an one-way ANOVA followed by a post hoc Fisher's test. A p value of <0.05 was considered to be statistically significant.

Results

GFAP-KO mice had a substantially greater behavioral score than WT mice after KA injection in a dose of 10 mg/kg ($P < 0.01$) (Fig. 1a). Few behavioral responses to KA in WT littermates were observed at 10 mg/kg KA (1.20 ± 0.45). However, there was a variable response to 10 mg/kg of KA injection with approximately all animals showing behavioral abnormalities (grade II to IV) in KO mice (3.16 ± 1.47).

Neuropathological results were consistent with behavioral analysis after KA administration. Damaged

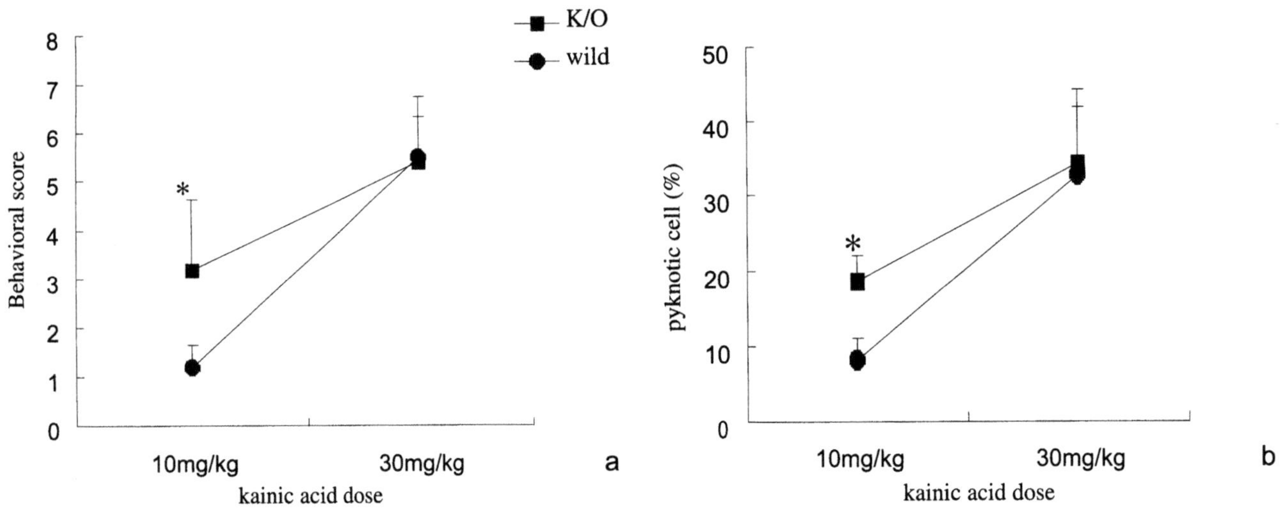

Fig. 1. Comparison of GFAP-KO mice and WT mice on Behavioral score (a) and the number of pyknotic CA3 neurons (b) after KA administration. * $p < 0.01$ compared to WT littermates at 10 mg/kg KA treatment

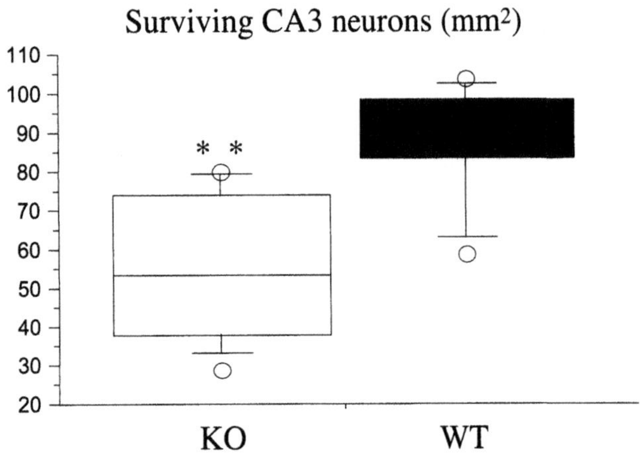

Fig. 2. Comparison of GFAP-KO mice and WT mice on the number of surviving CA3 neurons after cerebral trauma. ** $p < 0.05$ compared to WT littermates after cerebral trauma

pyknotic CA3 neurons increased in a dose-dependent fasion in both WT and KO mice. There were no significant differences between WT and KO mice after KA (30 mg/kg) administration. However, KO mice had a substantially higher number of pyknotic CA3 neurons (41.60 ± 7.16%) than WT (19.60 ± 7.19%) mice after KA (10 mg/kg) administration ($P < 0.01$) (Fig. 1b). On the other hand, KO mice had a substantially lower number of the surviving CA3 neurons than WT mice 72 hrs after impact acceleration brain injury ($P < 0.05$) (Fig. 2).

Discussion

GFAP is known to play an important role for nutritional and structural support for neurons. Under a variety of pathological condition, hypertrophy of astrocytes have been shown using immunohistochemistry with GFAP antibodies in selectively vulnerable lesions [1]. KA-induced seizure resulted in a generalized glial reaction characterized by upregulation of GFAP and vimentin [5]. In this study, we showed that GFAP-KO mice were more sensitive to KA-induced seizure or mechanical trauma than WT littermates. These results suggested that astrocytes normally provided a protective function after CNS insult.

Astrocytes regulate the ionic and neurotransmitter levels in the extracellular fluid, and secrete neurotrophic factors [2]. In addition, astrocytic endfeet closely contact capillary endothelial cells, thus they are well positioned to regulate capillary microcirculation [8]. Although the GFAP-KO mice appear normal with respect to gross morphology including the vascular anatomy [3], there are suggestions that their processes are shorter and thinner and that their BBB is defective [6]. In this manner, astrocytes likely influence the physiological properties of the adjacent neurons and perform a variety of functions in the CNS that contribute to the survival of neurons. Thereafter, it seems reasonable to hypothesize that GFAP might play an important role for neuroprotective function by affecting cerebral vasculature and supporting vessel permeability via the interaction between the endothelial cells and the astrocytic endfeet.

Our results show that GFAP-KO mice are highly sensitive to hippocampal neurodegeneration after traumatic or kainite-induced exitotoxicity. This finding provides a focus for continuing studies of this intermediate filament protein that should contribute to a better understanding the pathophysiology of the CNS and also might lead to improved prevention and treatment of traumatic or KA-induced brain injury.

References

1. Bjorklund H, Olson L, Dahl D *et al* (1986) Short- and long-term consequences of intracranial injections of the excitotoxin, quinolinic acid, as evidenced by GFA immunohistochemistry of astrocytes. Brain Res 371: 267–277
2. Brenner M (1994) Structure and transcriptional regulation of the GFAP gene. Brain Pathol 4: 245–257
3. Gomi H, Yokoyama T, Fujimoto K *et al* (1995) Mice devoid of the glial fibrillary acidic protein develop normally and are susceptible to scrapie prions. Neuron 14: 29–41
4. Hu RQ, Koh S, Torgerson T *et al* (1998) Neuronal stress and injury in C57/BL mice after systemic kainic acid administration. Brain Res 810: 229–240
5. Lenz G, Manozzo L, Gottardo S *et al* (1997) Temporal profiles of the in vitro phosphorylation rate and immunocontent of glial fibrillary acidic protein (GFAP) after kainic acid-induced lesions in area CA1 of the rat hippocampus: demonstration of a novel phosphoprotein associated with gliosis. Brain Res 764: 188–196
6. Liedtke W, Edelmann W, Bieri PL *et al* (1996) GFAP is necessary for the integrity of CNS white matter architecture and long-term maintenance of myelination. Neuron 17: 607–615
7. Nawashiro H, Messing A, Azzam N *et al* (1998) Mice lacking GFAP are hypersensitive to traumatic cerebrospinal injury. Neuroreport 9: 1691–1696
8. Sporbert A, Mertsch K, Smolenski A *et al* (1999) Phosphorylation of vasodilator-stimulated phosphoprotein: a consequence of nitric oxide- and cGMP-mediated signal transduction in brain capillary endothelial cells and astrocytes. Brain Res Mol Brain Res 20: 258–266

Correspondence: Naoki Otani, M.D., Department of Neurosurgery, National Defense Medical College, 3-2 Namiki, Tokorozawa, Saitama 359-8513, Japan. e-mail: grd1405@gr.ndmc.ac.jp

Acta Neurochir (2003) [Suppl] 86: 271–275

Antioxidant properties of the vasoactive endocannabinoid, 2-arachidonoyl glycerol (2-AG)

R. M. McCarron[1], **E. Shohami**[2], **D. Panikashvili**[2], **Y. Chen**[1], **S. Golech**[1], **A. Strasser**[1], **R. Mechoulam**[2], and **M. Spatz**[3]

[1] Resuscitative Medicine Department, Naval Medical Research Center, Forest Glen, MD
[2] Hebrew University, Jerusalem, Israel
[3] Stroke Branch, NINDS, NIH, Bethesda, MD

Summary

Reactive oxygen species (ROS) were shown to play a role in altering blood-brain barrier (BBB) permeability and formation of brain edema induced by trauma and/or ischemia. 2-arachidonoyl glycerol (2-AG), a novel, potent vasodilatory and cytoprotective endocannabinoid has been implicated to act as an antioxidative agent. This study examines: 1) the possible 2-AG modulation of BBB injury and edema formation induced by closed head injury (CHI); and 2) comparable effects between 2-AG and 4-hydroxy-2,2,6,6-tetramethylpiperidine-N-oxyl (TPL), a known antioxidant nitroxide on endothelial Ca^{2+} and cytoskeletal responses to H_2O_2 (ROS). 2-AG treatment reduced the CHI-induced increase in BBB permeability and brain edema. The endothelial H_2O_2-stimulated Ca^{2+} mobilization and cytoskeleton (vimentin) rearrangement was modified by either 2-AG or TPL. These findings provide evidence of 2-AG antioxidant activity and are consistent with the involvement of ROS in the pathomechanism of CHI-induced BBB injury and brain edema.

Keywords: Brain endothelium; endothelin-1; 2-arachidonoyl glycerol; H_2O_2; ischemia; reperfusion; cerebral blood flow.

Introduction

Brain edema has been well recognized as one of the important, early sequelae of ischemic and traumatic tissue injury. The event is touched off by perturbance in the cellular membranes. These disturbances lead to the release or activation of various cellular constituents [i.e., unsaturated fatty acids, reactive oxygen species (ROS), neurotransmitters, cytochrome c, kinases, endothelin-1 (ET-1) and nitric oxide] that normally are metabolized and utilized by brain cells to maintain homeostasis [8]. However, their excessive efflux, combined with either decreased or increased enzymatic activities (i.e., Na^+, K^+-ATPase or phospholipase A and C, respectively) contributes to a variety of noxious effects such as alteration of blood-brain barrier (BBB) permeability, brain edema, apoptosis and necrosis. Studies in vivo have shown that endogenously produced free radical species such as H_2O_2, OH^- and O^-, among other factors, play a major role in the pathomechanism of these phenomena. This conclusion was substantiated by antioxidative treatments that prevented or diminished the development of BBB and/or edematous tissue changes [1, 13, 18]. Among these studies, the nitroxide 4-hydroxy-2,2,6,6-tetramethylpiperidine-N-oxyl (TPL), a most efficient antioxidant was shown to reduce the altered BBB permeability and the production of edema induced by brain trauma or ischemia. It has also been demonstrated that formation of ROS or exposure of cultured endothelial cells (EC), the main constituent of BBB, to the individual members of ROS alters the permeability of these cells [9, 10]. The increased EC permeability was frequently associated with changes in intracellular Ca^{2+} and cytoskeleton arrangement [9]; permeability changes in the EC can also be prevented by treatment with similar antioxidant agents [7]. However, little information is available regarding the effect of antioxidants on ROS-induced Ca^{2+} mobilization and derangements of endothelial cytoskeleton.

Recently, it has been shown that 2-arachidonoyl glycerol (2-AG), the naturally occurring cannabinoid and potent vasodilator, is cytoprotective. It may also act as an antioxidant since it suppressed the endotoxin-induced production of TNFα and ROS [6]. Therefore, the focus of this study has been to: 1) determine the possible 2-AG prevention of BBB permeability changes in addition to the formation of edema reported in brain trauma; and 2) shed some light on 2-

AG action as an antioxidant by comparing the 2-AG effects with those of nitroxide, a known antioxidant, on intracellular Ca^{2+} and cytoskeletal changes induced in human brain EC by H_2O_2.

Materials and methods

Experiments in vivo

Animals. Male Sabra mice (Hebrew University strain) weighing 32–35 g were used for induction of trauma according to the method of Beit-Yannai [1]. Sham-operated mice were used as controls. All experiments reported herein were conducted according to the principles set forth in the "Guide for the Care and Use of Laboratory Animals," Institute of Laboratory Animals Resources, National Research Council, National Academy Press, 1966 and were approved by the Animal Care Committee of the Hebrew University.

Treatment. 2-AG was injected (0.1, 5, or 10 mg/kg, i.v.) 15 min after CHI, for the edema study, and 5 mg/kg for the effect on BBB.

Evaluation of edema. Cerebral edema was evaluated by determining the tissue water content in the injured brain, as previously described [2]. A tissue segment (about 20 mg), taken from the area bordering the lesion (no more than 1 mm from lesion), or from the contralateral hemisphere was weighed to yield wet weight (WW). After drying in a desiccator oven for 24 hr at 95 °C it was reweighed to yield dry weight (DW). The percentage of water in the tissue was calculated as: $\%H_2O = [(WW - DW) \times 100]/WW$. Edema was evaluated at 24 h, a time which was shown to exert maximal edema.

Evaluation of BBB permeability. The integrity of the BBB was evaluated using Evans blue dye (2% w/v in saline), injected i.v. 3 h after closed head injury (CHI) and circulating for at least 60 min. Then, under ether anaesthesia the intravascular dye was removed by perfusion through the left ventricle until a colorless perfusion fluid was obtained from the right atrium. The separate brain hemispheres were homogenized in 50% trichloroacetic acid (w/v) and centrifuged (10,000 rpm for 20 min). The supernatant was diluted 4-fold with ethanol and an aliquot was diluted 1-to-3-fold with the solvent (50% trichloroacetic acid/ethanol 1 : 3) for fluorescent measurement (excitation/emission wavelengths of 620/680 nm, respectively) on a Perkin-Elmer LS-5 fluoro-spectrophotometer [3]. Calculations were based on external standards in the solvent (100–500 ng/ml).

Experiments in vitro

Cell culture. Isolated brain microvessels and capillaries were used for the dissociation and cultivation of human brain endothelial cell (HBEC) as previously described [16]. The endothelial cell lines (derived from at least six different brains; passages 7–15) were >95% Factor VIII-positive. Cells were grown on gelatin (1%)-coated 24-well plates at 37 °C in humidified 5% CO_2/air for 48 hours.

Treatment. The HBEC response (Ca^{2+} measurements) was evaluated by exposing the cells to various concentrations of H_2O_2 (2.5–10 mM) for 30 sec. The antioxidant effect was tested with TPL (0.01, 0.1 and 1 mM for 60 min) or 2-AG (EC50 = 50 μM for 5 min) prior to the addition of H_2O_2 (10 mM).

$[Ca^{2+}]_i$ measurements. After washing (3×) in solution containing 137 mM NaCl, 5 mM KCl, 1 mM $MgCl_2$, 25 mM sorbitol, 10 mM HEPES, and 3 mM $CaCl_2$ (pH 7.0), HBEC were incubated with 2.5 μM fluorescent probe Fluo-3/AM for 90 min at 37 °C. Fluorescence was measured using a fluorescein filter pair (excitation, 485 ± 20 nm; emission, 530 ± 20 nm) as previously described [2]. Changes in $[Ca^{2+}]_i$ are expressed as fluorescence intensity ratio determined as: [Experimental fluorescence value (F_e) – basal fluorescence value (F_0)]/F_0 × 100%; the fluorescence change in presence of ET-1 alone was considered as 100% [3].

Immunocytochemistry. The expressions of actin and vimentin were determined by immunocytochemical analyses of HBEC (grown on coverslips) that were fixed (4% paraformaldehyde), permeabilized (0.1% Triton X-100) and stained with phalloidin or monoclonal antibodies to vimentin, respectively as previously described [3].

Statistics. Results were analyzed by Student's t test or one-way ANOVA followed by Fisher's protective least squares difference (PLSD) test.

Results

Brain edema and BBB permeability

At 24 h post CHI, the time for maximal edema, 2-AG (at all doses tested) significantly reduced water accumulation in the contused hemisphere by about 50% as compared to untreated animals. Also, BBB disruption was significantly attenuated after 2-AG treatment. In non-treated mice, extravasation of Evans blue into the contused hemisphere (241 ± 99 ng dye/g tissue) was 6-fold higher than was observed in the contralateral hemisphere (41 ± 19 ng dye/g tissue). Whereas, in 2-AG-treated mice, the level of extravasation of Evans blue into the contused hemisphere was only approximately 4-fold higher as compared to the contralateral hemisphere (142 ± 59 vs 33 ± 17, respectively). The values in the contused hemispheres of non-treated and 2-AG-treated animals were significantly different ($p = 0.04$).

Calcium and cytoskeleton

To elucidate the possible participation of HBEC in the observed altered BBB permeability and formation of edema induced by brain trauma or ischemia, experiments examined the effect of ROS (e.g., H_2O_2) on Ca^{2+} and cytoskeleton of these cells. H_2O_2 dose-dependently increased Ca^{2+} mobilization; the highest level was seen with 10 mM H_2O_2 (Fig. 1). This effect was the same in cells incubated in Ca^{2+}-free medium, but not in cells cultured in the

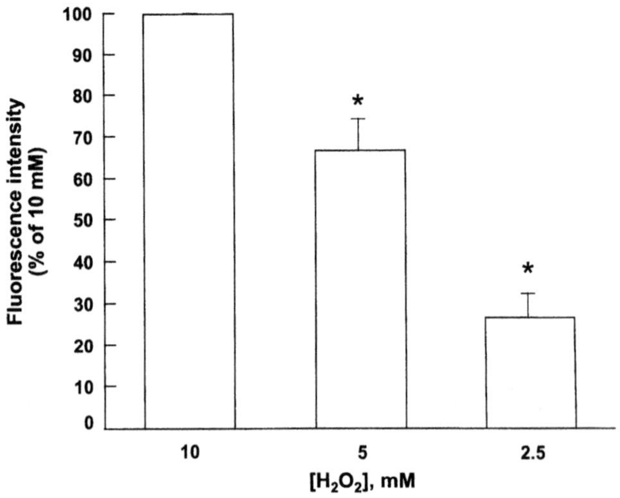

Fig. 1. Dose-dependent stimulation of Ca^{2+} mobilization in HBEC by H_2O_2. Ca^{2+} mobilization observed in the presence of 10 mM H_2O_2 is expressed 100 percent; *, significant difference from control ($p < 0.05$)

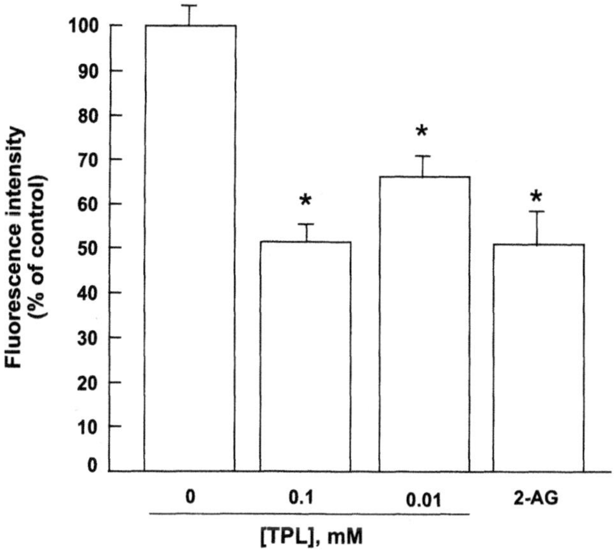

Fig. 2. Effect of TPL and 2-AG on H_2O_2-induced Ca^{2+} mobilization. Fluorescence intensity of HBEC exposed to 10 mM H_2O_2 alone are expressed as 100%. TPL (dose-dependently) and 2-AG (50 μM) reduced the H_2O_2-induced Ca^{2+} mobilization; *, significant difference from control ($p < 0.05$)

presence of BAPTA/AM. Pretreatment with TPL decreased the H_2O_2-induced mobilization in a dose-dependent manner (Fig. 2). A similar effect was seen with 2-AG pretreatment (50 μM 2-AG reduced the H_2O_2-stimulated Ca^{2+} mobilization).

Endothelial cytoskeletal changes were also observed in response to H_2O_2 treatment. These changes were manifested by increased thickness of vimentin filaments, but not actin fibers (see Fig. 3B). Pretreatment with either TPL or 2-AG reduced the H_2O_2 effect (Fig. 3C and D, respectively).

Discussion

The findings of this study unequivocally demonstrate that 2-AG decreased trauma-induced BBB permeability changes and edema formation in mice. These results are comparable with those reported after TPL treatment in the same model of trauma (i.e., CHI) in rats and in ischemia and reperfusion injury in both rats and gerbils [1, 13, 18, unpublished observation]. These results also clearly show that 2-AG or TPL are able to reduce Ca^{2+} mobilization and prevent cytoskeleton rearrangements (vimentin) induced by H_2O_2.

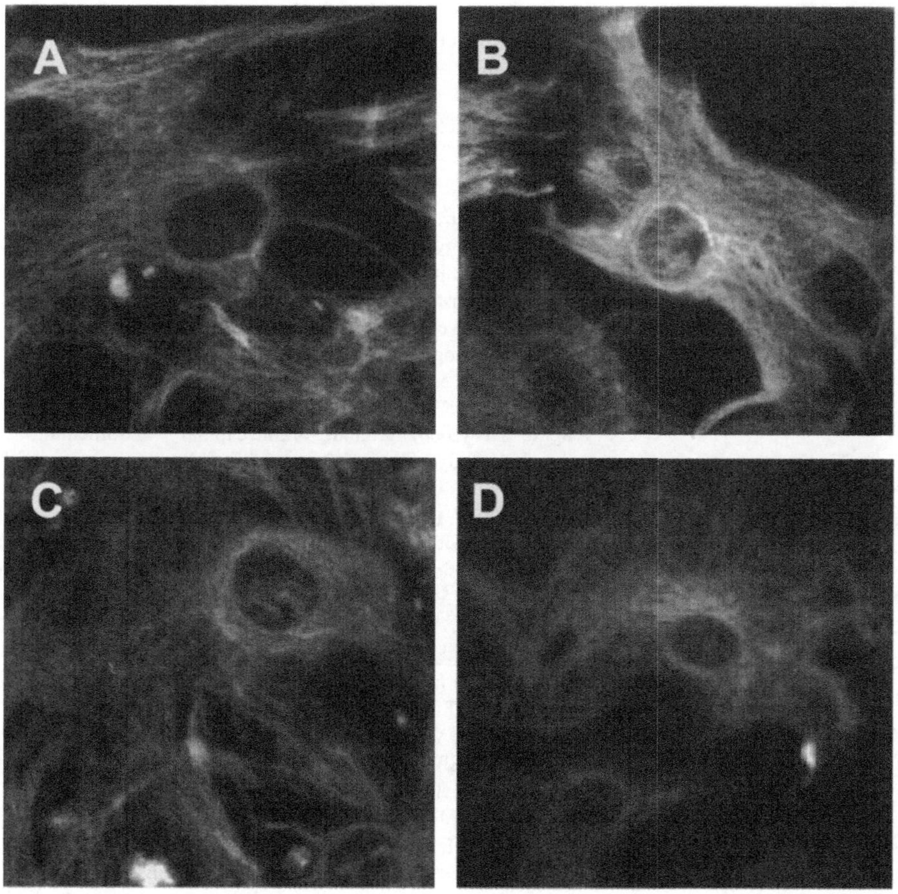

Fig. 3. Effect of H_2O_2 on vimentin filaments. HBEC were untreated (A) or treated with 5 mM H_2O_2 alone (B) or in the presence of either 1 mM TPL (C) or 50 μM 2-AG (D). All cells were treated with antibody to vimentin and examined with a Zeiss Axiovert S100 fluorescence microscope (magnification ×40)

2-AG is an endogenous cannabinoid produced in various organs (brain, gut) and cells [monocytes, platelets, endothelial cells (EC)] [4, 5, 11]. These substances elicit neuromodulatory and cardiovascular effects, which are mediated through cannabinoid (CB) receptors CB1, CB2 or vanilloid (VRA1) receptors [4, 11, 12, 14, 17]. The CB1 receptors have been localized in brain and peripheral organs; CB2 receptors were demonstrated in the periphery, especially immune cells; and VRA1 presence has been shown in human and rat brains. The endothelium, which synthesizes 2-AG, also expresses CB1 and CB2 receptors. Both substances have vasoactive (vasodilatory) and cytoprotective properties [4, 5, 11, 12, 14, 17]. 2-AG elicits a variety of responses in HBEC. It is capable, by itself, to stimulate Ca^{2+} uptake; this effect can be reduced by pretreatment with either Genistein or SR141716A (selective inhibitors of tyrosine kinase and CB1 receptors, respectively) [3, unpublished observation]. 2-AG was also demonstrated to abrogate the ET-1-induced Ca^{2+} mobilization and cytoskeleton (actin and vimentin) rearrangements in HBEC, which is partly mediated by CB1 receptor, G-protein, phosphoinositol signal transduction pathway, and Ca^{2+}-activated K^+ channels [3].

The TPL is a water-soluble molecule that readily crosses biological cell membranes including BBB. It has properties of a stable radical that can react with various ROS. TPL also has a low toxicity and mimics superoxide activity thereby protecting cells against oxidative stress. This substance has been shown not only to ameliorate the increased BBB permeability and brain edema, but also improve the postischemic hypoperfusion in both rats and gerbils [1, 13, 18, unpublished observation].

The observed H_2O_2-stimulated cytoskeleton (vimentin) effects are similar to those reported with ET-1, except that ET-1 also induced changes in actin [3]. The effect of H_2O_2 on vimentin, but not actin, is noteworthy since vimentin was recently reported to be linked with endothelial matrix adhesion in peripheral human microvascular endothelial cells [6]. H_2O_2 was also reported to preferentially oxidize thiol groups of vimentin in synovial cells [15]. The exact mechanism for H_2O_2 induction of Ca^{2+} mobilization associated with vimentin as well as their prevention by TPL or 2-AG remains to be clarified. At the present time there is also no clear understanding of the mechanisms involving Ca^{2+} mobilization and either increased cellular permeability or rearrangement of actin filaments induced by ROS in peripheral endothelium. The mitogen-activated protein kinase, tyrosine kinase, and GTPase have been implicated to play role(s) in the mechanism responsible for these phenomena [9].

The reported 2-AG inhibition of ROS, among other processes (i.e., inhibition of glutamate release, improved glucose utilization and counteracting vasoconstrictive effects of ET-1), has been implicated in cytoprotective effects [11, 12]. Thus, the comparable antioxidant effects between the TPL and 2-AG on HBEC subjected to H_2O_2 provide evidence of 2-AG antioxidant activity. In addition, the in vitro studies substantiate the participation of ROS in the mechanism responsible for CHI-induced BBB injury and formation of brain edema described here.

Acknowledgments

The authors wish to thank Joliet Bembry (Stroke Branch, NINDS, NIH) for culturing EC and Fred Lenz (The Johns Hopkins University School of Medicine) for brain tissue. This work was partially supported by Work Unit funding obtained from the Office of Naval Research (602233N.333.120.A0102). The opinions expressed in this paper are those of the authors and do not reflect the official policy of the Department of Navy, Department of Defense, of the U.S. Government.

References

1. Beit-Yannai E, Zhang R, Trembovler V, Samuni A, Shohami E (1996) Cerebroprotective effect of stable nitroxide radicals in closed head injury in the rat. Brain Res 717: 22–28
2. Chen Y, Constantini S, Trembovler V, Weinstock M, Shohami EJ (1996) An experimental model of closed head injury in mice: pathophysiology, histopathology, and cognitive deficits. J Neurotrauma 13: 557–568
3. Chen Y, McCarron RM, Azzam N, Reutzler C, Lenz FA, Spatz M (2000) Human brain capillary endothelium: 2-Arachidonoglycerol (Endocannabinoid) interacts with Endothelin-1. Circ Res 87: 323–327
4. Di Marzo V (1998) Endocannabinoids' and other fatty acid derivatives with cannabimimetic properties: biochemistry and possible physiopathological relevance. Biochim Biophys Acta 1392: 153–175
5. Gallily R, Breuer A, Mechoulam R (2000) 2-Arachidonylglycerol, an endogenous cannabinoid, inhibits tumor necrosis factor-alpha production in murine macrophages, and in mice. Eur J Pharmacol 406: R5–7
6. Gonzales M, Weksler B, Tsuruta D, Goldman RD, Yoon KJ, Hopkinson SB, Flitney FW, Jones JCR (2001) Structure and function of a vimentin-associated matrix adhesion in endothelial cells. Mol Biol Cell 12: 85–100
7. Haj-Yehia AI, Nassar T, Assaf P, Nassar H, Anggard EE (1999) Effects of the superoxide dismutase-mimic compound TEMPOL on oxidant stress-mediated endothelial dysfunction. Antioxid Redox Signal 1: 221–232
8. Lewan A, Matz P, Chan PH (2000) Free radical pathway in CNS injury. J Neurotrauma 17: 871–890

9. Lum H, Roebuck KA (2001) Oxidant stress and endothelial cell dysfunction. Am J Physiol 280: C719–C741
10. McCarron RM, Uematsu S, Merkel N, Long D, Bembry J, Spatz M (1990) The role of arachidonic acid and oxygen radicals on cerebromicrovascular endothelial permeability. Acta Neurochir (Wien) [Suppl] 51: 61–64
11. Mechoulam R, Spatz M, Shohami E (2002) Endocannabinoids and neuroprotection. Sci STKE 129: RE5
12. Panikashvili D, Simeonidou C, Ben-Shabat S, Hanus L, Breuer A, Mechoulam R, Shohami E (2001) An endogenous cannabinoid (2-AG) is neuroprotective after brain injury. Nature 413: 527–531
13. Rak R, Chao DL, Pluta RM, Mitchell JB, Oldfield EH, Watson JC (2000) Neuroprotection by the stable nitroxide Tempol during reperfusion in a rat model of transient focal ischemia. J Neurosurg 92: 646–651
14. Randall MD, Kendall DA (1998) Endocannabinoids: a new class of vasoactive substances. Trends Pharmacol Sci 19: 55–58
15. Rogers KR, Morris CJ, Blake DR (1991) Oxidation of thiol in the vimentin cytoskeleton. Biochem J 275: 789–791
16. Spatz M, Kawai N, Merkel N, Bembry J, McCarron RM (1997) Functional properties of cultured endothelial cells derived from large microvessels of human brain. Am J Physiol 272: C231–C239
17. Sugiura T, Kodaka T, Nakane S, Kishimoto S, Kondo S, Waku K (1998) Detection of an endogenous cannabimimetic molecule, 2-arachidonoylglycerol, and cannabinoid CB1 receptor mRNA in human vascular cells: is 2-arachidonoylglycerol a possible vasomodulator? Biochem Biophys Res Commun 243: 838–843
18. Yasuma Y, Strasser A, Ruetzel C, McCarron RM, Spatz M (1997) The effect of nitric oxide inhibition on ischemic brain edema. Acta Neurochir (Wien) [Suppl] 70: 202–205

Correspondence: Maria Spatz, M.D., National Institutes of Health, NINDS, Stroke Branch, 36 Convent Drive, MSC 4128, Bethesda, Maryland 20892-4128. e-mail: spatzm@ninds.nih.gov

Acta Neurochir (2003) [Suppl] 86: 277–279

Very early expression of vascular endothelial growth factor in brain oedema tissue associated with brain contusion

R. Suzuki[1], **N. Fukai**[2], **G. Nagashijma**[1], **J.-I. Asai**[1], **H. Itokawa**[1], **M. Nagai**[1], **T. Suzuki**[3], and **T. Fujimoto**[1]

[1] Department of Neurosurgery, Showa University, Fujigaoka Hospital, Japan
[2] Department of Cell Biology, Harvard Medical School, Boston, MA
[3] Laboratory of Histochemistry, Showa University, Fujigaoka Hospital, Japan

Summary

Background. Brain oedema associated with cerebral contusion can be life-threatening. Mechanisms of the development of brain oedema are still unclear.

Method. We investigated the expression of vascular endothelial growth factor (VEGF) and its receptor VEGFR-2 (KDR/Flk-1) in the contusional brain tissue obtained during neurosurgery from 5 patients.

Findings. VEGF is expressed in some but not all the astrocytes, and KDR/Flk-1 is expressed in vascular endothelial cells in the contusional tissue as early as 3 hours after onset.

Conclusion. The results suggested that the VEGF is induced in the contusional tissue in the very early period after onset, and that it increases capillary permeability via KDR/Flk-1 resulting in vasogenic type brain oedema.

Keywords: Cerebral contusion; Oedema; VEGF; KDR.

Introduction

Contusional brain damage can be life-threatening and may progress during the first few days. Mixed brain injury types such as primary brain damage, ischemia, and haemorrhage aggravate the tissue damage. Brain oedema also develops surrounding the damaged tissue and elevates the intracranial pressure causing a poor prognosis. Mechanisms of contusional brain oedema development are still unclear. However, mixed mechanisms including vasogenic, ischemic, and neurotoxic type of oedema may contribute [7]. Recent studies have proposed that vasogenic oedema may be related to the expression of vascular endothelial growth factor (VEGF) in various conditions [1, 8, 9, 10]. In the present study, we have studied the expression of VEGF and VEGF receptor-2 (Flk-1/KDR) in the oedematous tissue associated with brain contusion.

Method

We examined brain tissue samples obtained at surgery from 5 patients with brain contusion. Surgical procedures were for haematoma removal and the internal decompression of the contusional brain. Patients' conditions are listed in Table 1. The time period from onset to the operation varied from 3 hours to 72 hours. The tissue samples obtained from the lobectomized contusional brains were fixed with 4% paraformaldehyde. Four micron sections prepared from paraffin-embedded tissue blocks were mounted on silanized slides (Dako, Kyoto, Japan). The specimens were stained with haematoxylin and eosin (H&E), and immunohistochemistry, using a polyclonal antibody against the *N*-terminal region of human VEGF165 (Oncogene Science, Cambridge, MA) and the human VEGFR-2 (Flk-1/KDR) (R&D Systems Inc., MN), was performed.

Results

Histological studies confirmed by H&E staining showed brain contusion, micro haemorrhages, and many vacuolated spaces compatible with brain oedema in all cases. Immunohistochemistry showed VEGF positive staining in some astrocytes but not all of them, located in the oedematous tissue in 4 cases, and Flk-1/KDR positive staining in the vascular en-

Table 1. *Case presentation*

Case	Age/sex	GCS at onset	GCS at operation	Region	Hrs
1	48/F	8	7	Temporal	72
2	59/F	15	12	Frontal	5
3	23/M	8	3	Temporal	3
4	25/F	13	5	Frontal	48
5	18/M	5	4	Frontal	5

GCS Glasgow Coma Scale, *Region* regions from where the tissue was obtained, *Hrs* Time (h) at the start of the operation after onset.

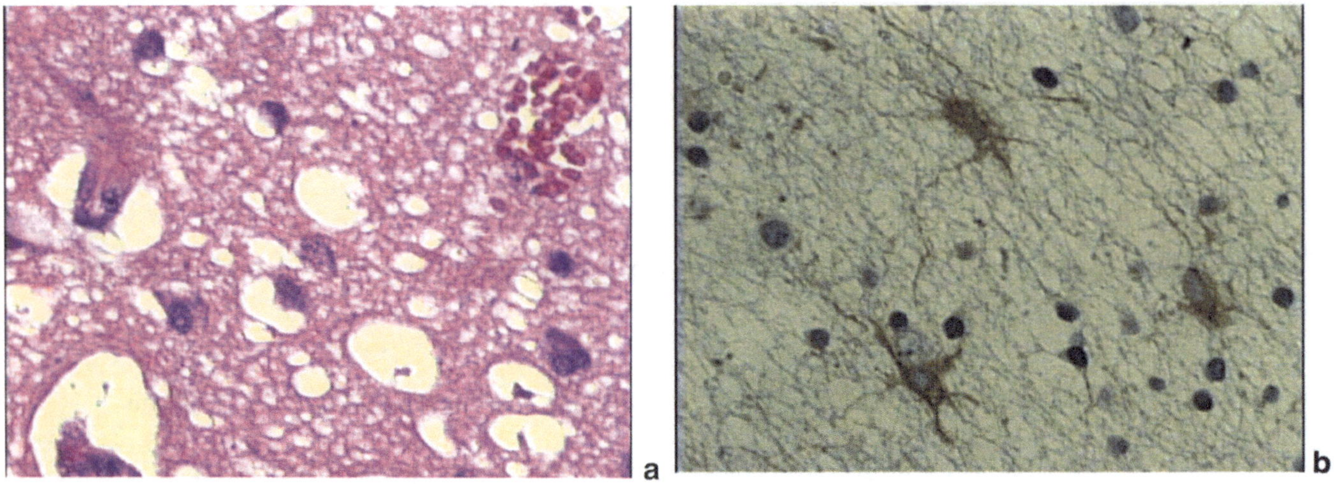

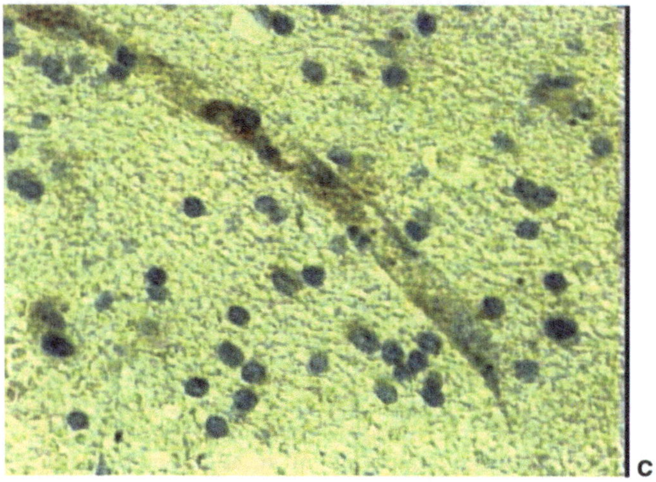

Fig. 1. Haematoxylin and eosin (H&E) staining (a), VEGF staining (b), and KDR/Flk-1 staining (c) in case 1. HE staining shows the vacuolated tissue compatible with vasogenic brain oedema (a). VEGF expressed in some astrocytes but not all of them in the oedematous tissue (b). KDR/Flk-1 expressed in vascular endothelial cells in the oedematous tissue (c)

dothelial cells located in the oedematous tissue in 4 cases (Fig. 1-A, B, C). In one case VEGF staining was negative in astrocytes and Flk-1/KDR staining was negative in endothelial cells. The results are shown in Table 2.

Table 2. *Results*

Case	Age/sex	Hrs	VEGF expression	Flk-1/KDR
1	48/F	72	positive in astrocytes	positive in VEC
2	59/F	5	positive in astrocytes	positive in VEC
3	23/M	3	positive in astrocytes	positive in VEC
4	25/F	48	negative	negative
5	18/M	5	positive in astrocytes	positive in VEC

Hrs Time (*h*) at the start of the operation after onset, *VEG* vascular endothelial cells.

Discussion

Primary brain damage due to the cerebral contusion progresses gradually during the first few days after injury. Brain oedema may add to further increase intracranial pressure resulting in a poor prognosis. Thus, the mechanism of brain oedema development needs to be clarified to obtain a better prognosis in patients with cerebral contusion.

VEGF is a potent mediator of angiogenesis and increase of vascular permeability [2]. VEGF has been considered to play an important role in the formation of vasogenic brain oedema, including peri-tumoural brain oedema and inflammatory brain oedema, and ischemic brain oedema due to its ability to increase vascular permeability [1, 8, 9, 10]. Endothelial cells express two related VEGF receptor tyrosine kinases, KDR (kinase-insert domain containing receptor, or VEGFR-2 also known as Flk-1) and Flt-1 (*fms*-like tyrosine kinase, or VEGF-1) [4, 5]. Hendrik recently demonstrated that KDR, but not Flt-1, up-regulation is responsible for the increase of vascular permeability by VEGF [4].

The immunohistological study revealed that VEGF expressed in some astrocytes located in the oedematous tissue as early as 3 hours after onset, and that KDR/Flk-1 is also expressed in the vascular endothelial cells of the same oedematous tissue. These results can provide an explanation for severe head injury causing primary brain damage resulting from tissue hypoxia or ischemia. Secondary hypoxia may induce VEGF in the astrocytes and up-regulate the induction of KDR/Flk-1 in endothelial cells located in damaged tissue. These reactions may increase vascular permeability resulting in the development of vasogenic brain oedema. Our results suggest that the permeability change occurs in the very early period after onset. This is consistent with clinical MRI data that showed gadolinium DTPA-enhancement in the peri-contusional tissue as early as 2 to 4 hours after onset [6].

Conclusion

We demonstrated the very early expression of VEGF and its receptor KDR/Flk-1 in contusional brain tissue in neurosurgical patients. VEGF appeared mainly in the astrocytes, and KDR/Flk-1 appeared in vascular endothelial cells. This suggests that expression of VEGF may play a role in the development of vasogenic brain oedema in contusional brain tissue.

References

1. Bruggen N, Thibodeaux H, Palmer JT, Lee WP, Fu L, Cairns B, Tumas D, Gerlai R, Williams SP, Campagne ML, Ferrara N (1999) VEGF antagonism reduces edema formation and tissue damage after ischemia/reperfusion injury in the mouse brain. J Clin Invest 104: 1613–1620
2. Connolly DT, Heuvelman DM, Nelson P, Olander JV, Eppley BL, Delfino JJ, Siegel NR, Leimgruber RM, Feder J (1989) Tumor vascular permeability factor stimulates endothelial cell growth and angiogenesis. J Clin Invest 84: 1470–1478
3. Hayashi T, Abe K, Suzuki H, Itoyama Y (1997) Rapid induction of vascular endothelial growth factor gene expression after transient middle cerebral occlusion in rats. Stroke 28: 2039–2044
4. Hendrik G, Kowalski J, Li B, LeCouter J, Maffat B, Zioncheck TF, Pelletier N, Ferrara N (2001) Analysis of biological effects and signaling properties of Flt-1 (VEGFR-1) and KDR (VEGFR-2). J Biol Chem 276: 3222–3230
5. Huiyan Z, Sanyal S, Mukhopadhyay D (2001) Tyrosine residues 951 and 1059 of vascular endothelial growth factor receptor-2 (KDR) are essential for vascular permeability factor/vascular endothelial growth factor-induced endothelial migration and proliferation, respectively. J Biol Chem 276: 32714–32719
6. Kushi H, Katayama Y, Shibuya T, Tsubokawa T, Kuroha T (1994) Gadolinium-enhanced magnetic resonance imaging of cerebral contusion. Acta Neurochir (Wien) [Suppl] 60: 472–474
7. Marmarou A (1994) Traumatic brain edema: an overview. Acta Neurochir (Wien) [Suppl] 60: 421–424
8. Provias J, Claffey K, delAguila L, Lau N, Feldkamp M, Guha A (1997) Meningiomas: role of vascular endothelial growth factor/vascular permeability factor in angiogenesis and peritumoral edema. Neurosurgery 40: 1016–1026
9. Vaquero J, Zurita M, Marales C (2001) Possible role for vascular permeability factor in the pathophysiology of vasogenic oedema association to brain abscess. Acta Neurochir (Wien) 143: 1039–1040
10. Vaquero J, Zurita M, Oya S (2001) Expression and significance of vascular permeability factor in tumour infiltrating lymphocytes of brain metastasis. Acta Neurochir (Wien) 143: 153–157

Correspondence: Ryuta Suzuki, Department of Neurosurgery, Showa University, Fujigaoka Hospital, Japan. e-mail: ryuta@med.showa-u.ac.jp

Acta Neurochir (2003) [Suppl] 86: 281–285

Changes of perivascular macrophages in the process of brain edema induced by cold injury

J. Wang[1,2], **K. Takeuchi**[1], and **S. Ookawara**[1]

[1] Department of Anatomy, Jichi Medical School, Tochigi, Japan
[2] Department of Anatomy, China Medical University, Shenyang, PR China

Summary

Perivascular macrophages are considered as cerebral scavenger cells under physical and pathological conditions. In this study, we tried to examine changes of perivascular macrophages, especially changes of the characteristic lysosomal inclusion bodies that are rich in hydrolytic enzymes, in the process of brain edema induced by cold injury. Wistar male rats aged 4 months were treated with dry ice for 20 minutes through a drilled hole at the parietal bone. At different time points after the cold injury, cerebral cortex was excised and the immunoreaction for cathepsin D, one kind of lysosomal protease, was examined by post-embedding immuno-electron microscopy. The reactions of cathepsin D were located in the inclusion bodies of perivascular macrophages. At 5 and 10 hours after cold injury, the reactions increased dramatically. Then the reactions inclined to decrease, and reached the minimum at 1 week after cold injury. The reactions seemed to recover at 2 weeks after cold injury. The changes of cathepsin D reactions suggest that the function of perivascular macrophages as scavenger cells were activated in the early phase of the process of brain edema, their later declines might be caused by severe pathological conditions.

Keywords: Perivascular macrophage; brain edema; cold injury; cathepsin D; immuno-electron microscopy.

Introduction

Perivascular macrophages, or Mato's fluorescent granular perithelial cells, are found within the central nervous system (CNS) in the perivascular space [7]. The cells express the phenotypes common to the macrophage lineage, including ED2 antigen [5], the scavenger receptor [10] and MHC-class II antigens [8]. The cells can take up horseradish peroxidase and ferritin administered via the femoral vein or cerebral ventricles [10]. Accompanying their function of pinocytosis, the cells have many large-sized lysosomal inclusion bodies. The inclusion bodies contain large quantities of lysosomal enzymes, such as acid phosphatase and protease. Moreover, the cells are activated in models of CNS inflammation and autoimmune disease [13]. The scavenger and immunoregulatory function is putting the perivascular macrophage on an important role in the CNS.

The cerebral cortical cold-injury model [11], one kind of cortical contusion, can induce dramatic blood-brain barrier breakdown and evident brain edema formation. The changes of perivascular macrophages in rat cerebral cortex of this model have been observed by transmission electron microscopy [3, 9]. Perivascular macrophages may remove the edema fluid by pinocytosis, and may break down protein molecules in the lysosomal inclusion bodies. However, the changes of lysosomal protease in the inclusion bodies have not been reported. The aim of this research is to reexamine the changes of perivascular macrophages in cold-induced brain edema. Especially, cathepsin D, one kind of lysosomal protease, was mainly observed by quantitative immunogold electron microscopic method. The functions of perivascular macrophages are also discussed.

Material and methods

A total of 40 Wistar male rats, aged 4 months, was used in this study. Dry ice was used for 20 minutes through a drilled hole at the parietal bone of rat. At 5 hours, 10 hours, 1 day, 2 days, 3 days, 1 week and 2 weeks after the operation, animals were perfused from the heart with a fixative containing 4% paraformaldehyde in 0.1 M phosphate buffer (pH 7.4). Cerebral cortex near the injury was excised, and was prepared for immunofluorescent and post-embedding immunogold labeling.

For dual-label immunofluorescent microscopy, the samples were immersed in a fixative containing 4% paraformaldehyde for 4 h at 4 °C, and coronal sections of 14 µm were cut in a cryostat. Sections were incubated by anti-cathepsin D (Santa Cruz Biotechnology, SC-

6487) and anti-ED2 (Serotec, MCA342R) for 1 hour. After rinsing, sections were incubated with Fluorescein Anti-Goat IgG (Vector Laboratories, FI-5000) and Texas Red Anti-Mouse IgG (Vector Laboratories, TI-2000) for 1 hour. Negative control was effected without primary antibodies. Sections were examined by OLYMPUS AX80TR fluorescent microscope.

For post-embedding immunogold labeling, the samples were immersed in a fixative containing 4% paraformaldehyde, 0.25% glutaraldehyde and 0.1% tannic acid for 4 h at 4 °C after 20 s microwave irradiation. After the procedure that has been reported by Berrymann and Rodewald [1], samples were embedded in LR Gold resin (London Resin Company), and were polymerized for 48 h with UV-light (360 nm) at −20 °C. Ultrathin sections cut by a microtome, were picked on collodium-coated nickel grids. The gold-particle-label procedures were performed as reported by Berrymann [1]. Sections were incubated subsequently by following solution: anti-cathpsin D, biotiylated anti-goat IgG (Vector Laboratories, BA-9500) and 15 nm gold EM Streptavidin (BBInternational, 4039). After being stained with 2% aqueous osmium tetroxide, and Reynold's lead citrate subsequently, sections were examined by JEM – 2000EX electron microscope. For negative controls the grids were incubated in dilution buffer instead of the primary antibody. More than 12 inclusion bodies of each group were taken randomly. The numbers of gold particles in inclusion bodies were counted. The average numbers of gold particles per μm^2 were calculated.

Results

Dual-label fluorescent microscopy

In the cerebral cortex of normal control, reactions of ED2 were observed in the perivascular areas of arterioles or venules, and were expressed in the perivascular macrophages. No reaction could be observed in the microglia in the brain parenchyma and the pericytes around capillaries (Fig. 1A). Strong reactions of cathepsin D were located in the ED2-positive cells,

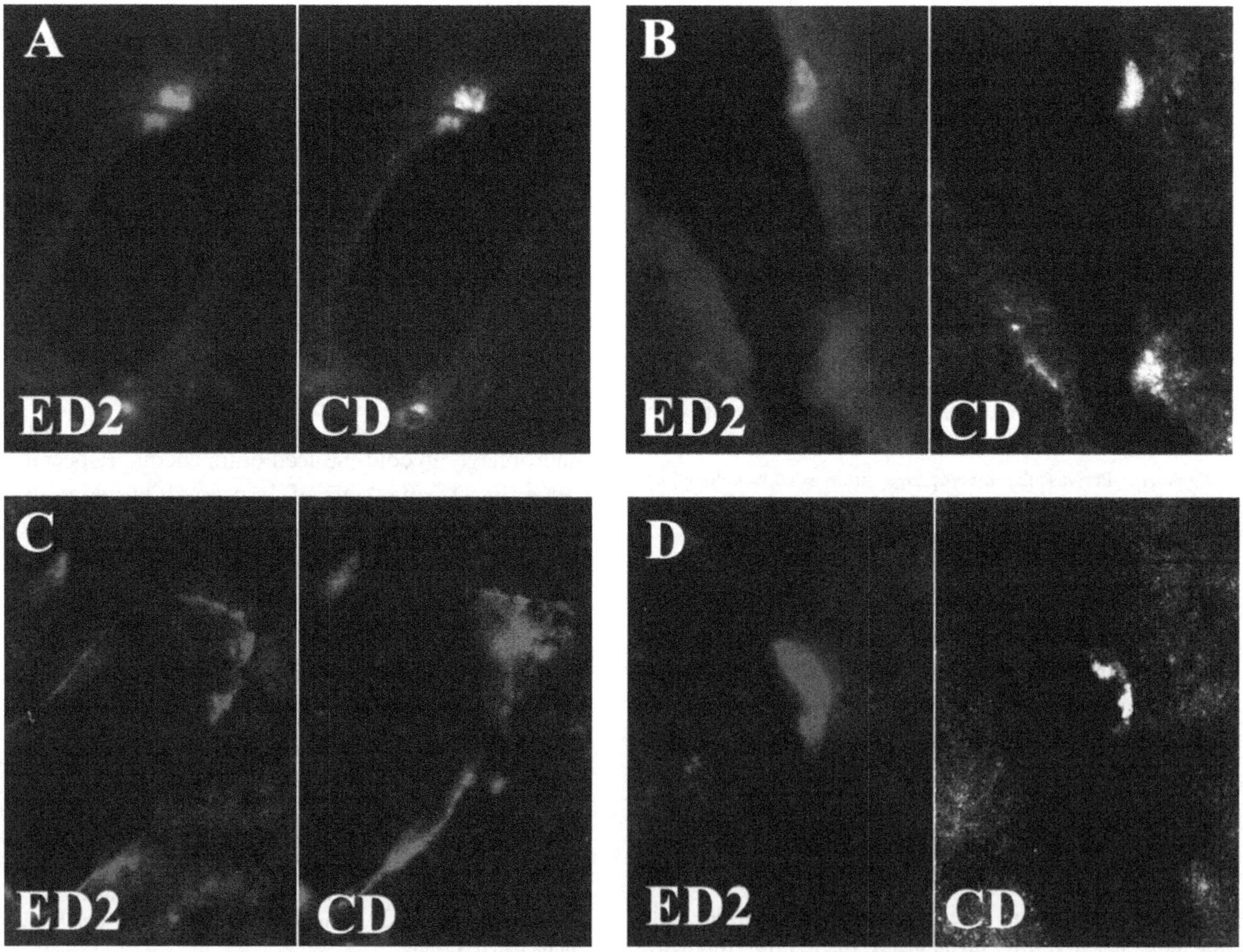

Fig. 1. Dual-label immuno-fluorescent microscopy: The perivascular macrophages were positive for ED2 and cathepsin D (CD). The reaction for cathepsin D appeared somewhat stronger at 10 hours than 2 days after cold injury. (A) Normal control. (B) 10 hours after cold injury. (C) 2 days after cold injury. (D) 2 weeks after cold injury. ×400

only very low reaction could be observed in some neurons or microglia, that were without ED2 reactions. Although in every time point of the process of brain edema, dual-positive cells were located only around arterioles or venules (Fig. 1B–D), the reaction for cathepsin D appeared somewhat stronger at 10 hours than 2 days after cold injury.

Immunogold electron microscopy

Although a small quantity of gold particle appeared in lipofuscins of neurons and microglia, most of reactions for cathepsin D were located in the inclusion bodies of perivascular macrophage (Fig. 2A), but did not appeare in pericytes. Changes of inclusion bodies and distribution of cathepsin D in perivascular macrophage are associated with the process of brain edema. At 5 and 10 hours after cold injury, some inclusion bodies fused each other, and reaction for cathepsin D increased (Fig. 2B). Then the reaction inclined to decrease. At 2 days after cold injury, membrane of inclusion bodies looked not clear, and reaction for cathepsin D decreased in the inclusion bodies. The reaction particles were also distributed in cytosole (Fig. 2C). At 1 week after cold injury, the reaction for cathepsin D reached to the minimum (Fig. 2D). Reaction for cathepsin D recovered at 2 weeks after cold injury (Fig. 2E). The result of quantitative analysis was summarized in Fig. 3.

Discussion

Mato *et al.* found perivascular macrophages to express class A scavenger receptor and to take up modified low density lipoprotein injected into the blood or cerebral ventricles [10]. Most recently, Linehan *et al.* found perivascular macrophages to express mannose receptor, which is important for endocytosis of glycoproteins. However, astrocytes and parenchymal microglia do not express mannose receptor in normal brain [6]. The large-sized lysosomal inclusion bodies may be helpful to digest proteins that were taken up into the cell. Moreover, in the present study, large quantities of Cathepsin D, one important kind of acid lysosomal protease, were located in the inclusion bodies of perivascular macrophages. All of the above findings suggest that the perivascular macrophage may be the most powerful scavenger in the brain. But what kind of role do the perivasular macrophages take in the process of cold-induced brain edema?

Immediately after cold injury, there was an increase of pinocytosis showed in perivascular macrophages (activation of taking-in ability) [3, 9]. In this research, from 5 hours to 10 hours after cold injury, an evident increase of immunoreaction for cathepsin D was observed. These findings suggest that the perivascular macrophages were activated in the early phase of cold-induced brain edema with an increased scavenger function. The activation of perivascular macrophage may be helpful to remove edema fluid including soluble protein molecules in the perivascular space.

The immunoreactions for cathepsin D in the inclusion bodies declined from 1 day to 1 week after cold injury, and needed at least another week to recover to normal level. Several possible pathways may contribute to the reduction of the enzyme in the inclusion bodies. First, the perivascular macrophages probably had taken up too much edema fluid composing large quantity of protein, which caused the enzyme exhaustion or dilution, but the enzyme could not get enough supply by composition. Second, cathepsin D infiltrated into cytoplasm from the inclusion bodies, as observed on the 2nd day after cold injury. Third, the enzyme may be secreted into the extracellular fluid. Thus, the reduction of cathepsin D reactions seems to be caused by severe pathological consequence, and suggests the functional fall and even degeneration of perivascular macrophages in the late phase of cold-induced brain edema.

As we know, oxygen radical mediated lipid peroxidation has been suggested to be one important factor in neuronal degeneration especially after head injury, and some reports have shown free radical and lipid peroxidation increase in brain cold injury models [2, 12]. On the other hand, Mato *et al.* found that feeding with chow containing a low content of the antioxidant vitamin E could result in the marked degeneration of perivascular macrophages [10]. Moreover, there was a research on cultured macrophage finding that oxidized low-density lipoprotein caused lysosomal destabilization and release of the cathepsin D to the cytosol [14]. Thus, in brain cold injury models it is very possible that the degeneration of perivascular macrophage and induction of cathepsin D may be caused by oxygen radical. Since melatonin, which has been suggested to be a free radical scavenger and an antioxidant, was found to be neuroprotective in instances of cold-injury-induced brain edema [4], we need more experiments to examine if antioxidant can be helpful for perivascular macrophage.

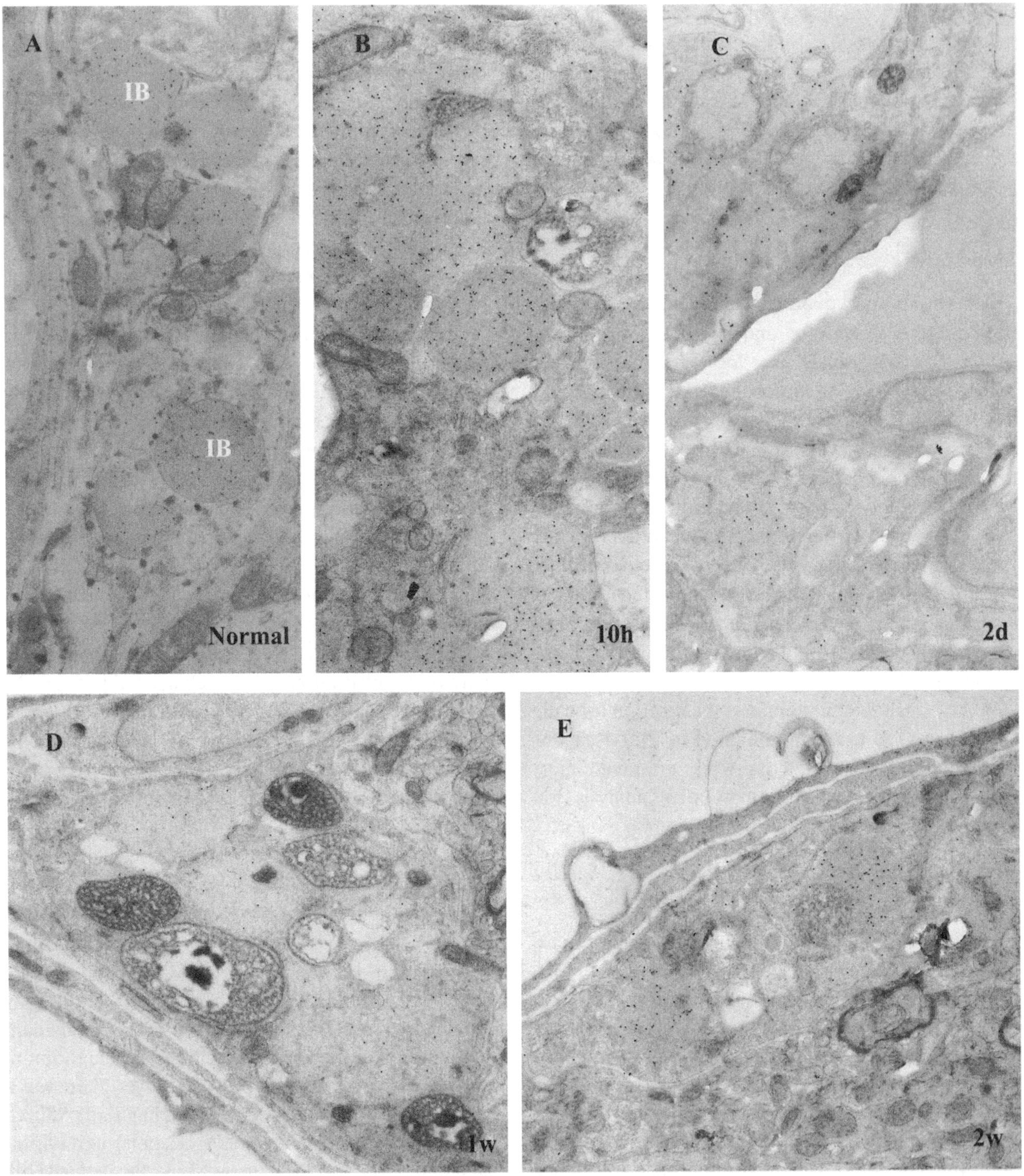

Fig. 2. Immunoelectron microscopy for cathepsin D: (A) Normal control. The perivascular macrophage contains several inclusion bodies (*IBs*) and mitochondria. IBs were labeled with many gold particles, which came from immunoreaction for cathepsin D. (B) 10 hours after cold injury. IBs show increase of the reaction for cathepsin D. Several IBs appear fusing each other. Several vacuoles can be seen with few gold particles. A honeycomb-like body shows diverse electron-dense and almost without immunoreaction for cathepsin D. (C) 2 days after cold injury. IBs show decrease of raction for cathepsin D. The membrane of IBs appears obscure. Big vacuoles can be seen with few gold particles. Several honeycomb-like bodies show no immunoreaction. (D) 1 week after cold injury. IBs show decrease of reaction for cathepsin D. (E) 2 weeks after cold injury. The reaction for cathepsin D in IBs appears recovered. ×14000

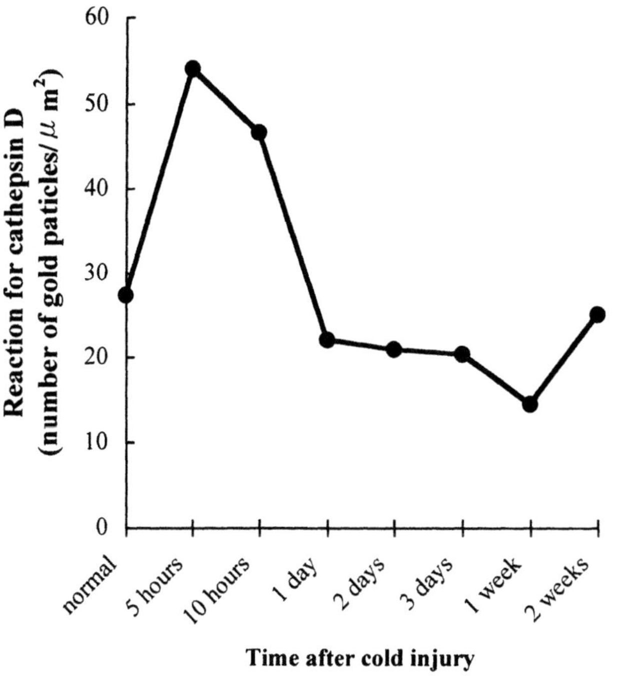

Fig. 3. Change of gold particle density: At 5 and 10 hours after cold injury reaction for cathepsin D increased. Then the reaction inclined to decrease. At 1 week after cold injury, the reaction for cathepsin D reached the minimum. Reaction for cathepsin D recovered at 2 weeks after cold injury

Acknowledgment

The authors thank Ms. Takeshi Inagaki for graphic assistance. This study was supported in part by the fund of the Ministry of Health, labor and welfare (Project of Longevity) of Japan.

References

1. Berryman M, Rodewald R (1990) An enhanced method for post-embedding immunocytochemical staining which preserves cell membranes. J Histochem Cytochem 38: 159–170
2. Erdincler P, Tuzgen S, Erdincler UD, Oguz E, Korpinar A, Ciplak N, Kuday C (2002) Influence of aging on blood-brain barrier permeability and free radical formation following experimental brain cold injury. Acta Neurochir 144(2): 195–200
3. Fukuda S, Mato M (1985) Studies on ultrastructural changes in perivascular cells (FGP) of small cerebral vessels at the initial stage after cold injury – chronological observations on FGP within 24 hours. No To Shinkei 37(4): 349–357
4. Gorgulu A, Palaoglu S, Ismailoglu O, Tuncel M, Surucu MT, Erbil M, Kilinc K (2001) Effect of melatonin on cerebral edema in rats. Neurosurgery 49(6): 1434–1442
5. Graeber MB, Streit WJ, Kreutzberg GW (1989) Identity of ED2-positive perivascular cells in rat brain. J Neurosci Res 22: 103–106
6. Linehan S, Pomares L, Stahl P (1999) Mannose receptor and its putative ligands in normal murine lymphoid and nonlymphoid organs: in situ expression of mannose receptor by selected macrophages, endothelial cells, perivascular microglia, and mesangial cells, but not dendritic cells. J Exp Med 189: 1961–1972
7. Mato M, Ookawara S (1979) A simple method for observation on capillary nets in rat brain cortex. Experientia 35: 501–503
8. Mato M, Ookawara S, Saito-Taki T (1986) Serological determinants of fluorescent granular perithelial cells along small cerebral blood vessels in rodent. Acta Neuropathol 72: 117–123
9. Mato M, Ookawara S, Hikishima H (1989) Role of perivascular phagocyte (FGP) of cerebral small blood vessel in cerebral edema. No To Shinkei 41(11): 1109–1117
10. Mato M, Ookawara S, Sakamoto A, Aikawa E, Ogawa T, Mitsuhashi U, Masuzawa T, Suzuki H, Honda M, Yazaki Y, Watanabe E, Luoma J, Yla-Herttuala S, Fraser I, Gordon S, Kodama T (1996) Involvement of specific macrophage-lineage cells surrounding arterioles in barrier and scavenger function in brain cortex. Proc Natl Acad Sci USA 93: 3269–3274
11. Nag S (1996) Cold-injury of cerebral cortex: immunolocalization of cellular proteins and blood-brain barrier permeability studies. J Neuropathol Exp Neurol 55: 880–888
12. Nag S, Picard P, Stewart DJ (2000) Increased immunolocalization of nitric oxide synthases during blood-brain barrier breakdown and cerebral edema. Acta Neurochir (Wien) [Suppl] 76: 65–68
13. Williams K, Alvarez X, Lackner AA (2001) Central nervous system perivascular cells are immunoregulatory cells that connect the CNS with the peripheral immune system. Glia 36(2): 156–164
14. Yuan XM, Li W, Brunk UT, Dalen H, Chang YH, Sevanian A (2000) Lysosomal destabilization during macrophage damage induced by cholesterol oxidation products. Free Radic Biol Med 28(2): 208–218

Correspondence: Koichi Takeuchi, M.D., Ph.D., Department of Anatomy, Jichi Medical School, Minamikawachi, Tochigi 329-0498, Japan. e-mail: koichi-t@jichi.ac.jp

Acta Neurochir (2003) [Suppl] 86: 287–289

Mitogen-activated protein kinases phosphorylation in posttraumatic selective vulnerability in rats

N. Otani, H. Nawashiro, N. Tsuzuki, H. Katoh, T. Miyazawa, and **K. Shima**

Department of Neurosurgery, National Defense Medical College, Tokorozawa, Japan

Summary

Mitogenic stimulation of the Mitogen-activated protein kinase (MAPK) pathway modulates the activity of many transcriptional factors leading to biological responses. Of these, three MAPK cascades are well characterized as extracellular signal-regulated protein kinase (ERK), c-Jun NH(2)-terminal kinase (JNK), and p38 pathways. The aim of this study was to investigate the topographic distribution and the role of activated MAPK pathways after fluid percussion injury (FPI) in rats. In the present results, FPI significantly induced ERK- and JNK-phosphorylation, but not p38-phosphorylation in the cortex and hippocampus at the injury site. The immunoreactivity for phospho-ERK was localized in the superficial neuronal layers, dentate hilar neurons, and the damaged CA3 neurons after 30 mins of FPI. Double immunostaining showed that phospho-ERK was prominent in astrocytes 6 hrs after TBI. The current results suggest that MAPK pathways are involved in signal transduction after FPI.

Keywords: Signaling pathway; MAPK; neuronal death; astrogliosis.

Introduction

Mitogen-activated protein kinases (MAPK), composed of extracellular signal regulated protein kinase (ERK), c-Jun NH(2)-terminal kinase (JNK), and p38 pathways, are activated by phosphorylation in response to a variety of mitogenic signals. On the other hand, Traumatic brain injury (TBI) causes neuronal degeneration and astrogliosis. However, little is known about mechanisms about mechanisms of these pathological findings. There has been no reports focusing on the expression and distribution of phospho-MAPKs after FPI. In the present study, we examined the activation of MAPK pathways after FPI in the rat brain using Western blot analysis and immunohistochemistry.

Material and methods

Adult male Sprague-Dawley rats weighing 300 to 400 g were subjected to lateral fluid percussion injury at a moderate severity (3.5 to 4.0 atmosphere) using a Dragonfly fluid percussion device as previously described [3]. For Western blot analysis, the rats were killed by decapitation at 5, 15, 30 mins, 1, 6, 24, and 72 hrs after injury ($n = 4$ per time point). The ipsilateral hippocampal or cortical tissues at the injured site was used for experiment. For immunohistochemistry, the rats were perfused transcardially ($n = 4$ per time point). The brains were embedded in paraffin after fixation in 4% buffered paraformaldehyde. 5-μm-thick serial coronal sections were prepared. The ipsilateral hippocampus of the injured hemisphere was used for Western blot analysis. After SDS-PAGE, the protein was transferred to PVDF membranes using an electrophoretic transfer system. The membranes were incubated with primary antibody against either polyclonal phospho-ERK1/2, phospho-JNK, and phospho-p38 antibodies (New England Biolabs, Beverly, MA, U.S.A). The liver tissue specimens of the rat undergoing ultraviolet treatment were used for positive staining to detect phospho-p38.

Immunohistochemistry was performed by the strept-avidin-biotin peroxidase complex method for immunostaining by polyclonal antibodies against phospho-ERK1/2, phospho-SAPK/JNK, and phospho-p38 (New England Biolabs). Double staining was performed with an EnVision system (DAKO, Japan) for GFAP, and a HISTOFINE SAB-PO (R) KIT (NICHIREI Co) for phospho-ERK1/2. The mean number of immunopositive CA3 neurons for phospho-ERK per millimeter was calculated. Cell counting was performed using a light microscope equipped with a 10× objective by independent observers in a blind manner.

All data are expressed as the mean ± SD. The densities of the immunoblotting band were quantified by using a NIH image analyzer. Statistical significance for Western blots was determined using an analysis of ANOVA followed by a post hoc Bonferroni/Dunn test.

Results

Our results showed that the phospho-ERK and phospho-JNK contents but not the phospho-p38 contents significantly increased in the rat brain after FPI. In hippocampus, the immunoreactivity for phospho-

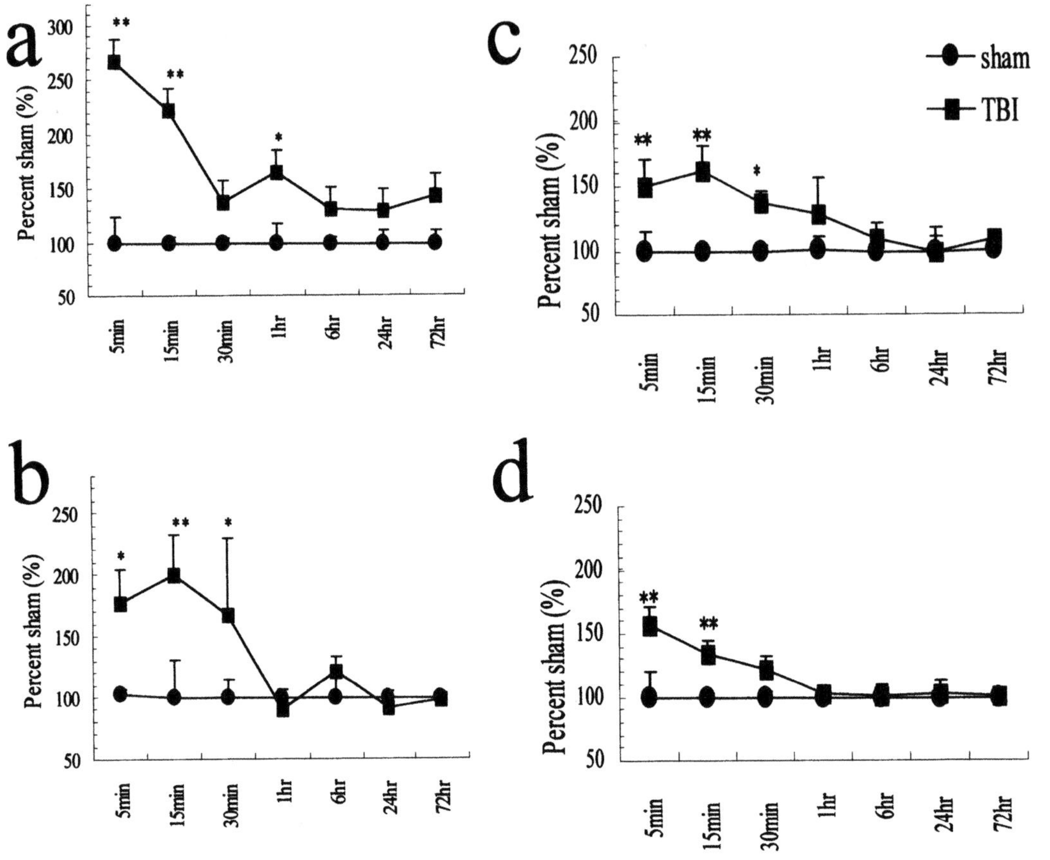

Fig. 1. FPI increases ERK-, and JNK-phosphorylation in the cortex (a and c) and hippocampus (b and d). Summary figure showing the time course of ERK- (a and b), and JNK-phosphorylation (c and d) from four independent experiments. Data are represented as percentage of the sham control (means ± SD). * $P < 0.05$, ** $p < 0.01$

JNK significantly increased at 5 mins, and thereafter gradually decreased to control levels at 1 hr after injury (Fig. 1d). On the other hand, the immunoreactivity for phospho-ERK increased at 5 mins, until then reached a maximum level at 15 mins, and then returned to control levels at 24 hrs after injury (Fig. 1c). In cortex, the immunoreactivity for phospho-JNK increased at 5 mins, until then reached a maximum level at 15 mins, and gradually decreased to control levels 1 hr after injury (Fig. 1b). On the other hand, the immunoreactivity for phospho-ERK significantly increased at 5 mins, and then returned to control levels at 24 hrs after injury (Fig. 1a). We could detect the immunoreactivity for phospho-p38 in the liver tissue specimens of the rats undergoing ultraviolet treatment, however, no discernible phospho-p38 protein was noted in the rat hippocampus at any time course after injury.

Immunohistochemical results showed that the immunoreactivity for phospho-p38 was not detected after injury in spite of being detected in rat brains after 72 hrs of transient forebrain ischemia. The immunoreactivity for phospho-JNK was maintained uniformly regardless of any regional selective vulnerability to FPI. The immunoreactivity for phospho-ERK was immediately induced in the superficial neuronal layers, dentate gyrus, mossy fibers, and CA3 neurons, but it was not detected in the CA1 neurons after injury. In addition, the induction of the phospho-ERK expression was localized in the dentate hilar neurons and the damaged CA3 neurons with the appearance of a small, shrunken, and dark nucleus at 30 mins after injury (Fig. 2). Cellular subtype distribution of immunopositive stained cells for phospho-ERK was almost exclusively observed in astrocytes after 6 hrs of FPI. The mean number of immunopositive CA3 neurons for phospho-ERK per millimeter was calculated. Cell counting was performed using a light microscope equipped with a 10× objective by independent observers in a blind manner.

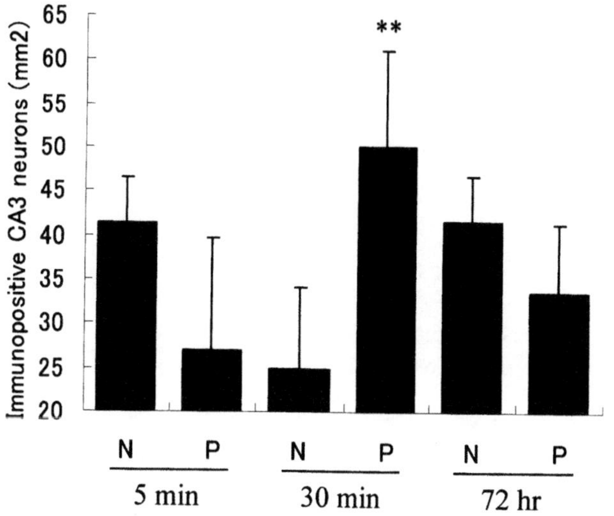

Fig. 2. The bar graph showing the number of immunopositive CA3 neurons for phospho-ERK after FPI. ** $p < 0.01$ compared to normal cell counts

Discussion

Previous studies have suggested that a selective vulnerability to FPI was observed in hippocampal CA3 neurons [1] and dentate hilar neurons [2]. However, the mechanism to explain the posttraumatic selective vulnerability of these neurons has yet to be fully elucidated. Several studies regarding the participation of MAPK cascades in transient forebrain ischemia have revealed that the JNK and p38 MAPK pathways play important roles in CA1 neuronal cell death during ischemia [5]. The distinct expression of phospho-ERK and phospho-JNK reflected pathophysiological differences between cerebral trauma and transient forebrain ischemia in the rat hippocampus.

Double immunostaining results showed that the immunoreactivity for phospho-ERK was prominent in astrocytes after 6 hrs of FPI. We speculated that the prolonged ERK-phosphorylation in astrocytes might thus play an important role in the pathophysiology of the late period of traumatic brain injury (TBI). The data presented in this article suggested that ERK- and JNK-, but not p38-phosphorylation, were associated with the molecular sequelae of TBI and that the discrepancy in the MAPK alterations reflected differences in selective vulnerability between the mechanical and ischemical events in the rat brain. The MAPK pathways might thus be important molecular targets for elucidating the mechanism following TBI.

References

1. Cortez SC, Mcintosh TK, Noble LJ (1989) Exerimental fluid percussion brain injury: vascular disruption and neuronal and glial alterations. Brain Res 482: 271–282
2. Lowenstein DH, Thomas MJ, Smith DH *et al* (1992) Selective vulnerability of dentate hilar neurons following traumatic brain injury: a potential mechanistic link between head trauma and disorders of the hippocampus. J Neurosci 12: 4846–4853
3. Matsushita Y, Shima K, Nawashiro H *et al* (2000) Real-time monitoring of glutamate following fluid percussion brain injury with hypoxia in the rat. J Neurotrauma 17: 143–153
4. Scheff SW, Baldwin SA, Brown RW (1997) Morris water maze deficits in rats following traumatic brain injury: lateral controlled cortical impact. J Neurotrauma 14: 615–627
5. Sugino T, Nozaki K, Takagi Y *et al* (2000) Activation of mitogen-activated protein kinases after transient forebrain ischemia in gerbil hippocampus. J Neurosci 20: 4506–4515

Correspondence: Naoki Otani, M.D., Department of Neurosurgery, National Defense Medical College, 3-2 Namiki, Tokorozawa, Saitama 359-8513, Japan. e-mail: grd1405@gr.ndmc.ac.jp

Acta Neurochir (2003) [Suppl] 86: 291–295

Treatment of cold injury-induced brain edema with a nonspecific matrix metalloproteinase inhibitor MMI270 in rats

N. Kawai, M. Kawanishi, and **S. Nagao**

Department of Neurological Surgery, Kagawa Medical University, Kagawa, Japan

Summary

Blood-brain barrier (BBB) disruption is a critical event leading to vasogenic brain edema and secondary brain damage after cold injury-induced brain trauma. Matrix metalloproteinases (MMPs) are implicated in BBB disruption in this model. The purpose of this study was to examine the effects of MMI270, a synthetic nonspecific MMP inhibitor, on cold injury-induced brain edema in rats. Treatment with MMI270, a bolus injection at a dose of 30 mg/kg, was started immediately after the induction of cold injury and was continued at a dose of 40 mg/kg/day using an intraperitoneal osmotic minipump. At 24 hours after the cold injury, the brain water content and the BBB permeability to Evans Blue (EB) were determined. The secondary brain lesion was assessed using hematoxylin and eosin (H-E) staining at 7 days after the cold injury. Compared with the untreated control group, treatment with MMI270 significantly reduced the brain water content in the ipsilateral core and intermediate areas and protected the BBB integrity to EB in the ipsilateral core area. The secondary lesion was significantly smaller in the MMI270-treated animals compared with the untreated animals. Our results indicate that treatment with MMI270 in rats exhibits protection in acute brain edema formation and secondary brain lesion by attenuating the BBB permeability after cold injury.

Keywords: Blood-brain barrier; brain edema; cold injury; MMI270; matrix metalloproteinase (MMP); MMP inhibitor.

Introduction

The blood-brain barrier (BBB) integrity contributes to the protection of the central nervous system from the environment. The tight junctions connecting the microvascular endothelial cells and the basement membrane maintain the integrity of the BBB. Matrix metalloproteinases (MMPs) are known as a family of calcium-requiring, zinc-containing proteolytic enzymes that degrade the molecules of the ECM. Two members of the MMP family, MMP-2 (72-kD gelatinase/gelatinase A) and MMP-9 (92-kD gelatinase/gelatinase B), specifically attack type IV collagen, fibronectin, laminin, which are the major components of the basement membrane around cerebral blood vessels. Increased expression of MMP-2 and MMP-9 has been shown in cerebral ischemia in animals [2, 5, 7].

Vasogenic brain edema is due to the dysfunction of endothelial cells, causing increased permeability of the BBB and leakage of the serum components into the brain tissue. Cold injury is well known and well characterized as a model of vasogenic brain edema. Previous studies suggest the possibility that the early induction and activation of MMPs may contribute to the brain edema and lesion formation after the cold injury [3]. In view of such findings, inhibition of MMPs has become an interesting target for brain edema therapy.

MMI270, *N*-hydroxy-2(R)-[(4-methoxysulfony)(3-picolyl)-amino]-3-metylbutaneamide hydrochloride monohydrate, is a novel synthetic hydroxamic acid derivative which is able to competitively bind the Zn^{2+} ion in the active site of a wide range of MMPs. In the present study, we sought to investigate the therapeutic potentials of MMI270 on the brain edema at 24 hours and secondary brain lesion at 7 days after the cold injury. In addition, the effect of MMI270 on the BBB permeability was also determined to clarify the protective mechanism on cold injury-induced brain damage.

Materials and methods

Experimental model

A total of 38 male Sprague-Dawley rats, each weighing 246 to 332 g, were used in this study. The rats were anesthetized with sodium pentobarbital (60 mg/kg, i.p.) and the body (rectal) temperature was maintained at 37 °C during the surgery using a feedback-controlled heating system. The rat was positioned in a stereotactic frame and the scalp was incised on the midline. The outer table of the right parietal skull was removed using a diamond bar. A copper probe, 4 mm in diameter and cooled with liquid nitrogen (−196 °C), was applied to the skull above the right parietal lobe for 30 sec.

Experimental protocols

This study was divided into three parts. The first part evaluated the effect of MMI270 on brain edema formation in the untreated (n = 7) and MMI270-treated (n = 7) rats at 24 hours after the cold injury. In the second part, Evans Blue dye extravasation for 24 hours after the cold injury was determined in the untreated (n = 6) and MMI270-treated (n = 6) rats to examine the BBB integrity. In the third part, tissue damage as the secondary lesion was determined using hematoxylin and eosin (H-E) staining in the untreated (n = 6) and MMI270-treated (n = 6) rats at 7 days after the cold injury.

Solutions containing MMI270 dissolved in water were injected intravenously at a dose of 30 mg/kg in the MMI270-treated rats. A mini-osmotic pump (ALZET 2001D or 2ML1, Alza Co., CA, USA) was subsequently implanted into the peritoneal cavity. This pump delivered MMI270 solution (dissolved in dimethyl sulfoxide and polyethylene glycol 300) continuously at a dose of 40 mg/kg/day. In the untreated group, rats received vehicle and were implanted with vehicle-containing osmotic pumps.

Measurements of brain water content

The rats were sacrificed by decapitation under deep sodium pentobarbital anesthesia at 24 hours after the cold injury. The brains were rapidly removed and separated into cortices and basal ganglia. Cortices were flattened on a piece of Parafilm, and using 5- and 10-mm-diameter cork borers, punches were made of the ipsilateral and contralateral cortices to obtain tissue samples from the core and intermediate areas of the cold injury. The tissue samples were immediately weighed on an electronic analytical balance to the nearest 0.1 mg to obtain the wet weight (WW). The tissue was then dried in an oven at 110 °C for 24 hours and weighed again to obtain the dry weight (DW). The formula (WW − DW)/WW × 100 was used to calculate the water content, which was expressed as a percentage of WW.

Blood-brain barrier permeability to Evans Blue

The integrity of the BBB was investigated using Evans Blue dye extravasation. Evans Blue (EB, 2% in saline, 1 ml/kg) was injected intravenously immediately after the induction of cold injury. At 24 hours, the rats were reanesthetized with sodium pentobarbital (60 mg/kg, i.p.). The chest was subsequently opened and the brain was transcardially perfused with 200 ml of saline through the left ventricle at 100 mmHg. After decapitation, the brain was removed and separated into cortices and basal ganglia. Punches were made of the ipsilateral and contralateral cortices to obtain tissue samples from the core and intermediate areas of the cold injury.

Samples were weighted and placed in 50% of trichloroacetic solution. Following homogenization and centrifugation, the extracted dye was diluted with ethanol (1 : 3), and its fluorescence was determined (excitation at 620 nm and emission at 680 nm) with a luminescence spectrophotometer (Hitachi 650-40, Tokyo, Japan). Calculations were based on the external standard (62.5–500 ng/ml) in the same solvent. The tissue content of EB was quantified from a linear standard line derived from known amounts of the dye and was expressed per gram of wet tissue.

Development of secondary lesion

We evaluated the development of the secondary lesion using H-E staining at 7 days after the cold injury. The brains were removed and frozen in isopentane. Cryostat coronal brain sections were obtained at 2-mm intervals including the center of the focal lesion at the parietal lobe. Secondary brain lesion appeared well demarcated: lesioned areas were unstained (white) tissue, which was easily contrasted from areas of viable tissue. The unstained area in the slice of maximum focal lesion was measured using a computerized image analysis system (Scion Image, Beta 4.02).

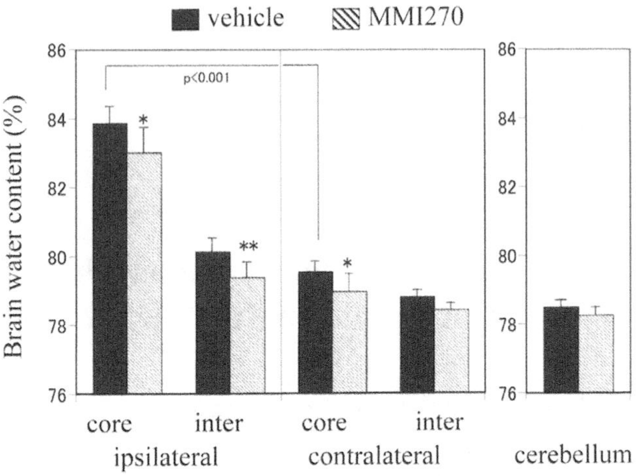

Fig. 1. Bar graph showing the brain water content in the core and intermediate (inter) areas of the ipsilateral and contralateral hemispheres of the cold injury in the untreated (n = 7) and MMI270-treated (n = 7) animals. Measurements were made 24 hours after the induction of cold injury. Values are mean ± SD. *, $p < 0.05$ and **, $p < 0.01$, statistically significant differences between the vehicle-treated and MMI270-treated groups

Results

Brain water content (Fig. 1)

Application of the copper probe cooled with liquid nitrogen on the parietal skull produced a significant increase in brain water content in the ipsilateral cerebral cortex compared with the contralateral cortex (83.9 ± 0.5 vs. 79.5 ± 0.3%; p < 0.001).

Brain water content observed in the cerebellum was not significantly different between the untreated and MMI270-treated animals (78.5 ± 0.2 vs. 78.3 ± 0.2%), indicating that the treatment with MMI270 per se did not influence the brain water content in the normal tissue. Treatment with MMI270 significantly reduced the brain water content when compared with the untreated animals in the ipsilateral core and intermediate areas of the cold injury (83.9 ± 0.5 vs. 83.0 ± 0.7%; p < 0.05 and 80.1 ± 0.4 vs. 79.4 ± 0.4%; p < 0.01, respectively). An increase in the brain water content was also significantly suppressed with MMI270 treatment in the contralateral hemisphere (79.5 ± 0.3 vs. 79.0 ± 0.5%, p < 0.05).

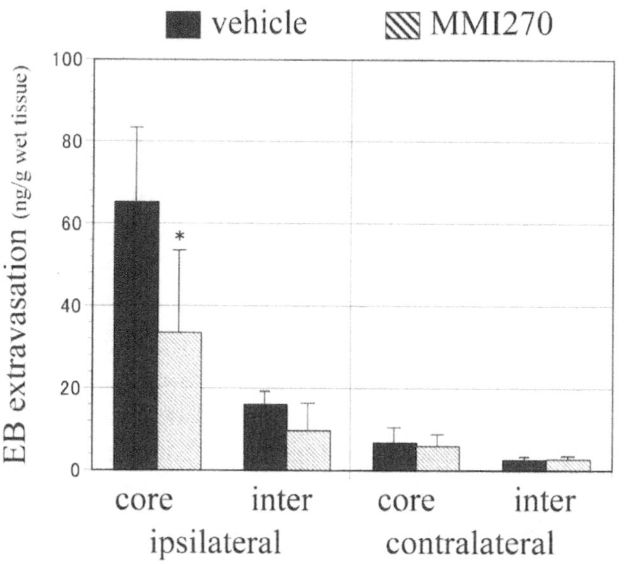

Fig. 2. Bar graph showing quantitative measurement of BBB permeability to EB in the core and intermediate (inter) areas of the ipsilateral and contralateral hemispheres of the cold injury in the untreated (n = 6) and MMI270-treated (n = 6) animals. Measurements were made 24 hours after the induction of cold injury. Values are mean ± SD. *, $p < 0.05$, statistically significant difference between the vehicle-treated and MMI270-treated groups

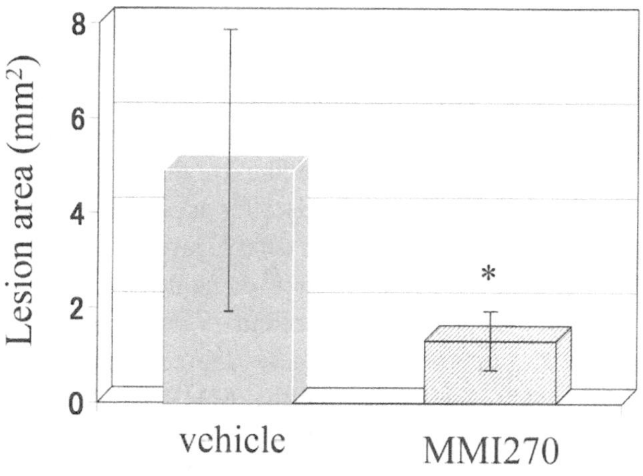

Fig. 3. Bar graph showing the secondary brain lesion after the cold injury using TTC staining in the untreated (n = 6) and MMI270-treated (n = 6) animals. Measurements were made 7 days after the induction of cold injury. Values are mean ± SD. *, $p < 0.05$, statistically significant difference between the vehicle-treated and MMI270-treated groups

Blood-brain barrier permeability (Fig. 2)

The area showing EB extravasation was obviously smaller in the MMI270-treated rat than it was in the untreated rat. The BBB permeability to EB was significantly reduced with MMI270 treatment when compared with the untreated animals in the ipsilateral core area (65.3 ± 18.1 vs. 33.5 ± 20.0 ng/g wet tissue; $p < 0.05$). In the ipsilateral intermediate area, an increase in BBB permeability was also decreased with MMI270 treatment in comparison with the untreated animals (16.1 ± 3.2 vs. 9.7 ± 6.6 ng/g wet tissue), but this did not reach statistical significance.

Secondary lesion following cold injury (Fig. 3)

The parietal cortex, where the cold injury was induced, showed a focal lesion with a relatively sharp demarcation at 7 days after the cold injury. We evaluated the development of the secondary lesion following cold injury using H-E staining at 7 days after the cold injury (Fig. 5). The unstained area was significantly smaller in the MMI270-treated animals compared with the untreated animals (4.9 ± 3.0 vs. 1.3 ± 0.6 mm^2, $p < 0.05$).

Discussion

Matrix metalloproteinase in cold injury

The blood-brain barrier (BBB) integrity contributes to the protection of the neuronal microenvironment. The tight-junctions and basement membrane maintain the normal BBB. When the integrity of the BBB is disrupted, inflammatory cells and fluid penetrate into the brain, causing brain edema and cell death. The basement membrane, which provides structural support of the vasculature, is composed of type IV collagen, fibronectin, laminin, various proteoglycans and heparan sulfates. A proteolytic cascade involving matrix metalloproteinases (MMPs) and plasminogen activators are important in the ECM disruption. MMPs are known as a family of calcium-requiring, zinc-containing proteolytic enzymes that degrade the molecules of the ECM. Two members of the MMP family, MMP-2 (72-kD gelatinase/gelatinase A) and MMP-9 (92-kD gelatinase/gelatinase B), are able to degrade the ECM components of the basement membrane around cerebral blood vessels. Increased expression of MMP-2 and MMP-9 has been shown in cerebral ischemia in animals [2, 5, 7]. Breakdown of the BBB with an associated increase in vascular permeability has also been shown in cerebral ischemia [2, 5, 7], suggesting the importance of MMP activation in the progression of brain damage by BBB disruption.

The cold injury model is one of the established models for the study of vasogenic brain edema and secondary brain injury. This model induces a reproducible lesion localized in the cortex, which does not extend much into the striatum and hippocampus under the cortical layer. In this model, the BBB permeability to EB, which is a protein-binding dye that binds to albumin, increased immediately after application of the injury, and it returned to the control level by 24 hours [4]. On the other hand, brain water content was maximized at 24 hours, and the secondary lesion gradually progressed up to 72 hours after the cold injury [4]. A recent study demonstrated that MMP-9 and MMP-2 were induced independently after the cold injury in mice. A marked induction of MMP-9 appeared as early as 2 hours after the cold injury and was maximized at 12 hours with mild induction of MMP-2, which preceded the peak of water content increase [3]. These results suggest the possibility that the early induction and activation of MMP-9 may contribute to the early opening of the BBB and brain edema formation following cold injury.

Inhibition of matrix metalloproteinase

Intraperitoneal administration of BB-94 (also known as batimastat), a synthetic broad-spectrum MMP inhibitor, at 30 minutes before and 3 hours after focal cerebral ischemia significantly reduced the ischemic lesion volume and neurological deficits at 24 hours after the ischemia in normal mild type mice [1]. Importantly, BB-94 did not reduce the ischemic lesion volume in MMP-9 knockout mice [1], suggesting that BB-94 protects the ischemic brain by inhibiting MMP-9 activity. Treatment with BB-1101, another synthetic broad-spectrum MMP-inhibitor, prevented the brain injury and reduced the BBB opening after focal cerebral ischemia in rats [6]. These results warrant that pharmacological modulations of MMP activity can offer therapeutic strategies for the treatment of experimental brain injuries.

MMI270, *N*-hydroxy-2(R)-[(4-methoxysulfony)(3-picolyl)-amino]-3-metylbutaneamide hydrochloride monohydrate, is a novel synthetic hydroxamic acid derivative. MMI270 is able to competitively bind the Zn^{2+} ion in the active site of a wide range of MMPs, inhibiting their activity at nM concentrations in vitro. In our experiment, blood samples were taken at 24 hours after the cold injury in the first part for the analysis of MMI270 concentrations. Mean plasma concentration of MMI270 was 540.9 ± 61.8 nM ($n = 7$). This level is $\geqq 10$ times the IC_{50} for the MMP enzymes, and therefore should provide near-complete inhibition of MMP activities. In the present study, we examined the effects of MMI270, a nonspecific MMP inhibitor, on cold injury-induced vasogenic brain edema and the BBB permeability in rats. We also quantitatively determined the BBB permeability for 24 hours after the cold injury in untreated and MMI270-treated rats using the EB fluorescence method. The long circulation time of 24 hours allowed the EB to spread far from the core area to the intermediate area. MMI270 treatment significantly reduced the EB leakage in the core area and tended to decrease in the intermediate area of the cold injury. This was evident by macroscopic examinations of cold-injured brains. Thus, an important finding of the present study, in addition to the reduction of brain edema, is the protection of BBB integrity with MMI270 treatment. Because the EB extravasation observed in the contralateral core area was not significantly reduced in the MMI270 rats compared with the untreated rats, the reduction of brain edema in the contralateral core area was thought to be caused by a decrease in the spread of edema fluid from the ipsilateral to the contralateral hemisphere through the corpus callosum. Severe brain edema increases intracranial pressure and causes a decrease of cerebral blood perfusion, resulting in secondary ischemia. It was demonstrated that the lesion size increased in a time-dependent manner, not only in the acute phase, but also in the delayed phase up to 72 hours after the cold injury in mice. Using the same model in mice, a previous study revealed that the activated form of MMP-9 was detected from 6 to 24 hours after cold injury, and the activated MMP-2 appeared at 24 hours, and was maximized at 7 days [3]. In view of the results of these studies, we evaluated the secondary lesion at 7 days after the cold injury, which showed a significant reduction in the brain lesion in MMI270-treated rats compared with untreated rats.

References

1. Asahi M, Asahi K, Jung J-C, del Zoppo GJ, Fini ME, Lo EH (2000) Role for matrix metalloproteinase 9 after focal cerebral ischemia: effects of gene knockout and enzyme inhibition with BB-94. J Cereb Blood Flow Metab 20: 1681–1689
2. Gasche Y, Fujimura M, Morita-Fujimura Y, Copin J-C, Kawase M, Massengale J, Chan PH (1999) Early appearance of activated matrix metalloproteinase-9 after focal cerebral ischemia in mice: a possible role in blood-brain barrier dysfunction. J Cereb Blood Flow Metab 19: 1020–1028
3. Morita-Fujimura Y, Fujimura M, Gasche Y, Copin J-C, Chan PH (2000) Overexpression of zinc superoxide dismutase in trans-

genic mice prevents the induction and activation of matrix metalloproteinase after cold injury-induced brain trauma. J Cereb Blood Flow Metab 20: 130–138
4. Murakami K, Kondo T, Yang G, Chen SF, Morita-Fujimura Y, Chan PH (1999) Cold injury in mice: a model of to study mechanism of brain edema and neuronal apoptosis. Prog Neurobiol 57: 289–299
5. Romanic AM, White RF, Arleth AJ, Ohlstein EH, Barone FC (1998) Matrix metalloproteinase expression increases after cerebral focal ischemia in rats: inhibition of matrix metalloproteinase-9 reduces infarct size. Stroke 29: 1020–1030
6. Rosenberg GA, Estrada EY, Dencoff JE (1998) Matrix metalloproteinase and TIMPs are associated with blood-brain barrier opening after reperfusion in rat brain. Stroke 29: 2189–2195
7. Rosenberg GA, Navratil M, Barone F, Feuerstein G (1996) Proteolytic cascade enzymes increase in focal cerebral ischemia in rat. J Cereb Blood Flow Metab 16: 360–366

Correspondence: Nobuyuki Kawai, M.D., Department of Neurological Surgery, Kagawa Medical University, 1750-1 Miki-cho, Kita-gun, Kagawa 761-0793, Japan. e-mail: nobu@kms.ac.jp

Acta Neurochir (2003) [Suppl] 86: 297–299

Prions prevent brain damage after experimental brain injury: a preliminary report

S. Hoshino[1], K. Inoue[2], T. Yokoyama[3], S. Kobayashi[4], T. Asakura[1], A. Teramoto[1], and S. Itohara[2]

[1] Department of Neurosurgery, Nippon Medical School, Tokyo, Japan
[2] Laboratory for Behavioural Genetics, Brain Research Institute, RIKEN, Wako, Saitama, Japan
[3] National Institute of Animal Health, Tsukuba, Ibaraki, Japan
[4] Department of Neurosurgery, Nippon Medical School, Chiba Hokusoh Hospital, Inaba, Chiba, Japan

Summary

The physiological function of the normal cellular form of prion protein (PrP^C) is not yet fully understood. In the current study we used prion protein gene knock-out mice (*Prnp*$^{-/-}$) to assess the role of PrP^C in traumatic brain injury. *Prnp*$^{+/-}$ and *Prnp*$^{-/-}$ mice were subjected to weight-drop contusional brain injury over the left parietal cortex. *Prnp*$^{-/-}$ mice manifested a significantly larger lesion volume and worse neuromotor scores than did their *Prnp*$^{+/-}$ littermates. IgG immunostaining revealed that in *Prnp*$^{-/-}$ mice the breakdown in the blood-brain barrier (BBB) was more extensive at 1 month after brain injury. Our results are in agreement with previous in vitro findings of the neuroprotective role of PrP^C and further support the hypothesis that functional loss of PrP^C plays a pathogenic role in prion diseases. We also suggest that PrP^C modulates BBB function.

Keywords: Blood-brain barrier; IgG; mouse; prion; traumatic brain injury.

Introduction

In prion diseases such as Creutzfeldt-Jakob disease in humans and spongiform encephalopathy in cattle, neurodegeneration is characterized by the accumulation of a mis-folded and pathological form of prion protein (PrP^C), termed PrP^{Sc} [10]. Since clinical symptoms may develop before or in the absence of PrP^{Sc} deposits, it has been proposed that functional loss of PrP^C plays a pathogenic role in prion diseases [1].

While the physiological function of PrP^C remains obscure, Brown *et al.* [5] reported that PrP^C exhibits superoxide dismutase-like activity. Moreover, a neuronal protective role of PrP^C has been uncovered in vivo [14] and in vitro [4, 6, 10, 11, 17]. It is also involved in signal transduction, neuronal adhesion, and neurite growth [6, 8, 17].

To further characterize the function of PrP^C in vivo, we inflicted traumatic brain injury on prion protein knock out mice (*Prnp*$^{-/-}$) and their *Prnp*$^{+/-}$ littermates.

Materials and methods

Generation of Prnp$^{-/-}$ *mice*

As did others we used homologous recombination to produce *Prnp*$^{-/-}$ mice with a C57/BL/6 background, as previously reported [7, 16]. The entire open reading frame of *Prnp* and the 3′-end of intron 2 were replaced with a *pgk-neo* gene cassette. The genotypes were determined by polymerase chain reaction analysis of DNA prepared from the tails of the mice.

Brain injury

Our contusion device consisted of a plastic, 50-cm long tube with a footplate (3 mm in diameter) at one end. By dropping the weight inside the tube, a force of 200 g/cm was produced, resulting in a 2-mm depression of the brain surface. After anesthetizing 8-week old male *Prnp*$^{-/-}$ and *Prnp*$^{+/-}$ mice ($n = 10$ in each group) with pentobarbital (100 mg/kg i.p.), we performed 5×5 mm square craniotomy over the left parietal bone with a dental drill and positioned the center of the footplate at 2.5 mm posterior and 2.5 mm lateral to the bregma. After brain injury induction, a craniotomied skull was applied with adhesives.

Neuroscore and lesion volume

An experienced investigator (SH) blinded to the genotype used previously described composite tests [13] to record posttraumatic neurological motor function. Evaluations were carried out at 48 hr and at 1, 2, and 4 weeks after injury. At 4 weeks post-injury, the mice were deeply anesthetized with pentobarbital and transcardially perfused with 4% paraformaldehyde. Paraffin sections (6 μm in thickness) were stained with H&E every 150 μm to determine volumetric tissue loss (lesion volume). Digital photomicrographs were processed on a personal computer. Differences between right and left brain-regions were determined with imaging software (NIH image version 1.61) to assess the lesion volume.

Immunostaining of IgG

To evaluate the integrity of the blood-brain barrier (BBB), we used immunohistochemical methods and endogenous IgG, a sensitive marker for brain- and neuronal damage [9]. We immunostained at the level of bregma – 1.82. After deparaffinization and rehydration, the sections were placed in 0.01 M citrate buffer and then in a microwave oven (95 °C, 10 min). Subsequently, they were immersed overnight in horse anti-mouse IgG (Vector, Burlingame, CA, USA; 1:100) in 5% normal horse serum. Bound antibody was detected with the avidin-biotin complex method (ABC Elite kit, Vector); 3,3′-diaminobenzidine was used as the chromogen. In the negative controls, normal serum was used instead of primary antibodies. The area of IgG-immunoreactivity was measured with imaging software (NIH image version 1.61). The staining intensity of the median eminence where the BBB is normally absent was used as the internal standard.

Results

Some of the $Prnp^{-/-}$ mice had a deformed tail. No mice died as a direct result of the brain trauma. At 2 weeks after injury, $Prnp^{-/-}$ mice exhibited significant neurological motor dysfunction compared with $Prnp^{+/-}$ mice (data not shown). Compared to their $Prnp^{+/-}$ littermates, $Prnp^{-/-}$ mice manifested a 43.82% increase in hemispheric tissue loss ($p < 0.05$, unpaired Student's t-test; Fig. 1). IgG immunostaining was observed in the hippocampi (especially fimbria), left lateral posterior thalamic nucleus, left parietal cortex, and the right optic tract with both group of mice, but $Prnp^{-/-}$ mice showed the larger and more intense immunoreactive area (Fig. 2). $Prnp^{-/-}$ mice manifested a 187.18% increase in the IgG immunoreactive area in the bilateral hemispheres ($p < 0.05$, unpaired Student's t-test; Fig. 3).

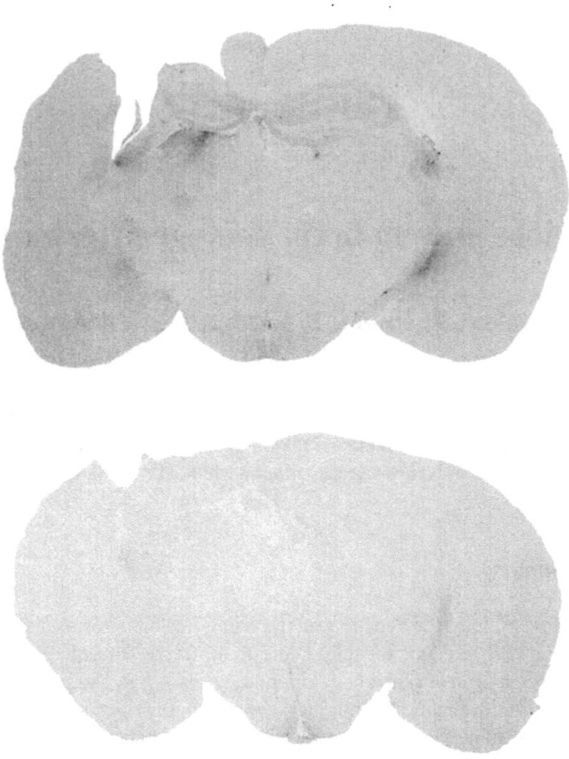

Fig. 2. Representative photomicrographs of IgG immunostaining in a $Prnp^{-/-}$ (upper) and a $Prnp^{+/-}$ (lower) mouse at 4 weeks after contusional brain injury

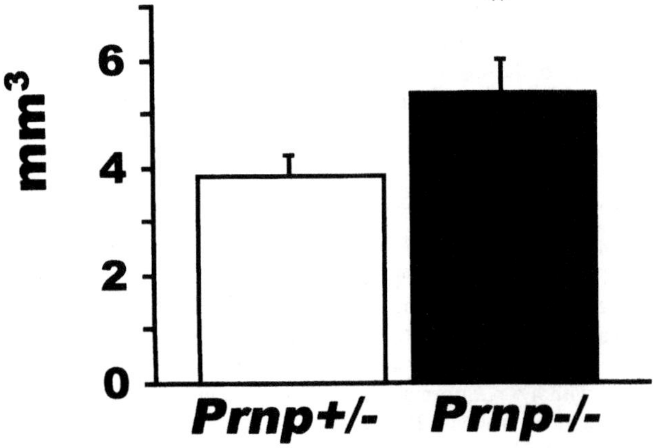

Fig. 1. Lesion volume in the left hemisphere of $Prnp^{+/-}$ and $Prnp^{-/-}$ mice at 4 weeks after contusional brain injury. * $P < 0.05$ (unpaired Student's *t*-test)

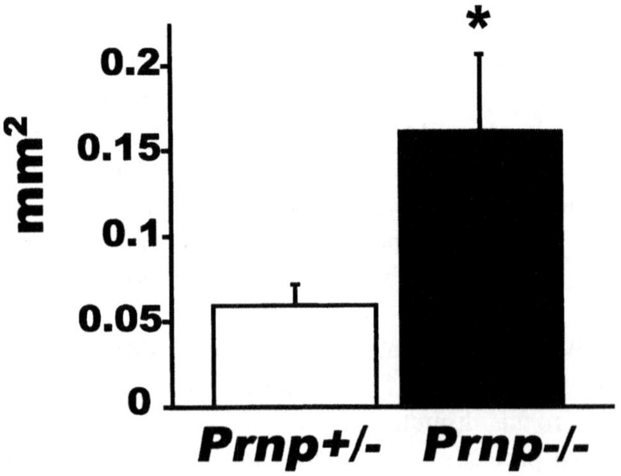

Fig. 3. Volume of IgG immunostained area in the bilateral hemispheres of $Prnp^{+/-}$ and $Prnp^{-/-}$ mice at 4 weeks after contusional brain injury. * $P < 0.05$ (unpaired Student's *t*-test)

Discussion

Although the abnormal isoform of PrP^C, PrP^{Sc}, is thought to be a main component in prion diseases [12], the function of PrP^C has not been fully elucidated [2, 11]. To prevent these diseases, and to develop effective therapies, the function of PrP^C must be clearly understood. While there is accumulating evidence for the neuroprotective function of PrP^C [2, 6, 10, 11, 14, 17], its in vivo role after brain injury has not been documented previously. We here report that in PrP^C knockout mice subjected to traumatic brain injury, PrP^C attenuated neurological motor dysfunction and the lesion volume. The observed neuroprotective effect coincides with previous findings that demonstrated decreased sensitivity in a mouse epilepsy model [14]. Several factors may play a role in the protective mechanism(s) of PrP^C. Excitotoxicity and oxidative stress are major candidates for the neuronal damage observed following traumatic brain injury. In vitro, PrP^C increased the glutamate uptake by astrocytes [3] and attenuated oxidative stress [15]. Therefore, in our model, the neuroprotective function of PrP^C may be attributable to both excitotoxicity and oxidative stress. Alternatively, PrP^C may prevent apoptosis [10]. The loss of neuroprotection upon the replacement of PrP^C by PrP^{Sc} may be involved in the neuronal damage observed in prion diseases. Furthermore, we speculate that PrP^C function may play a role in other chronic neurodegenerative diseases such as Alzheimer's disease.

We also found that in the presence of PrP^C, IgG immunoreactivity was decreased at 1 month after the induction of brain trauma. IgG immunostaining is a useful method that sensitively discloses both BBB disruption and cellular stress [9]. This result may reflect modulation of astrocyte function by PrP^C [3]. Moreover, binding of laminin, one of the components of the basement membrane, may affect endothelial cell barrier function [8]. Our results suggest that PrP^C plays a role in maintaining the BBB, and that its dysfunction may facilitate the propagation of PrP^{Sc}. Studies are underway to evaluate the BBB breakdown of *Prnp*$^{-/-}$ and *Prnp*$^{+/-}$ mice at the acute stage after contusional brain injury.

References

1. Aguzzi A, Weissmann C (1997) Prion research: the next frontiers. Nature 389: 795–798
2. Brown DR (2001) Prison and prejudice: normal protein and the synapse. Trends Neurosci 24: 85–90
3. Brown DR, Mohan CM (1999) Atrocitic glutamate uptake and prion protein expression. Glia 25: 282–292
4. Brown DR, Nicholas RS, Canevari L (2002) Lack of prion protein expression results in a neuronal phenotype sensitive to stress. J Neurosci Res 67: 211–224
5. Brown DR, Wong BS, Hafiz F, Clive C, Haswell SJ, Jones IM (1999) Normal prion protein has an activity like that of superoxide dismutase. Biochem J 344 pt 1: 1–5
6. Chiarini LB, Freitas AR, Zanata SM, Brentani RR, Martins VR, Linden R (2002) Cellular prion protein transduces neuroprotective signals. EMBO J 21: 3317–3326
7. Gomi H, Yokoyama T, Fujimoto K, Ikeda T, Katoh A, Itoh T, Itohara S (1995) Mice devoid of the glial fibrillary acidic protein develop normally and are susceptible to scrapie prions. Neuron 14: 29–41
8. Graner E, Mercadante AF, Zanata SM, Forlenza OV, Cabral AL, Veiga SS, Juliano MA, Roesler R, Walz R, Minetti A, Izquierdo I, Martins VR, Brentani RR (2000) Cellular prion protein binds laminin and mediates neuritogenesis. Mol Brain Res 76: 85–92
9. Hoshino S, Kobayashi S, Nakazawa S (1996) Prolonged and extensive IgG immunoreactivity after severe fluid-percussion injury in rat brain. Brain Res 711: 73–83
10. Kuwahara C, Takeuchi AM, Nishimura T, Haraguchi K, Kubosaki A, Matsumoto Y, Saeki K, Matsumoto Y, Yokoyama T, Itohara S, Onodera T (1999) Prions prevent neuronal cell-line death. Nature 400: 225–226
11. Martins VR, Linden R, Prado MA, Walz R, Sakamoto AC, Izquierdo I, Brentani RR (2002) Cellular prion protein: on the road for functions. FEBS Lett 512: 25–28
12. Prusiner SB (1991) Molecular biology of prion diseases. Science 252: 1515–1522
13. Scherbel U, Raghupathi R, Nakamura M, Saatman KE, Trojanowski JQ, Neugebauer E, Marino MW, McIntosh TK (1999) Differential acute and chronic responses of tumor necrosis factor-deficient mice to experimental brain injury. Proc Natl Acad Sci USA 96: 8721–8726
14. Walz R, Amaral OB, Rockenbach IC, Roesler R, Izquierdo I, Cavalheiro EA, Martins VR, Brentani RR (1999) Increased sensitivity to seizures in mice lacking cellular prion protein. Epilepsia 40: 1679–1682
15. Wong BS, Liu T, Li R, Pan T, Petersen RB, Smith MA, Gambetti P, Perry G, Manson JC, Brown DR, Sy MS (2001) Increased levels of oxidative stress markers detected in the brains of mice devoid of prion protein. J Neurochem 76: 565–572
16. Yokoyama T, Kimura KM, Ushiki Y, Yamada S, Morooka A, Nakashiba T, Sassa T, Itohara S (2001) In vivo conversion of cellular prion protein to pathogenic isoforms, as monitored by conformation-specific antibodies. J Biol Chem 276: 11265–11271
17. Zanata SM, Lopes MH, Mercadante AF, Hajj GN, Chiarini LB, Nomizo R, Freitas AR, Cabral AL, Lee KS, Juliano MA, de Oliveira E, Jachieri SG, Burlingame A, Huang L, Linden R, Brentani RR, Martins VR (2002) Stress-inducible protein 1 is a cell surface ligand for cellular prion that triggers neuroprotection. EMBO J 21: 3307–3316

Correspondence: Shigeru Hoshino, M.D., Ph.D., Heiwadai Hospital, 834-28, Fusa, Abiko, Chiba, 270-1101, Japan. e-mail: shigeru@nms.ac.jp

Acta Neurochir (2003) [Suppl] 86: 301–303

The effect of Cyclosporin A on brain edema formation following experimental cortical contusion

S. Fukui, S. Signoretti, J. G. Dunbar, and **A. Marmarou**

Division of Neurosurgery, Medical College of Virginia, Virginia Commonwealth University, Richmond, Virginia

Summary

Cyclosporin A (CsA) has been shown by our laboratory and others to be neuroprotective in the experimental animal model of traumatic brain injury. However, we found that the intrathecal administration resulted in a concomitant increase of brain edema. The aim of this study was to assess whether intravascular administration may also influence brain edema formation.

This project includes two independent series in which different doses of CsA were intravenously given to Sprague-Dawley rats of each group. In the first series, the animals were exposed to focal brain injury by a controlled cortical impact (CCI, 6 m/sec, 3 mm depth) and randomized into the following two groups: 20 mg/kg CsA and control vehicle. In the second series, animals were also injured by CCI and randomized into 35 mg/kg CsA and control vehicle. The intravenous continuous (1 h) infusion was begun 30 minutes after the insult. All animals were sacrificed at 24 hours post injury to assess the brain water content using the gravimetric method.

Intravenously-administrated CsA of either 20 or 35 mg/kg did not significantly change the brain water content. We therefore suggest that an intravascular route may be better for CsA administration because the intrathecal injection may exacerbate brain edema as found in our previous study.

Keywords: Cyclosporin A; intrathecal administration; intravenous administration; controlled cortical impact; focal brain injury; brain edema; gravimetric method.

Introduction

Recent evidence shows that Cyclosporin A (CsA) protects against various types of neuronal injury by its ability to prevent mitochondrial permeability transition during exposure to high levels of calcium or oxidative stress [1, 5–7]. CsA administered intrathecally or intravenously has been shown by our laboratory (unpublished data) and others to be neuroprotective in the experimental animal model of traumatic brain injury (TBI) as assessed by the blunting of N-acetylaspartate reduction [5]. However, our studies utilizing intrathecal administration resulted in a concomitant increase of brain edema, thus suggesting that there may be some toxic effect of intrathecally-injected CsA.

This is critical, since CsA is now in clinical trials for treatment of head injury and because edema is commonly believed to contribute to the high rate of secondary morbidity and mortality after TBI. The aim of this study was to assess whether intravascular administration of CsA may influence the brain edema following experimental cortical contusion.

Material and methods

All animals received humane care in compliance with the VCU Institutional Animal Care and Use Committee Regulations, according to NIH standards and guidelines. The present study includes two independent series in which different doses of CsA were intravenously given to the animals of each group. In the first series, 10 adult male Sprague-Dawley rats (350–400 g) were exposed to focal brain injury and randomized into the following two groups: intravenous injection of 20 mg/kg CsA ($n = 5$); intravenous injection of vehicle ($n = 5$). In the second series, 12 rats were also injured and randomized into the following two groups: intravenous injection of 35 mg/kg CsA ($n = 6$); intravenous injection of vehicle ($n = 6$).

All rats were given free access to food and water before surgery. General anesthesia was induced with 4% halothane. The animals were then intubated under direct vision and mechanically ventilated receiving a gas mixture of 70% nitrous oxide, 30% oxygen, and 1 to 1.5% halothane. A femoral vein was cannulated with polyethylene tubing for CsA or vehicle (cremaphor + 32% alcohol) infusion. The animals were placed in a stereotaxic frame to secure the head. A midline scalp incision was made and a craniotomy was performed in order to expose the right hemisphere of the brain to a controlled cortical impact (3 mm depth at a rate of 6 m/sec). After the insult, each animal underwent a cranioplasty and a skin closure.

An intravenous continuous (1 hour) infusion of either CsA or vehicle via the femoral vein was begun 30 minutes after the insult. After a removal of the cannula and a closure of the femoral wound, the animals were allowed to recover from the anesthesia. All animals were sacrificed at 24 hours post injury. The brain was immediately removed and cut into 3 mm-thick coronal sections. A 1.5 mm punch

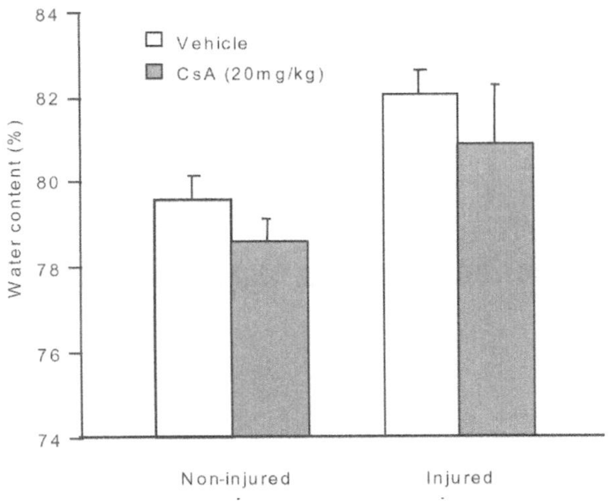

Fig. 1. Intravenously-administered CsA of 20 mg/kg/hr decreased the brain water content in both the contused and non-contused cortex, although the amelioration was not significant

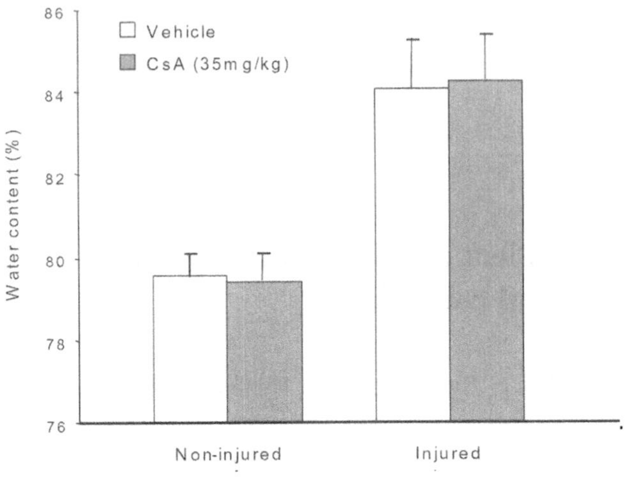

Fig. 2. The higher dose (35 mg/kg/hr) of CsA had no effect on the brain edema formation

was used to collect brain tissue in both the white and gray matter. Brain tissue water content was then measured by means of the gravimetric method (kerosene/bromobenze column) as described by Marmarou *et al.* [4].

All data were expressed as mean ± SD. Statistical analysis was performed by the Student's t test. Values of $p < 0.05$ were considered significant and all analyses were performed using the StatView Version 5.0 statistical package.

Results

In the first series, intravenously-administered CsA of 20 mg/kg decreased the brain water content in the contused cortex (80.8 ± 1.5%) as well as the contralateral non-contused cortex (78.6 ± 0.6%), although the amelioration was not significant compared to the control group (82.0 ± 0.5, 79.6 ± 0.5%, respectively, Fig. 1). In the second series, the higher dose of CsA (35 mg/kg) had no effect on brain edema formation (Fig. 2).

Discussion

Recent evidence suggests that CsA is neuroprotective against brain injuries by its ability to prevent mitochondrial permeability transition [1, 5–7], whereas it is originally used as an immunosuppressant drug in the clinical setting, especially in transplant patients. However, this drug may have a potentially critical side effect of neurotoxicity which may exacerbate brain edema formation [2, 3, 8]. It is therefore important to resolve this problem before this drug is applied in the clinical setting because edema is commonly believed to contribute to the high rate of secondary morbidity and mortality after TBI. Our previous research showed that an intrathecal injection of CsA (10 mg/kg) increased the brain water content, whereas it ameliorated the biochemical damages in the rat model of diffuse brain injury (unpublished data). Conversely, the present study has shown that the intravenous-administration did not exacerbate the brain edema formation at a dose of either 20 or 35 mg/kg while only the higher dose of CsA achieved neuroprotective effects as found in our intrathecal study.

These results suggest that an intravascular route may be better for CsA administration in both the laboratory setting and certainly in patients with traumatic brain injury. Although we have achieved neuroprotection without increased brain water at 35 mg/kg, further studies are needed to assess the optimal dose of an intravenous infusion of CsA.

Acknowledgment

The authors thank Angela M. Amorini, Ph.D. for her expert technical assistance.

References

1. Buki A, Okonkwo DO, Povlishock JT (1999) Postinjury cyclosporin A administration limits axonal damage and disconnection in traumatic brain injury. J Neurotrauma 16: 511–521
2. Gijtenbeek JM, van den Bent MJ, Vecht CJ (1999) Cyclosporine neurotoxicity: a review. J Neurol 246: 339–346
3. Hauben M (1996) Cyclosporine neurotoxicity. Pharmacotherapy 16: 576–583

4. Marmarou A, Tanaka K, Shulman K (1982) An improved gravimetric measure of cerebral edema. J Neurosurg 56: 246–253
5. Okonkwo DO, Povlishock JT (1999) An intrathecal bolus of cyclosporin A before injury preserves mitochondrial integrity and attenuates axonal disruption in traumatic brain injury. J Cereb Blood Flow Metab 19: 443–451
6. Okonkwo DO, Buki A, Siman R, Povlishock JT (1999) Cyclosporin A limits calcium-induced axonal damage following traumatic brain injury. Neuroreport 10: 353–358
7. Scheff SW, Sullivan PG (1999) Cyclosporin A significantly ameliorates cortical damage following experimental traumatic brain injury in rodents. J Neurotrauma 16: 783–792
8. Shah AK (1999) Cyclosporine A neurotoxicity among bone marrow transplant recipients. Clin Neuropharmacol 22: 67–73

Correspondence: Anthony Marmarou, Ph.D., Division of Neurosurgery, Medical College of Virginia, Virginia Commonwealth University, P.O. Box 950508, Richmond, Virginia 23298-0508, U.S.A. e-mail: marmarou@abic.vcu.edu

Acta Neurochir (2003) [Suppl] 86: 305–307

The glutamate AMPA receptor antagonist, YM872, attenuates regional cerebral edema and IgG immunoreactivity following experimental brain injury in rats

T. Atsumi[1], S. Hoshino[2], T. Furukawa[2], S. Kobayashi[3], T. Asakura[2], M. Takahashi[4], Y. Yamamoto[1], and A. Teramoto[2]

[1] Department of Emergency and Critical Care Medicine, Nippon Medical School, Tokyo, Japan
[2] Department of Neurosurgery, Nippon Medical School, Tokyo, Japan
[3] Department of Neurosurgery, Nippon Medical School, Chiba Hokusoh Hospital, Chiba, Japan
[4] Neuroscience Research, Pharmacology Laboratories, Institute for Drug Discovery Research, Yamanouchi Pharmaceutical Co. Ltd., Ibaraki, Japan

Summary

We previously reported the neuroprotective effects of the glutamate AMPA receptor antagonist YM872 on neurobehavioral motor function and cortical tissue loss (lesion volume) in a brain-injured rat model. Here we examined its effect on brain edema and the breakdown of the blood-brain barrier (BBB). Rats subjected to severe right lateral (parasagittal) fluid-percussion brain injury or sham injury received a 4-hr intravenous infusion of YM872 (20 mg/kg/hr, 20 mg/3 ml) or normal saline starting at 15 min post-injury. At 48 hr we removed their brains and evaluated the cerebral regional edema by the wet weight/dry weight method. Another group of rats was transcardially fixed with 10% formalin at 2 weeks after injury. Serial brain sections were immunostained for endogenous IgG and the extent and intensity of staining were evaluated. The administration of YM872 resulted in a significant reduction in regional cerebral edema in the injured parietal cortex and a markedly reduced area of IgG immunoreactivy in the injured cortex. Our results indicate that the post-traumatic administration of YM872 may be neuroprotective by reducing BBB breakdown and regional cerebral edema.

Keywords: AMPA receptor antagonist; blood-brain barrier; brain edema; glutamate; traumatic brain injury; YM872.

Introduction

The excitatory amino acid neurotransmitter glutamate is neurotoxic and is thought to mediate neuronal damage after traumatic brain injury (TBI) by excessive activation of glutamate receptors. Many reports demonstrated the neuroprotective effect of antagonists directed at both the NMDA and the AMPA glutamate receptors in experimental TBI; they point to secondary and delayed injury processes involving glutamate neurotoxicity.

We previously observed the neuroprotective effects of the AMPA receptor antagonist YM872. Compared to vehicle-treated animals, brain-injured, YM872-treated rats showed significant alleviation of neurobehavioral motor dysfunction and a reduction in cortical tissue loss (lesion volume). In the present study we examined the effect of YM872 on brain edema and the breakdown of the blood-brain barrier (BBB).

Materials and methods

Surgical procedures and brain injury

All experimental procedures used had the prior approval of the Animal Protection Committee of Nippon Medical School. Male Sprague-Dawley rats weighing 350–400 g were subjected to severe right lateral (parasagittal) fluid percussion (FP) brain injury. We modified our previously described procedure [2] to increase the severity of the traumatic brain injury (TBI) and used a device developed at the Department of Biotechnology, University of California, San Francisco, CA. Briefly, the rats were anesthetized with chloral hydrate (360 mg/kg, i.p.), and a polyethylene tube 50 was cannulated into the left jugular vein. Using a dental drill, a circular, 5-mm diameter craniectomy was made in the center of the parietal bone. A siliconized tube filled with normal saline was placed so as to contact the intact dura and fixed securely to the skull with dental acrylic cement. The FP injury device was connected to the skull via a Luer-Lok fitting and severe FP brain injury (5.0–6.0 atm) was inflicted. Throughout the surgical procedure, the rectal temperature was maintained at 37–38 °C with a thermostatically controlled heating pad.

Drug preparation and administration

YM872 (20 mg/3 ml; Yamanouchi Pharmaceutical) was dissolved in physiological saline and the solution was adjusted to pH 7.4 with 1 N NaOH [7]. At 15 min after FP injury, randomly selected rats were started on a 4-hr intravenous infusion of either YM872 (20 mg/kg/hr) or normal saline (NS; 3 ml/kg/hr). The rats were conscious and able to move freely during the infusion.

Regional cerebral edema

Animals that survived the TBI (brain-injured and YM872-treated, n = 6; brain-injured and NS-treated, n = 6; sham-injured and NS-treated, n = 6) were killed at 48 hr post-injury for the evaluation of post-traumatic cerebral edema. They were anesthetized and decapitated and a 3-mm-thick coronal slice was cut *in situ* at the parieto-occipital level. The brains were placed on an ice-cooled plate and the following sections were obtained from both hemispheres: the parietal cortex (at the impact site), the adjacent cortex (temporal cortex, injured side only), hippocampus, and thalamus. The degree of edema was calculated using the wet weight/dry weight technique. Briefly, fresh brain tissue was weighed on a piece of aluminum foil, dried for 24 hr at 100 °C, and re-weighed. Edema was defined as the percent of brain water content, where

$$\% \text{ water content} = [(\text{wet weight} - \text{dry weight})/\text{wet weight}] \times 100.$$

IgG immunostaining

At 2 weeks after surgery/injury, survived rats (brain-injured and YM872-treated, n = 10; brain-injured and NS-treated, n = 10; sham-injured and NS-treated, n = 2) were deeply anesthetized with pentobarbital (100 mg/kg i.p.) and transcardially perfused with 10% phosphate-buffered formalin. Their brains were embedded in paraffin and cut into 6 μm sections from 1.3 to 6.3 mm posterior to the bregma. Sections at the level bregma +3.3 were immunostained. After deparaffinization and rehydration, the sections were placed in 0.01 M citrate buffer and then in a microwave oven (95 °C, 10 min). Subsequently, they were immersed overnight in biotinylated goat anti-rat IgG (Vector, Burlingame, CA, USA; 1:100) in 5% normal goat serum. Bound antibody was detected with the avidin-biotin complex method (ABC Elite kit, Vector); 3,3′-diaminobenzidine was used as the chromogen. In the negative controls, normal serum was used instead of primary antibodies. The area of IgG-immunoreactivity was measured with imaging software (NIH image version 1.61). The staining intensity of the median eminence where the BBB is normally absent was used as the internal standard.

Statistical analysis

Regional cerebral edema results were compared with two-way ANOVA followed by Fisher's PLSD test. Comparisons of the IgG immunoreactive area between YM872- and saline-treated rats were performed with unpaired two-tailed Student's *t* test. These data were expressed as the mean ± standard error. Values of $p < 0.05$ were considered statistically significant.

Results

Regional cerebral edema

At 48 hr, a significant increase in tissue water content was observed in the injured parietal cortex ($p < 0.05$) and the thalamus ($p < 0.01$) of vehicle-treated rats compared to vehicle-treated sham-injured animals. The administration of YM872 resulted in a remarkable reduction in regional cerebral edema in the injured parietal cortex ($p < 0.01$). No significant increase in tissue water content was noted in the examined contralateral brain regions in either drug- or saline-treated brain-injured rats.

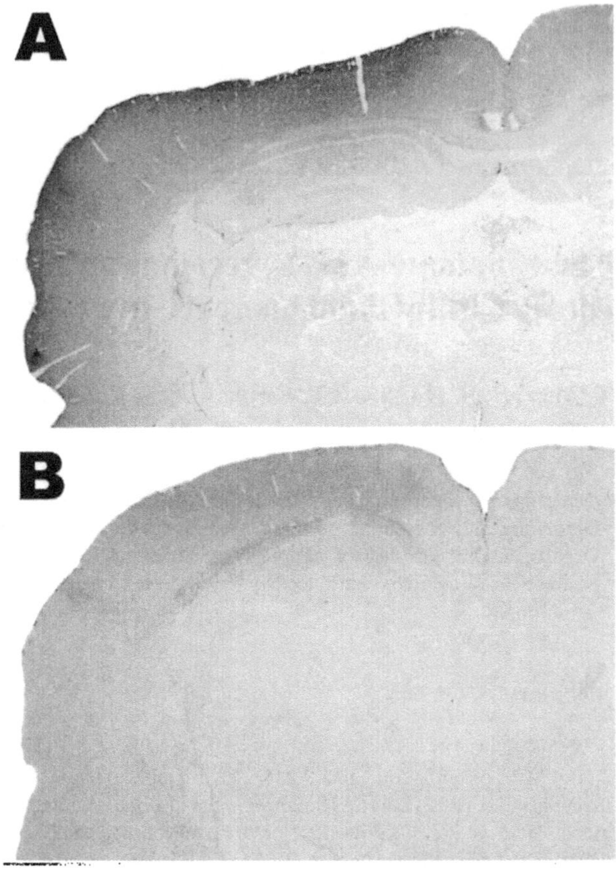

Fig. 1. Representative photomicrographs of IgG immunostaining in a normal saline-treated (A) and a YM872-treated (B) rat brain at 2 weeks after fluid-percussion brain injury

IgG immunoreactivity

In the NS-treated, brain-injured animals, intense IgG immunoreactivity was noted over the injured parietal cortex, and faint immunoreactivity was demonstrated in the injured hippocampus (Fig. 1). Post-injury treatment with YM872 resulted in a marked reduction of the area and intensity immunostained by anti-rat IgG ($p < 0.01$, Fig. 2).

Discussion

We found that YM872, a novel AMPA receptor antagonist, significantly reduced regional cerebral edema and BBB breakdown after severe FP brain injury in rats. Ours is the first report demonstrating the ability of the AMPA antagonist to alleviate regional

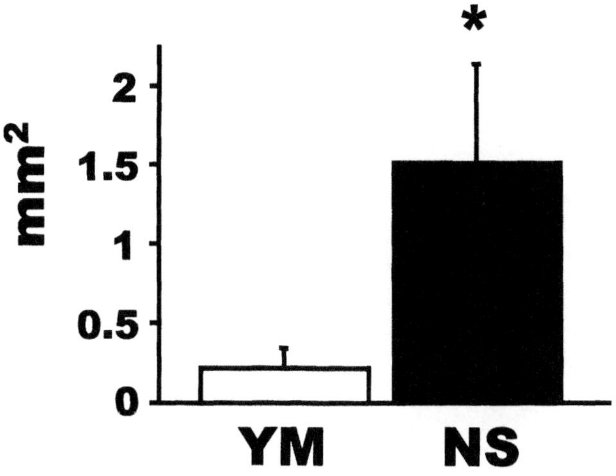

Fig. 2. IgG-immunostained area in the injured hemisphere of YM872-treated (*YM*) and normal saline-treated (*NS*) rat brain at 2 weeks after fluid-percussion brain injury. * $P < 0.01$ (unpaired Student's *t*-test)

cerebral edema and prolonged IgG immunoreactivity following experimental TBI.

Reinert *et al.* [5] reported an inverse relationship between increased extracellular potassium and intracranial pressure after severe TBI in humans. They proposed that prolonged ionic dysfunction, which could lead to massive swelling of astrocytes, is a major contributor to increased intracranial pressure and to the unfavorable outcomes seen in patients with severe TBI. Johansson *et al.* [3] also posited that the anti-edematous effect of NBQX, an AMPA antagonist, against BBB opening is related to cellular events within the brain rather than its ability to cross the BBB. Our findings that YM872 reduced post-traumatic edema in brain-injured rats may be attributable to a blockade of prolonged ionic dysfunction and the consequent cytotoxic edema.

In the present study, post-injury treatment with YM872 markedly reduced the area immunostained with anti-rat IgG. This observation strongly suggests that YM872 exerts a protective effect on the BBB after TBI. There have been many reports that show regulation of vascular permeability by glutamate. Based on their in vivo study, Mayhan and Didion [4] concluded that topical application of glutamate increases the permeability of the BBB and dilates cerebral arterioles via a nitric oxide-dependent mechanism. According to Collard *et al.* [1], extracellular glutamate decreases endothelial vasodilator-stimulated phosphoprotein by activating metabotropic glutamate receptors (mGluR1, mGluR4, and mGluR5), and increases endothelial permeability. Moreover, St'Astny *et al.* [6] recently reported that rat brain capillaries and albuminally adhering astrocyte processes carry functional NMDA and AMPA glutamate receptors and proposed that the effect of glutamate agonists and antagonists in the modulation of BBB function is mediated directly by cerebral microvessels. Considering these and our results, we posit that YM872 modulates the BBB by acting directly on endothelial cells and/or astrocytes. The current study suggests that YM872 is an effective neuroprotective agent following TBI and that its administration may reduce intracranial pressure in patients with severe head injury.

References

1. Collard CD, Park KA, Montalto MC, Alapati S, Buras JA, Stahl GL, Colgan SP (2002) Neutrophil-derived glutamate regulates vascular endothelial barrier function. J Biol Chem 277: 14801–14811
2. Hoshino S, Kobayashi S, Nakazawa S (1996) Prolonged and extensive IgG immunoreactivity after severe fluid-percussion injury rat brain. Brain Res 711: 73–83
3. Johansson BB, Westergren I (1994) Pharmacological reduction of brain edema induced by intracarotid infusion of protamine sulphate: a comparison between a free radical scavenger and an AMPA receptor antagonist. Acta Neurochir (Wien) [Suppl] 60: 128–131
4. Mayhan WG, Didion SP (1996) Glutamate-induced disruption of the blood-brain barrier in rats. Role of nitric oxide. Stroke 27: 965–969
5. Reinert M, Khaldi A, Zauner A, Doppenberg E, Choi S, Bullock R (2000) High level of extracellular potassium and its correlates after severe head injury: relationship to high intracranial pressure. J Neurosurgery 93: 800–807
6. St'Astny F, Schwendt M, Lisy V, Jezova D (2002) Main subunits of ionotropic glutamate receptors are expressed in isolated rat brain microvessels. Neurol Res 24: 93–96
7. Takahashi M, Ni JW, Kawasaki-Yatsugi S, Toya T, Ichiki C, Yatsugi S, Koshiya K, Shimizu-Sasamata M, Yamaguchi T (1998) Neuroprotective efficacy of YM872, an alpha-amino-3-hydroxy-5-methylisoxazole-4-propionic acid receptor antagonist, after permanent middle cerebral artery occlusion in rat. J Pharmacol Exp Ther 287: 559–566

Correspondence: Shigeru Hoshino, M.D., Department of Neurosurgery, Heiwadai Hospital, 834-28, Abiko, Chiba, 270-1101, Japan. e-mail: shigeru@nms.ac.jp

Acta Neurochir (2003) [Suppl] 86: 309–311

Effects of edaravone on experimental brain injury in view of free radical reaction

H. Nakamura[1], **M. Uzura**[2], **K. Uchida**[1], **H. Nakayama**[1], **Y. Furuya**[1], **T. Hayashi**[1], **H. Sekino**[1], **M. Ominato**[3], and **S. Owada**[3]

[1] Department of Neurosurgery, St. Marianna University School of Medicine, Kanagawa, Japan
[2] Department of Emergency & Critical Care Medicine, St. Marianna University School of Medicine, Kanagawa, Japan
[3] Department of Medicine, Akabane Central Hospital, Tokyo, Japan

Summary

The purpose of this study is to clarify the influence of a novel free radical scavenger edaravone on experimental brain injury. Male Wistar rats were anesthetized with 1–2% halothane. Brain injury was produced using a controlled cortical impact injury device. Experimental rats were divided into 2 groups. In the edaravone group, edaravone (3 mg/kg) was twice administered intravenously for 30 minutes. In the saline group, saline solution was administered in the same way. This administration of edaravone or saline solution made it possible to evaluate the relative effects of edaravone by assessment of free radical reaction and water content. As a result, the level of oxygen free radicals at 50 minutes after brain injury was significantly lower in the edaravone group than in the saline group. The water content in the injured brain at 180 minutes was significantly lower in the edaravone group than in the saline group. Therefore, we propose that edaravone may be effective for treatment in head injury.

Keywords: Controlled cortical impact injury; free radical reaction; ESR spectroscopy; edaravone.

Introduction

Recent studies have shown that oxygen free radicals are one of the important factors that induce secondary brain injury. Langfitt *et al.* emphasized 6 mechanisms which lead to secondary injury in brain tissue: direct mechanical injury of axons, brain ischemia and brain swelling, hemorrhage, edema, neurotransmitter failure, and toxic substances [4]. Free radicals are one of such toxic substances produced following traumatic brain injury. Recently the free radical scavenger, edaravone, has been used clinically and is reported to be effective for cerebral ischemic diseases. No new therapy or drug has made any remarkable advance in the treatment of head injury.

Therefore, we scheduled an experimental study using rats with brain injury in order to to evaluate the effects of edaravone in the injured brain tissue. And we discussed the time window for combating free radical reaction on traumatic brain injury in terms of administration of edaravone.

Materials and methods

Seventy two male Wistar rats, 8~9 weeks old, weighing 220~350 g, were used in this experiment. Standard diet and water were provided *ad libitum*, and the rats were kept under controlled conditions with respect to light, temperature, and humidity.

The experimental rats were anesthetized with 1~2% halothane* in a mixture of 50% oxygen and 50% air, and were allowed to breathe spontaneously. The rats were maintained at a rectal temperature of 37~38 °C using a heating pad. Teflon catheters were placed in the femoral artery and vein to continuously monitor blood pressure, to measure blood gas, and to administer edaravone or saline solution. The rats were mounted in a stereotaxic frame in the prone position, and a burr hole, 8 mm in diameter, centered 4 mm caudal from the coronal suture and 4 mm lateral from the midline was drilled through the right parietal cranium. Brain injury was produced using a controlled cortical impact injury device made by Virginia Commonwealth University [2] (conditions: impact velocity 3.2 m/sec, deformation 2 mm in depth, cylinder 5 mm in diameter). Each experiment was carried out after the scalp was sutured.

The experimental rats were divided into 2 groups. In the edaravone group, edaravone** (at 3 mg/kg in solution) was twice administered intravenously for 30 minutes, starting at 20 minutes and then at 80 minutes after the brain injury. In the saline group, saline solution was administered in the same way. This administration of edaravone or saline solution made it possible to evaluate the relative effects of edaravone.

Free radical reactions were measured using 300 MHz in vivo electron spin resonance (ESR) spectroscopy. ESR spectra were measured at 20, 50, 80, 110, and 180 minutes post injury (n = 6 for each point in time) in the edaravone and saline groups. The ESR spectrometer (JOEL, JES-RE3x) was equipped with a 300 MHz,

* Takeda Chemical Industries, Ltd. Osaka, Japan
** Mitsubishi Pharma Corp. Osaka, Japan

8.0 mW microwave power unit (JOEL, ES-LLBA21A) and a loop gap resonator. A sterilized solution of spin-labeled compound 140 mM (3-carbamoyl-2,2,5,5,-tetramethyl-pyrrolidine-1-oxyl: C-PROXYL*, pH 7.4, in solution in Sodium phosphate buffer) was injected into the femoral vein at 0.5 ml/100 g body weight. ESR spectra were measured for 10 minutes immediately after each scheduled injection of C-PROXYL. This technique is called the spin trapping method [7]. Signal intensity in the head implies the amount of C-PROXYL as a nitroxide radical spin probe. Graphed signal intensity by ESR spectroscopy shows the C-PROXYL curve on a line graph. The kinetic constants (/min) of spin clearance for C-PROXYL were used to calculate the inclination of the C-PROXYL curves and to assess the level of free radical reaction in the head.

We adopted water content to assess brain edema. At 180 minutes after the brain injury, the whole brain was removed within 60 seconds after decapitation. Peripheral damaged brain tissue from the site of impact was removed for samples. The water content at 180 minutes after the brain injury in the edaravone group (n = 6) and saline group (n = 6) was calculated as the percentage difference in tissue weight before and after drying for 1 day at 100 °C.

Data was presented as mean ± standard deviation. The data was compared using student's-t test between the edaravone group and saline group at the same point in time. A value of $p < 0.05$ was considered significant.

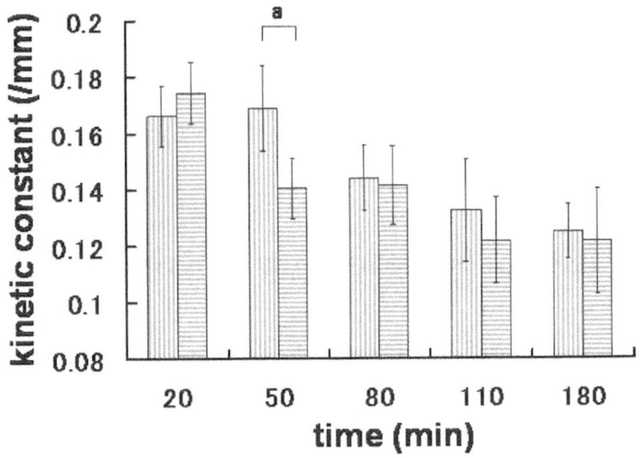

Fig. 1. Time course of the kinetic constants of spin clearance for C-PROXYL. Data was presented as mean ± standard deviation. (*a*) $p < 0.05$ vs saline group. ▥ saline; ▤ edaravone

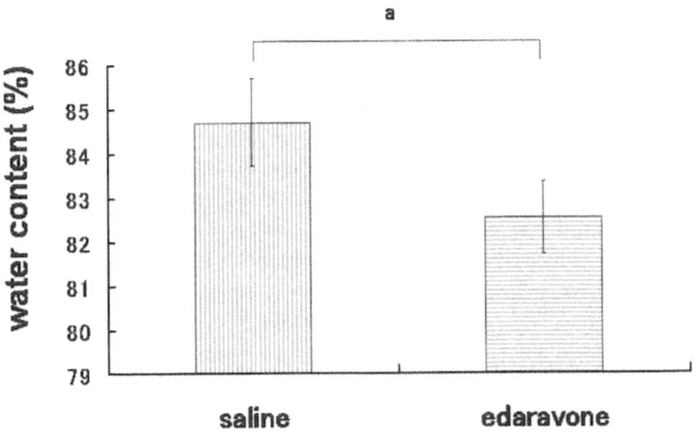

Fig. 2. Water content of the injured brain in the saline and edaravone groups. Data was presented as mean ± standard deviation. (*a*) $p < 0.05$ vs saline group

Results

In the course of this experiment, no significant difference was observed between the saline and edaravone groups in mean arterial blood pressure and arterial blood gas.

In the saline group, the kinetic constants of spin clearance for C-PROXYL increased up to 50 minutes and then gradually decreased up to 180 minutes. In the edaravone group, the kinetic constants of spin clearance for C-PROXYL increased up to 20 minutes and then gradually decreased up to 180 minutes. The most significant difference between the two groups was apparent at 50 minutes after brain injury ($p < 0.05$, Fig. 1).

The water content of the injured brain in the edaravone group was significantly lower at 180 minutes than in the saline group ($p < 0.05$, Fig. 2).

Discussion

There are several pathways in aerobic cells that can lead to the production of oxygen free radicals. The pathways consist of intracellular production from the mitochondria, auto-oxidation of catecholamines, the conversion of xanthine dehydrogenase to xanthin oxidase, activated neutrophils, and the arachidonic acid cascade [3]. The source of the oxygen free radicals observed in our study is not known. However, we measured oxygen free radicals generated by traumatic brain injury in addition to oxidative stress. At the time, we adopted 300 MHz in vivo ESR spectroscopy using a spin trapping method. Oxygen free radical reaction is short-lived in vivo, and the sequential univalent reduction of oxygen in the cerebral cells produces reactive products in sequence. C-PROXYL is a stable nitroxyl radical as a spin trapping drug. When C-PROXYL reacts to oxygen free radical generated traumatic brain injury, ESR spectra decrease as a result of the reaction with oxygen free radicals [7]. Therefore, in vivo ESR spectroscopy makes it possible to estimate free radical reactions in living animal non-invasively.

In this study, the kinetic constants of spin clearance for C-PROXYL were significantly lower at 50 minutes

* Sigma Chemical Co. St. Louis, USA

and the water content in the injured brain tissue at 180 minutes was significantly lower than in the saline group by edaravone was twice administered after the brain injury. Subsequent to our study in detail, there was cerebral tissue concentration of malondialdehyde (MDA) increased markedly from 80 to 180 minutes in the saline group [5]. And administration of edaravone twice delayed an increase in cerebral tissue concentration of MDA in the edaravone group. MDA is one of the end products of lipid peroxidation, and oxygen free radicals are one of the lipid peroxidation initiators. Therefore, lipid peroxidation measured by the level of MDA provides indirect evidence of oxygen free radicals. It has been reported that oxygen free radical reactions are very likely to play an important role in the pathogenesis of vasogenic brain edema [1, 6]. Free radical reactions induced and worsened the traumatic brain injury, and were one of the important factors which aggravated the brain condition following injury. These results indicate that edaravone attenuated not only the generation of oxygen free radicals induced by brain injury, but also lipid peroxidation lagging behind oxygen free radical generation, which together decreased the water content in the edaravone group by attenuating cell damage. Therefore, edaravone may be effective for treatment of head injury. There is a need for administration of free radical scavenger before the peak to cope effectively with free radical reactions. We propose that therapeutic window for combating free radical reaction with worsening condition of traumatic brain injury may lie within 50 minutes after brain injury at least. More detailed experiment will be needed.

References

1. Chan PH, Longar S, Fishman A (1984) Oxygen-free radical: potential edema mediators in brain injury. In: Inaba Y, Klatzo, Spatz M (eds) Brain edema. Springer Berlin Heidelberg New York Tokyo, pp 317–323
2. Dixon CE, Clifton GL, Lighthall JW, Yaghmai AA, Hayes RL (1991) A controlled cortical impact model of traumatic brain injury in the rat. J Neurosci Meth 39: 253–262
3. Ikeda Y, Long DM (1990) The molecular basis of brain injury and brain edema: the role of oxygen free radicals. Neurosurgery 27: 1–11
4. Langfitt TW, Gennarelli TA, Obrist WD, Bruce DA, Zimmerman RA (1982) Prospects for the future in the diagnosis and management of head injury: pathophysiology, brain imaging, and population-based studies. Clin Neurosurg 29: 353–376
5. Nakamura H, Sekino H (2002) Free radical reactions and effects of edaravone on experimental brain injury. St Marianna Med J 30 (preparation)
6. Olesen SP (1987) Free oxygen radicals decrease electrical resistance of microvascular endothelium in brain. Acta Physiol Scand 129: 181–187
7. Utsumi H, Ichikawa K, Takeshita K (1995) In vivo ESR measurements of free radical reactions in living mice. Toxicol Lett 82–83: 561–565

Correspondence: H. Nakamura, 2-16-1 Sugao, Miyamae-ku, Kawasaki-city, Kanagawa pref., Japan. e-mail: Homare@marianna-u.ac.jp

Acta Neurochir (2003) [Suppl] 86: 313–319

Antioxidant compounds EGB-761 and BN-52021 attenuate brain edema formation and hemeoxygenase expression following hyperthermic brain injury in the rat

H. S. Sharma[1], **K. Drieu**[2], and **J. Westman**[1]

[1] Laboratory of Neuroanatomy, Department of Medical Cell Biology, Biomedical Centre, Uppsala University, Uppsala, Sweden
[2] Institute Henri Beaufour-IPSEN-Pharma, Paris, France

Summary

Role of carbon monoxide (CO) in hyperthermic brain injury induced brain pathology was examined in a rat model using immunohistochemistry of the hemeoxygenase-2 (HO-2) enzyme. Exposure of rats to 4 h heat stress at 38 °C resulted in profound hyperthermia, breakdown of the blood-brain barrier (BBB), brain edema formation, cell damage and expression of HO-2 in several brain regions. Pretreatment with potent antioxidant compounds EGB-761 and BN-52021 markedly reduced the HO-2 expression, BBB breakdown, brain edema formation and cell damage without attenuating the hyperthermic response. This effect was most marked in animals treated with EGB-761. These observations suggest that upregulation of HO-2 representing generation of CO plays important roles in hyperthermic brain injury, and oxidative stress seems to be one of the most important signals in inducing HO-2 expression in hyperthermia, not reported earlier.

Keywords: Hyperthermia; hemeoxygenase-2; carbon monoxide; brain injury; EGB-761; BN-52021; immunohistochemistry; edema formation; blood-brain barrier; Evans blue; radioiodine.

Introduction

The pathophysiology of hyperthermic brain injury includes alteration in the blood-brain barrier (BBB) permeability, edema formation and cell injury in the central nervous system (CNS) [17]. Several neurochemical mediators such as serotonin, prostaglandin, histamine and nitric oxide (NO) appear to participate in the pathological changes following hyperthermia induced by heat stress [22].

Since no single chemical compound is solely responsible for all the pathological changes seen in hyperthermic brain injury, it appears that other potential neurochemicals and endogenous factors are involved in brain pathology of hyperthermia. In this investigation, we examined the involvement of recently described signal transduction molecule, carbon monoxide (CO) in the pathophysiology of hyperthermic brain injury.

CO is a free radical gas which influences neuronal communication by modulating signal transduction in the CNS [8, 21]. The CO is a by product of the reaction when heme is converted into the antioxidant biliverdin and iron [7]. The enzyme responsible for conversion of heme into the CO is known as hemeoxygenase (HO) [7, 15] that occurs in two isoforms [7, 21]. The constitutive isoform of HO (HO-2) is present in the few normal nerve cells in the CNS [8, 17]. The inducible isoform of HO (HO-1) can be seen following several noxious stimuli to the brain such as ischemia, hypoxia and trauma [for review, see 15]. Recently, upregulation of HO in the CNS is described following a variety of chemical, hormonal and biological stressors [7, 15, 19]. The functional significance of HO upregulation is not clearly understood.

There are reasons to believe that CO is involved in several biological processes in the CNS [15, 17]. Previous reports from our laboratory suggest that hyperthermia caused by heat exposure is able to induce an upregulation of HO-2 enzyme in the CNS [13, 15]. This observation indicates that hyperthermia has the capacity to induce an upregulation of HO-2. However, whether increased production of CO as evidences with HO-2 expression in hyperthermia is associated with cell injury or cell survival is still not clearly known.

Hyperthermia is capable to induce release of free radicals probably due to an increase in the oxidative stress [4, 5] that contributes to hyperthermia induced cell damage in the CNS [5, 8]. Since CO is a free radical gas, it seems likely that its release in hyperthermia will induce cell damage. Thus, antioxidant compounds capable to inhibit or scavenge the free radicals will induce neuroprotection in hyperthermia [18, 19] and thus influence cellular stress mechanisms [17].

The present investigation was thus undertaken to investigate the influence of two antioxidant compounds EGB-761 and Gingkolide B (BN 52021) [1, 2, 9, 10] on hyperthermia induced HO expression, brain edema formation, cell injury and cellular stress. We have focused our attention on the expression of HO-2 and cell damage in a few selected brain regions in the control and heat stressed rats with or without treatment with the EGB-761 or BN-52021.

Materials and methods

Animals

Experiments were carried out on 56 male Sprague Dawley rats (body weight 110–140 g, age 9–10 weeks) housed at controlled room temperature at 21 ± 1 °C with 12 h light and 12 h dark schedule. The rat food and tap water were supplied *ad libitum* before the experiments.

Heat exposure

Animals were exposed to heat in a biological oxygen demand (BOD) incubator maintained at 38 °C (relative humidity 45–50%; wind velocity 20–26 cm/sec) for 4 h [11, 12, 14, 17]. This experimental condition is approved by the Ethical Committee of Uppsala University, Uppsala, Sweden. Rats kept at normal room temperature for 4 h without food and water were served as controls [11, 12].

Drug treatments

Two potent antioxidant substances EGB-761, which is scavenger of the free radicals [2, 9, 18] and the Gingkolide B [9, 19] were administered as described earlier. EGB-761 (50 mg/kg/day, p.o.) is a standardised extract from Gingko biloba leaves and Gingkolide B (BN-52021, 2 mg/kg/day, p.o.) (IPSEN-Pharma, France) were administered daily for 5 days. One dose of these compounds (0.3 ml/rat in distilled water) was also administered 30 min before the onset of heat exposure [18, 19].

Stress symptoms

Changes in the core body temperature, occurrence of salivation and the behavioural prostration were recorded using standard protocol as described earlier [11] (Table 1).

HO-2 immunohistochemistry

The HO-2 immunostaining was examined on free floating Vibratome sections (40 µm thick) obtained from different brain regions using a commercial monoclonal HO-2 antiserum (Stress Gene Ltd., Canada) as described earlier [13–15, 17]. The HO-2 antiserum was diluted 1:5000 and applied at room temperature under continuous agitation for 48 h. The immune reaction was developed by using peroxidase-antiperoxidase technique. In few sections, the primary antibody was omitted and the reaction product was developed [15]. The control, heat stressed, drug-treated controls and drug-treated heat stressed rats were processed in parallel [15, 16].

Table 1. *Effect of EGB-761 and BN-52021 on stress symptoms and physiological variables in the control and 4 h heat stressed rats at 38 °C*

Type of Expt.	n	°C Rect. T	Salivation	Prostration	MABP torr	Arterial pH	PaO2 torr	PaCO2 torr	BBB permeability		Brain edema formation	
									EBA mg %	[131]Iodine %	Brain water %	% *f*
Control	5	–	nil	nil	110 ± 6	7.37 ± 0.03	81.34 ± 0.21	34.56 ± 0.33	0.24 ± 0.04	0.28 ± 0.03	74.34 ± 0.32	–
EGB-761 + control	5	−0.29 ± 0.05	nil	nil	108 ± 5	7.36 ± 0.08	80.98 ± 0.32	33.89 ± 0.76	0.18 ± 0.06	0.23 ± 0.05	74.18 ± 0.22	–
BN-52021 + control	5	+0.18 ± 0.08	nil	nil	112 ± 8	7.36 ± 0.04	81.02 ± 0.43	34.02 ± 0.44	0.26 ± 0.05	0.32 ± 0.08	74.47 ± 0.12	–
4 h HS	6	+3.48 ± 0.23***	++++	++++	78 ± 6***	7.34 ± 0.07	79.43 ± 0.98	32.89 ± 0.88	1.89 ± 0.24***	2.32 ± 0.28***	78.67 ± 0.34***	+18
EGB-761 + HS	5	+2.67 ± 0.18***a	++++	+++	82 ± 8***	7.33 ± 0.06	80.03 ± 0.76	33.04 ± 0.65	0.67 ± 0.11**a	0.89 ± 0.23**a	76.23 ± 0.12**a	+8
BN-52021 + HS	5	+2.89 ± 0.19***a	++++	++++	79 ± 6***	7.33 ± 0.09	79.89 ± 1.02	33.13 ± 0.87	0.86 ± 0.12**a	1.04 ± 0.18**a	76.98 ± 0.23**a	+12

Values are mean ± SD; *MABP* mean arterial blood pressure; *HS* heat stress; *nil* absent, ++++ severe, +++ mild; *** $P < 0.001$, ** $P < 0.01$ compared to control; *a* $P < 0.05$, compared to h HS, Student's unpaired t-test.

Blood-brain barrier permeability and edema formation

The BBB permeability to Evans blue and ^{131}I-iodine was examined using standard procedures described earlier [12]. After counting the radioactivity, tissue pieces were kept in an incubator at 90 °C for 72 h to determine their dry weight. The brain water content was calculated from the differences in the wet and dry weight of the samples as described earlier [13].

Cell injury

Tissue pieces from selected brain regions were processed for light and electron microscopy to study morphological changes using standard procedures [16].

Statistical analysis

Semiquantitative data were analysed using unpaired Student's t-test. A p-value less than 0.05 was considered as significant.

Results

Effect of EGB-751 and BN-52021 on stress response

Subjection of rats to 4 h heat stress resulted in profound hyperthermia (Table 1). Pretreatment with EGB-761 or BN-52021 slightly but significantly attenuated this hyperthermic response. However, the behavioural salivation or prostration following heat stress in these drug-treated rats were not affected (Table 1).

Effects of EGB-761 and BN-52021 on HO-2 expression

In normal animals only few cells showed mild HO-2 immunostaining distributed sporadically in the cortex, hippocampus, thalamus, hypothalamus in the brain stem (Fig. 1). Subjection of rats to 4 h heat stress resulted in massive upregulation of HO-2 expression in many brain regions. This was most extensive in the thalamus and hypothalamus. The nerve cell cytoplasm showed most immunostaining however, occasional nerve cell nuclei were also stained (Fig. 2). Pretreatment with EGB-761 or BN-52021 significantly reduced heat stress induced upregulation of HO-2 expression in several brain regions (Fig. 2). The influence of EGB-761 in attenuating heat stress induced HO-2 expression was most pronounced compared to the BN-52021 (Figs. 1 and 2).

BBB permeability and edema formation

Subjection of rats to 4 h heat stress resulted in marked extravasation of Evans blue albumin and radioactive iodine in the CNS (Table 1). This leakage of protein tracers was significantly attenuated by pretreatment with EGB-761 or BN-52021 (Gingkolide B) (Table 1). However, the effect of EGB-761 is far more superior in reducing the hyperthermia induced protein leakage in the brain compared to the Gingkolide B (BN 52021) alone (Table 1). These compounds are also able to thwart edema formation following heat stress (Table 1). This effect is most pronounced in the EGB-761 treated rats compared to the BN 52021 treatment (Table 1).

Cell injury in heat stress

Hyperthermia resulted in nerve cell distortion, edema, sponginess and general expansion of the brain (Fig. 2). These effects are significantly attenuated by pretreatment with EGB-761 or BN-52021 (Fig. 2). However, the effects of EGB-761 are far more superior than that of BN-52021 in reducing cell injury in the animals subjected to 4 h heat stress (Fig. 2).

Discussion

The present findings clearly show that heat stress induced hyperthermic brain injury has the capacity to induce an upregulation of HO-2 in several brain regions. This observation is in line with the idea that CO plays important roles in neuronal injury [13, 15]. There are several lines of evidences which suggest that upregulation of HO-2 is associated with increased production of CO [7, 21]. Thus, it seems likely that an abnormal production of CO during hyperthermia is responsible for cell injury. A marked reduction in HO-2 expression and cell injury in heat stress by prior treatment with EGB-761 and BN-52021 in present study further confirms this idea.

EGB-761 is a well known scavenger of free radicals [9, 18, 19]. On the other hand, Gingkolide B (BN 52021) one of the active constituent of the EGB-761 is a powerful antioxidant [9, 10]. Thus, EGB-761 and BN-52021 are able to attenuate heat stress induced generation of free radicals either by scavangering them or by attenuating the oxidative and/or cellular stress mechanisms [18–20]. This could be one of the most important mechanisms responsible for reduction in cell changes caused by these compounds in hyperthermia. Interestingly, these compounds are also able to attenuate HO-2 expression in heat stress. This observation suggests that the oxidative stress and associated free

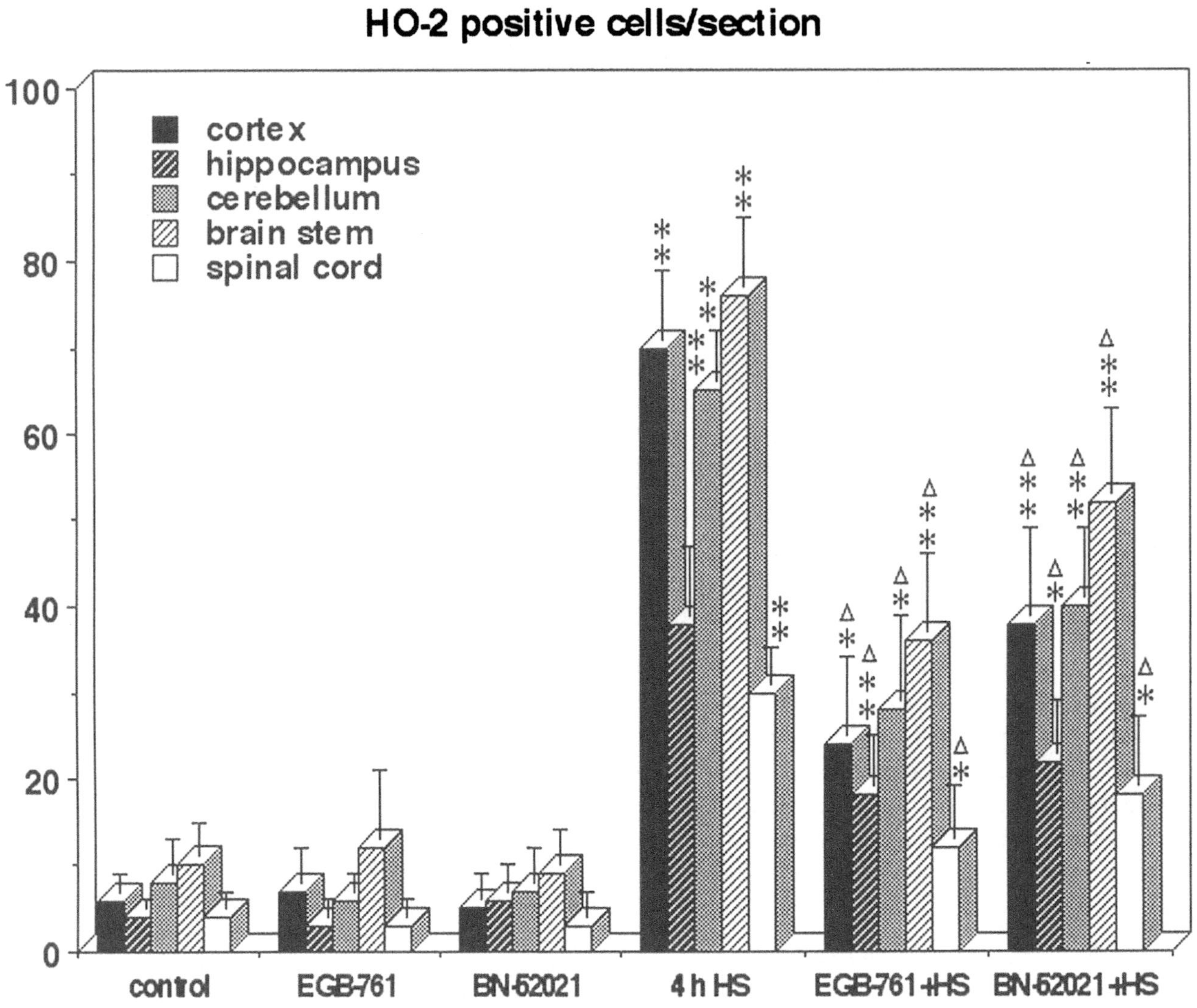

Fig. 1. Semiquantitative analysis of HO-2 immunostaining in control and heat stressed animals and their modification with EGB-761 or BN-52021. * = $P < 0.01$; ** = $P < 0.001$ compared with the control, D = $P < 0.01$ compared with 4 h heat stress (*HS*), Student's unpaired t-test

radical formation in hyperthermia is important in inducing HO-2 expression in the CNS.

Oxidative stress induces upregulation of nitric oxide synthase following trauma, ischemia or hypoxia [21]. Previous reports from our laboratory suggest that pretreatment with EGB-761 and BN-52021 reduce the upregulation of both constitutive and inducible isoforms of nitric oxide synthase [19]. This suggests that oxidative stress and free radical formation induce NOS upregulation in hyperthermia [18, 19, 21]. However, oxidative stress induced upregulation of HO is not well documented in the literature. Our results, probably for the first time using EGB-761 and BN-52021 in hyperthermia demonstrate that oxidative stress is somehow responsible for increased CO production as evident from an upregulation of HO-2 immunostaining. Taken together, these observations suggest that in hyperthermia NOS and HO seem to work in synergy [6] to induce cell damage.

Upregulation of CO as seen by increased immunostaining of HO-2 induces cell damage in spinal cord trauma [20]. Prevention of HO-2 upregulation in trauma is associated with neuroprotection. Since basic mechanisms of cell injury caused by trauma, ischemia or hypoxia appears to be quite similar in nature, it is likely that less intense immunoreaction of HO-2 in the EGB-761 or BN-52021 treated rats is related with mild cell damage in hyperthermia.

Hyperthermia is associated with severe cellular stress [4, 17] as seen by marked upregulation of heat shock proteins (HSP 72 kD) in the CNS [16, 17]. Cellular stress is caused by oxidative stress associated with

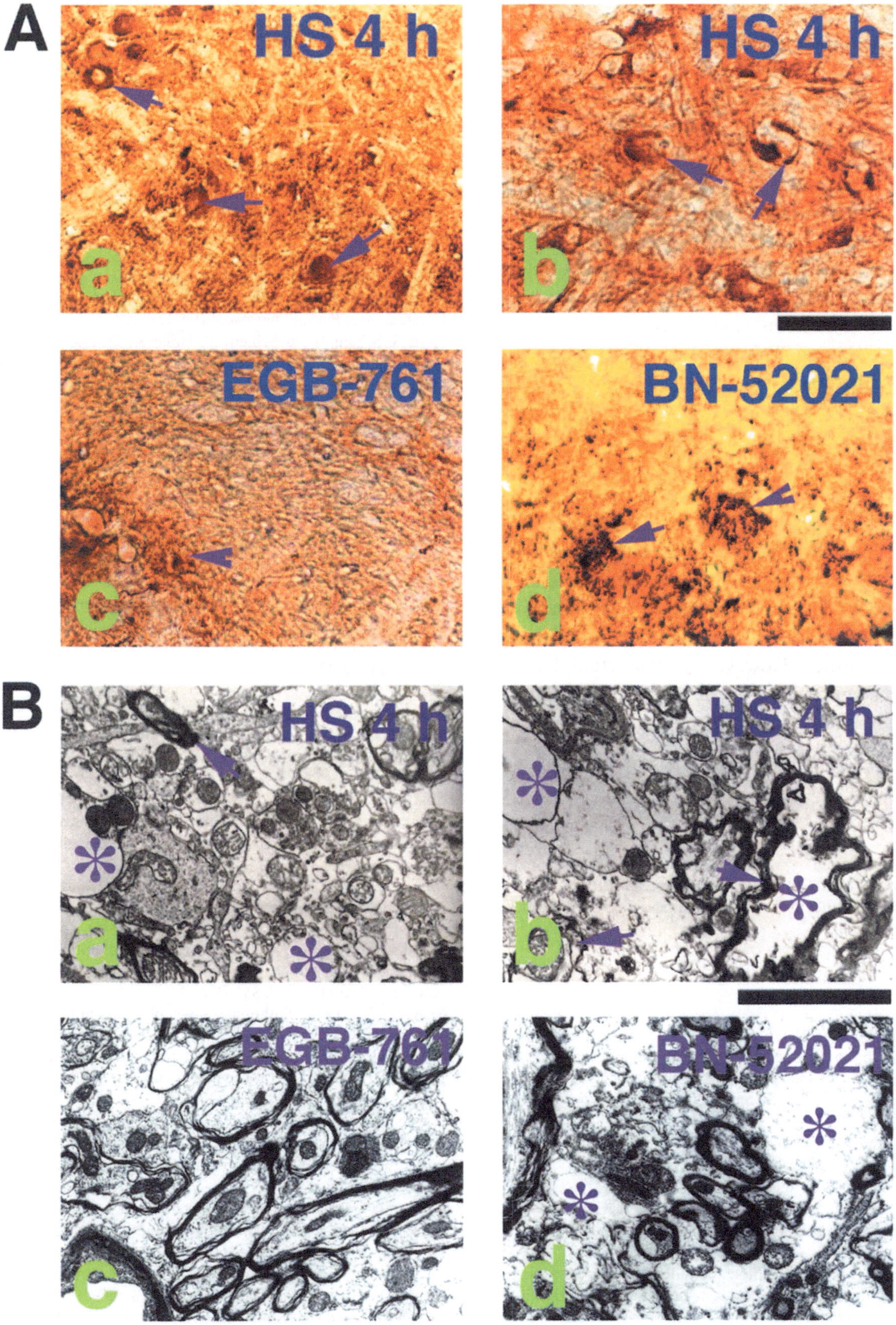

Fig. 2. (A) A representative example of HO-2 immunostaining in the spinal cord on one heat stressed rat (a, b) and its modification with EGB-761 (c) and BN-52021 (d) pretreatment. Bar = 50 μm. (B) Low power electron micrograph show pronounced cell injury in heat stressed rat (a, b) and its modification with EGB-761 (c) and BN 52021 (d) treatments. Marked neuroprotective effects are apparent in the EGB-761 treated stressed rat compared to the BN-52021 (bar –1 μm) (data modified after Westman *et al.*, 2000)

free radical formation [3, 5]. Since EGB-761 and BN-52021 are able to reduce HSP 72 kD expression in hyperthermia [22], it appears that these compounds are interfering with the molecular mechanims of cellular stress phenomena. Obviously, a less stress reaction in heat-stressed animals will result in mild release of free radicals.

Heat stress induced hyperthermia alone is not responsible for HO-2 expression or cell injury. Thus, EGB-761 or BN-52021 pretreatment resulted in a mild reduction in hyperthermia and no considerable reduction in the stress symptoms was observed in these drug-treated stressed group. This suggests that the magnitude of hyperthermia is not directly related with the severity of the heat stress induced behavioural symptoms. However, further studies are needed to clarify this point.

In conclusion, our study demonstrates that the antioxidants EGB-761 and its constituent Gingkolide B (BN 52021) has the capacity to attenuate heat stress induced HO-2 upregulation. This indicates that these compounds exert their neuroprotective effects in hyperthermic brain injury by influencing CO production, not reported earlier. However, further studies are needed to understand the mode of action of these compounds in attenuating HO-2 expression in hyperthermia.

Acknowledgments

This study is supported by grants from Swedish Medical Research Council (2710, JW, HSS) and IHB-Ipsen pharma, Paris, France. The skilful technical assistance of Kerstin Rystedt and Kärstin Flink are appreciated with thanks.

References

1. Bastianetto S, Zheng WH, Quirion R (2000) The Gingko biloba extract (EGb 761) protects and rescues hippocampal cells against nitric oxide-induced toxicity: involvement of its flavonoid constituents and protein kinase C. J Neurochem 74: 2268–2277
2. Drieu K, Vrancks R, Benassayad C, Haourigi M, Hassid J, Yoa RG, Rapin JR, Nunez EA (2000) Effect of the extract of gingko biloba (EGB 761) on the circulating and cellular profiles of polyunsaturated fatty acids: correlation with the anti-oxidant properties of the extract. Prostaglandins Leukot Essential Fatty Acids 63: 293–300
3. Flanagan SW, Mooseley PL, Buettner GR (1998) Increased flux of free radicals in cells subjected to hyperthermia: detection by electron paramagnetic resonance spin trapping. FEBS Lett 431: 285–286
4. Freeman ML, Spitz DR, Meredith MJ (1990) Does heat shock enhance oxidative stress? Studies with ferrous and ferric iron. Radiat Res 124: 288–293
5. Hedlund S, Sharma HS, Sjöquist P-O, Westman J (1999) A new antioxidant compound H-290/51 attenuates heat shock protein (HSP 72 kD) response, edema and cell injury following acute heat exposure. An experimental study using light and electron microscopy in the rat. J Therm Biol 24: 409–414
6. Kim CK, Rivier CL (2000) Nitric oxide and carbon monoxide have a stimulatory role in the hypothalamic-pituitary-adrenal response to physico-emotional stressors in rats. Endocrinology 141: 2244–2253
7. Maines MD (1997) The heme oxygenase system: a regulator of second messenger gases. Annu Rev Pharmacol Toxicol 37: 517–554
8. Miura T, Muraoka S, Fujimoto Y (1998) Temperature-dependent lipid peroxidation of rat brain homogenate. Res Commun Mol Pathol Pharmacol 100: 117–128
9. Pietri S, Seguin JR, d'Arbigny P, Drieu K, Culcasi M (1997) Gingko biloba extract (EGb 761) pretreatment limits free radical-induced oxidative stress in patients undergoing coronary bypass surgery. Cardiovasc Drugs Ther 11: 121–131
10. Sastre J, Millan A, Garcia de la Asuncion J, Pla R, Juan G, Pallardo, O'Conner E, Martin JA, Droy-Lefaix MT, Vina J (1998) A Gingko biloba extracts (EGb 761) prevents mitochondrial aging by protecting against oxidative stress. Free Radic Biol Med 24: 298–304
11. Sharma HS, Dey PK (1986) Probable involvement of 5-hydroxytryptamine in increased permeability of blood-brain barrier under heat stress. Neuropharmacology 25: 161–167
12. Sharma HS, Dey PK (1987) Influence of long-term acute heat exposure on regional blood-brain barrier permeability, cerebral blood flow and 5-HT level in conscious normotensive young rats. Brain Res 424: 153–162
13. Sharma HS, Alm P, Westman J (1997a) Upregulation of hemeoxygenase-II in the rat spinal cord following heat stress. Thermal physiology. In: Nielsen-Johanssen B, Nielsen R (eds) The August Krogh Institute, Copenhagen, 1997, pp 135–138
14. Sharma HS, Westman J, Cervós-Navarro J, Nyberg F (1997b) Role of neurochemicals in brain edema and cell changes following hyperthermic brain injury in the rat. Acta Neurochir (Wien) [Suppl] 70: 269–274
15. Sharma HS, Alm P, Westman (1998a) Nitric oxide and carbon monoxide in the pathophysiology of brain functions in heat stress. Prog Brain Res 115: 297–333
16. Sharma HS, Westman J, Nyberg F (1998b) Pathophysiology of brain edema and cell changes following hyperthermic brain injury. Prog Brain Res 115: 351–412
17. Sharma HS (1999) Pathophysiology of blood-brain barrier, brain edema and cell injury following hyperthermia: New role of heat shock protein, nitric oxide and carbon monoxide. An experimental study in the rat using light and electron microscopy. Acta Universitatis Upsaliensis 830: 1–94
18. Sharma HS, Drieu K, Alm P, Westman J (1999) Upregulation of neuronal nitric oxide synthase, edema and cell injury following heat stress are reduced by pretreatment with EGB-761 in the rat. J Therm Biol 24: 439–446
19. Sharma HS, Drieu K, Alm P, Westman J (2000a) Role of nitric oxide in the blood-brain barrier permeability, brain edema and cell damage following hyperthermic brain injury. An experimental study using EGB-761 and Gingkolide B pretreatment in the rat. Acta Neurochir (Wien) [Suppl] 76: 81–86
20. Sharma HS, Alm P, Sjöquist P-O, Westman J (2000b) A new antioxidant compound H-290/51 attenuates upregulation of constitutive isoform of heme oxygenase (HO-2) following trauma to the rat spinal cord. Acta Neurochir (Wien) [Suppl] 76: 153–157

21. Snyder SH, Ferris CD (2000) Novel neurotransmitters and their neuropsychiatric relevance. Am J Psychiatry 157: 1738–1751
22. Westman J, Drieu K, Sharma HS (2000) Antioxidant compounds EGB-761 and BN-52021 attenuate heat shock protein (HSP 72 kD) response, edema and cell changes following hyperthermic brain injury. An experimental study using immunohistochemistry in the rat. AminoAcids 19: 339–350

Correspondence: Hari Shanker Sharma, Dr. Med. Sci., Frödingsgatan 12:28, SE-75421 Uppsala, Sweden. e-mail: Sharma@medcellbiol.uu.se

Clinical trauma

Acta Neurochir (2003) [Suppl] 86: 323–327

Edema fluid accumulation within necrotic brain tissue as a cause of the mass effect of cerebral contusion in head trauma patients

Y. Katayama and **T. Kawamata**

Department of Neurological Surgery, Nihon University School of Medicine, Tokyo, Japan

Summary

The early massive edema caused by severe cerebral contusion results in progressive intracranial pressure (ICP) elevation and clinical deterioration within 24–72 hours post-trauma. Surgical excision of the necrotic brain tissue represents the only therapy, which can provide satisfactory control of the elevated ICP and clinical deterioration. In order to elucidate the mechanisms underlying the early massive edema, we have carried out a series of detailed clinical studies. Diffusion magnetic resonance (MR) imaging and apparent diffusion co-efficient (ADC) mapping suggest that cells in the central area of contusion undergo shrinkage, disintegration and homogenization, whereas cellular swelling is predominant in the peripheral area during the period of 24–72 hours post-trauma. The ADC values in the central and peripheral areas are maximally dissociated during this period. A large amount of edema fluid accumulates within the necrotic brain tissue of the central area beginning at approximately 24 hours post-trauma. We have found that fluid-blood interface formation within the central area does not represent an uncommon finding in various neuroimaging examinations of cerebral contusions, indicating layering of red blood cells within the necrotic brain tissue accumulating voluminous edema fluid. Intravenous slow infusion of gadolinium-DTPA and delayed MR imaging revealed that the central area of contusion can be enhanced at 24–48 hours post-trauma, implying that water supply from the blood vessels is not completely interrupted. Necrotic brain tissue sampled from the central area of contusion during surgery demonstrates a very high osmolality. It appears that the capacitance for edema fluid accumulation increases in the central area, whereas cellular swelling in the peripheral area elevates the resistance for edema fluid propagation. Combination of these circumstances may facilitate edema fluid accumulation in the central area. We also suggest that the dissociation of ADC values and high osmolality within the necrotic brain tissue may generate an osmotic potential across the central and peripheral areas and contribute to the early massive edema caused by cerebral contusion.

Keywords: Cerebral contusion; brain edema; diffusion image; diffusion co-efficient; magnetic resonance imaging.

Introduction

In cerebral contusions, at least two types of edema can be recognized. One is the early massive edema that occurs within the period of 24–72 hours post-trauma. This type of edema results in progressive elevation of the intracranial pressure (ICP) and clinical deterioration known as talk-and-deteriorate [9]. The other is the delayed peri-contusion edema, which is typically seen in the white matter adjacent to the cerebral contusion, at several days post-trauma by T2-weighted magnetic resonance (MR) imaging. This type of edema rarely causes ICP elevation.

Despite intensive medical therapy, the elevated ICP in patients with early massive edema is often uncontrollable and fatal. In such instances, surgical excision of the necrotic brain tissue is the only therapy, which can provide satisfactory control of the elevated ICP. The precise mechanism underlying such an early massive edema is not yet clearly understood. We review the results of our clinical studies [7–11], which have provided several lines of evidence to suggest that a large amount of edema fluid is accumulated in the necrotic brain tissue within the central area of contusion, and this contributes to the early massive edema.

Histopathology of cerebral contusion

The classical histopathological study of Lindenberg and Freytag [13] demonstrated the presence of two components of cerebral contusion; one is the central area in which cells undergo necrosis as the primary consequence of mechanical injury (contusion necrosis proper), and the other is the periperal area in which cellular swelling as a consequence of ischemia is observed. A clear demarcation line separates these two components. The area of contusion necrosis is histopathologically evident as early as at 3 hours post-trauma [5]. The cellular elements in the central area, both neuronal as well as glial cells, uniformly undergo shrinkage, and then disintegration, homogenation and cyst formation. In contrast, cell swelling is predominant in the peripheral area [13]. The ischemia in the peripheral area is largely attributable to microthrombosis. Numerous clinical studies have demonstrated a decrease in cerebral blood flow in the central as well as the peripheral areas of contusion [e.g., 1].

Contusion necrosis in the central (core) area

We have investigated the evolution of cerebral contusion by diffusion magnetic resonance (MR) imaging and apparent diffusion co-efficient (ADC) mapping in head trauma patients [10]. Following conventional T1- and T2-weighted MR imaging, diffusion MR images are obtained (TR: 0.8 ms, TE: 120 ms, matrix: 128 × 128, slice: 5 mm), and ADC mapping is computed from various b factor diffusion images. When contusion hemorrhage is observed on the T1- and T2-weighted images, the area of hemorrhage is carefully excluded from the analysis. The ADC ratio is calculated by dividing the ADC values in the region of interest by a reference ADC value, which was obtained from a normal brain area.

The diffusion MR images demonstrate a low intensity core in the central area, beginning at approximately 24 hours post-trauma. The ADC value within this central area clearly increases during the period of 24–72 hours post-trauma (Fig. 1) [10]. This elevated ADC value appears to represent the contusion necrosis

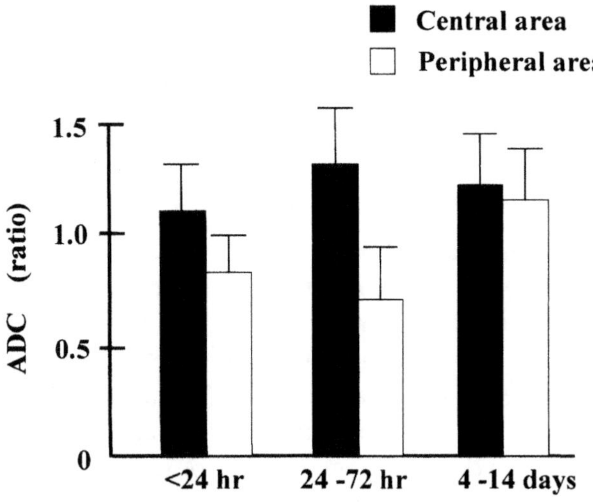

Fig. 1. Changes in apparent diffusion co-efficient of the central and peripheral areas of contusion after head trauma. The dissociation of the co-efficient during the period of 24–72 hours was statistically significant. See ref. 10 for details

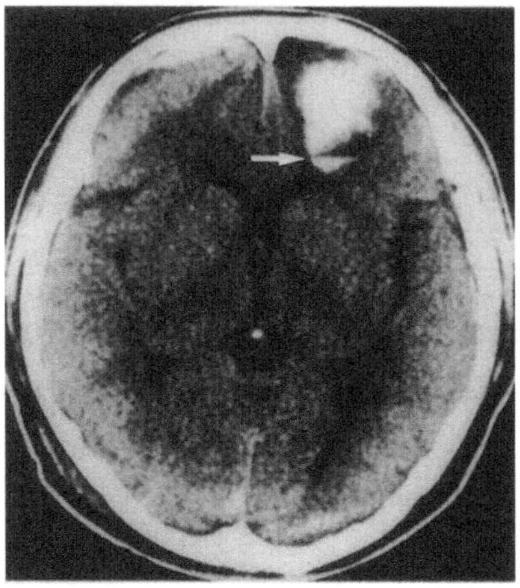

Fig. 2. Representative example of a small fluid-blood interface (arrow) seen at the bottom of a cerebral contusion with hemorrhages

proper, since cellular disintegration and homogenization in this area would result in an expansion of the extracellualr space.

Cell swelling in the peripheral (rim) area

The diffusion MR images demonstrate a high intensity rim in the peripheral area of contusion beginning at approximately 24 hours post-trauma, which corresponds with the timing of the elevated ICP and neurological deterioration. The combination of a low intensity core and a high intensity rim is a consistent finding in the cerebral contusion during this period, which can be termed a halo appearance. The ADC value decreases in the peripheral area during the period of 24–72 hours post-trauma (Fig. 1) [10]. This decreased ADC value appears to represent the cellular swelling, which would result in shrinkage of the extracellular space. The ADC values between the central and peripheral areas are maximally dissociated during the period of 24–72 hours post-trauma.

Peri-contusion edema within the adjacent white matter

The ADC value in the peripheral area shifts from a decrease to an increase after 72 hours post-trauma (Fig. 1), suggesting that the cellular swelling gradually resolves. At the same time, an increase in ADC value becomes evident in the adjacent white matter. This increase in ADC value in the adjacent white matter appears to represent the delayed peri-contusion edema, which can commonly be seen on T2-weighted MR images.

Increased cerebrovascular permeability

Since gadolinium (Gd)-DTPA, like plasma protein, does not normally cross the blood-brain barrier, enhancement with Gd-DTPA, if observed in association with cerebral contusion, implies an increased cerebrovascular permeability. In a previous study [12], Gd-DTPA was administered intravenously by bolus injection and MR imaging was undertaken soon after the administration. Such a procedure failed to detect any increase in cerebrovascular permeability associated with cerebral contusions during the initial few days post-trauma. Enhancement with Gd-DTPA has been reported to become detectable at 6–9 days post-trauma.

Furthermore, an immunohistochemical study of the post-mortem brain in such patients failed to reveal any plasma protein leakage around the contused brain areas during this early period [17]. Evaluations of the vascular permeability by 99mTc pertechnetate single photon emission tomography [3] failed to reveal any evidence of an increased vascular permeability within the area of contusion during the initial few days post-trauma. Such findings appear to contradict the widely held view that an increased cerebrovascular permeability is responsible for the development of contusion edema.

In general, increases in cerebrovascular permeability can be detected more clearly by intravenous slow infusion of contrast medium and delayed neuroimaging [6]. In order to detect an increased cerebrovascular permeability in the area of contusion in which blood flow is greatly reduced [e.g., 1], it appears necessary to maintain a sufficiently high blood concentration of Gd-DTPA for a longer period of time and to allow for delays caused by the slow penetration of Gd-DTPA into the edema areas. On this basis, we examined the changes in cerebrovascular permeability employing intravenous slow infusion of Gd-DTPA and delayed MR imaging [11]. This technique revealed that cerebral contusions can be enhanced at as early as 24–48 hours post-trauma. In MR imaging undertaken at 2 hours after Gd-DTPA administration, enhancement on T1-weighted images was observed in either the central area or the peripheral area, or both. It remains unknown whether the increased cerebrovascular permeability actually contributes to the early massive edema. However, it is at least evident that water supply from the blood vessels is not completely interrupted even in the central area of contusion.

Edema fluid accumulation in the central area

In approximately 50% of patients with cerebral contusions, a crescent-shaped zone of very high ADC value (ADC ratio = 1.38–1.61) develops at the border between the central and peripheral areas beginning at approximately 24 hours post-trauma [10]. This crescent-shaped zone was always located within the central area of contusion. No case demonstrated such a crescent-shaped zone within 24 hours post-trauma. The crescent-shaped zone still existed at 2 weeks post-trauma in some cases. The very high ADC value appears to represent edema fluid accumulation within the central area of contusion.

Fluid-blood interface within contusion necrosis

Edema fluid accumulation within the necrotic brain tissue is also suggested by the formation of a fluid-blood interface within cerebral contusions. While we have reported that fluid-blood interfaces associated with cerebral contusions do not appear to represent an uncommon finding if computed tomography (CT) scans or MR images are carefully examined (Fig. 2) [8], this phenomenon has rarely been discussed in the literature. The fluid-blood interface can be formed without a fluid cavity. When we carry out surgery, we find no fluid cavity but softened necrotic brain tissue is present [8].

Fluid-blood interfaces without a fluid cavity have been described in cases with intracerebral hematomas secondary to coagulation defects [4, 14], in which no plasma-blood interface was identified at autopsy. It has been suggested that this might be related to the brain fixation process or to evolution of the hematomas prior to death [14]. In cerebral contusion, however, no fluid cavity was identified at the time of surgery, indicating that a fluid cavity did not exist in any event. Within the contusion necrosis, hemorrhage undergoes enlargement within the initial 6 hours post-trauma [9], and red blood cells diffusely permeate the softened necrotic brain tissue [13]. The fluid-blood interfaces observed within the central area of contusion may represent layering of red blood cells in the softened necrotic brain tissue, which has accumulated voluminous edema fluid.

Osmotic potential of the contusion necrosis

We have demonstrated that a marked increase in tissue osmolality occurs within the area of contusion necrosis proper [7]. It is uncertain whether or not such a marked increase in osmolality is osmotically active and causes edema fluid accumulation. It appears that expansion of the extracellular space in the central area increases the capacitance for edema fluid accumulation. In contrast, shrinkage of the extracellular space in the peripheral area increases the resistance for edema fluid propagation or resolution. We hypothesize that the barrier formed by swollen cells in the peripheral area may prevent edema fluid propagation and also help to generate osmotic potentials across the central and peripheral areas. Since blood flow is greatly reduced but is not completely interrupted in the contused brain tissue, water is supplied from the blood vessels into the central area. We suggest that a combination of these events may facilitate edema fluid accumulation in the central area and contribute to the early massive edema of cerebral contusion.

Edema propagation to the white matter

The delayed peri-contusion edema has commonly been attributed to an increased cerebrovascular permeability. As mentioned above, enhancement with Gd-DTPA has been reported to become detectable at 6–9 days post-trauma. It is also possible, however, that resolution of the cellular swelling in the peripheral area of contusion might permit propagation of the edema fluid accumulated within the central area to the adjacent white matter. This would lead to edema appearance on CT and MR images which resembles that induced by an increased cerebrovascular permeability, i.e., vasogenic edema.

Edema fluid production reflected to cerebrospinal fluid formation rate

We have investigated the rate of formation of cerebrospinal fluid (CSF) in patients by CSF volume manipulation and ICP measurements via a ventricular catheter employing the method described by Marmarou *et al.* [15]. The CSF formation rate in patients with severe cerebral contusion is often elevated, suggesting that a large amount of edema fluid is being formed within the contused brain tissue.

Marmarou *et al.* [15] reported that a CSF formation rate in excess of 0.74 ml/min was observed in some of their severely injured head trauma patients. They believed that the high rate of CSF formation in such patients represented edema clearance through the CSF system [16]. We confirmed their observations and further noted that a high rate of CSF formation is frequently encountered in patients with cerebral contusions [9]. The increase in CSF formation rate in such patients is thus consistent with the findings of CT scans and MR imaging, suggesting the production of a large amount of edema fluid within the contusion necrosis.

Conclusion

There is at present no established medical treatment which can effectively inhibit edema fluid accumulation within cerebral contusions. The most effective therapy for ameliorating the potentially fatal edema is surgical excision of the necrotic brain tissue. The observed effect of such surgical excision of the necrotic brain tissue has commonly been accounted for on the basis of an increased space compensation for the mass lesions. It is possible, however, that excision of the necrotic brain tissue does mean the elimination of the cause of edema fluid accumulation. Arabi and Long [2] demonstrated that removal of freeze lesions can prevent the production and propagation of edema fluid in experimental animals. If the ICP is elevated by early massive edema due to cerebral contusions and the elevated ICP is medically uncontrollable, surgical excision of the necrotic brain tissue appears to represent the therapy of choice, regardless of the size of the associated hemorrhages.

References

1. Alexander MJ, Martin NA, Khanna M, Caron M, Becker DP (1994) Regional cerebral blood flow trends in head injured patients with focal contusions and cerebral edema. Acta Neurochir s60: 479–481
2. Arabi B, Long DM (1979) Dynamics of cerebral edema: The role of intact vascular bed in the production and propagation of vasogenic brain edema. J Neurosurg 51: 779–784
3. Bullock R, Statham J, Patterson D, Wyper D, Hadley D, Teasdale E (1990) The time course of vasogenic oedema after focal human head injury: Evidence from SPECT mapping of blood brain barrier defects. Acta Neurochir s51: 286–288
4. Dolinskas CA, Bilaniuk LT, Zimmerman RA, Kuhl DE (1977) Computed tomography of intracerebral hematomas. I. Transmission CT observations on hematoma resolution. AJR 129: 681–688
5. Eriskat J, Schurer L, Kempski O, Baethmann A (1994) Growth kinetics of a primary brain tissue necrosis from focal lesion. Acta Neurochir s60: 425–427
6. Ito U, Reulen H-J, Tomita H, Ikeda J, Saito J, Maehara T (1988) Formation and propagation of brain oedema fluid around human brain metastasis: A CT study. Acta Neurochir 90: 35–41
7. Katayama Y, Mori T, Maeda T, Kawamata T (1998) Pathogenesis of the mass effect of cerebral contusions: A rapid increase

in osmolality within the contusion necrosis. Acta Neurochir s71: 289–292
8. Katayama Y, Tsubokawa T, Kinoshita K, Himi K (1992) Intraparenchymal fluid-blood levels in traumatic intracerebral hematomas. Neuroradiology 34: 381–383
9. Katayama Y, Tsubokawa T, Miyazaki S, Kawamata T, Yoshino A (1990) Oedema fluid formation within contused brain tissue as a cause of medically uncontrollable elevation of intracranial pressure in head trauma patients. Acta Neurochir s51: 308–310
10. Kawamata T, Katayama Y, Aoyama N, Mori T (2000) Heterogeneous mechanisms of early edema formation in cerebral contusion: Diffusion MRI and ADC mapping study. Acta Neurochir s76: 9–12
11. Kushi H, Katayama Y, Shibuya T, Tsubokawa T, Kuroha T (1994) Gd-DTPA enhanced magnetic resonance imaging of cerebral contusions. Acta Neurochir s60: 472–474
12. Lang DA, Hadley DM, Teasdale GT, Macpherson P, Teasdale E (1991) Gadolinium DTPA enhanced magnetic resonance imaging in acute head injury. Acta Neurochir 109: 5–11
13. Lindenberg R, Freytag E (1957) Morphology of cortical contusions. AMA Arch Path 63: 23–42
14. Livioni JP, McGahan JP (1983) Intracranial fluid-blood levels in the anticoagulated patient. Neuroradiology 25: 335–337
15. Marmarou A, Maset AL, Ward JD, Choi S, Brooks D, Lutz H, Moulton RJ, Muizalaar JP, DeSalles A, Young HF (1987) Contribution of CSF and vascular factors to elevation of ICP in severe head-injured patients. J Neurosurg 66: 883–890
16. Reulen HJ, Tsuyumu M, Tack A, Fenske AR, Prioleau GR (1978) Clearance of edema fluid into cerebrospinal fluid. J Neurosurg 48: 754–764
17. Todd NV, Graham DI (1990) Blood-brain barrier damage in traumatic brain contusion. Acta Neurochir s51: 296–299

Correspondence: Yoichi Katayama, M.D., Ph.D., Department of Neurological Surgery, Nihon University School of Medicine, 30-1 Oyaguchi Kamimachi, Itabashi-ku, Tokyo 173-8610, Japan. Phone: 81-3-3972-8111, ext. 2480; FAX: 81-3-3554-0425; E-mail: ykatayam @med.nihon-u.ac.jp

Acta Neurochir (2003) [Suppl] 86: 329–331

Ultra-early study of edema formation in cerebral contusion using diffusion MRI and ADC mapping

T. Maeda[1,2], **Y. Katayama**[1], **T. Kawamata**[1], **S. Koyama**[2], and **J. Sasaki**[3]

[1] Department of Neurological Surgery, Nihon University School of Medicine, Tokyo, Japan
[2] Department of Neurological Surgery, Yokohama Chuo Hospital, Kanagawa, Japan
[3] Department of Radiology, Yokohama Chuo Hospital, Kanagawa, Japan

Summary

Objective. Our previous studies have reported that heterogeneous mechanisms exist in early edema formation in cerebral contusion, and cytotoxic edema plays an important role within 48 hours post-trauma. It is remains unclear, when edema begins to develop following injury. In order to determine the time course of edema development, diffusion imaging and ADC (apparent diffusion coefficient) mapping was performed in 10 patients within 24 hours post-trauma with cerebral contusion.

Methods. Diffusion imaging and ADC mapping were performed employing 1.0 T echo planar MRI. ADC values were indicated as a ration relative to the values of intact brain areas.

Results. Within 3 hours post-trauma, diffusion MRI showed no remarkable changes, and the ADC values were within normal limit (ADC ratio (=contused/normal brain) = 1.00 ± 0.21, (mean $\pm$ SD)). At 6 hours post-trauma, diffusion images demonstrated a low intensity core in the contusion proper and a high intensity rim in the peripheral area of contusion. The ADC value increased in the contusion proper (ADC ratio = 1.26 ± 0.13) and decreased in the peripheral area (ADC ratio = 0.58 ± 0.19).

Conclusions. These findings indicated that early cellular swelling in the peripheral area of contusion begins within 6 hours following injury. This delayed occurrence of contusion-induced cellular swelling suggests that the CBF does not decrease to ischemic level immediately following injury.

Keywords: Cerebral contusion; brain edema; diffusion MRI; ADC mapping.

Introduction

Brain edema following cerebral contusion is a common and serious complication of traumatic brain injury that is associated with poor neurological outcome [3]. The mechanisms leading to edema following contusion are not well understood. In the previous studies, we have reported that heterogeneous mechanisms exist in early edema formation in cerebral contusion [5, 6, 7]. Especially, cytotoxic edema plays an important role within 48 hours following injury [1, 2]. It remains unclear when edema begins to develop following injury. In the present study, we attempted ultra-early study of edema formation using diffusion imaging and ADC (apparent diffusion coefficient) mapping to determine the time-course of edema development in cerebral contusion.

Materials and methods

A total of 10 patients with cerebral contusion were investigated in diffusion MRI study within 24 hours. These patients were selected on the basis of the following criteria: GCS > 8, no hypoxic episodes, no massive ICP increase, no massive intracerebral hematoma, no severe multiple injury, and no systemic complications to avoid leading to severe secondary edema formation in cerebral contusion. Diffusion images were obtained and ADC mapping was computed from various b factor (=1000) diffusion images. Several regions of interest (ROI) for ADC were defined in the central area and peripheral area of cerebral contusion. When contusion hemorrhage was observed on the initial CT, the area of hemorrhage was excluded from the ROI. The ADC ratio (=ADC in contusion/ADC in intact brain) was calculated dividing the ADC values in the ROI by a reference ADC value that was obtained from intact brain tissue.

Results

Diffusion images

Within 3 hours, diffusion MRI showed no remarkable changes. At 6 hours, diffusion images demonstrated a low intensity core in the contusion proper and a high intensity rim in the peripheral area of contusion (Fig. 1).

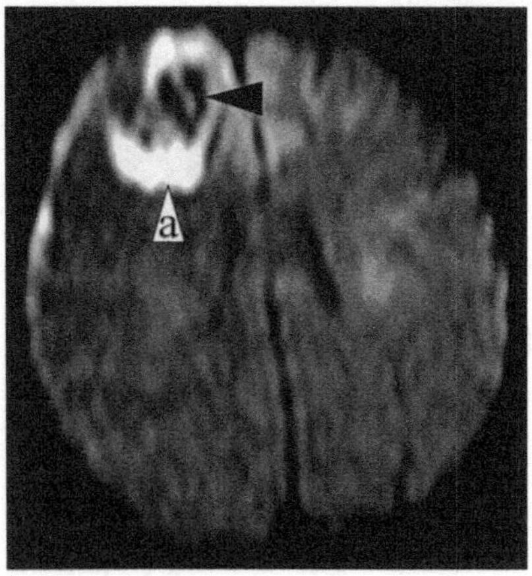

Fig. 1. At 6 hours post-trauma, diffusion images demonstrated a low intensity core in the contusion proper (black arrow) and a high intensity rim (white arrow) in the peripheral area of contusion

ADC ratio

All ADC ratios were within normal limit within 3 hours post injury. Within 6 hours post injury, the ADC ratio increased in the central area (1.26 ± 0.13 (mean ± SD)) and decreased in the peripheral area (0.58 ± 0.19) with in 6 hours post trauma. Interstitial edema extending along white matter fibers developed around the peripheral area of contusion at 24 hours post trauma (Fig. 2).

Discussion

Cerebral contusion is consisted of two components on histopathology; one comprises the area of contusion necrosis proper and the other is the surrounding area in which microthorombosis and cellular swelling is observed [8]. These two components are clearly separated by a line of demarcation. The central area of cerebral contusion that demonstrates high ADC values appears to correspond to the area of necrosis proper, since cellular disintegration and homogenization in this area would result in an increase in the extra-cellular space [4]. The peripheral area of cerebral contusion in which the ADC values decrease, appears to correspond to the surrounding area, since cellular swelling which is predominant in this area would result in a decrease in the extra-cellular space.

The present study shows that the ADC values were within normal limit in the central and the peripheral of cerebral contusion within 3 hours post-injury. At 6

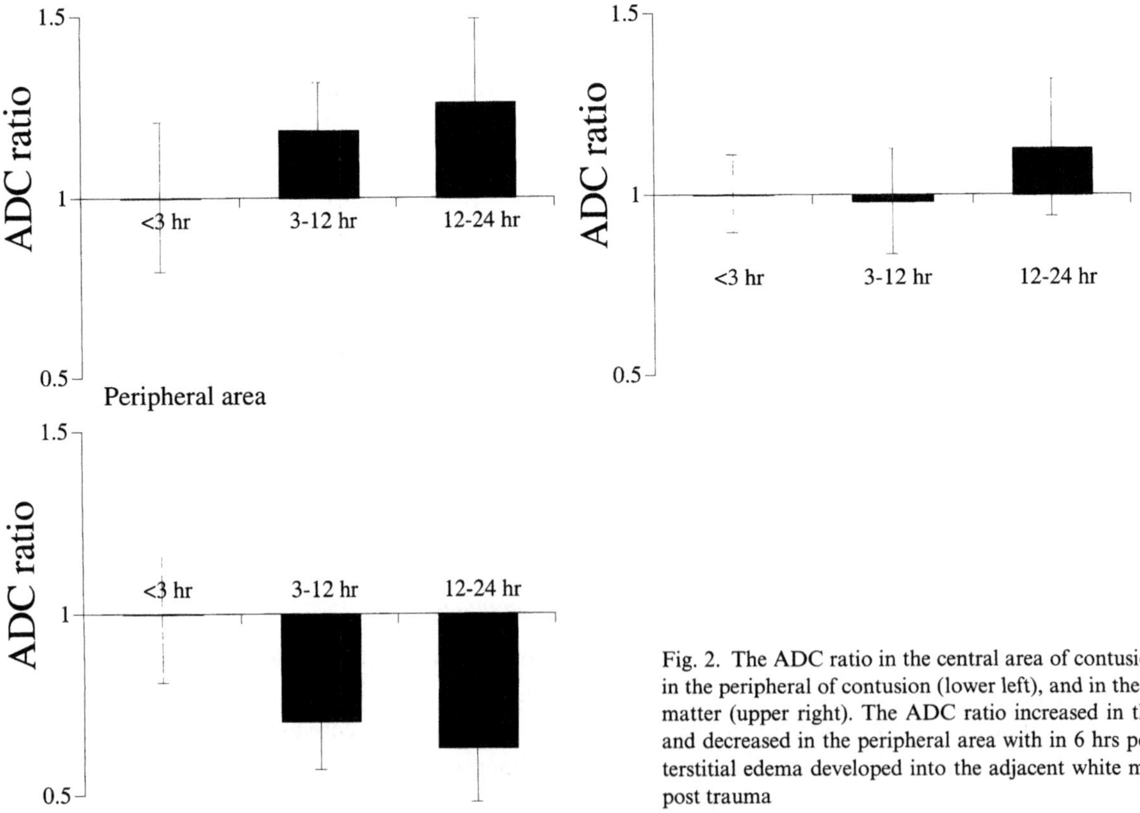

Fig. 2. The ADC ratio in the central area of contusion (upper left), in the peripheral of contusion (lower left), and in the adjacent white matter (upper right). The ADC ratio increased in the central area and decreased in the peripheral area with in 6 hrs post trauma. Interstitial edema developed into the adjacent white matter at 24 hrs post trauma

hours, the ADC ratio increased in the central area and decreased in the peripheral area. These results demonstrate that early cellular swelling in the peripheral area of contusion begins within 6 hours following injury. Our previous studies have reported that the tissue osmolality in the core of contusion proper (contusion necrosis proper) markedly increased approximately 25% in the normal brain [5, 7]. Such a marked increase in osmolality causes edema fluid accumulation. We hypothesize that the barrier formed by swollen cells in the peripheral of contusion may prevent edema fluid propagation and also helps to create an osmotic potential within the central area. Since blood flow is greatly decreased but is not ischemic at levels in the peripheral area, water is continuously supplied from the blood flow in to the central area [9]. These events facilitate edema fluid accumulation in the central area and contribute to the mass effect of early contusion edema.

In conclusion, diffusion images and ADC mapping are powerful tools for investigating spatially as well as temporally heterogeneous mechanisms of contusion edema in ultra-early studies following injury.

References

1. Barzo P, Marmarou A, Fatouros P, Hayasaki K, Corwin F (1997) Contribution of vasogenic and cellular edema to traumatic swelling measured by diffusion-weighted imaging. J Neurosurg 87: 900–907
2. Ito J, Marmarou A, Barzo P, Fatouros P, Corwin F (1996) Characterization of edema by diffusion-weighted imaging in experimental traumatic brain injury. J Neurosurg 84: 97–103
3. Katayama Y, Tsubokawa T, Miyazaki S, Kawamata T, Yoshino A (1990) Oedema fluid formation within contused brain tissue as a cause of medically uncontrollable elevation of intracranial pressure in head trauma patients. Acta Neurochir (Wien) [Suppl] 51: 308–310
4. Katayama Y, Tsubokawa T, Kinoshita K, Himi K (1992) Intra parenchymal fluid-blood levels in traumatic intracerebral hematomas. Neuroradiol 34: 381–383
5. Katayama Y, Mori T, Maeda T, Kawamata T (1998) Pathogenesis of the mass effect of cerebral contusions: rapid increase in osmolality within the contusion necrosis. Acta Neurochir (Wien) [Suppl] 71: 289–292
6. Kawamata T, Katayama Y, Aoyama N, Mori T (2000) Heterogeneous mechanisms of early edema formation in cerebral contusion: diffusion MRI and ADC mapping study. Acta Neurochir (Wien) [Suppl] 76: 9–12
7. Kawamata T, Katayama Y, Mori T, Aoyama N, Tsubokawa T Jr (2002) Mechanisms of the mass effect of cerebral contusion: ICP monitoring and Diffusion MRI study. Acta Neurochir (Wien) [Suppl] 81: 281–283
8. Lindenberg R, Freytag E (1957) Morphology of cortical contusion. Arch Path 63: 23–42
9. Maeda T, Katayama Y, Kawamata T, Aoyama N, Mori T (1997) Hemodynamic depression and microthrombosis in the peripheral areas of cortical contusion in the rat: role of platelet activating factor. Acta Neurochir (Wien) [Suppl] 70: 102–105

Correspondence: Takeshi Maeda, M.D., Ph.D., Department of Neurological Surgery, Nihon University School of Medicine, 30-1 Oyaguchi Kamimachi, Itabashi-ku, Tokyo 173-8610, Japan.

Acta Neurochir (2003) [Suppl] 86: 333–337

Mixed dishomogeneous hemorrhagic brain contusions. Mapping of cerebral blood flow

A. Chieregato[1], E. Fainardi[2], A. Tanfani[1], C. Martino[1], V. Pransani[1], F. Cocciolo[1], L. Targa[1], and F. Servadei[3]

[1] Neurorianimazione, Ospedale M. Bufalini, Cesena, Italy
[2] Neuroradiologia, Arcispedale S. Anna, Ferrara, Italy
[3] Neurotraumatologia, Ospedale M. Bufalini, Cesena, Italy

Summary

The aim of the study was to verify whether regional cerebral blood flow (rCBF) was distributed centrifugally in traumatic hemorrhagic contusions with multiple cores within an oedematous area. Seventeen traumatic brain contusions, from 14 patients with severe head injury (GCS < 9), were analyzed during 39 Xenon-enhanced computerized tomography (Xe-CT) studies. The CBF was measured in 3 concentric regions of interest (ROIs): the hemorrhagic core, the intracontusional oedematous low density area and a 1 cm rim of pericontusional normal-appearing brain tissue surrounding the contusion. Differences between rCBFs in the three ROIs were found ($p < 0.0001$). rCBF in both the hemorrhagic core (21.4 ± 19.4 ml/100gr/min) and the intracontusional low density area (28.4 ± 19 ml/100gr/min) were lower than rCBF in pericontusional normal-appearing area (41.9 ± 16 ml/100gr/min) ($p < 0.0001$). No significant differences were found between rCBF measured in the hemorrhagic core and intracontusional low density area ($p = 0.184$). Our study suggests that in the mixed density contusions with multiple hemorrhagic cores, the CBF is concentrically distributed, improving from the core to the periphery.

Keywords: Cerebral blood flow; ischemia; traumatic contusions; Xenon-CT.

Introduction

Traumatic contusions and intracerebral hematomas seem to be a heterogeneous group of lesions [13]. They include: a) *purely low density contusions*; b) contusions with a well defined hemorrhagic core and a surrounding concentric low density rim (*mixed homogeneous hemorrhagic contusions*); c) homogeneous areas of clotted blood with a diameter > 3 cm [21] called *traumatic hematomas*; d) contusions with multiples hemorrhagic cores distributed within a low density area (*mixed dishomogeneous hemorrhagic contusions*) [13]. Although it has been suggested that different subtypes of contusions may underline distinct mechanisms of tissue injury [13], accumulating evidences indicate that a reduction of regional cerebral blood flow (rCBF) may play a key role in the pathogenesis of these contusion forms. The rCBF seems to show a centrifugal distribution, improving from the centre to the periphery [5, 12, 15, 19], although these studies do not reach the same conclusions, which is probably due to the different timing of CBF investigations as well as topographic analysis and selection of the contusions. In the present paper we evaluated rCBF by means of Xenon-enhanced CT (Xe-CT) only in mixed dishomogeneous hemorrhagic contusions in order to verify whether a centrifugal rCBF profile also occurs in this subtype of contusions.

Material and methods

Patient summaries

From October 1999 to October 2001 the Neurointensive Care Unit of our hospital admitted 148 severely head injured patients (GCS ≤ 8). Of these, 52 patients underwent Xenon-CT studies according to the stability of physiological parameters and organizing facilities. Out of the thirty-eight patients with 59 traumatic contusions, our case material comprises 14 patients and 17 traumatic contusions with multiple hemorrhagic cores (larger than 1 cm in diameter), in a surrounding edematous area. In agreement with American and European guidelines, all patients were managed with a staircase protocol to maintain ICP below 20 mmHg and cerebral perfusion pressure (CPP) above 70 mmHg. The outcome at one year post-injury was measured by means of the Glasgow Outcome Scale (GOS).

Stable xenon-enhanced CT CBF measurement and analysis

CBF studies were conducted using a CT scanner (Picker 5000) equipped for Xe-CT CBF imaging (Xe/TC system-2™, Diversified Diagnostic Products, Inc., Houston, TX). Three different ROIs larger than 1 cm^2 were drawn manually on the diagnostic CT: 1) within the contusion hemorrhagic cores; 2) surrounding the intracontusional edematous low-density area, according to Schroder [15]; 3) within 1 cm of the normal-appearing brain tissue surrounding the contusion (pericontusional normal appearing area), according to McLaughlin [12]. An additional ROI, with a shape that corresponded to total contusional area (contusion hemorrhagic cores plus surrounding edematous low-density area), was located symmetrically in a normal-appearing brain tissue area of the contralateral hemisphere (contralateral contusional "mirror" area). When the contralateral contusional "mirror" area displayed an altered coefficient of density, this ROI was not analysed. In each ROI, rCBF values were calculated by dedicated software (Xe-CT System Version 1.0 w ©, 1998, Diversified Diagnostic Products, Inc., Houston, TX) and were expressed in ml/100gr/min.

Statistical analysis

Statistical analysis of the values of rCBF in the different ROIs was performed by means of ANOVA for repeated measures. Comparison between the ROIs was made using the paired T-test, and the Bonferroni adjustment was applied. Frequencies of rCBF values below 18 ml/100gr/min were compared applying the chi-square test. A value of p below 0.05 was considered significant.

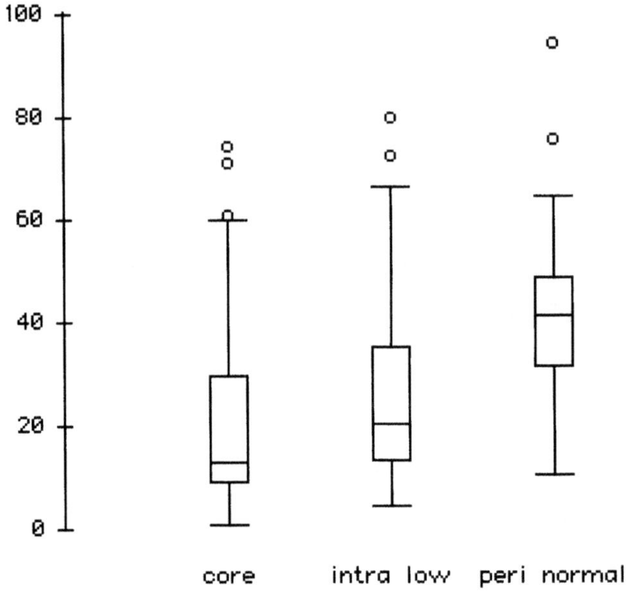

Fig. 1. Regional CBF (ml/100gr/min) in three different ROIs: Core (hemorrhagic core), intra low (intracontusional), peri normal (pericontusional normal appearing area). The centrifugal improvement of the rCBF from the center to the periphery can be noted

Results

The demographic and clinical parameters and the data concerning the CBF studies of our case material (14 patients) are reported in table 1 and table 2. We analyzed a total of 39 contusion studies as some of the 17 contusions were repeatedly studied by Xe-CT examinations during the acute phase.

As illustrated in Fig. 1, a centrifugal improvement of rCBFs in the three contusional ROIs was found ($p < 0.0001$). rCBF mean levels in both the hemorrhagic core (Table 1: 21.4 ± 19.4 ml/100gr/min) and the intracontusional low density area (Table 2: 28.4 ± 19 ml/100gr/min) were lower than in pericontusional normal-appearing area (41.9 ± 16 ml/100gr/min) ($p < 0.0001$). No significant differences were found between rCBF measured in the hemorrhagic core and intracontusional low density area ($p = 0.184$), nor between the values of rCBF in the contralateral contusional "mirror" area (mean $= 38.5 \pm 15$ ml/100gr/min) and the pericontusional normal-appearing area ($p = 0.4975$).

Discussion

Centrifugal distribution of rCBF

Our study aimed to investigate whether the contusional CBF may be associated with a progressive improvement of CBF from the core to the periphery as previously reported [5, 12, 15, 19]. To test this hypothesis, we employed a cumulative approach involving three different areas within and around the contusions. The intracontusional low-density area was designed as initially measured by Schröder [15] and called "edematous pericontusional tissue" or "perilesional tissue". The pericontusional normal-appearing area was first described by McLaughlin [12], while both the hemorragic core and the intracontusional low density area were included in a unique contusional ROI. Recently, Hoelper and co-workers [5] evaluated the hemorrhagic core, the intracontusional edematous area and the pericontusional normal-appearing area separately, in a subgroup of contusions (called "mixed contusions" by the authors).

As expected, we found a gradual improvement of rCBF values from the core to the periphery. Our data confirms results previously obtained in human CBF studies [5, 12, 15, 19] and experimental settings [10]. Our findings are also in agreement with those recently reported by Hoelper and colleagues [5] who measured rCBF mean values in mixed hemorrhagic contusions with a similar topographic approach.

Present study and generation of new hypothesis

Centrifugal distribution may be interpreted in two different ways: 1) the expression of a gradual disappearing of tissue damage from the centre to the

Table 1. *Characteristics of 14 severely head injured patients undergoing Xe-CT studies*

Patients	n		14
Age	mean ± SD	years	36 ± 18
Sex	n (%)	male	10 (71.4%)
		female	4 (28.5%)
Best initial GCS	median (IQR)		7.5 (6)
Best initial motor GCS	median (IQR)		4.5 (3)
Worst initial GCS	median (IQR)		4 (4)
Worst initial motor GCS	median (IQR)		2 (4)
Worst initial pupillary reactivity	n (%)	normal	6 (43%)
		one pupil dilated unreactive	8 (57%)
		bilaterally dilated unreactive pupils	0 (%)
Worst initial hypotension	n (%)	no hypotension	13 (93%)
		hypotension	1 (7%)
Worst initial hypoxia	n (%)	no hypoxia	6 (43%)
		hypoxia	8 (57%)
Worst CT	n (%)	DI II	2 (14.3%)
		DI III	2 (14.3%)
		EML	10 (71.4%)
		• EDH	1
		• SDH	6
		• Contusion	4
		NEML	0 (0%)
Outcome at discharge	n (%)	obeying	3 (21.4)
		not obeying	11 (78.6)
		dead	0 (0%)
Outcome at 6 months	n (%)	good recovery or moderate disability	11 (78.6%)
		severe disability	2 (14.3%)
		persistent vegetative state or death	1 (7.1%)

Table 2. *Details concerning the contusions and physiological parameters in 39 CBF studies*

Patients	n	14
Contusions	n	17
Xenon Studies	n	39
Time post injury of Xenon studies (hrs)	mean ± SD	130 ± 120
Contusions studied per time intervals (hours post injury)	24	3 (8%)
	24–96	18 (46%)
	post 96	18 (46%)
Hemorhagic core (mm^2)	mean ± SD	478 ± 329
Intracontusional low density area (mm^2)	mean ± SD	1072 ± 516
Pericontusional normal-appearing area (mm^2)	mean ± SD	1782 ± 396
Total contusional area (mm^2)	mean ± SD	1691 ± 635
ICP during the study (mmHg)	mean ± SD	17 ± 4
CPP during the study (mmHg)	mean ± SD	74 ± 12
PaCO2 during the study (mmHg)	mean ± SD	38.6 ± 4.5
Hb during the study (gr/dl)	mean ± SD	9.8 ± 1.2
Ht during the study (%)	mean ± SD	29 ± 3.6
SaO2 during the study (%)	mean ± SD	97 ± 2.6

periphery ("centrifugal static theory"), or 2) a "dynamic" (progressive over time) damage of the brain tissue surrounding contusional core ("centrifugal dynamic theory"). In this specific subtype of contusions (mixed dishomogeneous), the differences between the rCBF in the hemorrhagic cores and in the intracontusional low density area were not statistically relevant. Concomitantly, the rCBF in the pericontusional normal-appearing area seems to be markedly different in respect to the rCBF in the intracontusional low density area but not different from the contralateral normal-appearing brain tissue. This finding suggests that the normal-appearing area is not affected by any anticipatory reduction in rCBF. Even if our results generally confirm the hypothesis of a centrifugal distribution of rCBF, they suggest that the core and the intracontusional low density area may represent a part of the same initial insult ("centrifugal static theory"), rather than the result of a diffusion of the pathogenetic process from the centre to the periphery ("centrifugal dynamic theory"). In this scenario, the multiple small hemorrhagic cores might be the expression of a focal rupture of microvessels rather than a trigger promoting a progressive centrifugal involvement of the surrounding tissue.

The "centrifugal static theory" is supported by early histological studies. The central area of contusion ("contusion necrosis proper") [9] is histopathologically evident early, within 1–3 hours post-injury [3, 4]. The central area is due to a direct damage of neurons by a trauma-induced massive depolarization [14] and is associated to cell death and astrocytes swelling. The degree of impairment in the hemorrhagic core is so elevated that in experimental studies Sutton and co-workers showed that cortical necrosis evolve in a late cavitation [18]. Also biochemical evidence suggests that the centre of contusion is more primarily affected than the periphery since the initial concentration of glutamate is higher in the centre of the contusion than in the surrounding tissue [11].

Most of the "centrifugal dynamic theory" comes from suggestions obtained by studies on spontaneous intracranial hematoma. In fact the spontaneous intracranial hematoma may affect the surrounding tissue by various physiopathological mechanisms [1, 20], partially reversible [16]. Nevertheless, the comparison between traumatic and non-traumatic hematomas [17] revealed that the edema surrounding the core is more extensive in a traumatic hematoma than in a non-traumatic intracranial hematoma [17], suggesting that in head trauma a centrifugal "static" damage may be associated with other pathological processes. Recent studies indicate that the centrifugal dynamic distribution of edema may be independent from the presence of an hemorrhagic core and it can be driven by the osmotic forces produced by the central primary necrosis [6, 8]. The "centrifugal dynamic theory" is also supported by the demonstration that glutamate-mediated mechanisms involve the surrounding neurons and generate a progressive astrocyte swelling ("diffusion-reaction process") [7, 11] acting in combination with the well-known ischemic cytotoxic swelling [2].

Although we have described the two theories separately, it may be reasonable to believe that the "static" and "dynamic" mechanisms may coexist. Our findings involving only the rCBF, suggest that in mixed dishomogeneous hemorrhagic contusions, the primary injury and the "centrifugal static theory" may be predominant. Thus, further studies on basic physiology and temporal profiles of rCBF in each single subtype of contusions are required in order to understand to what extent "static" and "dynamic" processes contribute to the centrifugal distribution of rCBF.

Conclusion

The present study suggests that, in the mixed density contusions with multiple hemorrhagic cores, the CBF is concentrically distributed, improving from the core to the periphery. However, the differences between the core and the low density area seem less marked than expected. In this subtype of contusions, the intracontusional low density area is primarily and markedly involved by a post-traumatic reduction of CBF. In addition, our findings seem to support the hypothesis that the centrifugal distribution of rCBF is prominently orchestrated by "static" mechanisms produced by the initial traumatic impact rather than by "dynamic" processes originating in the hemorrhagic core.

References

1. Bullock R, Brock, Utne J, van Dellen J, Blake G (1988) Intracerebral hemorrhage in a primate model: effect on regional cerebral blood flow. Surg Neurol 29: 101–107
2. Bullock R, Maxwell WL, Graham DI, Teasdale GM, Adams JH (1991) Glial swelling following human cerebral contusion: an ultrastructural study. J Neurol Neurosurg Psychiatry 54: 427–434
3. Eriskat J, Schurer L, Kempski O, Baethmann A (1994) Growth kinetics of a primary brain tissue necrosis from a focal lesion. Acta Neurochir (Wien) [Suppl] 60: 425–427
4. Hekmatpanah J, Hekmatpanah CR (1985) Microvascular alterations following cerebral contusion in rats. Light, scanning and electron microscope study. J Neurosurg 62: 888–897
5. Hoelper BM, Reinert MM, Zauner A, Doppenberg E, Bullock R (2000) rCBF in hemorrhagic, non-hemorrhagic and mixed contusions after severe head injury and its effect on perilesional cerebral blood flow. Acta Neurochir (Wien) [Suppl] 76: 21–25
6. Katayama Y, Mori T, Maeda T, Kawamata T (1998) Pathogenesis of the mass effect of cerebral contusions: rapid increase in osmolality within the contusion necrosis. Acta Neurochir (Wien) [Suppl] 71: 289–292
7. Katayama Y, Tamura T, Becker DP, Tsubokawa T (1991) Inhibition of rapid potassium flux during cerebral ischemia in vivo with an exitatory amino acid antagonist. Brain Res 568: 294–298
8. Kawamata T, Katayama Y, Aoyama N, Mori T (2000) Heterogeneous mechanisms of early edema formation in cerebral contusion: diffusion MRI and ADC mapping study. Acta Neurochir (Wien) [Suppl] 76: 9–12
9. Lindenberg R, Freytag E (1957) Morphology of cortical contusion. Arch Pathol 63: 23–42
10. Madsen FF (1990) Regional cerebral blood flow in the pig after a localized cerebral contusion treated with barbiturates. Acta Neurochir (Wien) 106: 24–31
11. Maeda T, Katayama Y, Kawamata T, Yamamoto T (1998) Mechanisms of excitatory amino acid release in contused brain tissue: effects of hypothermia and in situ administration of CO2+ on extracellular levels of glutamate. J Neurotrauma 15: 655–664
12. McLaughlin MR, Marion DW (1996) Cerebral blood flow and vasoresponsivity within and around cerebral contusions. J Neurosurg 85: 871–876
13. Ribas GC, Jane JA (1992) Traumatic contusions and intracerebral hematomas. J Neurotrauma 9: S265–S268

14. Sachs F (1991) Mechanical transduction by membrane ion channels: a mini review. Mol Cell Biochem 104: 57–60
15. Schroder ML, Muizelaar JP, Bullock MR, Salvant JB, Povlishock JT (1995) Focal ischemia due to traumatic contusions documented by stable xenon-CT and ultrastructural studies. J Neurosurg 82: 966–971
16. Siddique MS, Fernandes HM, Wooldridge TD, Fenwick JD, Slomka P, Mendelow AD (2002) Reversible ischemia around intracerebral hemorrhage: a single-photon emission computerized tomography study. J Neurosurg 96: 736–741
17. Staham PFX, Todd NV (1990) Intracerebral haematoma: aetiology and haematoma volume determine the amount and progression of brain oedema. Acta Neurochir (Wien) [Suppl] 51: 289–291
18. Sutton RL, Lescaudron L, Stein DG (1993) Unilateral cortical contusion injury in the rat: vascular disruption and temporal development of cortical necrosis. J Neurotrauma 10: 135–149
19. von Oettingen G, Bergholt B, Gyldensted C, Astrup J (2002) Blood flow and ischemia within traumatic cerebral contusions. Neurosurgery 50: 781–790
20. Xi G, Wagner KR, Keep RF, Hua Y, de Courten-Myers GM, Broderick JP, Brott TG, Hoff JT (1998) Role of blood clot formation on early edema development after experimental intracerebral hemorrhage. Stroke 29: 2580–2586
21. Yamaki T, Hirakawa K, Tenjin H, Kuboyama T, Nakagawa Y (1990) Chronological evaluation of acute intracerebral haematoma. Acta Neurochir (Wien) 103: 112–115

Correspondence: Arturo Chieregato, Unità Operativa di Neurorianimazione, Ospedale M. Bufalini, viale Ghirotti 286, 47023, Cesena, Italy. e-mail: achiere@ausl-cesena.emr.it

Acta Neurochir (2003) [Suppl] 86: 339–342

Neuronal damage in pericontusional edema zone

H. Kushi, T. Saito, K. Makino, and **N. Hayashi**

Department of Emergency and Critical Care Medicine, Nihon University School of Medicine, Tokyo, Japan

Summary

In this study, we investigated the molecular biological and histopathological aspects of the etiological mechanisms for pericontusional edema zone (PEZ). The subjects were 5 patients with traumatic brain injury who underwent surgery to evacuate the resulting hematoma. The average age of the subjects was 52 $\pm$ 27.5 years. The GCS at the time of admission was 5–9. At operation apart from evacuating the hematoma, the PEZ was also excised and then examined histopathologically. Cerebrospinal fluid (CSF) levels of IL-6, IL-8, and IL-10 were measured at the time of admission and at 24 and 72 hours. Histological examination revealed large numbers of neutrophils accumulating within blood vessels in the PEZ, with some focal migration. IL-6: CSF levels at the time of admission and at 24, 72, and 72 hours were 550, 4350, and 878000 pg/ml, respectively (median values). IL-8: CSF levels were 715, 804, and 24900 pg/ml, respectively. IL-10: CSF levels were 15, 4, and 5 pg/ml, respectively. High levels of IL-6 and IL-8 were seen from an early stage, and became markedly higher with enlargement of the PEZ. The PEZ is thought to be due to microvascular disturbance by neutrophils stimulated by inflammatory cytokines, and neuronal damage from migrated neutrophils.

Keywords: Cerebral contusion; pericontusional edema zone; neuroinflammation; neutrophil.

Introduction

Cerebral contusions are known to cause a primary brain injury, and secondarily lead to neuronal necrosis in the pericontusional edema zone (PEZ), although the mechanism for the latter is unclear. Recent studies have placed the emphasis on the mechanism [24], etiological factors [17], histopathological findings [4, 17, 25], and outcome [8, 9] in cases of cerebral contusions, but much remains unexplained in this area. In a previous study, we measured cerebral blood flow (CBF) and cerebral blood volume (CBV) in the PEZ, finding both to be markedly reduced at 14.5 $\pm$ 2.5 ml/100g/min and 3.1 $\pm$ 0.3 ml/100g, respectively [16].

PEZ also developed as a secondary injury following neurotrauma. Secondary injury is mediated by a number of different mechanisms, including excitotoxicity, protease activation, metabolic failure, calcium accumulation, and inflammation. Cytokine are mediators of inflammation that are released by a variety of cell types and regulate a wide variety of metabolic, inflammatory, and regenerative processes [20].

Interleukin-6 (IL-6) is a cytokine that has proinflammatory and neurotrophic properties. Increased level of tumor necrosis factor (TNF) and IL-6 have been reported following fluid percussion injury in rats [27]. IL-6 has a neuroprotective effect through stimulation of nerve growth factor (NGF) [14], but in vitro models have shown that IL-6 impairs the blood brain barrier (BBB) function [7].

Interleukin-8 (IL-8) is a member of the cysteine-x-cysteine family of chemokines that has a potent neurotrophic cemoatactic and activating activity [3]. In a recent report, human astrocytes secrete IL-8 in response to pro-inflammatory cytokines in vitro, and IL-8 receptors are expressed by microglia and neurons in the normal brain [11]. In experimental animal models as well as in humans, traumatic brain injury (TBI) elicits an acute inflammatory response characterized by production of both pro- and anti-inflammatory cytokines [21].

Kossmann *et al.* [15] suggested that IL-8 mediates secondary injury after TBI. In their study IL-8 levels were increased in the cerebrospinal fluid (CSF) of adults for several days after severe TBI, and they postulated that IL-8 contributes to both injury and repair processes in the brain after TBI.

Interleukin-10 (IL-10), an 18-kDa polypeptide produced by T cells, B cells, monocytes and macrophage [15], is detected in the CSF and plasma after polytrauma [5], and is thought to globally suppress cytokine production. IL-10 has been found by immuno-assay to be present in the CSF of child [5] and adult patients [19] with severe head injuries. It was recently suggested that catecholamines released in response to brainstem compression injury elicit IL-10 secretion directly from native, non-stimulated peripheral blood monocytes [29]. In this study, we investigated the molecular biological and histopathological aspects of the etiological mechanisms for PEZ.

Materials and methods

Patient population

The subjects were 5 patients with closed traumatic brain injury, with cerebral contusion $\geqq$ 2 cm in length, who underwent surgery to evacuate the resulting hematoma. The average age of the subjects was 52 $\pm$ 27.5 years; there were 3 male and 2 female. The GCS at the time of admission was 5–9 points. Brain CT was performed at admission, and 24 and 72 hours later.

Management

After admission, neurological testing was performed by the neurosurgeons. Physiological parameters (ECG, blood pressure, arterial oxygen saturation, and core temperature) were also monitored.

Mannitol (2 mg/kg) was administered to prevent intracranial hypertension. Following CT scanning, the 4 patients with a GCS $\leqq$ 8 points underwent ventricular drainage and intracranial pressure (ICP) monitoring (intracranial pressure monitoring kit 110-4B: Intergra NeuroSciences, USA). The remaining patient had an admission GCS of 9 points, but later deteriorated reaching a GCS of 6 points, and ventricular drainage was performed 2 hours later. When the ICP rose over 25 mmHg, evacuation of the hematoma (including the contusion core and part of the PEZ) and external decompression, were performed. At operation, apart from evacuating the hematoma, the PEZ was also excised and then examined histopathologically (with the electron microscope). To prevent secondary brain damage, postoperative management followed to the following protocol: 1) After intubation, patients were ventilated to maintain a PaO_2 of 100 mmHg or more and a PCO_2 of 30 to 35 mmHg. 2) Albumin and crystalloid were infused to maintain the systolic blood pressure at 100 mmHg or more and the central venous pressure at 6 to 10 cm H_2O. 3) Blood transfusions were given when necessary, to maintain the hematocrit at 30% to 40%. 4) ICP was maintained at 20 mmHg or less through administration of a sedative (midazolam) and a muscle relaxant (pancuronium), raising of the head (20–30°), and mild hyperventilation ($PaCO_2$ of 25–30 mmHg). When the ICP remained above 20 mmHg for longer than 10 minutes despite the above, 20% mannitol was administered (10 to 15 minutes, 0.5–1 g/kg). Cerebral hypothermia was not performed, however, and normal body temperature was maintained.

Parameters measured

CSF samples were collected on three occasions (on admission and at 24 and 72 hours later) from each patient via the ventricular drain. After centrifugation, samples were stored for a maximum of one month at −70 °C. The IL-6, IL-8 and IL-10 concentration in the CSF was determined using a commercially available specific ELISA kit (Ohtsuka Pharma Co., Tokyo, Japan) according to the manufacturer's instructions.

Data analysis

All values for IL-6, IL-8 and IL-10 (in pg/ml) are expressed as mean $\pm$ SEM.

Results

The outcomes for the 5 patients were moderate disability in 3 cases and severe disability in 2 cases. Mean CSF levels of IL-6, CSF IL-8, CSF IL-10 are given below. CSF IL-6 levels at admission and 24 72 hours later, were 550 $\pm$ 950, 4350 $\pm$ 1679, and 878000 $\pm$ 152000 pg/ml, respectively. CSF IL-8 levels were 715 $\pm$ 320, 804 $\pm$ 420, and 24900 $\pm$ 10500 pg/ml. CSF IL-10 levels were 15 $\pm$ 6, 4 $\pm$ 0.5, and 5 $\pm$ 0.1 pg/ml. Brain CT showed an increasing extent of pericontusional edema after 24 hours post admission, with ICP exceeding 25 mmHg after 72 hours post admission. Evacuating of the hematoma and external decompression were performed between 80 and 100 hours after the trauma.

Histological examination (electron microscopic findings)

The electron microscopic findings of the PEZ were of intravascular accumulation and degranulation of neutrophils, leading to microcirculatory damage (Fig. 1). Some focal extravascular migration was also seen.

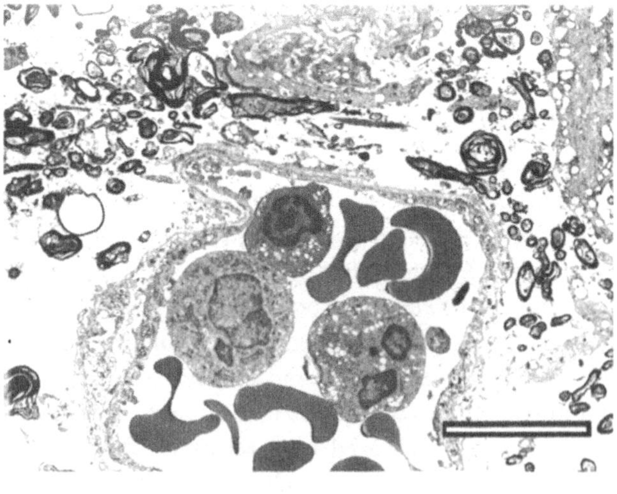

Fig. 1. Histological examination (electron microscopic findings). Histopathological examination of the PEZ revealed intravascular accumulation of neutrophils. Neutrophils contained vacuoles, and degranulation was recognized. No fibrin or platelet thrombi were seen, however

Discussion

Woiciechowsky *et al.* [30] reported that in cases of severe TBI, serum IL-6 decreased with time, but in our cases of cerebral contusion, CSF IL-6 and IL-8 were increased with time. IL-6 is a pleiotropic cytokine that is produced by a variety of cell types [26], and is a potent stimulator of B-cell differentiation and immunoglobulin synthesis [2]. In the CNS, IL-6 has also been reported to be produced by neuronal cells in response to glial cells and a variety of stimuli. IL-8 is generally produced by leukocytes of haematological origin. In recent years, synthesis of IL-8 by brain cells has also been reported [10].

Neuroinflammation is seen in cases of TBI, and is thought to be due to astrocytes, and activation of monocytes, macrophages, mocroglial cells [1]. Cerebral contusion is characterized by neuronal damage in the cortex, and increased levels of pro-inflammatory cytokines in the CSF indicates progression of neuroinflammation. The histopathological findings of intravascular accumulation of large numbers of neutrophils, with some focal extravascular migration, within the PEZ indicate continued neuroinflammation. These findings are in agreement with our previous report of decreased CBF and CBV in the PEZ [16]. Although cerebral contusion is a focal injury, there is progressive neuroinflammation, and the CT findings of enlargement of the PEZ with time suggest that the PEZ produces IL-6 and IL-8. Because it inhibits the release of inflammatory cytokines such as IL-1 and TNF-alpha, IL-10 is considered an anti-inflammatory cytokine. Following permanent focal and ischemia [23] and TBI [13] IL-10 displayed significant anti-inflammatory effects, including reduced TNF expression and enhanced neurological recovery after experimental TBI. Interestingly, the magnitude of the increase in IL-10 is small compared with the dramatic increase in IL-6 and IL-8. These findings are similar to those reported by Bell *et al.* [5]. High levels of IL-10 are detected in cases of bacterial meningitis in adults [28].

The endogenous increase in IL-10 has been reported to correspond to the decrease in synthesis of inflammatory cytokines in EAE mice [12], but in any cases, the source of IL-10 in the CNS is unknown. Astrocytes produce TGF-beta, but they clearly do not produce IL-10 [6]. It is possible that infiltrating monocytes produce IL-10 within the brain, and IL-10 is produced by human fetal microglia in cell cultures in response to TNF-alpha and IL-6 stimulation [22].

CSF levels of IL-6 and IL-8 rose with time in cases of cerebral contusion, but only a transient rise was seen in CSF IL-10 levels, and it failed to inhibit neuroinflammation. More investigation is required into the role and source of IL-10 in cases of cerebral contusion.

References

1. Aihara N, Hall JJ, Pitts LH, Fukuda K, Noble LJ (1995) Altered immunoexpression of microglia and macrophages after mild head injury. J Neurotrauma 12: 53–63
2. Akira S, Kishimoto T (1992) IL-6 and NF-IL6 in acute-phase response and viral infection. Immunol Rev 127: 25–50
3. Baggiolini M, Moser B, Clark-Lewis I (1994) Interleukin-8 and related chemotactic cytokines: Giles Filley Lecture. Chest 105: 95s–98s
4. Bell BA, Symon L, Branston NM (1985) CBF and time threshold for the formation of ischemic cerebral edema, and effect of reperfusion in baboons. J Neurosurg 62: 31–41
5. Bell MJ, Kochanek PM, Doughty LA, Carcillo JA, Andelson PD, Clark RSB, Wisniewski SR, Whalen MJ, Dekosky ST (1997) Interleukin-6 and interleukin-10 in cerebrospinal fluid after severe traumatic brain injury in children. J Neurotrauma 14: 451–457
6. Corsini E, Dufour A, Ciusani E, Gelati M, Frigerio S, Gritti A, Cajola L, Mancardi GL, Massa G, Salm (1996) Human brain endothelial cells and astrocytes produce IL-1 beta but not IL-10. Scand J Immunol 44: 506–511
7. de Vries HE, Blom-Roosemalen MC, van Oosten M, de Boer AG, van Berkel TJ, Breimer DD, Kuiper J (1996) The influence of cytokines on the integrity of the blood-brain barrier in vitro. J Neuroimmunol 64: 37–43
8. Eide PK, Trsnes OB (1992) Early and late outcome in head injury patients with radiological evidence of brain damage. Acta Neurol Scand 86: 194–198
9. Gennarelli TA, Spielman GM, Langfitt TW, Gildenberg PL, Harrington T, Jane JA, Marshall LF, Miller JD, Pitts LH (1982) Influence of the type of intracranial lesion on outcome from severe head injury. A multicenter study using a new classification system. J Neurosurg 56: 26–32
10. Glabinski AR, Ransofoff RM (1999) Chemokines and chemokine receptors in CNS pathology. J Neurovirol 5: 3–12
11. Horuk R, Martin AW, Wang Z, Schweitzer L, Gerassimides A, Guo H, Lu Z, Hesselgesser J, Perez HD, Kim J, Parker J, Hadley TJ (1997) Expression of chemokine receptors by subsets of neurons in the central nervous system. J Immunol 158: 2882–2890
12. Kennedy MK, Torrance DS, Picha KS, Mohler KM (1992) Analysis of cytokine mRNA expression in the central nervous system of mice with experimental autoimmune encephalomyelitis reveals that IL-10 mRNA expression correlate with recovery. J Immunol 149: 2496–2505
13. Knoblach SM, Faden AI (1998) Interleukin-10 improves outcome and alters proinflammatory cytokine expression after experimental traumatic brain injury. Exp Neurol 153: 143–151
14. Kossmann T, Hans VHJ, Imhof HG, Trentz O, Kossmann MCM (1996) Interleukin-6 released in human cerebrospinal fluid following traumatic brain injury may trigger nerve growth factor production in astrocytes. Brain Res 713: 14–152
15. Kossmann T, Stahel PF, Lenzlinger PM, Redl H, Dubs RW, Trentz O, Schlag G, Morganti-Kossmann MC (1997) Interleukin-8 released into the cerebrospinal fluid after brain injury is associated with blood-brain barrier dysfunction and

nerve growth factor production. J Cereb Blood Flow Metab 17: 280–289
16. Kushi H, Hayashi N, Saito T, Shibuya T (1996) Prospects of surgical removal of a contusional lesion and intracranial depressurization upon cerebral blood flow and brain tissue pH. Neurotraumatology (Jpn) 19: 58–62
17. Lindenberg R, Freytag E (1957) Morphology of cortical contusion. Arch Pathol 63: 23–42
18. Moor KW, O'Garra A, de Waal Malefyt R, Vieira P, Mosmann TR (1993) Interleukin 10. Annu Rev Immunol 11: 165–190
19. Morganti-Kossmann MC, Lenzlinger PM, Hans V, Stahel P, Csuka E, Amman E, Stocker R, Trentz O, Kossmann T (1997) Production of cytokines following brain injury: beneficial and deleterious for the damaged tissue. Mol Psychiatry 2: 133–136
20. Ott L, McClain CJ, Gillespie M, Young B (1994) Cytokines and metabolic dysfunction after severe head injury. J Neurotrauma 11: 447–472
21. Rothwell NJ, Strijbos PJ (1995) Cytokines in neurodegeneration and repair. Int J Dew Neurosci 13: 179–185
22. Sheng WS, Hu S, Kravitz FH, Peterson PK, Chao CC (1995) Tumor necrosis factor-α upregulate human microglia cell production of IL-10 in vitro. Clin Diag Lab Immunol 2: 604–608
23. Spera PA, Ellison JA, Feuerstein GZ, Barone FC (1998) IL-10 reduces rat brain injury following focal stroke. Neurosci Lett 251: 189–192
24. Stalhammer DH (1990) The mechanism of brain injuries. In: Braakman R (ed) Handbook of clinical neurology. Elsevire, Amsterdam, pp 17–41
25. Sutton RL, Liscaudron L, Stein DG (1993) Unilateral cortical contusion in the rat: Vascular disruption and temporal development of cortical necrosis. J Neurotrauma 10: 135–149
26. Taga T, Kishimto T (1997) Gp130 and the interleukine-6 family of cytokines. Annu Rev Immunol 15: 797–819
27. Taupin V, Toulmond S, Serrano A, Benavides J, Zavala F (1993) Increase in IL-6, IL-1 and TNF levels in rat brain following traumatic lesion. Influence of pre- and post-traumatic treatment with Ro5 4864, a peripheral-type (p site) benzodiazepine ligand. J Neuroimmunol 42: 177–185
28. Van Furth AM, Seijmonsbergen EM, Langermans JAM, Groeneveld PHP, de Bell CE, van Furth R (1995) High levels of interleukin-10 and tumor necrosis factor α in cerebrospinal fluid during the onset of bacterial meningitis. Clin Inf Dis 21: 220–222
29. Woiciechowsky C, Asadullah A, Nestler D, Eberhardt B, Platzer C, Schoning B, Glockneer F, Lanksch WR, Volk H-D, Docke W-D (1998) Sympathetic activation triggers systemic interleukin-10 release in immunodepression induced by brain injury. Nature Med 4: 808–813
30. Woiciechowsky C, Schöning B, Cobanov J, Lanksch R, Volk HD, Döcke WD (2001) Early IL-6 plasma concentration correlate with severity of brain injury and pneumonia in brain-injured patients. J Trauma 52: 339–345

Correspondence: Hidehiko Kushi, 1-8-13 Kanda, surugadai, Chiyoda-ku, Tokyo, Japan. e-mail: hkushi@med.nihon-u.ac.jp

Acta Neurochir (2003) [Suppl] 86: 343–346

Characteristics of parietal-parasagittal hemorrhage after mild or moderate traumatic brain injury

K. Kinoshita, H. Kushi, and **N. Hayashi**

Department of Emergency and Critical Care Medicine, Nihon University School of Medicine, Tokyo, Japan

Summary

The aim of this study is to clarify the clinical characteristics of parietal-parasagittal traumatic intracranial hemorrhage (TICH) after mild or moderate traumatic brain injury (TBI).

Methods. Subjects were 105 patients with mild or moderate TBI. The patients with parietal-parasagittal TICH were clinically analyzed based on the initial brain CT findings, hematoma sites and the clinical course as compared to those with TICH at other sites.

Results. Hematoma was located in the frontal or temporal lobes in 89.5% of the subjects and the parietal-parasagittal in 10.5%. Ten of the 11 patients suffering parietal-parasagittal TICH had skull fractures (7 depressed and 3 linear) but no depressed fracture observed in patients with frontal or temporal lobe hemorrhage. Neurological deterioration leading to a comatose state more frequently occurred in 63.6% of patients with parietal-parasagittal TICH than in those with frontal or temporal lobe hemorrhage (19.1%, $p < 0.01$). The incidence of hematoma growth was significantly higher in patients with parietal-parasagittal TICH (63.6%) than in those with frontal or temporal lobe hemorrhage (31.9%, $p < 0.05$).

Discussion. The incidence of parietal-parasagittal TICH is low, but the risk of neurological deterioration due to hematoma enlargement is significantly high. Parietal-parasagittal TICH may differ clinically from frontal-temporal TICH.

Keywords: Cerebral contusion; vascular damage; traumatic brain injury; venous thrombosis; vertical injury.

Introduction

Traumatic intracranial hemorrhage (TICH) due to focal injuries such as cerebral contusion, acute subdural hematoma (ASDH) or acute epidural hematoma may enlarge hematomas over time [7]. An enlargement of a mass lesion most frequently caused delayed neurological deterioration followed by death or severe disability. Early recognition and prompt surgical evacuation of the intracranial hemorrhage and large subdural hematomas usually prevent herniation and improve the outcome. These TICHs occur most frequently in the frontal and temporal lobes [1, 3], and are reported less frequently for the parietal-parasagittal region. The most important clinical issue is whether neurological deterioration or hematoma growth can be predicted in patients with suffering from traumatic brain injury (TBI). The clinical characteristics of TICH were examined, including the cause of injury, neurological deterioration and hematoma growth in the parietal-parasagittal region.

Materials and methods

Subjects were 105 patients with mild or moderate TBI who were transferred to Nihon University hospital. The patients sustained TICH and an initial Glasgow Coma Scale (GCS) score of 9 or more. Brain CT scans was conducted on admission and 3–6 hours after trauma or when neurological deterioration occurred. The patients with parietal-parasagittal TICH were clinically analyzed based on the initial brain CT findings, hematoma sites and the clinical course after hospitalization as compared to those with TICH at other sites. The relationship between the patients' ability to talk at some point before lapsing into a coma and the site of hematomas was evaluated.

Categorical data were evaluated using the Fisher's exact test. All statistical tests were performed using the StatView J-5.0 software package (SAS Institute Inc.). Differences were considered significant when the p value was <0.05.

Results

Patients' summary

Hemorrhage in the frontal and temporal lobes, such as cerebral contusion, was observed in 94 (89.5%) of the 105 subjects. Parietal-parasagittal intracranial hemorrhage was observed in 11 patients (10.4%), some of who had combined traumatic subarachnoid hemorrhage (TSAH) or ASDH (Table 1). All patients with parietal-parasagittal TICH demonstrated good verbal response immediately after injury.

Table 1. *Parasagittal-pareital traumatic intracerebral hemorrhage profiles*

Case no.	Age	Cause	Initial GCS	Neurological deterioration (hrs)	Skull fracture	Vascular damage	Combined hematoma	Hematoma growth	Surg.
1.	68	fall	15	4	depressed	sagittal sinus*1	non	yes	no*3
2.	60	fall	15	6	depressed	cortical vein*1 bridging vein*1	non	yes	yes
3.	24	fall	13	4	depressed	cortical vein*1	non	yes	yes
4.	47	blow to head	10	2	depressed	cortical artery*2 bridging vein*2	SDH	yes	yes
5.	67	fall	9	2	linear	cortical artery*2 bridging vein*2	SDH	yes	yes
6.	58	fall	10	2	depressed	bridging vein*2	SDH	yes	yes
7.	49	fall	11	1	depressed	cortical artery*2	SDH	yes	yes
8.	72	fall	13	none	depressed	unknown	TSAH	no	no
9.	38	TA	13	none	linear	unknown	TSAH	no	no
10.	55	TA	10	none	linear	unknown	TSAH	no	no
11.	49	TA	11	none	non	unknown	non	no	no

Case No Number of patients with parasagittal-pareital intracerebral hemorrhage, *TA* Traffic accident, *Neurological deterioration* Time until coma from injury in patients becoming comatose, *Depressed* Depressed skull fracture, *Linear* Linear skull fracture, *SDH* Acute subdural hematoma, *TSAH* Traumatic subarachnoid hemorrhage, *Surg.* Emergency craniotomy immediate after admission.
*1 Vascular thrombosis, *2 Vascular laceration, *3 Patient's family refused operative management.

Causes of parietal-parasagittal injury and presence of skull fracture

The cause of injury was falls in 7 of 11 patients with parietal-parasagittal injury. Case 4 received a blow to the head at the vertex in an assault. The contact point of the skull was the vertex in all cases. Seven of these patients had a depressed skull fracture, 3 had a linear fracture and in 1, there was no evidence of a skull fracture. Hemorrhage in the frontal and temporal lobes had linear skull fracture in 15 of 94 patients, but no depressed skull fracture. The incidence of depressed skull fracture was significantly higher for the patients with injury in the parietal-parasagittal region ($p < 0.0001$) than for those with frontal or temporal lobe injury.

Hematoma growth and neurological deterioration

Hematoma growth (major axis > 30 mm) or combined ASDH was observed in 30 (31.9%) of the patients with TICH located in the frontal and temporal lobe within 24 hours after trauma. Seven of the 11 patients had intracranial hematoma growth in the parietal-parasagittal region. The incidence of hematoma growth was significantly higher for the patients with injury in the parietal-parasagittal region ($p = 0.0488$) than for those with frontal or temporal lobe injury. In the patients with parietal-parasagittal TICH, 4 patients suffered from combined ASDH and there was no evidence of hematoma growth observed in brain CT scans in 4 other patients.

Eighteen of 94 patients (19.1%) with frontal or temporal lobe hematoma were fully alert or confused on admission, but subsequently exhibited neurological deterioration within 24 hours after injury. Seven of the 11 patients (63.6%) with parietal-parasagittal hematoma lapsed into a coma. Deterioration occurred after 4–6 hours in the patients with enlarged intracerebral hematoma and within 2 hours in the patients with combined ASDH following parietal-parasagittal TICH. The incidence of deterioration into a comatose state was significantly higher in the patients with parietal-parasagittal TICH after TBI ($p = 0.0034$) than in those with intracerebral hemorrhage in the frontal and temporal lobes, retrospectively.

Emergency craniotomy and cerebral angiography in patients with parietal-parasagittal TICH (Table 1)

Six of the 11 patients underwent emergency craniotomy immediately after admission. Non-operative management was chosen in Case 1 (Table 1) because patient's family refused operative management. Dura laceration was found in 4 patients. Vascular damages, such as vascular thrombosis or laceration, were observed in all operative cases. In combined ASDH, lacerations of the cortical veins, arteries and bridging veins caused by skull fracture were detected during surgery in 3 patients.

Discussion

TICH involves a high risk of neurological deterioration after a lucid interval (talk and deterioration) [6, 7, 8], and mainly occurs in the frontal and temporal lobes. However, the relationships between the TICH region and clinical courses are not well reported. The present study suggests that the TICH region is an important factor of hematoma growth and neurological deterioration following TBI.

Skull films indicated skull fracture for the 6 subjects with depressed skull fracture and 4 with linear skull fracture above the parietal-parasagittal TICH in this study. It is possible for the parietal-parasagittal region to sustain damage even in the absence of skull fracture [1]. In our study, all of the 11 patients hit the vertex directly, suggesting that parietal-parasagittal TICH is associated with a high incidence of a specific injury mechanism, which occurs when deformation or depression of the skull causes compressive strain exceeding the tolerance of the pial vessels and brain tissue [4].

Depressed or linear skull fracture itself may occlude or rupture the cortical surface vessels, such as cortical veins, cortical arteries, bridging veins and dural sinus. Damaged blood vessels may thrombose, leading to additional ischemic-hemorrhagic complications [2, 5]. Thrombosis is caused by a number of mechanisms. One is a high incidence of skull fracture, often depressed, that produces traumatic rifts occluding the sinus or damaging the endothelium, leading to thrombosis. In this study, vascular damages were detected in all surgical patients. Four patients suffered combined ASDH after parietal-parasagittal TICH. ASDH, which caused brain damage including vertical skull fracture, is often related to rupture of the bridging veins, and occasionally, rupture of the superficial cortical arteries or veins under the skull fracture (often depressed) [9]. The cerebral vertex has a rich variety of cerebral vessels, such as the superficial cerebral veins, cortical arteries, bridging veins and superior sagittal sinus, all of which are easily damaged by compression or directly by a depressed or injured skull. In some cases of ruptured bridging vein ASDH, there is little or no associated brain damage, and these patients may experience a brief lucid interval before becoming comatose, as in typical acute epidural hematoma. In this respect, parietal-parasagittal TICH may differ clinically from frontal-temporal TICH.

Once obstruction or laceration of the large vessels or superior sagittal sinus has occurred, symptoms depend largely on the exact site of vascular damage. If it is in the anterior superior sagittal sinus or in a nondominant transverses sinus, the patients may have no symptoms referable to sinus injury [2, 5]. If significant venous outflow is obstructed, however, venous engorgement results in increased intracranial pressure, cerebral edema and acute brain swelling. Propagation of thrombus from the sinus into the cortical veins produces venous infarction [2]. Patients may become comatose very rapidly. We recommend aggressive and early repeat CT scanning to detect hematoma growth, with goal of evacuating hematoma before patients suffer possible permanent damage.

Treatment of parietal-parasagittal intracerebral hemorrhage depends largely on the type of skull fracture and the presence of vascular damage. Depressed skull fracture should be treated as soon as possible to rearrange the cranial bone. If large vascular damage is present, neurological deterioration rapidly develops. Patients must be regularly and closely monitored in the neurosurgical intensive care unit, to ensure they are alert and awake. Operative indication must be evaluated as soon as possible.

Conclusion

The incidence of parietal-parasagittal TICH is very low in patients with mild or moderate TBI, but the high risk of hematoma enlargement and neurological deterioration must be recognized in these patients as compared to those with frontal or temporal hemorrhage after TBI. Parietal-parasagittal TICH may differ clinically from frontal-temporal TICH.

Patients must be regularly and closely monitored in the neurosurgical intensive care unit. Early identification of neurological changes and repeats CT scanning to detect hematoma growth might improve their prognosis.

References

1. Adams JH, Doyle D, Graham DI, Lawrence AE, McLellan DR, Gennarelli TA, Pastuszko M, Sakamoto T (1985) The contusion index: a reappraisal in human experimental nonmissile head injury. Neuropathol Appl Neurobiol 11: 299–308
2. Bousser MG, Chiras J, Bories J, Castaigne P (1985) Cerebral venous thrombosis – a review of 38 cases. Stroke 16: 199–213
3. Gurdjian ES (1976) Cerebral contusions: re-evaluation of the mechanism of their development. J Trauma 16: 35–51
4. Halliday AL (1998) Pathophysiology. In: Marion DW (ed) Traumatic brain injury. Thieme, New York, pp 29–38

5. Hesselbrock R, Sawaya R, Tomsick T, Wadhwa S (1985) Superior sagittal sinus thrombosis after closed head injury. Neurosurgery 16: 825–828
6. Humphreys RP, Hendric EB, Hoffmann HJ (1990) The head-injured child who "talks and dies." Childs Nerv Syst 6: 139–142
7. Lobato RD, Rivas JJ, Gomez PA, Castaneda M, Canizal JM, Sarabia R, Cabrera A, Munoz MJ (1991) Head-injured patients who talk and deteriorate into coma. Analysis of 211 cases studied with computerized tomography. J Neurosurg 75: 256–261
8. Marshall LF, Toole BM, Bowers SA (1983) The National Traumatic Coma Data Bank. Part 2: Patients who talk and deteriorate: Implications for treatment. J Neurosurg 59: 285–288
9. Vance BM (1950) Ruptures of surface blood vessels in cerebral subdural heamorrhage. Arch Surg 61: 992–1006

Correspondence: Kosaku Kinoshita, M.D., Department of Emergency and Critical Care Medicine, Nihon University School of Medicine, 30-1 Oyaguchi Kamimachi, Itabashi-ku, Tokyo 173-8610, Japan. e-mail: kosaku@med.nihon-u.ac.jp

Acta Neurochir (2003) [Suppl] 86: 347–350

L-8 is a key mediator of neuroinflammation in severe traumatic brain injuries

H. Kushi, T. Saito, K. Makino, and **N. Hayashi**

Department of Emergency and Critical Care Medicine, Nihon University School of Medicine, Tokyo, Japan

Summary

The subjects were 22 patients with severe head injury. The average age was 45 ± 18.3 years. There were 13 survivors and 9 fatalities. Samples of peripheral blood and cerebrospinal fluid (CSF) were taken four times, at the time of admission and at 24, 72, and 168 hours later. IL-6: For the survivor group, peripheral blood levels were 181, 105, 37, and 26 pg/ml, respectively (median values). CSF levels were 5376, 3565, 328, and 764 pg/ml, respectively. For the fatality group, peripheral blood levels were 102, 176, 873, and 3059 pg/ml, respectively, whereas CSF levels were 15241, 97384, 548225, and 366500 pg/ml, respectively. IL-8: For the survivor group, peripheral blood levels were 36, 15, 15, and 15 pg/ml, respectively, whereas CSF levels were 23736, 4074, 355, and 1509 pg/ml, respectively. For the fatality group, peripheral blood levels were 21, 28, 43, and 77 pg/ml, respectively, whereas CSF levels were 29003, 8906, 5852, and 8220 pg/ml, respectively. IL-6 and IL-8 levels were significantly higher after 72 hours in the fatality group. The fact that CSF IL-8 was 1000 times that in the peripheral blood at the time of admission, and decreased thereafter, indicates that IL-8 is a key mediator of neuroinflammation.

Keywords: Severe traumatic brain injury; neuroinflammation; interleukin-6; interleukin-8.

Introduction

Immunohistochemical studies in experimental and clinical settings have revealed that a number of immune mediators are activated by traumatic brain injury (TBI) [20]. A role of cytokines in the CNS with regard to infections, autoimmune diseases, degenerative diseases, and trauma has been reported [5, 19].

Cytokines are small peptides that regulate cellular functions and are excreted from almost all cells. Cytokines are the main regulators of the inflammatory response, since they initiate, perpetuate, and terminate this response. Cytokines are currently classified as pro-inflammatory cytokines and anti-inflammatory cytokines according to whether they promote or suppress immunoactivation. These molecules bind strongly to receptors and thus modulate intracellular communication [12, 23]. Interleukin 6 (IL-6) is a pleiotropic cytokine that is produced by many types of cells [30]. It is strong inducers of B-cell differentiation and immunoglobulin synthesis [2]. In the CNS, IL-6 is also produced by glial cells and neurons in response to various stimuli, and mRNA for IL-6 and the IL-6 receptor is always expressed in the normal rat brain [16, 27]. Promotion of IL-6 production has been reported in experimental models of neurological trauma in rodents and in clinical studies of TBI patients [17, 28]. TNF and IL-1 are the first mediators produced in response to invasive stimuli, but they are hard to detect in the blood. On the other hand, IL-6 can be easily detected at the time of invasion, making it a good marker of invasion [21, 26]. Interleukin-8 (IL-8) is a member of the cysteine-x-cysteine family of chemokines that has potent neutrophil chemotactic activity and a neutrophil-activating effect [4]. In principle, chemokines are produced by peripheral blood leukocytes. However, it was recently reported that chemokines are also produced in brain cells [7]. Development of neuroinflammation in TBI patients is considered to be due to the activation of monocytes, macrophages, and microglial cells or astrocytes [1]. In the presence of inflammatory leukocytes, IL-8 increases vascular permeability in vitro and in vivo [24]. Investigations in animal models of systemic inflammation have demonstrated the essential involvement of IL-8 in acute inflammation of diverse origins [31].

In the present study, we measured IL-6 and IL-8 levels in the serum and CSF after neurotrauma to find the key mediator of neuroinflammation and examined the relationship between the development of neuroinflammation and mortality.

Materials and methods

Patient population

The subjects were 22 patients with severe TBI patients who were transferred to our institution for management and had a score of 8 or less on the Glasgow Coma Scale (GCS). There were 15 men and 7 women. The mean age was 45 ± 18.3 years, and the mean GCS score on admission was 5 ± 0.4 points (4–8 points) (Table 1). The diagnosis was acute subdural hematoma in 12 patients, cerebral contusion in 7 and acute epidural hematoma in 3.

Treatment and postoperative management

After admission, neurological testing was performed by the neurosurgeons. Monitoring of physiological parameters (ECG, blood pressure, arterial oxygen saturation, and core temperature) was also performed. Under controlled ventilation with administration of midazolam, head CT scanning was performed. To prevent intracranial hypertension, mannitol (2 mg/kg) was administered. Surgery was generally performed as soon as possible. At operation, a sensor for intracranial pressure (ICP) monitoring (intracranial pressure monitoring kit 110-4B: Intergra NeuroSciences, USA) was inserted into the subdural space.

To prevent secondary brain damage, postoperative management was performed according to the following protocol: 1) After intubation, ventilation was done to maintain a PaO_2 of 100 mmHg or more and a PCO_2 of 30 to 35 mmHg. 2) Infusion of albumin and crystalloid was performed to maintain the systolic blood pressure at 100 mmHg or more and the central venous pressure at 6 to 10 cm H_2O. 3) Blood transfusion was done, when necessary, to maintain the hematocrit at 30% to 40%. 4) Administration of a sedative (midazolam) and a muscle relaxant (pancuronium), raising of the head (20–30°), and mild hyperventilation ($PaCO_2$ of 25–30 mmHg) were used to keep the ICP at 20 mmHg or less. When the ICP remained above 20 mmHg for longer than 10 minutes despite the above-mentioned therapy, 20% mannitol was administered (10 to 15 minutes, 0.5–1 g/kg). However, cerebral hypothermia was not performed and the body temperature was kept normal.

Parameters measured

The following items were compared between survivors and patients who died. Collection of cerebrospinal fluid (CSF) and serum CSF and sera samples were collected four times (on admission and at 24, 72, and 168 hours later.) from each patient via the ventricular drain and from a peripheral vein. After centrifugation, samples were stored for a maximum of one month at −70 °C. The IL-6 and IL-8 concentration in the CSF and serum was determined using a commercially available specific ELISA kit for IL-6 (Ohtsuka Pharma Co., Tokyo, Japan) according to the manufacturer's instructions.

Statistical analysis

Results are expressed as the mean ± standard error, and statistical analysis was performed by the Mann-Whitney U-test. A p value <0.05 was considered to indicate a significant difference.

Results

Thirteen patients survived and 9 died, with the cause of death being poor control of ICP. There were no significant differences between the survivor group and the fatality group in age, gender, or GCS.

Serum IL-6 level

The serum IL-6 level of the fatal group was 102 ± 10, 176 ± 24, 873 ± 395, and 3059 ± 982 pg/ml on admission and at 24, 72 hours, respectively. The corresponding serum IL-6 levels of the surviving group were 181 ± 36, 105 ± 15, 37 ± 14, and 26 ± 22 pg/ml, respectively. From day 3 onwards, the level was significantly higher in the fatal group.

CSF IL-6 level

The IL-6 level of the fatal group was 15241 ± 1456, 97384 ± 71234, 548225 ± 233777, and 366500 ± 234500 pg/ml on admission and at 24, 72 hours, respectively, while the corresponding values were 5376 ± 1456, 3565 ± 1163, 328 ± 173, and 764 ± 191 pg/ml in the surviving group. The level was significantly higher in the fatal group from day 3 onwards. The IL-6 level in the cerebrospinal fluid was more than 100 times higher than that in the peripheral blood.

Serum IL-8 level

The serum IL-8 level of the fatal group was 21 ± 7, 28 ± 8, 43 ± 9, and 77 ± 12 pg/ml on admission and at 24, 72 hours respectively. The corresponding serum IL-8 levels of the surviving group were 36 ± 19, 15 ± 2, 15 ± 1, and 15 ± 1 pg/ml, respectively. From day 3 onwards, the level was significantly higher in the fatal group.

CSF IL-8 level

The IL-8 level of the fatal group was 29003 ± 9605, 8906 ± 4744, 5852 ± 4769, and 8220 ± 3820 pg/ml on admission and at 24, 72 hours, respectively, while the corresponding values were 23736 ± 5330, 4074 ± 3025, 355 ± 207, and 1509 ± 542 pg/ml in the surviving group. The level was significantly higher in the fatal group from day 3 onwards. The IL-8 level in the cerebrospinal fluid was more than 1000 times higher than that in the peripheral blood.

Discussion

Recently, severe traumatic brain injury has been classified into primary and secondary brain injury.

Secondary brain injury occurs immediately after the initial injury and initiates a vascular, cellular, and biological cascade that induces increased intracranial pressure, ischemia, or swelling [3, 8]. However, it may take hours to days until these events are recognized clinically. Development of the state of secondary traumatic brain injury is complex, and is reported to involve immune mediators, excitatory amino acids, reactive oxygen intermediates, complement proteins, and ions, etc. [3, 18]. Changes of cerebral homeostasis occurring after primary brain injury become manifest clinically as increased intracranial pressure, decreased cerebral perfusion pressure, brain edema, and neurological deficits. Though rapid progress has been made in the treatment of patients suffering from traumatic brain injury, secondary brain injury has still not been controlled sufficiently [22]. The extent of brain injury, blood brain barrier (BBB) dysfunction, leukocyte infiltration, and the neurologic prognosis are all considered to by determined by harmful neuroinflammation that is caused by increases of immune parameters.

IL-6 after TBI

Since IL-6 is induced by pro-inflammatory cytokines, it increases later as compared with IL-1 and TNF. IL-6 was initially thought to be a pro-inflammatory cytokine, but recent findings suggest that IL-6 also has anti-inflammatory and immunosuppressive effects [15, 34]. While IL-6 is a cytokine that has direct or indirect neurotrophic activity [6], it was found to also exert a neurotoxic effect under certain conditions [9].

Kossmann *et al.* [13] reported that the IL-6 level in the CSF is markedly increased compared with the serum level and that IL-6 can be found in the CSF even after 3 weeks. In our study, the CSF level of IL-6 was 100 times the serum level. Woiciechowsky *et al.* [36] reported that the serum IL-6 level decreased with time in patients with severe TBI. We found that the serum IL-6 level decreased in survivors whereas it increased significantly in the group with a fatal outcome after 72 hours, suggesting that the changes of IL-6 differ between the dead and surviving groups of patients with severe TBI. IL-6 protein expression is up-regulated in animal models of TBI, such as fluid percussion, stab wound, closed head injury, and axonal injury [10, 28, 32, 37]. Furthermore, resident cells of the CNS product IL-6 after challenge with various stimuli [5]. In consideration of these reports and that fact that severe BBB dysfunction occurs at the site of cerebral damage, it is thought that a large amount of IL-6 is produced intrathecally, crosses the damaged BBB, and reaches the peripheral blood. Diffusion of IL-6 from the tissues of the brain into the CSF is considered to be caused by pressure-driven bulk flow, which drives soluble molecules into the spinal fluid [25].

IL-8 after TBI

There are only a very limited number of reported studies in which changes of the CSF IL-8 levels after TBI were determined. Whalen and colleagues [35] measured IL-8 levels in the CSF of children at one time and reported that the IL-8 level increased and was well correlated with mortality. Kossmann *et al.* [14] reported that in adult patients, there was no correlation between the CSF level of IL-8 and the Glasgow Coma Scale [33] or the Glasgow outcome scale. In our study conducted in adult patients, the CSF level of IL-8 on admission was 1000 times the serum IL-8 level in the dead group as well as in the survivors. The CSF level of IL-8 was highest on admission and decreased with time in both the dead and surviving groups. This pattern of change was different from that found with IL-6. In the dead group, the IL-8 level remained significantly increased after 72 hours. Similarly to the report by Kossmann *et al.* [14], our patients with severe TBI showed a markedly higher IL-8 concentration in the CSF than in the serum, suggesting the intrathecal production of this cytokine.

IL-8 is reported to induce neutrophil-dependent microvascular damage [11, 29] and is considered to be one of the causes of BBB dysfunction. Since increased BBB permeability aggravates brain damage by inducing secondary brain injury, the key mediator of neuroinflammation seems to be IL-8.

References

1. Aihara N, Hall JJ, Pitts LH, Fukuda K, Noble LJ (1995) Altered immunoexpression of microglia and macrophages after mild head injury. J Neurotrauma 12: 53–63
2. Akira S, Kishimoto T (1992) IL-6 and NF-IL6 in acute-phase response and viral infection. Immunol Rev 127: 25–50
3. Baethmann A, Marier-Hauff K, Kempski O, Unterberg A, Wahl M, Schurer L (1988) Mediators of brain edema and secondary brain damage. Crit Care Med 16: 972–978
4. Baggiolini M, Moser B, Clark-Lewis I (1994) Interleukin-8 and related chemotactic cytokines: Giles Filley Lecture. Chest 105: 95s–98s
5. Benveniste EN (1995) The role of cytokines in multiple sclerosis/autoimmune encephalitis and other neurological disorders. In:

Agarwal B, Puri R (eds) Cytokines, their role in research and therapy. Blackwell Science, Boston, pp 159–180
6. Frei K, Malipiero UV, Leist TP, Zinkernagel RM, Schwab ME, Fontana A (1989) On the cellular source and function of interleukin 6 produced in the central nervous system in viral diseases. Eur Immunol 19: 689–694
7. Glabinski AR, Ransofoff RM (1999) Chemokines and chemokine receptors in CNS pathology. J Neurovirol 5: 3–12
8. Graham DI, McIntosh TK, Maxwell WL, Nicoll JA (2000) Recent advances in neurotrauma. J Neuropathol Exp 59: 641–651
9. Gruol DL, Nelson TE (1997) Physiological and pathological roles of interleukin-6 in the central nervous system. Mol Neurobiol 15: 307–339
10. Hans VH, Kossmann T, Lenzlinger PM, Probstmeier R, Imhof HG, Trentz O, Morganti-Kossmann MC (1999) Experimental axonal injury triggers interleukin-6 mRNA, protein synthesis and release into cerebrospinal fluid. J Cereb Blood Flow Metab 19: 184–194
11. Harada A, Sekido N, Akahoshi T, Wada T, Mukaida N (1994) Essential involvement of interleukin-8 (IL-8) in acute inflammation. J Leukoc Biol 56: 559–564
12. Kitamura T, Ogorochi T, Miyajima A (1994) Multimetric cytokine receptors. Trends Endocrino Metab 5: 8–14
13. Kossmann T, Hans VH, Imhof HG, Stocker R, Grob P, Trentz O, Morganti-Kossmann C (1995) Intrathecal and serum interleukin-6 and the acute-phase response in patients with severe traumatic brain injuries. Shock 4: 311–317
14. Kossmann T, Stahel PF, Lenzlinger PM, Redl H, Dubs RW, Trentz O, Schlag G, Morganti-Kossmann MC (1997) Interleukin-8 released into the cerebrospinal fluid after brain injury is associated with blood-brain barrier dysfunction and nerve growth factor production. J Cereb Blood Flow Metab 17: 280–289
15. Lotz M (1993) Interleukin-6. Cancer Invest 11: 732–742
16. Marz P, Cheng JG, Gadient RA, Patterson PH, Stoyan T, Otten U, Rose-John S (1998) Sympathetic neurons can produce and respond to interleukin 6. Pro Natl Acad Sci USA 95: 3251–3256
17. McClain C, Cohen D, Phillips R, Ott L, Young B (1991) Increased plasma and ventricular fluid interleukin-6 levels in patients with head injury (see comments). J Lab Clin Med 118: 225–231
18. McIntosh TK, Smith DH, Meaney DF, Kotapka MJ, Gennarelli TA, Graham DI (1996) Neuropathological sequelae of traumatic brain injury: relationship to neurochemical and biomechanical mechanisms. Lab Invest 74: 315–342
19. Morganti-Kossmann MC, Otto VI, Stahel PF, Kossmann T (2000) The role of inflammation in neurologic disease. Curr Opin Crit Care 6: 98–109
20. Morganti-Kossmann MC, Rancan M, Otto VI, Stahel PF, Kossmann T (2001) Role of cerebral inflammation after traumatic brain injury: a revisited concept. Shock 16: 165–177
21. Murata A, Ogawa M, Yasuda T, Nishijima J, Oka Y, Ohmachi Y, Hiraoka N, Niinobu T, Uda K, Mori T (1990) Serum interleukin 6, C-reactive protein and pancreatic secretory trypsin inhibitor (PSTI) as acute phase reactants after major thoracoabdominal surgery. Immunol Invest 19: 271–278
22. Neugebauer E, Hensler T, Rose S, Maier B, Holanda M, Raum M, Rixen D, Marzi I (2000) Severe craniocerebral trauma in multiple trauma. An assessment of the interaction of local and systemic mediator responses. Unfallchirurg 103: 122–131
23. Otero GC, Merrill JE (1994) Cytokine receptors on glial cells. Glia 11: 117–128
24. Proost P, Wuyts A, van Damme J (1996) The role of chemokines in inflammation. Int J Clin Lab Res 26: 211–223
25. Rosenberg GA, Kyner WT, Estrada E (1980) Bulk flow of brain interstitial fluid under normal and hyperosmolar conditions. Am J Physiol 238: F42–F49
26. Sakamoto K, Arakawa H, Mita S, Ishiko T, Ikei S, Egami H, Hisano S, Ogawa M (1994) Elevation of circulating interleukin 6 after surgery: Factors influencing the serum level. Cytokine 6: 181–186
27. Schobitz B, Pezeshki G, Pohl T, Hemmann U, Heinrich PC, Holsboer F, Reul JM (1995) Soluble interleukin-6 (IL-6) receptor augments central effects of IL-6 in vivo. FASEB J 9: 659–664
28. Shohami E, Novikov M, Bass R, Yamin A, Gallily R (1994) Closed head injury triggers early production of TNF alpha and IL-6 by brain tissue. J Cereb Blood Flow Metab 14: 615–619
29. Smith WB, Gamble JR, Clark-Lewis I, Vadas MA (1991) Interleukin-8 induces neutrophil transendothelial migration. Immunology 72: 65–72
30. Taga T, Kishimto T (1997) Gp130 and the interleukine-6 family of cytokines. Annu Rev Immunol 15: 797–819
31. Taub DD, Oppenheim JJ (1994) Chemokines, inflammation and the immune system. Ther Immunol 1: 229–246
32. Taupin V, Toulmond S, Serrano A, Benavides J, Zavala F (1993) Increase in IL-6, IL-1 and TNF levels in rat brain following traumatic lesion. Influence of pre- and post-traumatic treatment with Ro5 4864, a peripheral-type (p site) benzodiazepine ligand. J Neuroimmunol 42: 177–185
33. Teasdale G, Jennett B (1974) Assessment of coma and impaired consciousness. A practical scale Lancet 2: 81–84
34. Tilg H, Dinarello CA, Mier JW (1997) IL-6 and APPs: anti-inflammatory and immunosuppressive mediators. Immunol Today 18: 428–432
35. Whalen MJ, Carlos TM, Kochanek PM, Wisniewsky SR, Bell MJ, Clark RSB, Dekosky ST, Marion DW, Adelson PD (2000) Interleukin-8 is increased cerebrospinal fluid of children with severe head injury. Crit Care Med 28: 929–934
36. Woiciechowsky C, Schöning B, Cobanov J, Lanksch R, Volk HD, Döcke WD (2001) Early IL-6 plasma concentration correlate with severity of brain injury and pneumonia in brain-injured patients. J Trauma 52: 339–345
37. Woodroofe MN, Sarna GS, Wadhwa M, Hayes GM, Loughlin AJ, Tinker A, Cuzner ML (1991) Detection of interleukin-1 and interleukin-6 in adult rat brain, following mechanical injury, by in vivo microdialysis: evidence of a role for microglia in cytokine production. J Neuroimmunol 33: 227–236

Correspondence: Hidehiko Kushi, 1-8-13 Kanda, surugadai, Chiyoda-ku, Tokyo, Japan. e-mail: hkushi@med.nihon-u.ac.jp

Acta Neurochir (2003) [Suppl] 86: 351–354

The risk factors for the occurrence of acute brain swelling in acute subdural hematoma

T. Saito, H. Kushi, K. Makino, and **N. Hayashi**

Department of Emergency and Critical Care Medicine, Nihon University School of Medicine, Tokyo, Japan

Summary

There are many cases of acute subdural hematoma (ASDH) in which acute brain swelling (ABS) occurs intra- or post-operative period and the control of intracranial pressure becomes impossible, resulting in death. Consequently, we investigated the risk factors for ABS in ASDH. Thirteen cases were used for this study. The average age was 53.8 years, and all the subjects were males whose Glasgow Coma Scale (GCS) was an average of 4.8 when hospitalized. We examined the relationship of blood gas values (pH, PaO_2, $PaCO_2$, HCO_3^-, and base excess) when hospitalized, mean arterial blood pressure (MAP) before surgery, and the presence of hypotension during the surgery with ABS. There were no cases, which showed evidence of hypoxia or hypotension when hospitalized or before the surgery. No statistical differences in blood gas values when hospitalized were observed between the 7 cases with ABS and the 6 cases without ABS. However, all of 7 cases with ABS showed MAP $\geq$ 140 mmHg, and 6 cases demonstrated intraoperative hypotension (systolic pressure < 90 mmHg). Conspicuous hypertension before the surgery was shown to be a possible risk factor for ABS. Therefore; MAP $\geq$ 140 mmHg may be a warning sign for the occurrence of ABS.

Keywords: Acute brain swelling; acute subdural hematoma; intraperative hypotension; preoperative hypertension.

Introduction

In spite of developments and improvements in emergency medical service systems, neuro-intensive monitoring, and treatment, acute subdural hematoma (ASDH) is a disorder with a very high mortality rate and extremely poor prognosis among traumatic brain injuries [8, 15, 17]. Extensive decompressive craniectomy is performed to treat ASDH surgically [4, 7, 12]. However, we have experienced many cases where acute brain swelling (ABS) occurs during surgery, making control of intracranial pressure (ICP) impossible thereafter, resulting in death [8, 10, 15, 17, 18]. Enlargement of the cerebrovascular bed due to a decrease in cerebrovascular tension is the nature of ABS. It has been reported that secondary brain insult due to hypoxia or hypotension is associated with the occurrence of ABS. Thus, we have examined the risk factors for ABS occurring during surgery for ASDH.

Materials and method

After those cases with multiple injuries were excluded from the total of 25 ASDH patients admitted and treated in our hospital during the period from April 2000 to March 2002, 13 patients who had undergone removal of hematoma and extensive decompressive craniectomy within two hours of being hospitalized were selected as the subjects of this study.

The patients were classified into two groups: one group (ABS group) consisted of 7 cases where acute brain swelling was noted during surgery and another group consisting of 6 cases where no brain swelling (non-ABS group) was observed. Blood gas values (pH, PaO_2, $PaCO_2$, HCO_3^- and base excess) at the time of hospitalization, and blood pressure upon hospitalization and during surgery were measured. The presence of preoperative hypoxia, preoperative hypotension, preoperative hypertension, and intraoperative hypotension was investigated; and then, the relation to the occurrence of ABS was examined.

Statistical analysis was performed using the Mann-Whitney U test and Fisher's exact probability test; in addition, a significance level of $p < 0.05$ was considered to be statistically significant.

Results

The average age of the 13 cases was 53.8 $\pm$ 17.3 years, and all cases were males with an average score of 4.8 $\pm$ 1.3 points on the Glasgow Coma Scale (GCS) when hospitalized. Average blood gas values on hospitalization were: pH: 7.33 $\pm$ 0.11, PaO_2: 248.9 $\pm$ 140.2 mmHg, $PaCO_2$: 37.6 $\pm$ 6.7 mmHg, HCO_3^-: 20.1 $\pm$ 2.8 mmol/l, and base excess: $-4.8 \pm$ 3.0 mmol/l, respectively. No case demonstrated hypoxia on hospitalization. The average value of the mean arterial blood pressure (MAP) on hospitalization was

Table 1. *Characteristics of patients with acute subdural hematoma*

No	Age	Sex	GCS	MAP	PH	PaO_2	$PaCO_2$	HCO_3^-	BE	IH	ABS	Outcome
1	26	male	5	100	7.410	462.0	25.8	16.0	−6.8	−	−	survival
2	48	male	4	147	7.381	156.0	37.0	21.4	−3.1	+	−	survival
3	48	male	7	80	7.308	239.0	26.9	20.0	−1.9	−	−	survival
4	28	male	4	131	7.355	207.0	39.9	21.8	−3.4	−	−	death
5	62	male	4	132	7.374	181.0	34.3	19.6	−4.9	−	−	survival
6	57	male	4	146	7.323	210.0	39.3	20.1	−5.4	+	+	death
7	74	male	5	151	7.318	232.7	44.2	21.9	−3.8	−	+	death
8	81	male	7	164	7.401	133.5	31.8	19.5	−4.7	+	+	death
9	65	male	3	187	7.257	173.5	48.2	21.0	−6.3	+	+	death
10	71	male	4	170	7.360	511.0	45.6	25.7	0.4	+	+	death
11	35	male	5	140	7.377	359.7	30.3	17.8	−5.6	+	+	death
12	58	male	7	97	7.019	359.7	40.7	15.6	−12.0	+	−	death
13	46	male	4	150	7.398	173.2	33.9	20.9	−2.5	+	+	death

MAP Mean arterial blood pressure; *BE* base excess; *IH* intraoperative hypotension; *ABS* acute brain swelling.

Table 2. *Comparison of blood gas values and blood pressure between two subgroup*

	ABS Group	Non-ABS Group	p Value*
pH	7.35 ± 0.05	7.31 ± 0.16	n.s.
PaO_2	247.3 ± 161.9	250.5 ± 130.5	n.s.
$PaCO_2$	39.0 ± 7.2	35.6 ± 6.0	n.s.
HCO_3^-	21.0 ± 2.5	18.9 ± 2.9	n.s.
base excess	−4.0 ± 2.3	−6.1 ± 3.6	n.s.
MAP	158.3 ± 16.3	114.5 ± 25.9	<0.01

MAP Mean arterial blood pressure; * According to Mann-Whitney's U test, mean ± SD.

138.1 ± 30.5 mmHg, and no case demonstrated hypotension. Eight cases lapsed into intraoperative hypotension (systolic blood pressure < 90 mmHg). Four cases survived but 9 cases died, resulting in a mortality rate of 69.2% (Table 1).

A statistically significant difference was not observed between the ABS group and the non-ABS group regarding blood gas values on hospitalization (pH: 7.35 ± 0.05/7.31 ± 0.16, PaO_2: 247.3 ± 161.9/250.5 ± 130.5, $PaCO_2$: 39.0 ± 7.2/35.6 ± 6.0, HCO_3^-: 21.0 ± 2.5/18.9 ± 2.9 and base excess: −4.0 ± 2.3/−6.1 ± 3.6). On the other hand, MAP on hospitalization in the non-ABS group was 114.5 ± 25.9 mmHg, whereas MAP in the ABS group was revealed to be an extraordinarily high value at 158.3 ± 16.3 mmHg. A statistically significant difference was observed between the two groups (Table 2).

Seven cases in the ABS group and one case in the non-ABS group, 8 cases in total, showed abnormally high MAP values: 140 mmHg on admittance to hospital, and MAP on hospitalization in 7 out of the 8 cases where intraoperative hypotension was noted showed a value of more than 140 mmHg. From these facts, the relationship of preoperative hypertension to intraoperative hypotension and ABS was examined, setting MAP 140 mmHg as a cut point, and a statistically significant relationship was demonstrated (Table 3, and Table 4).

Table 3. *Relationship between preoperative hypertension and intraoperative hypotension*

	IH (+)	IH (−)	Total
MAP ≥ 140	7	1	8
MAP < 140	1	4	5
Total	8	5	13

Fisher's exact probability P value was 0.0319.

Table 4. *Relationship between preoperative hypertension and acute brain swelling*

	ABS (+)	ABS (−)	Total
MAP ≥ 140	7	0	7
MAP < 140	1	5	6
Total	8	5	13

Fisher's exact probability P value was 0.0047.

Two out of the 6 non-ABS cases died, but all 7 cases in the ABS group died. A statistically significant relationship between ABS and outcome was observed. (Table 5)

Discussion

ASDH is a disorder with a very high mortality rate and extremely poor prognosis among traumatic brain injuries [5]. It is reported that in mild cases of ASDH

Table 5. *Relationship between acute brain swelling and outcome*

	Survival	Death	Total
ABS group	0	7	7
Non-ABS group	4	2	6
Total	4	9	13

Fisher's exact probability P value was 0.0210.

with a GCS of 11–15, conservative therapy produces more favorable results than surgery [4, 7, 8, 15, 17]. However, in severe cases of ASDH with a GCS < 8, surgery is frequently chosen since it is emergent. As a surgical treatment of ASDH, extensive decompressive craniotomy is performed for complete removal/arrest of the hemorrhage of the hematoma or control of ICP. There are many cases where ABS occurs suddenly during surgery and it becomes impossible to control ICP, resulting in death [8, 10, 15, 17, 18]. ABS is an enlargement of the cerebrovascular bed based on a decrease in cerebrovascular tonus. It is said that concomitant cerebral parenchyma injury, disorder of cerebral autoregulation, and secondary brain damage due to hypoxia or hypotension are involved in this occurrence.

Stocchetti *et al.* [19] reported that patients with severe cerebral injuries rescued by helicopter had a high rate of hypoxia or hypotension at the site of the accident. Chesnut *et al.* [3] warned of the possibility of neurogenic hypotension and the possibility and risk of iatrogenic hypotension by the use of mannitol or furosemide in patients with severe head injuries. In 1996, the resuscitation of blood pressure and oxygenation is stated as an item in Guidelines for the Management of Severe Head Injury compiled under the supervision of the Brain Trauma Foundation, which mentions the importance of prevention and correction of hypoxia and hypotension [1]. However, according to our study, no case demonstrated hypoxia or hypotension on hospitalization. The 13 cases in our study involved patients with isolated traumatic brain injuries not accompanied by multiple injuries, and no case demonstrated hypovolemic shock due to bleeding. The time of transfer of the patients from the accident site to the hospital was less than 30 minutes and hypoxia or hypotension was not noted in the paramedics' report on admittance to the hospital. It was considered that prehospital care was adequately performed by the paramedics. We did not use furosemide on traumatic brain injury patients with enhancement of ICP. Although mannitol was used, because the waiting time for surgery was less than 2 hours at most, there was no need for frequent administration and no risk of iatrogenic hypotension in the patients. After being hospitalized, endotracheal intubation was performed and assisted ventilation was carried out by ventilator with the help of sedatives and muscle relaxants. As the patients were monitored by puncturing a radial artery and continuously measuring arterial blood pressure, no possibility existed of overlooking hypoxia or hypotension.

Although it is well known that the secondary occurrence of trans-tentorial herniation disrupts brain stem function from ICP enhancement, the influence brainstem function disorder has on brain edema has become clear. Since Ishii [9] reported that the destruction of the dorsomedial hypothalamic nucleus caused acute brain swelling in monkeys, basic studies have been carried out on the relationship of abnormal brain-stem function with brain swelling and brain edema. It has been reported that the vasomotor center in the brain-stem plays an important role in the changes of cerebral blood volume and ICP [11, 14, 20]. It has also been revealed in clinical studies that disorders of brain-stem function due to brain injuries cause various levels of damage to the midbrain, pontine, and medulla oblongata, following shearing injury or diffuse brain swelling [2, 6, 13, 16]. In many cases of ASDH is often accompanied by cerebral contusion, and in severe ASDH cases, brain-stem function may be severely damaged by shearing injury and trans-tentorial herniation caused by ICP enhancement due to brain edema, so it is considered that abnormal hypertension tends to occur due to an excessive hyper-sympathetic condition. According to the results of our study, 8 out of the 13 cases showed abnormal hypertension after being hospitalized, and this condition persisted until the time of surgery. When hyper-sympathetic condition is suddenly released by a decrease in ICP as a result of an incision in the dura mater or removal of the hematoma by craniotomy, intraoperative hypotension often results. It is considered that when hypotension is resolved by volume resuscitation and administration of catecholamine, the vascular bed swells rapidly and extensively in the brain, resulting in a state of ABS, since cerebral autoregulation has been damaged. How should we then cope with this abnormal hypertension? There is no clear answer as to whether it is possible to treat it or whether it is even preferable to treat it. It is considered that an easy antihypertensive therapy causes a decrease in cerebral blood flow, thus further causing ischemic insult to the brain.

In conclusion, it has become clear that in patients where preoperative hypertension is observed, intraoperative hypotension is likely to develop, and ABS occurs at a high rate. Persistent prominent hypertension after hospitalization is considered to be a possible risk factor for ABS. Therefore, MAP 140 mmHg is considered to be a warning sign for the occurrence of ABS.

References

1. Bullock R, Chesnut RM, Clifton G, Chajin J, Marion DW, Narayan RK, Newell DW, Pitts LH, Rosner MJ, Wilberger JE (1996) Guidelines for the management of severe head injury. J Neurotrauma 13: 639–734
2. Britt RH, Herrick MK, Manson RT, Dorfman LJ (1980) Traumatic lesions of the pontomedullary junction. Neurosurgery 6: 623–631
3. Chesnut RM, Gautille T, Blunt BA, Klauber MR, Marshall LF (1998) Neurogenic hypotension in patients with severe head injuries. J Trauma 44: 958–964
4. Coplin WM, Cullen NK, Policherla PN, Vinas FC, Wilseck JM, Zafonte RD, Rengachary SS (2001) Safety and feasibility of craniectomy with duraplasty as the initial surgical intervention for traumatic brain injury. J Trauma 50: 1050–1059
5. Croce MA, Dent DL, Menke PG, Robertson JT, Hinton MS, Young BH, Donovan TB, Pritchard FE, Minard G, Kudsk KA, Fabian TC (1994) Acute subdural hematoma: nonsurgical management of selected patients. J Trauma 36: 820–826
6. Firsching R, Woischneck D, Diedrich M, Klein S, Rückert A, Wittig H, Döhring W (1998) Early magnetic resonance imaging of brainstem lesions after severe head injury. J Neurosurg 89: 707–712
7. Guilburd JN, Sviri GE (2001) Role of dural fenestrations in acute subdural hematoma. J Neurosurg 95: 263–267
8. Hatashita S, Koga N, Hosaka Y, Takagi S (1993) Acute subdural hematoma: severity of injury, surgical intervention, and mortality. Neurol Med Chir 33: 13–18
9. Ishii (1966) Brain-swelling; studies of structural, physiologic and biochemical aiterations. In: Caveness WH, Walker AF (eds) Head injury conference proceedings. Lippincott, Philadalphia Tront, pp 276–299
10. Kelly DF, Kodrestani RK, Martin NA, Nguyen T, Hovda DA, Bergsneider M, McArthur DL, Becker DP (1996) Hyperemia following traumatic brain injury: Relationship to intracranial hypertension and outcome. J Neurosurg 85: 762–771
11. Maeda M, Takachi T, Nakai M, Krieger AJ, Sapru HN (1994) Medullary cardiovascular center and acute brain swelling. Acuta Neurochir (Wien) [Suppl] 60: 168–170
12. Meier U, Zeilinger FS, Henzka O (2000) The use of decompressive craniectomy for the management of severe head injuries. Acuta Neurochir (Wien) [Suppl] 76: 475–478
13. Mitchell DE, Adams JH (1973) Primary focal impact damage to the brain stem in blunt head injuries. Lancet 2: 215–218
14. Nagao S, Kawauchi M, Ogawa T, Ohtomo T (1987) Effect of stimulation of the medullary reticular formation on cerebral vasomotor tonus and intracranial pressure. J Neurosurg 66: 279–287
15. Phuenpathom N, Choomuang M, Ratanalert S (1993) Outcome and outcome prediction in acute subdural hematoma. Surg Neurol 40: 22–25
16. Rosenblum WI, Greenberg RP, Seeling JM, Becker DP (1981) Midbrain lesions. Frequency and significant prognostic feature in closed head injury. Neurosurgery 9: 613–620
17. Sakas DE, Bullock MR, Teadale GM (1995) One-year outcome following craniotomy for traumatic hematoma in patients with fixed dilated pupils. J Neurosurg 82: 961–965
18. Shigemori M, Tokutomi T, Yuge T, Shojima K, Matsuo H, Moriyama T, Kuramoto S (1987) Treatment of acute subdural hematoma with low GCS score. No Shinkei Geka 15: 1305–1310
19. Stocchetti N, Furlan A, Volta F (1996) Hypoxemia and arterial hypotension st the accident scene in head injury. J Trauma 40: 764–767
20. Tsutsui T, Honma Y, Sunami N, Momma F, Tsuchimoto S, Nagao S (1984) The role of brain-stem vasomotor centers on the neurogenic control of cerebrovascular tonus part 1: in special reference to cerebral blood volume and intracranial pressure. No Shinkei Geka 12: 581–589

Correspondence: Takeshi Saito, 1-8-13 Kandasurugadai Chiyoda-ku, Tokyo, Japan. e-mail: tsaito@med.nihon-u.ac.jp

Acta Neurochir (2003) [Suppl] 86: 355–360

Volume-targeted therapy of increased intracranial pressure

C.-H. Nordström

Department of Neurosurgery, Lund University Hospital, Lund, Sweden

Summary

Fluid exchange across the intact blood-brain barrier (BBB) is counteracted by the low permeability to crystalloids (mainly Na^+ and Cl^-) combined with the high osmotic pressure (5,700 mm Hg) on both sides of the BBB. If the BBB is disrupted transcapillary water transport will be determined by the differences in hydrostatic and colloid osmotic pressure between the intra- and extracapillary compartments. Under these pathological conditions pressure autoregulation of cerebral blood flow is likely to be impaired and intracapillary hydrostatic pressure will depend on variations in systemic blood pressure.

The volume targeted "Lund concept" can be summarized under four headings: A. Reduction of stress response and cerebral energy metabolism; B. Reduction of capillary hydrostatic pressure; C. Maintenance of colloid osmotic pressure and control of fluid balance; D. Reduction of cerebral blood volume. The efficacy of the protocol has been evaluated in experimental and clinical studies regarding the physiological and biochemical (utilizing intracerebral microdialysis) effects and the clinical experiences have been favourable.

Keywords: Blood-brain barrier; brain oedema; microdialysis.

The "Volume targeted therapy" of increased intracranial pressure ("Lund concept") focuses on the physiological volume regulation of the intracranial compartments and unifies the principles behind both the surgical and non-surgical therapies [12].

Like other organs the volume regulation of the brain is mainly determined by mechanisms controlling the fluid exchange across the capillaries. Regarding these mechanisms the brain differs from other organs in its highly sophisticated semi-permeable capillary membrane function – the blood brain barrier (BBB). In addition to its other physiological functions the BBB is the most important regulator of cerebral volume [9]. Besides the volume regulation of the brain tissue itself the physiological regulation of cerebrospinal fluid (CSF), cerebral blood flow (CBF) and cerebral blood volume (CBV) determines the intracranial dynamics.

Blood-brain barrier permeability

The flux of water across a membrane or microvascular bed (J_v) is described by the equation

$$J_v = L_p \times A \times [\Delta P - \Sigma\sigma_s \times \Delta\Pi_s] \qquad \text{(Eq. 1)}$$

where L_p is hydraulic conductivity, A is the surface area available for fluid exchange, ΔP is the transcapillary hydrostatic pressure difference, $\Delta\Pi_s$ the transcapillary osmotic pressure difference and σ_s the reflection coefficient of each solute (s) of the system. The transcapillary water transport is thus determined by the difference in hydrostatic pressure (ΔP), the osmotic pressure gradient ($\Delta\Pi_s$), the endothelial component which will determine which solutes are reflected and will contribute to the osmotic pressure gradient (σ_s), and the hydraulic conductance of the capillary wall ($L_p \times A$). The two major solutes of biological fluids (Na^+ and Cl^-) have a blood-brain barrier reflection coefficient of 1.0. Water passing the BBB in any direction will thus be very dilute regarding crystalloids and an opposing osmotic gradient, which counteracts further fluid transport, will immediately be created.

According to eq. 1 the magnitude of the hydraulic conductance ($L_p \times A$) describes the rate by which the fluid is transferred across the capillary membrane whenever a driving force ($\Delta P - \Sigma\sigma \times \Delta\Pi$) is developed. This means that the only way of inducing transcapillary filtration or absorption is to disturb the balance between the hydrostatic and osmotic forces across the capillary membrane. For two reasons the magnitude of the CPP is of less importance for brain volume under physiological conditions. Firstly, intracapillary pressure and cerebral blood flow are physiologically tightly autoregulated and variations in systemic blood pressure are generally not transmitted to cerebral capillaries. Secondly, transcapillary fluid exchange is effectively counteracted by the low permeability to crystalloids combined with the high osmotic pressure of 5,700 mm Hg on both sides of the BBB [9]. This contrasts to most other capillary regions where the osmotic pressure force is mainly derived from the difference between plasma and interstitial colloid osmotic pressure approximately balancing the transcapillary hydrostatic pressure.

Regulation of cerebral blood flow and cerebral blood volume

The primary function of the tight regulation of cerebral blood flow is to provide a continuous supply of glucose and oxygen to secure the high metabolic demand. Two secondary effects of blood flow regulation have important consequences for the regulation of ICP: the intracapillary hydrostatic pressure and the intracerebral blood volume (mainly within the venous compartment). Several important principles within intracranial dynamics can be described in relation to the conventional physiological regulatory mechanisms of the cerebral blood flow: pressure autoregulation, CO_2-regulation, and metabolic regulation.

Pressure autoregulation implies that blood flow remains relatively constant despite variations in perfusion pressure within defined limits. The phenomenon is present in many organs and appears to be most prominent in the brain and in the kidneys. In physiological literature, autoregulation is usually described as a mechanism, which serves the purpose of keeping intracapillary hydrostatic pressure constant. In neurosurgical literature, autoregulation is often interpreted as a mechanism, which protects the brain from a fall in CBF when CPP decreases.

A rapid reduction of intracranial blood volume is conventionally obtained by controlled hyperventilation via the CO_2-regulation of CBF. The effect is obtained by a pH dependent constriction of precapillary resistance vessels. However, prolonged hyperventilation is probably of limited value or may even be harmful. Pronounced hyperventilation carries the risk of inducing focal ischemia and may have adverse effects on outcome. The reduction of CBF and CBV during hyperventilation is also transient in spite of preserved hypocapnia. Finally, in the most severely damaged patients hyperventilation is least effective in reducing CBV due to impaired cerebrovascular CO_2-reactivity [13, 23].

A lasting vasoconstriction and reduction of CBV can be achieved through reduction of cerebral energy metabolism and is often obtained by barbiturate infusion [18]. This mechanism is probably effective only in patients with preserved cerebrovascular CO_2-reactivity [18]. In addition, prolonged high-dose barbiturate treatment is often associated with pulmonary, cardiovascular and other serious complications [24].

Since about 70% of the intracranial blood volume is located within the venous compartment it would be possible to induce a lasting reduction of CBV with a minimal effect on CBF by pharmacological means. Dihydroergotamine (DHE) is known to primarily constrict venous capacitance vessels within the peripheral circulation. DHE has also been shown to decrease ICP in patients with severe head injuries [10] and this effect is obtained although CBF remains unaffected or increases [4, 16]. DHE has been shown to decrease ICP also in patients with impaired CO_2-reactivity [4] and in vitro studies have shown that DHE has a more pronounced constrictor effect on isolated human cortical veins than arteries [17].

Pathophysiological aspects of disturbed brain volume regulation

There are no techniques available for quantitative measurements of intracapillary hydrostatic pressure in the brain or the effects of hydrostatic pressure variations on transcapillary water transport. Some aspects of water transport across cerebral capillaries can however be studied in a model system utilizing a denervated cat skeletal muscle enclosed in a plethysmograph [1, 11, 12]. In this model the capillaries are permeable for water and most solutes but much less permeable for proteins ($\sigma \approx 0.8$–0.9) and pressure autoregulation of blood flow is impaired. The model thus may simulate the pathophysiological situation in a damaged brain.

The clinical implications of the data obtained in these experiments can be summarized as follows. If the injured brain has an impaired pressure autoregulation and a BBB with an increased permeability to crystalloids then transcapillary fluid exchange will be highly dependent on variations in systemic arterial blood pressure. Under conditions with increased BBB permeability to crystalloids a decrease in tissue pressure will cause a net transport of fluid from the capillaries into the interstitial space and, as discussed below, all surgical treatments of increased ICP are associated with a decrease in tissue pressure.

Surgical treatments of increased ICP – physiological aspects

Since the brain and its surroundings are enclosed within a rigid shell an increase in volume of one of the components must be at the expense of others:

$$V_{intracran} = V_{blood} + V_{brain} + V_{CSF} + V_{mass\ lesion} \quad \text{(Eq. 2)}$$

All surgical and non-surgical treatments of increased ICP are directed towards a change in volume of one or more of these components. The surgical treatments to decrease the intracranial volumes include evacuation of mass lesions and drainage of CSF. In selected cases decompressive craniectomy may be used to increase $V_{intracran}$.

It is well documented that, following initial resuscitation, early evacuation of focal mass lesions is the single most important treatment of head injured patients. The beneficial effects of CSF drainage in head-injured patients with sustained increase in ICP are not well documented but this treatment has been included in the recommendations by the US Brain Trauma Foundation [5]. The possibility to increase the $V_{intracran}$ by performing a craniectomy has been advocated as a last resort for treatment of increased ICP by several authors.

The various surgical treatments have one consequence in common: they all reduce tissue pressure. In a pathophysiological situation with increased permeability of the BBB to crystalloids a decrease in tissue pressure will lead to increased transcapillary water transport even if the BBB is not permeable to large molecules. The consequences are well known to all neurosurgeons: the gradual increase in ICP almost always observed after evacuation of focal lesions; the collapse of the ventricles often induced by ventricular drainage; the bulging of cerebral tissue through the bone defect after craniectomy. The physiological principles discussed above indicate that the rapid change in volume obtained by surgical treatment should always be combined with a non-surgical treatment aiming primarily at a slow and lasting reduction in brain water content [1, 2, 12].

Volume-targeted therapy of increased ICP

The volume-targeted "Lund concept" is based on the physiological considerations above and can be summarized under four headings.

Reduction of stress response and cerebral energy metabolism

Stress response is reduced by liberal use of sedatives (benzodiazepines) and analgesics (opioids). A further reduction of the stress response and catecholamine release is obtained by a continuous infusion of low-dose thiopental (0.5–$3\ mg \cdot kg^{-1} \cdot h^{-1}$) and fentanyl ($2$–$5\ \mu g \cdot kg^{-1} \cdot h^{-1}$) but also by treatment with the β_1-antagonist metoprolol and the α_2-agonist clonidine (see below). The dose of thiopental is kept low to avoid cardiac inhibition, pulmonary complications and other side effects [24].

Reduction of capillary hydrostatic pressure

Mean arterial blood pressure is reduced to the physiological level for the age of the individual patient with a combination of the β_1-antagonist metoprolol (0.2–$0.3\ mg \cdot kg^{-1} \cdot 24h^{-1}$ *i.v.*) and the α_2-agonist clonidine (0.4–$0.8\ \mu g \cdot kg^{-1} \times 4$–$6$ *i.v.*) [2, 3, 11, 12]. The antihypertensive treatment is initiated after evacuation of focal mass lesions when the patients are clearly normovolemic as obtained by red cell and albumin/plasma transfusions to normal albumin and hemoglobin values and to a normal central venous pressure. A CPP of 60–70 mm Hg is considered optimal but, if necessary to control ICP, a transient decrease in CPP to 50 mm Hg for adults and 40 mm Hg for children is accepted. Thiopental, as well as DHE, has a precapillary vasoconstrictor effect, which will contribute to lowering the intracapillary hydrostatic pressure.

Maintenance of colloid osmotic pressure and control of fluid balance

Red cell transfusions and albumin are given to achieve normal values (Hb/s 125–140 g/l, alb/s $\approx$ 40 g/l) to ensure normovolemia and to optimize oxygen supply. The albumin/plasma/blood transfusions also serve the purpose of obtaining a normal colloid osmotic pressure favoring transcapillary absorption. A balanced or moderately negative fluid balance is a part of the treatment protocol and is achieved by diuretics (furosemide) and albumin infusion. All patients are given a low calorie enteral nutrition (max energy supply 15–$20\ kcal \cdot kg^{-1} \cdot 24\ h^{-1}$).

Reduction of cerebral blood volume

Intracranial blood volume may be reduced both on the arterial side with thiopental [18] and DHE [4, 10, 16] and on the venous side with DHE [17]. DHE is only given when other treatments are insufficient and always at the lowest doses necessary to reduce intracranial pressure to values below 20–25 mm Hg. DHE should not be given for more than five days in order to minimize the risks of compromising peripheral circulation, in particular in patients with fractures of the extremities or renal insufficiency. The maximum doses of DHE used are: 0.8 $\mu g \cdot kg^{-1} \cdot h^{-1}$ day one, 0.6 $\mu g \cdot kg^{-1} \cdot h^{-1}$ day two, 0.4 $\mu g \cdot kg^{-1} \cdot h^{-1}$ day three, 0.2 $\mu g \cdot kg^{-1} \cdot h^{-1}$ day four, and 0.1 $\mu g \cdot kg^{-1} \cdot h^{-1}$ day five. In most patients intracranial pressure can be controlled at values below 25 mmHg with lower doses and an earlier discontinuation of DHE than five days.

Cerebral energy metabolism – relation to cerebral perfusion pressure

In patients with increased ICP a reduction of CPP might jeopardize cerebral energy metabolism especially in vulnerable regions such as the penumbra zones surrounding focal mass lesions. The intracerebral microdialysis technique permits bedside analyses of compounds reflecting energy metabolism (glucose, pyruvate, lactate) and indicators of excessive concentration of excitatory transmitters (glutamate) as well as cell membrane degradation (glycerol). The levels obtained during neuro intensive care can be compared to reference levels obtained in the normal human brain [20] and in patients with manifest cerebral ischemia [25]. By inserting multiple intracerebral catheters during open surgery it is possible to choose areas of interest *i.e.* the zones surrounding an evacuated focal lesion [26]. Figure 1 summarizes resent data, which support the view that in patients treated according to the "Lund concept" the lactate/pyruvate ratio (cytoplasmatic redox state) is not affected in the vulnerable penumbra zone until CPP below 50 mmHg [19].

Clinical results

Ideally all clinical therapies should be based on randomised controlled studies. This is often difficult to

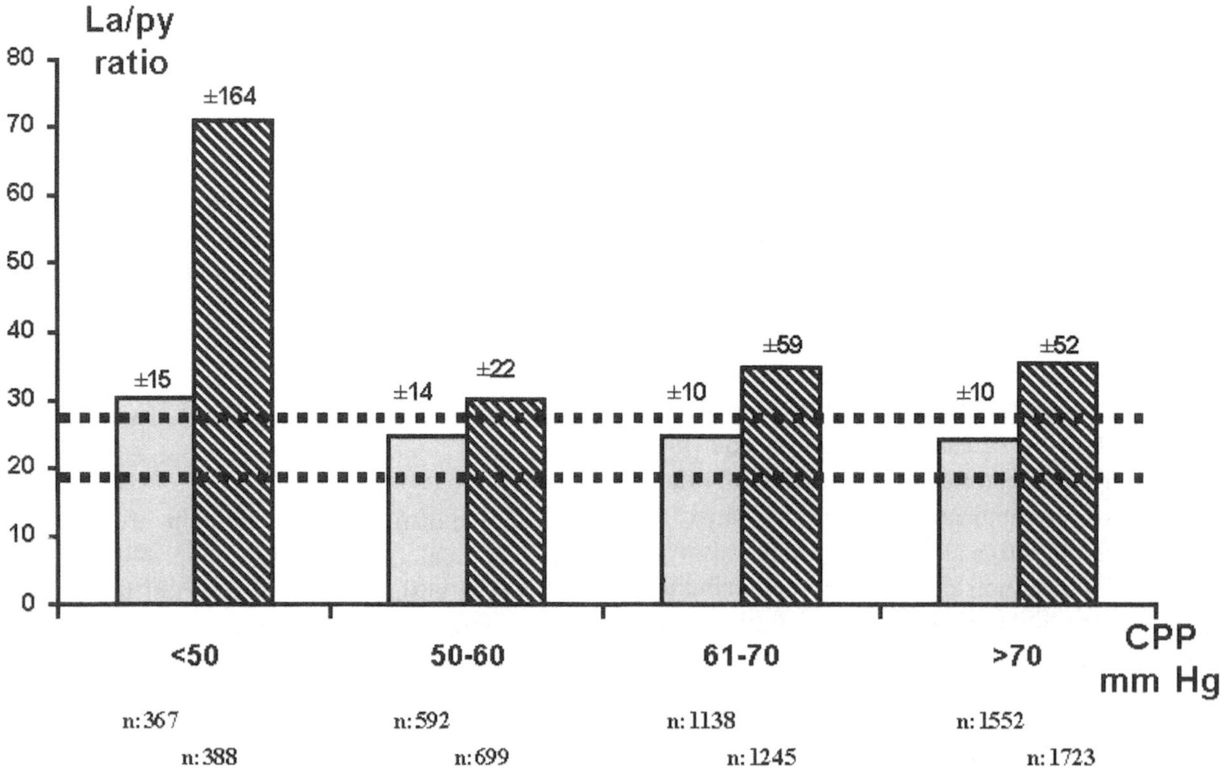

Fig. 1. Mean levels ± S.D. for the lactate/pyruvate (la/py) ratio (n: 7,704 measurements) in the "better" (gray) and "worse" (striped) positions in relation to four ranges of CPP. The broken lines indicate the range (mean ± S.D.; 23 ± 4) in normal human brain during wakefulness. Data from Nordström *et al.* [19]

attain in intensive care and there are virtually no such studies to support any specific treatment for increased ICP. The only randomised clinical trial that has compared the consequences of targeting different levels of CPP failed to show a long-term benefit of increasing CPP above 70 mmHg [22]. In a recent review on the management of cerebral perfusion pressure after head trauma it was concluded that it seemed likely that a CPP of 60 mmHg provided adequate perfusion for most patients with severe traumatic brain injuries [21].

Clinical outcome of volume-targeted therapy according to the "Lund-concept" has been reported from four Swedish neurosurgical centres [6, 8, 14, 15]. All four studies have shown a remarkably low mortality. In the original study, which included a selected group of patients with very severe traumatic brain lesion and ICP above 25 mmHg in spite of conventional treatment, the decrease in mortality was from 47% to 8% [6]. The significant decrease in mortality was associated with significant increases in the groups of "good recovery" and "moderate disability" but did not increase the number of patients in the groups of "severe disability" or "vegetative state". A comprehensive study of the long-term results in all patients in this center has recently been presented [7].

References

1. Asgeirsson B, Grände PO (1994) Effects of arterial and venous pressure alterations on transcapillary fluid exchange during raised tissue pressure. Intensive Care Med 20: 567–572
2. Asgeirsson B, Grände PO, Nordström CH (1994) A new therapy of post-trauma brain oedema based on haemodynamic principles for brain volume regulation. Intensive Care Med 20: 260–267
3. Asgeirsson B, Grände PO, Nordström CH, Berntman L, Messeter K, Ryding E (1995) Effects of hypotensive treatment with α2-agonist and β1-antagonist on cerebral hemodynamics in severe head injury. Acta Anasthesiol Scand 39: 347–351
4. Asgeirsson B, Grände PO, Nordström CH, Messeter K, Sjöholm A (1995) Cerebral hemodynamic effects of dihydroergotamine in patients with intracranial hypertension after severe head injury. Acta Anaesthesiol Scand 39: 922–930
5. Brain Trauma Foundation, American Association of Neurological Surgeons, Joint Section on Neurotrauma and Critical Care (1996) Guidelines for the management of severe head injury. J Neurotrauma 13: 641–734
6. Eker C, Asgeirsson B, Grände PO, Schalén W, Nordström C-H (1998) Improved outcome after severe head injury with a new therapy based on principles for brain volume regulation and improved microcirculation. Crit Care Med 26: 1881–1886
7. Eker C, Schalén W, Asgeirsson B, Grände P-O, Ranstam J, Nordström C-H (2000) Reduced mortality after severe head injury will increase the demands for rehabilitation services. Brain Injury 14: 605–619
8. Elf K, Nilsson P, Enblad P (2002) Outcome after traumatic brain injury improved by an organized secondary insult program and standardized neurointensive care. Crit Care Med 30: 2129–2134
9. Fenstermacher JD (1984) Volume regulation of the central nervous system. In: Staub NC, Taylor AE (eds) Edema. Raven Press, New York, pp 383–404
10. Grände PO (1989) The effect of dihydroergotamine in patients with head injury and raised intracranial pressure. Intensive Care Med 15: 523–527
11. Grände PO, Asgeirsson B, Nordström CH (1997) Physiological principles for volume regulation of a tissue enclosed in a rigid shell with application to the injured brain. J Trauma 42 [Suppl] 5: S23–S31
12. Grände PO, Asgeirsson B, Nordström CH (2002) Volume targeted therapy of increased intracranial pressure: the Lund concept unifies surgical and non-surgical treatments. Acta Anaesthesiol Scand 46: 929–941
13. Messeter K, Nordström CH, Sundbärg G, Algotsson L, Ryding E (1986) Cerebral hemodynamics in patients with acute severe head trauma. J Neurosurg 64: 231–237
14. Naredi S, Edén E, Zäll S, Stephensen H, Rydenhag B (1998) A standardized neurosurgical/neurointensive therapy directed toward vasogenic edema after severe traumatic brain injury: clinical results. Intensive Care Med 24: 446–451
15. Naredi S, Olivecrona M, Lindgren A, Östlund AL, Grände PO, Koskinen LOD (2001) An outcome study of severe traumatic head injury using the "Lund therapy" with low-dose prostacyclin. Acta Anaesthesiol Scand 45: 401–405
16. Nilsson F, Messeter K, Grände PO, Rosén I, Ryding E, Nordström CH (1995) Effects of dihydroergotamine on cerebral circulation during experimental intracranial hypertension. Acta Anaesthesiol Scand 39: 916–921
17. Nilsson F, Nilsson T, Edvinsson L, Björkman S, Nordström CH (1997) Effects of dihydroergotamine and sumatriptan on isolated human cerebral and peripheral arteries and veins. Acta Anaesthesiol Scand 41: 1257–1262
18. Nordström CH, Messeter K, Sundbärg G, Schalén W, Werner M, Ryding E (1988) Cerebral blood flow, vasoreactivity, and oxygen consumption during barbiturate therapy in severe traumatic brain lesions. J Neurosurg 68: 424–431
19. Nordström CH, Reinstrup P, Xu W, Gärdenfors A, Ungerstedt U (2002) Assessment of the lower limit for cerebral perfusion pressure in severe head injuries by bedside monitoring of regional energy metabolism. Anesthesiology. (accepted 2002)
20. Reinstrup P, Ståhl N, Hallström Å, Mellergård P, Uski T, Ungerstedt U, Nordström CH (2000) Intracerebral microdialysis in clinical practice. Normal values and variations during anaesthesia and neurosurgical operations. Neurosurgery 47: 701–710
21. Robertson CS (2001) Management of cerebral perfusion pressure after traumatic brain injury. Anesthesiology 95: 1513–1517
22. Robertson CS, Valadka AB, Hannay J, Contant CF, Gopinah SP, Cormio M, Uzura M, Grossman RG (1999) Prevention of secondary ischemic insults after severe head injury. Crit Care Med 27: 2086–2095
23. Schalén W, Messeter K, Nordström CH (1991) Cerebral vasoreactivity and the prediction of outcome in severe traumatic brain lesions. Acta Anaesthesiol Scand 35: 113–122
24. Schalén W, Messeter K, Nordström CH (1992) Complications and side effects during thiopentone therapy in patients with severe head injuries. Acta Anaesthesiol Scand 36: 369–377
25. Ståhl N, Mellergård P, Hallström Å, Ungerstedt U, Nordström CH (2001) Intracerebral microdialysis and bedside biochemical analysis in patients with fatal traumatic brain lesions. Acta Anaesthesiol Scand 45: 977–985

26. Ståhl N, Ungerstedt U, Nordström CH (2001) Brain energy metabolism during controlled reduction of cerebral perfusion pressure in severe head injuries. Intensive Care Med 27: 1215–1223

Correspondence: Carl-Henrik Nordström, Department of Neurosurgery, University Hospital, S-221 85 Lund, Sweden. e-mail: carl-henrik.nordstrom@neurokir.lu.se

Acta Neurochir (2003) [Suppl] 86: 361–365

Induced acute arterial hypertension and regional cerebral flow in intracontusional low density area

A. Chieregato[1], E. Fainardi[2], A. Tanfani[1], G. Sabia[1], C. Martino[1], R. Pascarella[3], F. Servadei[4], and L. Targa[1]

[1] Neurorianimazione, Ospedale Bufalini-Cesena, Italy
[2] Neuroradiologia, Ospedale S. Anna-Ferrara, Italy
[3] Neuroradiologia, Ospedale Bufalini-Cesena, Italy
[4] Neurotraumatologia, Ospedale Bufalini-Cesena, Italy

Summary

Traumatic brain contusions have been associated with regional ischemia. We aimed to measure the effect of induced supra-normal values of cerebral perfusion pressure (CPP) on regional cerebral blood flow (rCBF) in the intracontusional low density area surrounding the contusional hemorrhagic core. In 7 severely head injured patients (GCS ≤ 8) harbouring a contusion larger than 2 cm, the rCBF levels were measured, by means of Xenon-enhanced CT, in: 1) the intracontusional low density area; 2) contralaterally, in a normal brain symmetric area. CBF studies were performed at a baseline CPP of 65.3 mmHg $\pm$ 7 and after 20 minutes of norepinephrine-induced CPP supernormal values (88.3 mmHg $\pm$ 10.5) ($p = 0.0013$). A "paradoxical" reduction of rCBF levels was observed in both the intracontusional low density area ($p = 0.07$) and the contralateral "normal" area ($p = 0.08$). In particular, this decrease of rCBF in the intracontusional low density area (-25.7 ± 10 ml/100gr/min) ($p = 0.0009$) was present in only 4 cases, having a mean rCBF at baseline of 25 ± 16 ml/100gr/min. In the remaining 3 cases in which rCBF at baseline was abnormally low (12 ± 7 ml/100gr/min), rCBF values improved slightly (3.6 ± 2 ml/100gr/min) ($p = 0.61$). An acute increase of CPP seems to marginally affect rCBF in the intracontusional low density area having critically reduced initial values, but may greatly reduce rCBF in subjects starting from non-critical baseline values.

Keywords: Autoregulation; cerebral blood flow; cerebral perfusion pressure; contusion; ischemia; xenon CT.

Introduction

Traumatic contusion has been associated with regional ischemia. Low levels of regional cerebral blood flow (rCBF), below 18 ml/100gr/min [5], have been demonstrated in the intracontusional edematous low density area surrounding the contusional hemorrhagic core investigated by means of Xenon-enhanced CT (Xe-CT) [13]. Microscopic findings of cytotoxic edema and microvascular ischemia [13] were also detected in this area [10]. In the last decade, an increase of cerebral perfusion pressure (CPP) has been suggested as a therapeutic option for the control of elevated intracranial pressure (ICP) [12] and for the treatment of global and focal ischemia. When elevated CPP is used to improve ICP control, it benefits from a preserved autoregulation ("pressure active" pattern) which, in order to maintain CBF stability, induces a vasoconstriction and a reduction of cerebral vascular volume and, finally, ICP [12]. In a setting of focal and global ischemia, the pressure autoregulation is deranged and the rCBF passively follows the changes in CPP ("pressure passive" pattern) [11]. Experimental evidence suggests that while normal values of CPP may be associated with focal hypoperfusion [16] in traumatic contusions, an increase in CPP has been associated with raised rCBF [1, 7]. As far as we know, this concept has not yet been verified in a clinical setting. One of the current indications for surgery in severely head injured patients with focal parenchymal damage is the elevation of ICP, with the aim of removing the contusional necrotic/hemorrhagic tissue, whilst sparing any potentially viable surrounding brain matter [13]. However specific medical therapy might contribute to this objective and an increase of CPP, above the recommended values of 70 mmHg, could lead to an improvement of regional perfusion in intracontusional low density tissue surrounding the hemorrhagic core. The purpose of this study was to assess the effect of a briefly induced increase in CPP on intracontusional low density tissue rCBF.

Material and methods

Seven severely head injured patients (GCS ≤ 8) harbouring seven traumatic mixed density contusions larger than 2 cm. were included in the study. Patients with cardiovascular dysfunctions and those with unstable ICP were excluded. In agreement with guidelines [8], all patients were managed using a staircase protocol to maintain ICP below 20 mmHg and cerebral perfusion pressure (CPP) above 70 mmHg. The outcome at six months post-injury was measured by means of the Glasgow Outcome Scale (GOS) [4].

Stable xenon-enhanced CT CBF measurement

CBF studies were conducted by a wash–in protocol based on the inhalation of a mixture of 28% Xenon, 40% Oxygen and 32% room air using a CT scanner (Picker 5000) equipped for Xe-CT CBF imaging (Xe/TC system-2™, Diversified Diagnostic Products, Inc., Houston, TX). All patients were artificially ventilated, sedated, and paralysed.

CBF analysis

The rCBF was measured by means of Xenon-enhanced computerized tomography (Xe-CT) in two different ROIs drawn freehand on the CT scan: 1) the intracontusional low density area surrounding the contusional hemorrhagic core; 2) contralaterally, in a normal-appearing area corresponding to total contusional brain tissue (hemorrhagic core plus intracontusional low density area) located symmetrically to the contusion. According to previous authors [3, 9, 13], a value of CBF below 18 ml/100gr/min, traditionally considered to be the threshold indicating ischemia [5], was used to dichotomize the rCBF measurements.

Increase in CPP

Baseline rCBF values were obtained during an initial Xe-CT study. Following this, a progressive increase of mean arterial pressure (MAP) was induced and maintained for a total time of 20 minutes, by means of a continuous norepinephrine infusion. A stable increase of CPP, averaging 20 mmHg, was then obtained and a second Xe-CT measurement was carried out. Arterial blood gas analysis was performed before xenon administration, both at baseline and during hypertension phase. All efforts were made to maintain $paCO_2$ stable throughout both studies.

Statistical analysis

The baseline and post-hypertension values of physiological variables and rCBF were compared by means of the paired t-test. Frequencies of rCBF values below 18 ml/100gr/min were compared using the Fisher exact test. A value of p below 0.05 was considered significant.

Exploratory analysis

A relevant dispersion was noted in the mean value of rCBF changes in the intracontusional low density area during data analysis. Graphical analysis of data revealed that following the improvement of CPP, the intracontusional low density area rCBF increased (the rCBF follows passively the CPP, "pressure passive group") in three patients, whereas this did not occur in the remaining four patients ("pressure hyper-active group"). An exploratory analysis was applied with regard to the subgroups.

Results

The demographic, clinical and radiological data of our case material are reported in Table 1. Three cases had initial values of GCS above 8, but rapidly deteriorated during the acute phase.

Main results (Table 2)

A significant difference was found between MAP and CPP measured at baseline and after the induced increase with norepinephrine. No significant changes were observed for ICP, nor between $PaCO_2$ measured at baseline and during hypertension. After induced hypertension, the rCBF mean levels *decreased*, to level nearly statistically significant, both in the intracontusional low density area (-13.1 ± 17.2 ml/100gr/min) ($p = 0.07$) and the contralateral area (-11.3 ± 6.4 ml/100gr/min) ($p = 0.08$).

Exploratory analysis (Table 2)

A reduction of rCBF mean levels in the contralateral area was observed after the induction of arterial hypertension in all "pressure hyper-active" and "pressure passive" subjects (Fig. 1). In the "pressure hyper-active group", a significant reduction of intracontusional low density area rCBF mean levels (-25.7 ± 10 ml/100gr/min), below the threshold of 18 ml/100gr/min, was seen in all four patients. In this subset of patients, the baseline rCBF mean levels were above 18 ml/100gr/min. In the "pressure hyper active" group the reduction of rCBF was more relevant in the edematous area than contralaterally. In the "pressure passive group", rCBF mean values improved slightly (3.6 ± 2 ml/100gr/min) but without any statistical significance (Fig. 2). Two of the 3 patients had rCBF mean values at baseline below 18 ml/100gr/min and the remaining third patient showed an individual baseline CBF value of 20 ml/100gr/min.

Discussion

The pathophysiological background of the present study indicates that an improvement of rCBF levels in the intracontusional low density area, surrounding the hemorrhagic core of a brain contusion, may be obtained by means of an acute increase in CPP. This could be the consequence of a focal derangement of the

Table 1. *Characteristics of 7 severely head injured patients undergoing Xe-CT studies*

Patients (n)		7
Age (years)	mean ± SD	48.4 ± 20.4
Sex (n)	male/female	6/1
Best initial GCS	median (IQR)	10 (5)
Best initial motor GCS	median (IQR)	5 (1)
Worst initial GCS	median (IQR)	7 (5)
Worst initial motor GCS	median (IQR)	4 (2)
Worst initial pupillary reactivity	normal (n)	6 (86%)
	one pupil dilated unreactive (n)	1 (14%)
	bilaterally dilated unreactive pupils (n)	0 (0%)
Worst initial hypotension (n and %)	no hypotension	6 (86%)
	hypotension	1 (14%)
Worst initial hypoxia (n and %)	no hypoxia	5 (71%)
	hypoxia	2 (29%)
Worst CT and main lesions (n)	DI II	2
	EML	4
	• SDH	• 2
	• contusion	• 2
	NEML	1
	• contusion	• 1
Outcome at discharge (n)	obeying	1
	not obeying	6
	dead	0
Outcome at 6 months (n)	good recovery or moderate disability	5
	severe disability	1
	persistent vegetative state or death	1
Contusions	n	7
Intracontusional low density area (mm^2)	mean ± SD	1257 ± 378
Xenon studies	n	14
Time post injury of Xenon studies (hrs)	mean ± SD	133 ± 50
Contusions studied per time intervals (hours post injury)	24	0
	24–96	2
Hemoglobin	post 96	5
Hematocrit	gr/dl	10.4 ± 1.6
	%	30.6 ± 4.5

pressure autoregulation due to ischemic damage [11]. However the results did not support the hypothesis of the study, showing, on the contrary, that a supranormal CPP was associated with a *reduction*, instead of an improvement, of rCBF mean levels in the intracontusional low density area. Similar results were obtained for mean values of rCBF in the contralateral area. A possible explanation for the rejection of the study hypothesis may be that our case selection could not confirm the background of the study. In fact the intracontusional low density area rCBF was ischemic at baseline, below 18 ml/100gr/min [5], in only 2 patients. As a consequence the possible association between focal edema, ischemia and impaired autoregulation was not a common finding within the subjects studied. Moreover, contusion associated ischemia was found to be highly variable also in previous reports [3, 9, 13]. While the previous consideration excludes a diffuse derangement of pressure autoregulation in the lesions studied, it remains however to be explained why we observed, on average, an overregulation of rCBF following an increase of CPP. This "paradox" response had already have been reported in a similar experimental setting involving patients with subarachnoid hemorrhage and vasospasm [2], but the physiological reasons are unclear, even if a direct effect on brain microvessels of norepinephrine may be advocated [15].

Exploratory analysis

The exploratory analysis suggests that the induced increase in CPP could determine an improvement of rCBF in the low density tissue surrounding the hemorrhagic core only in a subset of patients. These "pressure passive" subjects, have pathological baseline values of rCBF, below or approximating the ischemia threshold [5]. In these areas the suggested association

Table 2. *Physiological variables (mean ± SD) in the three CBF phases*

		Baseline	During hypertension	p
All patients	n	7	7	
	intracontusional low density CBF (ml/100gr/min)	25 ± 16	12 ± 7	0.07
	individual subjects with intracontusional low density CBF below 18 ml/100gr/min (n)	2	6	0.10
	contralateral CBF (ml/100gr/min)	27.6 ± 12.6	16.3 ± 9.4	0.08
	individual subjects with contralateral CBF below 18 ml/100gr/min (n)	2	5	0.28
	MAP (mmHg)	94 ± 9	121 ± 14	0.0025
	CPP (mmHg)	66 ± 7	90 ± 10	0.0003
	ICP (mmHg)	26 ± 9	27 ± 12	0.88
	PaCO2 (mmHg)	35.8 ± 3.9	35.6 ± 4.2	0.94
	SaO2 (%)	97.3 ± 1.8	97.3 ± 2	0.93
Pressure passive patients	n	3	3	
	intracontusional low density CBF (ml/100gr/min)	12 ± 7	15.3 ± 9	0.61
	individual subjects with intracontusional low density CBF below 18 ml/100gr/min (n)	2	2	1
	contralateral CBF (ml/100gr/min)	18 ± 2	13 ± 3	0.0563
	individual subjects with contralateral CBF below 18 ml/100gr/min (n)	2	3	1
	MAP (mmHg)	90 ± 8	116 ± 21	0.12
	CPP (mmHg)	62 ± 4	84 ± 11	0.03
	ICP (mmHg)	28 ± 11	32 ± 16	0.77
	PaCO2 (mmHg)	35 ± 5	33 ± 5	0.7
	SaO2 (%)	95.7 ± 3	95.8 ± 3	0.7
Pressure hyper-active patients	n	4	4	
	intracontusional low density CBF (ml/100gr/min)	36 ± 13	10 ± 4	0.0009
	individual subjects with intracontusional low density CBF below 18 ml/100gr/min (n)	0	4	0.03
	contralateral CBF (ml/100gr/min)	35 ± 13	19 ± 12	0.12
	individual subjects with contralateral CBF below 18 ml/100gr/min (n)	0	2	0.42
	MAP (mmHg)	94 ± 11	117 ± 10	0.02
	CPP (mmHg)	71 ± 7	93 ± 9	0.005
	ICP (mmHg)	25 ± 7	23.5 ± 8	0.83
	PaCO2 (mmHg)	36.3 ± 3	37.4 ± 2.8	0.66
	SaO2 (%)	97.3 ± 1	97.4 ± 1	0.86

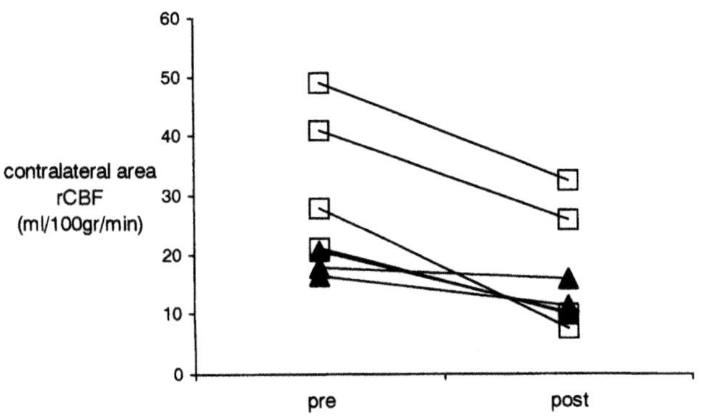

Fig. 1. Reduction of rCBF in the normal appearing contralateral brain (each line, a subject). Squares (pressure hyper-active subjects), triangles (pressure passive subjects)

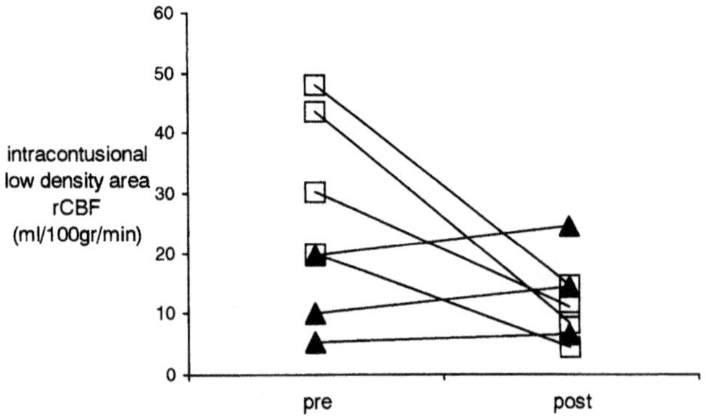

Fig. 2. Different patterns of rCBF in the intracontusional low density tissue (each line, a subject). Squares (pressure active subjects), triangles (pressure passive subjects)

between ischemia and deranged pressure autoregulation may be confirmed. However the change in rCBF seems to be, beside not statistically significant, also hardly biological relevant. Probably by the time we applied the elevated CPP, cytotoxic edema and microvascular ischemia were already well established and the microvessels in the intracontusional low density area may have been no more sensitive to reperfusion. In the experimental setting, the efficacy showed in restoring contusional CBF with cahecolamines may be in fact related to the early intervention post-injury [1, 7].

The exploratory analysis more clearly showed the extent of the paradoxical reduction of rCBF in the intracontusional low density area in the pressure "hyperactive" subjects. In these patients supra-normal CPP *inevitability* converted normal (for comatose patients) rCBF into ischemic values. A less pronounced reduction of rCBF was found in contralateral areas in the same subjects, suggesting that the damaged area may be relatively more sensitive to norepinephrine-induced arterial hypertension. The intracontusional low density area especially following some days [6] may be affected by vasogenic edema which does not involve a reduction of rCBF [14]. That may be the reason why we found, in the "pressure hyper-active" subjects, an association between preserved baseline rCBF and edema. In the case of the intracontusional edema in the damaged area, the edema was predominantly of the vasogenic type, and it may be associated with a blood brain barrier defect and the diffusion of neorepinephrine into the perivascular space. The present results suggest that intracontusional areas seem to behave, even in cases not associated with severe CBF disturbance, as in a penumbral area, more sensitive, than apparently normal brain, to para-physiological changes.

Conclusions

The present study failed to demonstrate that arterial hypertension induced a systematic improvement in rCBF in the intracontusional low density area and, on the contrary, showed an average reduction in rCBF. Although in an exploratory approach, a marginally relevant improvement in intracontusional rCBF was found in patients with critical rCBF values at baseline that needs more extensive research. More immediately relevant seems to be the detection of a specific vasoconstrictive response when the low density area was not associated with baseline ischemia. That suggests a further re-evaluation of the medical protocols based on standardized CPP elevation.

References

1. Cherian L, Chacko G, Goodman JC, Robertson CS (1999) Cerebral hemodynamic effects of phenylephrine and L-arginine after cortical impact injury. Crit Care Med 27: 2512–2517
2. Darby JM, Yonas H, Marks EC, Durham S, Snyder RW, Nemoto EM (1994) Acute cerebral blood flow response to dopamine-induced hypertension after subarachnoid hemorrhage. J Neurosurg 80: 857–864
3. Hoelper BM, Reinert MM, Zauner A, Doppenberg E, Bullock R (2000) rCBF in hemorrhagic, non-hemorrhagic and mixed contusions after severe head injury and its effect on perilesional cerebral blood flow. Acta Neurochir (Wien) [Suppl] 76: 21–25
4. Jennett B, Bond M (1975) Assessment of outcome after severe brain damage. A practical scale. Lancet 1: 480–484
5. Jones TH, Morawetz RB, Crowell RM, Marcoux FW, Fitz Gibbon SJ, DeGirolami U, Ojemann RG (1981) Thresholds of focal cerebral ischemia in awake monkeys. J Neurosurg 54: 773–782
6. Kushi H, Katayama Y, Shibuya T, Tsubokawa T, Kuroha T (1994) Gadolinium DTPA-enhanced magnetic resonance imaging of cerebral contusions. Acta Neurochir (Wien) [Suppl] 60: 472–474
7. Kroppenstedt SN, Sakowitz OW, Thomale UW, Unterberg AW, Stover JF (2002) Norepinephrine is superior to dopamine in increasing cortical perfusion following controlled cortical impact injury in rats. Acta Neurochir (Wien) [Suppl] 81: 225–227
8. Maas AIR, Dearden M, Teasdale GM, Braakman R, Cohadon F, Iannotti F, Karimi A, Lapierre F, Murray G, Ohman J, Persson L, Servadei F, Stocchetti N, Unterberg A (1997) EBIC-guideline for management of severe head injury in adults. Acta Neurochir (Wien) 139: 286–294
9. McLaughlin MR, Marion DW (1996) Cerebral blood flow and vasoresponsivity within and around cerebral contusions. J Neurosurg 85: 871–876
10. Muizelaar JP, Fatouros PP, Schroder ML (1997) A new method for quantitative regional cerebral blood volume measurements using computed tomography. Stroke 28: 1998–2005
11. Paulson OB, Strandgaard S, Edvinsson L (1990) Cerebral autoregulation. Cerebrovasc Brain Metab Rev 2: 161–192
12. Rosner MJ, Rosner SD, Johnson AH (1995) Cerebral perfusion pressure: management protocol and clinical results. J Neurosurg 83: 949–962
13. Schroder ML, Muizelaar JP, Bullock MR, Salvant JB, Povlishock JT (1995) Focal ischemia due to traumatic contusions documented by stable xenon-CT and ultrastructural studies. J Neurosurg 82: 966–971
14. Sutton LN, Greenberg J, Welsh F (1990) Blood flow and metabolism in vasogenic oedema. Acta Neurochir (Wien) [Suppl] 51: 397–400
15. Toda N (1976) Influence of dopamine and noradrenaline on isolated arteries of the dog. Br J Pharmacol 58: 121–126
16. Zhuang J, Schmoker D, Shackford SR, Pietropaoli JA (1992) Focal brain injury results in severe cerebral ischemia despite maintenance of cerebral perfusion pressure. J Trauma 33: 83–88

Correspondence: Arturo Chieregato, Unità Operativa di Neurorianimazione, Ospedale M. Bufalini, viale Ghirotti 286, 47023, Cesena, Italy. e-mail: achiere@ausl-cesena.emr.it

Acta Neurochir (2003) [Suppl] 86: 367–371

The importance of decompressive craniectomy for the management of severe head injuries

U. Meier and **A. Gräwe**

Department of Neurosurgery, Unfallkrankenhaus Berlin, Germany

Summary

Neurosurgical therapy aims to minimize the secondary brain damage after a severe head injury. This includes the evacuation of an intracranial space occupying bleeding, the reduction of intracranial volumes, in hematocephalus an external ventricular drainage, and the conservative therapy in order to influence an increased intracranial pressure (ICP) and a decreased $p(ti)0_2$. When conservative treatment fails to act a decompressive craniectomy might be successful in lowering ICP. From September 1997 until August 2001 we operated on 439 patients with severe head injuries. 50 patients (11%) were treated by means of a decompressive craniectomy.

The prognosis after decompression depends on the clinical signs and symptoms on admission, the patients age and the existence of major extracranial injuries. Our guidelines for an indication for decompressive craniectomy after failure of conservative interventions and evacuation of space occupying hematomas include a patients age below 50 years without multiple trauma, a patients age below 30 years in the presence of major extracranial injuries, a severe brain swelling on CT scan, the exclusion of a primary brainstem lesion or injury and the intervention before irreversible brainstem damage and secondarily while monitoring ICP and $p(ti)0_2$ in an interval up to 48 hours after the accident before irreversible brainstem damage or generalized brain damage has occurred.

Keywords: Severe head injury; brain oedema; decompressive craniectomy; intracranial pressure.

Introduction

Bergmann described in 1880 a decompressive craniectomy and Cushing published in 1908 a case report about a subtemporal decompressive craniectomy for relief of intracranial pressure. There is still a controversy going on about the value of operative decompression after severe head injuries with traumatic brain oedema [1, 2, 5, 6].

The aim of neurosurgical therapy after severe head injuries is the minimization of the secondary brain damage. General principles of neurosurgical therapy are the evacuation of space occupying hematomas, the reduction of intracranial volume, respectively the drainage of hematocephalus, and the conservative therapy focused on intracranial pressure (ICP), cerebral perfusion pressure (CPP) and brain tissue pO_2. In intractable intracranial hypertension, which is refractory to conservative interventions a decompressive craniectomy is indicated in a few patients. Indications for decompressive craniectomy, course of disease, prognostic criteria are analyzed and compared with the literature [16, 17, 18].

Patients and methods

All patients with a severe head injury at the Unfallkrankenhaus are treated by the neurosurgical service on the interdisciplinary ICU. The standard management includes an initial CAT-scan in the emergency room. Patients were included in our prospective study when we saw an indication for a neurosurgical operation. 439 patients with severe head injuries were operatively treated in the Department of Neurosurgery of the Unfallkrankenhaus Berlin between September 1997 to August 2001. The average age of the 356 male and 83 female patients was 39 years. 50 patients (11%) underwent decompressive craniectomy. In 36 patients craniectomy was performed additionally with removal of a space-occupying hematoma (subdural or epidural hematoma, contusion hemorrhage) because intraoperatively a generalized brain oedema was evident. The second group included 14 patients in whom conservative treatment on the intensive care unit was therapy-resistant under neuromonitoring with ICP-, CPP-, MAP- and $p(ti)0_2$-measurement revealed an intractable brain oedema. The average age of the group of patients who were decompressed was 30 years and the male/female ratio was 4:1 (41 male, 9 female). In 48 patients we performed an unilateral craniectomy (right side 22, left side 26) and in 2 patients a bilateral craniectomy was necessary.

Results

In accordance with the results of other groups [16] the majority of head injuries with indication for cra-

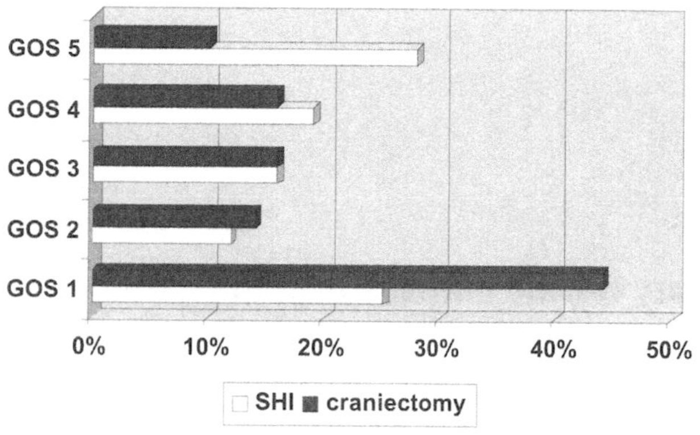

Fig. 1. Prognosis in severe head injuries with and without decompressive craniectomy

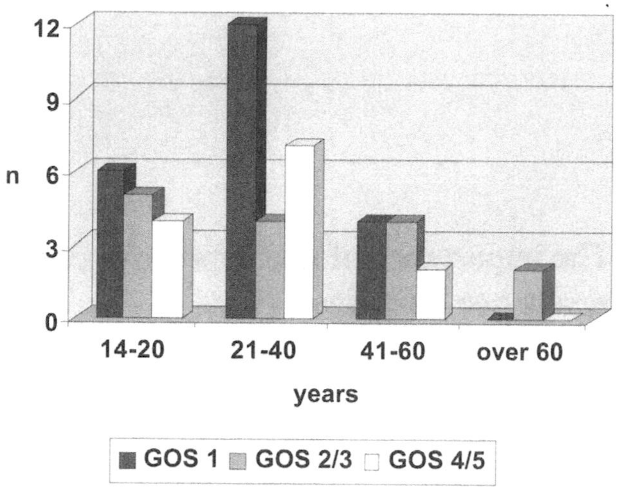

Fig. 2. Outcome versus patients age

niectomy was due to motor vehicle accidents (58%), followed by falls (22%), attempted suicide (8%) and violences (12%). 60 percent had a craniectomy because of a diffuse brain injury, 20 percent after an acute subdural hematoma and 12 percent with an open head injury. In patients with a diffuse brain injury the primary operative intervention was performed with the indication for evacuation of a space occupying contusion bleeding.

22 patients (44%) died despite decompression, 7 patients (14%) remained in a vegetative state, 8 patients (16%) had persistent severe neurological deficits, and 13 patients (26%) reached a GOS of 4 up to 5. The comparison of the course of disease between patients with a severe head injury and those who were decompressed is illustrated in Fig. 1. It is evident that patients with a decompressive craniectomy had a poorer outcome.

A comparison of the patients' age and the postoperative results is because of the small number of patients not sufficient (Fig. 2). Altogether, patients younger than 40 years have a better prognosis than the older ones. But also in the fourth and fifth decade satisfactory results are obtainable in 80 percent. According to our results a decompression is less favourable in patients over 60 years of age and our two patients in this group remained in a vegetative state.

There is evidence for an important influence on the outcome when major extracranial injuries are present in contrast to an isolated head injury (Fig. 3). We found an increased lethality of 63 percent when there was a multiple trauma in contrast to 32 percent with an isolated head injury. Furthermore the outcome is influenced by the existence of an extracranial injury with

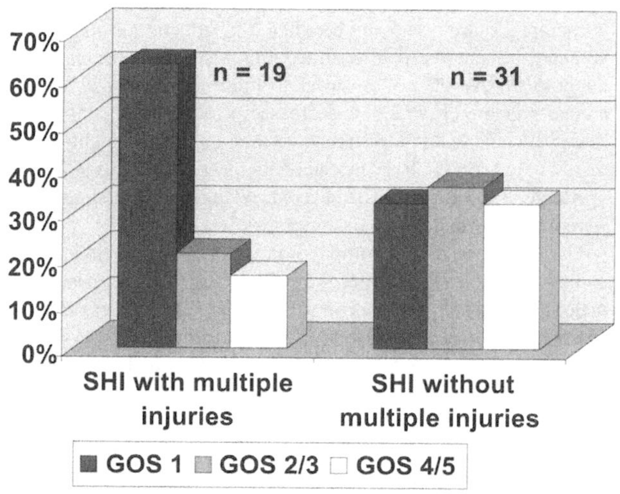

Fig. 3. Comparison between patients with and without major extracranial injuries

positive results in only 16 percent compared to 32 percent in isolated head injuries. In our small series we could not find a difference in the time span between accident and surgical decompression. This is not contradictory for pointing out that decompression should be performed early when indicated.

Discussion

The prognosis after severe head injuries depends on the clinical status on admission, the intracranial lesions, the patients age and the existence of accompanying injuries. Patients with epidural hematomas and open head injuries have a better prognosis than those with acute subdural hematomas. A poor prognosis is given in patients with brain contusions, diffuse brain injuries, traumatic injuries of the sinuses and/or less than 8 points on the Glasgow Coma Scale and an advanced age [16, 17, 18].

The pathophysiology of the posttraumatic primarily vasogenic and cytotoxic cerebral oedema is mainly a disturbance of the functional entity consisting of capillary, astroglia and neuron. The driving force for the formation of oedema is the pressure gradient across the injured capillary with loss of the blood brain barrier function and an accumulation of extracellular fluid. The fluid accumulation is not caused by diffusion but by a hydrostatic pressure gradient in the interstitium of the white matter. This global, mostly hemispherical blood brain barrier disturbance induces a generalized brain oedema which can only marginal be influenced by a conservative treatment. Up to a special degree an intracranial volume increase can be compensated by shifts in the cerebrospinal fluid or blood volume. According to the intracranial pressure-volume relationship, when the intracranial reserve areas are exhausted any further increase in volume is responsible for a considerable change in pressure. Exactly in this moment when the IVP increase decompensates, a decompressive craniectomy must be considered. The decompression changes the pressure-volume relationship and causes an increase of the compliance and therefore a shift to the right on the pressure-volume curve [1, 2, 6, 8, 9, 15].

In animal experiments it was found by other groups [1, 2, 4, 10] that after decompression the injured brain exhibits an ICP decrease and an increase of the pressure-volume-index (PVI) and furthermore a decrease of the interstitial fluid accumulation. There are contradictious statements in the literature concerning an improvement of the regional cerebral blood flow (rCBF) and of the cerebrovascular resistance [10, 24]. In clinical trials with PET, Xenon-CT and MR-spectroscopy Yamakami *et al.* [24] and Yoshida *et al.* [25] found an rCBF increase hemispheriel and in the thalamus and as well an increase of the cerebral metabolism after the craniectomy in comparison to the preoperative results. This requires an early decompression and the craniotomy must be wide enough and must include an opening and prolongation of the dura. The craniectomy should be performed when conservative treatment fails to influence an ICP between 25 up to 40 mm Hg and before the cerebral perfusion is irreversible disturbed [2].

There are different methods described in the literature. Clark *et al.* [83] prefers the circumferential craniotomy and other groups [13, 20, 23] advises the bifrontal craniotomy, Gerl *et al.* [6] recommend the bilateral craniotomy, and Gaab *et al.* [5, 12] recommend the wide unilateral or bilateral frontotemporoparietal craniectomy with dura opening and dura patch plasty. According to the recommendations of Gaab *et al.* [5, 12] we have performed the wide frontotemporoparietal craniectomy with dura plasty for which we used the temporal muscle and its fascia or a Neuropatch (Braun, Melsungen, Germany) as a graft. In 48 patients we made a unilateral decompression over the affected hemisphere and in one patient a frontal and in another patient a bilateral craniectomy. The bone flap is preserved by implantation subcutaneously into abdominal fat or by storage under sterile conditions at −80 °C. In accordance with Yoshida *et al.* [25] who found out that rCBF and cerebral metabolism is reduced in the area of craniectomy we perform the reimplantation after the regression of cerebral oedema and when a physiological ICP is present.

Beside decompressions in patients with diffuse brain injuries (Fig. 1) we performed a craniectomy primarily more often in patients with acute subdural hematomas when the development of cerebral oedema appeared during the operation and was resistant to conservative interventions.

The course of disease after craniectomy depends on the clinical status on admission (Fig. 2), and this was also found by other authors [5, 12, 21, 22]. In the literature the postoperative results are differing extremely. While analyzing the literature there are between 11 and 71 percent dead despite the decompression, 8 until 27 percent were in a vegetative state, 8 until 27 percent survived with severe neurological deficits, 0 until 21 percent survived with mild neurological deficits and 0 until 41 percent had good recovery [5, 6, 7, 11, 12, 14, 20]. In a large analysis Guerra *et al.* [12] found 49 percent of patient who had died, 13 percent survived with severe neurological deficits corresponding to a GOS of 2 and 3, and 32 percent had a favourable outcome with a GOS score of 4 and 5. Our results are mainly in accordance with those reported above but we had a slightly higher amount of patients in a vegetative state. In contrast to our results referring to the comparison of patients with severe head injuries and those after craniectomy (Fig. 3) Polin *et al.* [29] found significantly better results after decompression. This finding might be a fact of a reduction on patients with diffuse brain injuries who were not operated. In our comparable group there are all operatively treated patients after severe head injuries summarized.

The influence of the patients age on the outcome after severe head injuries and craniectomy (Fig. 4) was also found by other authors [5, 12, 14, 20, 21]. We critically have to re-evaluate the indication for decompressive craniectomy in those two patients over 60 years of age. Their postoperative results with a GOS score of 2 is one reason for our higher rate of patients in a vegetative state and we postulate that there should be an age restriction. Like Karlen *et al.* [11] the prognosis was worse when there were major extracranial injuries present (Fig. 5).

Under the influence of the results of Gaab *et al.* [5, 12, 21] we conclude the following guidelines for an indication for decompressive craniotomy after severe head injuries:

1. Patients age below 50 years without multiple trauma
2. Patients age below 30 years in the presence of major extracranial injuries
3. Severe brain swelling on CT scan
4. Exclusion of a primary brainstem lesion/injury
5. Intervention before irreversible brainstem damage
6. ICP increase up to 40 mm Hg and unsuccessful conservative therapy
7. Primarily for space-occupying hematomas with hemispheriel brain oedema
8. Secondarily while monitoring ICP and $p(ti)0_2$ in an interval up to 48 hours after the accident before irreversible brainstem damage or generalized brain damage has occurred.

References

1. Burkert W (1985) Zeitlicher Verlauf des posttraumatischen Hirnödems nach Ultraschalleinwirkung – Experimentelle Untersuchungen und Modellvorstellungen zum Zeitpunkt der operativen Dekompression und zur Größe der Trepanationsfläche. Dissertation B, Martin-Luther-Universität Halle
2. Burkert W, Plaumann H (1989) Der Wert der großen druckentlastenden Trepanation beim therapieresistenten Hirnödem. Tierexperimentelle Untersuchungen, erste klinische Ergebnisse. Zentralbl Neurochir 50: 106–108
3. Clark K, Nash TM, Hutchinson GC (1968) The failure of circumferential craniotomy in acute traumatic cerebral swelling. J Neurosurg 29: 367–371
4. Cooper PR, Hagler H, Clark WK, Barnett P (1979) Enhancement of experimental cerebral edema after decompressive craniectomy, implications for the management of severe head injuries. Neurosurgery 4: 296–300
5. Gaab MR, Rittierodt M, Lorenz M, Heissler HE (1990) Traumatic brain swelling and operative decompression, a prospective investigation. Acta Neurochir (Wien) [Suppl] 51: 326–328
6. Gerl A, Tavan S (1980) Die bilaterale Kraniektomie zur Behandlung des schweren traumatischen Hirnödems. Zentralbl Neurochir 41: 125–138
7. Gower DJ, Lee KS, Mc Whorter JM (1988) Role of subtemporal decompression in severe closed head injury. Neurosurgery 23: 417–422
8. Grände PO, Asgeirsson B, Nordström CH (1997) Physiologic principles for volume regulation of a tissue enclosed in a rigid shell with applications to the injured brain. J Trauma 42: 23–30
9. Hase U, Reulen HJ, Meinig G, Schürmann K (1978) The influence of the decompressive operation on the intracranial pressure and the pressure volume relation in patients with severe head injuries. Acta Neurochir (Wien) 45: 1–13
10. Hatashita S, Hoff JT (1987) The effect of craniectomy in the biomechanics of normal brain. J Neurosurg 67: 573–578
11. Karlen J, Stula D (1987) Dekompressive Kraniotomie bei schwerem Schädelhirntrauma nach erfolgloser Behandlung mit Barbituraten. Neurochirurgia 30: 35–39
12. Kleist-Welch Guerra W, Gaab MR, Dietz H, Mueller JU, Piek J, Fritsch MJ (1999) Surgical decompression for traumatic brain swelling, indications and results. J Neurosurg 90: 187–196
13. Kjellberg RN, Prieto A Jr (1971) Bifrontal decompressive craniotomy for massive cerebral edema. J Neurosurg 34: 488–493
14. Kunze E, Meixensberger J, Janka M, Sörensen N, Roosen K (1998) Decompressive craniectomy in patients with uncontollable intracranial hypertension. Acta Neurochir (Wien) [Suppl] 71: 16–18
15. Meier U (1985) Computertomographische Untersuchungen zum generalisierten Hirnödem nach Schädel-Hirn-Trauma und Subarachnoidalblutung. Dissertation A, Humboldt-Universität, Berlin
16. Meier U, Gärtner F, Knopf W, Klötzer R, Wolf O (1992) The traumatic dural sinus injury – a clinical study. Acta Neurochir (Wien) 116: 91–93
17. Meier U, Heinitz A, Kintzel D (1994) Operationsergebnisse nach schwerem Schädel-Hirn-Trauma im Kindes- und Erwachsenenalter – Eine Vergleichsstudie. Unfallchirurgie 97: 406–409
18. Meier U, Zeilinger F, Kintzel D (1996) Prognose und Ergebnisse der operativen Versorgung von schweren Schädel-Hirn-Traumen. Akt Traumatol 26: 1–5
19. Pasaoglu A, Kurtsoy A, Koe RK, Konts O, Akdemir H, Öktem IS, Selcuklu A, Kavuncu IA (1996) Cranoplasty with bone flaps preserved under the scalp. Neurosurg Rev 19: 153–156
20. Polin RS, Shaffrey ME, Bogaev CA, Tisdale N, Germanson T, Bicchicchio B, Jane JA (1997) Decompressive bifrontal craniectomy in the treatment of severe refractory posttraumatic cerebral edema. Neurosurgery 41: 84–94
21. Rittierodt M, Gaab MR, Lorenz M (1991) Decompressive craniectomy after severe head injury. Useful therapy in pathophysiologically guided indications. In: Bock WJ *et al* (eds) Advances in neurosurgery, vol 19. Springer, Berlin, Heidelberg, New York, Tokyo, pp 265–273
22. Shigemori M, Syojima K, Nakayama K, Kojima T, Watanabe M, Kuramoto S (1979) Outcome of acute subdural haematoma following decompressive hemicraniectomy. Acta Neurochir (Wien) [Suppl] 28: 195–198
23. Venes JL, Collins WF (1975) Bifrontal decompressive craniectomy in the management of head trauma. J Neurosurg 42: 429–433
24. Yamakami I, Yamaura A (1993) Effects of decompressive craniectomy on regional cerebral blood flow in severe head trauma patients. Neurol Med Chir 33: 616–620

25. Yoshida K, Furuse M, Izawa A, Iizima N, Kuchiwaki H, Inao S (1996) Dynamics of cerebral blood flow and metabolism in patients with cranioplasty as evaluated by ^{133}Xe CT and ^{31}P magnetic resonance spectroscopy. J Neurol Neurosurg Psychiatry 61: 166–171

Correspondence: Privat-Dozent Dr. med. Ullrich Meier, Head of Department, Department of Neurosurgery, Unfallkrankenhaus Berlin, Warener Strasse 7, D-12683 Berlin, Germany. e-mail: ullrich.meier@ukb.de

Acta Neurochir (2003) [Suppl] 86: 373–376

Importance of hemodynamics management in patients with severe head injury and during hypothermia

K. Kinoshita, N. Hayashi, A. Sakurai, A. Utagawa, and **T. Moriya**

Department of Critical Care and Emergency Medicine, Nihon University School of Medicine, Tokyo, Japan

Summary

Objective. To evaluate the hemodynamics in patients with traumatic brain injury (TBI) during therapeutic hypothermia.

Methods. Subjects were 25 patients with TBI (GCS; 8 or less). Mean arterial blood pressure (MAP), cerebral perfusion pressure (CPP), cardiac index (CI), systemic oxygen delivery (DO2), systemic vascular resistance index (SVRI), and pulmonary capillary wedge pressure (PCWP) were measured. Patients were retrospectively divided into 3 groups: normothermia ($n = 5$; NT), and survivors ($n = 14$; HT-S) and non-survivors ($n = 6$; HT-Non-S) after hypothermia, and hemodynamics were investigated for difference among groups at 24 hours from induction of normothermia or hypothermia.

Results. CPP target was above 70 mmHg, however, HT-Non-S could not maintain CPP above 70 mmHg. The low CPP was the result of elevated ICP, low MAP ($P < .05$), or both during hypothermia. In HT-Non-S, significantly high SVRI and low CI ($P < .05$) causing dehydration were observed during cooling. DO2 could not be maintained in HT-Non-S during hypothermia.

Conclusions. These results suggest that patients run the risk of impairing hemodynamics during therapeutic hypothermia. Hemodynamic management is essential during hypothermia. If dehydration occurs during hypothermia, MAP may be reduced due to inadequate sedation, analgesia, and excess use of diuretic agents.

Keywords: Severe head injury; hemodynamics; oxygenation; hypothermia.

Introduction

Clinical studies have documented the importance of intensive care unit (ICU) management of traumatically brain-injured patients in maintaining CPP [13]. It is important to maintain oxygen delivery meeting the oxygen demand in traumatized brain tissue [15]. Fluid management is also a critical factor during hypothermia in patients with traumatic brain injury (TBI) [4]. The main management goal in such patients is to provide an adequate oxygen supply and to deliver oxygen to damaged brain tissue to stabilize hemodynamics. A recent study reported that artificial cooling is associated with significant side effects, including significant fluid loss through hypothermia-induced diuresis and electrolyte depletion [1, 4, 11, 12, 14]. This is important because even brief episodes of hypovolemia can affect cerebral perfusion pressure (CPP) and cerebral blood flow and it has been shown that even very short episodes of decreased CPP can significantly affect neurological outcome [5]. However, changes in hemodynamics during cooling after TBI are not well understood. This study evaluates the hemodynamics in patients with TBI during cooling.

Materials and methods

All patients with severe closed head injury admitted within 1 hour of trauma were studied. Those for whom normothermia or hypothermia was indicated were examined to determine if they had closed head injury and a Glasgow Coma Scale (GCS) score of 8 or less upon admission ($N = 25$). Patients were excluded if they had hypotension (systemic blood pressure < 90 mmHg), serious other injuries. After initial brain CT, all patients underwent prompt surgical evacuation of mass lesions.

All cases were continuously monitored for mean arterial blood pressure (MAP), intracranial pressure (ICP), cardiac index (CI), pulmonary capillary wedge pressure (PCWP), and systemic vascular resistance index (SVRI). Cerebral perfusion pressure (CPP: MAP-ICP) was calculated. Oxygen consumption (VO2) and oxygen delivery (DO2) were calculated using Fick's equation. Patients underwent hypothermia immediately after surgery. All patients reached the target temperature (33–34 degree centigrade at the right jugular blood temperature) within 8 hours after injury using cooling blankets, and the target temperature was maintained for 48–72 hrs. After hypothermic therapy, the patients were gradually rewarmed at a rate of 1 °C per day.

Patients were retrospectively divided into 3 groups: normothermia ($n = 5$; NT), and survivors ($n = 14$; HT-S) and non-survivors ($n = 6$; HT-Non-S) after hypothermia, and hemodynamics (MAP,

ICP, CPP, CI, DO2, VO2, PCWP, and SVRI) were investigated for difference among groups at 24 hours from induction of normothermia or hypothermia.

The CPP target was above 70 mmHg. The patients were managed to have MAP above 90 mmHg and ICP below 20 mmHg under the ICP and CPP control protocol, according to which the patients were treated by cerebrospinal fluid drainage from the lateral ventricle and mannitol diuretic therapy. No patient underwent combined barbiturate coma therapy. Arterial PaCO2 was maintained at about 35 mmHg and that <30 mmHg was avoided.

The outcome was assessed at hospital discharge or 3 months after head injury based on the Glasgow Outcome Scale (GOS) score.

Statistical analysis

Data were expressed as mean values ± SD. One-way analysis of variance (ANOVA) was used to compare physiological data between the groups. Fisher's protected LSD analyses were performed to determine differences between each group. Differences were considered significant at $P < .05$.

Results

No statistically significant differences were seen among NT, HT-S, or TH-Non-S for age, GCS score on admission, MAP, ICP, or at hypothermia induction.

Overall outcome

All patients in the normothermia group survived. Their GOS scores indicated GR/MD in 2 patients and SD/PVS in 3 patients. Fourteen of the 20 cases survived after hypothermia. The survivors' GOS scores indicated that 9 patients had GR/MD and 5 patients had SD/PVS. Of the 6 non-survivors, 4 died from uncontrollable intracranial pressure during rewarming and 2 died of multiple organ failure after hypothermia.

Changes in systemic hemodynamics during cooling

Although no significant change in MAP (NT 101.0 ± 11.2, HT-S 98.2 ± 16.8, vs. HT-Non-S 96.7 ± 5.7 mmHg) was seen among the groups on ICU admission, MAP significantly decreased at 24 hrs after induction of hypothermia in HT-Non-S as compared to NT (NT 102.4 ± 11.8 vs. HT-Non-S 87.3 ± 7.1 mmHg, $P < .05$) (Fig. 1A). DO2I was lower in HT-Non-S (325.0 ± 71.3 ml/min/m^2) than in other groups ($P < .05$) at 24 hrs after normothermia or hypothermia (NT 563.3 ± 220.3, HT-S 505.9 ± 133 ml/min/m^2) (Fig. 1B). No significant change in VO2I was observed among the groups at 24 hrs after induction of normo or hypothermia (Fig. 1C). A significant ($P < .05$) decrease in CI (Fig. 1D) was seen at 24 hrs after injury in HT-Non-S (2.2 ± 0.5 ml/min/m^2) compared to NT (3.2 ± 0.7 ml/min/m^2). SVRI (Fig. 1E) increased statistically significantly ($P < .05$) in HT-Non-S compared to NT (HT-S 2574.1 ± 330.2 vs. HT-Non-S 3220.8 ± 692.1 dyne.s/cm5/m^2). HT-Non-S had a markedly decreased CI concomitant with an increase in SVRI during cooling. PCWP (Fig. 1F) also decreased significantly ($P < .01$) in HT-Non-S (2.3 ± 1.0 mmHg) compared to other treatment groups at 24 hrs after induction of normothermia or hypothermia (NT 6.6 ± 2.1, HT-S 5.4 ± 1.7 mmHg).

Changes in ICP and CPP

No statistically significant changes in ICP (NT 16.4 ± 2.7, HT-S 18.6 ± 7.3, vs. HT-Non-S 16.5 ± 5.4 mmHg) or CPP (NT 84.6 ± 11.3, HT-S 79.7 ± 14.0, vs. HT-Non-S 80.2 ± 8.2 mmHg) were seen at the start of normothermia or hypothermia. ICP decreased markedly (Fig. 1G) in HT-S (12.1 ± 2.5 mmHg) compared to other treatment groups (NT 21.4 ± 3.8, HT-Non-S 21.7 ± 2.5 mmHg, $P < .001$). CPP was significantly ($P < .05$) lower in HT-Non-S (65.7 ± 9.3 mmHg) than in other groups (NT 81.0 ± 14.2, HT-S 80.9 ± 12.1 mmHg) (Fig. 1H). Low CPP was the result of elevated ICP, low MAP, or both during hypothermia.

Discussion

These findings suggest that patients run the risk of impaired hemodynamics during therapeutic hypothermia. This may lead to compromised cerebral blood flow and oxygen supply, which in turn may adversely affect the final outcome [3, 15]. Recent clinical studies have shown that mild hypothermia for TBI can potentially limit the extent of secondary brain injury and may be clinically effective in patients with severe head injury [7, 8, 11]. Brain hypothermia, however, may also increase complications associated with high SVRI and intravascular volume depletion, leading to depression in cardiac output and to reduced systemic oxygen delivery. Major mechanisms for hemodynamic impairment during induced hypothermia are depression in cardiac output, a reduction in the intravascular blood volume [2, 6], and an increase in vascular capacity resulting from blood pooling [10]. During hypothermia, a body fluid shift to the interstitial space

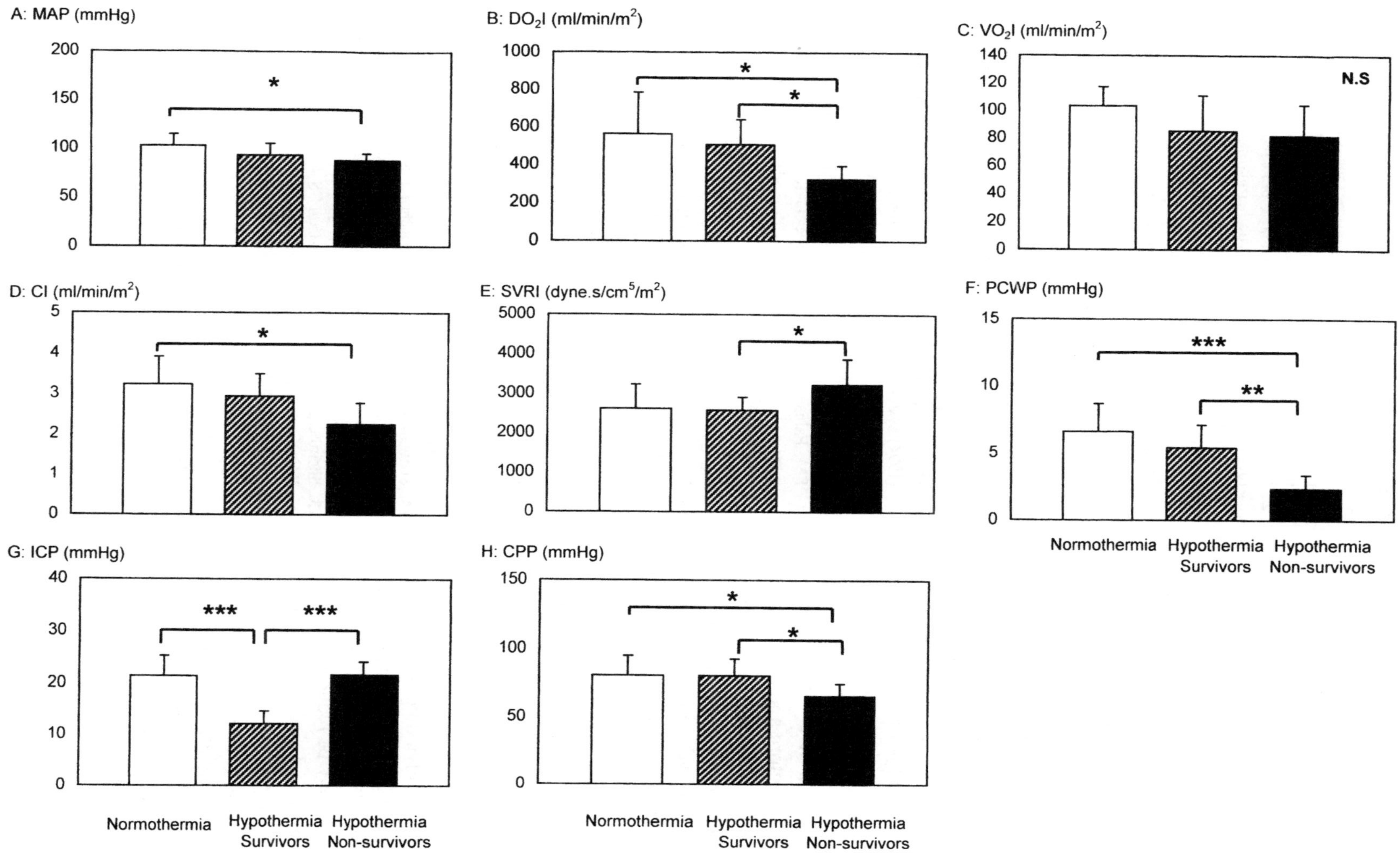

Fig. 1. Comparison of changes in hemodynamics at 24 hrs after induction of normothermia or hypothermia (33–34 °C). Normothermia: Patients treated in normothermia; Hypothermia Survivors/Non-Survivors: Patients who survived or did not survive after hypothermia. *MAP* Mean arterial blood pressure, *DO2* Oxygen delivery, *VO2* Oxygen consumption, *CI* Cardiac index, *SVRI* Systemic vascular resistance index, *PCWP* Pulmonary capillary wedge pressure, *ICP* Intracranial pressure, *CPP* Cerebral perfusion pressure (MAP-ICP), *N.S.* Not significant. $*P < .05$, $**P < .01$, $***P < .001$

caused by an increase in vascular permeability occurs, resulting in a decrease in intravascular plasma volume [2, 6]. Moreover, excessively increased SVR could decrease cardiac output leading to reduced systemic oxygen delivery, particularly disturbed microcirculation during hypothermia. Systemic disoxygenation leading to multiple organ failure may progress and worsen secondary ischemic brain injury. However, it is difficult to detect hypovolemia because MAP may be maintained due to increased SVR. Systemic intravascular volume depletion could accelerate due to fluid restriction or excess use of diuretic agents, such as mannitol, during hypothermia. Unexpected hypotension may occur under inadequate sedation, analgesia in such conditions. Urine volume increases due to the diuretic effect, and water appears to be sequestered in the capillaries during brain hypothermia [11].

Preventing and treating ischemia as soon as possible after injury is a major goal in trauma care, particularly in patients with head injuries. It is vital to restore and stabilize hemodynamics before and during hypothermia. Excess dehydration induces tissue hypoxia and microcirculation disturbance may be worsened by systemic oxygenation in dehydration during hypothermia, presumably having adverse effects in severely head injured patients during hypothermia. Further clinical studies are required to clarify the relationship between outcome and hemodynamics during hypothermia.

Conclusion

Knowledge of optimal hemodynamics management is essential, as fluid balance plays an important role in maintenance of MAP and CPP. If dehydration occurs during hypothermia, MAP may be reduced due to inadequate sedation, analgesia, and excess use of diuretic agents. Hemodynamic management is particularly important at the cooling stage during hypothermia. It is thus necessary to continuously monitor systemic hemodynamics. Maintaining MAP, CPP, and systemic oxygenation may then reduce serious complications in hypothermia in severe head injury.

References

1. Aibiki M, Kawaguchi S, Maekawa N (2001) Reversible hypophosphatemia during moderate hypothermia therapy for brain-injured patients. Crit Care Med 29: 1726–1730
2. Chen RYZ, Chien S (1977) Plasma volume, red cell volume, and thoracic duct lymph flow in hypothermia. Am J Physiol 233: 605–612
3. Chesnut RM, Marshall SB, Piek J, Blunt BA, Klauber MR, Marshall LF (1993) Early and late systemic hypotension as a frequent and fundamental source of cerebral ischemia following severe brain injury in the Traumatic Coma Data Bank. Acta Neurochir (Wien) [Suppl] 59: 121–125
4. Clifton GL, Miller ER, Choi SC, Levin HS (2002) Fluid thresholds and outcome from severe brain injury. Crit Care Med 30: 739–745
5. Ghajar J (2000) Traumatic brain injury. Lancet 356: 923–929
6. Green JF, Jackman AP (1979) Mechanism of the increased vascular capacity produced by mild perfusion hypothermia in the dog. Circ Res 44: 411–419
7. Hayashi N, Hirayama T, Utagawa A, Daimon W, Ohata M (1994) Systemic management of cerebral edema based on a new concept in severe head injury patients. Acta Neurochir (Wien) [Suppl] 60: 541–543
8. Marion DW, Penrod LE, Kelsey SF, Obrist WD, Kochanek PM, Palmer AM, Wisniewski SR, DeKosky ST (1997) Treatment of traumatic brain injury with moderate hypothermia. N Engl J Med 336: 540–546
9. Nose H (1982) Transvascular fluid shift and redistribution of blood in hypothermia. Jap J Physiol 32: 831–842
10. Paton BC (1991) Accidental hypothermia. In: Schonbaum E, Lomax P (eds) Thermoregulation pathology, pharmacology and therapy. Pergamon Press, Inc New York, pp 397–454
11. Polderman KH, Peerdeman SM, Girbes AR (2001) Hypophosphatemia and hypomagnesemia induced by cooling in patients with severe head injury. J Neurosurg 94: 697–705
12. Polderman KH, Peerdeman SM, Girbes ARJ (2001) Hypophosphatemia and hypomagnesemia induced by cooling in patients with severe head injury. J Neurosurgery 94: 697–705
13. Rosner MJ, Rosner SD, Johnson AH (1995) Cerebral perfusion pressure: management protocol and clinical results. J Neurosurg 83: 949–962
14. Schwab S, Georgiadis D, Berrouschot J, Schellinger PD, Graffagnino C, Mayer SA (2001) Feasibility and safety of moderate hypothermia after massive hemispheric infarction. Stroke 32: 2033–2035
15. Sheinberg M, Kanter MJ, Robertson CS, Contant CF, Narayan RK, Grossman RG (1992) Continuous monitoring of jugular venous oxygen saturation in head-injured patients. J Neurosurg 76: 212–217

Correspondence: Kosaku Kinoshita, M.D., Department of Emergency and Critical Care Medicine, Nihon University School of Medicine, 30-1 Oyaguchi Kamimachi, Itabashi-ku, Tokyo 173-8610, Japan. e-mail: kosaku@med.nihon-u.ac.jp

Acta Neurochir (2003) [Suppl] 86: 377–380

Changes in cerebrovascular response during brain hypothermia after traumatic brain injury

K. Kinoshita, N. Hayashi, A. Sakurai, A. Utagawa, and **T. Moriya**

Department of Critical Care and Emergency Medicine, Nihon University School of Medicine, Tokyo, Japan

Summary

Objective. Transcranial Doppler sonography (TCD) provides a rapid and noninvasive assessment of cerebral hemodynamics after traumatic brain injury. This study evaluates the change of cerebrovascular response at the rewarming stage of brain hypothermia using TCD.

Material and methods. Seventeen patients who had suffered from closed brain injury were investigated with daily TCD recordings and the changes in flow velocities were evaluated to determine whether they reflected the temperature during brain hypothermia. All patients who had treated brain hypothermia underwent continuous monitoring of SjO_2, mean arterial blood pressure, and intracranial pressure (ICP).

Results. No significant changes in ICP and cerebral perfusion pressure (CPP) were recorded in all cases. Of 17 patients, 5 had a significant increase in $SjO_2 > 75\%$ or more, retrospectively, with a robust increase in flow velocities of the middle cerebral artery ($P < .01$) at the rewarming stage. These cases marked a decrease in pulsatility index ($P < .05$) concomitant with an increase in SjO_2 values. The CT scan revealed acute brain swelling in these cases.

Conclusion. These data suggest cerebral vasoactivity could be altered at the rewarming stage. Hypothermia runs a risk of hyperemia at the rewarming stage, which induced a decrease in cerebral vasoresistance.

Keywords: Cerebral vasoactivity; hyperemia; hypothermia; transcranial Doppler; traumatic brain injury; autoregulation.

Introduction

Traumatic brain injury (TBI) often causes disturbances of the cerebrovascular response, which contribute to the infliction of secondary injury, although cerebral autoregulation is a major mechanism responsible for brain protection in circumstances in which cerebral perfusion pressure (CPP) may become unstable. Elevation in intracranial pressure (ICP), variations in mean arterial pressure (MAP), and other factors may lead to ischemic insults when autoregulation is defective. Transcranial Doppler sonography (TCD) provides a rapid and noninvasive means of assessment of cerebral hemodynamics after TBI. Recent studies have shown that mild-moderate hypothermia can potentially limit the extent of secondary brain injury and may be effective clinically in patients with severe head injury [5, 10, 15]. However, the changes in cerebrovascular response during brain hypothermia after TBI are not well known. This study examines the characteristic hemodynamics and investigates the changes in the cerebrovascular response using TCD during brain hypothermia of severe head injuries.

Materials and methods

All patients with severe closed head injury admitted to the Department of Critical Care and Emergency Medicine at Nihon University Hospital within 1 hour after trauma were studied. Those for whom brain hypothermia was indicated were examined to determine if they had closed head injury and a Glasgow Coma Scale (GCS) score < 8 or less upon admission to the emergency department. After initial computer tomography (CT) of the head, all patients underwent prompt surgical evacuation with external decompression of acute subdural hematoma. All patients received continuous monitoring of jugular oxygen saturation (SjO_2), MAP, and ICP. A total of 17 patients were investigated using daily TCD recordings (time-averaged maximum flow velocity; TAMX and pulsatility index; PI: [systolic flow velocity – diastolic flow velocity]/ mean flow velocity) of the middle cerebral artery (MCA) during hypothermia. The right MCA was insonated through the temporal window at a depth of 45 to 55 mm (Quantascope, Toshiba T2100, Vital Science, Amsterdam, The Netherlands) with a 2-MHz probe. Blood temperature at the right jugular bulb from the internal jugular catheters was controlled at 33–34 °C using compartment cooler blankets for 48–72 hours. After hypothermic therapy, patients were gradually rewarmed at a rate of 1 °C per day.

We divided brain hypothermia into 3 stages: induction (I), cooling (C), and rewarming (R) stages. Induction stage was defined as a jugular blood temperature of >34 °C. The cooling stage was defined as a jugular blood temperature (j-BT) under 33.9 °C. The rewarming stage was defined as when j-BT increased above 34 °C from the cooling stage.

Five cases were observed as having significantly increased $SjO_2 > 75\%$ or more, at the rewarming stage, retrospectively. Therefore, the patients were divided into 2 groups; high SjO_2 ($SjO_2 > 75\%$ or more; n = 5; H-group) and non-high SjO_2 ($SjO_2 < 75\%$ n = 12; Non-H-group) at the rewarming stage and were evaluated for difference in MAP, ICP, CPP, PI, and TAMX between the groups during hypothermia in this study.

General management during brain hypothermia

Patients with severe head injury were managed to have MAP above 90 mmHg and ICP below 20 mmHg under an aggressive ICP and CPP control protocol, according to which the patients were treated by cerebrospinal fluid drainage from the lateral ventricle and mannitol diuretic therapy. No patient underwent combined barbiturate coma therapy or active hyperventilation. Arterial blood gases ($PaCO_2$, PaO_2, and pH), hematocrit (Ht.), and hemoglobin (Hb.) were maintained during hypothermia treatment as follows: $PaCO_2 = 30–35$ mmHg, $PaO_2 = 120–150$ mmHg, pH = 7.35–7.45, Ht. = 32–35% and Hb. = 11–12 mg/dl.

Statistical analysis

Data were expressed as mean values ± SD. One-way analysis of variance (ANOVA) was used to compare physiological data between the groups. Scheffe's analyses post hoc test was performed to determine differences between the groups. Differences were considered significant if the P value was <0.05.

Results (Fig. 1)

There were no significant changes in MAP, ICP, or CPP between the groups during hypothermia. No significant changes in the TCD estimated PI or TAMX were observed from the induction to cooling stages between the groups. However, H-group was observed as having significantly increased TAMX (H-group; 100.3 ± 12.1 vs. Non-H-group; 40.2 ± 8.8 cm/sec, $P < 0.01$) in the MCA at the rewarming stage with a marked decrease in PI (H-group; 0.64 ± 0.12 vs. Non-H-group; 1.11 ± 0.16, $P < 0.05$) when compared to Non-H-group.

The CT scan, which indicated increased TAMX with high SjO_2 at the rewarming stage, revealed acute brain swelling.

Discussion

Our study demonstrated a markedly increased cerebral flow velocity and decreased pulsatility index estimating transcranial Doppler in patients with high SjO_2 during rewarming after TBI. Elevated transcranial Doppler flow velocities with high SjO_2 have been considered cerebral hyperemia after traumatic brain injury [11, 16]. The pulsatility index in the MCA fall on rewarming, providing evidence that the hyperemia is associated with a drop in distal cerebrovascular resistance related to the dilatation of cerebral arteries [11, 13, 16]. This study on a limited number of patients suggested patients run the risk of impairment of cerebral vasoactivity during therapeutic hypothermia, especially at the rewarming stage. This phenomenon is important, because it may lead to elevation in cerebral blood volume and ICP, which in turn may adversely affect the final outcome [7, 11, 16].

The remarkable ability of cerebral vessels to respond to changes in cerebral metabolism, perfusion pressure, and milieu interior, such as $PaCO_2$, are mediated by a number of cellular mechanisms. Although abnormal cerebral autoregulation, which includes CO_2 reactivity and vasoactivity, is seen after TBI [3, 7, 9] or hypothermia [14, 18, 19], the underlying mechanisms affecting this disturbance are not fully understood. Alternation of vasoactivity in rewarming may be caused by several mechanisms. The temperature changes could alter the balance of vasoconstrictive and vasodilator mediators [1]. Endothelial dysfunction [12] and release of free radicals [8] may be involved. Previous reports have also suggested that systemic cooling inhibits vasoactive mediators [17], such as nitric oxide [6]. We speculated that cooling might influence the vasomotor function through the modulation of neurotransmitter release, and vasodilatory mediators could

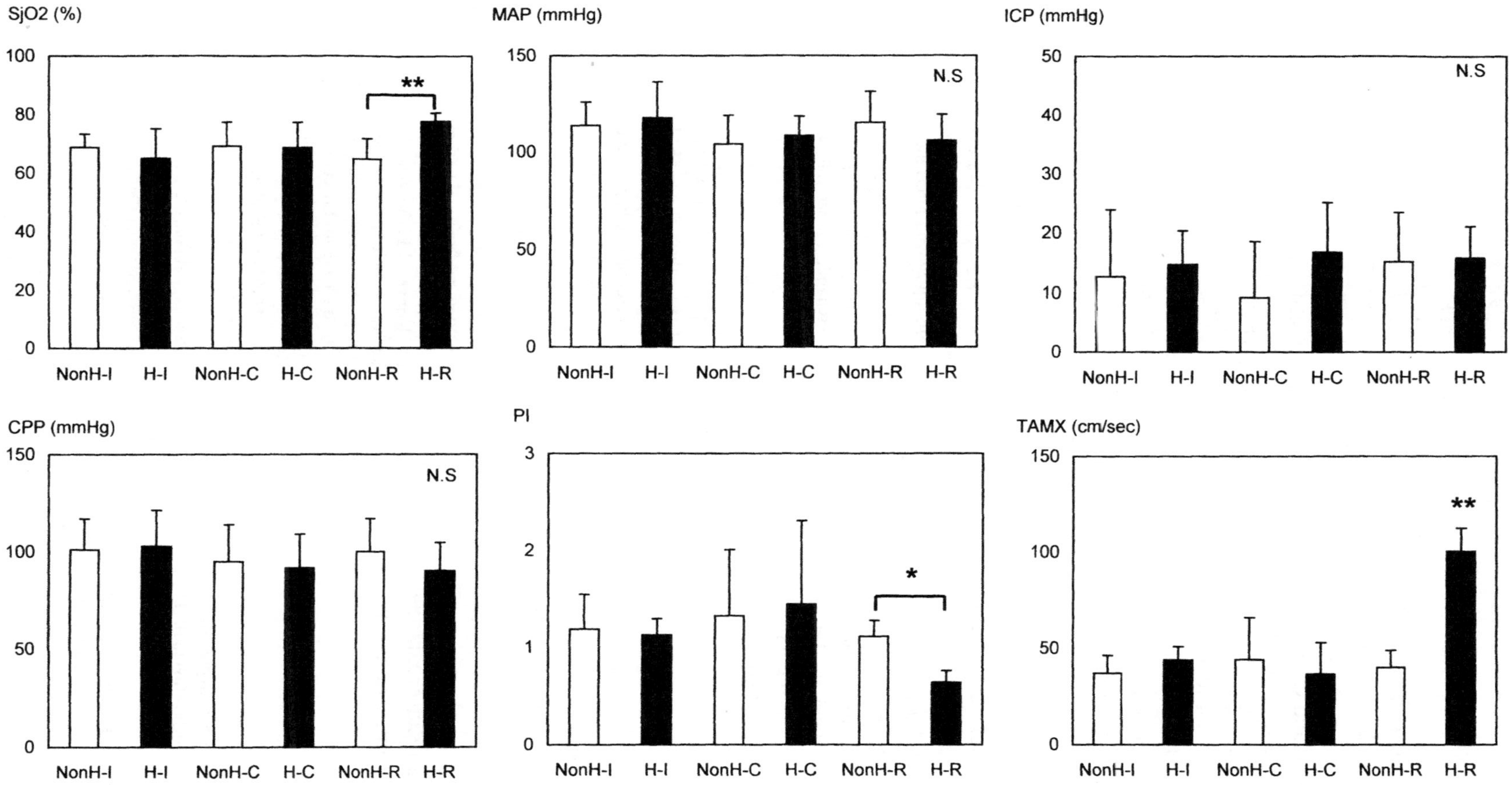

Fig. 1. The changes in physiological parameters during induction (*I*), cooling (*C*), or rewarming (*R*) stage in Non-H or H group. Non-H-I, C, or R: Induction, cooling, or rewarming stage in patients with non-high SjO_2 ($SjO_2 < 75\%$ at a rewarming stage). H-I, C, or R: Induction, cooling, or rewarming stage in patients with high SjO_2 ($SjO_2 > 75\%$ or more at a rewarming stage). (A) Saturation of jugular venous oxygen saturation (SjO_2; %); (B) Mean arterial blood pressure (MAP; mmHg); (C) Intracranial pressure (*ICP*; mmHg); (D) Cerebral perfusion pressure (*CPP*; mmHg); (E) Pulsatility (*PI*); (F) Time-averaged maximum flow velocity (*TAMX*; cm/sec). $^*P < .05$, $^{**}P < .01$

be re-produced at rewarming, in association with reduced cerebrovascular resistance, which might play a role in post-hypothermic hyperemia.

Hypothermia reduces the intravascular blood volume [2, 4], due to a body fluid shift to the interstitial space caused by the increase in vascular permeability that occurs during cooling. Since MAP and CPP were controlled during hypothermia, the changes in MAP and CPP were not significant between the groups during hypothermia in this study. When intravascular blood volume is restored at rewarming, hyperemia with intracranial hypertension may occur if normal cerebral vasoactivity was disturbed at rewarming caused by re-production of vasodilatory mediators.

A relation between raised intracranial pressure (ICP) and hyperemia has been suspected [11, 16], but reports have not been consistent, possibly due to dissociation between cerebral blood flow and cerebral blood volume in head-injured patients. Early identification of altered vasoactivity and aggressive treatment, including re-cooling, of these patients might improve their outcome. Further studies need to define the relation between changes in vasoactive mediators and temperature during hypothermia in patients with TBI.

Conclusion

This TCD-based study demonstrates marked changes in the alternation of vascular resistance at rewarming. Although the outcome of TBI after therapeutic hypothermia largely depends on the injury severity, TCD will lead us to accurately adjust the treatment in order to prevent and not exacerbate the deleterious effect of hyperemia with intracranial hypertension at rewarming.

References

1. Aibiki M, Maekawa S, Yokono S (2000) Moderate hypothermia improves imbalances of thromboxane A2 and prostaglandin I2 production after traumatic brain injury in humans. Crit Care Med 28: 3902–3906
2. Chen RYZ, Chien S (1977) Plasma volume, red cell volume, and thoracic duct lymph flow in hypothermia. Am J Physiol 233: 605–612
3. Czosnyka M, Smielewski P, Piechnik S, Steiner LA, Pickard JD (2001) Cerebral autoregulation following head injury. J Neurosurg 95: 756–763
4. Green JF, Jackman AP (1979) Mechanism of the increased vascular capacity produced by mild perfusion hypothermia in the dog. Circ Res 44: 411–419
5. Hayashi N, Hirayama T, Udagawa A, Daimon W, Ohata M (1994) Systemic management of cerebral edema based on a new concept in severe head injury patients. Acta Neurochir (Wien) [Suppl] 60: 541–543
6. Kader A, Frazzini VI, Baker CJ, Solomon RA, Trifiletti RR (1994) Effect of mild hypothermia on nitric oxide synthesis during focal cerebral ischemia. Neurosurgery 35: 272–277
7. Kelly DF, Kordestani RK, Martin NA, Nguyen T, Hovda DA, Bergsneider M, McArthur DL, Becker DP (1996) Hyperemia following traumatic brain injury: relationship to intracranial hypertension and outcome. J Neurosurg 85: 762–771
8. Kontos HA, Wei EP (1986) Superoxide production in experimental brain injury. J Neurosurg 64: 803–807
9. Lee JH, Kelly DF, Oertel M, McArthur DL, Glenn TC, Vespa P, Boscardin WJ, Martin NA (2001) Carbon dioxide reactivity, pressure autoregulation, and metabolic suppression reactivity after head injury: a transcranial Doppler study. J Neurosurg 95: 222–232
10. Marion DW, Penrod LE, Kelsey SF, Obrist WD, Kochanek PM, Palmer AM, Wisniewski SR, DeKosky ST (1997) Treatment of traumatic brain injury with moderate hypothermia. N Engl J Med 336: 540–546
11. Martin NA, Patwardhan RV, Alexander MJ, Africk CZ, Lee JH, Shalmon E, Hovda DA, Becker DP (1997) Characterization of cerebral hemodynamic phases following severe head trauma: hypoperfusion, hyperemia, and vasospasm. J Neurosurg 87: 9–19
12. McCulloh J, Edvinsson L (1984) Cerebrovascular smooth muscle reactivity: a critical appraisal of in vitro and in situ techniques. J Cereb Blood Flow Metab 4: 129–139
13. Muttaqin Z, Uozumi T, Kuwabara S, Arita K, Kurisu K, Ohba S, Kohno H, Ogasawara H, Ohtani M, Mikami T (1993) Hyperaemia prior to acute cerebral swelling in severe head injuries: the role of transcranial Doppler monitoring. Acta Neurochir (Wien) 123: 76–81
14. Niwa K, Takizawa S, Takagi S, Shinohara Y (1998) Mild hypothermia disturbs regional cerebrovascular autoregulation in awake rats. Brain Res 789: 68–73
15. Polderman KH, Peerdeman SM, Girbes AR (2001) Hypophosphatemia and hypomagnesemia induced by cooling in patients with severe head injury. J Neurosurg 94: 697–705
16. Romner B, Bellner J, Kongstad P, Sjoholm H (1996) Elevated transcranial Doppler flow velocities after severe head injury: cerebral vasospasm or hyperemia? J Neurosurg 85: 90–97
17. Takasu A, Yagi K, Okada Y (1996) Effect of mild hypothermia on ischemia-induced release of endothelin-1 in dog brain. Resuscitation 31: 59–64
18. Verhaegen MJ, Todd MM, Hindman BJ, Warner DS (1993) Cerebral autoregulation during moderate hypothermia in rats. Stroke 24: 407–414
19. Vigue B, Ract C, Zlotine N, Leblanc PE, Samii K, Bissonnette B (2000) Relationship between intracranial pressure, mild hypothermia and temperature-corrected $PaCO_2$ in patients with traumatic brain injury. Intensive Care Med 26: 722–728

Correspondence: Kosaku Kinoshita, M.D., Department of Emergency and Critical Care Medicine, Nihon University School of Medicine, 30-1 Oyaguchi Kamimachi, Itabashi-ku, Tokyo 173-8610, Japan. e-mail: kosaku@med.nihon-u.ac.jp

Spinal cord trauma

Acta Neurochir (2003) [Suppl] 86: 383–388

Neurotrophic factors attenuate microvascular permeability disturbances and axonal injury following trauma to the rat spinal cord

H. S. Sharma

Laboratory of Neuroanatomy, Department of Medical Cell Biology, Biomedical Center, Uppsala University, Uppsala, Sweden

Summary

Alterations of the blood-spinal cord barrier (BSCB) following spinal cord injury (SCI) and leakage of serum proteins induce vasogenic edema and cell damage. The possibility that two members of the neurotrophin family, BDNF or IGF-1 induce neuroprotection by attenuating the BSCB permeability following trauma was examined in a rat model. Repeated topical application of BDNF or IGF-1 (0.1 μg, 0.5 μg or 1 μg in 10 μl) onto the spinal cord 30 min before SCI or 2, 5, 10 or 30 min thereafter significantly attenuated BSCB permeability to Evans blue and iodine. In the neurotrophin treated rats, edema formation, degradation of MBP, and myelin vesiculation were much less frequent compared to the untreated traumatised rats. The protective effect of BDNF and IGF-1 was most pronounced at the high dose (1 μg in 10 μl) given either 30 min before or within 10 min after SCI. The observations suggest that early intervention with neurotrophins in high doses following trauma (within 10 min) attenuates disturbances of the fluid microenvironment of the spinal cord. This indicates that BSCB opening plays an important role in SCI induced myelin vesiculation and cord pathology.

Keywords: BDNF; spinal cord injury; edema formation; myelin vesiculation; blood-spinal cord barrier.

Introduction

Neurotrophins are a family of growth factors consisting of nerve growth factor (NGF), brain derived neurotrophic factor (BDNF), neurotrophin-3 (NT-3) and neurotrophin-4/5 (NT-4/5) [1, 7]. The other growth factors include insulin like growth factor-1 (IGF-1), glial-derived neurotrophic factor (GDNF), ciliary neurotrophic factor (CNTF), and transforming growth factor-β (TGF-β) [1, 3]. BDNF and IGF-1 are known to provide neuroprotection in various animal models of ischemic and traumatic injuries [see 7, 15]. Recently, neurotrophins are shown to reduce paraplegia or paralysis following SCI and to enhance axonal regeneration [7]. Axonal regeneration was experimentally achieved in animals by continuous application of neurotrophins, foetal tissue implants, or grafting of peripheral nerves following trauma [for review see 7]. However, details of the cellular and molecular mechanisms of neurotrophins in CNS injury with regard to axonal regeneration are still unknown.

The BSCB strictly regulates the fluid microenvironment of the spinal cord [18]. Breakdown of the BSCB allows entry of serum proteins into the spinal cord fluid microenvironment leading to vasogenic edema and cell injury [11, 15]. Thus, neurotrophins may have the additional potential to attenuate the microvascular permeability disturbances in SCI causing neuroprotection. The present investigation was carried out to determine, whether the neuroprotective properties of two potent members of the neurotrophin family, BDNF and IGF-1 are somehow related with their ability to reduce the breakdown of BSCB function in a well characterised rat model of SCI [9]. Furthermore, to evaluate the therapeutic utility of neurotrophins in clinically relevant conditions, BDNF and IGF-1 were administered at different time periods *after* trauma. Their neuroprotective efficacy against axonal damage was examined by using immunohistochemistry and electronmicroscopy.

Material and methods

Animals

Experiments were carried out in 48 male Sprague Dawley rats (body weight 250–300 g) under Equithesin (3 ml/kg, i.p.) anaesthesia. The animals were housed at controlled room temperature (21 ± 1 °C) with a 12 h light and 12 h dark schedule. Food and tap water were supplied ad libitum before the experiment.

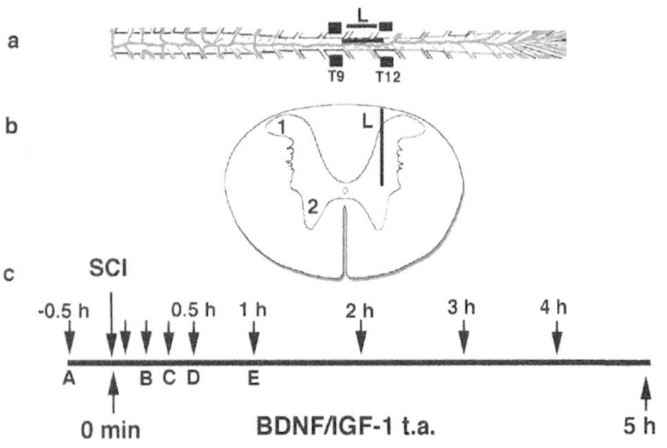

Fig. 1. Schematic representation of the experimental protocol of neurotrophin treatment in spinal cord injury (*SCI*). a = model diagram showing lesion (*L*) in the spinal cord and the tissue blocks (T9 and T12) used for the morphological and biochemical studies (bar = 5 mm). b = Cross section of the spinal cord segment T10–11 showing depth of the lesion (*L*) confined to Rexed's laminae VIII. Contralateral tissue blocks from dorsal (*1*) and ventral horn (*2*) were used for electron microscopy (bar = 2 mm). c = Topical administration (t.a.) of BDNF or IGF-1 was initiated at five different time points in separate groups (*A–E*). SCI was made at 0 min

Spinal cord injury

Under Equithesin anaesthesia (3 ml/kg, i.p.) a one-segment laminectomy was made at T10–11. After 30 min of stabilisation, focal injury was inflicted by incision into the right dorsal horn of the T10–11 segment using a scalpel blade [9] (Fig. 1). The deepest part of the lesion was close to the lamina VII–VIII (Fig. 1). The experimental protocol was approved by the Ethical Committee of Uppsala University, Uppsala, Sweden. Normal animals served as controls.

Treatment with neurotrophins

In separate groups of animals BDNF (n = 5) or IGF-I (n = 5) (0.1, 0.5 μg or 1 μg in 10 μl phosphate buffer saline) were topically administered 30 min before injury on the exposed spinal cord followed by repeated doses at 2, 5, 10 and 30 minutes after SCI (see Fig. 1). Thereafter, application of BDNF or IGF-1 was carried out at intervals of 1 h until sacrifice [14–17]. In another group, treatment with BDNF or IGF-1 was initiated 5 min, 10 min, 30 min or 1 h *after* SCI followed by repeated application of neurotrophins at 2, 3 and 4 hours post injury in an identical manner as described above (see Fig. 1). All animals were allowed to survive for 5 h after SCI (Fig. 1). Tissue pieces taken from T9 and T12 segments were processed for myelin basic protein (MBP)-immunostaining, and ultrastructural analysis of lanthanum extravasation and cell injury [10, 11, 13, 15, 18].

Parameters measured

The following parameters were examined in control, spinal cord injured, and neurotrophin treated injured animals.

Microvascular permeability and edema. The spinal cord water content and microvascular permeability for $^{[131]}$I-sodium was measured in the perifocal T9 and T12 segments as described earlier [8–10]. The tracers was injected into the right femoral vein 5 min before sacrifice [8]. The intravascular tracer was washed out by a brief saline rinse through the heart, and tissue pieces of T9, T10–11 and T12 were dissected out [8–10]. After counting the radioactivity, the tissue samples were dried in an incubator maintained at 90 °C for 72 h [9]. The water content was calculated as the differences between wet and dry weight [10].

Lanthanum extravasation. Passage of tracer across the BSCB was examined by assessing extravasation of lanthanum as described earlier [11].

Axonal damage. Axonal damage was assessed by myelin basic protein (MBP) immunostaining of paraffin embedded 3 μm sections obtained from the T9 segment of the cord [10]. Cell injury was examined at the ultratstructural level by transmission electronmicroscopy [8].

Statistical analysis of the data

ANOVA followed by Dunnet's test was applied to evaluate statistical significance of the data obtained for quantitative measurement of BSCB permeability and edema formation. The morphological scores were compared using non-parameteric Chi square test.

Results

Effect of neurotrophins on visible swelling

Traumatised rats had signs of haemorrhage and macroscopically visible swelling of the cord immediately after lesion. Swelling continued over the 5 h period of survival. Pretreatment with either BDNF or IGF-1 markedly attenuated the grossly visible swelling after injury without influencing microhaemorrhages (results not shown).

Effect of neurotrophins on edema formation

Spinal cord injured rats exhibited a profound increase in water content of the perifocal T9 and T12 segments (Fig. 2). Treatment with BDNF or IGF-1 in high doses either 30 min before or 5 and 10 min after SCI significantly attenuated the increase in spinal cord water content as compared to the untreated traumatised animals (Fig. 2). On the other hand, administration of neurotrophins 30 min *after* SCI did not reduce the tissue water content significantly (Fig. 2).

Effect of neurotrophins on BSCB permeability

SCI markedly increased the barrier permeability for Evans blue and ^{131}I-sodium in the adjacent T9 and T12 segments at 5 h. Application of neurotrophins in high concentrations prior to injury significantly attenuated the BSCB permeability to Evans blue and the iodine tracers (Fig. 2). Protection of the barrier offered

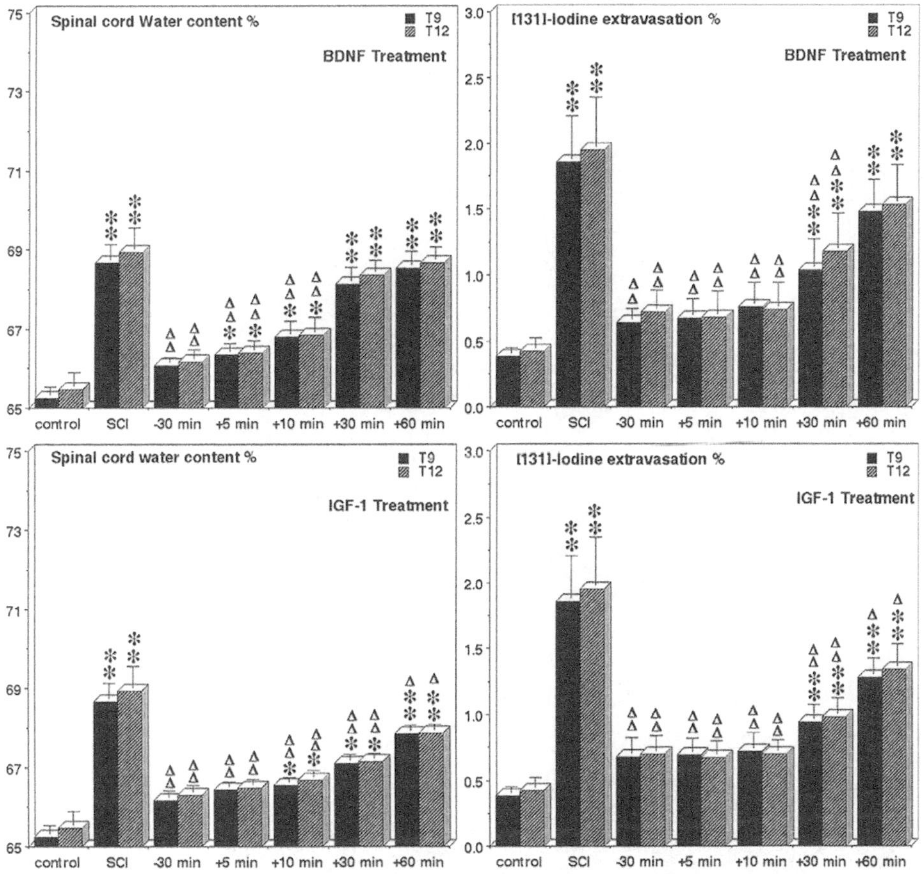

Fig. 2. Spinal cord water content and extravasation of radioiodine in the T9 and T12 segments following SCI and their modification by BDNF (upper panel) or IGF-1 (lower panel). ** = $P < 0.01$ compared from controls; DD = $P < 0.01$ compared from SCI, ANOVA followed by Dunnet's test for multiple group comparison

by neurotrophins was limited to their application within 10 minutes after SCI.

Effect of neurotrophins on cell changes

The spinal cord T9 and T12 segments of untreated rats had profound cell injury (Fig. 3), while these changes were considerably less severe in animals with pretreatment by high doses of BDNF or IGF-1 administered either 30 min prior to or within 10 min after SCI (Fig. 3).

Effect of neurotrophins on myelin basic protein degradation

Treatment with BDNF or IGF-1 significantly attenuated the SCI induced degradation of MBP in the spinal cord (Fig. 3). The most effective MBP preservation was seen in the spinal cord of animals with high doses of neurotrophin administered within 10 min of SCI (results not shown).

Effects of neurotrophins on axonal damage

The ultrastructural studies by electronmicroscopy also demonstrated that vesiculation of myelin and damage of axons was much less intense in traumatised animals receiving neurotrophins (Fig. 3) as compared to the untreated injured animals. The greatest therapeutical efficacy of neurotrophins was observed in animals administered with high doses within 10 minutes *after* SCI (results not shown).

Effects of neurotrophins on endothelial cell permeability

Application of BDNF or IGF-1 was also highly effective to limit extravasation of lanthanum across the BSCB, when given either 30 min before or within 10 min after SCI (Fig. 3). On the other side, onset of administration of neurotrophins at 30 min or 1 h *after* SCI was no more effective to protect the normal BSCB permeability to lanthanum (results nor shown).

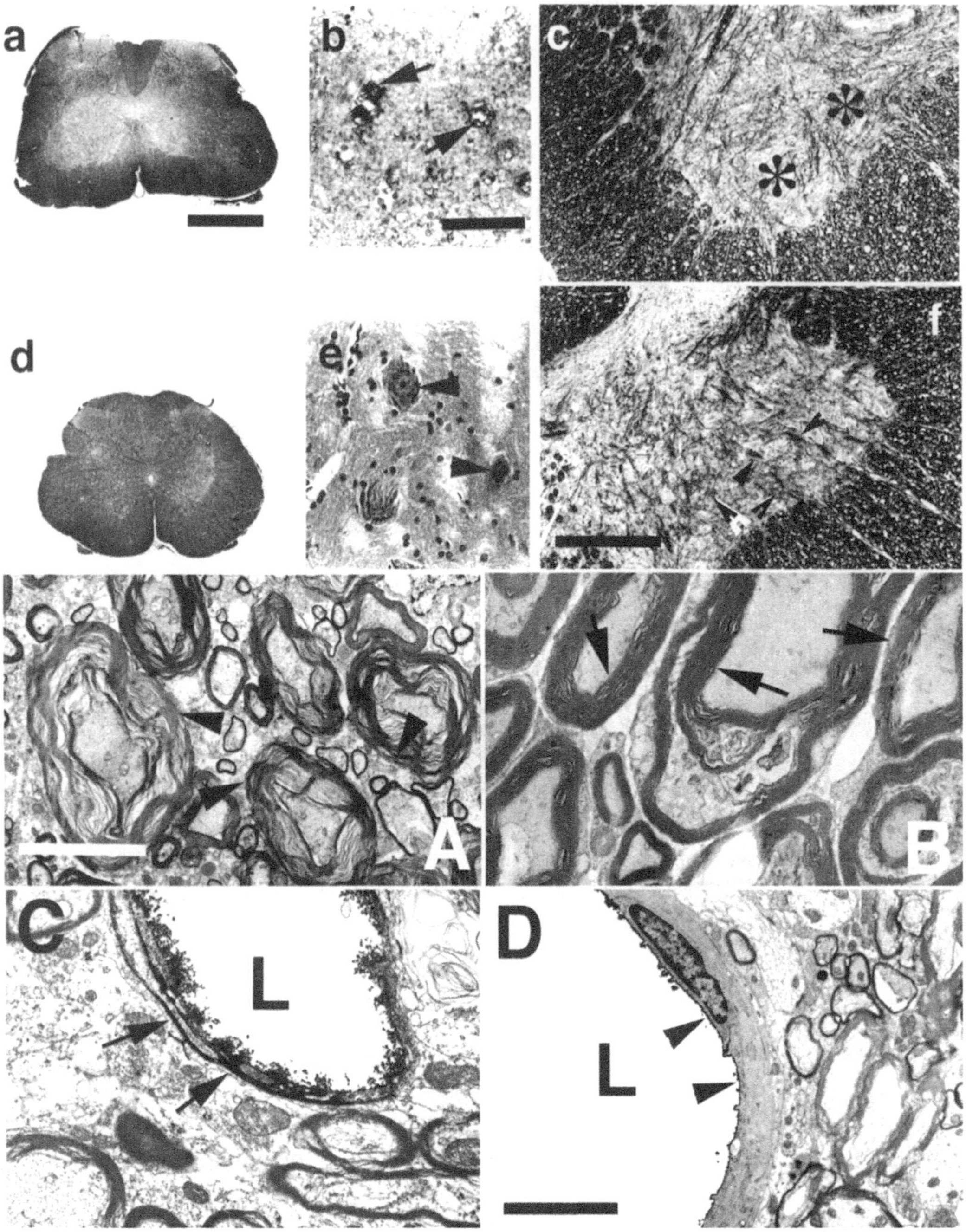

Fig. 3. Light (a–d) and electron (A–D) microscopical changes of the spinal cord T9 segment at 5 h *after* SCI and their modification by BDNF treatment. Swollen spinal cord segment (a) with damaged nerve cells, edema, sponginess (b) and loss of myelin basic protein (c, **) are seen in untreated injured rats. In BDNF treated traumatised animals, swelling of the cord (d), nerve cell changes in the ventral horn (e), and degradation of MBP (f) are much less pronounced. The ultrastructural morphology shows vesiculation of myelin and axonal damage in untreated injured rat (A) which was much less severe in the BDNF-treated and traumatised rat (B). Extravasation of lanthanum across a spinal cord microvessel in the ventral horn of the T9 segment of an untreated animal (arrows, C). In a BDNF treated traumatised rat, the lanthanum tracer was mainly confined to the lumen (L. arrowheads). Bars: a, d = 2 mm, b, e = 25 μm, c, f = 40 μm; A, B = 600 nm; C, D = 1 μm

Discussion

The salient new finding of the present study is that topical administration of neurotrophins onto the exposed traumatised spinal cord in repeated doses is able to induce marked therapeutical protection. This effect is dose-dependent and most pronounced when the neurotrophins are given either as pretreatment or within 10 min after SCI. The observations indicate that early intervention with neurotrophin therapy in high doses following trauma (within 10 min) has a neuroprotective potential.

Our studies have further shown that the neurotrophin administration limits the microvascular permeability disturbances following SCI. This suggests that the barrier disturbances and alterations of the fluid microenvironment of the spinal cord parenchyma following SCI play an important role in cell injury. Presumably, inhibition of the vascular and cellular reactions in the cord tissue following injury in animals with neurotrophins is likely to result in less cell and tissue damage [for review see 15, 17]. Most likely neurotrophins are able to attenuate the early mechanisms of secondary injury by protecting the fluid microenvironment of the cord, which has not been reported earlier.

Topically applied neurotrophins probably have a better access to the cord tissue within a short time period. Penetration of serotonin or nNOS antisera into spinal cord tissue following their topical application supports this idea [11–13]. Motoneurons and other neuronal populations of the spinal cord are important targets for neurotrophins [1, 3]. Most of the neuronal cells in the spinal cord express receptors for neurotrophins, such as trk-B, trk-C and the low-affinity neurotrophin receptor p75 [2, 7, 17]. Thus, exogenous supplementation of BDNF may elicit receptor mediated effects in the traumatised spinal cord underlying neuroprotection. The normal spinal cord is very rich in IGF, whereas its content is reduced following SCI [7, 17]. Exogenous supplementation of IGF-1 in SCI may, therefore, rescue nerve cells from damage [19, 20].

The mechanisms by which neurotrophins provide protection may be associated with their influence on intracellular cascades following trauma, e.g. by modifying cellular signal transduction, the intracellular Ca^{2+}-overload [5], and NOS upregulation, which all are injurious to the cord [4, 11, 19]. Restriction of cellular stress by modulation of the sensory information process by neurotrophins is also likely [1–3, 8]. Obviously, inhibition of the stress response following trauma or injury could be associated with neuroprotection. Further, neurotrophins may attenuate or block apoptotic cell death in SCI, an option which requires additional investigation.

Conclusion

Our results suggest that BDNF and IGF pretreatment is neuroprotective in spinal cord injury. The beneficial effect seems to be related with the capacity of neurotrophins to protect trauma induced BSCB permeability. Maintenance of a normal BSCB permeability in SCI might be instrumental in prevention of edema formation, neuronal injury, and axonal damage.

Acknowledgments

This study is supported by grants from Swedish Medical Research Council Nr. 2710; The University Grants Commission, New Delhi, India; The Indian Council of Medical Research, New Delhi, India; The Alexander von Humboldt Foundation, Bonn, Germany. Technical assistance of Elisabeth Scherer, Franziska Drum, Kärstin Flink, Kerstin Rystedt, Ingmarie Olsson and secretarial assistance of Aruna Sharma is highly appreciated.

References

1. Davies AM (1994) The role of neurotrophins in the developing nervous system. J Neurobiol 25: 1334–1348
2. Frisen J, Verge VMK, Cullheim S, Persson H, Fried K, Middlemas DS, Hunter T, Hökfelt T, Risling M (1992) Increased levels of trkB mRNA and trkB protein-like immunoreactivity in the injured rat and cat spinal cord. Proc Natl Acad Sci USA 89: 11282–11286
3. Hefti F (1994) Neurotrophic factor therapy for nervous system degenerative diseases. J Neurobiol 25: 1418–1435
4. Holtzman DM, Kilbridge J, Bredt DS, Black SM, Li Y, Clary DO, Reichardt LF, Mobley WC (1994) NOS induction by NGF in basal forebrain cholinergic neurons: evidence for regulation of brain NOS by a neurotrophin. Neurobiol Dis 1: 51–60
5. Mattson MP, Lovell MA, Furukawa K, Markesberry WR (1995) Neurotrophic factors attenuate glutamate-induced accumulation of peroxides, elevation of intracellular Ca^{2+} concentration, and neurotoxicity and increase antioxidant enzyme activities in hippocampal neurons. J Neurochem 65: 1740–1751
6. Novikov L, Novikova L, Kellerth J-O (1995) Brain-derived neurotrophic factor promotes survival and blocks nitric oxide synthase expression in adult rat spinal motoneurons after ventral nerve root avulsion. Neurosci Lett 200: 45–48
7. Schwab ME, Bartholdi D (1966) Degeneration and regeneration of axons in the lesioned spinal cord. Physiol Rev 76: 319–370
8. Sharma HS (1999) Pathophysiology of blood-brain barrier, brain edema and cell injury following hyperthermia: New role of heat shock protein, nitric oxide and carbon monoxide. An experimental study in the rat using light and electron microscopy. Acta Universitatis Upsaliensis 830: 1–94
9. Sharma HS, Olsson Y (1990) Edema formation and cellular alteration in spinal cord injury in the rat and their modification with p-chlorophenylalanine. Acta Neuropathol (Berl) 79: 604–610
10. Sharma HS, Olsson Y, Cervós-Navarro J (1993a) Early perifocal cell changes and edema in traumatic injury of the spinal cord are reduced by indomethacin, an inhibitor of prostaglandin synthesis. Acta Neuropathol (Berlin) 85: 145–153
11. Sharma HS, Olsson Y, Pearsson S, Nyberg F (1995b) Trauma induced opening of the blood-spinal cord barrier is reduced by indomethacin, an inhibitor of prostaglandin synthesis. Experimental observations in the rat using ^{131}I-sodium, Evans blue and lanthanum as tracers. Restor Neurol Neurosci 7: 207–215

12. Sharma HS, Westman J, Olsson Y, Alm P (1996a) Involvement of nitric oxide in acute spinal cord injury: an immunohistochemical study using light and electron microscopy in the rat. Neurosci Res 24: 373–384
13. Sharma HS, Westman J, Nyberg F (1997b) Topical application of 5-HT antibodies reduces edema and cell changes following trauma to the rat spinal cord. Acta Neurochir (Wien) [Suppl] 70: 155–158
14. Sharma HS, Nyberg F, Gordh T, Alm P, Westman J (1997a) Topical application of insulin like growth factor-1 reduces edema and upregulation of neuronal nitric oxide synthase following trauma to the rat spinal cord. Acta Neurochir (Wien) [Suppl] 70: 130–133
15. Sharma HS, Nyberg F, Gordh T, Alm P, Westman J (1998) Neurotrophic factors attenuate neuronal nitric oxide synthase upregulation, microvascular permeability disturbances, edema formation and cell injury in the spinal cord following trauma. Spinal cord monitoring. In: Stålberg E, Sharma HS, Olsson Y (eds) Springer, Wien New York, pp 181–210
16. Sharma HS, Nyberg F, Westman J, Alm P, Gordh T, Lindholm D (1998b) Brain derived neurotrophic factor and insulin like growth factor-1 attenuate upregulation of nitric oxide synthase and cell injury following trauma to the spinal cord. An immunohistochemical study in the rat. Amino Acids 14: 121–129
17. Sharma HS, Nyberg F, Gordh T, Alm P, Westman J (2000) Neurotrophic factors influence upregulation of constitutive isoform of heme oxygenase and cellular stress response in the spinal cord following trauma. An experimental study using immunohistochemistry in the rat. AminoAcids 19: 351–361
18. Sharma HS, Westman J (2003) The blood-spinal cord and brain barriers in health and disease. Academic Press, San Diego, pp 1–650 (in press)
19. Thoenen H (1995) Neurotrophins and neuronal plasticity. Science 270: 593–598
20. Yin QW, Johnson J, Prevette D, Oppenheim RW (1994) Cell death of spinal motoneurons in the chick embryo following deafferentation: rescue effects of tissue extracts, soluble proteins, and neurotrophic agents. J Neurosci 14: 7629–7640

Correspondence: Hari Shanker Sharma, Dr. Med. Sci., Frödingsgatan 12, app: 28; SE-75421 Uppsala, Sweden. e-mail: Sharma@medcellbiol.uu.se

Acta Neurochir (2003) [Suppl] 86: 389–394

Depletion of endogenous serotonin synthesis with p-CPA attenuates upregulation of constitutive isoform of heme oxygenase-2 expression, edema formation and cell injury following a focal trauma to the rat spinal cord

H. S. Sharma and **J. Westman**

Laboratory of Neuroanatomy, Department of Medical Cell Biology, Biomedical Center, Uppsala University, Uppsala, Sweden

Summary

The possibility that the upregulation of hemeoxygenase (HO) enzyme responsible for carbon monoxide (CO) formation in the spinal cord following trauma is involved in edema formation and cell damage was examined in a rat model. A focal trauma to the rat spinal cord by making an incision into the right dorsal horn of the T10–11 segment resulted in profound upregulation of HO-2 (the constitutive isoform of the enzyme) expression in the T9 and T12 segments 5 h after injury. In these segments a marked increase in edema formation, nerve cell damage, and expression of heat shock protein (HSP 72) were observed. Pretreatment with p-chlorophenylalanine (p-CPA, a serotonin synthesis inhibitor) significantly attenuated the trauma induced edema formation, cell injury, and HSP expression. Upregulation of HO-2 in p-CPA treated traumatised rats was considerably reduced. These observations suggest that (i) spinal cord injury has the capacity to induce an upregulation of HO-2 and HSP expression, (ii) abnormal production of CO as reflected by HO-2 expression is injurious to the cord, and (iii) that endogenous serotonin is involved in HO-2 expression in the cord.

Keywords: Spinal cord injury; heme oxygenase-2; carbon monoxide; edema; cell injury; trauma; serotonin; heat shock protein-72; p-chlorophenylalanine; immunohistochemistry.

Introduction

Trauma to the spinal cord is associated with alterations in fluid microenvironment of the cord and induce abnormal reactions in nerve cells, glial cells, and axons leading to severe long-term disability [4, 10, 19]. The outcome of injury is influenced by several mechanisms: (a) the severity of cell- and microfluid disturbances (edema, haemorrhage) at the site of the primary physical injury, (b) the degree of disruption and block of descending and ascending neuronal pathways including axonal and myelin alterations, (c) additional lesions in the segments cranial and caudal to the primary injury caused by edema, circulatory disturbances, and other factors, so-called perifocal or secondary injuries, and (d) repair mechanisms.

Proposed secondary injury factors include the products of phospholipid hydrolysis, such as polyunsaturated fatty acids, eicosanoids, free radicals, or neuropeptides, monoamines and changes of cations and amino acids [4, 10, 19]. Recently, nitric oxide (NO) and carbon monoxide (CO) which are newly discovered gases involved in neuronal communication, appear to play important roles in the pathophysiology of cell injury following spinal trauma [16, 17]. CO is a by-product of heme degradation and serves as a messenger molecule like NO [1]. There is evidence that CO mimics many actions of NO both in vivo and in vitro [3, 7, 20]. However, details of their biological function in spinal injury are not yet well understood.

CO is synthesised by the enzyme hemeoxygenase (HO) that exists in two different isoforms, i.e., HO-1 (inducible) and HO-2 (constitutive) [1, 17]. HO-1 is induced by many diverse agents, such as heme substances, histamine H_2-receptor antagonists, adrenaline, insulin, glucagon, bacterial toxins, heat shock, oxidative stress, and many other forms of cellular stress [see review by Sharma *et al.*, 17]. On the other hand, the role of HO-2 in cell injury is still not well understood. HO-2 is constitutively expressed and may participate in signal transduction in the central nervous system (CNS) [3, 20].

Since spinal cord injury (SCI) is associated with marked oxidative stress [13], HSP expression [15], edema formation, and cell damage [11], it seems likely that traumatic insult induces upregulation of HO-2

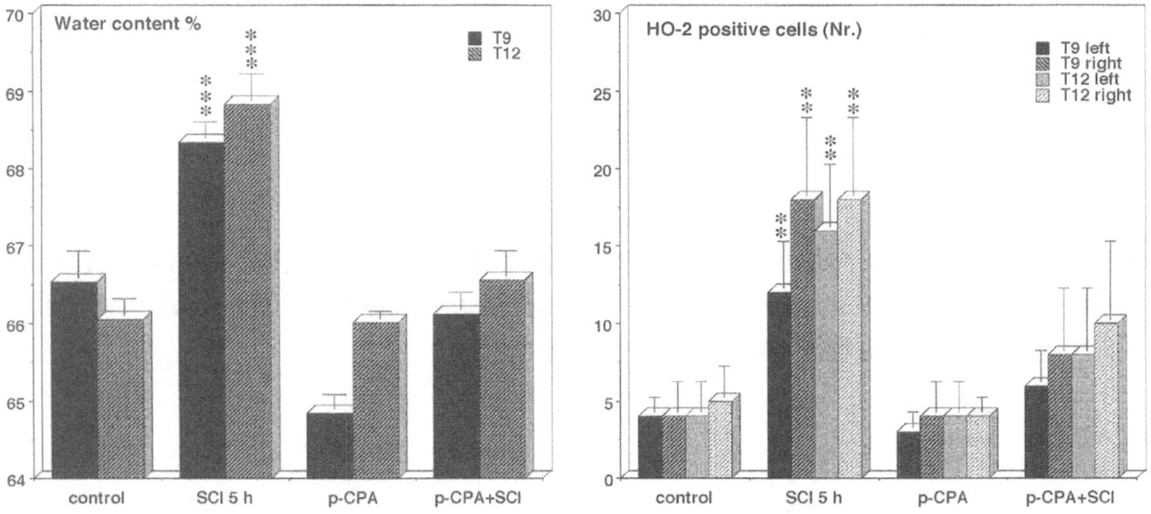

Fig. 1. Spinal cord water content (left panel) and number of HO-2 positive cells (right panel) in the spinal cord at 5 h following injury and its modification by p-CPA. *** $P < 0.001$, Student's unpaired t-test, ** $P < 0.01$, Chi Square test. SCI = spinal cord injury

expression. Previously, serotonin is shown to be a powerful mediator of spinal cord edema formation [14] and HSP response [15]. However, its interaction with the newly discovered neuromodulators NO and CO are not well understood. The present investigation was undertaken to examine, whether inhibition of endogenous serotonin synthesis prior to SCI influences upregulation of HO-2 expression. In addition, cell changes and HSP expression were also examined.

Material and methods

Animals

Experiments were carried out in 48 male Sprague Dawley rats (250 to 350 g) housed at controlled room temperature (21 ± 1 °C) with a 12 h light- and 12 dark schedule. Food and tap water were supplied ad libitum before the experiment.

Spinal cord injury

Under Equithesin anaesthesia (3 ml/kg, i.p.) laminectomy was carried out at the T10–11 segments. SCI was made by an incision into the right dorsal horn (about 2 mm deep and 4 mm long) using a sterile scalpel blade [11]. The wound was covered with cotton soaked in saline to avoid direct exposure of the cord to air. The level of anaesthesia was maintained by repeated administrations of a maintenance dose of Equithesin at regular intervals during the entire period of the experiments. The animals were allowed to survive 5 h after SCI. This experimental protocol was approved by the Ethics Committee of Uppsala University, Uppsala, Sweden. Equithesin anaesthetised normal animals were used as controls.

p-Chlorophenylalanine treatment

p-chlorophenylalanine (p-CPA-methyl ester, Sigma Chemical Co., USA) in powder form was dissolved in distilled water and administered intraperitoneally (100 mg/kg per day) for 3 days [11]. On the 4th day, SCI was made in the animals. One group of p-CPA treated animals were used as drug-treated controls.

Immunohistochemistry

5 h after SCI, the animals were perfused with a formalin based fixative (4% paraformaldehyde in 0.1 M phosphate buffered saline, pH 7.0 at 4 °C) preceded by a washout of blood with phosphate buffered saline [15]. The spinal cord segments from T9 and T12 were dissected out, and 40 μm multiple sections were cut on a Vibratome [15]. Immunostaining for HSP 72 and HO-2 were done on the free floating Vibratome sections according to a standard protocol described earlier [15, 17]. The dark brown reaction product was visualised by light microscopy and counted in a blinded fashion for semiquantitative analysis (see Fig. 1).

Spinal cord edema formation

The water content of the spinal cord of the T9 and T12 segments was determined in separate groups as described earlier [11]. The water content was calculated as the difference in wet and dry weight of the tissue sample as described earlier [13].

Spinal cord cell injury

The spinal cord tissue of the T9 and the T12 segments was processed for standard electron microscopy as described earlier [12]. In a group of rats, lanthanum chloride was added to the fixative to determine blood-spinal cord barrier permeability changes after trauma [11, 13].

Statistical analyses of the data obtained

Student's unpaired t-test was used to evaluate statistical significance of quantitative data obtained. The semiquantitative data were analyzed using non-parametric Chi-square test. A p-value less than 0.05 was considered significant.

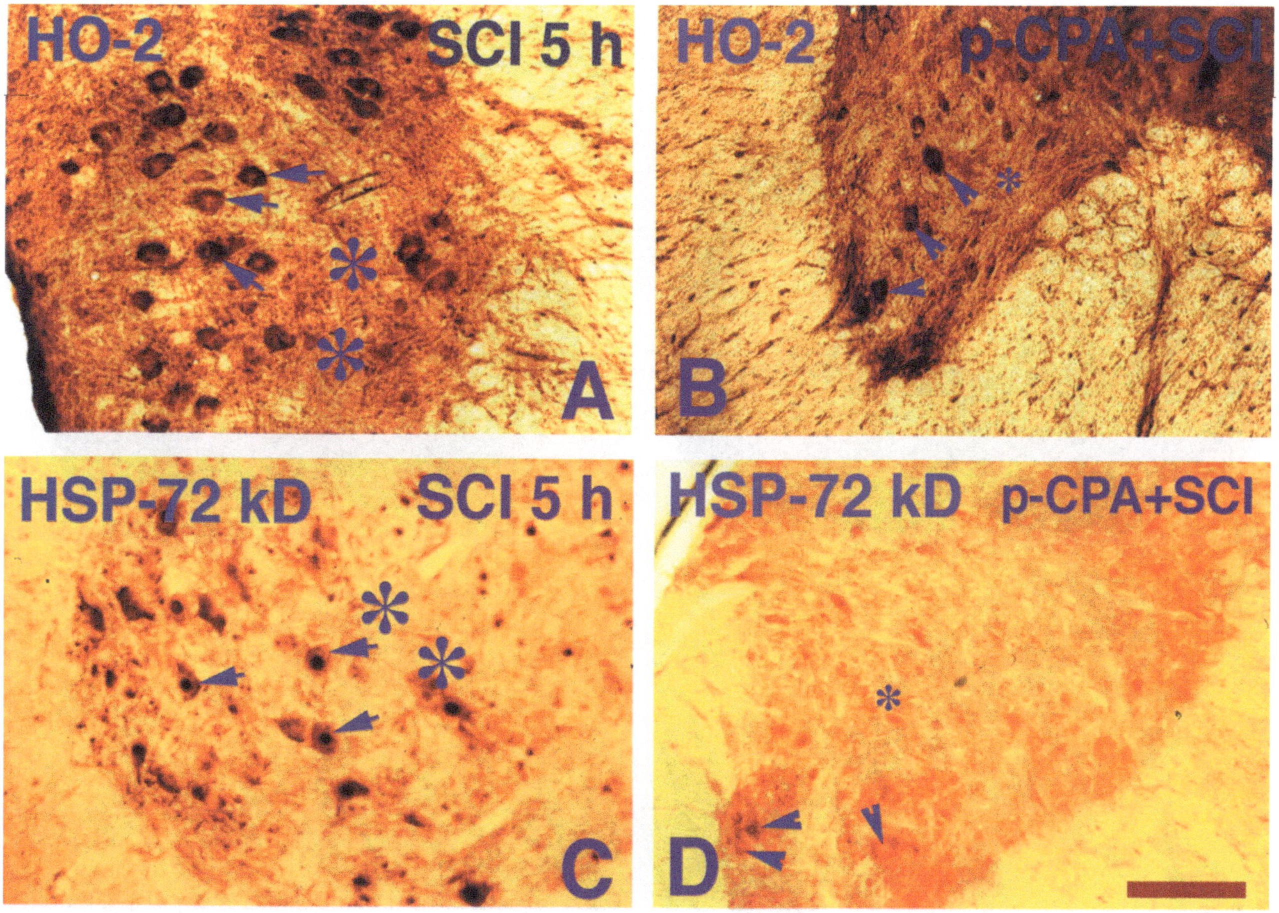

Fig. 2. Immunostaining of HO-2 (A,B) and HSP-72 kD (C,D) following spinal cord injury (*SCI*) and its modification by p-CPA. Many HO-2 (A) and HSP (C) positive cells can be seen following 5 h after SCI (arrows). Edema and sponginess are prominent (*) in these regions. Pretreatment with p-CPA markedly attenuated HO-2 expression (B) and HSP-72 kD upregulation (D) following SCI. In the treated group, only few positive cells (arrowheads) are present and signs of edema and sponginess are much less frequent. Bar: 25 μm

Results

Effect of p-CPA on HO-2 immunostaining

In normal animals, only few spinal cord nerve cells were positive for HO-2 immunostaining (Fig. 1). The focal trauma of the spinal cord induced a pronounced upregulation of HO-2 immunostaining in the T9 and the T12 segments (Fig. 1). The intensity of HO-2 immunostaining was most marked in the injured side of the cord as compared to the contralateral side (Fig. 1). In most cases, the nerve cytoplasm exhibited marked HO-2 immunostaining as compared to the cell nucleus. Pretreatment with p-CPA significantly attenuated the trauma induced HO-2 expression in the spinal cord (Fig. 1 and 2B). The effect of p-CPA on inhibition of HO-2 immunostaining was most marked in the contralateral side as compared to the ipsilateral cord (Fig. 1). In p-CPA treated traumatised animals, HO-2 expression was confined to only few sporadic nerve cells while in these regions, edema and expansion of the cord was markedly attenuated (Fig. 2B).

Effect of p-CPA on HSP immunostaining

The focal trauma was markedly increasing the number of HSP positive nerve cells in the spinal cord (Fig. 2C). HSP immunoreactivity was seen in both the nerve cell nucleus and the nerve cell cytoplasm. Cells exhibiting HSP immunoreactivity were often located in the edematous region of the cord. Pretreatment with p-CPA markedly reduced HSP expression in the traumatised cord (Fig. 2D). In this group of rats, the edematous expansion of the cord was considerably reduced. Thus, only few cells exhibited a HSP positive reaction with the cell nucleus remaining unstained (Fig. 2D).

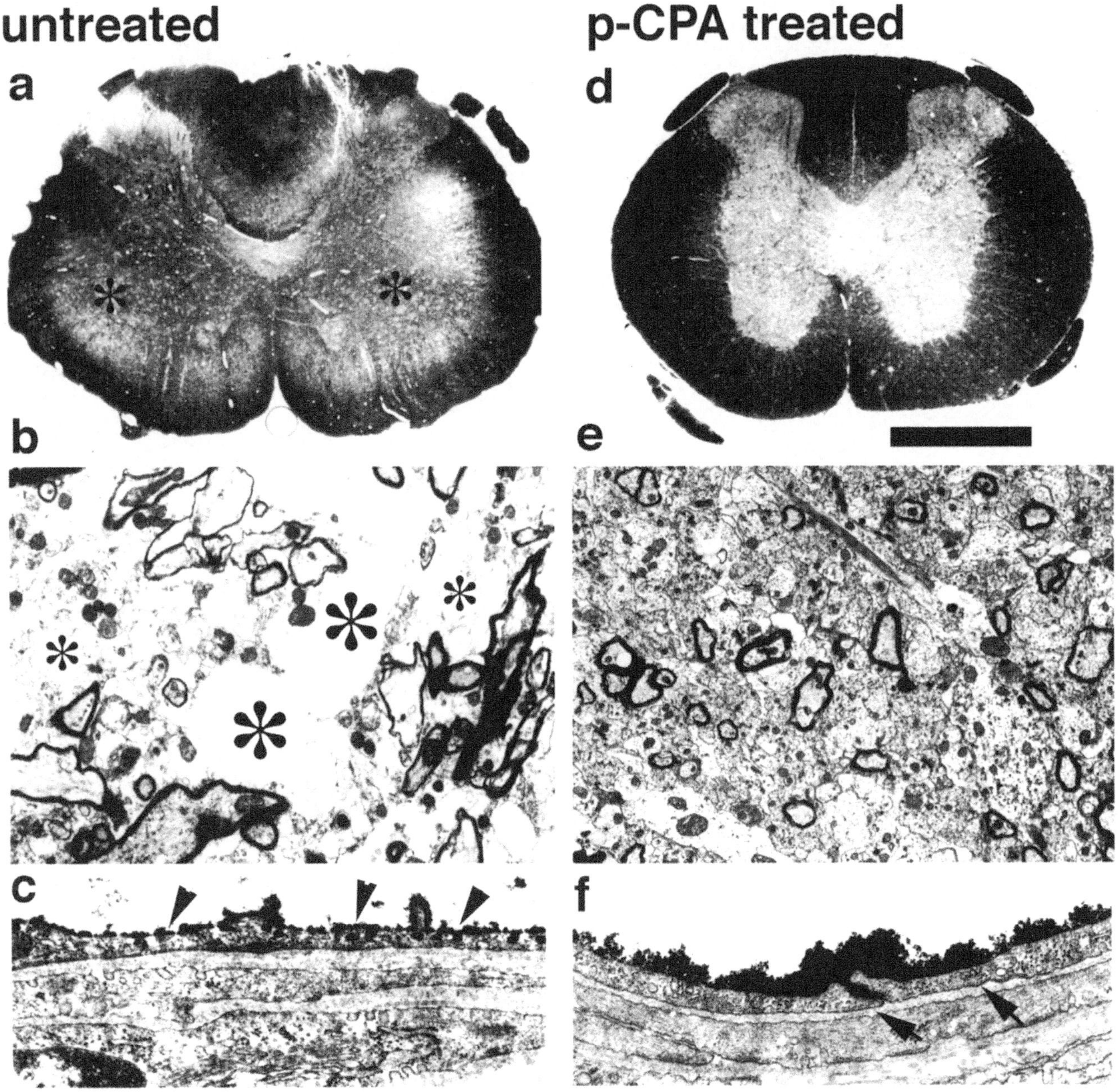

Fig. 3. Morphological alterations in the spinal cord and blood-spinal cord barrier following SCI and its modification by p-CPA. The gross pathology of the spinal cord following SCI shows profound expansion, of the cord, sponginess and edema (*) in the untreated injured animal (a). Pretreatment with p-CPA markedly attenuated the gross pathology of the spinal cord after trauma (d). The low power electron micrograph of the ventral horn of the spinal cord of one untreated rat exhibits membrane damage, vacuolization and edema (*) in the neuropil (b). These changes were much less apparent in the p-CPA treated traumatised rat (e). The high power electron micrograph of an endothelial cell of the dorsal horn of the spinal cord in an untreated injured rat shows several microvascular profiles loaded with lanthanum within the endothelial cytoplasm (arrowheads) (c). In a p-CPA pretreated injured animal, lanthanum is mainly confined to the vessel lumen, and its passage is stopped by the tight junction (arrow) (f). Bar: a,d = 5 mm; b,e = 1 μm, c,f = 600 nm (Modified after Sharma and Olsson 1990)

Effect of p-CPA on edema formation

Measurement of the water content of the spinal cord tissue revealed a profound increase of the T9 and T12 segments (Fig. 1). The edematous expansion of the cord is evident in the untreated traumatised animals (Fig. 3a), whereas pretreatment with p-CPA markedly attenuated the trauma induced edematous expansion indicative of swelling of the spinal cord (Fig. 3b).

Effect of p-CPA on cell injury

In the untreated traumatised rats, the breakdown of the blood-spinal cord barrier (BSCB) was quite widespread, as visualized by the lanthanum tracer in several regions of the T9 and T12 segments of the cord (Fig. 3c). Thus, exudation of lanthanum across the endothelial cell was quite common in the untreated traumatised rats, exhibiting edema, membrane disruption, vacuolization, and damage of the neuropil (Fig. 3b, c).

Pretreatment with p-CPA markedly attenuated extravasation of the lanthanum tracer (Fig. 3f), and damage of the neuropil (Fig. 3e). Signs of vacuolization, membrane damage, edema and distorted myelin were much less intense compared to the untreated traumatised rats (Fig. 3e).

Discussion

The new finding of this study is the pronounced increase of HO-2 expression in the perifocal segments of the cord at 5 h following SCI. This indicates that SCI induces upregulation of the constitutive isoform of the HO enzyme. At this time, breakdown of the BSCB, edema formation, and cell injury were prominent in the cord, indicating that secondary injury mechanisms, as leakage and spread of edema fluid in the cord contribute to HO-2 expression. The HO-2 immunoreactivity in the damaged and distorted nerve cells which were located in the edematous region of the cord supports this idea.

It appears that an increased expression of the constitutive isoform of HO is injurious to the cell, while it has been shown earlier that expression of the inducible isoform is neuroprotective [5]. Expression of HO-1 coincides with glial scar formation occurring during regeneration in the CNS [10]. An increased HO-2 expression at 5 h after SCI suggests that secondary injury factors, as the release of several neurotransmitters and cell injury play an important role. The spinal cord trauma induced microhaemorrhages can also be considered to enhance HO-2 expression.

Expression of HO-2 is associated with the production of CO that causes induction of LTP, a feature commonly known to mediate neuronal injury in many pathophysiological conditions [1, 7, 8]. Another possibility of HO/CO induced neuronal damage may be due to the release of free iron which acts as a catalyst for the production of free radicals thereby amplifying neuronal damage [8, 20]. Thus, generation of CO from over-expression or HO-2 may be associated with cell injury.

Oxidative stress, haemorrhage and/or ischemia are important factors in the generation of the HO response following fluid percussion brain injury in rats [1, 5, 9, 18]. Since SCI is markedly associated with ischemia, microhaemorrhages, and oxidative stress [13], these factors may somehow influence HO-2 induction in the cord following SCI.

In an *in vitro* model of anoxia and in fluid percussion injury, HO inhibitors metalloporphyrins (Sn-PP or Zn-PP) are neuroprotective. These inhibitors exert some anti-inflammatory effects which are apparent from the finding that Zn-PP attenuates infarct size and edema following cerebral ischemia [6]. In the present study, p-CPA treatment was markedly reducing SCI induced expression of HO-2 and cell damage. This supports that upregulation of HO-2 is associated with neurodestruction by generation of CO [for review see 17].

Trauma of the spinal cord induces a profound increase in plasma and cord tissue serotonin [14] indicating that an elevated level of tissue and blood serotonin influences HO-2 expression in the cord. In p-CPA treated traumatised rats, an elevation of the endogenous serotonin level was mainly absent [11]. In the absence of an endogenous serotonin elevation, the cellular and molecular mechanisms may not be operating causing overexpression of HO-2 in the cord. A direct inhibitory effect of serotonin on HO-2 expression in the spinal cord is unlikely, because depletion of serotonin with p-CPA in normal rats does not influence HO-2 expression.

The abnormal expression of HSP in the spinal cord following trauma suggests that cellular stress plays an important role in tissue damage. The absence of HSP expression in p-CPA treated and traumatised rats indicates that a serotonin dependent cellular stress response was somehow responsible for the HO-2 expression and cell injury. The reduction in HSP response following p-CPA treatment is indicative of a reduction of the trauma induced stress response, rather than of a direct inhibitory influence of p-CPA on HSP expression [12, 15, 21]. Thus, it seems likely that the SCI induced cellular stress plays a significant role in upregulation of HO-2 expression, a feature that requires additional investigation.

Conclusion

Our observations suggest that (i) an increased production of CO, as evident by upregulation of the HO-2 expression, is contributing to the spinal cord pathology and (ii) that endogenous serotonin is involved in the trauma induced overexpression of HO-2 in the cord, which has not been reported earlier.

Acknowledgments

This investigation is supported by grants from the Swedish Medical Research Council Nr. 2710; Göran Gustafsson Foundation, Sweden. The technical assistance of Kärstin Flink and Kerstin Rystedt is acknowledged with thanks.

References

1. Abraham NG, Drummond GS, Lutton JD, Kappas A (1996) The biological significance and physiological role of heme oxygenase. Cell Physiol Biochem 6: 129–168
2. Azmitia EC (2001) Modern views on an ancient chemical: serotonin effects on cell proliferation, maturation, and apoptosis. Brain Res Bull 56: 413–424
3. Dawson TM, Snyder SH (1994) Gases as biological messenger: nitric oxide and carbon monoxide in the brain. J Neurosci 14: 5147–5159
4. Faden AI, Salzman S (1992) Pharmacological strategies in CNS trauma. TiPS 13: 29–35
5. Fukuda K, Richmon JD, Sato M, Sharp FR, Panter SS, Noble LJ (1996b) Induction of heme oxygenase-1 (HO-1) in glia after traumatic brain injury. Brain Res 736: 68–95
6. Kadoya C, Domino EF, Yang GY, Stern JD, Betz AL (1995) Preischemic but not postischemic zinc protoporphyrin treatment reduces infarct size and edema accumulation after temporary focal cerebral ischemia in rats. Stroke 26: 1035–1038
7. Maines MD, Abraham NG, Kappas A (1977) Solubilization and partial purification of heme oxygenase from rat liver. J Biol Chem 252: 5900–5903
8. Maines MD (1988) Heme oxygenase: function, multiplicity, regulatory mechanisms and clinical applications. FASEB J 2: 2557–2568
9. Panizzon KL, Dwyer BE, Nishimura RN, Wallis RA (1996) Neuroprotection against CA1 injury with metalloporphyrins. NeuroReport 7: 662–666
10. Schwab ME, Bartholdi D (1996) Degeneration and regeneration of axons in the lesioned spinal cord. Phys Rev 76: 319–370
11. Sharma HS, Olsson Y (1990) Edema formation and cellular alteration in spinal cord injury in the rat and their modification with p-chlorophenylalanine. Acta Neuropathol (Berl) 79: 604–610
12. Sharma HS, Westman J (1997) Prostaglandins modulate constitutive isoform of heat shock protein (72 kD) response following trauma to the rat spinal cord. Acta Neurochir (Wien) [Suppl] 70: 134–137
13. Sharma HS, Sjöquist P-O (2002) A new antioxidant compound H-290/51 modulates glutamate and GABA immunoreactivity in the rat spinal cord following trauma. Amino Acids 217: 261–272
14. Sharma HS, Olsson Y (1990) Edema formation and cellular alterations following spinal cord injury in rat and their modification with p-chlorophenylalanine. Acta Neuropathol (Berlin) 79: 604–610
15. Sharma HS, Olsson Y, Westman J (1995) A serotonin synthesis inhibitor, p-chlorophenylalanine reduces the heat shock protein response following trauma to the spinal cord. An immunohistochemical and ultrastructural study in the rat. Neurosci Res 21: 241–249
16. Sharma HS, Westman J, Olsson Y, Alm P (1996) Involvement of nitric oxide in acute spinal cord injury: an immunohistochemical study using light and electron microscopy in the rat. Neurosci Res 24: 373–384
17. Sharma HS, Westman J, Nyberg F (1998) Pathophysiology of brain edema and cell changes following hyperthermic brain injury. Brain functions in hot environment. In: Sharma HS, Westman J (eds) Progr Brain Res 115: 351–412
18. Soltesz I (1996) Heme oxygenase inhibitors in neurotrauma. A commentary. NeuroReport 7: 385–386
19. Tator CH, Fehlings MG (1991) Review of the secondary injury theory of acute spinal cord trauma with emphasis on vascular mechanisms. J Neurosurg 75: 15–26
20. Verma A, Hirsch DJ, Glatt CE, Ronnett GV, Snyder SH (1993) Carbon monoxide: a putative neural messenger. Science 295: 381–384
21. Westman J, Sharma HS (1998) Heat shock protein response in the CNS following heat stress. Brain function in hot environment. In: Sharma HS, Westman J (eds) Progress in Brain Research 115: 207–239

Correspondence: Hari Shanker Sharma, Dr. Med. Sci., Frödingsgatan 12, App: 28, SE-75421 Uppsala, Sweden. e-mail: Sharma@medcellbiol.uu.se

Acta Neurochir (2003) [Suppl] 86: 395–397

High frequency electrical stimulation attenuates progressive necrosis and cavitation following spinal cord injury

M. Fujiki, H. Kobayashi, and **M. Isono**

Department of Neurosurgery, Oita Medical University, Oita, Japan

Summary

This study evaluates the influence of preconditioning by following electrical stimulation on primary and secondary lesion formation following spinal cord injury in rats. The dorsal surface of the spinal cord was stimulated (500 Hz, 10 pulses/train, inter train interval of 10 sec. for 2 hrs) at the T7 level 24 hrs before a right side hemisection, carried out immediately after injury and maintained every 24 hrs for 7 days. Preconditioning by electrical stimulation of the spinal cord activates reactive astrocytes and significantly attenuates edema and progressive necrosis and cavitation, concerning especially the primary (1, 3 weeks post injury) and secondary (24 hrs, 1, 3 weeks post injury) lesion volume. The results suggest that preconditioning by electrical stimulation prevents spinal cord secondary lesion formation after injury, and that the beneficial effect is provided by astroglial cells with regard to their ability to attenuate trauma induced cellular cascades.

Keywords: Spinal cord injury; secondary damage; electrical stimulation; preconditioning.

Introduction

An important approach to protect the spinal cord from injury is to reduce secondary tissue damage. In addition, preservation of the tissue and fiber tracts surviving the acute lesion is highly important. The secondary inflammatory response around the primary lesion site may determine the final outcome of tissue destruction. Although the cascade of tissue reactions and cell injury develops within days or weeks, cell death from spinal cord injury is most extensive within hours after trauma. This suggests that early intervention by therapeutic manipulations constitutes the most promising approach to rescue the spinal cord from further irreversible cell damage. Thus, influencing the early secondary injury processes around the lesion by modulating genes which control the upregulation of neurotrophic and/or neuroprotective factors seems to be highly relevant to minimize cord dysfunction from trauma. In the present study, we report on a novel non-invasive pre-treatment strategy of neuroprotection in spinal cord injury. We demonstrate that brief preconditioning by continuous electrical stimulation activates reactive astrocytes which substantially reduces the extent of secondary neuronal death of the injured cord following hemisection.

Material and methods

All experimental protocols were approved by the Ethical Committee of the Oita Medical University. Experiments were carried out in 30 adult female Sprague Dawley rats (body weight 300–350 g) housed at controlled room temperature with a 12 h light and 12 h dark schedule. Rat food pellets and tap water were provided ad libitum.

Preconditioning and treatment with electrical stimulation

The dorsal column of the spinal cord was electrically stimulated at the T7 segment using a bipolar flexible wire electrode (500 Hz at 2× threshold voltage of spinal cord evoked potential, 10 pulses/train 10 second inter train interval for 2 hours). Thus, repeated activation of the spinal cord was carried out 24 hours before (720 trains), at 0 min (at the time of injury; 720 trains) and thereafter every 24 h, i.e. 1 to 7 days post insult [20]. In control animals with spinal cord injury, the treatment schedule was exactly followed except that the animals under anesthesia with the electrode setting were not stimulated.

Spinal cord injury

The spinal cord was cut at the right side by inserting the tip of a No. 11 scalpel blade anteroventrally from the midline, sliding the blade to the lateral edge of the spinal cord. The procedure was repeated at the same side to ensure that all spinal tissue had been severed. Hemostasis was secured by topical application of cold sterile saline or Avitene Med. (Chem Products, Inc. Woburn, MA). Thereafter, the wound was closed by suturing the muscle with 6-0 silk and the skin with wound clips.

Quantitative histology

Animals were allowed to survive for 6 h, 24 h, 1, 3, 8 weeks after injury (n = 3 per time point). Animals of each time point and 24 hours after preconditioning stimulation alone (n = 3) were deeply anesthetized with sodium pentobarbital and perfused transcardially with 4% paraformaldehyde. The spinal cord was removed from the vertebral column and then immersed for 24 hours in the same fixative. The tissue was embedded in paraffin and sectioned horizontally (4 μm thickness), sections were collected on microscope slides, the paraffin was removed in xylene and the slides were re-hydrated through graded ethanols. Alternate sections were stained with H&E for histology.

Results

We have analyzed the sections obstained from animals sacrificed at 6, 24 hrs and 1, 3, 8 weeks post-injury. In the following the lesions are defined as total regions (primary and secondary lesion) containing hemorrhage, edema, inflammation, tissue necrosis and cysts or cavities.

Untreated control animals

By 1 week, the total primary and secondary lesion continued to enlarge in size. As a result of the phagocytic activity of macrophages and microglia, cavitation and necrosis in the primary and secondary lesion was progressive. The percent-area of cavitation of the total primary and secondary lesion was approximately (1.049/1.756 = 59.7%). By 3 weeks, progressive cavitation resulted in an enlargement of an empty lesion filled with fluid in the lesion center.

Treated animals

By 1 week, the total primary and secondary lesion remained unchanged in size, or tended only gradually to increase. The size of the total primary and secondary lesion at 1 week was significantly smaller in treated animals as compared to the controls. By 3 to 8 weeks, the total primary and secondary lesion had gradually increased in size and became statistically comparable to the size of untreated control animals. However, the primary lesion was completely filled by fibroblasts forming a continuous connective tissue matrix, which was similar to the well characterized matrix observed in spinal cord injury in mice (Fujiki *et al.*, 1996; Zhang *et al.*, 1996). The area of cavitation decreased gradually in size from 1 to 3 weeks, becoming significantly smaller in the treated animals as compared to the controls, even though the areas also tended to decrease.

Table 1. *Lesion area 1 day, 1 week after injury*

	6 hrs		1 day		1 week	
	Con.	ES	Con.	ES	Con.	ES
Non-cavity lesion	NS		NS			
Mean	0.131	0.141	0.201	0.159	0.707	0.434
SEM	0.024	0.049	0.047	0.072	0.109	0.055
Cavitation	NS		*$p < 0.05$			
Mean	0.189	0.176	0.722	0.432	1.049	0.196
SEM	0.016	0.044	0.106	0.101	0.139	0.116
Total lesion	NS		*$p < 0.05$			
Mean	0.319	0.316	0.922	0.591	1.756	0.631
SEM	0.041	0.005	0.152	0.101	0.231	0.165

The percent area of cavitation of the total primary and secondary lesion was approximately (0.196/0.631 = 31.06%), revealing that the huge cavitation surrounded by glial scar tissue was never replaced (Table 1).

Discussion

It is well known that electrical stimulation up-regulates the expression of a number of growth- and trophic factors, and also induces IEG transcription factors that could modulate the expression of other genes (Steward 1994; Steward *et al.*, 1991; Bonthius *et al.*, 1994). Any of these alterations in gene expression could play a role in the protection against secondary injury. It is important to note, however, that the time course of increase in growth factor expression does not correspond in any obvious way with the period of induced tolerance. Specifically, electrical stimulation causes an increase in growth factor expression that peaks within few hours after a seizure (Steward *et al.*, 1991; Steward 1994; Bonthius *et al.*, 1994). Nevertheless, it is possible that the key factor responsible for tolerance induction is expressed with a similar time course as the growth factors which were evaluated. Further studies are required to evaluate this possibility.

Altogether, our results indicate that the mechanisms by which electrical stimulation induces protective and/or inhibitory effects against secondary spinal cord injury are quite complex. Rather, multiple factors may play a role which contribute to neuronal and glial activation resulting in expression of specific genes for various neurotrophic factors (NGF, BDNF or GDNF, for example). Yet, we have not obtained so far any direct evidence on the nature of genes which are expressed and involved in the process of secondary injury. Further experiments must be made to answer this question. Nevertheless, the present findings have revealed that electrical stimulation of the spinal cord has a powerful therapeutical potential to afford injury tolerance specifically at frequencies of 500 Hz. Although, no evidence is so far available, e.g. on the details of genomic activation preventing the process of secondary necrosis formation and cavitation induced by this mode of electrical stimulation, we surmise that a protective cascade of mechanisms was induced, rendering the spinal cord more resistant against traumatic injury.

References

1. Bonthius DJ, Stringer JL, Lothman EW, Steward O (1994) Spreading depression and reverberatory seizures induce the upregulation of mRNA for glial fibrillary acidic protein. Brain Res 645: 215–224
2. Fujiki M, Zhang Z, Guth L, Steward O (1996) Genetic influences on cellular reactions to spinal cord injury: activation of macrophages/microglia and astrocytes is delayed in mice carrying a mutation (WldS) that causes delayed Wallerian degeneration. J Comp Neurol 371: 469–484
3. Fujiki M, Steward O (1997) High frequency transcranial magnetic stimulation mimics the effects of ECS in upregulating astroglial gene expression in the murine CNS. Brain Res Mole Brain Res 44: 301–308
4. Steward O, Torre ER, Tomasulo R, Lothman E (1991) Neuronal activity up-regulates astroglial gene expression. PNAS 88: 6819–6823
5. Steward O (1994) Electroconvulsive seizures upregulate astroglial gene expression selectively in the dentate gyrus. Brain Res Mol Brain Res 25: 217–224
6. Zhang Z, Fujiki M, Guth L, Steward O (1996) Genetic influences on cellular reactions to spinal cord injury: a wound-healing response present in normal mice is impaired in mice carrying a mutation (WldS) that causes delayed Wallerian degeneration. J Comp Neurol 371: 485–495

Correspondence: Minoru Fujiki, Department of Neurosurgery, Oita Medical University, 1-1, Idaigaoka, Hasama-machi, Oita, Japan 879-5593. e-mail: fujiki@oita-med.ac.jp

Acta Neurochir (2003) [Suppl] 86: 399–405

Low molecular weight compounds with affinity to melanocortin receptors exert neuroprotection in spinal cord injury – an experimental study in the rat

H. S. Sharma[1], **T. Lundstedt**[2], **M. Flärdh**[2], **J. Westman**[1], **C. Post**[2], and **A. Skottner**[2]

[1] Laboratory of Neuroanatomy, Department of Medical Cell Biology, Biomedical Centre, Uppsala University, Uppsala, Sweden
[2] Melacure Therapeutics AB, Uppsala, Sweden

Summary

The possibility that five new low molecular weight compounds with varying affinity and selectivity to the melanocortin receptors will exert neuroprotective effects in the spinal cord injury (SCI) induced edema formation and cell damage was examined in a rat model. A focal trauma of the rat spinal cord made by an incision into the right dorsal horn (T10–11) resulted in profound edema formation, leakage of Evans blue albumin and cell injury of the T9 segment at 5 h. Topical application of the Melacure compound ME10501 in high doses (10 μg in 10 μl) given 5 min after SCI resulted in most significant neuroprotection of the T9 segment of the cord compared to other compounds. Thus, marked reduction in water content, leakage of Evans blue albumin, and cell injury were observed in ME10501 treated traumatised rats. These observations suggest that the non-peptide compound ME10501 with affinity to the melanocortin receptor MC4 is capable to induce neuroprotection in the spinal cord following trauma not reported earlier.

Keywords: Melanocortin receptors; a-MSH; spinal cord injury; edema formation; blood-spinal cord barrier; cell injury; pharmacotherapy.

Introduction

Melanocortin peptides, such as α-melanocyte stimulating hormone (α-MSH) and adrenocorticotropin (ACTH) or its fragments are able to influence the neuronal development [19], to exert neurotrophic effects in damaged peripheral nerves [32] as well as functional recovery in injured peripheral and central nervous tissues [10, 23, 24]. Local administration of melanocortin peptides improves the functional recovery of peripheral and central nervous tissues [23, 24]. This indicates that melanocortin peptides are effective at the lesion site.

Local administration of ACTH4-10, α-MSH after a chemical lesion of the nucleus accumbens or in the septal regions accelerates functional recovery in rats [23, 24]. However, the molecular mechanisms behind such beneficial effects of melanocortin peptides are not well understood. It may be that melanocortins are acting on several melanocortin receptors identified recently to mediate such beneficial effects. In the rat brain, a MSH receptor (MC1-R), ACTH receptor (MC2-R), MC3-R, and MC4-R have recently been isolated and characterized [3, 5, 8, 11]. A fifth receptor, MC5-R has also been identified in the periphery and in the brain [2, 3]. It seems likely that these multiple melanocortin receptors play important roles in neurotrophic or neuroprotective effects in the brain as well as in the periphery following damage.

Recently, MC4-R has been identified in the rat spinal cord and the dorsal root ganglia [20]. An upregulation of spinal melanocortin receptors was shown in a rat model of neuropathic pain, which could be blocked by antagonists to MC4-R [22]. This suggests that MC4-R plays an important role in allodynia. The endogenous agonist of melanocortin receptors is α-MSH which is released in nociception-associated areas in the spinal cord [2, 4]. It has recently been suggested that α-MSH may have a therapeutic value for the treatment of brain injury [12]. However, its role in spinal cord injury is still not well known. It may be that MC4-R regulation is altered in spinal cord injury. Thus, alterations in melanocortin receptors may influence the pathophysiology of SCI.

Melacure has developed several low molecular weight, non-peptide compounds with varying affinity and selectivity for the melanocortin receptors [1]. In the present investigation, the efficacy of some of these compounds was evaluated in focal spinal cord trauma induced edema formation and cell injury in a well-established rat model [14].

Material and methods

Animals

Experiments were carried out on 136 male Sprague Dawley rats (200–260 g) housed at controlled room temperature (21 ± 1 °C) with a 12 h light and 12 h dark schedule. Food and tap water were supplied ad libitum before the experiment.

Spinal cord injury

Under Equithesin anaesthesia (3 ml/kg, i.p.), one-segment laminectomy was performed at the T10–11 level. Spinal cord injury (SCI) was inflicted by an incision into the right dorsal horn of the T10–11 segments by a sterilised scalpel blade [14]. The animals were allowed to survive 5 h after the injury. Equithesin anaesthetised normal animals were used as controls. This experimental condition is approved by the Ethics Committee of Uppsala University, Uppsala, Sweden.

Administration of low molecular compounds

Five different compounds identified as ME10092 (MS-1), ME10354 (MS-2), ME10393 (MS-3), ME10431 (MS-4) and ME10501 (MS-5) were used. The compounds had different affinities to the melanocortin receptors. Each compound was dissolved in saline and tested at 3 different dose-levels, i.e. 1 µg, 5 µg and 10 µg total dose using topical application to the spinal cord. The compounds were administered in a total volume of 20 µl 5 min after the injury, and the animals allowed to survive 5 h.

Histopathology of the spinal cord

The animals were perfused 5 h after SCI and the spinal cord was taken out, photographed for visible swelling and processed for histological evaluation using a standard protocol [15]. About 3 µm thick paraffin sections were cut of the T9 segment, stained with Haematoxyline and Eosin, and examined under a light microscope for sponginess, edema formation, cell injury, and cell death. For each parameter a rough score was assigned from 1 (least) to 4 (maximum) (Fig. 2). Analyses of structural changes were done in a blind fashion.

Spinal cord water content

The spinal cord water content was determined from the differences in wet and dry weight of the samples as described earlier [15, 16].

Blood-spinal cord barrier permeability

The blood-spinal cord barrier (BSCB) was examined using Evans blue albumin extravasation into the parenchyma as described earlier [16]. Evans blue albumin (2% of a 3 ml/kg solution) was administered into the right femoral artery 5 min before termination of the experiment [13]. The dye entering spinal cord tissue of the T9 and T12 segments was measured colorimetrically [13, 16].

Statistical analyses

ANOVA followed by Dunnet's test was used to evaluate statistical significance of the quantitative data obtained. A P value less than 0.05 was considered significant.

Results

Affinity and in vitro activity of the low molecular weight compounds to melanocortin receptors

The affinities of the five compounds were assayed in vitro using B16 mouse melanoma cells for the MC1R, and insect cells transiently transfected with the human MC3-, MC4- and MC5-receptors, respectively. The highest affinity for all compounds was shown to the MC1R (Ki = 0.4–4.5 µM). The compound ME10501 exhibited high affinity to MC4R (Ki = 0.9 µM), but no affinity for MC3R or MC5R. The affinity of other compounds which were tested was variable.

The *in vitro* activity of the compounds was assessed by measuring the increase in intracellular cAMP after incubation of either B16 mouse melanoma cells (for MC1R effects) or of transfected CHO cells expressing the MC3–5 receptors, respectively. Compounds ME10092, ME10354, ME10431 and ME10501 did not exert any significant agonistic effect with via MC1R, whereas ME10393 significantly stimulated cAMP formation in cells expressing MC1R. The stimulatory (agonistic) activity exerted via MC4R was observed with all compounds to a variable degree, except with ME10092.

Effect of the low molecular weight compounds on gross pathology of the spinal cord

The spinal cord of the untreated traumatised rats exhibited marked visible swelling at 5 h, microhaemorrhages, and edema around the lesion site (Fig. 2b). The perifocal regions of the spinal cord also appeared swollen in most of the cases (Fig. 2b). These pathological changes were considerably reduced in animals with topical application of ME10501 (10 µg in 10 µl) at 5 after SCI (Fig. 2). This effect of ME10501 was dosed dependent (results not shown). On the other hand, the other compounds ME10092, ME10354, ME10393 and ME10431 applied to the cord at 3 different doses had only a mild or no effect on the cord pathology (results not shown).

Effects of the low molecular weight compounds on edema formation and cell injury

Blind scoring of cell damage, cell loss, sponginess, and edema using an arbitrary unit from 1 (minimum)

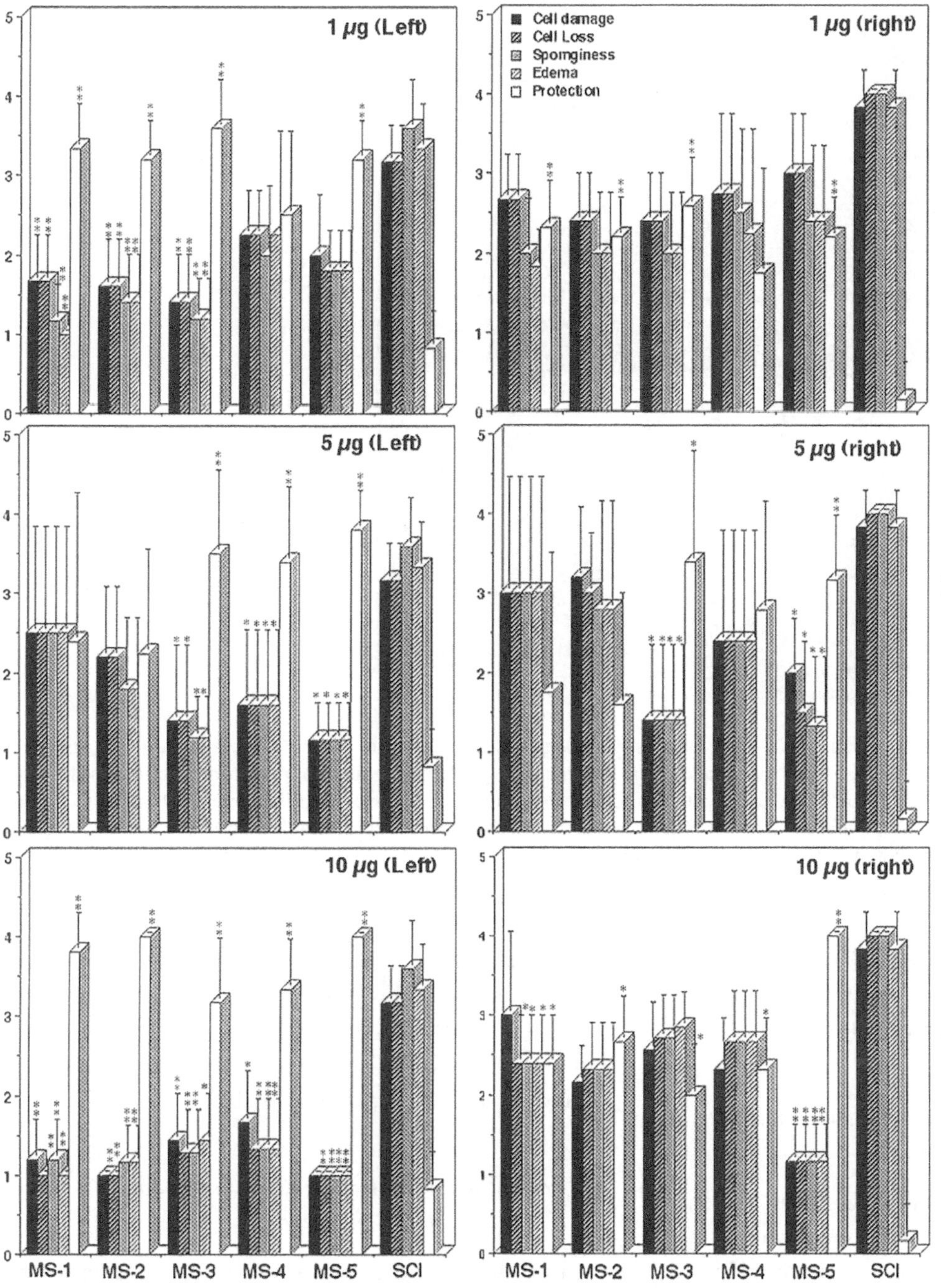

Fig. 1. Semiquantitative analysis of edema formation and cell changes in the ipsilateral (right) and contralateral (left) sides of the spinal cord following trauma and their modification with low molecular weight compounds (for details see text). * $P < 0.05$, ** $P < 0.01$, ANOVA followed by Dunnet's test for multiple group comparison

to 4 (maximum) revealed maximum cell damage at the right side of the T9 segment as compared to the left side in untreated animals of the traumatised group (Fig. 1). Interestingly, 10 μg of ME10501 were equally effective to afford neuroprotection in both the right and left side of the cord (Fig. 1). Neuroprotection by ME10501 (5 μg) was more effective in the contralateral side (Fig. 1). On the other hand, the low dose of ME10501 (1 μg) did not show any significant neuroprotection (Fig. 1) as compared to animals of the untreated group. The other compounds (ME10092, ME10354, ME10393 and ME10431) at high doses (10 μg) revealed moderate degrees of neuroprotection in the contralateral cord but not in the ipsilateral side (Fig. 1). ME10393, however, at 5 μg was able to provide mild neuroprotection in both sides of the cord

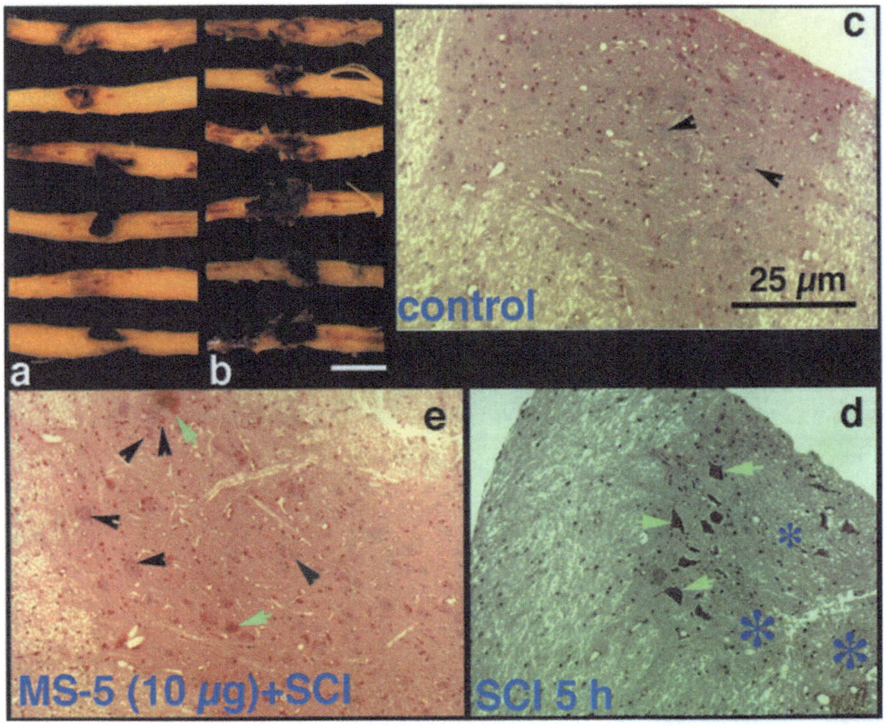

Fig. 2. Effect of ME10501 on spinal cord injury (*SCI*) induced gross pathology and cell changes in the ventral horn of the T9 segment compared to the untreated injured cords. Microhaemorrhages and visual swelling are significantly reduced in drug treated traumatised rats (a) compared to the untreated injured cords (b). Many dark and distorted neurones can be seen (arrows) in the untreated traumatised rat (d) compared to normal control group. (c). In ME10501 treated injured rats, nerve cell changes were considerably reduced (e) and signs of edema and sponginess (*) was less apparent. Bar: a, b = 5 mm; c, d, e = 25 μm

(Fig. 1). The low doses of ME10092, ME10354 and ME10393 (1 μg) exhibited a mild degree of neuroprotection in the contralateral cord (Fig. 1).

Effects of low the molecular weight compounds on fine structure of the spinal cord

Many dark and distorted cells were present in the left ventral horn of the T9 segment after 5 h SCI (Fig. 2d) compared to the control group (Fig. 2c). A general expansion of the cord, sponginess and edematous swelling of the neuropil was evident in the injured group (Fig. 2d). Topical application of ME10501 (10 μg) 5 min after injury markedly thwarted the edema development, cell damage, cell loss, and sponginess in the T9 segment of the cord (Figs. 1 and 2).

Effect of the low molecular weight compound ME10501 on spinal cord water content

Since ME10501 at the moderate (5 μg) and high (10 μg) dosis was able to afford the most pronounced neuroprotection in SCI, the spinal cord water content was measured in this group only. Treatment of animals with ME10501 significantly attenuated the cord edema in the T9 and T12 segments at 5 h. This effect was found to be dose dependent (Table 1).

Effect of the low molecular weight compound ME10301 on BSCB permeability

In order to understand the nature of edema, extravasation of Evans blue albumin in the cord tissue was examined in ME10501 treated animals with trauma. Topical application of ME10501 significantly attenuated extravasation of Evans blue into the T9 and the T12 segments of the cord in a dose dependent fashion (Table 1).

Discussion

The salient new finding of the present investigation is a pronounced neuroprotection of the new low-molecular weight compound ME10501 with high affinity to MC4-R, affecting edema, microvascular per-

Table 1. *Effect of ME10501 (MS-5) on spinal cord water content and BSCB permeability following spinal trauma in rats*

Expt. type	n	Spinal cord water %		Evans blue mg %	
		T9	T12	T9	T12
Control	6	65.46 ± 0.22	65.59 ± 0.15	0.28 ± 0.02	0.28 ± 0.03
SCI 5 h	5	68.23 ± 0.31**	68.63 ± 0.38**	1.15 ± 0.10**	1.18 ± 0.15**
MS-5 10 μg + SCI	6	66.27 ± 0.46*a	66.09 ± 0.21*a	0.56 ± 0.06*a	0.64 ± 0.08*a
MS-5 5 μg + SCI	6	66.99 ± 0.15**a	66.98 ± 0.31**a	0.76 ± 0.01**a	0.81 ± 0.04**a

Values are mean ± SD; ** = $P < 0.01$; * $P < 0.05$; Significantly different from Control; a = $P < 0.05$ significantly different from SCI; ANOVA followed by Dunnet's test for multiple group comparison.

meability disturbances, and cell injury following focal injury of the spinal cord. The other Melacure compounds at different doses had only mild or no effect on cell injury in the spinal cord. This observation suggests that compounds with affinity to MC4-R are neuroprotective in SCI, not reported earlier.

Application of ME10501 in high doses resulted in superior neuroprotection in SCI. Failure of other compounds to provide similar neuroprotection in SCI suggest that the affinity and specificity for melanocortin receptors are important for neuroprotection, which requires further investigation.

The neuroprotective effects of MC4-R receptor agonists in the present study were achieved by topical application of ME10501 onto the spinal cord. This observation is in line with previous studies in which local administration of melanocortins, α-MSH, and related peptides resulted in accelerated recovery of function following a sciatic nerve lesion in rats [9, 10, 18]. An improvement of the spinal cord function, both neurologically and electrophysiologically, in rats treated with α-MSH after experimental SCI further supports this hypothesis [20].

Another important observation of this study is that the low-molecular weight compounds when administered 5 minutes after SCI were able to thwart edema formation and cell injury in the spinal cord. This suggests that the early intervention with compounds active at melanocortin receptors seems to be beneficial. Further research using administration of potential melanocortin receptor agonists at different time intervals after trauma is needed to fully explore their therapeutic potential.

That melanocortin receptors are involved in brain injury is apparent from the fact that in hypoxic ischemia an upregulation of MC4-R mRNA occurs in the striatum of the non-injured hemisphere compared to controls [9]. Likewise, in the spinal cord expression of MC4-R occurs in animal models of neuropathic pain [22, 23]. This indicates that MC4-R receptors play an important role in disease processes in the spinal cord. The neuroprotective efficacy of the melanocortin receptor agonist ME10501 in spinal cord trauma in this study may suggest a possible upregulation of MC4-R expression under these conditions. Thus, it would be interesting, whether focal SCI influences expression of melanocortin receptors in our model.

The most pronounced neuroprotective effect of ME10501 was observed in the contralateral side of the spinal cord following injury. This suggests that melanocortin receptor agonists are more effective in attenuating the trauma induced spread of cell injury in the vicinity of a lesion. Alternatively, melanocortin receptors have the capacity to influence neuronal recovery in the uninjured side of the cord. Observations on the functional recovery in the contralateral brain in a model of unilateral hypoxic-ischemic brain injury using melanocortins are in line with this idea [9].

The significant reduction in Evans blue leakage across the BSCB following trauma by the MC4-R agonist compound ME10501 suggests that melanocortin receptors are somehow involved in the permeability disturbances of the spinal cord endothelial cells. The permeability of the BSCB is enhanced by several neurochemical mediators, such as serotonin, prostaglandins, histamine, bradykinin and other opioid peptides [13, for details see 13, 16, 17]. To our knowledge, the involvement of melanocortin receptors or melanocortin peptides in modulating BSCB permeability disturbances has not been reported earlier. Thus, this study appears to be the first to show an involvement of melanocortin receptors in modulating BSCB permeability and edema formation in SCI.

The significant inhibition of edema formation and cell injury following application of MC4-R agonist in SCI suggests that ME10501 has the capacity to attenuate BSCB permeability disturbances in SCI. Obviously, the less severe disruption of the BSCB permeability following trauma attenuates the abnormal alterations of the fluid microenvironment of the cord parenchyma, thereby reducing edema formation and cell damage [13, 15, 16].

The understanding of mechanisms involved in inhibiting the BSCB permeability changes by ME10501 is still speculative. It seems likely that cAMP mediated effects of melanocortins play a role [7]. In rat brain subcortical areas, ACTH has a biphasic effect on adenyl cyclase activity. In rat cortical slices, a dose dependent effect of ACTH on cAMP levels has been described [23]. Thus, it seems likely that ME10501 exerts its beneficial effects on BSCB permeability through cAMP dependent signal transduction pathways in the spinal cord endothelial cells. The fact that binding of endogenous ACTH and α-MSH to a putative ACTH/MSH receptor in certain brain regions and in spinal cord dorsal root ganglion activates specific signal transduction pathways involving cAMP as second messenger [4, 6] supports this idea. An increase in cAMP or cGMP levels of endothelial cells often is associated with an impaired microvascular permeability [see 7, 13, 16, 17].

In conclusion, our study provides new evidence that melanocortins are involved in the pathophysiology of spinal cord trauma and that topical application of the MC4-R agonist ME10501 exerts profound neuroprotection in spinal cord injury.

Acknowledgments

This study is supported by Grants from Melacure Therapeutics, Uppsala, Sweden and The Swedish Medical Research Council (2710). Expert technical assistance of Kärstin Flink is highly appreciated.

References

1. Andersson PM, Boman A, Seifert E, Skottner A, Lundstedt T (2001) Ligands to the melanocortin receptors. Expert Opin Ther Patents 11(10): 1583–1592
2. Barrett P, MacDonald A, Helliwell R, Davidson G, Morgan P (1994) Cloning and expression of a new member of the melanocyte-stimulating hormone receptor family. J Molec Endocr 12: 203–213
3. Chhajlani V, Wikberg JE (1992) Molecular cloning and expression of the human melanocyte stimulating hormone receptor cDNA. FEBS Lett 309: 417–420
4. Florjin WJ, Mulder AH, Versteeg DHG, Gispen WH (1993) Adrenocorticotropin/a-melanocyte-stimulating hormone (ACTH/MSH)-like peptides modulate adenylate cyclase activity in rat brain slices: evidence for an ACTH/MSH receptor mediated mechanism. J Neurochem 6: 2204–2211
5. Gantz I, Konda Y, Tashiro T, Shimoto Y, Miwa H, Munzert G, Watson SJ, DelValle J, Yamada T (1993) Molecular cloning of a novel melanocortin receptor. J Biol Chem 268: 8246–8250
6. Hol EM, Verhage M, Gispen WH, Bar PR (1994) The role of calcium and cAMP in the mechanism of action of two melanocortins: a MSH and the ACTH 4–9 analogue Org 2766. Brain Res 662: 109–116
7. Joó F (1985) The blood-brain barrier in vitro: Ten years of research on microvessels isolated from the brain. Neurochem Int 7: 1–25
8. Mountjoy KG, Mortrud MT, Low MJ, Simerley RB, Cone RD (1994) Localization of the melanocortin-4 receptor (MC4-R) in neuroendocrine and autonomic control circuits in the brain. Molec Endocr 8: 1298–1308
9. Mountjoy KG, Guan J, Elia CJ, Sirimanne ES, Williams CE (1999) Melanocortin-4 receptor messenger RNA expression is up-regulated in the non-damaged striatum following unilateral hypoxic-ischemic brain injury. Neuroscience 89: 183–190
10. Nyakas C, Veldhuis HD, De Wied D (1985) Beneficial effect of chronic treatment with ORG 2766 and a-MSH on impaired reversal learning of rats with bilateral lesions of the parafascicular area. Brain Res Bull 15: 257–265
11. Roselli-Rehfuss L, Mountjoy JG, Robbins LS, Mortrud MT, Low M, Tatro JB, Entwistle ML, Simerely R, Cone RD (1993) Identification of a receptor for g melanocortin and other proopiomelanocortin peptides in the hypothalamus and limbic system. Proc Natl Acad Sci USA 90: 8856–8860
12. Sarrafzadeh AS, Thomale U-W, Kroppenstedt S-N, Unterberg AW (2000) Neuroprotective effects of melatonin on cortical impact injury in the rat. Acta Neurochir (Wien) 142: 1293–1299
13. Sharma HS (2000) Bradykinin/B_2 receptor antagonist HOE-140 attenuates blood-spinal cord barrier permeability following trauma to the rat spinal cord. Acta Neurochir (Wien) [Suppl] 76: 159–163
14. Sharma HS, Olsson Y (1990) Edema formation and cellular alterations following spinal cord injury in rat and their modification with p-chlorophenylalanine. Acta Neuropathol (Berlin) 79: 604–610
15. Sharma HS, Olsson Y, Cervós-Navarro J (1993) Early perifocal cell changes and edema in traumatic injury of the spinal cord are reduced by indomethacin, an inhibitor of prostaglandin synthesis. Acta Neuropathol (Berlin) 85: 145–153
16. Sharma HS, Olsson Y, Pearsson S, Nyberg F (1995) Trauma induced opening of the blood-spinal cord barrier is reduced by indomethacin, an inhibitor of prostaglandin synthesis. Experimental observations in the rat using ^{131}I-sodium, Evans blue and lanthanum as tracers. Restorative Neurol Neurosci 7: 207–215
17. Sharma HS, Alm P, Westman J (1998) Nitric oxide and carbon monoxide in the pathophysiology of brain functions in heat stress. Brain functions in hot environment. In: Sharma HS, Westman J (eds) Progr Brain Res 115: 297–333
18. Strand FL, Kung TT (1980) ACTH accelerates recovery of neuromuscular function following crushing of peripheral nerve. Peptides 1: 135–138
19. Strand FL, Rose KJ, Zuccarelli LA, Kume J, Alves SE, Antonawich FJ, Garrett LY (1991) Neuropeptide hormones as neurotrophic factors. Physiol Rev 71: 1017–1046
20. van der Kraan M, Tatro JB, Entwistle ML, Brakkee JH, Burbach JP, Adam RA, Gispen WH (1999) Expression of melanocortin receptors and pro-opiomelanocortin in the rat spinal cord

in relation to neurotrophic effects of melanocortins. Mol Brain Res 63: 276–286

21. van de Meent H, Hamers FP, Lankhorst AJ, Joosten EA, Gispen WH (1997) Beneficial effects of the melanocortin alpha-melanocyte stimulating hormone on clinical and neurophysiological recovery after experimental spinal cord injury. Neurosurgery 40(1): 122–130
22. Verhaagen J, Edwards PM, Jennekens FGI, Schotman P, Gispen WH (1986) α-melanocyte-stimulating hormone stimulates the outgrowth of myelinated nerve fibres after peripheral nerve crush. Exp Neurol 92: 451–454
23. Wiegent VM, Dunn AJ, Schotman P, Gispen WH (1979) ACTH-like neurotrophic peptides: possible regulators of brain cyclic AMP. Brain Res 168: 565–584
24. Woolterink G, Van Zanten E, Van Ree JM (1990) Functional recovery after destruction of dopamine system in the nucleus accumbens of rats. IV. Delay by intra-accumbal treatment with ORG 2766 or a-MSH-antiserum. Brain Res 507: 115–120

Correspondence: Hari Shanker Sharma, Dr. Med. Sci., Frödingsgatan 12, app: 28; SE-75421 Uppsala, Sweden. e-mail: Sharma@medcellbiol.uu.se

Acta Neurochir (2003) [Suppl] 86: 407–413

Topical application of TNF-α antiserum attenuates spinal cord trauma induced edema formation, microvascular permeability disturbances and cell injury in the rat

H. S. Sharma[1], T. Winkler[2], E. Stålberg[2], T. Gordh[3], P. Alm[4], and J. Westman[1]

[1] Laboratory of Neuroanatomy, Department of Medical Cell Biology, Biomedical Center, Uppsala University, Uppsala, Sweden
[2] Department of Clinical Neurophysiology, University Hospital, Uppsala University, Uppsala, Sweden
[3] Department of Anaesthesiology, University Hospital, Uppsala University, Uppsala, Sweden
[4] Department of Pathology, University Hospital, Lund University, Lund, Sweden

Summary

The possibility that antiserum to tumour necrosis factor-alpha (TNF-α) is neuroprotective in spinal cord injury (SCI) was examined in a rat model. SCI was produced by making an incision into the right dorsal horn at the T10–11 segments. Topical application of TNF-α antiserum at three concentrations (1:10; 1:50 and 1:100) given 30 min before or 2 min, 5 min or 10 min after trauma resulted in marked reduction in visible swelling, edema formation, and leakage of radiolabelled iodine tracer within the T9 and T12 segments at 5 h in a dose dependent manner. This neuroprotective effect was most pronounced when the antiserum at the highest dose level (1:10) was applied 10 min after SCI. The TNF-α antiserum also reduced the SCI induced upregulation of neuronal nitric oxide synthase (nNOS) immunoreactivity in a concentration dependent manner. Taken together, these results suggest that local application of TNF-α antiserum is neuroprotective in SCI and that this effect is mediated through NOS regulation.

Keywords: TNF-α; spinal cord injury; nitric oxide synthase; edema formation; blood-spinal cord barrier permeability; TNF-α antiserum.

Introduction

Tumour necrosis factor alpha (TNF-α) is considered to be a cytotoxic cytokine in the central nervous system (CNS) [23]. Upregulation of TNF-α occurs following several kinds of traumatic, ischemic or hypoxic insults to the brain [6, 7, 23]. Thus, traumatic brain injury either caused by contusion or fluid percussion impact results in TNF-α production within 1 to 6 h after trauma [3, 23]. On the other hand, after focal cerebral ischemia, nerve cells may produce TNF-α from 6 h to several days, whereas microglial cells and macrophages express TNF-α 8 to 14 h after the insult [2, 9]. However, the role of TNF-α as neurotoxic substance in the spinal cord has not been examined in details.

Previous studies have shown that a focal trauma of the spinal cord induces upregulation of c-*fos* and TNF-α within 1 h after injury [4]. Exogenous supplementation of TNF-α may enhance neuroregeneration or exacerbate experimental autoimmune encephalomyelitis [5, 14]. These studies indicate that the effects of TNF-α on the spinal cord are time – and concentration dependent [13].

To understand the role of endogenous TNF-α in brain pathology antiserum to TNF-α has been used to neutralise its effect in vivo [for review see 22]. Thus, although the cytokine is produced following an insult to the CNS, its biological action is thwarted by its binding to the antibody. Based on this principle, it has been shown that chronic intravenous administration of TNF-α antiserum resulted in marked neuroprotection following experimental stroke in rats [6]. Similarly, in a mouse stroke model intraventricular administration of monoclonal TNF-α antiserum resulted in marked neuroprotection following middle cerebral artery occlusion with cerebral infarction [24]. These observations suggest that monoclonal anti-TNF-α antibodies are able to afford neuroprotection in ischemia and trauma. This indicates that TNF-α is neurotoxic in several pathological conditions. However, the detailed molecular and cellular mechanisms of neuronal damage caused by TNF-α are still not well known.

Recently, a marked increase in TNF-α immunoreactivity was reported in the cord following 4 h after compression injury [7]. The TNF-α upregulation is associated with an increased level of nitric oxide (NO), indicating that the neurotoxicity of TNF-α is mediated by formation of NO. This is supported by findings that application of TNF-α antiserum in this model significantly down regulates NOS expression and cell damage, indicating that neuroprotection by TNF-α antiserum is mediated by inhibition of NOS upregulation.

Previous reports of our laboratory have shown that a focal trauma of the rat spinal cord is able to induce profound upregulation of NOS at 5 h associated with marked edema formation, breakdown of the blood-spinal cord barrier (BSCB) permeability, and cell injury [19, 21]. These cell changes, BSCB permeability, edema formation and NOS upregulation were significantly attenuated by topical application of NOS antiserum [19], which indicates that NOS upregulation is an important factor in SCI.

In the present investigation we wanted to examine whether application of TNF-α antiserum in our model of SCI involving a lesion of the dorsal horn is capable to attenuate cell injury, and if so, whether the treatment is able to reduce edema formation and microvascular permeability disturbances as well.

Material and methods

Animals

Experiments were carried out in 66 male Sprague Dawley rats (260–280 g) housed at controlled room temperature 21 ± 1 °C with a 12 h light and 12 h dark schedule. Food pellets and tap water were supplied ad libitum before the experiment.

Spinal cord injury

SCI was induced in Equithesin anaesthetized rats at the T10–11 level by making an incision of the right dorsal horn. The animals were allowed to survive 5 h after the injury [17, 18]. Normal rats served as controls. The experimental protocol was approved by the Ethics Committee of Uppsala University, Uppsala Sweden.

Treatment with TNF-α antiserum

Antiserum to TNF-α (monoclonal, Sigma Chemical Co, USA) was applied in dilutions of 1:10; 1:20, 1:50 and 1:100 in phosphate buffered saline 30 min before, 2 min, 5 min and 10 min after injury onto the traumatized spinal cord (Total volume 10 µl). The animals were allowed to survive 5 h after the SCI (Fig. 1).

Parameters measured

The following parameters were measured in control, spinal cord traumatised, and TNF-a treated sham or injured animals.

Spinal cord edema formation

Edema formation was examined in the T9 and T12 segments by measuring spinal cord water content. The spinal cord water content was determined from the differences in the wet and dry weight of the spinal cord samples as described earlier [17].

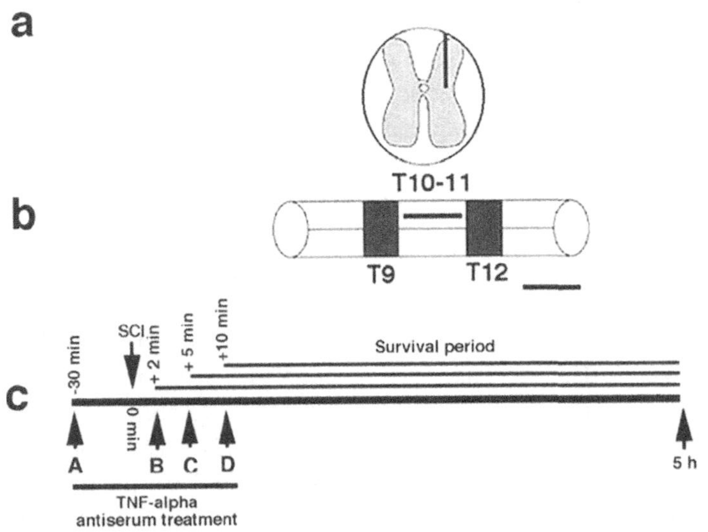

Fig. 1. Spinal cord injury (*SCI*) method and schedule of TNF-α antiserum treatment. a = SCI was produced by making an incision into the right dorsal horn (about 2 mm deep) of the T10–11 segments. b = Spinal cord tissue samples were taken from the T9 and T12 segments for analysis of edema formation and cell injury (bar = 5 mm). c = Schedule of TNF-α antiserum treatment. Topical application of TNF-α antiserum was made either 30 min before injury (*A*) or 2 min (*B*), 5 min (*C*) or 10 min (*D*) after SCI (for details see text)

Blood-spinal cord barrier permeability

The BSCB permeability in the T9 and T12 segments was examined by using two protein tracers, i.e., Evans blue albumin and [131]Iodine. These tracers were administered intravenously 5 min before sacrifice and the tracer distribution in the tissue was determined using standard procedures [15, 18].

NOS immunohistochemistry

NOS immunostaining was examined in the T9 and T12 segments using a monoclonal antiserum to the constitutive isoform of NOS [1] as described earlier [19, 21]. In brief, free floating 40 μm thick Vibratome sections were cut from the T9 segment and processed for NOS immunostaining. The immunoreactivity was visualized using a peroxidase-antiperoxidase technique as described earlier [19].

Spinal cord cell injury

The spinal cord cell injury was examined by standard light and electron microscopy. For high resolution light microscopy, 1 μm thick sections were cut and stained with toloudine blue and examined for gross cell changes in the spinal cord. The desired portions of spinal cord gray matter mainly from the contralateral dorsal and ventral horns of the T9 and T12 segments were cut on the ultramicrotome (LKB, Sweden), counterstained with lead citrate and uranyl acetate and examined using a Philips 400 electron microscope.

Statistical analysis

ANOVA followed by Dunnet's test was applied to evaluate the statistical significance of the quantitative data obtained. The semiquantitative data were examined by a non-parametric Chi-Square test. A p value less than 0.05 was considered to be significant.

Results

Effect of TNF-α on spinal cord edema formation

The focal trauma of the rat spinal cord induced marked edema formation as seen by the swelling of the spinal cord at 5 h. The visible swelling was markedly attenuated by the topical application of TNF-α antiserum in a dose- and time dependent manner. The most pronounced effect of TNF-α antiserum was obtained when the anti-TNF-α antibodies were applied in high concentration (1 : 10) 10 min after SCI (results not shown).

The observations on visible swelling were further confirmed by measurement of the spinal cord water content (Fig. 2). The application of TNF-α antiserum resulted in a concentration and time dependent attenuation of the spinal cord water content following SCI in both the T9 and the T12 segments (Fig. 2). The most marked effect of the TNF-α antiserum was seen at 10 min after application onto the traumatised spinal cord in high concentration (1 : 10) (Fig. 2).

Effect of TNF-α on BSCB permeability

The focal trauma resulted also in profound leakage of Evans blue into the spinal cord tissue as seen by visual inspection of the cord as well as by the colorimetrical measurements (results not shown). The leakage of Evans blue was considerably reduced by application of TNF-α antiserum in a time and concentration dependent manner. The most pronounced effect of the antiserum against TNF-α on Evans blue extravasation was seen at 10 min after application in high concentration (1 : 10) (results not shown).

The extravasation of radioiodine tracer into the cord parenchyma had a similar pattern in SCI (Fig. 2). Thus, application of TNF-α antiserum significantly reduced the uptake of radioiodine into the spinal cord T9 and T12 segments after SCI. The effect was again most marked when the antiserum was applied in high concentration (1 : 10) 10 min after the lesion (Fig. 2).

Effect of TNF-α on NOS immunohistochemistry

SCI resulted in marked upregulation of cNOS in the T9 and T12 segments (Figs. 2 and 3 A). The upregulation of NOS in the untreated injured animals was mainly confined to the edematous region of the spinal cord gray matter (Fig. 3 A). The topical application of TNF-α antiserum significantly attenuated the NOS upregulation in these cord segments after injury (Fig. 2). Thus, in TNF-α antiserum treated injured rats the intensity of NOS upregulation was considerably reduced (Fig. 3 A). This effect was most pronounced in rats receiving the antiserum 10 min after SCI in high concentration (1 : 10) (Figs. 2 and 3 A). In addition, in the TNF-α antiserum treated rats the expansion of spinal cord- and cell injury was also considerably reduced (Fig. 3 A).

Effect of TNF-α on spinal cord cell injury

SCI induced profound nerve cell and myelin damage as seen by light and electron microscopy (Fig. 3 B). Many dark and distorted nerve cells were seen in the T9 and T12 segments of untreated traumatised spinal cord (Fig. 3 B). The ultrastructural studies showed axonal damage, vesiculation of myelin, membrane disruption, and edema formation (Fig. 3 B:a,c). Treatment with TNF-α antiserum markedly reduced cell injury in the spinal cord and the nerve cell changes, as axonal damage and myelin vesiculation (Fig. 3 B).

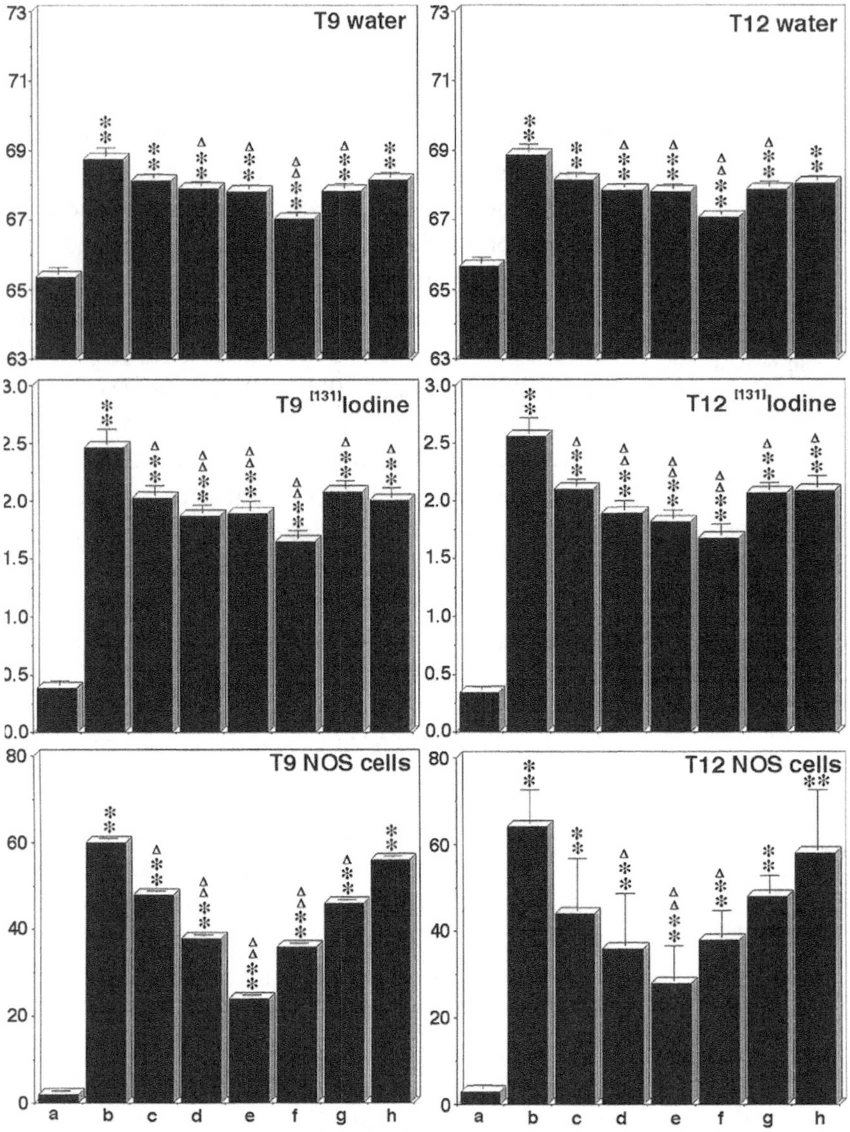

Fig. 2. Spinal cord water content (% upper panel), leakage of radiotracer across the blood-spinal cord barrier permeability (% middle panel) and upregulation of NOS activity (nr. lower panel) in the T9 (left side) and T12 (right side) segments of the spinal cord from untreated spinal cord injured rats and their modification with TNF-α antiserum application. a = control, b = 5 h SCI, c = −30 min, d = +5 min, e = +10 min TNF-α antiserum (1 : 10) application after SCI. TNF-α antiserum was applied 10 min after SCI at different dilution f = (1 : 20), g = (1 : 50) and h = (1 : 100) in phosphate buffered saline. Values are mean ± SD of 5–7 observations. ** $P < 0.01$ compared with control group (a). Δ $P < 0.05$, ΔΔ $P < 0.01$ compared with SCI (b) ANOVA followed by Dunnet's test for multiple group comparison

Thus, TNF-α antiserum given 10 min after SCI induced considerable neuroprotection in the spinal cord as seen by attenuation of the cellular damage when compared to the untreated traumatised rats (Fig. 3 B:b,d).

Discussion

The present findings show that endogenous TNF-α is neurotoxic in SCI (that contributes to neuronal and axonal injury. This idea is supported by the results that topical application of TNF-α antiserum remarkably reduced the pathological consequences of SCI. The effect of TNF-α antiserum was most marked 10 min after SCI indicating that it may have a therapeutic potential in spinal injury in the future.

Anti-TNF-α antibodies have been used in the past in neurological diseases in which TNF-α levels are elevated [see 22]. However, human trials using monoclonal murine TNF-α antibodies yielded mixed result [10]. The experimental results from this study further suggest that the usefulness of TNF-α antiserum

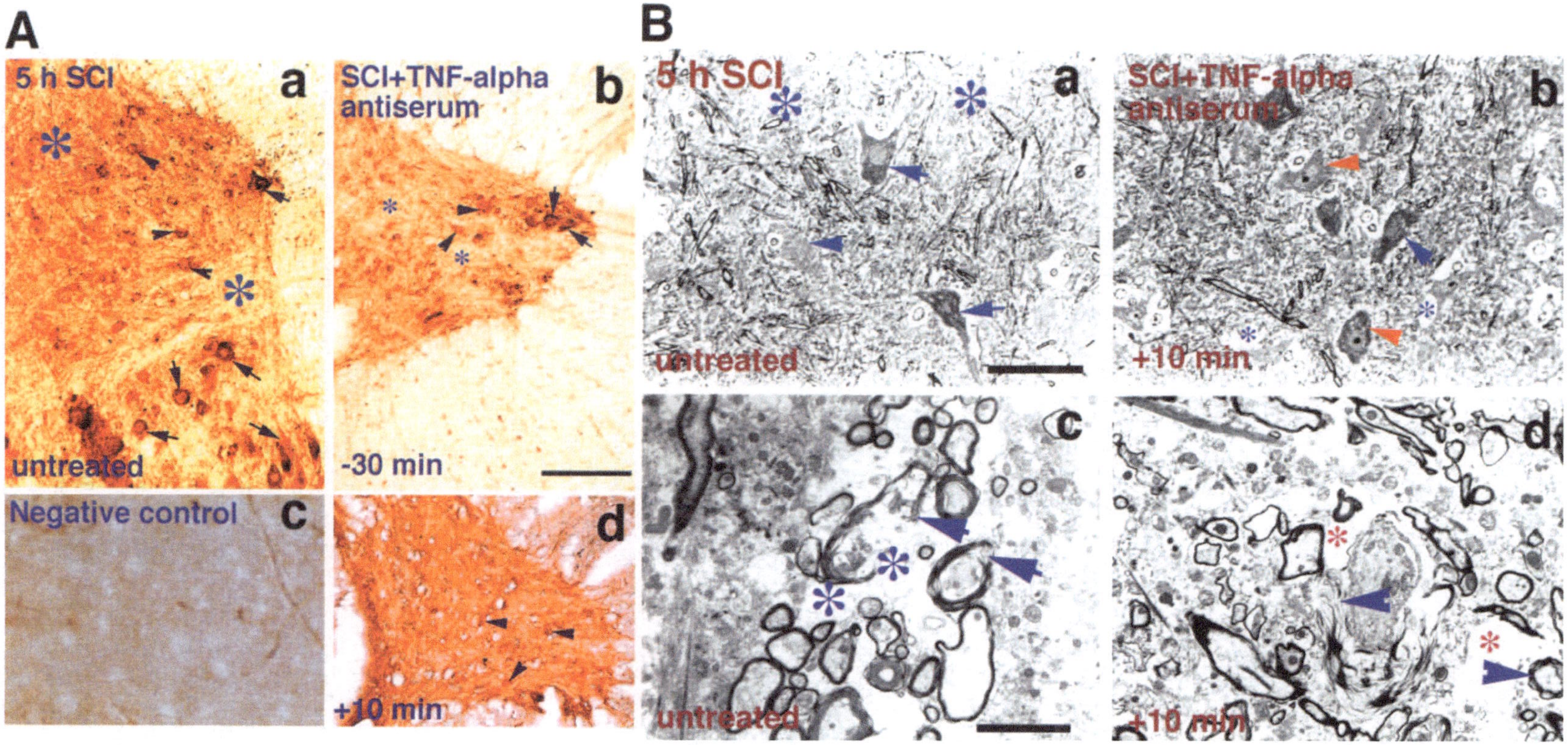

Fig. 3. NOS immunohistochemistry (A) and morphological changes (B) in SCI and their modification by TNF-α antiserum. Many NOS positive cells can be seen following 5 h SCI (A:a). Application of TNF-α antiserum (1 : 10) either 30 min before (b) or 10 min after (d) SCI markedly attenuated NOS activity. Negative control (c) exhibits some nonspecific background activity (bar = 20 μm). Low power light microscopy showed many dark nerve cells (arrows) in the ventral gray matter of the T9 segment in the untreated injured rats (B:a). Treatment with TNF-α antiserum 10 min after SCI markedly attenuated the number of damaged nerve cells (arrowheads) (B:b). The ultrastructural studies reveal widespread disruption of the neuropil in the untreated injured rats (B:c). Myelin damage (arrows) and edema (**) are quite prominent in animals of the untreated group. Topical application of TNF-α antiserum 10 min after SCI considerably reduced the ultrastructural changes in the neuropil (B:d) Bar: a, b = 30 μm, c, d = 1 μm

should be explored further in trauma or other neurodegenarative diseases.

Another important observation of this study is that TNF-α antiserum markedly reduces NOS upregulation following SCI indicating that this response is injurious to the cord. Our findings suggest that the expression of endogenous TNF-α is an important factor causing NOS upregulation in SCI.

Previously, upregulation of TNF-α was shown to occur at 1 h after compression injury of the cord [7]. An increased level of NO was also detected in this study at 4 h after SCI [7]. Administration of TNF-α antiserum several min after lesion into the center of the crushed spinal cord resulted in an attenuation of the NO level and of the number of apoptotic cells at 24 h after SCI [7, 8, 22].

Our study further suggests that the concentration of TNF-α antiserum and time of its application after SCI are important factors in reducing the pathological consequences of the SCI. The highest concentration of TNF-α antiserum (1:10) 10 min after SCI resulted in the most effective neuroprotection at 5 h, whereas the more diluted antisera (1:50 or 1:100) were not as effective to reduce the SCI induced cord pathologies and NOS upregulation. The properties of the antiserum to penetrate into the edematous cord tissue and its binding to the endogenous epitopes of TNF-α are likely to be instrumental in the antiserum induced neuroprotection [13].

In normal conditions, the transport of TNF-α through the BSCB is selective and restricted [11–13]. The transport system for TNF-α is upregulated after SCI [11, 12]. Moreover, SCI enhances TNF-α production both in spinal cord tissue as well as in the periphery [8]. Thus, it seems likely that antiserum against TNF-α given after SCI has a better access to its endogenous epitopes in the cord as compared to administration of the antibodies prior to injury.

It may be that the neutralization of TNF-α by the antiserum was also influencing downregulation of NOS. Obviously, upregulation of NOS associated with NO production is injurious to cells and membranes [21]. NO induces breakdown of the BBB permeability [for review see 16] and may thus induce vasogenic edema formation [20]. Leakage of serum proteins into the spinal cord is clearly evident after SCI, as seen by the uptake of Evans blue and radioiodine tracers. Protection of the normal BSCB permeability following TNF-α antiserum treatment suggests that endogenous TNF-α increases endothelial cell permeability through toxic actions on the cell membrane. Alternatively, a reduction in NOS upregulation by the TNF-α antiserum could also be responsible for the observed reduction in BSCB permeability.

An increased permeability of the spinal cord may result in an efflux and intra/extracellular translocation of several normally tightly controlled ions, as K^+ or Ca^{++}, neurochemicals as amino acids, and immunologically active material into the extracellular spinal cord fluid microenvironment. These factors together are likely to induce adverse cell reactions and profound damage of neuronal- and glial cells and myelin. The marked prevention of cell injury and of opening of the BSCB by TNF-α antiserum support this idea.

In conclusion, our observations suggest that TNF-α antiserum has the capacity to neutralise endogenous TNF-α *in vivo*, which causes neuroprotection. Furthermore, the resulting decrease in NOS upregulation is responsible for attenuating edema formation and microvascular permeability disturbances in SCI. However, additional research is needed to identify the optimal therapeutic window and dose level of the TNF-α antiserum to achieve neuroprotection in patients.

Acknowledgments

This investigation is supported by grants from Swedish Medical Research Council Nr. 2710 (JW/HSS); 11205 (PA/HSS); Alexander Humboldt Foundation, Bonn, Germany, The University Grants Commission, New Delhi, India and The Indian Council of Medical Research, New Delhi, India. Skilful technical assistance of Kärstin Flink and Kerstin Rystedt is acknowledged with thanks.

References

1. Alm P, Larsson B, Ekblad E, Sundler F, Andersson K (1993) Immunohistochemical localization of peripheral nitric oxide synthase-containing nerves using antibodies raised against synthesised C- and N-terminal fragments of a cloned enzyme from rat brain. Acta Physiol Scand 148: 421–429
2. Buttinin M, Appel K, Sauter A, Gebicke-Haerter PJ, Boddeke HW (1996) Expression of tumor necrosis factor alpha after focal cerebral ischemia in the rat. Neuroscience 71: 851–863
3. Fan L, Young PR, Barone FC, Feuerstein GZ, Smith DH, McIntosh TK (1996) Experimental brain injury induces differential expression of tumor necrosis factor alpha mRNA in the CNS Brain Res Molecular Brain Res 36: 287–291
4. Hayashi M, Ueama T, Nemoto K, Tamaki T, Senba E (2000) Sequential mRNA expression for immediate early genes, cytokines, and neurotrophins in spinal cord injury. J Neurotrauma 17: 203–218
5. Kuroda Y, Shimamoto Y (1991) Human tumor necrosis factor-α augments experimental allergic encephalomyelitis in rats. J Neuroimmunol 34: 159–164

6. Lavine SD, Hofman FM, Zlokovic BV (1998) Circulating antibody against tumor necrosis factor alpha protects rat brain from reperfusion injury. J Cereb Blood Flow Metab 18: 52–58
7. Lee YB, Yune TY, Bail SY, Shin YH, Du S, Rhim H, Lee EB, Kim YC, Shin ML, Markelonis GJ, Oh TH (2000) Role of tumor necrosis factor-α in neural and glial apoptosis after spinal cord injury. Exp Neurol 166: 190–195
8. Leskovar A, Moriarty LJ, Turek JJ, Schoenlein IA, Borgens RB (2000) Tha macrophage in acute neural injury: Changes in cell number over time and levels of cytokine production in mammalian central and peripheral nervous systems. J Exp Biol 203: 1783–1795
9. Liu T, Clark RK, McDonnel PC, Young PR, White RF, Barone FC, Feuerstein GZ (1994) Tumor necrosis factor-alpha expression in ischemic neurones. Stroke 25: 1481–1488
10. Marriot JB, Westby M, Dalgleish AG (1997) Therapeutic potential of TNF-α inhibitors old and new. Drug Discovery Today 2: 273–282
11. Pan W, Banks WA, Kastin AJ (1997) BBB permeability to ebiratide and TNF-α in acute spinal cord injury. Exp Neurol 146: 367–373
12. Pan W, Kastin AJ, Bell RL, Olson RD (1999) Upregulation of tumor necrosis factor-α transport across the blood-brain barrier after acute compressive spinal cord injury. J Neurosci 19: 3649–3655
13. Pan W, Kastin AJ (2001) Increase in TNF-α transport after SCI is specific for time, region, and type of lesion. Exp Neurol 170: 357–363
14. Schwartz M, Solomon A, Lavie V, Ben-Bassat S, Belkin M, Cohen A (1991) Tumor necrosis factor alpha facilitates regeneration of injured central nervous system axons. Brain Res 545: 334–338
15. Sharma HS (1987) Effect of captopril (a converting enzyme inhibitor) on blood-brain barrier permeability and cerebral blood flow in normotensive rats. Neuropharmacology 26: 85–92
16. Sharma HS, Alm P (2003) Role of nitric oxide in the blood-spinal cord and brain barriers following CNS injuries. In: Sharma HS, Westman J (eds) The blood-spinal cord and brain barriers in health and disease. Academic Press, San Diego, pp 191–229
17. Sharma HS, Olsson Y (1990) Edema formation and cellular alterations following spinal cord injury in rat and their modification with p-chlorophenylalanine. Acta Neuropathol (Berlin) 79: 604–610
18. Sharma HS, Olsson Y, Dey PK (1990) Early accumulation of serotonin in rat spinal cord subjected to traumatic injury. Relation to edema and blood flow changes. Neurosci 36: 725–730
19. Sharma HS, Westman J, Olsson Y, Alm P (1996) Involvement of nitric oxide in acute spinal cord injury: an immunohistochemical study using light and electron microscopy in the rat. Neurosci Res 24: 373–384
20. Sharma HS, Alm P, Westman J (1998) Nitric oxide and carbon monoxide in the pathophysiology of brain functions in heat stress. Brain functions in hot environment. In: Sharma HS, Westman J (eds) Progr Brain Res 115: 297–333
21. Sharma HS, Westman J, Nyberg F (1998) Pathophysiology of brain edema and cell changes following hyperthermic brain injury. Brain functions in hot environment. In: Sharma HS, Westman J (eds) Progr Brain Res 115: 351–412
22. Shohami E, Ginis I, Hallenbeck JM (1999) Dual role of tumor necrosis factor alpha in brain injury. Cytokine Growth Factor Rev 10: 119–130
23. Shohami E, Novikov M, Bass R, Yamin A, Gallily R (1994) Closed head injury triggers early production of TNF-α and IL-6 by brain tissue. J Cereb Blood Flow Metab 14: 615–619
24. Yang GY, Gong C, Qin Z, Ye W, Mao Y, Bertz AL (1998) Inhibition of TNF alpha attenuates infarct volume and ICAM-I expression in ischemic mouse brain. NeuroReport 9: 2131–2134

Correspondence: Hari Shanker Sharma, Dr. Med. Sci., Frödingsgatan 12, app: 28; SE-75421 Uppsala, Sweden. e-mail: Sharma@medcellbiol.uu.se

Acta Neurochir (2003) [Suppl] 86: 415–420

A new antioxidant compound H-290/51 attenuates spinal cord injury induced expression of constitutive and inducible isoforms of nitric oxide synthase and edema formation in the rat

H. S. Sharma[1], **P.-O. Sjöquist**[2], and **P. Alm**[3]

[1] Laboratory of Neuroanatomy, Department of Medical Cell Biology, Biomedical Center, Uppsala University, Uppsala, Sweden
[2] Department of Integrative Pharmacology, AstraZeneca, Mölndal, Sweden
[3] Department of Pathology, University Hospital, Lund University, Lund, Sweden

Summary

The role of oxidative stress in spinal cord injury (SCI) induced upregulation of constitutive or inducible isoforms of nitric oxide synthase (cNOS or iNOS) is not well known. The present investigation was undertaken to examine the influence of an antioxidant compound H-290/51 (Astra-Zeneca, Mölndal, Sweden) on SCI induced cNOS and iNOS upregulation in a rat model. SCI induced by incision into the right dorsal horn of the T10–11 segment resulted in marked NOS upregulation. Upregulation of cNOS was most prominent in the uninjured T9 and T12 segments. On the other hand, iNOS expression was most marked in the injured T10–11 segments. These NOS immunoreactivities were mainly confined to the injured cells located in the edematous regions of the cord exhibiting profound leakage of Evans blue and $^{[131]}$Iodine-sodium tracers. Pretreatment with H-290/51 markedly attenuated the trauma-induced cNOS and iNOS expression along with the microvascular permeability disturbances, edema formation and cell injury. These results suggest that (i) oxidative stress is involved in SCI induced induction of cNOS and iNOS, (ii) NO plays an important role in the cord pathology, and (iii) that the compound H-290/51 has a potential therapeutic value in SCI.

Keywords: Oxidative stress; spinal cord injury; nitric oxide synthase; constitutive and inducible isoform; edema; blood-spinal cord barrier permeability; H-290/51; antioxidant; cell injury.

Introduction

The involvement of nitric oxide (NO) in the pathophysiology of spinal cord injury (SCI) is still not well understood [1–3]. Previously, we have reported that topical application of an antibody raised against the neuronal isoform of NOS (nNOS, often called "constitutive" isoform, cNOS) is capable to attenuate blood-spinal cord barrier (BSCB) permeability, edema formation, and cell injury in the traumatised cord [16, 18–20]. The antiserum was also able to attenuate nNOS expression in SCI [16, 20–22] suggesting that NO is injurious to the cord following trauma [6, 11].

Upregulation of NOS is associated with an increased NO production and, thus, can contribute to cell injury in several brain diseases [1–3, 7, 19, 20]. One of the strongest evidence supporting the role of NO in neurotoxicity is based on the fact that nNOS null mice are quite resistant to ischemia induced cell damage in the brain [see 1–3]. However, the use of NOS inhibitors in ischemia or stroke with infarction yielded contradictory results [19, 20]. The blockade of the endothelial isoform of NOS (eNOS) aggravates ischemic injury, whereas inhibition of nNOS (cNOS) in pathological conditions seems to be beneficial [see 3]. Thus, the inhibition of particular isoforms of NOS is important in cell injury or survival. The understanding of the basic mechanisms of NOS upregulation in CNS disorders is still speculative [see 19].

SCI is associated with profound oxidative stress which may underly induction of cNOS. Lipid peroxidation, generation of free radicals and formation of NO alone or in combination can induce marked neuronal, glial or endothelial cell injuries [1–3]. Thus, antioxidant compounds could be potential therapeutic agents in the future to achieve neuroprotection in SCI. There are reports that SCI causes upregulation of the inducible isoform of NOS (iNOS) as well [11, 19]. Upregulation of iNOS results in a sustained production of NO in high quantities [2, 19] which may be more injurious to cells than upregulation of cNOS. However, the role of oxidative stress in inducing iNOS is not well characterised.

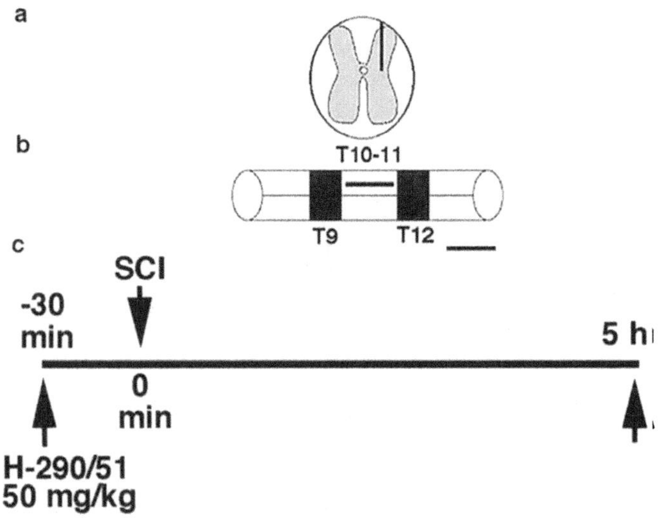

Fig. 1. Schematic representation of the experimental protocol of spinal cord injury of the T10–11 segment (a,b), and the specimens (T9 and T12) used for analysis of different parameters in this study, and treatment schedule of H-290/51 (c). bars: a = 2 mm; b = 5 mm

In the present investigation we have examined the influence of the known neuroprotective antioxidant H-290/51 (AstraZeneca, Mölndal, Sweden) on the upregulation of both cNOS and iNOS by using immunohistochemistry in our rat model of SCI. In addition, the effect of this compound on the SCI induced edema formation, BSCB permeability disturbances, and cell injury were also examined.

Material and methods

Animals

The experiments were carried out in 84 male Sprague Dawley rats housed at controlled room temperature 21 ± 1 °C with 12 h light and 12 h dark schedule. Food and tap water were supplied ad libitum.

Spinal cord injury

Under Equithesin anaesthesia (3 ml/kg, i.p.) laminectomy was performed at the T10–11 segment. Spinal cord injury (SCI) was produced by making a longitudinal incision into the right dorsal horn (about 2 mm deep and 4 mm long) [13]. The wound was covered with cotton soaked in saline to prevent direct exposure of the cord to air. The experimental protocol was approved by the Ethics Committee of Uppsala University, Uppsala Sweden. The animals were allowed to survive 5 h after injury. Equithesin anaesthetised normal rats served as controls.

Treatment with H-290/51

H-290/51 (50 mg/kg, p.o.) was administered 30 min before SCI in separate groups of animals. The traumatised animals were allowed to survive 5 h after injury [22] (Fig. 1). In one group of H-290/51 treated rats, no injury was inflicted and these animals were used as drug-treated controls.

NOS immunohistochemistry

Five h after spinal cord injury, the animals were perfused with about 200 ml of 4% paraformaldehyde fixative in 0.1 M phosphate buffer pH 7.4 at 4 °C preceded with a brief saline rinse [16]. After perfusion, the animals were wrapped in an aluminum foil and kept overnight at 4 °C in a refrigerator.

On the next day, the spinal cord samples were dissected out and placed in the same fixative at 4 °C. About 60 μm thick multiple vibratome sections were cut of the T9 and T12 segments and processed for NOS immunoreactivity using monoclonal antiserum to the constitutive isoform of neuronal NOS as described earlier (Sharma *et al.*, 1996a–c). Some tissue pieces were embedded in epon for routine light and electron microscopy for morphological examination by standard procedures [15, 17–19].

Blood-spinal cord barrier (BSCB) permeability

In separate groups of control, traumatized, or H-290/51 treated injured rats the BSCB permeability to Evans blue, [131]Iodine-sodium, and lanthanum was examined in the T9 and T12 segments as described earlier [14, 15].

Spinal cord edema and cell injury

The water content of the spinal cord was determined to evaluate edema formation [13]. Spinal cord cell injury was visualised by using light and electron microscopy of the T9 and T12 segments. Tissue pieces from the cord were embedded in paraffin or in Epon for light and electron microscopy, respectively using standard procedures [19]. About 3 μm thick paraffin sections and 1 μm thick epon sections were used for high resolution light microscopy to study edema and cell changes [19]. Neuronal, axonal, and endothelial cell changes were examined at the ultratstructural level as well [16–19].

Statistical analysis

ANOVA followed by Dunnet's test for multiple group comparison with one control was used to evaluate statistical significance of the data obtained. A p-value less than 0.05 was considered significant.

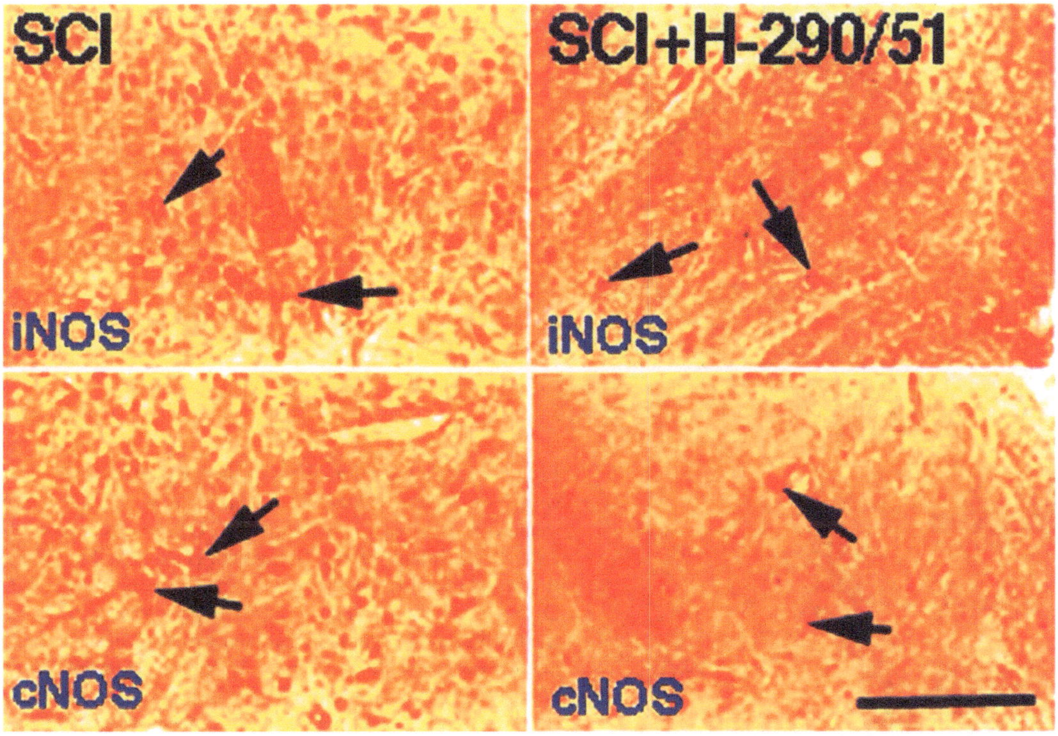

Fig. 2. Spinal cord injury induces profound upregulation of cNOS and iNOS in the spinal cord (arrowheads) of an untreated rat (left panel). Upregulation of NOS is mainly seen in the distorted cells located in the edematous region of the cord. Pretreatment with the antioxidant H-290/51 markedly attenuated iNOS and cNOS expression following spinal cord injury (right panel). Appearance of damaged cells is also reduced in the drug-treated traumatised rat. Bar = 100 μm

Results

Effect of H-290/51 on NOS upregulation

Focal trauma of the dorsal horn of the T10–11 segment resulted in a marked upregulation of cNOS and iNOS in the injured and the adjacent T9 and the T12 segments (Fig. 2). cNOS expression was most prominent in the T9 and T12 segments. On the other hand, upregulation of iNOS was mainly seen in the injured segments. These NOS immunoreactivities were mainly confined to the injured cells located in the edematous regions of the spinal cord.

Pretreatment with H-290/51 markedly attenuated the trauma-induced upregulation of cNOS and iNOS in the cord (Fig. 2). Thus, the H-290/51 treated injured rats exhibited much less cNOS or iNOS expression as compared to the untreated injured rats (Fig. 2).

Effect of H-290/51 on edema formation and cell injury

Many distorted nerve cells were observed in the spinal cord of untreated injured rats. Sponginess and edema were most prominent in untreated traumatised animals (Fig. 3). Pretreatment with H-290/51 resulted in marked neuroprotection in the spinal cord, following trauma (Fig. 3). In drug-treated traumatised rats the nerve cells were quite normal in appearance, and sponginess and edema were mainly absent (Fig. 3).

At the ultrastructural level, vacuolization, edema, membrane damage, and axonal injury were quite prominent in the untreated traumatised rats (Fig. 3). Myelin disruption and vesiculation were much more frequent in untreated traumatised rats (Fig. 3). Pretreatment with H-290/51 considerably reduced myelin vesiculation, membrane damage and edema in the spinal cord (Fig. 3).

The findings of the water content of the spinal cord were in good agreement with our morphological observations. Untreated traumatised rats had a profound increase in the spinal cord water content (T9 68.56 ± 0.23; T12 68.65 ± 0.28; $P < 0.01$) as compared to the control values (T9 65.36 ± 0.12; T12 65.43 ± 0.22). Pretreatment with H-290/51 significantly thwarted the spinal cord edema formation as shown by the markedly lower water content of the

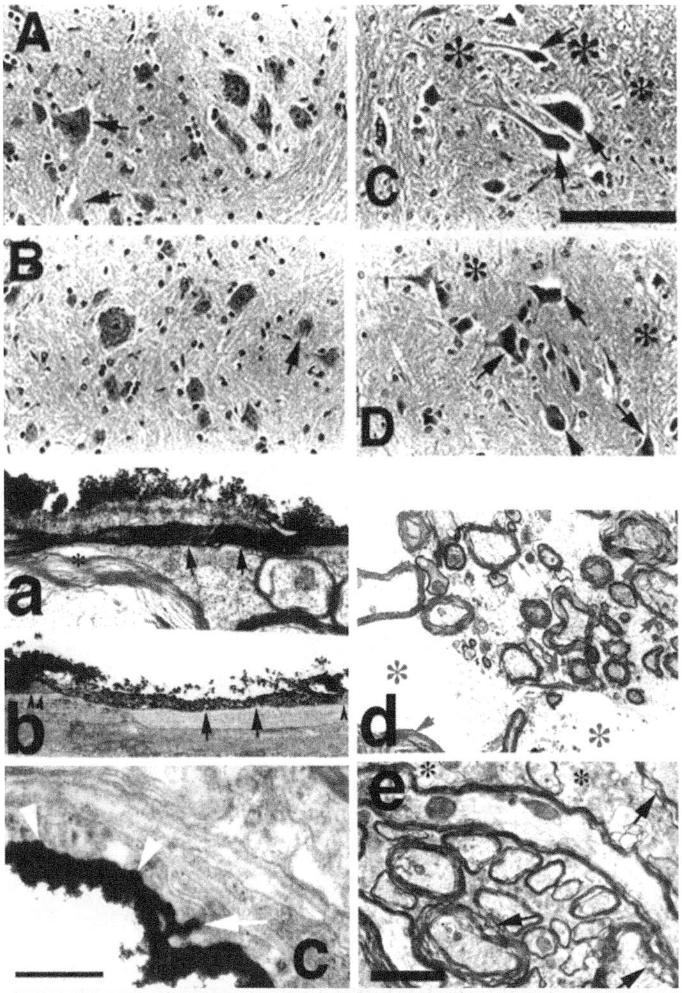

Fig. 3. Morphological analysis of the T9 segment of a spinal cord traumatised rat and its modification by H-290/51 pretreatment. By light microscopy, (A–D) paraffin embedded 3 μm thick sections after staining with Haematoxylin and Eosin, reveal many dark and distorted motoneurons from SCI (C, D). Sponginess and edema (*) are quite prominent in the untreated traumatised rat. Pretreatment with H-290/51 markedly attenuated motoneuron injury (A, B) and signs of edema and sponginess are mainly absent. Bar = 25 μm. The ultrastructural studies (a–e) exhibit profound extravasation of lanthanum into the basal lamina (a) and the endothelial cell cytoplasm (b) of an untreated spinal cord traumatised rat. Myelin vesiculation (d), vacuolization and edema are prominent in the untreated spinal cord injured rat (d). Pretreatment with H-290/51 profoundly reduced leakage of lanthanum across the endothelial cell layer (e) and myelin vesiculation, vacuolization and edema are much less evident (e) in the drug treated traumatised cord. Bar (a,b,d = 800 nm; c,d = 1 μm). Data modified after Winkler et al., 1998; Sharma *et al.*, 2001

traumatised spinal cord in this group (T9 66.48 ± 0.32; T12 66.74 ± 0.42; $P < 0.01$) as compared to the untreated injured rats.

Effect of H-290/51 on BSCB permeability

The BSCB permeability was significantly increased for Evans blue and radioactive iodine tracers in the spinal cord of traumatised animals (Evans blue T9 1.56 ± 0.16, T12 1.62 ± 0.16; radioiodine T9 2.14 ± 0.21, T12 2.28 ± 0.23, $P < 0.01$) as compared to the control group (Evans blue T9 0.24 ± 0.08, T12 0.28 ± 0.06; radioiodine T9 0.38 ± 0.08, T12 0.40 ± 0.10). Pretreatment with H-290/51 significantly reduced the extravasation of Evans blue and radioactive iodine into the T9 (Evans blue 0.48 ± 0.11; radioiodine 0.76 ± 0.14; $P < 0.01$) and the T12 (Evans blue 0.44 ± 0.14; radioiodine 0.81 ± 0.15; $P < 0.01$) segments as compared to the untreated traumatised rats.

At the ultrastructural level, the electron dense tracer lanthanum was mainly confined to the lumen of normal spinal cord microvessels (results not shown). SCI induced pronounced infiltration of lanthanum into the endothelial cell cytoplasm (Fig. 3) and occasionally to the basal lamina (Fig. 3). In H-290/51 treated injured rats, the lanthanum tracer was confined to the vessel lumen (Fig. 3), while infiltration of the tracer into the endothelial cytoplasm was absent.

Discussion

The new findings of this study show that SCI induces an upregulation of both constitutive and inducible isoforms of NOS in the edematous regions of the cord exhibiting cell damage. This indicates that upregulation of NOS is injurious to the cord tissue.

Our results further demonstrate that iNOS was confined mostly to the injured segments, whereas distribution of cNOS was more widespread. This suggests that iNOS upregulation is associated with intense cell injury. A prolonged and sustained release of NO by iNOS supports this idea [19]. Obviously, the intensity of adverse cell reactions, and spinal cord tissue damage was most prominent in the traumatised segments.

Expression of cNOS in remote regions suggests that a spread of injury factors or of edema fluid into adjacent segments of the spinal cord probably has triggered an upregulation of cNOS. Since normal spinal cord segments did not express cNOS, it is conceivable that cells expressing cNOS activity were more vulnerable to trauma induced cell death [3, 16].

The molecular mechanisms of upregulation of NOS in SCI are not clear. It appears that formation of free radicals is associated with the production of NO in SCI [3, 19–21, 24]. This is apparent from the fact that pretreatment with the antioxidant H-290/51 markedly reduced the expression of cNOS and iNOS in the cord following injury. H-290/51 is a chain-breaking antioxidant [23] and thus is capable to inhibit lipid peroxidation and the generation of free radicals [10, 23]. The compound as given in the present dose was capable to inhibit TBARS production [10] indicating that the drug is able to inhibit SCI induced lipid peroxidation [20]. These observations suggest that oxidative stress followed by lipid peroxidation and generation of free radicals in SCI contribute to NOS upregulation.

The source of free radical generation after SCI is not well known. Haemorrhage and extravasation of blood components, such as haemoglobin into cord parenchyma is one potent source of iron to catalyse formation of oxygen radicals and lipid peroxidation [6, 9]. Other factors include alterations in arachidonic acid metabolism and modification in the cyclooxygenase pathways, invading neutrophils undergoing oxidative bursts or auto-oxidation of catecholamines [8, 9]. Activation of phospholipase A_2 increases the level of platelet-activating factor (PAF) which seems to be related with the post-traumatic disturbances of the SCBF and the BSCB [4, 5]. Lipolysis by phospholipase A_2 (PLA_2) PLA_2 is also responsible for the generation of free radicals, the molecules which damage cell membranes due to their high random reactivity in the presence of unpaired electrons [see 1].

Oxidative stress and generation of free radicals contribute to spinal cord edema formation and membrane damage either directly or through indirect mechanisms. Direct membrane damage and increased BSCB permeability in the spinal cord allows for accumulation of water in the extra- and intracellular compartments, resulting in further membrane damage, vacuolization and cell swelling. Ultrastructural studies clearly show that accumulation of water occurs in both the extra- and intracellular compartments in the cord after trauma. Pretreatment with H-290/51 was associated with a considerable reduction in the BSCB permeability, edema formation, and cell injury. Obviously, an increased extravasation of serum proteins into the spinal cord due to disruption of the BSCB will enhance entry of fluid from the vascular compartment into the spinal cord fluid microenvironment from the shift in colloid osmotic pressure.

In most cases, endothelial cells exhibited currently a selective increase in lanthanum permeability leaving the adjacent cell completely intact. Activation of specific receptor mediated events and signal transduction mechanisms by free radicals in the endothelial cell membrane appears to be crucial for such a selective lanthanum infiltration. The prevention of lanthanum extravasation following SCI by H-290/51 pretreatment is in line with this assumption.

In conclusion, the salient new findings of the present experiments are that SCI induces upregulation of both cNOS and iNOS in the vicinity of cell injury, and that pretreatment with H-290/51 markedly inhibits NOS expression from trauma in the cord. These observations suggest that (i) NO plays an important role in spinal cord injury, (ii) oxidative stress is somehow related to the induction of cNOS and iNOS expression in the spinal cord following trauma, and (iii) that the antioxidant H-290/51 has a marked therapeutic potential to inhibit SCI induced cell injury and edema, a feature requiring further investigation.

Acknowledgments

This investigation is supported by Grants from Swedish Medical Research Council Nrs. 2710 (JW/HSS) and 11205 (PA/HSS) and Astra-Zeneca, Mölndal (HSS), Sweden. The technical assistance of Kärstin Flink is acknowledged with thanks.

References

1. Chiueh CC, Gilbert DL, Colton CA (1994) The neurobiology of NO and OH. Ann NY Acad Sci 738: 1–471
2. Choi DW (1993) Nitric oxide: foe or friend to the injured brain. Proc Natl Acad Sci USA 90: 9741–9743
3. Dawson VL, Dawson TM (1996) Nitric oxide neurotoxicity. J Chem Neuroanat 10: 179–190
4. Faden AI, Halt P (1992) Platelet-activating factor reduces spinal cord blood flow and causes behavioral deficits after intrathecal

administration in rats through a specific receptor mechanism. J Pharmacol Exp Ther 261: 1064–1070
5. Frerichs KU, Feuerstein GZ (1990) Platelet-activating factor; key mediator in neuroinjury? Cerebrovasc Brain Metab Rev 2: 148–160
6. Fukuda K, Richmon JD, Sato M, Sharp FR, Panter SS, Noble LJ (1996b) Induction of heme oxygenase-1 (HO-1) in glia after traumatic brain injury. Brain Res 736: 68–95
7. Kimura H, Steinbusch HWM (1996) The role of nitric oxide in the central nervous system. From basic to therapeutic aspect. J Chem Neuroanat 10: 179–322
8. McCord JM (1987) Oxygen-derivated radicals: a link between reperfusion injury and inflammation. Federation Proc 46: 2402–2406
9. Misra HP, Fridovich I (1972) The role of superoxide anion in the auto-oxidation of epinephrine and a simple assay for superoxide dismutase. J Biol Chem 247: 3170–3175
10. Mustafa A, Sharma HS, Olsson Y, Gordh T, Thorén P, Sjöquist P-O, Roos P, Adem A, Nyberg F (1995) Vascular permeability to growth hormone in the rat central nervous system after focal spinal cord injury. Influence of a new antioxidant compound H 290/51 and age. Neurosci Res 23: 185–194
11. Novikov L, Novikova L, Kellerth J-O (1995) Brain-derived neurotrophic factor promotes survival and blocks nitric oxide synthase expression in adult rat spinal motoneurons after ventral nerve root avulsion. Neurosci Lett 200: 45–48
12. Sharma HS (1987) Effect of captopril (a converting enzyme inhibitor) on blood-brain barrier permeability and cerebral blood flow in normotensive rats. Neuropharmacology 26: 85–92
13. Sharma HS, Olsson Y (1990) Edema formation and cellular alterations following spinal cord injury in rat and their modification with p-chlorophenylalanine. Acta Neuropathol (Berlin) 79: 604–610
14. Sharma HS, Cervós-Navarro J, Dey PK (1991) Rearing at high ambient temperature during later phase of the brain development enhances functional plasticity of the CNS and induces tolerance to heat stress. An experimental study in the conscious normotensive young rats. Brain Dysfunction 4: 104–124
15. Sharma HS, Olsson Y, Dey PK (1990) Early accumulation of serotonin in rat spinal cord subjected to traumatic injury. Relation to edema and blood flow changes. Neuroscience 36: 725–730
16. Sharma HS, Westman J, Olsson Y, Alm P (1996) Involvement of nitric oxide in acute spinal cord injury: an immunohistochemical study using light and electron microscopy in the rat. Neurosci Res 24: 373–384
17. Sharma HS, Westman J, Alm P, Sjöquist P-O, Cervós-Navarro J, Nyberg F (1997) Involvement of nitric oxide in the pathophysiology of acute heat stress in the rat. Ann NY Acad Sci 813: 581–590
18. Sharma HS, Nyberg F, Gordh T, Alm P, Westman J (1997) Topical application of insulin like growth factor-1 reduces edema and upregulation of neuronal nitric oxide synthase following trauma to the rat spinal cord. Acta Neurochir Suppl 70: 130–133
19. Sharma HS, Alm P, Westman J (1998) Nitric oxide and carbon monoxide in the pathophysiology of brain functions in heat stress. Brain functions in hot environment. In: Sharma HS, Westman J (eds) Progr Brain Res 115: 297–333
20. Sharma HS, Nyberg F, Gordh T, Alm P, Westman J (1998) Neurotrophic factors attenuate neuronal nitric oxide synthase upregulation, microvascular permeability disturbances, edema formation and cell injury in the spinal cord following trauma. Spinal cord monitoring. In: Stålberg E, Sharma HS, Olsson Y (eds) Springer, Wien New York, pp 181–210
21. Sharma HS, Westman J, Nyberg F (1998) Pathophysiology of brain edema and cell changes following hyperthermic brain injury. Brain functions in hot environment. In: Sharma HS, Westman J (eds) Progr Brain Res 115: 351–412
22. Sharma HS, Sjöquist P-O, Westman J (2001) Pathophysiology of the blood-spinal cord barrier in spinal cord injury. Influence of a new antioxidant compound H-290/51. In: Kobiler D, Lustig S, Shapra S (eds) Blood-brain barrier. Drug Delivery and Brain Pathology. Kluwer Academic/Plenum Publishers, New York, pp 401–416
23. Svensson L, Börjesson I, Kull B, Sjöquist P-O (1993) Automated procedure for measuring TBARS for in vitro comparison of the effect of antioxidants on tissues. Scand J Clin Lab Invest 53: 83–85
24. Winkler T, Sharma HS, Stålberg E, Westman (1998) Spinal cord bioelectrical activity, edema and cell injury following a focal trauma to the spinal cord. An experimental study using pharmacological and morphological approach. Spinal cord monitoring. In: Stålberg E, Sharma HS, Olsson Y (eds) Springer, Wien New York, pp 283–363

Correspondence: Hari Shanker Sharma, Dr. Med. Sci., Frödingsgatan 12, app: 28; SE-75421 Uppsala, Sweden. e-mail: Sharma@medcellbiol.uu.se

Acta Neurochir (2003) [Suppl] 86: 421–423

Src family kinase inhibitor PP1 improves motor function by reducing edema after spinal cord contusion in rats

C. Akiyama, T. Yuguchi, M. Nishio, T. Fujinaka, M. Taniguchi, Y. Nakajima, and **T. Yoshimine**

Department of Neurosurgery, Osaka University Medical School, Osaka, Japan

Summary

Following spinal cord injury vascular permeability increases around the area of injury, which possibly leads to secondary tissue damage. Vascular endothelial growth factor (VEGF) and Src which exists downstream of VEGF may contribute to edema formation. We here report that the Src family kinase inhibitor PP1 could reduce edema and the inflammatory response after spinal cord injury. In this study we have examined the effect of PP1 on motor function after mild spinal cord compression injury.

We utilized a mild spinal cord compression model in rats. PP1 or vehicle only was administered intraperitoneally after cord compression. The motor function of the hind limbs after injury was categorized into 7 grades. At 1, 3, 7 and 14 days after injury, the spinal cord was removed and the extent of edema formation and inflammation were examined using immunohistochemistry with an anti-IgG and anti-ED-1 antibody.

The immunohistochemical analysis revealed that the area of edema formation and inflammation was remarkably reduced in animals with PP1. The muscle function was flaccid in both groups immediately after injury. However, at 3 and 8 days after injury, a significant improvement was observed in the PP1 group. These results suggest that PP1 is a strong candidate for drug treatment of spinal cord injury.

Keywords: Spinal cord injury; edema; Src family kinase inhibitor; motor function.

Introduction

Following spinal cord injury, vascular permeability increases around the area of injury, which possibly leads to both edema formation and an inflammatory response, which in turn causes secondary tissue damage [9]. The control of vascular permeability, therefore, is essential in preventing secondary damage.

Vascular endothelial growth factor (VEGF), an endothelial mitogen as well as a potent mediator of vascular permeability [10] is reported to contribute to brain edema associated with brain tumors [7] and cerebral infarction [1]. In spinal cord injury, VEGF may contribute to edema formation. VEGF binding of the VEGF receptor (VEGFR) leads to recruitment of Src and stimulation of the catalytic activity of Src [8]. Activated Src may enhance vascular permeability [2], though underlying mechanisms are unclear.

PP1 (4-amino-5-(4-methylphenyl)-7-(t-butyl)pyrazolo[3,4-d]pyrimidine), which belongs to the pyrazolopyrimidine family, is a novel, potent and selective inhibitor of the Src family tyrosine kinases [3]. We have previously reported that PP1 reduces edema and the inflammatory response from spinal cord contusion. In the present study, we examine the effect of PP1 on the motor function after mild spinal cord compression injury.

Material and methods

Experimental model

Twenty five female Wistar rats (body weight = 200 g) were used for the experiments. They were anesthetized with pentobarbital (60 mg/kg, i.p.) and placed in the prone position. After midline skin incision, a single segmental laminectomy was perfomed at Th10, and the dura of the spinal cord was exposed. The spinal cord was compressed for 5 seconds through the dura using a Biemer vessel clip with a squeezing force of 25 g. The Src inhibitor PP1 (1.5 mg/kg) suspended in 30% DMSO with PBS was administered intraperitoneally 10 minutes after spinal cord compression (PP1 group). The control rats received vehicle only using the same method (control group).

At 1 day, 3 days, 7 days and 14 days after surgery, the rats were deeply anesthetized with pentobarbital (100 mg/kg) and the spinal cord was removed. The cords were quick-frozen, and sagittal sections were obtained (10 μm thickness).

All animal experiments were conducted in compliance with the Osaka University Medical School Guidelines for the Care and Use of Laboratory Animals. The surgical and experimental procedures were approved by the Osaka University Medical School Animal Care and Use Committee.

Evaluation of motor function

The motor function of hind limbs after injury was scored using a system which was reported by Kanellopoulos *et al.* [4]. A functional score was given to each rat according to the following criteria.

S0 = normal
S1 = walks normally but legs are weak (cannot pull the legs if one holds them)
S2 = assumes normal body posture on a flat surface and is able to walk but there is ataxia with or without spasticity
S3 = able to walk on the knuckles or walk on the feet but without proper stepping
S4 = drags legs but there is movement at the knees
S5 = drags legs without significant movement in the lower limbs and spasticity is present
S6 = drags legs without significant movement in the lower limbs and flaccidity is present

Spasticity was defined as continuous or intermittent tonic positioning of the hind limbs in extension and especially of the feet in plantar flexion. The degree of spasticity ranged from mild (usually present intermittently after provocation by lifting the tail or by stressing the animal) to conspicuous (continuous, even at rest and characterized by the assumption of a striking posture with hind limbs in extreme extension and occasionally with tonic upwards curvature of the tail). Flaccidity was defined as the absence or great reduction of the muscle tone during passive movement at the ankle joint.

Histopathological analysis

The expression of VEGF was examined by immunohistochemistry using a Vectastain ABC kit (Vector, USA); expression of IgG and ED-1 was also examined as measure of edema and infiltration of macrophages. The sections were fixed with 4% paraformaldehyde in 0.1 M phosphate buffer for 45 min and incubated with methanol at −20 °C for 10 min. After incubation with 3% hydrogen peroxidase for 30 min, the sections were added with normal horseserum for 20 min and with the primary antibody at 4 °C overnight (anti-VEGF monoclonal mouse IgG, 1:100, Santa Cruz, USA, or anti-ED-1 monoclonal mouse antibody, 1:100, Serotec, UK). Finally, the sections were administered with the biotinylated secondary antibody for 30 min and avidin-biotin complex for 30 min. The endproduct was visualized by using a diaminobenzidine solution and by counterstaining with hematoxylin. The procedure of using anti-IgG (rabbit, polyclonal antibody, 1:200, Vector, USA) differed slightly from the method described above. As this antibody was produced in rabbits and was already biotinylated, the sections were incubated with normal rabbit serum followed by the primary antibody at room temperature for 30 min and then the avidin-biotin complex for another 30 min. The other procedures were the same as described above.

Results

After recovery from anesthesia, all animals developed paraplegia. They dragged their legs, displaying no significant movement of the lower limbs, and flaccidity (S6) was confirmed. At 3 days after injury, spasticity or flaccidity was observed in the control group (S5–6), while walking on the knuckles or movement of the knees was observed in the PP1 group (S3–4). At 8 days after injury, only a slight movement of the knees was observed in the control group (S4), whereas an ataxic gait in the PP1 group (S2) (Fig. 1).

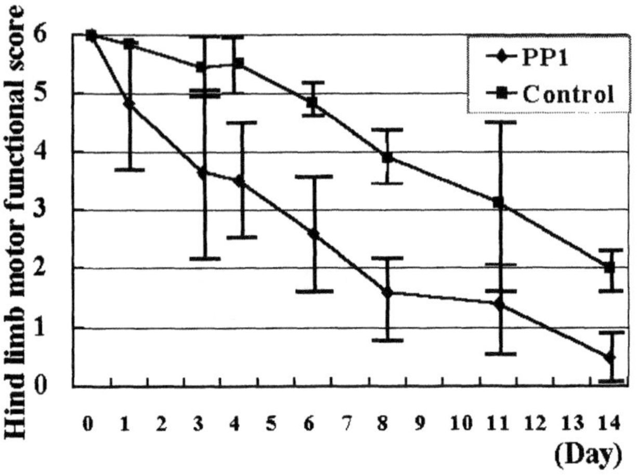

Fig. 1. Motor functional score in control and PP1 group during 14 days follow-up period

Immunohistochemistry

On day 3 and 7 after surgery, significant VEGF protein expressions was detected mainly in the glial cells around the contusion, reaching a peak on day 7. There was no remarkable difference between the PP1 and control group (data not shown). Edema was much less extensive in the PP1 group than in the control group on day one (data not shown). ED-1 positive cells were found to be restricted to the contusion area on day 3 in the PP1 group, while in the untreated controls ED-1 positive cells were found to be more widely dispersed. On day 7 and 14, the difference between both groups was obvious (Fig. 2).

Discussion

We have demonstrated a significant improvement of the motor function by intraperitoneal administration of PP1, and that although VEGF expression was similar in both groups on day 3 and day 7, the extent of edema formation and the inflammatory response were significantly reduced in the PP1 group. These findings may reflect a therapeutical efficacy of PP1, which inhibits Src kinase activity downstream of VEGF.

It has been reported that Src-deficient mice have a negligible vascular permeability response to VEGF and that they are protected against edema-associated damage from stroke [6]. Therefore, with regard to the currently observed mechanism by which edema was reduced, it is assumed that PP1 suppresses the function of Src kinases, thereby inhibiting the enhancement of vascular permeability, resulting in attenuation of edema.

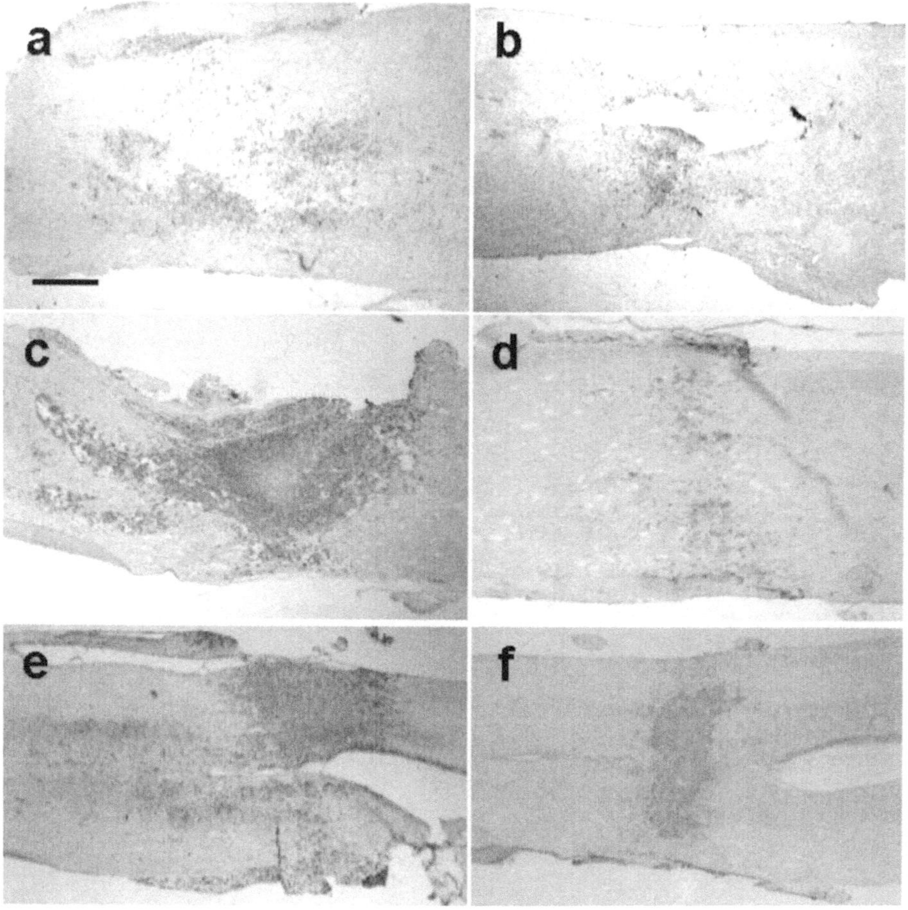

Fig. 2. Photomicrographs of ED-1 immunohistochemistry on day 3, 7 and 14. Control group (a, c, e), and PP1 group (b, d, f). ED-1 positive cells were found to be restricted to the area of the contusion on day 3 in the PP1 group (a), whereas in the control group (b) ED-1 positive cells were found to be more widely dispersed. On day 7 and 14, the difference between both groups was more apparent (c–f). Scale bars = 1 mm (a–f)

On the other hand, it has been reported that PP1 inhibits the production of TNF and of the inducible nitric oxide synthase (iNOS) in murine macrophages [5]. It is conceivable that because PP1 blocks the activity of macrophages, progression of inflammation and chemotaxis of macrophages was also inhibited in the PP1 group. We hope that PP1 is applicable for the medical treatment of patients with spinal cord injury in the future, requiring, however, that potentially adverse effects should be investigated beforehand.

References

1. Cobbs CS, Chen J, Greenberg DA *et al* (1998) Vascular endothelial growth factor expression in transient focal cerebral ischemia in the rat. Neurosci Lett 249: 79–82
2. Eliceiri BP, Paul R, Schwartzberg PL *et al* (1999) Selective requirement of Src kinases during VEGF-induced angiogenesis and vascular permeability. Mol Cell 4: 915–924
3. Hanke JH, Gardner JP, Dow RL *et al* (1996) Discovery of a novel, potent, and Src family-selective tyrosine kinase inhibitor. J Biol Chem 271: 695–701
4. Kanellopoulos GK, Kato H, Wu Y *et al* (1997) Neuronal cell death in the ischemic spinal cord: The effect of methylprednisolone. Ann Thorac Surg 64: 1279–1286
5. Orlicek SL, Hanke JH, English BK (1999) The Src family-selective tyrosine kinase inhibitor PP1 blocks LSP and IFN-gamma-mediated TNF and iNOS production in murine macrophages. Shock 12: 350–354
6. Paul R, Zhang ZG, Eliceiri BP *et al* (2001) Src deficiency or blockade of Src activity in mice provides cerebral protection following stroke. Nat Med 7: 222–227
7. Plate KH, Risau W (1995) Angiogenesis in malignant gliomas. Glia 15: 339–347
8. Schlessinger J (2000) New role for Src kinases in control of cell survival and angiogenesis. Cell 100: 293–296
9. Scoeld M, Cullheim S, Hammarberg H *et al* (2000) Induction of VEGF and VEGF receptors in the spinal cord after mechanical spinal injury and prostaglandin administration. Eur J Neurosci 12: 3675–3686
10. Senger DR, Galli SJ, Dvorak AM *et al* (1983) Tumor cells secrete a vascular permeability factor that promotes accumulation of ascites fluid. Science 219: 983–985

Correspondence: Chihiro Akiyama, M.D., Department of Neurosurgery, Osaka University Medical School, 2-2 Yamadaoka, Suita, Osaka 565-0871, Japan. e-mail: cakiyama@nsurg.med.osaka-u.ac.jp

Acta Neurochir (2003) [Suppl] 86: 425–432

An L-type calcium channel blocker, nimodipine influences trauma induced spinal cord conduction and axonal injury in the rat

T. Winkler[1], **H. S. Sharma**[2], **E. Stålberg**[1], **R. D. Badgaiyan**[2,4], **T. Gordh**[3], and **J. Westman**[2]

[1] Department of Clinical Neurophysiology, University Hospital, Uppsala, Sweden
[2] Laboratory of Neuroanatomy, Department of Medical Cell Biology, Biomedical Center, Uppsala University, Uppsala, Sweden
[3] Department of Anaesthesiology, University Hospital, Uppsala, Sweden
[4] Department of Psychiatry, Harvard Medical School and Department of Psychology, Harvard University, Cambridge, MA

Summary

The influence of the potent L-type $Ca^{[2+]}$ channel antagonist Nimodipine on spinal cord evoked potentials (SCEP) and axonal injury following trauma to the spinal cord was examined in a rat model. Spinal cord injury (SCI) was produced by an incision into the right dorsal horn of the T10–11 segments under urethane anaesthesia (1.5 g/kg, i.p.). SCEPs were recorded by epidural electrodes placed over the T9 (rostral) and T12 (caudal) segments after stimulation of the right tibial and sural nerves. SCI induced a pronounced decrease of the SCEP negative amplitude in the rostral (T9) recordings immediately after trauma. Axonal injury seen as degradation of myelin basic protein (MBP) immunostaining and myelin vesiculation at the ultrastructural level was most pronounced at 5 h. Continuous administration of Nimodipine (2 µg/kg/min, i.v.) from 30 min prior to injury until sacrifice markedly attenuated the changes in SCEP amplitude and latency. Axonal damage, loss of MBP, and myelin vesiculation were much less evident in the nimodipine treated traumatised rats. These observations suggest that $Ca^{[2+]}$ channels play an important role in the trauma induced alterations in SCEP and axonal injury, and indicate a therapeutic value of $Ca^{[2+]}$ blockers in SCI.

Keywords: Spinal cord evoked potentials; edema; spinal cord injury; $Ca^{[2+]}$ channels; nimodipine.

Introduction

A focal trauma to the spinal cord induces cell and tissue destruction leading to functional disturbances and/or permanent disability depending on the magnitude and severity of the primary lesion [29]. The final extent of tissue injury is determined by the primary physical insult followed by secondary or "perifocal" changes around the site of the lesion. An immediate loss of the spinal cord bioelectrical activity, an increase in extracellular potassium, and decrease in sodium-potassium ATPase activity after spinal cord injury (SCI) is a prominent finding [3, 4]. Even small variations in the concentrations of some ions and molecules greatly influence the neural excitability and synaptic transmission of the cord [8, 9].

Increased potassium levels in the extracellular space together with calcium depolarize neuronal populations in both gray and white matter which is mainly responsible for the conduction block in the damaged spinal cord segments [1, 10]. Chronic loss of calcium and intracellular potassium following trauma is related to disturbances in cell membrane function leading to cell injury or cell death [1, 17].

Calcium plays a crucial role in the regulation of sodium and potassium conductance of the cell membrane and is involved in neural excitability, regulation of many critical cellular enzymes, as well as storage and release of many neurotransmitters [10]. Balentine and Spector [1] were the first to demonstrate intra-axonal accumulation of calcium 30 min after spinal cord injury which is considered to be responsible for myelin vesiculation. The extracellular calcium store is decreased rapidly following trauma to the cord [37]. The total calcium level is increased 45 min after trauma reaching a peak at 8 h. This increase in the cord calcium level persists even for 1 week after injury [2, 14, 30].

Increased levels of intracellular calcium induce many adverse effects on cellular functions [10, 28]. One possible cause of cell death is activation of calcium-dependent phospholipase C and A_2 which causes a rapid breakdown of cellular membranes and production of arachidonic acid and its metabolites thromboxane, leukotrienes and free radicals which may directly or through inflammatory processes contribute to cell and membrane injury [5, 22, 30, 33, 36].

The degree of the initial impact is the major factor influencing magnitude and severity of the circulatory disturbances which occur around a primary traumatic injury of the cord. Alterations of microvascular permeability in the perifocal segments of the spinal cord are evident within minutes after injury [16, 25]. Edema of the cord is present as soon as 30 min after trauma [13, 20, 24–26]. The extracellular edema spreads in both longitudinal and transverse directions which is progressive with time [13].

Previously, we have used spinal cord evoked potentials (SCEP) to study the bioelectrical activity of the cord following trauma [26]. Our results have shown that early changes in the SCEP can be used to predict the later development of edema formation and other pathological changes in the cord [34].

Since spinal cord bioelectrical activity is greatly influenced by alterations of the intracellular calcium level [29], the present investigation was undertaken to examine the effect of the potent L-type calcium channel blocker, nimodipine on SCEP changes and edema formation in our well-established SCI model in rats [24, 25–27]. In addition, the influence of nimodipine on the spinal cord pathophysiology was examined.

Material and methods

Animals

The experiments were carried out in 28 male Sprague-Dawley rats (350–400 g, Alab, Stockholm) housed at a controlled room temperature (22 ± 1 °C), with a 12 h light and 12 h dark schedule. Food and water were given *ad libitum* before the experiment.

Spinal cord injury

Under urethane anesthesia (1.5 g/kg i.p.) SCI was inflicted by a longitudinal incision (2 mm deep and 5 mm long) of the right dorsal horn of the T10–11 segments leaving the white matter mainly intact (Fig. 1) [20, 25, 26]. The study was approved by the Ethical Committee of the Uppsala University.

Spinal cord evoked potentials

SCEP were recorded from electrodes (electrode resistance 0.11 Ω) placed in the epidural space of the T9 and T12 segments of the spinal cord (Fig. 1) after supramaximal stimulation (5–10 mA) of the right or left tibial and sural nerves (frequency 6 Hz; stimulus duration of 0.1 msec) as described earlier [37, 35].

Two hundred responses were collected and averaged by the EMG equipment (Medelec MS 20 "Mystro", filter 3 Hz–5 kHz, sweep duration 20 msec). The amplitudes and latencies of the maximal negative peak (MNP), and the maximal positive peak (MPP) were measured for statistical calculations [26, 47].

Physiological variables

The mean arterial blood pressure (MABP), arterial pH and blood gases were examined in separate group of animals using standard protocol [26].

Spinal cord blood flow

The spinal cord blood flow (SCBF) was measured using about 1 million carbonised microspheres (labelled with ^{125}Iodine; 15 ± 0.6 μm in diameter) which were injected into the left atrium [25, 26]. The reference blood flow sample (0.8 ml/min) was collected through right femoral artery. About 90 sec after microsphere administration, the spinal cord tissue was dissected out after decapitation, and the radioactivity was measured. The blood flow (ml/g/min) was calculated from the reference samples as described earlier [25].

Spinal cord edema

Five hours after lesion, the injured spinal cord segments (T10–11) were removed and weighed immediately to obtain the wet weight of the sample. The samples were then placed in an oven at 90 °C for 48 h to obtain the dry weight. The water content of the spinal cord was calculated from the difference between the dry and wet weights of the sample. An increase in water content was used as a measure of edema [25, 27].

Spinal cord cell changes

Spinal cord cell changes and axonal damage were examined using standard light and electron microscopical techniques [25, 27]. About 3 μm thick sections were cut from the paraffin embedded tissue blocks and processed for myelin basic protein (MBP) immunostaining as described earlier [27]. About 1 μm thick sections were cut from Epon embedded spinal cord tissues for high resolution microscopy. Ultrathin sections from the spinal cord dorsal or ventral horn were cut and processed for standard transmission electron microscopy [23, 27].

Influence of Nimodipine (Bayer AG, Germany): Intravenous infusion of nimodipine (2 μg/kg/min) was given continuously through an infusion pump starting 30 min before injury and continuing until 5 h after injury [23, 30].

Statistical analysis

In the calculations, the rostral and caudal MPP and MNP amplitudes and latencies in each group measured 30 min before injury were set to 100%. Comparisons were then made for each parameter between −2 min and each recording time, and for each recording time between groups. ANOVA (Fisher and Dunnett) and Wilcoxon's signed rank test were used in statistical calculations. P values less than 0.05 were considered significant.

Results

SCEP changes following SCI

The most pronounced changes in SCEP were seen rostrally after ipsilateral (right-sided) stimulation (Fig. 1). The MNP disappeared immediately (0 min) after injury and MPP increased in amplitude and latency. The depression of MNP showed some recovery in later

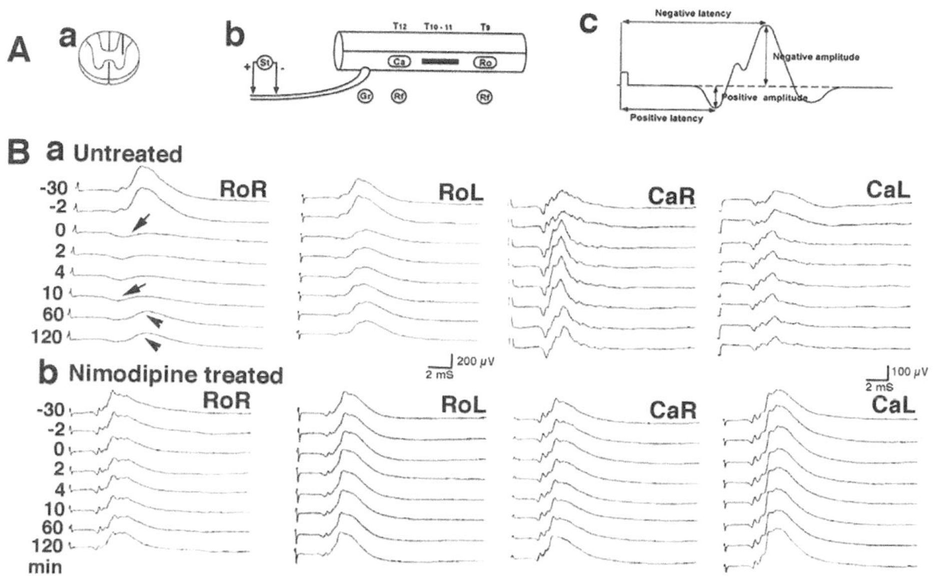

Fig. 1. (A) SCEP recording and methods. Spinal cord injury model (a), Recording of SCEP after stimulation of peripheral nerve (b), and analysis of SCEP amplitude and latency (c). (B) Representative records of SCEP in untreated traumatised rats (a) and their modification with nimodipine treatment (b). *RoR* Rostral right side stimulation, *RoL* Rostral left side stimulation, *CaR* caudal right side stimulation, *CaL* caudal left side stimulation. SCEP depression is evident in untreated injured group (arrows)

recordings. Contralateral (left-sided) stimulation also showed a depression of the rostral MNP. The caudal SCEPs showed only minor changes after SCI (Fig. 1).

Amplitudes

The rostral MNP showed a pronounced post-injury decrease in amplitude (Figs. 1 and 2), whereas the rostral MPP exhibited a marked increase in amplitude at 0 to 10 min after SCI (Figs. 1 and 2). The caudal MPP amplitude was much less altered compared to the rostral recordings (Figs. 1 and 2). After contralateral stimulation only the rostral MPP amplitude showed some increase (Fig. 1). Other amplitudes did not change significantly (Fig. 2).

Latencies

The rostral MNP latency after contralateral stimulation and caudal MNP latencies after ipsi- and contralateral stimulation increased somewhat after injury (Figs. 1 and 2). The rostral MPP latency after ipsilateral stimulation showed a pronounced postinjury increase throughout the recording period. The caudal MPP latency increased at 60 min after ipsilateral stimulation (Figs. 1 and 2). The rostral and caudal MPP latencies after contralateral stimulation showed only a slight increase (Fig. 2).

Effect of nimodipine on SCEP changes

Amplitudes

Pretreatment with nimodipine (2 µg/kg/min continued infusion) attenuated all postinjury SCEP changes following both ipsi- or contralateral stimulation in rostral and caudal recordings (Figs. 1 and 2). The rostral MNP amplitude showed some decrease after ipsilateral stimulation during the first 10 min but remained at a higher level throughout the postinjury recording period (Figs. 1 and 2). The caudal MNP amplitude was increased in all postinjury recordings after contralateral stimulation (Figs. 1 and 2).

Latencies

Nimodipine was found to be very effective in reducing latency changes. Thus, apparently no significant increase in latency was noted in the drug-treated rats after trauma as compared to the untreated group in both rostral or caudal recordings after ipsilateral stimulation (Figs. 1 and 2). The MPP latencies both

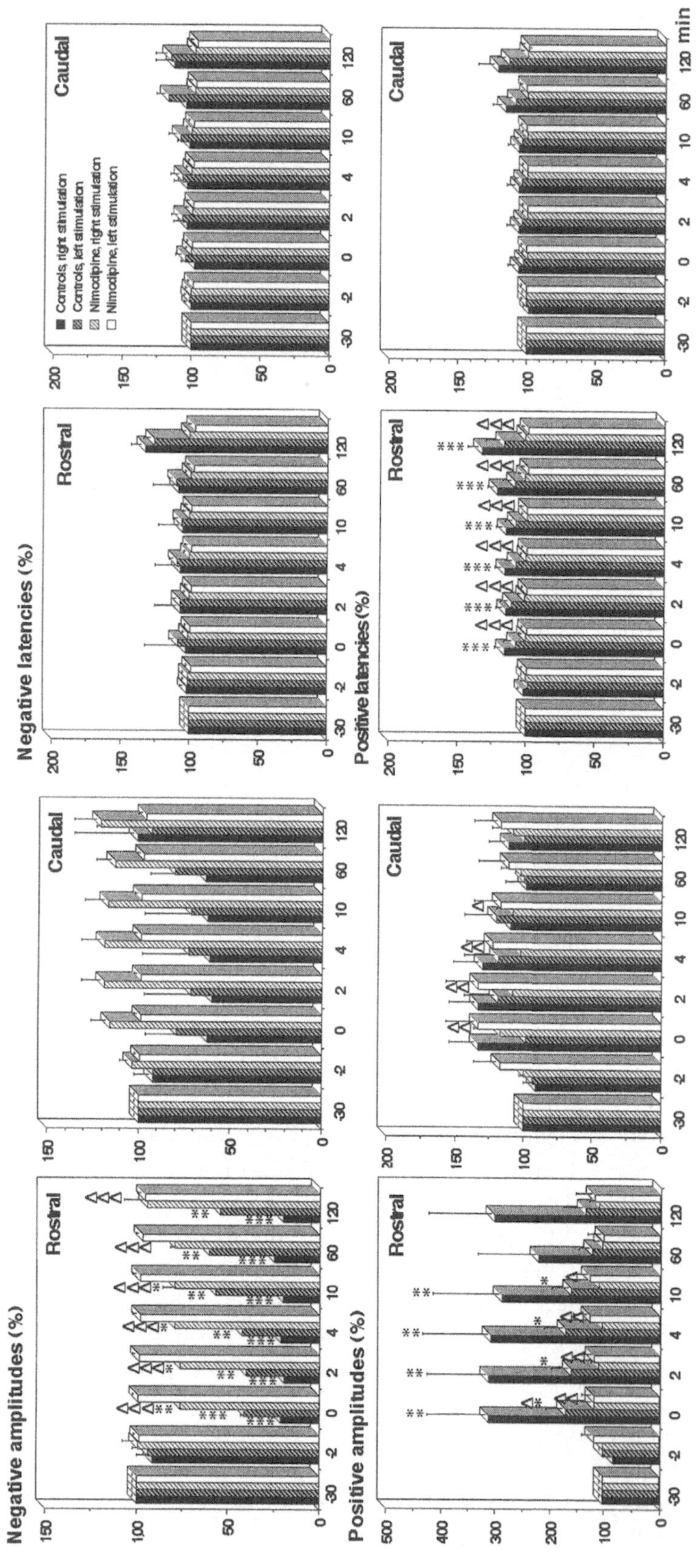

Fig. 2. Mean SCEP recording from rostral and caudal electrodes in untreated traumatised rats and their modification by the nimodipine treatment. *** P = <0.001, ** P = 0.01, * P = <0.05 from preinjury level, ΔΔΔ = P < 0.001; ΔΔ = P < 0.01, Δ = P < 0.05 from untreated traumatised rats

in rostral and caudal recordings following ipsilateral stimulation increased markedly after injury. Following contralateral stimulation there was no significant change in MPP latency in either rostral or caudal recordings (Figs. 1 and 2).

Effect of nimodipine on the physiological variables

Untreated spinal cord injured rats exhibited a marked reduction in the MABP at 5 h (90 ± 4 torr) compared to the normal animals (120 ± 8, $P < 0.01$). The decrease in MABP following 5 h injury was not significantly affected by the nimodipine treatment (84 ± 6 torr). Further, the arterial pH (7.38 ± 0.004) and blood gases (PaO2, 81.56 ± 0.34 torr; PaCO2, 34.47 ± 0.21 torr) were not significantly affected by trauma nor nimodipine treatment (results not shown).

Effect of nimodipine on spinal cord blood flow

SCI markedly reduced the SCBF in the T9 segment at 5 h (0.71 ± 0.08 ml/g/min) compared to normal rats (0.98 ± 0.06 ml/g/min, $P < 0.01$). The reduction in the local blood flow following SCI was significantly attenuated by the nimodipine treatment (0.88 ± 0.07 ml/g/min, $P < 0.05$).

Effect of nimodipine on spinal cord edema

The water content (68.56 ± 0.32%) of the injured spinal cord segments (T10–11) at 5 h was significantly higher than in normal rats (65.23 ± 0.21%, $P < 0.01$). Pretreatment with nimodipine significantly reduced the SCI induced increase in water content (66.67 ± 0.23%; $P < 0.01$) as compared to the untreated injured rats.

Effect of nimodipine on cell changes and axonal injury

The trauma to the spinal cord resulted in marked cell and tissue damage. Many nerve cells were dark and distorted in the gray matter of the T9 segment following 5 h injury (Fig. 3A.a). The spinal cord gray matter further showed a loss of nerve cells and general sponginess (Fig. 3A.a). In these rats, the immunostaining for MBP revealed a profound loss of myelin (Fig. 3A.c) as compared to the normal animals (results not shown). Pretreatment with nimodipine significantly attenuated the SCI induced cell loss, distortion of nerve cells, sponginess and edema in the spinal cord gray matter (Fig. 3A.b). In the nimodipine treated traumatised animals, the loss of MBP was much less pronounced (Fig. 3A.d) as compared to the untreated spinal cord injured rats (Fig. 3A.c).

Effect of nimodipine on the ultrastructural changes of myelin

At the ultrastructural level, a widespread damage of myelinated or unmyelinated axons was found in the spinal cord of the traumatised rats (Fig. 3B.a,b). Myelin vesiculation, edema and membrane damage were most common in the untreated traumatised rats (Fig. 3B.a,b).

Pretreatment with nimodipine markedly attenuated the axonal damage, membrane disruption, splitting of myelin and edema (Fig. 3B.c,d). Thus, in nimodipine treated traumatised animals, myelin vesiculation and membrane damage was much less frequent as compared to the untreated injured rats (Fig. 3B). Nimodipine treatment induced less distortion of the spinal cord, and the myelinated and non-myelinated axons were considerably less damaged (results not shown).

Discussion

The salient new findings of this study are that continuous low dose infusion of an L-type calcium channel blocker, nimodipine improves spinal cord conduction and causes marked inhibition in spinal cord ischemia, edema formation, and axonal injury without profound arterial hypotension. These observations suggest that calcium channel antagonists may be useful in SCI [8, 9, 22, 30].

Previously, ion channel blockers were found to improve conduction deficits from a trauma induced redistribution of ion fluxes across the cell membrane underlying an altered spinal cord conduction and cell injury. A potent blocker of the voltage dependent potassium channel is 4-aminopyridine (4-AP), which has beneficial effects by influencing the calcium influx and phosphorylation of membrane proteins [18, 19]. The drug is able to restore conduction in patients following chronic SCI [15] indicating its possible therapeutic use for the future.

The intracellular calcium influx is the final common pathway of cell death in the CNS [28]. Thus, the use of calcium channel antagonists appears to be beneficial in SCI, provided the associated arterial hypotension can be controlled by other supporting drugs. Tator's labo-

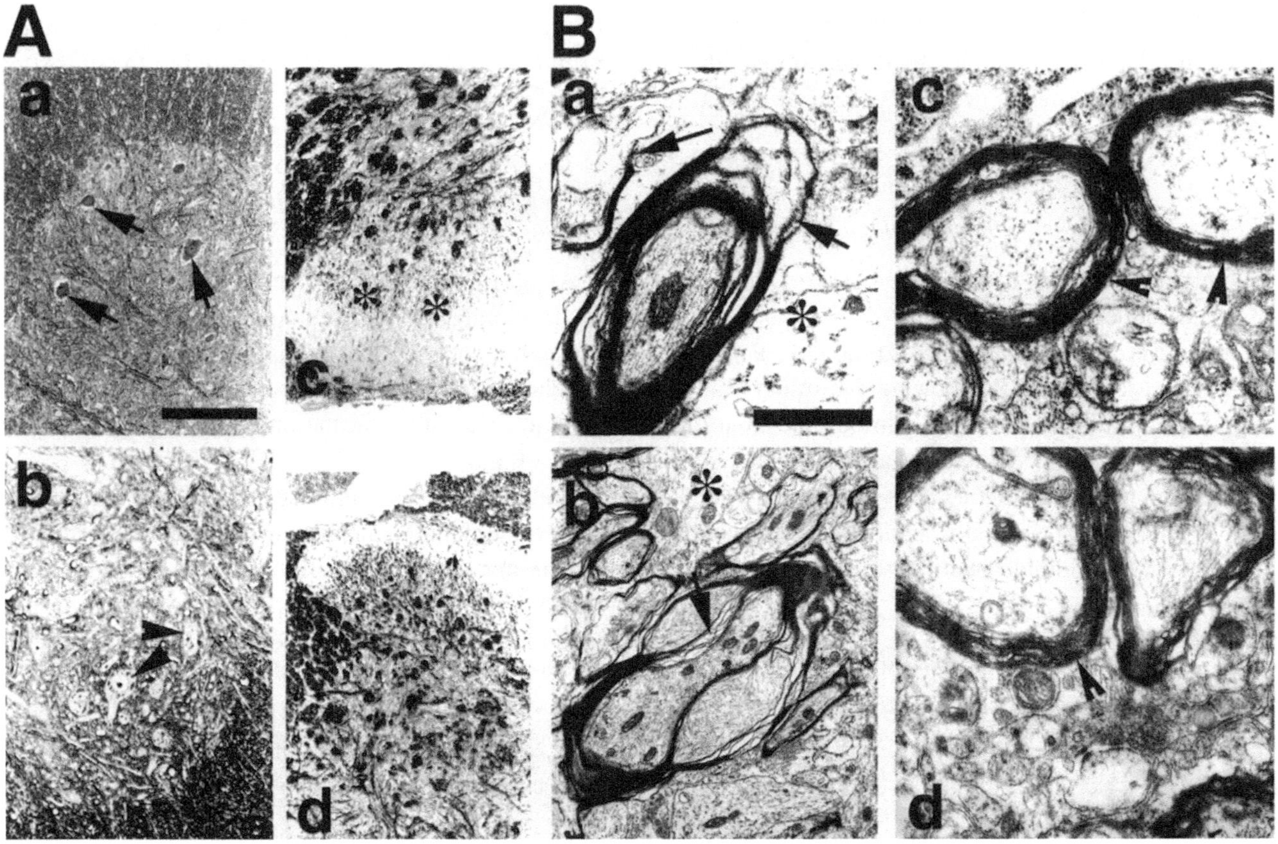

Fig. 3. Morphological alterations in the spinal cord following SCI and their modification with nimodipine treatment. (A) Toloudine stained 1 μm section from the ventral horn of an untreated rat 5 h after lesion (a). Many distorted nerve cells (arrows) can be seen after SCI (a). Pretreatment with nimodipine markedly attenuated these nerve cell changes (b). MBP immunostaining of 3 μm paraffin section from an untreated injured rat (c) and its modification with nimodipine treatment (d). Degradation of MBP is evident in untreated traumatised rats (**). (B) Ultratstructural changes in myelinated axons from an untreated traumatised rat (a,b) and their modification with nimodipine treatment (c,d). Myelin vesiculation (arrows) and edema (*) are prominent in the dorsal horn (c) and ventral horn (b) of the untreated injured rats. Nimodipine treatment markedly reduced the myelin vesiculation and edema in both the dorsal (c) and in ventral horn (d) after SCI. Bar: A = 25 μm, B = 500 nm

ratory was the first to show a beneficial effect of the calcium channel blocker nimodipine in spinal cord injury which reduces post-traumatic ischemia [see 30]. Since increased intracellular calcium is the common denominator of the action of many neurochemicals, the blockade of calcium channels by nimodipine should be neuroprotective in spinal trauma.

The spinal cord exhibits a wide area of negative field potentials within the dorsal horn up to lamina VII and adjacent dorsal columns [3, 11, 32]. A balance between synaptic excitation and inhibition is responsible for the normal spinal cord conductivity and SCEP response, which is greatly perturbed by a focal lesion of the dorsal horn [6, 21]. An increase in MNP in the caudal SCEP recording following lesion, represents a loss of synaptic inhibition [see 16].

The appearance of a broad positive potential peak appears to be similar to that of injury potentials or "killed end potentials" seen by various other workers in experimental SCI [see 7, 12]. According to the classical volume conduction theory, positive potentials represent a large positive 'killed end' signal, which is measured by a distal electrode [9]. The local ionic changes in the extra- or intracellular space at the site of the lesion may also contribute to this potential [12]. This injury potential suggests major cord damage [7, 12, 26].

The amplitude- and latency changes represent a combination of demyelination and axonal injury, respectively and constitute important parameters in the assessment of the severity of a spinal lesion [26, 33]. Nimodipine treatment significantly attenuated the

SCEP latency changes in this study, and was able to thwart myelin vesiculation. These observations suggest that nimodipine has a pronounced neuroprotective effect on vesiculation of myelin following SCI.

An increase in SCEP latency is one of the most sensitive indicators of cord injury underlying slowing of axonal conduction in motor tracts across the lesion site [3, 17]. The latency increase may be due to the demyelination of axons as observed by the loss of MBP immunostaining [12, 13, 27].

The effect of injury on the MPP latency was most pronounced in the rostral recordings after ipsilateral stimulation. Since the rostral MNP was virtually abolished after injury, the latency could not be calculated. However, the MNP latencies when present showed a similar increase after injury. An increase in latency is attributable to slower conduction by the tracts of the dorsal columns due to myelin damage as confirmed by the pathological findings in SCI [2, 4, 30]. An increased MPP amplitude is often combined with an increase of MPP latency.

Our investigation further demonstrates that the early SCEP changes occur in close association with the development of edema and cell damage of the injured cord. Edema is a primary response to all types of mechanical, metabolic and ischemic insults to the cord [9, 20, 25]. This indicates that SCEP could be of significance in evaluating edema formation in the cord after injury. This is further supported by the findings that in nimodipine-treated animals the postinjury decreases in MNP amplitude and increases in water content at 5 h were minimal or absent [23]. The MNP amplitude is also correlated with the local spinal cord blood flow [30]. In the nimodipine treated injured rats the SCBF was almost restored to normal values. This suggests that nimodipine has significant protective effects on SCI induced ischemia resulting in an improved SCEP activity in the traumatised cord.

The MNP amplitude in caudal recordings, however, showed a mixed effect. Thus, in some cases there was virtually an increase in the amplitude after injury. This could be due to the blockade of an inhibitory influence on SCEP [26]. A similar increase in SCEP potentials reported in monkeys and cats after SCI is in line with our observations in the rat model [31]. The less pronounced effect on SCEP amplitudes after contralateral stimulation suggests that the tracts of the intact side could still conduct action potentials. The occasional finding of a decrease in SCEP amplitude after contralateral stimulation may be due to the spread of spinal cord depression, e.g. by a release of secondary injury factors (neurochemicals) in the contralateral side [12, 26, 35].

Conclusion

In conclusion, our results demonstrate that pretreatment with nimodipine is effective in preventing postinjury changes of SCEP and spinal cord pathology, indicating a clinical usefulness of nimodipine for the treatment of acute spinal cord injuries. Our studies further indicate that the early spinal cord bioelectrical activity and pathology following trauma reflects an altered neurochemical homeostasis in the cord. Taken together, it appears that the SCEP changes are useful indicators of the development of spinal cord pathology after trauma.

Acknowledgments

This investigation is supported by grants from the Swedish Medical Research Council project nos. 2710 (JW/HSS), the Göran Gustafsson Foundation, Sweden, Alexander von Humboldt Foundation, Bonn, Germany, Bayer Pharmaceuticals, Leverkusen, Germany, and the University Grants Commission, New Delhi, India. technical assistance of Kärstin Flink, Kerstin Rystedt, Inga Hörte, and Madeleine Jarild is acknowledged with thanks.

References

1. Balentine JD, Spector M (1977) Calcification of axons in experimental spinal cord trauma. Ann Neurol 2: 520–523
2. Balentine JD (1988) Spinal cord trauma: in search of the meaning of granular axoplasm and vesicular myelin. J Neuropath Exp Neurol 47: 77–92
3. Bennet HM (1984) Correlations of ascending and descending conduction block following ischemic and compressive insult to the spinal cord of cat. Fundamentals and clinical application of spinal cord monitoring. In: Homma S, Tamaki T (eds) Saikon Publishing Co, Tokyo, pp 125–134
4. de la Torre, JC (1981) Spinal cord injury. Review of basic and applied research. Spine 6: 315–335
5. Demediuk P, Faden AI (1988) Traumatic spinal cord injury in rats causes increases in tissue thromboxane but not peptidoleukotrienes. J Neurosci Res 20: 115–121
6. Dimitrijevic MR, Dimitrijevic MM, Faganel J, Sherwood AM (1984) Suprasegmentally induced motor unit activity in paralyzed muscles of patients with established spinal cord injury. Ann Neurol 16: 216–221
7. Ducker TB, Brown RH (1989) Neurophysiology and standards of spinal cord monitoring. Elsevier, Amsterdam
8. Faden AI (1993) Experimental neurobiology of central nervous system trauma. Crit Rev Neurobiol 7: 175–186
9. Faden AI, Salzman S (1992) Pharmacological strategies in CNS trauma. TiPS 13: 29–35
10. Farber JL (1982) Biology of disease: membrane injury and calcium homeostasis in the pathogenesis of coagulative necrosis. Lab Invest 47: 114–123

11. Foreman RD, Kenshalo DR Jr, Schmidt RF, Willis WD (1979) Field potentials and excitation of primate spinothalamic neurones in response to volleys in muscle afferents. J Physiol (Lond) 286: 197–213
12. Fehlings MG, Tator CH, Linden RD (1989) The relationships among the severity of spinal cord injury, motor and somatosensory evoked potentials and spinal cord blood flow. Electroenceph Clin Neurophysiol 74: 241–259
13. Griffiths IR, Miller R (1974) Vascular permeability to protein and vasogenic oedema in experimental concussive injuries to the canine spinal cord. J Neurol Sci 22: 291–304
14. Happel RD, Smith KP, Banik NL, Powers JM, Hogan EL, Balentine JD (1981) Ca^{2+}-accumulation in experimental spinal cord trauma. Brain Res 211(2): 476–479
15. Hayes KC, Potter PJ, Wolfe DL, Hsieh JTC, Delaney GA, Blight AR (1994) 4-aminopyridine sensitive neurological deficits in patients with spinal cord injury. J Neurotrauma 11: 433–436
16. Homma S, Tamaki T (1984) Fundamentals and clinical applications of spinal cord monitoring. Saikon Publishing Co, Tokyo, Japan, pp 1–450
17. Levy W, McCaffrey M, York D (1986) Motor evoked potential in cats with spinal cord injury. Neurosurgery 19: 9–19
18. Lundh H, Thesleff S (1977) The mode of action of 4-aminopyridine and guanidine on transmitter release from motor nerve terminals. Eur J Pharmacol 42: 411–412
19. Molgo J, Lundh H, Thesleff S (1980) Potency of 3,4-diaminopyridine and 4-aminopyridine on mammalian neuromuscular transmissions and the effect of pH changes. Eur J Pharmacol 61: 25–34
20. Olsson Y, Sharma HS, Pettersson Å, Cervós-Navarro J (1992) Endogenous release of neurochemicals may increase vascular permeability, induce edema and influence on cell changes in trauma to the spinal cord. Progr Brain Res 91: 197–203
21. Schramm J, Krause R, Shigeno T, Brock M (1984) Relevance of spinal cord evoked potentials for monitoring. In: Homma S, Tamaki T (1984) Fundamentals and clinical applications of spinal cord monitoring. Saikon Publishing Co, Tokyo, Japan, pp 113–124
22. Schwab ME, Bartholdi D (1996) Degeneration and regeneration of axons in the lesioned spinal cord. Physiol Rev 76: 319–370
23. Sharma HS, Cervós-Navarro J (1990) Nimodipine improves cerebral blood flow and reduces brain edema, cellular damage and blood-brain barrier permeability following heat stress in young rats. Pharmacology of cerebral ischemia 1990. In: Krieglstein J, Oberpichler H (eds) CRC Press, Boca Raton, Florida, pp 303–310
24. Sharma HS, Olsson Y (1990) Edema formation and cellular alteration in spinal cord injury in the rat and their modification with p-chlorophenylalanine. Acta Neuropathol (Berl) 79: 604–610
25. Sharma HS, Olsson Y, Dey PK (1990a) Early accumulation of serotonin in rat spinal cord subjected to traumatic injury. Relation to edema and blood flow changes. Neuroscience 36: 725–730
26. Sharma HS, Winkler T, Stålberg E, Olsson Y, Dey PK (1991) Evaluation of traumatic spinal cord edema using evoked potentials recorded from the spinal epidural space. An experimental study in the rat. J Neurol Sci 102: 150–162
27. Sharma HS, Olsson Y, Cervós-Navarro J (1993a) Early perifocal cell changes and edema in traumatic injury of the spinal cord are reduced by indomethacin, an inhibitor of prostaglandin synthesis. Acta Neuropathol (Berlin) 85: 145–153
28. Siesjö BK (1988) Historical overview: calcium, ischemia, and death of brain cells. Ann NY Acad Sci 522: 638–661
29. Stålberg E, Sharma HS, Olsson Y (1998) Spinal cord monitoring. Basic principles, regeneration, pathophysiology and clinical aspects. Springer, Wien New York, pp 1–527
30. Tator CH, Fehlings MG (1991) Review of the secondary injury theory of acute spinal cord trauma with emphasis on vascular mechanisms. J Neurosurg 75: 15–26
31. Willis WD Jr (1984) Evoked spinal cord potentials in the cat and monkey: use in the analysis of spinal cord function. Fundamentals and clinical applications of spinal cord monitoring. In: Homma S, Tamaki T (eds) Saikon Publishing Co, Tokyo, pp 3–20
32. Willis WD Jr (1980) Spinal cord potentials. Spinal cord and its reaction to traumatic injury. In: Windle WF (ed) Marcel Dekker Inc, New York, pp 159–187
33. Winkler T (1994) Evaluation of spinal cord injuries using spinal cord evoked potentials. An experimental study in the rat. Acta Universitatis Upsaliensis, Comprehensive Summaries of Uppsala Dissertations from the Faculty of Medicine 467. Almqvist & Wiksell International, Stockholm, Sweden, 1–45 pp
34. Winkler T, Sharma HS, Stålberg E, Olsson Y, Nyberg F (1994a) Opioid receptors influence spinal cord electrical activity and edema formation following spinal cord injury. Experimental observations using naloxone in the rat. Neurosci Res 21: 91–101
35. Winkler T, Sharma HS, Stålberg E, Olsson Y (1993) Indomethacin, an inhibitor of prostaglandin synthesis attenuates alteration in spinal cord evoked potentials and edema formation after trauma to the spinal cord. An experimental study in the rat. Neuroscience 52: 1057–1067
36. Winkler T, Sharma HS, Stålberg E, Westman (1998) Spinal cord bioelectrical activity, edema and cell injury following a focal trauma to the spinal cord. An experimental study using pharmacological and morphological approach. Spinal cord monitoring. In: Stålberg E, Sharma HS, Olsson Y (eds) Springer, Wien New York, pp 281–348
37. Young W, Flamm ES (1982) Effect of high-dose corticosteroid therapy on blood flow, evoked potentials, and extracellular calcium in experimental spinal injury. J Neurosurg 57: 667–673

Correspondence: Hari Shanker Sharma, Dr. Med. Sci., Laboratory of Neuroanatomy, Department of Medical Cell Biology, Biomedical Centre, Uppsala University, S-751 23 Uppsala, Sweden. e-mail: Sharma@medcellbiol.uu.se

Acta Neurochir (2003) [Suppl] 86: 433–438

Multiple hyperbaric oxygenation (HBO) expands the therapeutic window in acute spinal cord injury in rats

L. Huang[3], M. P. Mehta[1], J. H. Eichhorn[1], A. Nanda[2], and J. H. Zhang[2]

[1] Department of Anesthesiology, University of Mississippi Medical Center, Jackson, Mississippi
[2] Department of Neurosurgery, Louisiana State University Health Sciences Center in Shreveport, Shreveport, Louisiana
[3] Division of Orthopaedic Surgery, The First Affiliated Hospital of Suzhou University, Suzhou, Jiangsu, China

Summary

Hyperbaric oxygenation (HBO) therapy has been reported to improve neurological recovery following spinal cord injury (SCI). In the present study, we examined whether multiple HBO expands the therapeutic window for acute SCI. Single HBO (2.8 ATA, 1 hour) treatment was used at 30 minutes, 3 hours, and 6 hours following SCI, and serial HBO treatment (once daily for 1 week) at 6 hours and 24 hours post-injury. Mild SCI was induced by adjusting the height for a weight drop insult (10 g) to 6.25 mm above the exposed spinal cord. The group of animals receiving a single HBO intervention beginning at 30 minutes and 3 hours, or serial HBO treatment starting at 6 hours following the injury had a significantly better neurological recovery than animals with SCI only. The results of this study demonstrate that multiple HBO expands the therapeutic window for acute SCI to 6 hours after injury, further that serial HBO administration is superior to single HBO therapy.

Keywords: Behavior; hyperbaric oxygen; spinal cord injury; therapy.

Introduction

Spinal cord injury (SCI), a catastrophic and disabling illness, occurs primarily in motor vehicle accidents, sports and recreational activities, accidents at work, falls, and diving. SCI afflicts approximately 8,000–12,000 patients per year in the United States [1, 2] with an annual rate of 20–40 per million in most other countries [3, 4]. The majority of affected patients are in the 15 to 29 years age group. SCI has permanent and devastating effects on the lives of these individuals. Although much experimental and clinical research has been carried out on the mechanisms of injury and on various treatment modalities in an effort to improve the neurological outcome, the success of the current treatment of SCI remains poor [5].

HBO has been advocated as therapy to improve the neurological recovery following brain injury [6] and cerebral ischemia [7–10] because it may improve oxygen delivery to the tissue. The basis of using HBO for SCI corresponds to the application of this therapy in brain injuries. Although HBO treatment for SCI has been investigated before, its effectiveness is still controversially discussed. So far, no data have been established on the therapeutic window of HBO. For the current study, we have used a clinically relevant model of SCI in rats to elucidate, whether multiple HBO treatment expands the therapeutic window.

Material and methods

All experimental procedures and animal care protocols were approved and conducted in accordance with the guidelines of the Animal Care and Use Committee at the University of Mississippi Medical Center.

Experimental groups

Seventy mature female Sprague-Dawley rats (weight range 300–350 g) used for this study were randomly divided into 7 groups (Fig. 1): Group 1 – Sham surgery (n = 5), Group 2 – SCI without treatment (n = 16), Groups 3–5 – Single HBO treatment 30 minutes (n = 10), 3 hours (n = 11), and 6 hours following SCI (n = 9), respectively, and Groups 6–7 – Serial HBO treatment 6 hours (n = 10) and 24 hours following SCI for 1 week (n = 9), respectively.

Animal model and surgical procedure

The mild NYU weight-drop SCI model was used in this study. A mild traumatic insult was preferred because too severe injury may result in complete paraplegia which would overshadow the effectiveness of the treatment [11]. Rats were anesthetized by an intraperitoneal injection of ketamine (80 mg/kg) and xylazine (10 mg/

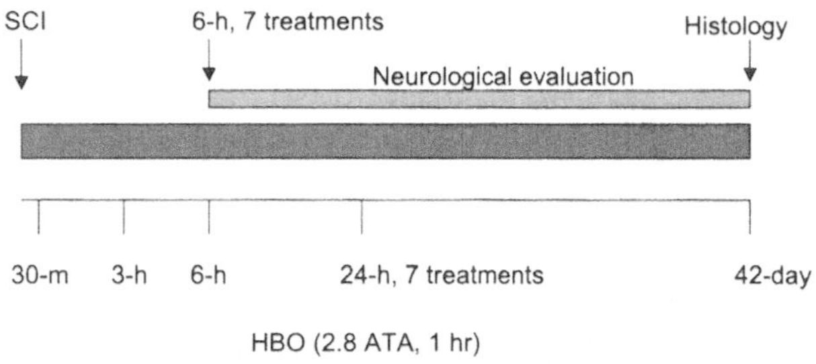

Fig. 1. Schematic representation of the experimental design. The red bar indicates the duration of the HBO treatment and the experimental observation period. Upon SCI, animals received a single HBO exposure either at 30 min (30-m), 3 hours, 6 hours (3-h and 6-h), or multiple HBO exposures at 6 hours (6-h) and at 24 hours (24 h). The green bar indicates the time period of neurological evaluation. Rats were sacrificed 42 days after SCI for histological examination

kg). A single-level laminectomy was performed at the vertebral T10 level to expose the dorsal aspect of the spinal cord without rupturing the dura. A mild SCI was induced by dropping a 10-gram rod from a height of 6.25 mm (6.25-gram centimeter force) directly onto the exposed spinal cord. The rod was immediately lifted thereafter. The surgical wound was closed then in layers by sutures.

Animals of group 3 to 7 were placed in a hyperbaric chamber at 30 minutes, 3 hours, 6 hours, or 24 hours following SCI according to the experimental protocol.

Neurologic assessment

The open field locomotor function was evaluated by using the BBB locomotor rating scale [4], which ranges from 0 – with no evidence of observable hind limb movement – to 21 in a normal rat. Examiners were blinded to the type of treatment.

Histopathological assessment

At 42 days post SCI, rats were deeply anesthetized with sodium pentobarbital for intracardiac perfusion fixation with 10% formaldehyde. The spinal cord segment containing the contusion epicenter, was embedded in paraffin. The tissue sections were stained with Luxol fast blue, hematoxylin, and eosin.

Statistical analysis

Data of the BBB open-field locomotor functional score were expressed as mean $\pm$ SEM. Statistical differences between the control (SCI, without HBO treatment) and other groups were assessed by using the one-way ANOVA. A p value of less than 0.05 was considered to be statistically significant.

Results

Neurologic assessment

The neurologic results are summarized in Fig. 2a–e. None of the rats in group 1 had a neurological deficit after sham-surgery, whereas all rats in the SCI group exhibited a pronounced hind limb locomotor deficit on the first testing day, deteriorating further on the third day. Eighteen days after injury, the locomotor performance scores of rats in group 3, 4, and 6 were significantly improved as compared to animals without treatment (group 2). The BBB scores of rats in groups 5 and 7 were not significantly different from those in group 2. This trend continued throughout the duration of the evaluation period.

Histopathological assessment

The results of the histopathological examination are shown in Fig. 3. The spinal cord sections obtained from sham-surgery animals appeared normal. The center of the injured spinal cord of animals of group 2 showed a characteristical necrosis, wide demyelination, and cavitation of the white matter in the dorsomedial region. It demonstrated also an almost complete loss of neurons in the gray matter as well as marked gliosis, infiltration of inflammatory cells, and vacuolization (Fig. 3a, A). The spinal cord of the animals of groups 5 and 7 had the same appearance as seen in group 2 (Fig. 3d, D, f, F). The histological sections of the injured spinal cord obtained from groups 3, 4, and 6 with HBO treatment beginning either at 30 minutes and 3 hours, or as serial administration beginning at 6 hours post-injury, contained more spared myelin (which was stained by Luxol Fast Blue in white matter) and more viable neurons in the gray matter (Fig. 3b, B, c, C, e, E) than in animals of group 2. The histological sections seen in Fig. 3 represent five samples of each group.

Discussion

The present study demonstrates that a single HBO exposure beginning at 30 minutes or at 3 hours post-injury is able to improve the neurological outcome from SCI. We report further that serial HBO treatment beginning at 6 hours following SCI has an even greater therapeutical efficacy for the neurological recovery than the single HBO exposure given at the same time. To the contrary, single HBO exposure beginning at 6 hours or serial HBO exposure starting only at 24 hours post-injury failed to provide an improvement of the locomotor function or of the histopathological findings when compared with the SCI-animals not receiving treatment.

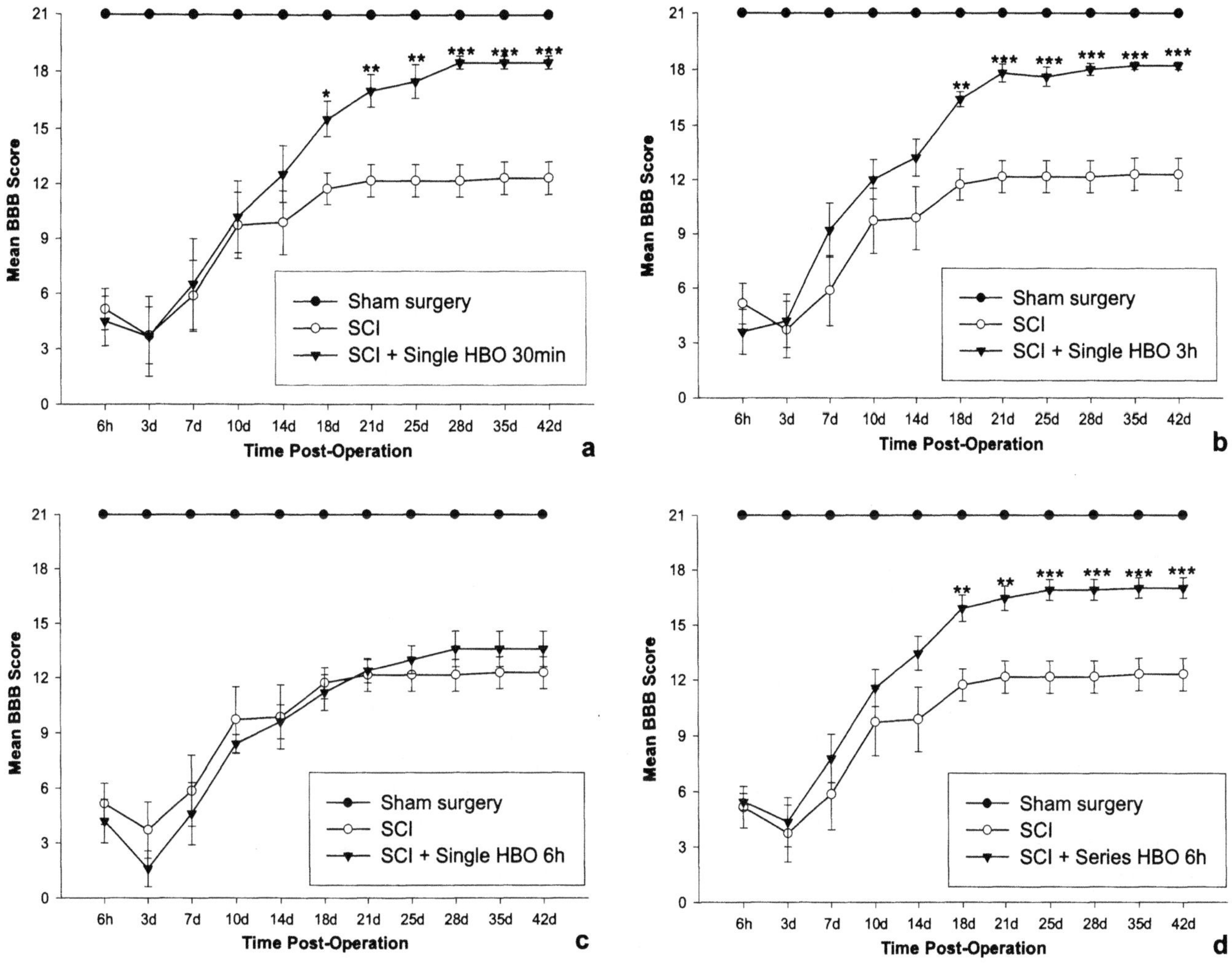

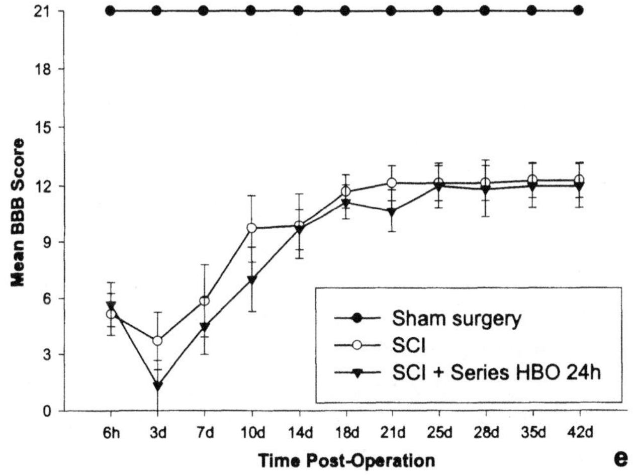

Fig. 2. Functional recovery and outcome. The open field locomotor scores of the BBB scale of SCI rats receiving a single HBO treatment beginning at 30 minutes post-injury became significantly better than that of the SCI rats without treatment at day 18 thereafter (a). The BBB scores of SCI rats receiving a single HBO exposure beginning at 3 hours post-injury were also significantly higher than that of the untreated SCI animals from day 18 on and later (b). On the other hand, there was no difference between SCI rats receiving a single HBO exposure beginning at 6 hours post-injury and the untreated controls (c). The BBB score of SCI rats which received serial HBO treatment beginning at 6 hours post-injury was also significantly better than that of SCI animals without treatment (d), while the scores did not show any differences between the SCI rats receiving serial HBO treatment with onset at 24 hours post-injury and the untreated controls (e). (*$p < 0.05$, **$p < 0.01$, ***$p < 0.001$)

Hypoxia and secondary injury after SCI

Many findings suggest that the spinal cord suffers not only from the primary mechanical insult, but also from secondary injury. Whereas the primary injury is immediate and irreversible, the secondary tissue damage evolves over time providing a window of opportunity for therapeutic interventions. In the current study, most of the experimental animals could move some of their hind limb joints extensively at 6 hours following

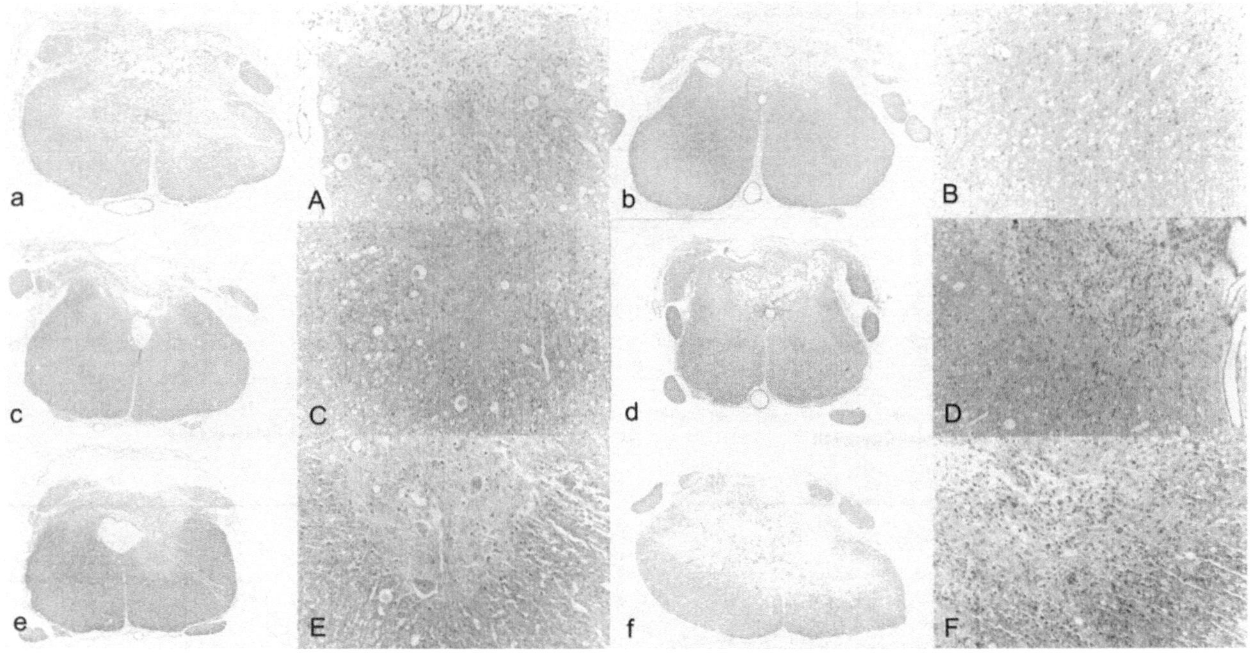

Fig. 3. Histopathological findings of the lesion center at 42 days after mild spinal cord injury in rats. The spinal cord sections obtained from animals of the SCI-only group had a characteristical necrosis, wide demyelination of the white matter in the dorso-medial region, and an almost complete loss of neurons in the gray matter with marked gliosis, infiltration of inflammatory cells, and vacuolization (a, A). The spinal cord of animals which received single HBO initiated at 6 hours or serial HBO beginning at 24 hours post-injury had a similar appearance (d, D, f, F). The sections obtained from rats with single HBO exposure beginning at 30 minutes or 3 hours, and serial HBO exposure starting at 6 hours following SCI retained more myelinated tissue and more viable neurons (b, B, c, C, e, E). The sections were stained with Luxol fast blue, hematoxylin, and eosin. Original magnification, ×50 (a, b, c, d, e, f), ×250 (A, B, C, D, E, F). The representative histological sections were obtained from five animals per group

mild SCI, while the motor function always appeared to have deteriorated on day 3 post-injury. This observation supports the concept of secondary damage following a primary mechanical insult.

Many investigators have shown that the secondary damage following traumatic SCI is associated with edema and ischemia [11, 12], which may be a consequence of vasoparalysis and of the direct injury of blood vessels of the spinal cord. Focal spinal cord ischemia is compounded by arterial hypotension and hypoxia, since autoregulation of blood flow is lost in SCI, which then passively follows the alterations of the systemic blood pressure [13]. Hypoxia develops from focal ischemia within and adjacent to the injured cord segments and/or from systemic cardiac and pulmonary dysfunction. According to various studies, hypoxia plays an important role in the pathogenesis of secondary spinal cord damage [14–16]. Based on this concept, it seems that the injured spinal cord would benefit from improving the microcirculation and the supply of oxygen to the injured parenchyma by HBO.

Mechanisms of HBO-induced protection

The present findings show that HBO initiated at an early stage following SCI improves the final neurological and histopathological outcome significantly. The rationale of the HBO treatment in SCI is that thereby the oxygen level in the hypoxic tissue is increased. Kelly and colleagues [17] have studied the tissue PO_2 of normal and traumatized spinal cords in dogs. They have observed that the tissue PO_2 of the injured cord dropped to near zero and did not respond to ventilation with 100% oxygen at ambient pressure. Administration of oxygen at 2 ATA, however, led the tissue PO_2 rise to high levels during a period of mechanical compression of the cord. Narayana [18] *et al.* have used magnetic resonance imaging (MRI) to assess the efficacy of HBO in experimental SCI in rats. The results indicate that the benefit of the HBO treatment in SCI is attributable to reducing edema and arrest of the spread of hemorrhage. Others [19] have found an enhanced regeneration of neuronal structures. Accordingly, damage of myelin and of axons was significantly less severe in animals with repetitive HBO treatment.

Therapeutic window of HBO

The therapeutic window of implementation of HBO *after* SCI remains controversial. Higgins [20] subjected SCI cats to HBO treatment at 2 ATA for a period of 3 hours. They found that a delay by 2 hours following insult abolishes the protection of long-tract neuronal conduction provided by HBO. Murakami [21] has observed that HBO initiated at 30 minutes after an ischemic insult was protective against spinal cord damage in rabbits, while a delay of 6 hours not. The majority of reports indicate that HBO in SCI has a therapeutical benefit only, if it is administered early after injury [11, 22, 23].

The present results confirm that SCI animals with HBO initiated within 3 hours post-injury have a better neurological recovery and histopathological outcome than those with a delay by more than 6 hours. The time period of 2 to 4 hours may represent the phase of "golden hours" [11] . Thereafter, the pathophysiological sequelae are progressing, resulting in permanent anatomical disruption of the initially functional cord. It is interesting to note in our experiments that the single HBO exposure at 6 hours *after* SCI had no beneficial effects, while serial HBO treatment initiated after the same time period not only improved the neurological recovery but also the histopathological outcome. Based on this result, we suggest that the therapeutic window for HBO in spinal cord injury is open for 6 hours post insult.

Acknowledgment

This work was partially supported by a grant from the American Heart Association National Grant-in-Aid, and a grant from the American Heart Association Bugher Foundation Award for the Investigation of Stroke to J.H.Z.

References

1. Yanagawa Y, Marcillo A, Garcia-Rojas R, Loor KE, Dietrich WD (2001) Influence of posttraumatic hypoxia on behavioral recovery and histopathological outcome following moderate spinal cord injury in rats. J Neurotrauma 18(6): 635–644
2. McVicar JP, Luce JM (1986) Management of spinal cord injury in the critical care setting. Crit Care Clin 2(4): 747–758
3. Yanagawa Y, Marcillo A, Garcia-Rojas R, Loor KE, Dietrich WD (2001) Influence of posttraumatic hypoxia on behavioral recovery and histopathological outcome following moderate spinal cord injury in rats. J Neurotrauma 18(6): 635–644
4. Kraus JF, Franti CE, Riggins RS (1975) Incidence of traumatic spinal cord lesions. J Chronic Dis 28: 471–492
5. Gamache FW Jr, Myers RA, Ducker TB, Cowley RA (1981) The clinical application of hyperbaric oxygen therapy in spinal cord injury: a preliminary report. Surg Neurol 15(2): 85–87
6. Rockswold SB, Rockswold GL, Vargo JM, Erickson CA, Sutton RL, Bergman TA, Biros MH (2001) Effects of hyperbaric oxygenation therapy on cerebral metabolism and intracranial pressure in severely brain injured patients. J Neurosurg 94(3): 403–411
7. Mink RB, Dutka AJ (1995) Hyperbaric oxygen after global cerebral ischemia in rabbits reduces brain vascular permeability and blood flow. Stroke 26(12): 2307–2312
8. Sunami K, Takeda Y, Hashimoto M, Hirakawa M (2000) Hyperbaric oxygen reduces infarct volume in rats by increasing oxygen supply to the ischemic periphery. Crit Care Med 28(8): 2831–2836
9. Veltkamp R, Warner DS, Domoki F, Brinkhous AD, Toole JF, Busija DW (2000) Hyperbaric oxygen decreases infarct size and behavioral deficit after transient focal cerebral ischemia in rats. Brain Res 853(1): 68–73
10. Badr AE, Yin W, Mychaskiw G, Zhang JH (2001) Dual effect of HBO on cerebral infarction in MCAO rats. Am J Physiol Regul Integr Comp Physiol 280(3): R766–R770
11. Sukoff M, Jain KK (1999) Hyperbaric oxygen therapy in neurosurgery. Textbook of hyperbaric medicine. Hogrefe & Huber Publishers, pp 351–371
12. Yanagawa Y, Marcillo A, Garcia-Rojas R, Loor KE, Dietrich WD (2001) Influence of posttraumatic hypoxia on behavioral recovery and histopathological outcome following moderate spinal cord injury in rats. J Neurotrauma 18(6): 635–644
13. Tator CH, Fehlings MG (1991) Review of the secondary injury theory of acute spinal cord trauma with emphasis on vascular mechanisms. J Neurosurg 75(1): 15–26
14. Osterholm JL (1974) The pathophysiological response to spinal cord injury. The current status of related research. J Neurosurg 40(1): 5–33
15. Holbach KH, Wassmann H, Linke D (1977) The use of hyperbaric oxygenation in the treatment of spinal cord lesions. Eur Neurol 16(1–6): 213–221
16. Lee MR, Sakatani K, Young W (1993) Interaction of hypoxia and hypothermia on dorsal column conduction in adult rat spinal cord in vitro. Exp Neurol 119(1): 140–145
17. Kelly DL Jr, Lassiter KR, Vongsvivut A, Smith JM (1972) Effects of hyperbaric oxygenation and tissue oxygen studies in experimental paraplegia. J Neurosurg 36(4): 425–429
18. Narayana PA, Kudrle WA, Liu SJ, Charnov JH, Butler BD, Harris JH Jr (1991) Magnetic resonance imaging of hyperbaric oxygen treated rats with spinal cord injury: preliminary studies. Magn Reson Imaging 9(3): 423–428
19. Haapaniemi T, Nylander G, Kanje M, Dahlin L (1998) Hyperbaric oxygen treatment enhances regeneration of the rat sciatic nerve. Exp Neurol 149(2): 433–438
20. Higgins AC, Pearlstein RD, Mullen JB, Nashold BS Jr (1981) Effects of hyperbaric oxygen therapy on long-tract neuronal conduction in the acute phase of spinal cord injury. J Neurosurg 55(4): 501–510
21. Murakami N, Horinouchi T, Sakurai M, Ejima Y, Matsukawa S, Kato M, Tabayashi K (2001) Hyperbaric oxygen therapy given 30 minutes after spinal cord ischemia attenuates selective motor neuron death in rabbits. Crit Care Med 29(4): 814–818
22. Gamache FW Jr, Myers RA, Ducker TB, Cowley RA (1981) The clinical application of hyperbaric oxygen therapy in

spinal cord injury: a preliminary report. Surg Neurol 15(2): 85–87
23. Sukoff MH (2001) Effects of hyperbaric oxygenation. J Neurosurg 95(3): 544–546

Correspondence: John H. Zhang M.D., Ph.D., Department of Neurosurgery, Louisiana State University Health Science Center in Shreveport, 1501 Kings Highway, PO Box 33932, Shreveport, LA 71130-3932. e-mail: Johnzhang3910@yahoo.com

Intracerebral and subarachnoid hemorrhage

Acta Neurochir (2003) [Suppl] 86: 441–443

The international surgical trial in intracerebral haemorrhage (ISTICH)

A. D. Mendelow on behalf of the Investigators and the Steering Committee

(Committees and Investigators from 107 centres in 27 countries will be named in the final publication of the unblinded results at the end of 2003)

Summary

At the XIth International Brain Oedema Symposium we reported that 208 patients had been randomized in this trial from 65 centres. At the time of submission of this manuscript we have randomized 985 patients from 107 centres including Japan. The trial will continue randomising patients until the end of February 2003. The objective is to determine whether "Early Surgical Intervention" or "Initial Conservative Treatment" is the best option in patients with spontaneous supratentorial Intracerebral haemorrhage (ICH). Patients with trauma, tumours, aneurysms and angiographically proven arteriovenous malformations are excluded. Ascertainment logs from the various centres have revealed substantial variation in the rate of operation for ICH from 2% in Hungary to 90% in Lithuania. The results will remain blinded until all 6 month follow-up information has been obtained. It is anticipated that these results will therefore be known towards the end of 2003. The pooled results have shown that patients with larger Haematomas and depressed consciousness are more likely to end up with a worse outcome.

Keywords: Intracerebral haemorrhage; prospective randomised controlled trial; surgery; stroke.

Introduction

Spontaneous supratentorial Intracerebral Haemorrhage accounts for about 20% of all strokes and is the major cause of morbidity and mortality. The role of surgical evacuation is controversial and a meta-analysis of 7 previous small Prospective Randomised Controlled Trials (PRCT's) has indicated that surgical treatment has no advantage over medical treatment [1]. Experimental [2] and clinical [3] evidence has indicated that there is a penumbra of functionally impaired, but potentially viable, tissue that surrounds the ischaemic zone, which immediately surrounds the ICH. Given that there is clinical equipoise about which patients require surgical removal of the ICH and the *potential* to restore function in the penumbra, a prospective randomised controlled trial of early surgical intervention was begun in 1995. This Surgical Trial in Intracerebral Haemorrhage became multicentre in 1998 and, because of the need to increase recruitment further, became International, hence the name ISTICH.

Methods and patients

In any patient with an ICH, the neurosurgeon is either uncertain about the indication for surgery or certain about what to do. If the neurosurgeon is uncertain, then the uncertainty principal is invoked and the patient is considered eligible for randomisation in the trial. Written informed consent is then obtained if the patient is able to understand or relatives' assent is obtained if the patient is confused or comatose. In each centre ethical approval was obtained from the Local Research Ethics Committee and, in the UK, the Multicentre Research Ethics Committee (MREC) of the Northern and Yorkshire Region approved the trial.

The inclusion criteria therefore are that the neurosurgeon should be uncertain about whether or not to operate, that there should be a supratentorial spontaneous intracerebral haemorrhage and that there should be written informed consent or assent. The randomisation must have taken place within 72 hours of the ictus. Guidelines for inclusion suggest that patients should have a Glasgow Coma Score (GCS) of 5 or more, that the minimum clot diameter on CT should be 2 cm and that the age should be >14 years.

Exclusion criteria are the suspicion of an aneurysm or an angiographically proven arteriovenous malformation. Patients with trauma or suspected tumours are excluded, as are patients with cerebellar haematomas.

The day to day management of the trial is undertaken in Newcastle by the Management Committee that meets on a weekly basis. A Steering Committee was set up according to the recommendations of the UK Medical Research Council (MRC) Guidelines for Good Clinical Practice (GCP) in Clinical Trials. These are strictly followed in all aspects of the trial. These guidelines are also followed in all the ISTICH committee structures and functions. The Oxford Clinical Trials Centre in England provides a 24 hour telephone randomisation service. After obtaining written informed consent or assent, a Randomisation Form is completed prior to this telephone call in each case and the Oxford Centre uses a minimisation technique to randomise the patient either to "Early Surgical Intervention" or "Initial Conservative Treatment". This process includes stratification for GCS, size and side of haematoma and age. A 2 week Follow Up or Death Form is then completed by the individual centre and returned to the Newcastle Coordinating ISTICH Office. All adverse events and all serious adverse events, including death, are reported to the Newcastle Coordinating ISTICH Office. This Newcastle Coordinating Office informs an independent Data Monitoring and Safety Committee (DMSC) of all adverse events and provides regular confidential safety reports to the DMSC. The DMSC devised stopping rules for the trial and advises the Steering Committee and the Northern and Yorkshire MREC about whether or not to continue randomising patients. At the time of writing this manuscript (November 2002), these committees have recommended continuation of the trial with randomisation scheduled to continue until the end of February 2003.

Once a patient is randomised to "Early Surgical Intervention", surgery should be performed within 24 hours of randomisation. If a patient, who is randomised to "Initial Conservative Treatment", deteriorates, then the surgeon may decide to undertake delayed surgical intervention as is often the routine practice in the management of patients with ICH. Nevertheless, such a patient would remain in the "Initial Conservative Treatment" limb of the trial because statistical analysis of outcome will be by intention to treat.

Outcome is assessed by postal questionnaire at 6 months using the validated 5 point and 8 point Glasgow Outcome Scores (GOS) unless prior notification of death is obtained from the 2 week or adverse event reporting processes described above. The primary outcome measure will be assessed with the Glasgow Outcome Score. Secondary Outcome measures include the Modified Rankin Score, the Barthel Activities of Daily Living Index, Length of stay and Mortality. An economic analysis using Length of stay and the Euroquol is undertaken in UK patients at 3 and 6 months.

Results

All investigators and ISTICH Committees, except the DMSC, remain blinded to the results and will remain so until the database is locked after all 6 month outcomes have been retrieved after August 2003. No results are therefore available about the effects of randomisation to one group or the other. Only pooled results of all the patients can be presented at this stage.

From 107 centres, 985 patients have been randomised. The majority were male and the median age was 63 years. Five hundred and twenty five were left sided and 460 right. Three hundred and ninety six were lobar, 415 in the basal ganglia and 170 occupied both areas. Most patients were not in coma (Fig. 1). By 6 months more than 16% had reached a favourable outcome on the GCS (Fig. 2). The median time from ictus to randomisation was 20 hours and 61% were randomised within 24 hours.

The results in the 2 groups of patients will only be released after the database is locked in October or November 2003.

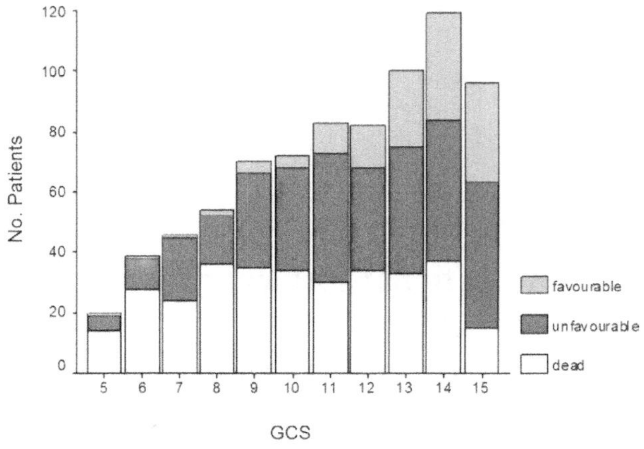

Fig. 1. Glasgow Coma Score at randomisation and outcome (Favourable indicates Good Recovery or Moderate Disability and Unfavourable indicates Severe Disability or the Vegetative state; Death is shown separately)

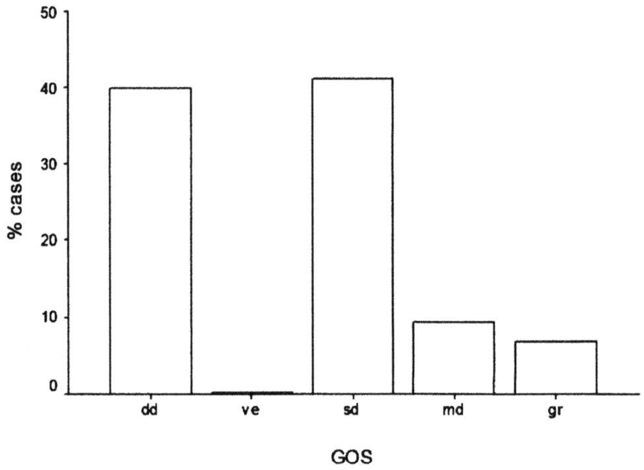

Fig. 2. Glasgow Outcome Score at 6 months

Discussion

There are fewer favourable outcomes at 6 months than was expected from the initial estimates. This may in part be explained by the age of the population and the use of the Glasgow Outcome Score which was designed for young head injured patients. In order to be regarded as independent on the 5 point Glasgow Outcome Scale (Moderate Disability or Good Recovery) patients must be able to use public transport and to shop independently. This may not be appropriate for the elderly population in whom independence within the home should be regarded as a "Favourable Outcome". For this reason, the 8 point Glasgow Outcome Scale has been used because Severe Disability is subdivided into upper and lower Severe Disability. Upper Severe Disability means that the patient is independent for many activities of daily living while Lower Severe Disability denotes dependent on others for many of these activities. For these reasons a dichotomous split between upper and lower Severe Disability would seem more appropriate for this age group.

Statistical methods to stratify the outcome are being developed at the University of Edinburgh and these studies may also provide a form of stratified outcome related to the patient's initial GCS. The precise method of analysis will be published by the Steering Committee before the database is closed and before the analysis takes place in 2003. The final results must therefore await these developments.

References

1. Fernandes HM, Gregson B, Siddique MS, Mendelow AD (2000) Surgery in intracerebral haemorrhage: the uncertainty continues. Stroke 31: 2511–2516
2. Mendelow AD, Bullock R, Teasdale GM, Graham DI, McCulloch J (1984) Intracranial haemorrhage induced at arterial pressure in the rat: Part 2. Short term changes in local cerebral blood flow measured by autoradiography. Neurol Res 6: 189–193
3. Siddique MS, Gregson BA, Fernandes HM, Barnes J, Treadwell L, Wooldridge TD, Mendelow AD (2002) Comparative study of traumatic and spontaneous intracerebral haemorrhage. J. Neurosurg 96: 86–89

Correspondence: Prof. A. D. Mendelow, Head of Department of Neurosurgery, Newcastle General Hospital, Newcastle upon Tyne, Westgate Road, England, NE4 6BE.

Acta Neurochir (2003) [Suppl] 86: 445–448

Chronological changes of perihematomal edema of human intracerebral hematoma

M. Inaji[1,2], **H. Tomita**[2], **O. Tone**[2], **M. Tamaki**[2], **R. Suzuki**[3], and **K. Ohno**[1]

[1] Department of Neurosurgery, Tokyo Medical and Dental University, Tokyo, Japan
[2] Department of Neurosurgery, Musashino Red Cross Hospital, Tokyo, Japan
[3] Department of Neurosurgery, Showa University, Fujigaoka Hospital, Kawasaki, Japan

Summary

Recent investigations have indicated the importance of secondary brain damage in the pathophysiology of intracerebral hemorrhage (ICH), which includes ischemic brain damage and edema formation around a hematoma. The purpose of the current study is to investigate chronological changes of perihematomal edema in patients with human ICH and also the correlation between volume of perihematomal edema and neurological status. Fourteen patients with medium-sized putaminal hemorrhage (29.4 ± 13.2 ml) without hematoma enlargement were included in this study. To investigate chronological changes of perihematomal edema, we performed CT scans prospectively on the day of hemorrhage and repeated them on days 1, 3, 7, 14, and 28. We evaluated the patients neurologically using the NIH stroke scale on the day a CT scan was performed. The volume of perihematomal edema in human ICH increased rapidly three days after hemorrhage and the score on the NIH stroke scale showed a deterioration. The volume of perihematomal edema then increased slowly until day 14 after hemorrhage, and decreased thereafter. Despite progression of perihematomal edema, the score on the NIH stroke scale improved gradually after day 3.

Keywords: Perihematomal edema; intracerebral hemorrhage; computed tomography; NIH stroke scale.

Introduction

Intracerebral hemorrhage (ICH) is a common disease with overall mortality of approximately 50% and morbidity approaching 75%. Nevertheless, various aspects of its pathophysiology are not well known. Injury due to ICH consists of two components: One is primary brain damage due to hematoma formation materializing as mechanical and immediate destruction of the neural structure. The other is secondary brain damage to which ischemia and edema formation of the perihematomal area arc major contributors. Some recent experimental studies have revealed the importance of secondary brain damage or edema formation by degrees. However, only a few clinical prospective studies have been reported, and the time course of clinical status and perihematomal edema formation is not clear. Edema formation of the perihematomal area is considered to play an important role in secondary brain damage. Thus, better understanding of edema formation will help to achieve better treatment for patients.

In the current study, we try to clarify the chronological changes of perihematomal edema in human ICH and to investigate the correlation between volume of perihematomal edema and neurological status.

Patients and methods

Fourteen patients with putaminal hemorrhage were included in this study. Eleven were males and three were females. Their mean age was 60.0 ± 11.4 years old (mean ± SD). All of the patients were admitted to our hospital within 24 hours after onset. All received administration of osmotic antidiuretics and other intensive medical treatments. The hematoma was more than 3 cm in diameter, and the mean volume of the hematoma was 29.4 ± 13.2 ml (mean ± SD). Hematoma enlargement was not observed in either case.

We prospectively performed axial CT scans using a GE9200 CT scanner on day of onset and repeated them on days 1, 3, 7, 14, and 28. We defined perihematomal edema as an area of lower density than the corresponding area in the contralateral hemisphere. Edema volume was calculated through a summation of all the measured volumes of low-density areas including hematomas in each CT slice multiplied by the thickness of each slice (1.0 cm). We expressed the perihematomal edema volume as increased volume % compared to the hematoma volume on day 0, which was expressed as $E + H/H0$ ratio (%) ($E + H$: volume of edema and hematoma, $H0$: volume of hematoma on day 0). We evaluated the patients neurologically using the NIH stroke scale on the day when a CT scan was performed.

Differences between the volumes of each point were evaluated using Fisher's protected least significance difference. All values are expressed as mean ± standard deviation. A probability value of $P < 0.05$ was considered significant.

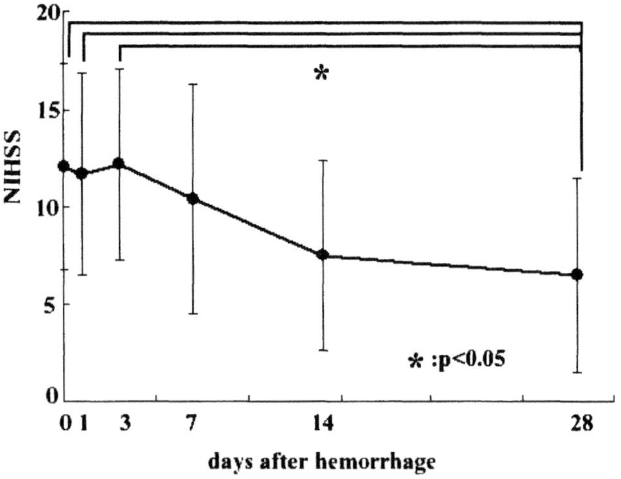

Fig. 1. CT scans at the level of hematoma on day of onset, day 1, day 3, day 7, day 14, and day 28. Edema increased gradually and reached the maximum at day 14, whereas NIH stroke scale scores improved after day 3

Results

Figure 1 shows a representative case followed up by periodical CT scans. This 45-year-old patient with sudden right-sided hemiparesis and aphasia was admitted to our hospital within 24 hours after onset. A CT scan demonstrated left putaminal hemorrhage. The volume of edema increased gradually and peaked on day 14. Then it decreased on day 28. Neurological status, which was expressed as a score on the NIH stroke scale, deteriorated during the three days after hemorrhage. Neurological status then improved gradually, despite a progression in the volume of edema (Fig. 1).

Of the fourteen patients in this study, mean value of E + H/H0 ratios on days 0, 1, 3, 7, 14, and 28 were 100%, $126 \pm 25\%$, $148 \pm 29\%$, $155 \pm 40\%$, $180 \pm 79\%$, $115 \pm 85\%$ respectively (Fig. 2). A significant difference in the mean value of the E + H/H0 ratios was observed between day 0 and day 3, day 0 and day 7, day 0 and day 14, day 14 and day 28. Mean scores on the NIH stroke scale at days 0, 1, 3, 7, 14, and 28 day were 12.1 ± 5.3, 11.7 ± 5.2, 12.2 ± 4.9, 10.4 ± 5.9, 7.5 ± 4.9, 6.5 ± 5.0 (Fig. 3). A significant difference in mean score was observed between day 0 and day 28, day 1 and day 28, day 3 and day 28.

The volume of perihematomal edema in human intracerebral hemorrhage rapidly increased during the first three days after hemorrhage, and then slowly increased until day 14 after hemorrhage. It decreased thereafter. The scores on the NIH stroke scale deteriorated during the three days after hemorrhage, then improved gradually after day 3 despite propagation of perihematomal edema. The propagation of perihematomal edema did not correspond with neurological deterioration.

Discussion

It is well known that a neurological deterioration similar to ischemic stroke is often observed in patients with ICH. According to the study by Meyer *et al.* [2], 33% of 46 noncomatose patients with ICH deterio-

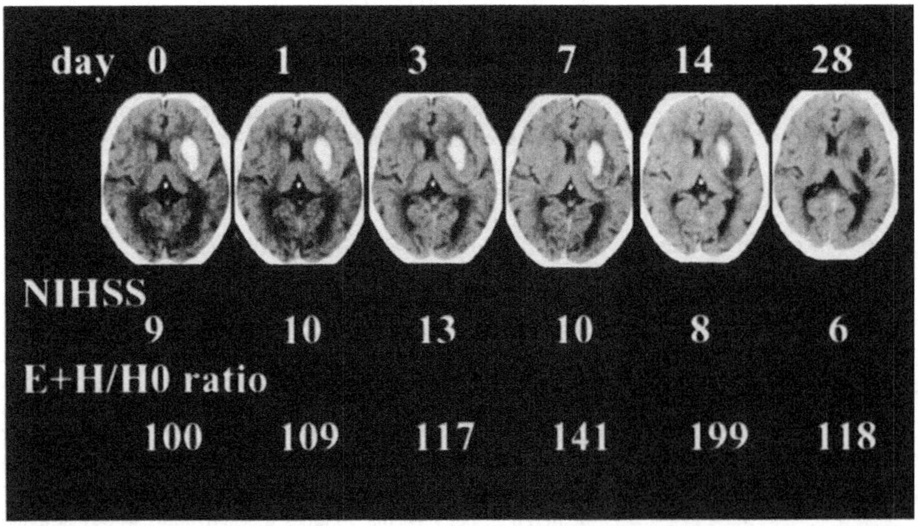

Fig. 2. Line graph shows mean value of E + H/H0 ratio on each day. A significant difference was observed between day 0 and day 3, day 0 and 7, day 0 and 14, day 14 and 28 ($p < 0.05$). Values are means $\pm$ standard deviation (bars)

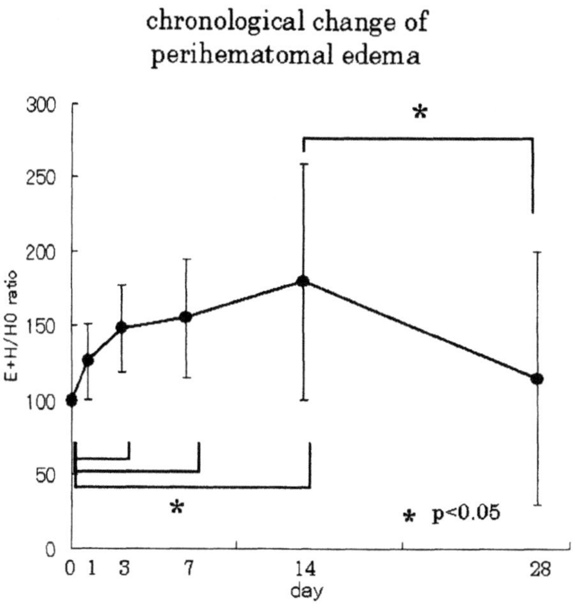

Fig. 3. Line graph shows mean score on the NIH stroke scale on each day. A significant difference was observed between day 0 and day 28, day 1 and day 28, day 3 and day 28 ($p < 0.05$)

rated after admission neurologically. Only 4 of 15 patients deteriorated because of hematoma enlargement. The size of the hematoma of the other 11 patients did not change. Therefore, this study suggests that neurological deterioration is caused by secondary brain damage after intracerebral hemorrhage. This mechanism mainly consists of ischemia and edema formation of the perihematomal area, though the pathophysiology of secondary damage has not been fully understood.

The pathomechanism for the development of brain edema associated with ICH seems to be quite different from that of cold lesion edema. Tomita *et al.* [5] reported chronological changes in brain edema induced by experimental ICH in cats. In their study, edema was mild 2 hours after the induction of ICH and was most severe 3 days after onset. It decreased from 7 days after onset. Comparing their results with those in a cold injury model, they concluded that perihematomal edema was prolonged. Their study suggests that the evolution and resolution of edema occurs more slowly after induction of ICH than the same processes after a cold injury.

Recent experimental studies have investigated the mechanisms of edema formation. Lee *et al.* [1] reported that solution of thrombin infused into the basal ganglia of rats induced blood-brain barrier disruption as well as parenchymal cell death, and that cell toxicity and blood-brain disruption by thrombin were triggering mechanisms for edema formation that followed ICH. Xi *et al.* [6] also showed that an intracerebral injection of lysed red blood cells into the caudate nucleus in a rat caused edema formation within 24 hours, whereas an injection of packed red blood cells caused brain edema formation 3 days later. These studies suggest that degradation products of blood play an important role in edema formation.

In the literature only a few clinical studies are reported. Most are retrospective studies, moreover including various locations and sizes of hematoma. Suzuki *et al.* [4] demonstrated in a study of patients with ICH that the mass effect due to perihematomal edema after putaminal hemorrhage peaked between 10 to 20 days after hemorrhage. In the present study, the volume of perihematomal edema was greatest at 14 days after the onset of ICH, which is in agreement with the observations by Suzuki *et al.* On the other hand, few reports have shown a relationship between volume of perihematomal edema and neurological status. Suga *et al.* [3] studied changes in the volume of the high intensity area on T2 weighted MRI scans in 7 patients with ICH, and showed that the volume of edema peaked at 2 weeks and declined at 4 weeks without a correlation with neurological status. Similarly no correlation between edema formation and neurological deterioration was observed in our study. The reason for this discrepancy remains unclear. Secondary brain damage forms a complicated cascade. It may be necessary as a next step to assess the mass effect of edema on the midline structure and to measure intracranial pressure. A further investigation is needed to clarify the clinical influence of perihematomal edema and the efficacy of various treatments to reduce perihematomal edema formation.

References

1. Lee KR, Kawai N, Kim S, Sagher O, Hoff JT (1997) Mechanism of edema formation after intracerebral hemorrhage: effect of thrombin on cerebral blood flow, blood brain barrier permeability, and cell survival in a rat model. J Neurosurg 86: 272–278
2. Meyer SA, Sacco RL, Shi T, Mohr JP (1994) Neurological deterioration in noncomatose patients with supratentorial intracerebral hemorrhage. Neurology 44: 1379–1384
3. Suga S, Sato S, Yunoki K, Mihara B (1994) Sequential change of brain edema by semiquantitative measurement on MRI in patients with hypertensive intracerebral hemorrhage. Acta Neurochir (Wien) [Suppl] 60: 564–567
4. Suzuki R, Ohno K, Hiratsuka H, Inaba Y (1985) Chronological changes in brain edema in hypertensive intracerebral hemorrhage

observed by CT and xenon-enhanced CT. In: Inaba Y, Klatzo I, Spatz M (eds) Brain edema. Springer, Berlin Heiderberg New York Tokyo, pp 613–620
5. Tomita H, Ito U, Ohno K, Hirakawa K (1994) Chronological changes in brain edema induced by experimental intracerebral hematoma in cats. Acta Neurochir (Wien) [Suppl] 60: 558–560
6. Xi G, Hua Y, Bhasin R, Ennis SR, Keep R, Hoff JT (2001) Mechanism of edema formation after intracerebral hemorrhage. Effect of red blood cells on blood flow and blood-brain barrier integrity. Stroke 32: 2932–2938

Correspondence: M. Inaji, 1-5-45 Yushima, Bunkyo-ku, and Tokyo, Japan. e-mail: inaji@nirs.go.jp

Acta Neurochir (2003) [Suppl] 86: 449–452

Thrombin preconditioning upregulates transferrin and transferrin receptor and reduces brain edema induced by lysed red blood cells*

G. Xi[1], J. Wu[1], Y. Jiang[1], Y. Hua[1], R. F. Keep[1,2], and J. T. Hoff[1]

[1] Department of Neurosurgery, University of Michigan, Ann Arbor, Michigan, U.S.A.
[2] Department of Physiology, University of Michigan, Ann Arbor, Michigan, U.S.A.

Summary

Pretreatment with a low dose of thrombin reduces brain edema after both hemorrhagic and ischemic stroke. We call this phenomenon thrombin preconditioning (TPC) or thrombin-induced brain tolerance. The present study examines whether TPC can attenuate the brain edema induced by lysed red blood cells (RBCs) to determine whether thrombin production early in an intracerebral hemorrhage (ICH) might alter potentially injurious events associated with clot resolution. It also examines whether TPC might be protective by altering iron handling within the brain, particularly through modulating transferrin (Tf) and transferrin receptor (TfR) levels. Brain edema was measured by wet/dry weight. Western blot analysis and immunohistochemistry were used for Tf and TfR measurements. We found that TPC reduces lysed RBC-induced brain edema and upregulates both Tf and TfR levels in the brain. Thrombin formation after an ICH may be part of a signaling cascade that acts to limit potentially injurious events associated with clot resolution through altering iron-handling proteins.

Keywords: Brain edema; thrombin; preconditioning; red blood cells; transferrin; transferrin receptor; cerebral hemorrhage.

Introduction

Thrombin production, erythrocyte lysis and hemoglobin toxicity result in brain edema after intracerebral hemorrhage (ICH) [5, 15, 17]. Hemoglobin-induced brain injury is, at least in part, iron-mediated [4]. Although a large dose of thrombin causes brain edema and cell death, low concentrations of thrombin are neuroprotective. We previously found that pretreatment with a low dose of thrombin attenuates the brain edema induced by thrombin or ICH, and significantly reduces infarct size and brain edema in a rat middle cerebral artery occlusion model [6, 14, 16]. We have termed this phenomenon thrombin preconditioning (TPC). The present study examines whether TPC can attenuate the brain edema induced by lysed red blood cells (RBCs) to determine whether thrombin production early in an ICH might alter potentially injurious events associated with clot resolution. It also examines whether TPC might be protective by altering iron handling within the brain, particularly through modulating transferrin (Tf) and transferrin receptor (TfR) levels.

* Supported by grants NS-39866 (GX), NS-17760 (JTH) from the National Institutes of Health and Grant-In-Aid 0255721N (GX) from American Heart Association.

Materials and methods

Animal groups

There were two sets of experiments in this study. In the first set, pentobarbital anesthetized male Sprague-Dawley rats received an intracerebral infusion of saline or one unit thrombin into the right basal ganglia. They received then a 30-μl lysed RBCs infusion one, three or seven days later. All rats were sacrificed after a further 24 hours for measurement of brain water content. In the second set of experiments, rats received an intracerebral infusion of either saline or one unit of thrombin. The rats were then killed 1, 3 or 7 days later for Western blot analysis and immunohistochemistry.

Animal preparation

Adult male Sprague-Dawley rats (275–325 g) were anesthetized with pentobarbital (40 mg/kg, i.p.). A polyethylene catheter (PE-50) was inserted into the right femoral artery in order to monitor arterial blood pressure and to obtain blood for analysis of blood gases, blood pH, hematocrit, blood glucose concentration. Body temperature was maintained at 37.5 °C by using a feedback-controlled heating pad.

Intracerebral infusion

Before intracerebral infusion, the rats were positioned in a stereotactic frame (Kopf Instrument, Tujunga, CA), the scalp was incised along the sagittal midline using a sterile technique. A cranial burr hole (1 mm) was drilled near the right coronal suture 4.0 mm lateral to the midline. A 26-gauge needle was inserted stereotaxically into the right basal ganglia (coordinates: 0.2 mm anterior, 5.5 mm ventral, and 4.0 mm lateral to the bregma). Thrombin, saline or lysed red blood cells were infused into the right basal ganglia using a microinfusion pump. After infusion, the needle was removed and the skin incisions were closed with sutures. Animals were allowed to recover.

Brain water content measurements

The rats were sacrificed by decapitation under deep pentobarbital anesthesia (60 mg/kg i.p.). The brains were removed immediately and a 3 mm thick coronal brain slice 4 mm from the frontal pole was cut. That slice was divided into four samples, ipsilateral and contralateral basal ganglia and ipsilateral and contralateral cortex. Cerebellum was obtained as a control. Tissue samples were weighed on an electronic analytical balance to obtain the wet weight (WW). The tissue is then dried in a gravity oven at 100 °C for more than 24 hours to determine the dry weight (DW). Tissue water contents (%) are calculated as $((WW - DW)/WW) * 100$.

Western blot analysis

The rat brains were perfused with saline, and a coronal brain slice was cut as described for brain water content measurements. The brain tissue was immersed in 0.5 ml of Western blot sample buffer and then sonicated for Western blot analysis [16]. Primary antibodies were rabbit anti-human transferrin antibody (DAKO, Glostrup, Denmark) and mouse anti-human transferrin receptor antibody (Zymed Laboratories Inc., South San Francisco, CA). The relative densities of Tf and TfR protein bands were analyzed using the NIH image software.

Immunohistochemistry

Immunohistochemical studies were performed as in our previous report [16]. The rats were reanesthetized with pentobarbital (40 mg/kg, i.p.) and brain was perfused with 4% paraformaldehyde in 0.1 M pH 7.4 phosphate-buffered saline (PBS). Removed brain was kept in 4% paraformaldehyde for six hours, then immersed in 25% sucrose for three to four days at 4 °C. The brains were embedded in O.C.T compound (Sakura Finetek U.S.A. Inc.) and sectioned on a cryostat (18 μm thick). Sections were incubated according to the avidin-biotin complex technique. Primary antibodies were rabbit anti-human transferrin antibody (DAKO, Glostrup, Denmark) and mouse anti-human transferrin receptor antibody (Zymed Laboratories Inc. South San Francisco, CA). Normal rabbit IgG was used as negative control.

Statistical analysis

All data in this study are presented as mean ± standard deviation. Data were analyzed with Student's t test. Significance levels were measured at $p < 0.05$.

Results

Mean arterial blood pressure, blood pH, PaO_2, $PaCO_2$, hematocrit, and blood glucose were controlled within normal ranges.

Intracerebral injection of lysed RBCs caused a marked increase in the water content of the ipsilateral basal ganglia (81.8 ± 0.5% vs. 77.8 ± 0.3% in the control). The time course showed that there was no protection when thrombin pretreatment was given one day before infusion of lysed RBCs. However, thrombin given three days before significantly reduced the amount of edema induced by lysed RBCs (79.4 ± 0.9% vs. 81.5 ± 1.4% in the control, mean ± SD, $n = 6$, $p < 0.01$; Fig. 1). Some protection was also found in rats given TPC seven days before infusion of lysed RBCs (80.8 ± 0.9% vs. 82.2 ± 1.0% in the control, $n = 6$, $p < 0.05$; Fig. 1), but this was less than seen in the three day group.

Western blot analysis indicated that both Tf and TfR protein levels were upregulated in the ipsilateral

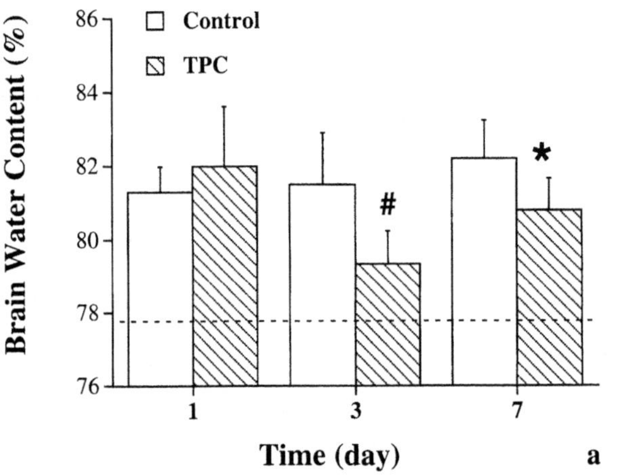

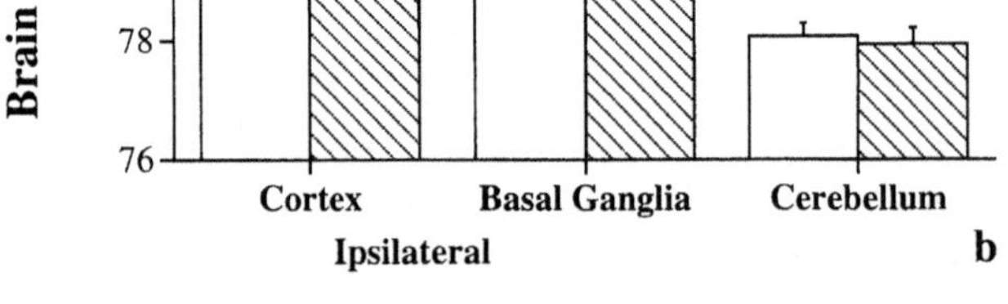

Fig. 1. (a) Brain water contents in the ipsilateral basal ganglia at 24 hours after intracerebral infusion of 30 μl lysed RBCs. The brains had been infused with either saline or 1 unit thrombin at 1, 3 or 7 days prior to the lysed RBCs. Values are expressed as mean ± SD, n = 5–6. *p < 0.05 vs. control, #P < 0.01 versus control. The hatched line shows the normal water content of the basal ganglia. (b) Brain water contents 24 hrs after intracerebral infusion of 30 μl lysed RBCs. The brains have been infused with either saline or 1 unit thrombin 3 days prior to the lysed RBCs. Values are expressed as mean ± SD, n = 5–6. *p < 0.01 vs. control

basal ganglia one day after thrombin infusion, peaked at day three (Tf: 100-fold increase; TfR: 10-fold increase, Fig. 2) and returned towards to the baseline levels at seven days.

Immunohistochemistry demonstrated an increase of immunoreactivities of Tf and TfR in the ipsilateral basal ganglia after one unit thrombin infusion. Most Tf and TfR positive cells were neurons (Fig. 3).

Discussion

This study provides evidence that intracerebral thrombin preconditioning reduces brain edema induced by lysed RBCs and upregulates brain Tf and TfR. All these data suggest that upregulation of iron handling proteins, Tf and TfR, may be responsible for thrombin-induced brain tolerance to RBC-induced brain injury.

Our previous experiments have indicated that lysis of RBC after an ICH results in delayed brain edema and that this is mediated by hemoglobin and, at least in part, its degradation products [4, 15]. In rat, intracerebral infusion of hemoglobin, hemin, bilirubin or $FeCl_2$ caused an increase in brain water content at 24 hours. Heme oxygenase inhibition by tin-protoporphyrin (SnPP) attenuated hemoglobin-induced edema.

Iron is essential for normal brain function, but iron overload may have devastating effects [1]. After erythrocyte lysis, iron concentrations in the brain can reach very high levels. Our recent study showed a 3-fold increase of brain non-heme iron after intracerebral hemorrhage (Wu *et al.*, unpublished data, 2002). In-

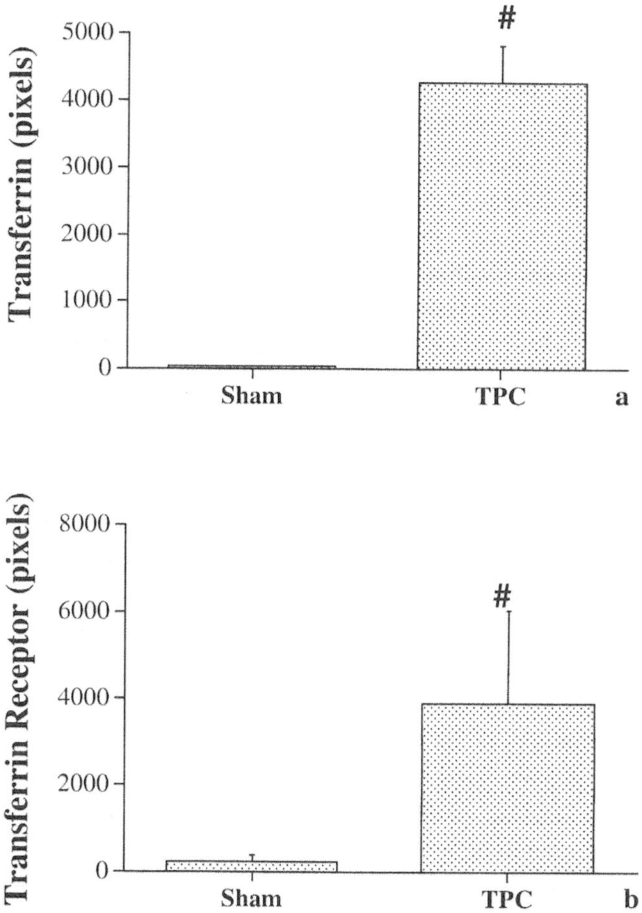

Fig. 2. Transferrin (a) and transferrin receptor (b) protein levels were measured by Western blot analysis in the ipsilateral basal ganglia 3 days after either saline or one unit thrombin infusion. Values are mean ± SD, #$p < 0.01$ vs. saline

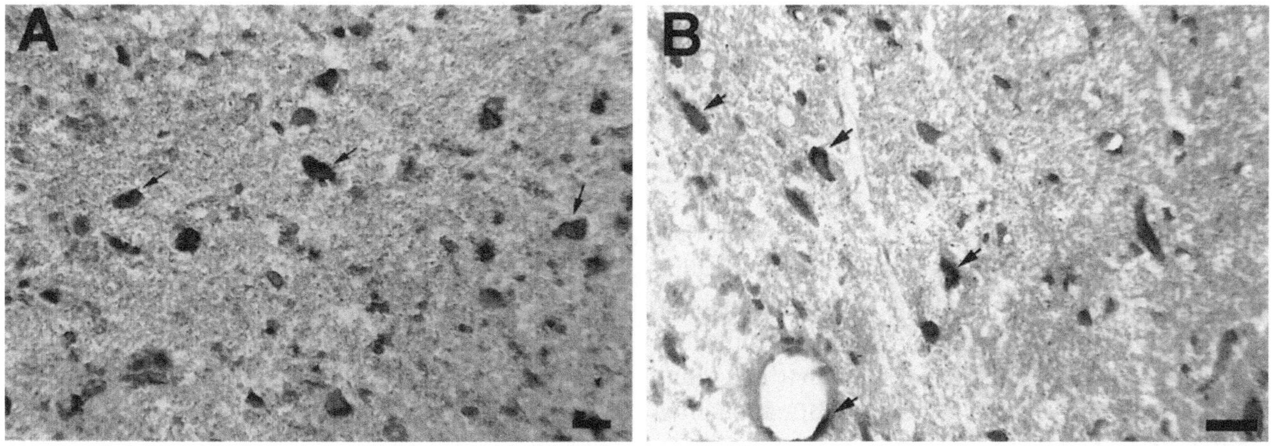

Fig. 3. Immunoreactivities of transferrin (A) and transferrin receptor (B) in the ipsilateral basal ganglia three days after intracerebral infusion of one unit thrombin. Arrows indicate the positive cells. Scale bars are 20 μm

tracerebral infusion of iron causes brain edema and an iron chelator reduces hemoglobin-induced edema, suggesting that iron plays an important role in edema formation after ICH [4]. Cortical iron injections cause focal epileptiform paroxysmal discharges [3, 13]. Iron and lipid peroxidation also have an important role in hemoglobin-induced brain injury. For example, a subpial injection of $FeCl_2$ induces brain edema and lipid peroxidation in the brain [12]. Iron can also stimulate the formation of free radicals leading to neuronal damage. It is known that ferrous (Fe^{2+}) and ferric (Fe^{3+}) iron react with lipid hydroperoxides to produce free radicals [9]. Deferoxamine, an iron chelator, reduces hemoglobin-induced brain Na^+/K^+ ATPase inhibition and neuronal toxicity [7, 8]. Furthermore, antioxidants block neuronal toxicity induced by hemoglobin and iron [7].

A number of proteins, including transferrin (Tf) and transferrin receptor (TfR), are involved in maintaining brain iron homeostasis. Transferrin, an 80-kDa protein, is a major iron distributor in the brain. Cellular uptake of transferrin-bound iron is achieved by binding to the TfR. In normal brain, TfR is expressed at very high levels in the blood-brain barrier where it is involved in iron uptake into brain. However, a recent report indicates that there is rapid efflux of Tf from brain to blood across the BBB [18] suggesting that Tf could contribute to iron clearance when there is brain iron overload. TfR is normally only expressed on brain parenchymal cells at low levels. Brain TfR can be altered under pathophysiological conditions. Decreases of brain TfR levels are associated with Alzheimer's disease and other neurodegenerative diseases [11], but TfR is upregulated by hypoxia [10]. Various neurodegenerative disorders alter brain Tf levels [11]. It should be noted that the brain can produce ferritin, the iron storage protein. Ferritin has two subunits, Heavy-ferritin (21 kDa) and Light-ferritin (19 kDa) [2]. We have found that thrombin preconditioning increase ferritin immunoreactivities in the brain (Xi *et al.* unpublished data, 2002).

In summary, our present results demonstrate that TPC reduces lysed RBC-induced brain edema. This protection may be via an upregulation in brain parenchyma Tf and TfR levels and/or enhanced iron efflux across the blood-brain barrier, events that may limit iron-related brain damage. Thrombin formation after an ICH may be part of a signaling cascade that acts to limit potentially injurious events associated with clot resolution.

References

1. Chiueh CC (2001) Iron overload, oxidative stress, and axonal dystrophy in brain disorders. Pediatric Neurology 25: 138–147
2. Connor JR, Menzies SL, Burdo JR, Boyer PJ (2001) Iron and iron management proteins in neurobiology. Pediatric Neurology 25: 118–129
3. Hammond EJ, Ramsay RE, Villarreal HJ, Wilder BJ (1980) Effects of intracortical injection of blood and blood components on the electrocorticogram. Epilepsia 21: 3–14
4. Huang F, Xi G, Keep RF, Hua Y, Nemoianu A, Hoff JT (2002) Brain edema after experimental intracerebral hemorrhage: role of hemoglobin degradation products. J Neurosurg 96: 287–293
5. Lee KR, Colon GP, Betz AL, Keep RF, Kim S, Hoff JT (1996) Edema from intracerebral hemorrhage: the role of thrombin. J Neurosurg 84: 91–96
6. Masada T, Xi G, Hua Y, Keep RF (2000) The effects of thrombin preconditioning on focal cerebral ischemia in rats. Brain Res 867: 173–179
7. Regan RF, Panter SS (1993) Neurotoxicity of hemoglobin in cortical cell culture. Neurosci Lett 153: 219–222
8. Sadrzadeh SM, Anderson DK, Panter SS, Hallaway PE, Eaton JW (1987) Hemoglobin potentiates central nervous system damage. J Clin Invest 79: 662–664
9. Siesjo BK, Agardh CD, Bengtsson F (1989) Free radicals and brain damage. Cerebrovascular Brain Metab Rev 1: 165–211
10. Tacchini L, Bianchi L, Bernelli-Zazzera A, Cairo G (1999) Transferrin receptor induction by hypoxia. HIF-1-mediated transcriptional activation and cell-specific post-transcriptional regulation. Journal of Biological Chemistry 274: 24142–24146
11. Thompson KJ, Shoham S, Connor JR (2001) Iron and neurodegenerative disorders. Brain Research Bulletin 55: 155–164
12. Willmore LJ, Rubin JJ (1982) Formation of malonaldehyde and focal brain edema induced by subpial injection of FeCl2 into rat isocortex. Brain Research 246: 113–119
13. Willmore LJ, Sypert GW, Munson JV, Hurd RW (1978) Chronic focal epileptiform discharges induced by injection of iron into rat and cat cortex. Science 200: 1501–1503
14. Xi G, Hua Y, Keep RF, Hoff JT (2000) Induction of colligin may attenuate brain edema following intracerebral hemorrhage. Acta Neurochir (Wien) [Suppl] 76: 501–505
15. Xi G, Keep RF, Hoff JT (1998) Erythrocytes and delayed brain edema formation following intracerebral hemorrhage in rats. J Neurosurg 89: 991–996
16. Xi G, Keep RF, Hua Y, Xiang JM, Hoff JT (1999) Attenuation of thrombin-induced brain edema by cerebral thrombin preconditioning. Stroke 30: 1247–1255
17. Xi G, Wagner KR, Keep RF, Hua Y, de Courten-Myers GM, Broderick JP, Brott TG, Hoff JT (1998) The role of blood clot formation on early edema development following experimental intracerebral hemorrhage. Stroke 29: 2580–2586
18. Zhang Y, Pardridge WM (2001) Rapid transferrin efflux from brain to blood across the blood-brain barrier. J Neurochem 76: 1597–1600

Correspondence: Guohua Xi, R5550 Kresge I, University of Michigan, Ann Arbor, Michigan 48109-0532, U.S.A. e-mail: guohuaxi@umich.edu

Acta Neurochir (2003) [Suppl] 86: 453–456

Effect of hypothermia on brain edema formation following intracerebral hemorrhage in rats

M. Kawanishi

Department of Neurological Surgery, Kagawa Medical University, Kagawa, Japan

Summary

Brain edema is an important clinical complication of intracerebral hematoma (ICH). Hypothermia has been employed as an effective neuroprotective treatment on cerebral ischemia and contusional brain injuries. In this study, we attempted to evaluate the effects of mild hypothermia (35 °C) on brain edema formation at 48 hours after ICH in the rat. Brain hypothermia started at 6, 12 and 24 hours after the induction of hematoma (HT6; n = 6, HT12; n = 11, HT24; n = 6). Brain water contents in the basal ganglia was significantly reduced in the rats treated with mild hypothermia compared with the normothermic rats (NT; 82.0 ± 0.04% vs. HT6; 78.6 ± 0.09%: $p < 0.01$, HT12; 79.7 ± 0.05%: $p < 0.01$, HT24; 79.7 ± 0.05%: $p < 0.01$). Differences in the brain water content were not significant among the hypothermic subgroups. The BBB disruption to Evan's blue was significantly reduced with hypothermia compared with the normothermic rats in the ipsilateral basal ganglia (42.3 ± 4.0 vs. 23.0 ± 5.2 ng/g wet tissue; $p < 0.05$).

Hypothermic treatment tended to inhibit the accumulation of polymorphonuclear (PMN) leukocytes in the lesion compared with the normothermic treatment (0.78 ± 0.13 vs. 0.51 ± 0.16 ΔAbs/mg tissue).

This study indicates that hypothermic treatment significantly reduces the brain edema formation after ICH in rats.

Keywords: Blood-brain barrier; brain edema; hypothermia; intracerebral hemorrhage; polymorphonuclear leukocyte.

Introduction

Brain edema is an important clinical complication after intracerebral hemorrhage [6]. Recent study demonstrates that an inflammatory response occurs around a blood clot following ICH in rats. Inflammatory response may also be involved in secondary injury following ICH [3, 7].

In this study, we attempted to evaluate the effects of mild hypothermia (35 °C) on brain edema formation and blood-brain barrier (BBB) disruption at 48 hours after the hematoma induction in rats. Accumulation of PMN leukocytes in the brain was also determined to clarify the protective mechanism of hypothermia on this model.

Materials and methods

Animal preparation

Adult male Sprague-Dawley rats weighing 250–400 g were anesthetized with intraperitoneal sodium pentobarbital (60 mg/kg). A polyethylene (PE-50) catheter was introduced into the femoral artery to monitor the arterial blood pressure and to obtain blood for analysis of blood gases, blood pH, hematocrit, and blood glucose. The rat's body (rectal) temperature was maintained at 37 °C during the surgery using a feedback-controlled heating system.

Experimental model

The rats were positioned in a stereotactic frame and the scalp was incised along the sagittal midline. Using a sterile technique, a 1 mm burr-hole was opened on the skull near the left coronal suture 3 mm lateral to the midline. A blunt 26-Gauge needle was inserted into the left basal ganglia under stereotactic guidance (coordinates: 0.2 mm anterior, 6 mm ventral, and 3 mm lateral to the bregma). Then, 100 µl of nonheparinized fresh arterial blood was infused at a rate of 20 µl/min (Eicom EPS-26, Kyoto, Japan). After completion of the infusion, the needle was withdrawn quickly and cyanoacrylate glue was placed around the burr hole. After the surgery, rats were divided into the normothermic and hypothermic groups. Normothermic rats (NT: n = 7) were housed in a room maintained at 25 °C for 48 h and hypothermic rats were housed in a cold room maintained at 5 °C. Hypothermia treatment was started at 6 (HT6: n = 6), 12 (HT12: n = 11), and 24 (HT24: n = 6) h after the induction of hematoma.

Experimental protocols

This study was divided into three parts. The first part evaluated the effect of temperature on brain edema formation in the normothermic and hypothermic rats. In the second and third parts, the BBB permeability to Evan's blue dye and the accumulation of PMN leukocytes were examined in the normothermic and hypothermic (HT6) rats to clarify the protective mechanism of hypothermia on brain edema formation.

Measurement of brain water content

The rats were killed by decapitation under deep pentobarbital anesthesia at 48 h after the injection of autologous arterial blood. The brains were removed rapidly and two coronal slices 3 mm from the frontal pole were cut 3 mm thick. The anterior slice (slice A) contained the needle trajectory and arterial blood injection site at the basal ganglia. The adjacent slice (slice B) was a section posterior to the arterial blood injection site. These sections of brain were divided along the midline and the cortex was separated from the basal ganglia bilaterally.

The tissue samples were immediately weighed on an electronic analytical balance to the nearest 0.1 mg to obtain wet weight (WW). The tissue was then dried in an oven at 110 °C for 24 h and weighed again to obtain the dry weight (DW). The formula (WW – DW)/WW × 100 was used to calculate the water content and expressed as percentage of WW.

Blood-brain barrier permeability to Evan's blue

The integrity of the blood-brain barrier was investigated using Evan's blue dye extravasation. Evan's blue (EB, 2% in saline, 2 ml/kg) was injected intravenously at 48 h after the arterial blood injection and allowed to circulate for 30 min under the anesthesia with i.p. sodium pentobarbital (60 mg/kg). The chest was subsequently opened and the brain was transcardially perfused with 200 ml of saline through the left ventricle at 100 mmHg pressure until colorless perfusion fluid was obtained from the right atrium. After decapitation, the brain was removed and two coronal slice 3 mm from the anterior pole were cut 3 mm thick. These sections of brain were divided along the midline and the cortex was separated from the basal ganglia.

Samples were weighed and placed in 50% trichloroacetic solution. Following homogenization and centrifugation, the extracted dye was diluted with ethanol (1 : 3), and its fluorescence was determined (excitation at 620 nm and emission 680 nm) with a luminescence spectrophotometer (Hitachi 650-40, Tokyo, Japan). Calculations were based on the external standard (62.5–500 ng/ml) in the same solvent. The tissue content of EB was quantified from a linear standard line derived from known amounts of the dye and was expressed per gram of tissue.

Measurement of myeloperoxidase (MPO) activity

Myeloperoxidase activity was used as an indicator of tissue PMN leukocytes accumulation. The brains were transcardially perfused with 200 ml saline through the left ventricle at 100 mmHg pressure and rapidly removed. A coronal slice 3 mm from the frontal pole was cut 5 mm thick and separated into the right and left hemispheres. Each hemisphere was immediately frozen in liquid nitrogen and stored at −70 °C for later biochemical analysis.

The procedure described by Bradley *et al.* [2], was used for the quantification of MPO activity with minor modifications. The brain samples were thawed on ice and the cortex of ICH site was separated from the basal ganglia. The wet weight in each sample was immediately measured. Each sample was homogenized in 20 w/v of 5 mmol/l potassium phosphate buffer (pH 6.0, 4 °C) and centrifuged at 30,000× g (30 min, 4 °C). The supernatant was discarded, and the pellet was washed again as described above. After decanting the supernatant, the pellet was extracted by suspension in 0.5% hexadecyltrimethylammonium bromide in 50 mmol/l potassium phosphate buffer (pH 6.0, 25 °C) for approximately 2 min at an original tissue weight-to volume ratio of 1 : 10. The samples were immediately frozen in liquid nitrogen. Three freeze/thaw cycles were performed with sonications (10 s, 25 °C) between cycles. After the last sonications, the samples were incubated at 4 °C for 20 min and centrifuged at 12,500× g (15 min, 4 °C). A 0.1-ml aliquot of supernatant was mixed with 2.9 ml of 50 mmol/l potassium phosphate buffer, pH 6.0, containing 0.167 mg/ml o-dianisidine dihydrochloride and 0.005% hydrogen peroxide. The change in absorbance at 460 nm during 2 min was measured with a spectrophotometer (Hitachi U-3300, Tokyo, Japan). One unit of MPO activity is defined as that which degrades 1 μmol of peroxidase per minute at 25 °C.

Results

Brain water content (Fig. 1)

Hypothermic treatment significantly reduced the brain edema formation compared with the normothermic group in the ipsilateral basal ganglia (NT; 82.0 ± 0.04% vs. HT6; 78.6 ± 0.09%: $p < 0.01$, HT12; 79.7 ± 0.05%: $p < 0.01$, HT24; 79.7 ± 0.05%: $p < 0.01$) in the slice of autologous blood injection (Fig. 1). An increase in brain water content was also significantly suppressed with hypothermia in the ipsilateral cortex of HT12 (NT; 81.3 ± 0.06% vs. *HT12;* 79.3 ± 0.02%: $p < 0.01$).

Blood-brain barrier permeability (Fig. 2)

The BBB was significantly protected from the breakdown with hypothermia compared with the normothermic rats in the ipsilateral basal ganglia (42.3 ± 4.0 vs. 23.0 ± 5.2 ng/g wet tissue; $p < 0.05$) in the slice of hematoma injection (slice A), but not in the ipsilateral cortex (53.3 ± 10.6 vs. 33.8 ± 10.4 ng/g wet tissue; n.s) (Fig. 2). In the adjacent slice B, the BBB disruption was significantly suppressed with hypothermia in the ipsilateral basal ganglia (55.5 ± 9.3 vs. 21.8 ± 3.2 ng/g wet tissue; $p < 0.05$) and cortex (34.3 ± 5.9 vs. 16.6 ± 3.5 ng/g wet tissue; $p < 0.01$).

Myeloperoxidase activity (Fig. 3)

Hypothermic treatment tended to inhibit the accumulation of PMN leukocytes in the lesion compared with the normothermic treatment (0.78 ± 0.13 vs. 0.51 ± 0.16 ΔAbs/mg tissue).

Discussion

Spontaneous ICH is a relatively common neurosurgical emergency associated with significant morbidity and mortality. The speculated mechanisms for brain edema formation after ICH are complicated. It

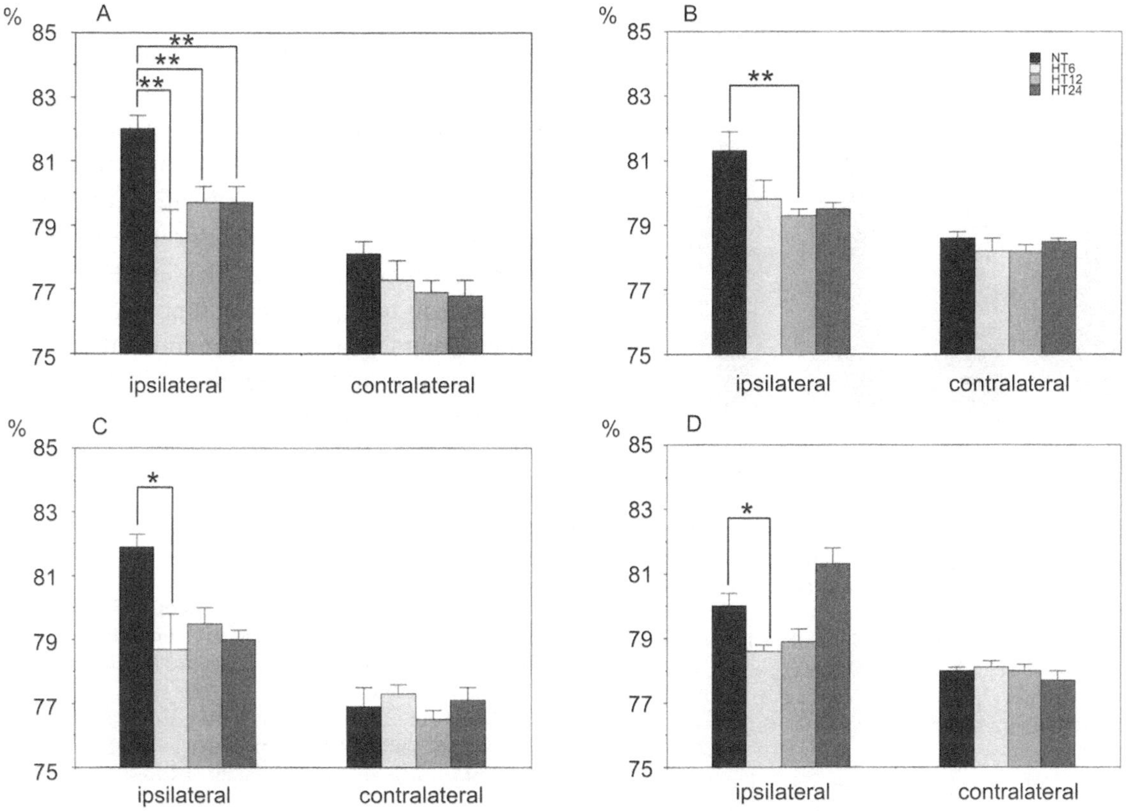

Fig. 1. Bar graph comparing the brain water content in the injection slice (A: basal ganglia, B: cortex) and in the adjacent slice (C: basal ganglia, D: cortex) in the normothermic (NT; n = 7), hypothermic (HT6; n = 6, HT12; n = 11, HT24; n = 6) rats. Brain slices were divided into the ipsilateral and contralateral hemispheres, and the cortex was separated from the basal ganglia bilaterally. Values are mean ± S.E.M. Hypothermia significantly reduced the brain water content in the ipsilateral basal ganglia in the slice of ICH (**$p < 0.01$). In the adjacent slice HT6 significantly reduced brain water content (*$p < 0.05$). Differences in the brain water content were not significant among the hypothermic subgroups (HT6, HT12, HT24)

has been believed that brain edema after ICH is due to local pressure compressing the microcirculation and producing ischemia in the region around the hematoma [5].

Attention has recently focused on the direct role of the biochemical substances like thrombin in the clot in generating brain edema. We previously demonstrated that injection of thrombin into basal ganglia produced a significant brain edema formation and hypothermic treatment reduced the brain edema in the basal ganglia at 24 h after the thrombin injection [7].

Moderate hypothermia has been shown to provide protection BBB permeability in rat model of traumatic brain injury [1, 4].

In our experiment, hypothermic treatment reduced the brain edema in the basal ganglia at 48 h after ICH. Hypothermia suppressed the BBB disruption to EB dye in a similar fashion, suggesting that hypothermia-induced decrease in brain edema formation observed in the present study is largely mediated by the protection of BBB integrity.

The recent study demonstrates that an inflammatory response occurs around a blood clot following ICH in rats. Inflammatory response may also be involved in secondary injury following ICH. Previously we revealed quantitative evidence that brain hypothermia significantly attenuates the accumulation of PMN leukocytes in the brain and the edema formation after the thrombin injection, however in this study differences in the myeloperoxidase (MPO) activity was not significant between ipsilateral basal ganglia in the normothermic group and that in the hypothermic group.

However, the hypothermic treatment has a tendency to inhibit the accumulation of PMN leukocytes into the brain compared with the normothermic treatment.

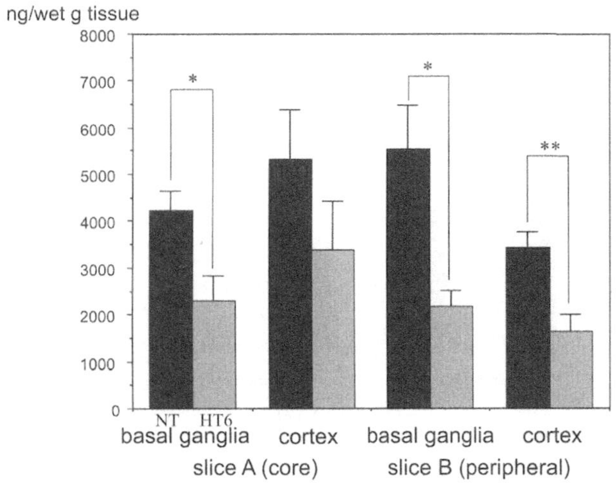

Fig. 2. Bar graph comparing the BBB permeability to Evan's blue in the injection slice (A) and in the adjacent slice (B) in the normothermic (NT; n = 6) and hypothermic (HT6; n = 6) rats. Evan's blue (EB) was injected intravenously at 48 h after ICH. Values are mean ± S.E.M. Hypothermia significantly reduced the BBB permeability to EB in the ipsilateral basal ganglia in slice A (*$p < 0.05$) and in the ipsilateral basal ganglia (*$p < 0.05$) and cortex (**$p < 0.01$) in slice B compared with the normothermic group

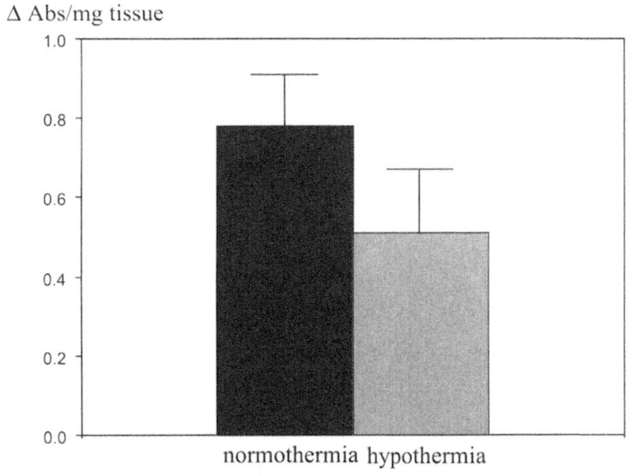

Fig. 3. Bar graph demonstrating the myeloperoxidase activity in the ipsilateral basal ganglia after 48 h following ICH. Hypothermic treatment tended to inhibit the accumulation of PMN leukocytes in the lesion compared with the normothermic treatment (0.78 ± 0.13 vs. 0.51 ± 0.16 ΔAbs/mg tissue)

Conclusion

This study indicates that hypothermic treatment significantly reduces the brain edema formation after ICH in rats. The decrease of brain water content was accompanied by a significant reduction in BBB permeability to Evan's blue dye. Inhibition of inflammation may be also responsible for the protective mechanism of hypothermia in this model. Hypothermic treatment may provide an approach to potentially reduce ongoing edema formation after ICH, detailed mechanism of hypothermia must be studied before applying in clinical settings.

References

1. Busto R, Dietrich WD, Gobus MY, Valdes I, Scheinberg P, Ginsberg MD (1987) Small differences in intraischemic brain temperature critically determine the extent of ischemic neuronal injury. Cereb Blood Flow Metab 7: 729–738
2. Bradley PP, Priebat DA, Christensen RD, Rothstein G (1982) Measurement of cutaneous inflammation: estimation of neutrophils content with an enzyme marker. J Inves Dermatol 78: 206–209
3. Gong C, Hoff JT, Keep RF (2000) Acute inflammatory reaction following experimental intracerebral hemorrhage in rat. Brain Res 871: 57–65
4. Clofton GL, Jiang JY, Lyeth BG, Jenkins LW, Hamm RJ, Hayes RL (1991) Marked protection by moderate hypothermia after experimental traumatic brain injury. J Cereb Blood Flow Metab 11: 114–121
5. Mendelow AD, Bullock R, Teasdale GM, Graham DI, McCulloch J (1984) Intracranial haemorrhage induced at arterial pressure in the rat. Part 2: short term changes in local cerebral blood flow measured by autoradiograhpy. Neurol Res 6: 189–193
6. Nayo NE, Neville D, Kirkland S, Ostbye T, Mustard CA Reeder, M Joffres, G Brauer, AR Levy (1996) Hospitalization and case fatality rates for stroke in Canada from 1982 through 1991: the Canadian collaborative study group of stroke hospitalizations. Stroke 27: 1215–1220
7. Kawai N, Kawanishi M, Okaushi M, Nagao S (2001) Effects of hypothermia on thrombin-induced brain edema formation. Brain Res 895: 50–58

Correspondence: Masahiko Kawanishi, 1750-1 Ikenobe, Miki-cho, Kita-gun, Kagawa, 761-0793 Japan. e-mail: mk@kms.ac.jp

Acta Neurochir (2003) [Suppl] 86: 457–461

Effect of delayed argatroban treatment on intracerebral hemorrhage-induced edema in the rat

T. Kitaoka[1,3], **Y. Hua**[1], **G. Xi**[1], **S. Nagao**[3], **J. T. Hoff**[1], and **R. F. Keep**[1,2]

[1] Department of Neurosurgery, University of Michigan, Ann Arbor, Michigan, U.S.A.
[2] Department of Physiology, University of Michigan, Ann Arbor, Michigan, U.S.A.
[3] Department of Neurological Surgery, Kagawa Medical University, Kagawa, Japan

Summary

Studies indicate that thrombin plays an important role in intracerebral hemorrhage (ICH) induced edema formation. However, the time window for administration of a thrombin inhibitor to reduce ICH-induced edema is unknown. Nor is it known whether this time window extends beyond the period when a thrombin inhibitor might exacerbate rebleeding. This study examines whether a thrombin inhibitor, argatroban, can reduce edema formation following intracerebral infusion of 100 µl of blood in the rat, the therapeutic time window for argatroban, and whether argatroban promotes rebleeding. Intracerebral injection of argatroban 3 hours after ICH caused a significant reduction in edema measured at 48 hours. The systemic administration of argatroban (0.9 mg/h) starting 6 hours after ICH also significantly reduced edema formation. There was no protection when the onset of argatroban administration was delayed to 24 hours after ICH. Argatroban did not increase collagenase-induced hematoma volume when given into the clot after 3 hours or given systemically at 6 hours. Our data suggest argatroban may be an effective therapy for ICH-induced edema.

Keywords: Intracerebral hemorrhage; thrombin; edema; argatroban.

Introduction

Direct intracerebral infusion of thrombin causes inflammation, edema, reactive gliosis, apoptosis and scar formation [5, 8, 13]. Thrombin is generated as blood clots and evidence indicates that thrombin plays an important role in intracerebral hemorrhage (ICH)-induced edema formation [9, 11]. Although thrombin formation occurs rapidly after ICH, enzymatically active thrombin may remain within a clot associated with fibrin [12], leading to a delayed release of thrombin into surrounding brain tissue. Thus, there could be a long therapeutic time window for a thrombin inhibitor. The goals of the present study were: 1) To determine whether delayed administration of argatroban (intracerebral or systemic) could reduce edema formation in a rat model of ICH involving injection of autologous blood into the caudate nucleus. 2) To examine whether argatroban administration would increase rebleeding in a collagenase model of ICH.

Materials and methods

Male Sprague-Dawley rats, each weighing 300–400 g, were used for all experiments. The animals were anesthetized with pentobarbital (50 mg/kg; intraperitoneal) and the right femoral artery was catheterized to monitor the arterial blood pressure and to sample blood for intracerebral infusion. Blood pH, PaO_2, $PaCO_2$, hematocrit, and glucose levels were monitored. Body temperature was maintained at 37 °C using a feedback–controlled heating pad.

Rats were then positioned in a stereotactic frame. A cranial burr hole (1 mm) was drilled near the right coronal suture 3.5 mm lateral to the midline. A 26-gauge needle was inserted stereotactically into the right basal ganglia (coordinates: 0.2 mm anterior, 5.5 mm ventral, and 3.5 mm lateral to the bregma). This needle was used for infusions of autologous blood or collagenase, the two ICH models used in this study.

The first set of experiments evaluated the effect of direct argatroban (provided by Mitsubishi Pharma Corporation) administration to an intracerebral hematoma on brain edema formation. Two groups of animals were studied. In each rats (n = 6) received a 100 µl infusion of autologous blood into the right basal ganglia at a rate of 10 µl/min. After infusion the needle was removed and skin incisions closed. Three hours later, the animals were reanesthetized and a Hamilton syringe was inserted using the same stereotactic coordinates. Twenty µl of saline or saline ± argatroban (0.5 µg/ml) was then infused over ten minutes. Forty-eight hours after ICH, animals were reanesthetized and decapitated. The brains were removed and a coronal brain slice (3 mm thickness) 4 mm from the frontal pole taken. The brain slice was divided into the two hemispheres and each hemisphere dissected into cortex and basal ganglia, a total of four samples. Samples were immediately weighed on an electronic analytical balance to obtain the wet weight. Brain samples were then dried at 100 °C for 24 hours to obtain dry weight and the water content determined as (Wet Weight – Dry Weight)/Wet Weight. The dehydrated samples were digested in 1 mL of 1 M nitric acid for 1 week prior to sodium and potassium content determination by flame photometer. Chloride was measured using a digital chloride meter. Ion content was expressed in milliequivalents per kilogram of dehydrated brain tissue (mEq/kg dry wt).

In the second and third sets of experiments, we investigated the effect of systemic argatroban administration on ICH-induced edema formation. For these experiments, argatroban was dissolved in physiological saline solution containing hydrochloric acid (pH 1.2–1.8). Physiological saline adjusted to the same pH with hydrochloric acid was used as vehicle. These solutions were used to fill osmotic minipumps (ALZET model 2ML1; Alza Corporation) that deliver 10 μl/h which were then implanted intraperitoneally.

All animals were sacrificed 48 hours after ICH or saline infusion for edema and ion measurements. In the second set, animals received an intracerebral infusion of 100-μl autologous blood. After six hours they were reanesthetized and implanted with the osmotic minipumps delivering vehicle or argatroban, (0.9 mg/h/rat). The animals were then allowed to recover. In the third set of experiments the osmotic minipumps were not implanted until 24 hours after the ICH. There were five and six animals in the vehicle and argatroban groups.

The last set of experiments examined whether argatroban would induce rebleeding in a collagenase model of ICH. Rats received a 2 μl injection (at 0.2 μl/min) of saline containing 0.4U bacteria collagenase (type VII, Sigma Chemical Co). Once the infusion was completed, the Hamilton syringe was left in place for five minutes. These animals were used for two sets of experiments. In the first set, the animals were reanesthetized after three hours when they received an intracerebral infusion of 20 μl saline or 20 μl saline + argatroban (0.5 μg/ml) over ten minutes ($n = 6$ per group). The stereotactic coordinates were the same as used for the collagenase injection. In the second set of experiments, the animals were reanesthetized after six hours and implanted with an osmotic minipump intraperitoneally delivering either saline (10 μl/h) or saline + argatroban. There were six rats per group. For both sets of experiments, the animals were decapitated 48 hours after the initial collagenase injection for determination of hematoma volume and brain hemoglobin content. Animals were decapitated and the brains rapidly removed and sectioned coronally at 2 mm intervals. Using a computerized image analysis system, the hemorrhage area for each section was measured. The total hematoma volume was calculated by summing the clot area in each section and multiplying by the distance between sections. For measuring hemoglobin content the ipsilateral cerebral hemisphere was collected from each animal. Distilled water (1 ml) was added to each hemisphere, followed by homogenization for 30 seconds, sonication on ice with an ultrasonicator for one minute, and centrifugation at 13,000 rpm for 30 minutes. The hemoglobin containing supernatant was collected and 80 μl of Drabkin's solution was added to a 20 μl aliquot. Fifteen minutes later the optical density (OD) of the solution at 550-nm wavelength was measured to assess hemoglobin content [2].

Data were analyzed using unpaired Student's t test. A two-tailed probability value of less than 0.05 was taken to indicate a significant difference.

Results

Intracerebral injection of argatroban 3 hours after ICH caused a significant reduction in perihematomal edema formation measured 48 hours post ICH. Thus, in the ipsilateral basal ganglia, water content was $80.9 \pm 1.0\%$ in the argatroban group compared to $82.6 \pm 0.8\%$ in the vehicle group (Fig. 1). The reduction of brain edema formation with argatroban was associated with reduced total ion accumulation in the treatment group (Fig. 1). Intraperitoneal administration (0.9 mg/h) of argatroban starting 6 hours after ICH also caused a significant reduction in perihematomal water content measured 48 hours post ICH. Thus, in the ipsilateral basal ganglia, brain water content was 80.3 ± 1.1 and $82.0 \pm 1.3\%$ in the argatroban and vehicle treated rats, respectively. Brain edema reduction in ipsilateral basal ganglia was associated with reduced total ion accumulation (Fig. 2). There was no reduction in edema formation when the onset of argatroban infusion was delayed to 24 hours post ICH.

Hematoma volume was assessed morphometrically 48 hours after ICH in rats receiving either argatroban or vehicle. Argatroban did not increase collagenase-induced hematoma volume when given into the clot at 3 hours or given systemically at 6 hours when compared to vehicle treated-rats (Table 1). Brain hemoglobin content was also used to assess hematoma mass. Again, there were no differences in hemoglobin content between argatroban and vehicle treated rats (Table 1).

Discussion

In previous rat studies, we have shown that injections of two thrombin inhibitors, hirudin and α-NAPAP, into an intracerebral hematoma within 5 minutes of blood injection can significantly reduce perihematomal edema formation [9, 11]. In contrast to normal blood, injection of anticoagulated blood (with heparin) does not cause perihematomal edema in rat or pig [17] and evidence indicates that anticoagulated blood is not edemagenic in humans as well [4]. Thrombin can induce cytotoxic effects on brain parenchymal cells, blood-brain barrier (BBB) disruption [10] and an inflammatory response [1]. These effects may contribute to edemagenic properties of thrombin. These previous studies have demonstrated that prior thrombin inhibition or thrombin inhibition immediately after ICH can reduce ICH-induced edema formation. The current study demonstrates that delayed intracerebral or systemic administration of a thrombin inhibitor, argatroban, can also reduce perihematomal edema formation. Although thrombin is formed almost immediately after an ICH, there are

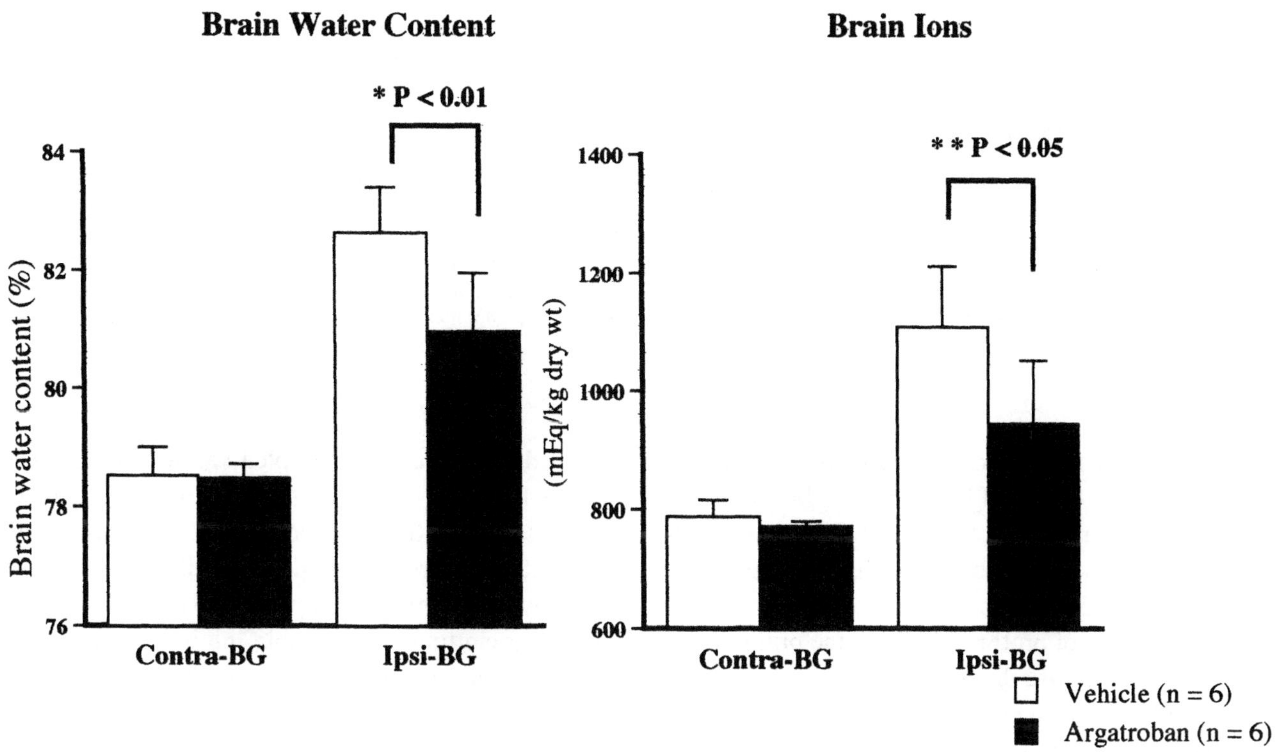

Fig. 1. Graph showing brain water content and brain ion contents 48 hours after intracerebral infusion of 100 μl of blood. Three hours after ICH, rats received an intracerebral infusion of argatroban (20 μl of 0.5 μg/ml) or saline. Values were expressed as the mean ± SD. *$P < 0.01$ compared to vehicle. **$P < 0.05$ compared to vehicle

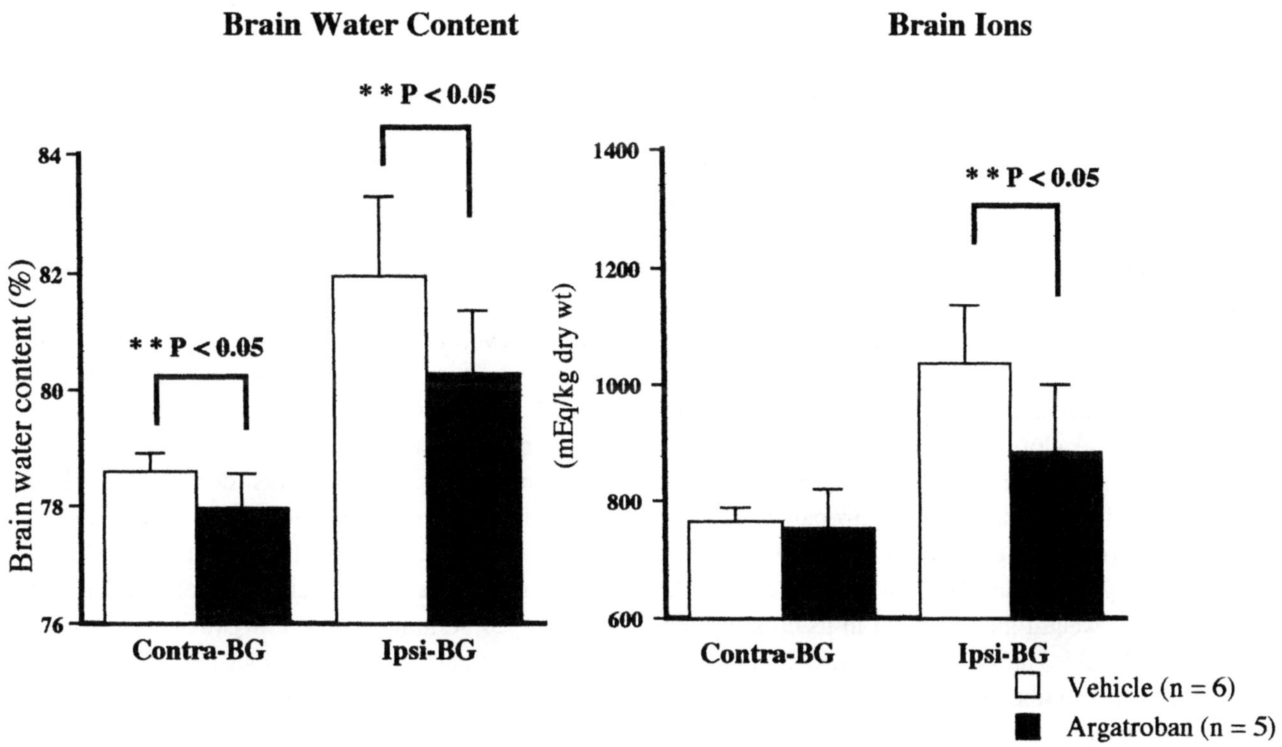

Fig. 2. Graph showing brain water content and brain ion contents 48 hours after intracerebral infusion of 100 μl of blood. Starting six hours after ICH, rats received an intraperitoneal infusion of argatroban or saline. Values were expressed as the mean ± SD. **$P < 0.05$ compared to vehicle

Table 1. *Hematoma volume and optical density (OD) for hemoglobin content in argatroban and vehicle treated rats*

		Vehicle	Argatroban
Intracerebral infusion of argatroban	hematoma volume (mm^3)	67.25 ± 6.99	65.12 ± 8.99
	OD at 550 nm	0.45 ± 0.07	0.39 ± 0.09
Intraperitoneal infusion of argatroban	hematoma volume (mm^3)	86.31 ± 17.96	76.56 ± 22.58
	OD at 550 nm	0.50 ± 0.06	0.49 ± 0.11

Table shows hematoma volume and brain hemoglobin (optical density) 48 hours after intracerebral infusion of collagenase. (Upper): three hours after ICH, rats received an intracerebral injection of argatroban (20 μl of 0.5 μg/ml) or saline. (Lower): six hours after ICH, rats received an intraperitoneal infusion of higher dose argatroban (0.9 mg/h/rat) or saline. Values are means ± SD; n = 6. There were no significant differences between argatroban and vehicle treated rats.

potential reasons why a delayed administration of a thrombin inhibitor would be effective. Thrombin can be bound within the clot [12] and only gradually released into the surrounding parenchyma. There may also be non-hematoma sources of thrombin (e.g. there may be an influx of prothrombin into the brain due to the ICH-induced BBB disruption).

Delayed systemic administration of argatroban, begun six hours after infusion of blood, also markedly reduced perihematomal edema and it was approximately equally effective as an intracerebral injection. In animals where systemic administration was delayed until 24 hours after the blood injection, there was no reduction in perihematomal edema formation. There is difficulty in translating animal data to the human. These studies employed an intracerebral injection of 100 μl of blood. Expressed per g of brain weight, this would correspond to an approximately 80 ml hematoma in the human. However, the absolute (rather than the relative) mass of the hematoma may affect the time window for argatroban treatment. In particular the time course for release of thrombin from the hematoma to the surrounding brain may be prolonged, if the hematoma mass is greater. Argatroban was chosen as the thrombin inhibitor for these studies for a number of reasons. Most importantly, it is able to inhibit both soluble and clot-bound thrombin [16], and the latter may be particularly important in relation to its effects in ICH. The ability of argatroban to inhibit clot-bound thrombin is probably a reflection of its low molecular weight compared to hirudin and heparin. Its low molecular weight may also aid in its crossing the disrupted BBB. Also importantly, argatroban does not directly affect thrombolytic enzymes [16] and it has a lower hemorrhagic potential in vivo than heparin or hirudin [6].

The primary ICH model used in this study involved direct injection of autologous blood into the caudate and it does not have a disrupted vasculature as basis for the ICH. To assess whether argatroban might induce rebleeding, the collagenase ICH model was used [15]. Although argatroban does not directly affect the fibrinolytic enzymes, it may exacerbate a rebleed by inhibiting clotting via thrombin inhibition and by reducing platelet adhesion to the vasculature [16]. In the collagenase model there was no evidence of rebleeding after argatroban administration (systemic or intracerebral) as assessed by measurement of hematoma volume or brain hemoglobin concentration. The doses used were the same as could reduce edema formation. It is difficult to translate safety data between rats and humans. The rat collagenase model has a disrupted vasculature as basis for the ICH but the form of that disruption is different from a normal human ICH. Although most patients stop bleeding shortly after a spontaneous ICH [14], spontaneous rebleeding may occur, most rehemorrhaging occurring within the first 24 hours [7]. Fujii *et al.*, reported early expansion of the hematoma in 14% of patients within 24 hours and suggested that early hematoma growth is related to coagulation abnormalities such as platelet aggregation and low levels of fibrinogen [3]. In view of this potential for spontaneous rebleeding, great care would be needed in determining the safety window for argatroban in humans.

In conclusion, these results demonstrate that a thrombin inhibitor, argatroban, can significantly reduce ICH-induced edema formation even when administration is delayed for six hours after ICH. In the rat, delayed administration of argatroban was not associated with an increase in rebleeding after ICH suggesting it may be a potential therapeutic agent for ICH-induced edema in humans. A number of mechanisms appear to contribute to ICH-induced edema, including erythrocyte lysis and hemoglobin-induced toxicity [18]. Thus, although argatroban is successful in reducing ICH-induced edema, it is possible that it would be even more effective as part of a combination therapy.

References

1. Carney DH, Redin W, McCroskey L (1992) Role of high-affinity thrombin receptors in postclotting cellular effects of thrombin. Sem Thrombosis Hemost 18: 91–103
2. Choudhri TF, Hoh BL, Solomon RA, Connolly ES, Pinsky DJ (1997) Use of a spectrophotometric hemoglobin assay to objectively quantify intracerebral hemorrhage in mice. Stroke 28: 2296–2302
3. Fujii Y, Tanaka R, Takeuchi S, Koike T, Minakawa T, Sasaki O (1994) Hematoma enlargement in spontaneous intracerebral hemorrhage. J Neurosurg 80: 51–57
4. Gebel JM, Brott TG, Sila CA, Tomsick TA, Jauch E, Salisbury S, Khoury J, Miller R, Pancioli A, Duldner JE, Topol EJ, Broderick JP (2000) Decreased perihematomal edema in thrombolysis-related intracerebral hemorrhage compared with spontaneous intracerebral hemorrhage. Stroke 31: 596–600
5. Gong C, Boulis N, Qian J, Turner DE, Hoff JT, Keep RF (2001) Intracerebral hemorrhage-induced neuronal death. Neurosurgery 48: 875–882
6. Hursting MJ, Alford KL, Becker JC, Brooks RL, Joffrion JL, Knappenberger GD, Kogan PW, Kogan TP, McKinney AA, Schwarz RP Jr (1997) Novastan (brand of argatroban): a small-molecule, direct thrombin inhibitor. Semin Thromb Hemost 23: 503–516
7. Kazui S, Naritomi H, Yamamoto H, Sawada T, Yamaguchi T (1996) Enlargement of spontaneous intracerebral hemorrhage: Incidence and time course. Stroke 27: 1783–1787
8. Lee KR, Betz AL, Keep RF, Chenevert TL, Kim S, Hoff JT (1995) Intracerebral infusion of thrombin as a cause of brain edema. J Neurosurg 83: 1045–1050
9. Lee KR, Betz AL, Kim S, Keep RF, Hoff JT (1996) The role of the coagulation cascade in brain edema formation after intracerebral hemorrhage. Acta Neurochir (Wien) 138: 396–400
10. Lee KR, Kawai N, Kim S, Sagher O, Hoff JT (1997) Mechanisms of edema formation after intracerebral hemorrhage: effects of thrombin on cerebral blood flow, blood-brain barrier permeability, and cell survival in a rat model. J Neurosurg 86: 272–278
11. Lee KR, Colon G, Betz AL, Keep RF, Kim S, Hoff JT (1996) Edema from intracerebral hemorrhage: the role of thrombin. J Neurosurg 84: 91–96
12. Mutch NJ, Robbie LA, Booth NA (2000) Human thrombi contain an abundance of active thrombin. Throm Hemost 86: 1028–1034
13. Nishino A, Suzuki M, Ohtani H, Motohashi O, Umezawa K, Nagura H, Yoshimoto T (1993) Thrombin may contribute to the pathophysiology of central nervous system injury. J Neurotrauma 10: 167–179
14. Ojemann RG, Mohr JP (1976) Hypertensive brain hemorrhage. Clin Neurosurg 23: 220–244
15. Rosenberg GA, Mun-Bryce S, Wesley M, Kornfeld M (1990) Collagenase-induced intracerebral hemorrhage in rats. Stroke 21: 801–807
16. Schwarz RP, Becker J-C, Brooks RL, Hursting MJ, Joffrion JL, Knappenberger GD, Kogan TP, Kogan PW, McKinney AA (1997) The preclinical and clinical pharmacology of Novastan (Argatroban): a small-molecule, direct thrombin inhibitor. Clin Appl Thrombosis/Hemostasis 3: 1–15
17. Xi G, Wagner KR, Keep RF, Hua Y, de Courten-Myers GM, Broderick JP, Brott TG, Hoff JT, Muizelaar JP (1998) Role of blood clot formation on early edema development after experimental intracerebral hemorrhage. Stroke 29: 2580–2586
18. Xi G, Keep RF, Hoff JT (1998) Erythrocytes and delayed brain edema formation following intracerebral hemorrhage in rats. J Neurosurgery 89: 991–996

Correspondence: Takuji Kitaoka, Department of Neurosurgery, Kagawa Medical University, 1750-1, Ikenobe, Miki-cho, Kagawa, 761-0793, Japan. e-mail: kitaoka@sky.quolia.com

Acta Neurochir (2003) [Suppl] 86: 463–467

Overexpression of interleukin-1 receptor antagonist reduces brain edema induced by intracerebral hemorrhage and thrombin

T. Masada[1], Y. Hua[2], G. Xi[2], G.-Y. Yang[2], J. T. Hoff[2], R. F. Keep[3], and S. Nagao[1]

[1] Department of Neurological Surgery, Kagawa Medical University, Japan
[2] Department of Surgery (Neurosurgery), University of Michigan, Ann Arbor, Michigan
[3] Department of Physiology, University of Michigan, Ann Arbor, Michigan

Summary

Recent studies indicate that inflammatory reaction occurs around hematoma after intracerebral hemorrhage (ICH). In this study the authors examine the hypothesis that overexpression of IL-1ra in the brain attenuate brain edema formation after ICH.

Adenoviruses expressing IL-1ra (Ad.RSVIL-1ra) or LacZ (Ad.RSVLacZ) or saline were injected into the lateral ventricle. On the fifth day after virus injection, 100 μl of autologous blood or 5 U thrombin was infused into the right basal ganglia. Rats with ICH were killed 24 or 72 hours later for measurement of brain water content. Thrombin-treated rats were killed 24 hours later for edema measurements and an assessment of polymorphonuclear leukocyte (PMNL) infiltration by myeloperoxidase (MPO) assay.

Compared with control groups, Ad.RSVIL-1ra treated rats had less brain edema formation in the ipsilateral basal ganglia 3 days after ICH (81.5 ± 0.3% compared with 83.4 ± 0.4% and 83.3 ± 0.5% in control animals). Ad.RSVIL-1ra treated rats had also less brain edema following thrombin injection. The reduction of brain edema induced by thrombin was involved in the reduction of PMNL infiltration in basal ganglia, as assessed by MPO assay.

Adenovirus-mediated overexpression of IL-1ra attenuated brain edema formation following ICH, perhaps by reduction of thrombin-induced brain inflammation.

Keywords: Intracerebral hemorrhage; brain edema; thrombin; interleukin-1 receptor antagonist; inflammation; rat.

Introduction

Brain edema induced by intracerebral hemorrhage (ICH) elevates intracranial pressure and causes brain herniation. Preventing the accumulation of edema is important for the management of ICH.

One of the mechanisms to produce brain edema after ICH is thrombin, which induces a variety of reaction to the brain tissue [7]. Thrombin upregulates inflammatory cytokines such as IL-1 and tumor necrotic factor-α, which are involved in the migration of white blood cells from blood to brain. Infiltration of inflammatory cells in the brain and release of cytokines such as IL-1β induce tissue injury and brain edema formation [3].

The interleukin-1 receptor antagonist (IL-1ra) inhibits several actions of IL-1, and administration of IL-1ra reduces ischemic and traumatic brain injuries [10]. Adenovirus-mediated overexpression of IL-1ra also attenuates ischemic brain injury [11]. The effect of IL-1ra on ICH has not been studied.

In this study, we examine whether adenovirus-mediated overexpression of IL-1ra can reduce the brain edema following ICH or thrombin instillation into the basal ganglia, and whether IL-1ra overexpression reduces the thrombin-induced inflammatory response.

Materials and methods

Animal preparation

The protocols for these animal studies were approved by the University of Michigan Committee on the Use and Care of Animals. Adult male Sprague-Dawley rats weighing 275–350 g were used in these experiments. The rats were given free access to food and water.

Adenovirus gene transduction

Two different recombinant virus vectors were used in this study: one containing the human IL-1ra complementary DNA, the other containing the Escherichia coli β-galactosidase gene. For both viruses, the RSV promoter was used to drive gene transcription. Ad.RSVlacZ virus was used as a control.

For administration of the virus, each rat was positioned in a stereotactic frame. Ten microliters of adenoviral suspension containing 10^{12} particles/ml were injected into the lateral ventricle.

Intracerebral infusion of blood and thrombin

Five days after administration of virus or vehicle, the rats were positioned in a stereotactic head frame. A stereotactically guided 26 gauge needle was inserted into the right basal ganglia (coordinate: 0.2 mm anterior, 5.5 mm ventral, and 3 mm lateral to the bregma). Whole blood from femoral artery or thrombin was infused into the right basal ganglia [8].

Experimental groups

In this study we compared rats that underwent injection of saline into the left ventricle, or saline containing Ad.RSVLacZ or Ad.RSVIL-1ra. After recovery for 5 days, the rats were used for one of three sets of experiments. In the first set, rats underwent injection of 100 µl of whole blood into the right basal ganglia, then they were killed 24 or 72 hours later to determine the degree of brain edema. In the second set, 5U of rat thrombin in 50 µl of saline was infused into the right basal ganglia, after which the rats were killed at 24 hours for determination of brain edema. In the third set, MPO activity was measured 24 hours after thrombin injection.

Measurement of brain water content

Animals were anesthetized and decapitated. The brains were removed, and a coronal slice approximately 3 mm thick and located 4 mm from the frontal pole was cut with a blade. Ipsilateral and contralateral basal ganglia were obtained from each brain. Brain samples were weighed to obtain the wet weight. The samples were dried in a gravity oven at 100 °C for 24 hours to obtain the dry weight and to determine the water content, which was calculated as (wet weight – dry weight)/wet weight.

Myeloperoxidase assay

Animals were anesthetized and decapitated. Bilateral basal ganglia were obtained as described earlier. The brain samples were immediately weighed and frozen in liquid nitrogen.

Tissue segments were homogenized in 4 ml of 50 mM Tris-HCl (pH7.4) at 4 °C. Samples were added to 20 ml of 5 mM phosphate buffer (pH6) at 4 °C, homogenized , and centrifuged at 30,000 G for 30 minutes. The supernatant was discarded and the pellet was washed again as described earlier. After the supernatant was decanted, the pellet was extracted by suspension for approximately 2 minutes in 0.5% hexadecyl trimethyl ammonium bromide in 50 mM phosphate buffer at 25 °C, at an original tissue weight/volume ratio of 1:4. The samples were frozen in liquid nitrogen, and freeze and thaw cycles were then performed three times, with 10-second sonications at 25 °C between cycles. After the last sonication, the samples were incubated at 4 °C for 20 minutes and centrifuged at 12,500 G for 15 minutes. The 100 µl supernatants were mixed with 100 µl of 50 mM phosphate buffer containing 0.001% hydrogen peroxidase (pH6) and 50 µl of 0.334 mg/ml σ dianisident dihydrochloride. Absorbance was measured at 460 nm at 15-second intervals over 3 minutes by using a spectrophotometer, and results were expressed as the relative change in absorbance per minute at 460 nm.

Statistical analysis

All data in this study are presented as the means $\pm$ the standard error. Statistical significance was accepted at probability values of less than 0.05.

Results

Brain edema formation

24 hours after ICH, water content in the ipsilateral basal ganglia was 80.8 $\pm$ 0.5%, compared with 77.9 $\pm$ 0.2% in the contralateral hemisphere. By 72 hours post ICH, the water content increased further in the ipsilateral hemisphere to 83.4 $\pm$ 0.4% (Fig. 1). Prior treatment with Ad.RSVLacZ did not change the amount of edema formation both 1 and 3 days after ICH. Rats that received prior treatment with Ad.RSVIL-1ra, however, had less edema formation in the ipsilateral basal ganglia than did the other two groups at 72 hours, but not at 24 hours (Fig. 1).

In rats that had received a ventricular infusion of saline or Ad.RSVlacZ, injection of 5U of thrombin into the right basal ganglia resulted in marked edema formation in that location (Fig. 2). In rats that had received prior treatment with Ad.RSVIL-1ra, edema formation was reduced.

Myeloperoxidase activity

In thrombin-injected rats, bilateral basal ganglia were assayed for MPO activity, which is an indicator of neutrophil accumulation. In the ipsilateral basal ganglia (the site of the thrombin injection) of Ad.RSVlacZ-transduced rats, there was a marked increase in MPO activity compared with the contralateral basal ganglia. This increase was inhibited in Ad.RSVIL-1ra–transduced rats (Fig. 3).

Discussion

In this study we demonstrated that adenovirus-mediated overexpression of IL-1ra reduces brain edema induced by ICH and thrombin. This protective effect of IL-1ra against brain edema formation may be mediated by a reduction of thrombin-induced inflammation.

It has been suggested that inflammatory reaction peaks 48 hours after ICH [2]. In our study, a protective effect of Ad.RSVIL-1ra on ICH-induced brain edema was found 3 days after ICH but not after 1 day. The lack of protective effect of IL-1ra at 24 hours after ICH may be due to the fact that inflammatory reaction does not peak 24 hours after ICH.

To explore a potential mechanism for the protective effect of IL-1ra overexpression on ICH-

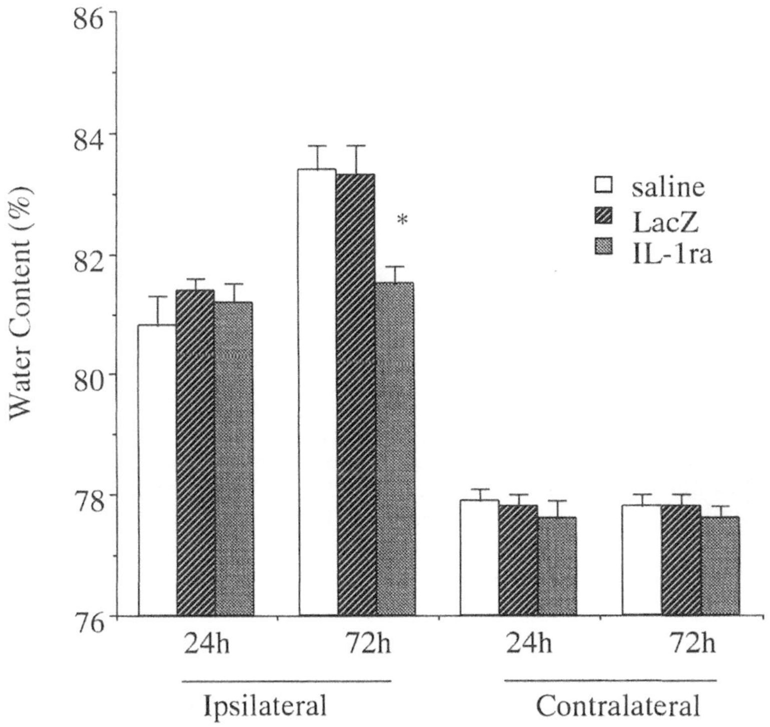

Fig. 1. Bar graph showing brain water content in the basal ganglia 24 hours or 72 hours after intracerebral infusion of 100 μl of blood. Five days before ICH, the rats received intraventricular injections of either Ad.RSVIL-1ra, Ad.RSVlacZ, or saline. Values are expressed as means ± SE. *$p < 0.05$ compared with saline and Ad.RSVlacZ. n = 5–6

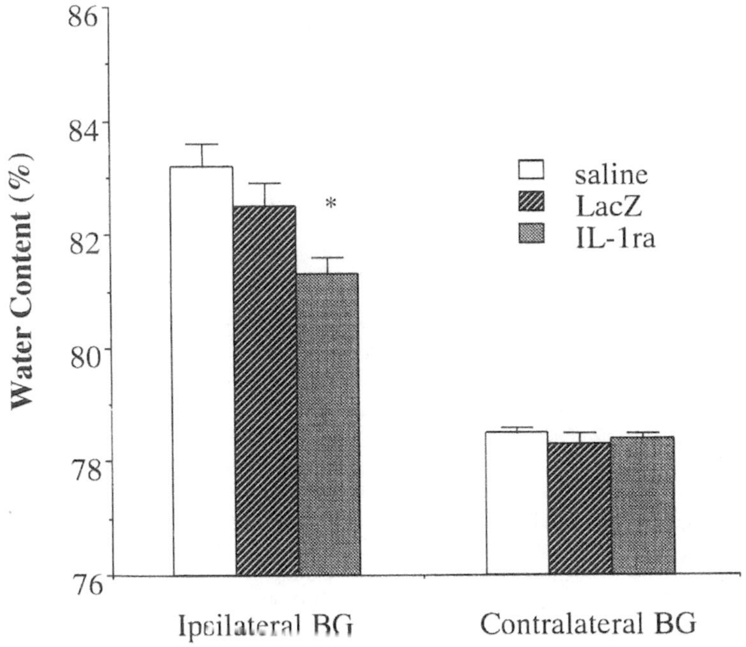

Fig. 2. Bar graph showing water content 1 day after injection with 5 U thrombin in the basal ganglia. The rats received intraventricular injections of either Ad.RSVIL-1ra, Ad.RSVlacZ, or saline 5 days before thrombin injection. *$p < 0.05$ compared with saline and Ad.RSVlacZ. n = 6

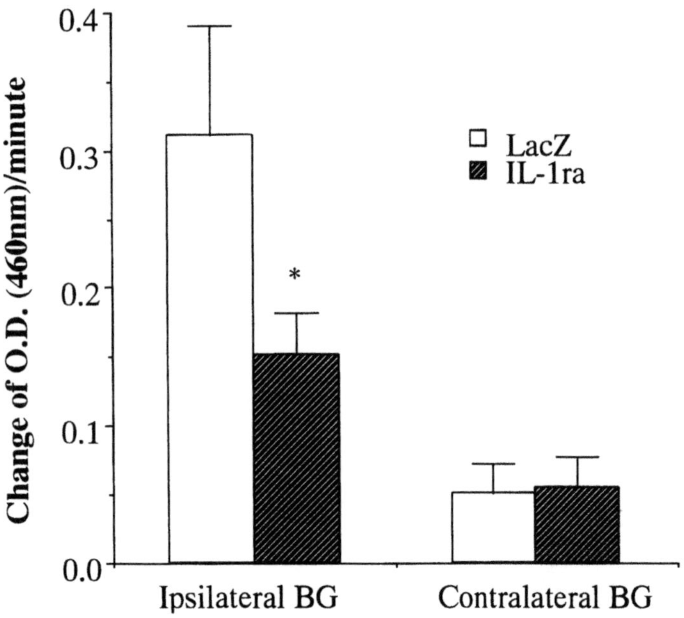

Fig. 3. Bar graph showing MPO activity 1 day after injection with 5 U thrombin. The rats received intraventricular injections of either Ad.RSVIL-1ra, AdRSVlacZ, or saline before thrombin injection. The MPO activity was measured in the ipsilateral and contralateral basal ganglia. *$p < 0.05$ compared with LacZ. OD = optical density. n = 8

induced brain edema, we also examined the effect of IL-1ra on thrombin–induced injury. Thrombin induced brain edema also attenuated after administration of Ad.RSVIL-1ra. Thrombin is one component of the hematoma that is responsible for edema formation after ICH [6]. Thrombin upregulates inflammatory cytokines such as IL-1 and tumor necrotic factor-α, which are involved in the migration of white blood cells from blood to tissue [3]. Infiltration of inflammatory cells in the brain induces tissue injury [5]. Therefore we examined whether reduction in thrombin–induced edema in Ad.RSVIL-1ra transduced rats is associated with a reduction in inflammation and PMNL infiltration. As observed on MPO assay, thrombin induced infiltration of PMNL was lower in Ad.RSVIL-1ra-treated rats.

One of the advantages of gene therapy is the potential for continuous production of a protective protein in the brain. Because of its size and short half-life, IL-1ra can be administered repetitively in the lateral ventricle to produce a protective effect [10]. On the other hand, one administration of adenovirus is enough to produce a protective effect [1]. In this study adenovirus was injected in the lateral ventricle 5 days before the ICH because earlier results had demonstrated that production of IL-1ra peaks at 5–7 days [11]. To clinically apply the gene therapy, we need an advance of the vectors to produce the protective protein earlier.

The contribution of IL-1 to ICH-induced injury has not been previously examined. The expression of IL-1 increases markedly after endotoxin infection and brain injury [9]. The source of IL-1 after brain injury is unclear, although there is evidence that it can be produced by a number of cell types, including microglia astrocytes, endothelial cells and invading macrophages and neutrophils [4]. IL-1 has multiple potentially harmful effects on the brain, including neurotoxicity, opening of the blood-brain barrier, induction of apoptosis, neutrophil infiltration, and activation of microglia [5]. To control IL-1 production will be an important strategy to reduce brain injury.

In conclusion, adenovirus-mediated overexpression of IL-1ra reduces brain edema formation after ICH. The attenuation of brain inflammation induced by thrombin may be one of the factors. These results may provide a new therapeutic target after ICH.

References

1. Betz AL, Yang GY, Davidson BL (1995) Attenuation of stroke size in rats using an adenoviral vector to induce overexpression of interleukin-1 receptor antagonist in brain. J Cereb Blood Flow Metab 15: 547–551
2. Del Bigio MR, Yan HJ, Buist R, Peeling J (1996) Experimental intracerebral hemorrhage in rats. Magnetic resonance imaging and histopathological correlates. Stroke 27: 2312–2320
3. Drake WT, Lopes NN, Fenton JW *et al* (1992) Thrombin enhancement of interleukin-1 and tumor necrotic factor-α induced polymorphonuclear leukocyte migration. Lab Invest 67: 617–627
4. Giulian D, Barker TJ (1985) Peptides released by ameboid microglia regulate astroglial proliferation. J Cell Biol 101: 2411–2415
5. Holmin S, Mathiesen T (2000) Intracerebral inflammation of interleukin-1β and induction of inflammation, apoptosis, and vasogenic edema. J Neurosurg 92: 108–120

6. Lee KR, Betz AL, Keep RF, Chenevert TL, Kim S, Hoff JT (1995) Intracerebral infusion of thrombin as a cause of brain edema. J Neurosurg 83: 1045–1050
7. Lee KR, Colon GP, Betz AL, Keep RF, Kim S, Hoff JT (1996) Edema from intracerebral hemorrhage: the role of thrombin. J Neurosurg 84: 91–96
8. Masada T, Xi G, Hua Y, Keep RF (2000) The effects of thrombin preconditioning on focal cerebral ischemia in rats. Brain Res 867: 173–179
9. Rothwell NJ, Relton JK (1993) Involvement of cytokines in acute neurodegeneration in the CNS. Neurosci Biobehav Rev 17: 217–227
10. Toulmond S, Rothwell NJ (1995) Interlekion-1 receptor antagonist inhibits neuronal damage caused by fluid percussion injury in the rat. Brain Res 671: 261–266
11. Yang GY, Zhao YJ, Davidson BL, Betz AL (1997) Overexpression of interleukin-1 receptor antagonist in the mouse brain reduces ischemic brain injury. Brain Res 751: 181–188

Correspondence: Tetsuya Masasda, Department of Neurological Surgery, Kagawa Medical University, 1750-1 Ikenobe Miki Kita Kagawa Japan 453-8511. e-mail: tmasada@kms.ac.jp

Acta Neurochir (2003) [Suppl] 86: 469–471

The role of haematoma aspiration in the management of patients with thalamic haemorrhage: a pilot study with continuous compliance monitoring

S. C. P. Ng[1], **W. S. Poon**[1], and **M. T. V. Chan**[2]

[1] Division of Neurosurgery, Department of Surgery, Prince of Wales Hospital, The Chinese University of Hong Kong, Shatin, Hong Kong
[2] Department of Anaesthesia and Intensive Care, Prince of Wales Hospital, The Chinese University of Hong Kong, Shatin, Hong Kong

Summary

Objective. Thalamic brain haemorrhage is a common disabling and potentially fatal condition. Management is mainly supportive and very rarely do neurosurgeons have to resort to evacuation of the haematoma. We hypothesised that haematoma aspiration may help to improve the outcome of the group of patients with thalamic haemorrhage.

Material and methods. Patients with thalamic haemorrhage were selected on clinical group (intraventricular haemorrhage, hydrocephalus and haematoma size > 15 ml) for intracranial pressure (ICP) monitoring and management in the intensive care unit (ICU). A Spiegelberg double lumen intraventricular balloon catheter was inserted as a fluid-filled catheter on the side of the haematoma. Data of ICP, arterial blood pressure (ABP), cerebral perfusion pressure (CPP) and compliance were collected at one-minute intervals. Hourly averages were used for analysis.

Results. 94,448 set of data were collected from eight patients with thalamic haemorrhage. Significant reduction in ICP and improvement in intracranial compliance were detected. In patients with haematoma aspiration, intracranial compliance was improved within 48 hours of the procedure whereas ICP did not.

Conclusions. This pilot study showed that haematoma aspiration in patients with thalamic haemorrhage could improve intracranial compliance. Whether this could be translated into clinical outcome benefits requires further study in a larger number of patients.

Keywords: Thalamic haemorrhage; intracranial compliance; intracranial pressure monitoring; intraventricular haemorrhage; hydrocephalus.

Introduction

Spontaneous intracerebral haemorrhage (ICH) is twice more common in Southern Chinese than in Caucasians. Thalamic haemorrhage accounts for 32% of these patients with ICH (Kneel 1991) and carries a 35% management mortality (Hsiang 1996).

In general, thalamic haemorrhage is complicated by intraventricular haemorrhage (IVH) with or without hydrocephalus. It causes more extensive and persistent metabolic depression and more pronounced cerebral blood flow reduction, and therefore a worse outcome (Tanaka 1996). Management is mainly supportive. In deteriorating patients with hydrocephalus, cerebrospinal fluid (CSF) drainage is necessary. However, in treating the deep-sited haematoma, there is no rigid guideline for surgical intervention.

In order to improve the outcome, apart from monitoring the intracranial pressure (ICP), we may need intensive monitoring to predict the development of hydrocephalus and to help clinical decision making in evacuating or aspirating the haematoma.

Continuous intracranial compliance monitoring provides information on the volume change resulting from unit change in pressure and helps to identify the process which leads to raised ICP. The Spiegelberg continuous compliance monitor consists of a double lumen intraventricular balloon catheter, which is inserted into the ventricle. Intracranial compliance is calculated from a an average of small ICP changes resulting from a sequence of measurements. This device produces a minute by minute measure of intracranial compliance.

In this pilot study with intracranial compliance monitoring, we hypothesized that haematoma aspiration may help to improve the outcome of patients with thalamic haemorrhage.

Material and methods

All patients with thalamic haemorrhage aged between 45 to 75 years were included. Patient selection for ICP monitoring and intensive care management was based on clinical background. They included impaired conscious levels (Glasgow Come Scale score

Table 1. *Comparison of ICP, ABP, CPP and intracranial compliance collected on day 1 and the last day in ICU (data are presented as mean ± s.d.)*

Parameters	Day 1	Last day in ICU	p-value
ICP (mmHg)	14.6 ± 2.5	9.7 ± 3.4	0.001*
ABP (mmHg)	95.6 ± 10.3	95.3 ± 9.3	0.892
CPP (mmHg)	81.0 ± 9.8	85.6 ± 9.7	0.188
Compliance (ml/mmHg)	0.41 ± 0.13	0.95 ± 0.27	0.001*

Table 2. *Stratified analysis based on the data of ICP, ABP, CPP and intracranial compliance collected on day 1 and the last day in ICU (data are presented as mean ± s.d.)*

	Day 1	Last day in ICU	p-value
No USG aspiration performed (n = 2)			
ICP (mmHg)	12.7 ± 2.6	7.4 ± 3.2	0.048*
ABP (mmHg)	98.2 ± 14.9	98.1 ± 16.0	0.925
CPP (mmHg)	85.5 ± 12.3	90.6 ± 12.8	0.180
Compliance (ml/mmHg)	0.40 ± 0.08	1.17 ± 0.25	0.096
USG aspiration performed (n = 6)			
ICP (mmHg)	15.3 ± 2.4	10.5 ± 3.3	0.014*
ABP (mmHg)	94.8 ± 10.0	94.4 ± 8.0	0.903
CPP (mmHg)	79.5 ± 9.6	83.9 ± 9.2	0.352
Compliance (ml/mmHg)	0.41 ± 0.15	0.88 ± 0.26	0.008*

G.C.S. < 13 points), evidence of IVH, hydrocephalus and haematoma volume greater than 15 ml.

ICP and intracranial compliance were measured on the side of predominant haemorrhage. Data of ICP, compliance, arterial blood pressure (ABP) and cerebral perfusion pressure (CPP) were collected at one-minute intervals. Hourly averages were used for analysis.

Results

Six male and two female patients age 55.0 ± 6.3 years (mean ± standard deviation, s.d.) were included. 94449 sets of valid data were collected from these patients. Median admission G.C.S. score for the patients in this study was twelve (range: 3–13 points). Mean ICU stay was nine days with a s.d. of 1.8 days. Six out of the eight patients were suggested to haematoma aspiration.

Table 1 shows the pooled results of all eight patients. Comparing data collected on day 1 and the last day in ICU, there was a significant reduction in ICP (14.6 ± 2.6 vs 9.7 ± 3.4 mmHg, p = 0.001) and an improvement in intracranial compliance (0.41 ± 0.13 vs 0.95 ± 0.27 ml/mmHg, p = 0.001). But no differences were found in ABP and CPP.

When the results were stratified by whether haematoma aspiration was performed (Table 2), reductions in ICP were observed in both groups (12.7 ± 2.6 vs 7.4 ± 3.2 mmHg, p = 0.048; 15.3 ± 2.4 vs 10.5 ± 3.2 mmHg, p = 0.014), however, improvement in compliance was only significant in the group of patients

Table 3. *Collected data of ICP, ABP, CPP and intracranial compliance are compared: (i) before aspiration versus 1 day, 2 days after aspiration, and the last day in ICU; (ii) 1 day after aspiration versus 2 days after aspiration and the last day in ICU (data are presented as mean ± s.d.)*

(i)	*Before*	*1 day after*	*2 days after*	*Last day in ICU*
ICP (mmHg)	16.8 ± 5.6	13.0 ± 4.2 (p = 0.126)	12.8 ± 4.3 (p = 0.093)	10.5 ± 3.3 (p = 0.015*)
ABP (mmHg)	92.5 ± 7.0	92.3 ± 7.1 (p = 0.660)	93.8 ± 6.1 (p = 0.773)	94.4 ± 8.0 (p = 0.982)
CPP (mmHg)	78.5 ± 6.6	79.3 ± 8.1 (p = 0.625)	81.0 ± 7.5 (p = 0.939)	83.9 ± 9.2 (p = 0.374)
Compliance (ml/mmHg)	0.39 ± 0.17	0.58 ± 0.26 (p = 0.005*)	0.83 ± 0.27 (p = 0.001*)	0.88 ± 0.26 (p = 0.001*)

(ii)	*1 day after*	*2 days after*	*Last day in ICU*
ICP (mmHg)	13.0 ± 4.2	12.8 ± 4.3 (p = 0.060)	10.5 ± 3.3 (p = 0.032*)
ABP (mmHg)	92.3 ± 7.1	93.8 ± 6.1 (p = 0.391)	94.4 ± 8.0 (p = 0.769)
CPP (mmHg)	79.3 ± 8.1	81.0 ± 7.5 (p = 0.108)	83.9 ± 9.2 (p = 0.239)
Compliance (ml/mmHg)	0.58 ± 0.26	0.83 ± 0.27 (p = 0.031*)	0.88 ± 0.26 (p = 0.769)

suggested to haematoma aspiration (0.41 ± 0.15 vs 0.88 ± 0.26 ml/mmHg, $p = 0.008$). There were no group to group differences in the four parameters, neither on day 1 nor the last day in ICU.

Table 3 shows the further analysed results of the data collected from the six patients with haematoma aspiration. By comparing data collected before and after aspiration (Table 3(i)), no significant changes were found in ICP, ABP and CPP whereas improvements in compliance were observed in a 48-hour interval (0.39 ± 0.17 vs 0.58 ± 0.26 ml/mmHg, $p = 0.005$; vs 0.83 ± 0.27, $p = 0.001$). By comparing data collected on day 1 and day 2 after aspiration (Table 3(i)), further improvement was observed in compliance (0.58 ± 0.26 vs 0.83 ± 0.27 ml/mmHg, $p = 0.031$). By comparing data collected on day 2 after aspiration and the last day in ICU (Table 3(ii)), no difference was found in compliance whereas ICP took a longer time to achieve significant improvement (12.8 ± 4.3 vs 10.4 ± 3.3 mmHg, $p = 0.032$).

Discussion

This pilot study showed that haematoma aspiration improves intracranial compliance in patients with thalamic haemorrhage, whereas ICP and CPP do not. Whether this can be translated into clinical outcome benefits, such as Glasgow outcome scale scores, the number of days requiring CSF drainage and the need for CSF diversion, requires further study in a larger number of patients.

Acknowledgment

This study was supported by a direct grant for research from the Chinese University of Hong Kong.

References

1. Hsiang JNK, Zhu XL, Wong LKS, Kay R, Poon WS (1998) Putaminal and thalamic haemorrhage in ethnic Chinese living in Hong Kong. Surg Neruology 50: 526–532
2. Kreel L, Kay R, Woo J, Wong Hy, Nicholls GM (1991) The radiological (CT) and clinical sequelae of primary intracerebral haemorrhage. Brit J Radiology 64: 1096–1100
3. Tanaka A, Yoshinaga S, Nakayama Y, Kimura M, Tomonaga M (1996) Cerebral blood flow and clinical outcome in patient with thalamic haemorrhages: a comparison with putaminal haemorrhages. J Neruo Sci 144: 191–197

Correspondence: Wai S. Poon, Division of Neurosurgery, Department of Surgery, The Chinese University of Hong Kong, Shatin, Hong Kong. e-mail: wpoon@surgery.cuhk.edu.hk

Acta Neurochir (2003) [Suppl] 86: 473–476

Acute decrease of cerebrocortical microflow and lack of carbon dioxide reactivity following subarachnoid haemorrhage in the rat

K. Jarus-Dziedzic[1], Z. Czernicki[1], and E. Koźniewska[1,2,3]

[1] Department of Neurosurgery, Medical Research Center, Polish Academy of Sciences, Warsaw, Poland
[2] Laboratory of Experimental Nuclear Medicine, Medical Research Center, Polish Academy of Sciences, Warsaw, Poland
[3] Department of Experimental and Clinical Physiology, Medical University of Warsaw, Warsaw, Poland

Summary

Experiments were designed to study the extent, duration and severity of the impairment of cerebrocortical microciroflow during acute ischemia following experimental SAH in the rat. Twenty five male, adult anesthetized and mechanically ventilated Wistar rats were used. SAH was induced by perforation of the middle cerebral artery (MCA) using intravascular filament. Cerebrocortical microflow (LDF) was recorded bilaterally in the territory supplied by the MCA at normocapnia before SAH and up to 180 min thereafter. Reactivity of microcirculation to CO_2 was also studied. In order to further explain mechanisms of post-SAH microcirculatory changes, L-arginine – a precursor of NO and 17-octadecynoic acid (17-ODYA) – an inhibitor of the enzymes of cytochrome P-450 family were administered.

SAH resulted in acute decrease of microflow on both sides although during the first 20 min this effect was much better pronounced on the side ipsilateral to ruptured MCA ($p < 0.05$). Pretreatment with L-arginine or 17-ODYA didn't improve microflow after SAH. On the contrary, in rats pretreated with 17-ODYA LDF on the side ipsilateral to bleeding significantly deteriorated ($p < 0.05$ vs. control group at all times beginning 10 min after SAH). Following SAH the impairment of CO_2 reactivity was also observed.

Keywords: Experimental SAH; cerebrocortical microflow; CO_2 reactivity; L-arginine; 17-ODYA.

Introduction

The main and extensively studied consequence of subarachnoid haemorrhage (SAH) is cerebral vasospasm which develops in a course of few days after intracranial bleeding. Despite many clinical and experimental efforts the mechanisms leading to vasospasm are far from being clarified. One of the causes of this delayed ischemic deficit may be connected with vascular changes provoked by acute, severe ischemia observed immediately after SAH.

Mechanisms of this acute ischemia have not been studied until recently. It has been postulated that the abrupt increase of intracranial pressure (ICP) results in the decrease of cerebral perfusion pressure and thus in lowering of cerebral blood flow. The dramatic increase of ICP immediately after SAH has been demonstrated both clinically and experimentally [3, 4]. However, it has been pointed out recently that ischemia after SAH lasts much longer than the increase of ICP and therefore active vasoconstriction has been proposed as a mechanism contributing to acute decrease of cerebral perfusion after SAH [3].

Active vasoconstriction in response to SAH may result from the withdrawal of endothelium-dependent vasodilatation, activation of endothelium-dependent vasoconstriction or activation of platelets with subsequent release of many vasoconstrictive substances which promote coagulation. Recently published data point to the decreased availability of NO [9, 10, 11] which is one of the most important endothelium-dependent vasodilators involved in the maintenance of baseline cerebral blood flow [6]. Besides a direct vasoconstriction due to withdrawal of NO-based dilatation, decreased availability of NO may result, also in disclosure of another important vasoconstrictor mechanism. Under physiological conditions NO inactivates the enzymes of cytochrome P-450 (CYP) family. One of them – CYP4A catalyzes transformation of arachidonic acid (AA) into strong vasoconstrictor hydroxyeicosatetraenoic acid (20-HETE). Thus, when NO is not available then vasoconstrictor activity of 20-HETE may come into play [1, 9].

Our present experiments were designed to study duration and severity of the impairment of cerebrocortical microflow during acute ischemia after SAH as well as contribution of decreased availability of NO and increased production of CYP metabolites of arachidonic acid to acute ischemia and to impairment of CO_2 reactivity of cerebrocortical microflow in the acute stage after SAH in rats.

Materials and methods

Twenty five male, adult Wistar rats weighing 280 g–350 g anaesthetized with 3,6% chloral hydrate (3,6 mg/100g body weight intraperitoneally followed by intravenous supplementation of 1.5 mg/100g every hour), tracheotomized and mechanically ventilated with small animal respirator (model 683, Harvard Apparatus, Kent, UK) were used for this study. Femoral arteries and veins were cannulated for mean arterial blood pressure (MABP) monitoring, blood sampling for gasometric analysis ($PaCO_2$, PaO_2, and pH), supplementation of anaesthesia, and drug administration. Body temperature was maintained around 37 °C with the rectal-thermometer-controlled heating pad.

Experimental SAH was induced using the endovascular suture model as originally described by Veelken *et al.* [14] and modified by Bederson *et al.* [2]. Briefly, after exposure of the right common carotid artery, the external carotid and pterygopalatine arteries were identified and ligated. Monofilament (Dermalon, 3.0) was advanced through the right common carotid and internal carotid artery to the level of the calvarium. The animals were placed in a sphinx position with the head mounted in a head holder. Before the perforation of the origin of the middle cerebral artery (MCA) the animals were prepared for bilateral LDF measurements. After midline skin incision the calvarium was exposed and two burr holes were drilled in the skull 5 mm lateral to the midline and 1 mm posterior to bregma over the MCA territory. LDF probes (DRT4, Moor Instruments Ltd., UK) were positioned carefully with the help of a micromanipulator to avoid large cortical vessels in their field of view. After at least 20 min of stabilization period and control of LDF/arterial blood pressure recordings, SAH was induced by pushing the suture deeper into the internal carotid artery and perforating MCA at its origin.

CO_2 reactivity test was performed before and 60, 90, 120, 150 and 180 min after SAH. Increase of arterial CO_2 partial pressure ($PaCO_2$) was achieved by decreasing tidal volume. $PaCO_2$ at the end of 5 min normoxic hypoventilation was not exceeding 55–60 mm Hg. Since during hypoventilation after SAH the increase of MABP was observed, the reactivity to CO_2 was calculated based on the change of cerebral vascular resistance (CVR) and not LDF. The latter parameter usually followed passively blood pressure changes.

For administration of 17-octadecynoic acid, the atlanto-occipital membrane was exposed by separating the overlying neck muscles in the midline and a 30 G needle attached to a PE-10 catheter was inserted into the cisterna magna.

Three groups of animals were studied. Group I (n = 12) – control was "pretreated" only with 0.9% NaCl. Group II (n = 6) obtained intravenous infusion of L-arginine (600 mg/kg in 1.5 ml of 0.9% NaCl) starting 5 min before SAH and lasting for 20 min. Group III (n = 7) was pretreated with 17-octadecynoic acid (17-ODYA, 500 ng in 50 µl of 0.9% NaCl) administered through the cisterna magna catheter 5 min before SAH induction.

Statistical analyses were performed by one-way analysis of variance with Student t-test for independent or paired data where appropriate. P values less than 0,05 were considered statistically significant. All data are presented as means ± SD.

Results

Hemodynamic parameters and blood gas analyses are included in Table 1. MABP was stable during the observation time in all groups within the range of 93 to 130 mm Hg. All animals were normocapnic (except for the CO_2 reactivity test) with PaO_2 above 95 to mmHg. Administration of L-arginine or 17-ODYA did not affect values of LDF, CVR, MABP and gasometric condition.

Table 1. *Physiological parameters during baseline and at selected time points following SAH*

Groups	MABP (mmHg)	pH	$PaCO_2$ (mmHg)	PaO_2 (mmHg)
Control (n = 12)				
Before SAH	102 ± 12	7,38 ± 0,02	38 ± 1	139 ± 13
30 min after SAH	102 ± 13	7,41 ± 0,04	37 ± 3	121 ± 9
1 hr after SAH	116 ± 15	7,38 ± 0,02	39 ± 2	138 ± 18
2 hr after SAH	105 ± 9	7,41 ± 0,02	39 ± 1	135 ± 7
L-arginine (n = 6)				
Before SAH	97 ± 9	7,39 ± 0,03	40 ± 1	158 ± 13
After SAH	95 ± 19	7,39 ± 0,06	38 ± 1	133 ± 16
1 h after SAH	102 ± 21	7,39 ± 0,05	37 ± 1	141 ± 22
2 h after SAH	93 ± 22	7,35 ± 0,07	38 ± 1	135 ± 21
17-ODYA (n = 7)				
Before SAH	113 ± 13	7,38 ± 0,03	38 ± 1	135 ± 16
After SAH	122 ± 10	7,37 ± 0,01	39 ± 1	133 ± 21
1 h after SAH	113 ± 19	7,36 ± 0,05	39 ± 2	127 ± 13
2 h after SAH	113 ± 16	7,38 ± 0,04	37 ± 2	137 ± 28

Data are presented as mean value ± SD.

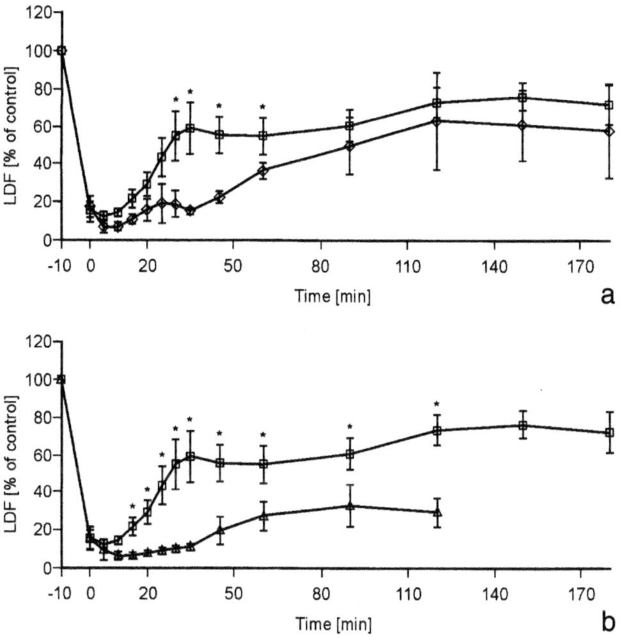

Fig. 1. Time course of ipsilateral LDF changes following SAH. Comparison between: (a) control (squares) and L-arginine pretreated rats (diamonds), (b) control (squares) and 17-octadecynoic acid pretreated rats (triangles). Asterisks denote statistical significance at $p < 0.05$

SAH resulted in the immediate decrease of LDF ($p < 0.05$) and increase of CVR ($p < 0.05$) on both sides and in all groups of animals. This decrease was much better pronounced on the side ipsilateral to SAH (LDF1). LDF1 dropped by 87% ± 21% in the control group (Fig. 1a and b), 81% ± 16% in animals treated with L-arginine (Fig. 1a), and 86% ± 9% in animals treated with 17-ODYA (Fig. 1b). In group 1 LDF1 was kept below microflow on the contralateral side for the first 20 minutes. Thereafter there was a tendency for LDF1 to recover and the difference between hemispheres disappeared. At the end of the experiment LDF1 in this group amounted to 73% ± 11% of pre-SAH value (Fig. 1a and b). The time course of recovery of LDF1 in L-arginine group (Fig. 1a) was similar whereas in animals pretreated with 17-ODYA, there was no recovery of LDF1 (Fig. 1b). Most of the animals in 17-ODYA group had flow arrest between 2 and 3 hours after SAH and therefore were not included.

Decrease of LDF2 (side contralateral to SAH) was less pronounced in all groups when compared with the side ipsilateral to SAH in each of them. This decrease amounted to 47% ± 7% (control), 38% ± 11% (L-arginine), and 60% ± 12% (17-ODYA) and did not differ between groups. There was also no difference in the recovery of LDF2 between groups.

Reactivity to CO_2 was comparable in all groups before SAH induction and amounted to 1% ± 0,3%, 0,93% ± 0,3% and 1% ± 0,5%, for control, L-arginine and 17-ODYA group, respectively. Following SAH CO_2 reactivity was generally diminished but the severity of impairment varied between animals, irrespective of the group and pretreatment. There was no statistically significant correlation between CO_2 reactivity test and time after SAH or microflow deficit in either group.

Discussion

There are two main new findings of this study: asymmetry of LDF decrease following SAH and significantly greater impact of SAH on ipsilateral LDF in animals pretreated with 17-ODYA.

The first result speaks in favor of active vasoconstriction of the ruptured vessel. If increase of ICP and decrease of CPP would be cause of ischemia, then microflow should decline symmetrically. Active vasoconstriction was suggested, as mentioned before, by few other authors who studied the mechanisms of acute ischemia after SAH [5, 8, 10, 13]. Some of them suggested that active vasoconstriction is caused by the decreased availability of NO and that NO donors can reverse acute vasoconstriction [10, 11]. According to our results, L-arginine-NO pathway is impaired after haemorrhage. The infusion rate of L-arginine used inthis study resulted in an increase of LDF by 10–15% 1 hour after the cessation of 20 min infusion in control rats (sham identical with SAH surgery, data not shown). In our present experiments there was no sign of LDF improvement in the L-arginine group after SAH on either side. This might be due to the effect of ischemia per se. It has been demonstrated that MCA occlusion in the rat for longer than 20 min results in the lack of the LDF response to L-arginine during reperfusion [12]. Moreover, according to the results of Schwartz *et al.* [10], who use the same model of SAH as we do, there is a biphasic pattern of NO availability after MCA rupture. NO-mediated vasodilatation is limited during the first 30 minutes of SAH and is restored 60 min after SAH. Thus, it seems that the lack of improvement of LDF in the rats pretreated with L-arginine in our model confirms the results of other studies.

Lack of NO availability was the inspiration for Kehl *et al.* [5] to study participation of 20-HETE as vasoconstrictor involved in the early ischemia after SAH. In a very elegant and convincing study these authors demonstrated that 20-HETE contributes in a very important way to the acute fall in regional cerebral blood flow following SAH in rats. Administration of 17-ODYA or more selective inhibitor of 20-HETE synthesis (HET0016) in their study resulted in the improvement of LDF after SAH by 40%. Our results clearly demonstrate that pretreatment with 17-ODYA results in deterioration and lack of LDF recovery following SAH on the side ipsilateral to MCA rupture where the decrease of LDF immediately after bleeding is below 80% of baseline. Although our results seem to contradict those of Kehl *et al.* we think that they are rather complementary. We used a different model of SAH than the one applied for the study of Kehl *et al.* and observed a much more dramatic decrease of LDF on the side ipsilateral to SAH. The inhibitor of CYP used in our study is not a selective one i.e. it irreversibly inhibits formation of 20-HETE – a vasoconstrictor as well as a product of arachidonic acid epoxygenation, epoxyeicosatrienoic acid (EET) – a vasodilator [9]. The deterioration of LDF after SAH in rats pretreated with 17-ODYA suggests that in a group without pretreatment, EET is released during SAH to control the degree of vasoconstriction. This part of our study requires further experimentation.

Lack of reactivity to CO_2 and diversity of the effects of L-arginine or 17-ODYA pretreatment on microflow responses to CO_2 after SAH is difficult to discuss at the moment in terms of the mechanisms involved. It also requires further studies.

In summary, our results indicate that 1) L-arginine – NO pathway is impaired during first hours following SAH, 2) reactivity of cerebrocortical microflow to changes in $PaCO_2$ is depressed during acute ischemia following SAH irrespective of the severity of ischemia. Moreover, it seems that during acute ischemia following SAH at the side of the ruptured artery, the mechanism counteracting vasoconstriction is also activated. This mechanism is related to the metabolism of AA via cytochrome P450 enzymes-dependent pathway.

References

1. Alonso-Galicia M, Hudetz AG, Shen H *et al* (1999) Contribution of 20-HETE to vasodilator actions of nitric oxide in the cerebral microcirculation. Stroke 30: 2727–2734
2. Bederson JB, Germano IM, Guarino L (1995) Cortical blood flow and cerebral perfusion pressure in a new model of subarachnoid hemorrhage in the rat. Stroke 26: 1086–1092
3. Bederson JB, Levy AL, Ding WH, Kahn R, DiPerna CA, Jenkins AL 3rd, Vallabhajosyula P (1998) Acute vasoconstriction after subarachnoid hemorrhage. Neurosurg 42: 352–362
4. Fisher CM (1975) Clinical syndromes in cerebral thrombosis, hypertensive hemorrhage, and ruptured saccular aneurysm. Clin Neurosurg 22: 117–147
5. Kehl F, Cambj-Sapunar L, Maier KG *et al* (2002) 20-HETE contributes to the acute fall in cerebral blood flow after subarachnoid hemorrhage in the rat. Am J Physiol 282: H1556–H1565
6. Koźniewska E, Osęka M, Styś T (1992) Effects of endothelium-derived nitric oxide on cerebral circulation during normoxia and hipoxia in the rat. J Cerebr Blood Flow Metab 12: 311–317
8. Nishizawa S, Chen D, Yokoyama T, Yokota N, Otha S (2000) Endothelin-1 initiates the development of spasm after subarachnoid haemorrhage through protein kinase C activation, but does not contribute to prolonged vasospasm. Acta Neurochir (Wien) 142: 1409–1415
9. Roman RJ (2001) P-450 metabolites of arachidonic acid in the control of cardiovascular function. Physiol Rev 82: 131–185
10. Schwartz AY, Sehba FA, Bederson JB (2000) Decreased nitric oxide availability contributes to acute cerebral ischemia after subarachnoid hemorrhage. Neurosurg 47: 208–214
11. Sehba FA, Ding WH, Chereshnev I, Bederson JB (1999) Effects of S-nitrosoglutathione on acute vasoconstriction and glutamate release after subarachnoid hemorrhage. Stroke 30: 1955–1961
12. Sporer B, Martens KH, Koedel U, Haberl RI (1997) L-arginine-induced regional cerebral blood flow increase is abolished after transient focal cerebral ischemia in the rat. J Cerebr Blood Flow Metab 17: 1074–1080
13. Thomas JE, McGinnis G (2002) Safety of intraventricular sodium nitroprusside and thiosulfate for the treatment of cerebral vasospasm in the intensive care unit setting. Stroke 33: 486–492
14. Veelken JA, Laing RJC, Jakubowski J (1995) The sheffield model of subarachnoid hemorrhage in rats. Stroke 26: 1279–1284
15. Zou AP, Ma YH, Sui ZH, Ortiz de Mentellano PR, Clark JE, Masters BS, Roman RJ (1994) Effects of 17-octadecynoic acid, a suicide-substrate inhibitor of cytochrome P450 fatty acid Omega-hydroxylase, on renal function in rats. J Pharmacol Exp Ther 268: 474–481

Correspondence: Ewa Kozniewska, Ph.D., Department of Neurosurgery, Medical Research Center, Polish Academy of Sciences, ul. Pawińskiego 5, 02-106 Warsaw, Poland. e-mail: ewak@cmdik.pan.pl

Acta Neurochir (2003) [Suppl] 86: 477–482

Heat shock proteins expression in brain stem after subarachnoid hemorrhage in rats

M. Satoh[1], **J. Tang**[2], **A. Nanda**[3], and **J. H. Zhang**[3]

[1] Department of Neurosurgery, University of Mississippi Medical Center, Jackson, MS
[2] Department of Physiology, Louisiana State University Health Sciences Center in Shreveport, LA
[3] Department of Neurosurgery, Louisiana State University Health Sciences Center in Shreveport, LA

Summary

The pathogenesis of brain damage after subarachnoid hemorrhage (SAH) especially at molecular or gene level remains unclear. We used complimentary deoxyribonucleic acid (cDNA) macroarray technique and compared gene expression in brain stem after experimental SAH in rats. The upregulation of several heat shock proteins (HSPs) demonstrated by cDNA array was further confirmed by Western blotting. The expressions of 9 genes were upregulated 30 minutes or 2 days after SAH. They included four upregulated HSPs: HSP90α, HSP60, HSP27, and HSP10. Western blotting demonstrated increases in the HSP27 and HSP10 proteins on Day 2. SAH enhanced the induction of several HSP mRNAs in the brainstems, even though the functions of these HSPs after SAH remain unclear.

Introduction

Several clinical and experimental studies have been directed to examine brain tissue damage reported after SAH, especially in the early stages [2]. Neuronal death, especially apoptosis, has been identified in the cortex and other brain regions and is probably associated with the blood lysate related cytotoxicity [10]. Other neurological damage and damage to other brain tissues caused by blood components, such as cerebral smooth muscle [12] were reported; however, the mechanisms, especially the signaling pathways at the gene level in SAH-induced brain tissue damage, remain to be explored. Recently, a complementary deoxyribonucleic acid (cDNA) array technique has been employed to detect gene expressions in several cerebral disorders that were caused by multiple factors, including cerebral vasospasm after SAH [20] and focal cerebral ischemia [16]. Consequently, there are few experimental studies of multiple gene expressions in brain tissue after SAH, especially in the regions of the brainstem that are thought to be affected by early microcirculation disturbance, an important factor for mortality and morbidity after SAH [13].

Material and methods

Subarachnoid hemorrhage model and experimental design

Male Sprague-Dawley (SD) rats (n = 30) weighing from 300 g to 330 g were intraperitoneally anesthetized with sodium pentobarbital (30 mg/kg). Single hemorrhage model of SAH was produced as described previously [11].

Thirty rats were randomly divided into three groups: the control group without SAH (n = 10), the group sacrificed at 30 min after blood injection (n = 10), and the group sacrificed 2 days after blood injection (n = 10). In each group, 2 rats were used for photograph, 2 rats were used for cDNA array, and 6 rats were used for Western blotting. All procedures were approved by the Animal Care and Use Committee at the University of Mississippi Medical Center (UMMC).

RNA preparation

The animals were sacrificed by an injection of 100 mg/kg sodium pentobarbital and perfused with 100 mL normal saline. The basilar arteries were removed after blood clots were observed around the brainstems. The brains were then transected to pick up the pons, immediately frozen with liquid nitrogen, and stored a temperature of −80 °C until use.

cDNA macroarray

Total RNA was extracted from the brainstems using the Atlas Pure Total RNA Labeling System (Clontech, Palo Alto, CA). Samples from two rats in each group were used for cDNA array, and the experiments were conducted using each individual samples. Poly(A)$^+$ RNA was isolated from 50 µg of total RNA and was then converted to ^{32}P labeled first strand cDNA using [α-^{32}P]dATP (3,000 Ci/mmol, Amersham, Piscataway, NJ). The cDNA probes were then hybridized to the ATLAS array membranes (Atlas^{TM} Rat Stress Array, Clontech) according to the manufac-

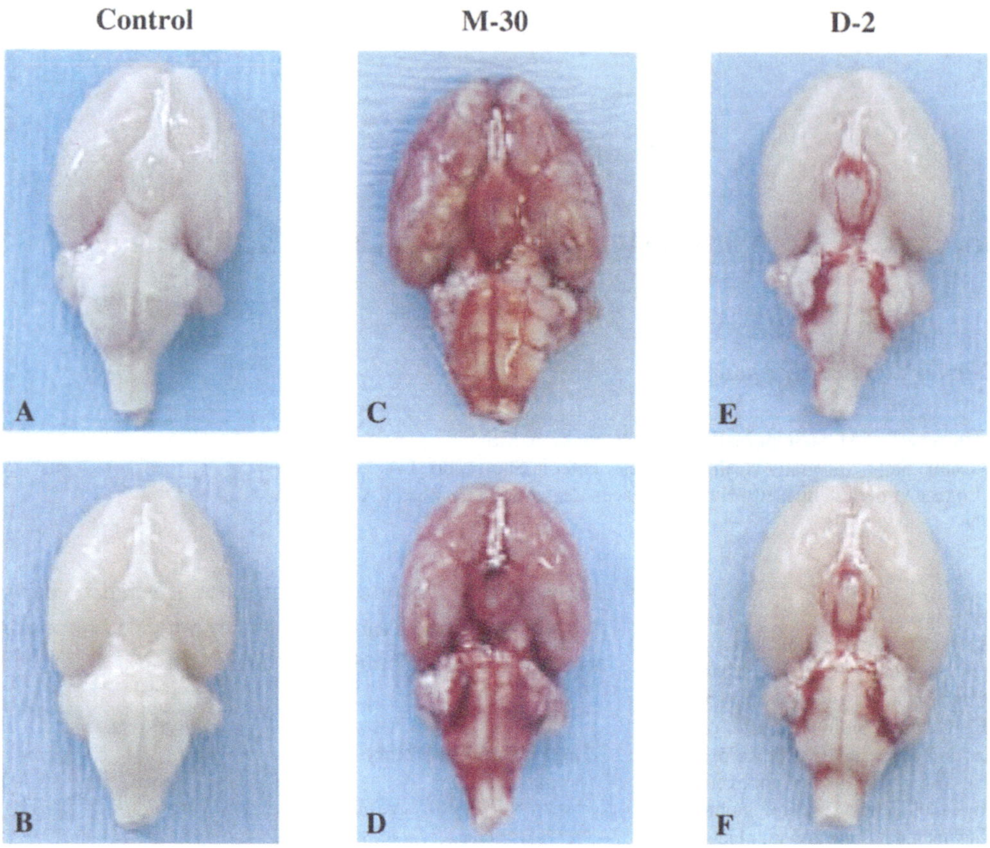

Fig. 1. Brain of a rat in control group (A, B), 30-minute group (C, D), and 2-day group (E, F). There were no clots found in the control group (A, B). In the 30-minute group, a diffused blood clot was recognized on the surface of the cerebral cortex or brainstem (C, D). In the 2-day group, blood clots were present only on the lateral sides, on the brainstem surface, and around the Circle of Willis (E, F)

turer's instructions. After six washings with SDS–SSC solutions, the membranes were exposed at room temperature to a phosphorimaging filter. Images were generated using a phosphorimaging system (Molecular Dynamics Typhoon 8600 variable mode imager, Amersham Pharmacia Biotech).

Quantification and analysis of array signals

The signals on each DNA array were quantified using AtlasImage® (Clontech). The intensity of the genes were averaged for each group and compared among the three groups after normalization against the sum of hybridization signals for all spots on the array [20]. The ratio of signals was calculated among the three groups and considered significant when its differences were more than 2-fold.

Western blotting

Western blotting was performed to confirm the results of cDNA array as previously described [15].

Statistical analysis

All data were expressed as the mean ± SEM. We used the one-way ANOVA followed by Scheffé's F post-hoc test to perform statistical comparisons among the three groups. A probability value was considered significant when $P < 0.05$.

Results

Subarachnoid hemorrhage model

There were no clots in the control group (Fig. 1). There was no mortality in the SAH groups. In all of the rats in the group sacrificed after 30 minutes, a diffused blood clot was recognized on the surface of the cerebral cortex or brainstem. The clot was thicker on the basilar artery, both lateral sides of the ventral brainstem, and around the Circle of Willis. In the group sacrificed after 2 days, most of the diffused clots were washed out, and the blood clots were present only on both lateral sides of the ventral brainstem and around the Circle of Willis.

cDNA array

The cDNA array nylon membrane included 207 known rat cDNA with 9 house keeping genes. Array hybridizations resulted in the detection of 32 genes

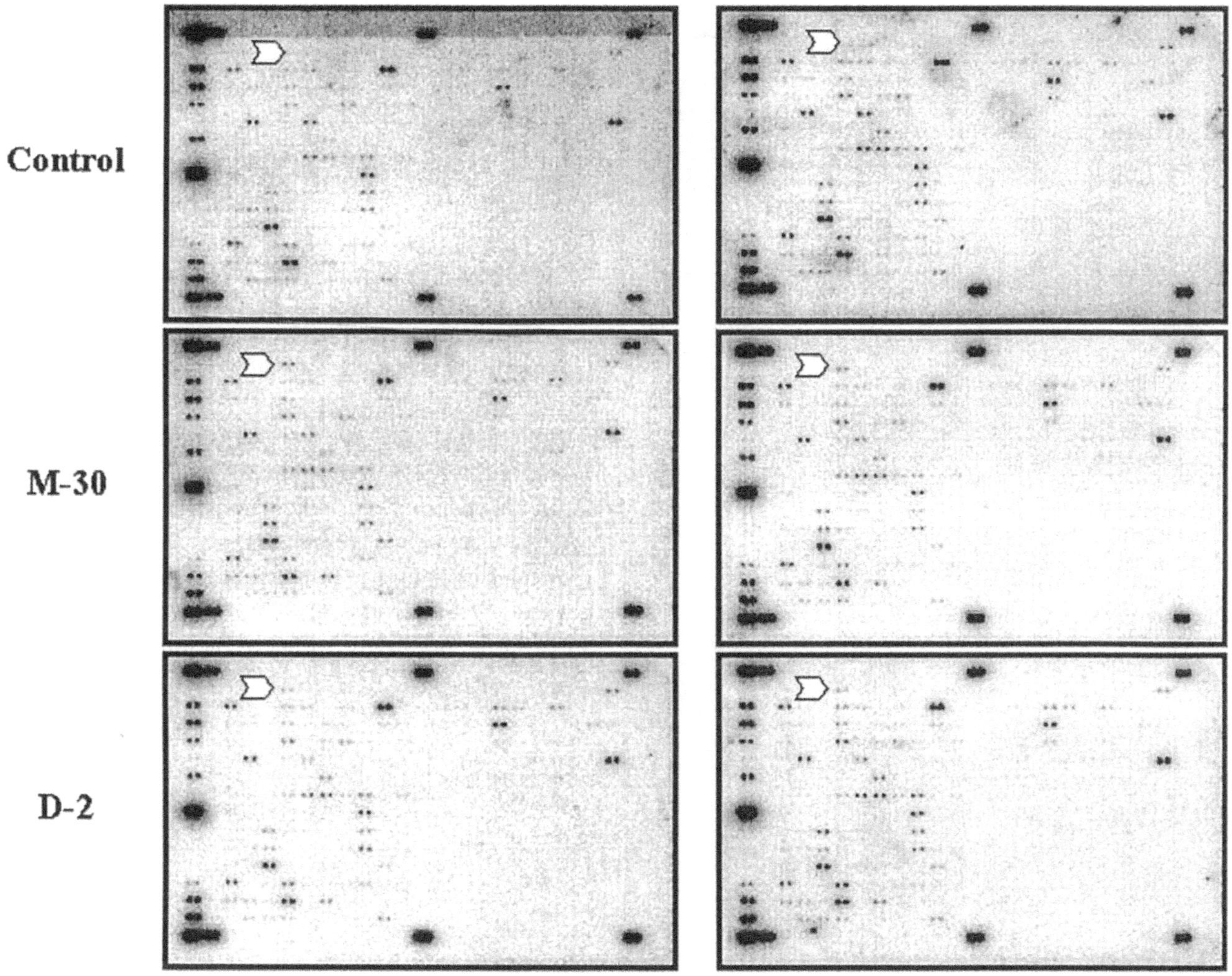

Fig. 2. Representative autoradiography of mRNA expression on cDNA array (n = 2 in each group). The gene of HSP60 is shown. Arrows point out the typical upregulated pattern

(15.5%) with adjusted intensities, which were two-fold above the background level (Fig. 2). There were no significant differences in each gene expression between the two membranes within each group. Nine out of 32 genes were upregulated in the brainstems with subarachnoid clots after SAH as follows: HSP90α, HSP60, HSP27, HSP10, 23-kDa progesterone receptor-associated protein, T-complex protein 1 α subunit (TCP1-α), endoplasmic reticulum stress protein 72 (ERP72), cell division control protein 25 B (CDC25B), and apurinic/apyrimidinic endonuclease (Table 1).

Western blotting

The time-course of four HSP expressions was confirmed at the protein level. HSP27 (Fig. 3) and HSP10 (not shown) were elevated ($P < 0.05$) in the brainstem 2 days after SAH. In addition, HSP27 protein decreased markedly in the group sacrificed after 30 minutes ($P < 0.05$, Fig. 3). Even though HSP10 decreased 30 min after the blood injection, it did not reach statistical significance ($P > 0.05$, not shown); however, HSP90α (not shown) and HSP60 (not shown) remained unchanged in both the group sacrificed 30 minutes after SAH and the group sacrificed 2 days after SAH ($P > 0.05$, ANOVA).

Discussion

In this study, we used a rat SAH model because it is easier to get several genes with known functions than it is in large animals such as dogs or monkeys. The rat single-hemorrhage model is known to show biphasic

Table 1.

Gene name	Average of intensities			Difference	
	Control (C)	M-30 (M)	D-2 (D)	Ratio	Time-course
1. HSP90α	24	86	32	1:3.6	C:M
2. HSP60	0	70	0	–	C:M
3. HSP27	17	28	72	1:2.6	M:D
4. HSP10	32	84	70	1:2.6	C:M
5. 23-kDa progesterone receptor-associated protein	0	52	0	–	C:M
6. T-complex protein 1 α subunit (TCP1-α)	0	45	15	–	C:M
7. Endoplasmic reticulum stress protein 72 (ERP72)	21	22	55	1:2.5	M:D
8. Cell division control protein 25 B (CDC25B)	0	65	53	–	C:M
9. Apurinic/apyrimidinic endonuclease	21	44	46	1:2.1	C:M

vasospasms in the acute phase (10 to 60 minutes) and in the late phase (2 days) [17]. Therefore, the changes of multiple gene expressions were examined at 30 minutes after SAH and at 2 days after SAH. The use of brainstem tissue seemed suitable for this experiment because SAH was induced through the cisterna magna. In fact, we observed many subarachnoid clots on the brainstem surface at 30 minutes after SAH and a few at 2 days after SAH, which can cause direct toxic effects. In addition, some authors have demonstrated that intraparenchymal arteries in the brainstem contract after SAH, which may induce cerebral ischemia in the early stage of SAH [13].

For the first time, we have demonstrated increases in the mRNAs and proteins of HSP27 and HSP10 in the brainstem after SAH. The HSP90α and HSP60 mRNA were raised after SAH, but the protein level remained unchanged. Interestingly, HO-1 (hemeoxygenase-1, HSP32) and HSP70 mRNA were not elevated in the brainstem, either at 30 minutes or 2 days

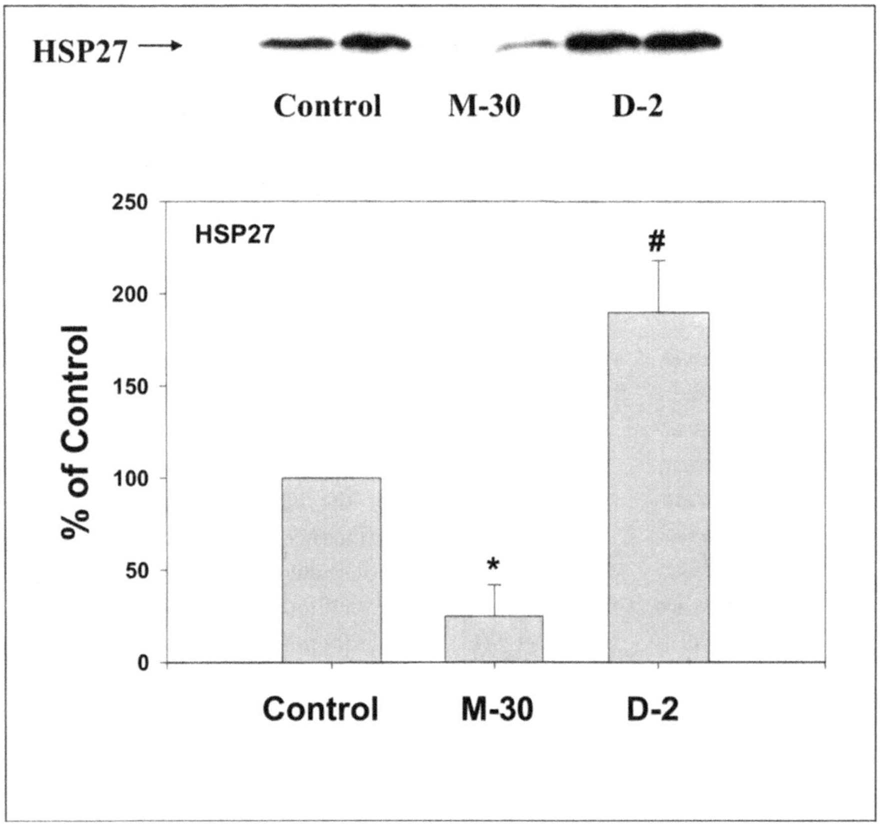

Fig. 3. A. The protein levels of HSP27 in the control group, 30-minute group, and 2-day group. HSPs in the 30-minute group and 2-day group were expressed as percentages of that in the control group. Lane 1 and 2, control group; Lane 3 and 4, 30-minute group; Lane 5 and 6, 2-day group

after SAH, when compared with normal controls, even though they (HSP32 and HSP70) were reported to increase in the cortex after SAH [10].

HSP27 increases during thermal stress and provides protection against heat shock. Recent studies have demonstrated that HSP27 elevated during cerebral ischemia and HSP27 mRNA expression increased dramatically 8 to 16 hours after focal cerebral ischemia in rats and was sustained at least 48 hours [5]. These reports are similar to our results in the present study. An increase in HSP27 mRNA expression was not found in the group sacrificed 30 minutes after SAH, but was found in the group sacrificed 2 days following SAH (Table 1). The protein level of HSP27 was also elevated on Day 2 (Fig. 3). HSP27 protein level, however, decreased significantly 30 minutes after SAH. This reduction may reflect a redistribution of HSP27 from the cytoplasm into the nucleus. Arrigo *et al.* [1] indicated that much of the HSP27 protein relocalized into the nucleus after heat shock, assembled into large aggregates, and was resistant to detergent extraction. Besides the thermal effect, HSP27 may have other functions. Leszczynski *et al.* [9] proposed that HSP27 might facilitate the development of brain cancer by inhibiting the cytochrome c/caspase-3 apoptotic pathway and cause an increase in blood-brain barrier permeability through stabilization of endothelial cell stress fibers. Thus, elevation of HSP27 may lead to both ischemic tolerance [4] and anti-oxidative stress [19] effects and other unwanted actions, such as instability of the blood-brain barrier after SAH [9].

HSP10, also known as chaperonin 10 (CPN10), and HSP60 are present in the mitochondria. Recent studies have indicated that the interaction between HSP10 and HSP60 plays an important role in mitochondrial protein folding and assembly. The simultaneous induction of both HSP10 and HSP60 mRNA has been reported from 4 to 24 hours following cerebral ischemia [6]. The present study also indicated the simultaneous induction of mRNA of both HSP10 and HSP60 30 minutes after SAH (Table 1). The induction of both mRNAs, however, appeared earlier after SAH than cerebral ischemia. This result may be due to the direct toxic effect of blood clots. Meanwhile, though the level of HSP10 protein did not increase in the group sacrificed 30 minutes after SAH, it did increase in the group sacrificed 2 days after SAH, much like the protein expression of HSP27 as shown in Fig. 3. Wagstaff *et al.* [18] suggested that there was a time-lag between the increase in mRNA levels in HSPs and the corresponding increase in protein levels. The unchanged level of HSP60 protein in the group sacrificed 30 minutes after SAH and the group sacrificed 2 days after SAH may also be connected to the time-lag, because HSP60 mRNA in the group sacrificed after 2 days had already returned to the control level (Table 1). The reason that the induction time was different between HSP10 and HSP60 is unclear. Even though HSP10 exists mainly in the mitochondria in a variety of tissues, including liver, heart, pancreas, kidney, anterior pituitary, salivary gland, thyroid, and adrenal gland, HSP10 was also identified in several extramitochondrial compartments similar to several other mitochondrial proteins (HSP60, tumor necrosis factor receptor-associated protein-1, P32 (gC1q-R) protein, and cytochrome c) [14]. Mitochondrial proteins HSP10 and HSP60 may be involved in mitochondrial-related apoptosis and genetic disorders. Indeed, blood components have been implicated in cell death especially apoptotic death in neurons [10] and HSP60 has an anti-apoptotic role in myocardial injury [8].

HSP90 has two isoforms, α and β. HSP90 is an abundant molecular chaperone, which functions in cellular homeostasis in prokaryotes and eukaryotes. It is well known that HSP90 plays a critical and indispensable role in regulating cell growth through the modulation of various signal transduction pathways. The induction of HSP90 during cerebral ischemia is controversial and, while HSP90α mRNA has been reported to be induced after transient global ischemia in gerbils [7], the level of HSP90 mRNA has been shown to be unchanged during focal cerebral ischemia in rats [5]. Considering that HSP90α mRNA was induced in the hippocampus at 1 hour and diminished 2 days after ischemia [7], the induction of HSP90α mRNA after SAH (Table 1) seemed to be similar to that after global ischemia. However, HSP90α protein was not remarkably different either 30 min after or 2 days after SAH, perhaps as a result of the time-lag. The function of HSP90 is not clear, especially since only a transient elevation was observed in the 30 min after SAH group. One recent study indicates that HSP90 is involved in the activation of endothelial nitric oxide synthase (eNOS). Stimulation of endothelial cells with vascular endothelial growth factors recruits eNOS and Akt to an adjacent region on the same domain of HSP90, thereby facilitating eNOS phosphorylation and enzyme activation [3]. This action of HSP90 might be involved in the microcirculation compromise in the acute stage of SAH, and eNOS might play a role in the regulation of vascular tone.

In conclusion, we have demonstrated the induction of several HSP mRNAs after SAH. Western blotting confirmed the over-expression of HSP27 and HSP10 2 days after SAH. Even though the function of most of the HSPs in SAH is unknown, HSP27 may be involved in ischemic tolerance and HSP10 may play a role in anti-apoptosis of the brainstem tissue after SAH.

Acknowledgment

This work was partially supported by a grant from the American Heart Association Bugher Foundation Award for Stroke Research to J.H.Z.

References

1. Arrigo AP, Suhan JP, Welch WJ (1988) Dynamic changes in the structure and intracellular locale of the mammalian low-molecular-weight heat shock protein. Mol Cell Biol 8: 5059–5071
2. Doczi T (1985) The pathogenetic and prognostic significance of blood-brain barrier damage at the acute stage of aneurysmal subarachnoid haemorrhage. Clinical and experimental studies. Acta Neurochir (Wien) 77: 110–132
3. Fontana J, Fulton D, Chen Y, Fairchild TA, McCabe TJ, Fujita N, Tsuruo T, Sessa WC (2002) Domain mapping studies reveal that the M domain of hsp90 serves as a molecular scaffold to regulate Akt-dependent phosphorylation of endothelial nitric oxide synthase and NO release. Circ Res 90: 866–873
4. Higashi T, Nishi S, Nakai A, Nagata K (1995) Regulatory mechanism of stress response in mammalian nervous system during cerebral ischaemia or after heat shock. Neuropathol Appl Neurobiol 21: 480–483
5. Higashi T, Takechi H, Uemura Y, Kikuchi H, Nagata K (1994) Differential induction of mRNA species encoding several classes of stress proteins following focal cerebral ischemia in rats. Brain Res 650: 239–248
6. Izaki K, Kinouchi H, Watanabe K, Owada Y, Okubo A, Itoh H, Kondo H, Tashima Y, Tamura S, Yoshimoto T, Mizoi K (2001) Induction of mitochondrial heat shock protein 60 and 10 mRNAs following transient focal cerebral ischemia in the rat. Brain Res Mol Brain Res 88: 14–25
7. Kawagoe J, Abe K, Aoki M, Kogure K (1993) Induction of HSP90 alpha heat shock mRNA after transient global ischemia in gerbil hippocampus. Brain Res 621: 121–125
8. Kirchhoff SR, Gupta S, Knowlton AA (2002) Cytosolic heat shock protein 60, apoptosis, and myocardial injury. Circulation 105: 2899–2904
9. Leszczynski D, Joenvaara S, Reivinen J, Kuokka R (2002) Non-thermal activation of the hsp27/p38MAPK stress pathway by mobile phone radiation in human endothelial cells: Molecular mechanism for cancer- and blood-brain barrier-related effects. Differentiation 70: 120–129
10. Matz P, Turner C, Weinstein PR, Massa SM, Panter SS, Sharp FR (1996) Heme-oxygenase-1 induction in glia throughout rat brain following experimental subarachnoid hemorrhage. Brain Res 713: 211–222
11. Meguro T, Clower BR, Carpenter R, Parent AD, Zhang JH (2001) Improved rat model for cerebral vasospasm studies. Neurol Res 23: 761–766
12. Ogihara K, Zubkov AY, Parent AD, Zhang JH (2000) Oxyhemoglobin produces necrosis in cultured smooth muscle cells. Acta Neurochir Suppl 76: 507–510
13. Ohkuma H, Ogane K, Tanaka M, Suzuki S (2001) Assessment of cerebral microcirculatory changes during cerebral vasospasm by analyzing cerebral circulation time on DSA images. Acta Neurochir Suppl 77: 127–130
14. Sadacharan SK, Cavanagh AC, Gupta RS (2001) Immunoelectron microscopy provides evidence for the presence of mitochondrial heat shock 10-kDa protein (chaperonin 10) in red blood cells and a variety of secretory granules. Histochem Cell Biol 116: 507–517
15. Satoh M, Parent AD, Zhang JH (2002) Inhibitory effect with antisense mitogen-activated protein kinase oligodeoxynucleotide against cerebral vasospasm in rats. Stroke 33: 775–781
16. Soriano MA, Tessier M, Certa U, Gill R (2000) Parallel gene expression monitoring using oligonucleotide probe arrays of multiple transcripts with an animal model of focal ischemia. J Cereb Blood Flow Metab 20: 1045–1055
17. Suzuki H, Kanamaru K, Tsunoda H, Inada H, Kuroki M, Sun H, Waga S, Tanaka T (1999) Heme oxygenase-1 gene induction as an intrinsic regulation against delayed cerebral vasospasm in rats. J Clin Invest 104: 59–66
18. Wagstaff MJ, Collaco-Moraes Y, Aspey BS, Coffin RS, Harrison MJ, Latchman DS, de Belleroche JS (1996) Focal cerebral ischaemia increases the levels of several classes of heat shock proteins and their corresponding mRNAs. Brain Res Mol Brain Res 42: 236–244
19. Wyttenbach A, Sauvageot O, Carmichael J, Diaz-Latoud C, Arrigo AP, Rubinsztein DC (2002) Heat shock protein 27 prevents cellular polyglutamine toxicity and suppresses the increase of reactive oxygen species caused by huntingtin. Hum Mol Genet 11: 1137–1151
20. Zhang ZD, Marton LS, Ono S, Komuro T, Weir B, Macdonald RL (2001) Complimentary deoxyribonucleic acid array analysis of cerebral arteries after experimental subarachnoid hemorrhage. Acta Neurochir Suppl 77: 17–19

Correspondence: John H. Zhang, M.D., Ph.D., Department of Neurosurgery, Louisiana State University Health Sciences Center in Shreveport, 1501 Kings Highway, PO Box 33932, Shreveport, LA 71130-3932. e-mail: johnzhang3910@yahoo.com

Acta Neurochir (2003) [Suppl] 86: 483–487

Apoptosis, blood-brain barrier, and subarachnoid hemorrhage

I. Gules[1], **M. Satoh**[1], **A. Nanda**[2], and **J. H. Zhang**[1,2]

[1] Department of Neurosurgery, University of Mississippi Medical Center, Jackson, Mississippi
[2] Department of Neurosurgery, Louisiana State University Health Sciences Center in Shreveport, Shreveport, Louisiana

Summary

This study was undertaken to investigate the role of apoptosis in the integrity of blood-brain barrier (BBB) in subarachnoid hemorrhage (SAH). BBB permeability changes were examined and found increased on day 7 in a double hemorrhage rat model using Evans blue dye. The BBB permeability increase is coincidental to brain microvascular endothelial cell apoptosis (expression of caspase-8 and -9) occurring on Day 7. However, caspase-8 and caspase-9 inhibitors failed to protect the BBB. Considering that treatment did not completely inhibit apoptosis in brain microvascular endothelial cells, higher doses, earlier and/or multiple applications, and, possibly, more potent caspase inhibitors may be needed.

Keywords: Blood-brain barrier; apoptosis; subarachnoid hemorrhage; rat, caspase.

Introduction

Numerous experimental studies using various animal models were performed to explore vascular permeability changes after SAH [2, 5, 10, 13]. It was reported that macro and microvascular permeability changes were remarkably evident following SAH [10]. Relatively few studies have examined the mechanisms contributing to the development of BBB disruption after SAH, and nothing is actually known about the existence of apoptosis in brain microvascular endothelial cells. Evidence of apoptosis has been seen in the cerebral arterial endothelium of a patient who died as a result of SAH-induced vasospasm [12]. Also, in vitro studies show that oxyhemoglobin has exhibited apoptotic effects on brain microvascular endothelial cells with increased levels of caspase 2, 3, 8, and 9 [7, 8]. In the present study, we specifically focused on exploring the potential role of apoptosis in the complex pathophysiology of BBB disruption after SAH.

Materials and methods

SAH animal model

The University of Mississippi Medical Center Committee on the Use and Care of Animals approved the protocol for the animals used in the experiment. Sprague-Dawley rats weighing 300 to 350 g were used in this study.

We divided 104 male Sprague-Dawley rats into the following groups: group I rats were non-operated (control), group II rats were sham-operated, and group III rats received one blood injection intracisternally. "Single-Hemorrhage Model" and the BBB permeability changes were measured 2 hours after initial induction of SAH and again 1 and 2 days after SAH. In group IV rats, which received two blood injections, "Double-Hemorrhage Model," and the BBB permeability changes were measured 3, 7, and 12 days later.

SAH was induced by a posterior craniocervical approach as described previously [8].

Treatment with caspase inhibitors

Caspase-8 (Z-IETD-FMK) and -9 (Z-LEHD-FMK) (Calbiochem, Inc.) inhibitors were dissolved in dimethyl sulfoxide (DMSO) solvent solution. The concentration of the stock solution prepared was 3 mM. The final concentration of DMSO in CM was approximately 0.3%. Caspase inhibitors were applied on Day 3. A single dose of 0.01 ml of either caspase-8 ($n = 10$) or -9 inhibitor ($n = 10$) or DMSO ($n = 10$) was administered intracisternally.

Evaluation of permeability changes of BBB

All animals were re-anesthetized at predetermined intervals. A jugular venous catheter was inserted, and the rats were injected with Evans blue dye intravenously (2%, 5 ml/kg) over two minutes. At the end of the 60 minute waiting period, trans-cardiac perfusion was performed with saline solution at 120-mm Hg pressure to clear the cerebral circulation of Evans blue dye until colorless perfusion fluid was obtained. The brain was immediately removed and weighed. Tissue samples were taken from the cerebral cortex, cerebellar cortex, and brainstem of the rats, as defined by the Paxinos and Vatson's atlas. The amount of extravasated Evans blue dye remaining in the brain was determined spectrofluorophotometrically (AMINCO-Bowman Series 2 Luminescence Spectrometer).

Immunohistochemistry for caspases

The frontoparietal cortexes were removed as 2-mm cubic blocks. The blocks were then dehydrated, embedded in paraffin, and sectioned, 5-µm thick, from the surface. Monoclonal anti-mouse caspase-8 (p-20) and caspase-9 (p-10) antibodies were purchased from Santa Cruz Biotechnology, Inc.

In situ apoptosis detection

In situ apoptosis detection was performed using the Apop Tag kit (Intergen Co.) following the manufacturer's protocol. This procedure detects the DNA using the terminal deoxynucleotidyl transferase (TdT) mediated d-uridine 5′-triphosphate (UTP) nick end labelling technique.

Statistics

All values are expressed as mean ± SEM. Statistical differences between the groups were compared using one way ANOVA as well as the Bonferroni test. Any probabilities less than 5% (P value < 0.05) were considered significant.

Results

BBB permeability changes

Figure 1 shows the permeability changes after SAH at six different time intervals: the second hour and Days 1, 2, 3, 7, and 12. SAH produced increased extravasation of Evans blue dye into the brain tissue, revealing significantly higher levels at Day 7 after SAH as compared to controls ($p < 0.05$). However, the amount of dye in the brain samples in all of the groups showed similar values at the rest of the time intervals.

Immunohistochemical staining

Both caspases in brain microvascular endothelial cells were visible under microscopic examination after being stained. Samples from day 3 and day 7 showed positive staining for caspase-8 and -9 antibodies. In most of the slides, immunoreactivity for caspase-8 was denser than caspase-9 (Fig. 2).

SAH-induced apoptosis in brain microvascular endothelial cells

Apop tag positive cells were rarely observed in all groups on Day 3, but Apop tag positive cells were considerably increased on Day 7 in the sample from the group with SAH only. Apop tag positively stained cells and the control group are presented (not shown). However, treatment with caspase-8 and caspase-9 inhibitors showed little effect on preventing apoptosis (Fig. 3).

Discussion

In the present study, the existence and importance of apoptosis in brain microvascular endothelial cells in SAH-induced BBB disruption was investigated in both Single-Hemorrhage and Double-Hemorrhage Models in rats. We provide the first evidence of SAH-induced apoptosis in brain microvascular endothelial cells in a rat in vivo model and this presence may lead to a marked increase in extravasation of Evans blue dye

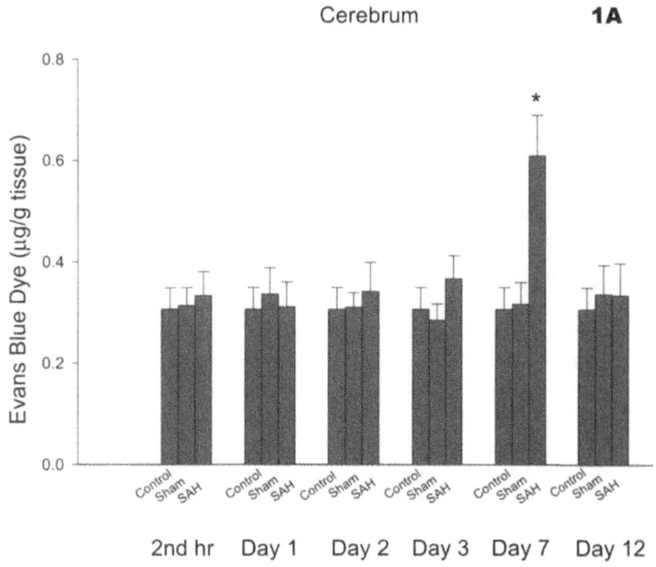

Fig. 1. The BBB permeability changes were measured in cerebrum by spectrofluorometric evaluation of Evans blue. Please note the significant increase in permeability on Day 7. *$p < 0.05$ vs. control (non-operated) and sham-operated. Each bar represents the mean ± SEM

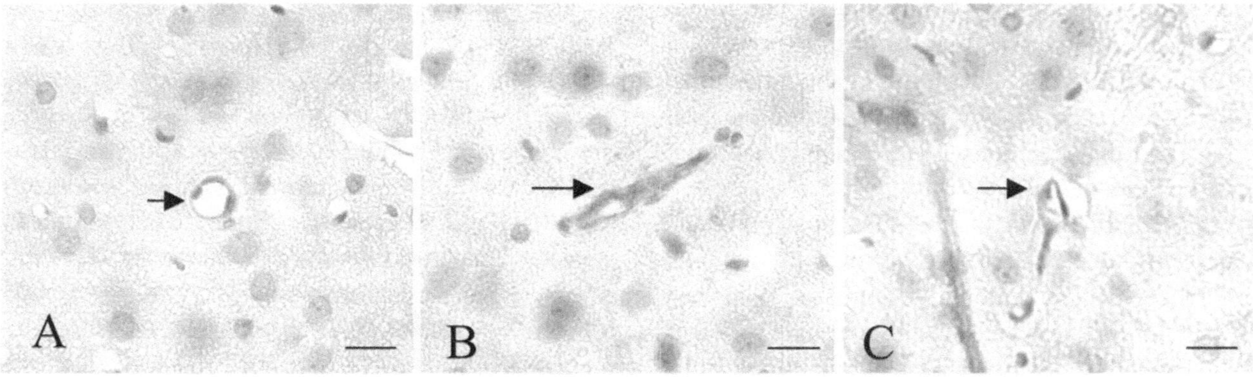

Fig. 2. Light photomicrographs illustrating immunoreactivity for caspases in paraffin embedded brain microvascular endothelial cells of cerebral cortex (A) without staining, (B) staining with caspase-8 antibody or (C) staining with caspase-9 antibody. Brain microvascular endothelial cells stained either with caspase-8 or caspase-9 antibody exhibited dense brown-stained areas in their cytoplasm. The samples were counterstained with hematoxylene. Scale bar = 25 μm

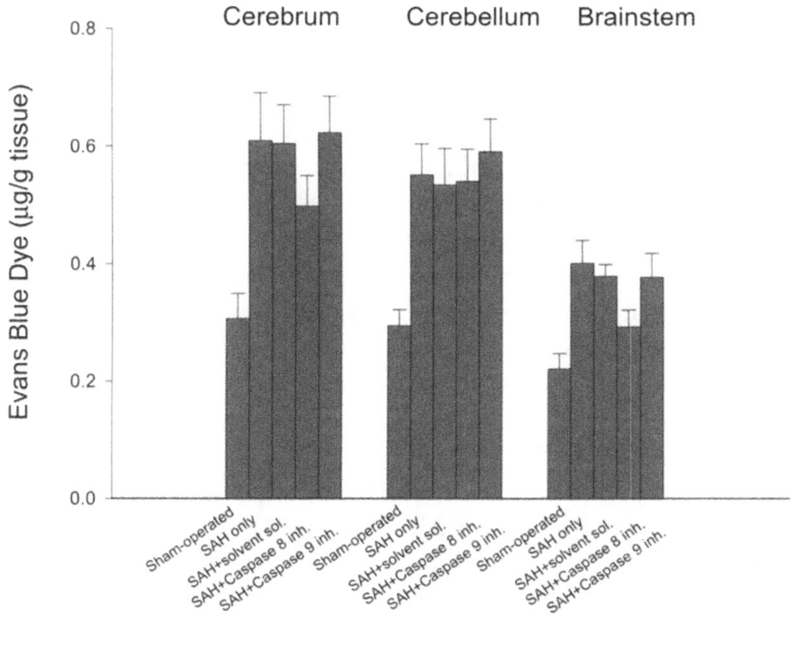

Fig. 3. The effect of caspase-8 and caspase-9 inhibitor on SAH-induced BBB disruption. Although caspase-8 inhibitor had a slight therapeutic effect on permeability increase in the cerebrum and brainstem, all treatment modalities exhibited no statistical therapeutic effect on SAH-induced BBB disruption. Each bar represents the mean ± SEM

into brain tissue on Day 7. However, therapy interventions with the peptide inhibitors caspase-8 and -9 resulted in failure.

The brain microvessels are formed by a specialized endothelium, which regulates the interchange of solutes between the blood and the brain. Endothelial cells are coupled by adherens and tight junctions (paracellular flux), and there are also very few endocystotic vesicles in brain microvascular endothelial cells (transcellular flux). A combination of those features constitutes the BBB. Any damage to intercellular junctions or activation of transcytosis across the endothelium causes alterations in permeability properties. Studies report that SAH causes increases in tight junctional and vesicular permeability [2]. The research revealed that there is a disruption in the BBB following

SAH. This disruption may be the underlying mechanism for raised intracranial pressure [2], hypertension (Doczi, 1985), the blood itself and/or the breakdown products, or the production of some substances such as free radicals, endothelin-1, arachidonic acid, and thromboxane-A_2 [3, 13].

The present study, for the first time, revealed permeability changes due to SAH in a rat Double-Hemorrhage Model. Among the studies that showed BBB disruption, there are some discrepancies, such as differences in time course and the magnitude of disruption, even in the studies that used rats [3]. The intraventricular administration of blood and the cortical model of SAH caused a permeability increase, which was evident as early as 3 hours [3]; the latter lasted for 48 hours. In Single-Hemorrhage Model, however, it began at the 36th hour and ended at the 60th hour [5]. Using different time intervals from the 2nd hour to Day 12 following SAH, we found a permeability increase occurred at Day 7.

Apoptosis is characterized by morphological features such as cell shrinkage, plasma membrane blebbing, chromatin condensation, and DNA fragmentation [6]. Every cell already contains all components of suicide machinery and is ready to engage in self-destruction. Caspase cascade can vary depending on cell type and stimulus. After an apoptotic stimulus is received, different apoptotic initiators can be activated in the endothelium. This phase is one of the earliest events in the apoptosis cascade. Initiator caspase-8 and -9 represent mitochondria-independent and mitochondria-dependent pathways, respectively. They are situated at pivotal junctions in this phenomenon. Apoptosis of the macrovascular endothelium has been reported in cases of chronic cerebral vasospasm after SAH in humans [12]. In vitro studies provided evidence that apoptosis occurred when endothelial cells were exposed to specific inducers, which are the degradation products of blood such as oxyhemoglobin (OxyHb) [7, 8], or vasoactive substances occurring in patients with SAH such as tumor necrosis factor-α (TNF-α) [1], tromboxane-A_2 [4], natriuretic peptides [11], including atrial natriuretic factors, brain natriuretic factors, and C-type natriuretic factors. It was also reported that a lesser number of selective inhibitors have been used to inhibit the function of caspase 2, 3, 8, and 9, which effectively protected endothelial cells from OxyHb-induced apoptosis [8].

After the first experiment was completed, a treatment dose that would be effective and not cause toxicity had to be chosen for both inhibitors. Both compounds are known as cell permeable and induce irreversible inhibition of caspases. Previous studies have shown that the application of caspase inhibitors prevented OxyHb-induced apoptosis at high concentrations (100 μM) but failed at low concentrations (10 μM) [7, 8]. Because the present study was an in vivo study and there was the expectation for long-term effects, a higher concentration of those agents was necessary. Considering that the tissue samples from the subjects with SAH had not produced apoptotic cells by Day 3, we decided to apply the treatment doses as 100 μM in the final intracisternal concentration on Day 3. Our primary hypothesis was that brain microvascular endothelial cell apoptosis might contribute to SAH-induced BBB disruption. In this study we, for the first time, demonstrated apoptosis in brain microvascular endothelial cells providing evidence of the expression of caspase-8 and -9. However, neither caspase-8 nor -9 inhibitors completely prevented brain microvascular endothelial cells apoptosis. Despite the limited permeability increase obtained in the group treated with caspase-8 inhibitor, both caspase inhibitors showed no effect on the prevention of BBB disruption after SAH.

Even though this study failed to show a positive result from the treatment using caspase inhibitors, it demonstrated the presence of apoptosis in brain microvessel endothelial cells and the rupture of BBB after SAH. The apoptotic death of endothelial cells shown in this study is consistent with our previous publication, which showed that apoptosis occurred in cerebral endothelial cells in a patient who died of cerebral vasospasm after SAH [12]. A better understanding of the detailed mechanisms of SAH-induced apoptosis in brain microvascular endothelial cells may potentially facilitate innovative therapeutic approaches to limit tissue injury.

Conclusion

Nothing has been known about brain microvascular endothelial cell apoptosis after SAH. This study reported a novel strategy designed to suppress caspase-8 and -9 pathways of apoptosis to prevent BBB disruption. These results supported the hypothesis that brain microvascular endothelial cell apoptosis induced by SAH might have a role in BBB disruption after SAH. However, either higher concentrations or other more potent apoptotic inhibitors are needed to prevent the integrity of BBB after SAH.

References

1. Dimmeler S, Breitschopf K, Haendeler J, Zeiher AM (1999) Dephosphorylation targets Bcl-2 for ubiquitin-dependent degradation: a link between the apoptosome and the proteasome pathway. J Exp Med 189: 1815–1822
2. Doczi T (1985) The pathogenetic and prognostic significance of blood-brain barrier damage at the acute stage of aneurysmal subarachnoid haemorrhage. Clinical and experimental studies. Acta Neurochir (Wien) 77: 110–132
3. Doczi T, Joo F, Adam G, Bozoky B, Szerdahelyi P (1986) Blood-brain barrier damage during the acute stage of subarachnoid hemorrhage, as exemplified by a new animal model. Neurosurgery 18: 733–739
4. Gao Y, Yokota R, Tang S, Ashton AW, Ware JA (2000) Reversal of angiogenesis in vitro, induction of apoptosis, and inhibition of AKT phosphorylation in endothelial cells by thromboxane A(2). Circ Res 87: 739–745
5. Germano A, D'Avella D, Imperatore C, Caruso G, Tomasello F (2000) Time-course of blood-brain barrier permeability changes after experimental subarachnoid haemorrhage. Acta Neurochir (Wien) 142: 575–580
6. Mallat Z, Tedgui A (2000) Apoptosis in the vasculature: mechanisms and functional importance. Br J Pharmacol 130: 947–962
7. Meguro T, Chen B, Lancon J, Zhang JH (2001a) Oxyhemoglobin induces caspase-mediated cell death in cerebral endothelial cells. J Neurochem 77: 1128–1135
8. Meguro T, Chen B, Parent AD, Zhang JH (2001b) Caspase inhibitors attenuate oxyhemoglobin-induced apoptosis in endothelial cells. Stroke 32: 561–566
9. Meguro T, Clower BR, Carpenter R, Parent AD, Zhang JH (2001c) Improved rat model for cerebral vasospasm studies. Neurol Res 23: 761–766
10. Peterson EW, Cardoso ER (1983) The blood-brain barrier following experimental subarachnoid hemorrhage. Part 1: Response to insult caused by arterial hypertension. J Neurosurg 58: 338–344
11. Suenobu N, Shichiri M, Iwashina M, Marumo F, Hirata Y (1999) Natriuretic peptides and nitric oxide induce endothelial apoptosis via a cGMP-dependent mechanism. Arterioscler Thromb Vasc Biol 19: 140–146
12. Zubkov AY, Ogihara K, Bernanke DH, Parent AD, Zhang J (2000) Apoptosis of endothelial cells in vessels affected by cerebral vasospasm. Surg Neurol 53: 260–266
13. Zuccarello M, Sasaki T, Kassell NF, Yamashita M (1988) Effect of intracisternal thromboxane A2 analogue on cerebral artery permeability. Acta Neurochir (Wien) 90: 144–151

Correspondence: John H. Zhang, M.D., Ph.D., Department of Neurosurgery, Louisiana State University Health Sciences Center in Shreveport, 1501 Kings Highway, PO Box 33932, Shreveport, LA 71130-3932. e-mail: johnzhang3910@yahoo.com

Acta Neurochir (2003) [Suppl] 86: 489–491

Focal brain edema and natriuretic peptides in patients with subarachnoid hemorrhage

S. Fukui, H. Nawashiro, N. Otani, H. Ooigawa, T. Toyooka, N. Tsuzuki, H. Katoh, S. Ishihara, T. Miyazawa, A. Ohnuki, and **K. Shima**

Department of Neurosurgery, National Defense Medical College, Tokorozawa, Japan

Summary

Central salt wasting syndrome may be caused by pathological increases in serum natriuretic peptides after subarachnoid hemorrhage (SAH). However, it is unclear as to why the serum concentration of atrial natriuretic peptide (ANP) or brain natriuretic peptide (BNP) increases in the subacute phase of SAH. The present study was designed to assess the correlation between focal brain edema and serum concentration of ANP or BNP in patients with SAH. Focal brain edema was found in 8 SAH-patients and peaked between days 4 and 7 of SAH. The mean serum ANP and BNP levels in patients with focal brain edema were significantly higher than those in patients without focal brain edema between days 4 and 14 of SAH. These results suggest that focal brain edema might correlate with increased levels of ANP and BNP in the subacute phase of SAH.

Keywords: Subarachnoid hemorrhage; atrial natriuretic peptide; brain natriuretic peptide; central salt wasting syndrome; brain edema.

Introduction

Central salt wasting syndrome (CSWS), which is characterized by excessive natriuresis regardless of hyponatremia, is often found in the subacute phase of subarachnoid hemorrhage (SAH) [1, 2, 4]. The excessive natriuresis observed during CSWS decreases the intravascular volume and exacerbates cerebral vasospasms thus leading to a deterioration in the outcome of SAH-patient [2]. A CSWS is commonly believed to be caused by the pathological increases of serum atrial natriuretic peptide (ANP) [4] or brain natriuretic peptide (BNP) [1, 2] in the subacute phase of SAH. However, the pathogeneses of these increases in the serum concentration of ANP or BNP in the subacute phase of SAH have not yet been fully investigated.

ANP is secreted mainly from right or left atrium and partially from left ventricle. Recent evidence has shown that BNP is also secreted mainly from heart, especially the left ventricle or the left atrium. Takahashi *et al.* [3] however, showed that there was some immunoreactivity of BNP also through the brains, including the cerebral cortex, thalamus, cerebellum, pons, and hypothalamus, thus indicating that BNP secretion may be induced by pathological processes involving these regions. In patients with SAH, focal brain edema, which may be caused by either an intracerebral hematoma associated with SAH or some surgical complications during a direct surgery for ruptured aneurysms, is occasionally found also during the subacute phase of SAH. We herein assessed whether focal brain edema has any relationship with serum concentration of ANP or BNP in patients with SAH.

Materials and methods

Subjects and clinical variables

We retrospectively studied 61 SAH-patients who were admitted within 72 hours after onset of SAH to the National Defense Medical College Hospital between January 1997 and November 2001, and whose blood samples, especially in the subacute phase of SAH, were available to assess the serum concentrations of ANP or BNP in the present study. Neurological condition on admission was evaluated according to the Hunt and Kosnik classification. The severity of SAH was assessed according to Fisher's classification.

Management of SAH

We mainly selected a direct surgery with aneurysmal clips as a first choice for the treatment of the SAH-patients in the acute phase of SAH, and partially performed an interventional procedure with Guglielmi detachable coil (GDC). During the direct surgery with a craniotomy, as a rule, both the cisternal and ventricular drainage were used for both avoiding a high intracranial pressure and the removal of clots in the subarachnoid spaces. The cisternal drainage was set at a pressure of 15 cm H_2O until Day 7 of SAH after the surgery and the ventricular drainage was set at a pressure of 20 cm H_2O until Day 3 of SAH after the surgery.

In the present study, there was no difference based on the presence of focal brain edema in the management of SAH including the intravenous fluid volume per day or the amounts of intravenous sodium intake per day (data not assessed).

Focal brain edema

Focal brain edema, which was assessed by means of consecutive CT examinations, was defined as low-density area on CT that had an obvious focal lesion such as an intracerebral hematoma associated with SAH, which belongs to Fisher's group 4, or a cerebral contusion due to the excessive brain retraction during the direct surgery for ruptured aneurysms.

Serum data

Serum levels of ANP, BNP, and antidiuretic hormone (ADH) were measured at 6 different time periods (Day 1, Day 2, Day 3, Days 4 to 7, Days 8 to 14, and Days 15 to 25) by means of a specific immunoradiometric assay. Serum levels of sodium, potassium, and glucose, and serum osmolarity were repeatedly measured through the clinical course. All serum data were measured, as a rule, at 7 a.m. after the second day of SAH.

Statistical analysis

The association between the focal brain edema and variable factors was analyzed by the chi square test, Student's t test, or Mann-Whitney U test. Values of $p < 0.05$ were considered.

Results

Focal brain edema

The consecutive CT examinations revealed focal brain edema in 8 SAH-patients in this series, which appeared on Day 1 or 2 and peaked between Days 4 and 7. In 3 patients, the focal brain edema was caused by the intracerebral hematoma associated with SAH, which belonged to Fisher's group 4. In 4 patients, the focal brain edema was caused by the surgical complications for the ruptured aneurysm, mainly the cerebral contusion due to the excessive brain retraction. In a patient, the focal brain edema was caused by the combination of the two different causes described above.

Factors associated with focal brain edema

There was a significant correlation between the presence of focal brain edema and Fisher's classification for CT ($p < 0.03$). The CT finding of Fisher's group 4, which was found in 9 patients, included 4 SAH with intracerebral hematoma and 5 SAH with intraventricular hematoma. All the 4 patients with an intracerebral hematoma associated with SAH had a focal brain edema around the hematoma, which gradually developed after the SAH.

The patients with focal brain edema had a poorer outcome (GOS, $p < 0.03$) compared with the patients without focal brain edema. There was a weak correlation ($p < 0.1$) but not significantly between the presence of focal brain edema and the symptomatic vasospasm. Both of the 2 patients with focal brain edema that had the poor outcome of either vegetative state or dead had symptomatic vasospasms in their clinical courses.

On the other hand, there was no correlation between the presence of focal brain edema and the patient age, gender, Hunt and Kosnik grade, aneurysm site, therapeutic interventions, hyponatremia during hospitalization, or the serum sodium or potassium levels on the any time after SAH.

ANP, BNP, and ADH

The mean serum ANP levels in the patients with focal brain edema increased between Days 4 and 14 of SAH. Between Days 4 and 7 of SAH, the mean serum ANP level in the patients with focal brain edema was significantly higher (131 ± 57 pg/ml, $p < 0.0001$) than that in the patients without focal brain edema (42 ± 32 pg/ml). Between Days 8 and 14 of SAH, the mean serum ANP level in the patients with focal brain edema was also significantly higher (79 ± 30 pg/ml, $p = 0.0008$) than that in the patients without focal brain edema (30 ± 31 pg/ml). On the other hand, there was no difference based on the presence of focal brain edema in the mean serum ANP levels either between Days 1 and 3 of SAH, or between Days 15 and 25 of SAH. In the patients without focal brain edema, the mean serum ANP levels were nearly equal to its upper limit of normal range (<43.0 pg/ml) through their clinical courses.

The mean serum BNP levels in the patients with focal brain edema also increased between Days 4 and 14 of SAH. Between Days 4 and 7 of SAH, the mean serum BNP level in the patients with focal brain edema was significantly higher (655 ± 335 pg/ml, $p = 0.0006$) than that in the patients without focal brain edema (97 ± 118 pg/ml). Between Days 8 and 14 of SAH, the mean serum BNP level in the patients with focal brain edema was also significantly higher (598 ± 276 pg/ml, $p = 0.0003$) than that in the patients without focal brain edema (91 ± 122 pg/ml). On the other hand, there was no difference based on the presence of focal brain edema in the mean serum BNP levels either between Days 1 and 3 of SAH, or between Days 15 and 25 of SAH.

The mean serum ADH levels were characterized by the prominent high value of the mean serum level on day 1 of SAH, which was common to both the patients with and without focal brain edema. The mean serum ADH level in the patients with focal brain edema (7.4 ± 8.9 pg/ml, $p < 0.03$) was significantly higher than that in the patients without focal brain edema (2.4 ± 1.9 pg/ml) between Days 8 and 14 of SAH.

Discussion

In this study, we found a significant correlation between the presence of focal brain edema and increased levels of natriuretic peptides following SAH. Because BNP could be produced in the brain itself, it is an attractive hypothesis that brain edema could elicit the production of BNP. In the present study, there was no difference based on the presence of focal brain edema in the management of SAH including the intravenous fluid volume per day or the amounts of intravenous sodium intake per day. However, we did not assess the hemodynamic parameters, such as cardiac output volume or pulmonary wedged pressure to rule out the possibility that the hemodynamic parameters are different among the patients. Furthermore, the level of BNP in the cerebrospinal fluid should be investigated to clarify the alternative origin of increased BNP after SAH. Further studies are required to investigate the pathophysiological basis for the increased BNP in the subacute phase of SAH.

References

1. Berendes E, Walter M, Cullen P, Prien T, Van Aken H, Horsthemke J, Schulte M, von Wild K, Scherer R (1997) Secretion of brain natriuretic peptide in patients with aneurysmal subarachnoid haemorrhage. Lancet 349(9047): 245–249
2. Sviri GE, Feinsod M, Soustiel JF (2000) Brain natriuretic peptide and cerebral vasospasm in subarachnoid hemorrhage. Clinical and TCD correlations. Stroke 31: 118–122
3. Takahashi K, Totsune K, Sone M, Ohneda M, Murakami O, Itoi K, Mouri T (1992) Human brain natriuretic peptide-like immunoreactivity in human brain. Peptides 13: 121–123
4. Wijdicks EF, Schievink WI, Burnett JC Jr (1997) Natriuretic peptide system and endothelin in aneurysmal subarachnoid hemorrhage. J Neurosurg 87: 275–280

Correspondence: Hiroshi Nawashiro, M.D., D.M.Sc., Department of Neurosurgery, National Defense Medical College, 3-2 Namiki, Tokorozawa, Saitama 359-8513, Japan. e-mail: nawa1957@me.ndmc.ac.jp

Tumor

Acta Neurochir (2003) [Suppl] 86: 495–498

Aquaporin 1 and aquaporin 4 expression in human brain after subarachnoid hemorrhage and in peritumoral tissue

J. Badaut[1,2], **J. F. Brunet**[1,2], **L. Grollimund**[1,2], **M. F. Hamou**[1], **P. J. Magistretti**[2], **J. G. Villemure**[1], and **L. Regli**[1]

[1] Department of Neurosurgery, Centre Hospitalier Universitaire Vaudois, Lausanne, Switzerland
[2] Institut of Physiology, University of Lausanne, Lausanne, Switzerland

Summary

Aquaporins (AQPs) are a protein family of water channels which facilitate the water flux through the plasmatic membranes. The expression of AQPs has been described in rat brain by several studies. Despite recent reports that have shown an over-expression of AQP1 and 4 in human tumoral cells, little is known about AQP expression in human brain. The purpose of this study was to investigate the expression of AQP1 and AQP4 in human brain after subarachnoid hemorrhage (SAH) and in peritumoral tissue by western blot and immunohistochemistry.

The results showed a marked increase of the expression of AQP1 and AQP4. This over-expression occurred on the astrocytic processes and polarization on astrocytic end-feet was lost. No expression was observed on neuronal cells.

This study is the first demonstration of the induction of AQP1 and AQP4 on reactive astrocytes in an acute brain injury, such as SAH. These results reinforce the hypothesis that AQPs may be involved in the dynamics of brain edema formation or resolution. Further studies are needed to understand their functional role.

Keywords: Aquaporins; brain; astrocyte; subarachnoid hemorrhage; human.

Introduction

Water movement across cell membranes is a fundamental property of life. The aquaporins (AQPs) are a growing family of specific molecular water channels, first identified in the early 90s', and presently composed by eleven sub-types, AQP0 to AQP10. They are membrane proteins, ubiquitously distributed in tissues and organs, challenging the concept of simple diffusion of water molecules across pure lipid bilayers. Recent data concerning AQPs have boosted the field of research with emergent interest in their physiological and physiopathological role [3]. To date, AQP1 and AQP4 have been clearly demonstrated and studied in rodent brain [3]. As suggested by the different studies, water-channels may play a physiological role in the homeostasis of the extracellular space. Different models of acute brain injury in rodents have been studied to try to understand the modulation of the AQP expression. These studies have described mainly AQP4 changes showing variable results depending on the experimental paradigm [2, 5, 6, 7, 11, 12].

Recent reports have evidenced an increase of AQP1 and AQP4 expression in human brain tumors [4, 8, 9, 10]. Such characterization of AQP expression in human pathological states is important to understand their possible implication in brain water movements. Brain swelling is observed in all acute brain disorders, and the expression and cellular distribution of AQP4 and AQP1 have not yet been studied in cerebral tissue from patients with brain diseases. The purpose of this study was to investigate the expression and distribution of AQP1 and AQP4 in human brain after acute subarachnoid hemorrhage (SAH) and peritumoral tissue compared to "normal" and tumoral tissue.

Materials and methods

Patients

Tissue samples were obtained during neurosurgical procedures from 12 patients that underwent operations for intractable epilepsy (n = 3, temporal lobectomies), for ruptured intracranial aneurysm (n = 3), for resection of a cerebral glioma grade IV (n = 3), and for resections of other glial tumors (n = 3 peritumoral tissue) at the University Hospital of Lausanne. All brain samples were from neocortical regions. The tissue from temporal lobectomies was also taken from the neocortex, distant from the epileptic focus, and is, therefore, considered to be "normal" control tissue. The study followed institutional guidelines of the ethics committee.

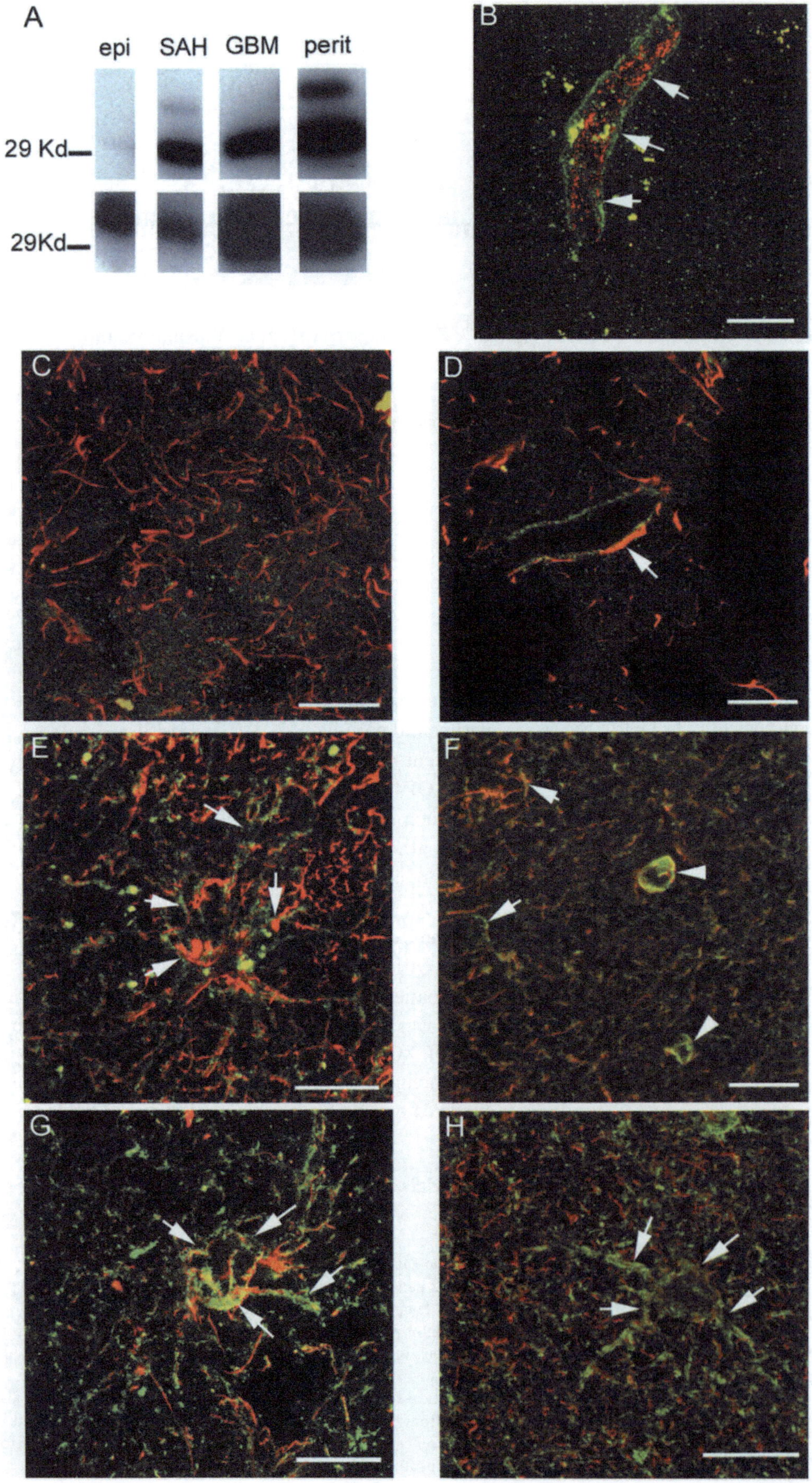
A
epi
SAH
GBM
perit
29 Kd
29Kd
B
C
D
E
F
G
H

Western blots

The tissue samples were frozen in nitreous liquid and cut in cryostat with 10 μm of thickness. The slices were put in a tube and gently homogenized on ice in medium as previously described [2]. Ten micrograms of total protein per lane were subjected to SDS polyacrylamide gel electrophoresis in 12% Tris-Glycine gels and transferred to the polyvinylidene fluoride membrane. The blot was probed with polyclonal antibodies against AQP1 (1/5000, Chemicon) and AQP4 (1/4000, Chemicon) in TBST with 3% non-fat milk for 4 hours at room temperature. After washing in TBST, the blot was probed with horseradish peroxidase conjugated secondary antibodies against mouse or rabbit IgG for two hours at room temperature, washed in TBST and processed with an enhanced chemiluminescence (ECL) Western blotting detection system (Calbiochem). The blots were exposed 30 s to film except for the blot of grade IV glioma and peritumoral tissue (5 s) (hyperfilm ECL, Amersham).

Immunohistochemistry

The tissue samples were frozen in nitreous liquid and cut in cryostat with 10 m of thickness. Samples were then fixed in paraformaldehyde 4% or in mix alcohol/acetone. All immunolabeling studies were carried out in PBS containing 0.1% Triton × 100 and 1% bovin serum albumin. Slices for immunofluorescence labelling were incubated overnight at 4 °C with one of the mix of primary antibodies: anti-AQP1 (1/200, Rabbit, Chemicon) with anti-GFAP (1/400, mouse, Sigma); anti-AQP4 with anti-GFAP (1/400, mouse, Sigma); anti-AQP1 (1/200, Rabbit, Chemicon) with anti-MAP 2 (1/400, mouse, Sigma); anti-AQP4 (1/200, Rabbit, Chemicon) with anti-MAP 2 (1/400, mouse, Sigma); anti-AQP4 (1/200, Rabbit, Chemicon) with anti-Factor VIII (1/400, mouse, Dako). After washing, 3 × 10 minutes in PBS, slices were incubated with the secondary antibody coupled with FITC (anti-rabbit-FITC, 1/200) or with CY3 (anti-mouse-CY3, 1/300) during two hours at room temperature in the dark and then rinsed 3 times and were coverslipped with Vectashield (Vector). Immunofluorescent preparations were examined under a confocal laser scanning microscope (Leica), equipped with a Leica optiphot microscope. Images from the preparations were processed using excitation at 488 nm, specific for FITC, and at 568 nm specific for CY3.

Results

Western blot

AQP1 and AQP4 proteins were detected by western blot at the expected band size (30 kDa) in the human tissue samples. AQP1 was not expressed in control tissue (epi, Fig. 1A). In all tissue samples of pathological states AQP1 was clearly over-expressed. The highest levels were seen in the tissue from gliomas grade IV and peritumoral tissue. Samples from SAH also showed over-expression of AQP1 (Fig. 1A). AQP4 protein was detected in all human tissue samples. Obvious over-expression was found in glioma grade IV and peritumoral tissue (Fig. 1A).

Immunohistochemistry

No specific AQP1 immunostaining was detected in control tissue (Fig. 1C), in contrast AQP1 was present on astrocytic processes of reactive astrocytes in SAH (Fig. 1E) and in peritumoral tissue (Fig. 1G) samples, as shown by double labeling with anti-GFAP (Fig. 1C,E,G). The highest labeling of AQP1 was detected on the tumoral cells of grade IV gliomas (not shown).

In control tissue, AQP4 labeling co-localized with GFAP staining and was polarized only around the cerebral microvessels (Fig. 1D), as confirmed by double immunolabeling with Factor VIII (Fig. 1B), indicating expression on the astrocytic endfeet. In SAH (Fig. 1F) and peritumoral tissues (Fig. 1H) AQP4 labeling was markedly increased and observed on all astrocytic processes (GFAP double labeling) (Fig. 1F,H). Grade IV glioma cells showed marked over-expression of AQP4. No AQP1 and AQP4 labeling was observed on neuronal cells (data not shown).

Discussion

Detailed studies of AQP distribution in rodent brain have shown that AQP1 is not detected on astrocytes, as opposed to AQP4 that shows a striking feature of polarization on astrocytic end-feet around cerebral microvessels [1, 3]. Our study showed similar results on human tissue samples. "Normal" control tissue showed no AQP1 expression in neocortex, but signifi-

◄

Fig. 1. Immunocytochemical analysis of AQP1 and AQP4 expression in human brain after subarachnoid hemorrhage (*SAH*), in peritumoral tissue (peri), in cerebral glioma grade IV (*GBM*) compared to "normal" brain tissue from peri-epileptical tissue (*epi*). (A) The immunoblot shows that the AQP1 and AQP4 reveals a specific band of approximatively 30 kDa, corresponding to the expected molecular weights of AQPs monomers. AQP1 is not detected in control tissue (epi). (B) Confocal photomicrograph of AQP4 immunolabeling (green) and Factor VIII immunostaining (red) in control tissue (epi) shows the expression on AQP4 around brain microvessels (arrows) revealed by Factor VIII labeling (in red). (C) Double labeling AQP1 (green) and GFAP (red) shows the absence of AQP1 labeling in control tissue (epi). (D) Double staining AQP4 (green) and GFAP (red) shows the expression of AQP4 on astrocyte endfeet revealed by GFAP labeling (arrows). (E) AQP1 immunostaining (green) in SAH tissues is expressed on astrocyte processes stained by anti-GFAP (arrows). (F) AQP4 labeling (green) in SAH is present around blood vessels (arrow heads) and also on astrocyte processes stained by anti-GFAP (red, arrows). (G) Double immunolabeling AQP1 (green) and (*GFAP*) in peritumoral tissue (perit) shows a high expression of AQP1 on astrocyte processes. (H) AQP4 staining is observed on astrocyte processes and cell body stained by anti-GFAP (red, arrows). Bar = 20 μm (B, C, D, E, F, G, H)

cant AQP4 expression. Moreover, immunohistochemistry showed co-localization of AQP4 with GFAP, and close expression around anti factor VIII labeling (although no co-localization). This pattern indicates that AQP4 expression is polarized on astrocytic end-feet around microvessels in human neocortex, just as demonstrated in rodent brain and recently reported by Sadoun *et al.* [9].

Although AQP expression changes after brain injury have been reported in rodents [2, 5, 6, 7, 11, 12], no similar data is available on AQP expression in human brain. Our study was able to demonstrate marked induction of astrocytic AQP1 and AQP4 expression in human neocortex after SAH. This increase of AQP4 expression was paralleled by an extension on the astrocytic processes and thus by the loss of polarization of AQP4 on perivascular astrocytic end-feet. Interestingly, induction of AQP1 and AQP4 was also observed in reactive astrocytes in peritumoral tissue samples. The expression of AQPs in a rodent model of SAH has not been reported yet. However, SAH from a ruptured aneurysm is a major injury to the brain and includes also an ischemic assault. Our results are in accordance with the induction of AQP4 observed after brain ischemia and traumatic brain injury in rodent models [5, 11, 12].

In conclusion, this study suggests that the induction of AQP1 and AQP4 seems to be a feature of reactive astrocytes after SAH as well as in the peritumoral gliosis. To date, the pathological role of the induction of these AQPs on astrocytes have not been determined. Injured brain parenchyma presents perturbations of homeostasis and can be accompanied by cytotoxic or vasogenic edema. Although the implication of these AQPs in the dynamic of edema formation or resolution can be suspected, further studies looking at the regulation of AQPs need to be done to understand there functional role.

Acknowledgment

This study was supported by a grant from "Fondazione per lo Studio delle Malattie Neurodegenerative" from Lugano, Swoizerland.

References

1. Badaut J, Nehlig A, Verbavatz J, Stoeckel M, Freund-Mercier M, Lasbennes F (2000) Hypervascularization in the magnocellular nuclei of the rat hypothalamus: relationship with the distribution of aquaporin-4 and markers of energy metabolism. J Neuroendocrinol 12: 960–969
2. Badaut J, Hirt L, Granziera C, Bogousslavsky J, Magistretti PJ, Regli L (2001) Astrocyte-specific expression of aquaporin-9 in mouse brain is increased after transient focal cerebral ischemia. J Cereb Blood Flow Metab 21: 477–482
3. Badaut J, Lasbennes F, Magistretti PJ, Regli L (2002) Aquaporins in brain: distribution, physiology, and pathophysiology. J Cereb Blood Flow Metab 22(4): 367–378
4. Endo M, Jain RK, Witwer B, Brown D (1999) Water channel (aquaporin 1) expression and distribution in mammary carcinomas and glioblastomas. Microvasc Res 58: 89–98
5. Ke C, Poon WS, Ng HK, Pang JC, Chan Y (2001) Heterogeneous responses of aquaporin-4 in oedema formation in a replicated severe traumatic brain injury model in rats. Neurosci Lett 301: 21–24
6. Kobayashi H, Minami S, Itoh S, Shiraishi S, Yokoo H, Yanagita T, Uezono Y, Mohri M, Wada A (2001) Aquaporin subtypes in rat cerebral microvessels. Neurosci Lett 297: 163–166
7. Manley GT, Fujimura M, Ma T, Noshita N, Filiz F, Bollen AW, Chan P, Verkman AS (2000) Aquaporin-4 deletion in mice reduces brain edema after acute water intoxication and ischemic stroke. Nat Med 6: 159–163
8. Markert JM, Fuller CM, Gillespie GY, Bubien JK, McLean LA, Hong RL, Lee K, Gullans SR, Mapstone TB, Benos DJ (2001) Differential gene expression profiling in human brain tumors. Physiol Genomics 5: 21–33
9. Saadoun S, Papadopoulos MC, Davies DC, Krishna S, Bell BA (2002) Aquaporin-4 expression is increased in oedematous human brain tumours. J Neurol Neurosurg Psychiatry 72: 262–265
10. Saadoun S, Papadopoulos MC, Davies DC, Bell BA, Krishna S (2002) Increased aquaporin 1 water channel expression in human brain tumours. Br J Cancer 87(6): 621–623
11. Taniguchi M, Yamashita T, Kumura E, Tamatani M, Kobayashi A, Yokawa T, Maruno M, Kato A, Ohnishi T, Kohmura E, Tohyama M, Yoshimine T (2000) Induction of aquaporin-4 water channel mRNA after focal cerebral ischemia in rat. Brain Res Mol Brain Res 78: 131–137
12. Vizuete ML, Venero JL, Vargas C, Ilundain AA, Echevarria M, Machado A, Cano J (1999) Differential upregulation of aquaporin-4 mRNA expression in reactive astrocytes after brain injury: potential role in brain edema. Neurobiol Dis 6: 245–258

Correspondence: Luca Regli, M.D., Department of Neurosurgery, Centre Hospitalier Universitaire Vaudois, Rue du Bugnon, 1011 Lausanne, Switzerland. e-mail: Luca.Regli@chuv.hospvd.ch

Acta Neurochir (2003) [Suppl] 86: 499–502

Aquaporin-1 expression in human glial tumors suggests a potential novel therapeutic target for tumor-associated edema

K. Oshio[1], D. K. Binder[1], A. Bollen, A. S. Verkman[2], M. S. Berger[1], and G. T. Manley[1]

[1] Brain Tumor Research Center, Department of Neurological Surgery, University of California, San Francisco, California, USA
[2] Department of Medicine, University of California, San Francisco, California, USA

Summary

Aquaporins are membrane proteins involved in water transport in many fluid-transporting tissues. The objective of this study was to investigate the expression of aquaporins in malignant glial tumors associated with cerebral edema. Eighteen human glial tumors were obtained from the UCSF Neurosurgery Tissue Bank. Aquaporin-1 (AQP1) expression was evaluated via Western blot and immunohistochemistry. Intense upregulation of AQP1 expression was found in all glioblastomas. The robust expression of aquaporins in glioblastomas suggests a pathologic role for these membrane water channels, and raises the possibility that selective AQP inhibition might offer a new therapeutic option for tumor-associated cerebral edema.

Keywords: Aquaporin; edema; brain tumor; gene array; swelling.

Introduction

The aquaporins (AQPs) are a family of hydrophobic intrinsic membrane proteins that function as "water channels" in many cell types involved in fluid transport. Aquaporin-1 (AQP1) was first identified from red blood cells and renal proximal tubular epithelium [2]. Humans without aquaporin-1, while phenotypically normal, have impaired urinary concentrating ability [4]. Lack of aquaporin-2 (AQP2) results in hereditary nephrogenic diabetes insipidus due to the inability to concentrate urine [1, 3, 10] and recent results from knockout mice implicate multiple physiological roles for aquaporins [12]. Recent studies imply a physiologic and pathologic role for aquaporins in the central nervous system [8]. For example, animals deficient in aquaporin-4 (AQP4), which is expressed in CNS glial cells lining ependymal and pial surfaces [7] show decreased cerebral edema and improved neurological outcome following water intoxication or ischemic challenge [5]. In addition, aquaporin-1 (AQP1) is expressed in the choroid plexus where it may play a role in CSF production [18]. In view of the marked peritumoral edema characteristic of glial tumors [9] and the probable role of aquaporins in cerebral water balance, we elected to investigate the expression of aquaporins in these tumors.

Materials and methods

Western blotting

Western blot analysis was performed on fresh frozen brain tumor tissue or on control brain (from resection for epilepsy). Samples were homogenized in buffer (250 mM Sucrose, 10 mM Tris·HCl pH 7.4, 20 μg/ml PMSF) and centrifuged at 1000 rpm for 10 minutes at 4 °C. The soluble fraction was dissolved in sodium dodecyl sulfate (SDS) (Pierce, Rockford, IL) and centrifuged at 10000 rpm for 10 minutes at 4 °C. Protein concentration was determined by Bradford assay (Bio-Rad, Hercules, CA). Ten μg of each sample was loaded onto 12% Tris-glycine gels (Invitrogen NOVEX, Carlsbad, CA) in Tris-glycine SDS buffer (Invitrogen NOVEX, Carlsbad, CA), and run at 90 V for 2 hours at room temperature. Then the gel was transferred to a PVDF membrane overnight at 4 °C. The membrane was rinsed with TBS-T buffer (0.02 M Tris, 0.225 M NaCl, 0.1% Triton and DW), and then placed in blocking solution (TBS-T with 3% nonfat milk (Sigma Chemical Co. Saint Louis, MO)) for 30 minutes before application of primary antibodies at room temperature. Then, affinity-purified rabbit polyclonal anti-rat aquaporin 1 (AQP1) antibody (CHEMICON International, Inc. Temecula, CA) was used. Primary antibody (1:1000) was incubated overnight at 4 °C in TBS-T buffer containing 1% BSA (SIGMA Chemical Co. Saint. Louis, MO), pH 7.6. After 3 rinses in TBS-T buffer, the membrane was incubated in horseradish peroxidase-linked donkey anti-rabbit IgG (Amersham Pharmacia, Piscataway, NJ) diluted 1:5000 in TBS-T buffer containing 1% BSA for 30 minutes at room temperature, pH 7.6. After rinsing in TBS-T buffer, the membrane was incubated with chemiluminescent detection reagents (ECL Plus; Amersham Pharmacia biotech, Piscataway, NJ) for 5 minutes and exposed to film (Hyperfilm™ ELC; Amersham Pharmacia biotech, Piscataway, NJ).

Immunohistochemistry

Fresh brain tumor tissue was fixed overnight in 10% formalin (Fisher Scientific, Pittsburgh, PA). The next day, the tissue was dehydrated in ethyl alcohol and xylene, paraffin-mounted and sectioned at 8 μm thickness. Sections were embedded on slides and stored at room temperature. For immunohistochemistry, slides were warmed to 60 °C for 10 min. Tissue was rehydrated and rinsed in 1X phosphate-buffered saline (PBS) (pH 7.4). Endogenous peroxidase activity was quenched by immersion in 3% H_2O_2 in methanol for 10 min. A nonspecific blocking procedure with 5% normal goat serum (Vector Laboratories, Inc., Burlingame, CA) was performed before application of primary antibodies. The slides were incubated overnight in aquaporin-1 (AQP1) antibody (1:1000) at 4 °C in PBS containing 5% normal goat serum, pH 7.4. The next day, sections were incubated in a biotinylated goat anti-rabbit IgG secondary antibody (1:500; Vector Laboratories, Inc., Burlingame, CA) for 1 hour at room temperature in PBS containing 5% normal goat serum, pH 7.4, followed by incubation in Vectastain ABC reagent (1:100; Vector Laboratories, Burlingame, CA) for 1 hour at room temperature in PBS containing 5% normal goat serum, pH 7.4. Diaminobenzidine (Sigma, St. Louis, MO) was used to develop the reaction product. After multiple rinses in PBS and a final rinse in distilled water, sections were counterstained with hematoxylin (Biomeda, Foster City, CA), dehydrated, cleared, and coverslipped.

Results

Western blotting demonstrated robust AQP1 protein expression in astrocytomas (Fig. 1). In glial tumors of various grades (Grade II, n = 3; Grade III, n = 3; Grade IV, n = 4), there was a dramatic upregulation of AQP1 protein expression (Fig. 1). In addition, it appeared qualitatively that higher-grade astrocytomas (III and IV) expressed more AQP1 protein than low-grade astrocytomas (Grade II) (Fig. 1). In nontumor brain (from epilepsy resection), there was a faint band at 28 kDa presumably reflecting low basal levels of expression of AQP1 or possible inclusion of a small amount of choroid plexus with the surgical specimen (N, Fig. 1). Protein from AQP1 −/− mouse spinal cord demonstrated no AQP1 band (−/−, Fig. 1). Protein from mouse and rat cortex also failed to demonstrate any AQP1 expression (data not shown).

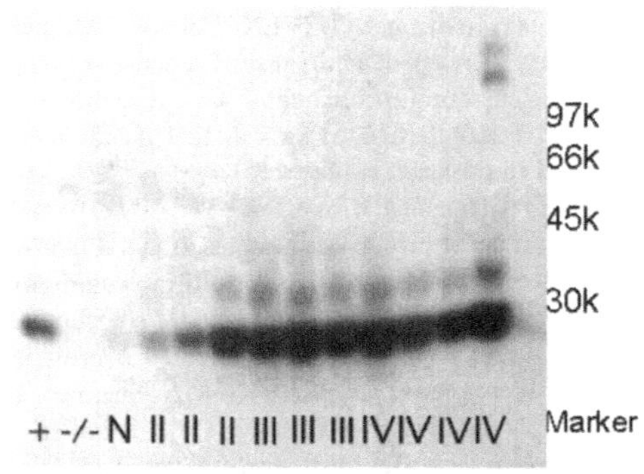

Fig. 1. Western blot analysis of aquaporin-1 protein in human glioma specimens. + positive control (oligoastrocytoma grade III). −/− negative control (AQP1 −/− mouse spinal cord). *N* Normal human brain (epilepsy patient). *II* astrocytoma (grade II). *III* anaplastic astrocytoma (grade III). *IV* glioblastoma multiforme (grade IV)

Immunohistochemical analysis with the same polyclonal AQP1 antibody used for Western blotting demonstrated robust AQP1 immunoreactivity in glioblastomas. Immunohistochemical analysis was performed on 18 tumors; representative data is shown from 2 tumors, both Grade IV (glioblastoma multiforme) (Fig. 2). Figure 2A shows typical Grade IV histology with anaplastic glial cells and foci of microvascular proliferation. The corresponding AQP1 immunoreactivity from the same tumor is shown in Fig. 2B. Strong AQP1 immunoreactivity is seen on the membranes of the neoplastic astrocytes. Interestingly, AQP1 immunoreactivity appeared to be localized to astrocyte membranes, and was not evident in areas of microvascular proliferation. Figure 2C shows a different Grade IV tumor with neoplastic astrocytes and multiple mitoses. Again, the corresponding AQP1 tumor immunoreactivity is localized primarily to glial membranes (Fig. 2D). AQP1 immunoreactivity in control brain was localized to the choroid plexus as previously described [8] (data not shown).

Discussion

We have demonstrated via Western blotting and immunohistochemistry that there is dramatic upregulation of the membrane water channel AQP1 in human glial tumors. Immunohistochemical data suggest that the increase in AQP1 protein is localized primarily if not exclusively to membranes of neoplastic astrocytes. No AQP1 expression was observed in cerebral endothelium. Western blot analysis of Grade II, III, and IV tumors also suggest qualitatively that lower-grade tumors may upregulate AQP1 to a lesser extent, but still dramatically more than control brain. However, a larger number of astrocytomas will need to be examined to determine if there are quantitative differences between different grades of tumors. It should be noted that Markert *et al.* [6] have also recently demonstrated expression of AQP1 in four glioblastomas using gene array analysis. Previous expression studies in rodents have failed to demonstrate AQP1 expression in normal astrocytes.

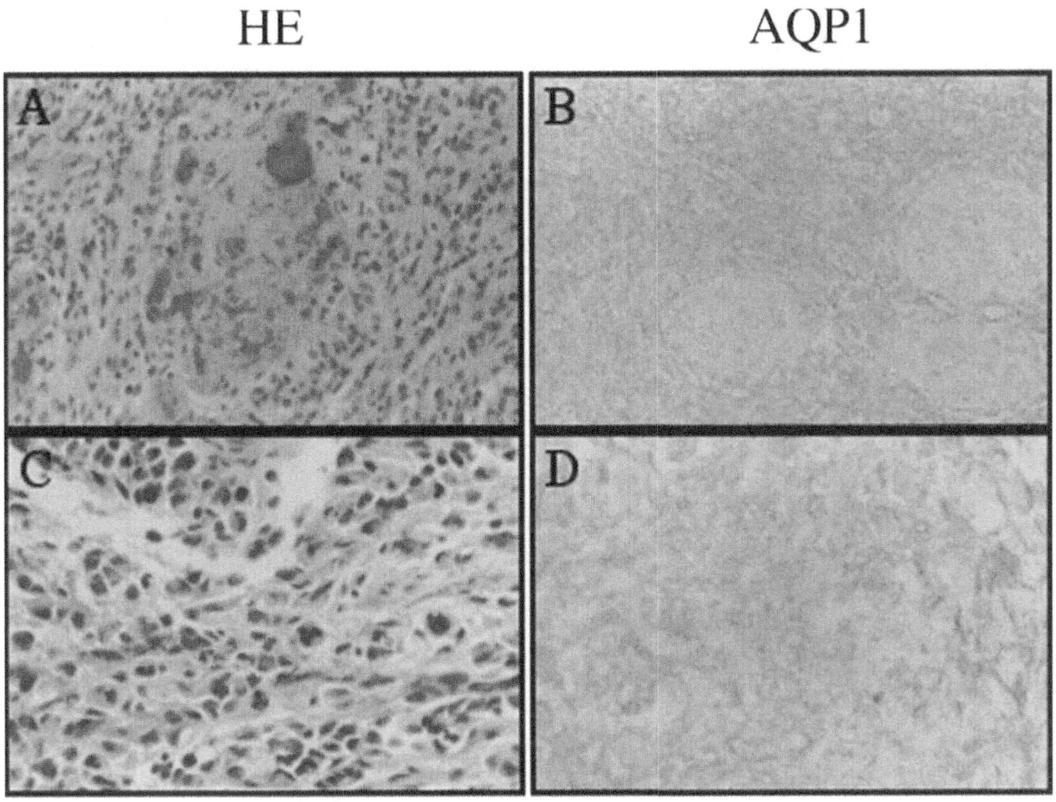

Fig. 2. Immunohistochemical analysis of aquaporin-1 protein expression in human glioma specimens. (A) This glioblastoma multiforme has multiple foci of microvascular proliferation (H&E, 250×). (B) The immunoperoxidase AQP1 stain (100×) demonstrates multiple negative foci of microvascular proliferation, but strongly positive immunoreactivity on the membranes of the neoplastic astrocytes. (C) This glioblastoma multiforme is composed of neoplastic astrocytes with multiple mitoses present (H&E, 250×). (D) The immunoperoxidase AQP1 stain (400×) demonstrates strong membrane immunoreactivity

The demonstrated role of aquaporins in transmembrane water transport and their marked upregulation in neoplastic glial membranes suggests a role in formation of tumor-associated edema. A common property of malignant glial tumors is marked edema in surrounding brain tissue (peritumoral edema), but the mechanism of this is uncertain [9]. It is possible that AQP1 upregulation allows excess water to flow out into surrounding brain interstitium, perhaps in combination with blood-brain barrier breakdown seen in glioma microvessels. Thus, abnormally high AQP1 expression in glial tumors may "re-set" the water balance in local areas of brain around the tumor, or prevent clearance of existing extracellular fluid. Tumor-associated edema carries significant clinical morbidity such as focal neurologic dysfunction, mass effect, and seizures. Inhibition of AQP expression or function may offer a new therapeutic option for tumor-associated cerebral edema. Future studies need to determine whether inhibition of AQP1 upregulation by genetic modification or direct inhibition of AQP1 protein using selective inhibitors will abrogate tumor-associated edema and improve clinical outcome.

References

1. Deen PM, Verdijk MA, Knoers NV, Wieringa B, Monnens LA, van Os CH, van Oost BA (1994) Requirement of human renal water channel aquaporin-2 for vasopressin-dependent concentration of urine. Science 264: 92–95
2. Denker BM, Smith BL, Kuhajda FP, Agre P (1988) Identification, purification, and partial characterization of a novel Mr 28,000 integral membrane protein from erythrocytes and renal tubules. J Biol Chem 263: 15634–15642
3. King LS, Agre P (1996) Pathophysiology of the aquaporin water channels. Annu Rev Physiol 58: 619–648
4. King LS, Choi M, Fernandez PC, Cartron JP, Agre P (2001) Defective urinary-concentrating ability due to a complete deficiency of aquaporin-1. N Engl J Med 345: 175–179
5. Manley GT, Fujimura M, Ma T, Noshita N, Filiz F, Bollen AW, Chan P, Verkman AS (2000) Aquaporin-4 deletion in mice reduces brain edema after acute water intoxication and ischemic stroke. Nat Med 6: 159–163
6. Markert JM, Fuller CM, Gillespie GY, Bubien JK, McLean LA, Hong RL, Lee K, Gullans SR, Mapstone TB, Benos DJ (2001) Differential gene expression profiling in human brain tumors. Physiol Genomics 5: 21–33

7. Nielsen S, Nagelhus EA, Amiry-Moghaddam M, Bourque C, Agre P, Ottersen OP (1997) Specialized membrane domains for water transport in glial cells: high-resolution immunogold cytochemistry of aquaporin-4 in rat brain. J Neurosci 17: 171–180
8. Nielsen S, Smith BL, Christensen EI, Agre P (1993) Distribution of the aquaporin CHIP in secretory and resorptive epithelia and capillary endothelia. Proc Natl Acad Sci USA 90: 7275–7279
9. Papadopoulos MC, Saadoun S, Davies DC, Bell BA (2001) Emerging molecular mechanisms of brain tumour oedema. Br J Neurosurg 15: 101–108
10. van Lieburg AF, Verdijk MA, Knoers VV, van Essen AJ, Proesmans W, Mallmann R, Monnens LA, van Oost BA, van Os CH, Deen PM (1994) Patients with autosomal nephrogenic diabetes insipidus homozygous for mutations in the aquaporin 2 water-channel gene. Am J Hum Genet 55: 648–652
11. Venero JL, Vizuete ML, Machado A, Cano J (2001) Aquaporins in the central nervous system. Prog Neurobiol 63: 321–336
12. Verkman AS, Mitra AK (2000) Structure and function of aquaporin water channels. Am J Physiol Renal Physiol 278: F13–28

Correspondence: Geoffrey T. Manley, M.D., Ph.D., Department of Neurosurgery, University of California, San Francisco, Room M-779, Box 0112, 505 Parnassus Ave., San Francisco CA 94143. e-mail: manley@itsa.ucsf.edu

Acta Neurochir (2003) [Suppl] 86: 503–506

A thrombin inhibitor reduces brain edema, glioma mass and neurological deficits in a rat glioma model*

Y. Hua[1], **R. F. Keep**[1,2], **T. Schallert**[1,3], **J. T. Hoff**[1], and **G. Xi**[1]

[1] Department of Neurosurgery, University of Michigan, Ann Arbor, Michigan
[2] Department of Physiology, University of Michigan, Ann Arbor, Michigan
[3] Department of Psychology and Institute for Neuroscience, University of Texas at Austin, U.S.A.

Summary

Although thrombin is a critical enzyme in the coagulation cascade, it has become apparent that it has many other effects. Thus, it may induce brain edema formation, angiogenesis and cell proliferation. Because of the importance of these three factors in the extremely poor prognosis of glioma patients, the present study examined the role of thrombin in that disease state. We found that thrombin activity is increased in a rat glioma model and thrombin positive cells were present in the tumor. Anti-thrombin treatment with argatroban reduced brain edema, tumor growth, and tumor-related neurological deficits. Our results suggest that thrombin is a new target for glioma treatment.

Keywords: Glioma; brain edema; thrombin; argatroban; neurological deficits; rats.

Introduction

Rapidly growing brain tumors, such as malignant gliomas, result in a high mortality within months. Brain edema contributes to the high mortality in glioma patients by causing herniation-related death. Our previous studies indicate that thrombin plays an important role in edema formation after intracerebral hemorrhage [6, 18]. Angiogenesis is essential for rapid glioma growth because of the need for oxygen and metabolites. Although many factors regulate angiogenesis, thrombin may play a key role [8]. In addition, thrombin is a potent mitogen, which enhances the synthesis and secretion of nerve growth factor in glial cells and stimulates astrocyte and tumor cell proliferation [1, 11]. Because of these multiple potential effects of thrombin in gliomas, the present study investigated whether a thrombin inhibitor, argatroban, can reduce edema formation, glioma mass and tumor-induced neurological deficits in a rat C6 glioma model.

* Supported by grants NS-17760 (JTH) and NS-39866 (GX) from the National Institutes of Health.

Materials and methods

Animal groups

Rat C6 glioma cells (6×10^5) were infused into the right caudate of adult male Fischer 344 rat. Two sets of experiments were performed. In the first set, rats were killed at day 12 for edema, glioma mass and thrombin activity measurements. Brain thrombin activity was measured using the thrombin-specific chromogenic substrate S2238. To further confirm the specificity of thrombin acitivity measurement, a specific thrombin inhibitor, PPACK was used to inhibit thrombin activity. In the second set, rats were implanted with an osmotic minipump to deliver a thrombin inhibitor (argatroban, 1 μg/μl, 0.5 μl/h) or saline into the right caudate at the time of C6 cell implantation. After 9 days, the rat brains were removed for determination of brain water content and glioma mass. Over the period of the experiment, the animals underwent behavioral testing [4, 15].

Glioma model

The rats were anesthetized with pentobarbital (40 mg/kg, i.p.). They then were placed in a Kopf stereotaxic frame. Body temperature was maintained at 37.5 °C using a rectal temperature probe and a feed-back regulator. A small burr hole was drilled in the skull overlying the right caudate nucleus. The tip of a 28 gauge stainless steel cannula was then lowered into the right caudate (0.2 mm anterior, 5.5 mm ventral, 3.5 mm lateral to the bregma) and 6×10^5 C6 glioma cells were infused.

Alzet osmotic mini-pump implantation

At the time of C6 glioma cell implantation (or saline injection), rats were implanted with an osmotic minipump to deliver a thrombin inhibitor (argatroban) or saline into the right basal ganglia. Utilizing the same burr hole as used for C6 cell infusion, the tip of a 28-gauge stainless steel cannula was lowered into the caudate. The cannula was fixed to the skull with dental acrylic and screws. An osmotic pump (Alzet) was connected to the cannula and implanted in the back of the neck via PE-60 tubing. The pumps were preloaded with or without drug and delivered 0.5 μl/h. The skin incisions were then closed and the animal was allowed to recover.

Behavioral tests

Animals were placed in a cylindrical enclosure to record preferential use of the non-impaired forelimb for weight shifting movements during spontaneous vertical exploration. The % independent use of the non-impaired forelimb (ipsilateral to the tumor), the % independent use of the contralateral forelimb, and the % use of both forelimbs together in rapid succession for stepping movements along the walls of the cylinder were assessed. A single score was then used to reflect forelimb use asymmetry: % ipsilateral limb use minus % contralateral limb use; low score = better function. In addition, a vibrissae-stimulated forelimb placing test (10 trials per side for each rat) was used to examine sensorimotor/proprioceptive capacity (high score = better function). These tests are highly correlated with extent of striatal injury without being influenced by repeated testing. All behavior was scored by experimenters who were blind to both neurological and treatment conditions.

Brain water content measurement

Rats were sacrificed by decapitation under deep pentobaribital anesthesia. The brains were removed immediately and cut into three parts: front part (frontal pole to 3 mm back), the middle part (3–7 mm from frontal pole) and the rear part (7 mm to occipital pole). The brain samples were then divided into cortex or basal ganglia (ipsilateral or contralateral). Tissue samples were weighed on an electronic analytical balance to obtain the wet weight (WW). The tissue was then dried in a gravity oven at 100 °C for more than 24 hours to determine the dry weight (DW). Tissue water contents (%) are calculated as $((WW - DW)/WW) * 100$.

Tumor mass

The weight difference between the ipsilateral (tumor side) and contralateral hemisphere was used to estimate the tumor mass.

Thrombin activity measurement

For thrombin activity measurements, rat brains were perfused transcardially with saline and sampled as described in the water content section. Brain samples were homogenized and thrombin activities were measured using the thrombin-specific chromogenic substrate S2238 (Chromogenix). The final concentration of S2238 was 0.3 mmol/L in PBS and the absorption at 405 nm of the supernatant was measured after one hour. A calibration curve of thrombin activity was made for the thrombin activity quantitation [16]. Measurements were made in the presence and absence of phenyl-propyl-arginyl-chloromethyl ketone (PPACK), a thrombin inhibitor, to ensure that the measurements were a reflection of thrombin activity.

Immunohistochemistry

Immunohistochemical studies were performed as in our previous report [17]. The primary antibody was sheep anti-human thrombin. Normal rabbit IgG was used as negative control.

Statistics

Data were analyzed by Student t test or Mann-Whitney U nonparametric test. Differences were considered significant at the $p < 0.05$ level.

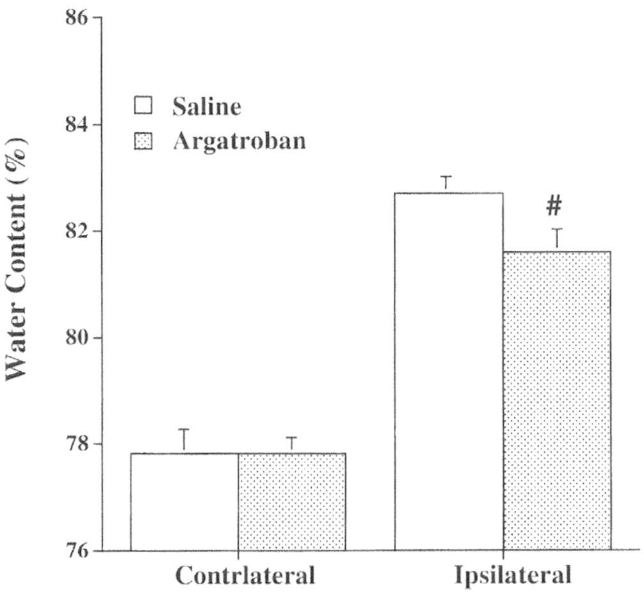

Fig. 1. Brain water contents in the caudate 9 days after intracerebral infusion of C6 glioma cells with or without argatroban treatment. Values are mean ± SD, n = 5, #$p < 0.01$ vs. vehicle, Student t test

Results

Intra-caudate implantation of C6 glioma cells caused brain edema and neurological deficits in the rat. Thrombin activity was increased in the glioma (340 ± 12 vs. 200 ± 40 mU/gram brain tissue in the contralateral side, $p < 0.05$). Immunohistochemistry also demonstrated the presence of thrombin-positive cells in and around the glioma.

Treatment with argatroban reduced brain edema (81.6 ± 0.5% vs. 82.7 ± 0.3% in control, n = 5, $p < 0.05$, Fig. 1), glioma mass (44 ± 21 mg vs. 106 ± 57 mg in the control, n = 4, $p < 0.05$) and neurological deficits (forelimb placing score: 85 ± 7% vs. 9 ± 4% in control, n = 10, $p < 0.05$, Fig. 2; Forelimb use asymmetry: −10 ± 3% vs. 25 ± 12% in control, n = 10, $p < 0.05$, Fig. 3) in rats at day 9.

Discussion

Our present findings demonstrate that thrombin (by activity assay and immunohistochemistry) is increased in the glioma. Anti-thrombin treatment reduces peri-gliomal edema formation, glioma mass and glioma-induced neurological deficits indicating that thrombin plays an important role in glioma growth.

As early as 1865, Trousseau found a connection between malignant cell growth and the coagulation. Much experimental data suggest a role for the coagu-

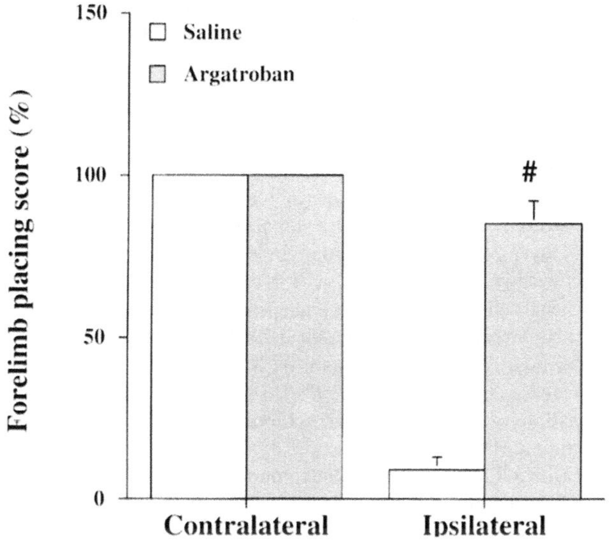

Fig. 2. Forelimb placing score in the rats after intracerebral infusion of C6 glioma cells. The rats were treated with argatroban or vehicle. Values are mean ± SD, n = 9–10, #p < 0.05 vs. vehicle, Mann-Whitney U rank test

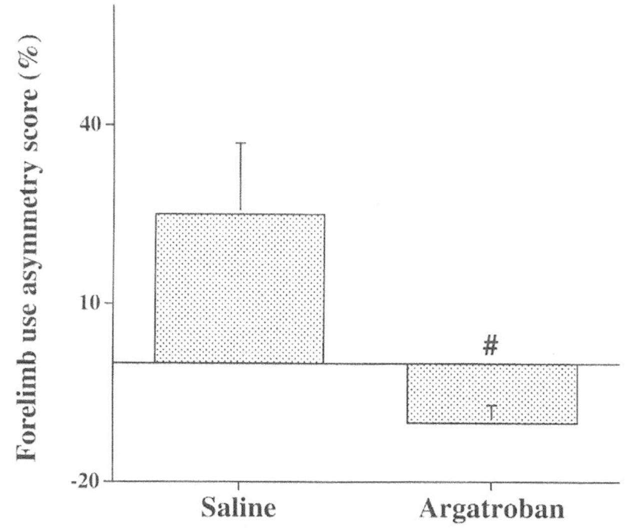

Fig. 3. Forelimb use asymmetry in the rats after intracerebral infusion of C6 glioma cells. The rats were treated with argatroban or vehicle. Values are mean ± SD, n = 9–10, #p < 0.05 vs. vehicle, Mann-Whitney U rank test

lation and fibrinolysis systems in tumor development, progression and metastasis. Also, some clinical data suggest that targeting the coagulation system might influence the course of malignant disease. The anti-coagulant drug warfarin prolongs the length of survival of patients with small cell lung cancer [19]. Heparin has a benefit in patients with several types of cancer [20]. Our current results suggest that the coagulation cascade (thrombin) may also be a therapeutic target in brain gliomas.

We chose this injection model for this glioma study because of several advantages. First, C6 glioma cells were injected into the caudate. This takes advantage of the fact that a number of behavioral tests have been devised to assess caudate damage induced by ischemia, hemorrhage or neurotoxins. Second, this model is stable and reproducible in terms of edema, tumor mass and neurological deficit measurements. Third, the properties of C6 glioma cell line can be retained for several years [5]. C6 gliomas have a disrupted blood-brain barrier suggesting that increased thrombin activity within the tumor might be due to an influx of prothrombin from blood. However, production of prothrombin by the tumor or adjacent cells can not yet be excluded.

Thrombin plays a major role in edema formation in the brain after intracerebral hemorrhage [6, 18]. Our current results suggest it may have a similar edemagenic role in tumors, although it is difficult to dissociate effects on tumor mass and edema. Thrombin is produced immediately in the brain after intracerebral hemorrhage or blood-brain barrier breakdown following many kinds of brain injury [3]. Direct infusion of large doses of thrombin into brain causes inflammatory cell infiltration, mesenchymal cell proliferation and brain edema formation [6, 13, 18]. Thrombin-induced brain edema results from a direct opening of the blood-brain barrier or a direct neurotoxicity [7].

The effect of thrombin on tumor mass may result from a direct effect on tumor cell proliferation. Thrombin enhances the synthesis and secretion of nerve growth factor in glial cells [11] and stimulates astrocyte proliferation [1]. In addition, thrombin may act as a growth factor for tumor cells [2] and induces proliferative response in T-47D mammary tumor cells [10]. In T98G and TM-1 human glioma cells, thrombin also induced proliferation. This mitogenic effect was abolished by hirudin, a specific thrombin inhibitor [14].

Thrombin is also a potent promoter of angiogenesis. Thrombin activates an angiogenic cascade through, at least in part, modulating hypoxia inducible factor-1, matrix metalloproteinases, vascular endothelial growth factor (VEGF) and the receptors of VEGF [8, 9, 12, 21, 22].

Traditional preclinical investigations of brain tumor therapies have focused primarily on inhibiting tumor cell proliferation or survival with little regard to behavioral assessment. However, in clinical trials, brain function has been an essential index of treatment benefit. Moreover, the literature on brain injury and functional recovery indicates that mechanisms of neural plasticity needed for optimal behavioral outcome in tumor patients might be vulnerable to some tumor treatments, including irradiation, pro-apoptotic drugs, or agents that adversely affect angiogenesis, mitosis or neurotrophic factor activity. A model that includes sensitive functional outcome measures would appear to represent a significant advantage in brain tumor research.

In summary, thrombin is present in the glioma and administration of a thrombin inhibitor (argatroban) results in less brain edema, glioma mass and neurological deficits. These results suggest that thrombin plays an important role in glioma growth. Modulating thrombin activity, either using thrombin inhibitors or upregulating endogenous thrombin inhibitors in the brain, may produce novel strategies for glioma treatment.

References

1. Cavanaugh KP, Gurwitz D, Cunningham DD, Bradshaw RA (1990) Reciprocal modulation of astrocyte stellation by thrombin and protease nexin-1. J Neurochem 54: 1735–1743
2. Chinni C, de Niese MR, Tew DJ, Jenkins AL, Bottomley SP, Mackie EJ (1999) Thrombin, a survival factor for cultured myoblasts. J Biol Chem 274: 9169–9174
3. Gingrich MB, Traynelis SF (2000) Serine proteases and brain damage – is there a link?. Trend Neurosci 23: 399–407
4. Hua Y, Schallert T, Keep RF, Wu J, Hoff JT, Xi G (2002) Behavioral tests after intracerebral hemorrhage in the rat. Stroke 33: 2478–2484
5. Lampson LA (2001) New animal models to probe brain tumor biology, therapy, and immunotherapy: advantages and remaining concerns. J Neuro Oncol 53: 275–287
6. Lee KR, Colon GP, Betz AL, Keep RF, Kim S, Hoff JT (1996) Edema from intracerebral hemorrhage: the role of thrombin. J Neurosurg 84: 91–96
7. Lee KR, Kawai N, Kim S, Sagher O, Hoff JT (1997) Mechanisms of edema formation after intracerebral hemorrhage: effects of thrombin on cerebral blood flow, blood-brain barrier permeability, and cell survival in a rat model. J Neurosurg 86: 272–278
8. Maragoudakis ME, Tsopanoglou NE (2000) On the mechanism(s) of thrombin induced angiogenesis. Advances Exper Med Biol 476: 47–55
9. Maragoudakis ME, Tsopanoglou NE, Andriopoulou P, Maragoudakis MM (2000) Effects of thrombin/thrombosis in angiogenesis and tumour progression. Matrix Biol 19: 345–351
10. Medrano EE, Cafferata EG, Larcher F (1987) Role of thrombin in the proliferative response of T-47D mammary tumor cells. Mitogenic action and pleiotropic modifications induced together with epidermal growth factor and insulin. Exper Cell Res 172: 354–364
11. Neveu I, Jehan F, Jandrot-Perrus M, Wion D, Brachet P (1993) Enhancement of the synthesis and secretion of nerve growth factor in primary cultures of glial cells by proteases: a possible involvement of thrombin. J Neurochem 60: 858–867
12. Nguyen M, Arkell J, Jackson CJ (1999) Thrombin rapidly and efficiently activates gelatinase A in human microvascular endothelial cells via a mechanism independent of active MT1 matrix metalloproteinase. Lab Investigat 79: 467–475
13. Nishino A, Suzuki M, Ohtani H, Motohashi O, Umezawa K, Nagura H, Yoshimoto T (1993) Thrombin may contribute to the pathophysiology of central nervous system injury. J Neurotrauma 10: 167–179
14. Ogiichi T, Hirashima Y, Nakamura S, Endo S, Kurimoto M, Takaku A (2000) Tissue factor and cancer procoagulant expressed by glioma cells participate in their thrombin-mediated proliferation. J Neuro Oncol 46: 1–9
15. Schallert T, Fleming SM, Leasure JL, Tillerson JL, Bland ST (2000) CNS plasticity and assessment of forelimb sensorimotor outcome in unilateral rat models of stroke, cortical ablation, parkinsonism and spinal cord injury. Neuropharmacology 39: 777–787
16. Stoop AA, Lupu F, Pannekoek H (2000) Colocalization of thrombin, PAI-1, and vitronectin in the atherosclerotic vessel wall: a potential regulatory mechanism of thrombin activity by PAI-1/vitronectin complexes. Arteriosclerosis, Thrombosis Vascular Biol 20: 1143–1149
17. Xi G, Keep RF, Hua Y, Xiang JM, Hoff JT (1999) Attenuation of thrombin-induced brain edema by cerebral thrombin preconditioning. Stroke 30: 1247–1255
18. Xi G, Wagner KR, Keep RF, Hua Y, de Courten-Myers GM, Broderick JP, Brott TG, Hoff JT (1998) The role of blood clot formation on early edema development following experimental intracerebral hemorrhage. Stroke 29: 2580–2586
19. Zacharski LR, Henderson WG, Rickles FR, Forman WB, Cornell CJ Jr, Forcier RJ, Edwards RL, Headley E, Kim SH, O'Donnell JF (1984) Effect of warfarin anticoagulation on survival in carcinoma of the lung, colon, head and neck, and prostate. Final report of VA Cooperative Study #75. Cancer 53: 2046–2052
20. Zacharski LR, Ornstein DL (1998) Heparin and cancer. Thrombosis Haemostasis 80: 10–23
21. Zagzag D, Friedlander DR, Margolis B, Grumet M, Semenza GL, Zhong H, Simons JW, Holash J, Wiegand SJ, Yancopoulos GD (2000) Molecular events implicated in brain tumor angiogenesis and invasion. Pediatric Neurosurg 33: 49–55
22. Zagzag D, Zhong H, Scalzitti JM, Laughner E, Simons JW, Semenza GL (2000) Expression of hypoxia-inducible factor 1alpha in brain tumors: association with angiogenesis, invasion, and progression. Cancer 88: 2606–2618

Correspondence: Ya Hua, M.D., R5550 Kresge I, University of Michigan, Ann Arbor, Michigan 48109-0532. e-mail: yahua@umich.edu

Acta Neurochir (2003) [Suppl] 86: 507–511

Tissue reconstruction process in the area of peri-tumoural oedema caused by glioblastoma – immunohistochemical and graphical analysis using brain obtained at autopsy

G. Nagashima, R. Suzuki, J.-I. Asai, M. Noda, M. Fujimoto, and **T. Fujimoto**

Department of Neurosurgery, Showa University, Fujigaoka Hospital, Fujigaoka, Aoba-ku, Yokohama-shi, Kanagawa-ken, Japan

Summary

Background. In the area of peri-tumoural oedema, proteolytic agents derived from the tumour cause tissue degradation, which promotes tumour cell invasion.

Method. We investigated the biological processes in the area of peri-tumoural oedema, using a brain obtained at autopsy from a patient who died from glioblastoma. Immunohistochemistry was performed to detect vascular endothelial growth factor (VEGF), c-myc, p53, paternally expressed gene-3 (PEG-3), transforming growth factor beta (TGFB), and tumour necrosis factor alpha (TNFA). The data were translated into colour graphics and the localization of these proteins was analyzed.

Findings. In the area of peri-tumoural oedema, Ki-67 and p53 positive cells were observed with TGFB expression. Moreover, c-myc, PEG-3, VEGF, and TNFA were also expressed strongly in the glial cells or extra-cellular spaces in the area of peri-tumoural oedema.

Interpretation. These data suggest that in the area of peri-tumoural oedema, tissue reconstruction processes take place with concomitant anti-tumour activities. The expression of c-myc, VEGF, and TNFA in the area of peri-tumoural oedema may indicate that these proteins are not utilized for tumour growth, but may be used to guard the brain against tumour invasion. Peri-tumoural oedema does not only indicate the tissue damage caused by tumour, but many tissue reconstruction processes take place in these areas against tumour cell invasion.

Keywords: Glioblastoma; invading front; TNFA; TGFB; PEG3.

Introduction

Neoplastic transformation of glial cells, tumour cell proliferation, neovascularization, degradation of surrounding matrices, and tumour cell invasion are major growth processes of glioblastomas. For the treatment of glioblastoma, selective therapeutic strategies against each of these processes are required. At present, the treatments for tumour cell proliferation or neovascularization, such as radiation therapy and chemotherapies, have limited effect against glioblastoma. Gene therapies using p53 or interferons, and treatment for neovascularization such as endostatin have also been shown to have a limited effect against glioblastoma. In these settings, the treatment of the tumour affects not only the tumour itself, but also damages the normal peri-tumoural tissues, facilitating tumour cell invasion [5, 6]. For these reasons, knowledge regarding the responses against the tumour in the area adjacent to the tumour has become a focus.

We have already reported that autoclave pretreatment as an immunohistochemical antigen-retrieval method, is effective in brains obtained at autopsy [6]. It was reported that tumour cells invaded far beyond the high-intensity lesion seen on T2-weighted magnetic resonance imaging (MRI T2WI), and that various proteins were expressed outside the tumour, at the invading front of the tumour, and in the area of peri-tumoural oedema. These proteins included vascular endothelial growth factor (VEGF), matrix metalloproteinase 2 (MMP2), membrane type matrix metalloproteinase (MT-MMP), interleukin 1 beta (IL1B), and IL6, which were expressed outside of the tumour, and were localized mainly in glial cells. It was postulated that these proteins expressed in the area of peri-tumoural oedema were induced against tumour cell invasion as an immune response against tumour [5]. These results suggest that a treatment against vascular proliferation in the tumour, such as endostatin, might also inhibit the tissue reconstruction processes in the area of peri-tumoural oedema around the tumour, and facilitate tumour cell invasion. In this study, we examined the immunohistochemistry of several other proteins, namely c-myc, paternally expressed gene 3 (PEG3), transforming growth factor beta (TGFB),

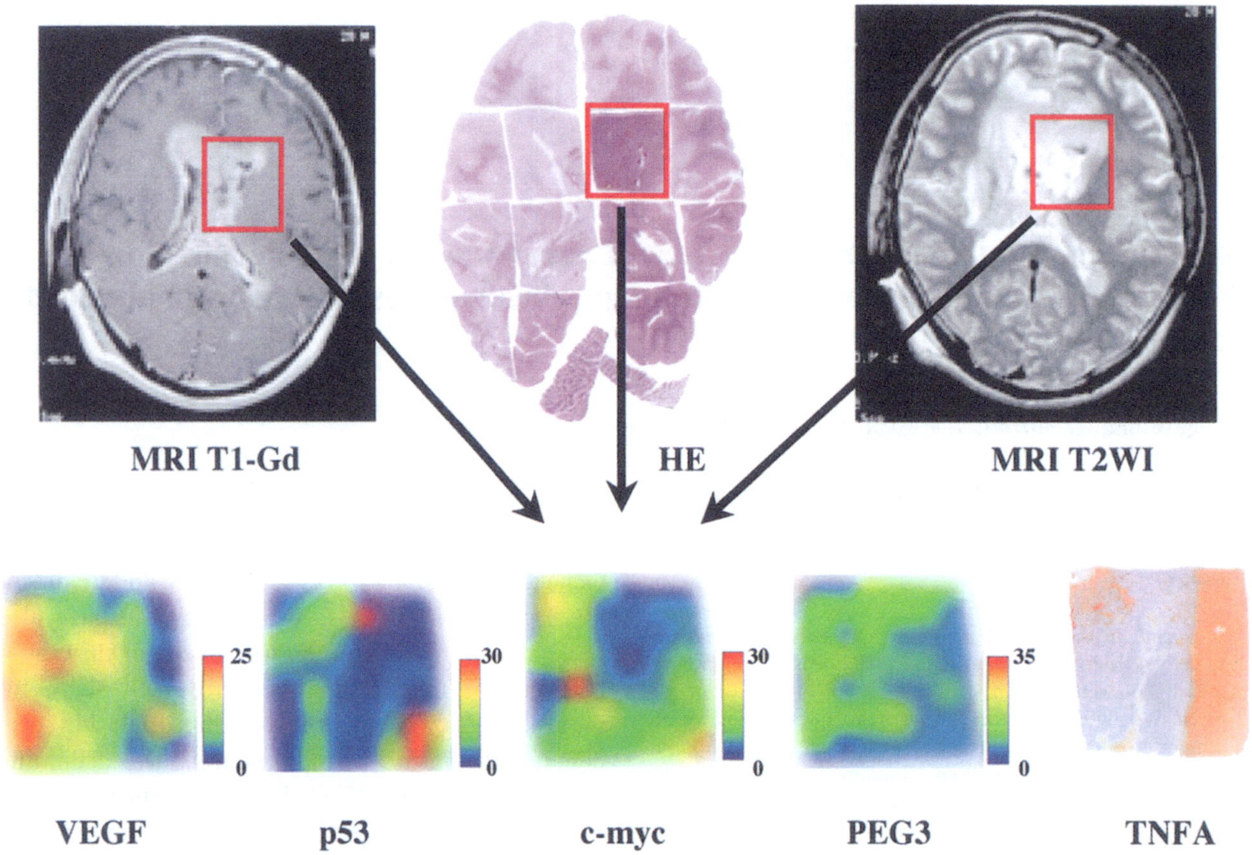

Fig. 1. The brain obtained at autopsy was cut parallel to the orbito-meatal line, and the 5 cm^2 area around the left lateral ventricle was selected where tumour was seen on gadolinium enhanced MRI T1WI. Invading front with peri-tumoural oedema seen on MRI T2WI, was included. The results of immunohistochemistry were translated into colour graphic maps. VEGF, c-myc, and p53 were stained outside the tumour at the invading front. The immunostaining of PEG3 was weak but localization was more prominent outside the tumour. TNFA was only expressed outside of the tumour where peri-tumoural oedema was recognized

tumour necrosis factor alpha (TNFA), and investigated the role of these proteins expressed in the invading front of the tumour with peri-tumoural oedema.

Materials and methods

The patient was a twenty-seven year-old male, referred to our institutes for his severe headache. Under the diagnosis of glioblastoma, radiation therapy of 60 Gy was given, followed by administration of nimustine hydrochloride and vincristine. The patient died of glioblastoma dissemination one year later. As demonstrated in Fig. 1, the brain, obtained at autopsy, was cut parallel to the orbito-meatal line, and an area was selected including the tumour and the peritumoural edema as the invading front of the tumour. Immunohistochemistry was performed using autoclave pretreatment as described previously [6]. MIB-1 monoclonal antibody (MIB-1 mAb, Immunotech S.A., Marseilles, France, ×100), p53 mAb (CM-1, Novocastra, Newcastle, UK, ×100), VEGF polyclonal antibody (VEGF pAb, Oncogene Science, Cambridge, MA, ×100), c-myc antibody (9E11, YLEM, Rome, Italy, ×100), PEG3 pAb (×100), TGFB mAb (Oncogene, ×100), TNFA mAb (Genzyme-Techne, ×100) were used as primary antibodies, and immunohistochemistry was performed using an ABC kit according to the manufacturers'-protocols (Dako, Tokyo, Japan). Colour graphic maps of the selected areas were prepared using the immunohistochemically positive cell number under randomly selected high power field (×400) for VEGF, c-myc, and PEG3. The rate of immunohistochemically positive cells among more than 500 cells was used for MIB-1 and p53 colour maps as described previously [6].

Results

Ki-67 (detected by MIB-1 antibody), p53, and TGFB were strongly expressed, not only in the tumour itself, but also in glial cells at the invading front where peri-tumoural oedema was recognized (Fig. 2). The immunostaining was entirely nuclear. The highest staining index of Ki-67 was 10% in the tumour itself, and 25% in the peri-tumoural oedema. The highest p53 staining index was 14% in the tumour, and 30% in the peri-tumoural oedema.

c-myc, PEG3, VEGF, and TNFA were also expressed outside the tumour. c-myc and PEG3 were also expressed in the nuclei of glial cells at the invading

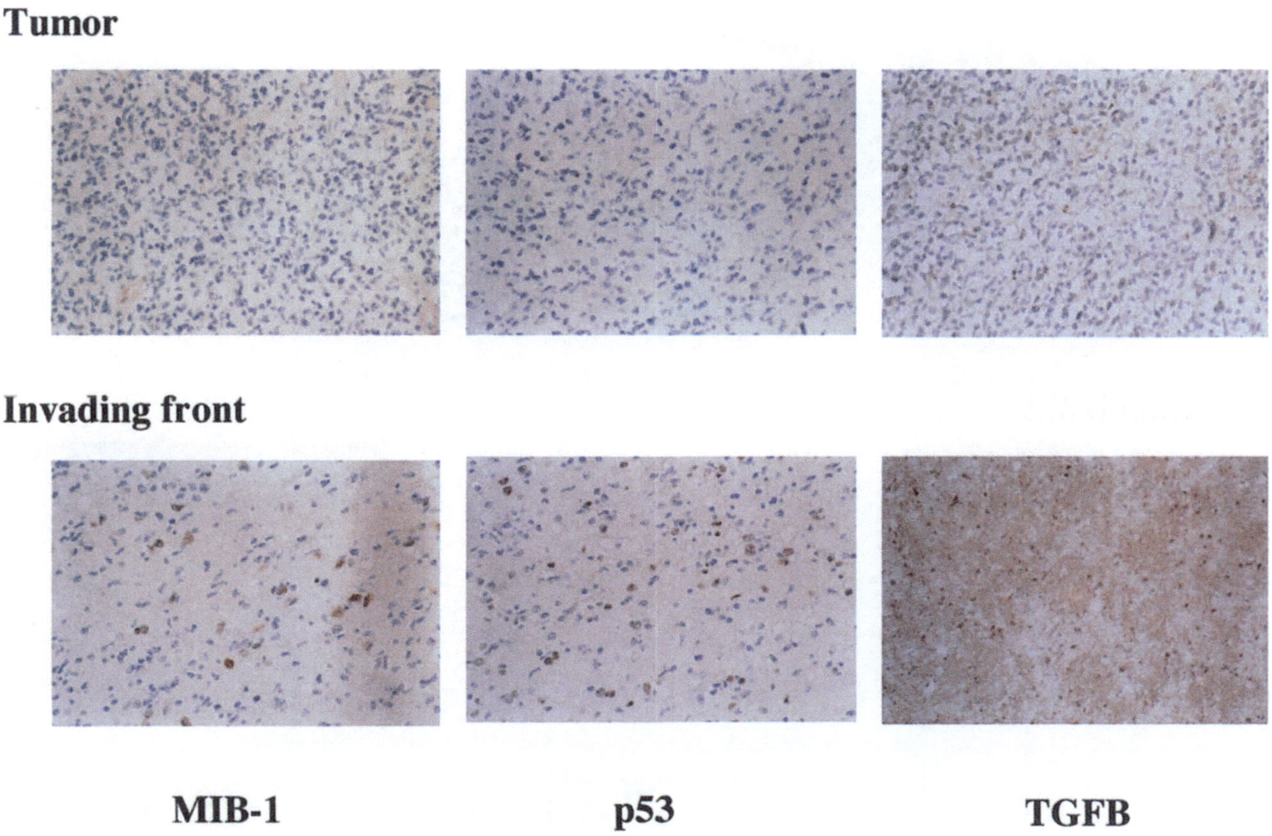

Fig. 2. Ki-67, p53, and TGFB were strongly expressed, not only in the tumour itself but also in the glial cells at the invading front, where peri-tumoural oedema was recognized. The staining was entirely nuclear (×200)

front, and TNFA was expressed mainly in the extracellular spaces (Fig. 3).

We translated these results into colour graphics in the 5 cm^2 area around the left lateral ventricle, where the tumour shown on gadolinium enhanced MRI T1WI and the invading front with peri-tumoural oedema on MRI T2WI were included (Fig. 1). The result of TNFA immunostaining was shown as it was on the slide glass, since its immunostaining was extracellular. As shown by the colour graphics, VEGF, c-myc, and p53 were all strongly stained in the tumour, and were also stained outside the tumour at the invading front. The immunostaining of PEG3 was weak, but the localization was more prominent outside of the tumour. TNFA was only expressed outside the tumour where peri-tumoural oedema was recognized.

Discussion

Ki-67 protein appears on the nuclear surface of proliferating cells, and its staining index is considered to be a reliable indicator of tumour cell proliferative potentials [8]. The nuclear protein, p53 is believed to play an important role in the progression of human malignancies, and a functional loss of p53 protein by genetic mutation or protein stabilization, may result in unrestricted tumour growth [4]. TGF is considered to be a tissue growth factor and is said to contribute to tumour cell growth [7]. In this setting, expression of MIB-1, p53, and TGFB in the area of the invading front of the tumour, indicates the active proliferation of invading tumour cells in these areas. These results demonstrate that tumour cells definitely proliferate in the area of peri-tumoural edema, that the normal mechanism of programmed cell death is disrupted, and that tumour cell growth was facilitated by several growth factors like TGFB.

In contrast, VEGF was expressed for the reconstructive process of angiogenesis [6], c-myc and PEG3 were expressed for tumour cell elimination through apoptosis [1, 3], and TNFA was expressed as inflammatory response against tumour [2].

Using similar methods, we previously demonstrated that proteins such as IL1B, IL6, MMP2, and MT-

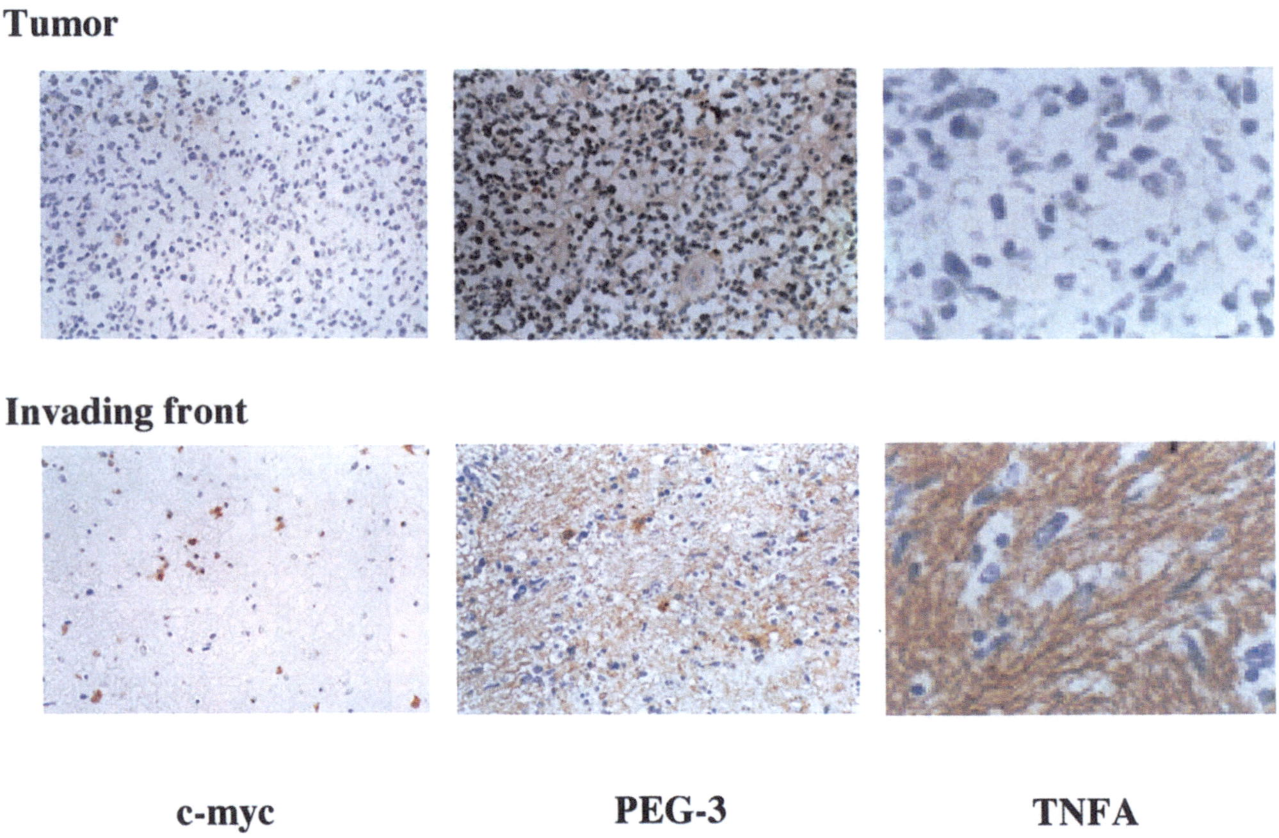

Fig. 3. c-myc, PEG3, VEGF, and TNFA were expressed outside of the tumour. c-myc and PEG3 were expressed in the nucleus of glial cells at the invading front, and TNFA was expressed mainly in the extracellular spaces (×200)

MMP, were expressed in the area of the invading front [5]. In this study, we further demonstrated that VEGF, c-myc, PEG3, and TNFA were also expressed in these areas mainly in reactive astrocytes. Taken together, VEGF, c-myc, PEG3, and TNFA were expressed concomitantly with IL1B, IL6, MMP-2, and MT-MMP, and these proteins expressed outside the tumour may inhibit the tumour cell invasion and help reconstruct the damaged tissues around the tumour. VEGF may work to repair the damaged tissues caused by oedema from tumour, c-myc, and PEG3 may facilitate tumour cell apoptosis as tumour suppressors, and TNFA may be expressed as an anti-tumour inflammatory cytokine against invaded tumour cells.

In conclusion, in the area of the invading front of the tumour, several factors that facilitate tumour cell growth, and other factors that inhibit tumour cell invasion and reconstruct damaged tissues, are simultaneously expressed. Further investigation is necessary to determine what kind of reactions take place at the invading front of the tumour, and to develop a new anti-tumour treatment that assists normal anti-tumour reactions.

References

1. Alitalo K, Koskinen P, Makela TP (1987) myc oncogenes: activation and amplification. Biochem Biophys Acta 907: 1–32
2. Duan L, Aoyagi M, Tamaki M (2001) Sensitization of human malignant glioma cell lines to tumor necrosis factor-induced apoptosis by cisplatin. J Neuro-oncol 52: 23–36
3. Khoda T, Asai A, Kuroiwa Y (2001) Tumor Suppressor activity of human imprinted gene PEG3 in a glioma cell line. Genes to Cells 6: 237–247
4. Louis DN (1994) The p53 gene and protein in human brain tumors. J Neuropathol Exp Neurol 53: 11–21
5. Nagashima G, Suzuki R, Asai J (2002) Immunohistochemical analysis of reactive astrocytes around glioblastoma: an immunohistochemical study of postmortem glioblastoma cases. Clin Neurol Neurosurg 104: 125–131
6. Nagashima G, Suzuki R, Hokaku H (1999) Graphic analysis of microscopic tumor cell infiltration, proliferative potential, and vascular endothelial growth factor expression in an autopsy brain with glioblastoma. Surg Neurol 51: 292–299
7. Platten M, Wick W, Weller M (2001) Malignant glioma biology: role of TGF-beta in growth, motility, angiogenesis, and immune escape. Microsc Res Tech 52: 401–410

8. Wakimoto H, Aoyagi M, Nakayama T (1996) Prognostic significance of Ki-67 labeling indices obtained using MIB-1 monoclonal antibodies in patients with supratentorial astrocytomas. Cancer 77: 373–380

Correspondence: Goro Nagashima, M.D., Department of Neurosurgery, Showa University, Fujigaoka Hospital, 1-30, Fujigaoka, Aoba-ku, Yokohama-shi, Kanagawa-ken, 227-8501, Japan. e-mail: goro-n@po1.dti2.ne.jp

Acta Neurochir (2003) [Suppl] 86: 513–517

Radiation necrosis and brain edema association with CyberKnife treatment

K. Sato[1], **Y. Baba**[1], **M. Inoue**[2], and **R. Omori**[2]

[1] Department of Neurosurgery, Okayama Kyokuto Hospital, Okayama, Japan
[2] CyberKnife Center, Okayama Kyokuto Hospital, Okayama, Japan

Summary

The CyberKnife (CK) is a frameless and image guided robotic controlled instrument for stereotactic irradiation. The authors studied CK treatment of glioma and glioblastoma, and analyzed frequency and risk factors of radiation necrosis. Of 61 patients with glioma and glioblastoma treated with CyberKnife, four patients showed symptomatic radiation necrosis. All of these patients were treated with stereotactic radiotherapy, varying from 3 to 6 fractions without previous radiation therapy. Two patients required necrotomy through craniotomy. Two patients were treated conservatively. Our four patients with radiation necrosis were not specific in terms of tumor volume and dose delivery. Glioma cells invade the normal brain tissue and over-radiation to this intermingling area is one of the risk factors for injury to normal endothelial cells. The homogeneity of the maximum dose area is an important factor to reduce over radiation to the normal brain parenchyma. The dose volume effect has been discussed in terms of risk factor; however, the number of fractions and dose per fraction should be considered to avoid radiation necrosis. We consider that conformal treatment with inverse algorism, fractionated stereotactic radiotherapy and precise anatomic targeting reduce the risk of radiation necrosis.

Keywords: Radiation; stereotactic irradiation; CyberKnife; Glioma; Glioblastoma; radiation necrosis; MR spectroscopy.

Introduction

Stereotactic irradiation (STI) has been shown to be an effective non-invasive treatment for a variety of intracranial conditions including benign and malignant tumors and AVMs. The primary risk associated with STI is delayed radiation necrosis, which typically occurs 3 months or more after treatment. The extreme goal of STI is high and accurate dose delivery to the lesion and to avoid radiation necrosis, which leads to satisfactory clinical results.

The CyberKnife (CK) (Accuray, Sunnyvale, CA. USA) is a frameless and image guided robotic controlled instrument for STI. The frameless STI enables painless STI and free choice of stereotactic radiosurgery (SRS) or stereotactic radiotherapy (SRT). The SRT is known to be radiobiologically advantageous for patients with malignant tumors. We started CK treatment at Okayama Kyokuto Hospital on June/1/2000. Seven hundred and seventy patients with intracranial and cervical lesions were treated with our CK over these twenty-nine months. In this group 61 cases of glioma and glioblastoma were included. The clinical results were satisfactory in contrast to historical data; however, four patients showed radiation necrosis. Two patients required necrotomy through craniotomy. After treatment, all of the four patients were alive and their Karnofsky performance status (KPS) was 80 and more. Based on this experience, we present our current management strategy for glioma and glioblastoma, and analyse the frequency and risk factors of radiation necrosis.

Materials and methods

The CyberKnife consists of small compact LINAC mounted with robotic manipulator with 6D freedom and Target Locating System (TLS). The robotic manipulator enables accurate conformal treatment with non isocenter dose deliver, or without multi-shot method which is applied by mechanical center STI modality, including gamma knife. TLS is checking real time head and cervical movement and enables frameless STI. We use UniFrame (MED-TEC, IA USA), thermoplastic soft shell, to immobilize the patient's head. The positioning repeatability and accuracy are reliable and the targeting error of our CK is less than 0.5 mm.

Our basic strategy for glioma and glioblastoma is that I) tumor volume (TV) is the high intensity area on T2WI or FLAIR of MRI, II) if the diameter of TV is more than 30 mm, fractionated SRT is applied, III) marginal dose is defined as the dose at 90% volume of dose-volume histogram, usually 80% isodoseline is applied, IV) conformal treatment and inverse algorism are applied and V) Linear-Quadrant formula [3, 8] is applied to calculate SRT dose to SRS dose, $\alpha/\beta = 10$ Gy applied.

Seven hundred and seventy patients were treated with our CK during these 29 months. Among them were 61 patients with glioma

Table 1. *Characteristics or forty six patients with glioblastoma and glioma*

Pathology	No. of cases (male/female)	Age (yo) (median)	KPS[1] (median)	Previous radiation (−/+)	Target volume (ml) (median)
Glioblastoma	24 (16/8)	11–81 (51.5)	40–100 (70)	8/16	1.5–182 (16.8)
Glioma	22 (8/14)	6–78 (56.5)	50–100 (90)	15/7	0.6–165 (9.6)

[1] Karnofsky performance status.

Table 2. *Summary of dose prescription*

Pathology	Previous radiation no. of cases	Marginal dose (Gy) (median)/no. of fractions	SRS (Gy) (median)[1]
Glioblastoma	(−) 8	30.0–42.0 (36.5)/3–8	19.0–23.7 (20.6)
Glioblastoma	(+) 16	8.6–39.1 (24.1)/1–6	7.1–24.7 (20.1)
Glioma	(−) 15	19.0–41.3 (24.4)/1–8	15.2–24.1 (20.0)
Glioma	(+) 7	18.4–31.4 (25.9)/1–5	14.5–21.7 (18.2)

[1] SRS was calculated with L-Q formula ($\alpha/\beta = 10$ Gy).

and glioblastoma. Forty six patients were pathologically confirmed and were analyzed in this study. Characteristics of these 46 patients are summarized in Table 1. Twenty three patients underwent previous radiation therapy, including conventional radiotherapy, gamma knife, and brachytherapy, showed local recurrent tumors.

The CK delivered 30.6 Gy of median marginal dose (range, 8.6–42.0 Gy) with SRS and 2–8 fractions of SRT, which is summarized in Table 2.

Results

Median survival period was 11 months in all glioblastoma patients. There was no statistical difference between CK alone group and prior radiation group in glioblastoma patients. Four patients presented symptomatic radiation necrosis at 4–18 months following CK treatment. All of these patients had no previous radiation therapy. Their clinical and dose prescription summary is shown in Tables 3 and 4. Two patients required necrotomy through craniotomy. All patients were treated with glycerin solution and steroid and could resume their previous work with a KPS score of 80 or more. We analyzed their dose volume relation in contrast to the non-necrotic group, which showed no significant difference (Fig. 1).

Illustrative case: case 1 (Figs. 2, 3)

A twelve year old girl presented with headache and dizziness in November 2000. MRI revealed cerebellar mass and she underwent craniectomy at the local hospital. Pathological examination revealed Gr. II of astrocytoma. She was referred to CK treatment for the residual cerebellar tumor. We prescribed 49.3/30.4 Gy (Maximum/marginal dose)/3 fractions of SRT for 10.3 ml of residual tumor. No early adverse event

Table 3. *Clinical summary of four patients with radiation necrosis*

Case no.	Age/sex	Pathology	Location	Symptom on radiation necrosis	Period to radiation necrosis	Treatment
1	12/F	astrocytoma Gr. II	Cerebellum	dizziness	6 mo	necrotmy
2	69/F	glioblastoma	Rt frontal lobe	hemiparesis	4 mo	necrotmy
3	36/F	astrocytoma Gr III	Lt medial frontal lobe	headache	18 mo	conservative treatment
4	50/F	astrocytoma Gr II	Rt medial temporal lobe	homonymous hemianopsia	7 mo	conservative treatment

Table 4. *Dose and clinical results of four patients with radiation necrosis*

Case no.	Target volume (ml)	Max/marginal dose Gy/no. of fractions	SRS Max/marginal dose (Gy)	Total follow up period (mo)	KPS at present
1	10.3	49.3/30.4/3 frs	31.4/20.2	19	90
2	26.9	56.9/42.0/6 frs	28.7/22.2	17	80
3	10.4	40.7/31.6/3 frs	26.4/21.0	24	100
4	5.3	38.5/29.0/3 frs	25.1/19.4	15	90

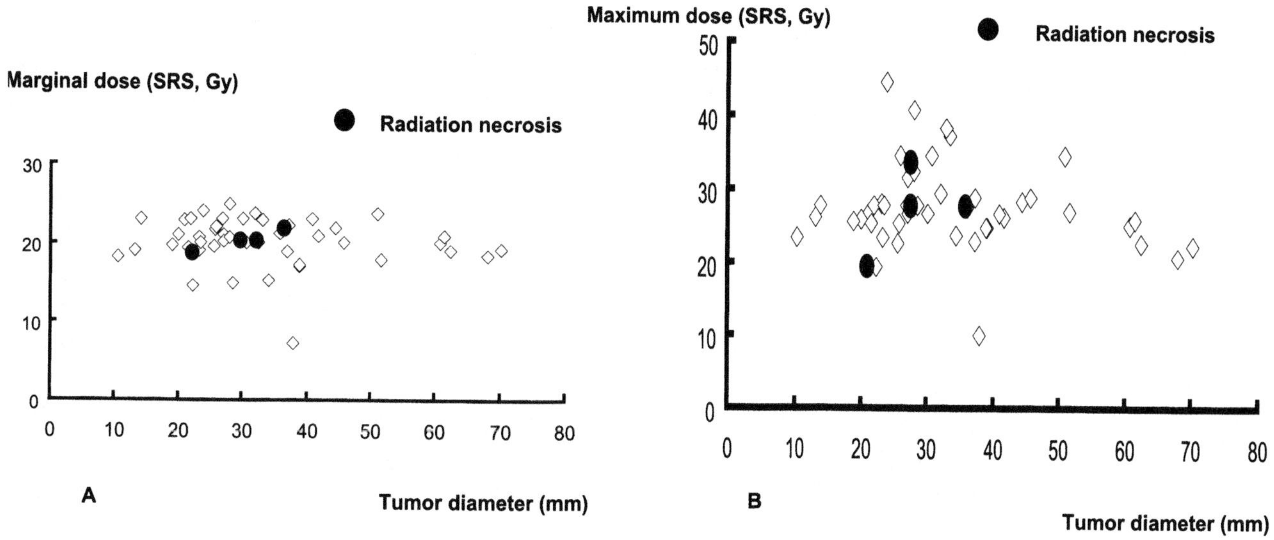

Fig. 1. Dose-volume relationship (A) Marginal dose vs tumor diameter, (B) Maximum dose vs tumor diameter. No statistical difference between radiation necrosis group and non necrosis group

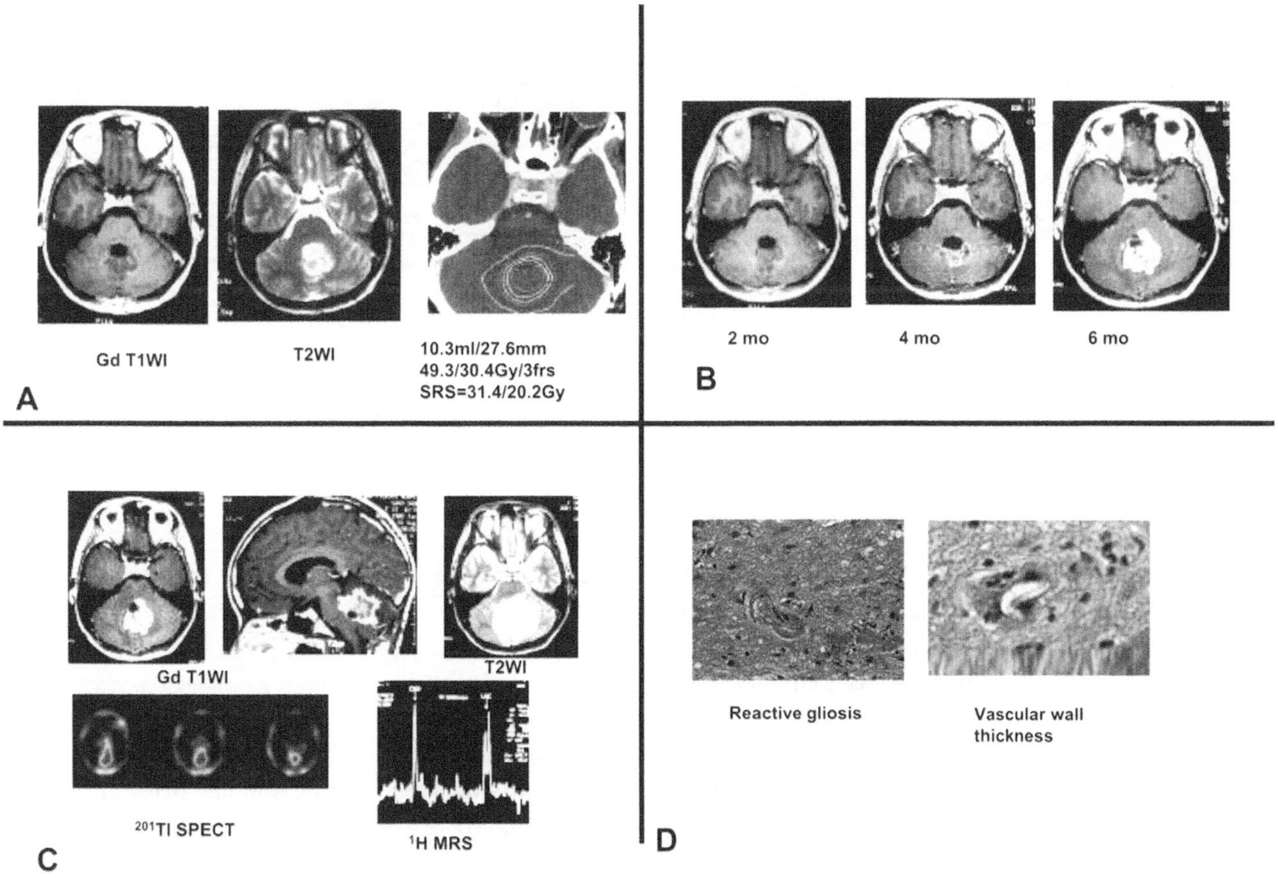

Fig. 2. Case 1. 12 yo girl with cerebellar Gr II astrocytoma. (A) Images at CK treatment, (B) Follow up MR Images (C) Image analysis at 6 months after CK treatment. Contrast enhancing mass increased in volume. Thallium SPECT revealed hot area in the target area. MRS showed choline and lactate peak. (D) Pathological examination revealed radiation necrosis. Reactive gliosis and vascular wall thickness were seen

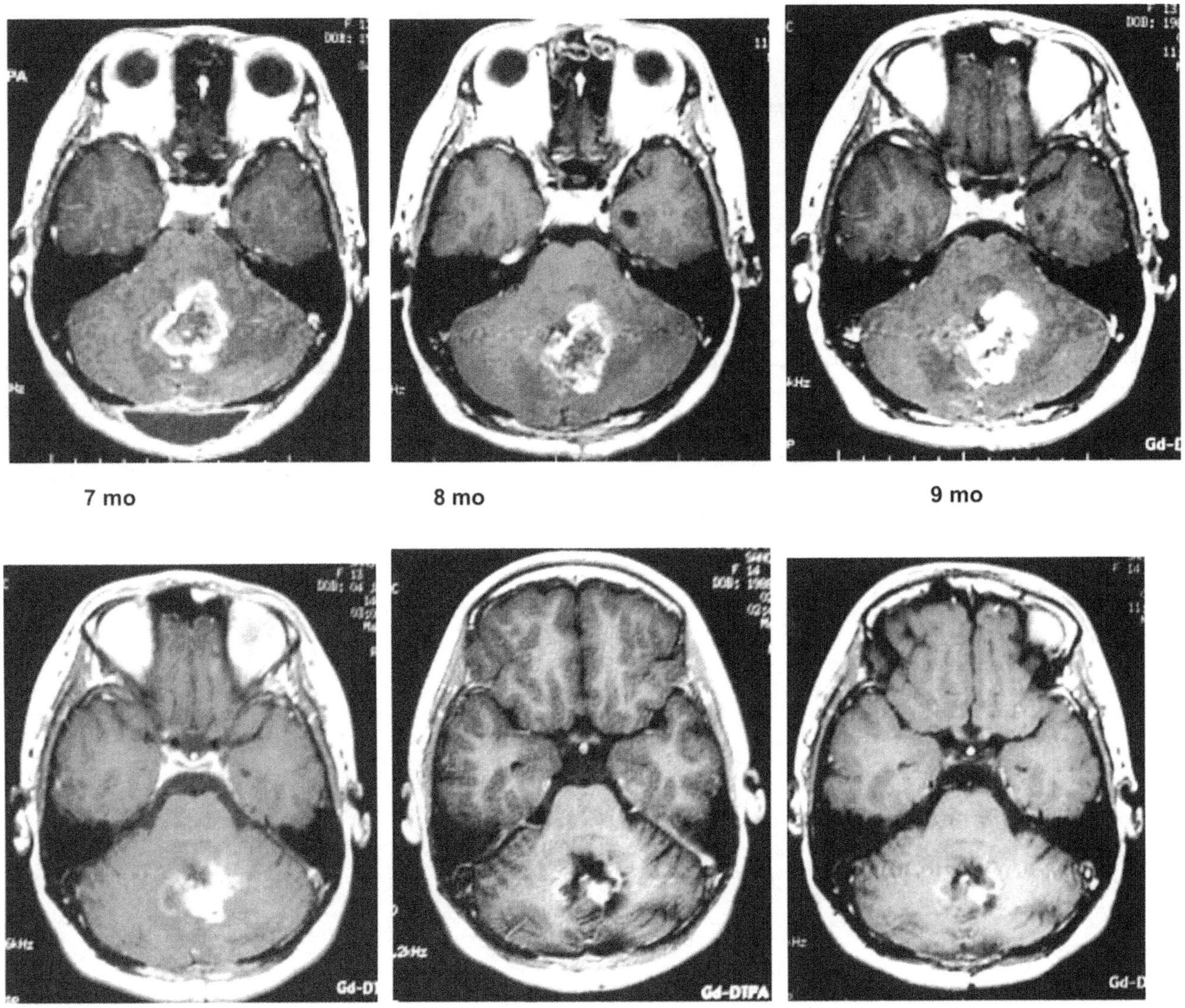

Fig. 3. Case 1. Follow up MRIs Necrotic mass increased in volume till 9 months and then decreased

occurred; however, follow up MRI at 4 months following CK treatment revealed a subtle enhancing lesion around the vermis. She complained of dizziness and unsteady gait at 6 months. At that time, MRI revealed enhancing mass lesion around the target area, ^{201}Thallium SPECT showed hot spot in the vermis and ^{1}HMRS revealed choline peak, which strongly suggested tumor recurrence. She underwent a second craniectomy, however, pathological examination revealed radiation necrosis. After the intervention, she was put under close MRI observation. Necrotic mass lesion increased in volume until 9 months and then decreased. She is going to junior high school with a KPS of 90.

Discussion

Radiation injury can be classified into acute reaction, early delayed reaction and late delayed reaction. Late delayed reaction occurs several months to years after radiation. Radiation necrosis is considered to be a late reaction due to vascular changes [7]. Glioma cells invade into normal brain parenchyma and the intermingling area exists peripherally. We consider over-radiation to these areaa as one of the risk factors for radiation necrosis, injuring normal endothelial cells. We use conformal treatment and inverse algorism of CK to reduce the maximum dose while maintaining an adequate marginal dose to avoid over radiation.

The dose volume effect is considered a factor of major importance for radiation necrosis. Kjellberg [2] and Flickinger [1] published 1–3% chance of radiation necrosis iso-effect dose-volume curve. Both curves were constructed using little clinical data and require verification with actual patient outcomes. Nedzi [4] reported that complications associated with the use of a LINAC-based radiosurgery system for treatment of brain tumors were more common with higher dosages, larger tumor volumes and tumor dose inhomogeneity. Shaw [6] also reported the same clinical data. However, our four patients with radiation necrosis were not specific in terms of tumor volume and dose delivery.

We initially delineated the target around the contrast enhancing area on T1WI of MRI; however peripheral recurrence was occasionally seen. Therefore we moved to high intensity area of T2WI or FLAIR as the target. Pirzkall [5] reported MR-Spectroscopy (MRS) guided target delineation for glioma. ^{1}H multivoxel MRS focusing on the ratio of choline to N-acetylaspartate is possible to demonstrate metabolically high area on MR images. In recent times we applied this method to delineation of the tumor volume. We believe that accurate anatomical targeting is also a key factor to reduce occurrence of radiation necrosis.

The CK is a frameless STI modality and easily applied both for SRS and SRT. The dose volume effect has been discussed with regard to risk factor; however, the number of fractions and dose per fraction should be considered to avoid radiation necrosis.

We consider that conformal treatment with inverse algorism, fractionated stereotactic radiotherapy, and precise anatomical targeting reduce the risk of radiation necrosis. Obviously longer term follow up will be needed to validate our strategy for glioma and glioblastoma.

References

1. Flickinger JC (1989) An integrated logistic formula for prediction of complications from radiosurgery. Int J Radiat Oncol Biol Phys 17: 879–885
2. Kjellberg RN (1988) Stereotactic Bragg peak proton radiosurgery results. In: Lunsford LD (ed) Modern stereotactic neurosurgery. Martinus Nijhoff, Boston, pp 463–470
3. Lea DE, Catcheside DG (1942) The mechanism of the induction by radiation of chromosome aberration in tradescantia. J Genet 44: 216–245
4. Nedzi LA, Kooy H, Alexander EIII (1991) Variables associated with the development of complications from radiosurgery of intracranial tumors. Int J Radiat Oncol Biol Phys 21: 591–599
5. Pirzkall A, McKnight TR, Graves EE (2001) MR Spectroscopy guided target delineation for high grade gliomas. Int J Radiat Oncol Biol Phys 50: 915–928
6. Shaw E, Scott C, Souhami L (1996) Radiosurgery for the treatment of previously irradiated recurrent primary brain tumors and brain metastases: initial report or radiation therapy oncology group protocol 90 05. Int J Radiat Oncol Biol Phys 34: 647–654
7. Sheline GE, Wara WM, Smith V (1980) Therapeutic irradiation and brain injury. Int J Radiat Oncol Biol Phys 6: 1215–1228
8. Thames HD, Peters LJ, Withers HR (1983) Accelerated fraction vs hyperfractionation: Rationales for several treatments per day. Int J Radiat Oncol Biol Phys 9: 127–138

Correspondence: Kengo Sato, 567-1 Kurata, Okayama, 703-8265, Japan. e-mail: kengosato@kyokuto.or.jp

Acta Neurochir (2003) [Suppl] 86: 519–521

Vascular endothelial growth factor expression in pituitary adenomas

S. Fukui[1], H. Nawashiro[1], N. Otani[1], H. Ooigawa[1], A. Yano[1], N. Nomura[1], A. M. Tokumaru[2], T. Miyazawa[1], A. Ohnuki[1], N. Tsuzuki[1], H. Katoh[1], S. Ishihara[1], and K. Shima[1]

[1] Department of Neurosurgery, National Defense Medical College, Tokorozawa, Japan
[2] Department of Radiology, National Defense Medical College, Tokorozawa, Japan

Summary

Vascular endothelial growth factor (VEGF) is known to be a mediator of angiogenesis and vascular permeability. A cystic component and hemorrhage are often found in pituitary adenomas. In the present study we assess the VEGF expression based on immunohistochemical examinations in 48 pituitary adenomas. All the adenomas showed some VEGF immunoreactivity mainly in the cytoplasm of tumor cells. Of the 48 adenoma-cases, 16 cases had a strong VEGF immunoreactivity, 26 cases had a moderate one, and 6 cases had a weak one. On the MR images, a cystic component was found in 16 cases (33.3%), and a hemorrhage was found in 18 cases (37.5%). The VEGF immunoreactivity had a significant relationship with the cystic component but neither the hemorrhage, size, recurrence, or HE classification. These findings suggest that VEGF might play a potential role in the pathogenesis of cystic formation in pituitary adenomas.

Keywords: Vascular endothelial growth factor; pituitary adenoma; cyst; hemorrhage.

Introduction

Vascular endothelial growth factor (VEGF), a heparin-binding dimeric glycoprotein, is a primary regulator of physiological angiogenesis and is also a major mediator of pathological angiogenesis such as tumor-associated neovascularization [9]. VEGF is considered to play an important role in the pathogenesis of tumor-associated hemorrhaging. In addition to its angiogenic capability, VEGF is the most powerful mediator among the known mediators of vascular permeability [2]. An astrocytoma with abundant cystic components and cystic fluid in glioblastoma contains a high amount of VEGF protein [7]. These findings suggest that VEGF also contributes to increased vascular permeability and subsequent cystic formation in astrocytic tumors.

Recent advances in radiological examinations, especially magnetic resonance (MR) imaging, have revealed that intratumoral hemorrhaging occurs most frequently in pituitary adenoma among all brain tumors [8], while cystic formation often occurs in pituitary adenoma [3]. However, a cystic component in both pituitary adenoma and astrocytic tumor is not a true cyst due to a lack of cyst-lining cells. Up to now, a cystic component of pituitary adenoma has usually been thought to be a cavity in which an intratumoral hemorrhaging occurred in the past [1, 5, 10], whereas a cystic component of astrocytic tumors such as a protoplasmic astrocytoma, is thought to be a fusion of many areas of microcystic degeneration [6]. In the present study, we assessed the VEGF expression based on immunohistochemical examinations in pituitary adenomas and also determined whether VEGF has any relationship with either the cystic component or hemorrhage on the MR images in pituitary adenomas.

Materials and methods

Patients

We retrospectively reviewed the cases of 80 patients with primary pituitary adenomas who were treated at the National Defense Medical College from 1989 to 2000. Fifteen of the 80 patients were thus excluded due to a poor condition of the surgical specimens. In addition, radiological examinations of MR imaging were not completely performed in 17 patients. Therefore, the association of the VEGF expression with the cystic component and hemorrhage on the MR images were discussed in the 48 cases in this series.

Radiological findings

The radiological features were examined on MR imaging before the initial operations. In the present study, a cystic component was defined as a lesion with both hypo-intensity on the T1 weighted image and hyper-intensity on the T2 weighted image. On the other hand, a hemorrhage was defined as a lesion either with a niveau formation or both with hyper-intensity on the T1 weighted image and hypo-intensity on the T2 weighted image. The Knosp radiological classification [4] was used to evaluate tumor invasiveness to the cavernous sinuses.

Histopathological examination

Specimens obtained at surgery were immediately fixed in 10% formalin and then embedded in paraffin. For histopathological examinations, the paraffin-embedded sections were cut in 4 μm slices, were heated at 55 °C overnight, and then were deparaffinized and hydrated through a series of xylene and alcohol. The sections were routinely stained with HE.

For the immunostaining of VEGF, antigen retrieval of the sections was performed in a microwave oven with a citrate buffer, pH 6. The tissue sections were then incubated 10 min with 3% hydrogen peroxide and rinsed in 0.01 M PBS. We used rabbit affinity-purified polyclonal antibody against VEGF (Santa Cruz Biotechnology, CA) at a dilution of 1 : 500 at 4 °C overnight.

Clinical follow up

Recurrence was defined as tumor regrowth of residual adenomas on the follow-up MR images, which needed a reoperation after an initial operation.

Statistical analysis

All data were statistically analyzed with the Student's t-test, the chi-square test, Spearman's correlation coefficient by rank, or Mann-Whitney U test. A p value of less than 0.05 was considered to be statistically significant.

Results

Clinical data

There were 18 males and 30 females in this study with a mean age of 44.3 years (range 16–70 years). Of the 48 patients, 8 patients had growth hormone secreting adenomas, 12 prolactinomas, 1 thyroid-stimulating hormone secreting adenoma, and 27 patients had nonfunctioning pituitary adenomas. These subtypes of pituitary adenomas were diagnosed according to the clinical symptoms or the hormonal data of serum. Recurrences were found in 7 patients in this series during follow-up periods ranging from 8 months to 134 months (mean: 57.3 months). Thirty-nine of the patients had no recurrent symptoms or radiological evidence of recurrence (median follow-up period: 54.5 months, range: 6–136 months). The recurrence rate was 14.6% (7 of 48 patients) in this series.

Radiological data

The average maximum diameter of the tumors was 27.3 mm (range 8.0–50.0 mm). Forty-seven of the tumors (97.9%) were macroadenomas (≥10 mm in size) and one tumor (2.1%) was a microadenoma. Gadolinium enhancement studies on MR imaging were performed in the 43 patients. The degree of gadolinium-enhancement of the tumors was noted as follows: none or weak 7 (16.3%), moderate 19 (44.2%), and strong 17 (39.5%). In the MR studies, cystic component in the tumors was noted in 16 cases (33.3%) and intratumoral hemorrhage was noted in 18 cases (37.5%). Of the 18 adenomas that were considered to have an intratumoral hemorrhage, 7 adenomas had a niveau formation and 11 adenomas had both a hyper-intensity on the T1 weighted image and a hypo-intensity on the T2 weighted image. In the Knosp grading system for adenoma invasiveness to cavernous sinuses, 4 cases were considered to be Grade 0, 13 Grade 1, 7 Grade 2, 14 Grade 3, and 10 cases were considered to be Grade 4.

Histopathological findings

In the classification based on the HE staining, 41 adenomas (85.4%) were diffuse type, 5 adenomas (10.4%) were sinusoidal type, and 2 adenomas (4.2%) were papillary type in this series. A microscopic hemorrhage was found in 9 adenomas. The immunoreactivity of VEGF in the tumor section was assigned one of three grades (weak, moderate, and strong) based upon the intensity of the visualized brown dots. Of the 48 adenoma-cases, 16 cases (33.3%) had a strong immunoreactivity of VEGF, 26 cases (54.2%) had a moderate one, and 6 cases (12.5%) had a weak one. The immunoreactivity of VEGF had a significant relationship ($p < 0.01$, Mann-Whitney U test) with the cystic component of the adenomas. The immunoreactivity of VEGF, however, showed no correlation with the hemorrhage, size, recurrence, the classification based on the HE staining, the degree of gadolinium-enhancement, or the Knosp Grade of the adenomas.

Cystic component and hemorrhage

The cases of adenoma with a cystic component showed a significantly stronger immunoreactivity of VEGF than those without a cystic component ($p < 0.01$, Mann-Whitney U test). There was no correlation (the chi-square test) between the cystic component and hemorrhage on the MR images.

Discussion

A cystic component of pituitary adenoma has usually been thought to be a cavity in which an intratumoral hemorrhaging occurred in the past [1, 5, 10]. There was, however, no correlation between the cystic component and hemorrhage in the adenomas on the MR images in this study. On the other hand, there was a significant relationship between the VEGF expression and the cystic component but not the hemorrhage in the adenomas. These findings suggest that VEGF plays a potential role in the pathogenesis of cystic formation in pituitary adenomas, in which there may be another pathogenesis of cystic formation different from that of the cavity in which an intratumoral hemorrhaging occurred in the past. There are, however, limitations to the conclusions that can be drawn, because we did not examine the serum concentration of VEGF whereas we examined the VEGF immunoreactivity in the cytoplasm of the tumor cells.

The VEGF immunoreactivity showed no correlation with the hemorrhage, size, recurrence, the degree of gadolinium-enhancement, or the lateral invasion of the adenomas. These findings suggest that VEGF may have little direct effect on either the tumorigenesis or angiogenesis of pituitary adenomas.

References

1. Cardoso ER, Peterson EW (1984) Pituitary apoplexy: a review. Neurosurgery 14: 363–373
2. Connolly DT (1991) Vascular permeability factor: a unique regulator of blood vessel function. J Cell Biochem 47: 219–223
3. Kannuki S, Bando K, Shirakawa N, Matsumoto K, Bando H, Saito S, Kusaka K (1993) [MRI findings and endocrinological dysfunction in hemorrhagic pituitary adenoma]. No Shinkei Geka 21: 1005–1012
4. Knosp E, Steiner E, Kitz K, Matula C (1993) Pituitary adenomas with invasion of the cavernous sinus space: a magnetic resonance imaging classification compared with surgical findings. Neurosurgery 33: 610–617
5. Kurihara N, Takahashi S, Higano S, Ikeda H, Mugikura S *et al* (1998) Hemorrhage in pituitary adenoma: correlation of MR imaging with operative findings. Eur Radiol 8: 971–976
6. Nishikawa R, Cheng SY, Nagashima R, Huang HJ, Cavenee WK, Matsutani M (1998) Expression of vascular endothelial growth factor in human brain tumors. Acta Neuropathol (Berl) 96: 453–462
7. Takano S, Yoshii Y, Kondo S, Suzuki H, Maruno T, Shirai S, Nose T (1996) Concentration of vascular endothelial growth factor in the serum and tumor tissue of brain tumor patients. Cancer Res 56: 2185–2190
8. Wakai S, Yamakawa K, Manaka S, Takakura K (1982) Spontaneous intracranial hemorrhage caused by brain tumor: its incidence and clinical significance. Neurosurery 10: 437–444
9. Xiong S, Grijalva R, Zhang L, Nguyen NT, Pisters PW, Pollock RE, Yu D (2001) Up-regulation of vascular endothelial growth factor in breast cancer cells by the heregulin-beta1-activated p38 signaling pathway enhances endothelial cell migration. Cancer Res 61: 1727–1732
10. Yoshioka H, Kurisu K, Arita K, Tominaga A (1998) A case of non-functioning pituitary adenoma accompanied by a large cyst. Acta Neurochir (Wien) 140: 293–294

Correspondence: Hiroshi Nawashiro, M.D., D.M.Sc., Department of Neurosurgery, National Defense Medical College, 3-2 Namiki, Tokorozawa, Saitama 359-8513, Japan. e-mail: nawa1957@me.ndmc.ac.jp

Hydrocephalus

Acta Neurochir (2003) [Suppl] 86: 525–528

Aquaporin-1 deletion reduces osmotic water permeability and cerebrospinal fluid production

K. Oshio[1], Y. Song[2], A. S. Verkman[2], and G. T. Manley[1]

[1] Department of Neurological Surgery, University of California, San Francisco, California, USA
[2] Department of Medicine, University of California, San Francisco, California, USA

Summary

Aquaporin-1 (AQP1) is a water channel that is strongly expressed at the ventricular-facing surface of choroid plexus epithelium. Using wildtype and AQP1 null mice, we developed novel methods to compare the water permeability in isolated choroid plexus, and cerebrospinal fluid (CSF) production in living mice. Osmotically-induced water transport was rapid in freshly isolated choroid plexus from wildtype mice as measured by a spatial-filtering optical method, and reduced by 5-fold by AQP1 deletion. CSF production, an isosmolar fluid secretion process, was measured by a dye dilution method involving fluid collections using a second microneedle introduced into the cisterna magna. CSF production in wildtype mice was (in μl/min) 0.37 ± 0.04 μl/min (control), 0.16 ± 0.03 μl/min (acetazolamide-treated) and 1.14 ± 0.15 μl/min (forskolin-treated), and reduced by up to 25% in AQP1 null mice. The impaired CSF production in AQP1 null mice provides direct functional evidence for the involvement of AQP1 in CSF formation.

Introduction

Cerebrospinal fluid (CSF) provides a mechanism support for the brain and a specialized extracellular environment for normal functioning of the central nervous system. Potential sites of CSF origin include the choroid plexus, brain parenchyma, and ependymal lining of the ventricles [2, 9, 12]. CSF produced by the choroid plexus and extra-choroidal tissues flow through the ventricles to sites of drainage in the subarachnoid space surrounding the brain and the spinal cord. The rates of CSF production and absorption must be equal in the steady state. When the balance between CSF production and absorption becomes disturbed, disorders such as hydrocephalus and increased intracranial pressure can occur. Most CSF appears to be actively secreted into the ventricles by the choroid plexus epithelium (CPE). However, the molecular mechanisms underlying fluid transport by this tissue remain largely unknown.

A family of molecular water channels called aquaporins (AQPs) has been identified in mammals [13]. AQPs are small integral membrane proteins (MW ~ 30,000) that provide a major pathway for water transport in the kidney, brain, secretory epithelia, and other organs. Several AQPs have been identified in regions of the central nervous system (CNS) that are thought to participate in the production and absorption of brain fluid [7]. Aquaporin-1 (AQP1) [10] is expressed in the ventricular-facing membrane of the choroid plexus epithelium (CPE). AQP1 functions primarily as a water pore to facilitate the transmembrane transport of water driven by osmotic gradients. The specific localization of AQP1 in the choroid plexus suggests a role in the production of CSF [4]. However, the function of AQP1 in CPE water transport and CSF production has not been elucidated. The purpose of this study was to define the role of AQP1 in CPE osmotic water permeability and CSF production. Novel methods were developed to measure CPE osmotic permeability of isolated CPE and CSF production in living mice. Comparisons were done in wild-type mice and mice lacking AQP1.

Materials and methods

Mice

AQP1 null mice were generated by targeted gene disruption in an outbred CD1 genetic background as originally described [5]. All experiments were performed on male wild type and AQP1-deficient littermates. Genotype analysis was done on all mice by PCR of tail

DNA. All protocols were approved by the University of California, San Francisco (U.C.S.F.) Committee on Animal Research.

Water permeability measurement in isolated choroid plexus

Choroid plexus was freshly isolated from the lateral and fourth ventricles of wild-type and AQP1 null mice by microdissection. Isolated choroid plexus tissue was immobilized on an 18-mm-diameter round treated with 0.01% poly-l-lysine. The cover-slip was mounted in a perfusion chamber (exchange time < 0.2 s) for measurement of osmotic water permeability by spatial filtering microscopy as described previously [3, 6]. Isolated choroid plexi were perfused with artificial CSF (aCSF: NaCl 120 mM, KCl 3 mM, NaH_2PO_4, 0.6 mM, $MgSO_4$ 0.8 mM, $NaHCO_3$ 18 mM, glucose 10 mM, $CaCl_2$ 1.1 mM, pH 7.4). The time course of transmitted light intensity was recorded in response to changing between perfusate osmolalities of 300 mOsm and 150 mOsm and 50% response times were calculated for the swelling and shrinking responses.

Cerebrospinal fluid production

The ventriculo-cisternal perfusion method originally described by Pappenheimer *et al.* [11] was modified for use in mice. After introduction of the micropipette in the lateral ventricle, a second micropipette held in a horizontal position by a stereotaxic manipulator was introduced into the cisterna magna. After visualizing CSF in the cisterna magna micropipette the ventriculo-cisternal perfusion was started in which the left lateral ventricle was perfused with aCSF containing 1 mg/ml tetramethylrhodamine isothionate-dextran (TRITC-Dextran) (MW 66.000) as a volume marker. For CSF production stimulation and inhibition experiments, the aCSF contained 1 mM forskolin or 0.1 mM acetazolamide (chemicals from Sigma, unless otherwise indicated) at 0.7 μl/min infusion rate. Fluid existing into the cisterna magna micropipette was collected every 20 min. The rate of CSF formation was calculated from, $V_f = r_i(C_i - C_o)/C_o$, where V_f = CSF formation rate (μl/min), r_i = infusion rate (μl/min), C_i = fluorescence of inflow solution, and C_o = fluorescence of outflow solution. TRITC fluorescence was measured in a fluorescence spectrometer.

Results

Choroid plexus osmotic water permeability

The role of AQP1 in CPE water permeability was determined from the kinetics of CPE swelling and shrinking in response to perfusion of the apical surface with hypotonic (150 mOsm) and isotonic (300 mOsm) artificial cerebrospinal fluid (aCSF). Relative CPE cell volume was measured using a spatial filtering light microscopy method. CPE was microsurgically isolated from the lateral ventricle and mounted in a perfusion chamber with the apical surface facing upward in direct contact with the perfusate. Fig. 1A shows a representative time course of transmitted light intensity responding to CPE swelling and shrinking. Changing the aCSF perfusate osmolality from 300 mOsm to 150 mOsm in wild-type (+/+) mice resulted in a rapid increase in cell volume that was reversibly upon perfusion with a 300 mOsm solution (Fig. 1B). In contrast, swelling and shrinking was remarkably slowed in CPE from AQP1 null (−/−) mice. The summary of results in different mice indicated 5-fold reduced osmotic water permeability in AQP1-deficient CPE ($p < 0.001$).

CSF production

The rate of CSF production was quantified using a modified Pappenheimer indicator dilution method. The average CSF production rate in wild-type mice at baseline was 0.37 ± 0.02 μl/minute (n = 8) (Fig. 2). This rate was significantly reduced by ~20% in AQP1 null mice (0.30 ± 0.01 μl/min; n = 8, $p < 0.01$). Acetazolamide, an carbonic anhydrase inhibitor of CSF formation reduced CSF production was reduced to similar levels in wild-type and AQP1 null mice (0.16 ± 0.03 μl/min vs. 0.15 ± 0.04 μl/min, respectively). The cyclic-AMP agonist forskolin increased CSF production in both wild-type and AQP1 null mice; however, the rate of increased CSF production AQP1 null mice was 25% less than wild-type mice ($p < 0.01$).

Discussion

The morphology of the choroid plexus and its location within the cerebral ventricles suggests that it is the principal site of CSF formation. The choroid plexus has many characteristics of fluid forming epithelium such as microvilli in the apical CSF side, infolding of the basal membrane, and tight junctions between cells. Hydrostatic pressure initiates the movement of water and ions from the choroidal capillaries to the interstitial fluid and choroid epithelium. Water and ions can then enter the ventricle in two ways: either moving from the basal membrane and then across to the apical membrane of the CPE, the trancellular route; or across the tight apical junction, the paracellular route. Water transport via either of these transmembrane routes is thought to be an active process dependent on ion transport [12].

Early studies of the choroid plexus demonstrated that active transport of sodium across the apical membrane of the CPE, leading investigators to conclude that local osmotic gradients are responsible for water transport [14]. It was also shown that the ratio of the osmotic to the diffusional water permeability was

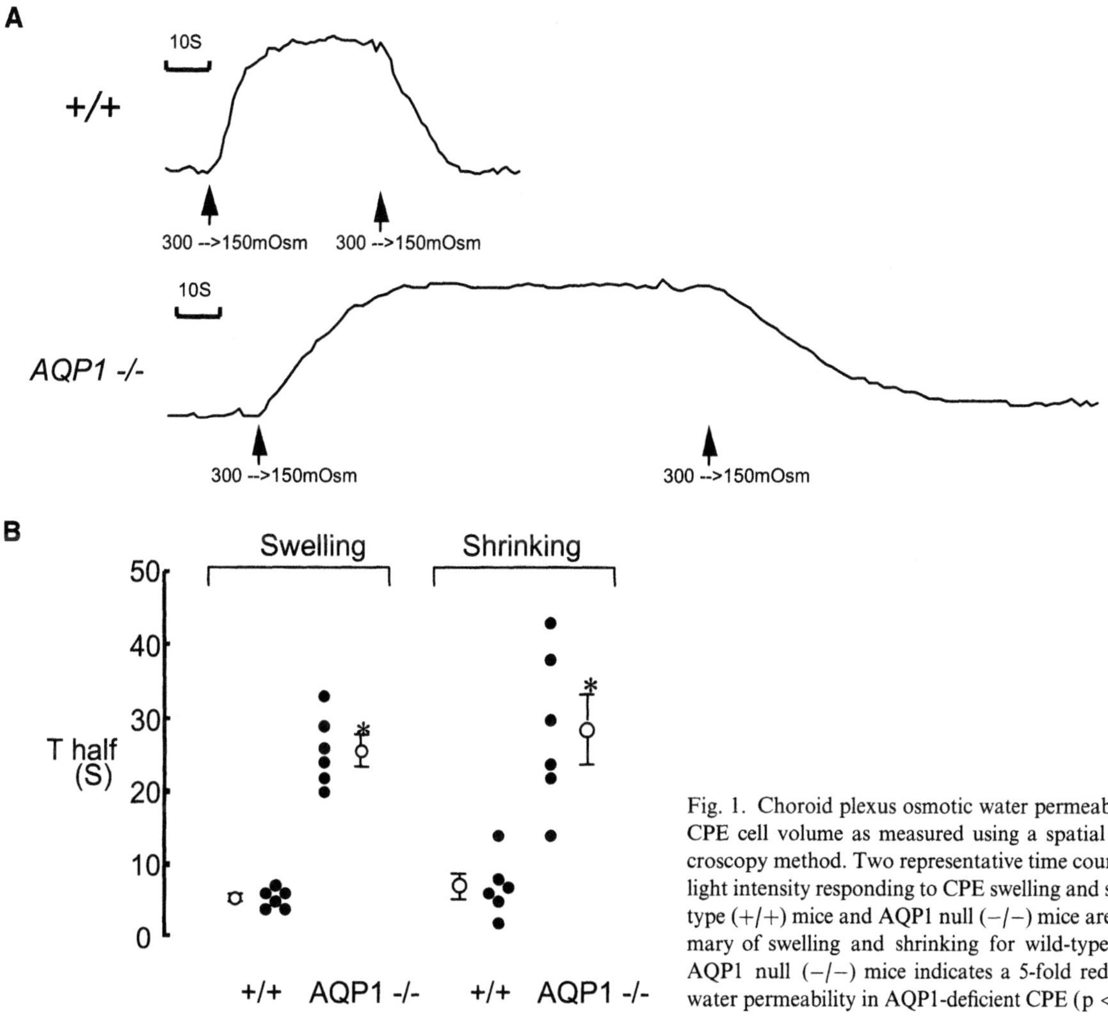

Fig. 1. Choroid plexus osmotic water permeability. (A) Relative CPE cell volume as measured using a spatial filtering light microscopy method. Two representative time courses of transmitted light intensity responding to CPE swelling and shrinking for wild-type (+/+) mice and AQP1 null (−/−) mice are shown. (B) Summary of swelling and shrinking for wild-type (+/+) mice and AQP1 null (−/−) mice indicates a 5-fold reduction of osmotic water permeability in AQP1-deficient CPE ($p < 0.001$)

very high. However, the reason for these high ratios was not clear. The location of the AQP1 water channel in the apical membrane of the choroid plexus provides an explanation for these earlier findings. As we have demonstrated, the AQP1 water channel is constitutively active and functions to move water through the apical surface of the CPE. Indeed, deletion of AQP1 resulted in a 5-fold reduction in osmotically-induced water transport.

While AQP1 deletion significantly reduced CPE osmotic permeability, the rate of CSF production in these transgenic mice was reduced by only 20% to 25%. There are several possible explanations for these results. The simplest explanation is that AQP1 facilitated transcellular water transport accounts for approximately one-quarter of CSF production, with the balance of water transport occurring through paracellular and non-AQP1 mediated transcellular routes. Another possibility is that the choroid plexus may not be the principal site of CSF production. Although the choroid plexus is thought to be responsible for 70% to 90% of CSF production, there are studies suggesting that extra-choroidal CSF production by the brain parenchyma may be significant. Milhorat reported that CSF production measured 2 to 6 months after removal of the choroid plexus in rhesus monkeys was only reduced by 31% [13]. Bering and Sato also reported that plexotomy reduced CSF formation by about half the rate observed in the normal animals [14]. Thus it is possible that in vivo, the choroid plexus may only account for 30% to 50% of CSF production. If this is correct, then water transport by AQP1 may in fact account for substantial percentage of choroidal CSF production. If however the choroid plexus is indeed responsible for the majority of CSF production, it is possible that our current results and those of Milhorat are due to some form of compensation in these animals. In both studies CSF production was measured in

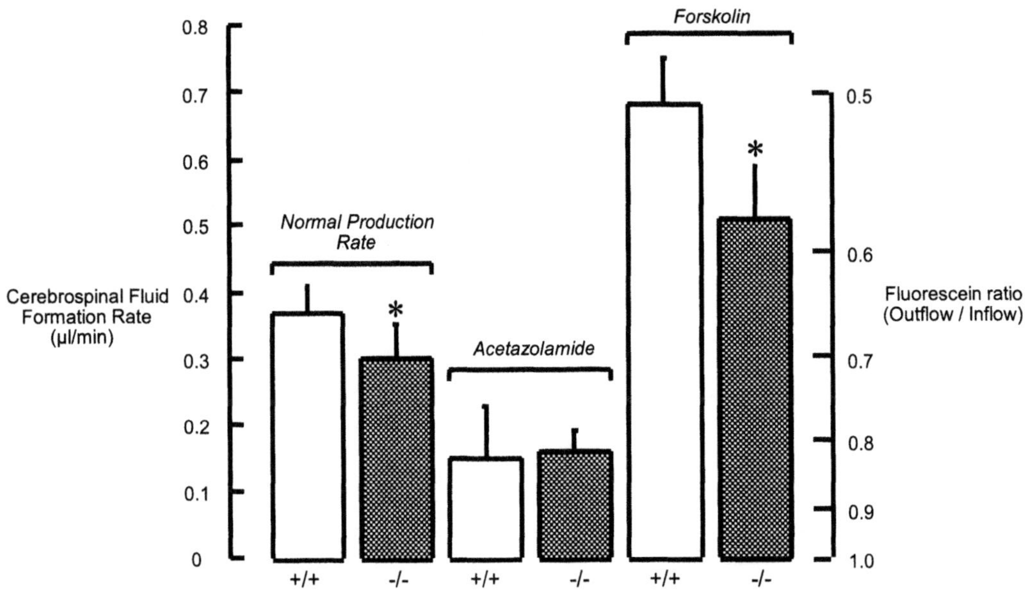

Fig. 2. CSF production rate. CSF production rate in wild-type mice at baseline was 0.37 ± 0.02 μl/minute and significantly reduced by ~20% in AQP1 null mice (0.30 ± 0.01 μl/min; $n = 8$, $p < 0.01$). Acetazolamide reduced CSF production was reduced to similar levels in wild-type and AQP1 null mice (0.16 ± 0.03 μl/min vs. 0.15 ± 0.04 μl/min, respectively). Forskolin increased CSF production in both wild-type and AQP1 null mice; however, the rate of increased CSF production AQP1 null mice was 25% less than wild-type mice ($p < 0.01$)

the chronic phase following plexotomy or AQP1 deletion. It is possible that extra-choroidal sources could have increased their percentage of contribution to CSF production over time.

In summary, we have found that AQP1 deletion in mice is associated with decreased CPE osmotic permeability and decreased CSF production. In addition, these studies provide the first extensive and quantitative description of mouse CSF dynamics. An important clinical implication of the reduced CSF production in AQP1-deficient mice is the potential application of AQP1 inhibitors for therapy of some forms of hydrocephalus and disorders of increased intracranial pressure.

References

1. Bering EA, Sato O (1963) Hydrocephalus: Changes in formation and absorption of cerebrospinal fluid within the cerebral ventricles. J Neurosurg 20: 1050–1063
2. Davson H, Hollingsworth G, Segal M (1970) The mechanism of drainage of the cerebrospinal fluid. Brain 93: 665–678
3. Farinas J, Kneen M, Moore M, Verkman AS (1997) Plasma membrane water permeability of cultured cells and epithelia measured by light microscopy with spatial filtering. J Gen Physiol 110: 283–296
4. Hasegawa H, Lian SC, Finkbeiner WE, Verkman AS (1994) Extrarenal tissue distribution of CHIP28 water channels by *in situ* hybridization and antibody staining. Am J Physiol 266: C893–C903
5. Ma T, Yang B, Gillespie A, Carlson EJ, Epstein CJ, Verkman AS (1998) Severely impaired urinary concentrating ability in transgenic mice lacking aquaporin-1 water channels. J Biol Chem 273: 4296–4299
6. Ma T, Song Y, Yang B, Gillespie A, Carlson EJ, Epstein CJ, Verkman AS (2000) Nephrogenic diabetes insipidus in mice deficient in aquaporin-3 water channels. Proc Natl Acad Sci USA 97: 4386–4391
7. Manley GT, Fujimura M, Ma T, Noshita N, Filiz F, Bollen AW, Chan P, Verkman AS (2000) Aquaporin-4 deletion in mice reduces brain edema after acute water intoxication and ischemic stroke. Nat Med 6: 159–163
8. Milhorat T (1972) Hydrocephalus and cerebrospinal fluid. Williams and Wilkins, Baltimore
9. Milhorat T, Hammock M, Fenstermacher J, Levin V (1971) Cerebrospinal fluid production by the choroid plexus and brain. Science 173: 330–332
10. Nielsen S, Smith BL, Christensen EI, Agre P (1993) Distribution of the aquaporin CHIP in secretory and resorptive epithelia and capillary endothelia. Proc Natl Acad Sci USA 90: 7275–7279
11. Pappenheimer J, Heisey S, Jordan E, Downer J (1962) Perfusion of the cerebral ventricular system in unanesthetized goats. Am J Physiol 203: 763–774
12. Segal M, Pollay M (1977) The secretion of cerebrospinal fluid. Exp Eye Res 25: 127–148
13. Verkman AS (2002) Physiological importance of aquaporin water channels. Ann Med 34: 192–200
14. Wright EM, Weidner G, Rumrich G (1977) Fluid secretion by the frog choroid plexus. Exp Eye Res 25: 149–155

Correspondence: Geoffrey T. Manley, M.D., Ph.D., Department of Neurosurgery, University of California, San Francisco, Room M-779, Box 0112, 505 Parnassus Ave., San Francisco CA 94143. e-mail: manley@itsa.ucsf.edu

Acta Neurochir (2003) [Suppl] 86: 529–532

Hysteresis of the cerebrospinal pressure-volume curve in hydrocephalus

M. Kasprowicz[1], M. Czosnyka[2], Z. Czosnyka[2], S. Momjian[2], P. Smielewski[2], H. Juniewicz[2], and J. D. Pickard[2]

[1] Division of Measurements and Measuring Systems, Faculty of Electronics, University of Technology, Wroclaw, Poland
[2] Academic Department of Neurosurgery, Box 167, Addenbrooke's Hospital, Hill Road, Cambridge, UK

Summary

The objective was to study the displacement of the cerebrospinal fluid pressure-volume curve during the descent relative to the ascent of intracranial pressure recorded during the cerebrospinal fluid constant rate infusion test. This phenomenon can be interpreted as the hysteresis of the pressure-volume curve.

The cerebrospinal fluid dynamics were tested in fifty-eight patients with clinical symptoms of hydrocephalus. After finished infusion, ICP was recorded until it returned to steady state level. Pressure-volume curves were plotted separately for ascending and descending phases of the test. The parameters of CSF compensation were estimated on the basis of mathematical mono-exponential model of CSF circulation.

The pressure-volume curve post-infusion was visibly shifted upward in 69% of tests. Those who demonstrated the upward shift of the pressure-volume curve had greater an elastance coefficient of the cerebrospinal space (with shift: $El = 0.26 \pm 0.14$; without shift: $El = 0.17 \pm 0.06$; $p < 0.05$). Magnitude of the shift was positively correlated with pulse amplitude of ICP ($r = -0.763$; $p < 0.0001$).

The accuracy of clinical examination of the pressure-volume compensatory reserve, which take into account both compression and decompression phase of the study, may be affected by this phenomenon.

Keywords: Hydrocephalus; pressure-volume curve; infusion test; mathematical modelling.

Introduction

The aim of this study was to analyse the displacement of the pressure-volume curve after a constant rate infusion test in patients with symptoms of hydrocephalus. The exponential nature of this relationship is well documented in the literature [6, 11]. The phenomenon of hysteresis of the relationship between ICP and the intracerebral volume was noticed in the past [5, 8, 10] but the displacement of pressure-volume relationship during and after infusion test has not been described quantitatively so far. It may play an important role in understanding of CSF flow into parenchyma and its interference with cerebrovascular function, often seen in patients suffering from hydrocephalus.

Materials and methods

A computerised CSF infusion studies were performed in fifty-eight patients (35 men and 23 women) with clinical symptoms of hydrocephalus. Mean age was 62 ± 22 years. None of the patients was previously shunted. Magnetic resonance imaging of the brain was obtained from 11 patients. Ventricular dilatation was confirmed radiographically in all of them.

The dynamics of cerebrospinal fluid was investigated using the computerised infusion test [3]. The tests were performed with a constant infusion rate (1.5 ml/min or 1.0 ml/min if the baseline pressure was higher than 15 mm Hg), using either lumbar puncture (33%) or a pre-implanted ventricular access device (67%). After finished infusion, ICP was recorded until ICP returned to the steady state level (i.e. plateau of the ICP time-trend) – see Fig. 1. The shape of the pressure-volume relationship was fitted to the mathematical model of a mono-exponential curve [6, 7, 12]:

$$P = (P_b - P_o)e^{E\Delta V} + P_o \qquad (1)$$

where: P – the intracranial pressure [mm Hg]; P_b – the baseline pressure [mm Hg], P_0 – the reference pressure [mm Hg], E – the elastance coefficient [1/ml], ΔV – the effective volume of fluid [ml].

Compensatory parameters were estimated on the basis of the physiological model of CSF circulation [1, 3, 7, 12] using computerised analysis. Calculations were repeated using ascending and descending phases of the CSF pressure recording during the test. If these two curves were displaced relative to each other, the test was classified as presenting with hysteresis. Three parameters ware used to describe this phenomenon: The hysteresis was quantified by a difference between pre- and post-infusion steady-state ICP. The area between both segments of the pressure-volume curve divided by the total volume effectively added during the study was used to describe of the magnitude of the hysteresis (Area Index, AI; this index has no units). Additionally a difference between the elastance coefficients for ascending (E1) and descending (E2) pressure volume curve ($\Delta E = E1 - E2$ [1/ml]) was calculated. The theoretical curves, describing the pressure rise and decrease observed within and after the constant rate infusion test (see Fig. 1), can be expressed by mathematical equations:

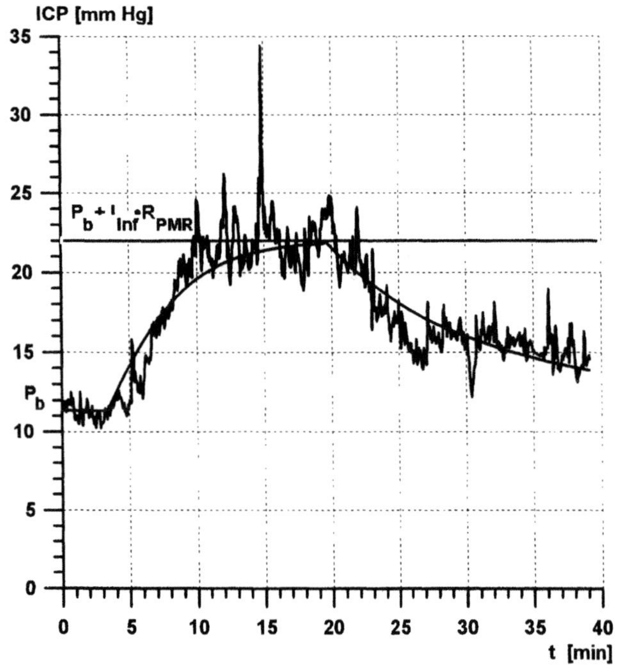

Fig. 1. Mean intracranial pressure (*ICP*) recorded during infusion test and the theoretical curves obtained from mathematical model

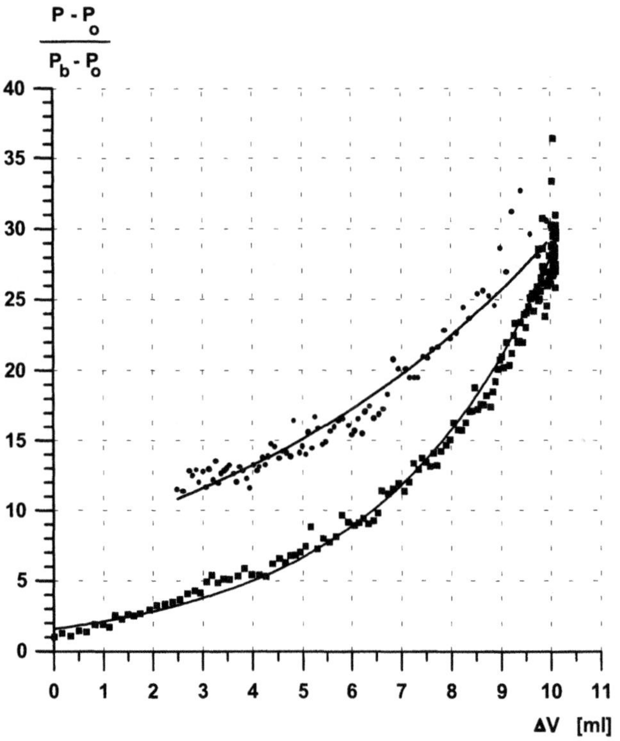

Fig. 2. The example of the pressure-volume curve hysteresis

$$P(t) = \frac{\left[I_{inf} + \frac{P_b - P_o}{R}\right] \cdot [P_b - P_o]}{\frac{P_b - P_o}{R} + I_{inf} \cdot e^{-E[(P_b-P_o)/R+I_{inf}]\cdot t}} + P_o \qquad (2)$$

$$P(t) = \frac{\left[I_{inf} + \frac{P_b - P_o}{R}\right] \cdot [P_b - P_o]}{\frac{P_b - P_o}{R} + I_{inf} \cdot [1 - e^{-E[(P_b-P_o)/R]\cdot t}]} + P_o \qquad (3)$$

where: I_{inf} – the infusion rate [ml/min], R – the resistance to CSF reabsorption or outflow [mmHg/(ml/min)].

Another description of the pressure-volume relationship can be derived from the relationship between time series of samples of mean CSF pressure and its amplitude of the pulsatile component (AMP) – so called AMP-P line [2]. It expresses changes in cerebral blood volume. The AMP-P curve was analysed for ascending and descending part of the pressure recording separately (slope (S), and intercept (P_0)) with the x (pressure axis) were studied. Under assumption that the rise of the blood volume after a heart contraction is equivalent to a rapid bolus addition of CSF fluid at the resting pressure P_b, the pulse amplitude (AMP) can be expressed as:

$$AMP = P_p - P_b = (P_b - P_o) \cdot (e^{E \cdot \Delta CBV} - 1) \qquad (4)$$

where: ΔCBV – the pulsatile cerebral blood volume.

Results

The hysteresis of the pressure-volume curve was identified in 40 infusion tests (69%). In 37 patients (92%) the descending curve decreased above the ascending one (Fig. 2). In 3 patients the opposite arrangement was observed. The average parameters describing hysteresis were summarized in Table 1.

Table 1. *Parameters of the hysteresis of the pressure-volume curve*

Parameter	Mean ± SD	Min	Max
ΔE (1/ml)	0.13 ± 0.12	0.02	0.55
ΔICP (mm Hg)	6.5 ± 4.4	0.5	18.1
AI	2.1 ± 3.5	0.1	20.5

AI (area index) – The area between both parts of the pressure-volume curve divided by the total volume effectively added during the study; *ΔICP* difference between pre- and post-infusion resting ICP; *ΔE = E1 – E2* difference between the elastance coefficient for ascending (E1) and descending (E2) pressure-volume curve.

When the groups of patients with and without hysteresis were compared manifestation of phenomenon was observed in patients with higher baseline cerebral elastance coefficient (with displacement: $E1 = 0.26 \pm 0.14$; no displacement: $E1 = 0.17 \pm 0.06$; $p < 0.05$). There were no statistically significant relationships between manifestation of hysteresis and clinical parameters: age of patients, ventricular size and type of hydrocephalus, neither with the site of volume addition (lumbar or ventricular). Parameters describing the cerebrospinal compensation measured using ascending and descending phases of the infusion study are summarised in Table 2 along with the results of statistical comparison. Contrary to the pressure-volume curve,

Table 2. *Parameters of cerebrospinal compensation*

Parameter	Increase of CSF pressure	Decrease of CSF pressure	p
E (1/ml)	0.24 ± 0.13	0.11 ± 0.06	<0.0001
ICP (mm Hg)	9.7 ± 4.4	16.2 ± 4.8	<0.0001
C (ml/mm Hg)	1.4 ± 1.7	3.2 ± 3.3	<0.0001
S	0.32 ± 0.11	0.32 ± 0.11	0.69
P_0 (mm Hg)	3.02 ± 5.06	3.56 ± 6.87	0.56
dCBV (ml)	1.7 ± 1.0	4.1 ± 3.0	<0.0001

E Cerebral elastance coefficient; *ICP* resting CSF pressure; *C* cerebral compliance; *S* slope of AMP-P regression line; P_0 reference pressure; *dCBV* pulsatile cerebral blood volume.

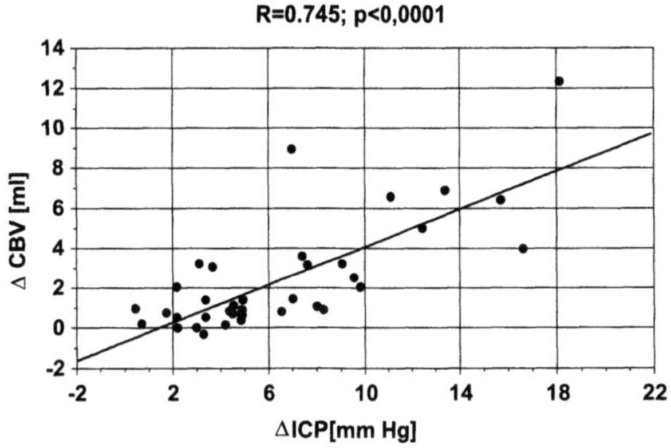

Fig. 3. Relationship between an increase of pulsatile fraction of cerebral blood volume detected after the infusion compared to the pre-infusion values (Y-axis) and the increase in the post-infusion ICP compared to the pre-infusion (baseline) ICP values (X-axis)

AMP-P regression line did not demonstrate any significant displacement.

The area index was positively correlated with elastance coefficient ($r = 0.605$; $p < 0.0001$), and inversely with the pulse amplitude of ICP ($r = 0.763$; $p < 0.0001$). The area index decreases when pulse amplitude of ICP grows ($r = -0.760$; $p < 0.0001$, reciprocal model). The increase in resting ICP level (pre-relative to post-infusion) was positively correlated with an increase in pulsatile cerebral blood volume ($r = 0.745$; $p < 0.0001$) – Fig. 3.

Discussion

We do not know the exact explanation of the hysteresis. It is likely that the different trajectory of the decrease of the pressure to the baseline level after the infusion is caused by the excessive accumulation of cerebrospinal fluid in the brain tissues during infusion, analogical to extracellular oedema. Therefore, after infusion, equivalent water volume stored in the brain tissue is greater than before. Time constant of displacement of this water is much longer than 10–15 minutes; an average time for observation of ICP trend after infusion.

Another possible explanation may touch 'vascular' factor of CSSF compensation. In accordance with the Monro-Kelly doctrine, the intracranial volume remains constant during the infusion test. The increase in CSF volume causes the displacement of predominantly venous blood from the intracranial space and an exponential rise of ICP. After the end of the infusion the recovery of the venous blood volume to the pre-infusion steady state initiates.

The resting ICP consists of two components: one derived from CSF circulation and one from pulsatile cerebral blood flow [4, 7]:

$$ICP = ICP_{CSF} + ICP_{vasc}$$
$$= R_{CSF} * CSF_{formation\ rate} + P_{ss} + ICP_{vasc}$$

It is likely that the process of changes in of CSF volume affected the vascular components of ICP (ICP_{vasc}). We do not possess direct evidence for such an explanation. An indirect one, however, exists: the pulsatile change in cerebral blood volume increases after the test (see Table 2), causing an increase in the vascular component of ICP, as it remains associated to pulsations of the cerebrovascular bed [4]. Indeed, the rise in the resting ICP after the test is significantly correlated with an increase in cerebral blood volume (see Fig. 3).

The different trajectories of the increase and decrease in the CSF pressure during and after the infusion study suggest differences between the values of CSF compensatory parameters calculated for the ascending and descending parts of the pressure-volume curve. Certain methods, which take into account both phases, may produce erroneous results [9]. Similarly, the calculation of the compensatory parameters from the bolus test can be biased.

References

1. Avazaat CJJ, van Eijndhoven JHM (1976) A new method of monitoring intracranial/volume relationship. Intracranial Pressure III: 308–313
2. Avezaat CJJ, van Eijndhoven JHM, Wyper DJ (1979) Cerebrospinal pulse-pressure and intracranial volume-pressure relationships. J Neurol Neurosurg Psychiatry 42: 687–700
3. Czosnyka M, Whitehouse H, Smielewski P, Simac S, Pickard JD (1996) Testing of cerebrospinal compensatory reserve in shunted and non-shunted patients: a guide to interpretation based on

observational study. J Neurol Neurosurg Psychiatry 60: 549–558
4. Czosnyka M, Richards HK, Czosnyka Z, Piechnik S, Pickard JD (1999) Vascular components of cerebrospinal fluid compensation. J Neurosurg 90: 752–759
5. Lesniak M, Clatterbuck R, Rigamonti D, Williams MA (2001) Low pressure hydrocephalus and ventriculomegaly: hysteresis, non-linear dynamics and benefits of CSF diversion. 3rd International Hydrocephalus Workshop, Abstract book, pp 1
6. Lofgren J, von Essen C, Zwetnow NN (1973): The pressure-volume curve of the cerebrospinal space in dogs. Acta Neurol Scand 49: 557–574
7. Marmarou A (1973) A Theoretical model and experimental evaluation of the cerebrospinal fluid system. Thesis, Faculty of Drexel University, Philadelphia, pp 1–132
8. Mchedlishvili G, Nikolaishvili L, Itkis M (1979) Pathophysiological mechanisms of brain edema development: role of tissue factors. Stroke 10(1): 52–57
9. Meier U, Zeilinger FS, Kintzel D (1999) Diagnostic in normal pressure hydrocephalus: A mathematical model for determination of the ICP-dependent resistance and compliance. Acta Neurochir (Wien) 141(9): 941–947
10. Shapiro K, Fried A, Marmarou A (1985) Biomechanical and hydrodynamic characterization of the hydrocephalic infant. J Neurosurg 63: 69–75
11. Shulman K, Marmarou A (1971) Pressure-volume considerations in infantile hydrocephalus. Dev Med Child Neurol 13(25): 90–95
12. Sliwka S (1980) A clinical system for the evaluation of selected dynamic properties of the intracranial system. PhD Thesis, Polish Academy of Sciences, Warsaw

Correspondence: Czosnyka Marek, Ph.D., D.Sc., Academic Department of Neurosurgery, Box 167, Addenbrooke's Hospital, Hill Road, Cambridge CB2 2QQ, UK. e-mail: mc141@medschl.cam.ac.uk

Acta Neurochir (2003) [Suppl] 86: 533–537

Is decreased ventricular volume a correlate of positive clinical outcome following shunt placement in cases of normal pressure hydrocephalus?

U. Meier[1], S. Paris[2], A. Gräwe[1], D. Stockheim[2], A. Hajdukova[1], and S. Mutze[2]

[1] Department of Neurosurgery, Unfallkrankenhaus Berlin, Germany
[2] Department of Radiology, Unfallkrankenhaus Berlin, Germany

Summary

It is well known that in patients with communicating hydrocephalus or normal pressure hydrocephalus, ventricular volume decreases following implantation of differential pressure valved shunts. Hydrostatic valves (Miethke dual-switch valves) were implanted in 60 normal pressure hydrocephalus patients at Unfallkrankenhaus Berlin between September 1997 and September 2001. One year postoperatively, these patients underwent CT scan, and their ventricular size was ascertained using the Evans index. Although 77% of these patients showed no postoperative change in ventricular volume, 65% nonetheless showed good to excellent clinical improvement, 13% satisfactory improvement and 22% no improvement. A moderate reduction in ventricular size was observed in 17% of the patients in our cohort. 40% of these patients showed good to excellent clinical improvement, 20% satisfactory improvement, and 40% unsatisfactory improvement. A marked reduction in ventricular size was observed in 6% of our patients. Of these latter patients, 50% showed good to excellent outcomes, while 50% had unsatisfactory outcomes. The favorable outcomes following implantation of a hydrostatic shunt in patients with normal pressure hydrocephalus did not correlate with decreased ventricular volume one year after operation. Better clinical outcomes were observed in patients with little or no alteration in ventricular size than in patients with a marked decrease in ventricular size. Postoperative change in ventricular volume should be assessed differently for patients with normal pressure hydrocephalus than in patients with hypertensive hydrocephalus.

Keywords: Normal pressure hydrocephalus; ventricular size; preoperative; postoperative; Evans index; shunt operation; hydrostatic valve.

Introduction

There is a great deal of disagreement among neurosurgeons, neuroradiologists and neurologists as to whether a clearly positive correlation exists between postoperative reductions in ventricular volume and clinical outcomes following shunt placement in patients with normal pressure hydrocephalus. It should be borne in mind, however, that all studies of clinical outcome following implantation of differential pressure valves have yielded different results. The advent of shunts with hydrostatic valves has only heightened this controversy, since prevention of overdrainage may be associated with an increased risk of relative underdrainage and with the related phenomenon of a reduced rate of ventriculomegaly regression [6, 3].

Material and method

Hydrostatic valves (Miethke dual-switch valve) were implanted in 60 normal pressure hydrocephalus patients at Unfallkrankenhaus Berlin between September 1997 and September 2001. The mean age of the 33 men and 27 women in the cohort was 64 years at the time of the study. Diagnoses were established and the decisions to perform shunt placement were made on the basis of the patients' histories as well as gait ataxia as a cardinal symptom, dementia and urinary incontinence as additional and late symptoms, and performance of an intrathecal infusion test [11, 13]. In addition, resistance to cerebrospinal fluid outflow R_{out} exceeding 13 Torr/ml/min was regarded as evidence of derangement of cerebrospinal fluid outflow in the craniospinal pathways [10, 12]. Following performance of an intrathecal infusion test, diagnostic siphoning of no less than 60 ml of cerebrospinal fluid was realized during the cerebrospinal fluid tap test, which was performed as an auxiliary diagnostic measure. The indications for shunt placement were as follows: gait ataxia, pathologically raised resistance to cerebrospinal fluid outflow, and improvement in clinical symptoms (particularly gait ataxia) following cerebrospinal fluid tap test. The hydrostatic Miethke dual-switch valve was implanted in all 60 patients [15, 16].

Design of the Miethke dual-switch valve

The dual-switch valve is composed of two pressure setting mechanisms that work independently of each other – one for when the patient is in the recumbent position and the other for the upright position. This design keeps intracranial pressure in the physiological range even when the patient is standing, while at the same time all problems of overdrainage are systematically eliminated. The requisite pressure is set by a tantalum ball, which, when the patient is in the upright position, prevents drainage of cerebrospinal fluid on the

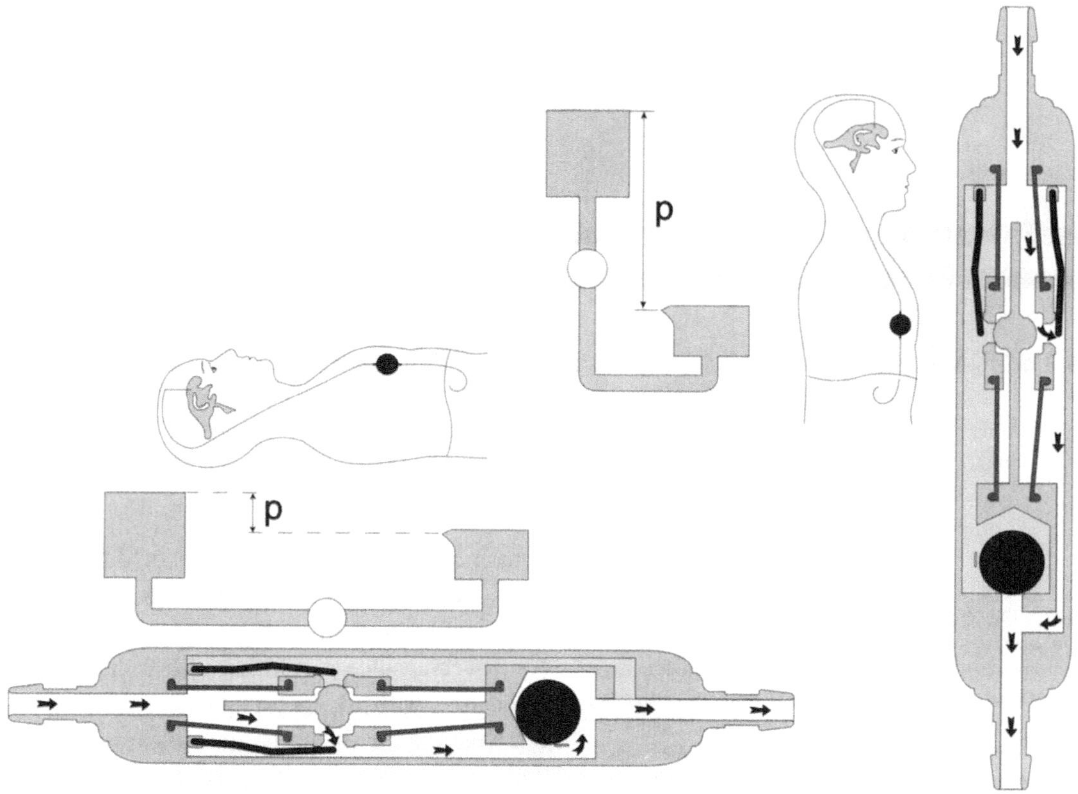

Fig. 1. Design principle of the Miethke dual-switch valve

low pressure side of the valve. Even if acute intracranial pressure levels are reached while the patient is in the upright position, the high pressure side of the valve opens, thereby preventing any further rise in pressure (Fig. 1).

Patient in recumbent position

As long as production and absorption of cerebrospinal fluid are in a state of equilibrium, the patient's intracranial pressure remains in the physiological range and the valve remains closed. But if the natural flow of cerebrospinal fluid is deranged, intracranial pressure rises. Inasmuch as the position switch (tantalum ball) is open, cerebrospinal fluid can be siphoned through the valve's low pressure chamber (Fig. 1).

Patient in upright position

When the patient moves into the upright position, the position switch (tantalum ball) closes, thereby preventing cerebrospinal fluid drainage that would otherwise be provoked by the additional hydrostatic pressure. If cerebrospinal fluid flow is disturbed and intracranial pressure increases further, the valve's high pressure chamber opens. This keeps intracranial pressure in the physiological range even when the patient is in the upright position, thereby systematically avoiding either overdrainage or an acute rise in intracranial pressure (Fig. 1).

CT studies were realized in all 60 patients pre-operatively and one year postoperatively. During all 120 of these CT scans, ventricular volume in the frontal horn was ascertained using the Evans index, and any postoperative changes were measured in accordance with this index as well (Fig. 2). Preoperative and postoperative results for all patients were evaluated, respectively, immediately prior to surgery and one year postoperatively using the Kiefer normal hydrocephalus grading scale [9, 14]. The investigation protocol also included an NPH recovery rate assessment [14] based on Kiefer clinical grading for NPH.

NPH recovery rate

$$= \frac{\text{NPH grading}_{\text{preoperative}} - \text{NPH grading}_{\text{postoperative}}}{\text{NPH grading after Kiefer [9]}_{\text{preoperative}}} \times 10$$

All graduated study results were classified into three patient groups: The group with good to excellent outcome: NPH recovery rate of ≥ 5 points; the group with a satisfactory outcome: NPH recovery rate of ≥ 2 points; the group with an unsatisfactory outcome: NPH recovery rate of <2 points.

Results

The clinical and CT findings of 60 patients with normal pressure hydrocephalus were evaluated following implantation of a hydrostatic valve. A reduction in the Evans index of ≥ 0.12 was defined as a clear or significant reduction in ventricular volume, while a reduction of ≥ 0.06 to 0.11 was classified as a mod-

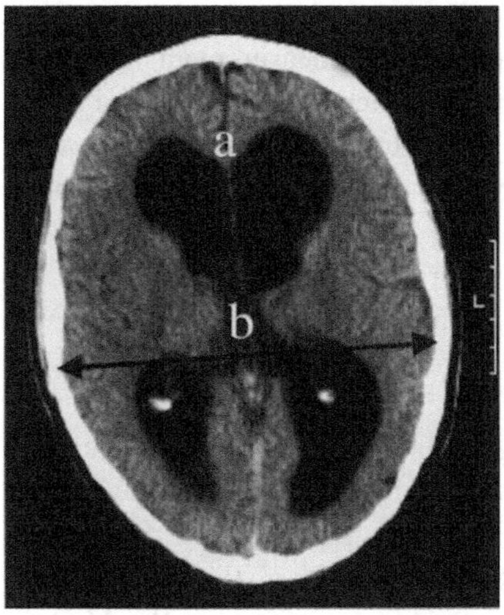

Fig. 2. Measurement of the Evans index = a/b

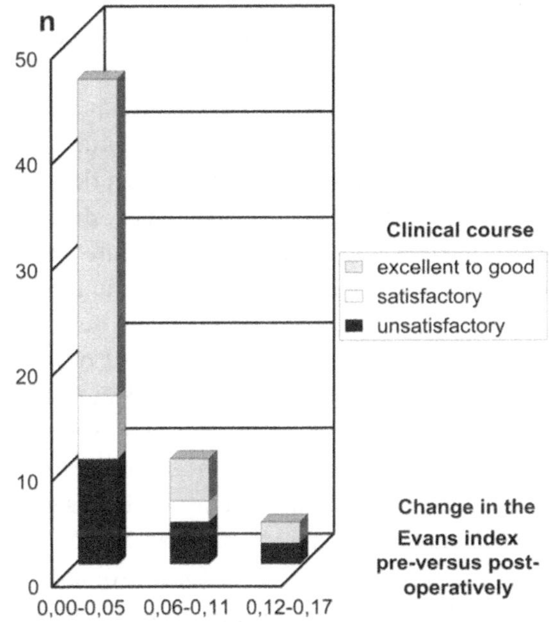

Fig. 3. Change in ventricular volume in relation to clinical course

erate improvement. Under this classification system, an Evans index differential of 0.05 constitutes little or no reduction in ventricular volume.

The following changes in ventricular volume were observed in shunted patients one year postoperatively: a slight reduction in 46 patients (77%), moderate reduction in 10 patients (17%), and marked reduction in four patients (6%). A comparison of outcomes one year postoperatively with changes in ventricular volume during the same period (Fig. 3) revealed the following: 65% of patients with a minimal reduction in ventricular volume showed good to excellent clinical outcomes, 13% showed satisfactory outcomes and 22% unsatisfactory outcomes; 40% of patients with moderate decreases in ventricular size had good to excellent clinical results, 40% had unsatisfactory results, and the remaining 20% showed satisfactory clinical outcomes; and 50% of patients with a marked reduction in ventricular volume had good to excellent clinical outcomes and the remaining 50% had unsatisfactory outcomes. No statistical analysis of these results could be performed due to the relatively low number of patients that showed a marked reduction in ventricular volume. Generally speaking, treatment outcome was characterized by only a minor reduction in ventricular size, accompanied in the vast majority of cases by good to excellent clinical improvement. The results of the CT scans (see Fig. 3) are surprising in that in 77% of cases, little or no reduction in ventricular volume (according to our definition of this parameter) was observed, notwithstanding the fact that good to excellent improvement was observed in a majority (68%) of cases.

Discussion

Hydrostatic valves were the first hydrocephalus valves to take account of the fact that hydrocephalus patients spend a great deal of their time sitting and standing and that patients have completely different needs with respect to cerebrospinal fluid dynamics in the upright versus the recumbent position [19]. Although hydrostatic valves have reduced the risk of overdrainage induced by valve design engineering [16, 20], problems of underdrainage have also become increasingly prevalent. Three (5%) of our patients underwent revision operations owing to clinical underdrainage that was confirmed using an intrathecal infusion test. Miethke dual-switch valves with a lower pressure setting for recumbent position were implanted in these patients. A review of the literature on underdrainage following shunt surgery reveals marked discrepancies and inconsistencies in the definitions and classifications of postoperative complications. Some authors calculate the overall number of complications, while others count only mechanical complications and exclude infections. Thus, for example, Sainte-Rose [21] fails to define underdrainage at all and characterizes overdrainage as a subdural hygroma, slit ventricle or isolated ventricle. Drake and colleagues [7] classify underdrainage and occlusion as a combination of complications, defining overdrainage as a subdural hydroma and a slit ventricle syndrome. Boon and colleagues [4] and Zemack [24] fail to define underdrainage and describe overdrainage as subdural hygroma and as a clinical symptomatic. In our view, underdrainage can only be said to exist when the pressure level of the valve is unduly high; when, due to a manufacturing defect, the actual valve pressure level is higher than the recommended level; when shunt function changes in vivo; or when cerebrospinal fluid outflow is reduced due to raised intraperitoneal pressure secondary to adiposis or insufficient cerebrospinal fluid resorption provoked by a pseudocyst. The UK Hydrocephalus Shunt Registry, which is comprised of over 9,000 cases, shows underdrainage to be a significant complication following shunt placement (52% of cases). Overdrainage provoked postoperative complications in only 3% of our patients, which stands in stark contrast to the results of some Scandinavian researchers who report that 80% of all postoperative shunt complications are attributable to overdrainage. It is difficult to carry out a comparison of underdrainage associated with other types of shunts for two reasons: first, only very few studies have reached any clear conclusions regarding this complication [21], and second, the majority of potentially comparable studies of patients with normal-pressure hydrocephalus fail to clearly define, or record clear data for, this complication [4, 5]. A positive correlation between ventricular volume reduction and clinical improvement following shunt placement has been reported by Anderson and colleagues [1], Benzel and colleagues [2], Pang and colleagues [17] and Tans and colleagues [22]. However, no such correlation was observed by either Greenberg and colleagues [8] or Yamashita and colleagues [23].

In the absence of a valve that is both adjustable [18] and authentically programmable – one that could continuously measure pressure and flow, and whose feedback mechanism would enable a regenerative coupling function – we will have to make the most of the technical capabilities offered by currently available valves as we strive to achieve optimal levels of intracranial pressure for individual patients. The valves available today lack the capacity to respond to either ventricular volume, cerebral blood flow, or alterations in cerebral metabolism; nor can they optimally regulate cerebrospinal fluid circulation. These valves only respond to the differential pressure between the region of the proximal catheter and the environment at the site of distal diversion. We must therefore continue to make efforts – through the now widely accepted radiological examinations, measurement of blood flow, measurments of intracranial pressure and differentiated infusion tests – to develop criteria that allow us to reliably predict the optimal pressure ratios needed by individual patients postoperatively in both the recumbent and upright positions.

Our results clearly indicate the following: in the vast majority of cases, good to excellent clinical results are accompanied by only a slight reduction in ventricular volume following implantation of a hydrostatic valve (Miethke dual-switch valve); a maximum reduction in ventricular volume is not a prerequisite for clinical improvement in patients with normal-pressure hydrocephalus. Five years of experience with the Miethke dual-switch valve have demonstrated that the design engineering of this gravitational, hydrostatic valve delivers dependable performance. The technical design principle consisting in a parallel two-chamber system containing large membranes [16] that minimize overdrainage, making the Miethe dual-switch valve greatly superior in this respect to diferential pressure valves and adjustable valves, as well as to the majority of other hydrostatic valves. The fact that the functionality of the Miethke dural-switch valve is determined neither by subcutaneous pressure nor by implantation height is an additional contributing factor here. In our view, the low overdrainage rate of 3% and the generally positive outcomes achieved with the Miethke dual-switch valve make it the therapy of choice for patients with normal-pressure hydrocephalus [15, 16].

References

1. Anderson RC, Grant JJ, de la Paz BA, Frucht S, Goodman RR (2002) Volumetric measurements in the detection of reduced ventricular volume in patients with normal-pressure hydrocephalus whose clinical condition improved after ventriculoperitoneal shunt placement. J Neurosurg 97: 73–79
2. Benzel EC, Pelletier AL, Levy PG (1990) Communicating hydrocephalus in adults: Prediction of outcome after ventricular shunting procedures. Neurosurgery 26: 655–660
3. Bergsneider M, Peacock WJ, Mazziotta JC, Becker DP (1999) Beneficial effect of siphoning in treatment of adult hydrocephalus. Arch Neurol 56: 1224–1229
4. Boon AJW, Tans JTJ (1998) Dutch normal pressure hydrocephalus (NPH) study. Randomised comparison of low and medium pressure shunts. Acta Neurochir (Wien) [Suppl] 71: 401–404
5. Di Rocco C, Marchese E, Velardi F (1994) A survey of the first complication of newly implanted CSF shunt devices for the treatment of nontumoral hydrocephalus. Childs Nerv Syst 10: 321–327

6. Drake JM, Silva M, Rutka JR (1993) Functional obstruction of an anti-siphon device of raised tissue capsule pressure. Neurosurgery 32: 137–139
7. Drake JM, Kestle JRW, Milner R, Cinnalli G, Boop F, Piatt J, Haines S, Schiff S, Cochrane D, Steinbok P, MacNeil N (1998) Randomized trial of cerebrospinal fliud valved shunt design in pediatric hydrocephalus. Neurosurgery 43: 294–303
8. Greenberg JO, Shenkin HA, Adam R (1977) Idiopathic normal pressure hydrocephalus – a report of 73 patients. J Neurol Neurosurg Psychiatry 40: 336–341
9. Kiefer M, Eymann R, Meier U (2002) Five years experience with gravitational shunts in chronic hydrocephalus of adults. Acta Neurochir (Wien) 144: 755–767
10. Meier U, Zeilinger FS, Kintzel D (1999) Diagnostic in normal pressure hydrocephalus: A mathematical model for determination of the ICP-dependent resistance and compliance. Acta Neurochir (Wien) 141: 941–948
11. Meier U, Zeilinger FS, Kintzel D (1999) Signs, symptoms and course of disease in normal pressure hydrocephalus in relation to cerebral atrophy. Acta Neurochir (Wien) 141: 1039–1048
12. Meier U, Bartels P (2001) The importance of the intrathecal infusion test in the diagnostic of normal pressure hydrocephalus. Europ Neurol 46: 178–186
13. Meier U, Kiefer M, Bartels P (2002) The ICP-dependency of resistance to cerebrospinal fluid outflow: A new mathematical method for CSF-parameter calculation in a model with H-Tx rats. J Clin Neurosci 9(1): 58–63
14. Meier U (2002) The grading of normal pressure hydrocephalus. Biomed Engineering 47: 54–58
15. Meier U, Kintzel D (2002) Clinical experiences with different valve systems in patients with normal-pressure hydrocephalus: evaluation of the Miethke dual-switch valve. Childs Nerv Syst 18: 288–294
16. Meier U, Kiefer M, Sprung C (2002) The Miethke Dual-Switch valve in patients with normal pressure hydrocephalus. Neurosurg Qu 12(2): 114–121
17. Pang D, Altschuler E (1994) Low-pressure hydrocephalic state and viscoelastic alterations in the brain. Neurosurgery 35: 643–656
18. Pollack IF, Albright AL, Adelson PD (1999) A randomized, controlled study of a programmable valved shunt versus a conventional valve for patients with hydrocephalus. Neurosurgery 45: 1399–1411
19. Pudenz RH, Foltz EL (1991) Hydrocephalus: Overdrainage by ventricular shunts. A review and recommandations. Surg Neurol 35: 200–212
20. Sainte-Rose C, Hooven MD, Hirsch JF (1987) A new approach in the treatment of hydrocephalus. J Neurosurg 66: 213–226
21. Sainte-Rose C (1993) Shunt obstruction: A preventable complication? Ped Neurosurg 19: 156–164
22. Tans JT, Poortvliet DCJ (1989) Does compliance predict ventricular reduction after shunting for normalpressure hydrocephalus? Neurol Res 11: 833–837
23. Yamashita N, Kamiya K, Yamada K (1999) Experience with a programmable valve shunt system. J Neurosurg 91: 26–31
24. Zemack G, Romner B (2000) Seven years of clinical experience with the programmable Codman Hakim valve: a retrospective study of 583 patients. J Neurosurg 92: 941–948

Correspondence: Ullrich Meier, M.D., Chairman, Department of Neurosurgery, Unfallkrankenhaus Berlin, Warener Strasse 7, D-12683 Berlin, Germany. e-mail: ullrich.meier@ukb.de

Acta Neurochir (2003) [Suppl] 86: 539–543

The ICP-dependency of resistance to cerebrospinal fluid outflow: a new mathematical method for CSF-parameter calculation in a model with H-Tx rats

U. Meier[1] and **M. Kiefer**[2]

[1] Department of Neurosurgery, Unfallkrankenhaus Berlin, Berlin, Germany
[2] Department of Neurosurgery, University of Homburg/Saar, Germany

Summary

The international accepted calculation methods concerning the cerebrospinal fluid dynamics proceed from a pressure independent resistance to cerebrospinal fluid outflow. In a new model we focus our attention on the pressure dependency of resistance. In our calculation model we are monitoring the complete pressure course p(t) over the time t during and after the infusion. The comparison of the pressure rise On(p) during the infusion and the descent Off(p) after the infusion in the same pressure level allows to construct all formulas for the compliance C(p) and resistance R(p).

The simultaneous measurement of the resistance and complications during a single investigation allows minimizing the patient's exertion. In contrast to the classical methods it is not necessary that the ICP reach a plateau. Our mathematical method diverges with the description of a pressure dependent slope of the function for the resistance from the static examination models. We proved our mathematical method by the use of a dynamic infusion test in ten H-Tx rats without hydrocephalus and five hydrocephalic H-Tx rats. For that we are able to take the non-linearity of the CSF resorption into consideration.

Keywords: Compliance; infusion test, normal pressure hydrocephalus, resistance, ICP, H-Tx rats.

Introduction

The computation of biosignals requires the description of physiological and pathophysiological events in the human organism by mathematical models, which represent a simplification of complex circuits. Contradictory results obtained from established methods and models for the examination of cerebrospinal fluid dynamics requires these methods and models to be revised. Restrictive assumptions in the kind of a pressure dependency of the parameters have to be excluded in a new model. The high variability of the craniospinal parameters has to be taken into account. In this paper we are investigating our own mathematical model in order to ascertain if a specific patient curve – the resistance to cerebrospinal fluid outflow R(p) in dependency on the intracranial pressure (ICP) – sufficiently characterizes the cerebrospinal fluid dynamics during the examination [14, 15]. Before putting the model into clinical operation its suitability was evaluated by means of experimental technical trials. It must be shown that this model is capable of describing the different conditions of the complicated structures in the human organism. These computer aided model trials with a pressure independent calculation for resistance serves as a comparison with the international accepted calculation methods, which assume a pressure independent value for resistance.

Material and methods

Ten nonhydrocephalic and five hydrocephalic H-Tx rats – a strain presumed to have a polygenic [10, 11] mode of hydrocephalus in a high percentage of offsprings – were used at an average age of 20 to 25 days. The results pertaining to morphology were demonstrated before the infusion-test for every animal in an MRI-scan. With the rats fixed in a stereotactic frame, placed on a thermostatically controlled heat pad and anaesthetised with halothane (0,5–2,5 vol%) vaporised with 100% O_2 (spontaneously breathing), a small bore hole was made 2 mm lateral to the middle line on each side at the level of the coronal suture. The dura was carefully opened and a parenchymal Codman® ICP sensor was inscried in the parenchyma at the depth of 3 mm (in nonhydrocephalic rats) or in the lateral ventricle (in hydrocephalic rats) of the right hemisphere. On the left hemisphere the lateral ventricle was punctured under stereotactic guidance with a 22-G needle. The boreholes were seated with cyanacrylate adhesive to avoid CSF leakage. The correct position for the needle within the ventricular system was confirmed by the visualisation of immediate CSF pulsation within the needles.

ICP values were monitored on-line and stored with PC-ICP [9] at a sampling rate of 100 Hz. The constant-rate infusion with 0,9% NaCl solution at 37 °C was performed using a ultra-precisc syringe pump (WPI, model; SP 120p) at flow rates between 0,5 and 2,0 ml/h in hydrocephalic rats, and 0,5 and 3,0 ml/h in normal rats. Initially the resting pressure was recorded over 20 min. Thereafter the first infusion test was started. The endpoint of the infusion was the achievement of pressure of a pressure. Have a pressure plateau or have an ICP of 50 mmHg. After the infusion was discontinued, the drop in ICP was recorded until the resting pressure was nearly recovered. Thereafter a new infusion test with a flow rate 0,25 ml/min higher than in the first test was started. Five to seven infusion test were performed in each animal. The CSF dynamics according to the algorithm for the "dynamic infusion test" were calculated off-line by a new self-designed software (Win-PCICP). The animals data for the animals own CSF production we used the data provided in the literature: 2,8 μl/min in healthy rats and 1,8 μl/min in hydrocephalic rats.

Results

After determining the formation rate I_o [Equation 1] the function for outflow resistance R(p) can be calculated with equation 2 and the function for compliance C(p) with equation 3 [14, 15]:

$$I_o = DV/Dp\, dp/dt \qquad \text{[Equation 1]}$$

$$R(p) = p/[I_o + I_{inf} Off(p)/[On(p) + Off(p)]] \qquad \text{[Equation 2]}$$

$$C(p) = I_{inf}/[On(p) + Off(p)] \qquad \text{[Equation 3]}$$

Resting pressure

The resting pressure of healthy and hydrocephalic rats was not very different. In hydrocephalic rats it was 13 ± 3,2 mmHg, while in normal rats it was 10 ± 4,3 mmHg. Spontaneous fluctuations of the baseline pressure with a frequency around 0,5 1/min and an amplitude of 3–5 mmHg were exclusively recorded in hydrocephalic rats.

$$102\ mmH_2O = 7{,}5\ mmHg = 7{,}5\ Torr = 1\ kPa$$

Accuracy of the dynamic infusion test

It must be emphasised that the algorithm fore the dynamic infusion test allows the calculation of CSF dynamics only between resting pressure and highest ICP value reached during the infusion test. Figure 1 shows the course of the pressure in five consecutive tests in an hydrocephalic animal. For example the first test was preceded with an infusion volume of 1 ml/h and was started at a resting pressure of 12 mmHg and ended at a plateau pressure of 28 mmHg. The fourth test in this animal was performed with an infusion rate of 1,78 ml/h, starting at the resting pressure of 18 mmHg and ending at the level of 38 mmHg. Consequently, the CSF dynamics were calculated at between 18 and 38 mmHg. Figure 2 shows on the right the results concerning the R_{out} of this animal in a summation figure. The overlaying of the results of the consecutive infusion tests in a given animal was, measured at the small values, excellent, with differences of <10% in a given animal for all calculated CSF dynamics (Fig. 3). This clearly proves the accuracy of the mathematical algorithm of the dynamic infusion test.

Resistance to outflow

The slope of R_{out} over a pressure range of 10 to 30 mmHg was the same in hydrocephalic and nonhydrocephalic animals. In the lower ICP levels, the R_{out} was high in both groups, but it was decreasing proportional to the increasing ICP and reached the lowest value at an ICP between 25 and 30 mmHg. From this lowest point the slope rose steadily with the increasing ICP. The R_{out} values at given ICP levels were clearly higher in hydrocephalic animals than in nonhydrocephalic rats. The lowest R_{out} was 1800 mmHg × min/ml on average in hydrocephalic rats, and 925 mmHg × min/ml on average in nonhydrocephalic rats (Fig. 2). While the R_{out} of nonhydrocephalic rats is between 2800 and 2600 mmHg × min/ml at an ICP < 10 mmHg or 34 mmHg, it is over 4800 mmHg × min/mlin hydrocephalic animals at the same ICP levels. The "V-shaped" slope of the curve is similar in both, regardless of the hydrocephalic state (Fig. 3), but the ICP range of the lowest R_{out} values is wider in healthy rats than in hydrocephalic.

Discussion

Only a few studies concerning the dynamics of CSF-flow in H-Tx rats have been performed until now. The methods used in the literature are quite different [1, 3, 12, 13, 17]. The classical method for the evaluation of the R_{out} is the constant rate infusion test, the Katzmann test [8]. Two facts must be emphasised: on the one hand this classical test provides data only on the R_{out}, and on the other hand the values it yields are true only for the ICP level of the plateau pressure, because it can be assumed that at this pressure level the CSF absorption is equivalent to the CSF produced by the living being itself and artificially added CSF [16, 18].

The values for the Resistance to outflow obtained in the study with H-Tx rats are correlating very well with the values reported in the literature [5, 6, 7]. For hy-

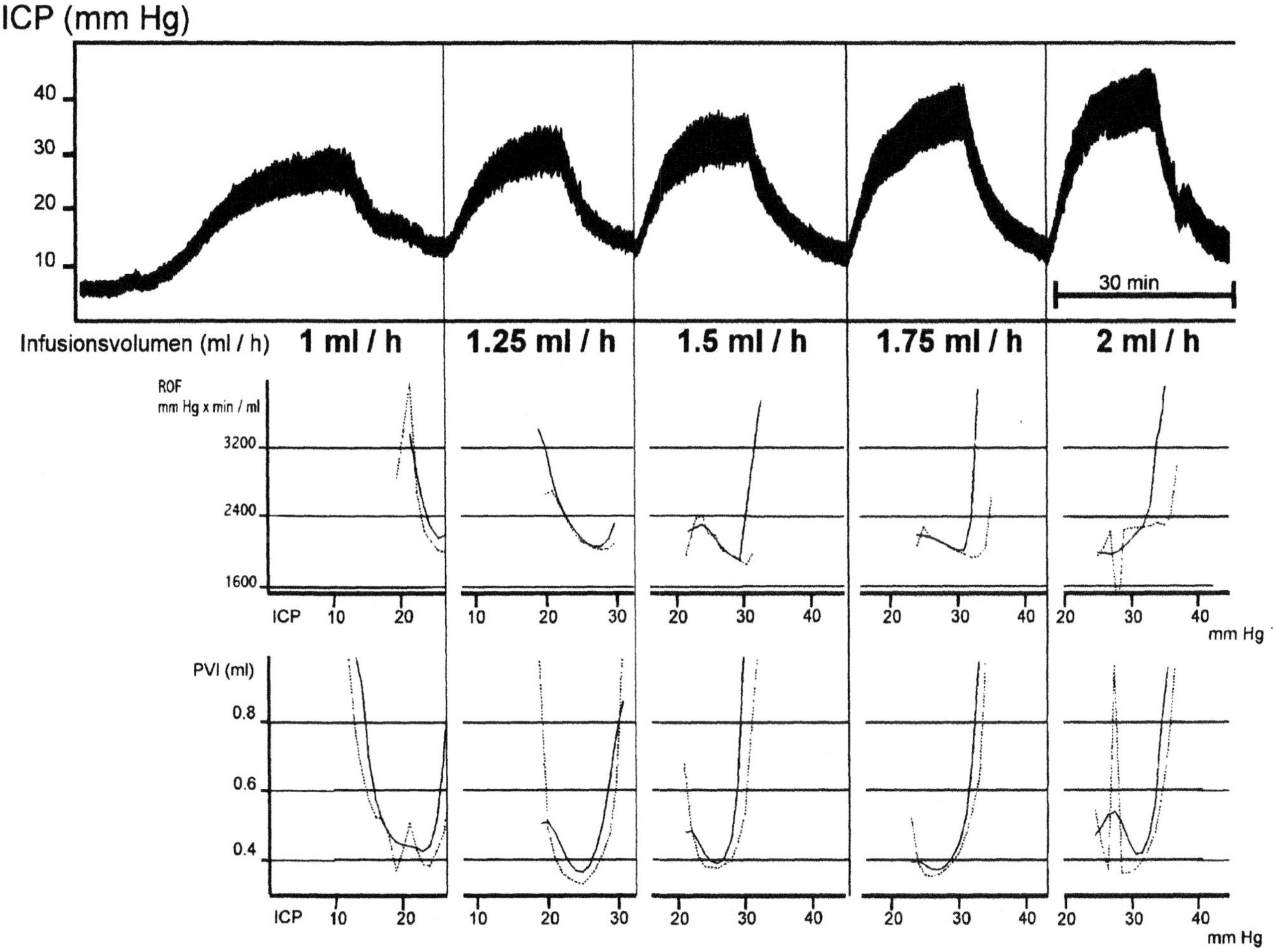

Fig. 1. Five consecutive infusion tests in a hydrocephalic rat. Upper row: original ICP record. Middler and lower row: results of each of the infusion according to the model of dynamic infusion test (middle: resistance to outflow R_{out}; lower: pressure volume index *PVI*)

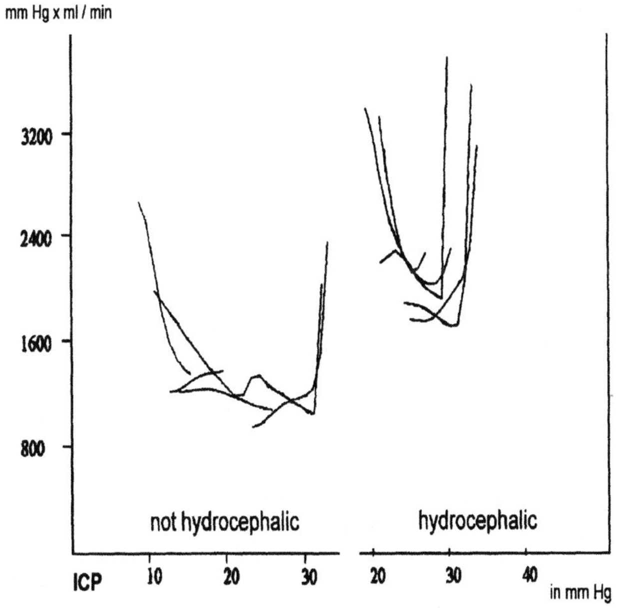

Fig. 2. Summary of the five infusion test results in one non-hydrocephalic (left) and one hydrocephalic (right) h-Tx rat. Note the good concordance of the different infusion tests in the same animal

drocephalic H-Tx rats, R_{out} values of 4500 mmHg × min/ml were reported [5]. The technique used by Jones and Bucknall [5] to evaluate the R_{out} was different from that in our study. Because they used the classical Katzmann test [8]. The plateau pressure levels ranged between 3 and 19 cmH_2O. The values for these ICP levels found in our study were similar. The R_{out} values on nonhydrocephalic rats ranged in our study between 900 and 2000 mmHg × min/ml. This is very well concordant with the values reported for normal pressure H-Tx rats. In comparison with other stocks [1, 6] these values are clearly high. However, it has to be emphasised that Jones and Bucknall [5] already found the R_{out} was somewhat higher in nonhydrocephalic H-Tx rats than in other stocks. The slope of the R_{out} may be divided in three parts. With the ICP levels < 25–30 mmHg the absorption of the CSF increases with the widening pressure differences between CSF pressure and the intraventricular pressure at the resorption

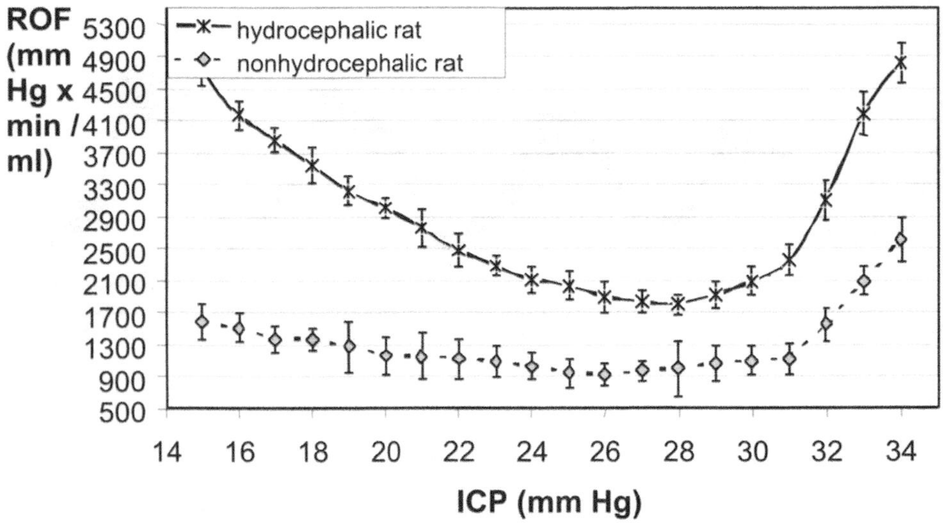

Fig. 3. The average R_{out} values of all animals, with SD bars. The ICP range illustrated here is smaller than the range in an individual animal, because the ICP level for which R_{out} values were only available from three different infusion tests in each animal were considered

sites. The infusion test promotes the expansion of the ventricular system, which results in a narrowing of the cerebral subarachnoidal space. Finally this process compensates a theoretically infinite decrease in R_{out} with a corresponding increase in ICP. This results primarily in a lowest R_{out} value at ICP levels between 25 and 30 mmHg. Beyond these ICP values, the effect of the ongoing reduction of the cerebral subarachnoidal space dominates over the facilitating effect of an increasing pressure difference CSF and venous blood pressure at the absorption sites. Therefore the R_{out} increases dramatically over ICP levels of 30 mmHg. This mechanism is the same in hydrocephalic and nonhydrocephalic animals. Accordingly, the slopes of the R_{out} curves are very similar in both groups. The definitive difference between hydrocephalic and nonhydrocephalic animals was – as expected – he absolute R_{out} value at a certain given ICP level. The R_{out} in hydrocephalic rats was at least twice as high as that in nonhydrocephalic H-Tx rats. With our model we proved for the first time the pressure dependency of CSF-flow in H-Tx rats.

Our method [14, 15] allows a simultaneous calculation of the discharge resistance and the compliance. A further advantage is the distinctly shortened examination time. In contrast to the classical method by Katzmann and Hussey [8], no ICP-plateau needs to form in order to calculate the parameter. Our mathematical computation deviates from the static approach. Using our model, a pressure dependent function course for the outflow resistance is obtained, thus enabling us to assess the system stability as well as to analyze the type of existing outflow disorder. Following this, it is possible to conduct a valve simulation using the patient characteristics curve in order to determine the optimal valve pressure region. Furthermore this method takes the non-linearity of the resorption behaviour into consideration and eliminates the assumption of a constant outflow resistance [2, 4, 13].

References

1. Bötel C, Brinker T (1993) Measurement of the dynamics of the cerebrospinal fluid system. J Exp Anim Sci 36: 78–83
2. Borgesen SE, Gjerris F, Sørensen SC (1979) Cerebrospinal fluid conductance and compliance of the craniospinal space in normal-pressure hydrocephalus. J Neurosurg 51: 521–526
3. Cserr H (1965) Potassium exchange between cerebrospinal fluid, plasma and brain. J Physiol 209: 1219–1226
4. Gjerris F, Børgesen SE (1992) Current concepts of measurement of cerebrospinal fluid absorption and biomechanics of hydrocephalus. In: Symon L, Calliauw L, Cohadon F, Lobo J, Loew F, Nornes H, Pásztor E, Pickard JD, Strong AJ, Yasargil MG (eds) Advances and technical standards in neurosurgery, vol 19. Springer, Wien New York, pp 145–177
5. Jones HC, Bucknall RM (1987) Changes in cerebrospinal fluid pressure and outflow from the lateral ventricles during development of congenital hydrocephalus in the H-Tx rat. Exp Neurology 98: 573–583
6. Jones HC, Deane R, Bucknall RM (1987) Developmental changes in cerebrospinal fluid pressure and resistance to absorption in rats. Dev Brain Res 33: 23–30

7. Kaiser G, Jones HC (1991) Cerebrospinal fluid pressure in 10-day-old rats with congenital hydrocephalus. Eur J Pediatr Surg [Suppl] I: 20–22
8. Katzmann R, Hussey F (1970) A simple constant-infusion manometric test for measurement of CSF absorption. Neurology 20: 534–544
9. Kiefer M (1995) Moderne Hirndruckdiagnostik. Minim Invasive Medizin 6(2): 78–86
10. Kiefer M, Eymann R, von Tiling S, Müller A, Steudel WI, Booz KH (1998) The ependyma in chronic hydrocephalus. Childs Nerv Syst 14: 263–270
11. Kohn DF (1981) Chinookoswong N, Chou SM: a new model of congenital hydrocephalus in the rat. Acta Neuropathol 54: 211–218
12. Mann JD, Butler AB, Rosenthal JE, Maffeo CJ, Johnson RN, Brass NH (1978) Regulation of intracranial pressure in rat, dog, man. Ann Neurol 3: 399–402
13. Marmarou A, Shulman K, LaMorgese J (1975) Compartmental analysis of compliance and outflow resistance of the cerebrospinal fluid system. J Neurosurg 43: 523–534
14. Meier U, Künzel B (1988) Rechnergestützte Modellversuche zur Liquordynamik des kraniospinalen Systems. Zentralbl Neurochir 49: 233–237
15. Meier U, Zeilinger FS, Kintzel D (1999) Diagnostic in normal pressure hydrocephalus: A mathematical model for determination of the ICP-dependent resistance and compliance. Acta Neurochir (Wien) 141: 941–948
16. Melton JE, Nattie EE (1984) Intracranial volume adjustments and cerebrospinal fluid pressure in the osmotically swollen rat brain. Am J Physiol 246: R533–R541
17. Nakamura S, Hochwald GM (1983) Spinal fluid formation and glucose influx in normal and experimental hydrocephalic rats. Exp Neurology 82: 108–117
18. Schuhmann MU, Thomas S, Hans VHJ, Beck H, Brinker T, Samii M (1998) CSF dynamics in a rodent model of closed head injury. Acta Neurochir (Wien) [Suppl] 71: 300–302

Correspondence: Privat-Dozent Dr. med. Ullrich Meier, Head of Department, Department of Neurosurgery, Unfallkrankenhaus Berlin, Warener Strasse 7, D-12683 Berlin, Germany. e-mail: ullrich.meier@ukb.de

Acta Neurochir (2003) [Suppl] 86: 545–547

Metabolic derangements in interstitial brain edema with preserved blood flow: selective vulnerability of the hippocampal CA3 region in rat hydrocephalus

T. Kawamata[1], **Y. Katayama**[2], **N. Tsuji**[1], and **H. Nishimoto**[2]

[1] Department of Neurological Surgery, Nihon University School of Medicine, Tokyo, Japan
[2] Saitma Children's Medical Center, Tokyo, Japan

Summary

Hydrocephalus induces interstitial brain edema, which causes neurological deficits, even if the intracranial pressure is maintained within the normal range, and the cerebral blood flow (CBF) does not decline to an ischemic level. The precise mechanisms underlying such edema-induced neuronal dysfunction remain unclear. In the present study, in an attempt to elucidate the metabolic derangements in brain tissue with interstitial edema, we evaluated the changes in CBF and oxidative/glucose metabolism using a rat model of kaolin-induced hydrocephalus. Hydrocephalus was produced in male Wistar rats by intrathecal injection of 0.1 ml aluminum silicate suspension (200 mg/mI) via the cisterna magna. CBF was determined by 14[C]-iodoantipyrine autoradiography. Oxidative metabolism was evaluated by cytochrome oxidase (CYO) histochemistry, and glucose metabolism by hexokinase (HK) histochemistry. CBF declined with the development of hydrocephalus, but did not reach an ischemic level. The CYO activity was diffusely depressed in both the cortex and hippocampus. The HK activity was preserved at the early stage of hydrocephalus. At the advanced stage, the HK activity was reduced in the hippocampal CA3 region first, and diffusely thereafter. In conclusion, interstitial brain edema impairs oxidative metabolism even at the early stage of hydrocephalus, and shifts the metabolism to anaerobic glycolysis despite a preserved CBF. Impairment of glucose metabolism was first observed in the CA3 region, suggesting that the CA3 is metabolically vulnerable, and CA3 dysfunction may contribute to the memory deficits seen in hydrocephalus.

Keywords: Hydrocephalus; interstitial edema; oxidative metabolism; glycolysis; cerebral blood flow; cytochrome oxidase, hippocampus, CA3.

Introduction

Interstitial edema is defined as increased water content in the extracellular space, as typically observed in the white matter of hydrocephalus. It is clear that interstitial edema clinically induces neurological deficits, even if the intracranial pressure is maintained within the normal range and the cerebral blood flow (CBF) dose not decline to an ischemic level [1]. However, precise mechanisms underlying such neuronal dysfunction remain unclear [4]. A previous study suggested that enlargement of extracellular space due to vasogenic brain edema may disturb diffusional delivery of oxygen from capillary vessels to neuronal cells, causing impairment of aerobic metabolism and resultant neuronal dysfunction [2, 7].

In the present study, in order to examine effects of the hydrocephalus-induced brain edema on oxidative and glucose metabolism, we measured the cytochrome oxidase (CYO) activity, a marker of mitochondrial respiratory function, and the hexokinase (HK) activity, a marker of glyeolysis, together with CBF employing iodoantipyrine autoradiography, in a rat model of kaolin-induced hydrocephalus.

Material and methods

Young Wistar rats (250–300 g) were anesthetized with mixture of 66% nitrous oxide, 33% oxygen and 1% halothane. The rectal temperature was maintained at 37–38 °C with a heating pad. The dura matter between occipital bone and C1 lamina was exposed and kaolin suspension (0.1 ml aluminum silicate suspension (200 mg/ml)) was injected into the cisterna magna. Saline injected animals were employed as a sham control. At 1, 2 and 4 weeks after the kaolin injection, the brain was perfused transcardially with 4% paraformaldehyde solution for CYO and HK histochemistry. The CYO and HK activity in hippocampus region were determined densitometorically from the histochemical staining, which is known to correlate directly with the biochemically quantitated CYO and HK activities [3, 7]. The CBF value was estimated autoradiographically [5] at the same time points. The ventricle size was measured on the coronal section crossing the septal nucleus, to classify the ventricular dilatation into 3 grades. Grade 1 was less than 300% ventricular dilatation as compared that of the sham operated animals. Grade 2 was 300–1000% ventricular dilatation. Grade 3 was more than 1000%. And then 3 groups were defined as follows: Group 1 animals had Grade 1 ventricular dilatation at 1 week after kaolin injection.

Group 2 had Grade 2 at 2 weeks, and Group 3 had Grade 3 at 4 weeks after the injection. Animals, which did not conform this classification, were excluded from the study.

Results

Brain water content

The water content in hippocampus was 81.3 + 2.72% in the sham animals, and 83.2 + 2.90%, 84.1 + 3.11%, 83.7 $\pm$ 2.13% in Group 1, 2, and 3, respectively, showing progressive edema formation along with the progression of hydrocephalus.

CBF in hippocampus

The %CBF in hippocampus was 95 $\pm$ 5.6%, 90 $\pm$ 4.6%, and 78 + 2.2%, in Group 1, 2, and 3, respectively. The reduction of CBF did not reach the ischemic level at any time points.

CYO activity and HK activity in hippocampus

The %CYO and %HK activity in hippocampus was obtained from the optical density of hisitochemistry in the hydrocephalic rats, divided by that in the sham animals [3, 7]. The %CYO was 79 + 4.7%, 72 $\pm$ 4.3%, 70 + 4.5% in Group 1, 2, and 3, respectively (Fig. 1), indicating that oxidative metabolism was significantly impaired even in the early stage of hydrocephalus ($p < 0.01$). The decrease in CYO activity in hippocampus was mainly observed in the pyramidal cell layer, and did not show significant difference among the area of CA1, CA3, and dentate gyrus.

The %HK activities in hippocampus was 92 + 5.8%, 85 + 4.8%, 78 + 3.8% in Group 1, 2, and 3, respectively, indicating that glucose metabolism was preserved at the near normal levels in the early stage of hydrocephalus (Group 1 and 2), and was significantly impaired in Group 3 ($p < 0.01$). The measurements of HK activities in CA1, CA3 and dentate gyrus revealed that CA3 region had lower activity as compared to CA1 and dentate gyrus (Fig. 2).

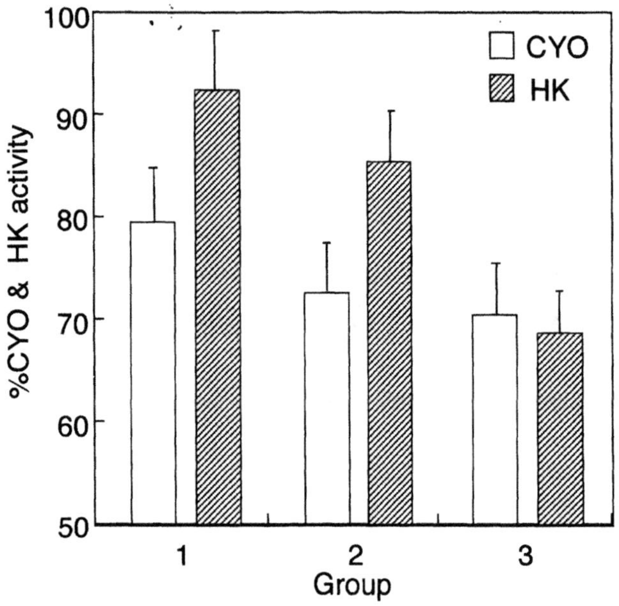

Fig. 1. The %CYO and %1-1K activity in hippocampus (n = 5 in each group). The CYO activity was decreased, while the 11K activity was preserved, in Group 1. The difference between CYO and 1-1K activities was still significant in Group 2 ($p < 0.01$). These findings indicate that oxidative metabolism is predominantly suppressed in the early stage of hydrocephalus

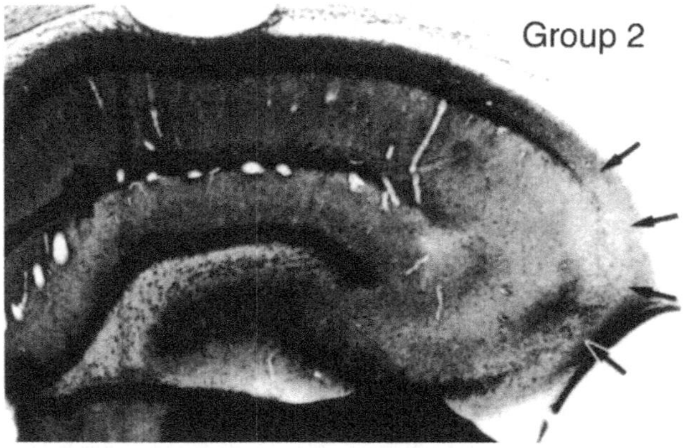

Fig. 2. A representative HK histochemistry of hippocampus in Group 2 clearly demonstrated the selective reduction of 11K activity in CA3 region (arrows), suggesting that CA3 is metabolically vulnerable

Discussion

One of the mechanisms underlying the neurological deficits observed in brain edema has been considered to be a decrease in regional CBF, resulting in impairment of axonal trasmission in the edematous white matter. However, the result of this study clearly demonstrated that the interstitial edema did not markedly influence on CBF and glucose metabolism, but did impair oxidative metabolism in the early phase of hydrocephalus.

Oxygen and glucose are utilized as energy sources to ensure maintenance of neuronal cell functions. For the delivery of these energy sources, two major routes exist between the capillary vessels and neuronal cells. One is diffusional delivery through the extracellular space, and the other is an active transport vial glial cell. Glucose is delivered by both these routes, while oxygen is delivered mainly by diffusion. Sutton *et al.* suggested that enlargement of the extracellular space more strongly disturbs the oxygen diffusion than the glucose transport [6]. If this is the case, edema formation could impair the oxidative metabolism predominantly [2], resulting in activation of anaerobic glycolysis, even under the condition with preserved CBF.

In Group 1 animals, the CBF and glucose metabolism were preserved, while oxidative metabolism was impaired. These findings are consistent with the hypothesis that edema formation disturbs the oxygen diffusion in the extracellular space. In the advanced stage, glucose metabolism was additionally impaired, resulting in aggravation of neuronal dysfunction. The impairment of glucose metabolism was fist seen in hippocampal CA3 region, suggesting that CA3 is metabolically vulnerable in hydrocephalus. Since no cell loss was observed on Nissl staining in the same animals, the reduction of HX activity was not due to the selective cell death in CA3 region.

In summary, the results of the present study indicate that hydrocephalus causes a unique metabolic derangements, involving impairment of oxidative metabolism despite preserved cerebral circulation, which can be regarded as a state of relative hypoxia. The hippocampal CA3 region is metabolically vulnerable, and the CA3 dysfunction may contribute to the memory deficits seen in hydrocephalus.

References

1. Adams RD, Fisher CM, Hakim 5 *et al* (1965) Symptomatic occults hydrocephalus with "normal" cerebrospinal fluid pressure: a treatable syndrome. N Engl J Med 273: 117–126
2. Kawamata T, Katayama Y, Kinoshita K *et al* (1994) Cytochrome oxidase and hexokinase activities in an infusion edema model with preserved blood flow. Acta Neurochir (Wien) [Suppl] 60: 268–270
3. Lawrence GM, Trayer IP (1984) Histochemical and immunohistochemical localization of hexokinase isoenzymes in rat kidney. Histochem J 16: 697–708
4. Ohta K, Marmarou A, Povlishock JT (1990) An immunocytochemical study of protein clearance in brain infusion edema. Acta Neuropathol (Ben) 81: 162–177
5. Sakurada 0, Kennedy C, Jehle J *et al* (1978) Measurement of local cerebral blood flow with iodol4[C] antipynine. Am J Physiol 234: 1159–1166
6. Sutton LN, Greenberg J, Welsh F (1990) Blood flow and metabolism in vasogenic oedema. Acta Neurochir (Wien) [Suppl] 51: 397–400
7. Wong-Riley MTT (1989) Cytochrome oxidase: an endogenous metabolic maker for neuronal activity. TINS 12: 94–101

Correspondence: Tatsuro Kawamata, M.D., Ph.D., Department of Neurological Surgery, Nihon University School of Medicine, 30-1 Oyaguchi Kamimachi, Itabashi-ku, Tokyo 173-8610, Japan. e-mail: kawamata@med.nihon-u.ac.jp

Acta Neurochir (2003) [Suppl] 86: 549–551

MRI analysis of hydrocephalus associated with acoustic neurinoma

K. Wada[2], **H. Nawashiro**[1], **A. Shimizu**[3], and **K. Shima**[1]

[1] Department of Neurosurgery, National Defense Medical College, Japan
[2] Undersea Medical Center, Japan Maritime Self Defense Force, Japan
[3] Japan Defense Force Central Hospital, Japan

Summary

We investigated the hydrocephalus in 24 patients associated with acoustic neurinoma. We found the high incidence of homo-lateral ventricular dilatation to the side of the acoustic neurinoma. Utilizing magnetic resonance imaging, the diameter of the tumor parallel to the pyramidal bone, diameter of the tumor perpendicular to the pyramidal bone, grade of the 4th ventricle deviation, and the shape of the tumor (round or oval) were analyzed. Ten (42%) of the 24 patients with acoustic neurinoma were found to have hydrocephalus. Seven (70%) of the 10 patients with hydrocephalus exhibited asymmetrical lateral ventricle dilatation: in all cases the lateral ventricle in the hemisphere homolateral to the acoustic neurinoma was larger than that of the contralateral side. The hydrocephalus was not related to the grade of the 4th ventricle deviation but rather to the diameter of the tumor parallel to the pyramidal bone ($p < 0.01$). The diameter of the tumor parallel to the pyramidal bone was also related to the asymmetrical lateral ventricular dilatation ($p < 0.05$).

Keywords: Hydrocephalus; acoustic neurinoma; MRI.

Introduction

Although the association of hydrocephalus and acoustic neurinomas is well known, the etiological factors and mechanisms are not fully understood [1, 3, 4]. Hydrocephalus is commonly related to the size of tumors and occurs more frequently in association with tumors greater than 3 cm in diameter [4]. We retrospectively investigated the incidence of hydrocephalus and the clinical presentations including the magnetic resonance imaging in 24 patients with acoustic neurinomas to clarify the potential etiological factors.

Patients and methods

We retrospectively studied 24 patients with acoustic neurinomas who were admitted to the National Defense Medical College Hospital between January 1988 and December 1995, and whose magnetic resonance imaging was available to morphometric analysis. The assessment of associated hydrocephalus depends on the ratio of the maximal width of the frontal horns to the maximal diameter of the inner table of the skull at the same level. If the ratio was more than 30%, the hydrocephalus was diagnosed irrespectively of clinical symptoms and signs. The parameters used were diameter of the tumor parallel to the pyramidal bone on axial MRI, diameter of the tumor perpendicular to the pyramidal bone on coronal MRI, grade of 4th ventricle deviation (classified into 3 grades), and the shape of the tumor (round or oval). The association between the hydrocephalus and asymmetrical dilatation of the lateral ventricles and variable factors was analyzed by the Mann-Whitney U test. Values of $p < 0.05$ were considered.

Results

Table 1 shows the incidence of the hydrocephalus and asymmetrical dilatation of the lateral ventricles related to the parameters on MRI in all patients. Ten (42%) of the 24 patients with acoustic neurinoma were found to have hydrocephalus (Figs. 1 and 2). Seven (70%) of the 10 patients with hydrocephalus exhibited asymmetrical lateral ventricle dilatation (Fig. 1): in all cases the lateral ventricle in the hemisphere homolateral to the acoustic neurinoma was larger than that of the contralateral side. The hydrocephalus was not related to the grade of the 4th ventricle deviation but rather to the diameter of the tumor parallel to the pyramidal bone ($p < 0.01$) and to the diameter of the tumor perpendicular to the pyramidal bone ($p < 0.05$). The diameter of the tumor parallel to the pyramidal bone was also related to the asymmetrical lateral ventricular dilatation ($p < 0.05$).

Discussion

In this study, we found a significant correlation between the presence of hydrocephalus and the size of the

Table 1. *Relationship between incidence of hydrocephalus and various parameters on MRI*

	Total	Hydrocephalus		Asymmetrical dilatation of lateral ventricles	
		+	−	+	−
Number of patients	24	10	14	7	17
Age	52.7 ± 11.2	54.7 ± 9.8	51.3 ± 12.3	55.0 ± 9.8	51.5 ± 12.3
Perpendicular size (cm)	2.6 ± 0.9	3.7 ± 0.6*	2.3 ± 1.0	3.9 ± 0.5	2.4 ± 1.0
Parallel size (cm)	3.0 ± 1.2	3.9 ± 0.5**	2.4 ± 1.0	4.2 ± 0.7*	2.5 ± 1.0
Shape – ovale	6	6	0	5	1
– round	18	4	14	2	16
4th ventricle deviation	1.8 ± 0.9	2.1 ± 0.9	1.6 ± 0.9	1.7 ± 0.8	1.9 ± 1.0

*: $p < 0.05$, **: $p < 0.01$

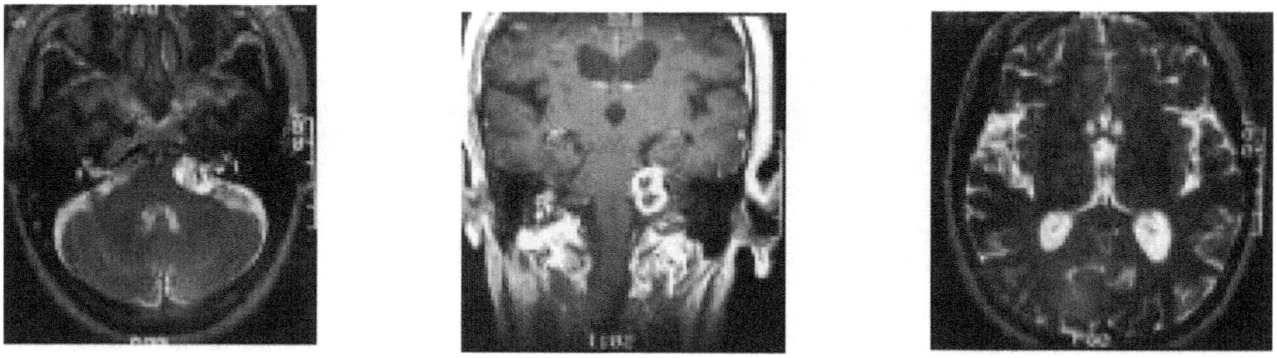

Fig. 1. A representative case with asymmetrical dilatation of lateral ventricles and its parameters on MRI

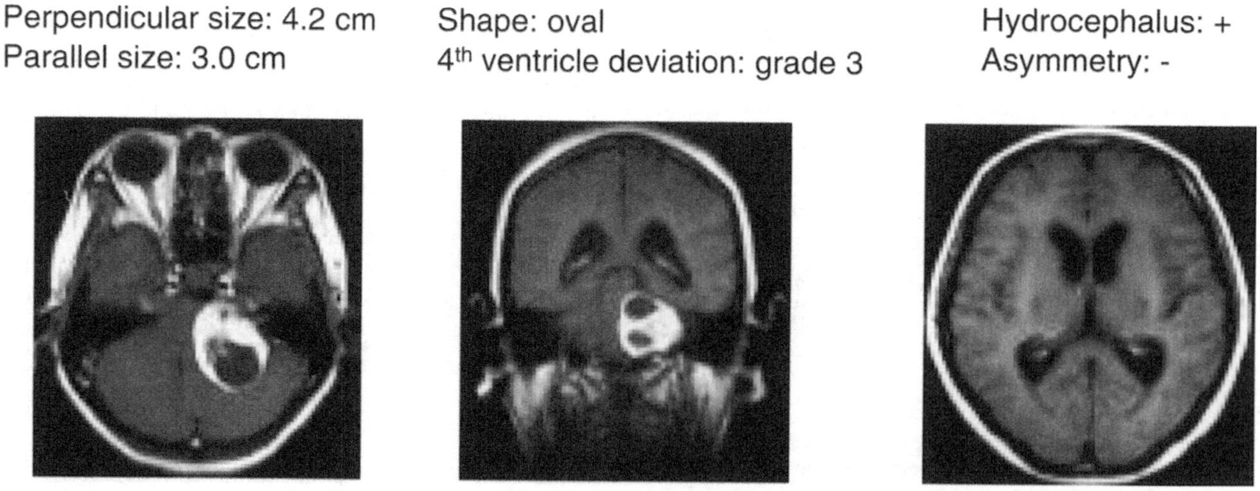

Fig. 2. A representative case with asymmetrical dilatation of lateral ventricles and its parameters on MRI

tumor, especially the size of the tumor parallel to the pyramidal bone on axial MRI. It has been reported 3.7% [2] and 15% [1] of patients with cerebello-pontine angle tumors develop hydrocephalus. High incidence (42%) of hydrocephalus in the present study may be related to the criteria to diagnose hydrocephalus. If the Evans' index (the ratio of the maximal width of the frontal horns to the maximal diameter of the inner table of the skull at the same level) was more than 30%, the hydrocephalus was diagnosed irrespectively of clinical symptoms and signs in the study. The most frequent clinical manifestations of hydrocephalus in these patients are similar to those observed in patients with normal pressure hydrocephalus. Typical symptoms and signs of increased intracranial pressure are relatively uncommon (20% in our series). Elevated total protein concentration of cerebrospinal fluid (CSF) is one of the etiologic factors of associated hydrocephalus. The asymmetrical lateral ventricular dilatation is a unique finding in our series. Especially the side of the asymmetrical dilatation was consistent with the side of the tumor. We speculate that the difference in CSF bulk flow around the cerebellopontine cisterns and around the cerebral cortices homolateral to the tumor might promote the pressure gradient between the lateral ventricles and convexity CSF space.

References

1. Atlas MD, Perez de Tagle JR, Cook JA, Sheehy JP, Fagan PA (1996) Evolution of the management of hydrocephalus associated with acoustic neuroma. Laryngoscope 106: 204–206
2. Briggs RJ, Shelton C, Kwartler JA, Hitselberger W (1993) Management of hydrocephalus resulting from acoustic neuromas. Otolaryngol Head Neck Surg 109: 1020–1024
3. Pirouzmand F, Tator CH, Rutka J (2001) Management of hydrocephalus associated with vestibular schwannoma and other cerebellopontine angle tumors. Neurosurgery 48: 1246–1254
4. Steenerson RL, Payne N (1992) Hydrocephalus in the patient with acoustic neuroma. Otolaryngol Head Neck Surg 107: 35–39

Correspondence: Hiroshi Nawashiro, M.D., D.M.Sc., Department of Neurosurgery, National Defense Medical College, 3-2 Namiki, Tokorozawa, Saitama 359-8513, Japan. e-mail: nawa1957@me.ndmc.ac.jp

Blood brain barrier, miscellaneous

Acta Neurochir (2003) [Suppl] 86: 555–558

Non-invasive opening of BBB by focused ultrasound

K. Hynynen, N. McDannold, N. Vykhodtseva, and **F. A. Jolesz**

Department of Radiology, Brigham and Women's Hospital, and Harvard Medical School, Boston, MA

Summary

Blood brain barrier (BBB) is a major barrier for delivering therapeutic agents in the brain. In this study we investigated the feasibility of open the BBB by using focused ultrasound. Rabbit brains were exposed to pulsed focused ultrasound while injecting ultrasound contrast agent containg microbubbles intravenously. The BBB opening was measured after the sonications by injecting MRI contrast agent IV and evaluating the local enhancement in the brain. Low ultrasound powers and pressure amplitudes were found to cause focal enhancement. Before sacrificing the animals trypan blue was also injected IV. After the sacrifice of the animals blue spots were found in the brain in the sonicated locations. This method may have potential for targeted delivery of macromolecules in the brain.

Keywords: Blood brain barrier; brain; ultrasound; drug delivery; therapy.

Introduction

In spite of the large number of very potent neurologically active substances and drugs, many CNS diseases are difficult to treat due the inability of mostly large molecular size agents to penetrate the blood-brain barrier (BBB) [1]. The BBB is a major limitation for utilizing neuropeptides, proteins (enzymes, antibodies) chemotherapeutic agents and genes for therapy [10]. After non-localized, diffuse opening of the BBB agents administered systematically or injected into the blood stream have undesired, often dose-limiting side effects due to their spread within the CNS. However, these substances could have more definitive therapeutic effect if their release in the brain would be localized. This implicates that the transient, reversible opening of the BBB is the most likely noninvasive technique which provides anatomically selective, targeted drug delivery. Image-guided monitoring and controlled opening of the BBB at a desired location thus establishes novel methods of treating CNS diseases such as brain tumors, seizure and movement disorders. Using large molecular size peptides, neuroactive proteins and various antibodies new innovative therapeutic interventions will be available for dealing with organic brain diseases and mental disorders.

The BBB prevents penetration of ionized water-soluble materials with molecular weight greater than 180 [7]. Chemical modification of the drugs to make them lipophilic or the use of other carriers such as aminoacid and peptide carriers are two ways to aid propagation through the barrier. Another option is to diffusely alter the function of BBB by temporarily opening the tight junctions which is possible with an increasing number of chemicals. Most of the clinical experience, however, has been confined to the BBB opening using intra-arterial injection of hyperosmotic solutions such as mannitol. This causes the endothelial cells to shrink and then open the tight junctions for a few hours [4]. Both the osmotic and chemical methods require intra-arterial catheterization and produce non-focal, transient BBB opening within the entire tissue volume supplied by the injected artery branch [1, 7].

A more localized drug delivery method can only be accomplished by injecting through a needle or catheter directly into the targeted brain area [7]. Such direct injections are invasive and require opening of the skull, penetration of non-targeted brain tissue, and carry the risk of neurological damage, bleeding, and infection.

It has been known for some time that pulsed ultrasound can sometimes induce focal BBB opening without damaging neurons in the sonicated area [12]. More recently it has been shown that by injecting an ultrasound contrast agent bolus containing the gas bubbles low power ultrasound exposures can be used to open the BBB [6]. Similarly, these agents have been shown to reduce the threshold for ultrasound-induced tissue damage in other tissues [2, 3, 9]. The purpose of this

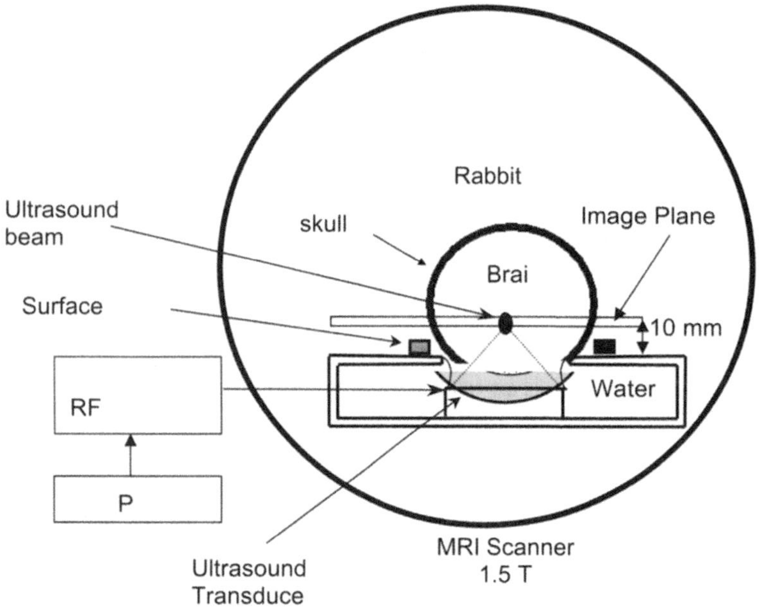

Fig. 1. A diagram of the sonication setup

study was to demonstrate the ability to use this mechanism for focal opening of open BBB for delivery of therapeutic agents into the brain.

Methods

New Zealand white rabbits (3–4 kg) were anesthetized. A piece of skull (approximately 20 × 20 mm) was removed and the skin was replaced over the bone window. The sonications were executed after the wound healed. The transducer had a diameter of 10 cm, a radius of curvature of 8 cm, and a frequency of 1.63 MHz. The transducer was moved within the MR table by a positioning device [5]. The MRI scanner was a standard clinical 1.5 Tesla system (General Electric Medical Systems) (Fig. 1). Four or six locations were sonicated with the focus aimed 10 mm deep in the brain at different peak acoustic power levels ranging from 0.2 to 11.5 W. This translated to acoustic temporal and spatial peak intensity range of 16 to 690 W/cm^2. The sonications were pulsed with the repetition frequency of 1 Hz. The duration of the whole sonication was 20 s. Approximately 10 s prior to the start of the sonication, a bolus of ultrasound contrast agent (Optison, Mallinckrodt Inc. St. Louis, MO) containing micro bubbles was injected in the ear vein.

The temperature dependent proton resonant frequency shift [8] was evaluated using a fast spoiled gradient echo sequence (FSPGR) with the following imaging parameters: repetition time (TR) = 50.6 ms, echo time (TE) = 20 ms, flip angle = 30°, bandwidth (BW) = 3.1 kHz, resolution 256 × 128, field of view FOV = 12 cm, slice thickness = 3 mm. The temperature elevation was calculated from the phase shift by using a coefficient 0.011 ppm/°C. T1-weighted images were obtained by using a Fast Spin Echo (FSE) pulse sequence (TR = 500 ms, TE = 17 ms, echo train length = 4, three data averages, field of view = 10 cm, matrix size = 256 × 256, slice thickness = 1.5 mm interleaved, BW = 16 kHz). These scans were repeated after a bolus of gadopentetate dimeglumine contrast agent (MAGNEVIST, Berlex Laboratories Inc, Wayne, NJ) was injected into the ear veins of the rabbits (dose 0.125 mmol/kg) to evaluate the blood brain barrier opening.

After the imaging Trypan blue was injected in the blood stream to evaluate the blood brain barrier opening for larger molecules.

Signal analysis

The amount of contrast enhancement was evaluated in the focal spot (location of the ultrasound beam focus) by averaging the signal intensity over a 3 × 3 voxels (1.17 × 1.17 mm) after normalizing to the baseline value before the contrast injection.

Animal sacrifice

The animals were sacrificed between 2 h and 7 days after the sonications. The animals on the first day were sacrificed typically at 4 h after the sonications. The brains were then immediately fixed in formalin, embedded in paraffin and serially sectioned at 6 micrometers (across the beam direction; parallel to the MRI slices). Every 50th section (interval of 0.3 mm) was stained with hematoxylin and eosin for histologic examination. The microscopy study was looking for areas with damaged neurons similar to those presented in [11]. The histology study was performed without the knowledge of the number, location, or power of the sonications.

Results

The focal signal intensity (SI) change after the MRI contrast agent injection increased with increasing

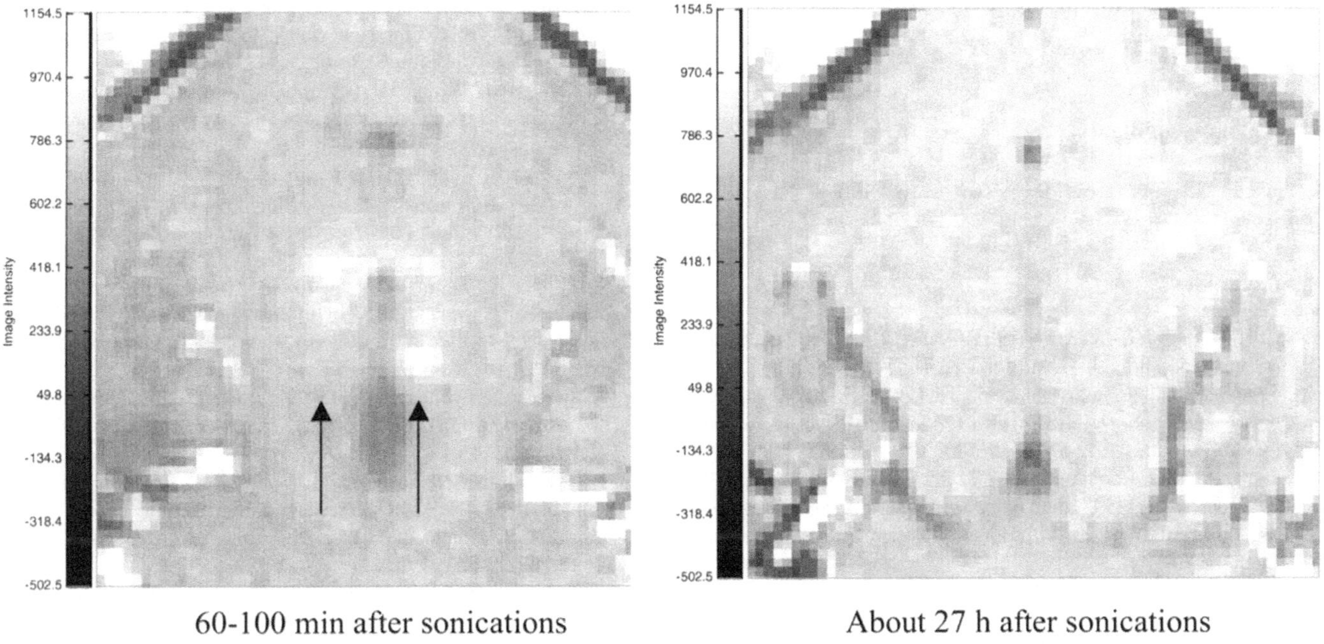

Fig. 2. T1-weighted image across the focal plane after sonications (three sonications on both hemisphere in a row – arrows) and contrast agent injection. The focal enhancement at five of the six focal locations is clearly seen after approximately 1 h post sonication (left) but not after one day (right)

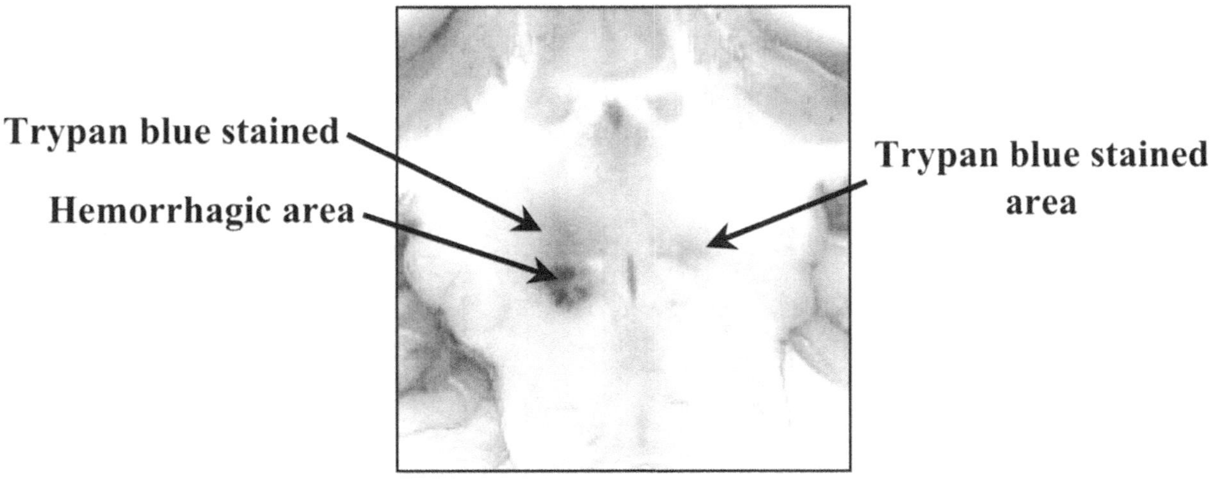

Fig. 3. Brain section at the focal plane across the focus demonstrating the diffusion of Trypan blue into the brain (2 areas) and one area of focal tissue damage and hemorrhage (different brain than in Fig. 2)

pressure amplitude (Fig. 2). The contrast agent diffused from the focal location to tissues around it. The trypan blue injected after sonications was focally visible in the post mortem brains (Fig. 3). The enhancement induced by a new contrast agent bolus injection at 6 h was reduced and not detectable at 24 h (Fig. 2).

Discussion

This study demonstrates for the first time that MRI guided focal BBB opening can be induced by using focused ultrasound. The results also demonstrate that on-line MRI can be used to monitor the magnitude of the opening so, on-line monitoring may be feasible.

This finding can have significant potential in targeted drug delivery.

Acknowledgments

This research was supported by NIH grants #46627.

References

1. Abbott NJ, Romero IA (1996) Transporting therapeutics across the blood brain barrier. Mol Med Today 2: 106–113
2. Dalecki D, Raeman CH, Child SZ, Cox C, Francis CW, Meltzer RS *et al* (1997) Hemolysis in vivo from exposure to pulsed ultrasound. Ultrasound Med Biol 23: 307–313
3. Dalecki D, Raeman CH, Child SZ, Penney DP, Mayer R, Carstensen EL (1997) The influence of contrast agents on hemorrhage produced by lithotripter fields. Ultrasound Med Biol 23: 1435–1439
4. Doolittle ND, Miner ME, Hall WA, Siegal T, Jerome E, Osztie E *et al* (2000) Safety and efficacy of a multicenter study using intraarterial chemotherapy in conjunction with osmotic opening of the blood brain barrier for the treatment of patients with malignant brain tumors. Cancer Feb 1, 88(3): 637–647
5. Hynynen K, Freund W, Cline HE, Chung A, Watkins R, Vetro J *et al* (1996) A clinical noninvasive MRI monitored ultrasound surgery method. RadioGraphics 16: 185–195
6. Hynynen K, McDannold N, Vykhodtseva N, Jolesz FA (2001) Noninvasive MR imaging guided focal opening of the blood brain barrier in rabbits. Radiology 220: 640–646
7. Kroll RA, Neuwelt EA (1998) Outwitting the blood brain barrier for therapeutic purposes: osmotic opening and other means. Neurosurgery 42: 1083–1099
8. Kuroda K, Chung A, Hynynen K, Jolesz FA (1998) Calibration of water proton chemical shift with temperature for noninvasive temperature imaging during focused ultrasound surgery. JMRI 8: 175–181
9. Miller DL, Thomas RM (1996) Contrast agent gas bodies enhance hemolysis induced by lithotripter shock waves and high intensity focused ultrasound in whole blood. Ultrasound Med Biol 22: 1089–1095
10. Qureshi NH, Chiocca EA (1999) A review of gene therapy for the treatment of central nervous system tumors. Crit Rev Oncog 10: 261–274
11. Vykhodtseva N, Sorrentino V, Jolesz FA, Bronson RT, Hynynen K (2000) MRI detection of the thermal effects of focused ultrasound on the brain. Ultrasound Med Biol Jun 1, 26(5): 871–880
12. Vykhodtseva NI, Hynynen K, Damianou C (1995) Histologic effects of high intensity pulsed ultrasound exposure with subharmonic emission in rabbit brain in vivo. Ultrasound Med Biol 21: 969–979

Correspondence: K. Hynynen, Department of Radiology, Brigham and Women's Hospital, and Harvard Medical School, 75 Francis Street, Boston, MA 02115.

Acta Neurochir (2003) [Suppl] 86: 559–563

Quantitative analysis of hyperosmotic and hypothermic blood-brain barrier opening

M. Ikeda[1], **T. Nagashima**[2], **A. K. Bhattacharjee**[1], **T. Kondoh**[1], **E. Kohmura**[1], and **N. Tamaki**[1]

[1] Department of Neurosurgery, Kobe University Graduate School of Medicine, Kobe, Japan
[2] Hyogo Children's Hospital, Kobe, Japan

Summary

Hyperosmotic opening of the blood-brain barrier (BBB) by mannitol is being used to enhance drug transport in human brains. Recently, cooling of the solution has been reported to have potential to open the BBB. However, the mechanism in barrier opening and closure remains elusive. We studied the rapid changes in cerebrovascular permeability after hyperosmotic and hypothermic BBB opening in rats, and then demonstrated that the Na^+/Ca^{++} exchange blocker (KB-R7943) prolongs opening.

BBB opening was attained by using intra-arterial infusion of hyperosmotic mannitol (1.6 M) and 1.1 M mannitol (which is less hyperosmotic than commonly used mannitol) at 4 degrees in Sprague-Dawley (SD) rats. To measure the changes in cerebrovascular permeability, perfusate-containing [^{14}C]-sucrose was infused intraarterially at different time points following hyperosmotic and hypothermic stress. Cerebrovascular permeability was then measured with the in situ brain perfusion technique.

1.6 M Mannitol produced opening of the BBB but the duration of the opening was less than 30 minutes. Use of 1.1 M Mannitol at 4 degrees indicated the same results. We then investigated the effect of a Na/Ca ion exchange blocker (KB-R7943) in both hyperosmotic and hypothermic BBB opening. KB-R7943 extended BBB opening up to 30 min without affecting the peak level of BBB permeability at 5 minutes.

Our findings represent important experimental information regarding pharmacological manipulation of BBB opening. The possibility of prolonging the transient opening of the BBB has major clinical implications.

Keywords: Blood-brain barrier; hyperosmotic opening; hypothermic opening; Na^+/Ca^{++} exchange blocker; cerebrovascular permeability.

Introduction

The blood-brain barrier (BBB) is a permeability barrier between the systemic blood stream and cerebral endothelial cells. It is a complex structure essential for maintaining homeostasis of the central nervous system. Since the discovery of the BBB, many experiments have been performed to determine its barrier integrity. The difficulty of crossing the BBB seen in many therapeutic compounds, such as antineoplastic agents and antibiotics, results in certain limitations for the treatment of pathological states in the central nervous system (CNS). Osmotic opening of the BBB by means of rapid intra-arterial infusion of hyperosmotic solutions has been used experimentally as well as clinically to increase the delivery of chemotherapy agents to intracerebral tumors [6, 8, 13, 14, 20]. Global delivery of genetic vectors to the CNS can be also enhanced by means of the BBB opening [10].

Mannitol of 1.1 M (20%) is the only hyperosmotic solution used in clinical practice. The safety and efficacy of this type of mannitol have been well established and it is generally used for reducing brain edemas. Experimental studies have demonstrated, however, that the osmolarity of mannitol at 1.1 M is not enough to achieve sufficient opening of the BBB, so that mannitol of 1.4 to 1.6 M has been commonly used instead [14, 16, 17]. Cooling of the solution for intra-arterial injection has been reported recently as an alternative approach to open the BBB [15]. Although the precise effect of cold solution on the BBB integrity has not been identified yet, this method is attractive in terms of prevention of complications since hypothermic conditions are well known to have a neuroprotective effect.

In the study presented here, we first examined the rapid changes in cerebrovascular permeability in rats after BBB disruption by intra-arterial infusion of hyperosmotic and cold mannitol and quantified the cerebrovascular permeability by using the *in situ* brain perfusion technique. We then demonstrated that the specific Na^+/Ca^{++} exchange blocker (KB-R7943) prolongs the barrier disruption induced by osmotic or hypothermic stress.

Materials and methods

Experimental animals, mannitol and perfusate preparation

Adult male Sprangue-Dawley (SD) rats weighing 388 ± 9 g were used for this study. The experiments were conducted in accordance with the Guidelines for Animal Experiments of the Kobe University School of Medicine. The rats were anesthetized with trychloroacetaldehyde (400 mg/kg, intraperitoneally, i.p.). Retrograde cannulation with a 10-cm polyethylene catheter (PE-50, Becton Dickinson and Company, Sparks, MD) into the right external carotid artery was performed under heparinization (100 IU/ml of normal saline). The rectal temperature of all the rats was maintained at 36.5 ± 0.3 °C with the aid of a heating pad. The brain temperature was monitored by means of a sensor needle inserted below the temporal muscle of the right cranium during the experiments.

Mannitol was dissolved in phosphate-buffered saline at a concentration of 1.1 M (20%) or 1.6 M (30%). In our study, 1.1 M mannitol was used at two different temperatures. Preparing the cold mannitol involved chilling with ice without precipitation. The temperature of the cold mannitol was 3.8 ± 0.2 °C immediately before the infusion. The mannitol at room temperature was used at 24.6 ± 0.3 °C.

For the perfusion fluid, [^{14}C]-sucrose (0.5 μ Ci/ml, NEN™ Life Science Products, Boston, MA) was added to PBS and allowed to reach equilibrium prior to infusion.

Experimental design

The rats in the control group were given intra-arterial infusion of 1.1 M mannitol at 24.6 ± 0.3 °C ($n = 12$). The rats in the hyperosmotic group were given intra-arterial infusion of 1.6 M mannitol at 24.6 ± 0.3 °C ($n = 21$). The rats in the hypothermic group were given intra-arterial infusion of 1.1 M mannitol at 3.8 ± 0.2 °C ($n = 28$). While those in the hyperosmotic and hypothermic group were first separated into two subgroups. In one group, hyperosmotic or hypothermic BBB opening was attained by means of the intra-arterial infusion of hyperosmotic ($n = 11$) or cold mannitol ($n = 14$). In the other group, hyperosmotic or hypothermic opening was brought about by intravenously injecting an Na^+/Ca^{++} exchange blocker (KB-R7943, Nippon Organon K.K., Osaka, Japan) prior to the infusion of hyperosmotic ($n = 10$) or cold mannitol ($n = 14$).

In these groups, cerebrovascular permeability for [^{14}C]-sucrose was determined with the *in situ* brain perfusion technique. Rats in the control group were infused with a perfusate containing [^{14}C]-sucrose to measure cerebrovascular permeability at 5 min ($n = 6$) and 30 min ($n = 6$) after the infusion of 1.1 M mannitol at 24.6 ± 0.3 °C. In the hyperosmotic group, 11 rats were infused with perfusate at 5 min ($n = 7$) and 30 min ($n = 4$) after the infusion of 1.6 M mannitol, and the other 10 rats were at first injected with KB-R7943 (i.v., 3 mg/kg) and then infused with perfusate containing [^{14}C]-sucrose at 5 min ($n = 5$) and 30 min ($n = 5$) after the infusion of 1.6 M mannitol. In the hypothermic group, 14 rats were infused with perfusate at 5 min ($n = 8$) and 30 min ($n = 6$) after the infusion of 1.1 M mannitol at 3.8 ± 0.2 °C, and the other 14 rats were at first injected with KB-R7943 (i.v., 3 mg/kg) and then infused with perfusate containing [^{14}C]-sucrose at 5 min ($n = 7$) and 30 min ($n = 7$) after the infusion of 1.1 M mannitol at 3.8 ± 0.2 °C.

BBB opening protocols

Mean blood pressure in the carotid artery in anesthetized rats after ligation of the pterygopalatine artery measured 86 ± 2 mmHg. To avoid hypertension during BBB opening, perfusion of mannitol was performed at a constant rate of 8.3×10^{-2} ml/s and at a perfusion pressure equaling 101 ± 4 mmHg into the right carotid artery for 30 s with an infusion pump (Model No. 100, KD Scientific Inc., Boston, MA). To measure permeability, the right common carotid artery was clipped and perfusate containing [^{14}C]-sucrose was infused for 60 s, after which the intravascular radioisotopes were washed out by infusion of normal saline for 30 s. Finally, the rats were decapitated. Immediately after removal of the brain, two regions, namely the ipsilateral frontal cortex, parietal cortex were dissected and placed in vials. The samples were then digested overnight at 40 °C in 1 ml of tissue solbidizer (BTS-450, Beckman Instruments Inc., Fullerton, CA). Finally, 10 ml of a scintillation cocktail for scintillation counting (Ready Organic, Beckman Instruments Inc., Fullerton, CA) and 70 μl of acetic acid to eliminate chemiluminescence were added. Counting was performed with a liquid scintillation counter (LS6000TA, Beckman Instruments).

The arterial blood gas of the experimental groups was measured by i-STAT (i-STAT Corporation, Princeton, NJ) at two time points, the first before and the second 5 or 30 min after the injection of mannitol. A 0.3 ml arterial blood sample was taken via the carotid artery for this purpose. Continuous temperature monitoring at the temporal muscle showed that the minimum level occurred at around one minute after the start of mannitol injection and returned to the previous temperature at around two minutes. The temperature before injection was 36.5 ± 0.3 °C, 34.4 ± 0.2 °C one minute after injection with room temperature mannitol and 33.1 ± 0.2 °C with cold mannitol.

Data analysis

Cerebrovascular permeability was measured in terms of [^{14}C]-sucrose, which is very slightly permeant at the normal BBB. The PA(s^{-1}) was calculated as brain [^{14}C] (dpm g^{-1})/perfusate [^{14}C] (dpm ml^{-1})/sec. All the values are expressed as mean ± S.E.. A two-sample Student's t test was employed for the comparison of two means. $P < 0.05$ was accepted as statistically significant.

Results

Hyperosmotic BBB opening by 1.6 M mannitol

In the control group, the PA ($\times 10^5$/s) was (6.3 ± 0.8) and (9.8 ± 1.9) for the frontal and parietal regions. Five minutes after the infusion of 1.6 M mannitol, the BBB was markedly opened (Fig. 1.a). The PA(s^{-1}) peaks for [^{14}C]-sucrose were $(34.9 \pm 9.5) \times 10^{-5}$ for the frontal and $(56 \pm 14.6) \times 10^{-5}$ for the parietal region, and these peaks were 5.5–5.7 times higher than the PA(s^{-1}) measured without osmotic opening. Immediately after osmotic opening, the process of BBB closure started and the absolute value for PA(s^{-1}) progressively declined. By 30 minutes, PA(s^{-1}) had decreased significantly but not returned to the pre-infusion level (Fig. 1.a).

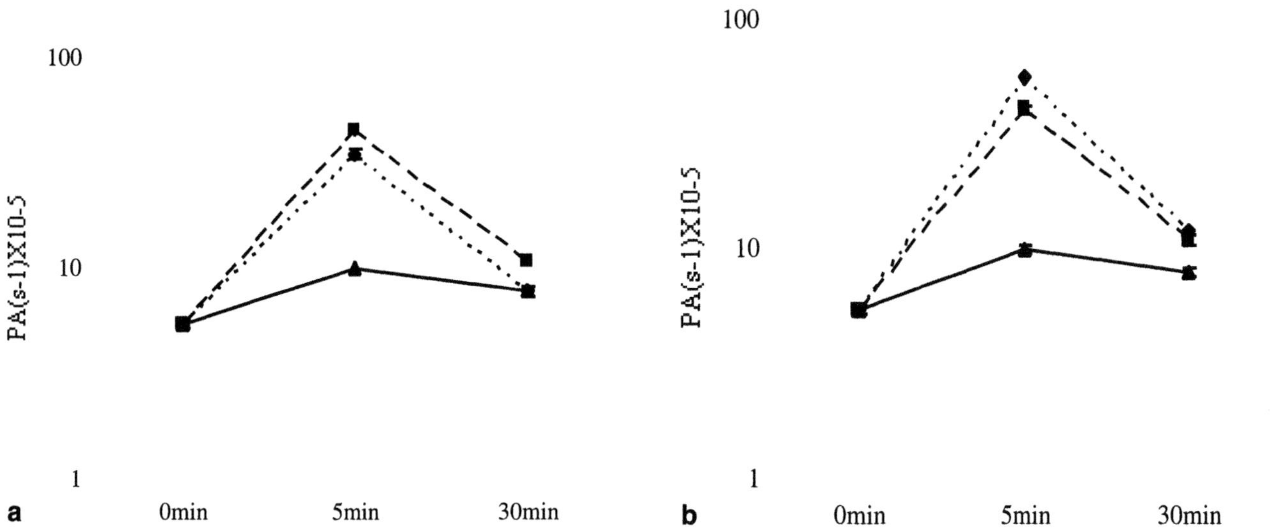

Fig. 1. (a) PA(s^{-1}) for [^{14}C]-sucrose in the frontal region. (b) PA(s^{-1}) for [^{14}C]-sucrose in the parietal region. Time course of cerebrovascular permeability, PA(s^{-1}) for [^{14}C]-sucrose after hyperosmotic or hypothermic opening of blood-brain barrier (BBB) by intra-arterial infusion of mannitol (1.6 mol/L) at 24.6 ± 0.3 °C or mannitol (1.1 mol/L) at 3.8 ± 0.2 °C. Five minutes after the infusion of hyperosmotic mannitol or cold mannitol, the PA(s^{-1}) demonstrated peaks. Thirty minutes after osmotic or hypothermic stress, PA(s^{-1}) had decreased significantly. ----◆--- hyperosmotic mannitol (1.6 M, 24.6 °C); — —■— — cold mannitol (1.1 M, 3.8 °C); ——▲—— control (1.1 M, 24.6 °C)

Hypothermic BBB opening by 1.1 M mannitol at 3.8 ± 0.2 °C

Five minutes after the infusion of 1.1 M mannitol at 3.8 ± 0.2 °C, PA(s^{-1}) significantly increased (Fig. 1.b). The PA(s^{-1}) peaks for [^{14}C]-sucrose were $(45.1 \pm 11.6) \times 10^{-5}$ for the frontal and $(40.7 \pm 12.0) \times 10^{-5}$ for the parietal region, and these peaks were 4.1–7.2 times higher than the PA(s^{-1}) measured without hypothermic opening. Thirty minutes after the infusion of 1.1 M mannitol at 3.8 ± 0.2 °C, the PA(s^{-1}) showed a significant decrease, so that there was no significant difference between the PA(s^{-1}) at 30 minutes after the hypothermic opening and the PA(s^{-1}) of the control.

Effect of Na^+/Ca^{++} exchange blocker on PA(s^{-1}) changes after hyperosmotic or hypothermic BBB opening

There was no significant difference in PA(s^{-1}) after five minutes between the rats treated with or without the Na^+/Ca^{++} exchange blocker. This indicates that the Na^+/Ca^{++} exchange blocker does not affect the degree of BBB opening itself (Fig. 2.a,b). Thirty minutes after the infusion of 1.6 M mannitol at 24.6 ± 0.3 °C or 1.1 M mannitol at 3.8 ± 0.2 °C, the increase in PA(s^{-1}) was preserved in the rats treated with the Na^+/Ca^{++} exchange blocker, which was significantly higher than the PA(s^{-1}) in rats without the blocker (Fig. 2.a,b). These findings indicate that the Na^+/Ca^{++} exchange blocker prolongs the duration of BBB opening.

Changes in physiological parameters before and after the infusion of mannitol

pH and HCO_3^- measured 5 min after the infusion of mannitol were significantly lower than those measured before the infusion, but other arterial blood gas parameters did not show any significant change. Arterial blood gas levels 30 minutes after the infusion of mannitol were not significantly different from those measured prior to infusion. These findings indicate that the rats suffered metabolic acidosis immediately after the infusion of mannitol, but had recovered by 30 minutes after infusion.

Discussion

In order to increase the barrier permeability at the ipsilateral cerebral hemisphere, hyperosmotic mannitol has been infused into the carotid artery of experimental animals and of humans [13, 14, 20]. One of the mechanisms of the resulting change in permeability is thought to be mediated via shrinkage of cerebrovascular endothelial cells caused by hyperosmolality,

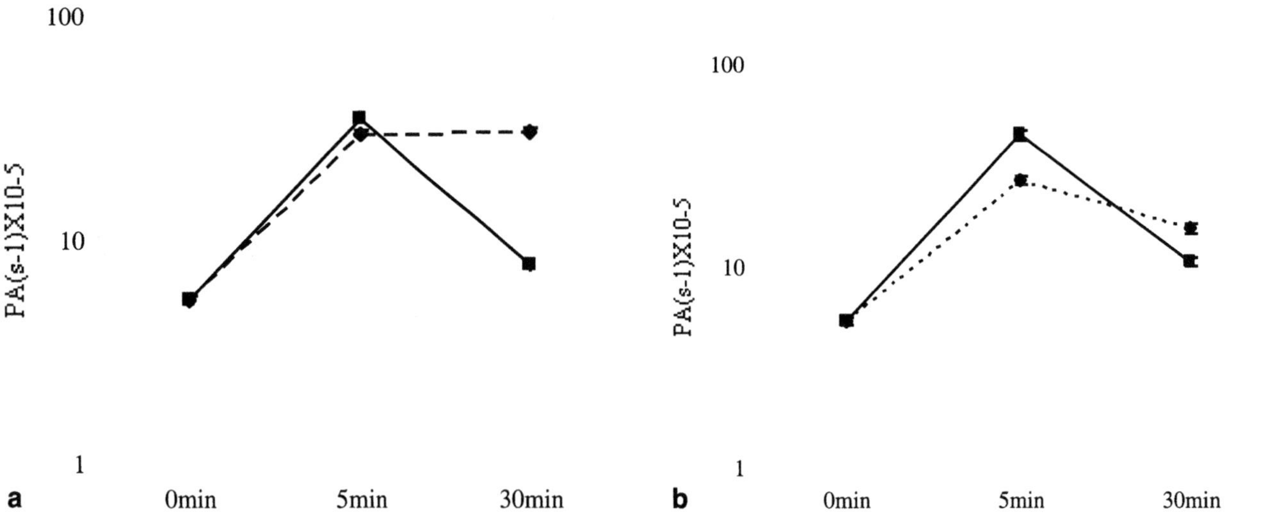

Fig. 2. (a) Effect of the Na^+/Ca^{++} exchange blocker on $PA(s^{-1})$ changes after osmotic stress in rats. ‒‒◆‒‒ KB-R7943+hyperosmotic mannitol; ——■—— hyperosmotic mannitol. (b) Effect of the Na^+/Ca^{++} exchange blocker on $PA(s^{-1})$ changes after hypothermic stress in rats. There was no significant difference in $PA(s^{-1})$ after five minutes between the rats treated with or without the Na^+/Ca^{++} exchange blocker. Thirty minutes after hyperosmotic or hypothermic BBB opening, the increase in $PA(s^{-1})$ was preserved in rats treated with the Na^+/Ca^{++} exchange blocker. ····◆···· KB-R7943+cold mannitol; ——■—— cold mannitol

which results in widening of the interendothelial tight junctions [17]. To achieve sufficient opening of the BBB, high osmolality obtained with 1.4 M to 1.6 M mannitol has been commonly used. A previous study of the concentration dependency of the effect of mannitol on BBB opening found that 1.0 M to 1.2 M mannitol did not result in opening of the BBB [17]. The data obtained by us with 1.1 M mannitol at room temperature are in agreement with those previously reported.

Osmotic opening of the BBB by mannitol has been reported in both animals and humans [7, 10, 18] but the time course in the early phase of opening could not be accurately identified due to indistinct time resolution [17]. In the study presented here, we demonstrated the precise time course of cerebrovascular reversibility with the earliest BBB opening occurring five minutes after the infusion of hyperosmotic mannitol in rats.

On the other hand, regarding the impact of the temperature on BBB permeability, a previous study has suggested that generalized hypothermia of the whole brain induced reduction in BBB permeability and that this effect was rather temperature-dependent with maximum reduction evident around 23 °C [9]. The nature of the mechanisms underlying this change in BBB permeability remains unexplained. By contrast, a recent study demonstrated that deep hypothermic saline infused into the carotid artery produced a transient BBB opening [15]. Opening of the BBB was demonstrated in all the animals receiving a hypothermic saline perfusion at 8 ± 1 °C. However, the details of the mechanism responsible for the effect of intra-carotid hypothermic saline infusion on BBB opening remains elusive. One explanation is that the very low temperature has a mechanical effect on the stability of the tight junctional complex and that below a postulated 'critical' temperature of about 7–8 °C some of the tight junctions may open [19]. It is clear that hypothermic opening of the BBB occurs at a threshold temperature, and in a pilot experiment for our study, no BBB opening was observed when 1.1 M mannitol was infused at temperatures higher than 8 °C (data not shown).

An other possible mechanism accounting for reversible BBB opening is that calcium mediated contraction of the endothelial cytoskeleton leads to opening of the inter-endothelial tight junctions [16]. KB-R7943 preserved the elevated intracellular calcium concentration of cerebral endothelial cells in vitro [11, 12] and has a prolonging effect on BBB opening in vivo [3, 4]. It is well known that KB-R7943 preferentially inhibits the reverse Na^+/Ca^{++} exchanger on both neuronal and non-neuronal cells and to have a protective effect on calcium-dependent cell death [2, 5].

By combining hyperosmotic or cooling of the mannitol and the application of KB-R7943, we could achieve prolonged BBB opening up to 30 minutes. In the study presented here, the synergistic effects of this combination indicate that this could be a promising modality for obtaining a safe reversible opening of the BBB. KB-R7943 did not affect the peak level of PA(s^{-1}) while it did prolong the opening time of the BBB produced by hyperosmotic and cold mannitol. There is an obvious need for investigating the possible mechanism of BBB opening following infusion of hyperosmotic and cold mannitol. Calcium is an important second messenger of intracellular signaling [1]. A rapid increase in intracellular calcium induced by hyperosmosis and hypothermia [11] and may cause drastic changes in cellular functions such as BBB integrity. Although changes in calcium levels were not measured in this in vivo study, similar changes in PA(s^{-1}) following hyperosmotic and hypothermic BBB opening imply that KB-R7943 exerts its effect via calcium dynamics.

Acknowledgments

This work was supported in part by a Grant-In-Aid for Scientific Research, (C)(2)(# 13671435), from the Ministry of Education, Science, Sports and Culture, Japan. The authors would like to thank Nippon Organon K.K., Osaka, Japan for providing the Na^+/Ca^{++} exchange blocker (KB-R7943) used in this study.

References

1. Adams DJ, Barakeh J, Laskey R, Van Breegman C (1989) Ion channels and regulation of intracellular calcium in vascular endothelial cells. FASEB J 3: 2389–2400
2. Arakawa N, Sakaue M, Yokoyama I, Hashimoto H, Koyama Y, Baba A, Matsuda T (2000) KB-R7943 inhibits store-operated $Ca^{(2+)}$ entry in cultured neurons and astrocytes. Biochem Biophys Res Commun 279: 354–357
3. Bhattacharjee AK, Nagashima T, Kondoh T, Tamaki N (2001) Quantification of early blood-brain barrier disruption by in situ brain perfusion technique. Brain Res Proto 8: 126–131
4. Bhattacharjee AK, Nagashima T, Kondoh T, Tamaki N (2001) The effects of the Na^+/Ca^{++} exchange blocker on osmotic blood-brain barrier disruption. Brain Res 900: 157–162
5. Bondarenko A, Chesler M (2001) Calcium dependence of rapid astrocyte death induced by transient hypoxia, acidosis, and extracellular ion shifts. Glia 34: 143–149
6. Doolittle ND, Miner ME, Hall WA, Siegal T, Hanson EJ, Osztie E, McAllister LD, Bubalo JS, Kraemer DF, Fortin D, Nixon R, Muldoon LL, Neulwelt EA (2000) Safety and efficacy of a multicenter study using intraarterial chemotherapy in conjunction with osmotic opening of the blood-brain barrier for the treatment of patients with malignant brain tumors. Cancer 88: 637–647
7. Gumerlock MK, Neuwelt EA (1990) The effects of anesthesia on osmotic blood-brain barrier disruption. Neurosurgery 26: 268–277
8. Kraemer DF, Fortin D, Doolittle ND, Neuwelt EA (2001) Association of total dose intensity of chemotherapy in primary central nervous system lymphoma (human non-acquired immunodeficiency syndrome) and survival. Neurosurgery 48: 1033–1041
9. Krantis A (1983) Hypothermia-induced reduction in the permeation of radiolabelled tracer substances across the blood-brain barrier. Acta Neuropathol (Berl) 60: 61–69
10. Kroll RA, Neuwelt EA (1998) Outwitting the blood-brain barrier for therapeutic purposes: osmotic opening and other means. Neurosurgery 42: 1083–1100
11. Nagashima T, Ikeda K, Wu S, Kondo T, Yamaguchi M, Tamaki N (1997) The mechanism of reversible osmotic opening of the blood-brain barrier. Acta Neurochir (Wien) [Suppl] 70: 231–233
12. Nagashima T, Wu S, Mizoguchi A, Tamaki N (1994) A possible role of calcium ion in osmotic opening of blood-brain barrier. J Auton Nerv Syst 49: 145–149
13. Neuwelt EA, Balaban E, Diehl J, Hill S, Frenkel E (1983) Successful treatment of primary central nervous system lymphomas with chemotherapy after osmotic blood-brain barrier opening. Neurosurgery 12: 662–671
14. Neuwelt EA, Frenkel EP, Diehl J, Vu LH, Rapoport SI, Hill S (1980) Reversible osmotic blood-brain barrier disruption in humans: implication for the chemotherapy of malignant brain tumors. Neurosurgery 7: 44–52
15. Öztas B, Küçük M (1995) Intracarotid hypothermic saline infusion: a new method for reversible blood-brain barrier disruption in anesthetized rats. Neurosci Lett 190: 203–206
16. Rapoport SI (2000) Osmotic opening of the blood-brain barrier: principles, mechanism, and therapeutic applications. Cell Mol Neurobiol 20: 217–230
17. Rapoport SI, Fredericks WR, Ohno K, Pettigrew KD (1980) Quantitative aspects of reversible osmotic opening of the blood-brain barrier. Am J Physiol 238: R421–R431
18. Remsen LG, Pagel MA, McCormick CI, Fiamengo SA, Sexton G, Neuwelt EA (1999) The influence of anesthetic choice, PaCO2, and other factors on osmotic blood-brain barrier disruption in rats with brain tumor xenografts. Anesth Analg 88: 559–567
19. Schindelmeiser J, Bergmann M, Lehmenkühler A, Kersting U (1987) Tracer permeability of rat cortical blood vessels during regional hypothermia. Acta Neuropathol (Berl) 73: 349–356
20. Siegal T, Rubinstein R, Bokstein F, Schwartz A, Lossos A, Shalom E, Chisin R, Gomori JM (2000) In vivo assessment of the window of barrier opening after osmotic blood-brain barrier disruption in humans. J Neurosurg 92: 599–605

Correspondence: Mitsuru Ikeda, M.D., Department of Neurosurgery, Kobe University Graduate School of Medicine, 7-5-1 Kusunoki Cho, Chuo-Ku, Kobe 650-0017, Japan. e-mail: mikeda@med.kobe-u.ac.jp

Acta Neurochir (2003) [Suppl] 86: 565–568

Chronic adrenomedullin treatment improves blood-brain barrier function but has no effects on expression of tight junction proteins

B. Kis[1,2], J. A. Snipes[1], M. A. Deli[3], C. S. Ábrahám[3], H. Yamashita[2], Y. Ueta[2], and **D. W. Busija[1]**

[1] Department of Physiology and Pharmacology, Wake Forest University, Winston-Salem, NC, USA
[2] Department of Physiology, University of Occupational and Environmental Health, Kitakyushu, Japan
[3] Institute of Biophysics, Biological Research Centre, Szeged, Hungary

Summary

We previously found that the production of adrenomedullin (AM) is one magnitude higher in cerebral endothelial cells (CECs) than in the peripheral endothelium and the AM concentration in the cerebral circulation is significantly higher than in other tested parts of the circulation. We also showed that CECs express AM receptors, and AM as an autocrine hormone is important to regulate the intracellular cAMP level in CECs. Further we reported that acute AM treatment has cAMP-like effects on specific BBB functions: AM decreased endothelial fluid phase endocytosis, activated the P-glycoprotein, increased transendothelial electrical resistance (TEER) and reduced endothelial permeability for sodium fluorescein, which suggests a tightening of intercellular junctions. In the present study, we found chronic AM exposure also increased TEER. In contrast, we could not detect significant effect of AM on the expression of tight junction proteins (claudin-1, occludin and zonula occludens-1). While not affecting expression of tight junction proteins, chronic AM treatment may influence the localization of these proteins which has been reported to correlate with functional changes of the BBB without a change in protein expression.

Keywords: Adrenomedullin; blood-brain barrier; cerebral endothelial cells; tight junction.

Introduction

The blood-brain barrier (BBB) is a physical and metabolic barrier between the central nervous system and the circulation and its primary role is to provide special ionic homeostasis for the proper functioning of neuronal circuits. Cerebral endothelial cells (CECs) are the anatomical and functional basis of the BBB characterized by the presence of tight junctions, as well as the low number of pinocytotic vesicles [4, 10]. Tight junctions are complexes of plasma membrane proteins that connect to the cytoarchitecture via membrane-associated accessory proteins [3, 10]. Tight junctions are located in the apical region of lateral plasma membranes and prevent paracellular diffusion of molecules. When the BBB is grossly disrupted, as sometimes happens following stroke, the resulting entry of proteins and ions into the central nervous system and the associated influx of water lead to brain edema [3].

Recently we described that the production of an endothelium-derived relaxing factor adrenomedullin (AM) is one magnitude higher in the CECs than in the peripheral endothelium [7]. We also found that the AM concentration in the cerebral circulation is significantly higher than in other tested parts of the circulation [7]. We showed that CECs express AM receptors, and AM as an autocrine hormone is important to regulate the intracellular cAMP level in these cells [7]. Moreover, we provided evidence that AM has cAMP-like effects on specific BBB functions in vitro [6]. Thus, exogenous AM decreased endothelial fluid phase endocytosis, activated the P-glycoprotein efflux pump in cultures of rat CECs [6]. AM also increased transendothelial electrical resistance (TEER) and reduced endothelial permeability for the low molecular weight sodium fluorescein, which suggests a tightening of intercellular junctions [6]. In the present study we examined the effect of chronic AM treatment on TEER and on the expression of tight junction proteins, claudin-1, occludin and zonula occludens-1 (ZO-1), which are major constituents of the tight junction assembly.

Methods

Cerebral endothelial cell culture

Primary rat CECs were isolated from 14-day-old Wistar rats as previously described [5] and were seeded onto collagen type IV and fibronectin coated dishes. The endothelial culture medium con-

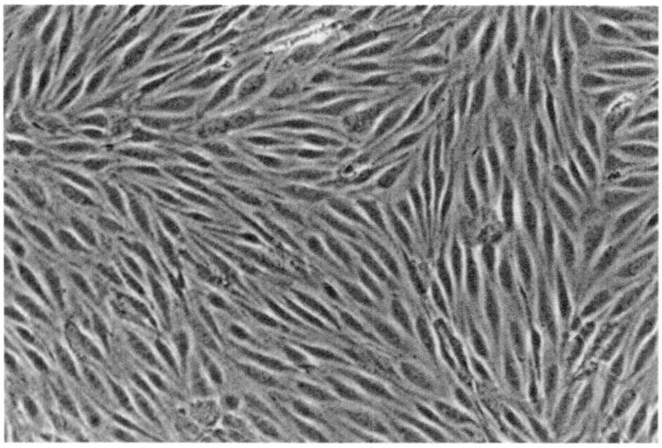

Fig. 1. Phase contrast micrograph of primary culture of rat cerebral endothelial cells

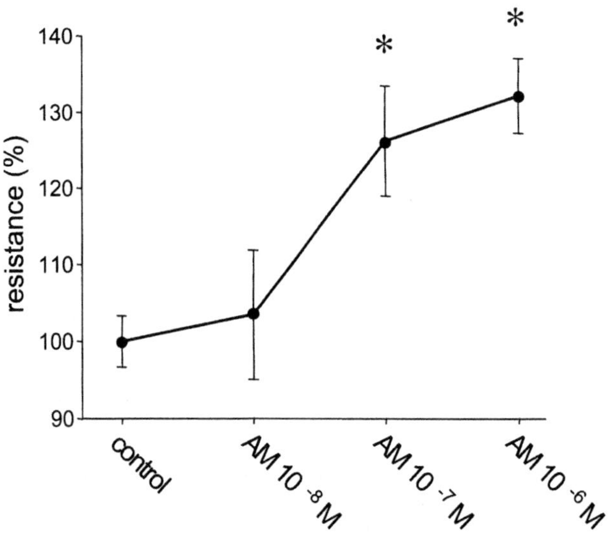

Fig. 2. Effect of chronic AM treatment on TEER. Primary cultures of rat cerebral endothelial cells were chronically treated with adrenomedullin (*AM*). TEER values were determined every day, and it showed continuous increase from day to day. The presented data were measured at day-in-vitro 5. TEER was expressed in the percent of the control value (mean $\pm$ S.E.M., n = 6–12, *$p < 0.05$ vs. control)

sisted of Dulbecco's modified Eagles medium (DMEM; Gibco BRL, Grand Island, NY) supplemented with 20% bovine plasma derived serum (Animal Technologies, Tyler, TX), 2 mM glutamine, 1 ng/mL basic fibroblast growth factor, 100 μg/mL heparin, 5 μg/mL vitamin C, and antibiotics. Confluent cultures (4–5th day in vitro) consisted of more than 95% of rat CECs (Fig. 1) verified by positive immunohistochemistry for von Willebrand factor, and negative immunochemistry for glial fibrillary acidic protein (GFAP) and α-smooth muscle actin.

TEER measurement

To study the effect of AM on TEER, rat CECs were grown on Transwell inserts (Corning Costar, Cambridge, MA) placed into plates containing confluent layers of rat cerebral astrocytes [7]. Cells were treated every day with different concentrations of AM (10^{-6}, 10^{-7} and 10^{-8} M) which were applied into the luminal compartment of the co-culture system. TEER was measured by an EVOM resistance meter (WPI, Sarasota, FL).

Western blotting

For western blot analysis, primary rat CECs were cultured in 35 mm dishes and were treated with different concentrations of AM (10^{-6}, 10^{-7} and 10^{-8} M) daily until the cultures reached confluency (day-in-vitro 4 or 5) when cells were harvested by scraping in ice cold NP40 lysis buffer supplemented with proteinase inhibitors (1 μg/ml aprotinin, 50 μg/ml phenylmethylsulfonyl fluoride, 1 μg/ml leupeptin). Lysates were centrifuged at 10,000 g for 10 min at 4 °C. The supernatants were collected and stored at −60 °C until assayed. Protein concentrations were determined with DC Protein Assay (Bio-Rad, Hercules, CA), and samples were mixed with 2× Laemmli sample buffer and heated at 95 °C for five minutes. An equal amount of protein for each sample was separated by 10% SDS-PAGE and transferred onto a polyvinylidine difluoride sheet (Polyscreen PVDF; Perkin Elmer Life Sciences, Boston, MA). Membranes were then incubated in a blocking buffer (Tris buffered saline, 0.1% Tween 20, 5% skimmed milk powder) for 1 hr at room temperature. Blots were subsequently incubated with rabbit polyclonal anti-occludin (1:500; Zymed, San Francisco, CA), or rabbit polyclonal anti-claudin (1:500; Zymed) or mouse monoclonal anti-ZO-1 (1:1000; Zymed) antibodies overnight at 4 °C. The membranes were then washed three times in Tris buffered saline with 0.1% Tween 20 then incubated for 1 hr in the blocking buffer with anti-rabbit IgG (1:50000; Jackson Immuno Research, West Grove, PA) or anti-mouse IgG (1:5000; Jackson Immuno Research) conjugated to horseradish peroxidase. The final reaction products were visualized using enhanced chemiluminescence (SuperSignal West Pico; Pierce, Rockford, IL) and recorded on X-ray film.

Results

Chronic AM treatment dose-dependently increased the TEER of primary rat CEC monolayers co-cultured with astrocytes (Fig. 2). 10^{-7} M and 10^{-6} M concentrations of AM significantly increased TEER by 26.1 $\pm$ 7.2% and 32.1 $\pm$ 4.9% (mean $\pm$ S.E.M.), respectively.

As Figure 3 shows, we did not find significant effects of the chronic AM treatment on the expression of the tight junction proteins claudin-1, occludin and ZO-1. With anti-ZO-1 antibody we detected two bands in rat CECs (Fig. 3). The two bands correspond to the two isoforms of ZO-1 which differ by the presence (α+) or absence (α−) of an 80 amino acid region known as 'motif-α' [11].

Discussion

Our previous results indicate that astrocyte-derived factors induce high AM production in CECs and the

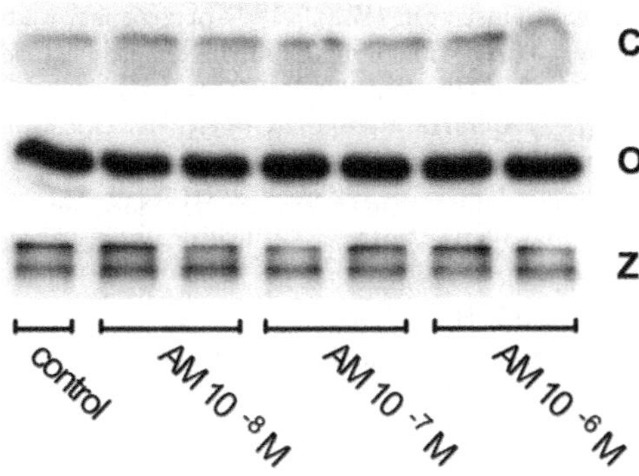

Fig. 3. Effect of adrenomedullin on the expression of tight junction proteins. Primary cultures of rat cerebral endothelial cells were chronically treated with adrenomedullin (*AM*), and then proteins were extracted and subjected to Western blot analysis. Representative Western blots for claudin-1, occludin and ZO-1 (n = 4, from two independent cultures)

produced AM as an autocrine mediator plays an important role in the maintenance of basal intra-endothelial cAMP level [7]. It has long been known that cAMP is implicated in the regulation of BBB functions. Elevation of intra-endothelial cAMP concentration induces high TEER and low BBB permeability for paracellular markers indicating a role for cAMP in tight junction assembly [4, 10]. We were the first to demonstrate that AM increases the barrier functions of endothelial cells [6] and very recently this was confirmed by the study of Hippenstiel *et al.* [1]. We have found that AM increases the characteristic BBB functions, including TEER, permeability and P-glycoprotein activity of primary rat CECs in vitro [6]. AM may be an autocrine inducer of BBB functions of cerebral endothelial cells *via* the activation of adenylate cyclase.

The effect of cAMP on BBB permeability is short-lived; as cAMP level returns to normal, junctional resistance decreases [3]. In the present study we found a chronic effect of AM in improving BBB function, which might be related to a cAMP-independent mechanism. Therefore we examined the expression of tight junction proteins, claudin-1, occludin and ZO-1, which might be influenced by chronic AM treatment.

Tight junctions in the BBB are composed of an intricate combination of transmembrane and cytoplasmic accessory proteins linked to an actin-based cytoskeleton of the CECs [3]. The transmembrane tight junction proteins consist of three integral protein families, claudins, occludin and junction adhesion molecules. Claudins bind homotypically to claudins on adjacent endothelial cells to form the primary seal of the tight junction and their short cytoplasmic section binds to occludin. Occludin connects claudins with tight junction cytoplasmic accessory proteins (i.e., ZO-1 and ZO-2) and the plasma membranes of adjacent cells [3]. The presence of occludin at the BBB is correlated with increased electrical resistance and decreased paracellular permeability of the BBB [2]. The cytoplasmic accessory proteins are necessary for structural support. ZO proteins interact with the C terminus of occludin and claudins. Binding of ZO proteins to actin indicates that one possible function of these molecules is to link the tight junction to the cytoskeleton [3].

With Western blot analysis we determined the expression of claudin-1, occludin and ZO-1. We could not detect significant effects of AM on the expression of these tight junction proteins which suggest the involvement of other mechanisms. AM may influence the translocation, phosphorylation state and assembly of tight junction proteins, which were reported to correlate with functional changes of the BBB without a change in protein expression [9]. Another possibility is that AM treatment leads to the disintegration of the microfilament system (stress fibers) of the endothelial cells which consist of F-actin and nonmuscle myosin. The disappearance of F-actin and myosin from stress fibers was reported to underlie the improved barrier function of endothelial cells [8]. The mechanism of the beneficial effect of chronic AM treatment on BBB properties needs further studies and is currently under investigation by our lab.

Acknowledgment

Supported by Grants-in-Aid for JSPS Fellows (S-02167 for B.K.) and for Scientific Research on Priority Areas (B) (No. 10218210 for Y.U.) from the Ministry of Education, Culture, Sports, Science and Technology, Japan and grants from NIH (HL-30260, HL-46558, HL-50587 for D.W. Busija).

References

1. Hippenstiel S, Witzenrath M, Schmeck B, Hocke A, Krisp M, Krull M, Seybold J, Seeger W, Rascher W, Schutte H, Suttorp N (2002) Adrenomedullin reduces endothelial hyperpermeability. Circ Res 91: 618–625
2. Hirase T, Staddon JM, Saitou M, Ando-Akatsuka Y, Itoh M, Furuse M, Fujimoto K, Tsukita S, Rubin LL (1997) Occludin as a possible determinant of tight junction permeability in endothelial cells. J Cell Sci 110: 1603–1613
3. Huber JD, Egleton RD, Davis TP (2001) Molecular physiology and pathophysiology of tight junctions in the blood-brain barrier. Trends Neurosci 24: 719–725

4. Joo F (1996) Endothelial cells of the brain and other organ systems: some similarities and differences. Prog Neurobiol 48: 255–273
5. Kis B, Szabo CA, Pataricza J, Krizbai IA, Mezei Z, Gecse A, Telegdy G, Papp JG, Deli MA (1999) Vasoactive substances produced by cultured rat brain endothelial cells. Eur J Pharmacol 368: 35–42
6. Kis B, Deli MA, Kobayashi H, Abraham CS, Yanagita T, Kaiya H, Isse T, Nishi R, Gotoh S, Kangawa K, Wada A, Greenwood J, Niwa M, Yamashita H, Ueta Y (2001) Adrenomedullin regulates blood-brain barrier functions in vitro. Neuroreport 12: 4139–4142
7. Kis B, Kaiya H, Nishi R, Deli MA, Abraham CS, Yanagita T, Isse T, Gotoh S, Kobayashi H, Wada A, Niwa M, Kangawa K, Greenwood J, Yamashita H, Ueta Y (2002) Cerebral endothelial cells are a major source of adrenomedullin. J Neuroendocrinol 14: 283–293
8. Langeler EG, van Hinsbergh VW (1991) Norepinephrine and iloprost improve barrier function of human endothelial cell monolayers: role of cAMP. Am J Physiol 260: C1052–1059
9. Mark KS, Davis TP (2002) Cerebral microvascular changes in permeability and tight junctions induced by hypoxia-reoxygenation. Am J Physiol Heart Circ Physiol 282: H1485–1494
10. Rubin LL, Staddon JM (1999) The cell biology of the blood-brain barrier. Annu Rev Neurosci 22: 11–28
11. Willott E, Balda MS, Heintzelman M, Jameson B, Anderson JM (1992) Localization and differential expression of two isoforms of the tight junction protein ZO-1. Am J Physiol 262: C1119–1124

Correspondence: Bela Kis, M.D., Ph.D., Department of Physiology and Pharmacology, Wake Forest University, School of Medicine, Medical Center Boulevard, Winston-Salem, NC-27157, USA. e-mail: bkis@wfubmc.edu

Blood brain barrier

Acta Neurochir (2003) [Suppl] 86: 571–573

Characteristic phosphorylation of the extracellular signal-regulated kinase pathway after kainate-induced seizures in the rat hippocampus

N. Otani, H. Nawashiro, A. Yano, H. Katoh, A. Ohnuki, T. Miyazawa, and **K. Shima**

Departments of Neurosurgery, National Defense Medical College, Tokorozawa, Japan

Summary

Extracellular signal-regulated kinase (ERK) pathways play a crucial role in cell growth and long-lasting neuronal plasticity. Several studies have shown that phosphorylated-ERK (p-ERK) significantly increases after kainic acid (KA) administration. However, little or no information is available about the spatial distribution of p-ERK after KA-induced seizures. We herein show that KA-induced seizures significantly increase p-ERK in both neurons and astrocytes in rat brain using Western blots and immunohistochemistry. A strong immunoreactivity for p-ERK was induced in the dentate hilar neurons and CA3 neurons 30 mins and 6 hrs after KA injection. In addition, immunoreactivity for p-ERK was seen in astrocytes 6 hrs after KA injection. 72 hrs after KA injection, all pyramidal neurons had died. These findings suggest that the ERK pathway participates in the KA-induced neurotoxicity in the rat hippocampus.

Keywords: Kainic acid; extracellular signal regulated kinase; hippocampus; seizure.

Introduction

Mitogen activated protein kinases (MAPK) are activated by phosphorylation in response to a variety of mitogenic signals. The MAPK cascades are composed of the extracellular signal-regulated protein kinase (ERK), c-Jun NH(2)-terminal kinase, and p38 pathways. Phosphorylated-ERK (p-ERK) plays an important role in the survival, proliferation, and differentiation of various cells.

Several studies have shown that kainic acid (KA)-induced seizures cause ERK phosphorylation in rat brain [5, 8]. However, up to now there have been no reports focusing on the spatial expression and distribution of p-ERK after KA administration. We herein show a characteristic activation of the ERK pathway in rat brain after KA-induced seizures. The findings suggest that a distinct ERK pathway might participate in the KA-induced neurotoxicity.

Material and methods

Adult male Sprague-Dawley rats weighing 150 to 200 g were used. KA dissolved in phosphate buffered saline was prepared for the induction of limbic seizures. Animals were subcutaneously injected with KA in a dose of 12 mg/kg ($n = 12$). For Western blot analysis, the rats were killed by decapitation under intraperitoneal anesthesia at 15, 30 min, 6, and 72 hrs after KA administration ($n = 6$ per time point). For immunohistochemistry, the rats were perfused transcardially with normal saline followed by 4% buffered paraformaldehyde at the same time points ($n = 6$ per time point). The brains were embedded in paraffin. Serial coronal sections (5 μm-thick) were prepared and stained with haematoxylin and eosin.

The bilateral hippocampus was used for Western blot analysis. After SDS-PAGE, the protein was transferred to PVDF membranes using an electrophoretic transfer system. The membranes were incubated with a primary antibody against either polyclonal p-ERK1/2 antibodies (New England Biolabs, Beverly, MA, U.S.A).

Immunohistochemistry was performed by the streptavidin-biotin peroxidase complex method for immunostaining by polyclonal antibodies against p-ERK1/2 (New England Biolabs). Double staining was performed with an EnVision system (DAKO, Japan) for GFAP, and a HISTOFINE SAB-PO (R) KIT (NICHIREI Co) for p-ERK1/2.

Results

The pattern of KA-induced seizures was reproducible. All animals survived the experimental procedure. The intraperitoneal injection of KA (12 mg/kg) caused a characteristic sequence of behavioral changes, such as staring posture, wet dog shakes, automatisms, and seizures.

The p-ERK protein content significantly increased after KA administration. The immunoreactivity for p-ERK was increased at 15 min, reached a maximum level at 30 min, and persisted until 6 hrs after KA administration. The immunoreactivity for p-ERK returned to control levels at 72 hrs after KA injection. The p-ERK protein contents in the sham control rats were not significantly different at any time point corresponding to the protocol of experimental animals with KA-induced seizures.

Yet, a few immunopositive cells stained for p-ERK were detected in sham control rats. Immunoreactivity for p-ERK was immediately induced in the dentate gyrus, mossy fiber bundles, and CA3 neurons. Furthermore, the p-ERK expression was prolonged in the dentate gyrus and all pyramidal neurons as seen at 6 hrs after KA injection. 72 hrs after KA administration all pyramidal neurons were lost in spite of a disappearance of the p-ERK immunoreactivity.

The double immunostaining showed that induction of p-ERK was observed in astrocytes surrounding the dentate neuronal cells and CA3 neurons at 6 hrs after KA injection. Thereafter, the immunostaining disappeared until termination of observation at 72 hrs after KA administration.

Discussion

Recent studies have shown that KA-induced seizures lead to expression of immediate early genes and of nerve growth factor in the rat hippocampus. Many neurotrophins are believed to promote cell survival and differentiation by activating intracellular signal transduction through Ras-ERK cascades [1]. On the other hand, several studies have revealed that the MAPK pathways were activated with a different temporal pattern in the rat hippocampus by KA-induced seizures [4, 8]. However, up to now there have been no reports focusing on the temporal and spatial distribution of the p-ERK after KA-induced seizures.

We have previously reported that p-ERK contents significantly increase in the ipsilateral hippocampus after TBI [9]. In this article, we have demonstrated that the immunoreactivity for p-ERK was localized in the dentate hilar neurons and the damaged CA3 neurons, a region which is selectively vulnerable to cerebral trauma [7]. In this study, immunoreactivity for p-ERK persisted in all pyramidal neurons at 6 hrs after KA-induced seizures. The difference in both magnitude and distribution of p-ERK expression reflects a pathophysiological distinction in pyramidal neuronal damage between cerebral trauma and KA neurotoxicity. We have, therefore, speculated that the remarkable and prolonged expression of p-ERK in pyramidal neurons may promote cell damage in vulnerable regions from KA-induced neurotoxicity.

Reactive astrogliosis is a most prominent response to diverse forms of damage from KA induced seizures. KA administration induces progressive glial scarring in the rat brain [3]. There is a close relationship between inflammation, cytokine production, and astrogliosis [6]. Double immunostaining results have shown that the immunoreactivity for p-ERK is almost exclusively localized in astrocytes at 6 hrs of KA administration. We speculate that the induction of p-ERK in astrocytes in the late period after KA-induced seizures plays an important role of astroglial activation leading to astrogliosis, which is beneficial for neuronal survival and repair of the damaged blood-brain barrier [2].

In this study, we have shown that immunoreactivity of p-ERK is significantly increased in areas which are selectively vulnerable to KA-induced seizures. We suggest that ERK-phosphorylation is associated with the molecular mechanisms elicited by KA administration, and that the differences in the ERK activation reflects differences in selective vulnerability between traumatic and neurotoxic injury to the rat hippocampus. However, the role of ERK phosphorylation after KA injection is still not fully understood. Further investigations using specific inhibitors will be necessary to elucidate the role of ERK phosphorylation after KA-induced seizure.

References

1. Bonni A, Brunet A, West AE *et al* (1999) Cell survival promoted by the Ras-MAPK signaling pathway by transcription-dependent and -independent mechanisms. Science 286: 1358–1362
2. Bush TG, Puvanachandra N, Horner CH *et al* (1999) Leukocyte infiltration, neuronal degeneration, and neurite outgrowth after ablation of scar-forming, reactive astrocytes in adult transgenic mice. Neuron 23: 297–308
3. Gramsberden JB, Berg KJ (1994) Regional and temporal profiles of calcium accumulation and glial fibrillary acid protein levels in rat brain after systemic injection of kainic acid. Brain Res 667: 216–228
4. Joen SH, Kim YS, Bae CD *et al* (2000) Activation of JNK and p38 in rat hippocampus after kainic acid induced seizure. Exp Mol Med 32: 227–230
5. Kim YS, Hong KS, Seong YS *et al* (1994) Phosphorylation and activation of mitogen-activated protein kinase by kainic acid-induced seizure in rat hippocampus. Biochem Biophys Res Commun 202: 1163–1168
6. Lieberman AP, Pitha RM, Shin ML *et al* (1992) Poly (A) removal is the kinase-regulated step in tumor necrosis factor mRNA decay. J Biol Chem 267: 2123–2126
7. Lowenstein DH, Thomas MJ, Smith DH *et al* (1992) Selective vulnerability of dentate hilar neurons following traumatic brain injury: a potential mechanistic link between head trauma and disorders of the hippocampus. J Neurosci 12: 4846–4853

8. Mielke K, Brecht S, Dorst A *et al* (1999) Activity and expression of JNK1, p38 and ERK kinases, c-Jun N-terminal phosphorylation, and c-jun promotor binding in the adult rat brain following kainate-induced seizures. Neuroscience 91: 471–483
9. Otani N, Nawashiro H, Fukui S *et al* (2002) Differential activation of the mitogen-activated protein kinase pathways following traumatic brain injury in the rat hippocampus. J Cereb Blood Flow Metab 22: 327–334

Correspondence: Naoki Otani, M.D., Department of Neurosurgery, National Defense Medical College, 3-2 Namiki, Tokorozawa, Saitama 359-8513, Japan. e-mail: grd1405@gr.ndmc.ac.jp

Acta Neurochir (2003) [Suppl] 86: 575–579

Highly polysialylated neural cell adhesion molecule (PSA-NCAM) positive cells are increased and change localization in rat hippocampus by exposure to repeated kindled seizures

K. Sato, M. Iwai, W.-R. Zhang, H. Kamada, K. Ohta, N. Omori, I. Nagano, M. Shoji, and **K. Abe**

Department of Neurology, Graduate School of Medicine and Dentistry, Okayama University, Okayama Japan

Summary

The highly polysialylated neural cell adhesion molecule (PSA-NCAM) is involved in migration of neural stem cells as well as in neural plasticity. Immunoreactive PSA-NCAM expression was examined in rats with repeated exposure to amygdaloid kindled generalized seizures (GS). The number of PSA-NCAM positive cells in the bilateral dentate gyrus (DG) increased significantly from GS. Although the total number of positive cells was not significantly different between animals with 3 times GS (3 GS) and 30 times GS (30 GS), in the latter group a greater number of positive cells was observed in the outer granule cell layer (GCL) and a marked extension of immunopositive dendrites to the molecular layer. These observations indicate that increased migration of newly generated cells as well as plastic changes of preexisting neural cells occur in response to recurrent GS. This may contribute to an abnormal reconstruction of the synaptic network in the hippocampus and, thus, epileptogenicity from kindling.

Keywords: Neural stem cell; migration; epilepsy; kindling; highly polysialylated neural cell adhesion molecule (PSA-NCAM).

Introduction

Repeated exposure of certain brain regions to a stimulus that triggers seizure activity results in progressively greater epileptiform responses, culminating in generalized seizures (GS), which is referred to as kindling phenomenon [8]. Kindling is an experimental model of temporal lobe epilepsy as well as neural plasticity [8, 16]. Mesial temporal lobe epilepsy is the most common form of epilepsy in adults, and the seizure attacks are often intractable and accompanied by selective vulnerability in the hippocampus [4]. In kindling, recurrent GS cause neural loss and reorganization of the hippocampal circuitry as in human epilepsy [2, 25, 26]. Such degenerative and regenerative structural changes have been demonstrated with repetition of GS, which may lead to a lowered seizure threshold culminating in spontaneous seizures [13, 15, 16].

Neural stem cells exist also in the adult mammalian brain especially in the dentate gyrus (DG) of the hippocampus [5, 7, 9, 12]. Neural stem cells have a potential of self-renewal and multidirectional differentiation [5, 7, 9]. Neurogenesis is regulated by internal molecular programs and the external epigenetic environment during normal development, but is also regulated under pathological conditions, such as ischemia and epileptic seizures [1, 3, 10, 11, 17, 18, 21, 22]. Proliferation (step 1) of neural stem cells is stimulated by prolonged seizure discharges in the DG. The newborn cells migrate (step 2), differentiate into certain neural cell species (step 3), and could make aberrant synaptic connections and neural networks [17, 18, 21].

The highly polysialylated neural cell adhesion molecule (PSA-NCAM) is expressed specifically at step 2 (migration) of the neural stem cell development. It plays an important role for neurite outgrowth in stem cell migration, synaptic remodeling, neuronal circuit formation, and cell-cell interaction [20, 23, 24]. The present study was made to investigate changes of PSA-NCAM expression in the DG, and to elucidate the role of migration of neural stem cells and synaptic plasticity on neural remodeling of the hippocampus from recurrent GS by kindling.

Material and methods

Tripolar electrodes were stereotactically implanted into the left basolateral amygdala of male Sprague-Dawley rats weighing 300–350 g under pentobarbital (50 mg/kg) anesthesia. After 14 days recovery, electrical stimuli (a 2 s train of 50 Hz, 1 ms rectangular waves at a current intensity of 200 μA) were given once a day. The animals developed a stage 5 GS [19] by the electrical stimulation. They were subjected to additional daily stimuli producing a total of 3 or 30 consecutive GS (n = 5 each of the 3 GS or 30 GS group, respectively). Age-matched animals with implanted electrodes without stimulation were used as controls (n = 5). The electroencephalogram was recorded by the remaining poles of the tripolar electrode and the skull screw electrode during all tests. The brain was rapidly removed then in deep anesthesia at 8 h after the last stimulation in the 3 GS and 30 GS group. The Animal Committee of the Graduate School of Medicine and Dentistry of Okayama University had approved the protocol of this experiment.

For immunohistochemical staining coronal 10 μm brain sections were cut at the dorsal hippocampal level (approximately 3.3 mm posterior to bregma) by using a cryostat at −20 °C and collected on glass slides. For identification of PSA-NCAM, the sections were fixed in acetone, and first incubated in 0.3% H_2O_2 for 30 min then blocked with 10% normal goat serum for 30 min. The sections were then incubated overnight at 4 °C with anti-PSA-NCAM mouse IgM monoclonal antibody (diluted 1:2000, Seki and Arai, 1993) [23]. Thereafter, the sections were incubated with biotinylated goat anti-mouse IgM (diluted 1:200; Vector laboratories, CA) for 1 h at room temperature. All sections were subsequently placed in a horseradish peroxidase/streptoavidin/biotin complex solution (Vectastain ABC Kit, Vector laboratories, CA) for 30 min, then incubated for 1.5 min in a peroxidase reaction solution (0.02% diaminobenzidine, 0.02% H_2O_2). Other sections were stained in a similar way without the primary antibody. Staining was examined by light-microscopy, positive cells in the DG were counted. Differences were statistically analyzed by one-way ANOVA (analysis of variance) followed by Fisher's Protected Least Significant Difference (Fisher's PLSD).

Results

As characteristic for kindling, there were no differences in the mean duration of the electroencephalographic afterdischarge upon the first electrical stimulation in the animals with either 3 GS or 30 GS. In rats with 3 GS or 30 GS, the mean duration of afterdischarge following the last electrical stimulation was 102.0 ± 26.4 s or 78.4 ± 23.4 s, respectively (n = 5, mean ± SD). The number of electrical stimuli needed to induce the first GS (stage 5) was 7.6 ± 1.8 or 8.6 ± 2.5 in animals with either 3 GS or 30 GS, respectively. The 3 GS and 30 GS animals were sacrificed at 9.6 ± 1.8 days or 37.6 ± 2.5 days, respectively after the initial electrical stimulation, respectively.

PSA-NCAM positive cells were located in the deepest portion of granule cell layer (GCL) of the ventral part of DG in control animals (Fig. 1a, d, filled arrowheads). In rats with GS granule cells in both the ventral (Fig. 1b, c, lower arrowheads) and dorsal (Fig. 1b, c, upper arrowheads) parts of the DG became positive for PSA-NCAM. The number of PSA-NCAM positive cells significantly increased bilaterally by 2 times in the 3 GS group (Fig. 1b, e, Fig. 2). Further increases were observed in the 30 GS group (Fig. 1c, f, Fig. 2), albeit with no significant differences between the 3 GS group. Although PSA-NCAM positive cells were still mainly located in the deepest part of the GCL, some positive cells were also observed in the outer GCL in the 3 GS animals (Fig. 1e, open arrowhead). In animals with 30 GS, a greater number of positive cells was located in the outer GCL (Fig. 1f, open arrowheads), as compared with the 3 GS group (Fig. 1e, open arrowhead). A minimal number of PSA-NCAM positive dendrites was found in the brain of controls (Fig. 1d, arrow), which increased slightly in animals with 3 GS (Fig. 1e, arrows), while markedly towards the molecular layer animals with 30 GS (Fig. 1f, arrows).

Discussion

The two key findings of this study are that the number of PSA-NCAM positive cells increased after 3 GS, while no further increase was observed in animals with 30 GS, and that repetition of GS markedly enhances the number of PSA-NCAM positive cells in the outer GCL and the extension of positive dendrites. PSA-NCAM is involved in migration of neural stem cells but also in synaptic plasticity [20, 24]. Thus, the current expression of PSA-NCAM should be considered from these points of view. The first finding indicates that in kindling induced GS, neural stem cell migration and synaptic remodeling are stimulated in the DG, and that this contributes to the reorganization of the neural network in the hippocampus. The lack of a further increase in the number of PSA-NCAM positive cells in animals with 30 GS shows that repeated GS do not lead to a further increment in neural stem cell migration and synaptic remodeling in the DG. Although repeated GS produce cell death in the DG, the electrical stimuli do not facilitate division of stem cells after GS were established [17]. It is possible that repeated GS increase the number of neurons involved in synaptic plasticity, rather than the number of neural stem cells which migrate following proliferation. Neural stem cell migration detected as increase of PSA-NCAM positively reacting cells in the DG is also enhanced by transient ischemia as well as in human temporal lobe epilepsy where hippocampal neurons die [6, 11, 14]. Currently, the repetitive GS might have been not strong enough to increase the number of neural stem cells migrating in the DG where neuronal damage was subtle.

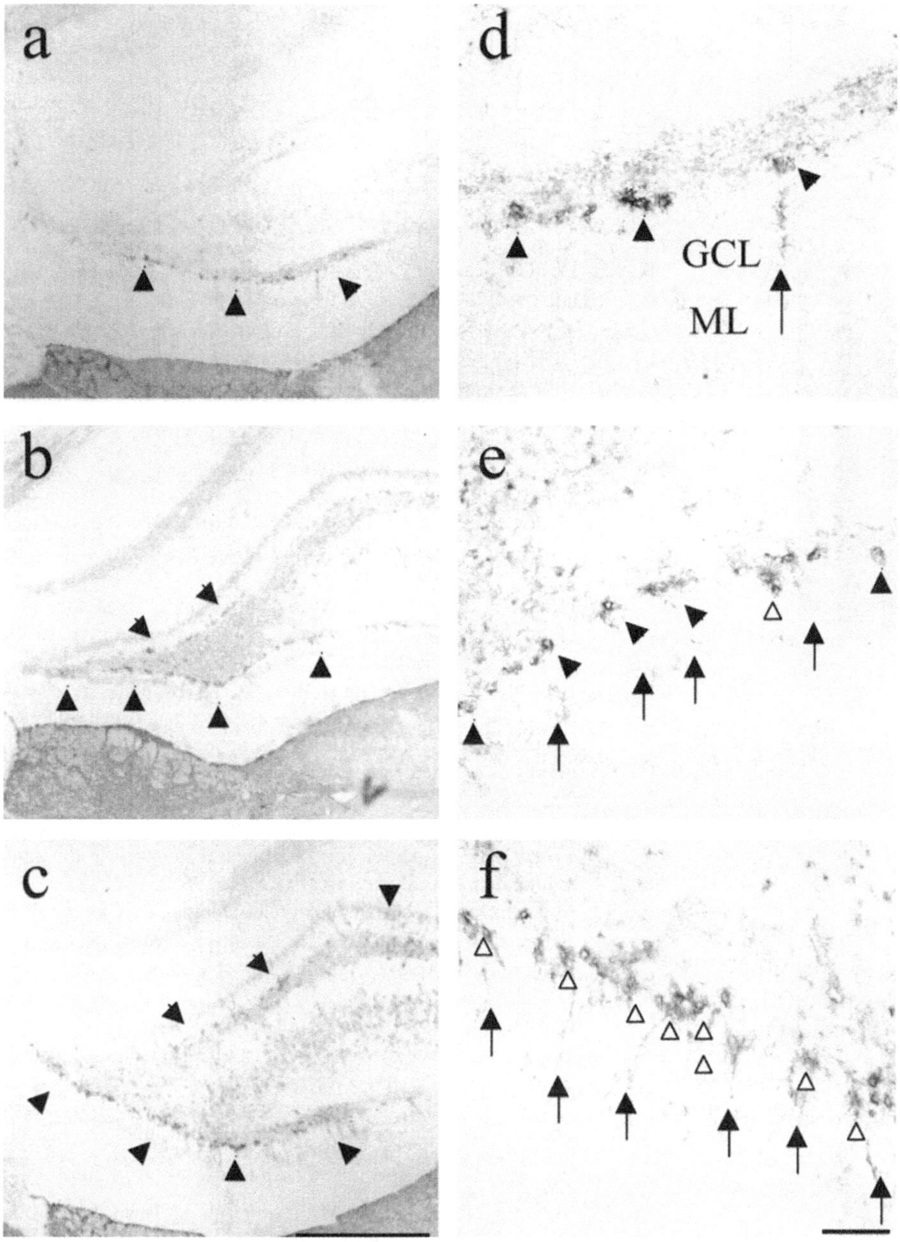

Fig. 1. Representative staining for immunoreactive PSA-NCAM in the dentate gyrus (*DG*) on the side of stimulation of sham-stimulated controls (a, d), or with 3 (b, e) or 30 generalized (c, f) seizures (3 GS and 30 GS, respectively). Note the stronger staining and increase of PSA-NCAM positive cells in the granule cell layer (*GCL*) in animals with 3 GS (b, e) or 30 GS (c, f). PSA-NCAM positive cells located in the deepest portion of the GCL are indicated by filled arrowheads in panels a–f, that in the outer GCL by open arrowheads in panel e and f. PSA-NCAM positive dendrites are indicated by arrows in panels d–f. Scale bar, 0.5 mm (a–c), 0.05 mm (d–f)

As for the other finding, recurrent GS are considered to enhance migration of neural stem cells from the deepest part of the GCL to the outer GCL, as well as the extension of PSA-NCAM positive dendrites. It is also possible that neurons of the outer GCL are involved in plasticity changes. The enhanced extension of PSA-NCAM positive dendrites in animals with GS might represent synaptic remodeling of both migrating neural stem cells and preexisting granule cells. Synaptic remodeling and neural stem cell migration from repeated GS may contribute to reorganization of the neural network with formation of abnormal synapses in the hippocampus. These changes are presumed to lower the seizure threshold culminating in spontaneous seizures in epileptic brain.

Although the *physiological* significance of an enhancement of neural stem cells is not understood so far, it occurs specifically in the DG and may contribute

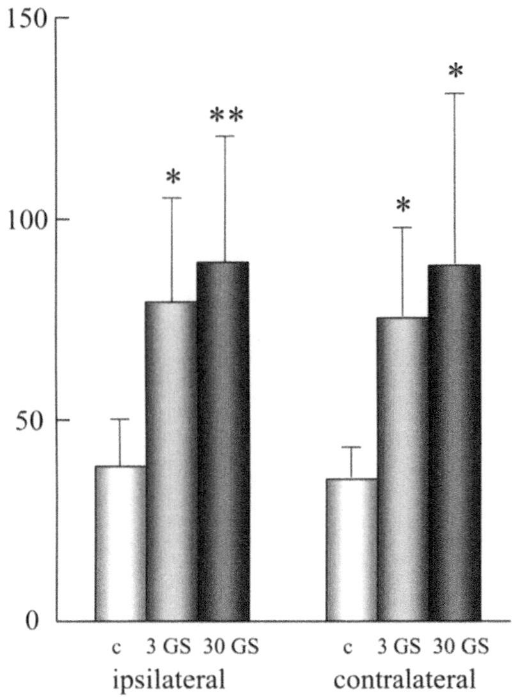

Fig. 2. Number of PSA-NCAM positive neurons in the DG. Note the increased number of PSA-NCAM positive cells both in animals with 3 GS and 30 GS as compared to the sham stimulated controls. Mean ± S.D., *: $p < 0.05$, **: $p < 0.01$, compared with control

there to an optimal functioning of the region [1, 3, 7]. Further studies are needed to elucidate the differentiation of neural stem cells and their anatomical, physiological, and chemical function in kindling. The understanding of the molecular basis of the present findings may offer key information on the mechanisms of acquisition of epileptogenicity.

Acknowledgments

We thank Dr. T. Seki for constructive criticism and technical advice for the manuscript. This work was partly supported by Grant-in-Aid for Scientific Research (B) 12470141 and (Hoga) 12877211 and National Project on Protein Structural and Functional Analyses from the Ministry of Education, Science, Culture and Sports of Japan, by grants (Itoyama Y, Kimura I and Kuzuhara S) from the Ministry of Health and Welfare of Japan, and by a grant from Japan Epilepsy Research Foundation.

References

1. Abe K (2000) Therapeutic potential of neurotrophic factors and neural stem cells against ischemic brain injury. J Cereb Blood Flow Metab 20: 1393–1408
2. Bengzon J, Kokaia Z, Elmer E, Nanobashvili A, Kokaia M, Lindvall O (1997) Apoptosis and proliferation of dentate gyrus neurons after single and intermittent limbic seizures. Proc Natl Acad Sci 94: 10432–10437
3. Bernabeu R, Sharp FR (2000) NMDA and AMPA/kainate glutamate receptors modulate dentate neurogenesis and CA3 synapsin I in normal and ischemic hippocampus. J Cereb Blood Flow Metab 20: 1669–1680
4. Blumcke I, Beck H, Lie AA, Wiestler OD (1999) Molecular neuropathology of human mesial temporal lobe epilepsy. Epilepsy Res 36: 205–223
5. Eriksson PS, Perfilieva E, Bjork-Eriksson T, Alborn AM, Nordborg C, Peterson DA, Gage FH (1998) Neurogenesis in the adult human hippocampus. Nat Med 4: 1313–1317
6. Fox GB, Kjoller C, Murphy KJ, Regan CM (2001) The modulations of NCAM polysialylation state that follow transient global ischemia are brief on neurons but enduring on glia. J Neuropathol Exp Neurol pp 132–140
7. Gage FH (2000) Mammalian neural stem cells. Science 287: 1433–1438
8. Goddard GVD, McIntyre C, Leech CK (1969) A permanent change in brain function resulting from daily electrical stimulation. Exp Neurol 25: 295–330
9. Gould E, Reeves AJ, Graziano MS, Gross CG (1999) Neurogenesis in the neocortex of adult primates. Science 286: 548–552
10. Gray WP, Sundstrom LE (1998) Kainic acid increases the proliferation of granule cell progenitors in the dentate gyrus of the adult rat. Brain Res 790: 52–59
11. Iwai M, Hayashi T, Zhang WR, Sato K, Manabe Y, Abe K (2001) Induction of highly polysialylated neural cell adhesion molecule (PSA-NCAM) in postischemic gerbil hippocampus mainly dissociated with neural stem cell proliferation. Brain Res 902: 288–293
12. Johansson CB, Momma S, Clarke DL, Risling M, Lendahl U, Frisen J (1999) Identification of a neural stem cell in the adult mammalian central nervous system. Cell 96: 25–34
13. Michalakis M, Holsinger D, Ikeda-Douglas C, Cammisuli S, Ferbinteanu J, DeSouza C, DeSouza S, Fecteau J, Racine RJ, Milgram NW (1998) Development of spontaneous seizures over extended electrical kindling. I. Electrographic, behavioral, and transfer kindling correlates. Brain Res 793: 197–211
14. Mikkonen M, Soininen H, Kalvianen R, Tapiola T, Ylinen A, Vapalahti M, Paljarvi L, Pitkanen A (1998) Remodeling of neuronal circuitries in human temporal lobe epilepsy: increased expression of highly polysialylated neural cell adhesion molecule in the hippocampus and the entorhinal cortex. Ann Neurol 44: 923–934
15. Milgram NW, Michael M, Cammisuli S, Head E, Ferbinteanu J, Reid C, Murphy MP, Racine RJ (1995) Development of spontaneous seizures over extended electrical kindling. II. Persistence of dentate inhibitory suppression. Brain Res 670: 112–120
16. Mody I (1999) Synaptic plasticity in kindling. Adv Neurol 79: 631–643
17. Nakagawa E, Aimi Y, Yasuhara O, Tooyama I, Shimada M, McGeer PL, Kimura H (2000) Enhancement of progenitor cell division in the dentate gyrus triggered by initial limbic seizures in rat models of epilepsy. Epilepsia 41: 10–18
18. Parent JM, Yu TW, Leibowitz RT, Geschwind DH, Sloviter RS, Lowenstein DH (1997) Dentate granule cell neurogenesis is increased by seizures and contributes to aberrant network reorganization in the adult hippocampus. J Neurosci 17: 3727–3738
19. Racine RJ (1972) Modification of seizure activity by electrical stimulation: II, Motor seizure. Electroenceph Clin Neurophysiol 32: 281–294
20. Rutishauser U, Landmesser L (1996) Polysialic acid in the vertebrate nervous system: a promoter of plasticity in cell-cell interactions. Trends Neurosci 19: 422–427

21. Scharfman HE, Goodman JH, Sollas AL (2000) Granule-like neurons at the hilar/CA3 border after status epilepticus and their synchrony with area CA3 pyramidal cells: functional implications of seizure-induced neurogenesis. J Neurosci 20: 6144–6158
22. Scott BW, Wang S, Burnham WM, De Boni U, Wojtowicz JM (1998) Kindling-indced neurogenesis in the dentate gyrus of the rat. Neurosci Lett 248: 73–76
23. Seki T, Arai Y (1993) Highly polysialylated neural cell adhesion molecule (NCAM-H) is expressed by newly generated granule cells in the dentate gyrus of the adult rat. J Neurosci 13: 2351–2358
24. Seki T, Arai Y (1993) Distribution and possible roles of the highly polysialylated neural cell adhesion molecule (NCAM-H) in the developing and adult central nervous system. Neurosci Res 17: 265–290
25. Sutula T, He XX, Cavazos J, Scott G (1988) Synaptic reorganization in the hippocampus induced by abnormal functional activity. Science 239: 1147–1150
26. Sutula T, Cascino G, Cavazos J, Parada I, Ramirez L (1989) Mossy fiber synaptic reorganization in the epileptic human temporal lobe. Ann Neurol 26: 321–330

Correspondence: K. Sato, M.D., Department of Neurology, Graduate School of Medicine and Dentistry, Okayama University, 2-5-1 Shikata-cho, Okayama 700-8558, Japan. e-mail: satokei@cc.okayama-u.ac.jp

Acta Neurochir (2003) [Suppl] 86: 581–585

Continuous assessment of cerebral autoregulation: clinical and laboratory experience

M. Czosnyka[1], P. Smielewski[1], Z. Czosnyka[1], S. Piechnik[1], L. A. Steiner[1], E. Schmidt[1], I. Gooskens[1], M. Soehle[1], E. W. Lang[2], B. F. Matta[1], and J. D. Pickard[1]

[1] Academic Neurosurgical Unit, Addenbrooke's Hospital, Cambridge, UK
[2] Department of Neurosurgery, University of Sydney, Westmead Hospital, Westmead, Australia

Summary

The method for the continuous assessment of cerebral autoregulation using slow waves of MCA blood flow velocity (FV) and cerebral perfusion pressure (CPP) or arterial pressure (ABP) has been introduced seven years ago. We intend to review its clinical applications in various scenarios.

Moving correlation coefficient (3–6 min window), named Mx, is calculated between low-pass filtered (0.05 Hz) signals of FV and CPP or ABP (when ICP is not measured directly). Data from ventilated 243 head injuries and 15 patients after poor grade subarachnoid haemorrhage, 38 patients with Carotid Artery stenosis, 35 patients with hydrocephalus and fourteen healthy volunteers is presented.

Good agreement between the leg-cuff test and Mx has been confirmed in healthy volunteers ($r = 0.81$). Mx also correlated significantly with the static rate of autoregulation and transient hyperaemic response test. Autoregulation was disturbed ($p < 0.021$) by vasospasm after SAH and worse in patients with hydrocephalus in whom CSF circulation was normal ($p < 0.02$). In head injury, Mx indicated disturbed autoregulation with low CPP (<55 mmHg) and too high CPP (>95 mmHg). Mx strongly discriminated between patients with favourable and unfavourable outcome ($p < 0.00002$).

This method can be used in many clinical scenarios for continuous monitoring of cerebral autoregulation, predicting outcome and optimising treatment strategies.

Keywords: Transcranial Doppler; autoregulation; head injury; hydrocephalus; subarachnoid haemorrhage; carotid artery stenosis.

Introduction

Autoregulation of cerebral blood flow (CBF) is the most potent intrinsic mechanism of cerebral self-protection against ischaemic insults. Failure of autoregulation has been associated with unfavourable outcome after head injury, subarachnoid haemorrhage, and stroke. The continuous assessment of cerebral autoregulation has been investigated by numerous, independently working groups over the past years [1, 5, 7, 9, 13, 14]. All methods converged to the similar principle of assessing interdependence between coherent slow fluctuations (starting from respiratory to 3-min period waves) in arterial blood pressure (ABP), or cerebral perfusion pressure (CPP), and CBF flow velocity (FV) monitored in the middle cerebral artery (MCA) using transcranial Doppler ultrasonography (TCD). Having a unique opportunity to haemodynamically assess patients suffering from severe head injury, subarachnoid haemorrhage, hydrocephalus, as well as healthy volunteers, we have previously published several reports regarding various aspects of cerebral autoregulation in clinical and laboratory practice [1–4, 8, 10, 11, 12]. Our intention is to summarize our own experience, to disseminate this useful methodology and encourage its use in various clinical scenarios.

Material and methods

Patients and volunteers

Fourteen healthy volunteers had autoregulation tested non-invasively on different levels (normo-, hypo-, and hyperventilation) of arterial partial pressure of CO_2 ($PaCO_2$). Two hundred and forty-three ventilated head injured patients after severe brain trauma or with a GCS ranging 9–13 at the scene, but who deteriorated further were monitored daily (ABP, ICP and TCD). We also included 15 ventilated poor grade patients after subarachnoid haemorrhage, 38 patients with common carotid artery stenosis (routine preoperative testing) and 35 patients suffering from hydrocephalus undergoing intraventricular infusion tests. In all patients, haemodynamic investigations using TCD were part of the routine clinical assessment and as such did not require informed consent. In the volunteers, informed consent was obtained.

Monitoring

Intracranial pressure was monitored continuously using micro-transducers in patients with head injury and SAH (Camino Direct Pressure Monitor, Camino Laboratories, San Diego, CA; or Codman MicroSensors, Johnson & Johnson Professional, Raynham, MA), inserted intraparenchymally into the frontal region. In hydrocephalus, ICP was monitored using external pressure transducer (Baxter) connected to the manometer line and a 25 G butterfly needle inserted into a pre-implanted Ommaya reservoir, connected to the ventricular CSF space. Arterial pressure was monitored directly from the radial or dorsalis pedis artery in head injury or in SAH (System 8000, S&W Vickers Ltd, Sidcup, UK or Solar 6000 System, Marquette, USA) or using a finger-cuff (Finapress) in patients with hydrocephalus, Carotid Artery stenosis and volunteers. The MCA was insonated daily on the side of the ICP bolt or bilaterally for a period of 20 minutes to 2 hours starting from the day of admission until discharge or day 8 following head injury, using the PCDop 842 Doppler Ultrasound Unit (Scimed, Bristol, UK) or Neuroguard (Medasonics, Fremona, CA). In other patients/volunteers MCA signal was monitored during the clinical test. Signals were recorded during periods of stable respiratory parameters, free from movement artefacts, physiotherapy, tracheal suction, and other disturbances.

Data series capture and analysis

Analog output from the pressure monitors and the TCD unit (maximal frequency envelope) were connected to the analogue-to-digital converter (DT 2814, Data Translation, Marlboro, USA) fitted into an IBM AT laptop computer (Amstrad ALT 386 SX, UK). Data were sampled, digitized and stored on the hard disk using software specifically designed for waveform recording and later processed using software developed in-house.

Time averaged values of ICP, ABP, CPP (CPP = ABP − ICP) were calculated using waveform time integration for 6 to 10-second intervals (integration period should be long enough to filter out the respiratory wave, alternatively, digital low-pass filtering can be used). Time averaged mean, systolic and diastolic values of FV were calculated after careful spectral filtration to reduce the influence of noise, and averaged within the same periods. A mean index (Mx) was calculated as a Pearson's correlation coefficient of 30 to 60 consecutive samples of CPP and mean FV, i.e. every 3–6 min. Period of calculation is not critical in this method, it should be long enough to cover the period of the longest visible slow waves in CPP and FV (usually the period is from 20 sec to 3 min). A positive association between CPP/ABP and FV (positive values of Mx) indicates passive dependence of blood flow on pressure, therefore defective autoregulation. Negative or zero values of Mx implicate active cerebrovascular responses to changes in CPP or ABP, i.e. preserved autoregulation.

Results

Volunteers

The rate of autoregulation (RoR) determined by the leg cuff test and Mx both significantly depended on $paCO_2$ (ANOVA, $p < 0.00001$, Fig. 1). At a high $PaCO_2$ autoregulation was disturbed (RoR < 0.17 [1/s] in all volunteers). This was in accordance with positive values of Mx (>0.4 in 86% of volunteers). Overall the relationship between RoR and Mx was confirmed to be linear ($R = 0.81$).

Head injury

The relationship between Mx and ABP or CPP has "u"-shape (Figure 2). Autoregulation was disturbed when CPP was too low (CPP < 65 mmHg) or too high (CPP > 90 mmHg) and in association with intracranial hypertension (ICP > 25 mmHg). Patients with an unfavourable outcome had more often disturbed autoregulation ($p < 0.005$) than those with favourable outcome.

Autoregulation was disturbed significantly (Mx > 0.2) during plateau waves of ICP. In 47 patients Mx was compared to the transient hyperaemic response test showing a significant correlation ($r = 0.49$; $p < 0.0001$). In 17 patients (Department of Neurosurgery, Christian-Albrechts University, Kiel) Mx was compared with the static rate of autoregulation. ABP was raised pharmacologically (norepinephrine) and a strong correlation between these two methods was confirmed ($r = 0.78$; $p < 0.05$).

With bi-lateral TCD, Mx was significantly ($p < 0.05$) worse on a side of contusion and on the side of brain expansion in patients presenting with a midline shift ($p < 0.05$), (Figure 3). Of those patients who died in hospital, significantly more presented hemispherical differences of Mx ($p < 0.05$).

Subarachnoid haemorrhage

Autoregulation during vasospasm was significantly worse than at baseline (Mx = 0.46 ± 0.32 versus Mx = 0.21 ± 0.24; $p = 0.021$). Mx correlated significantly with mean FV ($r = 0.577$, $p = 0.025$) and the Lindegaard ratio ($r = 0.672$, $p < 0.006$). Autoregulation was significantly worse ($p = 0.006$) on the side of vasospasm as compared to the contralateral side.

Common carotid artery stenotic disease

Impaired CO_2 reactivity was significantly correlated with worse pressure-autoregulation ($R = 0.34$; $p < 0.05$). In contrast to the data obtained in head injury, side-to-side differences in Mx failed to correlate with the degree of asymmetry of the Carotid Artery stenosis assessed with angiography. Mx indicated se-

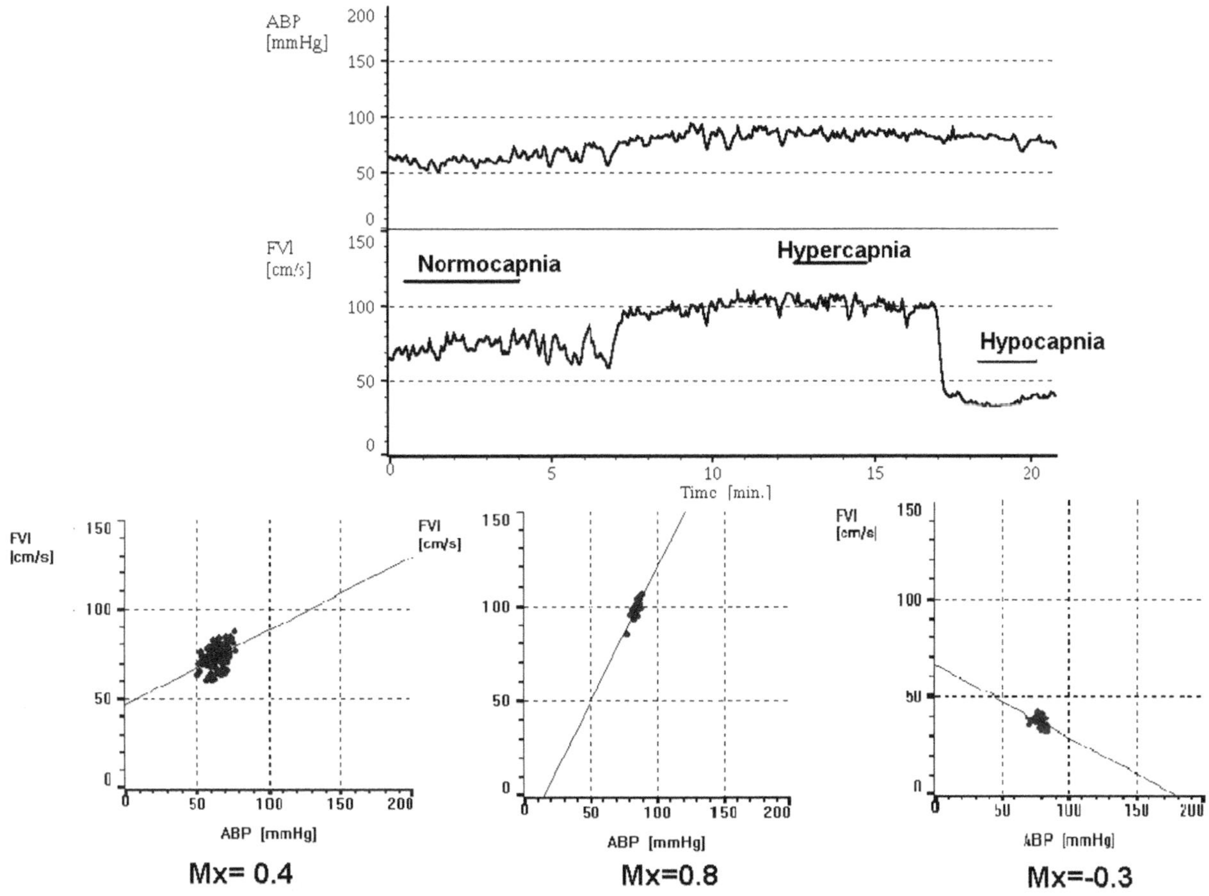

Fig. 1. Mean Flow Velocity (FVl- in this particular case form the left MCA) and arterial blood pressure (*ABP*) was recorded in volunteer during normo-, hyper- and hypo-capnia (upper graph). Regression plots of FVl versus ABP from these three periods reveals change in Mx according to CO_2 induced vasodilation/vasoconstriction (lower panels)

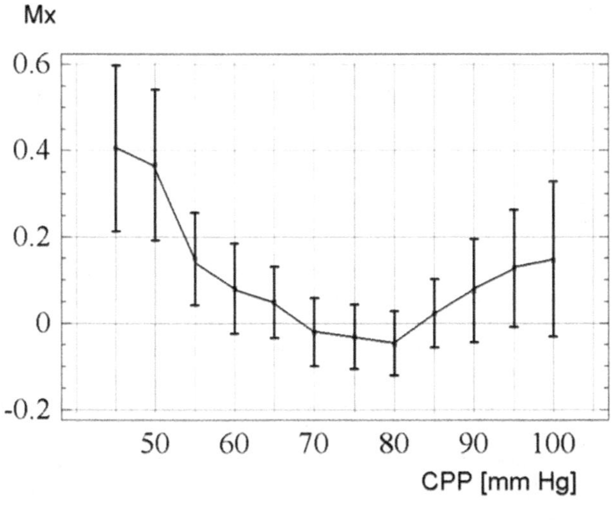

Fig. 2. Averaged MX versus CPP in 239 head injured patients indicate that either too low or too high CPP may be detrimental from the point of optimal autoregulation

vere disturbance in autoregulation in the presence of bilateral- in comparison to unilateral-stenosis ($p = 0.01$).

Hydrocephalus

Constant rate infusion tests were performed in 35 patients in whom CSF dynamics were measured before shunt implantation with simultaneous measurement of MCA FV. Mx was significantly correlated with the resistance to CSF outflow ($R = -0.41$; $p < 0.03$), indicating better autoregulation in those patients who had a *disturbed* CSF circulation.

Discussion

According to the classic theory, autoregulation interacts with $PaCO_2$ [6]. Indeed, Mx showed a strict relationship to $PaCO_2$, indicating that vasodilatation

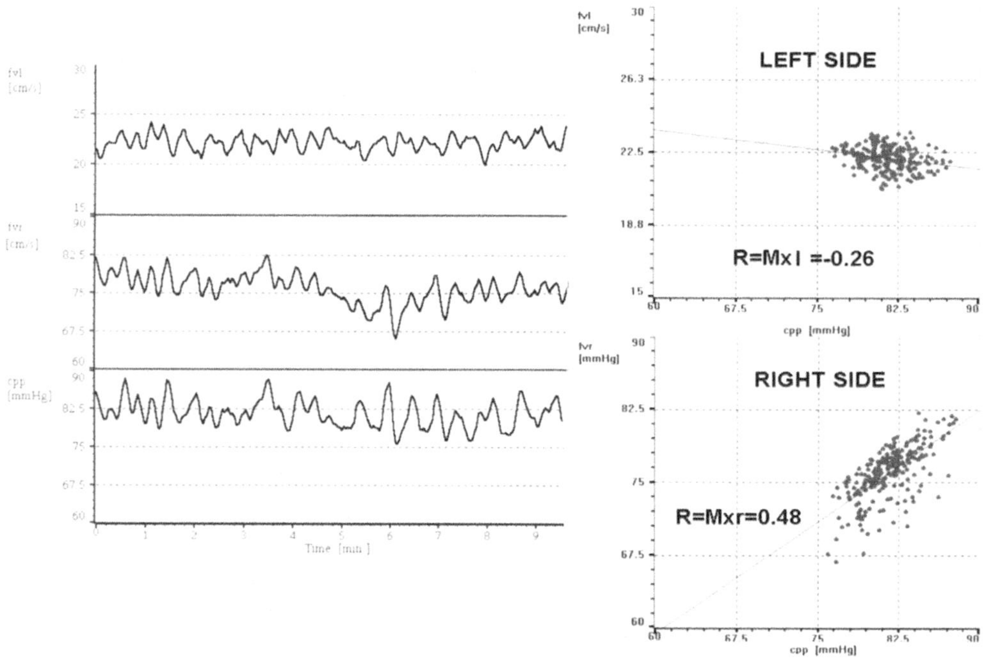

Fig. 3. Recording of Flow Velocity in left and right MCAs and CPP in the patient with right-to-left midline shift. While autoregulation on left side was good, Mx indicated heavily disturbed autoregulation at right side

caused by hypercapnia disturbs autoregulation even in healthy volunteers. This emphasises that autoregulation is a mechanism which can easily be disturbed, and is probably of transitional nature. Sitting in a not very well ventilated room, headaches (like these experienced during hangover, as our healthy volunteers discovered-unpublished data), commonly disturb autoregulation. Therefore, in clinical practice averaging of Mx over longer period (20–30 minutes minimum) is necessary.

Using of ABP instead of CPP to calculate Mx index seems to be reliable in all cases where ICP is assumed to be normal. This makes the assessment of autoregulation even more attractive: It can be done completely non-invasive. In head injured patients ($n = 187$) we calculated Mx using both ABP and CPP. The correlation between these two indices was close ($R = 0.82$), but there were patients in whom both indices differed to such an extent that false conclusion for autoregulation were possible. Interestingly, Mx calculated using ABP only was significantly higher than Mx calculated with CPP (on average 0.15). This should be taken into account, when comparison of autoregulation based on mixed data is to be made. In normoventilated head injured patients ($PaCO_2$ 4.5 to 5.5 kPa, CPP as an input function), the threshold of Mx indicating severely disturbed autoregulation seems to be 0.2. Mx less than 0 indicates good autoregulation. Values between 0 and 0.2 should be interpreted, to the best of our knowledge, as a grey zone.

The authors hope that the presented material proves the great potential of the continuous assessment of cerebral autoregulation in a wide range of clinical scenarios.

Acknowledgment

Supported by grant: Medical Research Council (UK) Acute Brain Injury.

References

1. Czosnyka M, Smielewski P, Kirkpatrick P, Menon DK, Pickard JD (1996) Monitoring of cerebral autoregulation in head-injured patients. Stroke 27: 829–834
2. Czosnyka M, Smielewski P, Piechnik S, Schmidt EA, Al-Rawi P, Kirkpatrick PJ, Pickard JD (1999) Hemodynamic characterization of intracranial pressure plateau waves in head-injured patients. J Neurosurgery 91: 11–19
3. Czosnyka M, Smielewski P, Piechnik S, Steiner LA, Pickard JD (2001) Cerebral autoregulation following head injury. J Neurosurg 95: 756–763
4. Czosnyka ZH, Czosnyka M, Whitfield PC, Donovan T, Pickard JD (2002) Cerebral autoregulation among patients with symptoms of hydrocephalus. Neurosurgery 50(3): 526–532

5. Diehl RR, Linden D, Lucke D, Berlit P (1998) Spontaneous blood pressure oscillations and cerebral autoregulation. Clin Auton Res 8: 7–12
6. Harper AM, Glass HI (1965) Effect of alterations in the arterial carbon dioxide tension on the blood flow through the cerebral cortex at normal and low arterial blood pressures. J Neurol Neurosurg Psychiat 28: 449–452
7. Hu HH, Kuo TB, Wong WJ, Luk YO, Chern CM, Hsu LC, Sheng WY (1999) Transfer function analysis of cerebral hemodynamics in patients with carotid artery stenosis. J Cereb Blood Flow Metab 19: 460–465
8. Lang EW, Mehdorn HM, Dorsch NW, Czosnyka M (2002) Continuous monitoring of cerebrovascular autoregulation: a validation study. J Neurol Neurosurg Psychiatry 72(5): 583–586
9. Panerai RB, White RP, Markus HS, Evans DH (1998) Grading of cerebral dynamic autoregulation from spontaneous fluctuations in arterial blood pressure. Stroke 29: 2341–2346
10. Piechnik SK, Yang X, Czosnyka M, Smielewski P, Fletcher SH, Jones AL, Pickard JD (1999) The continuous assessment of cerebrovascular reactivity: a validation of the method in healthy volunteers. Anesth Analg 89: 944–949
11. Schmidt EA, Czosnyka M, Matta BF, Gooskens I, Piechnik S, Pickard JD (2000) Non-invasive cerebral perfusion pressure (nCPP): evaluation of the monitoring methodology in head injured patients. Acta Neurochir (Wien) [Suppl] 76: 451–452
12. Smielewski P, Czosnyka M, Kirkpatrick P, Pickard JD (1997) Evaluation of transient hyperaemic response test in head injured patients. J Neurosurg 86: 773–778
13. Steinmeier R, Bauhuf C, Hubner U, Bauer RD, Fahlbusch R, Laumer R, Bondar I (1996) Slow rhythmic oscillations of blood pressure, intracranial pressure, microcirculation, and cerebral oxygenation. Dynamic interrelation and time course in humans. Stroke 27: 2236–2243
14. Zhang R, Zuckerman JH, Giller CA, Levine BD (1998) Transfer function analysis of dynamic cerebral autoregulation in humans. Am J Physiol 274: H233–H241

Correspondence: Dr. Marek Czosnyka, Academic Neurosurgical Unit, Box 167, Addenbrooke's Hospital, Cambridge CB2 2QQ, UK. e-mail: Mc141@medschl.cam.ac.uk

Acta Neurochir (2003) [Suppl] 86: 587–589

Spontaneous intracranial hypotension (SIH): The early appearance of urinary bladder activity in RI cisternography is a pathognomonic sign of SIH?

S. Ishihara, N. Otani, and **K. Shima**

Department of Neurosurgery, National Defense Medical College, Saitama, Japan

Keywords: Spontaneous intracranial hypotension; RI cisternography; early bladder visualization.

Introduction

Spontaneous intracranial hypotension (SIH) is a rare but increasingly recognized pathophysiological entity. It is a syndrome of low CSF pressure due to CSF leakage. This usually associates with postural headache which characteristically become worse when sitting or standing, and often ameliorates completely by lying flat. This could be caused by minor head trauma, vigorous exercise or a violent bout of coughing. In a clinical situation, it is sometimes difficult to differentiate it from other headache accompanying disease such as subarachnoid hemorrhage or chronic subdural hematoma because of its wide range of presenting symptoms. Characteristic neuroradiological findings have been well documented in the literature. MR imaging includes dural enhancement, subdural fluid collection and downward displacement of the cerebellar tonsils. Radioisotope (RI) cisternography, CT myelography and recently MR myelography are studies of choice to detect CSF leakage. Lumbar CSF pressure is usually below 60 mmH_2O and this can be assured by pressure monitoring.

Radioisotope (RI) cisternography findings are characterized by rapid clearance of the RI from the CSF space and early RI filling of urinary bladder. However, the changes of RI cisternography findings in the post-treatment state are still not fully understood. We herein report three cases who presented with severe postural headache diagnosed as SIH. In each case, symptoms and MR imaging findings resolved after treatment, except the early visualization of urinary bladder activity in RI cisternography. Therefore, we believe that this finding may not be pathognomonic in SIH.

Case presentation

Case 1

A 39-year-old woman developed a severe postural headache, which was diagnosed as muscle contraction headache at two other neurosurgical institutes and referred to our hospital 11 days after the acute onset of a severe headache. Lumbar puncture revealed an opening pressure of 50 mmH_2O with normal protein and glucose level. MRI myelography demonstrated CSF leakage from the left upper cervical nerve roots. The early bladder appearance was detected on RI cisternography at 4 hours post RI tracer injection (Fig. 1). Her symptoms resolved with treatment. The 3-year post-treatment follow up MRI presented no dural enhancement. Lumbar pressure was 150 mmH_2O. However the RI cisternography demonstrated early RI accumulation in the bladder as early as at 3 hours (Fig. 2).

Case 2

A 58-year-old woman with severe headache was transferred from another hospital presenting with vertigo. The headache was much less severe with reclining and was markedly worse if she stood. She presented with bilateral subdural hematomas on initial CT scan without any history of trauma. An MRI study showed diffuse dural enhancement with slight downward dis-

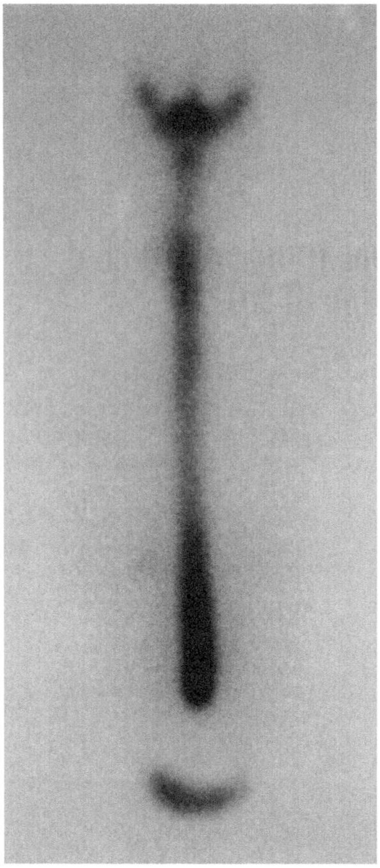

Fig. 1. The early bladder appearance was detected on RI cisternography at 4 hours post RI tracer injection. It shows the diffusion of the nucleotide into extra-arachnoid space in the right upper cervicothoracic junction

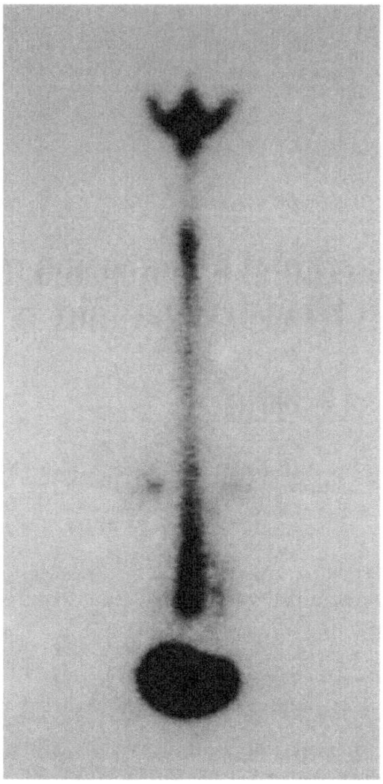

Fig. 2. RI cisternography demonstrated early RI accumulation in bladder as early as at 3 hours

placement of cerebellar tonsils. Then MRI myelography revealed multiple CSF retention at the nerve roots along the whole spinal axis, representing multiple meningeal diverticula. The RI cisternography demonstrated a single tracer uptake in the right lumbar region and early tracer accumulation in the bladder with less uptake in convexity region. Her symptoms recovered well in response to treatment. The follow up RI cisternography two years later, presented early bladder visualization with 90 mmH_2O lumbar pressure.

Case 3

A 53-year-old woman with postural severe cervico-occipital pain with subdural hematoma was referred to our hospital. Lumbar puncture revealed an opening pressure of 40 mmH_2O. Radio-isotope cisternography resulted in early bladder appearance and in delayed cortical tracer accumulation. Her symptoms disappeared completely with treatment. However, a two-month follow up study showed the early RI accumulation in the bladder without any detectable CSF leakage. For RI study, we used ^{111}DTPA (the diethlenetetramine pentaacetic acid chelate of indium 111).

Discussion

Spontaneous intracranial hypotension is a rare disease but is a very unique clinical entity for the differential diagnosis of headache. It was first described by Schaltenbrand in 1938 as "acute aliquorrhea". The presenting symptoms feature postural headache but occasionally neck pain, nausea, vomiting or vertigo may be associated as well as other cranial neuropathies. Low CSF pressure is a definitive confirmation of this disease and is usually lower than 60 mmH_2O as in our presented cases. CSF examination often demonstrates a high protein level, increased number of lymphocytes and red blood cells. The low pressure of CSF, especially if secondary to a leak from disruption of the meninges, may facilitate the passage of serum protein and blood cells into the subarachnoid space [10].

Characteristically, MRI studies in the patients with intracranial hypotension reveal diffuse enhancement of the meninges [2–12], associated subdural hematoma or hygroma [3, 9, 12], and caudal displacement of the cerebellar tonsils. Radio-isotope cisternography shows early accumulation of the nucleotide in the bladder and less activity over the cerebral convexities. These features are suggestive of an unusually rapid uptake of the tracer in the bloodstream [13]. Those presented cases demonstrated this early bladder visualization after resolution of symptoms with normal lumbar pressure. The sensitivity of RI cisternography is not high (60%), and spatial resolution of this study is rather poor [1]. Therefore one of the explanations of this finding is that early accumulation in the bladder does not necessarily indicate the presence of generalized CSF hyperabsorption [2]. And alternative explanation is the presence of a leak below the limits of resolution of the study [2]. However after treatment, all of those presented cases demonstrated normal lumbar pressure with complete resolution of diffuse dural enhancement on MRI study. They could not be diagnosed as intracranial hypotension at that stage. It is well known that besides arachnoid villi as a major site of CSF absorption, venous plexus, lymphatics along the spinal axis can act as alterative location of CSF reabsorption. Once the brain underwent the insult of intracranial hypotension, those alternative pathways may open and act as activated sites of CSF absorption. This may be a result of compensation for less CSF absorption from the arachnoid villi in those patients. Further study should be necessary to discover the CSF circulation in this pathophysiological condition.

Conclusion

We suggest that the early appearance of urinary bladder activity with RI cisternography is not pathognomonic for SIH. We presume that this interesting sign might reflect the up-regulated minor CSF pathway through the spinal arachnoid.

References

1. Benamor M, Tainturier C, Graveleau P *et al* (1998) Radionuclide cisternography in spontaneous intracranial hypotensionm. Clin Nusl Med 23: 150–151
2. Fishman RA, Dillon WP (1993) Dural enhancement and cerebral displacement secondary to intracranial hypotension. Neurology 43: 609–611
3. Good DC, Ghobrial M (1993) Pathologic changes associated with intracranial hypotension and meningeal enhancement on MRI. Neurology 43: 2698–2700
4. Hochman MS, Naidich TP, Kobetz SA, Maitin AF (1992) Spontaneous intracranial hypotension with pachymeningeal enhancement on MRI. Neurology 42: 1628–1630
5. Matsumura A, Anno I, Kimura H, Ishikawa E, Nose T (2000) Diagnosis of spontaneous intracranial hypotension by using magnetic resonance myelography. J Neurosurg 92: 873–876
6. Morki B, Parisi JE, Schethauer BW, Piepgras DG, Miller GM (1995) Meningeal biopsy in intracranial hypotension: meningeal enhancement on MRI. Neurology 45: 1801–1807
7. Niwa K, Yoshii F, Katayama M, Miyazaki H (1997) A patient with spontaneous intracranial hypotension: comparison between MRI findings and meningeal pathology. No To Shinkei 49: 541–546 (Japanese)
8. Pannullo SC, Reich JB, Krol G, Deck MDF, Posner JB (1993) MRI changes in intracranial hypotension. Neurology 43: 919–926
9. Rando TA, Fishman RA (1992) Spontaneous intracranial hypotension: report of two cases and review of the literature. Neurology 42: 481–487
10. River Y, Schwartz A, Gomori JM, Soffer D, Siegal T (1996) Clinical significance of diffuse dural enhancement detected by magnetic resonance imaging. J Neurosurg 85: 777–783
11. Schievink WI, Reimer R, Folger WN (1994) Surgical treatment of spontaneous intracranial hypotension associated with a spinal arachnoid diverticulum. J Neurosurg 80: 736–739
12. Schievink WI, Meyer FB, Atkinson JLD, Mokri B (1996) Spontaneous spinal cerebrospinal fluid leaks and intracranial hypotension. J Neurosurg 84: 598–605
13. Schievink WI, Morreale VM, Atkinson JLD, Meyer FB, Piepgras DG, Ebersold MJ (1998) Surgical treatment of spontaneous spinal cerebrospinal fluid leaks. J Neurosurg 88: 243–246

Correspondence: Shoichiro Ishihara, M.D., Department of Neurosurgery, National Defense Medical College, Address: 3-2 Namiki, Tokorozawa, Saitama, Japan 359-8513. e-mail: shopar72@me.ndmc.ac.jp

Acta Neurochir (2003) [Suppl] 86: 591–599

Roundtable discussion (Brain Edema 2002)

Moderators: J. T. Hoff (Ann Arbor), K.-A. Hossmann (Cologne), and **A. Marmarou (Richmond)**

Konstatin-A. Hossmann

Ladies and gentlemen, I am afraid that the most difficult part of the conference begins right now. We have been asked to summarize what we have heard during the last three days, to identify the most interesting topics and to make recommendations on what to do in the future. We sat together to identify a few topics which could be of interest for the general discussion, and Tony Marmarou will be happy to introduce this list to you.

Anthony Marmarou

First of all we would like to start with a series of disclaimers as it is impossible to review all of the studies: Firstly, everyone's work presented in this conference is important, and secondly, we hope that everyone will continue their work in the coming years. With consideration of the limited time available, we have divided the research into four basic sections: one, mechanisms; the second, gene profiling and regeneration; the third therapy; and the fourth, non-invasive imaging. Although it is almost impossible to list sufficient subtopics to capture all of the studies, we felt that thematically, this was the general sense of the work that you have brought to us. Let me start then with regard to mechanisms. There were several papers dealing with aquaporins, and it appears that aquaporins will remain as an integral part of the edema conference. Their role however, remains to be elucidated. Manley, in his work with regard to knockout mice, has clearly demonstrated that aquaporins are certainly involved in volume homeostasis. But at the same time he suggested that there may be alternate pathways and mechanisms that knockout mice may utilize in the resolution of edema fluid. So it is clear that additional work must continue with the knockout mice. At the same time one must tread carefully to ensure that one is not taken down a specific path or mechanism, knowing full well that there may be alternate pathways that are specific to the knockouts. We expect that this work would continue what would foster the continued development.

For the first time, some work that was presented with regard to the role of aquaporins in an in vivo model of brain injury. We believe that this is the direction that we must proceed. More specifically, in addition to the more basic work that was presented by several investigators showing markers tagging the epitopes so that we can see where these aquaporins lie, a parallel effort in elucidating the role of aquaporins in models of trauma and ischemia is necessary.

What can be said regarding the clearance mechanism of edema? Lets spend a moment discussing edema production. The question has become, whereas in past conferences, the blood brain barrier has truly been key in edema research, what is the role of aquaporins in the different types of edema that are now known to develop following traumatic injury. We have spent decades in exploring the blood brain barrier, and although these studies continue to be important we now realize that there may be a combination of edemas developing which involve the aquaporins. For example, in the development of cellular edema.

But consider for a moment that the aquaporins are also involved in clearance. Now how could that be? If we accept the notion that the reactive astrocytosis seen in many forms of injury as a glial response, allows the astrocytes to develop projections both to neurons and to capillaries which could then play a role in clearance? Although speculative, it entertains a new area of fertile investigation. Perhaps the next conference will provide some insight as to the contribution of barrier-produced vasogenic edema versus cellular edema as contributed by the aquaporins via the astrocytic response versus other forms or other causes of cellular edema such as will be covered by Dr. Hossmann.

These are the areas, that we find important, at least in the aquaporins, and what their relative role will be – both in traumatic injury and other disease processes. So for those investigators who have been working in these areas, particularly in knockout mice and particularly those that have shown the tagging of the epitopes to identify the upregulation, we believe this work is important and should continue.

If we move now to the extracellular matrix, Katayama and others had proposed that in the site of focal injury, the compaction of cells forms a barrier, and as result the normal propagation from the site of the lesion and the exudation of the edema fluid from that site would be limited. At the same time, the concept was put forth that perhaps a result of this compartmentalization, that there is an osmotic gradient that is built up, which then alters the resolution and of course the production of fluid.

What is the importance of understanding this relationship? Clinically it is extremely important because at the present time there is no known method by which one can determine if a contusion is likely to resolve or one to explode or blossom, eventually in some cases reaching a malignant form of refractory brain swelling.

So techniques which characterize the extracellular matrix surrounding lesions, would help us further identify the basis of the clearance mechanisms that have been viewed as operational the past several years – Are they intact? Do we visualize those as being intact when we know that there is now a cellular rim of edema that develops surrounding these lesions? And with regard to the classification – and this was brought up in our discussion – despite the fact that there may be a hydrostatically induced edema, it is in essence a hydrostatically induced vasogenic edema. And we feel that the original classification of vasogenic versus cellular still remains, as was pointed out by Prof. Baethmann. With regard to the third issue, the propagation, we believe the work, the seminal work, done by Hans Reulen exploring mechanisms of diffusion and bulk flow through the extracellular space have been clearly defined. Subsequent to these studies, we have found that as a result of diffusion that there can be a global or spherical resolution migrating to the surface of the brain and eventually into the subarachnoid space. Those mechanisms of edema propagation still exist in tumor, but it is in the contusion, and it is in diffuse injury where we still have a lot of work to do in order to be able to define what we mean in terms of propagation. What is the end result of this edema accumulation with regard to the development of intracranial pressure? These therapies will be discussed later by Dr. Hoff.

So we are left with still much work to do. More specifically to further elucidate the role of these aquaporins. Secondly, to more clearly define the role of the extracellular matrix as to its protection or perhaps its deleterious effect with regard to contusion. And finally, to determine what the role of the interaction of the differences in the extracellular matrix are to propagation of edema and the blossoming of lesions.

What I would like now is to ask Dr. Hossmann to discuss the gene profiling and regeneration. And we thought that as we finish these areas we may call upon one or two investigators or moderators in these area to give us their view as well.

Konstantin-A. Hossmann

Thank you Tony. The second issue is gene profiling and tissue regeneration. These two topics are seemingly unrelated, but we would like to propose them for this roundtable discussion because they are probably the two hottest research areas in neurobiology, and because they have been brought up for the first time at an Edema meeting.

High-throughput gene profiling is considered by some people as unscientific because it is a clear deviation from hypothesis-driven research. However, the unbiased screening of molecular abnormalities is probably the only way to liberate ourselves from trendy injury concepts that dominate the scientific thinking not because of evidence, but because they are mainstream. Glutamate toxicity, calcium toxicity, tPA toxicity or apoptosis are just a few examples of this kind of research which comes and goes. Gene profiling approaches such as the new possibilities of cDNA microarrays at the transcriptional level or of multi-Western blots at the translational level screen the whole genomic expression pattern of a complex disease and provide the chance to detect physiologically important molecules and mechanisms which are not in the focus of mainstream research. This new development has to be taken seriously, although it is still unclear how to handle the enormous amount of data generated by this approach.

The other new development which we want to bring

to the attention of this roundtable discussion is regeneration. For centuries, it has been believed that in the adult brain regeneration of neurons is not possible. However, with more sensitive techniques, in particular with the availability of proliferation markers, it suddenly comes to our attention that there is neurogenesis in the adult brain and that in various pathological conditions – stroke, trauma – newborn neurons migrate and settle down in the vicinity of the lesion. The big question is if they connect or not, and if it is possible to enhance this process, either by increasing the speed of neurogenesis or by transplanting neural progenitor cells into the brain. We had several presentations on regeneration, and we would like to ask to what extent neurogenesis contributes to the repair of a lesion. Is it of functional relevance or is it just an irrelevant side effect, which has no practical impact on the outcome of a disease? We will ask a few persons for their comments on this new and exciting research, and with that I would like to pass the microphone on to Dr. Hoff...

Julian Hoff

Thank you Peter. One of the nice things about this meeting to me as a clinician is that it is a melt of basic scientists and clinicians, which allows us to do or to pursue translational research. We have heard a lot about revolutions in neurobiology, but we also have heard about some of the frustrations clinicians have and are unable to solve by clinical trials. For example, the intracerebral hemorrhage, a clinical trial that is ongoing under the leadership of David Mendelow, is a great hope. But I think we need to be prepared even if the answer comes to us as clinicians that there is a therapy that is useful, that the bias that physicians have, and so on, has to be overcome as well. So we have to have hard irrefutable evidence that these studies can provide. That means the kind of studies that Dr. Mendelow is leading and hopes to pursue in additional directions. And so we will call on David later on in the discussion. We have also heard a fair amount about thrombin, and it seems to be sort of ubiquitous. It has become something that seems to affect practically everything. It is an inciter of edema, clearly, but has a role in the ischemic stroke as well as intracerebral hemorrhage, and it even possibly has a role in the inhibition or the propagation of tumors. With the thrombin story, one of the beauties of that is that we have pharmacological inhibitors that are currently available, and there are a lot of them that can be pursued to explore this thrombin story in greater detail. Where it will end I do not know, but it is certainly something that I think we need to continue to pursue in various directions.

One of the other interesting things that came out of this meeting for me was in the clinical trials arena dealing with decompression craniotomy that is currently being explored on both sides of the Atlantic. Also our Japanese colleagues are exploring decompression craniotomy for ischemic stroke, and particularly in young patients. Of course, the indications for and the outcomes of these trials are clearly unknown at the present time. There is a clinical trial on-going in the United States, but the enthusiasm for that too is marginal. We are not sure of the fate of this trial.

Dr. Umeo Ito (comments solicited after round table discussion)

We usually do a decompressive craniectomy with or without internal decompression, when the patient is threatened by unilateral transtentrial herniation and the operative result of the patients is not expected to be vegetative and/or severely disabled state. We never select this procedure as a first choice in case of increased ICP. Usually, the decision is necessary in an ambulance case of intracranial hematoma arriving within 30 minutes ~1 hour after the ictus or accident. During a careful observation in the NICU, when ipsilateral anisocoria with retarded light reaction occurs, we make the same decision. Some times, we open the burr hole at the emergency room or ICU bedside to evacuate the subdural or epidural hematoma, and/or perform a rapid drip infusion of mannitol. Immediately, we bring the patient to the operation room. A large craniectomy with enough subtemporal decompression and large radial dural incision is necessary in order that the bulged cortical surface to be trapped by bone edge of the cranial window.

Julian Hoff

There have been new therapies for clinical problems that we have heard about here – specifically, anti-inflammatory agents. It turns out, of course, that cytokines in the inflammatory process are at work in all of these issues that we are exploring associated with edema. Moreover, the use of inflammatories as therapeutic instruments has not been explored to the full extent, and should be. Particularly, notable are the spinal cord injury posters that were available a at this meeting.

Finally, the closed head injury story keeps on going – anti-inflammatory use there. The inhibition of cytokines and also the cyclosporin story that Tony Marmarou could possibly tell us more about. So I think from a therapeutic standpoint, we have a lot to do. At this point I will turn it back to Tony.

Anthony Marmarou

With regard to therapy, for those who have attended most of the poster work, in addition to a few of the oral presentations, it was amazing as to the number of neurotrophic factors which were effective, particularly by the Uppsala group in spinal cord. It's unfortunate, however, that in this work because of restrictions within Sweden that they are not able at the present time to evaluate functional outcome because of restrictions in animal use. Nevertheless, the pathophysiologic changes that were seen and of course were prevented were quite remarkable, so that this work certainly could continue. But at the same time one would hope that by virtue of collaborative efforts that might be fostered by meetings such as this that if certain studies could not be done in their laboratories perhaps they would combine with other investigators so that in three years from now when we meet that it would be possible to see some functional outcome of these very exciting applications.

The clinical trials, at least in traumatic brain injury, are also quite exciting, and that was based on the B2 receptor antagonist presented by Alex Baethmann. In this particular case, this trial is ongoing in the United States with a new bradykinin antagonist. The decompression trial that was mentioned by Dr. Hoff, is a big problem in the United States. When you look at the early works from our German colleagues with regard to the effectiveness of prophylactic decompression, or early decompression, it was rather positive. But to conduct this trial in the United States has truly been problematic. And again, there was an attempted randomization of patients, but to ask the surgeon to randomize patients based upon having a mass lesion, whether it be subdural or small contusion, and to randomize to complete decompression, prophylactic is bygone of the main issues. So we do not know what the outcome will be as to whether or not decompressive surgery will find its way to the U.S. in clinical trial. And for that reason we may call on David Mendelow later to give us an update as to what the prospects are conducting this trial in Europe as well. I would like at this point in time to hand it back to Peter so that he can talk about the application of non-invasive imaging to help guide perhaps some of the management of these patients.

Konstatin-A. Hossmann

Non-invasive imaging is a bridge between experiment and clinic, and if we have time we may go later into the details. But let us go to the first topic of our discussion, aquaporins. Do these channels contribute to the evolution of edema? Is there any evidence that influencing the water flux across these channels will be able to reduce, delay or even treat edema?

Anthony Marmarou

Yes I would like to invite comment by Dr. Manley with regard to extracellular matrix. And then I would like to call on Hans Reulen for the propagation.

Jeff Manley

The first time that the aquaporin was presented at this meeting three years ago in Newcastle, and since that time there's been a few more investigators that have come on board. I think that one of the things that we work against at this point is overcoming the idea that water's freely permeable across membranes. I believe that there are a number of examples that it's not. There's been a number of questions that have been raised regarding the fact that what we're looking at is a difference in rate of swelling, but not really what the end point is, and that everything swells to the same level based upon the ionic concentration. However, I would hold out that our experiments looking at water intoxication, a classic model of cellular edema that Dr. Klatzo used and part of the definition of this term shows a significant difference in outcome, and I think that is really important. As we look at these things we shouldn't look at genes going up and going down. We need models that show differences in outcome. In addition, we looked at permanent occlusion of the MCA. At 24 hours we saw a difference in outcome, neurological outcome, and we saw a difference in the swelling. So I think that what's incumbent upon us is to continue to do models which we're able to look at outcome to show that there is a difference. And as we've talked about with several other issues here, there may be something here in the rate of swelling which is affecting other systems besides just the water, such as potassium, glutamate and some other things that we need to look at, that could then possibly convince Dr. Hossmann that maybe the swelling is as important.

Konstantin-A. Hossmann

Can I ask Dr. Manley a question? I think you clearly pointed out that deletion of aquaporins retards, but does not prevent osmotic water equilibration. If it is the speed of water influx but not the final degree of swelling which can be influenced, what is the relevance for the severity of edema under steady state conditions? And what is the effect on metabolic water which is continuously generated by oxidation of glucose and which has to be transported out of the cell?

Jeff Manley

I cannot answer your question directly because I don't know, but I think that these things need to be looked at. For example, our poster downstairs deals with the influence of aquaporins on water transport across the choroid plexus and this leads to a number of interesting questions that will need to be addressed. One of the things that has been important for me at this meeting are the comments from folks like you who ask the difficult questions, so that we are not just going to bring up another trendy subject like programmed cell death or glutamate toxicity that people jump at. If aquaporins are important, fine, if they are not important, then we will go on to the next mechanism of edema that might result in improved patient outcome.

Marios C. Papadopoulos

My studies on aquaporins one and four have only been under observation using tissue from patients, and this is the beginning of a number of studies that I am planning to do. Next year I will be joining Jeff Manley's group to look at some of the possible functions of aquaporins – both aquaporin one and aquaporin four in edema, both in the context of tumors as well as brain injuries.

Anthony Marmarou

With regard to the work you had presented, as far as the presence of the aquaporins, do you see these markers as being truly effective? Do you think you are underestimating or overestimating?

Marios C. Papadopoulos

The data that I showed was up-regulational aquaporin one and aquaporin four in tumors associated with edema, and using immunohistochemistry. My colleague also found similar results. Whether those aquaporins are actually functional, and whether they actually serve to increase brain edema or resolve brain edema or have some other role, I do not know. I think the way forward would be transgenic mice that Jeff Manley has, and hopefully over the next few years the development of aquaporin channel blockers, which would allow experiments in bigger animals.

Anthony Marmarou

Let us leave the aquaporins for a moment and go back to the issue of edema propagation. Dr. Reulen may I have a comment from you with regard to mechanisms, specifically in clearance.

Hans-J. Reulen

With regard to propagation, the movement through tissue has been shown primarily in the cold injury model and then in the tumor. And I think, as we have shown in the tumor model later on, about 80 percent of the movement and of the clearance of the fluid are through ventricle and about 20–25 percent are to the surface, to the CSF space. It is probably the same in the hemorrhagic situation of the tumor. But for me this movement of edema fluid has become more important recently in the treatment of brain tumors, especially gliomas. Now it is extremely difficult, as we have seen with gene therapy, to bring the substances into the tissue at a level sufficient to reach a therapeutic effect. The gene therapy is like working with a sports car instead of having a big truck to bring the substances to the point and that it should work. Now there seems to be a new possibility in brain tumors, particularly in glial tumors. If you have the tumor removed, and then you have this so-called local regional therapy. You bring a substance, any substance which you wish that you believe is interesting, bring it into the cavity, and then with aid of the bulk flow or by diffusion moves in the tissue it may reach cells which are far outside of this area, of the original area, and may later become responsible for the recurrence. So I think this, in this context, bulk flow or diffusional movement may even become very important in the future. This is just to give a short idea what I think could be the future of this whole problem of edema propagation.

Konstantin-A. Hossmann

May I address a question to Dr. Reulen or anybody else in the audience who knows how extracellular matrix proteins might interfere with edema propagation. When you started to study bulk flow of edema fluid, extracellular matrix proteins were more or less unknown, but at this meeting two papers have been presented which may have something to do with this. One experiment showed that inhibition of matrix metalloproteinase inhibited the propagation of cold injury edema, and the other communication showed that it had no effect after cardiac arrest. As you know, cold injury but not cardiac arrest produces vasogenic edema. Could it be that matrix proteins determine the resistance of the extracellular matrix to fluid exchange, and that prevention of matrix protein degradation by inhibition of matrix metalloproteinase reduces bulk flow and hence propagation of vasogenic, but not cytoxic edema?

Hans-J. Reulen

I cannot answer this question and I do not know if anybody has looked at this. Close to the lesion, extracellular space is tremendously enlarged, and I would not expect that in this region matrix proteins interfere with bulk flow of edema fluid. However, this may be different in the border zone where extracellular space is not so much enlarged.

Konstatin-A. Hossmann

Should we then continue to proceed to the next topic, or do we have other questions coming up with mechanisms?

Anthony Marmarou

I would like Prof. Baethmann to comment on the influence of blood brain barriers to where we are at the present time, both as in its importance in stroke and its importance in traumatic brain injury.

Alexander Baethmann

Thank you Tony. Well in essence the opening of the barrier that takes place in massive lesions results in tissue destruction. This is not under dispute anymore. The question might be more subtle in cases where you do not have a gross traumatic, ischemic, necrotic lesions of brain, but you may have many small disseminated lesions, for example. Or maybe even some kind of ischemic insults which are not really producing gross opening but may be producing a disseminated subtle opening as in maybe minor embolic states, for example. My basic hypothesis is all is notwithstanding that for an opening of the barrier, the extravasation of vasogenic edema fluid in containing proteins, there must be at least some minor tissue necrosis. I do not believe that endothelial cells allow, except for some experimental, maybe even artifactual situations, allow for transport or movement of vasogenic fluid into tissue without additional necrobiosis of this tissue. You know there are situations as the administration of hypotonic fluid to parts of the brain where the barrier is open, but this is not producing vasogenic edema. It just allows for short time movement of edema fluid in the tissue. The tissue is stained, but if you measure the water content where it has not increased, this is not vasogenic edema. This is an opening of the barrier, which is reversible, and the cell elements seen really are not irreversibly damaged.

Anthony Marmarou

It's rather disappointing in terms of the number of years that we have been working with regard to the treatment of cellular edema, if you will, considering in the clinical arena that most of our therapies are targeted toward the removal of water from the extracellular space. It is my sense, with exception of the new neurotrophic factors and others which have shown to be edema preventers, in spinal cord, that it is only hypothermia which has been shown to be effective in some forms of cellular swelling. I would like Dr. Hayashi to at least give us some insight as to where he thinks the field is going in terms of hypothermia, as we have heard some very interesting papers today.

Nariyuki Hayashi

Thank you Dr. Marmarou. Many people are confused about the difference in the effectiveness of hypothermia for neuronal protection between laboratory studies and clinical trials. The reason of the difference is very simple. The mechanism of brain damage is different between experimental animal models and human brain. The insulin-resisted hyperglycemia, brain thermo-pooling phenomenon and hemoglobin dysfunction that occur in association with hypothalamus-pituitary-adrenal (HPA) axis stress reaction are major targets of management in the ICU. However, these stress-associated brain injury mechanisms are not observed in experimental animal models, probably due to the influence of anesthesia. Especially, hemoglobin dysfunction, which produces difficulty in the release of oxygen from binding hemoglobin in injured brain tissue. This means there is no effect of oxygen inhalation for ischemic human brain. Recently many pitfalls on the treatment of brain hypothermia have been recorded in severe brain injured patients. We need special consideration on HPA axis stress reaction and on the pitfalls for successful treatment of brain hypothermia. Two years later, February 5–6 in 2004, the International Brain Hypothermia Symposium 2004 will be held in Tokyo. You are welcome to the symposium.

Anthony Marmarou

In the United States I think everyone is aware than the hypothermia trial in traumatic brain injury was not effective, and it was stopped, possibly due to complications in terms of rewarming as well as futility. However, there was a subset of patients who came in cool and they had shown improvement in terms of Glasgow outcome score. So now the current clinical trial in hypothermia is structured as follows. If the patients come in cool you allow them to remain cool, but if they come in normothermic then of course they are in the controlled sector. So that is where we are with hypothermia, and it looks as if it is going to be with us for some time as we have seen papers presented here and also in poster, where combination therapies utilizing hypothermia perhaps to extend the therapeutic window then superimposing pharmacological treatment coincident with the hypothermia. Perhaps David Mendelow could comment on other clinical trials in progress.

David Mendelow

I think there are two types of decompression that we need to think about. First one is the internal decompression that I was talking about this morning, the stitch with the T-trial, which is removing the intracerebral haematoma, that's a kind of internal decompression, a standard craniotomy. And then there's the external decompression where you'd leave the bone out and allow the brain to swell. I think that those are two completely different types of trials, and we need to keep them separate. The external decompression really would be in two forms. One is for the stroke patients, the patients with the malignant middle cerebral swelling which is a unilateral decompression; and then there's the traumatic brain injury where we might have to be doing a bilateral decompressive craniectomy. And even in that situation, there are really two questions. One is the North American route which Tony Marmarou referred to. And there is the European approach, where the decompressive craniectomy would be reserved for the patient in the ITU, who gets refractory elevated ICP. Well you've exhausted all the medical treatments, you've gone along all the steps, you've used barbiturates, you've used hypothermia, you've used everything you can think of, and this patient's ICP continues to rise. The European proposal is that that sort of patient would then be randomized. Half would get surgery and half would get continuing best medical treatment.

As far as the European group is concerned, rather that starting the trial they decided to collect data and they have just completed a data gathering exercise, where all the EBIC centers have completed a data form and to try and put the trial into context. I don't think funding has been approved yet, but I think the Cambridge group with Peter Kirk, Patrick John Picard and Peter Hutchenson have got some preliminary funding to set up the external decompression trial in refractory elevated ICP.

I think it's mandatory that we do these trials; otherwise in 10 years' time we'll still be talking about the need for emergency decompressive craniectomy. And one has to remember that the first report of decompressive craniectomy was 100 years ago by Harvey Cushing. And 100 years later we still don't know whether this is the right operation to do or not. So I think we must make this a priority as clinicians that we need to undertake these trials and answer these questions as soon as possible.

Anthony Marmarou

Thank you David. I know that all of us are deeply involved with clinical trials. It would be nice to talk with someone that is outside of the box, if you will, looking in, and I know Dr. Vink has been paying a lot of attention to the new pharmacologic treatments that are upcoming. We would like to hear from as to how he sees many of these treatments that have been put forth, experimentally and as well as clinically.

Robert Vink

Well, you're right I am outside the box here, because I come from a totally different background. One of the things I'm working on currently is inflammation as you know, and I think people tend to restrict their view of these things, for example, arachidonic acid base cascades and cytokines. I think that we really should look at other fields. For example, in the peripheral work neurogenic inflammation has been known for a long time, and yet I don't believe anybody has really looked very closely at neurogenic inflammation. The thing about neurogenic inflammation involves C fibers, which contain many peptides that have been implicated in trauma. For example, dynorphin. If you inject dynorphin into the spinal cord you get edema. A dynorphin is contained in those same fibers. I think really you need to look at therapies being used in other areas, and try to apply that to some of our problems in neuroscience and in edema in particular.

Anthony Marmarou

There have also been advances in the neuromonitoring that have been taken from the laboratory setting to the clinical setting. Certainly the work of Nordstroem and the groups who are dealing with trying to describe the neuro-chemical changes that occur. I was wondering if we could hear as to what you think is the direction in terms of the application of the microdialysis techniques as far as patient care, and whether or not there is going to be a future online analysis in traumatic brain injury and in stroke.

Carl-Henrik Nordstroem

Well we discussed a lot about very complicated things, molecular biology and things like that, and then we jump to decompression and very simple surgical things. My main concern is about intensive care, and that is physiology. And I think that if we really take care of physiology and do really monitor ICP and CPP in all patients in all symptoms and look at these very simple physiological data then we would really in a very simple way improve the management of our patients. I don't think we so many new drugs unfortunately do really make a large improvement. Regarding microanalysis, this is a technique where we can see things we can't really detect with physiological measurements – that's for sure. We can detect what happens in penumbras pretty early on. But it is also quite complicated. And I feel confident that the best symptoms, they can really make further improvement if they adopt this technique. It will be quite difficult in the centers that are not that sophisticated. But the technique has great potential. Once we have this technique as a routine method you can really do a lot of clinical studies in addition. So I think there is a future in microanalysis.

Konstantin A. Hossmann

Coming close to the end of this roundtable discussion, I would like to raise a topic which goes beyond prevention or amelioration of damage. In many cases, such an approach is irrelevant because the tissue is dead and cannot be treated. Should we therefore dare a glimpse into the future and ask if it is possible that necrotic brain tissue is replaced by regenerated neurons? And if so, does regeneration restore lost function? I would like to call Dr. Dempsey, who was the one who raised this issue, to express his views where the future is heading.

Robert J. Dempsey

Thank you. I'm pleased that you've brought up gene profiling and regeneration, and even more so that this is being presented at a brain edema meeting, because I think this group is the therapeutic arm of neurobiology. It is a big advantage that the view of this group goes beyond that of a pure neurobiologist. As we look at these problems from the viewpoint of brain edema, we understand that these responses are hard wired into the genome and that they are there for a reason. This allows us to present hypotheses, which the gene profiling can answer rather quickly as opposed to a fishing trip of thousands of genes.

As regards repair, one should not overlook the fact

that only in the past decade we discovered that cells regenerate in the adult brain, and we don't understand why. There are some fascinating theories that this may have to do with the generation of memory, but we see regenerating cells also after brain injury, both trauma and ischemia: These new cells may be part of a large repair mechanism that may require supportive functions, such as the generation of neurotrophic factors. We also have to find out if we can make neurons connect. I think we will. We have the tools to look at that both in culture and in experimental models, such as the one we presented here today. We now understand how to separate regeneration from injury, we see that signaling is involved, and we can investigate these signals to learn more about the attempts of the CNS to restore function by regeneration.

This is a very exciting area, and it becomes even more exciting when people involved in therapies such as this group begin to investigate why the situation exists.

Konstantin-A. Hossmann

Thank you very much. I have been always very much impressed by the observation that in certain songbirds new song-related neurons are generated every year during the mating season, and then disappear when the season is over. This shows that in principle the brain is able to reorganize itself, even in the adult state. I am therefore optimistic that at some time in the future regeneration may help us to treat diseases.

We have two or three minutes left, and I wonder if we should not briefly touch the subject of non-invasive imaging. Here two issues have been raised: One is image-guided disease management, and the other one is injury prediction. Non-invasive image-guided management means that the individual pathophysiology should be determined in each individual patient, and that on this basis individual treatments should be designed for each particular case. To a certain degree this is already done, but in the future this approach will become much more sophisticated.

The other imaging aspect which is of practical interest right now is prediction of injury. Using so-called signatures of tissue injury, it is possible to differentiate between necrotic and non-necrotic tissue, but is it also possible to predict recovery of damaged tissue which has not yet undergone necrosis? I wonder whether the specialists in NMR or PET could give us their opinion.

Alexander Baethmann

I would rather make a more general statement. I am convinced that the more we know about head injured or stroke patients who are in critical condition, the better we will be able to predict recovery. Imaging is just one among several approaches to find out if the tissue is necrotic or if cells are just swelling. Does the barrier opening process increase? And what about these secondary processes leading to increasing destruction of tissue? The more information we acquire at the bedside, the more specific our treatment will be. An impressive example of this is intracranial pressure recording. Without such recordings, we never would know – as we heard today – that in head injury patients intracranial pressure may be refractory to treatment, and only with this information we can think about what to do and how this is influencing the management. The same is also true for many other complicating processes, such as secondary tissue destruction, secondary brain hemorrhage or others. The more we know, the more specific our management can be and the better the outcome of our treatment will be.

Anthony Marmarou

Thank you Alex. Well we are coming to a close. A little red light is flashing. And I'd like Dr. Kuroiwa to stand for a moment. Dr. Kuroiwa, you have worked us very hard. We want to thank you, Dr. Katayama and Dr. Ito and your entire staff for organizing – a meeting so that all of us can exchange our views, can discuss our differences and also plan for the future. So with that please accept our compliments and we look forward to the next meeting in Ann Arbor, Michigan.

Acta Neurochir (2003) [Suppl] 86: 601–602

Hakone best presentation award

The Hakone awards for the 3 best oral presentations and the 3 best posters were arrived at by a scientific committee chaired by the outgoing President of the International Brain Edema Society, Professor A. David Mendelow.

The committee consisted of Zbigniew Czernicki (Warsaw), Yukio Ikeda (Tokyo), Anthony Marmarou (Richmond), Maria Spatz (Bethesda), Ryuta Suzuki (Yokohama) and Tatsuya Nagashima (Kobe).

The committee first agreed on their own nominations based on their personal judgments. The moderators of the individual sessions had independently scored the oral presentations and posters from their own sessions and these were coordinated by the secretariat. These scores were remarkably well correlated with the judges who, taking everything into consideration, unanimously agreed to the following awards:

Oral presentations

Treatment of cold injury-induced brain edema with a nonspecific matrix metalloproteinase inhibitor MMI270 in rats

N. Kawai, M. Kawanishi, and **S. Nagao**
Department of Neurological Surgery, Kagawa Medical University, Kagawa, Japan

DNA microarray-based gene expression analysis for induced ischemic tolerance and delayed neuronal death following transient global ischemia in rats*

N. Kawahara[1], Y. Wang[1], T. Musaka[1], K. Furuya[1], T. Hamakubo[2], H. Aburatani[2], T. Kodama[2], and **T. Kirino[1]**
[1] Department of Neurosurgery, University of Tokyo, Tokyo, Japan
[2] Research Center for Advanced Science and Technology, University of Tokyo, Tokyo, Japan

Role of water channel proteins (Aquaporins) in brain tumor edema

M. C. Papadopoulos[1], S. Saadoun[2], D. C. Davies[3], S. Krishna[2], and **B. A. Bell[1]**
[1] Department of Neurosurgery, Atkinson Morley's Hospital, London, UK
[2] Department of Infectious Diseases, St. George's Hospital Medical School, London, UK
[3] Department of Anatomy, St. George's Hospital Medical School, London, UK

Poster presentation

Temporal and spatial profile of phosphorylated mitogen-activated protein kinase (MAPK) pathways following lateral percussion brain injury in rats

N. Otani, H. Nawashiro, S. Fukui, N. Tsuzuki, H. Katoh, T. Miyazawa, and **K. Shima**
Department of Neurosurgery, National Defense Medical College, Saitama, Japan

Recovery of apparent diffusion coefficient in a rat model of embolic stroke does not mean complete salvage from ischemic neuronal injury

Y. Tanaka[1], T. Kuroiwa[2], N. Miyasaka[3], F. Tanabe[3], T. Nagaoka[4], and **K. Ohno[1]**
[1] Department of Neurosurgery, Tokyo Medical and Dental University, Tokyo, Japan
[2] Department of Neuropathology, Tokyo Medical and Dental University, Tokyo, Japan
[3] Department of Comprehensive Reproductive Medicine, Tokyo Medical and Dental University, Tokyo, Japan
[4] Department of Physiology and Biophysics, Albert Einstein College of Medicine, New York, USA

I wish to thank all members of the selection committee and moderators for their cooperation.

* Papers by Dr. Nobutaka Kawahara was unfortunately withdrawn from this volume after the selection process by the author to prevent double publication.

Src family kinase inhibitor PP1 improves motor functions by reducing edema after spinal cord contusion in rats

C. Akiyama, T. Yuguchi, M. Nishi, T. Fujinaka, M. Taniguchi, Y. Nakajima, and **T. Yoshimine**
Department of Neurosurgery, Osaka University Medical School, Suita, Osaka, Japan

Professor A. David Mendelow
President of the International Brain Edema Society

Toshihiko Kuroiwa M.D.
Chairman, Organizing committee
for Brain Edema 2002

Author Index

Index of Keywords

Springer-Verlag
and the Environment

WE AT SPRINGER-VERLAG FIRMLY BELIEVE THAT AN international science publisher has a special obligation to the environment, and our corporate policies consistently reflect this conviction.

WE ALSO EXPECT OUR BUSINESS PARTNERS – PRINTERS, paper mills, packaging manufacturers, etc. – to commit themselves to using environmentally friendly materials and production processes.

THE PAPER IN THIS BOOK IS MADE FROM NO-CHLORINE pulp and is acid free, in conformance with international standards for paper permanency.

GPSR Compliance

The European Union's (EU) General Product Safety Regulation (GPSR) is a set of rules that requires consumer products to be safe and our obligations to ensure this.

If you have any concerns about our products, you can contact us on ProductSafety@springernature.com

In case Publisher is established outside the EU, the EU authorized representative is:

Springer Nature Customer Service Center GmbH
Europaplatz 3
69115 Heidelberg, Germany

Batch number: 10731592

Printed by Printforce, the Netherlands